ENCYCLOPEDIA OF
ANIMAL BEHAVIOR

ENCYCLOPEDIA OF
ANIMAL BEHAVIOR

EDITORS-IN-CHIEF

PROFESSOR MICHAEL D. BREED
University of Colorado, Boulder, CO, USA

PROFESSOR JANICE MOORE
Colorado State University
Fort Collins, CO, USA

AMSTERDAM • BOSTON • HEIDELBERG • LONDON • NEW YORK • OXFORD
PARIS • SAN DIEGO • SAN FRANCISCO • SINGAPORE • SYDNEY • TOKYO
Academic Press is an imprint of Elsevier

ACADEMIC
PRESS

Academic Press is an imprint of Elsevier
32 Jamestown Road, London NWI 7BY, UK
30 Corporate Drive, Suite 400, Burlington, MA 01803, USA
525 B Street, Suite 1900, San Diego, CA 92101-4495, USA

British Library Cataloguing in Publication Data
A catalogue record for this book is available from the British Library

Library of Congress Catalog Number: 2010922487

ISBN: 978-0-08-045333-0

For information on all Elsevier publications
visit our website at books.elsevier.com

10 11 12 13 14 10 9 8 7 6 5 4 3 2 1

Cover Photo: An iguana, Costa Rica, photograph by Michael D. Breed

Printed and Bound in Spain

In Memoriam

Christopher J. Barnard

Ross H. Crozier

PREFACE

Ancient drawings on the walls of caves speak for the ageless intrigue that animal behavior holds for human beings. In those days, the fascination was certainly motivated in part by survival; our ancestors were both predators and prey. There is some evidence that early humans also found animal behavior to be intrinsically interesting; the myths and stories that come down to us from prehistory contain elements of what animals do in the world and what they mean to people. These are the oldest statements of human relationship with the natural world and the living things that inhabit it.

Our ancestors would not recognize the far-flung universe of the modern science of animal behavior. Only 14 decades (approximately) have passed since Darwin first published *The Expression of Emotions in Man and Animals* – generally acknowledged as the starting point for the scientific study of animal behavior – and behavioral biologists now ask questions about topics ranging from the relationship of immunological phenomena and behavioral disorders in dogs, rats, and people to the integration of animal behavior and conservation. Experts in animal behavior provide commentary on the mating displays of rare primates, on television, where entire channels are devoted to the sensory worlds of insects and the ability of octopus to disappear in plain sight. In short, human fascination with animal behavior has produced a field that is rich beyond imagination, and frustratingly beyond the full embrace of any one person.

The almost hopelessly dispersed primary literature of animal behavior reflects the reticulated evolution of the field, which comes to us from field studies; from laboratory experiments; from our understanding of nerves, muscles, and hormones; and from our grasp of social interactions and ecology. It is difficult to think of a major area of biological inquiry that has not been touched by a behavioral tendril or two. A temptation exists to surrender to this fragmentation – allowing our intellectual landscape to reflect increasingly small and disjunct patches of thought and discovery.

Such surrender is, of course, distasteful to any scholar, but there is a more penetrating reason that makes it unacceptable: Anthropogenic change is occurring at a higher rate than ever before, and if we are to preserve our own habitat – the world that the ancients felt compelled to explain in their stories about animals – we must not fail in our attempts to understand its inhabitants. Those residents sustain our own habitat, and their requirements are varied, going far beyond calories and oxygen. They migrate and forage, choose mates, and defend territories, and all this behavior is influenced by hormones, external physical stimuli, trophic and social interactions, and eons of fitness outcomes. A fully integrated knowledge of animal behavior will be indispensable as scientists analyze changing populations, communities, ecosystems, and landscapes. Indeed, it will be indispensable for anyone who seeks to be an honest custodian of nature.

This encyclopedia offers over 300 authoritative and accessible synopses of topics ranging from dolphin signature whistles to game theory. As library reference material, the encyclopedia serves a public that is increasingly challenged to be aware of scientific advances. It is designed as a first stop for the curious advanced undergraduate or graduate student, as well as for the researcher desiring to learn about developments in fields related to his or her own study or to enter a new phase of inquiry.

In compiling this work, we contacted internationally known scientists in the broad array of fields that inform animal behavior. These accomplished men and women are the section editors, and they, in turn, invited some of the best scholars and rising stars in their subject areas to write for the encyclopedia. Thus, every contribution has been reviewed by experts. In short, the articles approach the best that our field has to offer, written by people whose passion for animal behavior is equaled only by their expertise.

Creating the list of sections was as daunting as it was enjoyable. Of course, we included traditional, major areas like foraging, predator–prey interactions, mate choice, and social behavior, along with endocrinology, methods, and neural processes, to name a few. You will see those and more as you survey these volumes. We also included areas that have recently captured the attention of an increasing number of behavioral biologists; those include infectious disease, cognition, conservation, and animal welfare. Looking to the future, we invited contributors from robotics and applied areas. We realized that we could not do the study of animal behavior justice without some exploration of the model systems – the landmark studies – that have molded and continue to guide the development of the field.

In general, you will not find human behavioral studies in this collection, although some articles are tangentially related to such work. That exclusion was a difficult decision, but was motivated not so much by some parochial commitment to a human/non-human divide as by the fact that the non-human literature itself is rich beyond description. Limiting the collection to non-human studies in no way removed the danger of intellectual gluttony.

We remember two eminent behavioral biologists in the dedication of this work – Professors Christopher J. Barnard and Ross H. Crozier. Both of these men played important roles in the creation of this work, and neither lived to see it come to fruition. Professor Barnard (1952–2007) was the first editor-in-chief of the encyclopedia and developed the initial overview of topics, but had to step back from the process because of the illness that eventually took his life. Professor Crozier (1943–2009) was the section editor of Genetics until his untimely death in late 2009. Their immense and varied contributions enriched our knowledge of animal behavior and are cataloged in numerous locations. Those contributions are remarkable in their scope and influence, but they are nonetheless dwarfed by the legions of students, friends, and family members who feel fortunate to have known these scientists and who will carry their legacy forward.

We are grateful to Dr. Andrew Richford, formerly the Senior Acquisitions Editor, Life Sciences Books, Academic Press, who guided us through the formative part of this project. His expertise in and enthusiasm for animal behavior provided significant momentum, not to mention some good fun. Simon Wood, Major Reference Works Development Editor, was indispensable to the project. He answered an amazing variety of questions from contributors and editors, kept the project organized and moving forward, and did all this without losing his fine sense of humor. Nicky Carter, Project Manager, guided us through the completion of the project, providing a pleasantly seamless interface between the scientific scribblers and other publishing professionals. We thank Kristi Gomez and Will Smaldon, also of Academic Press, for their roles in bringing the project to completion. Finally, working with the section editors (see pp. ix) was a real treat; their expertise and devotion to animal behavior is reflected in every page of this work. We particularly thank James Ha, Joan Herbers, James Serpell, and David Stephens for attending an organizational meeting to set the stage for the development of the project.

We are pleased to see this culmination of effort on the part of hundreds of authors and co-authors. Each article is the distillation of expert understanding, acquired over many years. We are excited to be part of such a remarkable collaboration, one that opens so many doors to the future of animal behavior for undergraduates and professionals alike.

Michael Breed, Boulder, CO
Janice Moore, Fort Collins, CO
August 2010

SECTION EDITORS

Bonnie Beaver

Bonnie is internationally recognized for her work in the normal and abnormal behaviors of animals. She has given over 250 scientific presentations to veterinary and veterinary student audiences on subjects of animal behavior, animal welfare, and the human–animal bond, as well as discussed many areas of veterinary medicine for the public media. In addition, she has authored over 150 scientific articles and has nine published books, including *The Veterinarian's Encyclopedia of Animal Behavior* (Blackwell Press), *Feline Behavior: A Guide for Veterinarians* (Saunders), and the newly released second edition of *Canine Behavior: Insights and Answers* (Saunders).

Bonnie is a member of numerous local, state, and national professional organizations and has served as president or chair of several organizations, including the American Veterinary Society for Animal Behavior, the American College of Veterinary Behaviorists, Phi Zeta, and the Texas Veterinary Medical Association. She is board certified by the American College of Veterinary Behaviorists and currently serves as its Executive Director. In addition, Bonnie is the President of the Organizing Committee for the American College of Animal Welfare.

Bonnie is a past president of the American Veterinary Medical Association and has served as Chair of the AVMA Executive Board. She has also served on several AVMA committees, including the Animal Welfare Committee, Council on Education, Committee on the Human–Animal Bond, and American Board on Veterinary Specialties.

In addition, she chaired the AVMA's Canine Aggression and Human–Canine Interactions Task Force, and the Panel on Euthanasia.

Professionally, Bonnie has been honored by being elected as a Distinguished Practitioner of the National Academies of Practice, named as the recipient of the 1996 AVMA Animal Welfare Award, awarded the 2001 Friskies PetCare Award in Animal Behavior, and received the 2001 Leo K. Bustad Companion Animal Veterinarian of the Year Award. She has been recognized for outstanding professional achievement in more than 150 editions of over 50 publications, including *Who's Who in America*, *The World Who's Who of Women*, *Who's Who in the World*, and *American Men and Women of Science*.

Michael Breed

After receiving his PhD from the University of Kansas in 1977, Michael came to Colorado to work as a faculty member at the University of Colorado, Boulder, where he has been ever since. He is currently a Professor in the Department of Ecology and Evolutionary Biology, and teaches courses in general biology, animal behavior, insect biology, and tropical biology. Michael's research program focuses on the behavior and ecology of social insects, and he has worked on ants, bees, and wasps. He has studied the nestmate recognition, the genetics of colony defense, the behavior of defensive bees, and communication, during colony defense. He was the Executive Editor of *Animal Behaviour* from 2006 to 2009.

Jae Chun Choe

After receiving his PhD from Harvard University in 1990, Jae became a Junior Fellow at the Michigan Society of Fellows. He then returned to his home country, Korea, to work in the School of Biological Sciences at Seoul National University. In 2006, he moved to Ewha Womans University to take the post of university chair professor and the director of its natural history museum. He served as the president of the Ecological Society of Korea and is currently serving as the co-president of the Climate Change Center. Since his return to Korea, he has been conducting a long-term ecological research of magpies while continuing to study insects. Quite recently, he began a field study of Javan Gibbons in the Gunuung Halimun-Salak National Park of Indonesia.

Nicola Clayton

Nicola is Professor of Comparative Cognition in the Department of Experimental Psychology at the University of Cambridge, and a Fellow of Clare College. She received her undergraduate degree in Zoology at the University of Oxford and her doctorate in animal behavior at St. Andrews University. In 1995, she moved to the University of California Davis where she gained her first Chair in Animal Behaviour in 2000. She moved to Cambridge and was appointed a personal Chair in 2005. She has 185 publications to her credit.

Nicola studies the development and evolution of intelligence. For example, she addresses the question of whether animals can plan for the future and what they remember about the past, as well as when these abilities develop in children. She is also interested in social and physical intelligence, such as whether animals can differentiate between what they know and what other individuals know. Nicola's work deals mainly with the members of the crow family (e.g., rooks and jays), and comparisons between crows, nonhuman apes, and young children.

Jeff Galef

After receiving his Ph.D. from the University of Pennsylvania in 1968, Jeff moved as an Assistant Professor to McMaster University in Hamilton, Ontario where, for 38 years, his research focused on understanding social influences on the feeding behavior of Norway rats and the mate choices of Japanese quail. Empirical work in his laboratory on social learning in animals has resulted in the publication of more than 100 scientific articles, (www.sociallearning.info) and his scholarly pursuits have produced three co-edited volumes (*Social Learning: Psychological and Biological Perspectives* (with TR Zentall), *Social Learning and Imitation: the Roots of Culture* (with CM Heyes), and *The Question of Animal Culture* (with KN Laland)) as well as a special issue of the journal *Learning & Behavior* (2004, 32(1) (with CM Heyes)). He was honored with the Lifetime Contribution Award of the Social Learning Group, St. Andrews University, Scotland, in 2005, and in 2009, was elected a Fellow of the Royal Society of Canada.

Sidney Gauthreaux

Sidney received his PhD in 1968 and did a post-doctorate at the Institute of Ecology at the University of Georgia in the following 2 years. He joined the zoology faculty at Clemson University in 1970 and retired as Centennial Professor of Biological Sciences in 2006. In 1959, he began working with weather surveillance radar at National Weather Service installations in an effort to detect, quantify, and monitor migrating birds in the atmosphere.

His research has focused on radar studies of bird migration across the Gulf of Mexico and over much of the United States in spring and fall. Since 1992, modern Doppler weather radar has 'revolutionized' the study of bird migration, and he has used it to monitor the flight behavior of birds in the surveillance areas of approximately 150 weather radar stations throughout the United States and explore the interrelationships of bird movements at different spatial scales in relation to geography, topography, habitat, weather, and climatic factors. Recent work with high-resolution surveillance radar (modified marine radar) and thermal imaging and vertically pointing radar (TI-VPR) has greatly enhanced his capability to work at small spatial scales and explore the behavior of migrating birds within 12 km of the radar.

Sidney was President of the Animal Behavior Society from 1987 to 1988 and was elected a Fellow in the American Association for the Advancement of Science in 1988. In October 2006, he received the William Brewster Memorial Award of the American Ornithologists' Union, and in April 2009, the Margaret Morse Nice Medal of the Wilson Ornithological Society.

Deborah M. Gordon

After receiving her PhD from Duke University in 1984, Deborah joined the Harvard Society of Fellows. She did her postdoctoral research at Oxford and at the Centre for Population Biology at Silwood Park, University of London. She came to Stanford in 1991 and is currently a Professor in the Department of Biology. She teaches courses in ecology and behavioral ecology. Deborah's research program focuses on the organization and ecology of ant colonies, and how colonies, without central control, use interaction networks to regulate colony behavior. Her projects include a long-term study of a population of harvester ant colonies in Arizona, studies of the invasive Argentine ant in northern California, and ant–plant mutualisms in Central America.

Patricia Adair Gowaty

Patricia is a Distinguished Professor of Ecology and Evolutionary Biology – UCLA and a Distinguished Research Professor *Emerita* of Ecology at the University of Georgia. After receiving her PhD in 1980, she supported herself with funding from NSF and NIH, until her first tenure track job as an Associate Professor of Zoology in 1993 at the University of Georgia. She studied social behavior, demography, and ecology of eastern bluebirds in the field for 30 years. She pioneered studies of extra-pair paternity in socially monogamous species. She studied fitness outcomes of reproduction under experimentally imposed social constraints in flies, mice, ducks, and cockroaches. Her theoretical work includes papers on the evolution of social systems, forced copulation, compensation, and sex role evolution. Currently, she is completing studies in the genetic mating system of eastern bluebirds, experiments on the fitness variation of males and females in the three species of *Drosophila,* and a book on reproductive decisions under ecological and social constraints. She was President of the Animal Behavior Society in 2001. She is a Fellow of the American Association for the Advancement of Science, the American Ornithologists' Union, the International Ornithologists' Union, and the Animal Behavior Society.

James Ha

James has a 1989 Ph.D. in Zoology/Animal Behavior from Colorado State University and has been on the faculty of the University of Washington since 1992. He is actively involved in research on the social behavior of Old World monkeys and their management in captivity, Pacific Northwest killer whales, local and Pacific island crows, and domestic dogs. He is also certified as an Applied Animal Behaviorist by the Animal Behavior Society and has his own private practice in dealing with companion animal behavior problems in the Puget Sound area.

Joan M. Herbers

Joan is a Professor of Evolution, Ecology, and Organismal Biology at The Ohio State University in Columbus Ohio. She has studied social evolution in ants for many years, with contributions to queen-worker conflict, sex ratio theory, and coevolution. She is currently serving as the Secretary-General of the International Union for the Study of Social Insects and also as the President of the Association for Women in Science.

Jeffrey Lucas

Jeffrey received a Ph.D. from the University of Florida in 1983, studying under Dr. H. Jane Brockmann. He then took a postdoc position in Dr. John Kreb's lab at Oxford University. After teaching at the College of William & Mary and Redlands University, he came to Purdue University in 1987, where he is currently a professor of Biological Sciences. Jeffrey teaches courses in ecology, animal behavior, sensory ecology, and animal communication. His research program focuses on the chick-a-dee call of chickadees and a comparison of auditory physiology in a variety of birds. He has worked on seed dispersal, antlions, and fish, and has published dynamic programming models of a number of systems. He is a past Executive Editor of *Animal Behaviour* and is a fellow of the Animal Behavior Society.

Constantino Macías Garcia

Constantino has been interested in animal behavior ever since he joined Hugh Drummond's laboratory to study the feeding habits of snakes for his BSc and MSc. His main research has been on sexual selection and the evolution of ornaments, which he has studied mainly in Goodeid fish. He was careless not to follow the early forays of his PhD supervisor, Bill Sutherland, into the hybrid field of behavior and conservation. But time, as well as the increasingly grim reality of Mexican fauna, has led him to investigate the links between behavior and conservation in fish, frogs, and birds.

Justin Marshall

Justin's interest in biology and the sea came from his parents, both marine biologists and keen communicators of the ocean realm. He was then fortunate to begin learning about sensory biology in aquatic life during his undergraduate degree in Zoology at The University of St Andrews. The Gatty Marine Laboratory and its then director, Mike Laverack, introduced him to the diversity of marine life and the challenges of different sensory environments under water. Enjoying the cold clear waters of Scotland, he also began to take interest in tropical biodiversity and traveled to Australia and The Great Barrier Reef toward the end of his undergraduate degree. Currently, he is the President of The Australian Coral Reef Society and lives in Australia working at The University of Queensland. He holds a position of Professor at The Queensland Brain Institute and is an Australian Research Council Professorial Research Fellow. Before moving into the superb sensory environment of Jack Pettigrew's Vision Touch and Hearing Research Centre, he did his D.Phil and spent his initial postdoctoral years at The University of Sussex in the UK., Mike Land and The University of Maryland's

Tom Cronin were his mentors during these years and Justin developed an enthusiasm for the amazing world of invertebrate vision only because of them. His work now focuses on the visual ecology of a variety of animals, mostly aquatic, and has branched out to include fish, reptiles, and birds. Animal behavior and questions, such as 'why are animals colorful?', form a large section of his current research.

Janice Moore

As an undergraduate, Janice was inspired by parasitologist Clark P. Read to think about the ecology and evolution of parasites in new ways. She was especially excited to learn that parasites affected animal behavior, another favorite subject area. Most biologists outside the world of parasitology were not interested in parasites; they were relegated to a nether world between the biology of free-living organisms and medicine. After peregrination through more than one graduate program, she completed her PhD studying parasites and behavior at the University of New Mexico. Janice did postdoctoral work on parasite community ecology with Dan Simberloff at Florida State University, and then accepted a faculty position at Colorado State University, where she has remained since 1983. She is currently a Professor in the Department of Biology where she teaches courses in invertebrate zoology, animal behavior, and the history of medicine. She studies a variety of aspects of parasite ecology and host behavior ranging from behavioral fever and transmission behavior to the ecology of introduced parasite species.

Daniel Papaj

After receiving his PhD from the Duke University in 1984, Daniel engaged in postdoctoral research at the University of Massachusetts at Amherst and at Wageningen University in The Netherlands. He joined the faculty of the Department of Ecology and Evolutionary Biology at the University of Arizona in 1991, where he has been ever since. His research focuses on the reproductive dynamics of insects, with special attention to the role of learning by the insect in its interactions with plants. Daniel's focal organisms have included butterflies, tephritid fruit flies, parasitic wasps, and more recently, bumble bees. Recent projects in the lab include the costs of learning in butterflies, the dynamics of social information use in bumble bees, the thermal ecology of host preference in butterflies, ovarian dynamics in fruit flies, multimodal floral signaling, and bumblebee learning. He teaches courses in animal behavior, behavioral ecology, and introductory biology.

Ted Stankowich

Ted grew up in suburban Southern California where opportunities to observe macrofauna in nature were few, but still found ways to observe and enjoy the animals that he could find in his own backyard. While his initial interests in biology were in biochemistry and genetics, after taking introductory courses at Cornell University, he quickly realized that these disciplines were not his calling. He developed interests in ecology and evolution after taking introductory courses and working in George Lauder's functional morphology lab for a summer at the University of California, Irvine, but he took an abiding interest in animal behavior after taking a course as a junior at Cornell and joined Paul Sherman's naked mole-rat lab, where he completed an honors thesis on parental pup-shoving behavior. Ted entered the Animal Behavior graduate program at the University of California, Davis to work with Richard Coss. He spent three field seasons working on predator recognition, flight decisions, and antipredator behavior, in Columbian black-tailed deer, and completed his dissertation in 2006. Ted served as the Darwin Postdoctoral Fellow at the University of Massachusetts, Amherst from 2006 to 2008, investigating escape behavior in jumping spiders. Since completing his tenure, he has continued to work as a postdoc and teach at UMass.

David W. Stephens

David received his PhD from Oxford University in 1982. Currently, David is a Professor at the University of Minnesota in the Twin Cities. His research takes a theoretical and experimental approach to behavior ecology. His research focuses on the connections between evolution and animal cognition, especially the evolutionary forces that have shaped animal learning and decision-making. His work makes connections with many disciplines within the behavioral sciences, and he has presented his work to groups of psychologists, economists, anthropologists, mathematicians, and neuroscientists. He is the author, with John Krebs, of the well-cited book *Foraging Theory*, and the editor (with Joel Brown and Ronald Ydenberg) of *Foraging: behavior and ecology*. He served as an editor of *Animal Behaviour* from 2006 to 2009.

John C. Wingfield

John's undergraduate degree was in Zoology (special honors program) from the University of Sheffield and he did his Ph.D. in Comparative Endocrinology and Zoology from the University College of North Wales, UK. Although John is trained as a comparative endocrinologist, he has always interacted with behavioral ecologists and has strived to integrate ecology and physiology down to cellular and molecular levels. The overarching question is how animals cope with a changing environment – basic biology of how environmental signals are perceived, transduced into endocrine secretions that then regulate morphological, physiological, and behavioral responses. The diversity of mechanisms is becoming more and more apparent and how these evolved is another intriguing question. He was an Assistant Professor at the Rockefeller University in New York and then spent over 20 years as a Professor at the University of Washington. Currently, he is a Professor and Chair in Physiology at the University of California at Davis.

Harold Zakon

Harold received a B.S. degree from Marlboro College in Vermont. He worked as a research technician at Harvard Medical School for 2 years and realized his love for doing research. He earned a Ph.D. from the Neurobiology & Behavior program at Cornell University, working with Robert Capranica, studying the regeneration of the frog auditory nerve. He did postdoctoral work at the University of - California, San Diego with Theodore Bullock and Walter Heiligenberg. There, he began working on weakly electric fish. He established his laboratory at the University of Texas in Austin, Texas where he has been studying communication in electric fish, and the regulation and evolution of ion channels in electric fish and other organisms. He was the first chairman of the then newly established Section of Neurobiology at UT. He has been Chairman for Gordon Research Conference on Neuroethology and organizer for the International Congress in Neuroethology. His hobbies include playing guitar and piano. He, his wife Lynne (mandolin), and son Alex (banjo), have a band called Red State Bluegrass. Their goal is to perform on Austin City Limits one day.

CONTRIBUTORS

J. S. Adelman
Princeton University, Princeton, NJ, USA

E. Adkins-Regan
Cornell University, Ithaca, NY, USA

J. F. Aggio
Neuroscience Institute and Department of Biology,
Atlanta, GA, USA

M. Ah-King
University of California, Los Angeles, CA, USA

I. Ahnesjö
Uppsala University, Uppsala, Sweden

J. Alcock
Arizona State University, Tempe, AZ, USA

L. Angeloni
Colorado State University, Fort Collins, CO, USA

B. R. Anholt
University of Victoria, Victoria, BC, Canada; Bamfield
Marine Sciences Centre, Bamfield, BC, Canada

C. J. L. Atkinson
University of Queensland, St Lucia, QLD, Australia

F. Aureli
Liverpool John Moores University, Liverpool, UK

A. Avarguès-Weber
CNRS, Université de Toulouse, Toulouse, France; Centre
de Recherches sur la Cognition Animale, Toulouse,
France

K. L. Ayres
University of Washington, Seattle, WA, USA

J. Bakker
University of Liège, Liège, Belgium

G. F. Ball
Johns Hopkins University, Baltimore, MD, USA

J. Balthazart
University of Liège, Liège, Belgium

L. Barrett
University of Lethbridge, Lethbridge, AB, Canada

A. H. Bass
Cornell University, Ithaca, NY, USA

D. K. Bassett
University of Auckland, Auckland, New Zealand

M. Bateson
Newcastle University, Newcastle upon Tyne, UK

G. Beauchamp
University of Montréal, St. Hyacinthe, QC, Canada

B. V. Beaver
Texas A&M University, College Station, TX, USA

P. A. Bednekoff
Eastern Michigan University, Ypsilanti, MI, USA

M. Beekman
University of Sydney, Sydney, NSW, Australia

J. A. Bender
Case Western Reserve University, Cleveland,
OHIO, USA

G. E. Bentley
University of California, Berkeley, CA, USA

A. Berchtold
University of Lausanne, Lausanne, Switzerland

I. S. Bernstein
University of Georgia, Athens, GA, USA

S. Bevins
Colorado State University, Fort Collins, USA

D. T. Blumstein
University of California, Los Angeles, CA, USA

C. R. B. Boake
University of Tennessee, Knoxville, TN, USA

R. A. Boakes
University of Sydney, Sydney, NSW, Australia

W. J. Boeing
New Mexico State University, Las Cruces, NM, USA

N. J. Boogert
McGill University, Montreal, QC, Canada

T. Boswell
Newcastle University, Newcastle upon Tyne, UK

A. Bouskila
Ben-Gurion University of the Negev, Beer Sheva, Israel

R. M. Bowden
Illinois State University, Normal, IL, USA

E. M. Brannon
Duke University, Durham, NC, USA

M. D. Breed
University of Colorado, Boulder, CO, USA

M. R. Bregman
University of California, San Diego, CA, USA

J. Brodeur
Université de Montréal, Montréal, QC, Canada

E. D. Brodie, III
University of Virginia, Charlottesville, VA, USA

A. Brodin
Lund University, Lund, Sweden

D. M. Broom
University of Cambridge, Cambridge, UK

J. L. Brown
University at Albany, Albany, NY, USA

J. S. Brown
University of Illinois at Chicago, Chicago, IL, USA

M. J. F. Brown
Royal Holloway University of London, Egham, UK

H. Brumm
Max Planck Institute for Ornithology, Seewiesen,
Germany

R. Buffenstein
University of Texas Health Science Center at San Antonio,
San Antonio, TX, USA

J. D. Buntin
University of Wisconsin-Milwaukee, Milwaukee, WI, USA

J. Burger
Rutgers University, Piscataway, NJ, USA

G. M. Burghardt
University of Tennessee, Knoxville, TN, USA

R. W. Burkhardt, Jr.
University of Illinois at Urbana-Champaign, Urbana,
IL, USA

N. T. Burley
University of California, Irvine, CA, USA

S. S. Burmeister
University of North Carolina, Chapel Hill, NC, USA

D. S. Busch
Northwest Fisheries Science Center, National Marine
Fisheries Service, Seattle, WA, USA

R. W. Byrne
University of St. Andrews, St. Andrews, Fife, Scotland, UK

R. M. Calisi
University of California, Berkeley, CA, USA

J. Call
Max Planck Institute for Evolutionary Anthropology,
Leipzig, Germany

U. Candolin
University of Helsinki, Helsinki, Finland

J. F. Cantlon
Rochester University, Rochester, NC, USA

C. E. Carr
University of Maryland, College Park, MD, USA

C. S. Carter
University of Illinois at Chicago, Chicago, IL, USA

F. Cézilly
Université de Bourgogne, Dijon, France

E. S. Chang
University of California-Davis, Bodega Bay, CA, USA

J. W. Chapman
Rothamsted Research, Harpenden, Hertfordshire, UK

J. C. Choe
Ewha Womans University, Seoul, Korea

J. A. Clarke
University of Northern Colorado, Greeley, CO, USA

N. S. Clayton
University of Cambridge, Cambridge, UK

B. Clucas
University of Washington, Seattle, WA, USA;
Humboldt University, Berlin, Germany

R. B. Cocroft
University of Missouri, Columbia, MO, USA

J. H. Cohen
Eckerd College, St. Petersburg, FL, USA

S. P. Collin
University of Western Australia, Crawley, WA, Australia

L. Conradt
University of Sussex, Brighton, UK

W. E. Cooper, Jr.
Indiana University Purdue University Fort Wayne, Fort
Wayne, IN, USA

R. G. Coss
University of California, Davis, CA, USA

J. T. Costa
Western Carolina University, Cullowhee, NC, USA;
Highlands Biological Station, Highland NC, USA

I. D. Couzin
Princeton University, Princeton, NJ, USA

N. J. Cowan
Johns Hopkins University, Baltimore, MD, USA

R. M. Cox
Dartmouth College, Hanover, NH, USA

J. Crast
University of Georgia, Athens, GA, USA

S. Creel
Montana State University, Bozeman, MT, USA

W. Cresswell
University of St. Andrews, St. Andrews, Scotland, UK

D. Crews
University of Texas, Austin, TX, USA

K. R. Crooks
Colorado State University, Fort Collins, CO, USA

J. D. Crystal
University of Georgia, Athens, GA, USA

S. R. X. Dall
University of Exeter, Cornwall, UK

D. Daniels
University at Buffalo, State University of New York, Buffalo, NY, USA

J. M. Davis
Vassar College, Poughkeepsie, NY, USA

K. Dean
University of Maryland, College Park, MD, USA

J. Deen
University of Minnesota, St. Paul, MN, USA

R. J. Denver
University of Michigan, Ann Arbor, MI, USA

C. D. Derby
Neuroscience Institute and Department of Biology, Atlanta, GA, USA

M. E. Deutschlander
Hobart and William Smith Colleges, Geneva, NY, USA

F. B. M. de Waal
Emory University, Atlanta, GA, USA

D. A. Dewsbury
University of Florida, Gainesville, FL, USA

A. Dickinson
University of Cambridge, Cambridge, UK

J. L. Dickinson
Cornell University, Ithaca, NY, USA

A. G. Dolezal
Arizona State University, Tempe, AZ, USA

D. Dollgoz
Université de Lyon, Villourbanne, France

R. H. Douglas
City University, London, UK

K. B. Døving
University of Oslo, Oslo, Norway

V. A. Drake
University of New South Wales at the Australian Defence Force Academy, Canberra, ACT, Australia

L. C. Drickamer
Northern Arizona University, Flagstaff, AZ, USA

H. Drummond
Universidad Nacional Autónoma de México, México

J. P. Drury
University of California, Los Angeles, CA, USA

J. E. Duffy
Virginia Institute of Marine Science, Gloucester Point, VA, USA

R. Dukas
McMaster University, Hamilton, ON, Canada

F. C. Dyer
Michigan State University, East Lansing, MI, USA

W. G. Eberhard
Smithsonian Tropical Research Institute; Universidad de Costa Rica, Ciudad Universitaria, Costa Rica

N. J. Emery
Queen Mary University of London, London, UK; University of Cambridge, Cambridge, UK

C. S. Evans
Macquarie University, Sydney, NSW, Australia

S. E. Fahrbach
Wake Forest University, Winston-Salem, NC, USA

E. Fernández-Juricic
Purdue University, West Lafayette, IN, USA

J. R. Fetcho
Cornell University, Ithaca, NY, USA

J. H. Fewell
Arizona State University, Tempe, AZ, USA

G. Fleissner
Goethe-University Frankfurt, Frankfurt, Germany

G. Fleissner
Goethe-University Frankfurt, Frankfurt, Germany

T. H. Fleming
University of Miami, Coral Gables, FL, USA

A. Florsheim
Veterinary Behavior Solutions, Dallas, TX, USA

E. S. Fortune
Johns Hopkins University, Baltimore, MD, USA

R. B. Forward, Jr.
Duke University Marine Laboratory, Beaufort, NC, USA

S. A. Foster
Clark University, Worcester, MA, USA

D. M. Fragaszy
University of Georgia, Athens, GA, USA

O. N. Fraser
University of Vienna, Vienna, Austria

P. J. Fraser
University of Aberdeen, Aberdeen, Scotland, UK

T. M. Freeberg
University of Tennessee, Knoxville, TN, USA

K. A. French
University of California, San Diego, La Jolla, CA, USA

A. Frid
Vancouver Aquarium, Vancouver, BC, Canada

C. B. Frith
Private Independent Ornithologist, Malanda, QLD, Australia

D. J. Funk
Vanderbilt University, Nashville, TN, USA

L. Fusani
University of Ferrara, Ferrara, Italy

C. R. Gabor
Texas State University-San Marcos, San Marcos, TX, USA

R. Gadagkar
Indian Institute Science, Bangalore, India

B. G. Galef
McMaster University, Hamilton, ON, Canada

C. M. Garcia
Instituto de Ecología, UNAM, México

S. A. Gauthreaux, Jr.
Clemson University, Clemson, SC, USA

F. Geiser
University of New England, Armidale, NSW, Australia

T. Q. Gentner
University of California, San Diego, CA, USA

H. C. Gerhardt
University of Missouri, Columbia, MO, USA

M. D. Ginzel
Purdue University, West Lafayette, IN, USA

L.-A. Giraldeau
Université du Québec à Montréal, Montréal, QC, Canada

M. Giurfa
CNRS, Université de Toulouse, Toulouse, France; Centre de Recherches sur la Cognition Animale, Toulouse, France

J.-G. J. Godin
Carleton University, Ottawa, ON, Canada

J. Godwin
North Carolina State University, Raleigh, NC, USA

E. Goodale
Field Ornithology Group of Sri Lanka, University of Colombo, Colombo, Sri Lanka

M. A. D. Goodisman
Georgia Institute of Technology, Atlanta, GA, USA

C. J. Goodnight
University of Vermont, Burlington, Vermont, USA

P. A. Gowaty
University of California, Los Angeles, CA, USA; Smithsonian Tropical Research Institute, USA

W. Goymann
Max Planck Institute for Ornithology, Seewiesen, Germany

P. Graham
University of Sussex, Brighton, UK

T. Grandin
Colorado State University, Fort Collins, CO, USA

M. D. Greenfield
Université François Rabelais de Tours, Tours, France

G. F. Grether
University of California, Los Angeles, CA, USA

A. S. Griffin
University of Newcastle, Callaghan, NSW, Australia

M. Griggio
Konrad Lorenz Institute for Ethology, Vienna, Austria

T. G. G. Groothuis
University of Groningen, Groningen, Netherlands

R. Grosberg
University of California, Davis, CA, USA

C. M. Grozinger
Pennsylvania State University, University Park, PA, USA

R. D. Grubbs
Florida State University Coastal and Marine Laboratory, St. Teresa, FL, USA; George Mason University, Fairfax, VA, USA

R. R. Ha
University of Washington, Seattle, WA, USA

J. P. Hailman
University of Wisconsin, Jupiter, FL, USA

I. M. Hamilton
Ohio State University, Columbus, OH, USA

R. R. Hampton
Emory University, Atlanta, GA, USA

I. C. W. Hardy
University of Nottingham, Loughborough, Leicestershire, UK

B. L. Hart
University of California, Davis, CA, USA

L. I. Haug
Texas Veterinary Behavior Services, Sugar Land, TX, USA

M. Hauser
Harvard University, Cambridge, MA, USA

L. S. Hayward
University of Washington, Seattle, WA, USA

S. D. Healy
University of St. Andrews, St. Andrews, Fife, Scotland, UK

E. A. Hebets
University of Nebraska, Lincoln, NE, USA

M. R. Heithaus
Florida International University, Miami, FL, USA

H. Helanterä
University of Sussex, Brighton, UK; University of Helsinki, Helsinki, Finland

J. M. Hemmi
Australian National University, Canberra, ACT, Australia

L. M. Henry
University of Oxford, Oxford, UK

J. M. Herbers
Ohio State University, Columbus, OH, USA

M. R. Heupel
James Cook University, Townsville, QLD, Australia

H. Hoi
Konrad Lorenz Institute for Ethology, Vienna, Austria

K. E. Holekamp
Michigan State University, East Lansing, MI, USA

R. A. Holland
Max Planck Institute for Ornithology, Radolfzell, Germany

A. G. Horn
Dalhousie University, Halifax, NS, Canada

L. Huber
University of Vienna, Vienna, Austria

M. A. Huffman
Kyoto University, Inuyama, Aichi Prefecture, Japan

H. Hurd
Keele University, Staffordshire, UK

P. L. Hurd
University of Alberta, Edmonton, AB, Canada

A. Jacobs
University of California, Riverside, CA, USA

V. M. Janik
University of St. Andrews, St. Andrews, Fife, Scotland, UK

K. Jensen
Queen Mary University of London, London, UK

C. Jozet-Alves
University of Caen Basse-Normandie, Caen, France

J. Kaminski
Max Planck Institute for Evolutionary Anthropology, Leipzig, Germany

L. Kapás
Washington State University, Spokane, WA, USA

A. S. Kauffman
University of California, San Diego, La Jolla, CA, USA

J. L. Kelley
University of Western Australia, Crawley, WA, Australia

A. J. King
Zoological Society of London, London, UK; University of Cambridge, Cambridge, UK

S. L. Klein
Johns Hopkins Bloomberg School of Public Health, Baltimore, MD, USA

M. J. Klowden
University of Idaho, Moscow, ID, USA

J. Komdeur
University of Groningen, Groningen, Netherlands

M. Konishi
California Institute of Technology, Pasadena, CA, USA

J. Korb
University of Osnabrueck, Osnabrück, Germany

I. Krams
University of Daugavpils, Daugavpils, Latvia

R. T. Kraus
Florida State University Coastal and Marine Laboratory, St. Teresa, FL, USA; George Mason University, Fairfax, VA, USA

W. B. Kristan, Jr.
University of California, San Diego, La Jolla, CA, USA

J. M. Krueger
Washington State University, Spokane, WA, USA

C. W. Kuhar
Cleveland Metroparks Zoo, Cleveland, OH, USA

C. P. Kyriacou
University of Leicester, Leicester, UK

F. Ladich
University of Vienna, Vienna, Austria

K. N. Laland
University of St Andrews, St Andrews, Fife, Scotland, UK

R. H. J. Lamberton
Imperial College Faculty of Medicine, London, UK

A. V. Latchininsky
University of Wyoming, Laramie, WY, USA

L. Lefebvre
McGill University, Montréal, QC, Canada

J. E. Leonard
Hiwassee College, Madisonville, TN, USA

M. L. Leonard
Dalhousie University, Halifax, NS, Canada

G. R. Lewin
Max-Delbrück Center for Molecular Medicine, Berlin, Germany

F. Libersat
Institut de Neurobiologie de la Méditerranée, Parc Scientifique de Luminy, Marseille, France

A. E. Liebert
Framingham State College, Framingham, MA, USA

C. H. Lin
University of British Columbia, Vancouver, BC, Canada

J. A. Linares
Texas A&M University, Gonzales, TX, USA

J. Lind
Stockholm University, Stockholm, Sweden

T. A. Linksvayer
University of Copenhagen, Copenhagen, Denmark

C. List
London School of Economics, London, UK

N. Lo
Australian Museum, Sydney, NSW, Australia; University of Sydney, Sydney, NSW, Australia

C. M. F. Lohmann
University of North Carolina, Chapel Hill, NC, USA

K. J. Lohmann
University of North Carolina, Chapel Hill, NC, USA

Y. Lubin
Ben-Gurion University of the Negev, Beer Sheva, Israel

J. Lucas
Purdue University, West Lafayette, IN, USA

S. K. Lynn
Boston College, Chestnut Hill, MA, USA

K. E. Mabry
New Mexico State University, Las Cruces, NM, USA

D. Maestripieri
University of Chicago, Chicago, IL, USA

D. L. Maney
Emory University, Atlanta, GA, USA

T. G. Manno
Auburn University, Auburn, AL, USA

S. W. Margulis
Canisius College, Buffalo, NY, USA

L. Marino
Emory University, Atlanta, GA, USA

T. A. Markow
University of California at San Diego, La Jolla, CA, USA

C. A. Marler
University of Wisconsin, Madison, WI, USA

P. P. Marra
Smithsonian Migratory Bird Center, National Zoological Park, Washington, DC, USA

L. B. Martin
University of South Florida, Tampa, FL, USA

M. Martin
North Carolina State University, Raleigh, NC, USA

J. A. Mather
University of Lethbridge, Lethbridge, AB, Canada

K. Matsuura
Okayama University, Okayama, Japan

T. Matsuzawa
Kyoto University, Kyoto, Japan

K. McAuliffe
Harvard University, Cambridge, MA, USA

E. A. McGraw
University of Queensland, Brisbane, QLD, Australia

N. L. McGuire
University of California, Berkeley, CA, USA

N. J. Mehdiabadi
Smithsonian Institution, Washington, DC, USA

R. Menzel
Freie Universität Berlin, Berlin, Germany

J. C. Mitani
University of Michigan, Ann Arbor, MI, USA

J. C. Montgomery
University of Auckland, Auckland, New Zealand

J. Moore
Colorado State University, Fort Collins, CO, USA

J. Morand-Ferron
Université du Québec à Montréal, Montréal, QC, Canada

J. Moreno
Museo Nacional de Ciencias Naturales, Madrid, Spain

K. Morgan
University of St. Andrews, St. Andrews, Fife, Scotland, UK

R. Muheim
Lund University, Lund, Sweden

C. A. Nalepa
North Carolina State University, Raleigh, NC, USA

D. Naug
Colorado State University, Fort Collins, CO, USA

D. A. Nelson
Ohio State University, Columbus, OH, USA

R. J. Nelson
Ohio State University, Columbus, OH, USA

I. Newton
Centre for Ecology & Hydrology, Wallingford, UK

K. Nishimura
Hokkaido University, Hakodate, Japan

J. E. Niven
University of Cambridge, Cambridge, UK; Smithsonian Tropical Research Institute, Panamá, República de Panamá

P. Nonacs
University of California, Los Angeles, CA, USA

A. J. Norton
Imperial College Faculty of Medicine, London, UK

B. P. Oldroyd
University of Sydney, Sydney, NSW, Australia

T. J. Ord
University of New South Wales, Sydney, NSW, Australia

M. A. Ottinger
University of Maryland, College Park, MD, USA

D. H. Owings
University of California, Davis, CA, USA

J. M. Packard
Texas A&M University, College Station, TX, USA

A. Pai
Spelman College, Atlanta, GA, USA

T. J. Park
University of Illinois at Chicago, Chicago, IL, USA

L. A. Parr
Yerkes National Primate Research Center, Atlanta, GA, USA

Y. M. Parsons
La Trobe University, Bundoora, VIC, Australia

G. L. Patricelli
University of California, Davis, CA, USA

M. M. Patten
Museum of Comparative Zoology, Cambridge, MA, USA

A. Payne
Tufts University, Medford, MA, USA

I. M. Pepperberg
Harvard University, Cambridge, MA, USA

M.-J. Perrot-Minnot
Université de Bourgogne, Dijon, France

S. Perry
University of California-Los Angeles, Los Angeles, CA, USA

K. M. Pickett
University of Vermont, Burlington, VT, USA

N. Pinter-Wollman
Stanford University, Stanford, CA, USA

D. Plachetzki
University of California, Davis, CA, USA

G. S. Pollack
McGill University, Montréal, QC, Canada

G. D. Pollak
University of Texas at Austin, Austin, TX, USA

R. Poulin
University of Otago, Dunedin, New Zealand

S. C. Pratt
Arizona State University, Tempe, AZ, USA

V. V. Pravosudov
University of Nevada, Reno, NV, USA

G. H. Pyke
Australian Museum, Sydney, NSW, Australia; Macquarie University, North Ryde, NSW, Australia

D. C. Queller
Rice University, Houston, TX, USA

M. Ramenofsky
University of California, Davis, CA, USA

C. H. Rankin
University of British Columbia, Vancouver, BC, Canada

F. L. W. Ratnieks
University of Sussex, Brighton, UK

D. Raubenheimer
Massey University, Auckland, New Zealand

S. M. Reader
Utrecht University, Utrecht, Netherlands

H. K. Reeve
Cornell University, New York, NY, USA

J. Reinhard
University of Queensland, Brisbane, QLD, Australia

L. Rendell
University of St Andrews, St Andrews, Fife, Scotland, UK

A. N. Rice
Cornell University, Ithaca, NY, USA

J. M. L. Richardson
University of Victoria, Victoria, BC, Canada

H. Richner
University of Bern, Bern, Switzerland

T. Rigaud
Université de Bourgogne, Dijon, France

R. E. Ritzmann
Case Western Reserve University, Cleveland, OHIO, USA

A. J. Riveros
University of Arizona, Tucson, AZ, USA

D. Robert
University of Bristol, Bristol, UK

G. E. Robinson
University of Illinois at Urbana-Champaign, Urbana, IL, USA

I. Rodriguez-Prieto
Museo Nacional de Ciencias Naturales, Madrid, Spain

B. D. Roitberg
Simon Fraser University, Burnaby, BC, Canada

L. M. Romero
Tufts University, Medford, MA, USA

T. J. Roper
University of Sussex, Brighton, UK

G. G. Rosenthal
Texas A&M University, College Station, TX, USA

C. Rowe
Newcastle University, Newcastle upon Tyne, UK

L. Ruggiero
Barnard College and Columbia University, New York, NY, USA

G. D. Ruxton
University of Glasgow, Glasgow, Scotland, UK

M. J. Ryan
University of Texas, Austin, TX, USA

R. Safran
University of Colorado, Boulder, CO, USA

W. Saltzman
University of California, Riverside, CA, USA

R. M. Sapolsky
Stanford University, Stanford, CA, USA

L. S. Sayigh
Woods Hole Oceanographic Institution, Woods Hole, MA, USA

A. Schmitz
University of Bonn, Bonn, Germany

H. Schmitz
University of Bonn, Bonn, Germany

J. Schulkin
Georgetown University, Washington, DC, USA; National Institute of Mental Health, Bethesda, MD, USA

H. Schwabl
Washington State University, Pullman, WA, USA

A. M. Seed
Max Planck Institute for Evolutionary Anthropology, Leipzig, Germany

M. R. Servedio
University of North Carolina, Chapel Hill, NC, USA

J. C. Shaw
University of California, Santa Barbara, CA, USA

S.-F. Shen
Cornell University, New York, NY, USA

B. L. Sherman
North Carolina State University, Raleigh, NC, USA

T. N. Sherratt
Carleton University, Ottawa, ON, Canada

D. M. Shuker
University of St. Andrews, St. Andrews, Fife, Scotland, UK

R. Silver
Barnard College and Columbia University, New York, NY, USA

B. Silverin
University of Göteborg, Göteborg, Sweden

A. M. Simmons
Brown University, Providence, RI, USA

S. J. Simpson
University of Sydney, Sydney, NSW, Australia

U. Sinsch
University Koblenz-Landau, Koblenz, Germany

H. Slabbekoorn
Leiden University, Leiden, Netherlands

P. J. B. Slater
University of St. Andrews, St. Andrews, Fife, Scotland, UK

C. N. Slobodchikoff
Northern Arizona University, Flagstaff, AZ, USA

A. R. Smith
Smithsonian Tropical Research Institute, Balboa, Ancon, Panamá

G. T. Smith
Indiana University, Bloomington, IN, USA

J. E. Smith
Michigan State University, East Lansing, MI, USA

B. Smuts
University of Michigan, Ann Arbor, MI, USA

E. C. Snell-Rood
Indiana University, Bloomington, IN, USA

C. T. Snowdon
University of Wisconsin, Madison, WI, USA

R. B. Srygley
USDA-Agricultural Research Service, Sidney, MT, USA

T. Stankowich
University of Massachusetts, Amherst, MA, USA

P. T. Starks
Tufts University, Medford, MA, USA

C. A. Stern
Cornell University, Ithaca, NY, USA

J. R. Stevens
Max Planck Institute for Human Development, Berlin, Germany

P. K. Stoddard
Florida International University, Miami, FL, USA

J. E. Strassmann
Rice University, Houston, TX, USA

C. E. Studds
Smithsonian Migratory Bird Center, National Zoological Park, Washington, DC, USA

L. Sullivan-Beckers
University of Nebraska, Lincoln, NE, USA

R. A. Suthers
Indiana University, Bloomington, IN, USA

J. P. Swaddle
College of William and Mary, Williamsburg, VA, USA

R. Swaisgood
San Diego Zoo's Institute for Conservation Research, Escondido, CA, USA

É. Szentirmai
Washington State University, Spokane, WA, USA

M. Taborsky
University of Bern, Hinterkappelen, Switzerland

Z. Tang-Martínez
University of Missouri-St. Louis, St. Louis, MO, USA

E. Tauber
University of Leicester, Leicester, UK

D. W. Thieltges
University of Otago, Dunedin, New Zealand

F. Thomas
Génétique et Evolution des Maladies Infectieuses, Montpellier, France; Université de Montréal, Montréal, QC, Canada

C. V. Tillberg
Linfield College, McMinnville, OR, USA

M. Tomasello
Max Planck Institute for Evolutionary Anthropology, Leipzig, Germany

A. L. Toth
Pennsylvania State University, University Park, PA, USA

B. C. Trainor
University of California, Davis, CA, USA

J. Traniello
Boston University, Boston, MA, USA

K. Tsuji
University of the Ryukyus, Okinawa, Japan

G. W. Uetz
University of Cincinnati, Cincinnati, OH, USA

M. Valentine
University of Vermont, Burlington, VT, USA

A. Valero
Instituto de Ecología, UNAM, México

J. L. Van Houten
University of Vermont, Burlington, VT, USA

M. A. van Noordwijk
University of Zurich, Zurich, Switzerland

C. P. van Schaik
University of Zurich, Zurich, Switzerland

S. H. Vessey
Bowling Green State University, Bowling Green, OH, USA

G. von der Emde
University of Bonn, Bonn, Germany

H. G. Wallraff
Max Planck Institute for Ornithology, Seewiesen, Germany

R. R. Warner
University of California, Santa Barbara, CA, USA

E. Warrant
University of Lund, Lund, Sweden

R. Watt
University of Edinburgh, Edinburgh, Scotland, UK

J. P. Webster
Imperial College Faculty of Medicine, London, UK

M. Webster
Cornell Lab of Ornithology, Ithaca, NY, USA

N. Wedell
University of Exeter, Penryn, UK

E. V. Wehncke
Biodiversity Research Center of the Californias, San Diego, CA, USA

M. J. West-Eberhard
Smithsonian Tropical Research Institute, Costa Rica

G. Westhoff
Tierpark Hagenbeck gGmbH, Hamburg, Germany

C. J. Whelan
Illinois Natural History Survey, University of Illinois at Chicago, Chicago, IL, USA

A. Whiten
University of St. Andrews, St. Andrews, Fife, Scotland, UK

A. Wilkinson
University of Virginia, Charlottesville, VA, USA

D. M. Wilkinson
Liverpool John Moores University, Liverpool, UK

S. P. Windsor
University of Auckland, Auckland, New Zealand

J. C. Wingfield
University of California, Davis, CA, USA

K. E. Wynne-Edwards
University of Calgary, Calgary, AB, Canada

D. D. Yager
University of Maryland, College Park,
MD, USA

R. Yamada
University of Queensland, Brisbane, QLD,
Australia

J. Yano
University of Vermont, Burlington, VT, USA

K. Yasukawa
Beloit College, Beloit, WI, USA

J. Zeil
Australian National University, Canberra, ACT,
Australia

T. R. Zentall
University of Kentucky, Lexington, KY, USA

E. Zou
Nicholls State University, Thibodaux, LA, USA

M. Zuk
University of California, Riverside, CA, USA

GUIDE TO USE OF THE ENCYCLOPEDIA

Structure of the Encyclopedia

The material in the Encyclopedia is arranged as a series of articles in alphabetical order.

There are four features to help you easily find the topic you're interested in: an alphabetical contents list, a subject classification index, cross-references and a full subject index.

1. Alphabetical Contents List

The alphabetical contents list, which appears at the front of each volume, lists the entries in the order that they appear in the Encyclopedia. It includes both the volume number and the page number of each entry.

2. Subject Classification Index

This index appears at the start of each volume and groups entries under subject headings that reflect the broad themes of Animal Behavior. This index is useful for making quick connections between entries and locating the relevant entry for a topic that is covered in more than one article.

3. Cross-references

All of the entries in the Encyclopedia have been extensively cross-referenced. The cross-references which appear at the end of an entry, serve three different functions:

i. To indicate if a topic is discussed in greater detail elsewhere
ii. To draw the readers attention to parallel discussions in other entries
iii. To indicate material that broadens the discussion

Example

The following list of cross-references appears at the end of the entry Landmark Studies: Honeybees

See also: Communication: Social Recognition; Invertebrate Social Behavior: Ant, Bee and Wasp Social Evolution; Invertebrate Social Behavior: Caste Determination in Arthropods; Invertebrate Social Behavior: Collective Intelligence, Invertebrate Social Behavior: Dance Language; Invertebrate Social Behavior: Developmental Plasticity; Invertebrate Social Behavior: Division of Labor; Invertebrate Social Behavior: Queen-Queen Conflict in Eusocial Insect Colonies; Invertebrate Social Behavior: Queen-Worker Conflicts Over Colony Sex Ratio.

4. Index

The index includes page numbers for quick reference to the information you're looking for. The index entries differentiate between references to a whole entry, a part of an entry, and a table or figure.

5. Contributors

At the start of each volume there is list of the authors who contributed to the Encyclopedia.

SUBJECT CLASSIFICATION

Anti-Predator Behavior

Section Editor: *Ted Stankowich*

Antipredator Benefits from Heterospecifics
Co-Evolution of Predators and Prey
Conservation and Anti-Predator Behavior
Defensive Avoidance
Defensive Chemicals
Defensive Coloration
Defensive Morphology
Ecology of Fear
Economic Escape
Empirical Studies of Predator and Prey Behavior
Games Played by Predators and Prey
Group Living
Life Histories and Predation Risk
Predator Avoidance: Mechanisms
Parasitoids
Predator's Perspective on Predator–Prey Interactions
Risk Allocation in Anti-Predator Behavior
Risk-Taking in Self-Defense
Trade-Offs in Anti-Predator Behavior
Vigilance and Models of Behavior

Applications

Section Editor: *Michael D. Breed* and *Janice Moore*

Conservation and Animal Behavior
Robot Behavior
Training of Animals

Arthropod Social Behavior

Section Editor: *Jae Chun Choe*

Ant, Bee and Wasp Social Evolution
Caste Determination in Arthropods
Collective Intelligence
Colony Founding in Social Insects
Crustacean Social Evolution
Dance Language

Developmental Plasticity
Division of Labor
Kin Selection and Relatedness
Parasites and Insects: Aspects of Social Behavior
Queen–Queen Conflict in Eusocial Insect Colonies
Queen–Worker Conflicts Over Colony Sex Ratio
Recognition Systems in the Social Insects
Reproductive Skew
Sex and Social Evolution
Social Evolution in 'Other' Insects and Arachnids
Spiders: Social Evolution
Subsociality and the Evolution of Eusociality
Termites: Social Evolution
Worker–Worker Conflict and Worker Policing

Behavioral Endocrinology

Section Editor: *John C. Wingfield*

Aggression and Territoriality
Aquatic Invertebrate Endocrine Disruption
Behavioral Endocrinology of Migration
Circadian and Circannual Rhythms and Hormones
Communication and Hormones
Conservation Behavior and Endocrinology
Experimental Approaches to Hormones and Behavior: Invertebrates
Female Sexual Behavior and Hormones in Non-Mammalian Vertebrates
Field Techniques in Hormones and Behavior
Fight or Flight Responses
Food Intake: Behavioral Endocrinology
Hibernation, Daily Torpor and Estivation in Mammals and Birds: Behavioral Aspects
Hormones and Behavior: Basic Concepts
Immune Systems and Sickness Behavior
Invertebrate Hormones and Behavior
Male Sexual Behavior and Hormones in Non-Mammalian Vertebrates
Mammalian Female Sexual Behavior and Hormones
Maternal Effects on Behavior
Memory, Learning, Hormones and Behavior

Cognition

Section Editor: *Nicola Clayton*

Communication

Section Editor: *Jeffery Lucas*

Conservation

Section Editor: *Constantíno Macías Garcia*

Decision Making by Individuals

Section Editor: *David W. Stephens*

Evolution

Section Editor: *Joan M. Herbers*

Túngara Frog: A Model for Sexual Selection and
    Communication
Turtles: Freshwater
White-Crowned Sparrow
Wolves
Zebra Finches
Zebrafish

Learning and Development

Section Editor: *Daniel Papaj*

Costs of Learning
Decision-Making and Learning: The Peak Shift
    Behavioral Response
Habitat Imprinting
Mate Choice and Learning
Play
Spatial Memory

Methodology

Section Editor: *James Ha*

Cost–Benefit Analysis
Dominance Relationships, Dominance Hierarchies and
    Rankings
Endocrinology and Behavior: Methods
Ethograms, Activity Profiles and Energy Budgets
Experiment, Observation, and Modeling in the Lab and
    Field
Experimental Design: Basic Concepts
Game Theory
Measurement Error and Reliability
Neuroethology: Methods
Playbacks in Behavioral Experiments
Remote-Sensing of Behavior
Robotics in the Study of Animal Behavior
Sequence Analysis and Transition Models
Spatial Orientation and Time: Methods

Migration, Orientation, and Navigation

Section Editor: *Sidney Gauthreaux*

Amphibia: Orientation and Migration
Bat Migration
Bats: Orientation, Navigation and Homing
Bird Migration
Fish Migration
Insect Migration
Insect Navigation
Irruptive Migration

Magnetic Compasses in Insects
Magnetic Orientation in Migratory Songbirds
Maps and Compasses
Migratory Connectivity
Pigeon Homing as a Model Case of Goal-Oriented
    Navigation
Sea Turtles: Navigation and Orientation
Vertical Migration of Aquatic Animals

Networks – Social

Section Editor: *Deborah M. Gordon*

Consensus Decisions
Disease Transmission and Networks
Group Movement
Life Histories and Network Function
Nest Site Choice in Social Insects

Neuroethology

Section Editor: *Harold Zakon*

Acoustic Communication in Insects: Neuroethology
Bat Neuroethology
Crabs and Their Visual World
Insect Flight and Walking: Neuroethological Basis
Leech Behavioral Choice: Neuroethology
Naked Mole Rats: Their Extraordinary Sensory World
Nematode Learning and Memory: Neuroethology
Neuroethology: What is it?
Parasitoid Wasps: Neuroethology
Predator Evasion
Sociogenomics
Sound Localization: Neuroethology
Vocal–Acoustic Communication in Fishes:
    Neuroethology

Reproductive Behavior

Section Editor: *Patricia Adair Gowaty*

Bateman's Principles: Original Experiment and Modern
    Data For and Against
Compensation in Reproduction
Cryptic Female Choice
Differential Allocation
Flexible Mate Choice
Forced or Aggressively Coerced Copulation
Helpers and Reproductive Behavior in Birds and
    Mammals
Infanticide

CONTENTS

VOLUME 1

A

B

C

D

E

F

VOLUME 2

G

H

I

K

L

M

VOLUME 3

Q

R

T

U

V

W

Z

A

Acoustic Communication in Insects: Neuroethology

G. S. Pollack, McGill University, Montréal, QC, Canada

Introduction

Until the advent of traffic, industry, and other noisy human activities, the dominant biogenic source of sound on earth was insects, and this remains the case in some parts of the world and, indeed, at some times of the day. It is perhaps not surprising that insects, being arthropods, are sound producers; almost any movement is likely to bang one part of the hard exoskeleton against another, with acoustical consequences. However, insects have gone far beyond these incidental sounds, evolving specialized mechanisms to amplify, tune, and broadcast acoustic signals both through the air and through the substrate, as well as sensory systems to detect them. There are two main reasons why insects produce sounds: to startle or warn off predators, and to communicate with other members of their species. This review focuses on the latter, focusing on what is being communicated, how the signals are produced, and how they are detected and analyzed by their recipients.

The familiar usage of the word 'sound' refers to a periodic variation in air pressure, which is detected by a pressure-sensitive receiver such as an eardrum or microphone diaphragm. Sound in this sense is indeed used for communication by insects, but they also use two other, related, types of signal. Close to a sound source, the molecules of air move back and forth coherently as the radiating sound source vibrates; in essence, this is rapidly oscillating wind. Because this only occurs close to the source (where 'close' means within a fraction of a wavelength), this is referred to as 'near-field' sound. Many insects communicate using near-field signals, which they detect with a variety of structures that are sensitive to the moving air mass. The coherence of air movement decays rapidly with distance, so communication with near-field signals is necessarily rather intimate, with the distance between the sender and the receiver typically only a few millimeters. Insects also communicate with substrate-borne vibrations that propagate from sender to the receiver through plant leaves and stems, as well as through the ground itself. This mode of signaling is covered in the article by Cocroft, and thus, will not be considered here.

The Functions of Acoustic Communication

Mate Attraction

The insect sounds that are most conspicuous to humans, such as the chirps of crickets, the rattlings of grasshoppers, and the squawks of cicadas, are love songs produced by individuals, mainly males, seeking mates. These signals are detectable at long distances from the singer, typically tens of meters and, in extreme cases, up to a kilometer. They declare the presence and location of a potential mate. The listener, typically a female, responds by walking, hopping, or flying toward the singer, a behavior known as phonotaxis. These sex-specific roles of sender and receiver are not universal. In some grasshoppers and katydids, potential mates sing a duet, in which a female answers a male's song with her own. One or both partners may then approach the other, guided by acoustical beacons.

Courtship

There is more to reproduction than simply finding a potential mate: actual mating is also required, and in many insects, this too depends on acoustical signaling. Male crickets, once having attracted a female from a distance, woo her with a distinct song that elicits copulation. Similarly, fruit-fly males court nearby females with a near-field song that they produce by vibrating a wing. Courtship signals operate at close range, after male and female have come together, and typically are only one component of a multimodal display that may include chemical, visual, tactile, and vibrational signals.

Territoriality

This has been studied most thoroughly in crickets and katydids, where the same long-range signal used to attract females also serves to adjust spacing between neighboring males. Crickets also engage in fights that include a distinct acoustic signal, aggression song.

Mechanisms of Sound Production

Insects produce sounds using a variety of mechanisms, including forcing air through specialized spiracles (whistling) in Madagascar giant cockroaches, rubbing series of cuticular 'teeth' against hardened cuticular scrapers in crickets, katydids, grasshoppers, and many other groups (stridulation), and buckling of ribs of cuticle in cicadas (think of 'clicker' toys). Although the acoustical mechanisms are diverse, all depend on the coordinated contraction of various muscles. In those cases that have so far been studied (crickets and grasshoppers), the timing and patterning of activation of motor neurons, and therefore of muscles, is determined by circuits of nerve cells within the central nervous system that are known as Central Pattern Generators (CPGs). These neurons are situated in the thoracic ganglia of the ventral nerve cord, which are the centers controlling movements of the legs and wings. The CPGs are turned on or off by specific nerve cells in the brain, called 'command neurons,' which communicate with the CPGs via axons that project from the brain to the thorax. When a cricket or grasshopper 'decides' to sing, it activates a command neuron that turns on the thoracic CPG. This type of hierarchical neural organization mirrors that found in a variety of animals, for other rhythmic motor behaviors such as walking and flying.

The Information Content of Acoustic Signals

Acoustic signals can inform their recipients of the species identity and fitness of the signaler, as well as of the signaler's intentions: for example, whether the signal is an invitation to mate or a threat of impending aggression. This information is represented by the physical structure of the signal, chiefly its temporal pattern; insect songs are generally not melodious, and the information they contain is carried by rhythm rather than tune. Songs consist of a series of discrete sound pulses; features such as the durations of the pulses and the intervals between them, and their higher-order groupings into chirps, trills, and phrases, tend to be species-specific. Recipients of the signals are 'tuned in' to the rhythmic parameters of their own species' songs, responding (e.g., with phonotaxis) only to stimuli that match the 'correct' signal parameters.

When communication is long-distance, the receiver has not only to decode the message but also to localize its source. Sound pressure is a scalar quantity that contains no information about the location of the source. Rather, this must be deduced from determining the direction in which sound is traveling, which in turn is based on the comparison of sounds at the two ears. The potential physical cues available for sound localization are binaural differences in amplitude and timing. Amplitude differs at the two ears because of the diffraction of sound around whatever separates them (for us, this is our heads), and the timing differs, because if the sound source is anywhere off the mid-saggital plane, the travel distance, and thus, the travel time, to the two ears will differ. Large organisms like us generate substantial binaural differences in both amplitude and timing, but this is difficult for insects because their small size limits both sorts of binaural difference. Insects have gotten around this constraint by evolving a variety of mechanical tricks that magnify the minute physical differences available at the ears. The details of how they do this are beyond the scope of this article, but the effectiveness of this strategy is demonstrated by the fact that a contender for the title of the world's best sound localizer is a fly, the two ears of which are separated by only 0.5 mm!

Neurobiology of Hearing

Ears

Hearing has evolved independently more than 20 times in insects, resulting in an astonishing diversity of ears. Pressure-sensitive hearing organs occur on the legs of crickets and tettigonnids, on the wings of lacewings, on the abdomens of grasshoppers, on the necks of beetles, on the throats of flies, on the chests of mantises, and on the chests, abdomens, or mouths of moths. Near-field-sensitive organs may be simple sensory hairs, or elaborate antenna. Despite the enormous variation in ear structure and location, the underlying cellular machinery is highly conserved. In all insect ears, except those consisting of sensory hairs, mechanical energy is transduced into neurophysiological activity by multicellular sensory structures known as scolopidia. Each scolopidium includes one or more sensory neurons that are in close association with several accessory cells that serve to anchor the scolopidium within the body and to couple it to the source of mechanical input, for example, an eardrum. Insect ears may contain as few as one scolopidium (in some moths), or as many as several thousand (in some cicadas).

The task of the sensory periphery is to translate behaviorally relevant features of acoustic signals, such as the temporal pattern of sound pulses, sound intensity and, in some cases, sound frequency, into sequences of action potentials in sensory neurons. Although many insects are tone deaf, others (crickets, grasshoppers, and cicadas) are equipped with ears in which different sensory neurons are 'tuned' to different sound frequencies, so that, in principle, the spectrum of the signal can be determined. This capacity allows crickets to discriminate between the relatively low sound frequencies in cricket songs and the ultrasonic frequencies of bat echolocation calls, as well as, in some species, spectral differences between long-range, mate attracting songs, and short-range courtship songs. Frequency discrimination also plays a role in communication between duetting grasshoppers, where song spectra differ between males and females. Another possible role of frequency-tuned neurons is capturing information about the distance to the singer. As sound travels through the environment, high frequencies are attenuated more severely than low frequencies because they can be reflected by smaller objects, such as blades of grass, and also because frictional loss of energy to the air increases with sound frequency. The extent of this environmental low-pass filtering, thus reflects how far the signal has traveled between the emitter and the receiver. Frequency tuning is also useful in tone-deaf insects. Although the different sensory neurons in their ears have similar tuning, they are nevertheless tuned, that is, they are more sensitive to some sound frequencies, usually those contained in behaviorally relevant signals, than to others. Thus, they serve as filters that selectively attenuate irrelevant sounds, thereby enhancing the signal-to-noise ratio for what matters.

As pointed out earlier, the information content of insect songs is encoded mainly in their rhythms, and this is generally represented faithfully in the responses of receptor neurons. Auditory receptor neurons tend to produce action potentials at a rather high rate, typically on the order of $300–500\,s^{-1}$, and in some cases, approaching $1000\,s^{-1}$. For the majority of insect sounds, in which the durations of sound pulses and of the intervals between them are on the order of tens of milliseconds, each sound pulse is answered by a burst of action potentials that lasts about as long as the stimulus, so that the duration and spacing of sound pulses can be read off rather easily from the responses of receptor neurons. Some insects, however, produce sound pulses at rates of $100\,s^{-1}$ or more (the individual pulses necessarily being brief). In these cases, each pulse may be marked by only a single action potential, so that information about pulse duration is lost. And, although receptor neurons respond to stimuli having a wide range of temporal patterns, the precision with which they do so may vary according to the structure of the stimulus. For example, receptor neurons of grasshoppers mark rapid sound-pulse onsets, such as occur in grasshopper songs, with remarkable precision. The timing of the first action potential with respect to the beginning of the sound pulse may vary, from trial to trial, by as little as 0.15 ms. As discussed below, the relative timing of responses at the two ears can serve as a cue for sound direction, providing a possible behavioral role for this exquisite temporal acuity.

In addition to capturing information about when sound pulses happen, the sensory periphery must encode the intensity of the signal for several reasons. First, sound intensity is the main cue for distance to the singer, a parameter that is of interest both to females, who might be willing to risk a short trip to a potential mate but not a long one, and to males, who might want to distance

themselves from potential rivals. Moreover, larger, stronger, individuals sing more loudly, so that the intensity can also be a clue to the fitness of the singer. Second, the difference in intensity at the two ears is the primary cue for sound location. As mentioned earlier, insect auditory systems embody mechanical specializations that can generate substantial interaural differences in energy input to the eardrums, despite the small size of the insect. Stimulus intensity is encoded by receptor neurons in three ways. First, the different receptor neurons typically differ in sensitivity; some are able to respond to weak stimuli, while others respond only to stronger stimuli. The number of neurons that respond to a signal, thus increases with increasing intensity as the less sensitive neurons are recruited. Second, once the threshold has been reached for a given neuron, the firing rate increases with sound intensity. Third, response latency decreases as intensity increases. Latency per se can be determined only by reference to an independent measure of when the stimulus occurs, which is not available to the insect (all it knows is what its receptor neurons tell it). However, interaural difference in latency can serve as a measure of interaural intensity difference, and thus, of sound direction. In behavioral tests (of crickets and grasshoppers) where the timing and intensity of sounds at the two ears are controlled separately, insects will turn toward the side where a sound stimulus is presented earlier (with no difference in intensity), and toward the side where the intensity is greatest (with no difference in timing). Thus, either the interaural difference in response timing or the difference in response magnitude (the number of responding neurons and their firing rates) can serve as a localization cue when it is presented alone. Under natural conditions, however, these two cues change in concert and thus, reinforce one another.

Extraction of Information in the Central Nervous System

Most of what we know about central processing comes from studies on three groups of insects in the order Orthoptera: crickets, grasshoppers, and katydids. These insects have garnered the lion's share of attention for a number of reasons. First, with the notable exception of cicadas, it is their songs that are most conspicuous to humans. Second, presumably because of this conspicuousness, they have been the focus of a large number of behavioral studies (which are less convenient with cicadas because of their arboreal life styles). Finally, these insects are relatively large, which is an advantage for neurophysiological studies.

Early processing: thoracic interneurons
Auditory receptor neurons terminate in the thoracic ganglia of the central nervous system, where they provide synaptic input to interneurons. Within a species, specific

interneurons, which are recognizable by the nature of their responses to sound stimuli and by their anatomy (which was revealed by injecting single neurons with dyes), can be found in specimen after specimen. Many of these 'identified neurons' can even be recognized in related species. For example, a number of specific interneurons have the same morphology and response properties in several grasshopper species, and the same is true for several cricket interneurons. A few interneurons can even be recognized as 'the same' across wider taxonomic divides, for example, in crickets and katydids, or grasshoppers and mantises.

The number of interneurons participating in this first stage of central processing varies from group to group, but in general, is rather small. In grasshoppers, about 20 distinct bilateral pairs of neurons have been identified; the corresponding number for crickets and katydids is about a dozen. Of course, the possibility that additional neurons are yet to be discovered, cannot be ruled out; however, all three insect groups have been studied intensively for more than 30 years, suggesting that most of the neurons have probably been described. In all three groups, some neurons seem specialized, by virtue of their sensitivity to ultrasound and large-diameter, rapidly conducting axons for mediating escape responses from echolocating bats, whereas others seem well suited for processing communication signals. In grasshoppers, different interneurons extract different behaviorally relevant features of the signal, dividing its information into parallel channels. For example, some neurons accurately mark the timing of sound pulses, but are not sensitive to stimulus direction, whereas others are highly directionally sensitive, but represent stimulus timing only poorly. Presumably, these separate channels are brought together in the brain. By contrast, in crickets, information about sound-pulse timing and stimulus direction is combined in the responses of a single bilaterally paired neuron that represents the only, or at least the principle, channel for transmitting information about communication signals to the brain. This difference in the 'strategy' for representing communication signals between grasshoppers and crickets, may reflect the different conditions under which hearing arose in these groups. In the grasshoppers hearing is believed to have evolved in the context of predator detection, whereas the primitive function of hearing in crickets is believed to have been communication.

A common feature of all of these insect groups is the enhancement of directional information through contralateral inhibition. Neurons that receive input from one ear are inhibited by one or more neurons that receive input from the other ear, thus amplifying through central neural circuitry the binaural differences that are generated peripherally.

The response properties of some of the thoracic interneurons show clear correlations with the behavioral requirements of communication. For example, a particular interneuron in grasshoppers (named AN4) responds to models of grasshopper song only if they do not contain

silent gaps of ~2 ms or more. Such gaps arise under natural conditions when a male has lost one of his two hind legs, presumably in an encounter with a predator. The song is produced by up-and-down scraping movements of the hind legs against the folded wings. A brief period of silence occurs with each reversal in leg direction, but this is obscured in intact males because the two legs are slightly out of phase, ensuring that periods of leg reversal do not coincide. In one-legged males, however, the gaps are revealed. Field studies show that one-legged males have poor mating success, and in laboratory tests, females fail to respond to gap-containing songs. Gap detection by females, thus allows them to avoid mating with males that are presumably less fit than those that were able to avoid contact with predators. The gap-sensitivity of females was quantified by measuring the probability of their entering into a duet with computer-generated test songs having gaps of various lengths. This parallels exactly the gap-sensitivity of AN4, strongly suggesting a role for this neuron in the female's assessment of song (and male) quality (**Figure 1**).

In crickets, specific interneurons (named AN1 and ON1) that respond to stimuli with cricket-like sound frequency do so for a wide range of temporal patterns. However, the timing of action potentials accurately reflects the temporal structure of the stimulus (which behavioral experiments have shown is the basis for song recognition) only for the narrow range of sound-pulse

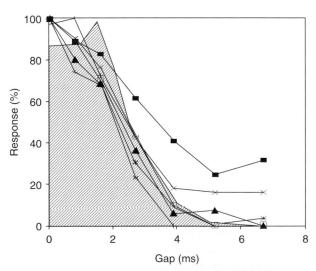

Figure 1 Gap-sensitivity in grasshoppers. The shaded area shows the probability of female responses to model songs containing gaps of varying durations. The lines show the normalized number of action potentials produced by AN4 in response to the same song models. The close correspondence between the neural and behavioral response functions suggests that AN4 plays a role in determining the female's response. Reproduced from Stumpner A and Ronacher B (1994) Neurophysiological aspects of song pattern recognition and sound localization in grasshoppers. *American Zoologist* 34: 696–705, with permission from Oxford University Press.

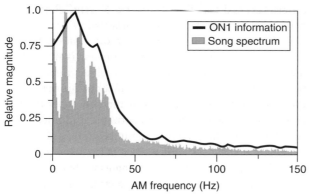

Figure 2 Temporal tuning in a thoracic interneuron of crickets. The shaded area shows the rates of amplitude modulation that occur in the songs of *Teleogryllus oceanicus*. The line shows the accuracy with which the pattern of amplitude modulation is captured by the timing of action potentials of the interneuron ON1. The close match indicates that the properties of the neuron are matched to the structure of the species' communication signals. Modified from Marsat G and Pollack GS (2004) Differential temporal coding of rhythmically diverse acoustic signals by a single interneuron.

rates that occur in the species' songs (**Figure 2**). Moreover, when response properties of ON1 were compared between species that sing with different tempos, the range of pulse rates that are coded accurately was higher in the faster-singing species (this has not yet been studied for AN1), again revealing a striking correlation between neuronal properties and behavioral requirements.

Correlations like those just described make a strong case for the involvement of specific neurons in behavior, but this can be shown conclusively only by manipulating the responses of neurons and studying the behavioral consequences of this. So far this has been only for crickets, where responses of AN1 and ON1 have been suppressed, during ongoing behavior, by manipulating their membrane potentials through intracellular current injection. Crickets were restrained in a manner that allowed them to turn an air-supported ball beneath their feet, and their turning behavior could be deduced from the movements of the ball. Under these conditions, they turn toward the side from which an attractive stimulus is played. However, when AN1 on one side was impaled with a microelectrode and negative current was injected to suppress its response, crickets turned toward the opposite side, no matter which side the sound was played from. For example, when the left AN1 was 'turned off' experimentally, crickets turned to the right even if the sound was played from the left. Similarly, suppressing the response of one of the ON1s biased walking direction toward the opposite side. Here, the effect was mediated through the removal of the contralateral inhibition that ON1 provides to the opposite AN1; turning off the left ON1 allows the right AN1 to respond more strongly, favoring turning to the right.

These experiments, thus demonstrate that information carried to the brain by AN1 is indeed instrumental in shaping the cricket's behavioral response.

Processing in the brain

The results of thoracic processing are relayed to the brain by the axons of ascending neurons. It is there that the 'decision' of whether or not to respond to a signal is made and, if the response is phonotaxis, in which direction to walk or fly. Although many brain neurons have been recorded, the circuits that they form, and the mechanisms by which they process acoustic information, are not well understood. Most of the emphasis has been on how selectivity for temporal pattern arises. Two main ideas are at the forefront, one supported by experiments on crickets, and the other by experiments on katydids. The first idea is that neurons in the brain function as rate filters; some, denoted low-pass, respond best only to low sound-pulse rates, and others, called high-pass, only to high pulse rates, and indeed neurons with these properties have been recorded in the brains of crickets. The cellular mechanisms underlying these filtering properties are not yet known, but standard rate-dependent processes, such as synaptic depression and facilitation, could, in principle, do the job. The most interesting class of brain neurons, called band-pass, respond well only to a middle range of pulse rates that corresponds to the pulse-rate selectivity of females in behavioral tests. In principle, these properties could arise if the band-pass neurons respond strongly only when receiving simultaneous input from both low-pass and high-pass neurons, the responses of which overlap only in the behaviorally relevant range of pulse rates (**Figure 3**). Whether this circuit actually occurs, and whether the band-pass neurons are indeed responsible for behavioral selectivity, remains to be shown experimentally. However, the close match between the filter properties of the neurons and behavioral selectivity is intriguing.

The second idea for pulse-rate selectivity posits a mechanism which does not require rate filters in the usual sense. Instead, selectivity is hypothesized to arise from interactions between acoustically driven excitation and intrinsic oscillation of a neuronal resonator. According to this model, arrival of a stimulus sets a neuron, or neural circuit, into oscillation, such that excitability alternately increases and decreases with a periodicity that matches the behaviorally preferred pulse rate (ionic and circuit mechanisms for producing intrinsic oscillations are well known in nervous systems). If auditory input arrives at the 'correct' pulse rate, then, input from successive sound pulses would coincide with successive cycles of the intrinsic oscillation, so that the summation of acoustically driven and intrinsic excitation would bring the oscillating neuron above the threshold. Input from improperly timed sound pulses would 'miss' the intrinsic peak in excitation, and thus, would be ineffective. The output of the system is maximal only when the input pulse rate

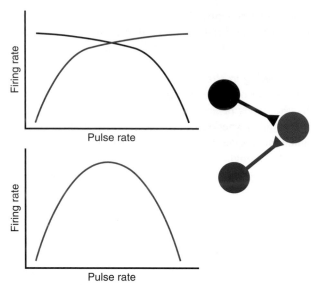

Figure 3 Temporal filters for pulse-rate selectivity. The red curve represents a high-pass neuron, that is, one that responds well only to high pulse rates, the blue curve represents a low-pass neuron, and the purple curve represents a band-pass neuron. The circuit at right is composed of the corresponding neurons; the purple neuron responds strongly only when it receives simultaneous input from the blue and red neurons. This happens only for the middle range of pulse rates, where their pass bands overlap. Reproduced from Pollack GS, Krahe R (2009) Signal identification: Peripheral and central mechanisms. In: Squire LR (ed.) *Encyclopedia of Neuroscience*, pp. 799–804. Oxford: Academic Press.

matches the intrinsic resonance frequency, in a manner analogous to the importance of timing when pushing a child on a playground swing. A prediction of the model is that the output should be rather high not only for the 'correct' stimulus rate, but also at integer submultiples of this. This is because pulses arriving at half the correct rate would coincide with every second cycle of the intrinsic oscillation; pulses arriving at one-third the correct rate with every third cycle, etc. (again, think of the swing analogy). Behavioral experiments with katydids showed that phonotaxis of females did indeed have response peaks for stimuli with one-half and one-third the species-typical pulse rate, a finding that cannot be explained by the rate-filter mechanism described earlier.

It is at first glance surprising that two quite different mechanisms for pulse-rate selectivity have been proposed

for crickets and katydids. These are closely related taxa, with similar methods of sound production (stridulation with the forewings) and similar auditory systems. In both crickets and katydids, for example, the ear is situated on the prothoracic tibia, and some identified interneurons (ON1, AN1) can be recognized in both groups. Some evolutionary biologists have proposed that acoustic communication arose in a common ancestor, but others have challenged this, holding that communication arose independently in these two groups. If the latter view is correct, then their use of different mechanisms for pulse-rate recognition might be less surprising than it appears.

See also: Insect Flight and Walking: Neuroethological Basis; Predator Evasion.

Further Reading

Hedwig B (2006) Pulses, patterns and paths: Neurobiology of acoustic behaviour in crickets. *Journal of Comparative Physiology A:* 192: 677–689.

Hedwig B and Pollack GS (2008) Invertebrate auditory pathways. In: Dallos P and Oertel D (eds.) *The Senses: A Comprehensive Reference: Audition,* vol. 3, pp. 525–564. Amsterdam: Elsevier.

von Helversen D and von Helversen O (1983) Species recognition and acoustic localization in acridid grasshoppers: A behavioral approach. In: Huber R and Markl H (eds.) *Neuroethology and Behavioral Physiology*, pp. 95–107. Heidelberg: Springer Verlag.

Hennig RM, Franz A, and Stumpner A (2004) Processing of auditory information in insects. *Microscopy Research and Technique* 63: 351–374.

Mason AC and Faure PA (2004) The physiology of insect auditory afferents. *Microscopy Research and Technique* 63: 338–350.

Pollack GS (2000) Who, what, where? Recognition and localization of acoustic signals by insects. *Current Opinion in Neurobiology* 10: 763–767.

Pollack GS and Imaizumi K (1999) Neural analysis of sound frequency in insects. *BioEssays* 21: 295–303.

Pollack GS and Krahe R (2009) Signal identification: Peripheral and central mechanisms. In: Squire LR (ed.) *Encyclopedia of Neuroscience*, pp. 799–804. Oxford: Academic Press.

Schildberger K (1994) The auditory pathway of crickets: Adaptations for intraspecific acoustic communication. In: Schildberger K and Elsner N (eds.) *Neural Basis of Behavioural Adaptations*, pp. 209–226. Stuttgart: Gustav Fischer.

Stumpner A and Helversen D (2001) Evolution and function of auditory systems in insects. *Naturwissenschaften* 88: 159–170.

Stumpner A and Ronacher B (1994) Neurophysiological aspects of song pattern recognition and sound localization in grasshoppers. *American Zoologist* 34: 696–705.

Yack JE (2004) The structure and function of auditory chordotonal organs in insects. *Microscopy Research and Technique* 63: 315–337.

Acoustic Signals

A. M. Simmons, Brown University, Providence, RI, USA

Introduction

Many species of animals use sounds to guide their behavior. Knowledge of the physical acoustic properties of the sounds made by animals in particular biological contexts can provide us with a window through which we can understand their behavior. Sounds convey such biologically relevant information as location (of food, of a mate, of a predator, of a prey, of an interesting object in the environment), identity (species, gender, individual), social status (dominant or submissive), motivation (fear, anger, willingness to mate), and even the animal's cognitive processes (in some species, different types of sounds are used to categorize different kinds of predators). The sounds used for these various purposes are typically complex in structure, but they usually can be described as being comprised of several simple sounds added together. Animals can control the information content of signals by actively changing their acoustic properties. The environment through which the sounds propagate also has a major influence on the properties of the sound as it arrives at the receiver. The environment can either aid transmission of the information carried by the sound or degrade the information by distorting the sound.

A great deal of the research on the use of sounds by animals attempts to uncover what specific properties of acoustic signals are used for guiding particular behaviors. Animals use a wide variety of sounds in their acoustic signals. Nonetheless, all sounds are governed by the same physical principles. The focus of this article is to provide descriptions of the important physical properties present in natural sounds, and how these properties are influenced by the environment.

Physical Properties of Simple Sounds

Sound is a physical disturbance in some medium (air or water) produced by the displacement of molecules as a result of mechanical action. In response to this physical disturbance, whether produced by a larynx, a loudspeaker, or a musical instrument, molecules in the medium are moved alternately closer together and farther apart around their equilibrium position (the resting air or water pressure). These cyclic patterns of inward and outward movement are called condensations (molecules move closer together) and rarefactions (molecules move farther apart). Very close to the source of the disturbance,

the molecules are physically displaced from their resting position in a flow called particle motion, which is the near-field component of sound. This is the 'wind' felt by sitting close to a large diameter bass loudspeaker in action. The cycles of condensations and rarefactions also propagate away from the mechanical source through the medium as a pressure wave, with no net flow. At distances from the source greater than about one to three wavelengths (see section 'Frequency, period, and wavelength'), the pressure wave, or far-field component of the sound, predominates over the near-field component. In ordinary circumstances, for acoustic communication at biologically useful distances, signaling is mediated by propagating pressure waves, whereas particle motion is usually thought of as a vibration that can be sensed only at short ranges. However, the relative contributions of pressure and particle motion depend on the nature of the organs for hearing. For most vertebrate animals, the 'ears' are sensitive primarily to the pressure component of sound, as are the ears of many insects. Many arthropods have sense organs that are sensitive primarily to particle displacement. Fishes, frogs, and toads have hearing organs that can detect both pressure and particle-motion components.

A simple sound wave can be visualized as a periodic or sinusoidal motion of instantaneous air or water pressure. A sine wave is graphed in **Figure 1** as cyclic pressure variations occurring over time. The peaks and troughs in the waveform represent the alternating cycles of condensation and rarefaction relative to the resting position (represented by the dashed line at zero amplitude in the graph or the zero-crossing point). There are four physical parameters that, taken together, define a sine wave uniquely. These are its (1) frequency, (2) amplitude, (3) phase, and (4) duration. For humans, frequency and amplitude are associated with two of the primary psychological percepts of sounds – pitch and loudness. The psychological percept of timbre, or sound quality (which makes an A note on an oboe sound different from an A note on a flute), depends on the relations among the amplitudes and phases of different frequencies present in a multiple-frequency sound, which excludes pure tones (single-frequency sinusoidal signals) from having a timbre. The duration of a sound has different effects on human auditory perception depending on its absolute magnitude. Very short sounds are perceived as getting louder as their duration increases, up to a critical time length that is often called the integration time of hearing. Then, as sounds become longer yet, they come to be perceived as

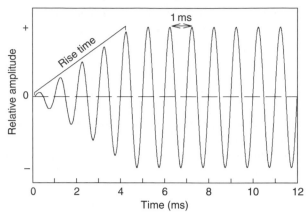

Figure 1 Amplitude–time display of the first 12 ms of a 1000 Hz sine wave, digitally generated by Adobe Audition v. 1.5. The horizontal line at relative amplitude of zero shows the zero-crossing point (ambient or resting pressure). The gradual onset of the sine wave is indicated by the rise time. Its period, as shown by the time interval between two consecutive peaks, is 1 ms.

longer, but not louder. It is important to appreciate that these psychological percepts, based on human auditory experience, are not simply equivalent to the physical parameters of frequency, amplitude, phase, and duration. For this reason, these psychological terms should be avoided when discussing animals' perceptions of their own sounds. Nonetheless, animals do behaviorally discriminate among sounds varying in these four physical parameters, and often the behavioral data obtained from experiments with animals resemble those obtained from human listening experiments where judgments of pitch, loudness, or timbre are made. One important aspect of research on animal auditory perception is to examine the potential equivalence of these behavioral similarities for evidence concerning the brain mechanisms involved in creating human auditory percepts.

Frequency, Period, and Wavelength

Frequency (f) is the number of times the sound repeats within a certain time interval. It is measured as Hertz, cycles per second (s), or as kiloHertz, thousands of cycles per second. When a sinusoid has completed one full cycle of vibration (it ends on the same point of displacement at which it has begun), it has traveled one complete cycle. For example, if a sinusoid has completed 1000 complete cycles in one second, it is said to have a frequency of 1000 Hz or 1 kHz. The amount of time taken to complete one cycle is called the period of the sound. Period (T) is the reciprocal of frequency ($T = 1/f$). The period is typically measured in milliseconds (ms), or thousandths of a second (1 s = 1000 ms). The relationship between frequency and period is important, because we do not know if an animal perceives sounds in terms of frequency (number of cycles) or period (cycle length), or both. The sine

wave in **Figure 1** has a period of 1 ms and a frequency of 1000 Hz.

The range of frequencies to which a species is sensitive is called its frequency range of hearing. Frequency range of hearing varies dramatically among different species. The human range of hearing lies between 20 and 20 000 Hz. Frequencies higher than the upper limit of the human audible range are called ultrasonic, whereas frequencies lower than the human range are called infrasonic. Cetaceans (dolphins and other toothed whales) and bats can hear higher frequencies than humans, up to about 150 000 Hz. These ultrasonic frequencies are used for biological sonar, or echolocation. Most species of songbirds cannot hear frequencies above 10 000 Hz, although the acuity of their hearing above 6000 Hz is usually too weak to use these higher frequencies for song. In contrast, owls typically can hear up to 12 000 Hz. Elephants can hear sounds at frequencies lower than the 20 Hz limit of human hearing.

The wavelength (λ) of a sound is its most important spatial feature, as distinct from the time features of the waveform shown in **Figure 1**. Wavelength is the distance in space spanned by a single cycle of a sound (i.e., from the maximum condensation or peak of a cycle to the maximum condensation or peak of the next cycle). Numerically, it is the ratio of the velocity of sound (m s)$^{-1}$ to the frequency (Hz). Because the speed of sound varies in different media, the wavelengths of sounds will also differ in these media. Thus, a single cycle of a sound wave at a particular frequency will travel farther in the ocean (where the speed of sound is approximately 1500 m s^{-1}, depending on depth, temperature, and salinity) than in air (where the speed of sound is approximately 343 m s^{-1}, depending on altitude, humidity, and temperature). The wavelength of sound in relation to the size and spatial separation of the ears has considerable impact on the ability of animals to locate sound sources in their environments.

Table 1 shows the relationship between frequency, period, and wavelength (in both air and water) of some sine waves that are audible to different species of animals.

Amplitude

Amplitude is the magnitude of the sound pressure change or particle displacement caused by the physical disturbance in the medium. After the sound has been picked up by a transducer such as a microphone or hydrophone, there are two different ways to express amplitude – peak-to-peak pressure or root-mean-square (RMS) pressure. The RMS pressure is obtained by squaring the numerical values of amplitude, which transforms the numbers from amplitude to power (power = pressure^2), and then taking the square-root, which turns the numbers back into amplitudes but also turns the positive and negative numbers of the cycles all into positive numbers. That is, the sine wave is rectified. The use of peak-to-peak

Table 1 Relationships between sound frequency, period, and wavelength for some simple sounds

Frequency (H$_Z$)	Period (ms)	λ (cm) Air	Seawater
100	10	343.3	1500
500	2	68.68	300
1000	1	34.3	150
2000	0.5	17.17	75
8000	0.125	4.29	18.75
12 000	0.08	2.86	12.5
20 000	0.05	1.717	7.5
50 000	0.02	0.6868	3
100 000	0.01	0.343	1.5

versus RMS amplitudes originated in the use of different electronic instruments for visualizing the sound – the oscilloscope for displaying successive cycles, and the sound level meter for registering a longer-term average amplitude. To a large extent, use of these instruments has been supplanted by widely available sound-display and analysis software in laptop computers, but these programs can be used to obtain both types of amplitude measures using the cursors and toolbars associated with their computer displays. The sine wave in **Figure 1** illustrates the utility of both measures. Individual cycles of the pictured waveform register their amplitude to the eye in terms of their height, which is the peak-to-peak sound pressure. This waveform shows the gradual increase in sound amplitude (called the rise time) from the beginning of the sound to its peak value (highest point). Biological and musical sounds usually have a corresponding fall time at the end of the sound, too. The duration of the sound's rise or fall time is related to the mechanics involved in producing the sound, such as laryngeal function. Because the segment of the sound shown in **Figure 1** changes in its amplitude over its duration, no individual cycle is a faithful reflection of the sound's amplitude. The RMS sound pressure represents an average amplitude over the sound's duration, and thus takes into account the gradual rise and fall in amplitude at the beginning and end of the sound. This point is important because few sounds in nature are short enough or stable enough in amplitude for peak-to-peak measurements to reflect the sound's effective amplitude to the receiver in communication. However, both types of amplitude measures are of value in animal bioacoustics. Analysis of the source's mechanics requires tracking these short-term changes in amplitude from peak-to-peak measurements, whereas evaluation of the perceived strength of the sound by the receiver more often involves making RMS measurements to summarize the sound's overall amplitude as a single number.

Numerical values of amplitude are obtained from voltages delivered by the microphone and converted into pressure units called Pascals ($1\,Pa = 1\,N\,m^{-2}$). However, the range of hearing in many animals from the weakest to the strongest biologically relevant sounds can span five or six orders of magnitude, which makes measurements in Pascals very cumbersome to use. For ease of expression, and convenience in thinking about how a sound is heard by the receiver, sound pressure is expressed on a logarithmic scale that goes approximately from zero (weakest sound) to 100 units (strongest sound). In this system, the amplitude of a sound wave is quantified as the ratio between the measured amplitude in Pascals and a reference value, also in Pascals, that represents the average threshold of human hearing at a frequency of 1000 Hz (where human hearing is very sensitive). This resulting unit is the decibel sound pressure level (dB SPL), a ratio of two sound powers, or pressures[2]. To calculate the amplitude of sounds on this scale, $dB = 20\log(p_1/p_2)$, where p_1 is the pressure of the sound to be expressed, and p_2 is the standard, or reference, pressure at the threshold of human hearing. (The factor of 20 squares the pressures to get power, and then divides each single logarithmic step into 10 manageable dB steps). For sounds in air, this reference pressure is 0.00002 Pa (20 µPa), which is an RMS value. A sound measured by the dB scale is stated as having an amplitude of so many dB SPL. Sound level meters typically display their measurements directly in dB SPL, whereas peak to peak measurements initially are in Pascals.

Along with their different frequency ranges of hearing, different animal species have different sensitivities to sound, and they communicate at different sound levels. The lowest sound amplitude audible to an animal at a particular sound frequency is called the threshold. For humans, thresholds lie around 0 dB SPL at their most sensitive frequencies of hearing. At certain frequencies of sounds, some animals (cats, for example) are even more sensitive than humans. Most animals communicate at levels well above their hearing threshold. Typical human conversations are in the range of 60–70 dB SPL, well above the thresholds for the frequencies present in speech sounds. Big brown bats emit their ultrasonic echolocation sounds at levels that can exceed 110–120 dB SPL; their thresholds of hearing at these frequencies are around 0–10 dB SPL. Bats emit loud calls because they need to overcome environmental constraints on sound propagation out to and then back from objects, and because little of the original emitted sound reflects off the small insect-sized objects they are trying to detect with echoes. Insects that can hear the echolocation sounds of approaching bats have hearing thresholds 20–30 dB higher than bats, but they only have to detect the sounds as they travel one way, out from the bat. Besides the intrinsic hearing sensitivity of animals, background sounds from the environment can cause reduced sensitivity by masking communication sounds. Animals, such as birds living near fast-running streams, use high-amplitude signals for communication to enable receivers to detect sounds against the background noise.

Phase

Phase refers to the location of a particular point in time along the condensation or rarefaction cycle, expressed relative to the sinusoidal wave rather than to absolute time. To fit different points at different frequencies into the same shape, phase is expressed as the angle along the sine cycle, not the time itself. Thus, a point at the start of a sine cycle, at an amplitude of zero with subsequent amplitudes going positive, is referred to as 0° phase, while a point half-way along the cycle, where the amplitude again is zero but with subsequent amplitudes going negative, is 180°. Thus, phase can be referred to as beginning- or onset-phase (the phase of the cycle at which the sound begins) or ongoing phase (the phase at some point during the sound with respect to some other event in time). In the sine wave in **Figure 1**, starting phase is 0°. Two different frequencies can have the same amplitude but different starting phases. Phase has biological meaning in two ways. The first concerns the relative phases of different frequencies that are in the same sound, particularly as these are affected by propagation from the source to the receiver. The second concerns differences in the time-of-occurrence of the same sound at two different receptors. In the case of vertebrates, with left and right ears, binaural time or phase differences are powerful cues for localization of sound. Phase differences in stimuli may also play a role in sound localization for animals, such as many fishes, that possess different types of receptors for sound particle velocity and sound pressure.

Physical Properties of Complex Sounds

Biologically produced sounds are typically not individual sine waves or pure tones, but are more often mixtures of tones with different frequencies at different amplitudes and phases. Sounds made up of multiple frequencies are called complex sounds. An example of a complex sounds, made up of five different frequencies (1000, 2000, 3000, 4000, and 5000 Hz), all with a starting phase of 0°, is shown in **Figure 2**. Although this sound has a more complicated structure than the sine wave in **Figure 1**, it still shows a recurring, or periodic, pattern, in which the same set of wave-shapes repeats. In this sound, the time interval, or period, between repeating sets is 1 ms, the same period as the 1000 Hz pure tone shown in **Figure 1**. Thus, although the sound waves shown in **Figures 1** and **2** have the same period, they have a different frequency structure, and they are perceptually distinct. For humans, the period of a complex sound gives a psychological sensation of periodicity pitch, related to the frequency of

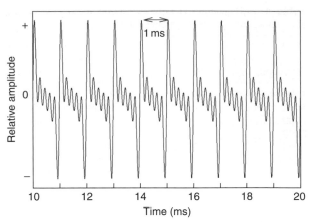

Figure 2 Amplitude–time display of 10 ms of a complex sound with a period of 1 ms. The onset of the sine wave is not shown.

the lower components. These two sounds are perceived as having the same low pitch, but they differ in their timbre. The fact that sounds with such different waveshapes have the same pitch is a perfect illustration of how the pitch of a sound is not simply the psychological equivalent of frequency. The actual wave shape of a complex sound depends on the phase angle of the individual frequency components. The specific structure (the frequencies, amplitudes, and phases of the individual components) of a complex sound is called its fine-structure. For humans, changes in fine-structure are perceived as changes in timbre. The psychological differences between periodicity pitch and timbre are easily explained in musical terms. The same note played by two different instruments can be identified as having the same pitch, even though the instruments can be identified as different because the notes sound different in timbre. The pitch of successive notes, nevertheless, carries the melody, not the identities of the instruments being played. Individual animals across a wide range of species can recognize each other from their vocal signals, suggesting that animals have percepts similar to periodicity pitch and timbre.

Frequency Analysis

Complex sounds can be mathematically described as being the sum of a series of component sine waves. Conversely, individual sine waves can be summed together to form a complex sound. The mathematical processes to describe these are called Fourier analysis and Fourier synthesis, respectively. Fourier analysis allows us to take a complex sound (or any continuous waveform) and decompose it into individual sine waves of specific frequencies, amplitudes, and phases. This process results in a frequency spectrum displaying the frequencies in the sound, together with their amplitudes and relative phases. Fourier analysis of recorded sounds is easily carried out by many widely available computer programs for characterizing sounds. Moveable cursors built into the displays of these programs permit selecting segments of sounds or whole sounds to determine their spectra. Animal sounds recorded in field conditions are most often characterized by their frequency–amplitude spectra, which display the relative strengths of different component frequencies in relation to the strengths of the same frequencies in background noise. This kind of display can be used, for example, to plot the salience above the background noise of communication sounds recorded at different distances from the calling animal. Phase spectra are less useful for examining vocal communication because propagation of sounds through complicated surroundings, which often contain vegetation and include reverberation from multiple objects, perturb the relative phases of frequency components in a manner that is difficult to relate to the receiver's perception of the sound, which depends chiefly on the amplitudes of the most critical frequencies.

By far, the most common and most useful display of the characteristics of communication sounds is the spectrogram. A spectrogram shows a running history of a sound's frequency content to show how its frequencies change over the sound's duration. It is created by dividing the sound into short overlapping segments and plotting the spectrum of each segment using Fourier analysis. Most computer programs for sound analysis have the capability of plotting spectrograms, and some can even display the spectrogram in real time as the sound progresses. A spectrogram does not, however, display the phases of different frequencies in the signal, but only the amplitudes. In a spectrogram, the relative amplitude of different frequency components of the sound is indicated by different colors or shades of gray.

The left plot in **Figure 3** shows the spectrogram of the 1000 Hz sine wave whose amplitude–time display was shown in **Figure 1**. The spectrogram shows one frequency band at 1000 Hz that is stable in its position over the entire duration of the sound. The right plot in **Figure 3** shows the spectrogram of the complex wave whose amplitude–time display is shown in **Figure 2**. The spectrogram of the complex sound shows that five frequency bands are present. These five bands are all of equal darkness in the plot, indicating that all five frequencies are of equal

amplitude. The five frequency bands are also horizontal and parallel, indicating that the frequency structure is stable across time. These frequency bands are separated by a fixed vertical interval of 1000 Hz. This means that the frequency components are in an integer relationship – the upper four frequencies are all integer multiples of 1000 Hz. This complex sound is said to be harmonic, or to have harmonic structure. In a harmonic sound, all frequency components are integer multiples of a base or fundamental frequency. In this example, 1000 Hz is the fundamental frequency, and the other components are harmonics of this fundamental frequency. The reciprocal of the fundamental frequency of a harmonic sound is the period of the entire sound. In the waveform shown in **Figure 3**, this period is 1 ms. This same period can be derived from the time interval between successive repeating units in the amplitude–time waveform, or from the frequency difference between the adjacent harmonics in the spectrogram.

Because of the nature of the vertebrate vocal tract, biologically produced complex sounds may not have their frequency components in exact integer ratios. For this reason, these sounds are more formally called quasi-harmonic sounds. **Figure 4** shows amplitude–time (top) and spectrogram (bottom) displays of one note in the advertisement call of the male bullfrog (*Rana catesbeiana*). This note (top left) is about 450 ms in duration and has a gradual onset (rise time) and offset (fall time). Expanding the time axis (top right display) shows that the note has a periodicity of about 122 Hz (8.2 ms). The spectrogram shows that the note contains 15 distinct frequency components, ranging from about 200 to about 1800 Hz. The individual harmonics in the note are not of equal amplitude, as shown by the unequal darkness of the individual frequency bands. The fundamental frequency of 122 Hz is not well-represented in the note's spectrum, but it can be calculated either from the frequency spacing between the harmonic bands, or from the repeating period of the waveform. Even when the fundamental frequency of a harmonic series is missing, male bullfrogs behave as if they can detect the period of their advertisement notes. Similarly, humans report that they detect a pitch at the missing fundamental frequency.

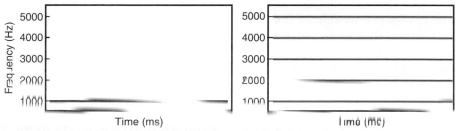

Figure 3 Spectrograms (displays of frequency over time) for the waves in **Figures 1** and **2**. The spectrograms were computed by Adobe Audition v. 1.5. The sine wave in **Figure 1** contains one frequency, 1000 Hz. The complex sound in **Figure 2** contains five frequencies. The fundamental frequency is 1000 Hz, and the harmonics are at 2000, 3000, 4000, and 5000 Hz (these are the second, third, fourth, and fifth harmonics of the 1000 Hz fundamental frequency).

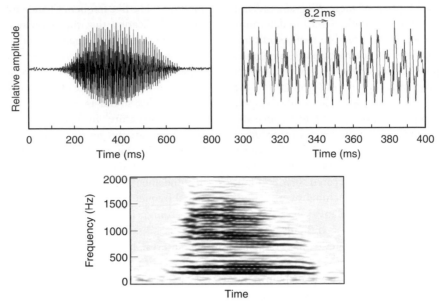

Figure 4   One note in the advertisement call of a male bullfrog, recorded in the frog's natural environment at a distance of 3 m away from the caller. The top left graph is the amplitude–time display of the entire waveform of the note. This note has a gradual rise time (onset) and fall time (offset), and has an overall duration of about 450 ms. The top right graph shows the waveform of a 100 ms segment of the note. The period of the note is 8.2 ms. The bottom graph shows the note's spectrogram. The note contains approximately 15 different frequencies, at a frequency spacing of about 122 Hz (the reciprocal of the period). The darkness of the frequency bands indicates the relative amplitude of the different frequency components. Most of the energy in the note is in the low-frequency range around 200–300 Hz.

Amplitude Modulation

Another note in the same male bullfrog's advertisement call is shown in **Figure 5**. This note fluctuates several times in its amplitude over its duration (top left graph). This pattern of change in amplitude is called amplitude modulation. Formally, there are a number of mechanisms that generate such periodic variations in amplitude. For example, one form of amplitude modulation results from the multiplication of two individual sine waves with a nonzero baseline pressure (as would result from air blown over a larynx). This process creates additional frequency components in the sound called sidebands. The bullfrog note in **Figure 5** still has the same 8.2 ms period (top right) as the unmodulated note in **Figure 4**, but its spectrogram differs by the relative darkness of some frequency bands, as well as gaps in the spectrum produced by the modulation process. The addition of amplitude modulation to a sound makes it perceptually distinct from an unmodulated call, and thus conveys additional information to the receiver of the call. Male bullfrogs behave as if they can discriminate between the modulated and unmodulated notes made by other bullfrogs. The advertisement and aggressive calls of the male green treefrog, *Hyla cinerea*, differ in their rates of amplitude modulation. Both male and female green treefrogs can behaviorally discriminate between these two types of calls, indicating that these animals detect the amplitude modulation as a cue.

Frequency Modulation

Other biological sounds vary in their frequency composition over time. This process is called frequency modulation. The left graph in **Figure 6** displays the amplitude–time waveform of the advertisement call of the male gray treefrog (*Hyla versicolor*). This call consists of a series of very short notes or pulses (20 successive call notes in the 1 s interval shown here), each of which has a sharp onset and offset. The spectrogram of a portion of this call, shown in the right graph, illustrates that each note contains several frequency components, and that these components are frequency modulated. The three main frequency components (around 1500, 2500, and 3500 Hz) sweep upward in their frequency over the note's duration. These three frequencies are in a quasi-harmonic relation, with the frequency spacing or period around 1000 Hz or 1 ms. Some species of insects and most vertebrates discriminate between sounds that differ in the direction and the extent of their frequency modulation. The addition of frequency modulation to a call can assist in its propagation through the environment.

Propagation of Sounds in the Environment

A sound pressure wave propagates or spreads outward from the source of the mechanical disturbance. As it propagates,

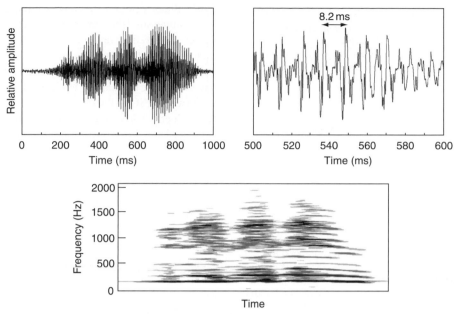

Figure 5   Another note in the advertisement call of the same male bullfrog as in **Figure 4**, also recorded at a distance of 3 m from the source. The amplitude–time display shows that this note is approximately 650 ms in duration and has a gradual rise time and fall time. The note is amplitude-modulated – it varies in its amplitude over its 650 ms duration. The period of the note is still 8.2 ms, but the spectrogram shows changes in the relative darkness of some frequency bands resulting from the modulation process.

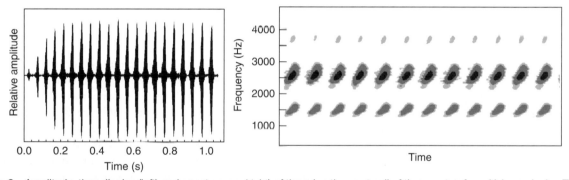

Figure 6   Amplitude–time display (left) and spectrogram (right) of the advertisement call of the gray treefrog, *Hyla versicolor*. The call was recorded in the frog's natural environment by Joshua J. Schwartz. This call is 1 s in duration, and contains 20 short notes, each with a sharp onset and offset. The spectrogram shows that each note contains three frequencies, each of which is frequency-modulated upward. The three frequencies in each note are in harmonic register.

it decreases in its amplitude as its wavefront spreads out to cover an ever-enlarging area of space, even as the total energy in the wave remains constant. Thus, the amplitude of a sound wave picked up at some point in space depends on the distance from the source. Eventually, the sound spreads out so much that it can no longer be detected at any one point. Moreover, in a natural environment, the propagating pressure wave is affected by the presence of other objects that interact with the sound to produce distortions in its waveform due to interference or reverberation. All of these affect the amplitude–time waveform of the signal, and they can affect the discreteness of frequency bands in the spectrogram. In cases where the presence and distinctiveness of closely spaced frequencies is critical for interpreting the message contained in communication

sounds, receivers located far from the source are less likely to be able to decipher the message even though the sound may still be detectable.

How signals propagate through the environment is important for understanding their biological function. Because of these environmental effects, within their range of hearing, different animal species use different sound frequencies for different communication purposes.

Propagation in an Ideal Environment

The amplitude, or pressure, of a sine wave in air decreases according to the inverse of the distance traveled. Specifically, its amplitude decreases by 6 dB (a halving of amplitude) for every doubling of the distance from the source.

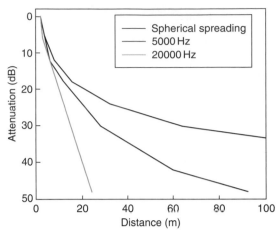

Figure 7 Attenuation of the amplitude of sound with distance as it propagates through the air. Attenuation is graphed as relative decibels, where 0 dB is the reference value. The solid dark line shows attenuation of sine waves as predicted by spherical spreading. The red line shows attenuation of a 5000 Hz sine wave, and the green line shows attenuation of a 20 000 Hz sine wave. Attenuation becomes more severe with higher frequencies, due primarily to atmospheric absorption. Data are replotted from Camhi JM (1984) *Neuroethology: Nerve Cells and the Natural Behavior of Animals*. Sunderland MA: Sinauer Associates.

This is called spherical spreading, and it describes the attenuation of (decrease in) sound amplitude with increasing distance (**Figure 7**). Spherical spreading is not dependent on the sound's frequency. Propagation of a sine wave over the horizontal surface of a body of water occurs by cylindrical spreading, or a decrease in amplitude by 3 dB for every doubling of distance, because the water prevents the formation of a spherical wavefront. Animals such as bullfrogs that call at the air–water interface thus experience less loss in the amplitude of their calls than if they were calling from a site on land. Underwater, the rate of sound propagation (whether spherical or cylindrical) depends on water depth and the physical distance of the sound source from the top (air–water interface) and bottom (seabed) boundaries.

Propagation in the Natural Environment

The rate of sound propagation in the natural environment is influenced by many variables, including temperature, humidity, turbulence (in air or in water), altitude (in air), depth (in water), presence of vegetation, and characteristics of the substrate (sand, mud, concrete). Sound waves also reflect from surfaces, resulting in a scattering or loss of energy. Moreover, in a natural environment, background noise, produced by other animals or by man-made objects, affects propagation and the fidelity of the sound at the receiver. Sounds can be distorted or undergo reverberation as they travel outwards from the source, and they can be reflected or scattered by objects or by boundaries. All of

these influence sound frequency, amplitude, and phase. In some situations, the environment can aid propagation. Sound waves refract or bend when they encounter a surface where the speed of sound changes. Some sounds travel better if they are emitted at the air–water interface; some sounds travel better if the animal is elevated above the substrate. Underwater (SOFAR) channels are found at particular ocean depths where refraction is maximal. Sounds in the SOFAR channel can travel long distances without attenuation. These phenomena can result in sound propagation as good as or even better than predicted by the inverse square law, even in the midst of environmental noise. These areas of increased sound propagation are called sound windows.

Outside of the ideal environment, sound attenuation occurs due to absorption of energy by the medium as well as by spreading losses. Such attenuation is affected by sound frequency (**Figure 7**). In effect, the propagating sound is robbed of energy that is absorbed by the medium, so the decrease in amplitude during propagation occurs faster than would be predicted by spherical spreading alone. Atmospheric absorption is a major constraint on the use of sounds for long-distance communication. High-frequency sounds attenuate more rapidly during propagation than do low-frequency sounds. This is because higher frequencies undergo more cycles of condensation and rarefaction per second, and in each cycle, acoustic energy dissipates as a result of thermal energy generated by the vibration of the molecules. Thus, high frequencies suffer more absorption than low frequencies. Lower frequencies in a complex sound are relatively more preserved as they travel through the environment. An example of this is shown in **Figure 8**. Here, the same note in a male bullfrog's advertisement call, as shown in **Figure 5**, where it was recorded at a distance of 3 m from the frog, is now recorded 40 m away from the frog. The waveform is, of course, at a lower amplitude, but the spectrogram plotted using the same digital settings as in **Figure 5** illustrates the effects of atmospheric absorption and other consequences of propagation through a natural environment. Comparing the spectrograms in **Figures 5** and **8**, only the low-frequency components of the call survive to be picked up at the 40-m recording site. The higher frequency components in the call have been absorbed, scattered, or attenuated by the environment.

The impact of the true environment on sound propagation means, for long distance communication, animals are more likely to use low-frequency than high-frequency signals because low frequencies travel farther and suffer less propagation loss. Territorial or aggressive calls that are used for communication between social groups tend to contain low frequencies. Advertisement or mating calls used by males to attract females to a breeding site also tend to contain low frequencies. Conversely, sounds that are used for communication over short distances, such as alarm calls directed toward members of the social group,

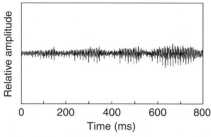 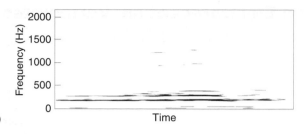

Figure 8 Amplitude–time display and spectrogram of the bullfrog note shown in **Figure 5**, this time recorded at 40 m distance from the calling male. Note that frequencies in the note above 400 Hz are now no longer present, showing that low frequencies suffer less propagation loss with distance than do the higher frequency components.

or contact calls between mothers and offspring, tend to contain high frequencies that will not propagate far from the source and possibly alert distant predators. The physical structure of animal signals may, thus, give clues as to the biological functions of these signals, and to the perceptual abilities of the receivers of these signals.

See also: Alarm Calls in Birds and Mammals; Anthropogenic Noise: Impacts on Animals; Hearing: Vertebrates; Mating Signals; Sound Production: Vertebrates.

Further Reading

Boatright-Horowitz SS, Cheney CA, and Simmons AM (1999) Atmospheric and underwater propagation of bullfrog vocalizations. *Bioacoustics* 9: 257–280.

Bradbury JW and Vehrencamp SL (1998) *Principles of Animal Communication*. Sunderland MA: Sinauer Associates.

Camhi JM (1984) *Neuroethology: Nerve Cells and the Natural Behavior of Animals*. Sunderland MA: Sinauer Associates.

Forrest TG (1994) From sender to receiver: Propagation and environmental effects on acoustic signals. *American Zoologist* 34: 644–654.

Gerhardt HC and Huber F (2002) *Acoustic Communication in Insects and Anurans*. Chicago: University of Chicago Press.

Marler P (1955) Characteristics of some animal calls. *Nature* 176: 6–8.

Waser PM and Brown CH (1984) Is there a 'sound window' for primate communication? *Behavioral Ecology and Sociobiology* 15: 73–76.

Wiley RH and Richards DG (1978) Physical constraints on acoustic communication in the atmosphere: Implications for the evolution of animal vocalizations. *Behavioral Ecology and Sociobiology* 3: 69–94.

Yost W (2000) *Fundamentals of Hearing*, 4th edn. New York: Academic Press.

Relevant Website

Discovery of Sound in the Sea – http://www.dosits.org/index.htm

Active Electroreception: Vertebrates

G. von der Emde, University of Bonn, Bonn, Germany

Introduction

Electroreception, that is, the detection of naturally occurring electric stimuli by animals with specialized electroreceptors in their skin, can be found only in animals that live in water and thus is always coupled to an aquatic medium. Many marine and freshwater fishes, with the important exception of most (but not all) teleosts, are electroreceptive. Most electroreceptive animals detect weak electric fields, which originate in the biotic or abiotic environment and stimulate their ampullary electroreceptors organs, a process called *passive electrolocation*. In contrast, animals that use *active electrolocation* actively emit electric signals and perceive them after they have been modified by the external world. In this case, objects are detected because they change the self-emitted signal in a way perceivable by the animal.

Active electrolocation is only used by weakly electric fishes that produce electric signals with specialized organs (electric organ discharges (EODs)) and perceive them with epidermal electroreceptor organs. This combination can be found only in the South American gymnotiforms (or Knifefishes) and the African mormyriforms (mormyrids). Despite its surprising similarity at several levels, the ability to actively electrolocate has evolved independently in South America and Africa.

While an EOD is emitted, an electrical field builds up around the fish in the water (**Figure 1**). For example, the field produced by the basically biphasic EOD of the mormyrid *Gnathonemus petersii* is an asymmetric dipole field with one smaller pole at the fish's tail and the other pole constituting the entire body of the fish anterior to the electric organ. Because water is a conducting medium, alternating electric current flows through the water and enters (or leaves) the fish's body mainly through the pores of the electroreceptor organs. The electroreceptor cells measure the electrical current flowing through them, which is proportional to the local electrical voltage between the inside and the outside of the fish.

If the fish approaches an object with electric properties different from those of the surrounding water, the electric field is distorted. The three-dimensional field distortions lead to a change in the voltage pattern within the 'electric image' which the object casts onto the fish's skin surface. Thus, the electric image is defined as the local modulation of the electric field at an area on the skin. In mormyrids, a typical electric image has a center-surround ('Mexican hat') spatial profile. For example, a good conductor (e.g., a

water plant, another fish, or a metal object) produces an image with a large center region where the local EOD amplitude increases, surrounded by a small rim area where the amplitude decreases. The image of a nonconductor such as a stone (or a plastic object) has an opposite appearance: in its center, local EOD amplitude decreases while it slightly increases in the surrounding rim area (**Figure 2**). In order to gain information about objects during active electrolocation, the fish has to scan the electric image with its electroreceptors, which are innervated by primary sensory afferent nerve fibers that project to the brain.

The Electric Organ Discharge

Electric fishes produce electric impulses by a muscle or nerve-cell-derived electric organs, which in the case of mormyrids lies in the caudal peduncle. In both Africa and South America, two basic types of EOD can be found: pulse-type EOD, where the interval between two EOD is clearly longer than the duration of a single EOD, and wave-type EOD, where discharges are produced one after another resulting in a quasisinusoidal wave signal (**Figure 3**).

In all cases, EOD are used for nocturnal orientation through active electrolocation and for electrocommunication. For both processes, EOD waveform plays a critical role. The waveform of an EOD depends on the morphology of the electric organ and on the hormonal state of the animal. The electroreceptor organs involved in electrolocation are tuned to the characteristics of the self-produced EOD and thus can detect object-induced modifications of the local EOD. Most objects in the environment of the fishes are mainly resistive, but animate objects also have capacitive properties, which lead to waveform shifts of the local EOD in addition to amplitude changes. By detecting these waveform changes, weakly electric mormyrids can detect and identify capacitive objects.

The Environment of Weakly Electric Fishes

Weakly electric fishes live in freshwater habitats of Africa and South America. The about 200 different species of Mormyrids and the more than 150 species of Knifefishes have conquered many diverse habitats from small creeks to smaller and larger rivers and lakes. Most of these waters have a rather low electrical conductivity and a temperature well above $20\,°C$. In waters of cooler or more arid areas, the number of electric fish species greatly diminishes. *G. petersii*, for example, lives in rivers and stream of the rain forests of central Africa. During the day, the animals hide in the vegetation or in cavities at the bank of the rivers. During the night, they become active, leave their hiding places and search for food at the ground of the river.

Different species of weakly electric fishes feed on a variety of food. Apparently most, if not all species are predators and insect larvae, such as chironomid larvae, constitute a high percentage of their diet, even for larger species. However, there are also fish predators, such as the mormyrids *Mormyrops anguilloides*, which grows up to a length of about 100 cm. In Lake Tanganyika, this species hunts in groups for sleeping cichlids at night, a behavior which is called 'pack hunting.' The great majority of weakly electric fishes are strictly nocturnal and in the absence of light, the major sense used for prey detection is the electric sense, in particular active electrolocation.

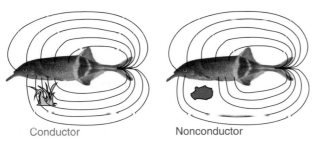

Figure 1 Schematic two-dimensional drawings of the electric fields of *G. petersii* distorted ventrally by a water plant (good conductor, left) or a stone (isolator, right). The fish is viewed from the side. Electrical field lines are drawn as thin lines. Modified after von der Emde G (1999) Active electrolocation of objects in weakly electric fish. *Journal of Experimental Biology* 202: 1205–1215.

Perception of Objects During Active Electrolocation

During active electrolocation, weakly electric fish not only detect and locate objects in their environment, but also identify several object properties. Fish can detect the electrical properties of the material an object is composed of. The electrical resistance of an object is determined by measuring the amplitude change imposed on the local EOD by the object. Mormyrids and gymnotiforms can also perceive capacitive object properties ('capacitance detection'). Especially living objects have complex electrical impedances consisting of capacitive and resistive components. *G. petersii* is able to measure both components independently and quantitatively, thereby being able to categorically discriminate between living and inanimate objects. Thus, living prey items (e.g., chironomid larvae) acquire an 'electrical color' and thus pop out of an inanimate, electrical gray background. Depending on the frequency compositions of the EOD, only a certain range of capacitive values can be detected by a fish. In several species of mormyrids, the detectable range of capacitances corresponds to the range of capacitive values of animated objects found in their natural habitat.

G. petersii can also localize nearby objects during active electrolocation. When presented with two objects at different distances, the fish can learn to choose the object located farther away than the alternative one. This ability is based only on distance, and is independent of the size or electrical properties of the object. The fish thus have a true sense of depth perception and perceive a three-dimensional electrical picture of their surroundings.

In addition to perceiving an object's material and location, *Gnathonemus* also perceive an object's shape. Fish can

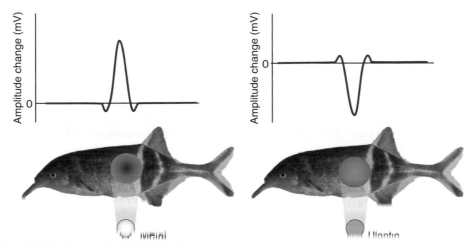

Figure 2 Electric images of a metal (left) or a plastic (right) object placed near the side of a *G. petersii*. The images on the fish's skin are color coded with local amplitude-increases depicted in red and amplitude-decreases shown in blue. Above each graph, a single one-dimensional transect through the image is shown, which plots the local EOD amplitude change versus horizontal location along the midline of the fish. Note that both objects project Mexican-hat like images, however, of an inverted sign.

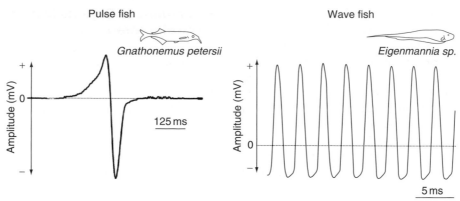

Figure 3 Electric organ discharges (EOD) emitted by a pulse fish (*G. petersii*) from Africa (left) and a wavefish (*Eigenmannia*) from South America. Note the different time scales. *G. petersii* emits single and brief pulses with long and variable pauses in between, while *Eigenmannia* emits a continuous sinusoidal signal.

recognize a free-standing object of a certain shape after it was moved within an arena. *G. petersii* quickly learn to recognize objects of various shapes and to discriminate them from differently shaped objects. Shape recognition persists even when the object is rotated in space, indicating a viewpoint-independent recognition of objects. In additional experiments, *G. petersii* demonstrated size constancy during object recognition, that is, they recognized an object of a certain shape even if its electric image appeared larger or smaller because of variations in distance. Fish also could identify the size of an object independent of its distance. For analyzing the shape or the size of a new object, fish have to perform so-called probing motor acts (PMA) (see later), that is, they have to swim around the object scanning it with their sensory surface from several viewpoints. This is in contrast to distance measurements, which can be achieved instantly and don't require scanning movements.

During active electrolocation, the electric images of two nearby objects will fuse in a nonlinear manner leading to a single, complex image. In spite of this effect, *G. petersii* is able to perceive the shapes of objects even when they are positioned right in front of a large background. This ability persists even if the object and the background are made from the same material. Fish are also able to detect small gaps in the millimeter range in a solid object.

These findings show that weakly electric fish have a remarkable ability to analyze complex three-dimensional scenes and can identify single object's properties even in a natural setting containing many bits and pieces of various sizes, shapes, and material.

Electromotor Behavior During Active Electrolocation

EOD are all-or-nothing events and their waveform cannot be modified by the animal on a short-term basis. However,

the fish can change the temporal pattern of EOD produced and thus influence the number of EOD emitted within a certain time window. In wave-type EOD, this results in different frequencies of the wave signal, while in pulse-type EOD, the sequence of pulse intervals (SPI) changes. In pulse fish, this SPI functions in electrocommunication by conveying different types of information between the fish. During active electrolocation, the SPI is important for regulating the flow of information about the environment to the animal.

In mormyrid pulse fish, typical SPI can be observed, when electrically inspecting an object or during foraging. The important parameter during active electrolocation of objects seems to be a *regular pattern* of interpulse intervals, as suggested by several authors for various species of mormyrids. All fish regularize their discharge activity during probing of an object, which is sometimes accompanied by PMA (see section 'Object Inspection: Probing Motor Acts'). These regular patterns contrast with the variable discharge rates during swimming and also during food search (see section 'Foraging'). Regularization during object inspection may serve to keep receptors and associated brain structures on a constant level of adaptation, which may be especially important because of a high degree of plasticity of electrosensory brain structures. Regular SPI ensure that all changes of spike activity in the brain are caused by the object under investigation and not by changed conditions of nerve or receptor cells.

Typical SPI during object probing can especially well be observed in fishes solving a conditioned object detection task. Under these conditions, the animals are mainly engaged in active electrolocation, and the aspect of electrocommunication has only a small influence on electric signaling behavior. **Figure 4** shows SPI of a *G. petersii*, which had to inspect an object in order to get access to an area in the aquarium that contained food. Identical training experiments were conducted with three other species of mormyrids which emitted either shorter or longer

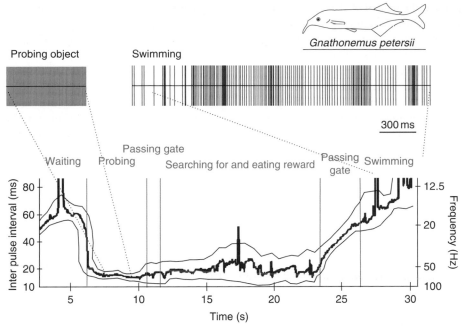

Figure 4 Sequence of pulse intervals (SPI) of a *G. petersii* during solving a conditioned electrolocation task. The graph shows interpulse intervals (IPI) versus time. Above, SPI during active probing (left) and during swimming (right) are shown in red as examples. The fish had to wait for the opening of a partition ('waiting') in a dividing wall of the experimental tank before it got excess to the electrolocation targets ('probing'). When it reached a decision, the fish swam through the partition ('passing gate') and searched for its food reward on the ground of the tank ('searching for and eating reward'). After this, it swam back through the partition and waited for the start of the next trial. Modified from Schwarz S and von der Emde G (2001) Distance discrimination during active electrolocation in the weakly electric fish *Gnathonemus petersii. Journal of Comparative Physiology A* 186: 1185–1197.

lasting EOD. Even though all fishes showed regularization during 'probing,' its level differed among members of different species. Fishes emitting longer EOD had the tendency to discharge at lower rates compared to fishes with shorter EOD.

Also, during food search mormyrid pulse fishes emit EOD at a high rate, which increases the amount of electrosensory input to the animal. However, no regularization behavior occurs. The significance of the lack of regularization during foraging remains to be examined, but one may assume that other factors (e.g., electrocommunication) play a role.

Locomotor Behavior During Active Electrolocation

In addition to emitting a certain temporal pattern of electric pulses, mormyrids perform certain stereotyped swimming movements when engaging in active electrolocation. Different locomotor behaviors can be observed during object inspection and during foraging.

Object Inspection: Probing Motor Acts

When investigating a novel object, mormyrids perform PMA, that is, characteristic behaviors composed of a

series of swimming maneuvers in close proximity to the object. Six types of PMA have been described, which all may serve to position the fish optimally for some aspect of active electrolocation. During 'lateral va-et-viens,' the fish scans the object with its lateral (or ventral) body surface. By doing so, it might be able to centrally compare inputs from receptors at different body locations. During 'radial va-et-viens,' a behavior performed mainly when a fish is investigating a potentially dangerous object, the fish slowly approaches the object backward while simultaneously displaying vehement lateral tail strokes to the left and right. This will modulate the input to all electroreceptors at one side of the body simultaneously. During the PMA called 'stationary wriggling,' the fish remains stationary lateral to the object and performs wriggling movements with the whole body, which continuously oscillates the distance between a fixed spot on the lateral body surface and the object. This allows the fish to compare inputs from the same electroreceptor organ at several distances to the object. When performing another type of PMA called 'stationary probing,' the fish rapidly approaches the object and suddenly stops when the head is only a couple of centimeters away.

While all species of mormyrids investigated so far perform the types of PMA just mentioned in a similar way, there exists one PMA which is only performed by *G. petersii*. During 'chin probing,' *G. petersii* brings its movable chin appendix, the so-called *Schnauzenorgan*, very close to the object, almost touching it. The fish then moves the Schnauzenorgan over the object, following its contours. This behavior resembles a haptic inspection of an object with the fingers of the human hand or the scanning movements of the fovea of the eye when looking at an object or inspecting a picture. The scanning of an object by these Schnauzenorgan movements will provide fine detailed electrical (and possibly touch) information about the shape of the object.

Foraging

When *G. petersii* is searching for small insect larvae (chironomidae) on the ground of the river, they never perform PMA. Nevertheless, characteristic and stereotyped behaviors occur in these situations, which help to optimize sensory input about the prey. *G. petersii* employs several senses to find their food: vision, olfaction, the mechanosensory lateral line, passive electrolocation and, most importantly, active electrolocation. This shows that food detection is a multisensory process, with several senses working together and being integrated by the brain.

During foraging, *G. petersii* employs a characteristic swimming posture: they swim at a constant angle of their body axis of $18° \pm 3.6°$ with their head toward the ground (**Figure 5**). With the tip of their Schnauzenorgan, they almost touch the ground, moving it in a stereotyped, rhythmic fashion from left to right while swimming forward. During these sweeping movements, the Schnauzenorgan scans an angle of about $110°$ during a full left-right cycle

(**Figure 5**), or $70°$, if only a half-cycle is performed. During exploratory and foraging behaviors, *Gnathonemus* can move its Schnauzenorgan with a high velocity of up to $800° \, s^{-1}$. These regular movements are often associated with high EOD emission frequencies of 55–80 Hz.

When prey or another object of interest is encountered, the scanning movements of the Schnauzenorgan stop abruptly, and it is brought in a twitching movement toward the object for further exploration. In the case of prey, exploration is very brief and the fish tilts forward to soak up the insect larva. In order to acquire an object buried in the soil, the Schnauzenorgan is used as a burrowing stick in order to dig out the prey up to a depth of 2 or 3 cm.

The described slanted swimming position during prey search probably serves an additional function. It ensures that the nasal region, the skin area above the mouth and between the nares at the fish's head, is held rather constant at an angle of about $50°$ relative to the ground (**Figure 5**). It thus points forward and slightly upward and might be in an optimal position to detect approaching objects such as obstacles, environmental landmarks, or swimming prey. When the fish approaches an obstacle, this object will project an electric image onto the nasal region and is thus detected and maybe identified. The nasal region, which contains an exceptionally high density of electroreceptors, is thus used like a fovea in the retina of the eye. The nasal region might also be used during catching of copepods suspended in the water. Because these prey items swim in the open water, they are usually not detected by the Schnauzenorgan, unless they happen to be very close to it. Instead, they will project an electric image on the nasal region, resulting in an orienting response of the Schnauzenorgan toward the prey, which then is followed by ingestion.

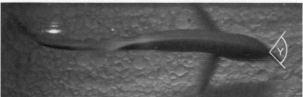

Figure 5 Photographs of a *G. petersii* during foraging taken from the side (above) and from above (below). The angle of the long axis of the fish toward the ground (blue, $\alpha = 18° \pm 3.6°$) and the angle of the surface of the nasal region toward the ground (red, $\beta = 50° \pm 5.8°$) are shown. During foraging, the Schnauzenorgan is moved left and right and thus sweeps over a wide angle (white, $\gamma = 108° \pm 44°$) over the ground.

The Electric-Fovea Hypothesis

The term 'fovea' (literally meaning 'small depression or pit') has been originally used for an area in the human retina of the eye containing a high density of only cone photoreceptors giving it a high spatial resolution. In addition, the fovea is strongly over-represented in the brain, with an overproportional number of neurons being devoted to process information coming from this retinal region. Behaviorally, the fovea is special, because during object inspection, eye movements let the fovea move over the object of interest and focus on important details. Lately, foveae have been reported to occur not only in visual systems but also in several other senses, including the acoustic fovea of echolocating cf-bats and the mechanosensory fovea of the star-nosed mole, which is located on 11th 'foveal' appendage.

Sensory systems can also contain double foveae. For example, pigeons have two specialized areas in each eye. One is used for long-range guidance, while the second is a shorter-range (food) detection system. The latter is a true foveal depression, which is located slightly below the retina's center. It is specialized for wide field monocular perception of the visual area around and lateral to each side of the bird and is presumably used for predator detection and flight control. The second specialized fovea (the 'area dorsalis') lacks a depression and is located in the upper temporal retina. This area serves the frontal region below the bird's beak and has presumably evolved for myopic foraging of food on the ground.

The idea that *G. petersii* possesses two separate electric foveae was developed, when behavioral, anatomical, and physiological results had revealed several similarities between certain electroreceptive skin regions of the elephantnose fish and the eyes of pigeons. Around the same time, a similar hypothesis was put forward for South American weakly electric fish, which have an electric fovea and a 'parafovea' around their mouth.

The two separate foveae in *G. petersii* constitute the Schnauzenorgan and the nasal region (**Figure 6**). Both regions independently fulfill the conditions of a fovea, because (1) the density of receptor elements in both regions is much higher than in the rest of the sensory epithelium; (2) both regions are over-represented in the brain of the fish, that is, there are more central neurons devoted to the processing of a single receptor element of the fovea compared to a receptor in the periphery; (3) there exist structural/morphological and physiological specializations of the receptors and accompanying structures within the foveae; and finally (4) the animals show behavioral adaptations for focusing a stimulus onto the fovea for detailed analysis.

It follows that *G. petersii* has two separate foveae, which both fulfill the premises of a real fovea, respectively. In addition, it becomes clear that like in the pigeon eye, the two foveae serve different functions: the nasal region is a long-range guidance system that is used to detect obstacles or other objects in front of and at the side of the animal. The Schnauzenorgan, on the other hand, is short-range movable (prey-) detection system that is used to find and identify prey on the ground or inspect details of objects (**Figure 6**). Like in the pigeon, both systems work simultaneously and ensure an optimal sensory inspection of the nocturnal environment of the fish.

The Novelty Response

African weakly electric mormyrid fish will respond to novel sensory stimuli that suddenly appear in their environment with a transient increase of the discharge rate of their electric organs (**Figure 7**). Similar 'novelty responses' can be found in the two unrelated groups of weakly electric pulse fishes from Africa and South America. This behavior will temporally increase the flow of sensory information to the fish allowing it to investigate in detail the new sensory environment and the cause of the change in sensory input. This function is backed by the association of the novelty response with several autonomic reactions, such as transient changes in heart and ventilatory rates. The novelty response can therefore be regarded as an 'orienting response,' first described by Pavlov and found in all vertebrates, where it facilitates sensory processing of important sensory information.

The novelty response of *G. petersii* occurs to all sensory modalities tested so far, that is, to mechanical, acoustical, electrical, and visual stimulation (**Figure 7**). A very effective stimulus is a sudden change in the electrical properties of an object close to the fish, which is detected by active electrolocation. In *G. petersii*, novelty response parameters such as duration, peak amplitude, and latency depend on stimulus intensity. In general, when stimulus intensity is high, the fish responds with a short latency, a high response amplitude, and a long-lasting novelty response. After repeated sensory stimulation, the novelty response will habituate, especially to nonsignificant, innocuous stimuli. Habituation of the novelty response follows a negative exponential function of the number of stimulus presentation, and is more pronounced the more rapid the frequency of stimulation and the lower the stimulus amplitude. Like a typical orienting response, the novelty response can be dishabituated by high-intensity stimuli of another modality.

The Schnauzenorgan Response

Both anatomical and behavioral evidence have shown that the moveable Schnauzenorgan is crucial for prey localization and object inspection. Recently, we observed another interesting reflex-like behavior of the Schnauzenorgan to nearby novel electrosensory stimuli. A sudden change

Figure 6 Schematic drawing of the posture of a *G. petersii* during foraging with the receptive beams of the two foveae indicated in yellow. The two foveae at the Schnauzenorgan (fovea 1) and at the nasal region (fovea 2) are sketched in red, the electric organ in the caudal peduncle in blue.

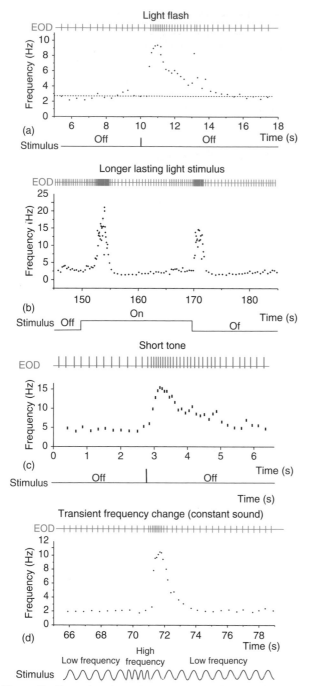

Figure 7 Examples of novelty responses of *G. petersii* to four types of sensory stimuli. Each graph shows the series of EOD of a single fish at the top with each EOD represented by a green vertical line. The middle diagram depicts the instantaneous frequency of EOD versus time. The third trace shows the occurrence of the stimulus. (a) The stimulus was a 10-ms light flash. (b) The stimulus consisted of a 20-s constant amplitude visual stimulus. Note that the fish responded with a novelty response both to stimulus on and off. (c) The stimulus was a short tone of *ca.* 750 Hz. (d) The stimulus consisted of a short-duration change in frequency from 500 to 600 Hz and back of an ongoing constant amplitude acoustical stimulus. Modified from Post N and von der Emde G (1999) The 'novelty response' in an electric fish: Response properties and habituation. *Physiology & Behavior* 68: 115–128.

in the properties of an object located close to the chin evoked one or several fast twitching movements of the Schnauzenorgan (**Figure 8**). These movements, called 'Schnauzenorgan response' (SOR), could be either evasive (movements away from the stimulus) or exploratory (movements toward the stimulus). When measuring the amplitude thresholds of this response, we could show that in contrast to the novelty response, the SOR only occurs to stimuli given next to the Schnauzenorgan or to a lesser degree near the head. In addition, SOR only occur reliably when stimuli are presented within about 3 mm of the fish's skin, whereas the novelty response occurs distinctly beyond this distance. The probability of evoking a SOR depended on the magnitude of the amplitude change of the electric input, with bigger changes eliciting SOR more

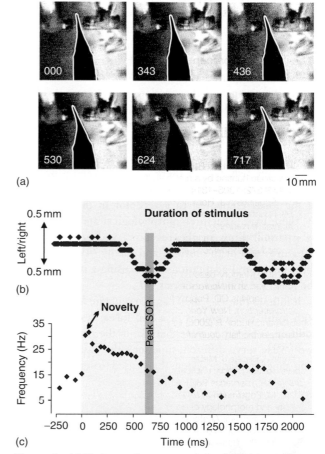

Figure 8 (a) Six frames from a sequence of a Schnauzenorgan movement filmed at 128 frames per second. In each frame, the time in ms is given. The resistance of the dipole-object at the left of the fish (yellow outline) was changed at time 0 h. The beginning of the SOR occurred at time 343 (second frame). The peak displacement of the SO happened at time 624, indicated by the red outline of the fish. (b) Tracking data of the Schnauzenorgan's displacement (72° s⁻¹). (c) Instantaneous EOD-frequency as measured during the sequence shown in (a) and (b). Note that the SOR peaks about 400 ms after the novelty response. In (b) and (c), the time of stimulation is indicated by the green background color; the time of maximal SOR is denoted by the orange vertical line.

reliably. Similarly, increasing the distance of the stimulus reduced the probability of the response.

While novelty responses are evoked by novel sensory stimuli of any modality (vision, audition, touch, electric, etc.), SORs are only evoked by electrolocation stimuli, which can, however, occur either during active or during passive electrolocation. Compared to the novelty response, the response latency of the SOR is much longer: about 300–500 ms versus *ca.* 40 ms of the novelty response (**Figure 8**).

The SOR appears to be a reflex-like behavior that is engaged in object detection and inspection, probably to quickly orient the Schnauzenorgan-fovea toward a suddenly emerging object during foraging. It appears to be mediated through a sensory motor loop from the receptors at the Schnauzenorgan (mormyromasts and/or ampullary electroreceptor organs) to the brain and back to the appropriate muscles of the lower jaw and the Schnauzenorgan.

See also: Electroreception in Vertebrates and Invertebrates.

Further Reading

Arnegard ME and Carlson BA (2005) Electric organ discharge patterns during group hunting by a mormyrid fish. *Proceedings of the Royal Society B* 272: 1305–1314.

Bacelo J, Engelmann J, Hollmann M, von der Emde G, and Grant K (2008) Functional foveae in an electrosensory system. *Journal of Comparative Neurology* 51(3): 342–359.

Bauer R (1974) Electric organ discharge activity of resting and stimulated *Gnathonemus petersii* (Mormyridae). *Behaviour* 50: 306–323.

Bell CC (2001) Memory-based expectations in electrosensory systems. *Current Opinion in Neurobiology* 11: 481–487.

Bullock TH, Hopkins CD, Popper AN, and Fay RR (2005) *Electroreception*. New York, NY: Springer.

Caputi AA and Budelli R (2006) Peripheral electrosensory imaging by weakly electric fish. *Journal of Comparative Physiology A* 192(6): 587–600.

Engelmann J, Bacelo J, Metzen M, et al. (2008) Electric imaging through active electrolocation: Implication for the analysis of complex scenes. *Biological Cybernetics* 98(6): 519–539.

Hollmann M, Engelmann J, and von der Emde G (2008) Distribution, density and morphology of electroreceptor organs in mormyrid weakly electric fish: Anatomical investigations of a receptor mosaic. *Journal of Zoology* 276: 149–158.

Hopkins CD (1999) Design features for electric communication. *Journal of Experimental Biology* 202(10): 1217–1228.

Kramer B (1990) *Electrocommunication in Teleost Fishes: Behavior and Experiments*. Berlin: Springer.

Maler L (2009) Receptive field organization across multiple electrosensory maps. II. Computational analysis of the effects of receptive field size on prey localization. *Journal of Comparative Neurology* 516(5): 394–422.

Moller P (1995) *Electric Fishes. History and Behavior*. London: Chapman & Hall.

Post N and von der Emde G (1999) The 'novelty response' in an electric fish: Response properties and habituation. *Physiology & Behavior* 68: 115–128.

Pusch R, von der Emde G, Hollmann M, et al. (2008) Active sensing in a Mormyrid fish – electric images and peripheral modifications of the signal carrier give evidence of dual foveation. *Journal of Experimental Biology* 211(6): 921–934.

von der Emde G (1992) Electrolocation of capacitive objects in four species of pulse-type weakly electric fish. II. Electric signalling behavior. *Ethology* 92: 177–192.

von der Emde G (2006) Non-visual environmental imaging and object detection through active electrolocation in weakly electric fish. *Journal of Comparative Physiology A* 192(6): 601–612.

von der Emde G, Amey M, Engelmann J, et al. (2008) Active electrolocation in *Gnathonemus petersii*: Behaviour, sensory performance, and receptor systems. *Journal of Physiology, Paris* 102(4–6): 279–290.

von der Emde G and Bleckmann H (1998) Finding food: Senses involved in foraging for insect larvae in the electric fish, *Gnathonemus petersii*. *Journal of Experimental Biology* 201: 969–980.

von der Emde G and Fetz S (2007) Distance, shape and more: Recognition of object features during active electrolocation in a weakly electric fish. *Journal of Experimental Biology* 210(17): 3082–3095.

Zupanc GKH (2002) From oscillators to modulators: Behavioral and neural control of modulations of the electric organ discharge in the gymnotiform fish, *Apteronotus leptorhynchus*. *Journal of Physiology, Paris* 96: 459–472.

Relevant Websites

http://biology4.wustl.edu/faculty/carlson/ – Carlson Lab: Behaviour and Communication.

http://www.nbb.cornell.edu/neurobio/Hopkins/Hopkins.html – Hopkins Lab: Communication and Evolution.

http://biology.mcgill.ca/faculty/krahe/ – Krahe Lab: Electrolocation and Communication.

http://www.neuromech.northwestern.edu/uropatagium/ – MacIver Lab with Focus on Robotics.

http://www.med.uottawa.ca/cellmed/eng/maler.html – Maler Lab: Theory of Electrolocation.

http://nelson.beckman.illinois.edu/ – Nelson Lab: Extensive Bibliography and Movies.

http://www.theangelsproject.org/tiki-index.php – Robotics.

http://www.fiu.edu/~efish/visitors/electric_field_animations.htm – Stoddard Lab.

http://www.zoologie.uni-bonn.de/NeuroEthologie/ – von der Emde Lab: Active Electrolocation and Bionics.

http://www.jacobs-university.de/directory/gzupanc/ – Zupanc Lab: Neural Mechanisms of Electroreception.

Adaptive Landscapes and Optimality

C. J. Goodnight, University of Vermont, Burlington, Vermont, USA

Fisher argued that selection acted to maximize the intrinsic rate of natural increase (r) of a population. The basis of this argument is that r is a predictor of the size of a population in the next time period. If two groups differ in their rate of increase, over time the group with the higher value of r would dominate the population. Similarly, those individuals with the highest lifetime reproductive rate (R_0) have the highest fitness. Thus, there is a clear idea that selection acts to maximize fitness as measured by r.

Maximizing fitness rarely means maximizing any one trait. Rather, under most circumstances, fitness is maximized when different aspects of the phenotype are at intermediate values that are compromises among the various selective forces affecting reproduction. That is, maximizing fitness equates to finding the best balance between tradeoffs. Thus, reproductive success is maximized within a set of external constraints that are imposed by the environment or the biology of the organism. The idea of maximizing fitness subject to constraints is known as optimization.

To study selection as an optimizing process, we must first identify constraints that most strongly produce trade-offs. One such constraint is the amount of energy available to an organism. In models that consider energy as a constraint, the total amount of energy available to an organism is fixed, and the model searches for the optimal partitioning of that energy among different functions in which the organism is engaged. For example, in its normal functioning an animal has to use energy for metabolic functions, growth, repair, and reproduction. The fixed amount of energy must be partitioned among these various functions, and the partitioning that maximizes the lifetime reproductive rate of the animal is by definition the optimal partitioning.

The actual partitioning of energy into these various components changes over the lifetime of an organism as it undergoes development and is also a function of the organism's ecology. Juveniles devote a relatively large amount of energy to growth and no energy to reproduction, whereas at maturity, energy devoted to growth may decrease or cease entirely, and energy devoted to reproduction increases. Similarly, ecological variables influence optimal energy partitioning. Large animals, including condors, whales, and presumably many of the Pleistocene megafauna have a life history that is typified by delayed maturation, low reproductive rates, and extended parental care. This life history maximizes lifetime reproduction if adults experience low predation because high adult survival counterbalances low reproductive rates. Conversely, mice are highly susceptible to predation throughout their lives and have quite different life histories. For mice, lifetime reproduction is maximized by early maturation with rapid and abundant reproduction at the expense of adult survivorship.

Lack presented an early application of optimization that examined brood sizes for animals, especially birds. Lack reasoned that the given amount of energy available for reproduction must be partitioned among the offspring. He envisioned a tradeoff between having many small offspring, each receiving a relatively small amount of energy (and therefore a fairly low probability of survival) versus producing a few large offspring, each receiving a larger share of the reproductive energy, and thus having a greater chance of survival. Lack showed that the optimal clutch size, the one that maximizes the number of fledged offspring, is intermediate. A simple numerical example illustrates this fundamental insight. Consider a situation in which the probability of survival to fledging declines linearly with the number of offspring. In this case, the number of fledged offspring will be:

Fledged offspring = Clutch size × Probability of an offspring surviving

Suppose the offspring produced from a clutch size of 1 gets 100% of the reproductive effort of the parents and has a probability of fledging of 90%. This decreases linearly such that with a clutch size of 10, each offspring gets 10% of the parental reproductive effort and has a 0% chance of survival (**Figure 1(a)**). In this example, the optimal clutch size is 5, which is expected to produce 2.5 offspring (**Figure 1(b)**). Below this optimum, individual offspring survive better, but because of the small clutch size fewer fledge. Above this clutch size, the number fledging decreases because survivorship is low.

It is important to note that what is optimal for the parents may not be optimal for the offspring. In the example given earlier, the optimal clutch size for the parents is 5 offspring, which maximizes the number or progeny that fledge. From the offspring's perspective, the optimal clutch size is one, which maximizes its own survival. This simple example is the foundation of a body of theory called 'parent–offspring conflict,' and illustrates that when there are interactions among individuals the optimal solution for one party may not be the optimal solution for other members of the interaction. Trivers provided the insight that as a result the final equilibrium may be a compromise among the different participants, and not optimal for any single individual.

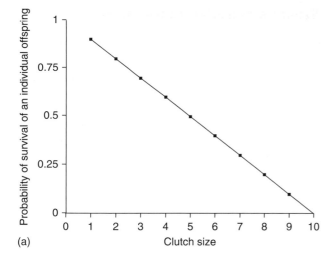

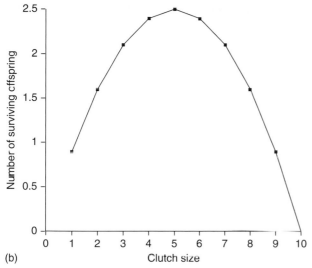

Figure 1 The probability of progeny survival as a function of clutch size in a hypothetical example in which the available parental effort must be partitioned among offspring. In this example, there is a linear relationship between clutch size and the energy available per offspring, and as a result a direct relationship between clutch size and probability of survival.

Foraging theory is another modeling tradition that uses an optimality approach. In its simplest form, optimal foraging focuses on energy, in the form of calories ingested. Thus, energy must be maximized, and the constraint that must be partitioned is time. That is, a foraging animal has a limited amount of time that must be partitioned between seeking food (e.g., prey), capturing it, and handling and processing it; the optimal diet is the one that maximizes the calorie intake per unit time. For example, a predator might focus on small, easily captured, but low energy prey, or instead focus on larger, harder to capture, but high-energy prey. More sophisticated models take into account complications such as nutritional value of the food and risk of being captured by a predator while foraging.

The important feature concerning both these modeling traditions is that they assume that there is some form of ecological tradeoff that cannot be relaxed. Thus, for clutch size models, the total amount of energy available for reproduction is fixed and must be partitioned among offspring, and for optimal foraging theory the energy and handling time for each food type is a fixed quantity. A second, often unstated assumption is that adequate genetic variation exists for selection to reach a fitness optimum.

Relating selective changes in phenotype to genetic changes requires the algebraic machinery of quantitative genetics. In quantitative genetic models, the change in phenotype due to selection can be shown to be equal to the additive genetic covariance between the trait and relative fitness:

$$\Delta \bar{z} = \text{cov}_A(z, \tilde{w})$$

where $\Delta \bar{z}$ is the change in the trait of interest, and $\text{cov}_A(z, \tilde{w})$ is the additive genetic covariance between the trait and relative fitness (Arnold and Wade, 1984). Thus, a trait changes as a result of natural selection only to the extent that it covaries with relative fitness.

It is generally true that selection always favors those individuals with the highest relative fitness. As a result, selection on relative fitness is always directional. However, the relationship between phenotypic traits and relative fitness is rarely linear, and the highest relative fitness is attained at intermediate values for most traits. As a result, most traits are under stabilizing selection for an intermediate optimum. This provides a genetic concept of optimality: The optimal phenotype is the phenotype that is at the joint selective value for a set of traits that maximizes relative fitness. Typically, this will be an intermediate value for most or all traits. Methods for studying stabilizing selection on one or more traits are well developed.

The classic example of stabilizing selection is human birth-weight. An early study in northern England identified an optimum birth weight of approximately 8 pounds, with infant mortality increasing with either higher or lower birth weights. Today, optimal birth weight varies strongly across human populations, primarily as a function of maternal nutrition and access to health care. The example of human birth weight illustrates that selection pressures above and below the optimum may be very different. Increased mortality for low birth weight babies reflect problems in early development and ability to thrive that are associated with premature and very small neonates; increased mortality among large birth weight babies reflect complications associated with difficult childbirth.

Even examples of strong directional selection eventually resolve into stabilizing selection for an intermediate optimum. Consider race times of thoroughbred horses. The fastest time for running the Kentucky Derby is currently held by Secretariat, a record set in 1973.

This record has stood for 35 years despite intensive selection on horses to win races. Failure to break this long-standing record probably reflects that faster horses tend to be more prone to injury. Thus, the directional selection for speed imposed by generations of horse breeders is now countered by natural selection against horses that are easily injured.

These examples raise the important point that although stabilizing selection can be modeled directly using a quadratic regression of phenotype on fitness, in most cases selection for an intermediate optimum involves 'correlational' selection. That is, the 'optimum' is typically a tradeoff between competing selective forces on different and correlated traits that maximizes the overall fitness of the individual. The theory of selection on correlated traits has been well developed, and in general emphasizes the point that when two traits have a negative genetic correlation the rate of evolution toward the joint optimum slows drastically.

Genetic correlations can arise in several ways. In genetic terms, they can be caused by pleiotropy, or by linkage. Pleiotropy occurs when one locus affects multiple traits. For example, a genetic locus that positively influences the running speed of a horse may negatively affect the robustness of its leg. The second, less common, cause of genetic correlations is linkage disequilibrium. Two traits may have a genetic correlation because alleles at two tightly linked loci have become nonrandomly associated by chance, selection, or through the mixing of populations. Genetic correlations due to linkage disequilibrium tend to be transient and are generally considered to be less important than pleiotropy.

Genetic correlations can also be defined from functional considerations. Genetic tradeoffs can result from fundamental physical, physiological, or phylogenetic constraints. As an example, consider body size in insects. Insects 'breathe' through a set of tubes called trachea that allow for passive gas exchange. Efficient gas exchange can only occur over short distances, and so the size of insects is limited by their tracheal system and its ability to deliver oxygen to their tissues. Breaking the genetic correlation that produces both trachea and small body size would require a fundamental change in the organism's physiology. Just as pigs cannot fly, insects cannot evolve very large body sizes. Yet other kinds of genetic correlations can be broken over evolutionary time, as was the case for resistance of *Escherichia coli* to T4 bacteriophage. In a classic study, Lenski showed that bacterial mutations conferring resistance to T4 substantially reduced competitive fitness in the early phases of selection. Over the course of 400 generations, however, resistant populations evolved to have competitive fitnesses approaching those of the sensitive populations even though they retained their resistance to the T4 virus. Similar amelioration of the deleterious effects of resistance to insecticides has been observed in insects

Wright's Adaptive Topography

A graphical representation of evolution in multivariate space was suggested by Wright. Wright envisioned an adaptive topography in which a set of 'horizontal' axes represented phenotypes in a population and a single 'vertical' axis represented fitness (**Figure 2**). Wright's adaptive topography model reflects three major generalizations about how genes contribute to fitness, which he inferred from decades of working on the genetics of coat color in guinea pigs:

1. *Multiple factor hypothesis:* The variations of most traits are affected by many loci
2. *Universal pleiotropy:* Allelic substitutions generally have effects on multiple traits
3. *Universal epistasis:* The effects of multiple loci on a trait generally involve a great many nonadditive interactions.

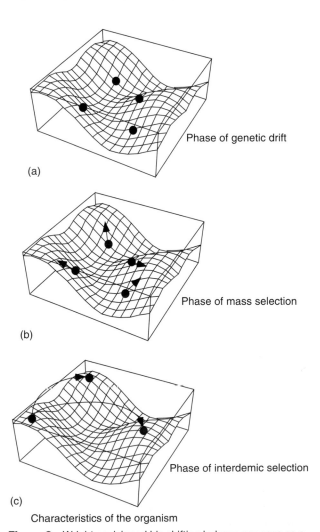

Characteristics of the organism

Figure 2 Wright envisioned his shifting balance process as a mechanism by which a species could shift from one adaptive peak to a second one. (a) Phase 1, the phase of genetic drift. (b) Phase 2, the phase of mass selection. (c) Phase 3, the phase of interdemic selection. See text for a full description.

Together, these generalizations led Wright to conclude that the adaptive topography has a complex shape with multiple adaptive peaks. In other words, *there typically is more than one solution to the problem of achieving high fitness.*

Also central to the concept of adaptive topography is the idea of 'adaptive gene complexes.' Adaptive gene complexes has never been well defined, but may be considered a suite of specific alleles at multiple loci that confer high fitness together, but interact in such a manner that high fitness is attained only when all these alleles are present in the genotype.

Gene Interaction, Adaptive Topographies, and the Shifting Balance Process

Traditional population genetics theory is built on the underlying assumption of additive gene action. This assumption that gene interaction has negligible effects on fitness is relaxed in certain cases. For example, the simplest form of gene interaction, dominance, is directly treated in many models. The most important class of gene interactions ignored by additive models is 'epistasis,' or interactions among different loci. Standard quantitative genetic models, often called 'infinitesimal models,' assume that populations are very large, and that traits are determined by a large number of loci, each with very small effect. Under these assumptions, and with random mating, the effect of an allele is adequately described by its average effect alone. That is, because each allele is found in all possible combinations with other alleles at other loci, and all genotypes are represented in proportion to the underlying frequencies of their constituent alleles, the epistatic interactions average out and can effectively be ignored. Infinitesimal models usually have a single optimum genotype that maps to a phenotype that is itself an optimal compromise among traits. In adaptive topography terms, this would be represented by an adaptive landscape with a single adaptive peak.

Yet few situations in nature fit the infinitesimal model. Wright focused on small populations that were subject to genetic drift. In such populations, interaction variance (dominance and epistasis) can be converted to additive genetic variance upon which selection can act by shifting the local average effects of alleles. The average effect of an allele can be thought of as the effect of that allele on the phenotype of an individual. The local average effect is the effect of an allele on the phenotype measured in a particular subpopulation. When there is gene interaction and genetic drift, the effect of the gene on the phenotype is no longer a function of the gene alone, but rather a function of the gene and the genetic background in which it is found. In most circumstances, that situation yields multiple solutions that maximize the fitness of the individual. These multiple genotypes with high fitness produce an adaptive topography with multiple adaptive peaks.

The infinitesimal model finds the optimum genotype simply, through mutation and selection. Populations starting at any point on the adaptive topography move through the landscape via mutation and selection, eventually climbing to the top of the single adaptive peak. By contrast, populations with multiple adaptive peaks may never reach the highest possible fitness. Mutation and selection alone lead a population to the nearest peak, the local optimum, rather than the highest peak, the global optimum. In many circumstances, the global peak may be separated from starting conditions for the population by a low fitness region of the adaptive topography. Crossing those low regions requires a different process altogether, one modeled by Sewell Wright as his Shifting Balance Theory (SBT).

The Shifting Balance Process

Wright realized that some process other than simple individual selection was needed if a population was to explore an adaptive landscape and arrive at a global optimum, rather than being stuck on a single local optimum. Wright's SBT starts with a population structured as a metapopulation, a set of relatively small subpopulations linked by occasional migration. Because the subpopulations are small, genetic drift is far more important than it would be if the populations were not subdivided. In these subdivided populations, Wright thought that movement between peaks would follow a three-phase process: The phase of genetic drift, the phase of mass selection, and the phase of interdeme selection.

Phase 1: The Phase of Genetic Drift

During this first phase, evolution in small populations is dominated by genetic drift (**Figure 2(a)**). Genetic drift is a function of population size: in very small populations, even selected alleles tend to behave as if they are neutral. Wright envisioned that this drift occurring in subpopulations allows each to move across the adaptive landscape independently. Indeed, some subpopulations potentially drift 'down hill' on the fitness slope and eventually cross an adaptive valley. Thus, genetic drift is the feature of the SBT allowing a subpopulation to escape the influence of one adaptive peak, move through an adaptive valley, and come under the selective influence of a new adaptive peak.

Phase 2: The Phase of Mass Selection

In this phase, subpopulations that drifted through the adaptive landscape come under the domain of influence of a new adaptive peak (**Figure 2(b)**). If selection becomes the dominant evolutionary force, subpopulations climb 'up' the fitness slope to the nearest adaptive peak. For selection to outweigh drift, either selection must become stronger, or the subpopulation sizes must increase. Such shifts are not guaranteed, however. Subpopulations that are small enough to be dominated by drift may remain too small to enter phase 2.

Phase 3: The Phase of Interdemic Selection

In this phase, the metapopulation is split over two or more local adaptive peaks, with some at their original adaptive peak, and some subpopulations on different adaptive peaks. Subpopulations at higher adaptive peaks experience higher fitness by definition, allowing them to grow; in time, they would tend to send out migrants. Subpopulations on the highest adaptive peaks become net exporters of migrants, whereas those on lower adaptive peaks become net importers of migrants (**Figure 2(c)**). Subpopulations that export migrants maintain their genetic integrity, while those that import migrants tend to have their gene complexes disrupted. The net result is to drive those subpopulations toward the higher adaptive peak. Over time, then, all subpopulations converge on the global optimum.

Modern Interpretations of Wright's Shifting Balance Process

Since Wright's first formulation, models incorporating epitasis have shown that drift changes the amounts of additive genetic variance within subpopulations and the local average effects of alleles. Subpopulations coming under the domain of influence of a different adaptive peak in Phase 1 experience shifts in the local average effects of alleles so that the relative fitness advantage conferred by an allele changes as well. Interpretation of phase 2 has been modified for finite populations as well. With gene interaction, selection changes gene frequency, and in the process changes local average effects, again changing which alleles are favored by selection. Thus, the effects of selection on an individual locus depend upon the total genetic background. Finally, phase 3, the phase of interdemic selection, is again influenced by gene interaction in that it is the 'adaptive gene complex' rather than individual genes that determine the fitness differences among subpopulations on different peaks.

Controversy Over the Shifting Balance Theory

The potential for the SBT to be an important model for evolutionary change has been a subject of considerable controversy. Recent theoretical and experimental studies have validated components of the SBT, and it remains an area of active research. Despite the controversy surrounding details of SBT, the concept of an adaptive topography has permeated modern thinking about evolutionary change, especially of complex phenotypes like behaviors.

Current research focuses on explicating relationships between phenotypes, genetic architecture, and fitness. The adaptive topography metaphor implies that the covariance between relative fitness and phenotypic traits of an organism will change in a different manner from the covariance between fitness and the underlying alleles. Numerous selection experiments have demonstrated that phenotypes often respond to selection according to predictions of infinitesimal models. However, molecular studies are confirming the primacy of gene interactions – supporting Wright's generalization of universal epistasis. Furthermore, theoretical studies show that the smooth and predictable behavior of phenotypes may not translate into the smooth and predictable behavior of the underlying genes. Thus, the adherence of phenotypic selection experiments to predictions from additive genetic models may be more apparent than real.

Conclusions

Phenotypic models, such as optimal clutch size and optimal foraging, seek to find a solution from a set of possibilities bounded by ecological, physiological, and evolutionary constraints. These phenotypic models share an underlying assumption that genetic limitations do not allow the organism to break the assumed constraints imposed on the model. Quantitative genetics provides a means of making these genetic assumptions explicit; they also add complexity, such as the assumption that genetic correlations are constant. One of the important features of these quantitative genetic models is that they invite us to view evolution as movement on an adaptive topography. Because phenotypic models often produce multiple optima, they appear to fit the SBT better than a Fisherian additive genetic paradigm. Resolving how behavior is optimized via natural selection will require a full explication of the interactions among behavioral phenotypes, fitness, and genetic architecture.

See also: Cost-Benefit Analysis; Levels of Selection; Optimal Foraging Theory: Introduction; Queen–Worker Conflicts Over Colony Sex Ratio; Trade-Offs in Anti-Predator Behavior.

Further Reading

Arnold SJ and Wade MJ (1984) On the measurement of natural and sexual selection: Theory. *Evolution* 38: 709–718.

Coyne JA, Barton NH, and Turelli M (1997) Perspective: A critique of Sewall Wright's shifting balance theory of evolution. *Evolution* 51: 643–671.

Fisher RA (1930) *The Genetical Theory of Natural Selection.* Oxford: Oxford University Press.

Goodnight CJ (1988) Epistasis and the effect of founder events on the additive genetic variance. *Evolution* 42: 441–454.

Lack D (1947) The significance of clutch size. *Ibis* 89: 302–352.

Lenski RE (1998) Experimental studies of pleiotropy and epistasis in *Escherichia coli*. II. Compensation for maladaptive effects associated with resistance to virus T4. *Evolution* 42: 433–440.

Trivers RL (1974) Parent–offspring conflict. *American Zoologist* 14: 249–264.

Wade MJ and Goodnight CJ (1998) The theories of Fisher and Wright: when nature does many small experiments. *Evolution* 54: 1537–1553.

Wright S (1931) Evolution in Mendelian populations. *Genetics* 16: 93–159.

Aggression and Territoriality

B. C. Trainor, University of California, Davis, CA, USA
C. A. Marler, University of Wisconsin, Madison, WI, USA

Introduction

A central component of theories of natural selection is that individuals are competing for limited resources. In many species, individuals use territories to maintain access to resources and mates, and aggressive behavior is frequently used to enforce the boundaries of territories. Although maintaining access to a resource such as food or a courtship site has obvious benefits, aggressive behavior has important costs (**Figure 1**). Aggressive displays are usually energetically expensive, and fighting increases the risk of injury or even death. In some vertebrate and invertebrate species, the decision to be territorial can be dependent on the density of conspecifics, food abundance, and distribution, as well as levels of stored energy. While aggression itself in the form of direct conflict between individuals is a primary mechanism for defending a territory, animals use a variety of other display behaviors to advertise their current ownership of a territory. Some species use brightly colored visual displays or advertise acoustically, while others deposit scent markings on the territory, particularly around the boundaries. When these signals are not sufficient to deter intruders, a territory holder may engage in physical aggression.

Aggression can be operationally defined as overt behavior that has the intention of inflicting physical damage on another individual. Examples include wrestling between horned beetles, biting in rodents, and darting flights by birds. In many species, there is a positive correlation between territory quality and reproductive success, suggesting there are important fitness consequences for winning aggressive encounters. Given the high costs of territorial aggression, it is not surprising, then, that aggressive behavior is tightly regulated with multiple levels of control that integrate information about the physical and social environment. Interestingly, it also appears that the same set of aggressive behaviors can be stimulated by different hormonal or neurobiological mechanisms under different environmental conditions.

Testosterone (T) is often a focus of studies examining hormonal mechanisms regulating aggression. Although it is usually assumed that T increases aggression, this relationship is much more complex and often depends on seasonal or social cues. In addition, T is a dynamic hormone that can change rapidly during a single aggressive encounter. Under some conditions, long-term baseline T levels do not correlate well with behavior, whereas a short-term increase in T may be closely associated with aggression and territory defense. Mechanistically, short-lived changes in T have tended to be ignored, possibly because it is usually assumed that steroids such as T and estrogens require several hours or days to exert a behavioral effect. However, there is increasing evidence that steroids can affect behavior rapidly. In addition, circulating T can be converted in the brain into dihydrotestosterone (a more potent androgen) or estradiol (a potent estrogen). Thus, differences in how T is metabolized within the brain can have important consequences for how behavior is affected by T.

Mechanisms of aggression have been studied in a wide variety of taxa, but due to space limitations, we will focus our discussions on studies conducted on rodents and free living birds. However, many interesting studies have identified mechanisms of aggressive behavior in other taxa, including fish, insects, reptiles, amphibians, and primates.

Studying Aggression in Birds and Rodents

The majority of studies on birds are conducted in field settings. An advantage of field studies is that aggressive behaviors can be observed in a complex environment, along with the fitness consequences of aggressive behaviors. A disadvantage is that because a field setting is less controlled, it is more challenging to conduct manipulations and physiological measurements. One of the most common methods for testing aggression in birds is the simulated territorial intrusion (STI), in which a caged male is placed near a resident male and a speaker is used to play songs. Typically, territory holders respond to STIs with a variety of aggressive behaviors, including producing song and darting at the intruder.

Almost all studies on rodents are conducted in the laboratory, and the most commonly used behavioral paradigm used is the resident–intruder test. The focal male (the resident) is housed in a cage for 2–5 days, and then an unfamiliar intruder is introduced into the resident's cage. In most species, male residents attack the intruder by biting the flanks or boxing with the forepaws. The frequencies of these behaviors can be a measure of the intensity of aggression. The motivation to fight can also be reflected by the latency to first attack. The resident–intruder test is designed to model a resident defending a territory, although it is only a rough approximation of

- Potential costs
  - Energy expenditure
  - Risk of injury
  - Exposure to predation

- Potential benefits
  - Access preferred food
  - Access to shelter
  - Mating opportunities

Figure 1 Potential costs and benefits to engaging in aggressive interactions. Distribution of resources and mating systems will influence the relative magnitude of the listed costs and benefits. Hawk photo by Steve Jurveston.

natural interactions. The main advantage to laboratory studies such as the resident–intruder test is the ability to conduct a wide variety of manipulations and measurements. Using implants, it is possible to measure the heart rate or the neurotransmitter release in real time. It is also possible to conduct precise hormone manipulations that would be difficult or impossible in a field situation.

Territoriality and Aggression in Seasonally Breeding Birds

In many passerine birds, breeding occurs in the summer and males defend breeding territories. This territorial aggression is usually associated with increased baseline plasma T. In many species, males provide parental care by feeding their chicks, and both T and territorial aggression decrease while males are provisioning their chicks. Hormone manipulation experiments in several species show that artificially increasing T with an implant during the parental phase can restore territorial aggression, but at the expense of parental behavior. In some cases, however, the negative relationship between increased T and paternal behavior has been dissociated. In species such as the rufous-collared sparrow, *Zonotrichia capensis*, increasing T does not inhibit paternal behavior. This insensitivity to T may have evolved because paternal care is essential, or because the breeding season is so short that it is impossible to breed late in the season.

Additional studies suggest that the relationship between T and aggression in birds is stronger during the breeding season. If a male song sparrow is removed from its territory, neighboring males compete to take over the recently vacated territory. If the experiment is conducted during the start of the breeding season, then the competing males have increased T. However, if the experiment is conducted outside the breeding season (autumn), then T is

not increased, even though competition over the vacated territory is intense. An STI conducted during the breeding season provokes an aggressive response by the resident as well as an increase in T. However, an STI conducted in the fall does not increase T, even though male residents respond aggressively. These studies suggest that T produced by the gonads is not essential for aggression outside of the breeding season, a hypothesis supported by observations that castration of male song sparrows does not reduce aggression during autumn STIs. Intriguingly, it appears that nonbreeding aggression is regulated by estrogens that are synthesized in the brain and not the gonads.

When male sparrows are treated with an aromatase inhibitor (to block synthesis of estrogens), aggression during the nonbreeding season is reduced. Interestingly, this effect is observed within 24 h, which is relatively fast for a steroid hormone manipulation. This is important because most changes in gene expression mediated by steroid hormones take several hours or days to occur. In contrast, physiological changes that do not depend on gene expression changes (so-called 'nongenomic' effects) can occur more rapidly. Subsequent studies suggest that the source of androgens for estrogen production may be the adrenal gland, specifically, dehydroepiandrosterone (DHEA). DHEA is not in itself an androgen, but is converted to androstenedione (an androgen) in the songbird brain. This is significant because plasma DHEA levels are elevated in nonbreeding males. Nonbreeding birds treated with DHEA implants increase singing behavior but do not increase aggressive behaviors, suggesting the possibility that a minimal threshold level of DHEA is necessary to support estrogen-dependent aggression. This hypothesis is supported by observations that DHEA levels decrease when males are molting feathers, a period when males are not aggressive.

Photoperiod and Aggression in Rodents

In many mammalian species, seasonal changes in behavior can be induced by light cycles, or photoperiod. For example, in many species of rodents, reproduction is inhibited in winter months and this inhibition can be induced by exposure to short days. In males, reproductive inhibition usually involves regression of the testes and a sharp decrease in T levels. Conventional thinking would then suggest that aggression levels should be reduced in winter-like short days. However, male aggression across a wide variety of hamsters and mice is increased in short days despite reduced T. Evidence from several species suggests that the increased aggression observed in short days may be independent of changes in T. For example, in Siberian hamsters, *Phodopus sungorus*, there is natural variation in the reproductive responses to photoperiod, and some individuals maintain large testes size and increased

T in short days. In a resident–intruder aggression test, these 'nonresponsive' individuals attack an intruder more quickly and more often than individuals housed in long days with equivalent testes sizes and T levels. Complementary evidence is seen in the California mouse (*Peromyscus californicus*), a species in which short days do not reduce testes size or T levels. Despite the absence of reproductive responses, male California mice are more aggressive in resident–intruder tests when housed on short days. These studies suggest that changes in T secreted by the testes cannot explain the effect of short days on aggression in rodents.

In hamsters, adrenal steroids play a role in plasticity in aggression. For example, removing the adrenal cortex of Siberian hamsters blocks the effects of short days on aggression. The hormone(s) affecting aggression from the adrenal glands are, however, unclear. In addition to producing glucocorticoids, the adrenal cortex can also synthesize DHEA, which could be converted into androgens in the brain. Siberian hamsters have increased DHEA levels in short days. Based on the importance of DHEA in regulating aggression in male sparrows outside the breeding season, it was hypothesized that short days might increase aggression by increasing DHEA. This hypothesis was not supported in a study demonstrating that DHEA implants did not increase aggression in hamsters housed in long days. Since removing the adrenal cortex blocks the effects of short days on aggression and increasing DHEA does not affect aggression, it may be that a minimal threshold of DHEA is required to promote aggressive behavior. It could also be that short days alter systems that were affected by DHEA or its metabolites. In birds, it appears that downstream estrogenic metabolites play a critical role in regulating aggression outside of the breeding system. A series of studies in the mice of the genus *Peromyscus* demonstrate that estrogens have important effects on male aggression, and that photoperiod plays an important modulating role.

In *Peromyscus*, estrogens increase male aggression in short days, as would be predicted based on results from sparrows. However, the relationship between estrogens and aggression in *Peromyscus* is complex. In old field mice (*P. polionotus*), estrogens decrease aggression in long days, but in short days, estrogens increase aggression. Intriguingly, estrogen receptor α (ERα) expression in the brain increases in short days, whereas estrogen receptor β (ERβ) expression in the brain increases in long days. These results appeared very relevant to aggressive behavior, because studies in estrogen receptor knock-out mice suggest that ERα increases aggression and ERβ decreases aggression. However, a different pattern of regulation was observed in *Peromyscus*. Drugs that selectively activate either ERα or ERβ decrease aggression when mice are housed in long days, and these same drugs increase aggression when mice are housed in short days. It appears that short days change how estrogens act in the

brain at a molecular level. When estrogen binds ERα or ERβ in the brain, these receptors can act as transcription factors, altering gene expression. This is a slow process, and most studies manipulating steroid hormones such as T or estradiol allow for at least 1 or 2 weeks for hormone manipulations to affect behavior.

To assess whether photoperiod influences how estrogens regulate gene expression, microarrays were used to measure the expression of genes that are estrogen dependent. In the bed nucleus of the stria terminalis (a brain region that regulates aggression), estrogen-dependent gene expression was up-regulated in mice housed in long days compared to short days. These data suggest that estrogens may decrease aggression in long days by promoting gene expression. In contrast, gene expression does not appear to be a central component of estrogen action in short days. In California mice and old-field mice, a single injection of estradiol increases aggression within 15 min if the mouse is housed in short days. If the mouse is housed in long days, an injection of estrogen has no effect on behavior. Fifteen minutes is considered too short for changes in gene expression to occur, so the rapid effect of estrogen on aggression in short day mice is most likely mediated by nongenomic mechanisms. Recent work has highlighted that steroids such as estradiol can phosphorylate kinases, regulate ion channels, or alter neurotransmitter release. All of these effects could contribute to rapid changes in aggressive behavior.

A final complicating factor is how studies conducted on rodents in short days compares with studies on nonbreeding birds. Although hamsters and many *Peromyscus* are reproductively suppressed in short days, there is important variability. For example, while most male Siberian hamsters have regressed testes under short days, some individuals are 'nonresponsive' and maintain their testes under short days. These 'nonresponsive' hamsters show high levels of aggression in short days, exactly like short-day hamsters with regressed testes. In addition, male California mice appear to be capable of reproducing throughout the year, yet expressed increased aggression levels in short days. A major unsolved question is why aggression in so many species of rodents is increased during short days when many individuals are not breeding.

Effects of Experience on Aggression

Within a season, individual variation in male territorial aggression may also occur. For example, in a contest over a territory, residents often have an advantage. The reasons vary with species, but can be related to traits intrinsic to the territory owner, such as fighting ability or size, or related to traits emerging from interactions between the territory owner and the physical environment, such as familiarity with the territory. Individual variation in territorial aggression in response to social stimuli can also be induced through a variety of mechanisms. For example, the behavioral response to an intruder can vary based on familiarity. The 'dear enemy' phenomena suggests that there is a lower aggressive response to a neighbor, but this may only be accurate as long as the boundaries remain stable. This effect has been described in both birds and frogs.

Another example of social influences is related to past experience such as the ability to win a contest with an intruder. The loser effect is well established and an individual that loses an encounter is more likely to lose future encounters as a result of long-lasting changes in the hypothalamic-pituitary-adrenal (stress) axis. Evidence for the winner effect has been established across a variety of species. *Peromyscus* mice have also become a model system for investigating this. The California mouse is strictly monogamous and defends territories year-round. In the laboratory, inexperienced male California mice are randomly assigned to win between 1 and 3 encounters as residents in the resident–intruder test. These tests were rigged so that the resident would always win. After this training phase, the residents were tested against a larger intruder. The residents accumulate an increased ability to win with an increased number of previous wins (winner effect). Because the residents were randomly assigned to experience different numbers of wins, it was demonstrated that the experience of winning (independent of intrinsic competitive ability) increased the probability of winning an aggressive encounter against a larger opponent. Interestingly, while California mice display a relatively robust winner effect, the strength of the winner effect in a familiar area can differ between species. In contrast to the California mouse, the white-footed mouse, which is promiscuous and significantly less territorial than the California mouse, displays a substantially diminished winner effect.

As described earlier, androgens play a significant role in facilitating aggressive behavior. Research related to the 'challenge hypothesis,' originally formulated from avian studies, has revealed that androgens often rise briefly after a male is challenged by another male and social stability has not been attained. Recent studies in California mice have investigated the function of this transient increase in testosterone and find that it modulates both future aggression and the ability to win. The ability of T to modulate future aggression appears to operate through androgen and not estrogen receptors in the brain (via conversion to estrogen through the enzyme aromatase). Furthermore, winning experience alone in males can induce an increased ability to win future encounters in an additive fashion with T. Interestingly, the white-footed mouse does not experience a testosterone surge at the same time after an encounter as the California mouse and, as stated earlier, it exhibits a much weaker winner effect. One variable that may influence

this species difference is the effect of residency, with the California mouse being a territorial species and the white-footed mouse (*P. leucopus*) being less territorial and using a roving strategy for finding females. The results thus far suggest that the separate effects of past winning experience and exposure to transient increases in T can induce changes in an individual's aggressive behavior. This relationship, however, is complex because of interactions between these two factors; T itself may influence future winning ability, but this influence is much stronger when it is coupled with the experience of winning a fight.

The species comparison between California mice and white-footed mice also raises the hypothesis that residency, a critical component to territoriality, may influence the development of the winner effect. In fact, California mice do not display a full robust winner effect unless they have the 'home advantage,' regardless of intrinsic fighting ability. Thus, interactions between the physical environment and the territorial resident can induce variation in the expression of the winner effect. For territorial animals, an individual's fitness depends on its ability to win aggressive disputes. During the establishment of a territory, frequent aggressive encounters may occur as individuals compete for or expand their territories. Because the costs of aggression can be high, a residency dependent winner effect might be adaptive because it would allow individuals to adjust their winning ability in a context-specific manner.

Although territoriality and aggression are typically overlooked in females, in some species they show high levels of aggressive behavior. This is especially prevalent in biparental species where males and females jointly defend territories. As in males, a role for androgens has been found in female territorial aggression, and seasonal context appears to be significant. In a neotropical songbird (the spotted antbird, *Hylophylax naeviodes*), females confronting an intruder show an increase in T during the prebreeding season, but no change in T during the breeding season. In the California mouse, male–female pairs defend territories together. During an aggressive encounter with another female, the dominant hormonal change is a rapid decrease in progesterone, similar in profile to the rapid increase in T observed during encounters in males. An identical decrease in progesterone during aggressive encounters was observed in female African cuckoos. In addition, female cuckoos treated with progesterone implants were less aggressive than females receiving empty implants, suggesting that a transient decrease in progesterone may indeed facilitate increased aggression. Additional studies in other species are necessary to determine whether the effects of progesterone on aggression in females are as widespread as the effects of T on male aggression appear to be.

There is a multitude of hormones and neurochemicals that influence territorial aggression. One with an intriguing influence on territorial behavior is the neuropeptide

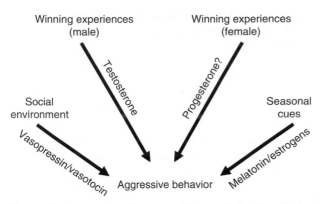

Figure 2 Hormones and neuropeptides mediating the effects of experience on aggressive behavior.

arginine vasopressin. In prairie voles, *Microtus*, variation in space use is related to the expression of the vasopressin 1a receptor in the areas of the brain implicated in spatial memory. Moreover, in estrildid finches, vasotocin (the nonmammalian homologue to vasopressin) neurons in the medial extended amygdala respond differently to social cues in gregarious versus solitary species. Space use is critical to the development of territorial behavior and can have a major impact on social systems. Vasopressin is also associated with aggression such that in the strongly territorial California mouse, the vasopressin receptor (V_{1a}) antagonist reduces resident–intruder aggression in male California mice, but not in the less territorial white-footed mice. Within this same species, paternal behaviors, such as retrieval of pups, have a long-term impact on adult aggression insofar as it increases aggression, as well as immuno-reactive vasopressin staining, in the brains of adult male offspring. Vasopressin antagonists also reduce dominance behaviors in golden hamsters. Overall, vasopressin and its homologues may exert a significant influence over territorial aggression (**Figure 2**). This may also be linked with androgens because these steroids often have a potent stimulatory effect on the vasopressin and vasotocin systems.

Future Directions

Territorial aggression is a major component of social systems that fluctuates seasonally as well as within breeding seasons. The plasticity in this behavior in response to a variety of physical and social aspects of the environment may reflect the multitude of selection pressures that can shape territorial behavior. Androgen manipulations have

revealed the costs in a variety of species, including field manipulations with the Mountain spiny lizard (*Sceloporus jarrovi*) and dark-eyed juncos (*Junco hyemalis*). In the appropriate contexts, territorial aggression is highly beneficial for gaining access to resources. The study of mechanisms controlling these behaviors is proving to be rewarding because of the striking plasticity of the behavior. New neural mechanisms are continually being discovered that reveal the complexity of control of aggression. Investigations into how different aspects of the environment influence territorial aggression through varying mechanisms and how this information is integrated at a neural level represents an opportunity for researchers interested in plasticity of behavior and neural mechanisms.

See also: Circadian and Circannual Rhythms and Hormones; Fight or Flight Responses; Hormones and Behavior: Basic Concepts; Stress, Health and Social Behavior.

Further Reading

Davis ES and Marler CA (2003) The progesterone challenge: Steroid hormone changes following a simulated territorial intrusion in female *Peromyscus californicus*. *Hormones and Behavior* 44: 185–198.

Demas GE, Polacek KM, Durrazzo A, and Jasnow AM (2004) Adrenal hormones mediate melatonin-induced increases in aggression in male Siberian hamsters (*Phodopus sungorus*). *Hormones and Behavior* 46: 582–591.

Goodson JL (2005) The vertebrate social behavior network: Evolutionary themes and variations. *Hormones and Behavior* 48: 11–22.

Hau M (2007) Regulation of male traits by testosterone: Implications for the evolution of vertebrate life histories. *BioEssays* 29: 133–144.

Oyegible TO and Marler CA (2005) Winning fights elevates testosterone levels in California mice and enhances future ability to win fights. *Hormones and Behavior* 48: 259–267.

Soma KK, Scotti MA, Newman AE, Charlier TD, and Demas GE (2008) Novel mechanisms for neuroendocrine regulation of aggression. *Frontiers in Neuroendocrinology* 29: 476–489.

Trainor BC, Bird IM, and Marler CA (2004) Opposing hormonal mechanisms of aggression revealed through short-lived testosterone manipulations and multiple winning experiences. *Hormones and Behavior* 45: 115–121.

Trainor BC, Finy MS, and Nelson RJ (2008) Estradiol rapidly increases short-day aggression in a non-seasonally breeding rodent. *Hormones and Behavior* 53: 192–199.

Trainor BC, Lin S, Finy MS, Rowland MR, and Nelson RJ (2007) Photoperiod reverses the effects of estrogens on male aggression via genomic and nongenomic pathways. *Proceedings of the National Academy of Sciences of the United States of America* 104: 9840–9845.

Wingfield JC, Hegner RE, Dufty AM, and Ball GF (1990) The challenge hypothesis – theoretical implications for patterns of testosterone secretion, mating systems, and breeding strategies. *The American Naturalist* 136: 829–846.

Agonistic Signals

C. R. Gabor, Texas State University-San Marcos, San Marcos, TX, USA

Introduction and Definitions

The process of communication involves information sent by a sender in the form of a signal that is detected by a receiver. The receiver, in making response decisions, uses the information content of the signal. The response of the receiver affects its own fitness, as well as that of the sender. In 'true communication,' both the sender and the receiver gain fitness benefits from their interaction. However, in some contexts, the sender suffers a fitness reduction (e.g., eavesdropping), or the receiver suffers a fitness reduction (e.g., deceit). More information on communication is discussed elsewhere. The focus of this article is on the mechanisms and the evolution of agonistic signals – those signals that are used in conflict resolution. Individuals often have conflict – especially over ownership of valuable resources such as food, territories used for foraging, territories used for breeding, and mates. Many animals use agonistic signals to minimize the costs of escalated violence. Such costs include the risk of injury, exposure to potential predators, as well as the energetic costs of fighting. Signals used in aggressive interactions function to resolve conflict and thus should benefit both the receiver and signaler, as escalated contests are costly to both senders and receivers. Signals used by senders in agonistic contests should be predictive of what the animal will do next if the intruder does not retreat. This could be an escalation of the intensity of the contest, or an impending attack. In the following sections, the design considerations of agonistic signals, followed by a discussion of the evolution of signal honesty in conflict resolution, are presented.

What Sensory Modalities Are Used in Agonistic Communication?

The type of signaling modality used by a sender is shaped by the costs and benefits associated with transmitting the signal. The costs and benefits of signal transmission are influenced by the type of information being transmitted, as well as by the abiotic and biotic environment in which the signal is transmitted. Because conflicts over resources are generally initiated and resolved with close distance between the actors, agonistic signals should be designed to travel short distances, be directed at individual rivals and be highly locatable, and reveal the age, body size, or social status of the signaler. They are usually short, forceful and conspicuous, as they need to send a clearly aggressive message. Threat signals often incorporate body parts and movements used in fighting into the ritualized display. Baring teeth in biting mammals and display of horns in fighting antelope are two such examples. Submissive signals often have the opposite states to aggressive signals and thus do not expose the signaler's weapons and reduce the chance that the signaler is attacked.

Signal modalities that are used in agonistic encounters include visual, acoustic, olfactory, electrical, and tactile. Because the signals are transmitted over short ranges, they do not need to travel long distances. Visual signals are generally limited to diurnal displays in open habitats that require the sender to always be present. For visual signals, individuals may use specific threat postures, coloration, or movement displays such as the territorial displays of *Anolis* lizards that include the extension of the brightly colored dewlap of males, as well as the headbob behavior. In addition, visual signals used in agonistic encounters are directed at the intended receiver (the rival) and are of short duration. Visual signals also convey information about the status or class (e.g., age, dominance) of the sender such as 'badges of status.' Badges of status are markings that may be used by animals to signal their size and dominance – they are indicators of rank. In many bird species, changes in color patterns or the development of badges of status are associated with an individual's aggressive tendencies. In house sparrows (*Passer domesticus*), there is a positive correlation between an individual's dominance level and the area of the status badge. Visual signals also convey information about an individual's body size, which can be enhanced with color patterns or striping patterns that enhance signal efficacy. In addition to indicating status or size, visual signals can convey information about variable levels of motivation or fighting ability via modulation of the degree of intensity of the signal. In some fish, for example, aggressive signals can be indicated by intensification of body coloration (bars) whereas suppression of coloration can signal defeat or subordination.

Acoustic signals are useful in defense of larger territories, because unlike visual signals, environmental barriers do not limit their use. They are effective over long distances and around corners. As such, the signaler does not need to be in direct visual contact to defend its territory from intruders. Acoustic signals are also localizable such that the receiver can usually determine where the signaler is located while calling. Acoustic signals are generally loud. Sound frequency can indicate body size, age, and sex. In addition, animals can modulate the intensity of acoustic signals by varying levels of call rate, call duration, frequency, and intensity. In frogs, acoustic signals are related to body size: call frequency and body size are negatively correlated in many species. It has also been shown that in many species of frogs, males alter their acoustic signals during interactions with other males. For example, male green frogs, *Rana clamitans*, significantly decrease the dominant frequency of their calls during aggressive territorial encounters and males differentially alter their behavior according to frequency of their opponent's calls, suggesting that they may use this information as an indication of a rival's body size.

Olfactory signals used in short range agonistic communications are usually volatile and rapidly diffuse. They are also transmitted to the receiver more slowly than visual and acoustic signals and are not directional. Some olfactory signals are liquid secretions, such as urine, that are used in territorial marking in many mammals. Olfactory signals are hard to localize, so substrate marking with long durable scents is most effective for delineating a territory via numerous marks around the territory. Olfactory signals can also be directly sprayed at rivals, as is found in skunks. Other olfactory signals are derived from maturation hormones, and thus contain information about the sender. For example, olfactory cues from testosterone can provide information about age and dominance status. For example, in five African cichlid species (*Neolamprologus pulcher, Lamprologus callipterus, Tropheus moorii, Pseudosimochromis curvifrons*, and *Oreochromis mossambicus*) androgen levels increase in males in response to territory intrusions. Olfactory signals are difficult to modulate and are often coupled with signals in other modalities for modulation of the threat.

Electrical signals consist of electric fields created by the electric organ discharge (EOD) that weakly electric fish use during agonistic encounters. Electric signals are localizable but only over very short distances. The frequency of these emissions can be modulated during encounters and can indicate aggression or submission. For example, male brown ghost knifefish, *Apteronotus leptorhynchus*, modulate the frequency of their electric organ discharge (EODF) such that winners use increasingly more abrupt EODF (signal of dominance and aggression) and more rapid frequency increases than losers.

Tactile signals occur via appendage movements of the sender that are detected by the receiver via nerve endings, such as pushing and pulling between rivals. For example, male thrips (*Elaphrothrips tuberculatus*) align their bodies in parallel and bat at each other with their abdomens. This tactile exchange of signals reveals the body size of the individuals engaged in the interaction. The likelihood of attack by either individual depends on their absolute size.

Some species use one signal to communicate in more than one context. For example agonistic signals used in territorial defense and signals used in mate attraction overlap, such as in many species of birds. These signals have a dual purpose: attracting conspecifics of the opposite sex and repelling conspecific rivals. In birds, both acoustic signals (calls and songs) and visual signals such as status badges may have this dual function. Often the signal itself is modulated between the functions where territorial songs may differ in length (usually shorter) than songs used in mate attraction (usually longer). An example of an honest signal of dominance or fighting ability that is sexually selected through male contest competition is found in red-collared widowbirds (*Euplectes ardens*). In this species, carotenoid coloration indicates status, with increasing aggressiveness correlated with larger redder collars.

For more detailed information on signal modalities, see other communication articles within this book.

How Many Signals Are Used?

Some animals use only a few displays in agonistic contexts, while others use many displays depending on the level of risk. Three hypotheses have been proposed for why multiple displays used in agonistic encounters have evolved and this is known as the 'intention signal controversy.' The first hypothesis, proposed by Tinbergen, is that displays serve slightly different functions and/or are used in different situations. Here, different displays could be used in conflicts over different types of resources such as females, food, and territories, or to indicate different types of opponents such as neighbor versus intruder or adult versus younger individuals. In addition, different displays could indicate higher versus lower probability of attack depending on the threat posed by the opponent. Alternatively, the mode of the display might vary on the basis of the transmission needs where acoustic displays might be more effective for distant threats and visual displays for closer intruders.

The second hypothesis, proposed by Andersson, is that threats lose their value over time and require the evolution of new threats. Andersson suggested that most threats evolve from intention movements – movements that indicate the animal is getting ready for an action. As these movements become ritualized the display may become decoupled from the subsequent action, and as a result, more bluffs occur until the point that the displays do not accurately predict what the sender will do next. Receivers are more likely to ignore these signals and focus on more reliable signals of the sender's future behavior. The outcome is that new threat signals evolve and the nonreliable displays stop being used such that at any given time, both predictive and nonpredictive displays are within a species' repertoire. The possibility of changes in sender and receiver strategies over time has been explored theoretically and empirically. Empirical evidence of changes in sender and receiver strategies over time as a result of inaccurate displays is still limited, as changes in displays across populations can be due to other factors such as environmental differences. Models of communication demonstrate that reliance on signals by receivers imposes intrinsic costs on senders that can only be made up if signals are honest and accurate. Theoretically, such conflict between senders and receivers should yield the evolution of new honest displays.

A third hypothesis, proposed by Enquist, for the evolution of multiple agonistic signals is that displays provide information on different intentions or aggressive motivation levels (termed 'motivational signaling'). Displays may serve to signal submission, offensive threat, defensive threat, dominance maintenance, victory, or ownership. Different displays may also be associated with different levels of fear or aggressive motivation. The effectiveness of the display in deterring opponents is correlated with the cost of performing the display, resulting in honest displays. An example of graded threat display is seen in the red-backed salamander (*Plethodon cinereus*). In this species, males use the all trunk raised (ATR) posture as a threat and the extent to which they raise their body off the ground is indicative of the intensity of the threat such that intruders show an increase in the rate of submissive response from ATR1 to ATR5 in **Figure 1**.

All three hypotheses probably work together, but evidence of signaling that are predictive of aggressive follow-up actions provides the strongest support for the third hypothesis of motivational signaling. To distinguish between the three hypotheses, future research must control for the receivers response to a signal when examining the subsequent response of the sender, as a threat signal sends the message that an attack will occur if the receiver does not retreat and vice versa if the receiver is highly motivated to escalate in the conflict.

Role of Agonistic Signals in Conflict Resolution

Conflicts occur over ownership of resources. Conflict resolution includes the exchange of threat and submissive signals between individuals. Such agonistic signals are hypothesized to have evolved from selection favoring the exchange of threat and submissive signals over escalated violence. Empirical and theoretical studies have demonstrated that there are at least four factors that determine the outcome of conflicts over resources (contest outcomes). The first is the relative fighting ability of a contestant (i.e., resource-holding potential, RHP). RHP is generally the main determinant of winning or losing agonistic encounters and is generally measured from the animal's relative body size, which provides an indication of the individual's relative fighting ability. However, other traits may also signal RHP, such as acoustic signals, that are correlated with body size. In addition, individuals of some species assess multiple components of RHP. For example, the acoustic agonistic signals of some frog species are honest indicators of body size as the signal frequency is constrained by body size. Visual cues that reveal RHP includes broadside threat displays that provide accurate information about relative body size, and weapons such as horn displays in some mammals (e.g., mountain sheep, red deer) that may reveal relative fighting ability.

Many studies of agonistic contests examine the ability of animals to assess and compare their own RHP with that of their rival, and to make decisions based on the estimated differences. If assessment in both directions is possible, then the individual with the lower RHP should terminate the contest immediately, thus reducing the time, energy, and risk of injury from an agonistic contest. As the difference in body size between opponents decreases, the average duration of contests and the variance in fighting duration should also increase. When size is not a good indication of RHP, repeated assessment is required to evaluate both resource value (RV) and relative fighting ability. In these cases, animals acquire more accurate information about the opponent with successive interactions, using repeated actions, because these reveal more accurate information about fighting ability.

The second type of trait that affects contest outcomes is RV, which arises from asymmetries in the quality of the

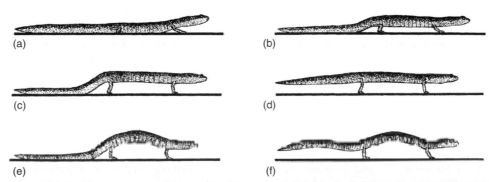

Figure 1 Gradational threat postures of *P. cinereus*. (a) A salamander rises from its resting posture to (b) ATR 1, low stance, to (c) ATR 2, high stance. (d) In ATR 3 the tail raised, or (e) in ATR 4 the back is arched, or (f) in ATR 5 both occur simultaneously.

contested resource (i.e., mate, food, nest, or territory). Fighting intensity varies with the quality of the resource as do the duration and the probability of victory. As the value of the resource increases, the contest duration should increase. When individuals value the resources differently, contests are likely won by the animal that places a greater value on the resource. While RHP alone can be an important factor in contest, many times RHP and RV both need to be evaluated and yield different predictive outcomes. For example, in some species such as hermit crabs (*Pagurus longicarpus*) and house crickets (*Acheta domesticus*), size (RHP) is predictive of the winner of contests, but motivation based on the signaler's assessment of their RV becomes the predominant factor when RV differs.

The third determinant of the outcome of agonistic interactions is related to the underlying aggressiveness of a given individual (called the aggressive syndrome). It has recently been demonstrated that there is variation between individuals in aggressive behavior and these differences lead to differences in aggressive behavior across contexts (e.g., similar aggressiveness in mating and agonistic encounters) that are not always related to RV or the RHP. These conserved 'personalities' across contexts are called behavioral syndromes. For example, behaviorally aggressive funnel web spiders, *Agelenopsis aperta*, attack both prey and conspecific territorial intruders more quickly than do less aggressive spiders.

Outside of consistently aggressive behavior, one main determinant of agonistic signal use is whether the main source of variation between contestants is in fighting ability or in RV. In species with large variation in body size or body condition, signals of fighting ability are expected. If the major source of variation is RV, then signals revealing aggressive motivation level should be used. This is expected in systems with little body size variation, in territorial systems with ownership asymmetries or those where prior contest outcomes affect subsequent behavior. This is so because signals indicating fighting ability via size are ineffective because of small variation in sizes.

A final determinant of the outcome of contests is prior ownership of a resource. Resident effects suggest that residents typically have higher probabilities of winning contests over territorial resources. Residents may have the advantage because they can base their decision on the true value of the resource and their relative fighting ability, whereas intruders can only base their decision on relative fighting ability. Studies of prior winner effect have found that prior fighting outcomes will affect subsequent fight outcomes where winners are more likely to win again. One study that examined why residents generally win contests found that in speckled wood butterflies (*Pararge aegeria*) intrinsically aggressive males are more likely to be residents, and continue to win because of a prior winner effect. Thus, residency does not serve as an arbitrary cue for contest settlement in this species; instead the likelihood of being able to acquire a territory is linked to the subsequent success in defending that territory.

Honesty or Deceit in Agonistic Signals?

Agonistic signals, including both threat and submissive signals, are expected to be honest and should reliably indicate different levels of fear or aggressive motivation, because of the cost of performing displays. Costs include physiological constraints, production expense, or risk of retaliation from the receiver. Agonistic signals can be classified as either 'performance signals' or 'strategic signals.' Performance signals (also referred to as unambiguous signaling, unbluffable signaling, assessment signaling, and revealing handicaps) are directly constrained by an individual's RHP and therefore must be honest. An example of an agonistic performance signal is found in fish that exhibit mouth wrestling such as in the cichlid, *Nannacara anomala*. These fish lock jaws and attempt to push each other backwards. The smaller individual in the interaction is physically more constrained in the force they can generate while pushing than the larger individual.

Strategic signals (also referred to as conventional signals) can be used by all senders and are not necessarily correlated with the quality of a resource. Therefore, strategic signals used in agonistic encounters can be deceitful, yet honesty could be maintained by costs, including production costs and the response of the receiver to the signal. One example of a strategic symbol is the status badge. In some species, receivers use badges to assess the agonistic abilities of strangers. In the wasp, *Polistes dominulus*, variable facial patterns function as badges and signal social status (**Figure 2**). Wasps assess these facial patterns and avoid opponents with badges that signal higher quality and challenge opponents that signal lower quality. Such responses ensure signal honesty while minimizing the costs of conflict.

The use of signals imposes on receivers intrinsic costs greater than the costs associated with signal processing alone. As a result, these costs can only be maintained if signals are sufficiently honest and accurate. Signals of intent in conflicts may provide information about the signaler's aggressive motivation or about what the signaler may do next, such as attacking. Signals of intent in conflict, however, are more susceptible to bluffing compared to those that are intrinsically constrained or costly to produce. If receivers impose costs via retaliation, then these signals can be stabilized. Individuals need to determine whether or not agonistic signals are correlated with the underlying quality of a contested resource. Once signals no longer convey dependable information, then

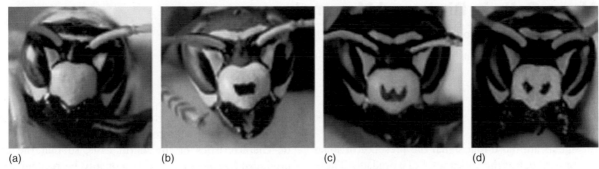

(a)   (b)   (c)   (d)

Figure 2 (a–d) Portraits of four *P. do minulus* paper wasps illustrating some of the naturally occurring diversity in the size, shape, and number of black facial spots. Wasps are arrayed from low advertised quality (0 spots, (a)) to high advertised quality (2 spots, (d)).

receivers should evolve to ignore these signals resulting in signalers ceasing to use them.

Despite the prediction that agonistic signals should be honest, some evidence has been found for deceitful agonistic displays (sometimes referred to as bluffs). One example of deceit in an aggressive display is the meral-spread display used by stomatopod shrimp of the genus *Gonodactylus*. Both newly molted individuals and intermolt individuals use this display but newly molted individuals cannot follow through on the aggressive behavior, whereas intermolt individuals are fully capable of attacking. This deceitful display by newly molted individuals may be maintained, because they are relatively rare in natural populations. Another example of deceptive signaling occurs in species where males adopt alternative strategies whereby some males display and exclude other males from their territory while also attracting receptive females, whereas other males adopt roaming/sneaking strategies. These males lack the adult male secondary sexual characteristics used in territorial and mating displays and thus look like females, resulting in essentially deceptive signaling toward territorial males. Despite these examples, deceit in agonistic signals is generally considered uncommon and is unlikely to persist over evolutionary time.

See also: Anthropogenic Noise: Implications for Conservation; Conflict Resolution; Deception: Competition by Misleading Behavior; Olfactory Signals; Punishment; Smell: Vertebrates.

Further Reading

Andersson M (1980) Why are there so many threat displays? *Journal of Theoretical Biology* 86: 773–781.

Bradbury JW and Vehrencamp SL (1998) *Principles of Animal Communication.* Sunderland, MA: Sinauer.

Bradbury JW and Vehrencamp SL (2000) Economic models of animal communication. *Animal Behaviour* 59: 259–268.

Dawkins MS and Guilford T (1991) The corruption of honest signaling. *Animal Behaviour* 41: 865–873.

Enquist M (1985) Communication during aggressive interactions with particular reference to variation in choice of behavior. *Animal Behaviour* 33: 1152–1161.

Enquist M and Leimar O (1987) Evolution of fighting behavior – the effect of variation in resource value. *Journal of Theoretical Biology* 127: 187–205.

Guilford T and Dawkins MS (1991) Receiver psychology and the evolution of animal signals. *Animal Behaviour* 42: 1–14.

Guilford T and Dawkins MS (1995) What are conventional signals? *Animal Behaviour* 49: 1689–1695.

Hurd PL and Enquist M (2005) A strategic taxonomy of biological communication. *Animal Behaviour* 70: 1155–1170.

Johnstone RA and Norris K (1993) Badges of status and the cost of aggression. *Behavioral Ecology and Sociobiology* 32: 127–134.

Kokko H, Lopez-Sepulcre A, and Morrell LJ (2006) From hawks and doves to self-consistent games of territorial behavior. *American Naturalist* 167: 901–912.

Parker GA (1974) Assessment strategy and evolution of fighting behavior. *Journal of Theoretical Biology* 47: 223–243.

Sih A, Bell A, and Johnson JC (2004) Behavioral syndromes: An ecological and evolutionary overview. *Trends in Ecology and Evolution* 19: 372–378.

Smith JM and Price GR (1973) Logic of animal conflict. *Nature* 246: 15–18.

Tinbergen N (1959) Comparative studies of the behavior of gulls (Laridae): A progress report. *Behaviour* 15: 1–70.

Alarm Calls in Birds and Mammals

C. N. Slobodchikoff, Northern Arizona University, Flagstaff, AZ, USA

Introduction

Alarm signals are those that signal the presence of some kind of threat, such as the appearance of a predator. These signals include vocalizations, acoustic signals such as foot drumming, alert postures, and olfactory signals such as alarm pheromones. Such signals are produced in response to a situation that elicits a fear response on the part of the signaler. Although this fear response is usually elicited by the presence of a predator, it can also be a result of an attack or other disturbance that is perceived by an animal as a potential threat. Alarm signals employ a number of different modalities, such as vision, acoustic, olfactory, and tactile. Visual signals can involve tail flagging, such as the tail waving found among some ground squirrels, or white rump patches of escaping deer or antelope, or tail feathers of birds that are exposed when the bird is fleeing, or head movements by waterfowl, or staring by primates at a predator. Olfactory signals tend to be composed of lower molecular-weight compounds, allowing for the rapid volatilization and fadeout of these signals. In some ant species, within an alarm pheromone, different compounds act differently depending on the distance from their release point. Some compounds attract ants at a distance, while other components of the alarm compounds induce the ants to attack and bite an intruder once the ants come nearer to the source of release of the alarm pheromone.

Most of the work on the evolution of alarm signals and on their possible information content comes from alarm vocalizations, or alarm calls. Therefore, the focus of this chapter is restricted to alarm calls.

Alarm Calls

A number of bird and mammals species are known to make alarm calls. Alarm calls can be elicited by an aerial predator flying overhead, such as a raptor, or they can be elicited by terrestrial predators, such as lions, leopards, coyotes, badgers, or snakes. Alarm calls can sometimes be elicited by nonpredatory situations, such as leaves being moved by the wind in an unusual fashion, or anything that is unusual for a particular animal species that can provoke a fear response.

Alarm calls have some acoustic characteristics common to a number of species. Some birds tend to have relatively high-pitched, pure tone alarm calls that can make them difficult to locate. Among some bird species, such as the reed bunting, blackbird, great titmouse, blue titmouse, and chaffinch, there appears to be a convergence in the acoustic characteristics of the alarm calls, so that the calls produced by one species can probably be recognized as alarm calls by other species as well. Although some predators may experience difficulty in localizing these high-pitched alarm calls, other species of predators are apparently not deterred by the ventriloquistic characteristics of such high-pitched alarm calls, and can locate their prey relatively easily by locating the source of the sound.

One of the benefits and drawbacks of high-pitched alarm calls is that the sound does not travel for very long distances. Higher-pitched sound frequencies tend to drop out rapidly as a function of distance, whereas lower-pitched frequencies can travel over longer distances. So high-pitched vocalizations can serve to alert other birds in the immediate vicinity of a predator, but they do not serve in alerting birds over longer distances. On the other hand, high-pitched alarm calls might not travel to the predator, and the predator might not be aware that it has been detected. Not all birds have high-pitched alarm calls. Some birds such as the scrubwren and fairy wren have broadband alarm calls that resemble those of terrestrial quadruped species.

Terrestrial species generally have two types of acoustically distinct alarm calls, though species differ in which one or both they use. One type of alarm call is a broadband buzz, whistle, or sound that covers a wide range of frequencies, while the other type of alarm call consists of a number of harmonics. Harmonics are frequencies that are multiples of a fundamental or base frequency. For example, if a string vibrates at a fundamental frequency of 300 Hz (or cycles per second), there can be harmonics at multiples of 300 Hz, at 600, 900, and 1200 Hz. These calls tend to be more localizable, both by the predator and by the conspecifics of the calling animal.

Evolution of Alarm Calling

The evolutionary origin of alarm calls appears to have been simple vocalizations that had no communicative function that result from a fear response. This vocalization presumably evolved into a true signal that serves to alert conspecifics to the presence of a predator. There is some controversy about the evolutionary selective pressures that result in the maintenance of alarm signals in populations; a number of hypotheses have been proposed to address this issue. The kin-selection hypothesis suggests that alarm calls serve to warn relatives. Although there is little documentation that the alarm caller is at risk by calling, the kin-selection hypothesis suggests that even if the caller is caught by a predator, the caller's relatives benefit by escaping, and consequently the genes that the caller and its relatives had in common receive a benefit from the caller's action. Another hypothesis suggests that alarm calls can have a manipulative function by flushing out animals that are unaware of the presence of the predator. This reduces the risk of the caller being caught, because the caller knows the location of the predator, while those animals that were flushed out by the call are not aware of the predator's location. A related hypothesis suggests that alarm calls synchronize the fleeing response of both the caller and other conspecifics, reducing the caller's risk through having more safety in numbers. This assumes that any given prey individual is as likely to be taken as any other. If this assumption is valid, the more prey a caller can flush, the less likely the caller is to be taken by the predator. All of these hypotheses assume that the caller is signaling to conspecifics, but this may not always be true.

Instead of signaling to conspecifics, alarm calls might have evolved as a pursuit-invitation or pursuit-deterrence. The pursuit-invitation or pursuit-deterrence hypotheses suggest that prey send a signal to the predator that they are capable of successful evasive action, and that there is no point in the predator investing its time and energy in pursuing them. For example, stotting gazelles tend to be pursued by cheetahs less frequently than nonstotting gazelles, and stotting is assumed to be a signal that a gazelle is vigorous enough to be able to evade the cheetah. In a number of species of rodents, species that are diurnal tend to give more alarm calls than nocturnal ones, regardless of whether the species are social or solitary. Diurnal conditions make it easier for a predator to evaluate the escape potential of an alarm-calling animal, and solitary species would not be expected to derive any evolutionary benefits from either kin-selection or from a synchronized fleeing response of conspecifics. Alternatively, the alarm vocalization of solitary species may simply be an expression of fear of a predator that is more easily detected by the alarm-calling animal under diurnal conditions.

Signal Information Content

Different levels of fear of the predator (part of what has been called a motivational or affective component) can generate information that other animals in a social group can find useful. This can take the form of response urgency, where an animal gives different signals depending on how urgently an escape response is required. For example, a diving hawk demands a more urgent response than a hawk that is merely circling overhead. Similarly, a predator who is running straight at a social group of animals demands a more urgent response than a predator who is passing by and showing no interest in hunting. With alarm calls, this often involves incorporating information about the distance that a predator is from a calling animal, as, for example, the response of marmots to predators.

Another level of information can be contained in referential communication, in which an animal produces a signal that refers to some aspect of a predator. A number of animals as diverse as chickens, many ground squirrels, and suricates produce two types of acoustically distinct calls, one for aerial predators and another for terrestrial ones. A few animals produce different calls for different predators. Vervet monkeys produce three types of calls, one for leopard-like predators, another for eagle-type predators, and a third for snake-type predators. Diana and Campbell's monkeys have two types of acoustically distinct calls, one for leopards and another for eagles. And prairie dogs have at least four different kinds of calls, one for humans, another for coyotes, a third for domestic dogs, and a fourth for red-tailed hawks (**Figures 1–4**)

Beyond the referential calls for different kinds or species of predators, some animals incorporate a greater level of description of a predator, in the form of descriptive labels. Black-capped chickadees incorporate information into their calls about the size and degree of threat of potential

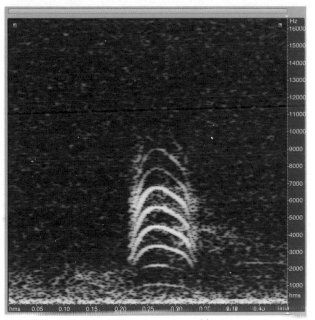

Figure 1 Alarm call of a Gunnison's prairie dog (*Cynomys gunnisoni*), elicited by a coyote (*Canis latrans*). Frequency in hertz is on the vertical axis, and time in seconds is on the horizontal axis. This call illustrates frequency-modulation with a series of harmonics.

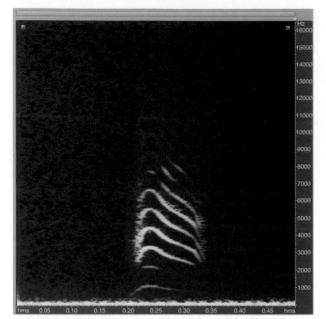

Figure 2 Alarm call of a Gunnison's prairie dog (*Cynomys gunnisoni*), elicited by a domestic dog (*Canis familiaris*). Frequency in hertz is on the vertical axis, and time in seconds is on the horizontal axis. This call illustrates frequency-modulation with a series of harmonics.

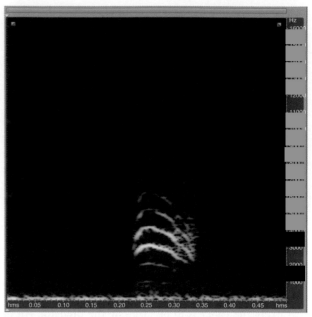

Figure 4 Alarm call of a Gunnison's prairie dog (*Cynomys gunnisoni*), elicited by a red-tailed hawk (*Buteo jamaicencis*). Frequency in hertz is on the vertical axis, and time in seconds is on the horizontal axis. This call illustrates frequency-modulation with a series of harmonics.

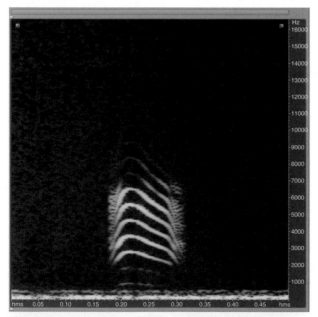

Figure 3 Alarm call of a Gunnison's prairie dog (*Cynomys gunnisoni*), elicited by a human (*Homo sapiens*). Frequency in hertz is on the vertical axis, and time in seconds is on the horizontal axis. This call illustrates frequency-modulation with a series of harmonics.

avian predators. Prairie dogs incorporate information about the size, shape, and color of humans and domestic dogs, as well as size and shape information about objects that they have previously never seen but could possibly be a threat.

Response-urgency and referential communication are not incompatible, as it is possible to incorporate both a motivational and a referential component into the same vocalization. Suricates have both a response-urgency component and a referential component in calls for aerial and terrestrial predators. Prairie dogs will speed up their call rate if a predator starts traveling faster and becomes more of a danger.

Signal Receivers

Although the calls serve the primary function of alerting conspecifics to potential danger, other species can benefit from the calls. A number of species recognize the alarm calls of other species, and respond appropriately. Nuthatches respond to the alarm calls of black-capped chickadees, superb starlings respond to the alarm calls of vervet monkeys, and Diana and Campbell's monkeys respond to each other's alarm calls, even though the acoustic structure of each species' calls might be quite different.

In the past, alarm calls were considered to be simple expressions of fear and nothing else. More recent research is showing that alarm calls can contain much more complex information than fear, and that animal communication systems might be far more sophisticated than previously imagined.

See also: Acoustic Signals; Co-Evolution of Predators and Prey; Defensive Avoidance; Ecology of Fear.

Further Reading

Caro, TM (1995). Pursuit-deterrence revisited. *Trends in Ecology and Evolution* 10: 500–503.

Charnov, EL and Krebs, JR (1974). The evolution of alarm calls: Altruism and manipulation. *American Naturalist* 109: 107–112.

Evans, CS, Evans, L, and Marler, P (1993). On the meaning of alarm calls: Functional reference in an avian alarm vocal system. *Animal Behaviour* 46: 23–38.

Kiriazis, J and Slobodchikoff, CN (2006). Perceptual specificity in the alarm calls of Gunnison's prairie dogs. *Behavioural Processes* 73: 29–35.

Manser, MB (2001). The acoustic structure of suricates' alarm calls varies with predator type and the level of response urgency. *Proceedings Royal Society of London, Series B* 268: 2315–2324.

Marler, P (1955). Characteristics of some animal calls. *Nature* 176: 6–8.

Owings, DH and Hennessy, DF (1984). The importance of variation in sciurid visual and vocal communication. In: Murie, JO and Michener, GR (eds.) *The Biology of Ground Dwelling Squirrels*, pp. 171–200. Lincoln: University of Nebraska Press.

Seyfarth, RM, Cheney, DL, and Marler, P (1980). Monkey responses to three different alarm calls: Evidence for predator classification and semantic communication. *Science* 210: 801–803.

Shelley, EL and Blumstein, DT (2004). The evolution of alarm calling in rodents. *Behavioral Ecology* 16: 169–177.

Sherman, PW (1977). Nepotism and the evolution of alarm calls. *Science* 197: 1246–1253.

Slobodchikoff, CN, Kiriazis, J, Fischer, C, and Creef, E (1991). Semantic information distinguishing individual predators in the alarm calls of Gunnison's prairie dogs. *Animal Behaviour* 42: 713–719.

Templeton, CN, Greene, E, and Davis, K (2005). Allometry of alarm calls: Black-capped chickadees encode information about predator size. *Science* 308: 1934–1937.

Warkentin, KJ, Keeley, ATH, and Hare, JF (2001). Repetitive calls of juvenile Richardson's ground squirrels (*Spermophilus richardsonii*) communicate response urgency. *Canadian Journal of Zoology* 79: 569–573.

Zuberbühler, K (2000). Referential labelling in Diana monkeys. *Animal Behaviour* 59: 917–927.

Zuberbühler, K (2001). Predator-specific alarm calls in Campbell's monkeys, *Cercopithecus campbelli*. *Behavioral Ecology and Sociobiology* 50: 414–422.

Alex: A Study in Avian Cognition

I. M. Pepperberg, Harvard University, Cambridge, MA, USA

Introduction and Background

When research with Alex, a Grey parrot, began in the late 1970s, animal–human communication and its use for studying cognition in the United States involved subjects with either close genetic relationships to humans (e.g., great apes), or large brains (e.g., dolphins); avian studies at the time generally employed pigeons, which scored significantly lower on intelligence tests than apes, monkeys, and often even rats. Birds seemed to lack a critical brain area – the cortical region responsible for humans' complex, cognitive processing – thus explaining avian failures. Furthermore, parrots were known for mindless mimicry: Prior studies to train meaningful speech, using standard psychological techniques (see section 'Training'), failed miserably. Therefore, studying a parrot, with a brain the size of a shelled walnut, whose closest relative to humans was a ∼280 Mya dinosaur, seemed ridiculous.

But research elsewhere, in ethology rather than psychology labs, suggested otherwise. Experiments in the 1940s and 1950s in places like Germany demonstrated that birds – particularly Grey parrots – exhibited advanced capacities. On tests involving number sense, or figuring out how to obtain food from various contraptions, parrots and jackdaws performed at levels comparable to non-human primates. Neurobiologists determined that birds had brain areas that did not look cortical but that functioned cortically, that the relative size of these areas compared to the entire brain correlated with intelligence – and that parrots and corvids had the greatest relative sizes. Scientists examining birds' song acquisition found similarities with ways in which children learn language; that is, en route to producing full song, birds engage in a child-like babbling period; that although a sensitive phase seemed to exist during which exposure to adult systems allowed species-specific song to develop most readily, social interaction could extend this learning period and enable birds to learn heterospecific songs, a bit-like human second-language learning; that some species learned multiple songs and appropriate contexts for their use (e.g., mate attraction vs. territorial defense); that parts of the avian brain were directly related to the learning, storage, and production of song and that these areas functioned in ways not unlike human brain areas responsible for language. And, notably, a German study showed that a particular training paradigm, the Model/Rival (M/R) technique, enabled Grey parrots to engage in apparently meaningful duets with humans. Might birds – at least Grey parrots – be appropriate subjects for studying non-human intelligence and communication after all?

Training: The M/R Technique, Referential Mapping

Because much of Alex's data resulted from M/R training, a brief explanation of the technique is necessary. Unlike operant conditioning, which involved starving a subject to 80% of its normal weight, placing it in a featureless box, and playing it tapes of human speech, with food rewards if some imitation occurred, M/R training uses social interaction to demonstrate the targeted vocal behavior and its use. Initially, labels are trained as requests for items, the items themselves being referential rewards; later, birds learn to separate labeling from requesting.

Sessions begin with a bird observing two humans handling an item desired by the bird. One human trains the second human (the model/rival; i.e., presents and asks questions about the item, 'What's here?,' 'What toy?'). The trainer rewards correct identifications by physically transferring this item (which thereby becomes an intrinsic reward), demonstrating referential and functional use of labels, respectively, by providing a 1:1 correspondence between label and item, and modeling label use to obtain the item. Training occurs with multiple exemplars of the items so the bird sees the label refer to a variety of related objects. The second human is a model for the bird's responses and its rival for the trainer's attention, and also enables demonstration of aversive consequences ensuing from errors: Trainers respond to a model/rival's intentionally garbled or incorrect responses with scolding, temporarily hiding the item. The model/rival is told to speak clearly or try again, thereby allowing a bird to observe corrective feedback. Unlike some other modeling procedures, here model/rival and trainer reverse roles to show how the communicative process is used by either party to request information or effect environmental change. After humans model the interaction several times, the bird is asked to label the item. Initially, any novel utterance the bird makes related to the target label (i.e., 'eee' for 'key') is rewarded; labels for other items or sounds used for other purposes are not. Humans then resume modeling. In subsequent sessions, the bird must approximate the targeted utterance more closely, being rewarded for successive approximations to a correct response; thus, training is adjusted to its level. Notably, if humans do not reverse roles during training, birds exhibit two behavior patterns inconsistent with interactive, referential communication: They respond only to the human who posed questions during training and do not learn both parts of the interaction. Thus role reversal promotes generalization of behavior.

Another technique, referential mapping, was used after Alex had learned numerous labels. Often, after acquiring a label, he engaged in sound play, recombining parts of this label and other known vocalizations in novel ways. If an utterance related to actual objects (e.g., after learning 'grey,' he produced 'grape,' 'grate,' 'grain,' 'chain,' 'cane') we immediately gave him the item (the fruit, a nutmeg grater, some seed, a paper clip chain, sugar cane) thus mapping the label to a referent; he immediately made the connection.

Other training techniques, used with additional Greys in my laboratory, did not engender referential communication. Exposure to audiotapes or videotapes in social isolation, exposure to videotapes or live video with human interaction but no modeling or referential reward, use of an LCD to avoid the disturbing flicker of CRT screens, use of a single trainer –all were unsuccessful.

Acquisition of Labels and Concepts

Label Acquisition

Alex eventually labeled over 50 exemplars (objects, foods, locations), seven colors (rose (red), green, blue, yellow, orange, grey (charcoal to black), purple), 6 shapes (1-, 2-, 3-, 4-, 5-, 6-corner), quantity to six (preliminary data exist for seven and eight), three categories (material, color, shape); he used 'no,' 'come here,' 'wanna go X,' 'want Y' (X, Y were appropriate location or item labels). His requests were intentional: Given Y after asking for X, he would toss Y back at the trainer. He combined these labels to identify, classify, request, or refuse ~100 items and alter his environment. Alex's abilities were tested rigorously, with all possible controls for various types and forms of inadvertent cuing; testing procedure details are in references below. Alex demonstrated competence comparable to apes and cetaceans similarly trained, and to children at the early stages of acquiring communication skills.

More importantly, he mapped human concepts, something critics did not expect. Of course, in nature, birds must have some conceptual understanding of their world. food/non-food, mate/non-mate, predator/not predator They likely categorize food sources, nest sites, and even mates in terms of relative quality (e.g., more/less, better/inferior). But Alex learned labels for various concepts, and many distinctions made by apes and human children.

Concepts of Category

Alex acquired both labels and concepts of category. For any given object, he labeled different hues, shapes or materials in response to vocal queries of 'What color?,' 'What shape?,' 'What matter?' (for some objects, 'What toy?'): An item could be 'green,' 'four-corner,' 'wood,' and 'block.' He understood that 'blue,' for example, is one instance of the category 'color,' and that, for any colored and shaped object, specific instances of these attributes (e.g., 'blue,' 'three-corner') represented different categories. He learned that different sets of responses – color, shape, and material labels – formed different hierarchical classes. These are higher-order class concepts, because individual color labels have no intrinsic connection to the label 'color'; likewise for 'shape' and 'matter.' The protocol, requiring categorization of one exemplar with respect different attributes, involved flexibility in changing the basis for classification. Such capacity for reclassification indicates the presence of abstract aptitude.

In further tests of these abilities, Alex was shown seven-member collections and asked to provide information about the specific instance of one category of an item uniquely defined by conjunction of two other categories (e.g., 'What object is color-A and shape-B?,' 'What shape is object-C and color-D?'). Other objects on the tray exemplified one, but not both, these defining categories. Specifically, each question contained several parts, the combination of which uniquely specified the targeted object; the complexity of the question was determined by its context (number of different possible objects from which to choose) and the number of its parts (e.g., number of attributes used to specify the target). Alex had to divide the query into these parts and use his understanding of each part to answer correctly. His accuracy, above 75%, matched that of marine mammals similarly tested and indicated he understood all elements in the questions.

Concepts of Same–Different, Absence

In the 1970s, same–different supposedly separated primates from other animals. Comprehension of this concept is complex, and is not tested by tasks such as matching-to-sample (matching stimulus A to sample A rather than B), nonmatching-to-sample, choosing the odd item from a set of two matching and one nonmatching items, or distinguishing homogeneity from nonhomogeneity. According to Premack, same–different requires use of arbitrary symbols to represent relationships of sameness and difference between sets of objects and to denote the attribute that is same or different. Animals would thus need symbolic representation – some elementary form of language – to succeed. Tasks listed above, in contrast, require showing only savings in the number of trials needed to respond to B and B as a match (or as a homogeneous field) after learning to respond to A and A as a match (and likewise by showing a savings in trials involving C and D after learning to respond appropriately to A and B as non-matching or nonhomogenous). Responses might even be based on novelty or familiarity (i.e., on the relative frequency A vs. B samples are experienced). Understanding same–different, however, requires knowing that two nonidentical red objects are related just as are two nonidentical blue objects – in terms of color – that the red objects relate to each other as do two nonidentical square items but via a different category, and, moreover, that this understanding immediately transfers to any attribute of an item; likewise, for difference.

Alex learned abstract concepts of same–different and, furthermore, to respond to their absence. When shown two objects that were identical or varied with respect to some or all of their attributes, and queried 'What's same/different?,' Alex uttered the appropriate category label (color, shape, matter), or 'none' if nothing was same or different. He responded equally accurately to pairs of novel objects, colors, shapes, and materials, including ones he could not label. Furthermore, he responded to the specific questions, and not merely based on training and physical attributes of the objects: He still responded above chance when, for example, asked 'What's same?' for green and blue wooden triangles. Had he ignored the question and responded instead based on prior training, he would have determined, and responded with the label for, the one anomalous attribute (here, 'color'). Instead, he uttered one of two appropriate answers (i.e., 'shape').

Alex's use of 'none' was significant: Understanding and commenting upon nonexistence or absence, although seemingly simple, denotes an advanced stage in cognitive and linguistic development. An organism reacts to absence only after acquiring a corpus of knowledge about the expected presence of events, objects, or other information in its environment, that is, only when it recognizes discrepancies between expected and actual states of affairs. (Such behavior differs qualitatively from learning what stimulus leads to absence of reward, wherein subjects learn what to avoid.) Many animal species, including Grey parrots, tested on absence using Piagetian object permanence, react to disappearance or nonexistence of specific items expected to be present; some songbirds react to absence of signs of territorial defense (e.g., song) from neighbors with positive acts of territorial invasion. Researchers in child development, however, suggest that both comprehension and verbal production of terms relating to nonexistence are necessary before an organism is considered to understand absence. Experimental demonstration of this concept thus can be difficult, even in humans, and Alex's capacities were notable (see sections 'Concept of Relative Size' and 'Numerical Concepts').

Concept of Relative Size

Alex understood categorical classes based on absolute physical criteria; what about relative concepts? Color and shape labels are symbolic and thus abstract but refer to concrete entities: Red is red. Relative concepts are more difficult; what is darker in one trial may be lighter in the next. Starlings, for example, preferred responding to stimuli on absolute rather than relative bases; to obtain the latter response, the former somehow had to be blocked. Might Alex learn to respond facilely to relative concepts, specifically to bigger/smaller? Such data would provide direct comparisons with marine mammal research.

After M/R training on 'What color bigger/smaller?' with a limited object set (yellow, blue, green cups, woolen felt circles, Playdoh rods), Alex was tested on different familiar and unfamiliar items, achieving scores of almost 80%; equivalent to that of certain marine mammals. He responded to stimuli outside of the training domain with 80% accuracy; by uttering the label for an attribute (i.e., color), he responded to novel objects of shapes, sizes, and colors not used in training with an accuracy of almost 80%. These objects were often of shapes or materials he could not label (e.g., styrofoam stars). We did not examine how close in size two objects must be before he could not discriminate a difference.

Notably, without any training, he responded 'none' when exemplars did not differ in size and answered questions based on object material rather than color. Thus he was not limited to responding within a single dimension, he was attending to our questions, and was able to transfer information learned in one domain ('none' from the same/different study) to another. Such transfer involves complex cognitive processing.

Numerical Concepts

Could Alex form a new categorical class involving quantity labels; that is, reclassify a group of, for example, wooden objects known until now as 'wood' or 'green wood' as 'five wood'? Success would mean understanding that a new set of labels ('one,' 'two,' 'three,' etc.), represented a novel class: a category based both on physical similarity within a group and the group's quantity, not only physical characteristics. He would need to generalize this new class of numerical labels to sets of novel objects, objects in random arrays, and heterogeneous collections, that is, develop a concept of number. German researchers had already demonstrated Grey parrots' sensitivity to number: Koehler's birds could open boxes randomly containing 0, 1, or 2 baits until they obtained a fixed number (e.g., 4). The number of boxes to be opened to obtain the precise number of baits varied across trials, and the number being sought depended upon independent visual cues:

black box lids denoted 2 baits, green lids 3, etc. Koehler did not state, however, whether colors actually represented particular quantities. Lögler studied Grey's number behavior with light flashes and flute notes, going from visual representations to sequential auditory ones. Could Alex, like chimpanzees, progress to using number as a categorical label?

Number labeling

Alex learned to recognize and label quantities of physical objects up to 6, later 7, and 8. The sets needn't have been familiar, nor placed in particular arrangements, such as squares or triangles (e.g., pattern recognition). Furthermore, for heterogeneous collections (e.g., red and blue balls and blocks) he appropriately quantified any subset uniquely defined by the combination of one color and one object category (e.g., 'How many blue blocks?'). His level was beyond that of young children who, if asked about subsets, generally label the total number of items in a heterogeneous set if, like Alex, they have been taught to label homogeneous sets exclusively.

Number comprehension and zero-like concept

Number label comprehension, crucial for assessing numerical competence, was also tested. Alex answered, without training, 'What color/object (number)?' for subsets of various simultaneously presented quantities (e.g., collections of 4 blue, 5 red, 6 yellow blocks or 2 keys, 4 corks, 6 sticks of one color). His accuracy, above 80%, was unaffected by array quantity, mass, or contour, demonstrating numerical comprehension competence comparable to that of chimpanzees and young children. Notable, however, was his unexpected behavior on one trial.

Here, Alex was asked 'What color 3?' for a collection of two, three, and six blocks. He replied 'five.' The questioner asked twice more; each time he replied 'five.' The questioner, unsure of his intent, finally said 'OK, Alex, what color 5?' Alex immediately responded 'none.' Remember, Alex had learned to respond 'none' in the same–different task and spontaneously transferred 'none' to the relative size study, but had never been taught the concept of absence of quantity or to label absence of an exemplar. The question was subsequently randomly repeated throughout other trials with respect to each possible number to ensure that this situation was not happenstance. On these 'none' trials, Alex's accuracy was over 80%. His one error was labeling a color not on the tray. Now, in Western cultures, labeling a null set, whether by 'zero' or 'none,' did not occur until the 1500s. But zero was represented in some way by a parrot, without training. Furthermore, not only had Alex spontaneously used 'none' to designate absence of a set of objects, but he also managed to manipulate the human into asking the question *he* wished to answer.

Addition, counting, and more on zero-like concepts

Alex's addition experiments were unplanned. Students and I had begun a sequential auditory number session (training to respond to, e.g., three clicks with the vocal label 'three') with another bird, saying 'Listen,' clicking (this time, twice), then asking 'Griffin, how many?' Griffin refused to answer; we replicated the trial. Alex, who often interrupted Griffin's sessions (e.g., 'Talk clearly' or to provide the answer), said 'four.' I shushed him, assuming his vocalization was not intentional. We repeated the trial yet again with Griffin, who remained silent; Alex now said 'six.' I thus decided to replicate the Boysen and Berntson study on chimpanzee addition as closely as possible, and to further study 'zero/none.'

Alex summed physical quantities 0–6, without explicit training. He answered, 'How many total X?' for two collections of variously sized items – hidden under cups, placed on a tray, that were separately, sequentially, raised and lowered so that total quantities were covered – with vocal English number labels. His 85% accuracy, independent of mass or contour, suggested addition abilities comparable to those of non-human primates and young children, particularly with respect to two sets of responses discussed next.

Alex had difficulty with $5 + 0$. He was statistically correct on $1 + 4$ and $2 + 3$ (i.e., fiveness was not the issue), but always erred on $5 + 0$ when given the usual 2–3s to respond, consistently stating '6.' However, given 10–15s to respond, his accuracy was 100%. He thus likely used different mechanisms to determine the answer in different situations. When given 2–3s for $X + 0$, $X = 1$ to 4, he might, like humans, have subitized (i.e., used a rapid visual recognition system), but for $X = 5$ may have had time only to perceive five as something large and, 'six' being his largest label, used it as a default for anything above four. Given 10–15s for $5 + 0$, may have used the additional time actually to count, possibly like humans, who subitize only up to four and then must count to be accurate.

Alex also responded interestingly to nothing under any cup. These trials were interspersed with standard addition trials. On five trials, he looked at the tray and said nothing; on three trials, he said 'one.' His failure to respond on five trials suggested he recognized a difference from the standard trials; that is, even if he did not understand what was expected, he knew his standard number answers would be incorrect. His response of 'one' on some trials matched that of Matsuzawa's chimpanzee Ai, who confused 'one' with 'zero' (i.e., answered from the low end of the number line). Alex thus could use 'none' to signify absence of an attribute (one set) in a collection, but not for absence of everything. This understanding of 'none' as a zero-like concept resembled that of young children, who lack full understanding until they are at least 4 years old.

Ordinality and inferential number abilities

Alex also combined various capacities – labeling the color of the bigger/smaller object in a pair, vocally quantifying sets, and training to vocally identify Arabic numerals 1–6 without associating these Arabic symbols with their relevant physical quantities – to demonstrate a more advanced level of number comprehension. Here he viewed pairs of Arabic numbers or an Arabic numeral and a set of objects and was asked the color of the bigger or smaller one. Alex's 75% accuracy showed he (1) understood physical number symbols as abstract representations of real-world collections, (2) inferred the relationship between the Arabic number and the quantity via use of the same English label for each entity, and (3) understood the ordinal relationship of his numbers. In a final study, he extended these abilities to 8.

Symbolic representation and number

Alex's considerable training on human number labels may, of course, have enabled him to use representational abilities otherwise inaccessible to non-humans. This enculturation factor is consistent with data on the human Pirahã tribe, who lack most number labels and whose numerical abilities (seemingly 'one,' 'two,' and 'many') appear to be far less complex than those of enculturated non-humans.

Phonological Awareness

Toward the end of his life, Alex began demonstrating skills that were somewhat linguistic. He was trained to associate wooden or plastic graphemes B, CH, I, K, N, OR, S, SH, T with their corresponding appropriate phonological sounds (e.g., /bi/ for BI). His accuracy, although above chance (about 50%, $p<0.01$, chance of 1/9), was never high enough (i.e., ~80%) to claim mastery; he never progressed to tests on sounding out untrained letter combinations like young children. But he began constructing targeted novel vocalizations from elements – trained grapheme sounds plus other labels – already in his repertoire. This combinatory behavior is a form of vocal segmentation, showing he understood that his existent labels were comprised of individual units that could be recombined in novel ways to create novel targeted vocalizations. Previous data (e.g., sound play where he produced rhymed strings) suggested, but could not substantiate, this behavior.

Notably, this behavior contrasted with my parrots' customary patterns of label acquisition. For example, when initially learning labels, Alex (and all my other, younger birds), progressed through specific steps – first a vocal contour, then introducing vowels, finally adding consonants; the latter might even require the esophagus to produce plosive sounds (imaging saying 'p' without lips). But now Alex demonstrated a different pattern. To produce 'spool,' he began using a combination of existing phonemes and labels:/s/(trained as noted above) and wool, to form 's'-(pause)-'wool'. He used this form for almost a year, at the end of which spontaneously produced 'spool,' perfectly formed, including appropriate adjustment of the vowel. Similarly, when learning 'seven,' he progressed from 's-(pause)-one' to 's-none' to 'seben.' Alex's ability was indeed a learned behavior, not uniquely human, and dependent upon having considerable experience with both English speech and sound-letter training. My younger birds, lacking such sound-letter training, do not engage in such behavior.

Conclusions

Over his lifetime, Alex demonstrated numerous complex cognitive capacities that were unexpected in a non-human, nonprimate, nonmammalian species. But what mechanisms did he use to learn what were clearly more than simple associations, to transfer knowledge learned in one domain to another, to make subtle phonological distinctions? Neurobiologists now suggest that the brain areas involved are derived from the same precortical tissue as those of humans, but such data do not explain how Alex used this cortical-like area in so many striking ways. Alex was not likely an exceptional individual; other birds in my laboratory give evidence of similar abilities. What defined Alex (and the other birds now in the laboratory) was the socially relevant, referential training he received; the 8–10 h day^{-1} of human interaction, of being treated in ways not unlike a human toddler. The results of this research not only show that parrots can learn to perform tasks once thought the exclusive domain of humans or non-human primates, but also that we must be open to the idea that animals that appear small-brained and that are evolutionarily remote from humans are nevertheless capable of complex, cognitive processing.

Acknowledgments

Research described in this paper was supported over the years by the National Science Foundation, the Harry Frank Guggenheim Foundation, a John Simon Guggenheim Foundation Fellowship, the MIT Media Lab, a Bunting Fellowship from the Radcliffe Institute, and many generous donors to *The Alex Foundation*.

See also: Animal Arithmetic; Categories and Concepts: Language-Related Competences in Non-Linguistic Species; Cognitive Development in Chimpanzees; Ethology in Europe; Psychology of Animals; Referential Signaling; Syntactically Complex Vocal Systems.

Further Reading

Boysen ST (1993) Counting in chimpanzees: Nonhuman principles and emergent properties of number. In: Boysen ST and Capaldi EJ (eds.) *The Development of Numerical Competence: Animal and Human Models*, pp. 39–59. Hillsdale, NJ: Erlbaum.

Herman LM (1987) Receptive competencies of language-trained animals. In: Rosenblatt JS, Beer C, Busnel M-C, and Slater PJB (eds.) *Advances in the Study of Behavior,* vol. 17, pp. 1–60. New York: Academic Press.

Hirsh-Pasek K, Golinkoff RM, and Hollich G (2000) An emergentist coalition model for word learning: Mapping words to objects is a product of the interaction of multiple cues. In: Marschark M (ed.) *Becoming a Word Learner: A Debate on Lexical Acquistion*, pp. 136–164. New York: Oxford.

Jarvis ED, Güntürkün O, Bruce L, et al. (2005) Avian brains and a new understanding of vertebrate evolution. *Nature Reviews Neuroscience* 6: 151–159.

Koehler O (1950) The ability of birds to 'count.' *Bulletin of the Animal Behaviour Society* 9: 41–45.

Pepperberg IM (1999) *The Alex Studies*. Cambridge, MA: Harvard.

Pepperberg IM (2006) Grey Parrot (*Psittacus erithacus*) numerical abilities: Addition and further experiments on a zero-like concept. *Journal of Comparative Psychology* 120: 1–11.

Pepperberg IM (2006) Ordinality and inferential abilities of a Grey parrot (*Psittacus erithacus*). *Journal of Comparative Psychology* 120: 205–216.

Pepperberg IM (2007) Grey parrots do not always 'parrot': Phonological awareness and the creation of new labels from existing vocalizations. *Language Sciences* 29: 1–13.

Pepperberg IM, Gardiner LI, and Luttrell LJ (1999) Limited contextual vocal learning in the Grey parrot (*Psittacus erithacus*): the effect of co-viewers on videotaped instruction. *Journal of Comparative Psychology* 113: 158–172.

Pepperberg IM and Gordon JD (2005) Numerical comprehension by a Grey parrot (*Psittacus erithacus*), including a zero-like concept. *Journal of Comparative Psychology* 119: 197–209.

Pepperberg IM, Sandefer RM, Noel D, and Ellsworth CP (2000) Vocal learning in the Grey parrot (*Psittacus erithacus*): Effect of species identity and number of trainers. *Journal of Comparative Psychology* 114: 371–380.

Pepperberg IM and Wilkes S (2004) Lack of referential vocal learning from LCD video by Grey parrots (*Psittacus erithacus*). *Interaction Studies* 5: 75–97.

Schusterman RJ and Krieger K (1986) Artificial language comprehension and size transposition by a California sea lion (*Zalophus californianus*). *Journal of Comparative Psychology* 100: 348–355.

Todt D (1975) Social learning of vocal patterns and models of their applications in Grey parrots. *Zeitschrift für Tierpsychologie* 39: 178–188.

Amphibia: Orientation and Migration

U. Sinsch, University Koblenz-Landau, Koblenz, Germany

Introduction

The orientation behavior of Amphibia concerns mainly the terrestrial part of their complex life cycle, as goal-oriented long-distance movements are restricted to this stage of life. There are usually two classes of long-distance movements: the postmetamorphic dispersal of mainly juveniles and rarely adults from the natal sites, and the return migrations of adults among different parts of the habitat such as breeding and hibernation sites, and daytime shelter. The spectacular mass migration of common toads *Bufo bufo* to their breeding pond in spring is a well-known example for return migrations because *B. bufo* as well as many other species show a high site fidelity to the natal pond. Still, the distances covered during movements are rather small-scaled compared with other vertebrates, ranging from a few meters in sedentary salamanders over about 15 km in the frog *Rana lessonae* during breeding migrations, to a maximum of nearly 50 km in dispersing cane toads *B. marinus*. Besides these active long-distance movements over ground, populations of some anuran species make use of passive displacement within streams. Common frogs *Rana temporaria* and common toads *B. bufo* cover distances of up to 10.5 km by drift.

Since long-distance movements usually force amphibians to cross unfavorable habitats with a high mortality risk due to desiccation and predation, fitness increases by minimizing the walking distance to the goal. If the time spent for random walking to a goal is reduced in favor of close to line-of-sight movements, orientation behavior is required. Consequently, homing of passively displaced individuals is observed in many amphibian species, that is, the ability to relocate to known sites using directional cues and orientation mechanisms.

In the following sections, I will briefly summarize the spatial range of migratory behavior, and introduce the environmental cues providing directional information for moving amphibians. The processing and integration of sensory information into different types of orientation mechanisms will form the final part of this article.

The Spatial Range of Migratory Behavior

Knowledge on the natural range of active movements of amphibians is necessary to address the question as to whether the homing ability of amphibians is restricted to the area of previous migratory experience. The appropriate answer is complicated to give because the migratory range of most members of a local population is small compared to that of the usually few dispersers which are responsible for the genetic exchange between neighboring populations and for the colonization of new breeding sites. Consequently, the migratory range may vary dramatically among the members of the same population, as recently shown for *B. fowleri* in Canada. Smith and Green reported in their recent analysis of long-term recapture data that irrespective of gender, most toads were recaptured within a range of 100 m distance from their initial capture site. Still, almost every year, some individuals moved further than 7.5 km and one individual was recaptured 34 km away after 1 year. A meta-analysis of migration studies on 90 amphibian species revealed that 7% of the anurans and 2% of the urodeles may migrate further than 10 km. These new data demonstrate that the spatial range of migratory behavior has probably been grossly underestimated. Therefore, long-distance homing of passively displaced individuals cannot be taken as a proof of navigational abilities, that is, map-based orientation from unfamiliar localities, not even in the case of the newt *Taricha rivularis* which homed following a 30-km displacement.

Brief History of Orientation Research

As early as in 1892, George F. Romanes suggested in his book *Animal Intelligence* that frogs have a map sense ('distinct idea of locality') and that they perceive moisture from a great distance to localize breeding ponds. Thus, two basic features of modern concepts on orientation in Amphibia were already mentioned: (1) spatial orientation requires environmental information in which direction an individual should move to approach a goal at the shortest distance possible, and (2) homing and navigation are processes which may include a cognitive map to determine the current location relative to the goal. Still, it has been a long way from Romanes ideas to the current view on the sensory basis of orientation and the strategies employed to move goal-oriented. While the importance of environmental information for orientation was recognized early, research was long hampered by misconceptions about the nature of the directional cues used for homing. Considering that research focussed on temperate zone amphibians which usually breed in ponds or streams, it seemed obvious that breeding-site orientation may be based on hygrotaxis, as proposed by Romanes. Experimental evidence suggests that moisture gradients are only perceived, if air humidity is below 50% RH. In the field, however, amphibians move within the almost water-saturated air close to the ground surface, and it is not surprising that neither Cummings in 1912 nor later studies provided any evidence for hygrotactic orientation. In the beginning of the twentieth century, Buytendijk and Czeloth were the first to recognize that visual and olfactory cues may provide the environmental information needed for the goal orientation of toads and newts.

The homing ability of amphibians remained widely enigmatic until the early 1960s, when several North American and European groups continued research on visual cues. Many species were reported to use celestial cues and visual landmarks. Some authors suggested the use of olfactory or acoustic cues, while others did not find supporting evidence. The main feature of this period of orientation research was the assumption that the homing ability of an amphibian species was based on the use of a single orientation cue. This paradigm was broken by Ferguson in 1971 who was the first to propose the still valid concept of a redundant, multisensory system which integrates all environmental directional information available. Since the first proposal, additional sensory capabilities such as magnetoperception and the detection of polarized light have been demonstrated experimentally.

To our current knowledge, Amphibia make use of acoustic, magnetic, mechanical, olfactory, and visual cues for homing. The species-specific and sometimes population-specific ranking of the sensory input obtained from the potential cues into a specific hierarchy optimizes the available information in every habitat. Individuals of the same species may vary in their preferred directional cues as demonstrated in the natterjack toad *B. calamita*. Keeping in mind that the availability and the reliability of environmental cues differs considerably among habitats, controversial data on preferred directional cues, even of conspecific populations, do not contradict the hypothesis that amphibians use modifications of the same basic system for orientation.

The last three decades of a century-lasting research on the homing ability of amphibians were mainly dedicated to the analysis of the orientation mechanisms. Specifically, the group of John B. Phillips focussed on the role of magnetoperception and provided convincing evidence that the earth's magnetic field may contribute information for both compass orientation and cognitive maps. Other studies emphasized the importance of olfactory homecoming and pilotage in homing amphibians. True navigation, that is, the ability of homing from unfamiliar sites outside the natural migratory range, has been demonstrated so far in a single salamander species (*Notophthalmus viridescens*).

The Directional Cues of Homing Amphibians

In this section, I will concentrate on the directional information provided by the environment and used by amphibians. Among the five classes of orientation cues, visual, olfactory, and magnetic cues, are often reported to play a major role in homing amphibians, whereas acoustic and mechanical cues seem to be restricted mainly to short-distance orientation.

The first sensory system known to be involved in orientation behavior is vision, as evidenced for toads by Buytendijk in 1918. It is used for the perception of environmental cues at all developmental stages of life history and mediates short- and long-distance orientation. Visual cues fall into two broad classes, the fixed landmarks and periodically 'moving' celestial cues. Examples for fixed landmark cues are shore lines for tadpoles and forest silhouettes for migrating adults, as already reported for newts by Czeloth in 1930. Their use for long-distance orientation seems limited. The celestial cues include sun, moon, and stars, reflecting or emitting visible light and also skylight polarization patterns. The use of celestial cues as directional references requires the compensation of their apparent movement by an endogenous clock. Two types of receptors are involved in the perception of visual cues, the lateral eyes for visible light and the extraocular photoreceptors of the pineal complex for polarized light. The importance of the extraocular receptors was recognized in the 1970s by derivation experiments reviewed in Adler (1982).

In his 1930 paper on the spatial orientation of newts *Triturus cristatus*, Czeloth also mentioned that besides visual cues, odors may provide directional information. Now, there is a bulk of evidence that many urodele and anuran species employ olfactory information to detect mates or prey at short distances, or for long-distance homeward orientation. Olfactory deprivation often leads to initial disorientation in migrating toads and newts. T-maze choice experiments demonstrate that amphibians can distinguish between specific odors which they had previously learned to recognize. The chemical nature of directional olfactory cues is completely unknown, but the multitude of sources of odors suggests that it is unlikely to assume that only a certain class of molecules is involved. Odors emanating from breeding ponds are common directional cues. The olfactory system of amphibians includes sensory epithelia which are located in the cavum principale connecting nares and buccal cavity, and in the vomeronasal organ (=Jacobson's organ). The sensory pathway involved in orientation behavior is still unknown.

One of the more recently discovered sensory modalities of major importance is magnetoperception, that is, the use of the earth's magnetic field for orientation. In 1977, Phillips demonstrated this capability for the first time in amphibians when studying the salamander *Eurycea lucifuga*. Later studies revealed that adults and tadpoles of several more amphibian species share the same magnetic sensitivity. Line of evidence for magnetoperception bases on deprivation experiments and on controlled alterations of magnetic field parameters in laboratory experiments. Deprivation of magnetic field information causes responses ranging from random directional choice to altered preferred directions under field conditions. The experimental variation of magnetic parameters predictably modifies directional choice in the laboratory. Phillips and coworkers demonstrated convincingly that inclination, that is, the vertical component of the magnetic field intensity is sensed by orientating amphibians and many other vertebrates. The receptors used to sense magnetic field parameters are still enigmatic. In the newt *Notophthalmus viridiscens*, magnetoperception may be located in the extraocular photoreceptors of the pineal complex, a trigeminal nerve system, or in ferromagnetic material (magnetite) in the head. In this species – the only one thoroughly studied with respect of magnetoperception – two distinct types of receptors seem to be present. One receptor is sensitive to changes in the wavelength of ambient light, coupled with the visual system, and related to compass orientation. The second one is light-independent, sensitive to the polarity of the magnetic field, and related to the map.

Acoustic cues seem to play a minor role in long-distance orientation and are mainly used for short-distance approaches within the mating behavior. Conspecific advertisement calls as orientation cues are exclusive features of anurans. Still, very few species produce calls with sound pressure levels which are audible at distances exceeding 100 m. Such an exception is the natterjack toad *B. calamita* in which males form loud choruses at potential breeding sites. Females were shown to move towards the less distant breeding chorus, that is, that with the highest sound pressure level from distances of up to 1 km. A recent study suggests that newts *Triturus marmoratus* use the advertisement calls of syntopic *B. calamita* for phonotactic homing, but it is not clear whether the newts' ear is sufficiently well-tuned to use the heterospecific calls for long-distance homing. Sound receptors are the external ears, often including a tympanic membrane. Frequencies above 1000 Hz are routed via the columnella to the papilla basilaris of the inner ear, and frequencies below 1000 Hz via the opercular complex and the papilla amphibiorum.

Mechanical cues such as water surface-waves and seismic signals are used exclusively for short-distance orientation to fixed references within the range of a few meters.

Orientation Mechanisms

The sensory information collected from the environment, either during active outward movements or following passive displacement at the release site, is processed and filtered in the central nervous system to control the course and distance of resulting movements. Amphibians accomplish goal-oriented long-distance movements using at least five basic orientation mechanisms: path integration (dead reckoning), beaconing, pilotage, compass orientation within an area of familiarity, and true navigation requiring both map and a compass steps for homing. It is a matter of definition, if rheotaxis (orientation based on the perception of water currents) is a special case of beaconing or an additional orientation mechanism, but its role is limited to the short-distance orientation of mostly newts. In this section, I shall briefly describe the current knowledge on the orientation mechanisms detected so far and on their relative importance for long-distance orientation.

The only orientation mechanism in which an active outward journey is mandatory is path integration (dead reckoning). Walking distance and direction is stored using kinesthetic senses. During return migration, the individual reverses the average walking direction, and based on the associated distance information, a straight homeward course is maintained. There is only one study on the newt *Taricha torosa* which suggests a role of path integration in homing, by Endler in 1970. It is still an open question whether the uninterrupted flow of information during the outward movement available to all active dispersers plays a significant role for long-distance homing of other amphibians.

If there is a direct sensory contact to the goal, home orientation becomes independent from outward journey information. The direct orientation along a gradient originating from the goal is termed beaconing and has been demonstrated in many urodeles and anurans. This simple type of orientation requires only the association of a gradually varying environmental cue with the goal and the ability to measure the intensity of the cue. Olfactory beaconing toward breeding pond odor is common in toads and newts. Acoustic beaconing (= phonotaxis) toward calling males is evident in female natterjack toads *B. calamita* and other anurans. Still, olfactory gradients are sensitive to wind influence and hardly stable over large distances, while advertisement calls are subject to environmental attenuation. Considering these constraints, it is reasonable to assume that beaconing plays only a minor role for long-distance homing.

Pilotage and compass orientation do neither require active outward migrations nor direct sensory contact with the goal, but familiarity with the territory from which goal-oriented movements are started. Amphibians dispersing from their natal sites acquire some kind of a cognitive map which enables them to move goal-oriented within the area of familiarity. Pilotage is one of the first orientation mechanisms discovered in amphibians and enables homing by following a sequence of fixed references within the landscape. Heusser found in 1969 that homing common toad *B. bufo* used fixed visual landmarks such as forest. It is not yet known whether pilotage may also rely on a mosaic map of local odors or sound sources. Previous studies on olfactory orientation have exclusively considered beaconing based on pond-emananting odors.

A more sophisticated level of orientation is the use of compasses to steer a straight course towards the goal. A compass system allows the amphibian to be aware of the direction in which it is moving. This type of orientation has been studied experimentally in individuals that have been passively displaced to familiar sites and deprived of outward journey information. Two compass systems have been analyzed so far, the celestial and the magnetic compass. The celestial compass was discovered first and related to the directional information provided by the positions of the sun, moon, and stars. As these celestial cues perform an apparent movement, an endogenous circadian clock is necessary to compensate for the cues' movement. In fact, the presence of a clock was evidenced by the predictable effects of clock-shifting on the directional choice of amphibians. When Taylor and Ferguson (1970) demonstrated that amphibians can directly perceive the axis (e-vector) of linearly polarized light using the extraocular receptors, it became uncertain if the position of the sun itself really serves as the reference of the sun compass or polarization pattern of the sky which is visible even on overcast days. About 10 years later, Phillips provided the first evidence that newts *N. viridiscens* use a magnetic compass for shoreward orientation. The peculiar compass system is sensitive to the inclination of the earth's magnetic field as that of many other vertebrates. However, it only works properly, if the newts were tested under either natural or short-wavelength light, suggesting that the receptors form part of the visual system. Exposure to long wavelengths (>500 nm) caused newts to steer courses which were 90° counterclockwise from the correct home direction. It remains to be studied whether both compass systems interact by means of calibration.

The most complex orientation mechanism is true navigation which is needed for homing if individuals are displaced to a site outside their natural migratory range. True navigation is thought to be a two-step process. The first step is to determine the individual's location relative to the position of the goal on a cognitive map. The map must rely on information which can be reliably extrapolated from the area of familiarity to unknown areas. Once the individual's position is known, the second step of orientation is to use a compass system to move straight toward home. There is only one amphibian species which has been experimentally shown to employ true navigation in homing, the eastern red-spotted newt *N. viridescens*.

Adult newts are aquatic with a rather small migratory range, but the terrestrial efts may collect spatial experience outside the adult's habitat. Phillips and coworkers focussed research on the nature of the map used by this newt for homing, specifically testing the magnetic map hypothesis. The most exciting results characterizing map features

were obtained in laboratory experiments at distances of about 40 km from the home pond. Alteration of local inclination by $\pm 2°$ produced predictable shifts in homeward orientation. Homeward orientation was observed on both the North–South and the East–West axes, suggesting a bicoordinate nature of the map. Finally, newts reverse the direction of homing orientation over a range of inclination of 0.5° spanning the home value, indicating that magnetic inclination (or its vertical and horizontal intensity components) is used to derive map information. These findings lent strong support to the magnetic map hypothesis.

Yet, what is the ecological significance of a supposedly magnetic map in the natural habitat? Amphibians usually stay within a small area of previous migratory experience in which they can choose among several types of orientation mechanisms for goal-directed movements. A map-compass system is clearly useful within the area of familiarity, too, but the reading of a magnetic map would require very sensitive magnetoreceptors. Considering a natural migratory range of about 1 km, a reasonable figure for most members of a population, an amphibian would be supposed to read the regular spatial variation of the earth's magnetic field which is, on average, only about $3–5\,\text{nT}\,\text{km}^{-1}$ (0.01%) with respect to the total intensity and on average $0.01°\,\text{km}^{-1}$ in inclination. Further complications for a small-scale magnetic map are magnetic storms and spatial irregularities. Thus, a determination of position using a magnetic map will often yield misleading or unreliable results at this scale and is probably backed up by alternative orientation mechanisms. On a larger spatial scale, it is easier to imagine that the neural hardware of amphibians enables the reading of reliable positions from a magnetic map. As new data aforementioned suggest that the migratory range of some individuals of a population has been underestimated considerably, a magnetic map might be particularly useful for a disperser. The next century of orientation research will reveal if the navigational ability is an evolutionary relict from wider-ranging ancestors or a sophisticated orientation mechanism which is also employed within the area of familiarity to reduce mortality risks during migrations.

See also: Bird Migration; Magnetic Orientation in Migratory Songbirds; Maps and Compasses; Pigeon Homing as a Model Case of Goal-Oriented Navigation; Sea Turtles: Navigation and Orientation.

Further Reading

Adler K (1982) Sensory aspects of amphibian navigation and compass orientation, *Vertebrata Hungarica* 21. 7–10.

Brassart J, Kirschvink JL, Phillips JB, and Borland SC (1999) Ferromagnetic material in the eastern red-spotted newt *Notophthalmus viridescens*. *Journal of Experimental Biology* 202: 3155–3160.

Endler J (1970) Kinesthetic orientation in the california newt (*Taricha torosa*). *Behaviour* 37: 15–23.

Ferguson DE (1971) The sensory basis of orientation in amphibians. *Annals of the New York Academy of Sciences* 188: 30–36.

Fischer JH, Freake MJ, and Phillips JB (2001) Evidence for the use of magnetic map information by an amphibian. *Animal Behaviour* 62: 1–10.

Phillips JB (1998) Magnetoreception. In: Heatwole H (ed.) *Amphibian Biology, Sensory Perception,* vol. 3, pp. 954–964. Chipping Norton: Surrey Beatty.

Phillips JB, Adler K, and Borland SC (1995) True navigation by an amphibian. *Animal Behaviour* 50: 855–858.

Russell AP, Bauer AM, and Johnson MK (2005) Migration in amphibians and reptiles: An overview of patterns and orientation mechanisms in relation to life history strategies. In: Elewa AMT (ed.) *Migration of Organisms*, pp. 151–205. Berlin: Heidelberg, Springer.

Sinsch U (1990) Migration and orientation in anuran amphibians. *Ethology Ecology & Evolution* 2: 65–79.

Sinsch U (1992) Amphibians. In: Papi F (ed.) *Animal Homing*, pp. 213–233. London: Chapman & Hall.

Sinsch U (2006) Orientation and navigation in Amphibia. *Marine and Freshwater Behaviour and Physiology* 39: 65–71.

Smith MA and Green DM (2005) Dispersal and the metapopulation paradigm in amphibian ecology and conservation: Are all amphibian populations metapopulations? *Ecography* 28: 110–128.

Smith MA and Green DM (2006) Sex, isolation and fidelity: Unbiased long-distance dispersal in a terrestrial amphibian. *Ecography* 29: 649–658.

Taylor DH and Ferguson DE (1970) Extraoptic celestial orientation in the southern cricket frog Acris gryllus. *Science* 168: 390–392.

Animal Arithmetic

J. F. Cantlon, Rochester University, Rochester, NC, USA
E. M. Brannon, Duke University, Durham, NC, USA

Non-humans Represent 'Numbers'

Early reports of animal numerical abilities argued that number is an unnatural dimension for the animal mind. These early studies by Davis and Perusse claimed that an animal can only conceive of numerical values after extensive training in a laboratory. For example, Davis and Perusse once argued that if a rat or pigeon is given a judgment task testing stimuli that vary in size and number, the animal will base its judgment on size rather than on number. In this view, a proclivity to use numerical concepts is a uniquely human phenomenon. However, since these early studies, many studies have established that animals other than humans represent numerical values, and in many cases, they do so spontaneously.

Animals as different as bees, fish, salamanders, birds, raccoons, rats, lions, elephants, and primates have been shown to make quantity discriminations. We can infer from this vast and diverse set of studies that all animals reason about quantities to some extent. Moreover, several studies such as those by Marc Hauser and Karen McComb suggest that animals attend to the quantitative attributes of their world naturally, spontaneously, and automatically. In the wild, groups of lions and chimpanzees naturally avoid unfamiliar groups of their conspecifics when they are outnumbered. Honeybees have been shown to use the number of landmarks (e.g., trees) that they pass along their foraging route to locate a feeding site. Salamanders preferentially choose feeding sites with large amounts of food (fruit flies) over small amounts of food. And female mosquitofish show a natural preference for joining a school with a larger number of fish in order to avoid sexually harassing males. Thus, animals of all types spontaneously use quantitative information to make adaptive decisions in their natural environments.

Laboratory studies by Cantlon and Brannon also indicate that animals have a spontaneous capacity for representing numerical values. In contrast to many prior naturalistic reports, these laboratory studies ensured that animals truly represent 'number' as opposed to some other quantitative dimension such as size. This level of control is important in order to determine whether animals are using pure numerical representations to make quantitative judgments instead of the total size or extent of the set (e.g., the cumulative surface area of the set). Under some circumstances, the number of items in a set is in quantitative conflict with the total size or extent of the set. For example, a group of six lions is numerically greater than a group of three elephants but, the group of three elephants takes up more space and has a greater cumulative surface area than the group of lions. This fact raises the question of whether animals use number and/or spatial extent to make quantitative decisions.

In a recent laboratory study, Cantlon and Brannon tested monkeys with and without prior numerical training on a numerical matching task to determine whether explicit training is necessary for animals to conceive of numerical values. Number-experienced and number-naïve monkeys were tested on a matching task in which they were allowed to freely choose the basis for matching from two dimensions: number, color, shape, or cumulative surface area. During this task, number was confounded with one of the other three alternative dimensions. For example, as shown in **Figure 1(a)**, in the shape and number condition, if a sample array contained two circles, the monkey would then have to choose between two circles (shape and number match) and four lightening bolts (shape and number mismatch) to find a match. The correct match could have been made on the basis of either number or shape, or both. Because number was always confounded with an alternative dimension during training, there was no explicit training for the monkeys to use number as the basis for matching. In fact, the monkeys could have completely ignored the numerical values of the stimuli and solved the task using the alternative dimension.

After each monkey could successfully solve the matching task, probe trials were introduced in which number was pitted against the alternative dimension (color, shape, or surface area) as shown in **Figure 1(b)**. The monkeys now had to choose which dimension they preferred as the basis for matching: number or color, shape, or surface area. During these probe trials, the monkeys were rewarded no matter which option they selected as the match so that they could freely indicate the dimension that guided their decisions. Remarkably, both number-experienced and number-naïve monkeys chose to match the stimuli on the basis of numerical value across a substantial proportion of the probe trials. These findings demonstrate that pure numerical value is a salient feature of the environment for monkeys, regardless of their prior training experience. Claims that 'number' is an unnatural dimension for the non-human animal mind are therefore false.

Taken together, studies of many different species using many different experimental protocols firmly indicate

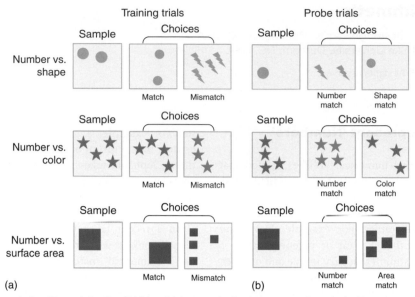

Figure 1 Monkeys were trained to match stimuli (a) in which numerical value was confounded with a second dimension of shape, color, or surface area Cantlon and Brannon (2007). Then, they were tested on the same task with probe stimuli (b) in which numerical value was in conflict with each of these three dimensions. Even though this training did not require monkeys to use number (because they could always use the alternative dimension of shape, color, or surface area), the data from probe trials indicated that they did use number to solve the matching task. Redrawn with permission from Cantlon JF and Brannon EM (2007). How much does number matter to a monkey? *Journal of Experimental Psychology: Animal Behavior Processes* 33(1): 32–41.

that non-human animals represent numerical values. Yet, such evidence does not imply that non-human animals are capable of 'counting' as adult humans do when they successively label elements with the verbal counting terms 'one,' 'two,' 'three,' etc. to precisely determine the total number of items in a set.

Non-human Animals Represent 'Numbers' Approximately

As alluded to earlier, non-human animals cannot represent numerical values precisely because they lack symbolic language. Symbols such as count words (one, two, three, etc.) or Arabic numerals (1, 2, 3, etc.) are required in order to conceive of precise numerical values because these symbols are discrete representations of the values for which they stand. That is, the Arabic numeral '3,' for example, always represents exactly three items. Non-human animals do not have a symbolic system for representing precise numerical values in this way. Instead of using discrete representations of numerical values, non-human animals represent numerical values approximately, which is akin to estimating. However, it is important to note that, like non-human animals, humans of all ages also represent numerical values approximately even after they learn to count and to use a precise numerical symbol system. Humans therefore possess both a precise and an approximate means of enumerating.

The main behavioral signature of approximate numerical representation is the numerical ratio effect: the ability to

psychologically discriminate numerical values depends on the ratio between the values being compared. This effect is known more broadly as Weber's law. An implication of the numerical ratio effect is that there is noise (i.e., error) in the psychological representation of each numerical value that is *proportional* to its value. Hence, larger values are noisier than smaller values. Numerical discriminations that exhibit a numerical ratio effect are approximate discriminations as opposed to precise discriminations because they are noisy.

There is compelling evidence that animals and humans rely on the same system for representing number approximately. When animals and humans are tested in the same nonverbal tasks, their performance is often indistinguishable. In one study, monkeys and adult humans were required to choose one of two arrays that contained the smaller number of elements (**Figure 2(a)**). Humans were instructed to respond rapidly, without verbally counting the elements. When monkeys and humans were tested on identical versions of this numerical comparison task, their patterns of performance were remarkably similar; both groups showed steady decreases in accuracy (**Figure 2(b)**) and increases in response time (**Figure 2(c)**) as the numerical ratio between the stimuli increases. Mathematically speaking, the larger the numerical ratio, the more similar two numerical values are to each other. In this study, as the numerical ratio approached one (a 1:1 ratio in numerical value), monkeys' and humans' performance approached chance accuracy (which was 50%), because the numerical values were too similar to be accurately discriminated by either group at larger ratios.

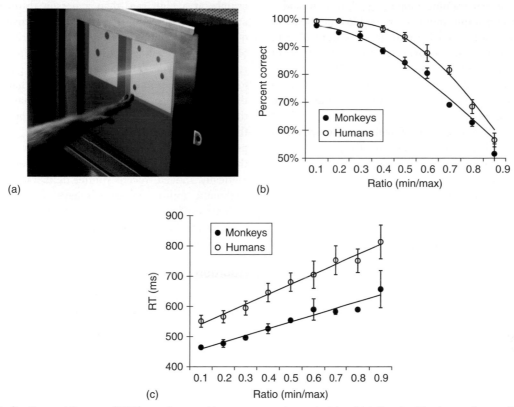

Figure 2 In Cantlon and Brannon (2006), monkeys and humans were given a task in which they had to choose the smaller number of elements from two visual arrays like those in panel (a). Humans were prevented from verbally counting during this task. Monkeys and humans performed very similarly on this task. Both groups performed significantly better than chance (chance = 50%), indicating that they could accurately compare the numbers. For monkeys and humans alike, accuracy decreased (b) and response time increased (c) as a function of the numerical ratio between the two numbers in a given pair (minimum number/maximum number in a given pair). This pattern of performance indicates that for monkeys and humans, numerical comparisons become more difficult as the numerical ratio between values because more similar (i.e., closer to 1 on this scale which represents identical numbers or, a 1:1 ratio in numerical value). Redrawn with permission from Cantlon JF and Brannon EM (2006) Shared system for ordering small and large numbers in monkeys and humans. *Psychological Science* 17(5): 401–406.

Similar parallels between human and non-human animal numerical abilities also have been reported for pigeons and rats on other numerical comparison and estimation tasks. It seems that numerical approximation is a widespread strategy for numerical discrimination throughout the animal kingdom.

A few studies such as those by Tetsuro Matsuzawa and colleagues have shown that animals can use discrete symbols, such as Arabic numerals, to represent numerical values. However, it is important to note that although non-human animals can be trained to associate a symbol with a particular numerical value, this association is not a precise numerical representation in these animals as it is in humans. For instance, macaque monkeys, chimpanzees, and pigeons can be trained to associate the Arabic numerals (e.g., the numerals 1, 2, 3, and 4) with their corresponding values (e.g., sets of 1, 2, 3, and 4 objects). However, after months and even years of training, the animals continue to represent these symbols approximately in that they exhibit a numerical ratio effect in their responses when they match the sets of objects to

their symbols. In contrast, adult humans who are given unlimited time to assign an Arabic numeral to a set of objects perform almost perfectly and do not display a numerical ratio effect in their accuracy, because they can verbally count the items to determine the precise numeral that corresponds to the set. Thus, when animals use symbols to compare numerical values, they are limited to approximate numerical representations, whereas adult humans can employ precise numerical representations by counting.

Despite the fact that animals cannot represent precise numerical values as humans do, they can appreciate the ordinal and continuous nature of numerical value. Studies testing animals' abilities to assess relative numerical value (e.g., choose the larger or smaller) have provided evidence that animals understand the ordinal relationships among quantities. There is clear evidence that when trained on one subset of numerical values presented nonsymbolically as arrays of elements (e.g., 1, 2, 3, and 4 elements), monkeys can transfer an ordinal rule (such as ordering from least to greatest) to novel numerical values that are outside that initial training range (e.g., 5, 6, 7, 8, and 9 elements). For example, monkeys trained to order boxes of 1, 2, 3, and 4 dots from least to greatest were able to spontaneously infer that 6 dots is less than 9 dots without being explicitly trained to order sets of 6 and 9 dots. Thus, when comparing quantities, animals appreciate numerical value as a continuum along which values can be ranked from least to greatest.

Ordering is a simple form of arithmetic computation that requires an individual to determine the proximity of a given value to the numerical origin (e.g., zero for humans). Evidence of this simple type of arithmetic ability in non-human animals raises the question of whether they can perform more complex arithmetic operations.

Non-human Animals Mentally Manipulate 'Numbers'

Arithmetic operations, such as addition, subtraction, division, and multiplication, require mental transformations over numerical values. For example, addition is an arithmetic operation that involves mentally combining two or more quantitative representations (addends) to form a new representation (the sum). That is, during addition, an individual has to mentally alter the information it is given (the addends) to create a new representation (the sum). The degree to which animals are capable of mental arithmetic therefore reflects their capacity to mentally transform numerical information.

Many models of foraging behavior assume that animals calculate the rate of return: the ratio of the number of food items or the total amount of food they obtain to the time it took to procure the food. For example, ducks are more likely to congregate around a person throwing bread crumbs at a high rate than a person throwing crumbs at a low rate, showing that they are sensitive to the rate of return of a feeding site. Additionally, great tits (a kind of bird) visit feeding sites in direct proportion to the relative abundance of food at that site (e.g., if a site has food 75% or the time, the birds will visit that site 75% of the time). This probability matching behavior indicates that animals are sensitive to the proportion of instances that an individual feeding site pays off. Such reports predict that animals not only represent 'number' but that they also manipulate numerical representations arithmetically. Indeed, recent studies deliberately testing the arithmetic abilities of animals have confirmed that animals are capable of manipulating their quantitative representations using arithmetic procedures such as addition, subtraction, and proportion (akin to division).

Addition

Several studies have tested animals' abilities to add numerical values together. Moreover, recent studies have begun to test animals' abilities to do mental arithmetic over large and complex ranges of addition problems. For instance, in one study, monkeys and adult humans were presented with two sets of dots on a computer monitor, separated by a delay. After the presentation of these two 'sample sets,' subjects were required to choose between two arrays: one with a number of dots equal to the numerical sum of the two sets and a second, distractor array that contained a different number of dots. The addition problems consisted of addends ranging from 1 to 16, tested in all possible combinations. Monkeys and humans (who were not allowed to verbally count the dots) successfully solved the addition problems, and the two species' accuracy (**Figure 3(a)**) and response times (**Figure 3(b)**) were

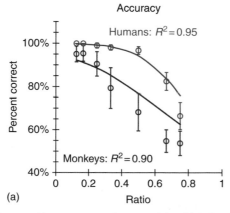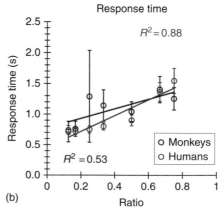

Figure 3 Monkeys and humans were given a task in which they had to add together two sets of dots that appeared successively and then choose the sum from two visual arrays. Humans were prevented from verbally counting during this task. Monkeys and humans performed very well on this task in that both groups performed significantly better than chance (chance = 50%). Moreover, both groups showed the same pattern of difficulty: decreasing accuracy (a) and increasing response time (b) as the numerical ratio between the two choice stimuli increased. Redrawn with permission from Cantlon JF and Brannon EM (2007) Basic math in monkeys and college students. *PLoS Biology* 5(12): e328.

similarly constrained by the numerical ratio between the choice stimuli, or Weber's law. A series of control conditions verified that monkeys' successful performance was not based on simple heuristics such as choosing the array closest to the larger addend. Like humans, monkeys performed approximate addition over the numerical values of the sets.

A series of studies by Michael Beran and colleagues tested non-human primates' abilities to choose adaptively among several food caches containing different amounts of food. These studies have demonstrated that non-human primates can reliably identify the cache containing the largest quantity of food, even when this requires tracking one-by-one additions to multiple caches over long periods of time. Thus, non-human primates are capable of maintaining separate running tallies of different food caches.

Hauser and Spelke have found that monkeys exhibit these kinds of arithmetic abilities even without prior training experience. For instance, when semifree ranging, untrained rhesus monkeys watched as two groups of four lemons were placed behind a screen, they looked longer when the screen was lowered to reveal only four lemons (incorrect outcome) than when the correct outcome of eight lemons was revealed. Monkeys' longer looking time to the incorrect arithmetic outcome can be interpreted as 'surprise.' Thus, as measured by their looking time, monkeys spontaneously form numerical expectations when they view addition-like events and are 'surprised' when an event violates their expectations.

Other studies have trained animals to associate symbols with specific numerical values and then tested the animals' ability to add the symbols. One study showed that pigeons reliably choose the combination of two symbols that indicates the larger amount of food. However, when the number of food items associated with the symbols was varied but total reward value (mass) was held constant, the pigeons failed to determine the numerical sum of the food items, suggesting that they performed the addition task by computing the total reward value represented by the two symbols, rather than by performing numerical arithmetic. Although these data do not demonstrate pure numerical arithmetic in non-human animals, they do indicate that pigeons can mentally combine 'amounts' to choose the larger sum of food. Moreover, this study shows that pigeons can compute total amount abstractly, using symbols to stand for, or represent, the different addends in the problems.

A particularly impressive test of symbolic numerical arithmetic in a non-human animal was conducted on a single chimpanzee by Boysen and Berntson. In this study, a chimpanzee was trained to associate the Arabic numerals 1–4 with their corresponding values. After the chimpanzee was proficient at identifying the value that corresponded to each Arabic numeral (and vice versa), she was tested on her ability to comprehend the arithmetic sum of two

Arabic numerals. Sets of oranges were hidden at various sites in a field. Each hidden set of oranges was labeled with two Arabic numerals the sum of which reflected the total number of oranges in the cache. The chimpanzee consistently chose the cache with a combination of Arabic numerals that corresponded to the greatest sum of hidden oranges. Additionally, this chimpanzee was able to view separate sets of oranges and then to identify an Arabic numeral that corresponded to its sum. These experiments demonstrate that non-human animals can use abstract symbols as representations of numerical values to compute approximate arithmetic outcomes.

Subtraction

While there is good evidence that animals can add, evidence that animals can subtract is very scarce. In fact, most studies report that animals struggle to compute the outcomes of subtraction problems. For instance, Hauser and Spelke tested semiwild monkeys' ability to subtract, using the looking time method described earlier that relies on animals looking longer at events that are surprising. In this study, subjects were shown an empty container and then a small number of eggplants were placed inside the container after which the contents of the container were revealed to the subjects. In one condition, two eggplants were placed inside the container, then one eggplant was removed ($2 - 1$ subtraction). When the contents of the box were revealed, the animals either saw one eggplant (not surprising) or two eggplants (surprising). In another condition, one eggplant was placed inside the container and then another was added to the container ($1 + 1$ addition). The contents of the box were revealed just as in the subtraction condition but, in this case, an outcome of one eggplant would be surprising whereas an outcome of two eggplants would be expected.

Monkeys that saw the addition condition looked (appropriately) longer at the unexpected outcome of one eggplant. However, the results were more ambiguous from the subtraction condition: although the majority of monkeys looked longer at the surprising outcome of two eggplants than at the unsurprising outcome, the magnitude of the difference in their looking time between surprising and unsurprising outcomes was not significantly different. One possibility is that monkeys found it significantly more difficult to predict the outcome of the subtraction event than the addition event.

Similar difficulties with subtraction problems relative to addition problems have been reported in chimpanzees. In a study by Michael Beran, chimpanzees watched as food items were added to or removed from two different containers. Then, they were allowed to choose from container to eat its contents. Chimpanzees chose adaptively, maximizing their food intake, when food items were added to containers, but they were less successful when items were subtracted from containers. However, although chimpanzees were not as good at subtractions as they are at additions, they were able to compute some simple subtraction outcomes. For example, when chimpanzees saw one food item removed from a container with anywhere from one to eight food items, they successfully chose the container with more items on the majority of trials. Thus, it was not the case that the chimpanzees failed to understand subtraction operations all together — they just failed to compute problems with large subtracted amounts.

A study by Hauser and Spelke testing the subtraction abilities of wild rhesus monkeys on the same type of subtraction task arrived at a similar conclusion. In that study, monkeys successfully computed the outcomes of simple subtraction problems involving three or fewer total food items but failed on problems involving larger operands. On the basis of these findings, the authors concluded that there might be an upper limit on the magnitude of the subtraction problems that animals naturally compute.

Taken together, findings from subtraction studies suggest that non-human animals are capable of computing subtractions, albeit with limited accuracy relative to addition. However, it may be the case that animals can perform better on subtraction problems in tasks that either do not present the elements of the problems as food items or that test the animals over a wider variety of problems with a more extended task exposure period. Due to the limited number of studies testing subtraction in animals, it is difficult to make a concrete conclusion about animals' capacities for subtraction.

Probabilities and Proportions

A variety of non-human animal species use the proportion and probability of food abundance to guide their feeding choices. Proportion and probability operations are similar to division in the sense that they all require partitioning one quantity as a function of a second quantity to derive a quotient. Two examples of this capacity, discussed earlier, described the use of rate of return and probability matching in foraging birds. These measures are employed not only by birds but can be observed also in the feeding behaviors of many mammalian species. Both the rate of return and probability matching require animals to calculate the total quantity of food divided by the total time foraging. Animals thus appear naturally sensitive to proportions such as rate and probability when these measures factor foraging time and food.

Animals are also capable of computing other forms of proportions. For instance, piranhas seem to use length proportions in order to identify their prey. Piranhas only attack and devour fish that have a 1:4 height–length ratio or greater. The 1:4 proportion rule prevents piranhas from attacking each other, as well as several other types of fish whose height–length ratios fall beneath 1:4. This is a different kind of proportion computation from rate of return and probability matching in the sense that it requires dividing one dimension by a second dimension within the same object.

Laboratory studies, such as those by Nieder and colleagues, have demonstrated that non-human animals can use proportions to reason about problems beyond those they experience in their natural environments. These controlled laboratory studies ensured that animals are truly capable of using 'proportion' to solve problems, rather than using an alternative cue such as absolute size. For instance, a recent study by Vallentin and Nieder showed that monkeys can match sets of lines on the basis of their length proportions.

They trained monkeys to look at a pair of lines, encode the ratio of the length of the first line to the second line, and finally, choose a pair of lines from among a few options that matched the initial pair in length ratio. Thus, if a monkey saw two lines in a 1:4 length ratio on a given trial, they should choose a pair of lines that was also in a 1:4 ratio as opposed to a 2:4, 3:4, or 4:4 ratio. Examples of the length ratios that the monkeys were tested with are shown in **Figure 4(a)**. Importantly, the absolute lengths of the lines were varied such that the animals had to encode the ratio of the lines to arrive at the correct answer; using the absolute length of either or both of the lines would lead to random performance. Monkeys' performance was not random, however, showing that they were capable of basing their matching choices on proportion. In fact, monkeys performed about as well as adult humans who were tested on an identical task (**Figure 4(b)**). Moreover, monkeys were subsequently tested with novel length ratios (3:8 and 5:8) and showed no decrement in performance on these novel ratios relative to the familiar ratios (1:4, 2:4, 3:4, and 4:4). Broadly speaking, this study demonstrates that monkeys are capable of calculating proportions flexibly, to solve novel tasks testing a range of problems.

Pigeons have been tested in a similar paradigm to this primate study. Jacky Emmerton has tested pigeons' abilities to calculate the proportion of red to green color within horizontal bars and arrays of squares. Half of the pigeons in this study were trained to choose stimuli with a greater proportion of green, whereas the other half chose stimuli with a greater proportion of red. Thus, unlike the previous primate study, this study did not require animals to identify a specific proportion (e.g., 1:4). However, the pigeons' accuracy indicated that they were sensitive to proportion: pigeons were much better at choosing the stimulus with the greater amount of their target color when the proportion was in a greater disparity (e.g., a 1:5 ratio of red to green was easier to discriminate than a 2:5 ratio). Furthermore, a series of control tests revealed that pigeons actually encoded proportion as opposed to absolute amount. Thus, the ability to use proportion flexibly may extend to nonprimate and even nonmammalian species.

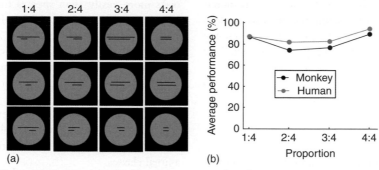

(a)   (b)

Figure 4 Vallentin and Nieder (2008) tested monkeys and humans on a length proportion matching task in which they had to match the length proportion of two lines to the proportion of another pair of lines (from a series of choices), regardless of the absolute sizes of the lines. As shown in (a), there were four different proportions (1:4, 2:4, 3:4, and 4:4) on which the animals were tested. Different exemplars of each proportion category are shown. Among proportion exemplars, the absolute lengths and positions of the lines vary. Monkeys and humans performed similarly on this task (b) in terms of their overall accuracy and in terms of which proportion categories they found most difficult. Redrawn with permission from Vallentin D and Nieder A (2008) Behavioral and prefrontal representation of spatial proportions in the monkey. *Current Biology* 18(18): 1420–1425.

Several studies have also investigated animals' abilities to make decisions on the basis of the probability of a reward pay-off. Recently, Yang and Shadlen demonstrated that monkeys are even capable of adding probabilities together to determine their sum. In this study, monkeys were shown different shapes that were each associated with a specific probability that one of two choice targets (a red circle or a green circle) would pay-off a reward (fruit juice). On each trial, the monkeys had to look at a shape and then choose either the red target or the green target. There were ten possible shapes whose probabilities of pay-off ranged from a 100% chance that the red target would pay off to a 100% chance that the green target would pay off. So, for example, a monkey might see a shape associated with a 70% chance that the red target would pay off and, in this case, he should choose the red target as opposed to the green target. Once the monkeys learned to choose targets appropriately on the basis of the probability of pay-off associated with each of the ten shapes, they were given a more complicated task.

In the more complicated version of the task, the monkeys were shown a combination of four shapes and were required to choose a target on the basis of the sum of the probabilities of the four shapes (**Figure 5**). For example, a monkey might see a shape associated with a 70% chance of red paying off, a second shape with a 90% chance of green paying off, a third shape with a 70% chance of green paying off, and a fourth shape with a 70% chance of red paying off. In this example, the sum of these probabilities results in a 20% chance that green will pay off and the monkey should choose the green target. This example is shown in Figure 5.

Across many trials, with many different combinations of shapes, the monkeys chose the correct target on the majority of trials on the basis of the cumulative probability of the shape combination. Of course, the monkeys computed these probabilities only approximately and thus, they made

errors when the difference between the sum of the red- and green-target pay-off probabilities was slight. However, it is impressive that monkeys chose the appropriate target on the majority of trials, given that 715 different combinations of shapes were tested. This large number of possible shape combinations would have made it impossible for the animals to learn or memorize the pay-off probabilities of the combined shapes. The animals therefore had to compute the sum of the probabilities across the four shapes to choose

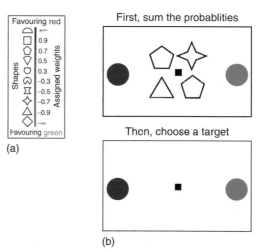

(a)

(b)

Figure 5 Each shape in panel (a) is associated with a probability that the red or green target will pay-off a juice reward. Monkeys were shown four shapes on a computer screen, and then they were required to choose either the red or green target (b). In order to choose the correct target, monkeys had to compute the probability that the red or a green target would pay-off by summing the probabilities from these four shapes for favoring the two targets. Using the scale in panel (a), the sum of the four shapes in panel (b) indicates that there is a 0.2 probability that the green target will pay-off and so, the monkey should choose the green target (Yang and Shadlen, 2007). Redrawn with permission from Yang T and Shadlen MN (2007) Probabilistic reasoning by neurons. *Nature* 447(7148): 1075–1080.

the appropriate target; this is analogous to performing a computation (addition) on a computation (probability). Evidence that non-human animals can compute complex calculations such as these raises the possibility that their minds are capable of computing a whole host of approximate arithmetic computations.

Conclusion

Quantitative thinking appears to be an inherent aspect of decision-making throughout the animal kingdom. Studies of the behavior of animals in their natural habitats and during controlled laboratory tasks have revealed a level of arithmetic sophistication in non-human animals that once may have been considered uniquely human. So far, there is evidence that animals can add, subtract, estimate a proportion or probability, and add probabilities. Unlike humans, non-human animals are limited to entering approximate quantitative representations into these operations. However, regardless of differences in the precision of human and animal representations, approximate arithmetic operations seem to function quite similarly in humans and other animals. Humans and non-human animals perform at comparable levels of accuracy on arithmetic tasks that force humans to use approximate numerical representations by preventing them from verbally counting or labeling units.

The parallels in human and non-human animal approximate arithmetic suggest an evolutionary link in their quantitative capacities. That is, the types of quantitative abilities described herein likely have been around for millions of years. Moreover, evidence of non-human animal arithmetic advances the hypothesis that quantitative reasoning is a component of a primitive cognitive system that exists even without language. Animals that do not use symbolic language to express their thoughts nonetheless possess the ability to perform arithmetic and quantitative computations. Together, these findings underscore the existence of extraordinary continuity in the thought processes of humans and other animals, despite the obvious differences between them.
The parallels in human and non-human

Although we have discussed some broad similarities in the quantitative capabilities of many animal species, the degree to which species subtly differ in their arithmetic abilities remains to be explored. For instance, animal species that naturally reason about length proportions in their environments (e.g., piranhas) may solve length proportion tasks more easily than species that do not. Studies that compare arithmetic capacities between species using comparable tasks are needed to address such questions. Additionally, in order to develop specific hypotheses for revealing differences among species, the degree to which animals face quantitative problems and use quantitative strategies in their natural environments needs to be explored in greater detail. This type of research would help to further delineate the evolutionary relationships among the quantitative abilities of different animal classes.

See also: Categories and Concepts: Language-Related Competences in Non-Linguistic Species; Cognitive Development in Chimpanzees; Time: What Animals Know.

Further Reading

Beran MJ and Beran MM (2004) Chimpanzees remember the results of one-by-one addition of food items to sets over extended time periods. *Psychological Science* 15(2): 94–99.

Biro D and Matsuzawa T (2001) Use of numerical symbols by the chimpanzee (*Pan troglodytes*): Cardinals, ordinals, and the introduction of zero. *Animal Cognition* 4(3–4): 193–199.

Boysen ST and Berntson GG (1989) Numerical competence in a chimpanzee (*Pan troglodytes*). *Journal of Comparative Psychology* 103(1): 23–31.

Brannon EM (2005) What animals know about numbers. In: Campbell JID (ed.) *Handbook of Mathematical Cognition,* vol. xvii, pp. 85–107. New York, NY: Psychology Press.

Cantlon JF and Brannon EM (2007) How much does number matter to a monkey? *Journal of Experimental Psychology: Animal Behavior Processes* 33(1): 32–41.

Davis H and Pérusse R (1988) Numerical competence in animals: Definitional issues, current evidence, and a new research agenda. *Behavioral and Brain Sciences* 11(4): 561–615.

Dehaene S (1997) *The Number Sense.* New York: Oxford University Press.

Dehaene S, Molko N, Cohen L, and Wilson AJ (2004) Arithmetic and the brain. *Current Opinion in Neurobiology* 14(2): 218–224.

Gallistel CR (1990) *The Organization of Learning.* Cambridge, MA: The MIT Press.

Gallistel CR and Gelman R (1992) Preverbal and verbal counting and computation. *Cognition* 44(1–2): 43–74.

Hauser MD and Spelke ES (2004) Evolutionary and developmental foundations of human knowledge: A case study of mathematics. In: Gazzaniga M (ed.) *The Cognitive Neurosciences III.* Cambridge: MIT Press.

Nieder A (2005) Counting on neurons: The neurobiology of numerical competence. *Nature Reviews Neuroscience* 6(3): 177–190.

Vallentin D and Nieder A (2008) Behavioral and prefrontal representation of spatial proportions in the monkey. *Current Biology* 18(18): 1420–1425.

Yang T and Shadlen MN (2007) Probabilistic reasoning by neurons. *Nature* 447(7148): 1075–1080.

Animal Behavior: Antiquity to the Sixteenth Century

L. C. Drickamer, Northern Arizona University, Flagstaff, AZ, USA

Early Humans

The scientific study of animal behavior has its origins in the late eighteenth and early nineteenth centuries. However, animal behavior was a constant in the lives of the early humans and information on animal behavior accumulated over many centuries, first by oral traditions and art and later through writing and art. Thus, it is entirely appropriate to begin this section on the history of animal behavior with topics that commenced with early humans and end with the Renaissance. A subsequent entry covers developments from this juncture up to the beginning of the twentieth century when the modern study of animal behavior first emerged.

Early humans made thoughtful observations of behavior and shared their knowledge about the natural world. Their survival depended on this knowledge. Archeological sites and associated evidence for fires, tools, art, and other cultural artifacts provide insight into the way early humans used their knowledge of animal behavior. The oral traditions of numerous indigenous peoples around the world, gathered from the 1600s onward, provide evidence of both the general and specific nature of these observations and their importance.

For daily living, humans required two basic types of information, consisting of observations of animals that they could use as food on the one hand, and knowledge of organisms that were potential predators on the other. Knowing aspects of behavior, such as when the animal is active, seasonal variations in presence and activity level, what it eats, where it sleeps, whether it lives alone or in social groups, and many other traits would enhance the ability to capture and kill organisms for meat. Indeed, the means by which capturing and killing a prey animal would be dependent on this knowledge and some period of trial and error with snares, traps, spears, and other implements.

Louis Leakey told a story about how early humans might have captured a rabbit. The method is based on the knowledge that when a rabbit feels threatened, it initially freezes, but when the source of threat moves too close, the rabbit will generally break to the left or right from its frozen position. So, if a hunter wants to catch and kill the rabbit he would find a suitable stone, and then look for a rabbit. Once spotted, and with the animal now frozen in hopes of not being discovered, the hunter moves steadily toward the rabbit. When the critical distance is reached, the rabbit will bolt in one direction or the other. The hunter makes a calculated guess and goes left or right.

If correct, he will be nearly on top of the rabbit and can kill it with his stone. If he guessed incorrectly, perhaps he will turn and locate the rabbit for a second attempt.

Having observed prey and their behavior, humans were able to select from several different strategies. One involves stalking the animal, moving behind some form of barrier, using camouflage, and working from the appropriate wind direction to reach a point where it could be captured or killed. This distance would depend on whether the prey was to be dispatched by hand, with a stone, using a spear, or by some other means. Clearly, knowing about the animals in the region where they lived would enable early humans to provide a supply of meat for themselves and a family or similar group.

A second strategy, termed sit and wait, was used for animals that frequent a particular location such as a water hole. Remaining hidden near such a location, a hunter was able to use information gleaned from numerous previous visits to know when certain types of animals were most likely to return to the water hole. He would then be ready to kill or capture the animal.

For peoples who lived near water and used fish as a steady source of meat and protein, similar observational knowledge was critical. Knowing the types of fish and their different swimming patterns, social groupings, and particularly their habitat, would make capture possible, using nets for example, or spearing bigger fish.

Survival was also dependent on avoiding becoming a prey item for some other animal. Obviously, this concern was more relevant in some locations than others. There are a few large mammals that can kill and consume humans; bears are one example as are many of the large cats. Pack-dwelling canids can pose a threat as could large herds of grazing animals if they were startled and stampeded. In the marine environment, sharks would be a primary threat. In some locations, reptiles such as snakes and crocodilians would pose a hazard. Thus, knowledge of the habits and habitats of each of these potential predators would be necessary on a daily basis.

Various forms of artistic expression that came to be part of the cultural heritage of early humans also demonstrate their knowledge about animal behavior. Drawings in caves and the many of the petroglyphs and pictographs found throughout the world depict animals. Some of these artworks are of prey, some are of potential predators, and others appear to have a religious or spiritual connection. In all cases, there are examples where the art is true to life, with very accurate characteristics of the animal depicted, and other examples where the representation is a caricature of some form, recognizable as a particular animal, but drawn with special features or only as a rough sketch. Some art of these types depicts canids or, in a few cases, hoofed animals, that were subjects of early attempts at domestication. In addition, at some locations, we find artifacts, such as small animal sculptures, exhibiting characteristics that show a firm knowledge of the anatomy and behavior of the animals in their world.

Various early forms of communication were tied, in part, to information gathered about non-human animals and shared with others. The aforementioned artwork is a good example. Also, by some accounts, aspects of language originated from mimicking animal sounds; this is called onomatopoeia. Such oral communication could be used, for example, to lure prey closer and, of course, as a set of signals about prey locations and behaviors while the hunt was on.

Finally, early humans were students of their own behavior. Most animals acquire a set of social skills and means of communication that derive, in part, from observations of others. In order to function in a social group, early humans learned generally accepted norms as for example, their place in the progression of food sharing. They also used observation to gather useful information about individuals in their sphere, including personality quirks, habits, and preferences.

Agriculture and Domestication

As humans evolved in terms of their physical traits, so did they evolve culturally. Among the events that transpired, two in particular are dependent on knowledge of animal behavior. The rise of agriculture was a driving force in the appearance of towns and then cities; humans could congregate because food supplies were now produced in large quantities. These efforts to domesticate plants, such as corn or squash, required many decades of experimentation, some almost certainly accidental, but accompanied by a growing knowledge of how to reproduce desired traits in subsequent generations.

Animal behavior is important to these efforts in terms of a variety of pest organisms, during both the growing phase and when grains and other agricultural products were stored for use during nongrowing times of the year. Just as today, there are numerous insects, birds, mammals, and other organisms that can harvest the fruits of our labor before we do. Knowing the habits of these creatures would enable at least some attempts to prevent damage to the growing plants and fruit. However, as we know quite well, many of these issues are still with us today, whether it is insects like the bark beetle destroying forests in western North America or rodents consuming rice and other grains in the growing fields of the Philippines or countries of southeast Asia.

Humans have been more successful in terms of storing grains and other products. The use of various kinds of containers or vessels generally prevents damage from rodents and birds, though insects can sometimes still be a problem. Placing foodstuffs in granaries that are hard to reach, as in the case of storage structures on cliff faces or in caves worked well for protecting the food supply. These means of successfully storing foods required a knowledge of what types of animals posed threats to the stored goods, when during the day or year the animals were active, and possibly exploring various means of discouraging the attempts to steal food. The latter involved, in some instances, plant compounds that acted like deterrents. Also important in this regard were domesticated canids that could warn of or help ward off larger predators that came to raid the fields or storage locations.

Coincident with some early humans, several types of animals were domesticated. There are at least four rationales for domestication: (1) companionship and protection; (2) as a food supply; (3) to provide skins and fur for clothing, bones for utensils, intestines for water storage, and other animal parts used for various cultural artifacts; and (4) as an aid to transportation and travel. In all the cases, having a thorough knowledge of the habits of the animals would have been very necessary to have any degree of success with the sequence of steps involved in domestication.

The most well known of the companion animals are the canids, primarily the wolf. Animals domesticated for food include large species such as cattle and swine, and many smaller species such as rabbits and guinea pigs. Among the species domesticated for clothing (and other purposes) were sheep and musk ox, though all of the animals eaten for food served as a supply of skins, bones, and other useful items. Animals used for transportation included horses, llamas, and camels. Each of these domesticated stocks derives from one or more wild ancestors. Humans in different locations around the globe often domesticated local animals for similar purposes, as exemplified by those used for transport. Observing the wild animals over many generations provided an understanding of the behavioral traits, diet, and particularly the social and reproductive biology of the target species. In addition, early peoples likely determined that occasional introduction of wild stock animals into the breeding program would help maintain genetic heterozygosity, avoiding potential problems associated with too much inbreeding.

Domestication likely began as some form of close association between humans and the wild mammals. As time passed, the symbiosis with humans increased and some controlled breeding, in captivity, followed. As with the domestication of crops, experimentation resulted in animals with traits that made them more suitable for human uses. Much later, as human interest became more concentrated and economic in nature, different breeds of the domestic stocks were bred for specific traits and locations with varying climates. This process continues even today with most of our livestock. Indeed, we often use our knowledge of the behavior and genetics of the organism to facilitate breeding programs. A few mammals and some birds have also been domesticated. These include rabbits, guinea pigs, chickens, and a variety of animals we now use for laboratory research, many of them being rodents such as mice and rats.

Greeks and Romans

Both Greeks and Romans made substantial contributions to the general understanding of animals, including commentaries on behavior. Among the major developments during this period extending from early Greeks (*c.* 2700 BCE) to the Byzantine Empire (*c.* 600) with respect to animal behavior were (1) a better understanding of anatomy and some physiology, (2) the beginnings of true natural history at the time of Aristotle and his development of a classification scheme for plants and animals, and (3) the application of science to agriculture and domestic livestock.

Beginning by about 650 BCE, dissections of humans and other animals have led to some basic understanding of anatomy, particularly in terms of bones and muscles. This has led, in turn, to an early, but incomplete knowledge of how these systems worked, as for example in locomotion. Accompanying the anatomy was a basic understanding of the five senses and the manner in which we humans sampled our environment. In the Aristotelian view, the human mind was, at birth, a blank slate and the senses were a primary source of input and to aid in formulating the rules for processing this information. The attribution of human qualities to gods and to mythical creatures continued to be a major part of the spiritual life of the Greeks. Together these lines of evidence indicate that the Greeks were cognizant of the qualities of non-human animals with which they shared plant Earth.

Aristotle (384–322 BCE) and his followers are the first major influence on what we would recognize as the scientific study of animal behavior. At least three lines of evidence support this statement. Systematic, recorded natural history observations provided a basis for understanding both the immediate world surrounding them and, with time, the discoveries made in what were then distant places. His use of consistent methods in observing and recording became the basis for natural history for many centuries. His many hours spent watching marine organisms like starfish, mussels, and fishes produced the first ethograms, including daily activities, modes of reproduction, and development from egg, to larva, to adult.

The use of the comparative method, both in anatomy and in natural history began during this period. Aspects of animal reproduction were a common theme in many of Aristotle's writings. Last, Aristotle organized what was then known about animals into a classification scheme, a basic ladder of the living organisms. Because he had limited exposure to land animals, his classification consists mainly of marine and fresh water aquatic organisms.

The history of animal behavior during Roman times centered on applied topics, particularly with domestic stocks, which included, in addition to the farm animals with which we are familiar, treatises on birds and bees. Pliny the Elder (AD 23–79) compiled a 37-volume encyclopedia of the natural world. He relied both on his own observations and a collation of written works from his many predecessors. Work by Galen (AD 129–200) extended the knowledge of anatomy through a series of dissections. These findings could then be extended to the understanding of body functions, like locomotion, feeding, and reproduction.

The Middle East and Asia

Peoples of Asia and the Middle East have produced extraordinary work in areas of astronomy, physics, medicine, and mathematics, but what little evidence we have suggests that there was some work done on domestic animals and little on observations of behavior in wild animals. Much care and thoroughness, and early development of the scientific method, including empirical data and experimental techniques, have characterized work on astronomy, but these same procedures were not used for animal study, or at least no records of this type of work have survived. Depictions of animals, particularly birds and mammals, in artwork and descriptions of the behavior of these animals in myths and stories indicate that some species were held in high regard. Art showing animals catching prey, feeding, nursing, copulating, and engaging in other behaviors is found at numerous sites from many cultures in this large region.

Domestic animals included livestock, beasts of burden, and a strong affinity for cats and dogs. Cats in particular are depicted in a number of locations and situations; they were revered animals in some cultures, particularly in Egypt. This was particular true of house cats, domesticated from wild ancestors, used for capturing mice and rats, and then, over several millennia, elevated to the status of gods. Drawings, figurines, and other depictions of wild animals, including raptors and hoofed animals like ibex and water buffalo, show prey animals captured for food and skin and bone.

Throughout this vast region, and extending, via Native American migrations to North and South America, there is a strong connection between animals, their behavior, and religion. A variety of deities were defined in terms of a particular animal and were assumed to have the observed traits and features of that animal. In most of this region, animals like camels, elephants, and some horses were used for military purposes. To work with these animals, people using them needed to understand their behavior through direct experience and observation, and passed to future generations in oral or written form.

In India, where Hinduism has long been the dominant religion, much work centered on animals used for commerce, either as transportation or for trade, hunting, staged fights, and various sacrificial ceremonies. As a result, there are written works on topics like training and taming elephants, along with early schemes of classification that used, among other traits, the behavior of animals as a criterion.

Several precursors of much more modern ideas in life science and ecology that have strong relevance to animal behavior appeared during this period. These were a product of what is sometimes called the Arab Agricultural Revolution, which transpired during the European Middle Ages. Among these were (1) the notion of food chains, (2) the struggle for existence among animals, and (3) the effects of environmental determinism. The last is a topic that recurs in animal behavior up to the present.

Precursors of Animal Behavior in Europe

From *c.* AD 200 onward, the entire scientific enterprise in Europe entered a long decline and disintegration that lasted until the late 1300s. This period was characterized, among other things, by a return to the use of mythology to explain the behavior of humans and other animals. There was little interest in natural history. Religion was the dominant force and most knowledge was framed and characterized in terms of religious beliefs and traditions.

With the beginning of the Renaissance (late fourteenth century through the sixteenth century) a return to basic and then experimental science resulted in renewed interest in animal behavior. An early beginning to this period, in the late twelfth century was curtailed by the black or bubonic plague that descended on Europe during the 1300s. As natural history observations became more thorough and sophisticated, information was organized by animal type, location, habitats, and other criteria. Astronomy and other physical sciences, and physiology, anatomy, and medicine dominated this period.

However, there were important developments in areas of science that bear on what we now know as the study of animal behavior. In the early Renaissance, considerable information was collected and added to the general knowledge about the natural history of a variety of animals from both Europe and neighboring regions such as

northern Africa and western Asia. These findings were, in large measure, a function of increased trade, with visits made to an ever-widening circle of nations. The Age of Exploration began in the fifteenth century, resulting in voyages to many locations around the globe and observations of wild animals and, in some instances, the collection of live specimens for the examination and viewing of diverse activities. The Linnaean classification scheme, with the notion of immutable species and a hierarchy of levels to fully characterize each organism's place in the grand scheme emerged and became universal. Two further developments had enormous impact on human history: the printing press and methods for illustration. The latter was particularly relevant for animal behavior in that depictions of animals engaging in various activities became part of the permanent record.

As the Renaissance progressed, several advances bore directly on viewing and interpreting behavior. Descartes (1596–1650) promulgated the mechanistic notion of life. Animal motions and activities were all driven by a vital spirit. His theory encompassed all living matter and was the dominant view adopted by most scientists for the subsequent several centuries. Science was starting to expand, the Enlightenment, which began in the late 1500s, brought on a new era for science, with diverse approaches to many areas of knowledge and new methods to employ to study topics like animal behavior. At this time, we first begin to see the splitting of natural philosophy into various specific disciplines; areas like botany, biology, medicine, physiology, etc., that we recognize today.

Conclusion

As we know from history, the 1000-year period from the end of the Roman Empire (late fifth century) until the late fourteenth century is labeled as the Dark Ages followed by the Middle Ages. While it is certain that knowledge about animal behavior accumulated in this period, no major treatises on behavioral topics appear to have been written and what work was done with animal behavior was in conjunction with domesticated stocks and breeding programs. As travel to distant lands emerged in the fifteenth century, new varieties of wild animals were encountered and stocks of domesticated animals were transported between distant locations.

Humans were involved with animal behavior from very early times. Many of the areas of behavior that were of interest to these early peoples remain important to us today. In various parts of the world, people still depend on wild game for portions of their diet and are keen observers of behavior patterns of their potential prey. Domestic animals abound, with more than 100 species of mammals and birds, and breeding for specific traits continues. In addition, as time passed and philosophy became subdivided, with modern science as one outcome, the planned study of animal behavior emerged. Between the early humans and those scientific studies, a long period of virtually no systematic investigation or descriptions of animal behavior was followed by world-wide travel, exploration, and accumulation of enormous information about animals and their actions.

See also: Animal Behavior: The Seventeenth to the Twentieth Centuries.

Further Reading

Drickamer LC, Vessey SH, and Jakob B (2002) *Animal Behavior: Mechanisms, Ecology, Evolution,* 5th edn, Chapter 2. New York, NY: McGraw-Hill.

Farber PL (2000) *Finding Order in Nature: The Naturalist Tradition from Linnaeus to E.O. Wilson.* Baltimore, MD: Johns Hopkins University Press.

Huff TE (1993) *The Rise of Early Modern Science, Islam, China, and the West.* New York, NY: Cambridge University Press.

Nordenskiold E (1928) *The History of Biology.* New York, NY: Alfred A. Knopf.

Relethford J (2007) *The Human Species: An Introduction to Biological Anthropology.* New York, NY: McGraw Hill.

Singer C (1959) *A History of Biology to About the Year 1900.* New York, NY: Abelard-Schuman.

Washburn SL (ed.) (1970) *The Social Life of Early Man.* Chicago, IL: Aldine.

Relevant Websites

http://www.historyworld.net/wrldhis/PlainTextHistories.asp?historyid=ac22 – History of Biology.

http://www.bioexplorer.net/History_of_Biology/ – History of Biology.

http://www.press.uchicago.edu/presssite/metadata.epl?mode=synopsis&bookkey=24142 – Paleolithic Art.

http://en.wikipedia.org/wiki/History_of_science_in_Classical_Antiquity – History of Science in Classical Antiquity.

http://www.normalesup.org/~adanchin/history/Antiquity.html – History of Biology.

Animal Behavior: The Seventeenth to the Twentieth Centuries

L. C. Drickamer, Northern Arizona University, Flagstaff, AZ, USA

The Seventeenth Century

The view of biological phenomena during the first half of this century and beyond was dominated by the mechanistic perspective promulgated by Descartes (1596–1650) and his followers. All actions were thought to be the result of some vital force. His fondness for and deep understanding of the subject led to a conception of life as entirely driven by principles of mathematics and statistics. One of his most significant works (*Discourse on Method*) contains an essay outlining his views on the proper approach to science. Among these principles are several that remain instructive today for scientists in the field of animal behavior.

Among his tenets are the following. (1) Divide the problem under investigation into as many separate parts as possible and then work on each of these parts individually. (2) Using this method, conduct the investigation in a stepwise fashion, to build up the larger answers pertinent to the entire problem. (3) All information used in conjunction with the problem and its constituent parts must be factual and obtained objectively.

Although the study of animal behavior was an integral part of natural history, there is no separate or comprehensive treatment of the subject. Rather, it is important to understand that a series of developments in the seventeenth century had significant consequences for the development of more specific studies of animal behavior during the eighteenth and nineteenth centuries.

Several naturalists contributed to the understanding of animal behavior during the seventeenth century. Accurate descriptions of behavior, rather than the too often myth-like depictions of previous centuries, became the norm. Explorations of more distant lands and varied habitats provided much new information, expanding the ability to make comparisons, producing more general conclusions tying together form, function, and behavior. And the earliest recorded experimental work on the behavior of animals occurred during the seventeenth century.

John Ray (1627–1705), an Englishman, was likely the most significant of this group that also included his student Francis Willughby (1635–1672), Jan Swammerdam (1637–1680), who elucidated insect metamorphosis, Rene Antoine Reaumur (1683–1757) and his work on ants spanning two centuries, and Maria Sibylla Merian (1647–1717), who provided some of the earliest life history information on insects in Europe and South America.

Ray produced plenty of work dealing with natural history, a classification scheme, and he and Willughby performed experimental work showing the rise of sap in trees. Ray and Willughby traveled extensively in Europe, gathering information on flora and fauna. Ray's work centered on plants with later work on mammals and reptiles. Willughby completed the treatises on birds and fishes prior to his early death. The descriptions of animal life included notes on life history and behavior. Ray wrote about instinct, noting that each bird species made nests with the same materials, even when hand reared with no prior access to the materials. At this time, religion still held strong influence over science as evidenced by Ray's volume *The Wisdom of God in Creation*. The number of different species was thought to be ordained by God and the task of the naturalist was to decipher and understand the wonders of nature.

Other scientific developments during the century had indirect effects on the understanding of animal behavior. The work by Harvey on circulation and elucidation of other aspects of physiology provided the groundwork for later discoveries related to the interrelationships (causation) of behavior. The discovery, by van Leeuwenhoek, of the potential for using the microscope expanded the known world to include microbes and many smaller organisms not readily observed by the unaided eye. Much later, studies of behavior involving these small creatures would produce findings relevant to disease, pollution, and the roles of detritivores in ecosystems.

During the seventeenth century, several scientific societies emerged and grew; many of them served as places for presentations and discussions of scientific work. These societies emerged from efforts to circulate abstracts, primarily in England and countries of Europe. Most of these learned societies involved just presentations, but a few, such as the Royal Society of London, had a component that involved conducting experiments. During the later half of the century, several societies began publications on a wide variety of scientific topics. Articles written for publication initially took the form of reviews of existing information in various disciplines, but shortly, articles that involved ongoing investigations were included as well. Only in the nineteenth century would specialized journals emerge and in the case of animal behavior, it was well into the twentieth century before journals specific to the field appeared. There was a publication, *Journal of Animal Behavior* in the first decade of the twentieth century but it lasted only for 5 years.

Although a few collections of animals, museums containing specimens, and scientific libraries existed at the start of the 1600s, all these types of institutions expanded in number and scope during the century. Since anatomy was a longstanding area of work, some of the earliest museums, dating back more than a millennium, were of anatomical specimens. Formal museums, containing animal, plant, and mineral specimens, developed from individual collections generated by the scientists in the course of their work. Among the earliest of these was the British Museum in London, where these personal collections were combined, catalogued, and made available for wider use. The wide-ranging travels, both by sea and on land, resulted in acquisition of many previously unseen organisms. These exotic animals were often added to the collections (zoos or menageries) at country estates of nobility. Perhaps the earliest zoological garden was the Menagerie due Parc established at Versailles in the 1660s. Zoos as we know them, with many different functions, and open to the public, did not emerge until the nineteenth century.

A key feature of science is that information is stored in written form where it can be accessed by others, debated, and retested as needed. The scientific journals already mentioned were circulated to some individuals who had small libraries of their own. As volumes written on topics such as natural history were published, they too became part of these collections. Some colleges and universities had libraries, making materials available to scholars during the course of their work. Some of the early societies also maintained libraries of relevant books and journals. Eventually, some of these individual collections and small libraries coalesced to form true libraries open to scholars and students. A good example of this is the Bodleian Library at Oxford University, which opened in 1602.

The Eighteenth Century

The early decades of the century are best characterized by the work of individuals like Linnaeus (1707–1778), Buffon (1707–1788), and Lamarck (1744–1829) expanding on the naturalist tradition of previous centuries. Also Erasmus Darwin (1731–1802), the grandfather of Charles Darwin was a physician, botanist, and naturalist. His writings include mention of the idea of evolution and the interrelatedness of all living forms, as well as a poem about evolution. Thomas Malthus (1766–1834) is best known for his thoughtful combination of economics and demography, leading to the assumption that population growth was a potential threat to continued human existence. Gilbert White (1720–1793) wrote *The Natural History of Selbourne*, a compilation of his observations of animals and plants. He was among the first to explore bird song as a means of differentiating related species. John Bartram

(1699–1777), an explorer and the first great American botanist, wrote about animals he encountered in the southeastern states. His son, William Bartram (1739–1823), continued in this tradition, becoming well known as an ornithologist and artist of natural history.

Carolus Linnaeus was a Swedish biologist, equally adept at both plants and animals. His comprehensive knowledge of both living forms and the considerable attempts by others who preceded him to find a system to classify all of nature provided the basis for his hierarchical scheme for organizing plants and animals. His introduction of the binomial nomenclature served as the foundation for his scheme. Organisms were divided into distinct species on the basis of observable characteristics, including external features, anatomy, and for animals, some aspects of their behavior. His work on plants in particular provided some of the foundation for what we know today as ecology.

Georges-Louis Comte de Buffon, responsible early in his career for key developments in mathematics related to probability theory, spent the bulk of his scientific career as a natural historian. At a young age, he moved to Paris and was appointed as the director of what is today the Jardin des Plantes, which he transformed from a king's garden into a scientific establishment with plants from many locations around the world. For those interested in animal behavior, his most significant work was in the area of natural history. He compiled and wrote a 44-volume encyclopedic coverage of all that was known to that time about plants and animals. These writings contain a great deal about the behavior of animals, most of which was based on factual observations and reports, though with some errors and misconceptions. His work also included, for the first time, a systematic approach to the distributions of various animal and plants types; the forerunner of modern biogeography. The latter is important for animal behavior in that it becomes the basis for the comparative approach used, for example, to contrast animals with various adaptations to particular habitats such as deserts or tropical rain forests or to compare the dietary habits of marmots and marmot-like animals living on several different continents.

Jean-Baptiste Lamarck is best known for the idea that animals could pass along to the next generation (inherit) characteristics acquired during their lifetime. His major focus was on plants, though he also provided insights into animals, with a major treatise on invertebrates. Where Buffon is credited with mentioning processes similar to evolution, Lamarck provided the first truly comprehensive notion of how change in form and function could occur over time. His work spans the last decades of the eighteenth century and the early portion of the nineteenth century. A major tenet of his evolutionary theory was that organisms adapted to their environment. Thus, his claims had strong relevance to animal behavior in terms of the form, function, and action sequence.

For defining the history of animal behavior, a little known Frenchman named Charles G. Leroy (1723–1789) played a major role. At mid-century, Leroy was the gamekeeper at the menagerie maintained at the Versailles Palace. His book on animal intelligence was based on many years of keen observations of the animals, primarily mammals and birds that could be readily seen on the palace grounds and in neighboring regions. In the book he describes the need for accurately listing and defining behavior and he describes some traits of what we now know as an ethogram, a complete rendering of all of the habits and life history traits of an organism. Many sections of his book deal with comparisons between species, for example herbivores and carnivores. Further, he argued that animals were not completely mechanistic in their behavior; foreshadowing the longstanding debate about nature and nurture. He alludes to both the earlier claims, but mixes this with an understanding that animal actions can depend on differing motivations and needs.

During the eighteenth century, collections of animals, called menageries initially and later zoos or zoological gardens, became important for the observation and study of animal behavior. In the ensuing two centuries, zoos built on three major functions: scientific study, conservation, and education. In ancient times, some small menageries were maintained for pleasure, as in China, where deer herds were kept, for curiosity and novelty, and as happened with emperors, kings, and other nobility, and for use in the arena for example in staged fights involving bulls, bears, lions, many other animals, and humans. While some observations were undoubtedly made on these captive animals, the primary emphasis was on standard husbandry practices. By the mid-eighteenth century, zoos could be found in many cities, particularly in Europe. Most notable of these was the menagerie at Versailles outside Paris, France. The early collections were for aristocrats only, but a few such as the Tiergarten Schonbrunn in Vienna opened to the public in the later decades of the eighteenth century. Zoological gardens had not existed in America until the later half of the nineteenth century, that is, the decades after the Civil War.

The First Half of the Nineteenth Century (1800–1850)

Several key developments, ongoing at the close of the eighteenth century, continued into the first decades of the nineteenth century, including the work by Lamarck. His theory on inheritance of acquired characteristics was accompanied by views that species did not originate from special creation and that transformations in form and function occurred over time. Many of his examples, such as the length of the neck for giraffes, are based, in part, on observations of behavior, as in feeding on vegetation high up in bushes and small trees.

One of the most important events of this period involved a great debate between Georges Cuvier (1769–1832) and Geoffrey Saint-Hilaire (1772–1844). The debate between the two contemporaries was emblematic of a key controversy: were species and their traits fixed or were there ongoing transformations occurring in nature. Cuvier, working primarily in the laboratory, held that function was the result of form, whereas Saint-Hilaire proposed that, based on observations in nature as well as laboratory work, function was at least in part a manifestation of activity. A considerable portion of the evidence used was from comparative anatomy, but animal behavior was also considered as part of the evidence for the differing points of view. Saint-Hilaire can be credited with providing some of the earliest ideas about the concept of homology, the idea that traits with common functions may be the result of some degree of common ancestry. Cuvier felt that common functions did not signal common ancestry. As with some earlier differences, this sounds a bit like the never-ending debate about nature and nurture, which has been an important stimulus for work in animal behavior for several centuries, including experimental research, which began in the last half of the nineteenth century.

Ongoing advances in other areas of life sciences during the early nineteenth century were important for understanding animal behavior and remain keys to our interpretation of animals in their natural world today. Growing from the emphasis on medicine and comparative anatomy, which characterized much of life science during the preceding several centuries, physiology became more involved with the synergy of form and function, one result being observed behavior. Principles, like the uniformatarianism, put forward by Charles Lyell (1797–1875), for understanding Earth's geological history were based on the notion that all of nature involved slow, but gradual changes; these forces were thought to be still in effect and would remain so into the future. This idea is important to the thinking of Darwin later in the century.

Ideas, some with roots extending back 1000 years, provided the basis for ecology, the study of interactions between organisms and their environment. These included notions of populations, communities, and various types of organismal interactions, involving both animals and plants. Much of this work, though in its infancy, set the stage for areas within animal behavior that emerged during the twentieth century. Animals and their behavior are at the middle of a ladder of life that extends from biochemistry to the biosphere. Thus, exploration of factors that control and influence behavior is complemented by the understanding of the population and community importance of various behavioral actions.

The Second Half of the Nineteenth Century (1851–1900)

The writings of Darwin (and Wallace) dominated thinking in natural history for the last half of the nineteenth century. As scientists grasped the full meaning of evolution by natural selection and grappled with the consequences of this view, a process was set in motion that cast new light on observed behavior patterns. This, in turn, led to further field observations with more specific intent, for example, details of courtship activities in birds and to experiments involving behavior. Brief synopses of work done by a variety of individuals over the course of the last half of this century provides insights into the shift from mostly natural observations to more experimental work with research conducted in both laboratory and field settings. Among the scientists active during this period were Douglas Spalding (1841–1877), George John Romanes (1848–1894), Charles O. Whitman (1843–1910), and C. Lloyd Morgan (1862–1936). Too often we think of animal behavior as starting in about the 1930s with the work of individuals like Tinbergen and Lorenz. But, as these examples show, there were many ideas, considerable research, and healthy discussion beginning at least 60 years earlier. As has been noted already, we can trace some of today's work on animal behavior to these early roots.

Spalding, from Great Britain, helped pioneer the experimental approach to behavior studies. He worked primarily with birds, performing tests of flight in fledglings. His conclusions, based on the research, supported the idea that observed behavior was a mixture of instinct, the idea that animals perform actions with no prior experience, and developmental experiences. Another way to phrase this is that instinct guides learning. He is credited with being the first scientist to properly describe imprinting in birds.

Romanes, a Canadian, worked primarily with invertebrates, but his thinking and writing extended much further. He was a good friend of Darwin and generally supported the new ideas on evolution by natural selection. His books include topics involving insights regarding thought and how mental processes evolved: *Animal Intelligence* and *Mental Evolution in Animals*. He also published many of his observations on invertebrate physiology and related behavior and several volumes concerning Darwin's theory and its corollaries.

Whitman, an American who helped established the Marine Biological Laboratory at Woods Hole, Massachusetts, studied pigeons, moving his subjects each summer from his position at the University of Chicago to the laboratory on Cape Cod. He is remembered for his work to establish zoology as an independent discipline and for his work on the evolutionary bases for animal behavior. He retained a view that life was a progression of stages, ending with humans as the top species.

Finally, Morgan, from Great Britain, was a major figure in launching the field of animal behavior. His work had importance for both comparative psychology and ethology. His book, *Animal Behaviour*, could be considered as the first 'textbook' in the field. A key tenet that he put forward is known as Morgan's Canon. It reads: 'in no case may we interpret an action as the outcome of the exercise of a higher psychical faculty, if it can be interpreted as the outcome of the exercise of one which stands lower in the psychological scale.' This notion was directed at anecdotal observations and conclusions by contemporaries as well as a tendency toward anthropomorphism. The simple-sounding phrase significantly impacted on empirical research and theoretical work in animal behavior. Morgan's own work focused on several topics, and among them are the relationship between the animal and human mind and experiments on bird migration and nest building. He also stressed that both the objective and subjective approaches offered value in terms of explaining behavior.

Others whose work during this period is important to the founding of animal behavior as a discrete subfield within both biology and psychology include Jacques Loeb (1859–1924), who worked with animal movements, particularly tropisms. Jakob von Uexkull (1864–1944) formalized the concept of Umwelt, the notion that each animal is surrounded by a series of events and characteristics that give it a unique perspective. Thus, to properly and fully understand the behavior of an animal, it is necessary to look at its world from its perspective. William Morton Wheeler (1865–1937) is best known for his extensive work with ants, particularly their social life. In 1902, he wrote a short piece for *Science* in which he discussed zoology and ecology, and proposed that ethology was the best term for describing that segment of zoology involved in "... their physical and psychical behavior towards their living and inorganic environment ..." Jean Henri Fabre (1823–1915) championed the study of insects, including descriptions of and experiments on behavior.

By the end of the century, the three threads that eventually merged in the last half of the twentieth century to become modern animal behavior were in nascent stages of development. These three threads are sometimes divided along geographical lines with respect to their main locus of thoughts and research. However, a bit of reading in the history of science provides a picture of emergence of the three threads across both Europe and North America. The three threads are (1) a tradition of animal observation centered primarily in Europe that led to the field of ethology, where explanations are sought concerning the evolution and functional significance of behavior, (2) a tradition of psychology, beginning with attempts to understand the human mind and how we think, and extending to comparisons with and among non-human animals that emerged in America, and (3) a physiology-zoology tradition that involved examination of underlying body functions as they affected observed behavior. Today, ethology and animal behavior, as terms used to define a subfield, are used interchangeably.

The discipline of study we call animal behavior or ethology has its origins with the earliest humans. As a scientific enterprise, various pieces of work in anatomy and natural history, and early formulations concerned with the different roles of genetics and experience affecting behavior, the manner in which animal interact with each other and their environment, and the classification of organisms – all affected the emergence of animal behavior in the later half of the nineteenth century. Of course, the major stimulus, both in his initial book on the origin of species, and in a number of subsequent volumes on topics that encompass many areas of animal behavior, Charles Darwin can properly be viewed as a principal founder of this field of study.

By the end of the 1800s, several developments characterize the study of behavior. First, the diversity of organisms examined in field and laboratory was greatly increased. Second, the use of the experimental method became the acceptable way to explore behavior, though certainly the gathering of information via observation and natural history remained quite significant. Third, the topics being investigated were more diverse than in previous times. For example, there were books on topics like play behavior, and entire books devoted to the behavior of invertebrates.

Transition to the Twentieth Century

Growing from the critical, initial progress of the late 1800s, several key developments in the history of animal behavior occurred during the first two decades of the twentieth century. In the area of European ethology, major figures, in addition to those who overlapped between the nineteenth and twentieth centuries, include Heinroth, Thorpe, Baerends, von Frisch, Lorenz, and Tinbergen. For comparative psychology, a similar list includes, Thorndike, Lashley, Watson, Skinner, and Yerkes. In the case of American zoology, individuals who were part of the infant stages of animal behavior include, Allee and Noble. With

reference to the point made earlier that this sort of history does not always lend itself to exact categorization, there are some individuals who made significant contributions to the progress in animal behavior, but who cannot be said to have an exact fit with the three threads. Notable among these would be Erich von Holst, who worked on neural mechanisms, Sewall Wright, who made efforts in genetics and evolutionary theory, and Curt Richter, who worked at the intersection of genetics, biology, and psychology.

The reader is encouraged to explore the entries on (1) developments in animal behavior from the early twentieth century to the 1960s and (2) the history of comparative psychology.

See also: Behavioral Ecology and Sociobiology; Comparative Animal Behavior – 1920–1973; Darwin and Animal Behavior; Ethology in Europe; Future of Animal Behavior: Predicting Trends; Integration of Proximate and Ultimate Causes; Neurobiology, Endocrinology and Behavior; Psychology of Animals.

Further Reading

Burkhardt RW, Jr (2005) *Patterns of Behavior.* Chicago, IL: University of Chicago Press.
Dewsbury DA (1989) A brief history of the study of animal behavior in North America. In: Bateson PPG and Klopfer PH (eds.) *Perspectives in Ethology,* vol. 8. New York, NY: Plenum.
Elliott RM (ed.) (1957) *A History of Experimental Psychology.* New York, NY: Appleton-Century-Crofts.
Gardner EJ (1972) *History of Biology.* Minneapolis, MN: Burgess.
Locy WA (1908) *Biology and Its Makers.* New York, NY: Henry Holt.
Nordenskiold E (1928) *The History of Biology.* New York, NY: Alfred A. Knopf.
Thorpe WH (1979) *The Origins and Rise of Ethology.* London: Praeger.

Relevant Websites

http://books.google.com/books – Reader's Guide to the History of Science.
www.britannica.com/EBchecked/topic/492392/John-Ray – Encyclopedia Britannica Online.
www.fact-archive.com/encyclopedia/Douglas_Spalding (see similar entries for other individuals) – Fact Archive – Encyclopedia.
http://en.wikipedia.org/wiki/Erasmus_Darwin (see similar entries for other individuals) – Wikipedia.

Ant, Bee and Wasp Social Evolution

R. Gadagkar, Indian Institute Science, Bangalore, India

Introduction and Definitions

Ants, bees, and wasps belong to the Hymenoptera, a 220-My-old monophyletic order, which is among the largest and most diverse in the class Insecta (**Figure 1**). Among the characters that are common to all Hymenoptera, perhaps the one that is of greatest interest to the topic of this article (even if it eventually turns out to be only of historical interest) is haplodiploidy, a term that indicates that males are haploid on account of developing parthenogenetically from unfertilized eggs and females are diploid on account of developing from fertilized eggs. In spite of the size of this order (more than 250 000 described species), only a small fraction (<10%) of the species is social, and thus of interest to us here. Even among those that are social, there is a very large variation in the degree of sociality. To focus attention on the phenomena that give rise to the most sophisticated forms of sociality, it has become customary to demarcate the eusocial species from all 'lesser' forms of sociality. To qualify as eusocial, a species must exhibit cooperative brood care, differentiation of colony members into fertile reproductive castes (queens) and sterile nonreproductive castes (workers), and an overlap of generations such that offspring assist their parents in brood care and other tasks required for colony maintenance. The discussion in this article thus refers to the evolution of eusociality in ants, bees, and wasps.

It is also customary to distinguish two subgroups within the eusocial species. Some species have relatively small colonies (<100 individuals) and their queens and workers are not morphologically differentiated. These species are often labeled primitively eusocial. In other species by contrast, colony sizes are large (>100 individuals, often in the thousands or hundreds of thousands, sometimes over a million) and queens and workers are morphologically differentiated. These species are often labeled advanced or highly eusocial on the argument that they have acquired more traits that are unlikely to have been present in their solitary ancestors compared to the primitively eusocial species. There is one more remarkable fact about hymenopteran sociality that needs to be mentioned. Feminine monarchy, a phrase used by the cleric Charles Butler in 1634 to describe the honeybee society, applies to all hymenopteran societies, as they are all headed by one or a small number of fertile queens. Indeed, males eclose, usually leave the nest, mate, and die, playing no role in colonial life. As queens do little more than lay eggs, all tasks connected with nest building, brood care, and colony maintenance, are performed by the workers, and this leads to a society of female subjects headed by a female monarch.

The Paradox of Altruism

Queens mate and lay both haploid, unfertilized eggs and diploid, fertilized eggs. Workers are, by and large, sterile. In some ant genera, such as *Pheidole*, *Pheidologeton*, and *Solenopsis*, workers have altogether lost their ovaries. In others, they have much smaller ovaries compared to their queens and in some species, workers have lost the ability to mate. In any case, workers never or very rarely lay eggs when their queen is alive so that they usually spend their whole lives working to assist their queen to reproduce, while they themselves die without leaving behind any offspring. Sometimes, the sacrifice of the workers on behalf of their colonies is more dramatic. Honeybee workers defend their colonies by stinging their marauders but such stinging results in the death of the bee, as her sting is armed with barbs pointing outwards, and once inserted, cannot be withdrawn. She flies away leaving behind her sting, the poison gland, and a part of her digestive system only to die within a few minutes. The poison gland continues to pump venom into the body of the victim for up to 60 s, making this a very efficient venom delivery mechanism, but for the owner of the sting, it is an act of suicide. In the Brazilian ant *Forelius busillus*, a small number of workers perform preemptive self-sacrifice every day as they lay down their lives, because the last workers involved in closing the nest entrance in the evening remain outside and die.

Kin Selection

How can such sterility and sacrifice, seen in tens of thousands of species, arise and persist in nature? Why does natural selection not eliminate these altruistic traits as they do not appear to contribute to the Darwinian fitness of their bearers? These questions remained unanswered and often remained unrecognized as unanswered questions for over a 100 years after Darwin had proposed the theory of natural selection. But today, we have a powerful theoretical framework on account of which the evolution of altruism by natural selection is no longer a paradox, at least to the theoreticians, although empiricists

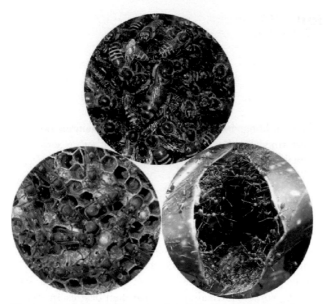

Figure 1   A portion of the nest of the Asian dwarf bee *Apis florea* showing the morphological differentiation between the single, large queen and the many small workers surrounding her (top), a portion of the primitively eusocial wasp *Ropalidia marginata,* showing a portion of the paper carton nest and the absence of morphological caste differentiation (bottom left), and a portion of leaf nest and workers of the Asian weaver ant *Oecophylla smaragdina*. Photos: Dr. Thresiamma Varghese.

may have to wait a bit longer before they can claim to have demystified the evolution of altruism.

The necessary theoretical framework was developed by W.D. Hamilton in 1964 and has come to be known popularly as kin selection. Hamilton argued that self-sacrificing altruism is not necessarily paradoxical if it is directed toward close genetic relatives who share genes with the altruist. This concept is known as inclusive fitness, implying that fitness can be gained not only by producing offspring (direct fitness) but also by aiding genetic relatives (indirect fitness). Hamilton showed that an altruistic trait can be favored by natural selection if the inequality

$$b/r - c > 0$$

is satisfied, where b is the benefit to the recipient, c the cost to the altruist, and r the coefficient of genetic relatedness between the altruist and recipient. This has come to be known as Hamilton's rule. There is now good evidence that this is a theoretically robust idea, but the hard part has been to show that ants, bees, and wasps, behave as if they obey Hamilton's rule. There was initially a fascinating red herring. On account of haplodiploidy, a hymenopteran female can be more closely related to her full sister ($r = 0.75$) than she would be to her own offspring ($r = 0.5$). This means that Hamilton's rule can be rather easily satisfied and thus altruistic sterility can evolve rather more easily in the Hymenoptera than in diploid organisms. This haplodiploidy hypothesis had a great

appeal for some time because it seemed to have strong empirical support – eusociality was known to have originated at least eleven times independently in the Hymenoptera compared to a single origin (termites) in the rest of the animal kingdom. The euphoria was unfortunately short-lived.

Testing the Theory

A major problem for the haplodiploidy hypothesis was that although workers can be related to their full sisters by 0.75, they are related to their brothers by 0.25, bringing the average relatedness to siblings back to the diploid value of 0.5. This problem can, in principle, be surmounted if workers invest more in sisters than in brothers. But this would lead to a conflict between workers that would prefer a female-biased investment and their mothers that would prefer equal investment in sons and daughters. How this conflict is resolved and whether its resolution affirms or negates the role of haplodiploidy in social evolution are still being vigorously researched and debated. A more serious problem for the haplodiploidy hypothesis is the increasing evidence for reduction in relatedness among the workers themselves because of both multiple mating (polyandry) by the queens and parallel or serial polygyny (multiple queens). It is now widely accepted that the asymmetries created by haplodiploidy are, by themselves, inadequate to explain the evolution of eusociality in the Hymenoptera.

The demise of the haplodiploidy hypothesis by no means weakens Hamilton's rule, which has been often tested. Unfortunately, most tests ignore the benefit and cost terms, and test the limited prediction that altruism should be directed at close rather than distant relatives. This effort, combined with increasingly powerful DNA-based molecular techniques to measure genetic relatedness, has spawned a number of efforts to measure intracolony genetic relatedness. These values tend to be quite variable, often below 0.75 and even below 0.5. Social insect colonies are thus sometimes composed of rather distantly related or even unrelated individuals. Some specific phenomena are elegantly explained by the observed variability in relatedness values. For example, a comparison of intracolony genetic relatedness in honeybees and stingless bees explains why daughter queens inherit the nest and mother queens leave to found their own nests in multiply mated honeybees, while the mother stays and the daughter leaves in singly mated stingless bees. Another well-known example is found in worker policing. In colonies with singly mated queens, workers should prefer to rear nephews ($r = 0.375$) rather than brothers ($r = 0.25$), while in species with multiply mated queens, workers should prefer to rear brothers ($r = 0.25$) rather than nephews ($r = 0.125$). There is now good empirical evidence that in stingless bees which mate singly, worker oviposition is common and oophagy by other workers is not common. Conversely, oviposition is rare and worker policing is common in honeybees whose queens mate multiply.

Nevertheless, the low values of genetic relatedness are somewhat embarrassing for the general idea of kin selection, but some investigators now argue that lifetime monogamy is a fundamental condition for the evolution of eusociality, and processes such as polyandry and polygyny that lead to lower relatedness are later elaborations. Others have given a near burial to kin selection itself by arguing that kin selection is only a weak force and that high genetic relatedness is more likely to be a consequence of eusociality rather than a factor in its origin. A potential problem is that Hamilton's rule is seldom tested in its entirety, by simultaneously measuring b, r, and c and when that is done, it does appear to have impressive explanatory power. It is another matter though that such tests point to the greater importance of b and c, over r, which is tantamount to greater importance for ecology and demography over genetic relatedness (**Figure 2**).

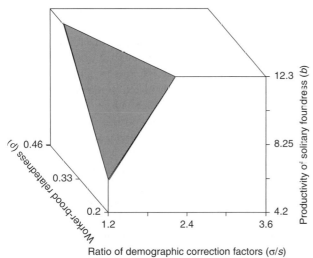

Figure 2 A graphical illustration of a unified model showing the parameter space where worker behavior is selected (unshaded) and the missing chip of the block where solitary nesting behavior is favored in the primitively eusocial wasp *Ropalidia marginata*. The model simultaneously considers the benefit, cost, and relatedness parameters in Hamilton's rule and therefore incorporates genetic, ecological as well as demographic factors in the evolution of eusociality. The unified model predicts that about 95% of the wasps in the population studied should opt for the altruistic, sterile worker role and only about 5% of the wasps in the population should opt for the selfish solitary nest founding role. In striking confirmation of the predictions of the model, empirical field data indicate that about 92–96% of the wasps choose to nest in groups (where most of them will end up as altruistic sterile workers) and only about 4–8% of the wasps choose to nest solitarily and reproduce. Reprinted with permission from Gadagkar R (2001) *The Social Biology of Ropalidia marginata – Toward Understanding the Evolution of Eusociality*. Cambridge, MA: Harvard University Press. Copyright © 2001 by the President and Fellows of Harvard College.

The Proximate Causation of Social Behavior

Debates about the relative importance of understanding the so-called ultimate (evolutionary) factors and proximate (causation) factors in the evolution of eusociality have waxed and waned much as they have in many other fields of animal behavior. By and large, investigations relating to the proximate causation of social behavior have lagged behind evolutionary studies, especially since the advent of kin selection. There is, however, a strong attempt to redress the imbalance in the past decade or so.

Division of Labor

The remarkable ecological success of social insects is attributed primarily to their sociality and in turn largely to the division of labor that social insects achieve in organizing their work. Division of labor is therefore a topic that has successfully rivaled kin selection in attracting the attention of social insect researchers. A major early interest concerned how the morphologically differentiated subcastes among some species of ant workers increase the ergonomic efficiency of the colony by pursuing parallel rather than serial processing of tasks. Later, attention shifted to how a worker decides what it needs to do at any given time. There is considerable evidence that division of labor, especially in species that lack morphologically differentiated worker subcastes such as honeybees, is based on age polyethism. Thus, there is a strong correlation between what a bee does and its age – young bees begin their adult careers as cleaner bees and gradually move through the tasks of building the nest, nursing the larvae, unloading and processing the food, guarding the nest, and finally go out of the nest in search of food. In the 1990s, there developed an interesting debate about whether internal physiological factors such as age and hormone levels drive task allocation of the workers or whether it is governed by external factors such as prior experience and work availability at any given time. Supporters of the latter idea came up with the interesting phrase 'tasks allocate workers' to bring out the contrast with the idea that workers allocate themselves to different tasks based on their physiology. Like so many debates, this one too appears to have died down with time, without settling many of the interesting issues raised during the heat of the debate. Instead, discussion has moved on to other, perhaps more productive, topics.

A theme that has gained currency and has produced a large body of extremely interesting, and indeed, practically useful knowledge, relates to the self-organization of work in insect colonies. It is now recognized that workers in social insect colonies accomplish rather complex tasks not by any top-down control by a leader or foreman but by bottom-up control. In this scenario, individual workers follow simple local rules but their collective labor leads to the emergence of complex patterns such as an architecturally sophisticated nest or the choice of the shortest foraging path. This has come to be known as swarm intelligence, also known as distributed or collective intelligence. Swarm intelligence is flexible, robust, and self-organized. Flexible because the locally available workers can quickly respond to a change in their local environment without waiting for the central authority to perceive that change or without a potential conflict with some physiologically programmed universal algorithm. Robust because even if some individuals fail to perform, there are others who can substitute for them. And self-organized because there is no need for supervision. The concept of swarm intelligence has led not only to a better understanding of how social insects achieve the remarkable feats they do but also to a surprising degree of practical application in the context of the performance of human tasks. Based on our understanding of how ants, bees, and wasps utilize swarm intelligence, so-called ant colony optimization algorithms (AOL) are in regular use in the telecommunication industry, on the Internet, and in the cargo industry, and the number of such applications is growing.

Regulation of Reproduction

From a physiological point of view and especially for someone interested in controlling insect reproduction, the fact that a single queen maintains reproductive monopoly in colonies consisting of hundreds or thousands of potentially reproductive workers, is even more remarkable than anything we have discussed so far. Unfortunately, we do not understand the mechanism behind this feat in any degree of detail. Traditionally, it has been thought that queens in small colonies of primitively eusocial species achieve reproductive monopoly by suppressing worker reproduction through physical aggression and intimidation. Workers are thought to succumb to such suppression even if they might get more fitness by laying a few of their own eggs because they have no choice – they are physically too weak to fight back and leaving the nest is worse than staying on and attempting to get indirect fitness. Queens in large colonies of highly eusocial species cannot obviously physically aggress against every worker and hence behavioral dominance is not an option for them. They are known in many cases to produce pheromones that might serve the same purpose. In imitative language, queens of highly eusocial species have long been said to suppress worker reproduction by means of pheromones. In a thoughtful essay, Laurent Keller and Peter Nonacs (see Further Reading) pointed out that this idea is untenable. It is hard to imagine how queens can suppress worker reproduction against their interests by means of pheromones because workers can fight back by evolving enzymes or other chemical weapons that would neutralize the queen pheromone. Hence, it must be assumed that it is in the evolutionary interest of the workers themselves to refrain from reproduction and strive to increase the productivity of their colonies. The direct fitness they thus lose would be small, as they are no match to their large physogastric queens in terms of egg laying.

This has led people to be cautious of the language they use, but even more importantly, it has led to the idea that the queen pheromone must be an honest signal not only of their superior fertility but also of their health and vigor at any given time. This has in turn spawned a plethora of studies attempting to detect and understand these signals. While honeybee queen pheromones were long thought to be volatile compounds produced by the queen's mandibular glands, ant and wasp researchers have now drawn attention to cuticular hydrocarbons (CHC), mostly linear or branched long-chain hydrocarbons, present adsorbed to the wax coating on the cuticles of the insects. The primary function of the CHCs appears to be to protect them from dehydration and since they are highly variable, they are thought to have been co-opted to serve the function of signals. Each individual may have a unique CHC profile that has led to the phrase 'cuticular hydrocarbon signature.'

Ironically, it is not the honeybees or the ant species with large-sized queens and large numbers of workers that have been at the forefront of the CHC research. Instead, queenless ponerine ants in which mated workers (gamergates) serve as the sole egg layers of their colonies, bumble bees in which the queens only modestly outsize their workers, and even primitively eusocial wasps without any morphological caste differentiation, have led this research from the front. This has had two consequences. First, CHCs have also been implicated in nestmate recognition, a function of crucial importance to all social insects (see section 'Kin and Nestmate Discrimination'). Second, honest signaling of fertility is also being attributed to the queens of primitively eusocial species without morphologically differentiated queens. The whole field of CHC research is in its infancy; and there is rather scanty evidence yet that the insects themselves perceive the diversity in the CHC cocktail to a degree of precision and sophistication that can begin to match the increasingly sophisticated gas chromatograms and multivariate statistical analysis tools that researchers now use to detect the compounds and discriminate different individuals. On the other hand, it might well be that the true suppression of worker reproduction by physical aggression and intimidation, even in the small colonies of primitively eusocial species, may be a myth, and the regulation of reproduction in all species of social insects may depend on CHCs and other similar honest chemical signals. It must be admitted that there is really no direct experimental evidence that physical aggression and intimidation are necessary and sufficient to suppress worker reproduction. Future research in this area is eagerly awaited.

Kin and Nestmate Discrimination

Low average intracolony genetic relatedness, if accompanied by high variance, is not really a difficulty for kin selection if there is good intracolony kin recognition so that altruism can be selectively directed to close kin. Hence, there has been an earnest search for the evidence of kin recognition. But that search has yielded nothing but disappointment. There are really no good examples of incontrovertible evidence for intracolony kin recognition in any ants, bees, or wasps. On the other hand, social insects have very well-developed abilities for nestmate discrimination. This suggests that keeping away nonnestmates and thus maintaining colony integrity are more important in the daily lives of social insects than to pursue intracolony nepotism. For some 15 years after Hamilton proposed the idea of kin selection, there was no attempt to test whether animals had direct kin recognition abilities. Then the deluge began and the first pieces of evidence for nestmate recognition were enthusiastically welcomed as evidence for kin recognition and as further vindication of the kin selection theory. This error in judgment was soon realized and fortunately, it did not dampen the enthusiasm for extending the studies of nestmate discrimination to scores of other species of social insects. Today, CHCs mentioned earlier in connection with the regulation of reproduction, are also thought to mediate nestmate discrimination. Whether the same set of molecules can simultaneously mediate both the discrimination of nestmates and nonnestmates and queens and workers within a colony is still a matter of debate. Perhaps, the evidence today for the role of CHCs in nestmate discrimination is stronger than the evidence for their role in reproductive regulation.

The Social Behavior Toolkit

Theoreticians modeling the origin and evolution of social behavior posit a gene (allele) for altruism or other social traits and investigate how they might fare against competition with their selfish or other ancestral counterparts under the action of natural selection. The implied idea of an allele for altruism should perhaps remain a metaphor. But it is not uncommon for empiricists to take this concept literally and begin to look for genes for altruism, sterility, dominance, etc. To help disengage from this trend, Mary Jane West-Eberhard explicitly suggested what has come to be known as the Ovarian Groundplan hypothesis. The idea she emphasized is that apparently new traits shown by social insects may be a result of co-opting phenotypically plastic traits already existing in their solitary ancestors. This has now been developed into the Diapause Groundplan hypothesis, which argues that the worker and gyne castes of a primitively eusocial wasp such as *Polistes* arise from the developmental pathway already present in bivoltine solitary insects. More generally, investigators refer to the Reproductive Groundplan Hypothesis, which can encompass species that do not diapause.

Gyne is the term used for wasps eclosing in the fall in the annual colonies of social wasps such as *Polistes*, which overwinter and found new colonies in the following spring. In solitary bivoltine species, there are two generations of females produced per year: a first generation of females (G1) that reproduce soon after eclosion and a second generation of females (G2) that undergo diapause before reproduction. This hypothesis leads to the prediction that *Polistes* workers (who do not reproduce) correspond to G1 females who are programmed to reproduce and the gynes (who are the future reproductives) correspond to G2 females that have their reproduction turned off (**Figure 3**). This prediction at first seems counterintuitive because workers are generally thought of as sterile and gynes as fertile. But these predictions are testable and there is now considerable evidence to support the diapause ground plan hypothesis. The idea that the social behavior toolkit of the ants, bees, and wasps are borrowed with some modification (and that modification is made possible because of phenotypic plasticity) from their solitary ancestors is not only powerful but also one that suggests many new lines of investigation.

How Does Social Behavior Develop?

Debates between the practitioners of the proximate and ultimate questions have by no means died down, and indeed they sometimes threaten to go out of hand. One way to reduce conflict is to introduce more players into the ring. Help to do precisely that and thus to channel these debates into more productive directions may come from an unexpected quarter – ethology's Nobel laureate Niko Tinbergen. In an influential paper in the early 1960s, Tinbergen argued that we should simultaneously be asking four different kinds of questions concerning any behavior: What is the current adaptive value of the behavior? What are the proximate factors that cause the behavior? How does the behavior develop in the life time of an individual animal? What is the evolutionary history of the behavior? The first two questions correspond to the ultimate and proximate questions we have already discussed. Tinbergen's third question, which concerns the ontogeny of behavior, has only very recently begun to be asked in the context of social behavior in insects. Not surprisingly, ontogenetic questions have first and foremost been applied to understand how some individuals in insect societies come to develop and behave as fertile queens, while others come to develop and behave as sterile workers. It has long been assumed that caste determination is not genetic but entirely environmental.

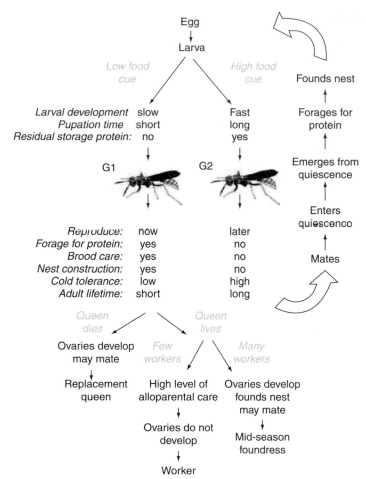

Egg
↓
Larva

Low food cue   *High food cue*

Founds nest
↑
Forages for protein
↑
Emerges from quiescence
↑
Enters quiescence
↑
Mates

Larval development	slow	Fast
Pupation time	short	long
Residual storage protein:	no	yes

G1   G2

Reproduce:	now	later
Forage for protein:	yes	no
Brood care:	yes	no
Nest construction:	yes	no
Cold tolerance:	low	high
Adult lifetime:	short	long

Queen dies   *Queen lives*

Few workers   *Many workers*

Ovaries develop may mate
↓
Replacement queen

High level of alloparental care
↓
Ovaries do not develop
↓
Worker

Ovaries develop founds nest may mate
↓
Mid-season foundress

Figure 3 The *Polistes* life cycle incorporates the fundamental elements of the bivoltine ground plan, larvae respond during development to a food cue and diverge onto one of two trajectories. Scanty provisioning leads to the G1 pathway, which is signaled by slow larval development (due to low nutrient inflow), short pupation time, and no storage protein residuum in emerging adults. More abundant provisioning leads to more rapid larval development, longer pupation time, and residual storage protein in emerging G2 adults. G1 females have a 'reproduce now' phenotype, and they forage for protein, care for the brood, and construct nests. The expression of these behaviors is conditional, as indicated by branching points in the G1 sequence. If the queen is lost, a G1 female can develop her ovaries, mate if males are present, and become a replacement queen. If a queen is present but the number of workers is low, a G1 female will alloparentally express maternal behaviors (i.e., nest construction, nest defense, brood care, and foraging) as a worker at her natal nest. Finally, if a queen is present and the number of workers is high, a G1 female may depart from the natal nest and found a satellite nest in midseason. Because the cold tolerance of G1 females is low, they do not survive quiescence, and lifetimes are short. In contrast, G2 females have a 'reproduce later' phenotype. They express no maternal behaviors the first year, but after emerging from quiescence, they break reproductive diapause and shift to the reproduce now phenotype. Reprinted with permission from Hunt JH and Amdam GV (2005). Bivoltinism as an antecedent to eusociality in the paper wasp genus Polistes. *Science* 308: 264–266.

And there is plenty of evidence that environmental factors, especially the nutrition in the early larval stage, influence the future caste of the individual. What is surprising, however, is that more than a negligible number of cases of genetic determination of caste, or at least genetic influences on caste ratios, are being thrown up when genetic, and more recently, molecular genetic techniques, are applied. These genetic influences remain poorly understood at the molecular or physiological level. But the mechanisms involved in nutritional influence on caste have recently begun to be unraveled in an impressively sophisticated experimental paradigm. It is well known that honeybee larvae fed with royal jelly develop into queens, while those denied royal jelly develop as workers. Gene expression profiles determined using microarray analysis have shown that queen- and worker-destined larvae differ greatly in which genes they upregulate and which they downregulate, paving the way for tracing the ontogenetic development of caste-specific morphology, physiology, and behavior. A similar study but one that used adults of the primitively eusocial wasp *Polistes canadensis* also helps identify genes that are involved in producing caste-specific phenotypes in queens and workers. A more recent microarray study shows that gene expression in the brains of worker-destined wasps is similar to that of nest foundresses, both

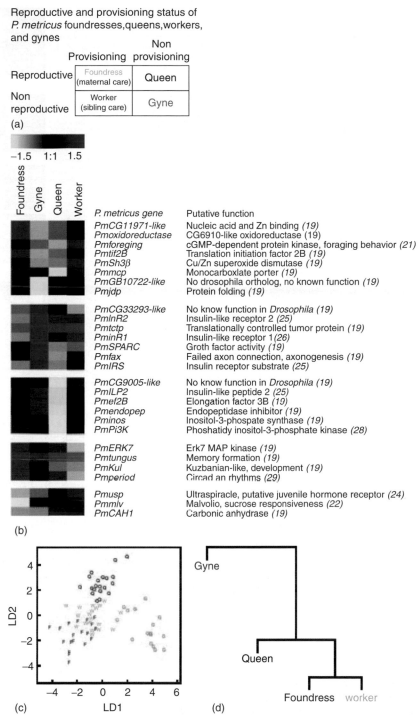

Figure 4 *Polistes metricus* wasp brain gene expression analysis tests the prediction that maternal and worker (eusocial) behavior share a common molecular basis. (a) Similarities and differences in reproductive and brood provisioning status for the four behavioral groups analyzed in this study: foundresses, gynes, queens, and workers. Each individual wasp (total of 87) was assigned to a behavioral group on the basis of physiological measurements (b–d). Results for 28 genes selected for their known involvement in worker (honeybee) behavior. (b) Heatmap of mean expression values by group and a summary of analysis of variance (ANOVA) results for each gene. Genes were clustered by K-means clustering, those in red show significant differences between the behavioral groups. *P. metricus* gene names were assigned on the basis of orthology to honeybee genes, putative functions were assigned on the basis of similarity to *Drosophila melanogaster* genes. (c) Results of linear discriminant analysis show that foundress and worker brain profiles are more similar to each other than to the other groups. (d) Results of hierarchical clustering show the same result (based on group mean expression value for each gene). Reprinted with permission from Toth AL, Varala K, Newman TC, et al. (2007) Wasp gene expression supports an evolutionary link between maternal behavior and eusociality. *Science* 318: 441–443.

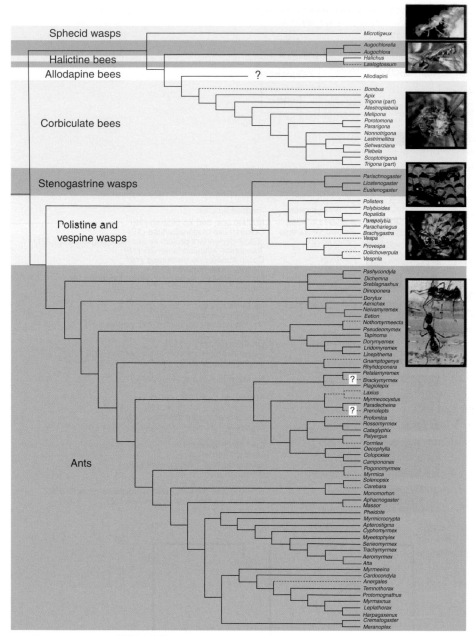

Figure 5   Phylogeny of genera of eusocial Hymenoptera (ants, bees, and wasps) for which female mating frequency data are available. Each independent origin of eusociality is indicated by alternately colored clades. Clades exhibiting high polyandry (<2 effective mates) have solid red branches, those exhibiting facultative low polyandry (<1 but <2 effective mates) have dotted red branches, and entirely monandrous genera have solid black branches. Mating frequency data are not available for the allodapine bees. Reprinted with permission from Hughes WOH, Oldroyd BP, Beekman M, and Ratnieks FLW (2008) Ancestral monogamy shows kin selection is key to the evolution of eusociality. *Science* 320: 1213–1216.

of which specialize in maternal behavior, compared to queens and gynes, which do not display maternal behavior (see **Figure 4**). These kinds of studies are expected to usher in a new era in the study of insect sociality that, if integrated into the more classical approaches involving ultimate and proximate causation rather than pursued in isolation as a more modern and more correct alternative approach, should revolutionize the study of insect sociality.

The Phylogeny of Social Behavior

The good news is that we have also begun to put Tinbergen's fourth question to work for us. In the last decade or two, DNA sequencing has become routine, and powerful statistical techniques to construct molecular phylogenetic trees are being developed. This has made it possible to reconstruct the phylogeny of social insect lineages and trace the evolutionary history of social behavior to a level of sophistication that Tinbergen could hardly have imagined. It is gradually becoming fashionable to investigate the salience of different traits to the origin of eusociality in the context of a phylogenetic tree. The first such study made on eusocial wasps identified nesting, oviposition into an empty nest cell, progressive provisioning of larvae, adult nourishment when they feed larvae, and inequitable distribution of food among nestmates, as the most important traits in the origin of vespid eusociality. Interestingly, no role was obvious for the asymmetries in genetic relatedness created by haplodiploidy, although haplodiploidy was found to be important as a mechanism that permitted females to choose the sex of their offspring and produce an all-female brood. In contrast, a similar study that includes many more eusocial lineages in addition to the wasps says nothing explicit about haplodiploidy but shows that high relatedness expected from monogamy is ancestral to all the eusocial lineages considered. It therefore suggested a high salience for genetic relatedness in the origin of eusociality (**Figure 5**). That these studies may, in the beginning, yield a contradictory and confusing picture is a minor problem compared to the expected long-term benefits of taking this approach. A more impressive and less controversial study of attine ants has helped trace the origin of five different agricultural systems of ants. There can be little doubt that these studies will multiply rapidly in the near future. What is less likely to happen automatically, which may therefore require some special effort, is the preservation of the climate required for a simultaneous pursuit of all of Tinbergen's four questions and a meaningful integration of the knowledge gained for these diverse approaches.

See also: Division of Labor; Kin Selection and Relatedness; Recognition Systems in the Social Insects; Social Evolution in 'Other' Insects and Arachnids; Spiders: Social Evolution; Subsociality and the Evolution of Eusociality; Termites: Social Evolution.

Further Reading

Bourke AFG and Franks NR (1995) *Social Evolution in Ants*. Princeton, NJ: Princeton University Press.

Crozier RH and Pamilo P (1996) *Evolution of Social Insect Colonies – Sex Allocation and Kin Selection*. Oxford: Oxford University Press.

Detrain C, Deneubourg J-L, and Pasteels JM (eds.) (1999) *Information Processing in Social Insects*. Basel: Birkhäuser Verlag.

Gadagkar R (2001) *The Social Biology of Ropalidia marginata – Toward Understanding the Evolution of Eusociality*. Cambridge, MA: Harvard University Press.

Gadau J and Fewell JE (2009) *Organization of Insect Societies – From Genome to Sociocomplexity*. Cambridge, MA: Harvard University Press.

Hölldobler B and Wilson EO (1990) *The Ants*. Cambridge, MA: Harvard University Press.

Hölldobler B and Wilson EO (2009) *The Superorganism – The Beauty, Elegance, and Strangeness of Insect Societies*. New York, London: W.W. Norton & Company.

Hunt JH (1999) Trait mapping and salience in the evolution of eusocial vespid wasps. *Evolution* 53: 225–237.

Hunt JH (2007) *The Evolution of Social Wasps*. Oxford: Oxford University Press.

Keller L and Nonacs P (1993) The role of queen pheromones in social insects: Queen control or queen signal? *Animal Behaviour* 45: 787–794.

Ross KG and Matthews RW (eds.) (1991) *The Social Biology of Wasps*. Ithaca and London: Cornell University Press.

Toth AL and Robinson GE (2007) Evo-devo and the evolution of social behavior. *Trends in Genetics* 23: 334–341.

Toth AL, Varala K, Newman TC, et al. (2007) Wasp gene expression supports an evolutionary link between maternal behavior and eusociality. *Science* 318: 441–443.

West-Eberhard MJ (2003) *Developmental Plasticity and Evolution*. New York, NY: Oxford University Press.

Wilson EO (1971) *The Insect Societies*. Cambridge, MA: The Belknap Press of Harvard University Press.

Anthropogenic Noise: Impacts on Animals

H. Slabbekoorn, Leiden University, Leiden, Netherlands

Introduction

The world is rapidly becoming more and more noisy. Most of us get used to it without realizing the every day cacophony around us. If you pay attention and listen, there is an amazing level and diversity of noise, when we are in traffic, in shopping malls, at sporting events, or just when we open the front door. The fact that we habituate to the rise of noise levels and that we have acknowledged the presence of many loud sounds associated with locations and context does not mean that noise is not harmful. Noise pollution is a serious threat to human health and well-being. Traffic noise levels, especially, in combination with being annoyed by them are, for example, significantly correlated to the probability of ischemic heart disease. Children also show impaired cognitive development at schools when they are exposed to repetitive noise from overflying aircraft or nearby road or rail traffic. Animals have not received much attention until recently, but are also affected by noisy conditions.

The impact of noise on animal health and well-being is often hard to assess. Correlational studies between noise levels and animal presence or condition can be suggestive, but the disturbing role of noise often remains speculative as many things may vary in concert with noisy conditions in urban and industrial areas or along highways. Therefore, there are still many questions that are not yet or only partly answered. For example, what impact does noise have on animals? Can they be annoyed like humans and can they also experience heart problems? Or what would the consequence of noise be in restricting the use of species-relevant sounds? And is there no way of getting around such problems of masking noise? Are all animal species equally affected, irrespective of which habitat they occupy? Or are some species better suited than others in their ability to cope with anthropogenic noise?

To address these questions, I will first briefly touch on the role that sounds play in the life of animals and to what extent efficient hearing of relevant sounds may contribute to their survival and reproductive success (**Figure 1**). The fundamental features of sound that explain the dominant role of acoustic communication across the animal kingdom will be highlighted (**Box 1**), while the use of sounds in the variety of media in which animals live is also reviewed. The subsequent discussion of natural signal-to-noise ratios serves to provide insight into the potential harm of anthropogenic noise levels. The last part deals with what we actually know now about causes and consequences of artificially raised noise levels related to human activities. A case study on an avian urban survivor provides, probably, the best insight, to date, into why some species can cope with extreme noise levels and why others fail.

The Importance of Sounds to Animals

Many animal species produce sounds which may play an important role in communication among conspecifics. Correlations between acoustic variation in a signal and sender qualities, motivational state, or environmental conditions may broadcast a message that may be detected and perceived as meaningful by one or more receivers. A sound may provide information about the presence of a male individual of a specific species, or the distance between sender and receiver. Variation in acoustic signals within an individual may reflect condition or motivational state: whether an animal is in the mood of chasing others out of his territory, or whether it is in the mood of receiving a female partner at a freshly built nest. Acoustic signals or cues may also convey information about the presence of food or predators. In this context, sounds used for intraspecific communication or sounds generated by activities such as eating or moving around may be picked up by receivers of another species.

There are several examples in which one species has specialized on exploiting the signals of another; for example, when bats find frogs by their mating calls or when parasitic flies home in on calling crickets. Bats are, of course, also well known for their use of echo-location: reflections of their own sounds that tell something about distance and shape of objects in their environment.

Not all heterospecific use of signals is detrimental to the senders. Several newt species, which produce no or few sounds themselves, seem to listen for the calls of frog species which are indicative of the location of a suitable breeding pond, suitable for both the frogs and the newts. This phenomenon is called 'soundscape orientation' and entails the use of any biotic or abiotic sound from the environment for orientation, which allows animals to find important resources such as feeding or breeding places or to pick out a suitable direction for dispersal or migration.

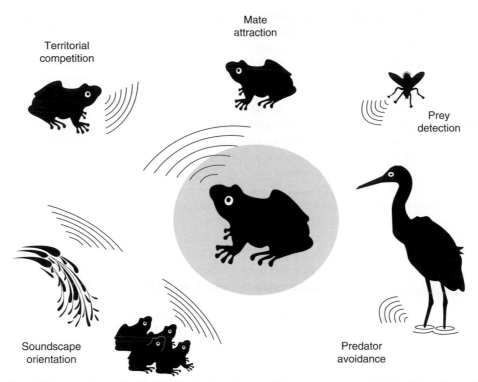

Figure 1 Illustration of the role of sound in the life of animals using a male frog as an example. Typically, male vocalizations serve to defend resources against competitors and to attract potential partners to mate. Sound is also used to detect and localize prey and to detect and avoid predators. In general, all sounds in the environment, such as calling conspecifics or heterospecifics or a waterfall can be used for soundscape orientation.

Box 1: Fundamental Features of Sound

The importance of sounds to animals in acoustic communication or soundscape orientation is closely related to features that are specific to sound. First of all, sound can be transmitted over long distances and is not necessarily obstructed by obstacles in the landscape between senders and receivers. Sound does not rely on the presence of light and can be heard during day and night. Sound is also multidirectional, both from the perspective of the sound source, or the sender of a signal, as well as from the sound receiver. Acoustic signals or noises radiate away from a point source (a bird singing or an engine running) or a line source (the surf at the coast or a highway) in all directions. Similarly, receivers of sounds hear signals and noises coming in from all directions and do not need to face the source direction to detect its presence acoustically. This directionality varies with the frequency of sound: high frequencies are more directional than low frequencies and, therefore, high twitters or shrieks will more rapidly loose power in directions other than the one faced into by the sender. In addition, high-pitched signals attenuate relatively fast over distance as sounds of shorter wave lengths are more prone to medium absorption and scattering compared to sounds of longer wavelengths. Nevertheless, if the amplitude level at a receiver is above the detection threshold, high-pitched signals are more easily used for locating the signaller as shorter wavelengths allow the use of interaural differences in arrival times to assess the angle to the sound source.

Sound Use Through Different Media

Most well-known examples and well-studied model systems of acoustic communication concern species transmitting signals through air. Roaring red deer, howling monkeys, singing birds, calling frogs, and chirping insects all have to deal with similar properties of attenuation and degradation inherent to sound transmission through air. And we should not forget: whether we whistle, clap our hands, speak, or honk a horn, we are also typical users of air-borne sounds ourselves. Some important properties of air that may lead to variation in sound transmission properties are temperature, humidity, and turbidity. Higher temperatures and lower humidity levels lead to higher sound energy absorption rates. So, if the target of an acoustic message is far away, it makes sense to sing or call when the air is cold and humid, such as during the night and early morning. Air turbidity leads to more signal attenuation as well as degradation. The result is that few animals spend much time trying to get their message across during very windy conditions.

Although we are most aware of sounds in the air, not all acoustic signals and cues in the animal kingdom concern air-borne sounds. As a matter of fact, water is a much better medium for sound transmission than air. There is less attenuation and less degradation of sound transmitted through water than for sound transmitted over the same distance in air, which can be explained by the higher molecular density of water. Many aquatic animal species are known to rely on sounds in a variety of contexts. Whales and dolphins are famous for their often elaborate vocal repertoire and intricate relationship between acoustic similarities among individuals and their degree of social affiliation. Many fish species are also known to produce courtship sounds which are used to attract and seduce potential mating partners. Furthermore, fish are thought to depend heavily on their accurate hearing ability for finding prey or detecting predators, while soundscape orientation in general is very likely a widespread phenomenon under water.

Sounds may also transmit through solid matter and the soil, although with much higher attenuation and degradation compared to water and air. Although soil is generally not an obvious medium for communication, some events are better audible by sounds radiating through the soil than by the air-borne component. You may think, for example, of native Americans listening for running bison beyond eye-sight with their ear against the ground, but also elephants can sense far-away rumbles of conspecifics through soil vibrations conducted through their toes. Male mole crickets call from a hole in the ground to attract overflying females, but these sounds also reach the holes of competing males through the substrate. Furthermore, many rodents use species-specific drum patterns in the context of territorial threat displays and predator alarm, while mole rats bang their heads against the ceiling in their underground burrows to reach next-door neighbors.

Finally, vibrational communication through twigs and leaves is especially common among insects, such as lacewings, cicadas, and spiders, but is also known from chameleons. The communicative vibrations can be generated by vibrating legs, abdominal shaking, or through the production of air-borne sounds which resonate in the plant substrate. These vegetation-borne signals do not often transmit over large distances, but they may reach intended receivers without attracting unwanted receivers with air-borne signals. Nevertheless, the vibrations can be detected by some predators such as birds, lizards, and frogs, and even humans may exploit them via special devices, for example, to assess pest insect infestation in agricultural crops or timber trees.

Signal-to-Noise Ratios in the Natural World

Natural environments are noisy. Successful information transfer among communicating animals is, therefore, not only dependent on signal amplitude, receiver sensitivity, and sound transmission properties, but also on signal interference through masking. More or less continuous, broadband sounds from rustling leaves, ocean surf, wave or wind turbulence, and fast-flowing streams and torrents may cause signal detection problems though masking. The same is true for the accumulation of sounds produced by the vocal animal community, especially, where seasonal and diurnal timing, as well as spectral range, overlap among community members. Detection of acoustic cues and conspecific signals implies the ability of extraction of relevant sounds from a background of detected but irrelevant sounds. The subsequent perception of meaningful variation in an extracted acoustic signal also depends heavily on suitable signal-to-noise ratios at the receiver's location.

Both biotic and abiotic noise levels may fluctuate considerably through relatively long-term or short-term cycles. Incoming waves in tidal rock-pools may render courtship calls of male fish inaudible to female for large parts of the day. Deep inside a forest, grunting monkeys may be sheltered from strong winds and thereby have better noise conditions than at the forest edge. And especially at the edge, wind can come in gusts, can gradually grow in power, but also suddenly diminish. Although wind may be the driving force of a dominant noise component in many open habitats, the very presence of trees often provides the noise-generating substrate: two birds may sing equally loud, but only one may hear the other well if one is on a bare rock and the other on a tree among rustling leaves.

Noisy signaling conditions for vibrational communication through soil or vegetation are not conceptually different from the signal-to-noise ratios in air and water. Wind and rain drops can generate quite noisy conditions in the soil and vegetation, while air-borne noise can also be transferred into plant matter and interfere as such with vibrational signals. Wind generates movement, turbulence, and vibration in tree trunks, twigs, and leaves, for which low-frequency noise will be transmitted through trunk conduction into the soil. Wind and rain will deteriorate signaling conditions in a wide frequency range with often dramatic peaks at low frequencies between 10 and 50 Hz. Transmission properties are equally bad for signals and noise, although specific leaf size or stalk lengths may result in frequency-dependent resonance favoring transmission of some frequencies over others.

The Rise and Nature of Anthropogenic Noise

Anthropogenic noise levels have risen globally and are still on the rise. The human population is growing rapidly and most people rely heavily on motorized transport and all sorts of machinery for their well-being and luxury.

Truck and car traffic is one of the most dominating noise sources in cities, on highways, and far beyond. Therefore, we can derive the timing of noise development roughly from the numbers of cars in a region. Personal car ownership in Europe and the United States has grown significantly since the early 1920s and has led to the current conditions of one car per two persons in Europe and three for every four persons in the United States. The ongoing population growth and the rising number of cars per person still resulted in a 40% increase over the last 15 years to, for example, over 250 million passenger cars currently driving around in Europe.

Besides cars, there is an amazing variety of noise sources that ranges from leaf blowers, lawn mowers, and ghetto-blasters, to air conditioners, pumping systems, car alarms, church bells, and other calls to prayer. Busses, trains, and airplanes pass on the ground, underground, and overhead. There is hammering and sawing at constructions sites, running engines and other noisy activities in industrial areas, container shipping and drilling in the ocean, and dredging and piling in and around river systems. Irregular amplitude peaks are associated with outdoor activities, such as concerts, festivities, manifestations, and large sporting events. Many recreational activities are also motorized and particularly noisy, such as with the use of motorbikes, all-terrain vehicles, speedboats, water scooters, or snow mobiles. Extensive windmill farms at sea, along dikes, or on top of windy hills have recently been added to this already spectacular soundscape of our modern globe.

Most anthropogenic noise is heavily biased towards low frequencies. Car engines and most machinery generate wide-band noise with the majority of sound energy below 1.0 kHz, and gradually declining amplitudes at higher frequencies. Beeps, whistles, and sirens can have dominant frequencies that may be slightly higher but typically do not exceed 2.0 kHz. Some sources generate noise at a constant and predictive level, while others fluctuate in time and vary acoustically. There are sources of moderate levels and others that are extremely loud such as piling, sonar and seismic guns under water, or sonic booms when a jet fighter flies through the sound barrier. Like the abiotic and biotic sounds of nature, human-generated sounds also exhibit seasonal and diurnal cycles. Despite a growing awareness of the potentially detrimental impact on human health and well being, noise pollution is not expected to diminish any time in the near future. The economic values typically overrule ecological values, and people are not willing to leave their cars at home.

Impact of Noise on Animals

There are several ways in which noise may be harmful to animals (**Figure 2**). First of all, very loud sounds may cause physical damage depending on amplitude, but also on duration and repetition of exposure. It also matters whether sound reaches the ear through water or air, and not all species and age classes are equally sensitive. Under water, extremely loud sounds from, for example, seismic air guns or piling may even cause ruptures and internal bleeding in organs like swim bladders or eyes, which may induce death. Military sonar has been suggested to lead to the stranding of whales either due to hearing damage or decompression sickness after rapid surfacing in panic. In general, temporary or permanent hearing loss will lead to significant fitness loss, as hearing is critical to many natural activities that benefit survival and reproduction.

Less, but potentially still, serious damage concerns noise-related stress. Domestic animals, several species of ungulates, as well as birds, have been shown to have a raised heart rate in response to noise of overflying air planes. Also, fish can exhibit increased secretion of the

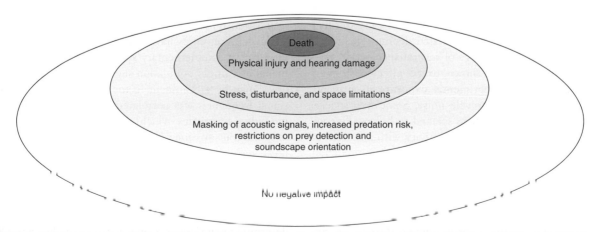

Figure 2 Schematic representation of the negative impact of anthropogenic noise on animals. At close proximity and extreme sound levels, noise may cause death. From the center of the figure outwards, noise exposure is less extreme and the severity of impact levels off; from physical injury and hearing damage to masking of communicative and other biologically relevant sounds.

stress-hormone cortisol in response to exposure to boat engine noise. Other indicators of stress are evasive behavior, panic, or temporary interruption of activities. Repeated exposure can lead to habituation and thereby to a reduction in overt stress, but there are also examples of noise-related reductions in growth and rate of reproduction in zoos and on farms. Especially, noises which are loud, low, and fluctuating unpredictably, seem to have a negative impact on health and reproductive success of animals in captivity. This may also be true for animals in the wild.

Disruption of behavior or an increase in evasive behavior may be an indicator of stress, but can also imply a restriction on habitat availability and on the time and energy budget of animals. Many bird, bat, and frog species are known to avoid otherwise suitable habitat alongside highways, and noise may be the reason. Anecdotal evidence on great reed warblers revealed that territory density in a reed bed alongside a road went up in 2 years of road closure and went down again in the spring of reopening. Congruently, when given a choice, bats preferred to spend more time in a silent than in a noisy compartment of an experimental arena. Foraging efficiency of these bats was also negatively affected by traffic noise, which is ascribed to the masking of the rustling sounds of their prey. Many primates, birds, and fish make similar use of sounds while hunting and may be affected in the same way. For example, foraging efficiency of chaffinches also goes down with rising noise levels: they were shown to shift towards more visual scanning for predators at the expense of eating time, as auditory detection of danger became less reliable.

Masking of acoustic communication is another threat to social relationships, reproductive success of individuals, and the survival of local populations. The continuous presence of cargo boat noise in the ocean has, for example, reduced the historical signal range of blue whales from thousands of kilometers down to about 50 km. A study on ovenbirds defending territories at artificial forest gaps in Canada, either close to noisy generators or close to silent wells, showed dramatic noise-related declines in success rate of mate attraction. Frogs of several species and geographical regions were shown to be affected in their calling behavior by experimental exposure to airplane, highway traffic, or motorcycle noise. Some frog species stop or reduce calling, others seemed to be vocally triggered by noise and raise calling activity with noise. All of which may be detrimental to signal function, predation risk, and optimal allocation of energy and time. Little is known yet, but male fish, calling through water to attract female partners, will likely suffer from noise-related breakdown of the signal detection range. The same will be true for mole crickets emitting signals through the soil to competing neighbors. For vibratory communication through vegetation, we also know that leafhopper females

are less attracted to signaling males when there is spectral overlap between signal and noise. The wideband noise of highway traffic will, therefore, likely render roadside vegetation less suitable to vibratory insect courtship.

Insights from Urban Survivors

Noise is not a novel feature of natural environments, and animals for which sounds play an important role in life have evolved several strategies to overcome the problems of interference and masking. On the production side, animals may, for example, raise signal amplitude in response to a rise in noise levels (the Lombard effect). They may also produce longer calls or sing more repetitions to stand a better chance to get the message across through some gaps of relative silence. Redundancy of the same message in multiple acoustic characteristics may also allow receivers to extract the meaning under challenging noise conditions. On the receiver side, animals may have highly specialized feature detectors, perceptually tuned to the properties of conspecific acoustic signals, and adapted to filter out noise spectra typical for their natural habitat. Several Asian frog species that live in an incredibly noisy environment with broadband river noise have evolved the ability to generate and hear ultrasonic sounds to get into an agreeable frequency channel for communication.

Given that many species have specially evolved capacities to cope with naturally noisy conditions, it should not be surprising that there are animal species able to resist noisy conditions of anthropogenic origin. The great tit (*Parus major*) is one bird species successful in noisy urban areas, while still common in natural forest areas. The secret behind their success may be related to their singing style. Great tits in urban populations across Europe were found to sing, on average, with a higher minimum frequency than the populations in nearby forests (**Figure 3**). This pattern may have emerged because sound transmission through dense vegetation favors low frequency use; cities have more open space than forests. However, a complementary study within urban environments in a single population suggests that environmental noise may be a dominant factor. Individual variation in signal frequency was correlated to local noise levels in individual territories, which were highly similar in terms of sound transmission properties. Male great tits inhabiting noisy territories sang with a higher minimum frequency than the nearby birds in relatively quiet territories. Given that cities are heterogeneous but still consistently noisier than forests as a result of low-pitched traffic noise, masking avoidance by pitch shift may be the individual strategy determining the noise-related patterns at the individual and population level. A recent study with experimental noise exposure revealed the behavioral mechanism

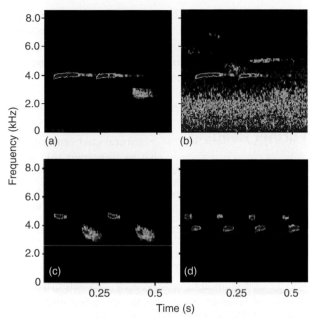

Figure 3 Sonographic illustration of great tit songs. In the upper part of the figure, the same three-note song type is depicted without (a) and with (b) anthropogenic noise. Comparing these two sonograms, it will be clear that the low-frequency noise band typical of anthropogenic sound sources has the most dramatic masking impact on the lowest notes in great tit songs. In the lower part of the figure, two different two-note song types depict a typical example of a relatively slow song (two repetitions in half a second) with a low minimum frequency from forest habitat (c) and a typical example of a relatively fast (four repetitions in half a second) with a high minimum frequency from urban habitat (d). The yellow lines demarcate the spectral divergence in these examples. Listen to sound files 3(a) great tit in quiet forest, 3(b) great tit in noisy city, 3 (c) forest song type SLOW and LOW, and 3(d) city song type FAST and HIGH, Recorded by the author.

underlying the masking avoidance: great tits extend the bout duration of song types that do well under current noise conditions. They stop singing songs with low-frequency notes sooner when they hear loud traffic noise, and keep on going for longer if they happen to sing a song type without such low notes.

These data on great tits draw special attention to the fact that masking is dependent on spectral overlap and that flexibility in frequency use may allow some animals to escape detrimental effects. However, many species may not be capable of such flexibility. As a result, masking of acoustic signals important to resource defense and mate attraction may be harmful to many birds, frogs, and fishes. Taking the typical bias towards low frequencies into account for anthropogenic noise, species-specific frequency use, and associated hearing range will provide a rough insight into which animals are more likely to suffer than others. Humans and many other mammals hear and use low-frequencies down to 60 or 40 Hz, but we also detect much higher frequencies (young persons up to 20 kHz).

The sensitivity of dogs goes up higher to 40 kHz, while cats hear up to 60 kHz. Many bats and mice are not only sensitive to even much higher frequencies, but they can also be completely insensitive to low frequencies; for example, below 1 kHz. We can expect most vulnerable species to be birds that depend on relatively low pitch (<4 kHz), and also, especially, frogs and fish, as they typically rely exclusively on low-pitched signals and are most sensitive in the range between 50 Hz and 2.0 kHz, overlapping dramatically with typical anthropogenic noise. Notably, the insights into spectral overlap and sensitivity can be important to safeguard or conserve, but the repellent effects of sound can also be applied in the opposite way (**Box 2**).

Concluding Remarks

In conclusion, increasing amounts of human activities generate artificially loud low-frequency noise in nearly all habitats around the world. Noise affects animals across taxa and in all environments, including air, water, soil, and within vegetation. Animals close to extremely loud sounds are affected most dramatically and, especially under water, high noise levels may cause physical damage and even death. Not only amplitude, but also duration and repetition rate of noise exposure are critical to the damage. Although relatively little is known, more moderate noise levels can lead to hearing damage, stress, and disturbance, leading to restrictions in available habitat, or to a negative impact on feeding or breeding activities. Here, predictability will play an important role in determining the strength of the impact, besides noise amplitude and the overlap with species-specific auditory sensitivity.

The use of acoustic signals for communication is ubiquitous across the animal kingdom and is spread across all habitats. Masking of acoustic signals through anthropogenic noise is, therefore, a potentially widespread detrimental factor, which may reduce individual fitness and may harm population health. The need to extract communicative signals from a background of noise is not an evolutionary novel phenomenon and some species have evolved coping strategies that do well with anthropogenic noise. Data on the flexibility of such survivors in the cacophony of our modern society can provide insight into why other less-flexible species may not cope and flee or fail in noisy areas.

Our world is not likely to become quieter any time in the near future. On the contrary, anthropogenic noise is one of the exceptional pollutants which is expected to remain on the rise. Hopefully, the increasing awareness about the detrimental effects of noise will not only lead to more advanced repellents, but also to more sensible control of anthropogenic noise, to the benefit of animals and ourselves.

Acknowledgments

I thank Hans-Joerg Kunc, Marine Danek-Gontard, and Jeffrey Lucas for helpful comments on earlier drafts of the manuscript. The work on songs of urban great tits was supported by grants from the Netherlands Organization for Scientific Research (NWO, ALW-projects 831.48.006 and 817.01.003).

See also: Acoustic Signals; Anthropogenic Noise: Implications for Conservation.

Further Reading

Bouton N, van Opzeeland I, Coers A, ten Cate C, and Slabbekoorn H (submitted for publication) Anthropogenic noise impact on fish: A neglected conservation issue.

Brumm H and Slabbekoorn H (2005) Acoustic communication in noise. *Advances in the Study of Behavior* 35: 151–209.

Popper AN and Hasting MC (2009) The effects of human-generated sound on fish. *Integrative Zoology* 4: 43–52.

Slabbekoorn H, Bouton N, van Opzeeland I, Coers A, ten Cate C, and Popper, AN (in revision) A noisy spring: the impact of globally rising underwater noise levels on fish. Trends in Ecology & Evolution.

Slabbekoorn H and Halfwerk W (2009) Behavioural Ecology: Noise annoys at community level. *Current Biology* 19: R693–R695.

Slabbekoorn H and Ripmeester EAP (2008) Birdsong and anthropogenic noise: implications and applications for conservation. *Molecular Ecology* 17: 72–83.

Tyack P (2008) Large scale changes in the marine acoustical environment and its implications for marine mammals. *Journal of Mammalogy* 89: 549–558.

Warren PS, Katti M, Ermann M, and Brazel A (2006) Urban bioacoustics: it's not just noise. *Animal Behaviour* 71: 491–502.

Anthropogenic Noise: Implications for Conservation

H. Brumm, Max Planck Institute for Ornithology, Seewiesen, Germany

Introduction

Anthropogenic disturbance is a major cause of declines in biodiversity all over the world. To tackle the loss of species richness observed in many ecosystems, it is crucial for conservation biologists to understand the implications of manmade disturbance for animal populations. Anthropogenic noise pollution is an increasing problem that has received relatively little attention in the past, but awareness of its detrimental effects on animal populations is growing.

The urbanization of our planet is ever-increasing and the amount of local and global traffic continues to rise, with more and more people and goods being moved from one location to another. Both urbanization and traffic, including road transport, air traffic and shipping, are projected to increase further in the next decades. As a result, anthropogenic noise levels are likely to rise even further and an increasing number of areas will be affected by noise pollution.

In humans, noise at high levels and over prolonged periods is harmful to health, for instance, by causing hearing damage. However, even at low levels, environmental noise can have severe negative effects, such as sleep disturbance and induction of chronic stress, which in turn leads to heart diseases and psychiatric problems. Many animals are affected in similar ways by anthropogenic noise; in addition, some species also suffer impairment to their communication. From insects to mammals, a whole range of different taxa use sound to exchange vital information, for instance to find mating partners or to defend territories. This close relationship between acoustic communication and reproduction means that in many species variation in signaling efficiency is likely to have major fitness consequences.

As the environmental noise level rises, a listening animal finds it increasingly difficult to detect and recognize acoustic signals. Above a certain threshold, acoustic signals can no longer be recognized and thus communication breaks down. The degree to which an acoustic signal is masked by anthropogenic noise depends on the spectral and temporal overlap between signal and noise as well as the hearing sensitivity of the receiver. We know much about the hearing sensitivity of many species from laboratory studies that used comparably simple stimuli such as pure tones. However, much less is known about the effects of different types of anthropogenic noises on the perception of species-specific signals in the wild. Such data are necessary to provide an accurate representation of noise effects on animal communication, since it is critical to understand how noise reduces the maximum distance at which one animal can detect, discriminate, and identify the acoustic signals of another in its natural habitat. Few studies have addressed this challenging task to date and much more are needed to better inform management policies.

As we shall see, many species from various taxa have been found to be affected by the masking of their signals or the acoustic cues used in predator–prey interactions. However, we are only beginning to understand the implications anthropogenic noise has for animal populations and how noise pollution is linked to the loss of biodiversity. In the following section, I will review some notable examples of how animal communication is impaired by environmental noise.

Impairment of Signal Transmission

Communication is the foundation upon which all social relationships in animals are built. Many species use acoustic signals to exchange information, and the messages conveyed may function, for instance, in finding a mate, synchronizing courtship, competing over resources, parent–offspring recognition, and group cohesion. Therefore, disturbance of signal transmission by masking noise can affect the fitness of individuals and, in the long term, may have negative effects on whole populations.

Insects

Several neuroetholgical studies have investigated the capacity of insect auditory systems to detect and recognize acoustic signals in noise. These studies provide a good starting point for estimating the effects of anthropogenic noise on the communication abilities of insects, and thus are potentially very useful for conservation purposes if combined with data on the noise and sound degradation characteristics of particular insect habitats. A very good example of how auditory perception can be linked to environmental acoustics is that of Römer and Lewald, who used the activity of an identified neuron in the auditory pathway of bushcrickets as a 'biological microphone' and measured signal transmission distances in the natural habitat.

Very recently, Samarra and coworkers investigated how environmental noise affects reproductive behavior in fruit flies. Male fruit flies produce a courtship song that is a prerequisite for mating to occur and which females use to select mates. Female sexual behavior was disrupted by an impairment of signal recognition, as females showed significantly decreased responses to courtship song in the presence of high levels of noise within the same frequency band as the courtship song. In this way, environmental noise can affect mate choice and thus potentially have a negative effect on the viability of insect populations in the long term.

Fish

Underwater noise pollution is a growing environmental problem, and in freshwater habitats the sounds produced by shipping are of particular concern. It has been shown that underwater ship noise in rivers and lakes causes stress in fish, as indicated by increased cortisol levels. Chronic physiological stress has detrimental effects on growth, immunological functions, and reproduction in fish. As a consequence, shipping noise may affect the conservation status of populations. In addition to physiological stress, impairment of communication may also be a negative effect of anthropogenic underwater noise. Several fish species use sounds to communicate and masking noise may impair signal exchange. Data from laboratory tests on hearing sensitivity in fish show that prolonged exposure to high-intensity noise can cause temporary hearing loss, which decreases the detection of short transient signals and the temporal resolution ability. However, without more information, we have only a very crude idea of the effect of environmental noise on the perception of species-specific acoustic signals in nature.

Amphibians

The global decline of amphibian populations is a long-standing concern. In the past, conservationists have focused their attention mainly on habitat destruction, climate change, and chemical pollution, and it was not until recently that the effects of anthropogenic noise have also moved into the focus of attention. Noise pollution will particularly affect frogs and toads which make up nearly 90% of all extant amphibians and are a very vocally active group. Males attract female mating partners by calling and females use male calls in mate choice. Thus, acoustic masking of anuran calls may impair their sexual behavior and breeding success which can potentially lead a decrease in populations over time.

Male European tree frogs (*Hyla aborea*) decrease their calling behavior in the presence of traffic noise played back to them at realistic levels, which shows that anthropogenic noise may interfere with anuran communication and thus potentially have deleterious consequences on their reproduction. A similar decrease in calling activity has been found in a tropical mixed-species assemblage exposed to airplane and motorcycle noise. However, the suppression of calling behavior in a whole set of species in the assemblage also stimulated calling in one particular other species. Recently, Bee and Swanson used a phonotaxis assay to present female gray tree frogs (*Hyla chrysoscelis*) with male advertisement calls that were masked by realistic levels of road traffic noise. They found that in the presence of the noise, females took a longer time until they responded, they had considerably increased response thresholds, and a decreased orientation toward the target signal. This suggests that traffic noise can disturb frog communication in two ways, by suppressing male calling activity and, at the same time, by decreasing the females' ability to locate calling males.

Previous conservation studies on anurans have looked at the effects of roads in terms of habitat fragmentation and increased mortality due to road kills during spring migration to breeding ponds. Moreover, the latest evidence from bioacoustic studies indicates that the sound of roadway traffic may also be relevant. Future studies are needed to directly link levels of anthropogenic noise to reproductive success to develop reasonable action plans for noise abatement.

Birds

The effect of anthropogenic noise on avian communication is receiving increasing attention from behavioral biologists and conservationists alike. A few years ago, it was found that urban birds exposed to high levels of traffic noise increase their song amplitude and sing at a higher pitch compared to conspecifics in less noisy areas. A noise-related increase in vocal amplitude has also been shown for begging calls of nestling birds. In playback experiments with feeding parents, white noise eliminated parental preferences for higher begging rates of their young, but the preference was restored when call amplitude was increased to the level that nestlings normally produce in response to noise of that level.

These findings suggest that birds are able to, at least partly, mitigate acoustic masking by adjustments of signal structure and performance. Certainly, the scope of such mitigation is limited, and effective communication will break down as soon as these limits are crossed. For instance, it has been suggested that environmental noise pollution may affect the mating of songbirds that rely on acoustic signals to establish and maintain pair bonds. Indeed, it has been found that the pairing success of ovenbirds (*Seiurus aurocapilla*) was lower in areas around noisy industrial sites compared with areas around noiseless facilities. Moreover, it is conceivable that noise-related changes in song structure (e.g., pitch or song syntax) could lead to population differences that may also affect female preference functions or even species recognition, which is clearly of conservation concern.

Lohr and colleagues used operant conditioning to test the capacity of birds to detect and discriminate natural vocal signals in the presence of masking noise including traffic noise. They found that the thresholds for discrimination between different calls are higher than those for detection, and that these thresholds are crucially affected by the level, spectrum, and spectral shape of the ambient noise. This study also provides a model to estimate distances over which acoustic signals may be effective and to which extent this active signal space will be constrained by increased noise levels. Psychoacoustic studies like these are very relevant to the ecology of birds, as they provide a more accurate representation of noise effects on communication abilities.

Mammals

Over the past two decades, the impacts of anthropogenic noise on marine mammals have been a matter of concern, particularly the long-term effects on cetacean populations. In addition to physiological stress responses and possible tissue damage caused by high-intensity sound sources (such as air guns used for seismic exploration and military sonar), cetaceans are also vulnerable to interference of their communication. The main anthropogenic noise in the seas stems from the propulsion of ships, with most of the sound energy concentrated below 200 Hz. Such low-sound frequencies propagate particularly far, and in certain places shipping has elevated underwater noise of up to 100 times in this frequency band. As Tyack pointed out, many whale species use the same low-frequency band for acoustic long-range signals, and concern has been raised that the intense anthropogenic noise in the seas may interfere with their ability to communicate effectively and thus may ultimately be detrimental on normal cetacean behavior and breeding biology.

Impairment of Cue Recognition

Animals use the sounds produced by others in various ways to extract information about the environment. In addition to acoustic signals discussed in the previous section, animals may also extract meaning from acoustic cues that do not function primarily in communication. Such acoustic cues play a particular role in predator–prey interactions.

On one hand, species that detect their prey acoustically undergo reduced foraging success in noisy areas, and on the other, many species potentially suffer increased predation in places where they cannot hear an approaching predator.

Finding Food

From insects to mammals, many animals find their food by listening to the sounds their prey produce, a capacity referred to as 'passive listening.' If environmental noise masks the acoustic prey cues, passive listening can be impaired, thereby reducing the foraging success of the predator. Such a noise-related impairment of foraging success has been demonstrated in greater mouse-eared bats (*Myotis myotis*). This species belongs to a group of bats that find their food by listening to the rustling sounds their arthropod prey produces while moving on the substrate. In a choice experiment with two foraging compartments, Schaub and colleagues investigated the influence of different types of background noise, including traffic noise, on foraging effort and foraging success. They found that the bats spent less time in the noisy foraging compartment and preferred to search for prey in the quiet area, and the animals tested also caught considerably less prey in the compartment where noise was broadcast. This finding indicates how areas exposed to high levels of traffic noise could be less suitable as foraging areas for animals that use passive listening to locate their prey.

Similar detrimental effects of anthropogenic noise on foraging success are also possible in species that use echolocation rather than passive listening to locate their prey, as some man-made noise extends well into the ultrasonic range (**Figure 1**). Echolocating animals include whale and bat species that are of conservation concern, but the possible interference of their prey localization abilities by anthropogenic noise has not received much attention yet. If the returning echoes of the echolocation signal cannot be discriminated from the background noise, prey detection (and orientation in general) will become increasingly difficult. A recent experimental study on free tailed bats (*Tadarida brasiliensis*) showed that these animals change the structure and performance of their echolocation calls to reduce masking from environmental noise. However, the scope for compensation is surely limited, and where large areas are subject to high-intensity noise pollution during significant times, it is conceivable that the impairment of foraging may affect populations.

Avoiding Predators

One potential negative effect of human disturbance on animal populations is increased predation in areas where acoustic predator cues are masked by anthropogenic noise. It is conceivable that predation risk rises with increasing noise levels, as an animal's perceptual threshold to acoustically detect an approaching predator will be higher, the higher the level of masking noise is. However, some species may compensate the increased predation risk in noisy areas by an increase of vigilance behaviors.

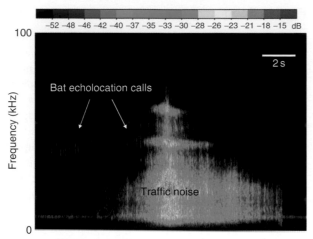

Figure 1 Acoustic masking of Galápagos red bat (*Lasiurus brachyotis*) echolocation calls by traffic noise. In this rather extreme example, the bat was foraging above a small street in approx. 4 m height while a motorbike was passing by beneath the flying bat. Note that the sounds produced by the motorbike extend well into the ultrasonic range and overlap with the frequency band of the bat's echolocation calls. Depending on the sound level and spectral properties of the traffic noise, it may impair signal recognition in hunting bats and thus reduce foraging success. In this way, anthropogenic noise pollution may reduce habitat quality and thus affect distribution patterns of species. Recording made by the author on Santa Cruz Island, Galápagos.

Such noise-related changes in vigilance have been shown for foraging chaffinches (*Fringilla coelebs*) that were exposed to white noise. During the noise treatment, the birds had shorter interscan intervals spending shorter time with their head down compared to periods when no noise was played back to them. Moreover, no evidence for habituation to the noise was found, as the change in antipredator vigilance did not vary within or between successive trials. However, the increased vigilance came at a cost to the birds: during the noise treatment the birds made fewer pecks when feeding, which led to a reduction in food intake rate, suggesting that compensating for the increased predation risk could lead indirectly to a fitness cost. So, noise pollution has the potential to affect population viability either way, directly through increased predation or indirectly by reduced food intake due to higher vigilance.

Animals Avoid Noisy Areas

It has been reported that marine mammals avoid certain human underwater sound sources at ranges of many kilometers, which raised the concern about displacement from important habitats. Often, researchers find a correlation between environmental noise levels and changes in species composition or abundance. In particular, several studies have shown lower species diversity and density of

breeding birds along motorways, and the negative effect of roads on bird populations has often been linked to noise pollution. However, as far as roads are concerned, it is difficult to pinpoint the impact of traffic noise because other factors, such as visual disturbance, road kills, edge effects, light pollution during night, chemical pollution, etc. could have similar detrimental effects on animal abundance. These factors can be excluded in controlled experiments that investigate the behavior of individual animals. In such an experimental setting, it has been demonstrated that foraging bats avoid traffic noise and prefer to forage at quiet sites where they perform much better at finding prey (see earlier).

Recently, Bayne and colleagues investigated the abundance of songbirds in a boreal forest in which noise from energy-sector industries is rapidly increasing. They found that areas near noise-producing gas compressors (that serve to maintain the flow of gas through pipelines) had a significantly lower bird density than areas near noiseless energy facilities. Interestingly, the noise did not affect abundance in all the 23 species investigated, and it would be worthwhile to test whether the differences in noise vulnerability were related to the acoustic structure of the species-specific songs. Such a relationship between acoustic masking and abundance has been shown by Rheindt, who found that species with lower-pitched songs declined more strongly in abundance toward a motorway. Most likely, this was because higher-pitched songs (well above the low-frequency traffic noise) make birds less vulnerable to noise pollution. In general, birds that sing in a similar frequency band as that masked by environmental noise are more likely to be affected by noise than those whose songs overlap less. These species differences in vulnerability to noise are an important topic critical for conservation action.

Overall, the current evidence indicates that the abundance of some species is negatively affected by manmade noise and – among other things – impairment of acoustic communication is one of the reasons for the decline. More studies measuring the effect of anthropogenic noise on the viability of populations are crucial for species conservation and the advice of future action plans.

See also: Anthropogenic Noise: Impacts on Animals; Conservation and Behavior: Introduction; Hearing: Vertebrates.

Further Reading

Bayne EM, Habib L, and Boutin S (2008) Impacts of chronic anthropogenic noise from energy-sector activity on abundance of songbirds in the boreal forest. *Conservation Biology* 22: 1186–1193.
Bee MA and Swanson EM (2007) Auditory masking of anuran advertisement calls by road traffic noise. *Animal Behaviour* 74: 1765–1776.

Bermúdez-Cuamatzin E, Ríos-Chelén AA, Gil D, and Macías Garcia C (2009) Strategies of song adaptation to urban noise in the house finch: Syllable pitch plasticity or differential syllable use? *Behaviour* 146: 1269–1286.

Brumm H (2004) The impact of environmental noise on song amplitude in a territorial bird. *Journal of Animal Ecology* 73: 434–440.

Brumm H and Slabbekoorn H (2005) Acoustic communication in noise. *Advances in the Study of Behavior* 35: 151–209.

Dooling RJ (2004) Audition: Can birds hear everything they sing? In: Marler P and Slabbekoorn H (eds.) *Nature's Music – The Science of Birdsong*, pp. 206–225. San Diego, CA: Elsevier Academic.

Francis CD, Ortega CP, and Cruz A (2009) Noise pollution changes avian communities and species interactions. *Current Biology* 19: 1415–1419.

Habib L, Bayne EM, and Boutin S (2006) Chronic industrial noise affects pairing success and age structure of ovenbirds *Seirus aurocapilla*. *Journal of Applied Ecology* 44: 176–184.

Lengagne T (2008) Traffic noise affects communication behaviour in a breeding anuran, *Hyla arborea*. *Biological Conservation* 141: 2023–2031.

Leonard ML and Horn AG (2005) Ambient noise and the design of begging signals. *Proceedings of the Royal Society of London, Series B* 272: 651–656.

Lohr B, Wright TF, and Dooling RJ (2003) Detection and discrimination of natural calls in masking noise by birds: Estimating the active space of a signal. *Animal Behaviour* 65: 763–777.

Nemeth E and Brumm H (2009) Blackbirds sing higher pitched songs in cities: Adaptation to habitat acoustics or side effect of urbanization? *Animal Behaviour* 78: 637–641.

Quinn JL, Whittingham MJ, Butler SJ, and Cresswell W (2006) Noise, predation risk compensation and vigilance in the chaffinch *Fringilla coelebs*. *Journal of Avian Biology* 3: 601–608.

Rabin LA, McCowan B, Hooper SL, and Owings DH (2003) Anthropogenic noise and its effects on animal communication: An interface between comparative psychology and conservation biology. *International Journal of Comparative Psychology* 16: 172–192.

Reijnen R, Foppen R, and Meeuwsen H (1996) The effects of traffic on the density of breeding birds in Dutch agricultural grasslands. *Biological Conservation* 75: 255–260.

Rheindt FE (2003) The impact of roads on birds: Does song frequency play a role in determining susceptibility to noise pollution? *Journal für Ornithologie* 144: 295–306.

Römer H and Lewald J (1992) High-frequency sound transmission in natural habitats: Implications for the evolution of insect acoustic communication. *Behavioral Ecology and Sociobiology* 29: 437–444.

Samarra FIP, Klappert K, Brumm H, and Miller PJO (2009) Background noise constrains communication: Acoustic masking of courtship song in the fruit fly *Drosophila montana*. *Behaviour* 146: 1635–1648.

Schaub A, Ostwald J, and Siemers BM (2008) Foraging bats avoid noise. *Journal of Experimental Biology* 211: 3174–3180.

Slabbekoorn H and den Boer-Visser A (2006) Cities change the songs of birds. *Current Biology* 16: 2326–2331.

Slabbekoorn H and Ripmeester EAP (2008) Birdsong and anthropogenic noise: Implications and applications for conservation. *Molecular Ecology* 17: 72–83.

Sun JWC and Narins PM (2005) Anthropogenic sounds differentially affect amphibian call rate. *Biological Conservation* 121: 419–427.

Tressler J and Smotherman MS (2009) Context-dependent effects of noise on echolocation pulse characteristics in free-tailed bats. *Journal of Comparative Physiology A* 195: 923–934.

Tyack P (2008) Implications for marine mammals of large-scale changes in the marine acoustic environment. *Journal of Mammalogy* 89: 549–558.

Wysocki LE, Dittami JP, and Ladich F (2006) Ship noise and cortisol secretion in European freshwater fishes. *Biological Conservation* 128: 501–508.

Wysocki LE and Ladich F (2005) Effects of noise exposure on click detection and the temporal resolution ability of the goldfish auditory system. *Hearing Research* 201: 27–36.

Antipredator Benefits from Heterospecifics

E. Goodale, Field Ornithology Group of Sri Lanka, University of Colombo, Colombo, Sri Lanka
G. D. Ruxton, University of Glasgow, Glasgow, Scotland, UK

Introduction

If it were possible to mark all the individuals of all the species of animals in an area with different colors of incandescent light so that their positions relative to each other could be observed from on high, it would be clear that the spacing between animals of different species is not random. Although spatial variation in food and water are often responsible for clumped distributions, predation is also an important factor: clumped distributions may be caused by animals sharing refuges where predators can not attack them, or actively joining other species in ways that reduce predation.

This article reviews how animals use heterospecifics in predation contexts, proceeding from situations in which animals live in the same area but rarely meet, to examples of species that spend virtually their entire lives together. This article first discusses animals that *eavesdrop* on the *information* about predators available from heterospecifics in the same habitat; this information transfer does not effect the positions of the species relative to each other. Second, the article describes *mobbing*, a behavior in which animals come together temporarily to confront predators. Third, other kinds of mixed-species groups are discussed – ranging from aggregations that form around food sources, to nesting colonies, to stable groups that reform everyday – in which many of the benefits of group-living are related to reducing predation. Finally, the article concludes by examining pairs of species that are highly dependent on each other for protection and often live together for life.

The progression of the article is thus one of increasing complexity and co-evolution of species interactions, starting with *commensalisms*, such as eavesdropping on heterospecific information, and ending with well-studied examples of *mutualism*. Mobbing and other kinds of mixed-species groups are intermediate on this gradient because species that tend to lead the groups expend greater costs and received fewer benefits than other species that follow them. We hope to explain why some species play leading roles in mixed-species groups by examining their antipredatory behavior. To introduce this subject, we first discuss two kinds of antipredation benefits that can be derived from heterospecifics: information and protection.

Heterospecific-Derived Information About Predators

In a provocative article, Juanne-Tuomas Seppänen and colleagues suggest that in some cases, information obtained from heterospecifics can be more *valuable* than information obtained from conspecifics (**Figure 1**). Specifically, at short distances between individuals, competition between conspecifics can be high. In contrast, a nearby heterospecific may offer information that is relevant because of its spatial proximity, with less competition.

Another potential benefit to heterospecific information is that the quality of the information may have some characteristics that are superior to conspecific information. Some characteristics of information that might vary among species are as follows:

1. *Detection*: Species may vary in the kinds and breadth of stimuli that they detect, and the probability of detection of each stimulus type. These differences may arise because of differences among species in their sensory capabilities, in the number of individuals per group, or in their different habitat preferences or spatial positions. Species that are preferentially targeted by predators and/or have relatively poor escape performance should be expected to invest more heavily in their ability to detect threats.
2. *Reliability*: Species may make 'false alarms' because (1) they inaccurately detect a predator when none is present, or (2) even though they do not detect a potential predator, they mistakenly make the signal anyway, or (3) they do not detect a predator but make a signal to manipulate the response of the receivers. Reliability may also depend on the species making and receiving the information: if species A is predated by only some of the same predators as species B, the reliability of species B's calls for A will depend on what percent of B's signals provide information about A's predators.
3. *Production*: Species may vary in the percentage of detected stimuli for which they will produce signals, due to the conspecific audience toward which they are directing the call. For example, an individual of a species that lives in groups of related individuals may be likely to give alarm calls, even at the risk of attracting the predator's attention to itself.

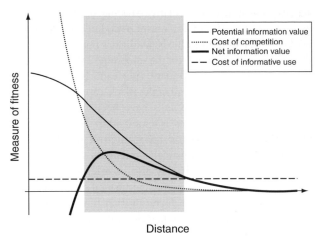

Figure 1 Seppänen et al. (2007) argue that the value of information, measured by the effects the information has on fitness, is a function of the distance from the signaler to the receiver. Closer signalers produce more valuable information, but competition between them and the receivers is high. Information becomes adaptive for the receiver at some distance (shaded area). Hence, at some close distances the information produced by heterospecific signalers, which are assumed to compete less with the receiver because of differences in their ecological niches, may be more valuable than information from a conspecific.

4. *Transmission:* Detectability of heterospecific signals may vary because of species differences in their capacity to produce signals that travel long distances.

Defensive Services Offered by Heterospecifics

Particularly in those situations where heterospecifics live in close proximity, they may provide a variety of benefits, listed here, and referred to in later sections. Two different kinds of benefits can usefully be delineated: complementary services are provided by species that are radically different from each other, and similarity-based services are those that are best used by species that are alike.

Complementary services:

1. Animals can consume predators of another species. Large, aggressive species can remove the predators of smaller species, and mobile animals can consume the predators of immobile animals that are unable to protect themselves.

2. Animals can serve as a refuge against predators, particularly large species providing structural cover for smaller species.

Similarity-based services:

3. Species can associate with distasteful or toxic species which they imitate (i.e., Müllerian or Batesian mimicry).

4. Species can dilute the risk of predation in a group, particularly if they appear similar to other group members.

Responses to Heterospecific Alarm Signals and Cues

The first category of heterospecific interactions that involve predation occur between generally noninteracting species that inhabit the same area. This is part of the larger phenomenon of *'public information'*: information that may be created by animals for their own function or to benefit a conspecific, but is accessible by eavesdropping to heterospecifics, as has been reviewed by Sasha Dall and colleagues. Such interactions are also known to occur in plants, which emit volatile chemicals when they are attacked by herbivores; these chemicals can be detected by other plants and used to initiate defensive mechanisms, or can be detected by parasitoids of the herbivore and used to locate a potential host.

Chemicals are also used by aquatic animal species as indications of alarm. For example, many fish have specialized club cells in the surface layer of the epidermis that rupture when attacked, releasing chemicals (originally dubbed 'Schreckstoff' or 'fear substance' by Karl von Frisch) that fish rapidly respond to, including fish of other species. Amphibians and aquatic invertebrates have also been shown to respond to the chemicals released by heterospecifics in a predation event. Different chemicals are given out at different stages of the predation process: for example, *disturbance chemicals* are released when a predator has been detected, whereas *damage-released chemicals* are emitted by animals that have been injured or killed by a predator. Because release of these chemicals appears involuntary, and because it is not clear how the animal that releases the chemicals benefits (although in some cases the chemicals may attract another predator that can attack the original one), such chemicals might be considered *cues* rather than *signals*. Indeed, research by Douglas Chivers, Brian Wisenden, and colleagues has concluded that club cells evolved as a generalized defense against parasites and ultraviolet radiation, with fish evolving secondarily the ability to respond to the presence of the chemicals when released by predation.

In terrestrial systems, alarm signals are usually acoustically based (*note:* but chemical alarm systems do exist in terrestrial species, e.g., alarm cues emitted by rats and deer). It has been repeatedly shown that species that share the same predators and live in the same environment can often recognize and respond to each other's acoustic alarms. For example, species of the same family that live in the same area respond to each other's alarms, as shown by a ground squirrel (*Spermophilus lateralis*) and a marmot (*Marmota flaviventris*) in North America, or by the Madagascar primate species *Propithecus verreauxi* and *Eulemur fulvus*. Cross-taxa alarm recognition has also been found, starting with the work of Robert Seyfarth and Dorothy Cheney, who studied how vervet monkeys (*Chlorocebus pygerythrus*) respond to the alarm calls of the superb starling (*Spreo superbus*). Since that pioneering work, examples of cross-taxa alarm recognition include squirrels responding to the alarm calls of jays, monkeys responding to alarm calls of deer, and antelope responding to an especially vigilant bird known as the 'go-away bird' (*Corythaixoides leucogaster*). The recognition of acoustic alarms is not limited to birds and mammals or even to animals that are able to produce acoustic signals: Maren Vitousek and colleagues demonstrated that a marine iguana in the Galapagos can respond to the alarm calls of a mockingbird that inhabits the same habitats.

The amount of information that animals can extract from heterospecific alarms is also impressive. For example, in their study of vervet monkeys discussed earlier, Seyfarth and Cheney showed that the monkeys are able to distinguish between two types of superb starling alarm calls: those made specifically to raptors and those made to a wide variety of predators. In the taxonomic reversal, a bird species, the Yellow-casqued hornbill (*Ceratogymna elata*), responds appropriately to referential information encoded in the alarm calls of Diana monkeys (*Cercopithecus diana*). Diana monkeys, in turn, show further sophistication, in being able to discern a syntactical rule in the alarm calls of the Campbell's monkey (*Cercopithecus campbelli*) that indicates that the alarm is of low risk. Finally, a nuthatch (*Sitta canadensis*) is able to use information encoded in the mobbing calls of chickadees (*Poecile atricapillus*) to judge the level of risk posed by a predator.

The question of whether the recognition of heterospecific alarms is learned or innate remains a controversial one, and answers often differ between taxa. For example, the chemicals released by fish are highly conserved among related species, which may explain some heterospecific recognition; however, some fish and amphibian species are able to learn predator-associated chemical cues. Bird alarms calls are also often similar to each other, perhaps due to convergence around certain acoustic characteristics – narrow frequency bandwidth and relatively high frequency – that make the calls difficult to localize for predators. Yet in some of the examples of heterospecific response seen earlier, particularly those between distantly related taxa or those that include a large quantity of information, it is likely that the species learn to associate heterospecific information with danger.

Heterospecific Mobbing

Mobbing is an interesting phenomenon from the perspective of heterospecific groups because it is a predation-related

activity that affects the relative spacing of individuals of different species, if only for a brief period of time. Mobbing has been best studied in birds, and it has been long known that birds will respond to the mobbing calls of heterospecifics. For example, so many species in North America respond to the mobbing call of the Black-capped Chickadee (*P. atricapillus*) that these calls have been used by André Desrochers and colleagues as a method to census breeding birds. While heterospecific recognition of mobbing may be due to general acoustical similarities between most species' mobbing calls, it may also be a learned behavior: Eberhard Curio and colleagues performed elegant experiments in the 1970s, demonstrating that birds could learn to mob a model if they watched other birds mob it, even if the model was harmless.

Attracting heterospecifics may give benefits to a caller. First, it may reduce the level of risk to the caller simply by dilution, and animals may have less to lose (in kin or mates) if heterospecifics are attracted. Second, attracting heterospecifics may lead to an enhanced level of mobbing that may be more effective in driving off the predator. Indeed, some heterospecifics may be attracted that are aggressive or predators of the predator. Finally, Indrikis Krams and Tatjana Krama have argued that callers may also benefit if other species reciprocate by engaging in mobbing and calling at later times, and have suggested that species in stable communities (as opposed to transient, migratory communities) would be able to monitor reciprocation.

Mobbing and other similar forms of group defense also occur in mammals, especially primates, sciurids, and ungulates, and has been rarely reported in fish. To our knowledge, however, there have not been formal descriptions of heterospecifics associating during such mobbing in these taxa, although it is suspected to occur (Tim Caro, pers. comm.; Grant E. Brown, pers. comm.), and further study of this matter is warranted.

Mixed-Species Groups in Relation to Predation

Unlike mobbing, where the stimulus that induces the behavior is clearly the predator, mixed-species groups are not adaptations to predation solely. The adaptive functions of mixed-species groups are multiple, nonexclusive of each other, and similar to single-species groups, with the important exception that competition among individuals inside the group is generally assumed to be less, since different species will overlap less in their foraging preferences than conspecifics would. Increased foraging efficiency, in particular, is an important benefit to many such groups. However, the evidence is strong that predation is an important, if not predominant, driver behind the formation of mixed-species groups: for example,

Jean-Marc Thiollay and colleagues have shown that species that are most vulnerable to predation tend to be found in mixed-species flocks of birds, and that more flocks are found where predators are most dense. Here we will ignore the other factors that lead to mixed-species group formation (e.g., foraging related benefits), and concentrate on factors that reduce predation. These factors apply to groups that vary in their stability: mixed-species assemblages that form transitorily over sources of food, mixed-species roosts formed at night, mixed-species colonies formed during the nesting season, and mixed-species foraging flocks that can often consist of the same individuals day after day. Such groups can be found in a variety of taxa, including spiders, amphibians, and fish, although the phenomenon of mixed-species grouping has been best studied in mammals and birds.

Perhaps the most basic benefit to group living is dilution of risk, both in terms of a predator finding the group and in terms of the risk for one individual to be predated once the group is found. This advantage holds only if larger groups do not attract more predators; large, persistent assemblages of species may indeed attract predators. In some cases, large assemblages of prey species that reproduce quickly (e.g., frogs) may be able to 'swamp' out predators, increasing their numbers more quickly than the predators can, in a manner similar to the way in which masting plants swamp seed predators. In addition, group-living animals that scatter at the approach of a predator may be able to confuse the predator (known as the 'confusion effect') and inhibit its ability to select an individual to catch. These potential benefits are found in single-species groups, and may also occur in mixed-species groups, especially if the species resemble each other. If, however, the species do not resemble each other, odd members of the group might be easier to capture; such an 'oddity effect' has been demonstrated experimentally, especially with fish. Finally, one kind of possible dilution benefit to grouping applies only to mixed-species groups: Clare Fitzgibbon, who worked on mixed-species ungulate groups, has suggested that individuals of one species could benefit from such groups if predators preferred to prey upon the other species.

A second benefit to grouping, and heterospecific grouping in particular, may be related to mobbing or other kinds of physical protection against predators. This aggressive function of heterospecific groups would be most effective if large or aggressive animals were included in them. For example, in some monkey assemblages, it has been reported that smaller species join larger ones, perhaps for protection against some predators. Aggression is particularly applicable to nesting colonies of birds. Birds in the center of a nesting colony have been shown to be well protected from predators because of mobbing. Birds will also nest close to biting or stinging social insects or perhaps even crocodilians, as well as aggressive or predatory birds.

Indeed, in a review in 2005, Tim Caro listed 30 studies that have documented a less aggressive species nesting with a protective bird species. These relationships can be mutualistic (if the less aggressive species contributes toward mobbing, or is consumed by the protector species at a low rate), commensal (e.g., stinging insects and birds), or even parasitic. As an example of the latter case, Martha J. Groom reported that nighthawks obtain greater fledging success when nesting near terns that aggressively mob predators. However, terns that had nighthawks nesting close to them spent more time in antipredatory defense and had reduced fledging success.

A third predator-related benefit from mixed-species grouping is increased vigilance. As in the case of alarm calling among animals that live in the same locality, animals that participate in flocks often make alarm calls and respond to each other. The simple amalgamation of information from species that occupy different spatial positions in the flock may benefit individual group participants, as first argued by the ornithologist E. O. Willis. Also, if multiple species call at once and they differ in the qualities of their information discussed earlier, a member of a mixed-species flock would have more information about predators than a member of a single species flock, as argued by E. Goodale and S. W. Kotagama (**Figure 2**).

One pattern found in many mixed-species groups is that they form around species that are particularly vigilant. For example, the Stonechat (*Saxicola torquata*), a heathland bird, was found by P. W. Greig-Smith to be very vigilant about predators and to take flight at a large distance from a predator, and thus was followed by less vigilant species; a similar argument has been made for several different species of shorebirds that are associated with less vigilant or wary species. In mammals, other species are suspected of joining Diana monkeys because they are especially vigilant, and other dolphin species may follow the highly vigilant spinner dolphins (*Stenella longirostris*).

Some leading species are particularly vigilant because their foraging ecology makes them good detectors of danger. For example, Neotropical bird flocks are led by antshrikes and shrike-tanagers, and drongos play important roles in south Asian flocks. These species all 'sally' for prey – taking insects on the wing – and thus seem to be particularly vigilant because of their frequent scanning for insects; hence, they are often referred to as 'sentinel species.' Saddleback tamarins (*Saguinus fuscicollis*) spend much time looking toward the ground and therefore are highly vigilant for ground predators, whereas moustached tamarins (*S. mystax*) spend more time looking up at the canopy and are better detectors of aerial threats. Differences among species in their vigilance may also be due to purely morphological traits: for example, Thomson's gazelles (*Gazella thomsoni*) may join Grant's gazelles (*G. granti*) because Grant's gazelles are taller and detect cheetahs at greater distances.

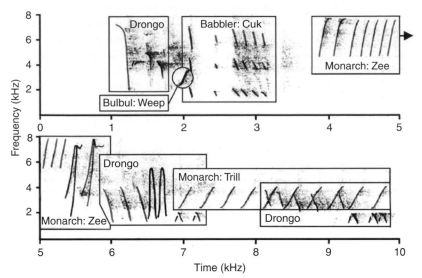

Figure 2 An example of alarm calls made by multiple species, from Goodale and Kotagama's (2005) study of mixed-species bird flocks in Sri Lanka. Goodale and Kotagama argue that species have different characteristics of their alarm calls and that hence the information available to participants is richer in a mixed-species as opposed to single-species flock system. Species calls are manually traced on the spectrogram for emphasis. Species included: Drongo – Greater Racket-tailed Drongo (*Dicrurus paradiseus*), Bulbul – Yellow-browed Bulbul (*Iole indica*), Babbler – Orange-billed Babbler (*Turdoides rufescens*), Monarch – Black-naped Monarch (*Hypothymis azurea*).

Animals may also benefit by associating with species that have a high propensity to produce signals about predators because they live in groups of kin or associate closely with their mates. Many '*nuclear species*' in mixed-species bird flocks – so called because they seem to play important roles in flocks and lead them – are gregarious and also tend to be cooperative breeders: traits that may make them more likely to participate in kin-selected behaviors such as alarm calling. Alarm calls can then be eavesdropped on by other species, as was shown in the classic studies of Kimberly Sullivan on vigilance by woodpeckers that follow gregarious groups of Black-capped chickadees; woodpeckers following chickadees scan less, eat more, and do not make their own alarm calls.

Before concluding this section on mixed-species groups, a phenomenon called '*social mimicry*' should be discussed, in which species of similar appearance group together. Martin Moynihan described several examples of this pattern in Neotropical bird flocks and explained the resemblances among species as adaptations to maintain group cohesion. C. J. Barnard argued against Moynihan's interpretation, suggesting that similarity among group members may be due to selection against the oddity effect (see above, second paragraph under 'Mixed-Species Groups in Relation to Predation') or that group members may be mimicking species that were distasteful or toxic. In relation to the toxicity hypothesis, a very striking case of social mimicry was described in bird flocks of Papua New Guinea, in which the experienced researcher H. L. Bell took several years in order to fully distinguish between species. Interestingly, J. P. Dumbacher and colleagues later discovered that

several of the nuclear species of that system are toxic, suggesting that Batesian mimics attempt to associate with the models that they imitate.

Mutualisms Based on Antipredation Benefits

In the discussion of some aspects of mixed-species groups, we touched on certain species (e.g., highly vigilant or aggressive species) that are particularly important to the organization of groups, or lead them. Thus, mixed-species groups may be viewed in part as commensalisms, with less vigilant or aggressive species following or attempting to associate with vigilant or protecting species (it should be acknowledged, however, that dilution benefits likely accrue to all parties). In contrast to such commensalisms as usually involve multiple species, here we discuss mutualisms that occur between two coevolved species partners. We see two basic categories of such mutualists: those that benefit each other by sharing information about predators and those that physically defend each other.

Information is usually traded between very dissimilar organisms. A well-studied case is that of goby fish and shrimp; about 100 species of gobies and 20 species of alpheid shrimp are involved in this mutualism (Karplus, 1987). The shrimp build burrows that the goby fish also dwell in; in turn, the fish warn the shrimp of predator presence through tactile communication. Andrew R. Thompson has shown that fish and shrimp are both negatively effected when either one is removed. An analogous mutualism that occurs under very different ecological conditions is the interaction between mongooses and hornbills, studied by O. Anne Rasa. Here the hornbills warn of predators by giving alarm calls, and in exchange, forage on insects that are disturbed by the mongooses. This system is particularly interesting because of the degree of co-evolution between the species; hornbills appear to give alarm calls to predators that are dangerous to the mongooses but not to themselves.

In other interactions between species in close proximity, physical protection is part of the exchange. Similar to the well-known phenomenon of ants defending plants, ants also protect sap-feeding insects, both hemiptera and lepidopterans, and receive nutritional benefits in exchange. Both in ant–plant relationships and in ant–insect relationships, variations abound in how facultative or obligate the relationship is, and there are several cases of interactions where mutualism has been replaced by parasitism. Other mutualistic partners provide protection in the form of a structural refuge. For example, corals provide crabs with protective structures as well as nutrition; in exchange, crabs consume potential competitors like algae. Another well-studied example is that of damselfish that seek protection among anemones and in exchange, consume specialized anemone predators. Such relationships represent the extreme of different species (large vs. small, mobile vs. immobile) giving diverse, complementary benefits to each other that similar species could not provide.

Conclusion

In reviewing the variety of associations between prey species that are related to predation, an interesting tension can be seen between mechanisms that appear to maintain similarity between species and mechanisms that facilitate very different species exchanging different kinds of benefits. In general, the kinds of benefits that maintain similarity (e.g., dilution, the confusion effect, the oddity effect, and mimicry between toxic and nontoxic species) apply to large assemblages or flocks in which there are many individuals and often many species. In contrast, mutualistic species tend to be diverse in their taxonomy, morphology, and behavior, and to complement each other in the services that they provide. Such a separation is not a rule, however; we have also argued here that the multiple characteristics of species in groups could make the overall information available in the group greater, and that nonvigilant species associate with vigilant ones.

There is much to learn about the role of heterospecifics in antipredator behavior. To better explore the idea of complementarity, we hope that species can be categorized by the 'functional role' that they play in heterospecific interactions and the services they offer heterospecifics. For example, alarm-calling species would compose one functional group, and food-disturbing species another. Studies are then needed to measure the dependency of species on each other, because while interdependencies are often known for mutualisms, they have not been well measured in mixed-species groups and in mobbing interactions. Having set such a foundation, we could then systematically investigate the conditions under which similar species are favored to interact and the contrasting conditions when species with different functional roles, such as alarm-calling species and food-disturbing species, are favored to associate together.

See also: Defensive Chemicals; Defensive Coloration; Group Living; Interspecific Communication; Parasitoids; Risk-Taking in Self-Defense.

Further Reading

Barnard CJ (1982) Social mimicry and interspecific exploitation. *American Naturalist* 120: 411–415.

Chivers DP and Smith RJF (1998) Chemical alarm signaling in aquatic predator–prey interactions: A review and synthesis. *Ecoscience* 5: 338–352.

Dall SRX, Giraldeau L-A, Olsson O, McNamara JM, and Stephens DW (2005) Information and its use by animals in evolutionary ecology. *Trends in Ecology & Evolution* 20: 187–193.

Fitzgibbon CD (1990) Mixed species grouping in Thomson's and Grant's gazelles: The antipredator benefits. *Animal Behaviour* 39: 1116–1126.

Goodale E and Kotagama SW (2005) Alarm calling in Sri Lankan mixed-species bird flocks. *Auk* 122: 108–120.

Haemig PD (2001) Symbiotic nesting of birds with formidable animals: A review with applications to biodiversity conservation. *Biodiversity and Conservation* 10: 527–540.

Karplus I (1987) The association between gobiid fishes and burrowing alpheid shrimps. *Oceanography and Marine Biology* 25: 507–562.

Krause J and Ruxton GD (2002) *Living in Groups*. Oxford: Oxford University Press.

Mönkkönen M, Forsman JT, and Helle P (1996) Mixed-species foraging aggregations and heterospecific attraction in boreal bird communities. *Oikos* 77: 127–136.

Morse DH (1977) Feeding behavior and predator avoidance in heterospecific groups. *Bioscience* 27: 332–339.

Moynihan M (1968) Social mimicry: Character convergence versus character displacement. *Evolution* 22: 315–331.

Rasa OAE (1983) Dwarf mongoose and hornbill mutualism in the Taru desert, Kenya. *Behavioral Ecology and Sociobiology* 12: 181–190.

Seppänen J-T, Forsman JT, Mönkkönen M, and Thomson RL (2007) Social information use is a process across time, space and ecology, reaching heterospecifics. *Ecology* 88: 1622–1633.

Seyfarth RM and Cheney DL (1990) The assessment by vervet monkeys of their own and another species' alarm calls. *Animal Behaviour* 40: 754–764.

Sullivan KA (1984) Information exploitation by downy woodpeckers in mixed-species flocks. *Behaviour* 91: 294–311.

Apes: Social Learning

A. Whiten, University of St. Andrews, St. Andrews, Fife, Scotland, UK

Introduction

The great apes include the chimpanzee (*Pan troglodytes*), the bonobo (*Pan paniscus*), and the gorilla (*Gorilla gorilla*) of Africa, and their more distant relative, the Asian orangutan (*Pongo pygmaeus*). Because the closest living relative of the chimpanzee is our own species (*Homo sapiens*), some authors also refer to humans as great apes. In this article, the focus is on non-human ape species, although some key comparisons with humans will be made. So little is known of social learning in the lesser apes, the gibbons (*Hylobates* spp.), that no mention will be made of them here. 'Apes' will refer to the great apes: chimpanzee, bonobo, gorilla, and orangutan.

Apes have played a particularly prominent role in the study of social learning in animals, as a result of their long-standing reputation for learning from others ('to ape' has long been a turn of phrase), as well as the fact that they are the closest relatives of humans, for whom culture is massively important in shaping behavioral repertoires. The importance of studies of apes in the history of studies of social learning probably reflects the complexity of their social learning. However, progress has also been made because of the need to develop methods to meet the challenges inherent in studying these intelligent, dexterous, and socially sensitive animals.

Experimental studies with captive apes began early in the twentieth century. However, I focus first on studies in the wild, which, although they started later, describe the important natural context to which apes' capacities for social learning are likely to be adapted.

The Study of Ape Culture in the Wild

The first serious field studies of chimpanzees began in the 1960s, followed by similar studies of gorillas and orangutans. The numbers of chimpanzee study sites gradually multiplied, with a subset of them being maintained for decades, providing unique opportunities to fully document the behavioral repertoires of different communities.

As such records accumulated, local or regional differences in behavior began to be noticed. Authors began to collate reports of such local variability in behavior and to treat such reports as providing circumstantial evidence of cultural variation in the behavioral repertoires of chimpanzee populations. However, relying solely on researchers' published descriptions of behavior is not, for several reasons, a satisfactory method for establishing the range of behavioral differences between populations. For example, authors do not necessarily publish complete catalogs of the behavior patterns common at their field site, and in particular, they do not necessarily document behaviors that are well-known elsewhere but absent in the community that they study.

In the 1990s, a consortium of chimpanzee researchers therefore agreed to pool both their published and unpublished observations spanning several decades, permitting the first systematic analysis of regional variations in chimpanzee behavior. The resulting catalog identified a surprisingly large number of behavioral variations that were commonly seen in at least one community of chimpanzees, but were absent in at least one other, without any obvious genetic or environmental explanations, such as local absence of a crucial raw material. Accordingly, researchers attributed the differences to social learning, an interpretation supported, in many cases, by youngsters intently observing adults' expertise in a local skill, and subsequently practicing that skill for long periods without reward (e.g., attempting to crack nuts using hammer stones). These 39 putative cultural variants included a great diversity of types of behavior: foraging skills, forms of tool use, social behavior, grooming styles, and courtship gambits (**Figure 1**).

The interpopulation variation in behavior was remarkable in two ways. First, the sheer number of cultural variants identified was unprecedented in studies of non-ape species. Second, each chimpanzee community had its own suite of behavioral variations, making its members culturally unique. In humans, cultures are defined by vast suites of traditions. That a similar phenomenon is recognizable in chimpanzees, though on a more modest scale, suggests that the common ancestor of humans and chimpanzees, living 6–7 Ma, is likely to have relied on a chimpanzee-like level of cultural learning.

Studies of wild orangutans have revealed an array of putative cultural variations almost as numerous as those seen in chimpanzees. Like the traditions of chimpanzees, these cultural variations distinguish local communities. If one knows enough about the behavioral repertoire of a wild orangutan or chimpanzee, just as in humans, their local origin can be determined from their cultural profile alone. Accordingly, van Schaik and colleagues, who pooled long-term records of orangutan behavior to arrive at this picture, inferred that a level of culture marked by multiple traditions spanning different modes

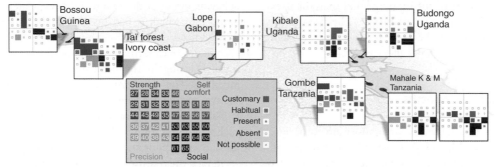

Figure 1 The putative cultures of wild chimpanzees. Reproduced from Whiten A (2005) The second inheritance system of chimpanzees and humans. *Nature* 437: 52–55. 'Customary' acts are those typical in a community, 'habitual' are those less frequent yet consistent with social learning. Each community displays its own profile of such local behavioral variants, providing evidence of a unique culture for each locality. Numbers identify behavior patterns in the catalog attached to Whiten et al. (1999) and illustrated at http://culture.st-and.ac.uk/chimp. Whiten A (2005) The second inheritance system of chimpanzees and humans. *Nature* 437: 52–55.

of behavior, such as tool use and social behavior, is likely to have originated at least 14 Ma in the ancestor of all the great apes.

The excitement of these recent discoveries of putative traditions in great apes has to be tempered, however, by the difficulty of establishing compelling evidence through observations alone that the variations in behavior, so comprehensively and carefully documented in free-living animals, truly all depend on social learning. To be sure, youngsters often intently watch skilled adults performing actions, such as tool-aided termite fishing, and later master the art themselves. It has even been found that young female chimpanzees, which are much more assiduous than their male peers in observing their mother's termite fishing, are much faster to perfect the skill and even adopt their mother's preference for particular lengths of tool, whilst the less-observant males do not. Many scientists believe such correlations provide compelling evidence of social learning, especially when seen within the totality of the observational evidence for cultural variation in the wild. Others are more skeptical, noting the possibility of undetected environmental or genetic factors that could generate the behavioral differences under discussion.

A salutary lesson relevant to this issue was provided by studies of ant-dipping by Humle and Matsuzawa. Earlier work at Gombe in East Africa had shown that chimpanzees strip a long wand of leaves and wipe this over swarming safari ants (**Figure 2**). They then use the other hand to wipe the ball of ants off the wand and into their mouths. In the Taï Forest of West Africa, chimpanzees instead use a short stick to dip for safari ants and transfer the ants direct to their mouths from the stick. That this difference in behavior persists despite the much greater efficiency of the two-handed version of ant dipping was one reason that each local variant was assumed to be culturally maintained. However, working at Bossou in Guinea, West Africa, Humle and Matsuzawa observed chimpanzees using both techniques. The chimpanzees used the longer

Figure 2 Ant-dipping using long wands at Gombe (Photo courtesy of D. Bygott).

wand to harvest a species of ant with a particularly painful bite, suggesting that the two techniques of ant dipping resulted from individual learning about avoiding bites rather than from social learning of arbitrary traditions. Intriguingly, further work comparing local communities around Bossou and in other parts of West Africa has suggested that aspects of ant-dipping other than the use of long and short tools do vary culturally. The present picture of ape culture in the wild has become far more complex than it was just a decade ago.

Diffusion Experiments

Experimentally, it is relatively straightforward to design a robust test for social learning. Subjects in an experimental condition are allowed to view an individual (the 'model' or 'demonstrator') already proficient in a novel task. Subjects in a second group see no model, providing a control condition that permits individual learning but not social learning. If the novel behavior is acquired only by observers in the experimental condition, then social learning has clearly occurred. This is precisely the degree of certainty that field studies lack because they do not have a powerful control condition to check that the behavior of interest is learned from others and not through individual learning. Consequently, laboratory social learning experiments are complementary to field studies. Each plays a crucial role in the study of social learning. Field studies are essential to map out the potential range of cultural phenomena in the wild. Laboratory experiments rigorously test for social learning and, as discussed further below, can even determine the particular type of social learning involved.

Because traditions and cultures are group-level phenomena defined by the spread of novel behaviors across populations of animals, a particular kind of experiment, the diffusion experiment, has been used to examine evidence for cultural transmission. In diffusion experiments, the spread (or failure to spread) of a novel behavior is systematically examined in a group of animals. This contrasts with the bulk of social learning experiments that typically permit only a single observer to witness the actions of a model before being tested alone to determine how well it has acquired the behavior that it observed. Only eight controlled diffusion experiments with primates have been completed, and of these, six used chimpanzees as subjects. Although, in principle, such experiments could be executed in the field, the logistics are extremely challenging, and diffusion studies have been conducted only with captive apes.

In the first of these diffusion studies, a high-ranking female chimpanzee was temporarily separated from her group and taught to use a stick to lift a blockage out of a foraging device (the 'pan-pipes') thus releasing trapped food. A female from a different group was shown how to use the stick in a completely different way, inserting it through a small flap to poke the blockage backwards so that the food fell and rolled towards the chimp. Once each model had attained proficiency in her respective technique, she was reunited with her group.

The first question was whether, relative to control groups whose members saw no demonstrations of either behavior, either of the two techniques would spread in the group into which it had been seeded. In fact, no control group solved the task. By contrast, all but one member of each of the 16 experimental groups did, clearly implicating social learning in the acquisition of solutions by group members. The more critical question was whether each of the two techniques would spread in the groups into which it had been seeded, or whether corruption would occur such that both groups behaved similarly, either both opting for one solution or showing a mixture of both solutions.

In fact, the traditions of lifting and poking did spread in the groups into which they had been introduced. However, some corruption in traditions occurred. Half the lift group, in particular, discovered, and increasingly used the poke method, at least initially. Intriguingly, however, over the next 2 months, there was a significant tendency for animals that had 'strayed' from the behavioral norm of their group to return to this norm, a pattern of behavior that the investigators interpreted as evidence of conformity, defined as preferring an option just because a majority of one's group-mates exhibit it. The experiment showed, consistent with fieldworkers' interpretations of regional differences in the behavior of wild chimpanzees, that chimpanzees will not only sustain different traditions by social learning, but they also show signs of the strong social learning tendency known as conformity.

Further diffusion experiments have built on this early work, demonstrating other aspects of social learning. In one experiment, highly arbitrary actions that involved the relatively bizarre posting of a rubber 'token' into either of two different receptacles to gain food, spread in two groups, with only a single individual rejecting the local tradition. In another experiment, researchers introduced foraging tasks, in which the alternative actions needed to acquire food, that were more complex than those involved in retrieving food from the pan-pipes. These experiments required the chimpanzees to engage in a sequence of actions to gain access to food. Once again alternatives spread differentially in two groups. Even two further groups of chimpanzees that watched the activities of the founder groups acquired the behaviors of the group they had watched, as did a third pair of groups that watched the second group (**Figure 3**). The striking fidelity in copying may have been a result of the complexity of the actions involved, making it unlikely that individuals would acquire either behavior by individual trial and error learning.

All of these experimental designs can be described as 'open diffusion' designs because which members of a group watch or acquire the actions of a founder individual is not controlled, and identifying how much transfer is directly from a founder to the rest of its group or is, instead, stepwise through other group members, is difficult to determine. To examine the multiple transmission events that would be necessary for long-term, intergenerational preservation of traditions, experimenters use a diffusion

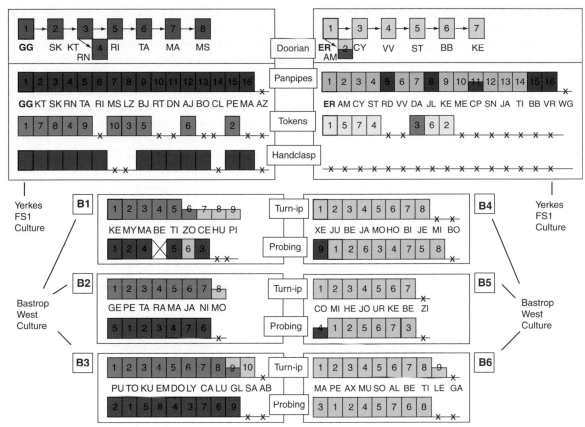

Figure 3 The results of diffusion experiments in captive chimpanzees. Reproduced from Whiten A, Spiteri A, Horner V, et al. (2007) Transmission of multiple traditions within and between chimpanzee groups. *Current Biology* 17: 1038–1043. Each *rectangle* represents a chimpanzee with two-character ID codes. Tasks, named in the center, were available in both local populations named on either side, but different techniques, color coded here, were seeded in one individual, marked here as No. 1, each population. The Doorian experiment was run as a transmission chain, as indicated by the *arrows*; all other experiments involved open diffusion, with no predetermination of potential order of transmission. At Bastrop, transmission extended from group B1 to B2 and B3, and from group B4 to B5 and B6. Handclasp grooming spread spontaneously in the FS1 population. Numbers represent order of acquisition for each task. For further explanation, see text. These studies demonstrate the capacity of chimpanzees to sustain multiple-tradition cultures, consistent with the interpretation of regional variations among wild chimpanzees summarized in **Figure 1**.

(or transmission) chain design. Here, model A is observed by individual B. Once B masters the task, B becomes the model for C, and so on along a potential chain of culture-bearers. Results can be compared with both a no-model control condition and with a chain that begins with a model demonstrating an alternative solution to the task at hand. Less than a half-dozen such experiments have been completed with non-human animals and only one with chimpanzees. In this experiment, the alternative options involved obtaining food either by simply sliding a door to one side or by lifting a smaller door set in the sliding door. Despite the potential for corruption between these fairly subtle alternatives, fidelity of transmission was maintained over six experimental generations (**Figure 3**), which would correspond to about 90 years of mother–daughter cultural inheritance in wild chimpanzees.

In sum, diffusion experiments have demonstrated that chimpanzees have the capacity to sustain different traditions through social learning.

A further question that arises from the discovery of a capacity for tradition in chimpanzees concerns the learning mechanisms that underlie their traditional behaviors. Because of the 'two-action' alternatives built into the diffusion experiments described above, we know that such mechanisms must be sufficiently powerful to permit replication of behavioral alternatives. However, the ability to copy behavioral alternatives does not necessarily require reliance on imitation, in the strict sense often used in social learning research. To see why, I next focus specifically on the mechanisms involved in chimpanzee social learning.

Social Learning Mechanisms

To 'ape' means to copy or, in common parlance, to imitate. Both anecdotal reports and results of experimental studies carried out in the early twentieth century tended to provide information consistent with the view that apes imitated. However, more recently, there has been considerable skepticism as to apes' imitative competence, and acceptance of the hypothesis that apes can imitate has waxed and waned over recent decades. Fortunately, the controversy concerning apes' imitative abilities led to increasingly rigorous scientific methodologies for examining imitative abilities.

Imitation and Emulation

An important change in perspective in ape research that has also influenced the field of social learning more generally, began with experiments by Tomasello, Call and colleagues. In an early experiment, chimpanzees witnessed a model using a stick as a tool to recover food that was out of reach. The chimpanzees failed to copy a skilled maneuver that facilitated success, although they did learn something basic about using the stick to rake in food that a control group that did not see a model using a stick to rake in food failed to discover. Tomasello concluded from this finding that chimps did not necessarily imitate in the strict sense of copying a skilled model's behavior. Rather, chimpanzees learned about environmental affordances that the model's behavior made salient, in this case, that the stick had potential as a rake. Tomasello labeled this mode of learning 'emulation,' describing it as intermediate between imitation and mere local enhancement. In local enhancement, observers simply have their attention focused on relevant items or locations by the actions of a model. Emulation involves more than this.

Subsequent studies led Tomasello and Call to the conclusion that, in contrast with human children who are active imitators, apes are better characterized as emulators. The result of emulation may often superficially resemble imitation, thus explaining apes' misleading reputation for being habitual imitators.

Recognition of emulation has enriched the study of animal social learning, particularly in primates, and has fuelled continuing controversies about the nature of apes' social learning. This controversy has been complicated by the tendency for 'emulation' to be used to refer to any lack of imitative fidelity (typically, in comparison to children) when the behavioral match between model and observer is limited to the final outcome of the action of interest, for example in the raking in of food described above. There is a consensus in the research literature that, compared to children, apes are relatively emulative. Human children have an extraordinary tendency to imitate any action they see performed. Apes are more likely to avoid imitating if they can instead use a more economical approach of their own devising to solve a task at hand, reaching an outcome they have learned about by observing the behavior of another, but using alternative means to achieve the same

end. Consequently, apes may often complete observed tasks more efficiently than children, who on seeing an inefficient method demonstrated will often copy it with undue fidelity, a phenomenon described as 'over-imitation.'

However, when emulation is defined more rigorously as learning about the environmental results of action rather than about the action itself, results of several recent experiments contradict the contention that apes are primarily emulators. In such 'ghost' experiments, a model is removed from the scene and objects that the model would normally manipulate are manipulated by devices such as pulling on a fine fishing line or manipulation of video displays, so objects move just as they would if a model had interacted with them. Observers now see just the components of a display that would normally be caused by the actions of a model, but no model is in view, so no behavior is to be seen. Experiments of this kind, in which the blockage in the pan-pipes described above was made to move as if by a ghostly chimpanzee, showed that observer chimpanzees were unable to learn from observing this ghost control, whilst they could learn from observing a chimpanzee model. Perhaps chimpanzees are not so neatly described as emulators. Recalling Thorndike's concise nineteenth-century definition of imitation as 'learning to do an act from seeing it done,' chimpanzees could be seen as imitators insofar as to learn effectively they needed to see another ape perform an operation.

Other experiments show, however, that such an interpretation would be an over-simplification. Although involvement of an ape as a model is important to observer apes, once such a model is in place, learning does involve processing information about aspects of the environment that change as a result of what the model does. For example, in an experiment where young chimpanzees watched a familiar human obtain a food reward from a box, subjects in one condition saw a stick tool first rammed into the top of an opaque box, then later used to extract food from a central tunnel (**Figure 4**). Chimpanzees in a second condition saw exactly the same thing, except the box was transparent. Consequently, when the stick was rammed into the top of the box, it was evident that it did no useful work, merely hammering on a barrier above the tunnel with the food in it. When allowed their own attempt, chimpanzees given the transparent box were less likely to perform the initial ramming action, than were chimpanzees exposed to the opaque box. The chimpanzees appeared to be sensitive to perceptual cues about causality within the box. With the opaque box they were relatively imitative, copying the sequence of actions performed by the model, whereas with the transparent box they were more emulative, omitting part of what they witnessed, and focusing more on final results. These apes thus seem to have a portfolio of social-learning approaches that they employ appropriately in different situations. Such flexibility may explain some past controversies about whether apes are 'really' imitators or are merely emulators. Apparently, they can be either, depending on circumstances.

Other studies have identified situations in which apes learn more from observing changes in the environment in the absence of a demonstrator than they did in the ghost pan-pipes study described above. In a ghost experiment in which the task was simply to push a hatch to one side or another to reveal food, chimpanzees did show an initial, if fleeting, tendency to match what they saw, implicating emulation. In another experiment, chimpanzees observed an 'artificial fruit' that had been taken apart in either of two ways to reveal an edible core. These subjects later recreated the outcome they had observed, an emulative response that would misleadingly have appeared imitative if the subjects had watched a model perform only one way of opening the artificial fruit and subsequently did the same.

To accommodate such results relating to the issue of whether chimpanzees imitate or emulate, it can be theorized that chimpanzees, and perhaps other apes as well, have a tendency to emulate when they can. Emulation will tend to occur when the task is relatively simple and solutions are therefore relatively readily available within the ape's repertoire. More complex or novel tasks may require greater reliance on copying what a model actually does, and imitation may then become a strategy.

Copying Sequential and Hierarchical Action Structure

Byrne and Russon have drawn a distinction between imitation at the action level (e.g., 'lift tool') and what they call program-level imitation, in which the overall structure of a complex action sequence is copied, with or without its attendant details. These authors speculated that complex actions ranging from feeding on difficult-to-process foods by wild gorillas to apparent attempts at fire-making by orangutans in a rehabilitation camp, might exemplify copying at the program level. This idea remains little tested, but chimpanzees have been shown to copy a laboratory task with its constituent elements in different sequential orders, either ABCD or CDBA.

'Do-as-I-Do'

A very different way to test for imitation other than the naturalistic approaches outlined above is to explicitly train a participant ape to copy in response to the request 'do this!' using a set of training actions, then testing subjects on a battery of relatively novel actions different from those on which they were trained. The advantage of the 'do this' approach is that it allows exploration of what a species can or cannot copy.

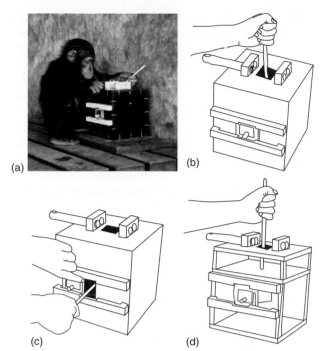

Figure 4   Testing for imitation versus emulation. Reproduced from Whiten A (2005) The second inheritance system of chimpanzees and humans. *Nature* 437: 52–55. (a) Chimpanzee working on top part of task after watching model; (b) model probing in top of opaque box; (c) extraction of food from the opaque box; (d) model probing in top of transparent box, where it can be seen that this action is ineffectual, merely hitting a barrier.

One thorough test with an orangutan elicited 58% full imitations and 36% partial imitations of a battery of 48 actions. Tests with chimpanzees have also shown significant but less impressive copying that included copying of actions involving parts of the body not visible to the subject, such as touching the back of one's head. Interestingly, attempts to train monkeys to participate in 'do-this' tests have failed, leading to the hypothesis that among primates, only apes are able to grasp the concept of imitation necessary for success in 'do this' tasks.

Conclusions

The history of ape social-learning research over the last few decades has been one marked by much controversy as to whether apes can imitate, and consequently by considerable fluctuation in the attribution of imitative and cultural capacities ascribed to the higher primates. It is probably true that the pendulum has recently swung towards the end of the continuum that recognizes apes as having capacities for both imitation and cultural transmission unmatched by any other non-human species, but that still leaves them far short of the cultural capacities and inclinations of humans. Progress in understanding the ability of apes to acquire behavior through social learning has involved, in part, discovery of the context dependence of their social-learning strategies. Such context dependence can cause great variation in apes' copying fidelity. We now know an enormous amount about social learning, traditions and culture in our nearest relatives, probably more than about social learning in any other non-human animal. This article has been able to do no more than outline some central findings in what has by now become a very large literature.

See also: Avian Social Learning; Imitation: Cognitive Implications.

Further Reading

Buttelmann D, Carpenter M, Call J, and Tomasello M (2008) Enculturated chimpanzees imitate rationally. *Developmental Science* 10: 31–38.

Byrne RW and Russon AE (1998) Learning by imitation: A hierarchical approach. *Behavioral and Brain Sciences* 21: 667–721.

Call J (2001) Body imitation in an enculturated orangutan (*Pongo pygmaeus*). *Cybernetics and Systems* 32: 97–119.

Call J, Carpenter M, and Tomasello M (2005) Copying results and copying actions in the process of social learning: Chimpanzees (*Pan troglodytes*) and human children (*Homo sapiens*). *Animal Cognition* 8: 151–163.

Heyes C, Huber L, Gergely G, and Brass M (eds.) (2009) Evolution, development and international control of imitation. *Philosophical Transactions of the Royal Society B* 364 (1528): 2291–2443.

Hopper LM, Lambeth SP, Schapiro SJ, and Whiten A (2008) Observational learning in chimpanzees and children studied through 'ghost' conditions. *Proceedings of the Royal Society of London B* 275: 835–840.

Humle T and Matsuzawa T (2002) Ant-dipping among the chimpanzees of Bossou, Guinea, and some comparisons with other sites. *American Journal of Primatology* 58: 133–148.

Laland KN, and Galef BG, Jr (eds.) (2009) *The Question of Animal Culture.* Cambridge: Harvard University Press.

Lonsdorf E, Ross S, and Matsuzawa T (eds.) (2008) *The Mind of the Chimpanzee.* Chicago: Chicago University Press.

Mobius Y, Boesch C, Koops K, Matsuzawa T, and Humle T (2008) Cultural differences in army ant predation by West African chimpanzees? A comparative study of microecological variables. *Animal Behaviour* 76: 37–45.

van Schaik CP, Ancrenaz M, Borgen G, et al. (2003) Orangutan cultures and the evolution of material culture. *Science* 299: 102–105.

Whiten A (2005) The second inheritance system of chimpanzees and humans. *Nature* 437: 52–55.

Whiten A, Goodall J, McGrew WC, et al. (1999) Cultures in chimpanzees. *Nature* 399: 682–685.

Whiten A, Horner V, and de Waal FBM (2005) Conformity to cultural norms of tool use in chimpanzees. *Nature* 437: 737–740.

Whiten A, Horner V, Litchfield CA, and Marshall-Pescini S (2004) How do apes ape? *Learning and Behavior* 32: 36–52.

Whiten A, Spiteri A, Horner V, et al. (2007) Transmission of multiple traditions within and between chimpanzee groups. *Current Biology* 17: 1038–1043.

Aplysia

J. F. Aggio and C. D. Derby, Neuroscience Institute and Department of Biology, Atlanta, GA, USA

General Biology

Aplysia and its close relatives are sea hares, so named because their hunched back and earlike rhinophores give them a superficial resemblance to 'real' hares. They are actually snails, members of the order Anaspidea and class Gastropoda, part of the phylum Mollusca, which also includes squids and octopuses (Class Cephalopoda) and clams and oysters (Class Bivalvia). As molluscs, their body plan includes a head with a pair of eyes, tentacles that carry the chemical sense organs, a mouth, a foot, a visceral mass, and a mantle that protects the visceral mass and secretes the shell. A distinctive feature of sea hares is parapodia, which are lateral outgrowths of the mantle that protect the internal organs, create a respiratory current over the gills, and are used for swimming by some species. Another characteristic of sea hares is a greatly reduced and internalized shell, which may enhance certain forms of locomotion but removes one of the forms of defense typical of gastropods. This loss of a physical defense is compensated by a wealth of chemical defenses, which make these animals fascinating subjects for researchers interested in marine chemical ecology.

Sea hares are represented by over 40 species in 10 genera, most belonging to *Aplysia*. They are distributed throughout the world's oceans. Most are endemic, but a few are cosmopolitan, such as *Aplysia dactylomela*, *A. parvula*, and *A. juliana*. Almost all live in warm shallow waters, an exception being *A. punctata,* which also occurs inside the Arctic Circle. Local habitats are varied, even for individuals of the same species. For example, *A. dactylomela* can live in rocky shores, tidal pools, eelgrass beds, sand, and other habitats.

Sea hares are herbivores, feeding mainly on green and red algae in various combinations and proportions. One notable exception is *A. juliana*, which eats only green algae and among these greatly prefers *Ulva*. Their choice of diet is extremely important, not only for nutritional reasons but also because they absorb algal secondary metabolites and utilize them as chemical defenses. They spend several hours each day feeding and may eat up to one third of their body weight. This is a prodigious amount for a cold-blooded animal and translates into a very rapid rate of growth. This explains why sea hares can be so large, despite brief life spans – the largest recorded *A. californica* tipped the scales at 6.8 kg!

Reproduction and Life Cycle

Copulation

Sea hares are simultaneous hermaphrodites, meaning that each adult individual possesses both male and female reproductive structures that are mature at the same time. Although this could allow two individuals to reciprocally exchange sperm, this is relatively rare, as is self-fertilization. Instead, in a mating pair of sea hares, one animal donates sperm and is thus a functional male, and the other animal donates eggs and is thus a functional female. The factors influencing which sexual role an animal assumes are not clear, but it is known that in *A. juliana* older animals tend to assume one role or the other while younger ones show no preference. Actual mating is preceded by a brief courtship in which the animal that will donate sperm generally takes a more active role, approaching its partner while waving its head side-to-side in a movement that is very similar to that performed before feeding. Upon making contact, and for a brief period of time, the active individual will crawl over and around the inactive one before copulation begins.

Copulation generally involves several to many individuals aggregating and forming chains in which one individual donates sperm to another one while receiving it from a third. In some cases, the first and last links of the chain copulate, thus forming a ring. These aggregations contain individuals that are copulating as females (which can simultaneously lay eggs), individuals that are laying eggs, individuals that are neither copulating nor laying eggs, and recently laid eggs.

Egg Laying

As mentioned earlier, sea hares in the female role are capable of simultaneously copulating and laying eggs, but egg laying can also follow copulation after several hours or even days, since accumulated sperm is viable for up to 15 days. Sea hares attach the eggs in strings, called cordons, to the substratum, using three stereotypical head movements: up-and-down, possibly to prepare the substratum; sideways, to distribute the egg cordon; and in-and-out, to attach the egg cordon to the substrate. The choice of the substratum itself is also important, and different species have different preferences, with a common theme being that they tend to avoid laying eggs in the seaweed species that they prefer as food. In addition, individuals within a breeding aggregation tend to lay eggs on top of preexisting egg cordons, creating accumulations of unknown functional advantage. The amount of eggs laid can be enormous, as exemplified by an individual *A. californica* that was documented as laying 478 million eggs in 27 egg-laying episodes in a little over a month.

Egg laying is under the control of the egg-laying hormone (ELH) and other peptide hormones released from bag cells when stimulated by as-yet unidentified neurons located in their central nervous system. ELH acts directly on the gonads to induce egg release and on neurons in the central nervous system to inhibit feeding. Other peptide hormones depolarize the bag cells as part of a positive feedback loop. Although ELH cannot depolarize bag cells, it can elicit a full-fledged bout of egg laying, which suggests that the performance of certain aspects of the egg-laying behavior influences other aspects of the same behavior.

Development

Egg development leads to the release of planktonic veliger larvae. These larvae live and grow in the water column for approximately 1month, after which they are competent to settle and metamorphose. However, this process will not occur unless the appropriate chemical signal is present. This signal varies from species to species but its source is typically the seaweed that constitutes the preferred adult food. If this is present, the competent larvae stop swimming, reach the algae, attach to it, and metamorphose. The sea hares then begin to graze on the seaweed, and further develop into juvenile and adult phases.

Pheromone Control of Reproductive Behavior

Pheromones are important in controlling several aspects of the reproductive behavior of *Aplysia*. During the mating season, non-egg-laying individuals are attracted to each other, thus providing a mechanism for the initiation of the breeding aggregations. This attraction is chemically mediated, but the responsible compounds have yet to be identified. The mating and subsequent egg laying result in the release of a set of pheromones that increase the attraction of other *Aplysia*, thus forming breeding aggregates. The laid egg cordons themselves contribute to this attraction. The albumen gland, which fabricates the egg-packaging materials, laces egg cordons with at least four peptide pheromones with wonderfully descriptive names: Attractin, enticin, seductin, and temptin. These pheromones enhance the attraction of sea hares to each other and induce copulation. Thus, the result of breeding, namely eggs, attracts more breeders to the site. Interestingly, these protein pheromones share similarities in sequence and structure with the mammalian fertilin system, which is involved

in egg–sperm interactions, thus raising the possibility of a common origin.

Defenses

Chemical

Sea hares lack a well-developed shell to afford them mechanical protection from predation and they are sluggish, so they are unable to use armor or speed to avoid fast predators such as crabs or fish. They have evolved other means of protection, including crypsis, large size, and an impressive array of chemical defenses (**Figure 1**). Chemical defenses can be passive or active, depending on the involvement of the animals' nervous system. Passive defenses are always present, consist mainly of the accumulation of toxic and/or unpalatable compounds in the animals' tissues, and do not involve active behaviors – except the choice of algal diet and their constituent defensive compounds. Active defenses include one of the most striking behaviors in sea hares, called 'inking.' Inking typically involves the simultaneous release of both ink and opaline, each from a separate gland. Ink is typically a dark purple secretion. Its color is mainly due to the compound aplysioviolin, which is derived from phycoerythrobilin, a major photosynthetic pigment in the red algae favored in the hare's diet. Opaline is a clear and viscous secretion. Both ink and opaline are complex mixtures, and although of different chemical composition, both are effective chemical defenses against predators of sea hares such as fish, crabs, lobsters, and anemones. Ink and opaline are released to the mantle cavity, mixed there for a few seconds and then expelled via the siphon in the direction of the attacker. The fact that both secretions are mixed before reaching the attacker is significant, because components in them interact to generate aversive compounds that

Figure 1 A sea hare, *Aplysia californica*, releasing its defensive secretion of ink and opaline. The purple ink is clearly noticeable from its color, and the whitish opaline is less conspicuous. Photo courtesy of Genevieve Anderson.

are not present in either secretion alone. An example of this is the 'escapin pathway.' Escapin, an L-amino acid oxidase, is an enzyme in ink the preferred substrates of which are L-lysine and L-arginine, present in large quantities in opaline. This enzymatic reaction quickly gives rise to millimolar concentrations of the corresponding α-keto acids and hydrogen peroxide ('intermediate products'), which then recombine spontaneously to form carboxylic acids ('end products'). Hydrogen peroxide is a potent irritant to spiny lobsters and a feeding deterrent to blue crabs. In addition, intermediate products alone or in combination with hydrogen peroxide, but not end products, are mild feeding deterrents to wrasse.

Although it was initially thought that inking was a high-threshold, all-or-nothing behavior, more detailed studies have shown that neither of these affirmations is accurate. Threshold is variable and, not unexpectedly, depends on factors such as the environment (animals living in turbulent environments have higher thresholds), fullness of the gland (animals with severely depleted but not empty glands tend to have very high thresholds), and relevance of the stimulus (lower thresholds for brief contact with an anemone tentacle and higher thresholds for relatively intense electrical shock). Regarding its all-or-nothing aspect, research has shown that the first suprathreshold stimulus received by a sea hare with full ink glands induces the sea hare to release approximately half of its ink, and each subsequent stimulus causes it to release 30–50% of its remaining ink. This implies that sea hares optimize the use of a limited resource and try to conserve it, which is important since a sea hare needs approximately 2 days to replenish the contents of its ink gland, provided it has access to the adequate food supply.

Locomotion and Escape

The normal mode of locomotion in sea hares is crawling on the substrate by undulating their muscular foot. When attacked or disturbed, they will attempt to escape, generally by turning away from the source and locomoting either by crawling, galloping, or swimming. The prevalence of swimming is somewhat controversial, as is the nature of the mechanisms employed. The two known species of the genus *Notarchus*, *N. punctatus* and *N. indicus*, swim by means of jet propulsion. The parapodia of these sea hares are fused around the mantle cavity except for a small anterior opening, which allows the water in the mantle cavity to be expelled with sufficient force to elevate the animal above the substrate. This results in an awkward, bursting form of locomotion that is not suitable for translocation from one place to another, but rather to remove the animal quickly from a perceived danger or disturbance. In the swimming species that lack fused parapodia, swimming is achieved by the rhythmical flapping of the parapodia, and sculling has been proposed as

the mechanism by which they achieve movement. Animals that swim in this fashion may do so to escape a predator, but also to locomote from one point to another in search of food or mates, and there have been documented swimming episodes lasting more than an hour. Finally, members of a primitive family, Akeridae, use a slightly different mechanism in which they open and close their parapodia much like an umbrella and swim with their bodies in a roughly vertical position with the heavy shelled part on the bottom, sometimes in large swarms of many individuals. However, little is known about these animals and it is by no means clear that the swimming behavior is executed to escape from predators.

One exclusively escape-related mode of locomotion is known as galloping, and it involves an inchworm-like saltatory behavior that consists in detaching the leading edge of the foot from the substrate and extending the body before reattaching it, at which point the tail is detached and brought forward without contacting the substrate. Galloping is clearly faster than normal crawling and thus can enhance escape. But galloping also produces a discontinuous slime trail, which could aid sea hares in avoiding predators such as the snail *Navanax* that locate their prey by following them.

The chemical defensive and locomotive means of avoiding predators can be linked behaviorally. Ink and opaline contain chemical alarm signals that cause neighboring sea hares to turn and gallop away. This antipredatory behavior is also elicited by the ink of other sea hares and even of cephalopods, suggesting that the mechanism may be widespread among mollusks.

Aplysia as a Model for Learning and Memory

Aplysia californica has been used extensively as a model to study the neural basis of learning and memory, because it presents two advantages for this kind of study: clear and easily elicited behaviors and an accessible and relatively simple nervous system. Studies of *A. californica* have contributed enormously to our understanding of how organisms acquire, store, and eventually use their experiences. This work led to the 2000 Nobel Prize in Medicine or Physiology being co-awarded to Eric Kandel. The recent sequencing of its genome promises to allow scientists to use *A. californica* to explore deeper into the molecular and genetic bases of behavior. The following section is a brief summary of our current understanding of the mechanisms underlying learning and memory on the basis of studies of *A. californica*.

A. californica withdraws its gill and siphon when lightly touched, and this behavior can be modified by three well-known forms of learning. These include one form of associative conditioning – classical conditioning – and two forms of nonassociative learning – sensitization and habituation. Sensitization is defined as an enhanced response to a stimulus due to experience with another stimulus. In the example of gill–siphon withdrawal behavior of *Aplysia*, application of an electric shock to the animal's head or tail increases, or sensitizes, the intensity of the gill–siphon withdrawal to a standard touch to the siphon, and this enhancement lasts for several minutes after the shock. The sensitization survives longer if a series of electric shocks spaced over several days is used. Although it may seem similar to classical conditioning because two stimuli are involved, no temporal pairing of the two stimuli is needed to elicit sensitization. Habituation is defined as a diminished response to a stimulus due to periodic and repetitive presentation of that stimulus, because the animal learns to ignore it rather than through sensory adaptation. This diminution of response can also survive for long periods of time under the correct training conditions. The third form of learning, classical conditioning, involves a temporal pairing of a conditioned stimulus (CS), which does not produce a certain response, and an unconditioned stimulus (US), which produces a response, eventually leading to the CS by itself producing the response. In the example of gill–siphon withdrawal, a weak tactile stimulus to the siphon that does not elicit gill–siphon withdrawal (CS) is temporally paired with an electric shock to the body that does elicit the gill–siphon withdrawal (US), causing the CS to take on the ability to cause the gill–siphon withdrawal. This learned response can be retained for long periods of time, again depending on the training conditions. This classical conditioning, a form of associative learning, is different from sensitization and habituation, forms of nonassociative learning, in that in the latter two the stimuli are not temporally paired or associated (sensitization) or there is only one stimulus (habituation). However, some authors contend that two stimuli are effectively involved in habituation, one being the stimulus itself and the other the context in which it is applied.

A major strength of the *Aplysia* model is that the neural circuit underlying these behaviors is relatively simple, and it involves cells that are relatively invariant and identifiable from animal to animal and that are large enough for their electrical and biochemical properties to be studied. The siphon is innervated by 24 mechanosensory neurons that make both direct and indirect synaptic contacts with gill motor neurons. The direct contact is monosynaptic, and the indirect contact is mediated by both excitatory and inhibitory interneurons. Kandel and colleagues showed that this circuit is responsible for all three behaviors and is a crucial part of the three types of learning. They showed that memory is stored as modifications in the sensory-motor synapses, thus providing an important link between synaptic plasticity and memory. This change in synaptic strength can be decremental as in habituation or incremental as in sensitization or their classical conditioning paradigm. Furthermore, each type of learning is mediated by different changes and processes. Habituation is a homosynaptic process, in which activity of the sensory and/or motor neurons decreases the synaptic strength, which is mirrored behaviorally by a reduction in the response to the stimulus. Sensitization is mediated by a heterosynaptic process, in which the activity of one or more interneurons activated by the electric shock to the body enhances the sensory-motor synaptic strength by increasing the amount of neurotransmitter released by the sensory cells. Finally, classical conditioning, which involves pairing of touch of the siphon and electrical shock to the body, is explained by a combination of homosynaptic and heterosynaptic processes. Since animals can be trained to exhibit both short- and long-term memory, the molecular and biochemical mechanisms mediating both types of memory have been studied. Short-term memory is maintained by covalent modifications of preexisting proteins, while long-term memory requires protein synthesis, dramatic morphological changes in existing synapses, and formation of new synapses.

Ecological Context of the Withdrawal Reflexes

Although traditionally the gill and siphon withdrawal reflexes have been considered defensive, some authors question this assertion. The rationale behind this is that the siphon is used by sea hares to direct the defensive secretions at an attacking predator, and thus it would be advantageous not to withdraw it when facing attack. In the same vein, the gill is normally protected by the mantle and parapodia, and the withdrawal could be a way to remove it from the path of ink and opaline. In support of this hypothesis is the fact that the presence of ink inhibits siphon withdrawal.

Concluding Remarks

We have given a brief panorama of what is known on the behavioral ecology of sea hares. Since some topics are better studied than others, they have been treated here in more detail. There is still much to learn about all of these behaviors. Of particular interest are the complexity of the ink and opaline secretions and the many different ways in which they affect predators, the mechanism of propulsion during swimming, and the nature of the interplay between the different pheromones involved in mate attraction and mating.

See also: Neuroethology: What is it?.

Further Reading

Aggio JF and Derby CD (2008) Hydrogen peroxide and other components in the ink of sea hares are chemical defenses against predatory spiny lobsters acting through non-antennular chemoreceptors. *Journal of Experimental Marine Biology and Ecology* 363: 28–34.

Carefoot TH (1987) *Aplysia*: Its biology and ecology. *Oceanography and Marine Biology: An Annual Review* 25: 167–284.

Cummins SF, Nichols AE, Schein CH, and Nagle GT (2006) Newly identified water-borne protein pheromones interact with attractin to stimulate mate attraction in *Aplysia*. *Peptides* 27: 597–606.

Derby CD (2007) Escape by inking and secreting: Marine molluscs avoid predators through a rich array of chemicals and mechanisms. *Biological Bulletin* 213: 274–289.

Donovan DA, Pennings SC, and Carefoot TH (2006) Swimming in the sea hare *Aplysia brasiliana*: Cost of transport, parapodial morphometry, and swimming behavior. *Journal of Experimental Marine Biology and Ecology* 328: 76–86.

Johnson PM and Willows AOD (1999) Defense in sea hares (Gastropoda, Opisthobranchia, Anaspidea): Multiple layers of protection from egg to adult. *Marine and Freshwater Behaviour and Physiology* 32: 147–180.

Kandel ER (1979) *Behavioral Biology of Aplysia*. San Francisco: W.H. Freeman and Company.

Kandel ER (2003) The molecular biology of memory storage: A dialog between genes and synapses. In: Jörnvall H (ed.) *Nobel Lectures in Physiology or Medicine 1996–2000*, pp. 392–439. Singapore: World Scientific Publishing Company.

Kicklighter CE, Germann M, Kamio M, and Derby CD (2007) Molecular identification of alarm cues in the defensive secretions of the sea hare *Aplysia californica*. *Animal Behaviour* 74: 1481–1492.

Kriegsin AR (1977) Stages in the post-hatching development of *Aplysia californica*. *Journal of Experimental Zoology* 199: 275–288.

Nolen TG and Johnson PM (2001) Defensive inking in *Aplysia* spp.: Multiple episodes of ink secretion and the adaptive use of a limited chemical resource. *Journal of Experimental Biology* 204: 1257–1268.

Painter SD, Cummins SF, Nichols AE, et al. (2004) Structural and functional analysis of *Aplysia* attractins, a family of water-borne protein pheromones with interspecific attractiveness. *Proceedings of the National Academy of Sciences USA* 101: 6929–6933.

Susswein AJ and Nagle GT (2004) Peptide and protein pheromones in molluscs. *Peptides* 25: 1523–1530.

Tomsic D, Pedreira ME, Romano A, Hermitte G, and Maldonado H (1998) Context-US association as a determinant of long-term habituation in the crab *Chasmagnathus*. *Animal Learning & Behavior* 26: 196–209.

Vonnemann V, Schrodl M, Klussmann-Kolb A, and Wägele H (2005) Reconstruction of the phylogeny of the Opisthobranchia (Mollusca: Gastropoda) by means of 18S and 28S rRNA gene sequences. *Journal of Molluscan Studies* 71: 113–125.

Relevant Websites

http://www.seaslugforum.net/ – Sea Slug Forum.

http://en.wikipedia.org/wiki/Sea_hare.

http://tolweb.org/ – Tree of Life Web Project. 2008. Heterobranchia. *Version 09 February 2008 (temporary)*.

Aquatic Invertebrate Endocrine Disruption

E. Zou, Nicholls State University, Thibodaux, LA, USA

Introduction

The notion of endocrine disruption was first introduced at the first World Wildlife Federation Wingspread Conference in 1991, which was convened in the light of numerous reports on the adverse effects of various environmental chemicals, including some pesticides, industrial chemicals, and pharmaceuticals, on hormonally mediated functions in various groups of vertebrates and humans. A new field of scientific inquiry, commonly referred to as endocrine disruption, was born thereafter. The original focus of endocrine disruption research was on sex steroids-regulated functions in vertebrates, since the initially documented adverse effects of anthropogenic chemicals on endocrine systems included phenomena such as demasculinization and feminization of male fish and birds, masculinization of female fish, decline in male/female sex ratio in reptile and bird populations, and health problems in the children of women who had taken diethylstilbestrol (DES), a potent synthetic estrogen. One of the mechanisms for the estrogenic effects of environmental chemicals is estrogen-mimicry. Some chemicals, such as the plasticizers bisphenol A (BPA) and certain phthalates, render their estrogenic effects through mimicking the endogenous estrogen by binding to and activating the estrogen receptor, thereby turning on estrogen-sensitive genes in spite of the apparent lack of structural similarity between environmental chemicals and natural estrogens.

Given the fact that the estrogen-like signaling is a pervasive phenomenon, even found in the legume-rhizobial bacteria system where the estrogen-like signaling, vital for the symbiotic relationship between leguminous plants and *Rhizobium* bacteria, is also prone to disruption by environmental estrogens, McLachlan proposed the notion of environmental signaling to describe the signaling effects of environmental chemicals that ultimately lead to alterations in the expression of estrogen-regulated genes. The term environmental signaling, in my opinion, has a broad connotation and should encompass any signaling effects of environmental chemicals that directly or indirectly result in alterations in gene expression. To avoid confusion, it must be noted that environmental signaling in the context of environmental endocrine disruption does not include the signaling effects of physical environmental parameters such as photoperiod and temperature. After all, the field of vertebrate endocrine disruption has expanded to also include the disruption of thyroid hormone-regulated functions by the signaling of environmental chemicals. Environmental chemicals are also known to induce behavioral changes through disrupting chemical communication, which plays an important role in the mating and social interactions of many animals, or through altering endocrine and neuroendocrine processes. Therefore, the signaling effects of environmental chemicals that result in behavioral changes should also pertain to the realm of environmental signaling.

Invertebrates constitute over 95% of known animal species. Because of the ecological importance of these animals and the commercial value of many invertebrate species, research attention is increasingly being paid to environmental signaling in these animals, especially among aquatic invertebrates. However, owing to the deficiency in the understanding of endocrine, neuroendocrine, and chemical communication mechanisms of many aquatic invertebrates, much of environmental signaling research in aquatic invertebrates thus far is merely at the stage of defining the disrupting effects of environmental chemicals on endocrine functions, such as growth, development, and reproduction, as well as on behaviors, such as mating behavior, predator avoidance, and prey-capture behaviors, and swimming and burrowing behaviors. Only in the cases of gastropod imposex, a sexual abnormality which causes male sex organs to grow in a genetic female due to exposure to organotin, is the research currently at the level of unraveling the underlying molecular mechanisms. The material presented here summarizes the current status of research on endocrine and behavioral disruption owing to environmental signaling in selected aquatic invertebrates, including cnidarians, rotifers, molluscs, crustaceans, and echinoderms. Insects are omitted in view of the fact that most insects are terrestrial and endocrine disruption is commonly used as a mechanism for insecticide development.

Cnidarians

Endocrinology

Cnidarians, including hydrozoans, anthozoans, and scyphozoans, represent the earliest invertebrates that possess a nervous system. The cells that make up such a primitive nervous system are multifunctional, with the combined properties of sensory, motor, and interneurons. Strictly speaking, the term 'endocrinology' is not applicable to cnidarians since these primitive animals have neither a true circulation system nor defined endocrine glands. Information on endocrine-like signaling in cnidarians is sporadic. Although cnidarians lack a defined endocrine gland, certain cells in the nervous system are known to produce neuropeptides, which are implicated in the regulation of metamorphosis and peristaltic contractions. Cnidarians often have a life cycle consisting of planula larva, polyp, and medusa stages. A neuropeptide, pGlu–Gln–Pro–Gly–Leu–Trp–NH$_2$, induces metamorphosis of the marine hydroid from a hydroid planula larva to a polyp. Gonadotropin-releasing hormone (GnRH)-like peptides have been discovered in endodermal neurons of the sea pansy and the sea anemone. These GnRH-like factors inhibit the amplitude and frequencies of peristaltic contractions in the sea pansy, suggesting that these neuropeptides may play a role in the modulation of neuromuscular transmission. Besides, the cnidarian nervous system synthesizes serotonin, which appears to function as a signaling molecule, being capable of stimulating rhythmic muscular contraction and spawning, and inducing metamorphosis of hydrozoan larvae.

The vertebrate steroids estrone and 17β-estradiol (E$_2$) have been detected in the tissue of the scleractinian coral, *Montipora verrucosa*. Tissue concentrations of estrone and E$_2$ in *M. verrucosa* reportedly vary during the year, with peaks for estrone occurring in April, a time of rapid gamete growth, and in early July, prior to spawning, while E$_2$ peaks preceded peaks for estrone, suggesting that estrogens may play a role in the regulation of coral gametogenesis and spawning. It is not clear whether these steroids are synthesized in the coral, derived from dietary sources or sequestered from the overlying water. While steroid signaling in cnidarians has not been confirmed yet, several nuclear receptors, including homologs to chicken ovalbumin upstream promoter transcription factor (COUP-TF) and retinoid X receptor (RXR), have been identified. The natural ligands for these receptors are still unknown.

Disruption of Reproduction, Regeneration, Metamorphosis, and Symbiosis

Although the pathway of steroid signaling has yet to be delineated, the effects of estrogenic compounds in cnidarians have been documented. The treatment of coral with exogenous estradiol prior to spawning causes the release of fewer egg–sperm bundles. Synthetic estrogen 17α-ethinylestradiol can reduce both the number of testes and the time for which the sperm remains active. This synthetic estrogen can also decrease the number of oocytes and shorten the time the oocyte is attached to the polyp. Besides, exposure to 17α-ethinylestradiol or BPA results in tentacle damage and regeneration inhibition in the hydroid. Additionally, high doses of BPA elicit adverse effects on sexual and asexual reproduction in a

hydroid, which are believed to be due to the general toxicity of BPA, rather than its estrogenic signaling.

Organic pollutants, including lower alcohols, aliphatic and aromatic hydrocarbons, thiophenes, tributyltin (TBT) and crude oil, inhibit induced metamorphosis (the transformation of the freely swimming larvae into the polyps) in the colonial hydroid, with more hydrophobic agents exhibiting greater inhibitory potency. Since metamorphosis in the hydroid appears to be regulated by neuroendocrine factors, it is likely that these organic agents could render their antagonizing effects through interfering with the signaling of neuroendocrine factors.

Some anthozoans have a symbiotic relationship with protists (algae). The signaling between the anthozoan and its endosymbiont is crucial for both organisms. Exposure of the sea anemone, *Aiptasia pallida*, to the antifouling agent, TBT, not only disrupts symbiosis, via expulsion of the symbiosis partner zooxanthellae, but also produces other effects, such as increased mucus secretion, thickening of the pedal disc ectoderm, and a decrease in the number of undischarged nematocysts. However, it is not known whether this disruption of symbiosis in *A. pallida* by TBT is due to interference with the signaling between the host and its symbiotic partner.

Rotifers

Endocrinology

Rotifers are a group of free-living, planktonic pseudocoelomates characterized by possessing a wheel of cilia called a corona at the anterior end. There is an alternation of parthenogenic and sexual reproduction in the life cycle of rotifers. At the beginning of the growing season, diploid parthenogenic females hatch out from the resting eggs of the previous season. Parthenogenic females reproduce unisexually by laying diploid eggs, which develop into females. Upon receiving appropriate environmental cues, rotifers switch the mode of reproduction from parthenogenic reproduction to sexual reproduction, wherein diploid sexual females are produced. Diploid sexual females then produce haploid eggs through meiosis, which develop into haploid males or resting eggs if fertilized by males.

The minute size of rotifers makes it impossible to use the conventional injection/implantation techniques to study rotifer endocrinology. Much of the available information on rotifer reproduction control is obtained from aqueous exposure experiments. Exposure of the marine rotifer, *Brachionus plicatilis*, to several vertebrate and invertebrate hormones and neurotransmitters results in changes in reproduction. γ-Aminobutyric acid (GABA), growth hormone (GH), human chorionic gonadotropin (HCG), and 5-hydroxytryptamine (5-HT) significantly increase population growth, whereas E$_2$ decreases population growth. Juvenile hormone (JH), 20-hydroxyecdysone or triiodothyronine (T3) has no effect on population growth. An increase or decrease in the population growth of the rotifer, *B. plicatilis*, treated with an exogenous hormone, is obviously mediated by stimulatory or inhibitory actions of such an agent on parthenogenic reproduction. JH, 5-HT, GH, E_2, GABA, and 20-hydroxyecdysone stimulate the production of sexual females, whereas neither T3 nor HCG has an effect on the production of sexual females. Furthermore, treatment of maternal rotifers with JH enhances the production of sexual females in the second and third generations, but has no effect on the production of sexual females in the first generation. Of these agents, GABA and 5-HT are endogenous to rotifers, suggesting that these two factors may be involved in the regulation of cyclic reproduction of rotifers. Recently, the sexual reproduction-inducing signal has been identified. It is a 39 kDa protein released from the parthenogenic females under the crowding condition. This signaling molecule is believed not to interact directly with the oocyte before extrusion, but binds to a receptor on the mother, thereby initiating an internal signal that is transduced to the oocyte. Now, the question is how the simple signaling molecules, such as GABA and 5-HT, and the signaling protein coordinate to regulate the initiation of sexual reproduction in rotifers. Presumably, environmental cues, such as crowding, trigger the release of GABA and 5-HT from the nervous system, which directly or indirectly result in synthesis and release of this signaling protein that then ultimately leads to the differentiation of haploid males and the production of sexual eggs.

A sex pheromone has also been identified in the marine rotifer, *B. plicatilis*. A glycoprotein, 29 kDa in molecular weight, bound to the body surface of the female rotifer, acts as a contact-mating pheromone. Recognition of this glycoprotein by the chemosensory receptors in the male corona initiates mating behavior.

Disruption of Reproduction

Endocrine and pheromone signaling plays a critical role in the life cycle of rotifers. Although the details of the signaling pathways have yet to be elucidated, these signaling processes are likely susceptible to perturbations by environmental chemicals, thereby resulting in disruption of reproduction. Pentachlorophenate (PCP) and chlorpyrifos do not affect parthenogenic reproduction of the rotifer *Brachionus calyciflorus*, but reduce sexual reproduction. This reduction in sexual reproduction is due to the inhibition of sexual female production. Chlorpyrifos also inhibits the production of males by sexual females. Additionally, the pesticides, diazinon, fenitrothion, isoprothiolane, and methoprene, trigger resting egg production in *B. plicatilis*, but reduce the hatchability of resting eggs, the most sensitive parameter for detecting the effects of pesticides in rotifers. It is not known whether these

effects arise from the disruption of hormonally regulated or pheromone-mediated processes.

For the purpose of exploring the freshwater rotifer, *B. calyciflorus*, to be used as a screen for endocrine disruptors, the effects of a suite of hormonally active compounds on the reproduction of rotifers have been assessed. Flutamide (an androgen antagonist), testosterone, and nonylphenol significantly inhibit fertilization of sexual females by decreasing the percentage of fertilized sexual females. Besides, testosterone reduces the proportion of unfertilized sexual females, while methoprene (JH agonist), precocene (JH antagonist), flutamide, or nonylphenol, elicits no effect on the percentage of unfertilized females. Since fertilization and resting egg production in rotifers are likely to be regulated by endocrine mechanisms, it has been suggested that rotifer reproduction assay may be an effective screen for the detection of endocrine disrupting activities in aquatic invertebrates. In a similar study aimed to obtain suitable endpoints for the screening of endocrine disruptors, Radix and coworkers investigated the effects of endocrine disruptors and steroid hormones on the reproduction and the ratio of sexual/parthenogenic females of the rotifer, *B. calyciflorus*. Nonylphenol, ethinylestradiol, and testosterone can decrease the intrinsic rate, an indicator of reproduction, of population increase of *B. calyciflorus* in a dose-response manner. Because of a clear dose–response relationship, these investigators concluded that the intrinsic rate of population increase is a suitable endpoint for the screening of endocrine disruptors. However, it is not specified as to whether this endpoint is specifically for the screening of endocrine disruption in rotifers or for the detection of endocrine disrupting activities in aquatic invertebrates in general. In my opinion, any idea of using rotifer reproduction as a screen for vertebrate endocrine disruptors is premature until there is concrete evidence for the existence of steroid signaling in rotifers. A recent report shows that progesterone, a vertebrate steroid, appears to be active in the rotifer, *Brachionus manjavacas*. However, the progesterone effects, including enhanced resting egg production, increased male copulating behavior, and increased resting egg hatchability, are elicited at ppm (parts per million) concentrations of progesterone, and therefore could well simply arise from the general toxicity of this vertebrate steroid.

Molluscs

Endocrinology

Mollusca consists of a diverse group of animals, including polyplacophores, gastropods, bivalves, and cephalopods. Because of the disparities in morphology and life history of various groups of molluscs, there are great variations in endocrine mechanisms among these discrete mollusc groups. Since gastropods and bivalves have received most attention in studies of environmental endocrine disruption, only the endocrinology of gastropods and bivalves is presented.

Neurosecretory centers of gastropods and bivalves are located in the central nervous system, which consists of four ganglia, namely, the cerebral, pleural, pedal, and abdominal ganglia. Neuropeptides play an important role in the regulation of growth and reproduction. The egg-laying hormone (ELH), a neuropeptide secreted from a bag cell cluster in the abdominal ganglion, regulates gonad maturation and ovulation in *Aplysia*, while caudodorsal cell hormone (CDCH), produced from the cerebral ganglion, controls the development of female accessory sex organs, gonad maturation, and ovulation in *Lymnaea*. Penis morphogenic factor (PMF), produced from the pedal ganglion, induces differentiation of the penis, while a lysis neurohormone from the pedal ganglion controls the regression of the penis after the mating season is over. The chemical nature of the PMF has yet to be determined. Ala–Pro–Gly–Trp–amide (APGWamide), a neuropeptide produced in the pedal ganglion, appears to be a good candidate of the PMF. APGWamide is known to regulate male sexual activities in *Lymnaea stagnalis* and treatment with APGWamide induces the development of the penis in the mud snail, *Ilyanassa obsoleta*.

Whether steroids are hormonally active in gastropods and bivalves is still not resolved. There have been reports suggesting that sex steroids are active in gastropods and bivalves. The vertebrate sex steroids E_2, testosterone, and progesterone have been found to be endogenous to gastropods and bivalves. Enzymatic systems responsible for the synthesis and metabolism of sex steroids also exist in gastropods and molluscs. Changes in the levels of sex steroid correlate with the process of sexual maturation in several bivalve species, suggesting a role for steroids in the control of reproduction. Administration of the exogenous steroids can induce responses concerning sexual development and reproduction in bivalves. E_2 injection to the oyster, *Crassostrea gigas*, in the early stages of sexual development reverses sex from male to female. The administration of testosterone induces the development of male sex organs in gastropods. The injection of E_2, testosterone, or progesterone stimulates oogenesis and spermatogenesis, while E_2 exposure triggers vitellogenesis and serotonin-induced egg release. However, the results of recent molecular studies do not support the notion that E_2 is active in the mussel, *Mytilus edulis*. Using the expression of estrogen receptor and vitellogenin genes as endpoints, Puinean and coworkers did not observe any significant induction of the mRNA of either gene in the mussel exposed to aqueous E_2.

It is generally known that the signalling of steroid hormones is mediated by receptors, whether they are located intracellularly or on plasma membranes. Thus far, molluscan estrogen receptors have been characterized in three gastropods, including *Aplysia californica*, *Thais clavigera*, and *Marisa cornuarietis*, and one bivalve, *C. gigas*. Additionally, two partial sequences of estrogen receptor genes of the mussel, *M. edulis*, have also been described. But the molluscan estrogen receptors, all constitutively active when transfected into mammalian cells, are not responsive to and have no affinity for E_2. Therefore, it is unlikely that the molluscan estrogen receptors can mediate intracellular estrogen signaling leading to transcriptional activation. It has been theorized, though, that estrogen may act as a nongenomic signal inducer in molluscs and that estrogen signaling could be mediated by the membrane-bound receptor. The inconsistencies in the data regarding the effects of sex steroids in gastropods and bivalves warrant further research for ultimate clarification of the hormonal roles of steroids in these animals, considering the fact that these animals are increasingly attracting research attention due to the disruption of sexual development and reproduction by environmental chemicals.

Disruption of Sexual Development and Reproduction

The best defined case of invertebrate endocrine disruption is the imposex of neogastropods. Neogastropods, a group of prosobranchs, are normally gonochoristic. Imposex is a type of sexual abnormality wherein male sex organs, such as the penis and vas deferens, develop on the genetic female neogastropods as a result of exposure to low levels (ppt) of organotin, the active ingredient of antifouling paints. Since the male organs are superimposed upon the female, this kind of sexual anomaly has been termed the imposex. The imposex phenomenon was first reported by Blaber in 1970 in the snail, *Nucella lapillus*. This sexual anomaly has now been described in over 140 neogastropod species from locations worldwide. In some species, such as *I. obsoleta* and *Nassarius reticulates*, the development of male sex organs does not appear to cause reproductive problems to the affected female, whereas in other species, such as *Nassarius lapillus* and *Ocenebra erinacea*, imposition of male sex organs can cause sterility of the female snail. The sterility of the affected females is caused by the blockage of the genital pore by convolutions of the vas deferens. Occurrences of imposex have been linked to the extinction of gastropods in some areas of high shipping and boating activities.

As far as the environmental etiology of this sexual aberration is concerned, there are three theories, each backed up by experimental evidence. The first one, known as the testosterone theory, attributes the masculinization of the female snail after exposure to organotin to an elevation in testosterone titer in the snail. This theory is based on the fact that exposure of the female snail to organotin at an environmentally realistic concentration results in the occurrence of a penis, concurrent with an elevation in testosterone levels, that the exogenous testosterone treatment of the female snail induces imposex, and that both antiandrogen and estrogen are capable of negating the imposex-inducing effects of organotin. An increase in the testosterone level in the female snail after exposure to organotin is caused by inhibition by the organotin of aromatase-mediated metabolism of testosterone to E_2, and by the suppression of testosterone elimination. In spite of the evidence for the testosterone theory, Oberdörster and McClellan-Green argued that increased testosterone titers are not the cause for imposex induction because of the fact that organotin and testosterone induce the penis and vas deferens in as little as 2 weeks, with maximum induction occurring after 1 month, and that testosterone titers do not become elevated until 2 months after organotin exposure. Injection of the neuropeptide APGWamide, a possible candidate for the PMF, into the female mud snail, *I. obsoleta*, induces imposex development, which has led these investigators to propose an alternative theory, or the neuropeptide theory, for imposex etiology. According to this theory, organotin can cause the abnormal release of APGWamide from neurosecretory cells, triggering the development of the male sex organs, which then release androgens, possibly testosterone, to maintain the male sex organs and spermatogenesis. Whether the actions of testosterone in gastropods are mediated by the receptor, as in the case with vertebrate systems, is still unknown. To date, no such receptor has been discovered yet in any molluscs. Both the testosterone and neuropeptide theories are somewhat discredited by the fact that the penis length induced by the exogenous testosterone or APGWamide is much shorter than that of the female snails inhabiting organotin-contaminated environments or of female snails exposed to organotin in the laboratory. Recently, the third theory, or the RXR theory, has been proposed to explain the imposex phenomenon. Given the data that organotin has affinity for the snail RXR and that 9-cis retinoic acid (9-cis RA), a natural ligand for RXR, induces imposex in the snail, *T. clavigera*, Nishikawa and his colleagues proposed that RXR mediates the signaling of organotin that leads to imposex development. The latest evidence is supportive of this RXR theory. The imposex-inducing effect of 9-cis RA and its affinity with the snail RXR have been corroborated in another neogastropod, *N. lapillus*. It must be noted that the size of 9-cis RA-induced penis on the female snail is similar to that of the penis induced by organotin, but much bigger than the size of the imposex penis induced by either the exogenous testosterone or APGWamide peptide.

Neogastropods are not the only molluscs adversely impacted by organotin. Both the giant abalone, *Haliotis madaka*, and Roe's abalone, *Haliotis roei*, inhabiting organotin-contaminated waters are masculinized with an ovo-testis.

The ovaries of the affected females have a small amount of testicular tissue in the process of spermatogenesis. Besides, the male/female sex is increased in the soft-shell clam, *Mya arenaria*, living in the tributyltin-impacted environment.

Environmental contaminants other than organotin have also been found to cause sexual abnormalities in gastropods and bivalves. Treatment of the freshwater snail, *M. cornuarietis*, with environmental estrogens, BPA and octylphenol, induces a type of sexual abnormality called the 'superfemale,' where affected females are characterized by the formation of additional female organs, for example, a second vagina with a vaginal opening to the mantle cavity and/or an enlargement of the pallial accessory sex glands. Oviduct malformation and an enhancement of spawning mass production were also observed in superfemales. Similar sexual abnormalities were also observed in the marine snail, *N. lapillus*, treated with BPA or octylphenol, where accessory pallial sex glands were enlarged and oocyte production increased. The 'superfemale' phenomenon also occurs in the freshwater mud snail, *Potamopyrgus antipodarum*, when exposed to estrogenic compounds, such as BPA, octylphenol, and nonylphenol. The 'superfemale' phenomenon has been linked to the estrogenicity of BPA and alkyl phenolic compounds. BPA reportedly acts as an estrogen receptor agonist since the super-feminization effect of BPA can be nullified by antiestrogens, such as tamoxifen and ICI 182,780, and estrogen-specific binding sites appear to exist in *M. cornuarietis*.

Disruption of Behaviors

Besides affecting endocrine signaling, organotin contamination also appears to be capable of disrupting pheromone-mediated communication in gastropods. The attraction between sexually active male and female *I. obsoleta* is mediated by sex pheromones since it has been demonstrated in the laboratory that male snails are attracted to the water conditioned by sexually active females, and female snails attracted to the water conditioned by sexually active males. The chemical nature of the presumed snail pheromones remains unknown. In assays using snails from the high imposex site with severe organotin contamination, Straw and Rittschof found that neither normal females nor imposex females respond differentially to waters containing male pheromone and those with female pheromone, suggesting that the chemical communication ability of organotin-impacted snails is compromised. These investigators also noted that there is apparent absence of breeding activity and egg deposition among morphologically normal snails at the high imposex site. The laboratory behavioral results, together with the field observation, imply that behavioral disruption may be enough to compromise snail sexual reproduction and that the sex attraction behavior could be a sensitive measure of the toxic effects of organotin on snail reproduction.

In bivalves, because of their limited locomotion capability, burrowing behavior is generally believed to be adaptive and allows the animal to escape predation. Medetomidine, a novel antifouling agent, decreases the burrowing responses of the burrowing bivalve, *Abra nitida*. The presence of naphthalene, a petroleum contaminant, can not only stimulate mucus secretion but also inhibit the burrowing behavior of the tropical acrid clam, *Anadara granosa*, inhabiting an intertidal marine environment. The failure to burrow in the presence of naphthalene could make the clams more susceptible to predation.

Crustaceans

Endocrinology

Crustaceans are one of the larger animal groups. There are over 66 000 known crustacean species, all aquatic, except some exclusively terrestrial isopods. Since most studies on crustacean endocrine disruption are concerned with sexual development, reproduction, and molting, only the endocrinology for these processes is presented.

Sexual development and reproduction

Crustaceans consist of diverse animal groups. There are two types of reproduction employed by these animals. Primitive crustaceans, such as cladocerans, artemians, notostracans, and most ostracods, have both parthenogenic and sexual reproduction in their life cycles, while most copepods and all cirripedes and malacostracans undergo only sexual reproduction.

For crustaceans that are capable of both parthenogenic and sexual reproduction, the mode of reproduction depends upon environmental conditions. For instance, cladocerans under favorable conditions, such as right temperature, enough food, long day length and no crowding, undergo parthenogenic reproduction. However, when environmental conditions turn unfavorable, that is, extreme temperature, food scarcity, short day length, and crowding, cladocerans switch the mode of reproduction to a sexual one, with just males being produced initially. After receiving appropriate environmental cues, the parthenogenic female deposits male-producing eggs into the brood chamber. The crustacean terpenoid hormone methyl farnesoate (MF), the unepoxidated form of insect juvenile hormone III (JH III), has been shown to be the determinant for male sex differentiation in *Daphnia*. Presumably, the reception of environmental cues leads to elevation in MF level in *Daphnia*, which results in the ovary to lay male-producing eggs. The critical period in sex determination in *Daphnia* is about 1 h before egg deposition, suggesting that whether male- or female-producing eggs are laid depends on the environmental cues received

during a very short time interval, and that the MF must complete its action rather quickly. The short time interval for *Daphnia* sex determination may be the reason that mixed broods containing both male and female daphnids have been observed by several investigators.

Much of the knowledge on endocrine control of sexual development and reproduction has been obtained from malacostracan crustaceans because of their large body sizes. There is evidence suggesting the existence of a genetic sex-determining mechanism in malacostracans. In some species, such as the isopod, *Asellus aquaticus*, males possess a heteromorphic sex chromosome pair, while in other species, such as the prawn, *Macrobrachium rosenbergii*, and the crayfish, *Cherax quadricarinatus*, males are homogametic. In malacostracan crustaceans, the development and maintenance of male secondary sexual characteristics is solely controlled by the androgenic gland hormone (AGH), which is a protein. The AGH of the isopod, *Armadillidium vulgare*, is a 11–13 kDa peptide, while a similar peptide, 17–18 kDa in molecular weight, has been reported for the prawn, *M. rosenbergii*. The activity of the androgenic gland is regulated by the gonad-inhibiting hormone (GIH) produced in the X-organ-sinus gland complex and the gonad-stimulating hormone (GSH) produced in the brain and in the thoracic ganglia, both of which are peptides. The rudiments of androgenic glands have been found in second instar juveniles of both sexes of the amphipods, *Orchestia cavimana* and *O. gammarella*, but only in males do these rudiments develop further. The development of female secondary sexual characteristics, such as the ovigerous setae and the brood pouch, is controlled by the ovarian hormone produced by the ovary. The chemical nature of this hormone is still not characterized.

Both GIH and GSH regulate gonadal maturation. In females, GIH and GSH act directly on the ovary, which then secretes the ovarian hormone, while in males, GIH and GSH act on the androgenic gland. As mentioned above, the AGH not only directly controls the secondary characteristics of males but it also controls the development of the male reproductive system.

Additionally, there is evidence that MF, produced by the mandibular organ, enhances ovarian maturation and stimulates the testicular development in malacostracan crustaceans. MF also controls male morphogenesis in the crayfish, *Procambarus clarkii*. However, the receptor for MF or its insect counterpart, JH III, has not been definitively recognized. There is evidence pointing to crustacean RXR being a candidate for MF receptor.

Molting

For growth to occur, crustaceans must periodically molt their rigid, confining exoskeleton. Molting in crustaceans is regulated by a multihormonal system, but is under the immediate control of steroid hormones called ecdysteroids.

In decapods, ecdysteroids are produced in the Y-organ whose activity is held in acquiescence during the intermolt stage by the molt-inhibiting hormone (MIH), which is a peptide, from the X-organ-sinus gland complex. When the animal enters the premolt stage, this inhibition of Y-organ activity by the MIH stops and ecdysteroidogenesis in the Y-organ intensifies. The ecdysteroid titer in the hemolymph is, therefore, elevated. However, this generally accepted model regarding the MIH-ecdysteroids dynamics during the molting cycle may need revision in light of the recent data obtained with the crab, *Carcinus maenas*. MIH transcription and translation in *C. maenas* are somewhat invariant during the molting cycle of *C. maenas*, and the most important step in controlling ecdysteroid synthesis appears to lie in intracellular signaling pathways within the Y-organ. Ecdysteroids regulate gene activities at the transcriptional level by interacting with the ecdysteroid receptor (EcR), which then heterodimerize with crustacean RXR. This EcR/crustacean RXR dimer binds to the DNA response elements of target genes. Besides, the MF may also be involved in molting control, since injection of mandibular organ extract accelerates molting in crustaceans. The molt-promoting effects of the MF could be mediated by the crustacean RXR, a putative MF receptor.

Endocrine Disruption

Disruption of sexual development and reproduction

Concerns were expressed over the possibility that the same chemical agents capable of disturbing the male development of vertebrates can also interfere with the male differentiation of cladocerans. The proportions of male *Daphnia* in Lake Mendota, Wisconsin, have declined rather drastically over several decades. The decrease in the male/female sex ratio was hypothesized to be caused by some of the same estrogenic agents capable of disrupting the male development of vertebrates. Such field data have generated several investigations to examine whether estrogenic agents can interfere with male differentiation in *Daphnia*. DES, endosulfan, E_2, TBT, dichlorodiphenyltrichloroethane (DDT), or methoxychlor has no effects on male differentiation in *Daphnia*, whereas nonylphenol promotes male differentiation only at a low food concentration. The absence of effects of various estrogenic chemicals on sex differentiation of *Daphnia*, especially the lack of response to the natural estrogen E_2 and the potent synthetic estrogen DES, strongly suggests that the strength of a chemical's estrogenicity is not a significant determining factor for the shift of sex determination in *Daphnia*. This is not surprising, since the determinant for male sex differentiation in *Daphnia* is the crustacean hormone MF. MF and its analogs, such as fenoxycarb and pyriproxyfen, are all capable of inducing male differentiation in *Daphnia*. The decline in the proportions of male *Daphnia* in Lake Mendota, Wisconsin, over several decades must be attributed to the disruption of MF signaling in *Daphnia*, instead of resulting from the estrogenicity of environmental chemicals in the lake.

Abnormal sexual characteristics of copepods have also attracted much attention. Harpacticoid copepods normally have separate sexes. However, intersexuality, an occurrence of an individual carrying sexual characteristics of both sexes, has been recorded in benthic harpacticoids in the vicinity of a large sewage discharge near Edinburgh in the United Kingdom. No evidence has been found for a correlation between intersex frequency and proximity to the sewage discharge. E_2, estrone, or 17α-ethynylestradiol has no effect on the female/male sex ratio in the copepod in laboratory studies. The field evidence, together with laboratory data, suggests that estrogenicity is not a factor behind the intersex problem in copepods and that the endocrine control for sexual development in copepods is different from that for vertebrates.

Malacostracans are mostly gonochoristic, but intersexuality is common among certain parasitic isopods, caridean shrimps, and anomurans. Among these malacostracan crustaceans, two types of hermaphroditism have been recognized. One is called sequential protandrous hermaphroditism, in which individuals first mature as sperm-producing males, and then with increasing size, switch sex to become breeding females, and the other is called simultaneous hermaphroditism where an intersex individual can function as either a male or a female at any given time.

Neither of these two types of hermaphroditism has been found in brachyurans. However, several forms of sexual abnormalities have been reported for brachyuran decapods. The fiddler crab, *Uca pugilator*, exhibits striking sexual dimorphism, with males characterized by a pair of asymmetrical chelipeds and a narrow abdomen, and females having two small, identical claws and a broad abdomen that covers the entire sternum. Four different forms of sexual aberrations have recently been discovered among the crabs purchased from the Gulf Specimen Marine Laboratories, Inc. of Panacea, Florida (**Figure 1**). In one form, the sexually abnormal crabs had a wide abdomen, female-type pleopods, and a pair of asymmetrical chelipeds. In another form, the specimen possessed a pair of symmetrical, enlarged claws and a female abdomen, along with the female-type pleopods. A specimen with reduced asymmetry in chelipeds, an intermediate abdomen, male-type pleopods (gonopods), and female-type apertures (female gonopores) was discovered. Representing the fourth form is a specimen possessing a pair of uneven claws with reduced asymmetry, female gonopores, and an intermediate abdomen with both gonopods and female-type pleopods. Whether an environmental etiology underlies these sexual anomalies in *U. pugilator* is unknown. Sexually abnormal crabs, with the external

Normal male   Normal female

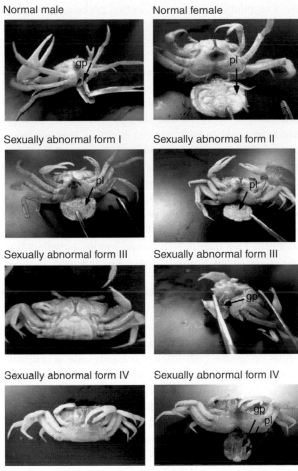

Sexually abnormal form I   Sexually abnormal form II

Sexually abnormal form III   Sexually abnormal form III

Sexually abnormal form IV   Sexually abnormal form IV

Figure 1   Forms of sexual abnormalities in the fiddler crab, *Uca pugilator*. gp, gonopod; pl, female pleopod.

appearance of a complete male but having female genital openings, were at frequencies of 8–32% among the freshwater crab, *Geothelphusa dehaani*, from ten rivers in the Nagasaki prefecture of Japan. This form of sexual aberration is induced not at the early stages of male development but after sexual maturation, presumably, in response to long-term exposure to environmental chemicals. Male *C. maenas* inhabiting areas in the United Kingdom, known to render high estrogenic responses in vertebrates, bear morphometric abnormalities, such as reduced pleopod-length ratios and enlarged abdomen width, and have enhanced ecdysteroid levels and the presence of vitellin-like proteins in the hepatopancreas of male crabs. While the exact environmental etiology for these morphological and physiological abnormalities remains nebulous, estrogenicity does not appear to be a contributing factor since there is plenty of evidence that crustacean reproduction is not responsive to vertebrate sex steroids. The feminization observed in male *C. maenas* is most likely due to direct or indirect alterations of AGH and MF signaling by certain environmental agents that have yet to be identified.

The amphipod, *Gammarus pulex*, collected from an unpolluted reference site, had a significantly higher male/female ratio than did those living below sewage treatment works (STW). The male/female size differential was significantly reduced in *G. pulex* collected from one of the polluted sites. Additionally, the allometric relationships of the male secondary sexual characteristics, gnathopod and genital papillae, to body size for *G. pulex* living below STWs, were different from those from the reference site. Moreover, abnormal oocytes were observed in *G. pulex* collected from one of the polluted sites that is known to elicit estrogenic responses in vertebrates, while no significant difference was found in the gonadal structure of males from the two polluted sites. These changes in male reproductive morphology were interpreted as being caused by interference with androgenic hormone signaling by substances in the sewage effluents, while the low number and aberrant distribution of yolk bodies in the abnormal oocytes are likely induced by steroid mimics in the effluents capable of interacting with ecdysteroid signaling. A laboratory investigation demonstrated that continuous exposure of the amphipod, *Corophium voluntator* (age < 5 days), to nonylphenol does not significantly affect the sex ratio, but induces elongation of the second antenna, which is a male characteristic, and enhances fertility, that is, the number of juveniles per female. The stimulating effect of nonylphenol on the second antenna is likely due to nonylphenol triggering hypersecretion of the AGH that controls the development of male secondary sexual characteristics. The large second antenna would be a selective disadvantage to the males because the large antennae make the males more vulnerable to predation in the field, which could lead to a skewed sex ratio and the subsequent population decline.

The possibility that sexual and reproductive aberrations can be induced in malacostracans, particularly among decapods, by exposure to environmental pollution needs to be extensively investigated, considering the ecological and economic importance of many of the malacostracan species. Both field and laboratory studies should be done. The laboratory component needs to focus on alterations in the physiology of relevant hormonal cascades that regulate sexual development and reproduction.

Disruption of molting

Because of industrial and agricultural activities, aquatic environments are increasingly contaminated with various kinds of pollutants. Through feeding and direct uptake from water and sediments, these contaminants accumulate in crustaceans living in various aquatic environments. Because of the generally high lipophilicity and refractory chemical nature of organochlorine compounds, these organic contaminants, such as polychlorinated biphenyls (PCBs), DDTs, hexachlorocyclohexanes (HCHs), hexachlorobenzene (HCB), heptachlor epoxide, dieldrin, endosulfan, and chlordane, can readily accumulate in the fatty tissue, such as hepatopancreas, of crustaceans. Like organochlorines, polycyclic aromatic hydrocarbons (PAHs), the pollutants from the activities of petroleum and smelter industries, are also highly lipophilic and can easily accumulate in crustacean tissues.

Several laboratory studies have demonstrated that organochlorine compounds, including Aroclor 1242 (a PCB mixture), 2,4,5-trichlorobiphenyl (PCB29), endosulfan, methoxychlor, kepone and heptachlor, and aromatic hydrocarbons, including benzene and dimethylnaphthalene, inhibit molting of crustaceans. The stimulatory effects of xenobiotics on crustacean molting have also been reported. The nonsteroidal ecdysone mimic RH 5849 and the pesticide emamectin benzoate are capable of accelerating crustacean molting. Since the adverse effects of environmental contaminants on crustacean molting are not readily visible in the field, the disruption of crustacean molting by xenobiotics has been termed the invisible endocrine disruption.

As mentioned above, molting in crustaceans is regulated by a multihormonal system, but is under the immediate control of ecdysteroids. An elevation in ecdysteroid titer in the premolt stage triggers secretion of molting fluid by the epidermis. This molting fluid contains hydrolytic enzymes necessary for degradation of the chitinous exoskeleton that will be shed. In theory, any step in the cascade of endocrine events, for example, ecdysteroidogenesis, metabolism of ecdysteroids, and EcR binding, could be the target of xenobiotics. Should a chemical agent be capable of assaulting this hormonal system, the effect would be manifested at the terminal step, the activity of the exoskeleton-degrading enzymes. Chitobiase, also known as N-acetyl-β-glucosaminidase, is one of the two chitinolytic enzymes found in the molting fluid, which are required for complete digestion of exoskeletal chitin. This enzyme is a product of the gene regulated by the molting hormone. The exposure of U. pugilator to molt-inhibiting xenobiotics decreases chitobiase activity. This inhibition of epidermal chitobiase activity can at least partly account for the slowing of molting caused by the molt-inhibiting xenobiotics because chitobiase is indispensable for the degradation of exoskeletal chitin. Since chitobiase is a biomarker for the actions of the molting hormone, inhibition of this enzymatic activity by molt-inhibiting chemicals strongly suggests that exposure to these xenobiotics disturbs the Y-organ-EcR axis.

Disturbance of the Y-organ-EcR axis could arise from the actions of xenobiotics on ecdysteroid signaling in the epidermis through direct binding to the EcR or its heterodimerization partner RXR and through ligand-independent activation of these nuclear receptors. It has been shown that diethyl phthalate, an industrial chemical capable of inhibiting crustacean molting, produces antagonistic activity in an in vitro assay utilizing an ecdysteroid-responsive insect cell line. The inhibitory effects of diethyl phthalate on ecdysteroid signaling can at least partly explain the inhibition of crustacean molting by this agent.

The disturbance to the Y-organ-ecdysteroid receptor axis may also result from alterations in ecdysteroidogenesis in the Y-organ and/or changes in ecdysteroid excretion. Very little is known about the effects of environmental chemicals on ecydysteroid homeostasis. But attempts have been made to look into the effects of xenobiotics on the enzymatic system responsible for ecdysteroid metabolism. A new cytochrome P450 member, CYP45, which appears to be involved in ecdysteroid metabolism, has been found in the hepatopancreas of the lobster, *Homarus americanus*. Exposure of *H. americanus* larvae to heptachlor results in modulations in CYP45 expression and ecdysteroid levels. This shift in ecdysteroid profile is attributed to the slowing of larval molting caused by this pesticide.

Disruption of Behaviors

Mating in the shore crab, *C. maenas*, coincides with the female molt and copulation takes place shortly after the ecdysis of female crab when the exoskeleton is still soft. Sex pheromone appears to mediate the attraction of male crabs to precopula female crabs. It has been demonstrated that male *C. maenas* typically responds to an inanimate object, such as a stone tainted with the water conditioned by precopula females, by grasping the stone and testing its hardness with its claws. Male *C. maenas* collected from sites around the coast of the United Kingdom where water is known to elicit estrogenic activity in vertebrates exhibited reduced responses to sex pheromone from the precopula females in comparison to males from reference sites. The disruption of sex pheromone communication is concurrent with morphological and physiological feminization of male crabs from contaminated sites. The failure of males to display the stereotypical behavior to females, along with morphological and physiological abnormalities found among males, may be linked to exposure to endocrine disrupting chemicals. However, whether the estrogenicity of the environmental agents plays a role is unknown. Environmental chemicals could disrupt sex pheromone communication through direct suppression of pheromone reception capability of the male or via retarding male sexual development, thereby delaying male sexual readiness.

A sexually mature female amphipod is only available for mating immediately after ecdysis when the shell is flexible enough to allow the eggs to be released into the brood pouch through the gonopores. There is only a brief window of opportunity for the male to deposit sperm into the brood pouch. Some amphipod species, such as *G. pulex* and *Hyalella azteca*, exhibit a type of male precopulatory behavior called mate-guarding behavior, wherein a mature male locates a mature female prior to the female's ecdysis and guards her against other males until the female sheds her exoskeleton and can be fertilized. There are two forms of mate guarding. In some species, such as *G. pulex*, the male grasps and holds the female until she molts, while in other species, such as *Ampithoe lacertosa*, the male does not carry the female but stays in close proximity. This amphipod mate-guarding behavior appears prone to disruption by environmental contamination. A field study by a group of British scientists has shown that *G. pulex* living near STW exhibit significantly less mate-guarding behavior than those from a reference site. The mate-guarding behavior of male amphipods can be reduced by laboratory exposure to lindane, esfenvalerate (a pyrethroid insecticide), Aroclor 1254 (a PCB mixture), or bleached kraft pulp mill effluent. Whether this behavioral disruption by environmental pollution is pheromone-based awaits further investigations.

Environmental pollution can also disrupt behaviors not directly associated with reproduction, including locomotor, prey capture, and burrowing behaviors. Locomotor behavior of the mysid, *Neomysis integer*, and the brine shrimp, *Artemia salina*, has been shown to be susceptible to disruption by exposure to organophosphate pesticides, with the chemically treated animals exhibiting compromise swimming ability. It is not unexpected that organophosphates can cause disruption to the locomotor behavior of crustaceans considering that these pesticides inhibit choline esterase in the nervous system, resulting in impairment to neural signal transmission. Grass shrimps, *Palaemonetes pugio*, inhabiting an environment impacted by landfills are less capable of capturing preys than those from a relatively un-impacted site. The presence of PAHs in sediments reduces burrowing activities of the estuarine copepod, *Schizopera knabeni*.

Echinoderms

Endocrinology

Echinoderms, including sea stars, sea lilies, sea urchins, and brittle stars, are deuterostomes. Evolutionarily, echinoderms and vertebrates belong to the same lineage. It is not surprising, therefore, that vertebrate steroids are active in echinoderms. This is for three reasons. First of all, vertebrate sex steroids, such as testosterone, androstenedione, progesterone, estrone, and estradiol, are endogenous to echinoderms, and these animals possess enzymatic systems for steroid synthesis and metabolism. Second, sex steroid levels in the gonads vary during the reproductive cycle of echinoderms, with the peak levels corresponding to specific reproductive events. For example, testosterone and estradiol levels in the ovaries of *Asterias rubens* are highest at the onset of oogenesis. Third, reproductive functions of echinoderms are responsive to exogenous steroids. Dietary administration of androstenedione or estrone stimulates

gonadal growth of the male sea urchin, *Pseudocentrotus depressus*, while dietary treatment with E_2 promotes ovarian growth but inhibits oocyte growth, and testosterone stimulates oocyte growth in *Lytechinus variegates*. Although the receptors for vertebrate steroids have not been characterized, specific estrogen and androgen binding sites have been reported in echinoderms.

Neural factors also play a role in the regulation of echinoderm reproduction. A gonad-stimulating substance (GSS), secreted from the radial nerves and also called the radial nerve factor (RNF), is known to be involved in the control of maturation and spawning of oocytes in echinoderms. The GSS stimulates the follicular cells that envelop each primary oocyte to produce the maturation-inducing substance 1-methyladenine. 1-Methyladenine then acts on the primary oocyte to stimulate its maturation and eventual spawning. An extrafollicular source of the maturation-inducing substance that mediates the actions of the GSS in an echinoderm has recently been reported.

Echinoderms are well known for their striking regenerative abilities. Three classes of regulatory molecules have been proposed as possible candidates for growth-promoting factors in regeneration. These molecules are monoamines, neuropeptides, such as substance P and SALMFamides 1 and 2, and growth factor-like molecules, such as nerve growth factor, transforming growth factor beta, and basic fibroblast growth factor.

Disruption of Reproduction, Development, and Regeneration

Information on endocrine disrupting effects of environmental chemicals on echinoderms is scarce. Exposure of the sea star, *Asterias rubens*, to Clophen A50, a PCB mixture, reduces ovarian growth and impairs the quality of the offspring. These adverse effects of the PCB mixture are related to a compromised ability of steroid synthesis in the pyloric caeca and gonads, since treatment with Clophen A50 inhibits steroid synthesis rates in the pyloric caeca and gonads of *A. rubens*. Treatment of the adult sea urchins, *Anthocidaris crassispina*, with phenol lowers sperm quality and reproductive success. While reproductive consequences are unknown, exposure of the crinoid, *Antedon mediterranea*, to triphenyltin (TPT) or *p,p'*-DDE increases the testosterone level and triphenyltin exposure decreases the estradiol level. TBT, TPT, and PCBs can lead to reduced fertilization success and development malformations in the sea urchin, *Paracentrotus lividus*, while 4-octylphenol, BPA, and TBT stunt the development of sea urchin embryos. Additionally, BPA suppresses juvenile growth in sea urchins.

Regeneration in echinoderms can be adversely affected by PCBs. Aroclor 1260, a PCB mixture, causes abnormal arm growth, characterized by an accelerated growth of the regenerate, massive cell migration/proliferation, hypertrophic development of the coelomic canals, and extensive rearrangement of differentiated tissues of the stump, in *A. mediterranea*, whose three arms were amputated at the level of the autonomy plane. Whether the actions of Aroclor 1260 are mediated by steroid hormones or the growth-promoting factors remains unknown. TPT and fenarimol, a chlorinated pesticide, adversely affect regenerative growth in *A. mediterranea*, and induce malformations in the internal anatomy and histopathological anomalies involving cell proliferation, migration, and differentiation which are likely regulated by steroid hormones. Interestingly, although TPT and fenarimol elicit the same morphological and histological alterations, the underlying biochemical mechanisms may be quite different since TPT elevates the testosterone level while fenarimol enhances estradiol concentration.

Disruption of Behaviors

The presence of sex pheromones that are produced in the ovary and induce the spawning of conspecific males has been reported in several echinoderms. But the information on the effects of environmental pollution on pheromone-mediated communication is nonexistent.

Asteroid echinoderms are foraging predators in the marine environment. Asteroids locate their prey through chemoreception, being capable of detecting the source of prey odor with chemoreceptors concentrated in terminal podia. Laboratory tests have shown that petroleum contaminants can disrupt chemoreception in asteroids, thereby impairing the ability of these animals to capture preys.

Perspectives

Invertebrates, which make up the vast majority of known animal species, have, in recent years, attracted much research attention to the possibility that environmental chemicals capable of perturbing the endocrine functions of vertebrates can also disrupt endocrine signaling or endocrine-like signaling processes in invertebrates. Although it is generally believed that growth, development, metamorphosis, and reproduction in invertebrates are regulated by hormones or hormone-like factors, our knowledge of the endocrinology of these diverse organisms remains fragmental or even nonexistent for many invertebrate groups. This deficiency in the understanding of invertebrate endocrinology has impeded the efforts to define the disrupting effects of environmental chemicals on signaling processes that underlie growth, development, metamorphosis, and reproduction of aquatic invertebrates. Among the invertebrates covered in this article,

the understanding of crustacean endocrinology is most advanced. The hormonal control of molting and the reproduction of malacostracan crustaceans is well understood, but little is known about the endocrine mechanisms for crustacean metamorphosis, a process that is possibly vulnerable to environmental pollution. The molecular aspects of the actions of crustacean hormones involved in the regulation of molting and reproduction still need to be elucidated. Information on the endocrinology of small-sized crustaceans, such as copepods and cladocerans, is scarce. For future investigations, more efforts should be directed toward the investigation of basic endocrinology of various invertebrate groups.

Based on the available information on invertebrate endocrinology, there appears to be only limited overlap between the endocrinology of invertebrates and that of vertebrates. First, vertebrate sex steroids appear to play a hormonal role in the regulation of echinoderm reproduction. This is not unexpected, since echinoderms and vertebrates belong to the same evolutionary lineage, the deuterostomes. Second, the RXR, an ancient nuclear receptor, appears to mediate intracellular hormonal signaling in both invertebrates and vertebrates. Other than these two commonalities, profound differences exist between the endocrine mechanisms of nondeuterostome invertebrates and vertebrates. Solid evidence that vertebrate sex steroids play a hormonal role in nondeuterostome invertebrates, is still lacking. Although the orthologs of vertebrate estrogen receptors have been found in molluscs, none of these molluscan estrogen receptors are responsive to the natural estrogen. Therefore, caution must be exercised when extrapolating to nondeuterostome invertebrates the results of endocrine disruption studies with vertebrates. Conversely, it is inappropriate to suggest that an endocrine-disrupting effect in a nondeuterostome invertebrate can be used as a marker to monitor environmental endocrine disruption in vertebrates.

Pheromone-mediated chemical communication plays an important role in the mating behavior of many invertebrates. The presence of sex pheromones has been evidenced in several invertebrate groups. Available data suggest that pheromone-mediated signaling is prone to disruption by environmental pollution, leading to reduced reproductive success. Environmental chemicals could disrupt pheromone signaling through inhibiting pheromone release or blocking pheromone reception. Pheromone signaling disruption could also be elicited by adversely impacting endocrine processes, thereby delaying the recipient's physiological readiness. This latter effect would link endocrine disruption to behavioral disruption. Future studies should be directed toward chemical characterization of invertebrate pheromones, along with their receptors, and delineation of action pathways of environmental chemicals.

See also: Invertebrate Hormones and Behavior; Vertebrate Endocrine Disruption.

Further Reading

Blaber SJM (1970) The occurrence of a penis-like outgrowth behind the right tentacle in spent females of *Nucella lapillus* (L.). *Proceedings of the Malacological Society of London* 39: 231–233.

Crews D and McLachlan JA (2006) Epigenetics, evolution, endocrine disruption, health, and disease. *Endocrinology* 147: S4–S10.

Hotchkiss AK, Rider CV, Blystone CR, et al. (2008) Fifteen years after "Wingspread"-environmental endocrine disrupters and human and wildlife health: where we are today and where we need to go. *Toxicological Sciences* 105: 235–259.

Lye CM, Bentley MG, Clare AS, and Selfton EM (2005) Endocrine disruption in the shore crab *Carcinus maenas*-a biomarker for benthic marine invertebrates? *Marine Ecology Progress Series* 288: 221–232.

McLachlan JA (2001) Environmental signaling: What embryos and evolution teach us about endocrine disrupting chemicals. *Endocrine Reviews* 22: 319–341.

Nishikawa J-I, Mamiya S, Kanayama T, Nishikawa T, Shiraishi F, and Horiguchi T (2004) Involvement of the retinoid X receptor in the development of imposex caused by organotins in gastropods. *Environmental Science & Technology* 38: 6271–6276.

Oberdörster E and McClellan-Green P (2000) The neuropeptide APGWamide induces imposex in the mud snail, *Ilyanassa obsoleta*. *Peptides* 21: 1323–1330.

Olmstead AW and LeBlanc GA (2002) Juvenoid hormone methyl farnesoate is a sex determinant in the crustacean *Daphnia magna*. *Journal of Experimental Zoology* 293: 736–739.

Puinean AM, Labadie P, Hill EM, et al. (2006) Laboratory exposure to 17β-estradiol fails to induce vitellogenin and estrogen receptor gene expression in the marine invertebrate *Mytilus edulis*. *Aquatic Toxicology* 79: 376–383.

Radix P, Severin G, Schramm K-W, and Kettrup A (2002) Reproduction disturbances of *Brachionus calyciflorus* (rotifer) for the screening of environmental endocrine disrupters. *Chemosphere* 47: 1097–1101.

Straw J and Rittschof D (2004) Responses of mud snails from low and high imposex sites to sex pheromones. *Marine Pollution Bulletin* 48: 1048–1054.

Sugni M, Mozzi D, Barbaglio A, Bonasoro F, and Carnevali MDC (2007) Endocrine disrupting compounds and echinoderms: New ecotoxicological sentinels for the marine ecosystem. *Ecotoxicology* 16: 95–108.

Thornton JW, Need E, and Crews D (2003) Resurrecting the ancestral steroid receptor: Ancient origin of estrogen signaling. *Science* 301: 1714–1717.

Zou E (2003) Current status of environmental endocrine disruption in selected aquatic invertebrates. *Acta Zoologica Sinica* 49: 551–565.

Zou E (2005) Impacts of xenobiotics on crustacean molting: The invisible endocrine disruption. *Integrative and Comparative Biology* 45: 33–38.

Avian Social Learning

L. Lefebvre and N. J. Boogert, McGill University, Montréal, QC, Canada

Introduction

In 1921, birds described as tits (Paridae) were first seen to open milk bottles in the small, southern English town of Swaythling. Over the next 25 years, observations of birds opening milk bottles were reported from hundreds of other sites all over Great Britain, Ireland, and continental Europe. The first scientific article on the phenomenon was published in 1949. A short discussion of bottle opening by birds is a good introduction to the topic of avian social learning because the questions asked about milk-bottle opening are indicative of those that have governed almost all subsequent research on avian social learning: How did this behavior originate? Did its appearance in blue tits (*Parus caeruleus*) have anything to do with the cleverness or boldness of this species? Was the rapid spread of bottle-opening over many areas of Great Britain and Ireland due to cultural transmission? If so, were the birds imitating one another or was something simpler going on? Given that eleven species of birds were found to open bottles, did we see transmission between as well as within species? (**Figure 1**).

Cultural Transmission of Foraging Behavior

First of all, why did the new behavior appear in tits? In England, home delivery of bottled milk, which had started in the years before the First World War, was interrupted during that conflict, to be resumed shortly after its conclusion. The fact that the instances of milk bottle opening by birds were first recorded after the resumption of home delivery of milk bottles suggests that the innovation might be less surprising than often thought. Tits, and in particular, the species thought to have originated the new behavior, blue tits, are relatively tame, urbanized, and inquisitive birds, easily attracted to winter feeders and other sources of food provided by humans. One of the blue tits' normal food searching behaviors is to peck and peel bark from trees to look for insects, a technique very similar to the one they use to open bottle caps.

At the taxonomic level of the family, tits are also large-brained, ranking above all others in the parvorder Passerida (3500 species) in terms of the brain size corrected for body size. Several tit species, especially the great tit (*P. major*) and the blue tit, are known for other novel or unusual feeding behaviors, such as piercing the base of

flowers to drink nectar, eating the brain of a pied flycatcher, or folding paper to store food in it.

Was the increase in bottle opening observed over the years due to cultural transmission? Culturally transmitted behaviors are sometimes said to 'spread like wildfire.' This description implies three things: rapid spread, a vast spatial scale, and a temporal pattern that starts off slowly and then spreads rapidly, until the spread eventually ceases for lack of new material to spread to. In mathematical terms, such a pattern of increase is called a logistic function and is characterized by an S shape. Milk bottle opening shows all the three features of 'wildfire': In 25 years, it spread from a single site to nearly 400 locations, with the number of milk bottle-opening birds presumably increasing from one to several hundred thousand; the spatial scale went from a single doorstep in one small town to several countries; the mathematical function that describes the spread of bottle opening shows two of the phases of a logistic, S-shaped curve, the slow start (1921–1936) followed by a sharp acceleration (1937–1947). The absence of the final slowdown phase of the logistic is probably due to the fact that, in 1947, the last year surveyed, there were still many places to which the new behavior could spread. However, the spread of bottle opening does not show one feature of mathematical models of cultural transmission. If an innovation originates in a particular place and spreads elsewhere through a kind of 'wave of advance,' then sites close to Swaythling should show early dates of bottle opening and sites progressively farther and farther away should show progressively later dates. This is not what we see. There is no clear relationship between distance from Swaythling and the time that has elapsed since the presumed origin of bottle opening in 1921.

This exception might lead us to think that something besides cultural transmission was behind the increase seen in bottle opening over time. What could that something be? Could many different birds all over the British Isles have discovered how to open milk bottles independently? Tits are very inquisitive. Consequently, it is quite possible that many of them could have invented the new behavior on their own. This possibility suggests a crucial control test that needs to be incorporated in any study of social learning. One or several observers placed in front of a knowledgeable demonstrator might very well adopt the new behavior, but they could be doing so on their own by trial-and-error learning, without actually needing demonstrations. In other words, tits might just be so exploratory that they easily discover by themselves that

Figure 1 Blue tit opening a milk bottle. Photo courtesy of BBC Devon.

a bottle top can be pierced. Back in Britain, they might also have stumbled upon a bottle that had already been opened by another bird and drunk some leftover cream (it is the fat from the cream that the tits can digest, not the carbohydrates from the milk), without witnessing the bottle being opened by another tit. Even if naïve observers profited from watching demonstrators, the social information they acquired might have been vague and served only as a basis for individual perfection of the complete technique of milk-bottle opening. Researchers in Canada and Austria have set up laboratory analogs of bottle opening with captive tits and chickadees. Independent spontaneous discoveries, learning by feeding on an open bottle, and social learning, all occurred in these experiments, supporting the view that bottle opening probably spread in the wild via independent innovations as well as several learning processes, both social and nonsocial.

Imitation of Foraging Behavior

Our discussion of the spread of milk-bottle opening in birds suggests that individuals can obtain many types of information from each other. Social learning is a very general term used to describe *any* process through which one individual (the 'demonstrator') influences the behavior of another individual (the 'observer') in a way that makes the observer more likely to learn the behavior in which the demonstrator engages. Imitation has always been the most popular social learning process to study. Milk-bottle opening would be an instance of imitation if an observer bird learned to copy the precise technique used by its demonstrator. Birds that invented milk-bottle opening on their own, without a demonstrator, or birds that obtained only vague social information about what to fiddle around with might use any technique, including piercing the bottle cap with sharp downward pecks, or ripping it with a sideways motion. In contrast, an imitator would copy the precise technique that it saw demonstrated.

To understand how you could demonstrate imitation in birds experimentally, we can examine in some detail a study Fawcett and colleagues conducted with starlings (*Sturnus vulgaris*). The study has two important features, a 'two-action method' and a 'ghost control.' The two-action method ensures that we can separate the effects on an observer of seeing the behavior of its demonstrator from the effects of seeing a demonstrator act on a particular object. The 'ghost control' ensures that the information an observer obtains is truly social, that is, that the information comes from observing the behavior of the demonstrator itself and not from observing the effects of its demonstrator's behavior on the objects with which the demonstrator interacts.

In the 'two-action' method, each observer sees a demonstrator use one of two actions directed toward exactly the same portion of the environment. In the case of the study with starlings, the demonstrators were trained to remove a plug from a box to gain access to food either by pushing the plug downwards into the box (**Figure 2(a)**) or by pulling it upwards out of the box (**Figure 2(b)**). In the 'ghost control' condition, the plug was pushed or pulled via a fishing line controlled by the researchers, independent of the actions of the bird in the demonstrator compartment.

As is usual in such social learning experiments, the demonstrator and the observer were in adjacent but separate cages during training, so that the observer could watch the demonstrations but not interact with the demonstrator or the box during the experiment. In addition, the observer did not have access to a box of its own during the demonstration, so it needed to memorize the information it saw for later use.

After numerous demonstrations, the demonstrator was removed and the apparatus was presented to the observer. In Fawcett's experiment, observers that had seen a 'Pull demonstrator' were more likely to open the box by pulling the plug; those that had seen a 'Push demonstrator' were more likely to open the box by pushing the plug. Observers that had seen ghost control pushes and pulls were equally likely to push or pull. The fact that the observer starlings moved the plug *in the same way* as the demonstrator cannot be explained by any social learning process other than imitation. For example, observers were not simply attracted to the plug (stimulus enhancement) or to the same location as the demonstrator (local enhancement).

Learning from Other Species

As we mentioned before, several bird species learned to open milk bottles in Great Britain, Ireland, and continental Europe. Did they learn by watching blue tits, the first and most frequent openers? One of the usual assumptions of social learning is that copying members of one's own species should be more likely than copying members of

Figure 2   Typical response topographies of (a) a Push demonstrator and (b) a Pull demonstrator. Note the widely gaping beak of the Push demonstrator. Reprinted from Fawcett TW, Skinner AMJ, and Goldsmith AR (2002) A test of imitative learning in starlings using a two-action method with an enhanced ghost control. *Animal Behaviour* 64: 547–556, Copyright (2002), with permission from Elsevier Ltd.

other species. Why? Because the goals of other members of one's own species (food, mates, nesting sites, and predator avoidance), as well as the sensory equipment and the motor capabilities available to reach those goals, are more similar within than between species. If both a tit and a starling see a tit ripping open a milk-bottle top, the adequacy of the beak as a ripping instrument, the propensity to approach bottles or other man-made objects, and the motivational value of cream are all greater for an observing tit than for an observing starling. Despite all these reasons favoring social learning within species, we see a surprising amount of social learning between species.

On the tropical island of Barbados in the West Indies, Carib grackles (*Quiscalus lugubris*) and Zenaida doves (*Zenaida aurita*) often form mixed foraging flocks with bullfinches (*Loxigilla noctis*), Shiny cowbirds (*Molothrus bonariensis*), and Ground doves (*Columbina passerina*). Amid the many hotels, restaurants, and parks along the Barbados coast, members of all the five species readily join with others and feed together on food remnants left by humans.

Although these five bird species are all opportunistic feeders, they belong to two distinct avian orders that differ greatly in cognitive ability. Zenaida and Ground doves are Columbiformes that show less innovative behavior, perform worse on learning tasks in captivity, and have smaller brains than do Passeriformes, the order to which the grackles, bullfinches, and cowbirds belong. The Carib grackle is particularly innovative, belonging to the genus with the second highest number of reported innovations of all passerines in North America.

Carib grackles forage in small mobile flocks and are usually among the first birds to arrive when food becomes available. Although they boldly exploit feeding opportunities, grackles remain constantly vigilant for predators.

As soon as one grackle in a flock detects danger, such as an approaching mongoose, monkey, dog, or threatening human, it gives alarm calls that induce flight and even more alarm calls in nearby grackles. Zenaida doves, in contrast, are rarely the first to discover a food source, feed alone or with their mates, and have no alarm calls of their own. They are territorial over most of Barbados and vigorously chase away other Zenaida doves that represent a threat to both mate and territory. However, Zenaida doves are rarely aggressive toward grackles.

The assumption that learning from members of your own species is easier than learning from members of other species implies that doves will learn from doves. However, the feeding ecology of doves and grackles suggests otherwise. Grackles are useful informers about food and danger, whereas doves are territorial competitors that fight with one another when they meet. In accordance with this ecological and social scenario, only one of eleven Zenaida dove observers learned the solution of a feeding task from a dove demonstrator, whereas the majority of doves learned the solution when it was demonstrated by a grackle.

Grackles, on the other hand, do not defend foraging territories against other grackles, and learned as readily from grackle as from dove demonstrators, copying the precise technique that each of their demonstrator species used: closed beak pecking by doves, and open beak probing by grackles. Grackles also learned to treat a previously innocuous stimulus (a painted pigeon decoy) as a potential predator (a dog decoy) when grackle alarm calls were paired with it.

Social Learning About Predators

Although most research on avian social learning concerns song and feeding, some of the most elegant work on avian social learning both within and between species has been done on habitat choice (see section 'Social Learning About Habitats and Nest Sites') and predator recognition. Curio and colleagues conducted the pioneering experiments on social transmission of predator recognition with European blackbirds (*Turdus merula*). In the presence of a predator, these birds emit mobbing calls that summon nearby individuals to cooperate in attacking the predator.

As a result of co-evolution between predator and prey, predator avoidance behaviors may become heritable, making individual trial-and-error learning about coevolved predators by their natural prey unnecessary. However, species change their ranges over time, predators' diets change with changes in predator and prey abundance, and humans introduce nonnative species. Thus, birds may be preyed upon by evolutionarily unfamiliar species that they do not innately recognize as predators. Because a bird might not get a second chance after an encounter with a predator, social learning to recognize predators could be very valuable.

For their experiments, Curio and colleagues devised an ingenious apparatus containing observer and demonstrator blackbirds in opposite compartments, separated by a hallway containing a presentation box (**Figure 3**). The compartment of the presentation box facing the observer contained an object that the blackbirds had never seen before, while the compartment facing the demonstrator contained a little owl (*Athene noctua*), a familiar predator that triggered vigorous mobbing in demonstrators. Because the demonstrator and observer could not see the compartment of the apparatus that the other was able to see the observer was tricked into perceiving the demonstrator as mobbing the novel object.

Curio used two novel objects similar in size to the little owl to test for cultural transmission of predator recognition: a dummy of an Australian honeyeater (*Philemon corniculatus*), a bird unfamiliar to wild blackbirds, and a multicolored plastic bottle. Observation of the demonstrator mobbing the owl led the observers to mob whichever novel object was presented in their compartment of the presentation box. Furthermore, the observers also mobbed that novel object when it was presented 2 h later in the absence of the demonstrator, showing that the observers had learned to treat that novel object as a predator. However, observers showed a stronger mobbing response toward the honeyeater than toward the bottle. This last result suggests that learning about danger is influenced by characteristics of the 'dangerous' stimulus.

After the original blackbird demonstrator, who mobbed the owl, trained a first blackbird observer to mob the honeyeater, that observer was used as a demonstrator for a second observer. This second observer learned to mob the honeyeater and then became the next demonstrator for a naïve blackbird, and so on.

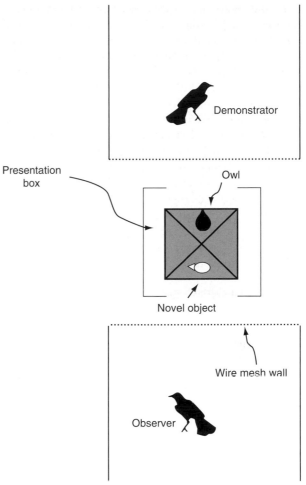

Figure 3 Experimental apparatus to test for social learning of predator recognition in blackbirds. The owl and novel object are positioned to elicit mobbing behavior in the observer towards the novel object through the demonstrator's mobbing of the owl. Reproduced from Vieth W, Curio E, and Ernst U (1980) The adaptive significance of avian mobbing. III. Cultural transmission of enemy recognition in blackbirds: Cross-species tutoring and properties of learning. *Animal Behaviour* 28(4): 1217–1229, Copyright (1980), with permission from Elsevier Ltd.

Information about the honeyeater passed along a chain of six blackbirds without any noticeable decrease in response strength.

Social Learning About Mates

The fitness of an animal depends not only on finding food and avoiding predators, but also on choosing a mate likely to maximize its reproductive success. If sampling candidate mates and comparing their quality are costly in terms of energy and time, then learning from the mate choice of others might provide the most efficient way to obtain a good match.

In some polygynous species such as black grouse (*Tetrao tetrix*), sage grouse (*Centrocercus urophasianus*), and white-bearded manikins (*Manacus trinitatis*), males gather at a 'lek,' an area where each male has his own tiny territory to display to the female audience. Often, only a very small number of the many males displaying on a lek acquire the great majority of matings.

Researchers studying black grouse in the field could not find any physical or behavioral traits that were consistently associated with the few successful males, while the pattern of female visits to leks suggested that females may simply copy the choices of females that mated before they did. If so, a male that was able to mate with a female that came to a lek early might well become overwhelmingly popular. In the field, however, it is difficult to separate social and individual factors governing mate choice, something that can be accomplished much more easily in laboratory experiments.

In the laboratory, female Japanese quail (*Coturnix japonica*), given a choice between an unmated male and one that they had seen courting and mating with another female, preferred the previously successful male. Furthermore, females that had watched a nonpreferred male mate with another female laid more fertilized eggs after mating with him than did females that did not see a nonpreferred male mate with another female.

While domesticated Japanese quail will court and mate whenever the opportunity arises, zebra finches (*Taeniopygia guttata*) form pair bonds that often last for life. One might expect mate choice copying to be rare in this and other monogamous species as compared to the polygynous quail and grouse species. However, in the dry areas of Australia where zebra finches abound, the time available for mate choice and reproduction is constrained; zebra finches have to be ready to reproduce whenever the unpredictable rainfalls provide necessary resources for rearing young. It might therefore benefit inexperienced females to copy the mate choices of others rather than spend time appraising mates for themselves. Indeed, laboratory experiments show that female zebra finches, like female Japanese quail, tend to copy the mate choices of other females of their species. More importantly for a monogamous bird, female zebra finches transfer their socially acquired preference to males similar to the one they had seen with a female. If a male seen mating wore a white leg band, virgin females presented with a pair of unfamiliar males preferred the one with a white leg band over the one with an orange leg band.

Social Learning About Habitats and Nest Sites

Imagine a young migratory bird, for instance, a collared flycatcher (*Ficedula albicollis*), arriving later than most of its fellows at a breeding area. Young birds have little prior breeding experience and need to find a nest site and start laying eggs as soon as possible. They do not have time to explore an area for a couple of weeks to pick the best site for a nest. In these conditions, a young bird might rely on the information provided by birds that have already settled in an area. These settled birds might be individuals of its own species or they might be birds of a resident species that stays in the area all year round. Researchers from both France and Finland have shown that flycatchers use cues both from resident birds of other species and by monitoring the breeding success of birds of their own species that have settled in an area before them to choose their nesting sites.

Doligez, Danchin, and colleagues manipulated the apparent breeding success of flycatchers by adding or removing chicks from a set of nests. Areas with added chicks were settled by a greater number of incoming flycatchers, while the opposite was true of areas where chicks were removed; control areas where chicks were simply taken from and put back in their original nest showed no change in the number of incoming flycatchers choosing to settle in them.

In a conceptually similar experiment, Seppänen and Forsman put nest boxes in four 5–12 ha. forest patches, two in the Swedish island province of Gotland and two in the Finnish city of Oulu. Once resident great and blue tit species had started building nests in these boxes, the researchers painted white circles around the nest box entrances in one forest patch in Gotland and Oulu, and white triangles in the other forest patch. They placed an empty nest box with the opposite symbol on the nearest tree similar to that containing the occupied nest box, to create the impression that the nesting tits in the patch had all chosen nest boxes with a particular geometric symbol.

The first males from the two migratory flycatcher species under study (Gotland: collared flycatchers *F. albicollis*; Oulu: pied flycatchers *F. hypoleuca*) arrived in the forest patches after the resident tits had started to nest. The researchers placed additional pairs of empty boxes, one box with a triangle, the other with a circle, 25 m from the nearest tit nest (**Figure 4**). Female flycatchers arriving and laying their eggs early did not have a preference for either symbol. However, as the breeding season progressed, female flycatchers started to match the nest box 'preference' shown by the tits, and more than 75% of the last third of females arriving at the breeding area chose a nest box with the same symbol as that on the tits' nest boxes.

Nest site choice used to be considered an innate and inflexible behavior. This field study shows, however, that migratory birds can copy the nest site choice of resident birds when the date of arrival at the breeding area imposes time constraints on individual learning.

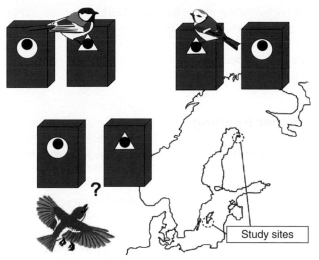

Figure 4 Experimental design to test for nest site copying between species. Once the resident great and blue tits had initiated nests, either a circle or a triangle was painted at the nest box entrance of all nests in a given forest patch. An empty box with the opposite symbol was placed 2–6 m away. Arriving migrant flycatchers were forced to choose between an empty nest box with the symbol 'preferred' by the resident tits and an empty nest box with the other symbol. Reprinted from Seppänen J-T and Forsman JT (2007) Interspecific social learning: Novel preference can be acquired from a competing species. *Current Biology* 17: 1248–1252, Copyright (2007), with permission from Elsevier Ltd.

Is Avian Social Learning Rare and Limited to Large-Brained, Social Species?

Blue tits are as good at individual learning as they are at social learning. They are far faster at both types of learning than are, for instance, Marsh tits (*P. palustris*, who are best at spatial learning). Across all species and individuals that have been studied in captivity, social and individual learning show positive correlations. Such positive correlations led most researchers to view social and individual learning as the same process, rather than treating them as separate cognitive modules that are adaptively specialized to different lifestyles.

Do social and individual learning in birds also correlate with brain size? Social learning does correlate with brain size in primates: the species that show the most social learning are also the ones that have the largest cortex. In birds, the trends are not as clear. In the field, the great majority of purported cases of avian social learning occur in a single suborder, the Oscines (songbirds). This bias precludes a fair analysis of overall trends throughout the class Aves.

Birds also differ from mammals in that a species' ability to imitate does not seem to vary much with brain size. Apes, for example, are far better imitators than are monkeys, and apes also have a larger cortex than any monkey species; similarly, large-brained dolphins are thought to be capable of imitation, whereas small-brained horses seem to be incapable of even the simplest form of social learning. In birds, even small-brained species such as Japanese quail and pigeons (*Columba livia*) seem to be able to pass the two-action test. At the other extreme of brain size, social learning in the kea (*Nestor notabilis*), an opportunistic, omnivorous, and large-brained parrot, has proved remarkably limited in two field studies, and keas also showed little evidence for imitation in laboratory experiments.

Inhibitors of Avian Social Learning

Researchers have identified several behavioral mechanisms that block or slow down social learning in birds: scrounging, bystanding, and territoriality. In many species, the discovery of a new food source, whether by social or individual learning, attracts the attention of others that often join the discoverer and feed with it. Joining can lead to theft or to simple scrounging. In scrounging, joiners unaggressively consume part of the food that the producer discovered rather than taking food away from its discoverer. In many cases, joining allows close observation of a producer's food finding behavior and seems to favor social learning by a scrounger. In other cases, however, scrounging can actually prevent social learning. In pigeons, competing with other scroungers, as well as identifying and following a given producer to its food discoveries, seems to interfere with observation of the producer's food-finding technique. Scrounger pigeons that have followed a producer to hundreds of its food discoveries will not perform the food-finding technique themselves after the producer is removed.

Like scrounging, the presence of bystanders seems to interfere with social learning in pigeons. If a caged demonstrator showing a food-finding technique is surrounded by several caged birds doing nothing but pacing in their cages, the observers' social learning is worse than in situations where only a single observer watches a demonstrator.

A final situation in which social learning appears to be impeded involves territorial boundaries. You may recall that in experiments on wild-caught birds, territorial Zenaida doves did not learn well from other doves. In the field, this implies that an innovation that occurs on one territory would not spread to adjacent territories. Unless grackles spread a new behavior, an innovation occurring on a dove territory would stay there. There are two examples where feeding innovations are known to have remained localized in a territorial species. As we have seen, winter flocks of tits learn socially to feed on milk bottles. However, flower piercing, an innovation performed in the spring when tits defend territories, has not spread. A similar localized pattern characterizes the opening of sugar packets by bullfinches in Barbados. Rather than observing and learning from an intruder opening sugar packets on its territory, a naïve territory holder aggressively attacks the intruder, preventing the innovative sugar-packet opening behavior from spreading beyond a restricted area. Packets offered to bullfinches foraging only a few hundred meters from the site where other bull-finches routinely open sugar packets are ignored.

Conclusion

Research on avian social learning has come a long way since the publication, 60 years ago, of the pioneering article on milk-bottle opening by tits. Many questions remain, however, concerning (1) the rarity of social learning in birds compared to primates and (2) the fact that all birds tested so far seem to pass the two-action test for imitation, regardless of their brain size. Among the most promising directions today are field experiments on breeding sites and predator recognition.

In tests of social learning about food in the wild, often, as we have seen in the case of bottle opening, it is impossible to separate individual and social learning processes. Only in controlled experiments can the effects of social and environmental cues about food be distinguished. In contrast, social and individual information can be manipulated separately when avian social learning tests involve alarm calls, mobbing calls, and the presence or success of others at breeding sites. In the coming years, field experiments in well-studied ecological settings on behaviors other than feeding should lead to rapid advances in research on avian social learning.

See also: Apes: Social Learning; Imitation: Cognitive Implications; Vocal Learning.

Further Reading

Curio E (1988) Cultural transmission of enemy recognition by birds. In: Zentall TR and Galef BG Jr. (eds.) *Social Learning: Psychological and Biological Perspectives*, pp. 75–97. Hillsdale, NJ: Lawrence Erlbaum Associates.

Doligez B, Danchin E, and Clobert J (2002) Public information and breeding habitat selection in a wild bird population. *Science* 297: 1108–1170.

Fawcett TW, Skinner AMJ, and Goldsmith AR (2002) A test of imitative learning in starlings using a two-action method with an enhanced ghost control. *Animal Behaviour* 64: 547–556.

Fisher J and Hinde RA (1949) The opening of milk bottles by birds. *British Birds* 42: 347–357.

Galef BG Jr (2008) Social influences on the mate choices of male and female Japanese quail. *Comparative Cognition & Behavior Reviews* 3: 1–12.

Hinde RA and Fisher J (1951) Further observations on the opening of milk bottles by birds. *British Birds* 44: 392–396.

Lefebvre L and Bouchard J (2003) Social learning about food in birds. In: Fragaszy DM and Perry S (eds.) *The Biology of Traditions: Models and Evidence*, pp. 94–126. Cambridge: Cambridge University Press.

Seppänen J-T and Forsman JT (2007) Interspecific social learning: Novel preference can be acquired from a competing species. *Current Biology* 17: 1248–1252.

Swaddle JP, Cathey MG, Correll M, and Hodkinson BP (2005) Socially transmitted mate preferences in a monogamous bird: A non-genetic mechanism of sexual selection. *Proceedings of the Royal Society of London B* 272: 1053–1058.

Avoidance of Parasites

D. W. Thieltges and R. Poulin, University of Otago, Dunedin, New Zealand

Introduction

Infection by parasites or pathogens usually puts a high burden on animals. Infected hosts generally achieve lower fitness: they either show lower growth rates, reduced reproductive output, or even lower survival than uninfected members of their species. Therefore, animals are better off if they can avoid becoming infected with parasites and pathogens in the first place. Not surprisingly, natural selection has favored many adaptations reducing the risk of infection. To avoid contact with parasites and pathogens, animals have developed numerous precontact avoidance measures (**Figure 1**). This first line of defence against parasites and pathogens is mainly behavioral. If the first line of defence fails, animals have a second chance to resist infections. This second line of defence consists of postcontact defensive measures of a behavioral, physiological, and/or immunological nature (**Figure 1**). Here, we focus our discussion on precontact avoidance measures, excluding mate choice. These involve a wide spectrum of behavioral adaptations serving to prevent infection, ranging from where to live or what to eat, all the way to whether or not to associate with other animals.

Avoidance Is Not Free

It is often said that there is no such thing as a free lunch: everything has a hidden cost. Thus, we cannot expect that avoidance of parasites is free. Whatever measures they use to prevent infections by parasites and pathogens, animals have to pay a cost in terms of reduced fitness (this also applies to postcontact defence measures, of course). For example, behavioral actions initiated to avoid parasites and pathogens may consume host energy that will subsequently no longer be available for other purposes. Alternatively, avoidance behaviors may take time away from other important activities such as searching for food or mates. The cost of avoidance may be small, but that does not mean that it is insignificant. In theory, in the complete absence of parasites and pathogens, these fitness costs result in a lower total fitness of resistant animals compared to susceptible ones (**Figure 2**). However, in the presence of parasites and pathogens, resistant animals achieve a higher total fitness compared to susceptible ones (**Figure 2**). The investment in avoidance measures pays off for potential hosts when the costs of avoidance are lower than the costs of infection. Only in this case will

natural selection favor the evolution of avoidance mechanisms. In situations where the parasite is relatively benign, that is, it has hardly any effect on host fitness, and where avoiding it comes at some cost to the animal, we would not expect animals to bother trying to avoid infection.

Note that in many cases avoidance mechanisms are inducible, that is, they are only triggered after parasites or pathogens are detected, such that their cost is only incurred when necessary. This is generally the case with behavioral avoidance measures: the animal only starts to behave in a defensive mode when the presence of parasites is detected. Thus, the full cost of avoidance is not paid when there are no parasites around: this means that in **Figure 2**, the magnitude of the cost of avoidance would be even lower compared to the cost of infection. Inducible behavioral defences are also advantageous because by chance alone, many animals would never encounter parasites anyway. Parasites are never distributed evenly among all individual hosts in an animal population, and thus it pays off to possess a defence mechanism that can only be turned on if and when needed. Behavioral avoidance is therefore a cheap method of preventing infection, one that should be greatly favored by selection in most animals.

Ben Hart and others have stressed that two criteria have to be met before any particular avoidance measure can be accepted to function as an effective defence against parasites and pathogens. First, the parasites and pathogens in question have to exert detrimental effects on the host's fitness. This may not always be as easy to demonstrate as one would think. Only in few cases are parasites and pathogens exerting dramatic effects on their hosts' fitness, such as causing gross pathology or mass mortality. In the majority of cases, the effects of parasites and pathogens on their hosts are much more subtle. A healthy, well-fed, and nonstressed host may not suffer noticeably from moderate parasite and pathogen burdens. However, this might change when the host experiences periods of starvation and other environmental or social stressors, or when the host has to fight against conspecifics or has to escape from predators. In such cases, parasites and pathogens may act as additional stressors and cause a reduced fitness of infected hosts. Hence, seemingly benign parasites and pathogens may nevertheless have an important effect on their hosts under certain circumstances. Second, the avoidance measure in question has to be effective in preventing or reducing contact with parasites and pathogens. Although this is a logical criterion, the effectiveness of a

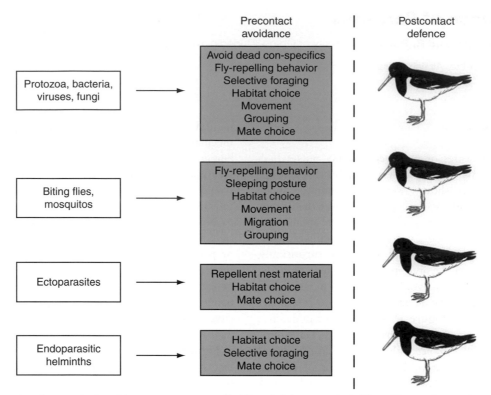

Figure 1 Examples of precontact avoidance measures as a first line of defence against different types of parasites and pathogens in birds. These are mainly behavioral measures to avoid contact with parasites and pathogens in the first place. Once a host has come into contact with parasites and pathogens, there is a second line of defence in the form of postcontact defensive measures that can be either behavioral, physiological, or immunological.

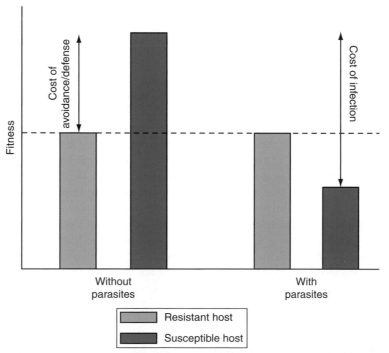

Figure 2 Mechanisms of parasite and pathogen avoidance or defence are costly and thus reduce the fitness of resistant hosts compared to susceptible hosts in the absence of parasites. However, in the presence of parasites, the fitness of resistant hosts is higher than that of susceptible hosts with the latter suffering more from the costs of an infection. Precontact avoidance and postcontact defensive measures can only evolve if the cost of infection exceeds the cost of avoidance and defensive measures.

particular behavior is often not rigidly tested and many reports of avoidance strategies are rather anecdotal. However, circumstantial evidence generally supports the effectiveness of many avoidance behaviors, and there are also studies that quantitatively observed or experimentally tested the effectiveness of various avoidance measures. The ideal test of these two criteria would involve the experimental manipulation of parasite burdens to see whether animals make the expected adjustments in their investment in avoidance measures, or the experimental impairment of avoidance mechanisms to see whether impaired hosts acquire more parasites than unimpaired controls, or some other rigorous experimental demonstration. In the following section, various types of avoidance measures are discussed that animals have developed against parasites and pathogens, though it is pointed out right now that rigorous demonstrations of their adaptive nature is often lacking.

Camouflage Against Infection

Perhaps the simplest way to avoid infection is to avoid being detected by parasites. Just as prey species use shape and color to camouflage themselves from predators, potential hosts might use a similar approach to avoid attacks by parasites and pathogens. A particularly intriguing, though not properly tested, example of camouflage against parasites and pathogens involves the conspicuous pattern of black and white stripes of zebras. Jeff Waage has suggested that this peculiar fur coloration pattern may be a means to avoid bites by tsetse flies. Biting arthropods are generally attracted by large uniform surfaces, a behavior that is thought to help flies find large animals that serve as hosts. The patchy color pattern of zebras is thought to camouflage their actual size by breaking up the regular outline and shape of the animals, and may thus help to avoid fly bites. There is circumstantial evidence in support of this hypothesis. First, it seems that zebras are indeed rarely bitten by tsetse flies. And second, it appears that the color pattern is most pronounced in parts of Africa where the flies are most abundant. Hence, this avoidance measure seems to be effective and could provide a benefit for zebras as it reduces the number of annoying and blood-letting bites as well as the risk of acquiring sleeping sickness or other pathogens transmitted by tsetse flies.

No other examples of camouflage against parasites are known. However, it is easy to imagine that several aquatic animals could use chemical camouflage to avoid detection or recognition by the larval stages of digenean trematodes. The free-swimming infective larvae (miracidia or cercariae) of these parasites use chemical cues produced by their hosts to locate and identify animals such as snails or fish that they must penetrate and parasitize in order to survive and complete their life cycle. Trematode infection comes

at a huge cost: parasitized snails are typically castrated permanently following infection, whereas other animals like fish that are infected by several trematode cercariae often incur greater risks of predation. Since trematodes use specific components of host mucus to find and recognize their hosts, one can speculate that natural selection would favor any host producing mucus with a different biochemical profile that would go unnoticed by the parasites. There may indeed be many more cases of camouflage, either visual or chemical, against parasite infection that have escaped the attention of biologists.

Moving away from Parasites or Repelling Them

Another way to avoid contact with parasites and pathogens is simply to move away from them. Cattle sometimes literally run away when approached by biting flies, a behavior referred to as gadding. The same behavior can be observed in other large grazing mammals like mule deer and elk that, by doing so, avoid not only annoying and blood-letting bites but also potential additional infections by pathogens transmitted by the flies. Fish can also avoid infections by moving away from infectious agents. Rainbow trout avoid eye fluke (trematode) infections by moving away from the infective cercarial stages of these debilitating parasites that use trout as second intermediate hosts in their life cycle. The parasite encysts in the eyes of the fish and heavy infections cause cataract that impairs the vision of infected fish. This in turn reduces the feeding ability and growth of the fish, as well as increasing their risk of predation. Avoiding infections with eye flukes by moving away from infective stages is thus of obvious advantage for trout.

Besides moving away from parasites and pathogens, staying away from potentially infected individuals is another avoidance strategy. In particular, avoiding contact with dead conspecifics may be a successful strategy because their death may have been caused by parasites or pathogens. Wild geese provide an example of such behavior as they avoid close contact with birds that died of avian cholera and thus avoid contact with the contagious stages of the pathogen. More generally, staying away from sick-looking individuals would also be of great benefit. Who among us would get close to or willingly embrace a friend with a runny nose and a persistent cough?

If moving away from parasites and pathogens is not an option, potential hosts can still avoid contact with some of them by various body movements. For example, there are various fly-repelling behaviors that effectively reduce the number of fly attacks. In ungulates, ear twitching, head tossing, leg stamping, muzzle flicking, muscle twitching, and tail switching are common ways to avoid attacks by flies. These behaviors help to avoid or reduce the pain and loss of blood caused by biting flies, and also reduce the number of eggs deposited by parasitic warble or bot flies. They may also help against fly-borne pathogens. Small mammals and birds show similar avoidance measures like tail and ear flipping, face rubbing, foot stamping, bill snapping, head shaking, and wing flapping. These body movements protect them from loss of blood to mosquitoes and other flies as well as mosquito-borne pathogens like malaria. A less demanding way of avoiding parasites and pathogens may simply involve changes in body posture. The sleeping posture of birds, for instance, in which a bird often sticks its head under its plumage and stands on just one leg, can do the trick. This posture reduces exposure of the feather-free parts of the bird's body to mosquitoes. As mosquitoes transmit serious pathogens like avian malaria, this behavior protects potential hosts from infection. Native Hawaiian birds reportedly slept with their heads and legs exposed and not covered in their plumage before the arrival of mosquito-borne avian malaria with introduced birds. Today, the sleeping posture of Hawaiian native birds is similar to that of birds from other localities where avian malaria is common, indicating that this behavior might be an effective pathogen-avoidance measures that has been adopted by surviving native species.

Habitat Choice, Habitat Modification, and Migration

If parasites and pathogens are associated with certain habitats, animals can avoid infections by choosing a different microhabitat within the area, modifying their habitat to make it less hospitable to parasites, or moving to another geographically distant habitat. As an example of microhabitat selection on a small scale, consider the oviposition behavior of female mosquitoes. It has been shown experimentally that female mosquitoes avoid water containing heavily parasitized mosquito larvae when they lay their eggs. Although the trematode parasite infecting the larvae is not transmitted from larva to larva, infected larvae indicate the presence of the first intermediate snail host from which the cercariae that infect the larvae originate. Since trematode infection can be fatal to a mosquito larva, the mother's decision to lay her eggs in a particular site can have a devastating consequence for her offspring, and the ability to use chemical cues to detect the presence of the parasite is very advantageous. By avoiding habitats with the snail present, female mosquitoes in search of a pool of water in which to lay their eggs can thus reduce the infection risk for their offspring. Gray tree frogs show a similar avoidance behavior when depositing their eggs, that is, they use chemical cues to avoid sites where the snail intermediate host of a debilitating trematode parasite is present. The right choice of habitat can protect offspring from parasites and pathogens,

but it may also be an important avoidance measure for adult animals. For example, when horse flies are abundant, hippopotamuses avoid foraging on land and remain submerged in water for longer periods.

In birds, the choice of nesting habitats can help to avoid parasites and pathogens. Several field experiments, mostly using next boxes in which parasites can be added or removed, have shown conclusively that birds are sensitive to the risk of infection when selecting a nest site. For example, great tits consistently choose parasite-free nests and avoid nests where ectoparasitic fleas are present. Similarly, cliff swallows are able to recognize old nests infected with ectoparasitic flies when they return to their nesting areas after migration, and then select only the cleanest nests. Sometimes the reuse of old nest sites will invariably result in high infection risk; in these situations, it pays to either modify these sites prior to reusing them or to create new ones. Many passerine birds do not bother to identify clean nests and prefer to build new, uninfected nests and thus avoid contact with ectoparasites by creating their own parasite-free habitat. An even more sophisticated way of ensuring a parasite-free habitat is the use of green plant material for nest construction that repels parasites and pathogens due to secondary plant metabolites. This practice has been particularly well documented in European starlings. These birds regularly include green leaves among the twigs making up their nest. The leaves are not just a random sample of leaves from local trees; instead, they are selected only from one or two specific tree species in which the leaves exude chemicals that either repel or even kill arthropod ectoparasites such as lice or mites. This good example of habitat modification leads to a parasite-free space in which the birds can rear their offspring.

Habitat choice in relation to the presence of parasites and pathogens can also occur on a larger scale in the form of animal migration. For example, the seasonal migrations of reindeer may be, at least in part, related to a parasitic fly that lays eggs under the reindeers' skin. The larvae leave the skin after 3–4 months and drop to the ground to pupate. With the emerging adults the cycle starts again, and the number of pupae on the pasture determines the infection levels experienced by reindeer. In migratory herds, fewer parasitic fly larvae are found in the skin, which probably results from lower numbers of pupae dropped within their summer pastures by migratory herds compared to the higher numbers on pastures where animals have been present all year round. Migrating to habitats with a lower parasite and pathogen pressure is a strategy adopted also by caribou. During their summer grazing, they migrate to higher altitudes, where mosquitoes are less common compared to lower altitudes. Another particularly intriguing example of migration related to parasite and pathogen presence comes from Hawaiian birds. With the introduction of avian malaria, many native birds became extinct at the beginning of the last century. In contrast to the introduced birds, the native species were highly susceptible to the pathogen. While the invaders subsequently took over the lowland forest areas, some native birds survived by moving to higher mountainous zones, where avian malaria is absent. The birds now migrate downwards to lower areas during the day to feed while the mosquitoes are largely inactive. They then return to their high-altitude roosts in the evening before the mosquitoes become active again. Animals are therefore not at the mercy of parasites within their habitat: they can move, on small or large scales, to habitats that present lower risks.

Choosing What to Eat

You are what you eat; or at least you harbor the parasites that you eat. In all cases where parasites and pathogens are acquired during foraging and food consumption, potential hosts can avoid infections by feeding selectively on safe food items. For grazers, a very simple strategy is to avoid foraging on patches of grass contaminated with the feces of other grazers. The feces of animals often contain eggs and larvae of parasites, and other animals can become infected by feeding on contaminated pasture. Intestinal nematodes are often transmitted in this manner, and they can be particularly costly in terms of host fitness (i.e., reduced growth) when they reach high numbers in a host. In order to avoid becoming infected with these parasites or other pathogens, horses, cattle, sheep, and presumably other ungulates avoid grazing or browsing on forage in close proximity to recently dropped feces. It is probably the odor of feces that helps to avoid contaminated patches. Horses are reported to be even more restrictive in their feeding behavior, as they feed in certain areas and defecate in others.

It is not just herbivores feeding on contaminated pasture that are at risk: carnivorous animals can also become infected with parasites and pathogens via infected prey. Many parasites use complex life cycles including intermediate hosts, and transmission to final hosts is often achieved by predation. In this case, a good avoidance measure for the predator serving as final host should be to avoid consuming infected prey. Infected prey can be identified either directly by, for instance, visual or olfactory clues, or by indirect estimates of the severity of infection using some kind of proxy. It must be pointed out here that prey serving as intermediate hosts for parasites often display aberrant behaviors or appearance, in what appears to be a manipulation of the host's phenotype by the parasite. The manipulated hosts are more visible, and often more susceptible to predation by the parasite's final host. Why is it that if a parasitized prey is visually very distinct from a nonparasitized one, predators still

choose the parasitized prey? Should natural selection not favor picky predators that discriminate against parasitized prey and thus avoid acquiring parasites? Well, no, according to a cost–benefit analysis of these situations based on optimal foraging principles. There are costs associated with prey capture, such as time and energy spent searching for and handling the prey. Presumably, the more visible parasitized prey are cheaper in that sense, being easier to see and capture. If the cost that comes with acquiring a parasite from these prey is very small, that is, smaller than the total cost of prey capture, then the predator actually benefits from selectively feeding on parasitized prey. Thus, in their coevolutionary arms race with hosts, many parasites avoid discrimination against their intermediate hosts by predatory definitive hosts by being not harmful to the latter.

The situation can be more complex, of course, as illustrated by the following example. Oystercatchers have been reported to reject their clam prey when it is infected with a particular trematode parasite, probably because they can notice the parasites because of their bright coloration. In this case, the predator can see the parasite directly, but otherwise parasitized prey are not easier to detect and capture than nonparasitized ones, and so they should be avoided. In cases where the direct detection of parasites is not possible, oystercatchers follow a different strategy. Their favorite prey are cockles, abundant bivalves infected with small larval stages of trematodes that utilize the birds as final hosts. The infective stages are so small that visual inspection is not possible, and thus the oystercatchers employ an indirect avoidance strategy. In general, the energy intake to oystercatchers resulting from ingesting a single cockle is positively correlated with the cockle size, but with increasing size cockles also contain higher loads of the larval trematodes (**Figure 3**). Although they would be more profitable in terms of energy intake, oystercatchers avoid large cockles and preferably prey upon smaller sizes. This is considered to be an avoidance mechanism against acquiring high parasite loads. However, total avoidance is not possible in this case as all cockle size classes are infected with parasites to some degree.

There may also be food-related avoidance mechanisms in humans. For example, disgust as a human emotion might have evolved to protect us from the risk of disease by preventing us from ingesting potentially infectious food items. It might also work with nonfood acquired parasites, as disgust responses in test persons are also common to other objects in the environment that represent potential threats in terms of infection, like images of wounds, sick people, and ectoparasites. Another way humans avoid contact with infective agents might be use of spices in cooking. When the frequency of spice use in meat-based recipes was compared among 36 countries and correlated with the local temperature regimes, an

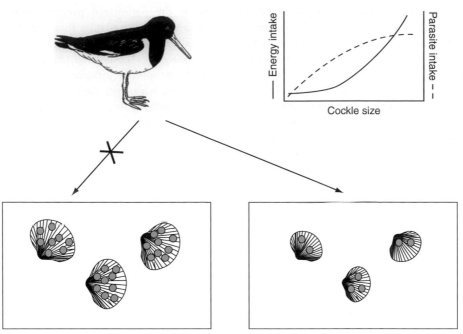

Figure 3 Selective prey selection can help to avoid acquiring parasites and pathogens. Oystercatchers take smaller cockle prey individuals with suboptimal energy gains. Smaller cockles harbor fewer infective stages of trematode parasites (green dots), which utilize the birds as final hosts, compared to larger cockles. By preferring smaller cockles the birds thus reduce their parasite intake and avoid acquiring high loads of parasites. Total avoidance of parasites is not possible in this case because all cockle sizes are infected to some extent. Schematic graph based on Norris K (1992) A trade-off between energy intake and exposure to parasites in oystercatchers feeding on a bivalve mollusc. *Proceedings of the Royal Society of London Series B* 266: 1703–1709.

interesting pattern emerged. As mean annual temperatures (being a proxy for the relative spoilage rates of unrefrigerated foods) increased, the proportion of recipes containing spices as well as the number of spices per recipe and the total number of spices used increased (Billing and Sherman, 1998). As spices often contain antimicrobial secondary metabolites (with garlic, onion, oregano, thyme, cinnamon, tarragon, cumin cloves, lemon grass, bay leave, capsicums, and rosemary being effective against >75% of bacteria tested), the use of spice might be an effective prophylaxis against pathogen contact. Cannibalism taboos might serve a similar function and prevent contact with parasites and pathogens. Such aversion to cannibalism can also be found in various animals and might be a very general avoidance measure against parasites and pathogens.

Joining Others to Form Groups

Group-living is often thought as mostly an adaptation against predation: think of a school of fish or a herd of wildebeests. Forming groups can also help to avoid parasites and pathogens. In fish, schooling can protect individual hosts from free-swimming, blood-sucking crustacean parasites. In experimental situations, sticklebacks exposed to the highly mobile crustacean ectoparasite *Argulus canadensis*

formed larger schools than control fish kept in identical conditions but without parasites. In addition, the attack success of the ectoparasite decreased as the size of the school it attacked increased. Similarly, in birds and mammals, individual hosts experience reduced fly bites with increasing group size (dilution) as long as the larger group does not attract more flies *per capita* due to its higher visibility (encounter). This has been coined the encounter–dilution effect. A good example of this phenomenon is seen in heifers, which usually graze in a normally dispersed manner on the pasture. However, when heifers are attacked by horse flies, they form grazing lines and continue grazing by moving along parallel to each other with the dominant animals well protected in the center. If the horse fly attacks become more severe, the heifers will stop grazing and form bunches. With this behavior, the heifers reduce the number of bites per individual and thus protect themselves from annoying bites and also from transmission of arthropod-borne pathogens. However, this dilution effect resulting from grouping only works against attacks of biting flies that satiate after biting one or two hosts. Otherwise grouping might actually increase the risk of becoming bitten. Besides cattle, caribou, reindeer, horses, and primates are known to form larger groups when biting fly intensity is high, and some studies have shown that denser grouping actually reduces the number of bites per host. Grouping is also an effective avoidance measure against parasites and pathogens in birds. Black grouse in Finland are often harassed by black flies during summer. By grouping together, grouse reduce their individual risk of getting bitten by these flies. As black flies transmit the two most common blood parasites of grouse, this behavior is an effective avoidance measure and a convincing example for the encounter–dilution effect in birds (**Figure 4**).

To protect themselves even more, individual hosts can position themselves in the center of a large group and thus further reduce the risk of contact with parasites and pathogens. This is called the selfish-herd effect; it can help individual hosts to avoid fly bites and potential subsequent infections with pathogens if they manage to position themselves in the center of a group where exposure is lower than at the periphery (**Figure 4**). While there are only a few tentative examples from birds, this avoidance strategy has been well documented in mammals. In reindeer calves parasitized by warble flies, individual parasite load not only decreases in larger groups because of the encounter–dilution effect, but also decreases with body

mass of individual calves. In reindeer, body mass is itself positively correlated with fitness and social status, and heavier individuals are thus better able to occupy the best central positions within the herd that provide protection from fly bites via the selfish-herd effect.

The Evolutionary Ecology of Avoidance

Avoidance is only the first line of defence, and natural selection has favored other safeguards in case avoidance fails. This second line of defence ranges from mechanical and physiological barriers, behavioral defences like grooming and preening, all the way to immunological defenses. We seem to know much more about the function and costs of these postcontact defensive measures, especially immune responses, than we do about avoidance behaviors. Although some behaviors, like nest site selection in birds, have been thoroughly studied using an experimental approach, many other apparently efficient avoidance mechanisms are only supported by anecdotal or circumstantial evidence. These suggest that much of what an animal does is aimed at avoiding parasites, and yet several important questions about the evolution and ecology of parasite avoidance remain unanswered. Most of these questions could be tackled using either an experimental or a comparative approach. What is the actual cost of avoidance behaviors? How have avoidance behaviors evolved, and were they originally serving a different purpose before being co-opted for defence against parasites? Is the diversity of avoidance behaviors shown by an animal, or the time and energy invested in their expression, roughly proportional to the number of different parasite species, their virulence, or their local abundance, that this animal faces? Many more questions come to mind, but the main one concerns the effectiveness of avoidance behaviors. Clearly, parasites are doing well: some estimates suggest that more than half of living species are parasitic and that all the remaining free-living animals have parasites. Within any vertebrate population, it is almost impossible to find a single individual that does not harbor at least some parasites. It is easy to argue that many individuals escape predation and hence that antipredation behaviors must work. But no one escapes parasitism. Still, it may be that antiparasite defences are very efficient and that the average number of parasites per host would be several times higher without these defences.

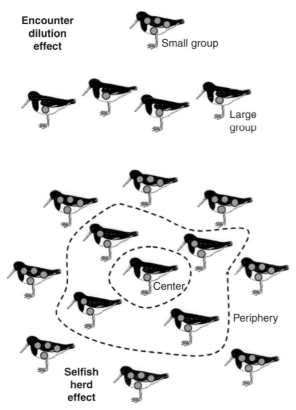

Figure 4 Forming groups can help to avoid parasites and pathogens. In birds and mammals, individual hosts suffer from reduced fly bites (green dots) per capita with increasing group size (dilution), as long as the larger group does not attract more flies due to a higher visibility (encounter). Positioning themselves at the center of a large group can also help individual hosts to avoid fly bites and potential subsequent infections via the selfish-herd effect.

See also: Intermediate Host Behavior; Parasites and Sexual Selection; Threespine Stickleback.

Further Reading

Billing J and Sherman PW (1998) Antimicrobial functions of species: Why some like it hot. *Quarterly Review of Biology* 73: 3–49.

Clark L and Mason JR (1985) Use of nest material as insecticidal and anti-pathogenic agents by the European straling. *Oecologia* 67: 169–176.

Curtis V, Aunger R, and Rable T (2004) Evidence that disgust evolved to protect from risk of disease. *Biology Letters* 271: S131–S133.

Folstad I, Nilssen AC, Halvorsen O, and Andersen J (1991) Parasite avoidance: The cause of post-calving migrations in *Rangifer*? *Canadian Journal of Zoology* 69: 2423–2429.

Hart BL (1994) Behavioural defences against parasites: Interaction with parasite invasiveness. *Parasitology* 109: S139–S151.

Hart BL (1997) Behavioural defence. In: Clayton DH and Moore J (eds.) *Host-Parasite Evolution: General Principles and Avian Models*, pp. 59–77. Oxford: Oxford University Press.

Karvonen A, Seppälä O, and Valtonen ET (2004) Parasite resistance and avoidance behaviour in preventing eye fluke infections in fish. *Parasitology* 129: 159–164.

Lowenberger CA and Rau ME (1994) Selective oviposition by *Aedes aegypti* (Diptera: Culicidae) in response to a larval parasite, *Plagiorchis elegans* (Trematoda: Plagiorchiidae). *Environmental Entomology* 23: 1269–1276.

Moore J (2002) *Parasites and the Behavior of Animals*. New York: Oxford University Press.

Mooring MS and Hart BL (1992) Animal grouping for protection from parasites: Selfish herd and encounter-dilution effects. *Behaviour* 123: 173–193.

Norris K (1992) A trade-off between energy intake and exposure to parasites in oystercatchers feeding on a bivalve mollusc. *Proceedings of the Royal Society of London Series B* 266: 1703–1709.

Sheldon BC and Verhulst S (1996) Ecological immunology: Costly parasite defences and trade-offs in evolutionary ecology. *Trends in Ecology and Evolution* 11: 317–321.

Waage JK (1981) How the zebra got its stripes: Biting flies as selective agents in the evolution of zebra coloration. *Journal of the Entomological Society of South Africa* 41: 351–358.

Barn Swallows: Sexual and Social Behavior

R. Safran, University of Colorado, Boulder, CO, USA

An inhabitant of most of the Holarctic (with the exception of Greenland and Iceland), the barn swallow is the most widespread species of the swallow family, *Hirundinidae*. The extensive breeding range of the barn swallow is believed to be due to their close association with human populations. On the basis of human colonization patterns across Eurasia and recent studies of the colonization of swallows, it appears that this close association with humans has persisted for millennia. Indeed, nearly everywhere you find a barn, building, or bridge, especially if these are situated near water and fields, you find the swallow's mud cup nest tucked in under the eaves or constructed along beams and planks. As such, the barn swallow, or simply the swallow as it is called throughout much of its range, is well known. Swallows are also well-loved as evidenced by hundreds of examples of their portraiture in fine and folk art. Some of the earliest discoveries of barn swallow art date back to the Bronze Age: Recent discoveries of cave paintings from the ancient society, Thera, feature swallows in flight fighting over feathers used as nest lining – a behavior that persists even today!

Formal studies of barn swallow behavior began with publications in the early part of the last century and number well into the high hundreds. These studies represent a tremendous amount of breadth ranging from classic ethological studies of parental behavior to sophisticated molecular studies of physiology and reproductive biology. Much of the detail of this research is covered in one of two academic books published on barn swallows. The first, written by Anders Møller in 1994, focuses on sexual selection, with an emphasis on the author's incredible long-term data set on swallows. The second, published very recently (in 2006, entitled *The Barn Swallow*) by Angela Turner is a comprehensive review, ranging from conservation status to taxonomy. In this short review, it is difficult to even scratch the surface of the incredible wealth of knowledge accumulated on this 17–20 g bird, and I will first focus on a trait that these birds are arguably the most famous for: their tail streamers. Indeed, much research on the barn swallow revolves around this trait

(**Figure 1**), as their streamers impact mate-selection, nest construction, flight aerodynamics, parental care (**Figure 2**), and physiology. Moreover, the tale of tail streamers has as many interesting twists and turns as this trait causes its bearer to make in flight; the tail of the swallow appears to be constantly evolving and changing in different ways depending on where one studies it. Even within populations, there is much debate on the kind of information this trait conveys to conspecifics. In the latter half of this review, I summarize what is known about the fascinating variation in the social behavior of swallows.

For Fancy, or Flight, or Both: The Controversy About Tails

One of the best-known articles on barn swallows was written by Anders Møller in 1988. This article, published in *Nature* and cited over 300 times, employed techniques of tail manipulation pioneered by Malte Andersson to examine the relationship between the streamer length of male barn swallows and their mating success. The elegant experimental design involved looking at the pairing dates of males randomly assigned to four treatment groups: males whose streamers were artificially elongated by 20 mm, males whose streamers were artificially shortened by the same length, control males whose streamers were cut and reglued, and yet another control group of males whose streamers were not manipulated at all.

That males with elongated streamers attracted mates earlier than their neighbors with short-streamers was the first demonstration of a causal relationship between male tail length and female mate choice. Indeed, since that article and the dozens that have followed it, tail streamers in the European population of the barn swallow *H. rustica rustica* have become a textbook example of sexual selection. Experimental and correlational studies show that females prefer males with the longest tail streamers, and among paired individuals, female tail length is positively correlated with male streamer length, providing evidence

Figure 1   A North American swallow (*H. rustica erythrogaster*) in flight. In North American populations of barn swallows, ventral plumage color is typically darker compared to the European nominate subspecies (*H. rustica rustica*), whereas tail streamers are shorter. Photo credit: © Marie Read.

Figure 2   A North American swallow attending young nestlings. Both males and females participate in feeding of shared offspring, although paternity is typically mixed within each brood. Interestingly, the degree to which males participate in incubation varies among subspecies of swallows, males participating to a greater degree in North America compared to Europe. Photo credit: © Marie Read.

for assortative mating on the basis of this trait. Long-tailed males produce the most offspring (in their first clutch and total number of young per season) each year because they pair and breed earlier and successfully fledge more broods than males with shorter tails.

Of course, merely counting the number of chicks in the nest that a male is provisioning is not enough to truly understand his evolutionary fitness. Barn swallows, like so many other social animals, have complicated sex lives. They form a cooperative social pair bond that can last throughout an entire breeding season, or longer, but they also pursue extra-pair mating strategies on the side. Indeed, molecular parentage analyses provide the only definitive way to measure the reproductive activities of a

male; use of these methods allows for more accurate assessment of the amount of sexual selection associated with streamer lengths.

To confirm the correlation between a male's streamer length and his social mating success, researchers in Europe also showed that males with the longest streamers enjoy a significantly greater share of paternity in their nests and the nests of others, relative to their short-streamered neighbors. In fact, Saino and collaborators replicated Møller's classic tail manipulation experiment to look for paternity differences among males in the four treatment groups and as predicted, found that males whose streamers were elongated sired more offspring in their nests and those of others compared to males in the shortened and control groups. Studies of extra-pair mating strategies in other populations of barn swallows (throughout Europe and North America) found that the percentage of broods with extra-pair young ranges from 33% to 50%. As shown in the Geographic Variation section of this article, the relationship between extra-pair mating success and ornamental traits becomes very important when comparing the role of sexual selection for shaping male appearance both within and among populations of swallows.

But tail streamers are also critical to barn swallow flight performance, as they need to function efficiently for these acrobatic aerial insectivores. For evidence that tails are important outside of mating, one needs to look no further than female and juvenile barn swallows – they too exhibit extensively forked tails. Research has also shown that males with the longest tail streamers pay costs associated with bearing this trait. A year after publishing his first experimental paper on tail streamer manipulations, Møller demonstrated lower survival for males carrying elongated streamers, suggesting that these traits are cumbersome in flight.

If longer streamers impose a burden, this trait could convey honest information about a male's ability to bear the costs of his long tail and also to maintain a high-quality nest location. Interestingly, rather than appearing to be solely under directional sexual selection (as would be predicted if long streamered males were always chosen as the favorite mates), this trait appears to be an interesting balance of both sexual and natural or survival-based selection. Previous studies suggest that individuals with longer streamers suffer from impaired aerodynamic performance that may result in lower foraging efficiency. Swallows with too short a set of streamers also suffer from reduced flight skills. The balance between too long and too short implies that natural selection already shaped the morphology of this species to accommodate elongation and sexual dimorphism of tail streamers. It appears that tail streamer lengths represent a tug of war consequence between sexual and natural selection; how much of each form of selection has contributed to the evolution of this trait has generated great controversy, stirred by the elegant aerodynamic performance studies of Matthew Evans and colleagues since the 1990s.

Recently, a novel set of experiments conducted by Jakob Brø-Jorgensen, Rufus Johnstone, and Matthew Evans utilized an individual-based approach to identify the extent to which variation in the length between a male's streamers either reflects differential ability to withstand the costs of 'too long' streamers, as predicted by sexual selection, or represents the individual-specific match between body size and tail streamer length to optimize flight and foraging performance, as predicted by survival-based natural selection. Through the analysis of aerodynamic performance in a flight maze after a series of manipulations of the same individuals' tail lengths, these researchers, working in a Scottish population of swallows, worked out the relative importance of natural and sexual selection contributing to the variation in the length of the tail streamer.

The conclusions of this article are surprising as they found no evidence to support the prevailing view that the sexually selected component of this trait reflects individual variation in some aspects of male quality which would serve as advertisements to choosy females or competitive males. Instead, the authors suggest that the optimal streamer length for flight varies significantly among males, but that the additional component of the streamer – assumed to be caused by sexual selection – does not. The conclusion, which counters the patterns predicted for variable sexdimorphic traits under sexual selection, is that the naturally selected – and not the sexually selected – component of the streamer conveys information about a male's flight and foraging performance, leaving open the question of why streamers are elongated past this optimal value. To interpret their findings, Evans and colleagues speculate that tail streamer lengths may simply serve to signal the age and sex of the individual (adult male vs. female or juvenile). Further experimental studies that adopt this highly powerful within-individual experimental approach with additional treatments related to mate-selection may provide a definitive test for understanding the likely contributions of both sexual and natural selection on this trait.

Geographic Variation in Phenotypes

The pursuit of whether natural selection, sexual selection, or likely both cause streamer elongation is far from over. Intriguing phenotypic differences in tail streamer length and plumage color exist among the six most well-known subspecies of barn swallows (it is speculated that there are several more subspecies throughout the enormous breeding range of swallows). Combinations of tail and color are not correlated, that is, dark color does not imply longer streamers and statistically, variation in one trait does not at all predict variation in the other. Looking at the average phenotypes of males from throughout the

Holarctic region, one sees nearly all possible pair-wise combinations of color and streamer length. Males of European *H. rustica* subspecies swallows have nearly the palest ventral color and the most exaggerated tail streamer lengths of all of the barn swallows, while swallows from the North American populations are substantially more colorful, with streamers that are among the shortest of all subspecies (**Figures 1** and **2**). Intriguingly, populations from the two Middle Eastern subspecies (*savignii* along the Nile and *transitiva* throughout Israel, Lebanon, Jordan, and Syria) have combinations of dark plumage coloration with streamer lengths that are almost the same as those of the European subspecies. Populations in northern Asia (*H. rustica tyleri*) possess intermediate values of streamer lengths and feather color relative to their conspecifics while *H. rustica gutturalis*, which occurs throughout much of Asia, has among the least exaggerated features of all, with the palest ventral color and shortest tail streamers. Though differentially sexually dimorphic with respect to both streamer lengths and color, differences in female morphology are highly concordant with differences in males throughout the entire range of this species complex. Ongoing research is focused on determining the underlying causes of these fascinating phenotypic differences.

Differential Sexual Selection?

What causes differences in the phenotypic variation among the subspecies of barn swallows? Three ecological variables are likely to play a key role. First, most populations are migratory but the Middle Eastern populations are not. Second, there are interesting differences in the extent to which males participate in parental care. Finally, latitudinal differences in streamer length (longer in the north for the most widespread populations in Europe and North America, though not a sweeping generalization for the species complex as a whole) are the rule, though this pattern remains unexplained. Sexual selection is also likely to be playing a role, since many of the phenotypic differences among populations are seen in sexually dimorphic traits. The hypothesis that sexual selection operates differently on streamer length and color among various populations is under current study in three subspecies: *H. rustica rustica*, *H. rustica erythrogaster*, and *H. rustica transitiva*, for which phenotype manipulation experiments will likely reveal interesting differences in the role of mateselection decisions related to these traits.

Sexually selected traits are often sexually dimorphic, predict patterns in mate-selection, and show a relationship with indices or measures of individuality present. While streamer lengths are sexually dimorphic in North America (though to a lesser extent than the dimorphism of streamers in western Europe), studies of the sexual selection of tail streamers of North American populations of barn swallows have yielded mixed results, with an overall impression that sexual selection is at the very least a lot weaker on this trait in North America. For example, streamer variation in males and females does not predict patterns of assortative pairing in *H. rustica erythrogaster*, as is the case in European populations. Male streamer length is not a predictor of many measures of seasonal reproductive success in most correlational data sets, with the exception of a paternity study conducted by Oddmund Kleven and colleagues that I describe later in this article. Collectively, these results may indicate reduced or absent sexual selection on this trait in this continental population.

Some other interesting comparisons between studies of males in North America and Europe are also noteworthy; Nicola Saino and colleagues reported a significant positive association between streamer length and the proportion of offspring sired in first breeding attempts in a northern Italian population of barn swallows, whereas Colby Neuman and others found no association between these two variables in North America using the same test statistic. Likewise, Anders Møller and colleagues report a significant linear relationship between the proportion of offspring sired by the resident male of the nest in relation to his streamer length from a population near Milan, Italy, whereas no such relationship was found in males from Ithaca, New York, using the same type of data analyses. Anders Møller and Håkan Tegelström, in the late 1990s, reported a negative correspondence between the proportion of broods being sired by extra-pair males and the streamer length of the male nest owner in a population in Denmark, indicating that longer-tailed males are less likely to be cuckolded. However, using the same statistical data analyses, Rebecca Safran and colleagues found no such correspondence between a male's streamer length and his probability of being cuckolded.

To date, only one study has experimentally manipulated the streamer length of North American barn swallows. Unfortunately, this experiment, designed to replicate Møller's 1988 study, is difficult to interpret because of the small sample sizes. Though Hendrik Smith and collaborators found that males whose streamers were experimentally elongated attracted social mates earlier in the breeding season than those whose streamers were shortened, these long-tailed males received less paternity from their social mates, compared to males with shortened tails. Interestingly, Oddmund Kleven went back to the same study sites in Ontario, Canada, nearly 15 years after the original experiment was published to conduct a large paternity study. Kleven and colleagues report that males with naturally long streamers received extra-pair benefits from females outside their social pair bond, but not within-pair benefits from their own mates, compared to their shorter-streamered neighbors. It is difficult to reconcile these results with others conducted on this subspecies; and

as previously mentioned, large-scale tail manipulation experiments in North American populations of barn swallows currently underway, are sorely needed.

The Color of Feathers

If tail streamers do not drive mate choice, what does? Recently, Rebecca Safran and Kevin McGraw found that ventral coloration, not streamer length, is correlated with patterns of pairing and seasonal reproductive success in a population of North American barn swallows. Experimental manipulations of male coloration demonstrated that individuals use this trait to assess male quality. Feather color in barn swallows is derived from melanin-based pigments; these are produced by the birds and, as such, do not reflect an individual's diet directly, as is the case with the beautiful pink feathers of the flamingo or the bright red beak of the zebra finch. Though we know little about why females might favor the use of color for mate-selection in one population and streamer length in another, Safran and colleagues recently demonstrated a causal relationship between coloration and testosterone, a sex steroid often linked with aggressive and sexual behavior. Darker males with higher levels of circulating testosterone in the early part of the breeding season may be more competitive for high-quality nesting territories. Further studies on the underlying production costs of streamers and ventral color would be particularly illuminating.

Explanations for Geographic Variation in Tail Streamers

There is mounting evidence that the function of elongated streamers varies between European and North American populations, and perhaps others including *H. rustica transitiva* in Israel.

Interestingly, despite latitudinal variation in streamer lengths in European populations so that males in Denmark have longer streamers compared to males in Italy, the function of streamers, in terms of the benefits of social and genetic reproductive success, does not vary tremendously between these two populations. Although the breeding latitude of males in North America most closely corresponds to males in the Italian study areas, there were no similarities in the benefits from elongated streamers in a population in New York compared to males in the intensively studied population near Milan. In the Italian population benefits associated with this sexual signal are apparent, while they are not in New York. Considered in concert, the results of studies in North America demonstrate that the pattern of sexual selection on tail streamers varies geographically. Hendrik Smith and Robert Montgomerie suggest that this geographic variation may relate to differences in male behavior during the incubation period, as male barn swallows in North American spend ~12% of daylight hours on the nest during the incubation stage of the breeding cycle, while males in the European population do not participate in incubation. It is possible that the longer-tailed males in North America may be at a higher risk of tail streamer breakage during incubation at nests as streamers often brush against a wall or roof. The resulting broken streamers may be shorter than the aerodynamic optimum, thereby decreasing the fitness of the bird.

This explanation is not entirely compelling because the average length of male streamers in North America is equivalent to those of females in Europe. The females' streamers would be even more subject to abrasion during entry into the nest as the female is the sole incubating parent in that population. Potentially, males in North America have less time to forage because of their incubation duties, and therefore they must be more efficient flyers. The additional time constraint of incubation may be sufficient enough to select against those individuals whose tails are beyond the aerodynamic optimum. Consistent with this explanation, previous studies have found that only a small distal region of the tail streamer (~10–15 mm) in the European population appears to be under sexual selection, while the majority of the tail streamer length has evolved to a naturally selected aerodynamic optimum that is very similar to the shorter mean streamer length in the North American population.

Because male ventral coloration predicts patterns of social and genetic reproductive success, in addition to influencing his mate's rate of parental care (females feed more to shared offspring when paired to darker males, **Figure 2**), feather coloration may be a more reliable signal of male quality than tail streamer length in North American populations. As mentioned earlier, the jury is still out as to why this trait might be more informative than streamer lengths.

Sociable Swallows

The physical appearance of barn swallows is not the only highly variable feature of this fascinating species. In fact, early studies of this species by Barbara Snapp in Ithaca, New York, and later by Anders Møller in Denmark and William Shields in a separate population in northern New York focused on variation in the sociality of barn swallows. Throughout their extensive breeding range, barn swallows breed in solitary pairs or with groups of conspecifics; they are not obligately social breeders. Typically, colony sizes range from 2 to 200 breeding pairs, with the majority of individuals breeding either solitarily or in groups ranging from 9 to 35 pairs.

Early studies demonstrated few benefits and many costs for group breeding for barn swallows. Barbara Snapp's pioneering studies of social behavior found none of the benefits to group-breeding barn swallows that are typically found in other highly social organisms. Barn swallows in her study area near Ithaca, New York, received no benefits from social foraging or collective predator defense. Snapp concluded that barn swallows breed in groups as a function of limited nest sites. Similarly, Anders Møller, working in Denmark, found no net social foraging benefits to group breeding, yet he did detect slightly shorter reaction times in larger colonies to the experimental presentation of a potential nest predator.

Møller concluded that group breeding in barn swallows may be beneficial to older males and unpaired males. These males gain extra-pair mating opportunities in social groups, but this does not explain why females or younger males tolerate the costs of sociality. Møller defined the costs in terms of competition for food, infanticide, nest parasitism, and parasite transmission. Another long-term study of barn swallow sociality in New York by Shields and colleagues generated an overall assessment of group breeding that was similar to Snapp's – that ideal nest sites are limiting. As a consequence, these researchers developed the *traditional aggregation hypothesis*, which predicts that group breeding is related to nest-site selection behavior.

Overall, research on group living in barn swallows has shown either a negative relationship or no relationship at all between average reproductive success and group size, leaving open the question of why individuals breed socially.

A distinctive attribute of many species in swallow family (Hirundinidae) is the persistent use of mud nests across breeding seasons. The reuse of old nests is a predominant nest-site selection strategy of barn swallows across their extensive breeding range. Anywhere from 45% to 82% of pairs reuse old nests for their first breeding attempts. Once constructed, nests can persist in the environment for decades, and the majority of breeding pairs at a site attempt to refurbish or reuse these structures instead of constructing new ones. Pairs settling in old nests for first breeding attempts lay eggs earlier and have greater numbers of fledged young compared to pairs that construct new nests at the start of the breeding season, regardless of their previous breeding experience. A primary benefit from reusing old nests is that these pairs breed earlier than those that construct new nests at the start of the season. Evidence also suggests that individuals avoid the costs associated with ectoparasites by selectively avoiding old nests with remnant mite populations. Because nests and nest scars are only rarely completely removed from sites between breeding seasons, it is logical to assume that these nests offer important information to individuals making decisions about where to breed.

A fascinating consequence of nest reuse is that the number of old nests at a breeding site strongly predicts the number of breeding pairs that settle there. Because site fidelity is the rule in barn swallows with prior breeding experience (natal philopatry – the return to the birth site in a following season – is incredibly low), group breeding persists even in the absence of old nests, suggesting strong benefits of site familiarity. In order to truly demonstrate that group size is a function of individuals searching for old nests, a critical experiment tested for a relationship between the number of immigrants that settle at sites and the number of old nests at the site at the start of the season. In the same breeding population as Barbara Snapp's studies but nearly three decades later, Rebecca Safran compared the return and immigration rates of adults at sites where all old nests had been experimentally removed and sites where old nests remained untouched between breeding seasons. That the proportion of immigrants was significantly lower during removal years and the number of immigrants was positively related to the number of old nests collectively provided compelling evidence that group size is strongly influenced by the number of new breeders at a site. In turn, the number of immigrants was experimentally shown to be related to the number of old nests at a site at the start of the breeding season. This strong relationship between the number of old nests and the number of immigrants settling at a site suggests that not only do immigrants use old nests as a cue for settlement decisions, but they also settle with a probability that is proportional to the number of old nests at a breeding location. Experiments designed to analyze further the benefits of site fidelity *per se* in the absence and presence of old nests would provide further resolution on the relationship between group size and the number of old nests present at a site.

The Past, Present, and Future

Having been featured so prominently in the biological literature, it is difficult to leave out the dozens of other reasons why barn swallows are wonderful subjects for studies related to animal behavior. Besides being tractable, easy to handle, robust to manipulation both during and after handling, and fairly common, they are highly variable in so many morphological and behavioral dimensions. I have mentioned a few here and Further Reading is offered to provide more details. Angela Turner's recent book will be extremely helpful to those who want more information.

Sadly, it is common these days to conclude an article like this with the bad news. Like so many other species on our planet, barn swallow populations appear to be declining. Formal demographic studies throughout Europe and anecdotal stories from elsewhere are providing sobering evidence that this once hugely abundant species is dwindling throughout its range. Though it is still common enough to observe swallows in flight almost everywhere you look, changes in agricultural practices and the move

toward metal and concrete over the use of wood for barn construction, and the usual detrimental effects related to human population growth appear to be taking their toll. One can purchase artificial nests or provide wooden ledges within buildings that might otherwise prove inhospitable to these beautiful birds. Reduced pesticide use will also help boost the populations of aerial insects upon which these birds rely.

Two comprehensive reviews of barn swallows have been published and are recommended here. The first is a treatment of sexual selection in European barn swallows published in 1994 by Anders Møller; as such, I provide suggestions related to sexual selection that were published after this book or that deal with sexual selection in North American populations. Many of the references given in the Further Reading are studies that were published after Angela Turner's wonderful synthesis of recent literature on barn swallows in 2006.

See also: Mate Choice in Males and Females; Social Selection, Sexual Selection, and Sexual Conflict; Visual Signals.

Further Reading

Brø-Jorgensen J, Johnstone RA, and Evans MR (2007) Uninformative exaggeration of male sexual ornaments in barn swallows. *Current Biology* 17: 850–855.

Brown CR and Brown MB (1999) Barn swallow (*Hirundo rustica*). No. 42. In: Poole A and Gill F (eds.) *The Birds of North America*. Philadelphia, PA: The Birds of North America, Inc.

Evans MR (1998) Selection on swallow tail streamers. *Nature* 394: 233–234.

Kleven O, Jacobsen F, Izadnegahdar R, Robertson RJ, and Lifjeld JT (2006) Male tail streamer length predicts fertilization success in the North American barn swallow (*Hirundo rustica erythrogaster*). *Behavioral Ecology and Sociobiology* 59: 412–418.

Møller AP (1987) The advantages and disadvantages of coloniality in the swallow *Hirundo rustica*. *Animal Behaviour* 35: 819–832.

Møller AP (1994) *Sexual Selection and the Barn Swallow*. Oxford: Oxford University Press.

Neuman CR, Safran RJ, and Lovette IJ (2007) Male tail streamer length does not predict apparent or genetic reproductive success in North American barn swallows. *Journal of Avian Biology* 38: 28–36.

Safran RJ (2004) Adaptive site selection rules and variation in group size of barn swallows: Individual decisions predict population patterns. *American Naturalist* 164: 121–131.

Safran RJ, Adolman J, McGraw KJ, and Hau M (2008) Sexual signal exaggeration affects physiological state in a social vertebrate. *Current Biology* 18: R461–R462.

Safran RJ, Neuman CR, McGraw KJ, and Lovette IJ (2005) Dynamic paternityallocation as a function of male color in barn swallows. *Science* 309: 2210–2212.

Shields WM and Crook JR (1987) Barn swallow coloniality: A net cost for group breeding in the Adirondacks? *Ecology* 68: 1373–1386.

Smith HG and Montgomerie R (1991) Sexual selection and tail ornaments of North American barn swallows. *Behavioral Ecology and Sociobiology* 28: 195–201.

Snapp BD (1976) Colonial breeding in the Barn swallow and its adaptive significance. *Condor* 78: 471–480.

Turner AK (2006) *The Barn Swallow*. London: T & AD Poyser.

Zink RM, Rohwer PA, and Drovetski SV (2006) Barn swallows before barns: Population histories and intercontinental colonization. *Proceedings of the Royal Society of London B* 273: 1245–1251.

Bat Migration

T. H. Fleming, University of Miami, Coral Gables, FL, USA

Introduction

Migration is an essential feature of the life history of a substantial fraction of the world's animal fauna. Among vertebrates, migration, which can be defined as a seasonal, usually two-way, movement from one habitat to another to avoid unfavorable climatic conditions and/or to seek more favorable energetic conditions, is common in fish and birds. It is less common in amphibians, reptiles, and mammals. Among volant vertebrates, it is much more common in birds than in bats, and birds migrate much longer distances, on average, than bats. Nonetheless, a considerable number of bats, including both temperate and tropical species, undergo significant seasonal movements between habitats. As I discuss in this article, these movements have important population and conservation consequences. In a world of global climate change and increased extinction risk for an estimated 25% of the world's mammals, migratory species are especially vulnerable to extinction risks and are of major conservation concern. These concerns include the effect of wind turbine farms on migratory bats and their roles as reservoirs for emerging diseases that can be fatal to humans.

An Overview of Bat Migration

Temperate Bats

Until recently, migration in bats had mostly been studied in temperate regions of North America and Europe where migratory behavior is closely associated with hibernation. Results of those studies indicate that temperate bats exhibit three broad patterns of spatial behavior: (i) sedentary (non-migratory) behavior in which bats breed and hibernate within a 50-km radius or less; (ii) regional migration in which bats migrate 100–500 km between summer and winter roosts; and (iii) long-distance migrants in which bats migrate 1000 km or more between seasonal roosts. Examples of European sedentary taxa include species of *Eptesicus*, *Plecotus*, and *Rhinolophus* and certain species of *Myotis*. Their North American counterparts include *Eptesicus fuscus*, *Corynorhinus rafinesquii*, and *Antrozous pallidus*. European regional migrants include several species of *Myotis*, and North American taxa include several species of *Myotis* and *Pipistrellus (or Perimyotis) subflavus*. European long-distance migrants include several species of *Nyctalus* as well as two species of *Pipistrellus* and *Vespertilio murinus*. Although they migrate relatively long distances between summer and winter roosts, they do so within, rather than between, continents, unlike many European migratory birds which are intercontinental migrants. In North America, long-distance migrants include species of *Lasiurus*, *Lasionycteris noctivagans*, and the subtropical/tropical seasonal migrants *Leptonycteris curasoae*, *L. nivalis*, *Choeronycteris mexicana*, and *Tadarida brasiliensis*. With the possible exception of *Lasiurus* species, these taxa are also intracontinental migrants. Unlike other temperate migrants, these species do not hibernate.

As is the case in birds, temperate zone long-distance migrant bats tend to differ morphologically from more sedentary species and are adapted for rapid, energetically efficient flight. They have wings with high aspect ratios (i.e., they are long and narrow), pointed wing tips, and high wing loading. As a result, most insectivorous long-distance migrants forage in uncluttered air space away from vegetation where slow, highly maneuverable flight is not needed. In addition, many long-distance migrants (e.g., the European species plus *Lasiurus* and *Lasionycteris* in North America) roost in small colonies (or are solitary) in trees and buildings rather than in large numbers in caves, as is often the case in regional migrants and sedentary species. Exceptions to this include North American *T. brasiliensis* and *L. curasoae* which are highly gregarious cave dwellers year round.

Not all individuals of species of temperate migratory bats undergo seasonal migrations. Partial migration occurs in a number of species, including long-distance migrants. In Europe, partial migrants include species of *Nyctalus* and *Pipistrellus* and *V. murinus*. In North America, partial migrants include *Lasiurus cinereus*, *L. noctivagans*, *L. curasoae*, and *T. brasiliensis*. In Europe, sedentary populations of *P. nathusii* share hibernation caves with migratory populations, whereas sedentary and migratory populations of *V. murinus* have geographically separate ranges year round. A situation similar to that of *V. murinus* is seen in North American *T. brasiliensis* in which western and southeastern populations in the United States are sedentary and seasonal populations in the south-central United States overwinter in Mexico.

Regardless of the distances involved, most temperate zone migratory bats undergo a characteristic annual physiological and reproductive cycle that is closely tied to hibernation. This cycle includes hyperphagia and fat deposition in the fall and mating in the fall or winter. Unlike birds, which sometimes increase their body mass by 50% by fat deposition prior to migration, bats increase their mass by only 12–26% by fat deposition. Also, unlike birds, bats must use most of the fat they deposit prior to or during migration as a fuel source during hibernation. Conservation of stored fat for use during hibernation is an important reason why many temperate bats migrate relatively short distances. Finally, the annual reproductive cycle of temperate bats usually involves mating in the fall prior to hibernation. In some species (e.g., *N. noctula*, *P. nathusii*), mating occurs along female migratory pathways with males defending roost sites which are visited by females for mating. In other species (e.g., *Myotis* species in North America), individuals of both sexes form 'swarms' at the entrances of hibernation or other caves, probably for mating. After mating, females of hibernating species store viable sperm in their oviducts during the winter and ovulate and undergo fertilization in the spring prior to migrating to their summer maternity roosts. In nonhibernating long-distance migrants such as *Lasiurus borealis*, *L. curasoae*, and *T. brasiliensis*, mating and fertilization occur simultaneously prior to or during spring migration.

Tropical Bats

Migratory behavior is much less common in tropical bats than in temperate bats and is never associated with hibernation. Whereas temperate bats migrate and hibernate to avoid habitats that are energetically and physiologically unfavorable during winter, tropical, and subtropical bats usually migrate along food resource gradients or among seasonally ephemeral resource patches. A clear example of this is the nectar-feeding, long-distance migrant *L. curasoae* in Mexico and the southwestern United States. During the fall and winter, populations in western Mexico live in tropical dry forest and visit flowers produced by trees and shrubs during an annual flowering peak. After mating, many pregnant females (but few males) migrate 1000 km or more to the Sonoran Desert of northwestern Mexico and Arizona along a 'nectar corridor' of blooming columnar cacti. Once in the Sonoran Desert, they form maternity colonies and feed at a super-rich source of nectar and pollen produced by several species of spring-blooming columnar cacti. Populations of this species living in central or southern Mexico are more sedentary because their floral resources are available year round. In Africa, several species of fruit-eating pteropodid bats, including *Eidolon helvum*, *Myonycteris torquata*, and *Nanonycteris veldkampi*, migrate up to 1500 km away from equatorial forests to savanna woodlands to feed on seasonal bursts of fruit. In eastern Australia, the pteropodid bat *Pteropus poliocephalus* contains sedentary coastal populations that feed on fig fruits year round and inland populations that migrate hundreds of kilometers between ephemeral but rich patches of flowering eucalypt trees. Finally, several species of insectivorous rhinolophid and vespertilionid bats undergo short-distance migrations between coastal and inland habitats in East Africa in response to seasonal changes in food availability.

Methods for Studying Bat Migration

Obtaining precise quantitative data on the distances that bats migrate and their migratory pathways has been technologically challenging. From the 1930s until recently, placing numbered aluminum bands on bats was the predominant method used to study bat (and bird) migration. This method involves banding bats at one roost (either a summer or winter roost), and then attempting to recapture banded bats (or recover their bands) somewhere else. This method is very labor intensive and inefficient because only a tiny fraction (=1%) of banded bats are usually ever recovered away from their original banding sites. For example, over 400 000 *T. brasiliensis* were banded in winter roosts in Mexico and summer roosts in the southern United States in the 1950s and 1960s, and only a handful were ever recovered away from their banding sites.

In addition to banding, methods that are currently being used to study bat migration include analyses involving DNA and stable isotopes and tracking studies using radio or satellite transmitters. Control region mitochondrial DNA is potentially very informative about genetic connections among distant roosts from which migratory connections can be inferred. A study of *L. curasoae* using this technique, for example, was able to identify two pathways along which females moved from south-central Mexico into the Sonoran Desert and southeastern Arizona, respectively, using data from only 49 individuals captured at a total of 13 roosts. Since this species is federally endangered in the United States and Mexico, large-scale banding operations were not feasible and other methods were needed to determine the scale of its migratory movements. Genetic analysis proved to be a very efficient method to obtain this information and to identify roosts in Mexico of special conservation concern (e.g., mating roosts). Other species whose migratory behavior has been studied using genetic techniques include *Miniopterus schreibersii*, *N. noctula*, and *T. brasiliensis*.

Another analytical technique that has provided important new insights into the migratory behavior of hard-to-study bats is stable isotope analysis. Stable isotopes of carbon and hydrogen are especially useful for this. The ^{13}C stable isotope of carbon allows one to determine whether herbivorous animals are feeding on plants that use the CAM, C4, or C3 photosynthetic pathway. The first two pathways are used by succulent plants (e.g., cacti and agaves) or tropical grasses (e.g., corn), respectively, and are enriched in ^{13}C compared with the more common C3 plants. By analyzing carbon stable isotopes in muscle tissue taken from museum specimens collected throughout its geographic range in Mexico and the southwestern United States, Fleming and colleagues found that individuals of *L. curasoae* use CAM plants (primarily columnar cacti in the spring and agaves in the fall) as food sources during migration and C3 plants during the winter. Knowing that columnar cacti occur in the Pacific coastal lowlands of Mexico and that agaves occur in upland portions of the Sierra Madre, they were able to identify the likely 'nectar corridors' along which these bats fly during migration.

Deuterium (D), the stable isotope of hydrogen, is useful for determining the migration distances of solitary roosting insectivores such as *L. cinereus* and *L. noctivagans*. Values of D vary inversely with latitude, elevation, and distance from coasts and can indicate approximately where bats were living when new tissue such as hair was produced. Since the aforementioned bats tend to molt before migrating in the fall, an analysis of hair samples from bats caught during migration or in their winter locations can be used to determine how far they migrated after molting. This technique has been used to determine migration distances of 1800–2600 km for both males and females of *L. cinereus* in North America. Some bats molting during the summer in Canada were captured in Mexico in the winter. The researchers concluded that this technique holds considerable promise for studying long-distance migration in bats.

Radiotagging and satellite tracking are two methods for directly studying the migratory behavior of bats. Radio transmitters weighing <1 g are readily available and have been used to study the foraging and roosting behavior of many species of bats, including species weighing <10 g. Except for the Australian *Pteropus poliocephalis*, which can carry transmitters with large batteries because of their large size, they have not yet been used to study bat migration for at least two reasons: (i) the battery life of these transmitters is short (about 2 weeks) and limits the amount of data that can be gathered from individual bats; (ii) since they migrate at night and sometimes at substantial altitudes (up to 2400 m), following radiotagged bats during migration involves potentially dangerous night time airplane flights over unknown terrain. Less dangerous but more costly in terms of transmitters (which currently cost US$1–3K) and daily or weekly downloads, satellite transmitters offer great potential for studying migration in bats, as it does for birds. Solar-powered satellite transmitters now weigh 12 g and have been used to study foraging and migration movements of one species of pteropodid bat that roosts in tree canopies (rather than in caves) during the day. Richter and Cumming studied the movements of four individuals of the 300 g frugivore *Eidolon helvum* that they tagged in central Zambia, Africa. These bats foraged up to 59 km from their day roost and traveled 878–1975 km over a period of two or more weeks when flying north to the Democratic Republic of Congo. During the return trips, these bats averaged 90 km per night.

Satellite transmitters equipped with conventional batteries (total package weight = 33–40 g) have been placed on two young males of *Pteropus poliocephalus* weighing 790 and 857 g in south-eastern Australia. Over the course of about a year, these bats made round trips of >2000 km spanning over 4° latitude as they moved up to 400 km among roosts in response to changes in the local availability of eucalypt blossoms.

Population and Genetic Consequences of Migration

Migration can have strong effects on the population and genetic structure of bats. In temperate bats, males and females typically hibernate in the same caves and are largely trophically inactive during winter. During the trophically active season, the behavior of males and females of migratory species often differs with respect to distances they migrate and locations where they spend the summer. In many species, females migrate longer distances and form larger summer (maternity) colonies than males. As a result, the sexes are often geographically separated at a variety of spatial scales during the summer. An extreme example of this is *Lasiurus cinereus* in which males spend the summer in the mountains of western North America and females roost in north-central and northeastern United States and Canada. Sex biased migration and seasonal spatial segregation of males and females also occur in *L. curasoae* and *T. brasiliensis* in North America. In both species, females migrate north from south-central Mexico to form large maternity colonies in the Sonoran Desert and south-central United States, respectively.

Seasonal movements can also have community consequences whenever regional and long-distance migrants move into and out of habitats containing resident species. In West Africa, for example, three species of frugivorous pteropodid bats migrate from equatorial forests to more northern savanna woodlands where they join a resident community of several species of frugivorous bats and birds. Differences in habitat and fruit preferences reduce potential competition among these species. A similar situation is seen in East Africa where several species of migrant insectivorous bats join a community of resident insectivores during periods of increased insect availability. Differences in morphology and foraging behavior again allow these species to coexist. Finally, seasonal influxes of migrant nectarivores and insectivores increase the species richness of bats in Sonoran Desert communities during the summer. As in the other two examples, differences in diets and foraging behavior minimize competition between residents and migrants.

As might be expected given their high mobility, the genetic structure of migrant species also differs from that of nonmigrants in a predictable way: a lower degree of genetic subdivision and larger effective population sizes in migratory species. An extreme example of a sedentary species is the Australian megadermatid *Macroderma gigas* in which widely separated colonies seldom exchange genes. As a result, Wright's index of subdivision F_{st} for this species is about 0.87 (out of 1.0 for complete subdivision; a value of 0.0 represents complete panmixia). At the other extreme is the long-distance migrant *N. noctula* whose F_{st} is 0.006 and whose region of panmixia in central Europe has a diameter of about 3000 km. Other species in both mobility classes have less extreme levels of F_{st}, but values for nonmigratory species are generally >0.10 whereas those of migrant species are <0.10.

Conservation Consequences of Migration

Because their annual ranges often encompass substantial geographic areas that usually cross different federal or international boundaries, the conservation of migratory bats, like that of migratory birds, can be challenging. Consequently, conservation efforts need to be geographically and politically broad in scope. This conservation must involve protecting a variety of different roost sites, including those used for mating, migration, and maternity, as well as the foraging habitats around critical roost sites. In addition, habitats used en route during migration, including stopover habitats where bats can refuel, need protection. Plant-visiting bats such as the nectarivore *L. curasoae* and the frugivore *E. helvum* likely migrate along specific food corridors that also need protection. Based on their satellite-tracking results, for example, Richter and Cumming noted that only a fraction of the migratory pathway along which *E. helvum* flies between the Democratic Republic of the Congo and Zambia is currently protected. Loss of forest habitat containing fruiting trees along this pathway could seriously disrupt its annual migration. Similarly, destruction of parts of the columnar cactus 'nectar corridor' along the Pacific coast of Mexico would have a strong negative effect on migrating pregnant females of *L. curasoae*. Based on levels of fat that these bats deposit prior to and during migration, Fleming has estimated that the maximum flight range of these bats is about 550 km. If the average distance between rich patches of cacti exceeds this value, then the migration of thousands of bats could be disrupted. And because intact populations of fall-blooming agaves are also needed by this species to complete the return leg of its migration, habitat protection over a large portion of western Mexico is needed. Migration of insectivorous bats is usually much more diffuse geographically than that of plant visitors, but they also need intact foraging habitat, as well as protected stopover roost sites. A landscape that is devoid of safe caves, intact forests, and unpolluted lakes and streams is just as threatening to the existence of migrating insect bats as a landscape devoid of flowering and fruiting plants is to plant visitors.

In addition to the usual litany of threats to bats and other wildlife (e.g., habitat destruction, pollution, and specifically for bats, malicious destruction of their colonies in caves and other roosts), a new threat to their conservation has emerged recently – wind turbine farms. In an effort to tap alternate sources of energy, wind farms have increased markedly in number and size in Europe, Australia, and North America in recent years. While these establishments clearly have positive value for energy production, they can have negative value for wildlife because they kill migratory birds and bats. In North America and Europe, for example, peak bat fatalities occur in late summer and fall and are heavily concentrated in long-distance migrants such as species of *Nyctalus* and *P. nathusii* in Europe and species of *Lasius* in North America. In the United States, wind farms located in forested parts of the east coast experience higher kill rates than those located in the Rocky Mountains and Pacific Northwest. Why lasiurine bats, which migrate in flocks despite being solitary roosters during the summer, are more vulnerable to fatal interactions with wind turbines is not yet known. Fatalities are most common on nights with low wind speed ($<6\,\mathrm{m\,s^{-1}}$) and before and after the passage of storm fronts when large numbers of bats (and birds) are likely to be migrating. Bat fatalities occur only when turbines are spinning, not when they are stationary. After reviewing available data, Arnett and colleagues concluded that the number of bat fatalities at wind farms could be reduced substantially by temporarily stopping turbines at night at certain times of the year and under certain climatic conditions.

My final topic – one that has important conservation implications, especially for migrant bats, as well as health implications for humans – is bats as reservoir hosts for emerging viruses. It has long been known that bats are important reservoirs for rabies virus and that they sometimes (but rarely) transmit it to humans. According to a review by Calisher and colleagues, bats are known to harbor a substantial number of viruses only a few of which are known to be pathogenic when transmitted to mammals, including humans. In addition to rabies, these include other lyssaviruses (Family Rhabodviridae) as well as Hendra and Nipah viruses (Family Paramyxoviridae) and possibly SARS-Coronavirus-like viruses (Family Coronaviridae). Hendra virus has been found in species of *Pteropus* in Australia; Nipah virus has been found in species of *Pteropus* in South and Southeast Asia and SARS coronavirus occurs in species of *Rhinolophus* in Eurasia. Both Hendra and Nipah viruses have been found in humans via transmission from intermediate host mammals (e.g., pigs). Ebola virus RNA has been found in three species of African pteropodids, including *E. helvum*, but not the virus itself.

Because they are geographically wide ranging, migrant bats have the potential to spread pathogenic viruses over wide areas. Outbreaks of rabies virus in Europe, for example, have occurred along the migration routes of *P. nathusii*, and different geographic variants of this virus have been found in two North American migrants, *P. subflavus* and *T. brasiliensis*. Migrant bats in general have a classic metapopulation structure featuring discrete populations (roosts) interconnected by dispersal or migration; between-colony movements can expose resident as well as migrant populations to new variants of rabies or exchange virus variants among colonies. Calisher and colleagues suggested that this kind of population structure has the potential for seasonal virus transmission, annual outbreaks of viral diseases, and periodic outbreaks among spatially separate populations. Geographically discrete outbreaks of rabies in the (nonmigratory) vampire bat *Desmodus rotundus* or outbreaks of Hendra virus in migratory Australian *Pteropus* bats may reflect this. Finally, the long-distance migrant *E. helvum* has the potential for spreading infectious diseases over large areas of sub-Saharan Africa.

The conservation implications of the fact that bats harbor pathogenic organisms are enormous. A major reason why bats in general are persecuted throughout Latin America is the fear of 'vampiros y la rabia.' The association between rabies and bats, in vampires or otherwise, is well known throughout the world, and bats tend to be maligned worldwide as a result. Because of this association, the ecologically beneficial 'services' provided by bats such as control of injurious insect populations (e.g., *T. brasiliensis* and cotton-boll worms), and the broad dispersal of pollen and seeds (e.g., by *L. curasoae* and *E. helvum*) tend to be overlooked. In truth, the positive benefits of migratory (and nonmigratory) bats far outweigh their negative aspects. Migratory bats play an important role in many ecosystems around the world, and their conservation is essential.

Conclusions

Many species of bats are migratory and serve as 'mobile links' between geographically separate habitats and ecosystems. They move energy and nutrients among ecosystems, help to control insects on a broad scale, and serve as wide disseminators of pollen, seeds, and pathogens. Like their avian counterparts, migratory bats have many morphological, physiological, and behavioral adaptations for 'life on the move.' Because of their mobile lifestyles, these bats have special conservation needs that must be addressed politically at the national or international level. Although they sometimes harbor pathogenic organisms, their positive attributes far outweigh their negative attributes. Increased public awareness of the lives of these fascinating bats worldwide is the key to their conservation.

See also: Bats: Orientation, Navigation and Homing; Bird Migration.

Further Reading

Arnett EB, Brown WK, Erickson WP, et al. (2008) Patterns of bat fatalities at wind energy facilities in North America. *Journal of Wildlife Management* 72: 61–78.

Calisher CH, Childs JE, Field HE, Holmes KV, and Schountz T (2006) Bats: Important reservoir hosts of emerging viruses. *Clinical Microbiology Reviews* 19: 531–545.

Cleveland CJ, Betke M, Federico P, et al. (2006) Economic value of the pest control service provided by Brazilian free-tailed bats in south-central Texas. *Frontiers in Ecology and the Environment* 5: 238–243.

Cryan PM, Bogan MA, Rye RO, Landis GP, and Kester CL (2004) Stable hydrogen isotope analysis of bat hair as evidence for seasonal molt and long-distance migration. *Journal of Mammalogy* 85: 995–1001.

Cryan PM and Diehl RH (2009) Analyzing bat migration. In: Kunz TH (ed) *Ecological and Behavioral Methods for the Study of Bats*, vol. 2, pp. 477–488.

Federico P, Hallam TG, McCracken GF, et al. (2008) Brazilian free-tailed bats as insect pest regulators in transgenic and conventional cotton crops. *Ecological Applications* 18: 826–837.

Fleming TH (2004) Nectar corridors: migration and the annual cycle of lesser long-nosed bats. In: Nabhan GP (ed.) *Conserving Migratory Pollinators and Nectar Corridors in Western North America*, pp. 23–42. Tucson, AZ: University of Arizona Press.

Fleming TH and Eby P (2003) Ecology of bat migration. In: Kunz TH and Fenton MB (eds.) *Bat Ecology*, pp. 156–208. Chicago: University of Chicago Press.

Fleming TH, Nunez RA, and Sternberg LSL (1993) Seasonal changes in the diets of migrant and non-migrant nectarivorous bats as revealed by carbon stable isotope analysis. *Oecologia* 94: 72–75.

Hutterer R, Ivanova T, Meyer-Cords C, and Rodrigues L (2005) Bat migrations in Europe: A review of banding data and literature. In: *Naturschutz und Biologische Vielfalt*, vol. 28, 180 pp. Federal Agency for Nature Conservation in Germany.

Richter HV and Cumming GS (2008) First application of satellite telemetry to track African straw-coloured fruit bat migration. *Journal of Zoology* 275: 172–176.

Bat Neuroethology

G. D. Pollak, University of Texas at Austin, Austin, TX, USA

Introduction

Bats belong to the their own mammalian order, Chiroptera, meaning 'wing handed,' and are distinguished from all other mammals by their ability for sustained flight. Their most intriguing attribute, however, is their sophisticated use of biosonar in which they emit ultrasonic sounds and by listening to the returning echoes that are reflected from objects they 'see' their world through sound. This form of biosonar, known as 'echolocation,' enables bats to orient in complete darkness and to hunt insects in the night sky. In the following sections, we first consider the various ecological niches that bats occupy together with the variety of echolocation signals different species use and why the signals emitted by each species are suited to the habitat in which it hunts. In the second part of this article, we turn to some of the pronounced adaptations in their cochleae that correlate closely with their echolocation calls. The third section deals with adaptations in the central auditory system. The theme here is that bats are mammals and nature did not invent a new auditory system for bats. Indeed, the bat's auditory system is similar to the auditory system of all other mammals, with the same structures, wiring, and mechanisms for processing information that are possessed by all other mammals. What distinguishes the auditory system of bats are not novel mechanisms, but rather that some common mechanisms and features are far more pronounced in their auditory systems than in other mammals. Moreover, the features and mechanisms that were emphasized were highly adaptive and allowed the various species of bats to compete successfully for food resources in a wide range of different habitats.

The Habitats of Bats

The adaptations that allowed bats to fly and 'see in the dark' with echolocation were among the most successful nature ever created. Bats were so successful that they invaded every region on earth, tropical and temperate, except for the polar ice caps. Bats may well be the most successful mammal on earth, in terms of the number of species (about 900 species of bats are known) and absolute numbers of individuals. There are, for example, about 20 million bats in Bracken Cave in central Texas, and the vast majority of the Bracken bats are one species, Mexican free-tailed bats.

As bats invaded the four corners of the earth, they exploited any type of food that was plentiful and different species focused their hunting strategies on a food resource that was plentiful in their habitat. Several examples are shown in **Figure 1**. Many species hunt insects in the night sky, others catch large ground-dwelling insects and arachnids, such as scorpions. Other species, the infamous vampire bats of Central and South America, even feed exclusively on blood. Others use their echolocation to orient in the jungle to exploit the rich supply of tropical fruits, while others seek night-flowering plants for their offerings of sugar and pollen. Yet others are even carnivorous, such as the frog-eating bats of Central and South America. Fishing bats even use their sonar to detect small fish swimming near the surface of a pond or lake.

Each group tailored the physical features of their bodies, their tongues, teeth, noses, or claws to enhance their ability to exploit an abundant food supply. They not only evolved the physical structures for capturing or exploiting a particular source of food, but they also tailored their echolocation calls to suit one or another ecological niche. Thus, each species employs more or less unique echolocation calls characterized by their frequency content (spectral composition), call duration, how frequencies are emitted over time (temporal features of the calls), the frequencies that are emphasized in each call, and the intensities of their echolocation cries.

Bats Emit Three Major Types of Echolocation Calls

Although there are a wide variety of echolocation calls emitted by bats, the various types fall into one of three general categories. The first are loud, frequency modulated (FM) calls (**Figure 2**). These are brief FM signals with durations of about 1.0–5.0 ms that sweep downward about an octave in frequency over the duration of the call and are typically emitted at 100 dB SPL or more. The calls are rich in frequencies as they are emitted with one or more harmonics. I shall refer to bats that emit these signals as loud FM bats.

The second signal type is soft FM calls. These are also downward FM sweeps that are spectrally similar to the loud FM calls but are emitted at a substantially lower intensity, typically about 70 dB SPL. The bats that emit these signals are sometimes called 'whispering bats,' and the justification for distinguishing the two types of FM

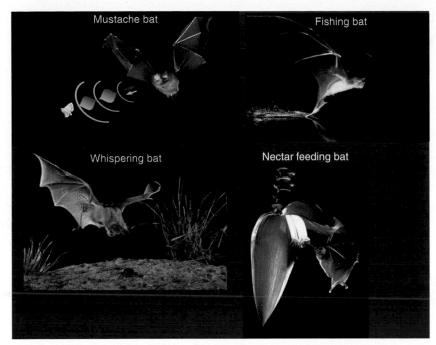

Figure 1 Four bats displaying different hunting strategies. Reproduced with permission from Bat Conservation International.

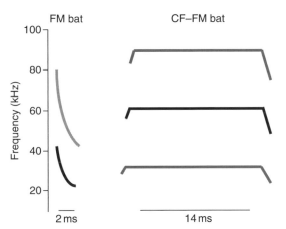

Figure 2 Sound spectrographs of echolocation calls showing how frequency varies over time. The call on the left illustrates the brief, downward sweeping frequency modulated (FM) calls emitted by both loud FM and whispering FM bats. The call on the right is for CF–FM bats. Echolocation calls are emitted with one or more harmonics. The harmonic with the greatest energy is shown in black while gray indicates lesser energy.

bats is that whispering bats occupy different habitats from those of the loud FM bats and use the calls in ways that are somewhat different from those of the loud FM bats (see later). The third signal type is also emitted at a high intensity (a common set) and is characterized by an initial 'long' constant frequency (CF) component whose duration ranges from about 15 ms up to about 80 ms, depending upon the species (**Figure 2**). These calls are also

emitted with one or more harmonics. The CF component of the mustache bat, for example, has a fundamental frequency of 30 and harmonics at 30, 60, and 90 kHz (**Figure 4**). However, most of the energy, by far, is in 60 kHz, the second harmonic of the calls. The end of each call always has, a brief, downward sweeping FM component, similar to the signal emitted by loud FM bats. Bats that emit these calls are known as CF–FM bats.

Echolocation Calls and Habitat

Loud FM Bats

Natural selection tailored their biosonar systems, as well as other features of their bodies, to enable each species to exploit one or another type of food supply in their environment. In general, bats that emit loud FM signals hunt insects in the night sky where there is little or no clutter from echoes of other objects (**Figure 3**). Some loud FM bats hunt in the open sky, away from trees and other objects, while others hunt insects that occur around the edges of trees or other foliage. The calls of all of these bats are loud FM chirps, but the spectral composition of the calls emitted by bats that hunt in the open sky is slightly different from those that hunt around the edges of foliage. Specifically, the FM echolocation calls of the bats that hunt around edges sweep over a larger frequency range (have a broader bandwidth) than those that hunt closer to foliage. Even those differences in spectral composition are adaptive in that the broader bandwidth emitted by some

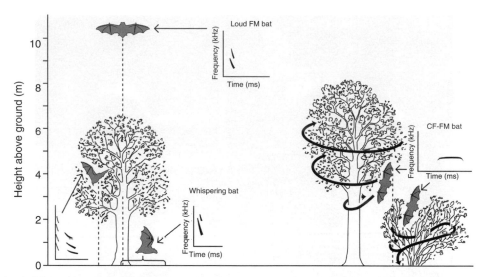

Figure 3   Hunting habitats of bats. Loud FM bats hunt either in the open sky (bat at top left) or around edges of vegetation (bat shown at bottom on far left). CF–FM bats hunt within vegetation and are specialized to detect the echoes of fluttering insects among the echo clutter reflected from trees, bushes and other objects in their foraging areas. Whispering bats hunt close to the ground where they typically listen passively for prey-generated noise. Adapted from Neuweiler G (1984) Foraging, echolocation and audition in bats. *Naturwissenschaften* 71: 446–455, with permission.

species enhance their abilities to detect targets in the more cluttered environment in which they hunt.

The Whispering FM Bats

Whispering bats, in contrast to loud FM bats, do not hunt insects in the open air. Many species prey upon terrestrial insects or even terrestrial vertebrates while others forage on fruits or flowering plants (**Figures 1** and **3**). Some whispering bats are gleaners in that they hover over tree trunks and walls and use their low-intensity biosonar signals to scan for moving prey, but others utilize echolocation primarily for orientation rather than for prey capture. Others feed on nectar and pollen, where they pollinate night-flowering plants that have structural adaptations designed to attract the bats. Yet other whispering bats have what is in essence 'two separate' auditory systems, one designed for orientation via echolocation and the other for detecting and locating the sounds made by prey by listening passively for prey-generated sounds. Three examples are especially interesting. One is the frog-eating bat, *Trachops cirrhosis*. These neotropical whispering bats orient with echolocation and simultaneously listen for the advertisement calls of tungara frogs, one of their favorite prey items. Frog-eating bats not only detect the mating calls of tungara frogs, but they also distinguish those calls from the advertisement calls of poisonous sympatric species. Another example is the infamous vampire bat, *Desmodus rotundus*, a species that feeds exclusively on the blood of large mammals. Vampire bats are found

throughout Central and South American and also orient through the forest with echolocation. However, they identify their prey not with echolocation but rather from the breathing rhythms of the mammals they prey upon. Finally, there is the pallid bat, *Antrozous pallidus* (**Figure 1**). These bats are found in the deserts of the southwestern United States, and like the vampire and frog-eating bats, pallid bats use echolocation for detecting obstacles and orientation and listen for prey-generated noise that they use to detect, locate, and capture their prey. The sounds of interest to pallid bats are the noisy sounds arthropods make as they scurry over leaves and other foliage on the desert floor. In a later section, we consider the neurophysiological studies that have revealed an elegant partitioning of the auditory system where one portion is specialized for the processing of the FM signals used in echolocation and the other specialized to respond to prey-generated noise.

The CF–FM Bats

Bats that emit CF–FM signals include one species found in the new world, the mustache bat, *Pteronotus parnellii* (**Figure 4**), and numerous species found only in the old world, the horseshoe bats and the hipposiderid bats. These bats use echolocation for both orientation and hunting. They hunt under the forest canopy where the echoes from bushes, trees, and other objects generate a massive clutter that the bats have to distinguish from echoes generated by the insects flying among the foliage.

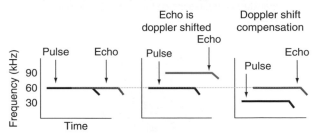

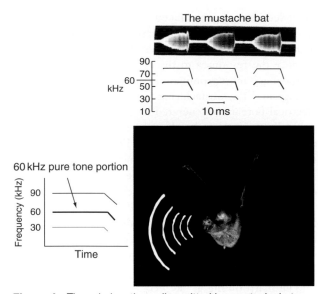

Figure 4 The echolocation calls emitted by mustache bats.

Figure 5 Schematic of the 60-kHz emitted CF component and Doppler-shifted echoes in three successive pulses to illustrate Doppler-shift compensation by the mustache bat. The first pulse is emitted by a stationary (nonflying) bat in which the echo has no Doppler shifts and returns at the same frequency as the emitted pulse. The second pulse is emitted by a flying bat that receives a Doppler-shifted echo. The third pulse is emitted at a lower frequency that compensated for the Doppler shift in the previous echo. Consequently, the Doppler-shifted echo returns at a frequency that is almost the same as the echo frequency received when the bat was not flying.

The key feature for detecting and recognizing prey under those conditions is the CF component and the way these bats manipulate the frequency of the CF components they emit during echolocation.

CF–FM Bats and Doppler-Shift Compensation

CF–FM bats exhibit a unique behavior called 'Doppler-shift compensation,' a behavior illustrated for the mustache bat in **Figure 5**. The frequency of the CF component emitted by a bat that is not flying is almost constant, varying by only 50–100 Hz from pulse to pulse. As they fly under the forest canopy, the CF component of the echo is shifted upward in frequency due to Doppler shifts caused by the relative movement of the bat toward the stationary background. The amount of Doppler shift is determined by the bat's flight velocity relative to the stationary background. Thus, the CF echoes returning to the bat's ears are replicas of the emitted CF component, basically pure tones, but are higher in frequency and lower in amplitude. These bats compensate for the Doppler shifts in the echoes they receive by lowering the frequencies of their emitted CF components by an amount nearly equal to the upward frequency shift in the echo. Consequently, the bat 'clamps' the echo CF component and holds it within a narrow frequency band that varies only slightly from pulse to pulse.

Although the flight velocity, the relative movement of the bat toward stationary background, is the feature that creates the Doppler shifts, relative flight velocity, is *not* the acoustic feature of interest to the bats. Rather, the bats are interested in the flutter caused by the beating wings of

an insect that happened to fly among the background foliage in the bat's acoustic space. The acoustic cues generated by flutter, that is, the up and down motion of the insect's wings are periodic frequency (FM) and amplitude (AM) modulations. The frequency modulations are imposed on the echo CF by the Doppler shifts created by the motion of the insect's wings and amplitude modulations are generated by the changes in the reflective surface area as the wings move up and down (**Figure 6**). The echo CF therefore has both AM and FM, very much like the signals received by a radio. Radios have a tuner that the listener adjusts to receive the particular frequency that carries the AM or FMs. In a comparable way, CF–FM bats tune into a carrier frequency by Doppler-shift compensation, whereby the bat adjusts its voice, the emitted CF, to hold or tune into the CF component of the echo, the frequency that carries the information generated by the beating wings of an insect.

Cochlear Specializations

The cochleae of bats exhibit a number of specializations that enhance and overrepresent the frequencies that each species depends most heavily upon for its survival. The specializations that have been examined primarily concern features of the basilar membrane, the membrane that transforms frequency into a place of maximal vibration along its length, the structures that couple the basilar membrane to the organ of Corti, the structure that supports and contains the hair cells, and the density of neural innervation along the cochlear partition. Not surprisingly, the bats with the most specialized behaviors have the most prominent specializations.

Fluttering wings of an insect generate frequency and
amplitude modulations on the echo CF component

Figure 6 Drawing of a flying mustache bat and small insect
to illustrate the periodic frequency modulations that are imposed
on the echo CF component by the moving wings of the insect.
Reproduced from Pollak GD and Casseday JH (1986) *The Neural
Basis of Echolocation in Bats*. New York, NY: Springer-Verlag,
with permission.

The cochleae of the loud FM bats are typically mammalian and exhibit the least pronounced specializations. One example is the Mexican free-tailed bat, a loud FM bat. The frequencies in its echolocation calls sweep downward from about 30 to 15 kHz, but the fundamental frequencies of its communication calls are often lower, extending down to 10 kHz or below. In general, the basal portion of the basilar membrane, representing frequencies from about 80 to 30 kHz, is typically mammalian with no exceptional features. The specializations are seen in the region representing 30–13 kHz, which is expanded, in that an equivalent range of higher frequencies (in terms of octaves) occupies a smaller stretch of membrane. This is the range of frequencies to which the bat is most sensitive, both behaviorally and neurophysiologically, and at which neurons exhibit the sharpest tuning (i.e., neurons most sensitive to frequencies between 30 and 10 kHz respond to a narrower range of frequencies than other neurons and thus are more selective for frequency). In addition, there are morphological specializations in the basilar membrane in the region representing 30–13 kHz that enhance sensitivity and tuning sharpness. In summary, although there are some specializations of the basilar membrane in the portion representing 30–13 kHz, the specializations are subtle and are not nearly as dramatic as those seen in the CF–FM bats, bats that possess the most specialized cochleae of any mammal.

The cochleae of both the mustache bat, *Pteronotus parnellii*, and the greater horseshoe bat, *Rhinolophus ferrumequinum*, have been the most thoroughly studied and both bats have features that greatly emphasize the responses to frequencies that correspond to the echo CF component of their echolocation calls. In both bats, the region of the cochlea representing the CF component, 60 kHz in mustache bats and 83 kHz in horseshoe bats, is greatly

expanded, has a pronounced innervation density, and expresses a number of specializations in the basilar membrane and tectorial membranes that both sharpen and enhance the responses to the echo CF component. The sharpening is most prominent in the cochlear microphonic audiogram of the mustache bat. Cochlear microphonics are electrical potentials generated by the responses of outer hair cells to sound and reflect the mechanical motion of the basilar membrane. As shown in **Figure 7**, the cochlear microphonic audiogram of the mustache bat is sharply tuned to 60 kHz and is by far the most sharply tuned cochlear microphonic audiogram of any known animal. The exact frequency to which the cochlea is tuned varies from about 60 to 63 kHz among individual bats, but each bat is sharply tuned to one frequency, the frequency at which it clamps the CF of the echo during Doppler-shift compensation.

The sharp tuning at '60' kHz is due to a pronounced resonance in the ear that is generated in large part by the specialized morphological features on the basilar and tectorial membranes. The resonance greatly enhances the amplitude of the cochlear microphonic response evoked by 60 kHz and acts to create auditory nerve fibers with tuning curves that are also sharply tuned. Indeed, the sharpness of the tuning curves of auditory nerve fibers that innervate the 60-kHz portion of the basilar membrane, and thus are most sensitive to 60 kHz, is 1–2 orders of magnitude sharper than the tuning curves of fibers that innervate other regions along the basilar membrane or of auditory nerve fibers in other bats and other mammals. Not only are the auditory nerve fibers that are most sensitive to '60' kHz sharply tuned, they comprise about one-third of the total population of the entire auditory nerve. In other words, the range of frequencies to which the population is most sensitive ranges from about 10 kHz to over 120 kHz, but one-third of entire population is most sensitive to one frequency, the frequency of the echo CF component.

Horseshoe bats also have a greatly expanded cochlear region with multiple specializations at the region representing 83 kHz, specializations that produce a disproportionately large population of auditory nerve fibers sharply tuned to 83 kHz. Since mustache and horseshoe bats evolved independently, horseshoe bats in the old world and mustache bats in the new world, the specializations in the two species are different, although the different specializations converged on the same result in the two groups. The common result is that the sharply tuned regions of the cochleae in these bats' act as an acoustic fovea. Just as foveation in vision, where we move our eyes to keep an object of interest focused on the fovea of the retina that has densely packed receptors with small receptive fields, CF–FM bats actively manipulate the frequency of their voices through Doppler-shift compensation so that the echo CF always falls on the region of their cochlea

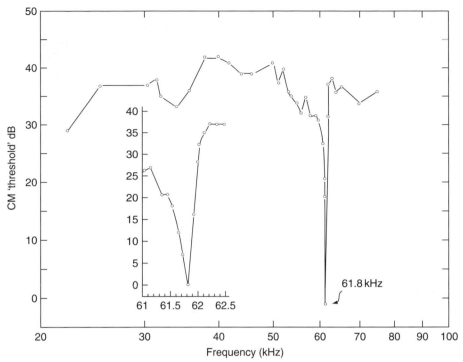

Figure 7 Audiogram showing the sensitivity of cochlear microphonic responses (CM) evoked from the cochlea of a mustache bat by various tonal frequencies. The audiogram is sharply tuned to 60 kHz, the same frequency at which the bat clamps the frequency of the echo CF component when compensating for Doppler shifts. Reproduced from Pollak GD and Casseday JH (1986) *The Neural Basis of Echolocation in Bats*. New York, NY: Springer-Verlag, with permission.

that is innervated a large population of sharply tuned neurons. As described in the following section, the large populations of sharply tuned neurons provide the neural substrate for analyzing the fine structure of the their targets, just as comparable features of retinal ganglion cells in the fovea allow for the resolution of fine details of objects that we are looking at.

Processing in the Auditory Pathway in the Central Nervous System

The purpose of the cochlea is to transduce, that is, change, the mechanical motion along the basilar membrane into the language of the nervous system, trains of action potentials evoked by sound. The information generated by each frequency is then conveyed into the central nervous system where the spike trains of the auditory nerve fibers are initially processed in the cochlear nucleus (CN), the first synaptic center in the ascending auditory pathway (**Figure 8**).

The principal effect of processing is that a particular feature or features of the sound that is encoded by the incoming spike trains are transformed. The transformation is expressed as a difference in the vigor of responding, the timing of the discharge patterns or more importantly, as response selectivity, whereby the incoming spike trains

are evoked by a wide variety of stimuli while the neuron's output, its discharge train, is evoked by only a select group of stimuli, whereas other stimuli evoke no action potentials. Before turning to the types of transformations expressed by auditory neurons in different centers along the auditory pathway, a brief review of the flow of information and the major organizational features of the mammalian auditory system are given in the following section, so that the adaptations seen in bats can be placed in context.

Principles of Organization in the Mammalian Auditory System

The central auditory pathway is composed of a number of nuclei and complex pathways that ascend within the brainstem and end in the cortex (**Figure 8**). The pathways are even for experts, terribly complex and the details of the various connections are, for our purposes, not important. The pathway into the brain begins with auditory nerve fibers that project from the cochlea into the brain where they first make synaptic connections with other neurons in the CN. The axons from the neurons in the CN project deep into the medulla where they make synaptic connections with a group of nuclei called 'the superior olive complex.' The axons of neurons in the superior olive

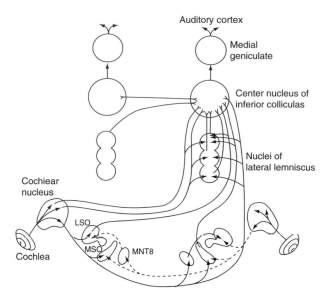

Figure 8 The basic wiring of the mammalian auditory system. Auditory nerve fibers coming from the cochlea branch to innervate the various cell groups in the cochlear nucleus (CN). The CN cells send their axons into the brain as a series of parallel pathways. One pathway innervates the neurons in the superior olivary complex, shown as the medial nucleus of the trapezoid body (MNTB), the medial superior olive (MSO), and the lateral superior olive (LSO). The MSO and the LSO are the first regions of the auditory system that receive innervation from the two ears. The projections from the superior olive, as well as from other groups, shown as the nuclei of the lateral lemniscus, then converge at a common site in the inferior colliculus (IC), the nexus of the auditory system. The projections from the IC then innervate the medial geniculate (auditory thalamus) and projections from the medial geniculate provide the major innervation to the auditory cortex (AC). Reproduced from Pollak GD and Park TJ (1995) The Inferior Colliculus. In: Popper A and Faye RR (eds.) *Hearing by Bats, Springer Handbook of Auditory Research*, pp. 296–367. Berlin-Heidelberg: Springer-Verlag, with permission.

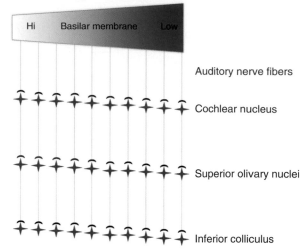

Figure 9 Projection patterns that underlie the tonotopic organization of the auditory system.

in turn project to yet other nuclei. Ultimately all the projections from the lower nuclei terminate in the auditory midbrain, the inferior colliculus (IC). The outputs of the IC are sent to the medial geniculate body, the auditory thalamus, and then to the auditory cortex (AC) in the temporal lobe.

The most fundamental organizational feature of the auditory system is the topographic remapping of the sensory (cochlear) surface upon each succeeding higher neural center. **Figure 9** shows that the orderly representation of frequency in the cochlea, a feature called 'tonotopic organization' or 'tonotopy,' is preserved in the projections to the CN, the first auditory nucleus in the central auditory system, and in the projections to every higher auditory region, from CN to AC. Stated differently, the same tonotopic organization that was first established in the cochlea is imposed on each auditory nucleus in the ascending auditory system.

The representation of a particular frequency in the CN (and other, higher regions) is not a single neuron but rather is composed of a sheet of neurons, all having the same 'best frequency,' the frequency to which the neuron is most sensitive, and which reflects the point on the cochlea from which the neuron received its innervation (**Figure 10**). This sheet of isofrequency cells is 'the unit module' because it represents the total processing that occurs for that frequency in each nucleus.

As illustrated in **Figure 10**, each isofrequency sheet in the CN is composed of a variety of cell types. Each cell type extracts some aspect of the information contained in the spike train of auditory nerve fibers, and the information extracted by that particular neuronal type is different from the information extracted by the other neuronal types. Each CN cell type, in turn, sends its axonal projections into the brain stem as a series of parallel projections, where each projection innervates one or a few of the various auditory nuclei in the brainstem in a tonotopic fashion. Interestingly, the cells are fairly homogeneous in each nucleus above the CN, and thus each nucleus extracts some aspect of information that is different from the information extracted by each of the other lower nuclei. The neurons in each of the lower nuclei send their axonal projections to terminate in corresponding isofrequency sheets of the IC in the midbrain. Thus, the IC receives the convergent projections from almost all lower nuclei and is, therefore, the nexus of the ascending auditory system. The output of the IC, in turn, is sent to the medial geniculate body and then to the AC, where the projections terminate tonotopically.

The remainder of this article will be concerned with only two regions of the auditory system: (1) the auditory

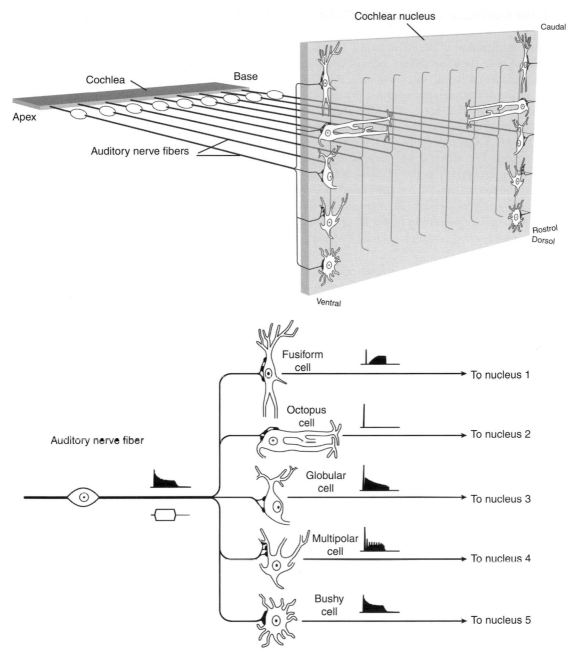

Figure 10 Schematic representation of projections of auditory nerve fibers to CN cell groups as they occur for different frequencies across the cochlear surface. Lower panel shows the response pattern of an auditory nerve fiber to a tone burst at the frequency to which the fiber is most sensitive (its best frequency). Notice that although each of the neurons in the CN is innervated by the same auditory nerve fiber, the response patterns of each of the neurons are different from that of the auditory nerve fiber that innervated all the neurons. Reproduced from Kaing NYS (1975) Stimulus representation in the discharge patterns of auditory neurons. In: Tower DB (ed.) *The Nervous System, Human Communication and Its Disorders*, vol. 3. New York, NY: Raven Press.

midbrain or IC because the IC receives the convergent projections from almost all lower auditory nuclei; and (2) the AC because it is the highest center in the ascending auditory system. Focus will be on a comparison of two features of the central auditory system: (1) the tonotopic organization of their auditory nuclei; and (2) the processing of frequency modulations, the signals that are so prominent in both the echolocation and the communication signals

emitted by bats. The two neural features will be illustrated for three bats: (1) the big brown bat, which is a loud FM bat; (2) the mustache bat, which is a CF–FM bat; and (3) the pallid bat, a whispering bat. This comparison is not comprehensive but will serve to illustrate the ways that basic organization and response features were adapted in different types of bats to suit the particular habitat in which they hunt.

The Inferior Colliculus in Loud FM Bats

The tonotopic organization of the IC in loud FM bats is similar to that of other less specialized animals and thus is typically mammalian. Like all other mammals, the IC is composed of sheets of isofrequency neurons, that is, neurons tuned to the same frequency, where the sheets are stacked one on top of the other. The sheets representing low frequencies are located dorsally with sheets representing progressively lower frequencies located ventrally. The frequency representation in the IC however is not uniform in that the frequencies of greatest importance to each species are overrepresented, as they are in the cochlea. The tonotopic organization in the IC of the big brown bat, *Eptesicus fuscus*, a prototypical loud FM bat is shown in **Figure 11**.

One of the most important features of some neurons in the IC is that they express selectivity for the direction of FM sweeps, whether it sweeps upward or downward. In lower regions of the auditory system, neurons generally respond to both upward and downward FM sweeps with equal vigor. Neurons in the IC, in contrast, respond only or far more strongly to FM signals that sweep in one direction than in the other direction. Thus, FM directionality is one of the major transformations that occur in the IC. It should be emphasized that selectivity for FM direction is not a feature unique to the IC to bats but rather is a feature seen in the IC of all mammals that have been studied. Loud FM bats, however, emphasize this selectivity and display a preference for downward FM sweeps, the types of FM sweeps in their echolocation calls, compared to preferences for upward FM sweeps.

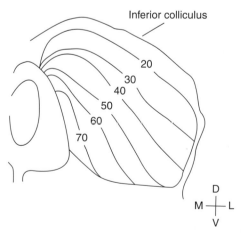

Figure 11 Tonotopic organization of the IC of the big brown bat. Each line indicates a sheet of isofrequency cells. Reproduced from Pollak GD and Casseday JH (1986) *The Neural Basis of Echolocation in Bats*. New York, NY, Springer-Verlag, with permission.

The Auditory Cortex of Loud FM Bats

The AC of big brown bats, like the auditory corticies of all mammals, is tonotopically organized (**Figure 12**) and also has a population of FM directionally selective cells.

Many cortical neurons however also display an adaptation called 'delay tuning,' an adaptation that is especially well suited for echolocation. Delay-tuned neurons respond weakly or not at all to a single FM signal but rather respond vigorously to two signals presented with a particular temporal period between the first and second signals (**Figure 13**). The first signal can be thought of as the emitted pulse and the second signal as the echo. Thus, the neurons respond best to a particular 'pulse-echo' delay. Different neurons respond best to different delays. The delay between the emitted pulse and the returning echo is determined by the distance between the bat and its target, and pulse-echo delay is the cue bats use to determine the distance to their targets. Thus, many investigators consider these neurons to code for target distance or range.

One problem with this interpretation is that delay-tuned neurons are broadly tuned and respond to a wide range of pulse-echo delays that differ by several milliseconds. For example, one of the neurons in **Figure 13** fires at 80% of its maximal discharge rate to a 7 ms range of pulse-echo delays. Bats can resolve target distances that differ by as little as 1.0 cm, which represents a pulse-echo difference of about 75 µs between the two signals; that is, if the bat were presented with two targets where one is closer to the bat by 1.0 cm, the bats can resolve that difference in range. How neurons that are so broadly tuned for delay can distinguish between two targets in which the pulse-echo delays differ by only 75 µs is difficult to explain.

Dear et al. (1993) have offered an alternative hypothesis. Their hypothesis holds that what the delay-tuned neurons are doing is to partition distance into smaller segments, where a population of neurons tuned to shorter delays process information from closer targets and the population of neurons tuned to longer delays process information from targets located at progressively longer distances. Thus, the acoustic scene, all targets in the bat's acoustic space, are each processed by a different population of neurons where nearer targets are processed by the population of neurons tuned to shorter delays, while the farther targets are processed by the population tuned to longer delays. In other words, the neurons do not provide information about the exact distance of each target, but rather break the scene up where each population 'looks' at a particular depth in the acoustic field.

The mechanism that underlies delay tuning is a feature called 'paradoxical latency shift.' Auditory neurons almost always respond to softer stimuli with longer latencies and louder stimuli with shorter latencies. Neurons with

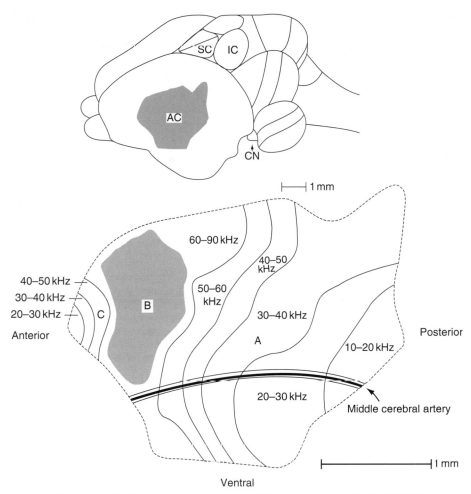

Figure 12 Brain of the big brown bat. Auditory areas include the CN, IC, and superior colliculus (SC). Shaded area defines AC. The lower panel shows the AC in greater detail and its tonotopic organization. Adapted from Dear SP, Fritz J, Haresign T, Ferragamo M, and Simmons JA (1993) Tonotopic and functional organization in the auditory cortex of the big brown bat, *Eptesicus fuscus*. *Journal of Neurophysiology* 70: 1988–2009, with permission.

paradoxical latency shifts, in contrast, respond to softer stimuli with shorter latencies and to louder sounds with longer latencies. The idea then is that the louder emitted pulse (first signal) evokes a long latency discharge and that the softer echo evokes a short latency discharge (**Figure 14**). The delay-tuned neuron is constructed by receiving innervation from neurons with paradoxical latency shifts, where the neuron's best delay (BD) is determined by the difference in latencies evoked by neurons that have a longer latency to the louder pulse and those with shorter latencies evoked by the softer echo. If that latency difference is 10 ms, then the excitation arriving from both neurons will occur for a delay of 10 ms and the neuron will fire most vigorously at that delay. For other neurons, the difference in latencies will be either shorter or longer, thereby generating the variety of delay-tuned neurons in the cortex.

Paradoxical latency shifts, however, are not unique to bats. Indeed, they have been seen in the auditory systems of frogs. Whether neurons in other mammals combine different delays to generate delay-tuned neurons is unclear since that feature has not been evaluated in other mammals. Nevertheless, it seems reasonable to assume that the *particular ways* by which bats combine the innervation from neurons with paradoxical latency shifts to create a population of delay-tuned neurons having delays that correspond closely to the pulse-echo delays they actually receive in nature is almost certainly a unique adaptation of bats.

Whispering Bats

As mentioned previously, whispering bats comprise a diverse group of bats that occupy a wide variety of habitats and feed upon an equally wide variety of food sources, but they do not prey upon flying insects as do the loud FM and CF–FM bats. Many whispering bats

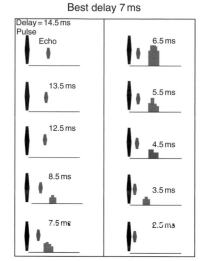
Best delay 7 ms

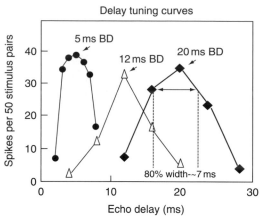

Figure 13   Delay-tuned neurons in the AC of the big brown bat. Left panel shows stimuli used to evaluate delay tuning. Two stimuli are presented, the first (pulse) was louder than the second (fainter echo). The delay between the pulse and echo was varied and the discharges evoked by each pulse-echo pair are shown. For this neuron, the best delay (BD) was about 6.5 ms. Adapted from Sullivan WE (1982) Possible neural mechanisms of target distance coding in auditory system of an echolocating bat. *Journal of Neurophysiology* 48: 1033–1047. Right panel shows plots of delay tuning for three cortical neurons, each having a different BD. The delay tuning is broad and encompasses several ms in each neuron. For the neuron with a BD of 20 ms, the delays that evoked firing rates that were 80% of the maximal rate spanned 7 ms. Adapted from Dear SP, Simmons JA, and Fritz J (1993) *Nature* 364: 620–623, with permission.

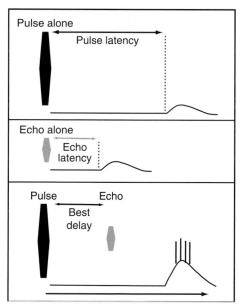

Figure 14   Paradoxical latency shifts that underlie delay tuning. The subthreshold response to the loud emitted pulse (top panel) has a longer latency than the subthreshold response to the fainter echo (middle panel). When the echo is delayed such that its response coincides in time with the louder pulse, the two subthreshold events summate to evoke a large response that evokes a strong series of discharges (bottom panel).

use echolocation primarily for orientation and hunt by passively listening to the sounds made by their prey. Although the processing of acoustic information in whispering bats has received less attention than in loud FM

or CF–FM bats, several neural adaptations in the pallid bat are especially interesting.

Recall that pallid bats are found in the deserts of the southwestern United States, and use echolocation for orientation while simultaneously listening for the noisy sounds made by insects or other small animals as they scurry on the desert floor. The echolocation calls are FM signals that sweep downward from about 60 to 30 kHz over a duration of 2–4 ms. Prey-generated noise, in contrast, is composed of frequencies from about 5 to 35 kHz. The remarkable feature of the IC and AC of pallid bats is that both structures are functionally divided into two regions, one adapted for processing the FM components in their echolocation calls and the other for processing the noise produced by their prey.

The Inferior Colliculus of the Pallid Bat

The IC of the pallid bat has medial and lateral divisions, as shown in **Figure 15**. The medial division is tonotopically organized in a manner similar to the IC of loud FM bats. Low frequencies are represented by sheets of isofrequency neurons located dorsally, while high frequencies are represented in sheets in more ventral locations. Neurons tuned to low frequencies, which are located in the dorsal isofrequency sheets, are less specialized than those tuned to the high frequencies in the ventrally located sheets. The high frequencies, from about 30 to 60 kHz, correspond to the frequencies in the pallid bats'

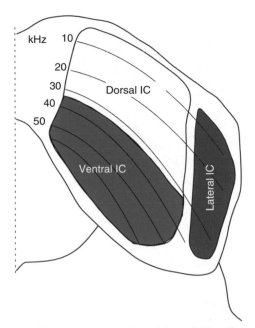

Figure 15 Tonotopic organization of the pallid bat IC. Isofrequency sheets together with the frequencies to which each sheet is tuned are shown. The nucleus is also functionally partitioned into dorsal, ventral, and lateral areas. The ventral area is tuned to the frequencies in the bat's echolocation calls and most neurons are sensitive to the direction the FM sweep. Neurons in the lateral area are tuned to lower frequencies and are specialized for responding to prey-generated noise. Courtesy of Zoltan Fuzessery.

echolocation calls. Low-frequency neurons respond (discharge) to many types of signals, which include tones and FM sweeps, but generally are unresponsive to noise. These neurons may play a role in the processing of their social communication calls that often have frequencies considerably lower than those in the echolocation calls. High-frequency neurons, in contrast, respond poorly to tones and noise but respond vigorously to FM sweeps. Moreover, they respond preferentially to downward FM sweeps that have particular sweep rates, features that correspond closely with the sweep direction and sweep rates of the FM signals pallid bats use for echolocation. Thus, the high-frequency neurons in the ventral isofrequency sheets in the medial portion of the IC are specialized for processing echolocation calls. It should also be noted that the mechanisms that generate the FM directional selectivity in the high-frequency neurons are the same mechanisms that generate FM directional selectivity in other bats and other mammals.

The lateral portion of the IC is markedly different than the medial portion of the IC. Frequencies represented in the lateral region are biased toward low frequencies, from about 5 to 40 kHz, frequencies that are prevalent in prey-generated noise but are lower than the frequencies in the pallid bat's echolocation calls. Neurons in the lateral

portion respond weakly to tones and FM signals but respond strongly to noise, the signals that are generated by the terrestrial insects these bats feed upon. Thus, the IC of the pallid bat is functionally divided into three major regions: (1) low-frequency neurons in the dorsal medial portion of the IC; (2) high-frequency neurons in the ventral medial portion of the IC that are specialized for processing echolocation calls; and (3) neurons in the lateral portion of the IC that are specialized for processing prey-generated noise.

The species-specific features that emerge in the IC are then imposed on the AC of pallid bats. Its AC contains a tonotopic map, as in the AC of all other mammals. As in the IC, the high-frequency representation, frequencies between 30 and 60 kHz, is dominated by neurons sensitive to FM sweeps, and thus the neurons in this region are important for processing echolocation calls. The neurons tuned to lower frequencies have not been as intensively studied but many low-frequency neurons respond to FM and tones whereas others respond most vigorously to noise and thus are important for the detection and localization of prey-generated sounds.

The Inferior Colliculus of the Mustache Bat

Ethological influences on the central auditory system are perhaps most elegantly and dramatically expressed in the tonotopic organization in the IC of CF–FM bats. The organization reflects the overrepresentation of the frequencies in the cochlea that comprise the bat's acoustic fovea and re-emphasizes the importance that Doppler-shift compensation plays in the acoustic behavior of the animals. These features are illustrated in the IC of the mustache bat (**Figure 16**), which shows that one-third of the entire volume of the mustache bat's IC is devoted to one frequency, '60 kHz,' the frequency at which the bat clamps the frequency of its echo CF component during Doppler-shift compensation. The overrepresentation distorts the normal tonotopic arrangement seen in all other mammals, where sheets of cells representing frequencies from low to high are stacked dorsoventrally as in the IC of the loud FM bat. The greatly expanded representation of 60 kHz literally pushes the sheets of cells representing frequencies lower than 60 kHz forward while simultaneously pushing the sheets of cells representing frequencies higher than 60 kHz more medially.

The cells tuned to '60 kHz' are not only overrepresented, but they are also much more sharply tuned than cells tuned to other frequencies. That is, 60-kHz cells respond only to a very narrow range of frequencies that can be as small as 100–200 Hz (**Figure 17**). Both the sharp tuning of 60-kHz cells and their overrepresentation are features that were first created in the cochlea and are

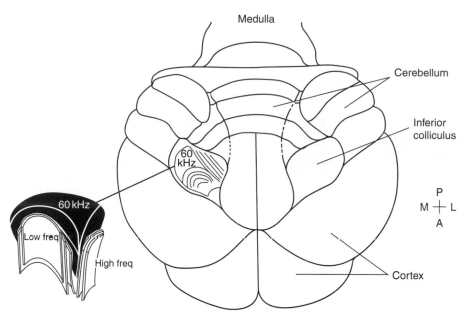

Figure 16 Schematic drawing of a dorsal view of the mustache bat's brain to show the location of the 60-kHz region in the IC. The hypertrophied inferior colliculi protrude between the cerebral cortex and the cerebellum. In the left colliculus are shown the isofrequency sheets as determined in anatomical and physiological studies. Low-frequency sheets, representing an orderly progression of frequencies from about 59 to 10 kHz and high-frequency sheets from about 59 to 120 kHz, are shown on either side of the greatly enlarged 60-kHz isofrequency region that occupies about one-third of the entire IC. Reproduced from Pollak GD and Casseday JH (1986) *The Neural Basis of Echolocation in Bats*. New York, NY: Springer-Verlag, with permission.

simply imposed on the IC due to the tonotopic projections of the ascending auditory pathway. The sharp tuning, moreover, endows 60-kHz cells with exceptional abilities to encode the frequency modulations on the echo CF component that are imparted by the fluttering wings of an insect, as described previously. Even the smallest frequency excursions generate discharge trains that are tightly locked to the wing-beat frequencies. As shown in **Figure 18**, frequency modulations as small as ±10 Hz evoke discharges in 60-kHz neurons that are locked to each modulation cycle. The locking of discharges is evoked by the frequency modulations as frequencies move in and out of a 60-kHz neuron's sharp tuning curve. This creates a unique pattern of discharge activity across the population of 60-kHz cells and provides the neural substrate for insect recognition.

Many neurons in the mustache bat's IC, including those tuned to 60 kHz, have an unusual feature; they respond much more vigorously to two tones that have a specific frequency relationship and temporal relationship than they do to a single tone at their 'best frequency,' the frequency to which they are most sensitive. These neurons are called 'combination-sensitive neurons' because they are far more responsive to combinations of tones than to single tones (**Figure 19**). These cells are similar to the delay-tuned cells of loud FM bats, described previously, except that combination-sensitive cells require both a specific temporal interval between the two tones *and*

a specific frequency difference. As illustrated in **Figure 19**, most of these cells fire best or most vigorously when the frequency of the first tone corresponds to the fundamental frequency of either the CF component or FM sweep of mustache bat's echolocation calls, about 30 kHz, and the second tone to the frequency of either the CF or FM sweep in the second or third harmonic of the calls, 60 or 90 kHz.

One of the functional attributes assigned to combinatorial neurons is for coding the range or the distance between the bat and its target. The rationale is that the temporal intervals that many, although not all, combination-sensitive neurons respond with facilitated responses vary from about 1 to 20 ms. These intervals would be generated by target distances of about 0.3–3.0 m and correspond to the timing differences between the pulse and the echo that the bats receive during echolocation. While ranging is an appealing hypothesis for these neurons, it should be noted that, like the delay-tuned neurons in loud FM bats, the delay tuning of these neurons is also broad, sometimes encompassing pulse-echo delays of up to 10 ms. It is also significant that combinatorial neurons are not unique to CF–FM bats but also have been found in the IC of mice. In mice, and in bats as well, the combinatorial neurons are most likely important for the processing of communication calls, which are broadband signals that have specific temporal features. The requirement that specific frequencies be combined and that the various frequencies must also have a specific

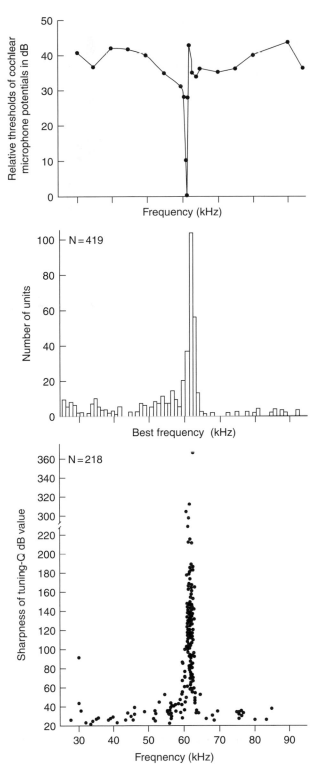

timing relationship is important for generating response selectivity where neurons respond more vigorously to certain calls and less vigorously or not at all to other calls, even though the frequency composition of the various calls would share common frequencies.

Auditory Cortex of the Mustache Bat

The AC of the mustache bat has been intensively studied by Nobuo Suga and his students and is the most elegantly organized cortex of any mammal. The organization of the mustache bat cortex is a highly elaborated version of the features described previously for the IC, but in the cortex, each of the features in the IC is spatially segregated, in that specific features are represented in their own functional regions (**Figure 20**).

Three regions of the mustache bat AC are especially prominent: one tonotopically organized region and two others that are composed entirely of combinatorial neurons, but the type of combinatorial neuron in the two regions is different and distinctive (**Figure 20**). The tonotopically organized region is similar to the tonotopic organization in the corticies of other mammals, except that 60 kHz is greatly overrepresented, as it is the mustache bats' IC. Indeed, Suga and his colleagues refer to the '60' kHz representation as the 'Doppler-shifted constant frequency' (DSCF) area, because all the neurons are tuned to the frequency at which the bat clamps the 60-kHz CF component of echo while Doppler compensating.

The most interesting and unique features of the mustache bat cortex are two other regions composed of combinatorial neurons, each of which lie near to but are separate from each other and from the DSCF area, the tonotopically organized region. One region, the so-called CF–CF area, contains neurons that are only responsive to the CF components of the echolocation calls, and require that particular combinations of CF harmonics to evoke a response. Thus, the neurons in one region require that the signal be composed of a 30- and a 60-kHz tones (the 30-kHz tone is the fundamental frequency, while the 60-kHz tone is the second harmonic frequency of the bat's calls), whereas others are sensitive to 30- and 90-kHz combinations. The exact frequency combinations of 30- and 60-kHz and 30- and 90-kHz differ slightly from neuron to neuron, and thus these neurons are probably responding to Doppler-shift magnitude. The other combinatorial region, the FM–FM region, is even more interesting. Here, the neurons require two FM sweeps, one having the frequencies of the

Figure 17 Cochlear microphonic audiogram (top panel), distribution of best frequencies (middle panel), and tuning sharpness of neurons tuned to various frequencies (lower panel) recorded from the mustache bat's IC. Tuning sharpness is shown as Q values defined as the range of frequencies that evoke discharges at 10 dB above the threshold defined by the neuron's best frequency. The higher the Q value, the sharper the tuning. A disproportionate number of neurons have best frequencies that correspond to the sharply tuned notch of the cochlear

microphonic audiogram and those neurons have much higher Q values (and are more sharply tuned) than neurons with best frequencies either below or above 60 kHz. Reproduced with permission from Pollak GD and Casseday JH (1986) *The Neural Basis of Echolocation in Bats*. New York, NY: Springer-Verlag.

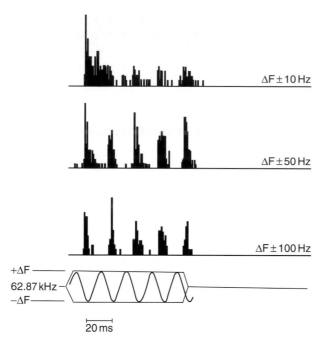

Figure 18 A neuron in the mustache bat IC sharply tuned to 60 kHz responds to each cycle of sinusoidally FM signals. The neuron displays discharges to each cycle of modulated signals in which the frequency varies by only ±10 Hz around a carrier frequency of 62.87 kHz. Reproduced from Pollak GD and Park TJ (1995) The inferior colliculus. In: Popper AN and Fay RR (eds.) *Hearing by Bats. Springer Handbook of Auditory Research.* New York, NY, with permission.

fundamental (i.e., from about 30 to 23 kHz) and the other the frequencies of either the second harmonic (i.e., from about 60 to 45 kHz) or third harmonic (i.e., from about 90 to 70 kHz) of the echolocation calls. Moreover, the neurons are delay tuned and thus also require a specific temporal interval between the first harmonic FM and the other harmonic that ranges from a few milliseconds to about 20 ms. Thus, these neurons respond selectively to echo delay, similar to the delay-tuned neurons in the mustache bats' IC. What these features show is that the cortex of the mustache bat is specially adapted for processing the acoustic features that it normally receives during echolocation.

Conclusions

This brief tour of the neuroethology of bats illustrates how the auditory systems of different bats are adapted to enable each species to exploit one or another type of food supply in the environment in which they hunt. The adaptations in their auditory systems, like the adaptations of their bodies, did not involve new inventions but rather took basic features present in all mammals and emphasized one or more features that were adaptive for a particular habitat. Thus, the adaptations in the auditory system

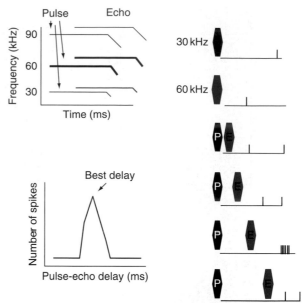

Figure 19 Combination-sensitive delay-tuned neuron in the mustache bat IC. Left panel shows the various harmonics in a mustache bat echolocation call and a Doppler-shifted echo that returns several milliseconds after the emitted pulse. The right panel shows the weak responses to two tones, one at a frequency corresponding to the fundamental frequency of the emitted pulse (30 kHz) and the other at a frequency in the second harmonic of the echo (60 kHz). The neuron responds weakly to each signal presented alone. When the 30-kHz signal (pulse) is presented first and the 60-kHz (echo) signal is presented shortly thereafter, the responses are still weak except at a particular pulse-echo delay that evokes a strong response. The number of spikes evoked at each delay is plotted in the graph in the lower left and graphically shows that the neuron responds weakly at all pulse-echo delays except for the BD, at which it responds strongly.

are primarily, although not exclusively, a matter of quantity, where a species expresses certain features that are shared by other species but to a greater degree or in a more pronounced form, rather than expressing wholesale qualitative changes in the mode of processing. The idea that each species evolved neural adaptations suited to their habitat is a universal feature of all sensory systems in all animals, and is especially well illustrated in the auditory systems of bats because of the diverse habitats in which these animals live.

See also: Bat Migration; Bats: Orientation, Navigation and Homing; Hearing: Vertebrates.

Further Reading

Barber JR, Razak KA, and Fuzessery ZM (2003) Can two streams of auditory information be processed simultaneously? Evidence from the gleaning bat, *Antrozous pallidus. Journal of Comparative Physiology A* 189: 843–855.

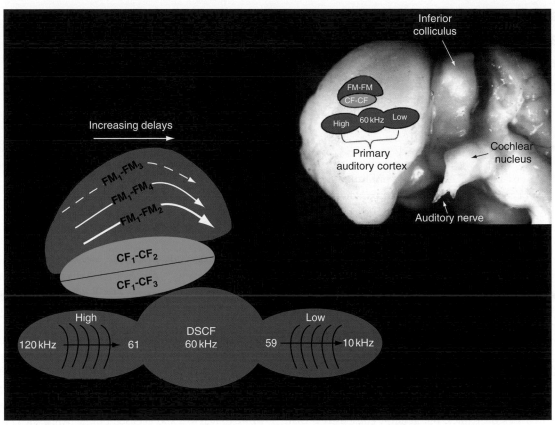

Figure 20   AC of the mustache bat showing the tonotopically organized region and the specialized combinatorial regions, the CF–CF and FM–FM regions. The tonotopically organized region has a disproportionately large representation of 60 kHz, the DSCF region. Based on Suga N (1990) Biosonar and neural computations in bats. *Scientific American* 262: 60–68.

Dear SP, Simmons JA, and Fritz J (1993) A possible neuronal basis for representation of acoustic scenes in auditory cortex of the big brown bat. *Nature* 364: 620–623.

Goldman LJ and Henson OW Jr (1977) Prey recognition and selection by the constant frequency bat, *Pteronotus parnellii*. *Behavioral Ecology and Sociobiology* 2: 411–419.

Griffin DR (1958) *Listening in the Dark*. New Haven, CT: Yale University Press.

Neuweller G (1990) Auditory adaptations for prey capture in echolocating bats. *Physiological Reviews* 70: 615–641.

Pollak GD and Casseday JH (1986) *The Neural Basis of Echolocation in Bats*. New York, NY: Springer-Verlag.

Popper AN and Fay RR (eds.) (1995) *Hearing by Bats. Springer Handbook of Auditory Research*. New York: Springer.

Schnitzler H-U, Moss CF, and Denzinger A (2003) From spatial orientation to food acquisition in echolocating bats. *Trends in Ecology and Evolution* 18: 386–394.

Schuller G and Pollak GD (1979) Disproportionate frequency representation in the inferior colliculus of Doppler-compensating greater horseshoe bats: evidence for an acoustic fovea. *Journal of Comparative Physiology A* 132: 47–54.

Suga N (1990) Biosonar and neural computation in bats. *Scientific American* 262: 60–68.

Sullivan WE (1982) Possible neural mechanisms of target distance coding in auditory system of an echolocating bat. *Journal of Neurophysiology* 48: 1033–1047.

Thomas JA, Moss CF, and Vater M (eds.) (2004) *Echolocation in Bats and Dolphins*. Chicago: University of Chicago Press.

Ulanovsky N and Moss CF (2008) What the bat's voice tells the bat's brain. *Proceedings of the National Academy of Science* 105(25): 8491–8498.

Von der Emde G and Schnitzler H-U (1990) Classification of insects by echolocating greater horseshoe bats. *Journal of Comparative Physiology A* 167: 423–430.

Bateman's Principles: Original Experiment and Modern Data For and Against

Z. Tang-Martínez, University of Missouri-St. Louis, St. Louis, MO, USA

Introduction

In 1948, Angus Bateman, a botanist, published a review destined to become the most important paradigm in most studies of sexual selection for the next 50 years; he never again wrote on sexual selection. By a paradigm, I mean here a particular world view, or set of interconnecting ideas, that comes to permeate a field and is accepted as self-evident by most researchers in a discipline. Bateman's paradigm, for example, has dominated much of the thinking on sexual selection theory, parental investment theory, evolution of mating systems, and the evolution of parental care, among other areas of study.

Because of the influence of a 1994 study by Steve Arnold, parts of Bateman's ideas have in modern times become known as Bateman's principles. Although there is still some confusion about what exactly should be included under Bateman's principles, the general consensus is that they include the following predictions: (1) male RS increases with number of mates, while female RS does not; (2) male RS will show greater variance than female RS; and (3) the sex with the greater variance in RS (i.e., males) will be more greatly affected by sexual selection. The acceptance of these ideas has led to the conclusion that males have greater variance in number of mates and in RS and, consequently, that they undergo more intense sexual selection when compared with females. The only exceptions to this would be found in so-called sex-role-reversed species in which females compete for males and males take care of the young, and in which the number of matings and RS of females is potentially higher than those of males.

In effect, Bateman's principles have become a cornerstone of sexual selection theory and have influenced numerous areas of behavioral ecology. Here, I review Bateman's original experiments and recent criticisms of this research. Additionally, I provide an overview of modern data that support Bateman's ideas, as well as of data that contradict, or otherwise throw into question, some of Bateman's assumptions and conclusions.

In summary, Bateman's research has been responsible for an impressive body of work and his principles appear to be valid in some species. However, an increasing literature, both theoretical and empirical, suggests that not all of Bateman's ideas can be supported as universal principles. Rather, it is likely that much variation exists from species to species and that, in many species, at least some aspects of Bateman's principles will need to be abandoned or recharacterized as hypotheses.

A Historical Perspective

In 2004, Donald Dewsbury reviewed the history of Bateman's study, 'Intra-sexual Selection in *Drosophila.*' The first point to make is that Bateman saw his study on *Drosophila* as a test of Charles Darwin's ideas on sexual selection. While Darwin's theory of natural selection had been relatively widely accepted, his suggestion that a different process, sexual selection, could explain sexual dimorphism in animals had been, for the most part, met with skepticism. In fact, even Alfred Russel Wallace, the codiscoverer of the concept of natural selection, once wrote to Darwin and accused him of being 'un-Darwinian' because of the latter's advocacy of sexual selection.

As envisioned by Darwin, sexual selection consisted of two different processes: (1) intrasexual selection, which involved direct combat among members of the same sex (usually thought to be more common in males) for access to members of the opposite sex and (2) intersexual selection, which required individuals of one sex (usually thought to be more common in females) to choose among individuals of the opposite sex. Thus, the traits selected by intrasexual selection tended to be size, vigor, strength, and fighting ability, traits believed by Darwin to be mainly characteristic of males. Intersexual selection resulted in the evolution of traits that made their owners more 'attractive' to the opposite sex. For example, male ornaments such as long tails, brightly colored plumage, or complex songs or calls were all considered to have evolved because females preferred males that exhibited these traits as mates. These two forms of sexual selection came to be known as male–male competition and female choice, respectively (although Darwin knew and acknowledged that, in a some species, females are the more ornamented sex and may physically compete for males, while males are less showy and do the choosing of mates). Relying on the fact that males produce large numbers of small sperm and females produce relatively few, costly eggs, Darwin implied that females can be expected to be more sexually passive, while males should be sexually active in their search for mates. In 1972, these ideas were further elaborated by Robert Trivers, and by Geoff Parker and his collaborators. These authors took Darwin's comments on gamete size and numbers, and his comments on passive females and profligate males, and reified them into a modern cost–benefit analysis that 'explained' the differences between the sexes: a female's eggs are few and costly and she should, therefore, guard them carefully and be coy, mating only with the best male (in terms of male quality) available to her. She could afford to mate with only one male because that male would produce enough sperm to fertilize all of her eggs. The male, on the other hand, could produce unlimited numbers of cheap sperm; he could afford to mate with all and any females because he incurred no serious costs by doing so.

Bateman's experiments and his interpretations (see section 'Bateman's Research') concurred with Darwin's ideas and later with those of Trivers and Parker. This is because Bateman proposed that, in the case of males, RS is a function of the number of mates a male can obtain. On the other hand, Bateman concluded that the peak of female RS is reached after mating with only one male and that additional matings will not increase a female's RS. This led him to propose that females should be sexually cautious and very discriminating, and that males should behave promiscuously.

An interesting aspect of the Bateman history is that his ideas lay dormant for ~26 years, until Trivers revived them when he published his now classical study 'Parental Investment and Sexual Selection,' which relied heavily on Bateman's work. Prior to the Trivers' publication, Bateman's study had been cited a total of 24 times. Yet, between 1973 and 2002, a period of 29 years, Bateman's work was cited 601 times, and this upward trend has continued unabated until the present with ~1000 total citations.

In summary, what could be called the Darwin–Bateman–Trivers–Parker paradigm has been, and among many researchers continues to be, the guiding principle and cornerstone for much of sexual selection theory. Up until very recently, the unquestioned assumptions underlying the study of sexual selection have been that eggs are expensive while sperm are unlimited and cheap, that males should therefore be promiscuous while females should be very choosy and should mate with only the one best male, and that there should be greater reproductive variance among males (as compared to females) because it is males that compete for females and mate with more than one female. Since females are, presumably, mating with only one male, this means that some males may mate with many females, while others may mate with few or none. This reproductive variance is then responsible for the sexual selection of traits possessed by the more successful males.

Bateman's Research

Bateman conducted six series of experiments using fruit-flies, *Drosophila melanogaster*. He placed equal numbers of male and female flies (3–5 of each sex) in bottles and allowed them to mate for a variable number of days (3 or 4). In some experiments, he transferred the flies (and eggs and larvae) to a new bottle every day, while in others he left the flies in the same bottle throughout the experiment. Additionally, in different experiments, the flies varied in age from 1 to 6 days old. The flies were from several inbred strains that contained mutant markers that he could then use to identify the parentage of the progeny (both the mother and the father). In some tests, some of these flies were backcrossed to other inbred strains for several generations, in an attempt to increase heterozygosity. However, as pointed out by graduate student Brian Snyder and his advisor, Patty Gowaty in 2007, even then, the flies would have been nearly genetically identical except for the mutations.

With regard to behavior, Bateman never actually observed copulations by his flies but, rather, inferred the number of mates based on the number of young sired by different males in any one female's offspring. For example, if a female's offspring all had the same father (based on the genetic mutation markers), Bateman concluded that the female had only mated with one male. If her offspring were sired by two males, he concluded she had mated with two males. Thus, matings that, for any reason, did not produce offspring were not detected or counted by Bateman in any of his analyses. In other words, if a female mated with three males, but only the offspring of one male survived, Bateman had no way of knowing that she had actually mated with three males.

The six series, including 4–9 replicates of each series, were analyzed as two separate sets of combined data (**Figure 1**). The first set consisted of series 1–4. The results, presented as Graph 1 in Bateman's study, showed that RS of males ('relative fertility' in Bateman's graph) increased steadily with the number of mates (meaning, of course, the number of females that a male fathered offspring with) up to three mates, then declined somewhat at four mates. Females' RS also increased, but slightly less steeply, up to three mates, and no females had young sired by more than three mates. The second set consisted of series 5 and 6 and the results are presented in Graph 2 of Bateman's work. In this case, the results are strikingly different. Male RS climbs continuously with each mating up to four mates. However, female RS peaks at one mate and then plateaus. That is, females gain nothing in RS by mating with more than one male. Snyder and Gowaty pointed out that Bateman's stated reason for presenting his data in two different graphs was that the results obtained in series 5 and 6 looked different from those obtained in series 1–4.

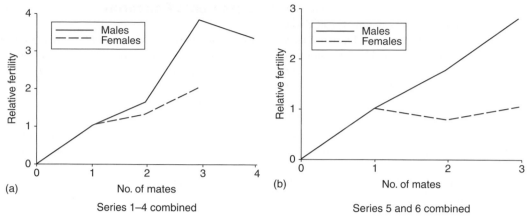

Figure 1 The two graphs produced by Bateman in his two series of experiments. Dashed lines represent females while solid lines represent males. See text for details. Reprinted from Bateman AJ (1948) Intra-sexual selection in Drosophila. *Heredity* 2: 349–368, with permission from Macmillan Publishers Ltd.

It is interesting that, although Bateman presents both graphs and all of his data in the paper, in his discussion he emphasizes primarily the results of the second set of experiments. Consequently, he concludes that females will be choosy and mate with only one male because they have nothing to gain by mating with more than one male. Males, on the other hand, have everything to gain by mating promiscuously with many females. Moreover, he presents evidence of male reproductive skews and concludes that the slopes of the lines in his graphs (now known as 'Bateman Gradients' or 'sexual selection gradients') represent the strength of sexual selection for each sex.

Perhaps more remarkable is what then happened to Bateman's two graphs. With a few exceptions, most subsequent researchers presented and relied only on the data from Bateman's series 5 and 6 (Bateman's second graph). General discussions of sexual selection, and even textbooks in animal behavior, almost always presented only the second graph and the discussion was limited to these results, usually as an explanation of why males are promiscuous and females coy and choosy. The results of series 1–4, and any discussion of increases in female RS as a function of the number of males the female mated with, for all practical purposes disappeared from the literature. The only exceptions appear to be a few theoretical models by Steve Arnold and his colleagues who re-examined Bateman's findings.

Methodological and Statistical Problems

Certain problems inherent in Bateman's methodology were discussed in a review by Zuleyma Tang-Martinez and graduate student T. Brandt Ryder. First, the use of inbred strains raises the possibility that the mating behavior of the mutants was not normal. Second, the age of the females may have been a factor. Female *D. melanogaster* do not

mature sexually until they reach 4 days of age (males are generally mature at 1 day of age). Consequently, females that were less than 4 days of age at the beginning of the experiments could be expected to have mated fewer times, as compared to older females or males. Third, it is not clear whether there was food limitation for the larvae in those experiments in which the flies were not transferred to a new bottle every day. Food limitation could have meant that larvae carrying certain mutations might have survived less well because they were less competitive in the face of a limited resource. Fourth, *Drosophila* males may have produced pheromones, or ejaculated secondary compounds that either inhibited the receptivity of the females with which they mated (causing them to become less attractive to other males), and/or resulted in the females having longer remating latencies. Fifth, the length of each experiment (number of days the experiment lasted) could have influenced how many females mated more than once. This is because *D. melanogaster* females can store sperm for a variable number of days (up to 2 weeks) and may not need to remate and replenish sperm supplies in experiments that lasted only 3–4 days. Lastly, anything that would have affected offspring mortality in a systematic way could have biased the results because only surviving offspring were included in the analysis. Examples of factors that could have affected mortality include lethal mutations (and most of Bateman's mutations were homozygous lethals), mutations that decrease survivorship for various reasons (e.g., heterozygous effects on viability, reduced competitive ability possibly as a result of food limitation), and maternal effects that affect survivorship of larvae as a function of maternal age.

Bateman recognized some of the concerns summarized in the preceding paragraph and attempted to control for them. Based on analyses of variance, he concluded that age, mutations, and population differences in the various experiments and bottles did not affect his results.

Despite the widespread acceptance of Bateman's principles, there has been a number of significant theoretical critiques of his research. For example, in 1985, William Sutherland devised a model that looked at the handling time of matings and their aftermath, searching time for new mates, and random rates of mating encounters. He concluded that Bateman's sexually dimorphic differences in behavior could be explained by chance alone. In 1987, Steve Hubbell and Leslie Johnson extended Sutherland's model to include two constraints: effects of survival probability over time (taking into account that everyone eventually dies) and also that, in diploid, sexually reproducing species, mean mating success for males and females must be equal among all mating individuals. They then examined the effects of these variables on lifetime mating success. They concluded that the variance in lifetime RS can be due to stochastic and nonheritable factors that affect time allocation budgets for mating. These results brought into question some of the accepted conclusions that form part of Bateman's principles.

Gowaty and Hubbell emphasized in 2005 that all these factors, alone or in combination, can affect choosiness or indiscriminate sexual behavior regardless of sex. Consequently, their model emphasizes that mating strategies can be flexible, that under certain conditions males may be choosy and females indiscriminate, and that individuals may be able to switch sex roles. Moreover, it highlights the importance of random forces and nonheritable factors that can affect variance in RS and mating success without influencing sexual selection.

Recently, a very substantive critique of Bateman's methodology has been presented by Snyder and Gowaty. In addition to addressing several of the confounds and difficulties already mentioned, they did an extensive reanalysis of Bateman's data. Among the problems they identified are:

1. Bateman's sampling was likely biased. In many cases, Bateman was able to assign partial parentage to either the mother or the father of an offspring, but not both. As a result, in 5 out of the 6 series, his males and females did not have equal mean RS, an impossibility for diploid species with equal sex ratios, when sampling is complete and unbiased. Moreover, because means and variances in RS are correlated to one another, this methodology may well have affected the reliability of his interpretations, throwing into doubt his conclusion that males have higher variance in RS than do females.

2. There were mathematical mistakes in Bateman's calculations of variance in number of mates. Snyder and Gowaty found that due to either rounding errors or mathematical miscalculations, three of the calculations of variance in number of mates for both males and females were incorrect. Moreover, a statistical analysis of these data found significant differences in only two of nine recalculated ratios of the variance in number of mates. Bateman had presented these data in a table but had not subjected them to statistical tests. Given that the putative sex differences in variance in number of mates is a crucial component of Bateman's principles, the failure to find consistent significant differences in this measure raises serious doubts about Bateman's conclusions.

3. Bateman's analysis of variance in RS suffers from pseudoreplication. An examination of the data and analyses presented by Bateman demonstrates that he calculated the degrees of freedom as $N-1$, where $N=$ the total number of adult flies in each treatment, rather than the total number of populations in each treatment, which constituted pseudoreplication. The appropriate analysis would have been to use $N-1$, where $N=$ the number of populations. Again, the fact that males are supposed to have higher variance in RS as compared to females is a central canon of Bateman's principles. Modern statistical techniques consider pseudoreplication completely inappropriate (although this was not the case at the time Bateman analyzed his data). Consequently, his use of pseudoreplication and his use of variance in RS as a characteristic of individuals, rather than populations, raise additional questions about the validity of his interpretations.

4. A reanalysis of Bateman's data shows that RS of both males and females increases as a function of number of mates. Snyder and Gowaty attempted to do a combined probability analysis of the data presented in Bateman's two graphs but were unable to do so because the relevant data (individual reproductive success) were not included in the review. Instead, they used the next best analysis, a regression of the pooled mean number of offspring with the number of mates. Results showed that the slope for females is slightly lower, but very close to the slope for males. Moreover, the fact that means are used and that there were sampling errors (see point 1, this section) suggests that the actual difference in RS of males and females as a function of number of mates may have been even smaller. Graph 2 is widely used to support Bateman's conclusions about the effect of number of mates on RS in males versus females. The reanalysis described here shows that there likely has been a misguided over-reliance on this graph.

5. Arbitrary decisions made by Bateman could have biased his results. As discussed earlier, Bateman decided to split his data into two different graphs for reasons that can only be considered arbitrary (i.e., that the two series yielded results that he said 'looked different') Snyder and Gowaty argue that if Bateman had grouped his data differently, for example by analyzing series 4 together with series 5 and 6, he might have obtained different results and reached different conclusions.

Overall, Snyder and Gowaty conclude that the problems with Bateman's analyses and conclusions are so serious that the tenets of Bateman's principles are questionable and unreliable. They recommend instead that investigators consider Bateman's 'principles' as 'hypotheses' awaiting further testing. It is particularly surprising that, given the central place of Bateman's principles in modern behavioral ecology and in the study of sexual selection, there have been few attempts to replicate his results until very recently.

Modern Studies in Support of Bateman

Despite the criticisms reviewed, there have been a number of theoretical and empirical studies that support all or part of Bateman's findings and conclusions. For example, based on a mathematical model and selection theory from quantitative genetics, Steve Arnold and David Duvall found strong support for the principle that the relationship of mating success (number of mates) and RS influences the strength of sexual selection. Thus, they concluded that Bateman's gradients are critical to understanding the operation and intensity of sexual selection. Sexual selection will be stronger in whichever sex has the steeper partial regression slope of fecundity plotted as a function of number of mates.

In a subsequent theoretical model, Arnold tested whether Bateman's prediction that males should always show stronger selection as compared to females (based on the male and female sexual selection gradients obtained in Bateman's experiments) is applicable to all other species. This was an important question because this relationship and prediction had been assumed to be a universal characteristic of all sexually reproducing species. The results showed that the relationship between fecundity and mating success demonstrated by Bateman's fruitflies cannot be considered a universal attribute. In fact, under some conditions and in certain mating systems, females may undergo sexual selection more intensely than males.

Both these studies were important in examining Bateman's hypotheses at greater depth. Interestingly, while one study supported the importance of Bateman gradients in predicting the intensity of sexual selection, the other found that the assumption that males would invariably be subject to more intense sexual selection could not be supported (see section 'Modern Research').

More recently, sex-role-reversed species have featured prominently in attempts to test Bateman's perspectives. Sex-role-reversed species are ones in which the male takes care of the young and the females compete for mates and experience stronger sexual selection. If Bateman was correct, then one would expect that the females in these species would have steeper Bateman gradients as compared to the males. Adam Jones and colleagues tested this prediction in the pipefish, *Syngnathus typhle*, a species in which males are choosy, carry the eggs and embryos in a pouch throughout 'pregnancy,' and give birth. Females mate multiply and compete aggressively for mates. Because of the reproductive dynamics (the father fertilizes the eggs once they are in the pouch), certainty of paternity was 100% and could be assigned unequivocally. Maternity was assigned by the use of microsatellite loci. The investigators found that the females in this species have steeper Bateman gradients. That is, there is a stronger correlation between number of mates and RS in the females, results that are consistent with Bateman's predictions.

In a second study, Jones and his collaborators tested Bateman's principles in the rough-skinned newt, *Taricha granulosa*, a species which is not sex-role-reversed. Female and male newts were captured in the field after copulations but prior to egg laying. The offspring were sampled genetically to determine paternity (in this case, the mothers were always known because the females were housed individually before they began laying eggs). Morphological measurements suggested that there was strong sexual selection in males but not in females. Analyses showed that males had greater variance in both mating success and RS, and that Bateman gradients were significantly steeper for males as compared to females. All these findings are in accordance with Bateman's principles and predictions (**Figure 2**).

A particularly interesting case involves the redback spider, *Latrodectus hasselti*, a species that exhibits 'terminal investment' by males. In this species, the male typically mates only once and is usually killed (and eaten) by the female (**Figure 3**). However, a few males do survive and manage to mate a second time. Maydianne Andrade and Michael Kasumovic tested Bateman's principle by allowing females to mate with 1, 2, or 3 males; males were rescued after the first mating and allowed to mate with a second female (males are incapable of mating more than twice). The sexual selection gradients concurred with Bateman's prediction that males should have steeper gradients. Despite the very high reproductive investment of these males (usually their lives under natural conditions), males were highly competitive and their RS (number of offspring) increased markedly with number of mates (but note that the maximum number of mates was only 2, and that, under natural conditions, most males only mate once). Females, on the other hand, showed only a weak increase in RS as their number of mates increased from 1 to 3 (**Figure 2**).

A final and notable example is a recent study by Adam Bjork and Scott Pitnick which seems to be the first attempt to partially replicate Bateman's experiments. These investigators generated Bateman gradients using several species of *Drosophila*, including *D. melanogaster*, the same species employed by Bateman. They were particularly interested in the effects of anisogamy, the difference

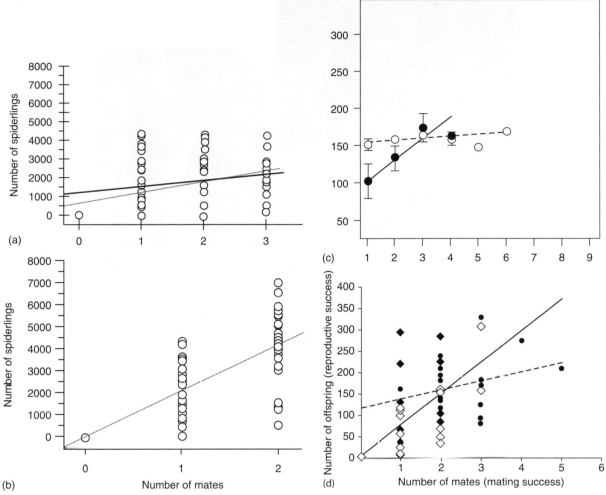

Figure 2   Bateman gradients for species reported to be consistent with Bateman's predictions: (a) female redback spider, *Latrodectus hasselti* (dashed line represents laboratory matings; solid line represents matings in the field adjusted by estimate of females that do not mate in nature) and (b) male redback spider (Andrade and Kasumovic, 2005; Courtesy of Oxford University Press). (c) Fruitfly *Drosophila melanogaster*; dashed line represents females and solid line represents males (Bjork and Pitnick, 2006; Courtesy of the authors). (d) The rough-skinned newt, *Taricha granulosa*; dashed line is for females and solid line is for males (Jones et al., 2002).

in size between male and female gametes. They predicted greater sexual differences in Bateman gradients when eggs and sperm were very different in size (as in *D. melanogaster*) and little or no difference when eggs and sperm were relatively more isogamous (i.e., more similar in size, as is the case in *D. bifurca*). *D. melanogaster* used in this experiment were outbred but had been held in the laboratory for more than 250 generations. *D. bifurca* were descendants of flies that had been trapped in the field and brought into the laboratory. For each species, groups of four males and four females were placed in experimental vials, copulations were recorded, and the number of eggs that hatched was counted. In *D. melanogaster*, sexual selection gradients were steeper for males as compared to females. On the other hand, in *D. bifurca*, selection gradients of males and females were not significantly different. These results and the degree of the differences in the slopes of the selection

gradients, apparently as a function of the relative size of eggs and sperm, are consistent with Bateman's findings and conclusions.

Modern Research That Is Not Consistent with Bateman's Hypotheses

Up until the advent of new genetic technologies that allow for parentage analyses, few studies ever attempted to rigorously determine the RS of males and females as a function of number of mates. In fact, in most cases, it was simply assumed that males increased their RS with an increase in number of mates and that females did not. A number of studies have now demonstrated that this is not always the case (**Figure 4**). For example, studies of juncos (*Junco hyemalis*) and brown-headed cowbirds

Figure 3   Species with sexual selection gradients that are consistent with Bateman's predictions: (a) Redback spider (*Latrodectus hasselti*) female with dead male in the foreground and a second male below (photo by Ken Jones; copyright MCB Andrade); (b) Fruitfly *Drosophila virilis* (drawing by A. J. Patterson, courtesy of Scott Pitnick); (c) Pipefish, *Syngnathus typhle*, a sex-role-reversed species (photo by Anders Berglund).

(*Molothrus ater*) found that sexual selection gradients for males and females are essentially the same. More recently, Andrew DeWoody and colleagues reported that in the small-mouthed salamander (*Ambysoma texanum*) not only is there no difference in the slope of the Bateman gradients between the sexes, but females engage in multiple matings significantly more than males and have greater standardized variances in mating success. They obtained similar results in tiger salamanders (*A. tigrinum*), although in this case males had greater variance in mating and reproductive success. In the katydid (*Conocephalus nigropleurum*), Patrick Lorch and coworkers reported that females have gradients that are steeper than those of males (**Figure 5**). Additionally, in the yellow-pine chipmunk (*Tamias amoenus*), the slope of the sexual selection gradient for males was only marginally higher than that for females (RS in both sexes increases with the number of mates). The RS of female Gunnison's prairie dogs (*Cynomys gunnisoni*) and mealworm beetles (*Tenebrio molitor*) also increases as a function of number of mates. Thus, there is now ample evidence that in some species female sexual selection gradients, acting through variance in number of mates, are as steep, or almost as steep, as those of males. In fact, in recent years, the number of studies reporting on species in which female sexual selection gradients are nearly identical to those of males appears to be increasing.

A special case that challenges (or at least modifies) Bateman's predictions occurs in cooperatively breeding species in which one pair reproduces, while the remaining adult members of the group are reproductively suppressed

and help in the care of the breeding pair's young. Mark Hauber and Eileen Lacey conducted a meta-analysis of such species among birds and mammals and found that females typically have greater variability in RS as compared to males. The obvious inference based on these findings is that females in these species are likely subject to more intense sexual selection as compared to males.

Lastly, Bateman's hypotheses may not always be applicable to externally fertilizing species. For example, Don Levitan examined a sea urchin, *Strongylocentrotus franciscanus*, and concluded that the intensity of sexual selection on the two sexes is dependent, at least in part, on mate density. Additionally, variance in RS was lower in males than in females, contrary to Bateman's predictions.

Modern Data on Corollaries and Assumptions

One of the basic, underlying assumptions of Bateman's principles is that sperm are cheap and produced in unlimited numbers, while eggs are costly and produced in limited numbers. Based on this, the prediction is that males should be promiscuous and indiscriminate, always attempting to mate with as many females as possible (note that this exactly parallels Bateman's gradients: as a male's number of mates increases, so does his RS). The concomitant prediction is that females will be sexually uninterested, and very choosy, in most cases mating only with one 'best' male who produces enough sperm to fertilize all the female's eggs. Again, this idea reinforces

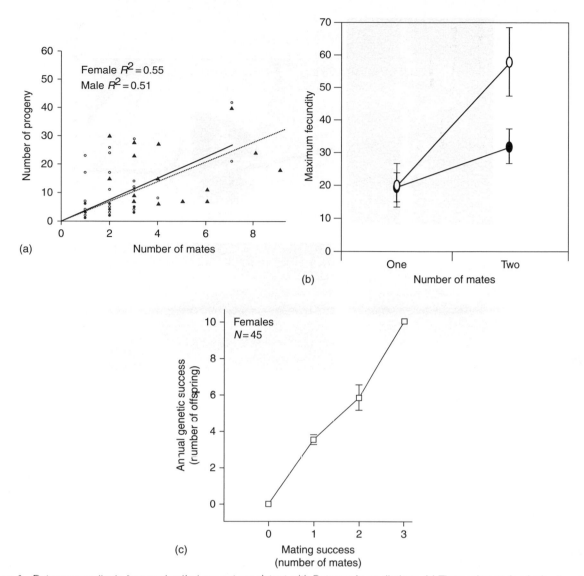

Figure 4 Bateman gradients for species that are not consistent with Bateman's predictions. (a) Tiger salamander *Ambystoma tigrinum*; dashed line and dark triangles represent females while solid line and open circles represent males (Williams and DeWoody, 2009; Courtesy of Springer). (b) Katydid *Conocephalus nigropleurum*; line with open circles represents females and line with solid circles represents males (Lorch et al., 2008; Courtesy of Brill). (c) Female dark-eyed juncos, *Junco hyemalis* (abstracted from Ketterson et al., 1998; Courtesy of Allen Press and the American Ornithologists' Union).

Bateman's conclusion that females will achieve peak RS after only one mating.

Modern data have revealed that the assumptions about cost of sperm and the corollaries that predict differences in male and female sexual behavior, may not be always valid, a point that was emphasized by Tang-Martinez and Ryder. Specifically, it is now clear that the comparison of the costs of one sperm with one egg makes very little sense. In order to fertilize even one egg, males typically produce millions of sperm as well as accessory gland secretions that constitute semen. In some cases, males must ejaculate multiple times in order for fertilization to occur. Therefore, a more meaningful comparison would

be that between the cost of one egg and the cost of the total number of ejaculates needed to fertilize that egg.

Moreover, there is now evidence from several species that demonstrate that sperm production can be quite costly. In the nematode *Caenorhabditis elegans*, mutants that do not produce sperm but are otherwise normal, have a 65% longer life span as compared to males that do produce sperm. Similarly, male adders, *Vipera berus*, lose as much weight during quiescent periods of sperm production as during periods during which they actively search for mates. Furthermore, it is now well established that sperm depletion is a problem faced by males of many species – the quantity of sperm that can be produced in

Figure 5 Species with sexual selection gradients that are inconsistent with Bateman's predictions. (a) Katydids, *Conocephalus nigropleurum* (photo by Michelle Prytula); (b) Brown-headed cowbird, *Moluthrus ater* (photo by Adrian Aspiroz); (c) Flour beetles, *Tenebrio molitor*, mating (photo by Bradley Worden); (d) Small-mouthed salamander, *Ambystoma texanum* (photo by Andrew DeWoody).

any given interval of time is limited. As emphasized by Wedell and colleagues and Tang-Martinez and Ryder, this means that males will often carefully allocate sperm (i.e., tailor the size of their ejaculates) based on aspects of a female's quality or previous mating status. This allocation of sperm is a form of male choice and discriminating sexual behavior that would not be expected if sperm were very inexpensive and produced in unlimited numbers.

The widespread existence of polyandrous behaviors in females further draws into question the predictions regarding sexual passivity and coyness in females. Females of many species are now known to engage in multiple copulations with the same or different males. Moreover, extra-pair copulations and extra-pair fertilizations are extremely common in species that are socially monogamous. As examples, multiple paternity of a female's offspring has been found in many species of birds, mammals, and insects. What is even more telling is that, in many cases, females are known to play an extremely active role in seeking and soliciting multiple males for copulations. Thus, Bateman's predictions about female behavior are not in accordance with current information on the active role played by females in sexual interactions. In fact, polyandrous behavior by females is so common that many different hypotheses have now been formulated to explain how females might benefit from multiple matings.

Leks have been considered classical examples of mating systems in which there is very high variance in male RS and very little variance in female RS. The reason is that, based on behavioral observations, it appeared that one or a very few males in the lek aggregation obtained all or most of the copulations, while the majority of males obtained none. Females, on the other hand, were believed to enter the lek, choose the highest quality male to mate with, then leave the lek. The assumption was that she mated only with the one, most attractive male in the lek. Recent evidence turns these assumptions and predictions on their head. Females in a growing list of lekking species (primarily in birds) are now known to mate multiply, sometimes with more than one male within a lek, and sometimes with males in different leks. One of the most interesting examples involves the buff-breasted sandpiper, *Tryngites subroficollis*, studied by Richard Lanctot and colleagues. Behavioral observations over a 2-year period yielded results that closely adhered to the traditional view of lekking behaviors: females appeared to be mating with only one male, and variance in male RS appeared to be extremely high because most males in the leks were never observed to mate. However, this traditional pattern completely broke down when parentage was determined using microsatellite DNA analyses. Genetic results demonstrated that multiple paternity was quite common in egg clutches (i.e., that females were mating and having young with multiple males), that most males were mating and siring young, and that most males were fathering offspring in only one clutch. These results were totally contrary to the long-standing and time-honored view of leks. In retrospect, it seems likely that the true dynamics of many lekking systems were not understood in the past because little attention had been given to the behavior of females, and because without the new molecular genetic technologies, it had not been possible to examine and accurately determine paternity of clutches.

Conclusions

This review has attempted to highlight the evidence that favors Bateman's paradigm as well as the criticisms of Bateman's ideas and experiment. It should be mentioned, however, that despite its many problems, at the time it was conducted, Bateman's experiment was imaginative, used up-to-date-technology, and included quantification and statistical analyses, something that was relatively uncommon in the 1940s. Moreover, some of the problems that have been identified more recently were not recognized in Bateman's time. A case in point is pseudoreplication, which was not acknowledged as a serious problem in statistical analyses until the 1980s. There can also be no doubt that Bateman's principles and ideas have provided a framework for modern studies of sexual selection and have spawned a very large body of research that has attempted to elucidate the dynamics of sexual selection.

So, where do we stand with Bateman's Principles? Probably the one aspect of Bateman that is the least controversial is that the slopes of Bateman gradients (i.e., sexual selection gradients) are indicative of the intensity of sexual selection. Much more problematic is Bateman's 'universal' prediction that males will have more steep gradients than females and, consequently, that males will undergo stronger sexual selection than females. Certainly, there are species which fit Bateman's principles and predictions. However, an increasing number of studies have found species in which female RS increases as a function of the number of mates, and in which there is little or no difference between the Bateman gradients of males and females. In such species, the parsimonious assumption is that females are also undergoing intense sexual selection.

There are, however, frequently neglected problems in many studies that have attempted to test Bateman's hypotheses. For example, one common criticism is that most of these studies do not test the null hypothesis to determine whether the observed results are due to chance, as was first suggested by Sutherland as early as 1985. A failure to test the null hypothesis leaves otherwise convincing studies open to debate. Specifically, given that sex differences in the variance of lifetime mating and RS can be due to stochastic and nonheritable factors,

and not exclusively to sexual selection, the results of studies that do not control for these factors may be questionable. To accurately measure the effects of sexual selection, it is necessary to partition out the effects due to sexual selection from those due to random and nonheritable factors, a point first noted by Hubbell and Johnson in 1987, and later reemphasized by Gowaty in 2004 and by Gowaty and Hubbell in 2005.

A related and important issue is that studies that examine male and female sexual selection gradients to compare the intensity of selection on the two sexes implicitly assume that sexual selection in females is mediated primarily through number of mates. It is possible, however, that other factors, such as the quality of mates, have a stronger impact on the intensity of sexual selection in females.

Lastly, there can be no doubt that Bateman and Darwin's predictions about female sexual behavior were incorrect. Females of many species are actively engaged in sexual solicitation and mate with multiple males. This has led some investigators to conclude that the most common mating system is probably polyandry or complete promiscuity (i.e., by both males and females). As mentioned earlier, in at least some cases, it has been well established that females that mate with multiple males have increased RS. Although many interesting hypotheses have been proposed to explain the benefits of multiple matings in females, no consensus has been reached and it is probable that the relevant selective pressures will vary from species to species.

In summary, Bateman's ideas have been extremely influential for the last half century and will likely continue to be so in the future. Nevertheless, we are currently experiencing a thorough re-evaluation of these ideas, including their assumptions, predictions, and corollaries. This reassessment may lead to a shift in paradigms and in our understanding of sexual selection and other forces that undoubtedly influence male and female lifetime RS.

See also: Helpers and Reproductive Behavior in Birds and Mammals; Mate Choice in Males and Females; Monogamy and Extra-Pair Parentage; Social Selection, Sexual Selection, and Sexual Conflict; Sperm Competition.

Further Reading

Andrade MCB and Kasumovic MM (2005) Terminal investment strategies and male mate choice: Extreme tests of Bateman. *Integrative and Comparative Biology* 45: 838–847.

Arnold SJ (1994) Bateman's principles and the measurement of sexual selection in plants and animals. *American Naturalist* 144: S126–S149.

Arnold SJ and Duvall D (1994) Animal mating systems: A synthesis based on selection theory. *American Naturalist* 143: 317–348.

Bateman AJ (1948) Intra-sexual selection in *Drosophila*. *Heredity* 2: 349–368.

Bjork A and Pitnick S (2006) Intensity of selection along the anisogamy–isogamy continuum. *Nature* 441: 742–745.

Darwin C (1871) *The Descent of Man and Selection in Relation to Sex.* London: John Murray.

Dewsbury DA (2005) The Darwin-Bateman paradigm in historical context. *Integrative and Comparative Biology* 45: 831–837.

Gopurenko D, Williams RN, and DeWoody JA (2007) Reproductive and mating success in the small-mouthed salamander (*Ambystoma texanum*) estimated via microsatellite parentage analysis. *Evolutionary Biology* 34: 130–139.

Gowaty PA (2004) Sex roles, contests for the control of reproduction, and sexual selection. In: Kappeler P and van Schaik C (eds.) *Sexual Selection in Primates*. 37–54 Cambridge, UK: Cambridge University Press.

Gowaty PA and Hubbell SP (2005) Chance, time allocation, and the evolution of adaptively flexible sex role behavior. *Integrative and Comparative Biology* 45: 931–944.

Hauber ME and Lacey EA (2005) Bateman's principle in cooperatively breeding vertebrates: The effects of non-breeding alloparents on variability in female and male reproductive success. *Integrative and Comparative Biology* 45: 903–914.

Hoogland JL (1998) Why do female Gunnison's prairie dogs copulate with more than one male? *Animal Behaviour* 55: 351–359.

Hubbell SP and Johnson LK (1987) Environmental variance in lifetime mating success, mate choice, and sexual selection. *American Naturalist* 130: 91–112.

Jones AG, Arguello JR, and Arnold SJ (2002) Validation of Bateman's principles: A genetic study of sexual selection and mating patterns in the rough-skinned newt. *Proceedings of the Royal Society of London B* 269: 2533–2539.

Jones AG, Rosenqvist G, Berglund A, Arnold SJ, and Avise JC (2000) The Bateman gradient and the cause of sexual selection in a sex-role-reversed pipefish. *Proceedings of the Royal Society of London B* 267: 677–680.

Ketterson ED, Parker PG, Raouf SA, Nolan V, Ziegenfus C, and Chandler CR (1998) The relative impact of extra-pair fertilizations on variation in male and female reproductive success in dark-eyed juncos (*Junco hyemalis*). In: Parker PG and Burley NT (eds.) *Avian Reproductive Tactics: Female and Male Perspectives Ornithological Monographs*, vol. 49, pp. 81–101. Lawrence, KS: Allen Press.

Lanctot RB, Schribner KT, Kempenaers B, and Weatherhead PJ (1997) Lekking without a paradox in the buff-breasted sandpiper. *American Naturalist* 149: 1051–1070.

Levitan DR (2005) The distribution of male and female reproductive success in a broadcast spawning marine invertebrate. *Integrative and Comparative Biology* 45: 848–855.

Lorch PD, Bussière L, and Gwynne DT (2008) Quantifying the potential for sexual dimorphism using upper limits on Bateman gradients. *Behaviour* 125: 1–24.

Parker GA, Baker RR, and Smith VGF (1972) The origin and evolution of gamete dimorphism and the male–female phenomenon. *Journal of Theoretical Biology* 36: 529–553.

Schulte-Hostedde AI, Millar JS, and Gibbs HL (2004) Sexual selection and mating patterns in a mammal with female-biased sexual size dimorphism. *Behavioral Ecology* 15: 351–356.

Snyder BF and Gowaty PA (2007) A re-appraisal of Bateman's classic study of intrasexual selection. *Evolution* 61: 2457–2468.

Sutherland WJ (1985) Chance can produce sex differences in mating success and explain Bateman's data. *Animal Behaviour* 33: 1349–1352.

Tang-Martínez Z and Ryder TB (2005) The problem with paradigms: Bateman's worldview as a case study. *Integrative and Comparative Biology* 45: 821–830.

Trivers RL (1972) Parental investment and sexual selection. In: Campbell B (ed.) *Sexual Selection and the Descent of Man*, pp. 136–179. Chicago, IL: Aldine Publishing.

Wedell N, Gage MJG, and Parker GA (2002) Sperm competition, male precedence, and sperm-limited females. *Trends in Ecology & Evolution* 17: 313–320.

Williams RN and DeWoody JA (2009) Reproductive success and sexual selection in eastern wild tiger salamanders (*Ambystoma t. tigrinum*). *Evolutionary Biology* 36: 201–213.

Woolfenden BE, Gibbs HL, and Sealy SG (2002) High opportunity for sexual selection in both sexes of an obligate brood parasitic bird, the brown-headed cowbird (*Molothrus ater*). *Behavioral Ecology and Sociobiology* 52: 417–425.

Worden BD and Parker PG (2001) Polyandry in grain beetles leads to greater reproductive success: Material and genetic benefits? *Behavioral Ecology* 12: 761–767.

Bats: Orientation, Navigation and Homing

R. A. Holland, Max Planck Institute for Ornithology, Radolfzell, Germany

Introduction

Bats represent one of the most diverse orders of mammals with nearly 1000 extant species. They are most famous for their nocturnal lifestyle and ability to orient and catch prey, using echolocation. Echolocation most probably evolved in bats initially for spatial memory, allowing them to find their way in the dark. Even relatively small-scale orientation in the dark presents a number of challenges that bats are able to achieve with apparent ease. Bats are also one of the only two vertebrate taxa to have evolved flight. This gives them the possibility to range over a wide area at a rapid speed. Such ability means that the animal can quickly travel beyond an area that it is familiar with, out of range of familiar visual or echo-acoustic landmarks. Indeed, some bats are known to make large seasonal migrations in excess of 1000 km, and other bats may forage over large distances. Bats, therefore, face the challenge of orientation and navigation at many spatial scales, from a few centimeters to thousands of kilometers.

Many migrating bats return to the same roost every year, and resident bats are often faithful to a single roost. If a bat needs to return to a familiar roost, then it must have a mechanism by which to achieve this from an unfamiliar place, in effect, something akin to our Global Positioning System (GPS). The biological equivalent of a GPS is true navigation, the ability to locate a final goal without using cues that emanate from it, or that are learned en route. Only cues available at their current, unknown position need to be used. It has been suggested that bats and birds are the only animals capable of performing such a feat. A true navigation ability is highly adaptive as, if an animal possesses a global scale navigation system, then it will never truly be lost. Displacements by climate or geographical barriers can be corrected by such a system. While there is evidence that birds are able to perform true navigation, the situation is less clear in bats, mainly through a lack of evidence.

If we compare a Web of Science search for navigation AND bats with navigation AND birds, the former brings up 72 records, whereas the latter brings up 1182 records. Navigation AND crustaceans brings up twice as many records as bats. Navigation in bats is, thus, one of the most poorly understood topics in animal behavior. Echolocation AND bats brings up 1015 records, highlighting the difference between our understanding of the short-range orientation behavior of these animals with their long-distance

navigation abilities. This article therefore focuses on the latter, in an attempt to stimulate further research on long-distance orientation and navigation in bats.

Why do we know so little about how bats navigate over long distances, compared to birds and other animals? The problem lies in the challenge of studying them, notably in the lack of suitable model species or techniques. In birds, there are two lines of research that have facilitated 50 years of study of the mechanisms of navigation in this taxon. First, researchers have used the homing pigeon, a semidomesticated species that is highly motivated to return to a home loft as a field-based model for orientation and navigation. When released from unfamiliar places, homing pigeons are very quickly oriented in the direction of their home loft, often only minutes after their release. By recording the bearing at which pigeons vanish from sight through binoculars, scientists can measure the orientation performance of a group of pigeons. Second, migratory passerine birds display a syndrome known as migratory restlessness when kept in a cage at the time of their normal migration. They will become highly active, and if the direction of this activity is measured, it normally corresponds to their species' normal direction of migration. This provides a system whereby the orientation of migratory birds that normally move thousands of miles can be measured by watching them hop around in a small area of a few centimeters. Using these two model systems, a number of important discoveries on the orientation and navigation of birds have been made.

In bats, no such model systems exist. There are no semi-domesticated species, and to our knowledge, captive bats do not display migratory restlessness. This has left the only option of studying the behavior of wild caught animals in a natural setting. This represents a challenge and is possibly one of the most important directions for research on animal behavior in the twenty-first century: the study of wide-ranging animals in a natural setting. Bats are small, fast-moving animals that are active at night. This makes observing them in the wild difficult and goes a long way to explaining the relative lack of research on their orientation and navigation abilities. Nevertheless, a number of techniques do exist for the study of such animals, and this article explains how orientation and navigation in bats can be studied, the breakthroughs that have been made, and future directions for research on these incredible animals in their natural setting.

Techniques

How can the movement of animals that are small, cryptic, and faster than the observers (if on foot) be studied in the wild? A number of techniques have been developed to study the movement of animals and all have been used in an attempt to learn more about the ability of bats to navigate.

Banding

Banding is a technique that has been used extensively to learn about the movement of both migratory bats and birds. In this, a small metal band is placed on the forearm of the bat and it can then be identified upon recapture. Banding is most commonly used to inform about the seasonal movement of migratory animals, but in the case of bats, was used extensively to study the homing ability of these animals in the 1950s and 1960s. However, concerns about the impact of the bands on the health of bats led to a cessation of such study in the 1970s. Banding is also limited in that it indicates only start points and subsequent points of capture. If an animal is not captured in a homing task, does this mean that it did not home or that it just evaded capture? Banding studies are thus an important first step in studying the movement of wild animals but not sufficiently detailed to make a full investigation of the cues and mechanisms involved in navigation.

Tracking

To study the movement of animals and thus the decisions they make in response to behavioral challenges, it would be ideal to monitor their movement over all parts of their journey. The past 30 years have seen great advances in technology to allow us to use such methods to track animals, but with tracking devices, there are limitations on the extent to which this technology can be used depending on how heavy the animal is. To track an animal, it is necessary to attach a device that allows the researcher to record the animal's position. Ideally, this device should not affect the behavior of the animal, and the general rule for vertebrates is that if the device weighs <5% of the animal's body weight, it should not do so. There are two types of tracking technology available, satellite tracking, which allows the animal to be tracked on a global scale, and radio tracking, whereby the researcher must maintain contact with the transmitter attached to the animal to locate it.

Satellite tracking

It is now possible to track animals globally, using satellite transmitters. These devices calculate their position, using the GPS satellite system and then relay the data, using the ARGOS satellite back to a storage server. Once the animal has been tagged, it is then possible to track its position without having to leave the comfort of an office. Two species of migratory fruit bat have been successfully tracked by these means. However, satellite transmitters currently weigh in excess of 5 g; this limits the size of the animal that can be tracked to 100 g. All temperate-zone bats fall below this cut-off, and many of these animals make the longest seasonal migrations. Tracking these animals globally using satellite transmitters is thus currently out of reach for a majority of bats.

Radio tracking

Animal radio tracking was developed in the 1960s by Bill Cochran of the Illinois Natural History Survey. It requires the attachment of a transmitter to the animal. The transmitter emits a radio pulse that can be detected using a radio receiver as long as the observer maintains the line of sight and range to the animal (**Figure 1**). Therefore, unlike satellite tracking, the researcher must maintain active contact with the animal to locate its position. While in theory it is possible to lose contact and later relocate the animal, this becomes difficult in practice for animals that move over large distances. With the advent of satellite tracking, radio tracking has come to be viewed as a poor relation, because of the labor-intensive nature of collecting the data. However, for small animals (<100 g) such as insectivorous bats, it is still the only technology small enough to allow the movement of the animal to be tracked with any detail. Several recent breakthroughs in the study of bat navigation, which are discussed later in this article, have used radio tracking to study the movement of the animals. As the transmitters can now be as small as 0.2 g, they are well within the 5% limit of weight carried. Tracking with this technique is still relatively labor intensive, but if plans to create a satellite to detect these transmitters from low Earth orbit come to fruition, then the field of orientation and navigation in bats will blossom.

Theory and Practice of Animal Navigation

The theory of how animals achieve true navigation has remained relatively unchanged since the proposal by Gustav Kramer in the 1950s that animals must use a system that corresponds to our map and a compass. In effect, therefore, animals must be able to locate their position with respect to their goal and then take up the direction needed to return to it. These two steps must be achieved using the cues available in the environment that can be detected by the animal's senses. The cues available for use as a map and compass have been reviewed many times and have most recently been covered comprehensively by Wallraff in his excellent book 'Avian Navigation: pigeon homing as a paradigm'. For the purposes of this article, it is important to have an idea of what cues have

Figure 1 Animals can be radio-tracked on foot (a), by motor vehicle (b), or by aircraft (c). For monitoring long-distance movement by migratory flying animals, an aircraft gives the best chance of maintaining contact with the signal from the radio transmitter, but both vehicle and foot-based tracking have been successfully used to investigate navigation behavior in bats and birds.

been demonstrated to play a role in true navigation to discuss their possible role in bats' orientation and navigation. The cues used to provide direction (compass) have been shown to include the sun, the stars, and the Earth's magnetic field. How animals are able to locate their position is more controversial, mainly because, despite a number of proposed cues, there is no clear evidence that any one of them solves the mystery of how animals can locate their position with respect to their goal in an area they have never previously visited. If the animals were in a familiar place, then it would be a simple matter of recognizing the local landmarks and working out where that placed them with respect to home. In an unfamiliar place, there are no learned landmarks to rely on and so, according to theory, the animals must extrapolate on the basis of their

experience from a familiar area. If the animal can recognize that if at least two environmental cues or gradients vary predictably in its familiar area, one on a north–south axis (latitude) and one on an east–west axis (longitude), then if these cues continue to vary predictably outside the animal's familiar area, then they can be used to locate the animal's position with respect to its goal in a familiar area. This has been called a bi-coordinate gradient map and is the best explanation we have for how animals are able to return to a familiar roost, wintering, or breeding ground after their displacement (**Figure 2**).

In birds, recent experiments have shown that even after displacements of over 3500 km, migrating adults can correct and head back to their known wintering area from a place that they cannot have been before. Despite the soundness of

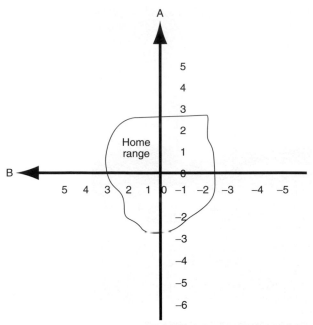

Figure 2 In a bi-coordinate gradient map, animals learn that at least two cues, ideally intersecting at 90°, vary in strength within the home range, and the animal assumes by extrapolation that they continue to vary in this way outside the home range. In the case of many long-distance migrants, these cues would need to vary consistently on a global scale, but at the very least on the basis of current evidence, on a continental scale. In the schematic shown, if the animal finds itself at A5, B5, then even though it has never encountered these values in its home range, they are both increasing values. As gradient A increases northward and gradient B increases westward within the home range, this means that it is north and west of its home range and must fly south east to return.

the framework provided by the bi-coordinate gradient map, there is actually little evidence that animals use such a map. There are a number of ways in which the latitude could be determined, including the altitude and arc of the sun, and the intensity or inclination of the magnetic field. No animal tested so far has been shown to use the sun's arc or altitude, however. The Earth's magnetic field has been proposed to be the most likely parameter using which animals detect their latitude, but although there is evidence that animals can detect the intensity of the Earth's magnetic field, proof that it is used as part of a bi-coordinate gradient map is still lacking. How animals determine longitude is even more of a mystery, and was a challenge that humans only solved in the eighteenth century.

Since animals had been making long migratory journeys successfully for thousands if not millions of years, it must be assumed that they had solved the problem much earlier, but we still do not know what cues they use. The possibility that they use the stars, as humans did to solve the longitude problem, has not yet been ruled out. It is also possible that they use the Earth's magnetic field, which varies in intensity

longitudinally in some (but not all) places. One of the biggest mysteries of animal navigation, however, is that the best evidence that exists for the cues used by animals to navigate comes from homing pigeon navigation. Research in the 1970s demonstrated that homing pigeons, which are normally able to return to their home loft from unfamiliar places, are unable to do so when their sense of smell is removed. A vast array of follow-up studies has confirmed these findings. This does not seem to fit into the bi-coordinate system, however, as it has been shown that pigeons can still successfully navigate without magnetic cues.

While the theory of how animals navigate is sound, this section indicates that in some respects, our knowledge of bat navigation is not that far behind other animals, if only because the question of how animals can locate their position in an unfamiliar place is yet to be answered. What evidence exists for how bats orient and navigate? Thanks to some old experiments and some recent breakthroughs, the cupboard is not entirely bare, and there is much to be optimistic about for future research.

Evidence for Orientation and Navigation in Bats

Are bats true navigators? While we cannot yet be certain of that fact, evidence from a number of displacement experiments in the 1950s and 1960s using recapture of banded animals, indicates that bats certainly have some impressive homing abilities. Summarized in an excellent review by Davis, in general, performance in these studies was measured in terms of the percentage of bats returning. It appeared that both the distance of release and experience were factors in the return rate; in general, bats returned more often when released close to their home roost, and adults were more likely to return than juveniles. A few of the studies were performed on species that are long-distance migrants, and none during migration. The longest return was performed by a nonmigratory species (*Eptesicus fuscus*), of 450 miles. This compares with the maximum normal range of movement of 142 miles for this species. The homing percentage was very low (4.6% of 155 bats), and by today's standards, would possibly not be considered as evidence of navigational ability. However, it is hard to see how an animal could locate its home roost by random search from such a distance.

In general, the studies of homing after displacement provide inconsistent results, but do occasionally show relatively high return percentages from within 100 miles. The return performance is a relatively crude measure of navigational ability as it relies on recapture of the individuals. When an animal did not return, it is unknown whether it was simply not detected or had failed to home. Roost switching is known in many bat species, and so, bats may have headed to a different roost when not recorded at the site of capture. These experiments do indicate that at least some bats are capable of homing, and certainly can return from outside an area that would be considered familiar to them. This suggests that bats might be capable of true navigation.

What sensory cues do bats use for navigation? Early experiments, again based mainly on homing performance, indicated that vision plays a crucial role, at least beyond 12 km. Studies in which displaced bats were blindfolded and then displaced from 5 km up to 100 km from their roost indicated better performance within 12 km than at any other distance. Deafening bats reduced performance within this region, which suggests that echolocation was used within this area to return home. This may not initially seem surprising, but given that the range of echolocation is at best 30 m, it is actually an incredible feat to become familiar with a 10 km area using this sense alone. Beyond this distance, it seemed that vision was essential for successful homing, although again, the use of banding to determine homing performance did not allow firm conclusions on the relative success of blindfolded versus sighted bats. Given the importance of the visual system to motivation, it is difficult to know whether these experiments represented disruption of a visual based navigation system or simply reluctance to fly from an area they did not recognize from their echolocation system.

These questions suggested follow-up investigations, but fears about the detrimental effect of banding all but stopped the study of bat orientation and navigation at the beginning of the 1970s. Little new research on bat navigation was performed in the next 30 years. However, recently, there have been new breakthroughs, based both on radio-tracking of animals in the wild and lab studies. Along with coresearchers, I have investigated the possibility that bats orient using a magnetic compass and/or the stars. These are known to be among the primary reference cues used by birds for orientation. A recent experiment indicating that migratory thrushes calibrate the magnetic compass by the sunset was used as a model for the investigation of these cues. It was observed that big brown bats, *E. fuscus*, rapidly return to their home roost if displaced 20 km north of it. This gave a baseline to investigate the possible role of the stars and the Earth's magnetic field in their orientation. By exposing the bats to an altered magnetic field at sunset and until only stars were visible, it was possible to test whether bats used the Earth's magnetic field or the stars as a compass (**Figure 3** for predictions). Bats were deflected according to the prediction expected for the use of a magnetic compass calibrated by the sunset (**Figure 4**), indicating that bats have a magnetic compass for orientation and that like birds, they may calibrate it to the direction of sunset. This would allow them to correct for the difference between true north and magnetic north (declination), and also to calibrate for anomalies in the magnetic field.

A subsequent experiment by Wang and coworkers investigated the way in which bats use the magnetic field to indicate direction. It is possible to derive directional information from the Earth's magnetic field in two ways (**Figure 5**). The polarity can be used, which indicates the direction of the magnetic north pole, or alternatively, the inclination (i.e., the angle the magnetic field forms with the earth, ranging from 90° at the poles to 0° at the equator) could be used to indicate whether the animal is flying toward the pole (increasing inclination) or toward the equator (decreasing inclination). Current evidence suggests that birds use the inclination of the magnetic field to orient, and they are unaffected by artificial changes in polarity when tested in orientation cages, whereas the

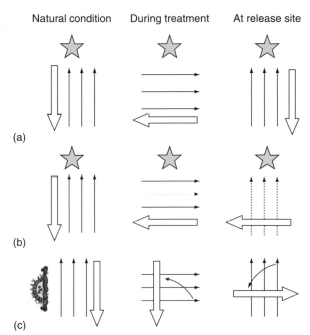

Figure 3 Schematic diagram to illustrate the predictions for direction of orientation after exposure to a magnetic field shifted clockwise (CW) through sunset depending on the interaction between magnetic and celestial cues. The three lined arrows in each diagram represent the magnetic field direction and the solid arrow represents the perceived direction of the home site for bats displaced north of home. (a) If there is a simple dominance of the magnetic compass or a star compass, with no calibration, then when released in a natural magnetic field at the release site, the bats should fly south regardless of the direction of the field during treatment. (b) If the magnetic field calibrates a star compass, then during treatment, the field is shifted CW, making the bat perceive north as east and therefore, south as west. When released, a bat trying to fly south to reach home would thus fly west. (c) If sunset cues calibrate a magnetic compass, then when the field is shifted east at sunset, the bat will perceive that geographical south is 90° CW to the magnetic field direction. When released in a normal magnetic field, flying at 90° will result in the bat flying east. Reproduced from Holland R, Thorup K, Vonhof M, Cochran W, and Wikelski M (2006) Bat orientation using Earth's magnetic field. *Nature* 444: 653.

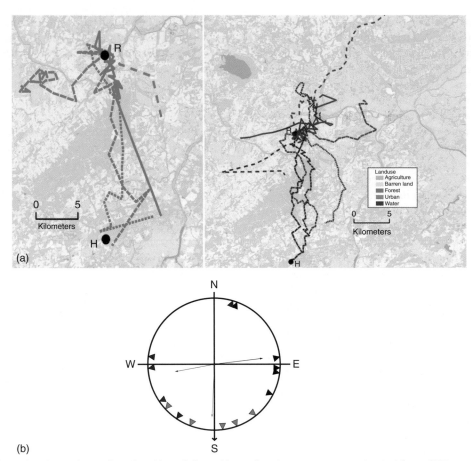

Figure 4 (a) Control and experimental tracks of bats followed by radio telemetry, as reconstructed from GPS waypoints. Red = counter clockwise rotation of the magnetic field at sunset (CCW), blue = clockwise (CW) rotation of the magnetic field at sunset, and green = control. R = release site, H = home. (b) Circular diagram of headings all groups released at the site 20 km north of home. Red = counter CCW rotation of the magnetic field at sunset, blue = CW rotation of the magnetic field at sunset, and green = control bats. The single-headed arrow outside the circle indicates the home direction (South). Reproduced from Holland R, Thorup K, Vonhof M, Cochran W, and Wikelski M (2006) Bat orientation using Earth's magnetic field. *Nature* 444: 653.

only mammal tested so far, the mole rat, appears to use the polarity of the magnetic field.

While investigating their roosting behavior, it was discovered that captive *Nyctalus plancyi*, the Chinese noctule bat, preferred to roost at the north end of its cage. When these bats were tested for their roosting preference in artificial magnetic fields that had either the polarity or the inclination reversed, changing only the polarity changed the roosting preference of the bats to the south end of the cage. Thus, unlike birds, which appear to use the inclination of the magnetic field, that is, fly pole wards or equator wards, these bats use the polarity of the magnetic field, in the same way as our needle compass works.

How do bats, or indeed, any animal, detect the magnetic field? There are no obvious sensory organs to do so, and initially there was resistance to the idea that animals could actually use the magnetic field, given the lack of evidence for a way to detect it. However, a large body of evidence was gathered, especially from birds, indicating that manipulation of the magnetic field could alter the orientation direction of animals, and so it became clear that a way to detect it must exist. There are now two theories as to how animals can detect the magnetic field, and they are often set up as competing, although it may be that in some animals, they are actually complementary systems. The first theory is based on the discovery of iron oxide particles called magnetite in the cells of nearly all organisms that have been studied. It has been proposed that animals can use these like tiny compass needles to indicate the direction of the magnetic field, and can also use the movement (possibly measured by torque) of these particles in the magnetic field to detect its strength. This is the magnetite hypothesis. The second theory is based on the knowledge that certain chemicals react to the magnetic field, which changes their chemical state. Certain light-sensitive molecules (such as cryptochromes) involved in vision appear to behave in this way, and it has been proposed that these reactions could be detected by the visual system, allowing animals to effectively see the magnetic field. Since removing certain wavelengths of

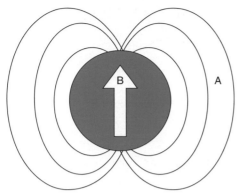

Figure 5 The Earth's magnetic field can be used to indicate the direction in two ways. The inclination angle of the field lines in relation to the Earth's surface (A) indicates the direction in terms of whether flying towards (decreasing inclination) or away from (increasing inclination) the equator. The polarity of the magnetic field (B) can also be used to indicate the direction, in the same way as a magnetic compass needle works. Birds appear to use inclination, they are unaffected by changes in polarity, whereas the reverse is true in the only species of bat tested so far.

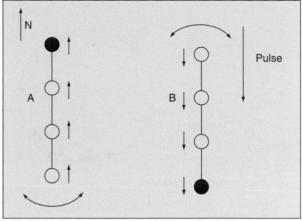

Figure 6 A schematic representation of the theoretical behavior of magnetite chains in a sensory cell if they are free to rotate. Normally, the chain aligns to the Earth's magnetic field as in (A). One end is 'painted red' to indicate the direction of the magnetic field. How the cell signals this in practice is still unknown. The strength of oscillations around this direction is one possible way in which the intensity of the magnetic field might be measured. In (B), if a strong, short-duration magnetic pulse is applied to the magnetite when aligned with a biasing field stronger than the earth's, if the magnetite is free to rotate, it will flip its orientation, and the 'compass needle' will now be pointed south instead of north, causing the animal to reverse its orientation.

light appears to disrupt the ability of some animals to use the magnetic compass, as predicted by this mechanism, this has been labeled the light-dependent hypothesis.

The evidence for these two systems is reviewed in a number of recent papers (see Further Reading), but in birds, it appears that both the systems may be in operation. They seem to use the light-dependent mechanism for orientation, using their compass, but additional studies indicate that magnetite plays a role in orientation in the absence of light as well as in the orientation of adult migrating birds.

A recent experiment has indicated that magnetite plays a role in the navigation behavior of bats. It tested for the use of magnetite to detect the Earth's magnetic field, using a technique that was first used in magneto-tactic bacteria, that of pulse remagnetization (**Figure 6**). Bacteria, which contain magnetite and are normally passively north seeking because of this magnetic material within their cell, can have their orientation reversed by changing the magnetic moment of the magnetite. This was done by exposing the bacteria to a brief and strong magnetic pulse in the opposite direction of the normal magnetic field. If animal cells contain magnetite and this is used for detecting the magnetic field, then a similar brief pulse should disrupt the orientation behavior of the animal. In birds, such a pulse disrupts their ability to orient, and changes their direction, but it does not change their magnetic compass use, suggesting that the magnetite is being used in the map, not the compass. In the experiment, one group of bats was exposed to a strong pulse 5000 times that of the Earth's magnetic field in the direction opposite to that of the normal magnetic field, that is, antiparallel. This was strong enough to overcome the

natural magnetization of magnetite and reverse its magnetic direction. Another group of bats received the same pulse but in the same direction as the Earth's magnetic field, that is, parallel. In bacteria, this pulse had no effect, but in birds, it disrupted their orientation in the same way as an antiparallel pulse.

When this strong magnetic pulse was applied to Big Brown Bats (*E. fuscus*) and they were then released 20 km north of home, the parallel-pulsed group flew south, whereas the antiparallel-pulsed group was split: half flew north and half flew south (**Figure 7**). This difference in behavior between the two groups indicated that as there was a difference between the parallel and antiparallel groups (control bats also flew south), the bats must have been using magnetite to detect the Earth's magnetic field and were using that information to take up a direction. If all the bats in the antiparallel group had flown north, this would have been a clear indication of the magnetite rotating in the same way as a compass needle and, therefore, that the magnetite was being used to indicate the compass direction. However, the split behavior of this group means that the interpretation is still open; the pulse disrupted the magnetite, but this may mean that the bats' ability to detect north was changed or that their ability to detect magnetic intensity was altered. Nevertheless, these experiments demonstrate at least one of the mechanisms by which bats detect the Earth's magnetic field.

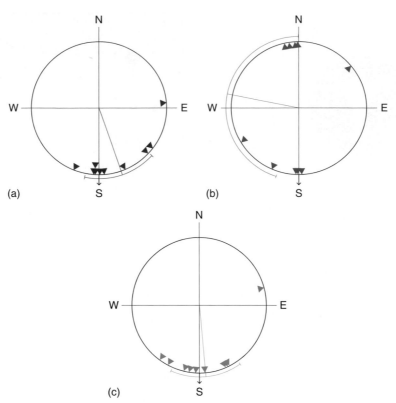

Figure 7 Circular diagram indicating the orientation of bats subjected to a strong magnetic pulse. (a) Control, (b) antiparallel pulsed, and (c) parallel-pulsed bats. The mean direction and 95% confidence interval are shown. The arrow on the edge of the circle indicates the home direction. Reproduced from Holland RA, Kirschvink JL, Doak TG, and Wikelski M (2008) Bats use magnetite to detect the Earth's magnetic field. *PLoS ONE* 3: e1676.

Future Directions

The three recent experiments on navigation in bats indicate possible directions of research in this taxon. We now have a field-based method to study the orientation and navigation of bats, and a laboratory-based method to study their compass orientation. Together, these two techniques should combine to reveal much more about the mysteries of navigation in bats. There are a number of open questions to be addressed and they are summarized as follows in the hope of stimulating further empirical research on these animals.

1. Are bats true navigators? In birds, the ability of migrating animals to correct for massive displacements to places they cannot have previously visited, with appropriate responses to this displacement, is an indication of their true navigation ability. In a recent experiment, Kasper Thorup and coworkers displaced adult and juvenile white-crowned sparrows (*Zonotricia leucophrys*) from Seattle to Princeton (i.e., from east to west coast of the United States). They found that by 25 km from the release site, juveniles were heading south, as their species would during normal migration on the west coast, but adults were heading west south west, toward the

wintering grounds of the species in California. The conclusion, which confirmed an experiment by Perdek nearly 50 years earlier, was that adults used a true navigation map to correct for the displacements, whereas juveniles used an innate, population-specific compass direction. There are bats that migrate over 1000 km, but little is known about their ability to correct for displacement, or whether juveniles migrate alone using an innate compass direction or follow adults in social groups. Displacement experiments have been performed before but never on bats that migrate.

2. What are the cues used in the map? This is an unanswered question not just for bats but also for birds. How do animals detect their longitudinal and latitudinal position? The demonstration that bats use magnetite in homing behavior indicates the potential for them to use magnetic intensity to locate their latitudinal position, although it has not yet been conclusively demonstrated that they can actually detect the intensity of the magnetic field. Birds have been shown to use the intensity or inclination of the magnetic field as a signal to start refueling, but it has not yet been conclusively demonstrated to play a role in their map. Olfactory cues have also been indicated to play a role in true navigation in homing pigeons. They have additionally

been shown to be used in homing by migratory birds but not yet been demonstrated to play a role in navigation during migration.

3. Do bats use only magnetite to detect the Earth's magnetic field? The experiments discussed in the previous section together indicate a polarity-based magnetic compass. As the proposed light-dependent mechanism is not thought to be able to detect polarity, this suggests that the mechanism must be magnetite, but the lab-based alignment behavior shown by Wang and colleagues has not been confirmed in a field-based navigation task. And the experiment demonstrating magnetite as a detection mechanism was not conclusive in determining whether the compass or a possible map cue was affected when pulses disrupted the orientation of Big Brown Bats.

4. To what extent do bats use echolocation to remember their location in flight? The experiments in the 1960s suggest that it is possible that they could use this sense to home from as much as 12 km from their roost, but this is at odds with a sensory system that has a range of 30 m at best. If bats can really remember their location over this distance, using echolocation, then further investigation is needed to discover how they achieve this remarkable feat.

5. What other compass cues do bats use for orientation? The experiment by me and my colleagues indicated that bats that were misdirected by changed magnetic fields could correct for this faulty orientation eventually and return home. They must have been able to switch to another compass mechanism. The stars are the most obvious candidate but bats may possess a system hitherto unknown.

These questions are just a beginning. Bat navigation research is 50 years behind bird navigation, but the methods are now available to discover how these most fascinating of animals are able to orient and navigate during their nightly foraging trips, or their seasonal migrations. The next 50 years should bring a wealth of data to redress the shortfall.

See also: Bat Migration; Behavioral Endocrinology of Migration; Bird Migration; Magnetic Orientation in Migratory Songbirds; Maps and Compasses; Pigeon Homing as a Model Case of Goal-Oriented Navigation; Spatial Memory.

Further Reading

Bingman VP and Cheng K (2006) Mechanisms of animal global navigation: Comparative perspectives and enduring challenges. *Ethology Ecology and Evolution* 17: 295–318.

Davis R (1966) Homing performance and homing ability in bats. *Ecological Monographs* 36: 201–230.

Freake MJ, Muheim R, and Phillips JB (2006) Magnetic maps in animals: A theory comes of age? *Quarterly Review of Biology* 81: 327–347.

Gagliardo A, Ioale P, Savini M, and Wild JM (2006) Having the nerve to home: Trigeminal magnetoreceptor versus olfactory mediation of homing in pigeons. *Journal of Experimental Biology* 209: 2888–2892.

Holland R, Thorup K, Vonhof M, Cochran W, and Wikelski M (2006) Bat orientation using Earth's magnetic field. *Nature* 444: 653.

Holland RA (2007) Orientation and navigation in bats: Known unknowns or unknown unknowns? *Behavioral Ecology and Sociobiology* 61: 653–660.

Holland RA, Thorup K, and Wikelski M (2007) Where the wild things go. *Biologist* 54: 214–219.

Holland RA, Kirschvink JL, Doak TG, and Wikelski M (2008) Bats use magnetite to detect the Earth's magnetic field. *PLoS ONE* 3: e1676.

Neuweiler G (2000) *The Biology of Bats*. New York: Oxford University Press.

Thorup K, Bisson I, Bowlin M, et al. (2007) Migration routes of adult and juvenile white-crowned sparrows differ after continent-wide displacement during migration. *PNAS* 104: 18115–18119.

Ulanovsky N and Moss CF (2008) What the bat's voice tells the bat's brain. *Proceedings of the National Academy of Sciences of the United States of America* 105: 8491–8498.

Wallraff HG (2005) *Avian Navigation: Pigeon Homing as a Paradigm*. Berlin: Springer.

Wang Y, Pan Y, Parsons S, Walker MM, and Zhang S (2007) Bats respond to polarity of a magnetic field. *Proceedings of the Royal Society B-Biological Sciences* 274: 2901–2905.

Wikelski M, Kays RW, Kasdin NJ, et al. (2007) Going wild: What a global small-animal tracking system could do for experimental biologists. *Journal of Experimental Biology* 210: 181–186.

Wiltschko R and Wiltschko W (2006) Magnetoreception. *Bioessays* 28: 157–168.

Behavioral Ecology and Sociobiology

J. L. Brown, University at Albany, Albany, NY, USA

Introduction

Behavioral ecology and sociobiology are relatively new scientific disciplines. They rose to prominence in the 1960s and 1970s as parts of other disciplines, and then solidified their status as new disciplines in the 1980s and 1990s as new mathematical and molecular tools became available.

What Are Behavioral Ecology and Sociobiology?

Behavioral ecology and sociobiology are terms that designate an approach to the study of behavior in which inquiry typically focuses on how behavior has evolved by natural selection, especially how behavioral differences among species have arisen in relation to their natural environments. As sociobiology, which is limited to social behavior, is essentially included within the term behavioral ecology, custom has come to favor use of the more inclusive term behavioral ecology. The dominant integrative theme of behavioral ecology, namely the role of natural selection, is unfortunately not explicit in either of these terms. Nevertheless, both involve the close integration of ideas from ecology, evolution, and ethology.

Recognition of a New Field

The First Books and Journals

The origins of behavioral ecology as a recognized field may be documented using the earliest published documents that employ the term. The first book to use 'behavioral ecology' in its title was a slim textbook in 1962 by Klopfer containing treatment of selected subjects **emphasizing behavioral aspects of community ecology**. Klopfer's books contained examples of behavioral ecology, but did not develop the field in the manner that was popular in later literature. The term 'sociobiology' entered the public attention in 1975 with Wilson's tome by that name. He intended it as an expansion of his 1971 book on 'The Insect Societies' to include all animals. The first of a series of textbooks that focused explicitly on behavioral ecology with comprehensive coverage appeared in 1978. These were at two levels, graduate (Krebs and Davies, 1978) and undergraduate (Krebs and Davies, 1981).

Other early surveys appeared in 1980 by Morse and 1985 by Sibly and Smith. The social behavior of birds received much attention in all these books, especially as most of the authors and editors had worked on birds to some extent.

Origin of the International Society for Behavioral Ecology

The field's first journal, *Behavioral Ecology and Sociobiology*, began publication in 1976 but was too expensive even for some libraries. An affordable journal for individuals had to await a new scientific society.

Despite the growth and influence of behavioral ecology as a scientific field, as late as the early 1980s there was no society that gave primary consideration to this field. Therefore, there were few opportunities for many people interested in behavioral ecology to come together to listen to each other's research contributions. Consequently, the bird people went to ornithological meetings, the insect people to entomological meetings, and so on. The societies in North America that might have brought the different taxonomic varieties of behavioral ecology together were the Animal Behavior Society, the Ecological Society of America, American Society of Naturalists, the International Society for the Study of Evolution, and a few others, but these represented many other interests, and in some ways, these societies were too specialized. Theorists, geneticists and empiricists still tended to go to different meetings, so far as behavior was concerned, and this impeded exchange of ideas among scientists interested in behavioral ecology.

In October 1986, the first International Behavioral Ecology Meeting was held in Albany, NY, USA. The organizers were faculty members and students at the State University of New York at Albany; they had no formal relationships to any previous society. Twenty-two speakers were invited from five countries to cover a wide range of topics of general interest to behavioral ecologists and to bring together people who might not otherwise attend the same meeting. The meeting was attended by 370 participants and was considered to have been very successful. By the next meeting in October 1988, the fledgling society had a name, a constitution, officers, a newsletter, and plans for the new journal, which became known as Behavioral Ecology. Membership in the society increased to about 1350 by 1998 and since then has remained close to that level (http://www.behavecol.com/pages/society/welcome.html).

Origins of the Ideas and Approaches in the 1960s and 1970s

Here we examine the development of the major ideas that preceded the recognition of behavioral ecology as a field of its own. Behavioral ecology had its roots in the 1960s when cost–benefit studies of social behavior became interesting and when foraging theory and Hamilton's kinship theory were introduced. It was at first largely expressed in studies of the population biology of social behavior: a trend that solidified with books by Alcock, Brown, and Wilson published in 1975. Later treatments by Krebs and Davies unified the field in 1978 by joining foraging behavior with social behavior under the general framework of behavioral ecology.

In the 1960s ethology, the study of animal behavior, made major contributions and was honored by a Nobel Prize awarded to three of its most prominent scientists, Lorenz, von Frisch, and Tinbergen While ethologists deserved their honors, there were great gaps in the ethology of those years (elaborated in Brown, 1994).

The Population Biology of Behavior

One major gap in ethology, as practiced by Lorenz, von Frisch, and Tinbergen in the decades leading up to the 1960s was the lack of a population biology of behavior, which developed rapidly later in the 1960s (see below). No doubt Tinbergen and his students were early contributors in this area, but in my opinion, the principal impetus to behavioral ecologists among Americans, such as Orians, Fretwell, MacArthur, and I, and Europeans, such as Hamilton and Maynard Smith, came from population biology, and not from ethology.

Reflecting this foment, in 1975, the three books already mentioned, by Alcock, Brown, and Wilson, appeared. These emphasized a new perspective among textbooks on behavior, namely that of natural selection. Previous texts had emphasized the developmental and physiological *Mechanisms of Animal Behavior* (Marler and Hamilton) or had attempted a 'synthesis of ethology and comparative psychology' (Hinde). In contrast, the 'central unifying theme of biological evolution' characterized Brown's text; I wrote that 'the central concepts in this book are concerned with populations.' Although the primary message of Wilson's *Sociobiology* was interpreted by social scientists to mean an emphasis on nature rather than nurture (an argument summarized by George Barlow), for ecologically minded behaviorists, it was a heavy dose of population biology. Alcock's textbook (1975 and later editions) also emphasized natural selection rather than physiological or developmental mechanisms. Thus, the 1970s were a period of major synthesis in which bridges between behavior study and population biology were built – at some cost to the study of mechanisms and development.

Although these three books had some influence of their own, they clearly reflected an already substantial existing trend. Behavioral ecology and sociobiology did not originate in these books. Even without these textbooks, the trend toward fusion of population ecology, genetics, and behavior was booming and would have continued.

An unexpected new dimension in this rather obscure scientific field was the media. The unprecedented front-page promotion of *Sociobiology* in the 28 May 1975 issue of the *New York Times* before the book had even been seen by most interested scientists (including myself) and the Sunday-supplement treatment by the *Times* (12 October 1975) made this scholarly book a media event and consequently a political issue. Actually, anthropology had already absorbed lessons from the fusion of ethology and ecology as highlighted by Tiger. The animal nature of 'the naked ape' had intrigued the common man well before Wilson's tome and had been a particular point of interest to Lorenz and other popular authors.

What then were the origins of these trends, if we cannot attribute them to textbook writers? I would nominate two sources, the population ecology of David Lack and the population genetics of W. D. Hamilton. Certainly these writers influenced many ornithologists and entomologists for decades. Two continuing themes in Lack's writings from 1947 to 1968 were the ecology of populations of birds and the evolution of reproductive rates by individual rather than 'group' selection (an idea with a long history before Lack). Although Lack had already convinced most ornithological readers of the supremacy of individual selection, the opposing view favoring population- or deme-level selection was presented conspicuously by Wynne-Edwards in his controversial 1962 book. The theme was not original with Wynne-Edwards, but he published a lavish and ponderous elaboration of it. Squid-like, this book propelled the population biology of social behavior backward in a cloud of black ink.

Reaction was strong and immediate. Although some sophisticates chose to ignore the crude reasoning of Wynne-Edwards because it was so obviously wrong as a general explanation, others attempted to refute his arguments; and still others felt that this was the time to elaborate the ways in which social behavior could be influenced by individual selection. For example, territoriality was Wynne-Edwards' prime example of a population-limiting behavior, but I presented a general model that showed how various kinds of territorial behavior could evolve by individual selection – without the need for population-level selection. Similarly, Orians advanced a graphical model to explain the evolution of mating systems on the basis of individual selection. Similar arguments on behalf of individual selection were made by others for a variety of avian social systems. Many of the papers of this era were concerned with spacing behavior and its effects on populations. Together these led to a comparative ecology of social systems. An even more global approach was taken by George C. Williams, who attempted to outline how adaptations in general could evolve by individual rather than group selection. These items of work and many others established a consensus position on the side of individual selection. They created an atmosphere that was hostile to any mechanism that seemed to differ from 'old-fashioned' individual selection. This conservative view proved to be an impediment to the acceptance of some exciting new ideas.

The most influential new idea in the realm of selection thinking was the theory of inclusive fitness launched by Hamilton (Hamilton, 1963, 1964). To prevent this kind of selection from being confused with 'group selection,' John Maynard Smith coined the term kin selection, including both direct and indirect components of inclusive fitness. While it was not spelled out explicitly, confusion of group and kin selection was implicit in some of the early literature attacking the use of the inclusive fitness theory as part of an explanation for helping behavior in birds. Whether for this reason or for others, such as recoiling from the term altruism, the inclusive-fitness thinking met a hostile reception that was based partly on the misunderstanding of the concept, as shown by Dawkins in his 1979 article, and partly on old-fashioned conservatism. It would take three decades for facts and reason to overcome this resistance.

Thus, the principal origins of what we now know as sociobiology were on the one hand from the tremendous impetus that Hamilton's theory gave to the study of social insects and sociality in general in the 1960s and, on the other hand, from what I have termed the comparative ecology of social systems that developed in the 1960s among ornithologists. This fusion of population genetics and the work of avian field ecologists in the tradition of David Lack provided at least part of the wave of interest that resulted in the syntheses of 1975 and certainly was a strong stimulus to the study of avian social behavior in the 1970s.

Behavioral Ecology Emerges

Add optimal foraging theory to sociobiology and you have behavioral ecology – or close to it. Optimal foraging theory arose from theories of niche exploitation written by the messiah of American ecology, Robert MacArthur and one of his prominent students Eric Pianka and by John Emlen. MacArthur's followers were influential, and they quickly expanded this approach into a flourishing field. Although foraging is usually not considered to be social behavior, ideas from optimal foraging theory provided the best-developed applications of optimality methods to behavior, and these methods came to be applied also to conventional social behavior. For example, Caraco showed how the transition between flocking and territoriality in juncos could be predicted on the basis of

foraging and predation hazard. Caraco and Wolf's paper on optimal group size in lions spawned a series of interesting papers on optimal group size in birds. Optimality theory was further developed for the social insects by George Oster and E. O. Wilson, and it later applied to helping behavior in birds; but in general, it has been more useful for foraging than for social behavior. Perhaps this was so because foraging behavior is more susceptible to simple models and experiments.

Territorial behavior and resources

Because of its relevance to population stability, the territorial behavior of birds received considerable attention in the 1960s. This subject had been considered at length both by Lack in the 1950s and by Wynne-Edwards in his book. With the focus of interest turned to cost–benefit theories, however, attention shifted to the energetic consequences of territorial behavior. An influential study of costs and benefits of territorial behavior in nectar-feeding birds in Africa was the culmination of a productive research program led by Wolf on nectar-feeding birds in North America. These studies generally agreed with the expectations raised by cost–benefit theories.

Studies of the transition between territorial and flocking behavior in the nonbreeding season were pursued in a variety of species. Both experimental and optimality methods were employed. Many predictions of a cost–benefit nature were tested and the theory was further elaborated with respect to avian social behavior.

The ideal free distribution and resources

When territories are compressible, increasing density of breeders may depress reproductive success. This depression lowers the value of a territory in a good habitat so that a newly arriving bird might have better success by breeding in a poorer habitat with lower density than a better habitat with a higher density. This tradeoff between the habitat quality and density was first formalized into an optimality model now known as the ideal free distribution (IFD) and tested with data by Brown. A similar tradeoff, although complicated by additional factors, is inherent for females in Orian's model. The density–habitat tradeoff was again formalized using elementary algebra by Stephen Fretwell and Lucas, who named it the IFD of competitors. Fretwell, an ornithologist, promoted the global applicability of this concept in a 1972 book. Ornithologists at first paid little attention to the IFD but after a convincing experimental demonstration of it in fishes by Milinski, it became a popular topic for modelers and lab tests in behavioral ecology. A study of 'ideal free ducks' demonstrated the tradeoff in free-living birds. Its relevance to community ecology has been developed at length by Rosenzweig's group using hummingbirds among their test animals (reviewed in Rosenzweig, 1991).

Mating systems

As cost–benefit modeling became popular in the 1960s, attention turned to mating systems. Early work done on North American Icteridae by Orians in 1961 led to a graphical polygyny threshold model. In it a decisive role was assigned to evaluation by females of territory quality and other conditions that affect female reproductive success. The classification of Emlen and Oring, in contrast, named and explained avian mating systems on the basis of the male's behavioral response to various kinds of environment. More recently, from the 1980s to 1990s, research on avian mating systems has focused on the role of sexual conflict and the variability of mating systems.

Sexual selection

A major change in the way ornithologists viewed sexual selection occurred in the 1980s. An anecdote from my own experience illustrates it. In 1975, I gave modest space in an article on sexual selection to Fisher's ideas and raised the unpopular possibility of the importance of female choice in birds. In 1978, at a meeting of behavioral ecologists at Ann Arbor, this passage was cited in a negative way; for the dogma then among ornithologists was that sexual selection in birds was caused only by aggressive competition among males, which could be easily seen, as in LeCroy's explanation of sexual selection in birds of paradise, and not at all by competition among males through attraction and persuasion of females. Aggression is more easily documented objectively than persuasion.

The landmark study in this area and one that killed the old dogma with a single blow was a carefully controlled field experiment by Andersson on the Long-tailed Widow (*Steganura paradisaea*). The data allowed the hypothesis of aggressive competition to be rejected, leaving competition by persuasion and attraction as the remaining alternative. This empirical study together with Zahavi's reshuffling of Fisher's and John Emlen's ideas into the 'handicap' theory stimulated much work on mate choice in birds.

Accepting that female choice had to be taken seriously for birds, modelers turned their attention to the reasons why females preferred particular traits under conditions in which resources were not at stake. Did females prefer males with exaggerated signals because they identified genes that would make sons superior at attracting females, or did females prefer such males because they identified genes that would make both sons and daughters more viable, or both? Models of the former situation allowed the initial stage of the male trait 'before' selection to be entirely neutral. The eminent geneticist and statistician, R. A. Fisher, however, had earlier thought that the process would begin with traits that were correlated with general good condition and that females would choose on the basis of male condition. In other words, he combined the two

processes. Many authors have chosen to present these theories as alternatives, thus polarizing the field and tending to delay compromises. It seems possible to me that many sexual signals in birds identify males whose progeny will be both more viable and more sexually attractive. There should be a continual tendency for selection to carry condition-sensitive traits 'too far,' with their 'dishonest' character only being selected against after some delay.

Aid-giving behavior

A good example of behavioral ecology as a new and exciting discipline before it was formally recognized as such was provided by the study of aid-giving behavior in the 1960s. Traditionally, it was accepted that parents aid their offspring because the offspring carry genes of their parents. Such aid is known as parental care. It had long been known, however, that in some species some individuals behaved as parents toward offspring that were not their own, as in the workers of social insects and the helpers in some species of bird and other vertebrates. Without a discrete theory to explain such anomalous observations, they received little attention. In 1963, in the first of W. D. Hamilton's many major contributions to behavioral ecology, he revolutionized the study of aid-giving by viewing it in relation to relatedness but in a broader sense than simple parent–offspring relationships. Undoubtedly, this single contribution was a major stimulus to sociobiology.

Science works best when there are clear alternative hypotheses. In this case, an alternative was provided by Trivers in what came to be called reciprocal altruism, an obvious idea that became formalized by Axelrod and Hamilton as the game called Prisoner's Dilemma. The significance of these theories is still being evaluated.

New Tools, New Directions

Research in behavioral ecology from the 1960s has been mainly theory driven and led most conspicuously by the many ideas of W. D. Hamilton. In this vein, the field of behavioral ecology has been enriched relatively recently by a new theoretical orientation derived from game theory but adapted to the context of natural selection. During the same period, tools derived from the rapidly developing field of molecular biology began to be employed to answer questions about paternity and evolution. As a result, behavioral ecology became more method driven, with theoretical insights being informed by detailed knowledge of genealogical relationships among animals in populations.

See also: Cooperation and Sociality; Kin Recognition and Genetics; Levels of Selection; Optimal Foraging Theory: Introduction; Wintering Strategies: Moult and Behavior.

Further Reading

Alcock J (1975) *Animal Behavior. An Evolutionary Approach,* 1st edn. Sunderland, MA: Sinauer Associates, Inc.

Brown JL (1964) The evolution of diversity in avian territorial systems. *Wilson Bulletin* 76: 160–169.

Brown JL (1969) The buffer effect and productivity in tit populations. *American Naturalist* 103: 347–354.

Brown JL (1975) *The Evolution of Behavior.* New York, NY: Norton.

Brown JL (1994) Historical patterns in the study of avian social behavior. *Condor* 96: 232–243.

Dawkins R (1979) Twelve misunderstandings of kin selection. *Zeitschrift fur Tierpsychologie* 51: 184–200.

Emlen ST and Oring LW (1977) Ecology, sexual selection, and the evolution of mating systems. *Science* 197: 215–223.

Fisher RA (1915) The evolution of sexual preference. *Eugenics Review* 7: 184–192.

Fisher RA (1930) *The Genetical Theory of Natural Selection.* Oxford: Clarendon Press.

Fretwell SD and Lucas HL (1970) On territorial behaviour and other factors influencing habitat distribution in birds. *Acta Biotheoretica* 19: 16–36.

Hamilton WD (1963) The evolution of altruistic behaviour. *American Naturalist* 97: 354–356.

Hamilton WD (1964) The genetical evolution of social behaviour. I and II. *Journal of Theoretical Biology* 7: 1–52.

Hamilton WD (1996) *Narrow Roads of Gene Land: The Collected Papers of W.D. Hamilton. Vol. 1 – Evolution of Social Behaviour.* San Francisco, CA: W.H. Freeman/Spektrum.

Hamilton WD (2001) *The Collected Papers of W.D. Hamnilton: Narrow Roads of Gene Land,* vol. 2. New York, NY: Oxford University Press.

Klopfer PH (1962) *Behavioral Aspects of Ecology.* Englewood Cliffs, NJ: Prentice-Hall, Inc.

Krebs JR and Davies NB (eds.) (1978) *Behavioural Ecology. An Evolutionary Approach,* 1st edn. Sunderland, MA: Sinauer Associates, Inc.

Krebs JR and Davies NB (1981) *An Introduction to Behavioural Ecology,* 1st edn. Sunderland, MA: Sinauer Associates.

Lack D (1954) *The Natural Regulation of Animal Numbers,* 1st edn. Oxford: Clarendon Press.

MacArthur R and Pianka E (1966) On the optimal use of a patchy environment. *American Naturalist* 100: 603–609.

Marler P and Griffin DR (1973) The 1973 Nobel prize for physiology and medicine. *Science* 182: 464–466.

Maynard Smith J (1982) *Evolution and the Theory of Games.* Cambridge: Cambridge University Press.

Orians GH (1969) On the evolution of mating systems in birds and mammals. *American Naturalist* 103: 589–603.

Schoener TW (1987) A brief history of optimal foraging theory. In: Kamil AC, Krebs JR, and Pulliam HR (eds.) *Foraging Behavior*, pp. 5–67. New York, NY: Plenum Press.

Tinbergen N (1953) *Social Behaviour in Animals,* 1st edn. London: Methuen.

Wilson EO (1971) *The Insect Societies.* Cambridge, MA: Belknap Press of Harvard University.

Wilson EO (1975) *Sociobiology.* Cambridge: Harvard University Press.

Wynne-Edwards VC (1962) *Animal Dispersion in Relation to Social Behavior.* New York, NY: Hafner.

Relevant Websites

http://www.behavecol.com/pages/society/welcome.html – International Society for Behavioral Ecology.

Behavioral Endocrinology of Migration

M. Ramenofsky, University of California, Davis, CA, USA

Introduction

When you search online for the term migration, a multitude of definitions appear regarding the movement of everything, from large charismatic mega fauna – whales and elephants – to the repositioning of atoms within a molecule and from the monumental treks of ancient human populations across continents to the recent heart-wrenching tales of people seeking political asylum. The underlying thread of all these definitions is the movement from one location to another. Migration, as intended here, is the movement of organisms between distinct geographical locations. Such regular migrations, in general, involve outward and return journeys, one for breeding and the other for nonbreeding, each offering seasonal resources that increase the overall fitness of an individual. The migration life history stages (i.e., outward and return) have evolved multiple times across phyla, creating diverse forms and striking parallels, all providing a rich platform for investigating multiple pathways for this adaptation and the endocrine mechanisms underlying it.

Historical Perspective of Migration

Knowledge of the major movements of organisms has been appreciated since ancient times. Seasonal appearance of prey is a vital cue for predators to time breeding. In fact, there are biblical references to the predictable return of the delectable European Quail (*Coturnix coturnix*) to the Middle East in autumn. The Paleolithic cave paintings of the Ardèche region of France depicting herds of mammals suggest a keen awareness of predictable animal movements representing a seasonal food source for primitive peoples. However, not all migratory movements are predictable or constructive. The sporadic appearance of plague species such as the migratory locusts (*Schistocerca gregaria, Chortoicetes terminifera*), African armyworms (*Spodoptera exempta*), Brown Plant hoppers (*Nilaparvata lugens*), aphids (*Toxoptera graminum*), and the avian Redbilled Quelea (*Quelea quelea*) can denude fields of planted crops at a devastating speed and present serious problems to agriculture and the livelihood of people.

Seasonal movements did not evade such keen observers of nature as Aristotle and Linnaeus. Their writings noted that swallows, namely, *Hirundo rustica*, would appear in early spring, flying low over open water, hunting insects. At the end of the summer, the birds vanished mysteriously. Little did the observers realize that at departure, the birds were embarking on a long-distance migration that would take them from breeding grounds in Europe to wintering sites in South Africa, some 7000 km away. Rather, it was thought by some that the birds had dove into the water and were hibernating in the mud at the bottom of the ponds and lakes where they were last seen hunting at the close of summer. To accomplish this, the birds would have to transform to an alternative state allowing them to withstand the cold and anoxic conditions. These ideas held sway over some naturalists, but the writings of Reverend Gilbert White in his treatise, the *Natural History of Selbourne* in the eighteenth century, describe how he actually observed birds leaving for the south and never saw any diving into the water. In spring, they would return flying from the south. Such astute observations by White and others herald the beginning of the study of seasonal movements – migration.

Migration: A Response to Living in a Changing Environment

Organisms that live in locations with distinct seasonal changes alter their behavior, physiology, and morphology to minimize mortality and maximize fitness under diverse sets of conditions. To be successful, organisms must be able to perceive and respond to changes in the environment. These changes fall into three basic categories. First, predictable or seasonal changes include the phenological fluctuations of resources, including food, water, and shelter. Second, the unpredictable changes in weather, predator numbers, and ecological features may occur at any time of the year. Third, the impact of social interactions, that is, competition (dominant-subordinate relationships), can affect access to valuable resources, that is, food and shelter, for certain members of a group. All the three environmental conditions can influence individuals by forcing them to move to breed or to survive. In the discussion that follows, each of these three conditions serves as selective agents molding the diversity of migratory patterns across taxa.

Considering Migration as Life History Stages of the Life Cycle

The annual cycle of species that live for more than 1 year is made up of a repeating sequence of unique stages of specific activities, each representing adaptations for the environmental conditions that exist at a location at a specific time of the year. Such species are referred to as iteroparous signifying the annual cycle reiterates or cycles. For example, the life history stages for a migratory bird can include wintering, prealternate molt, spring migration, breeding and prebasic molt, and autumn migration stages (**Figure 1**). Each of the stages may be considered an alteration in the expression of morphological, physiological, and behavioral traits representing phenotypic flexibility throughout the annual cycle. Upon closer inspection, each stage is composed of three phases – development, mature expression, and termination – that involve the differentiation of cells, tissues, and organs resulting in specific behaviors and physiology. Factors that regulate the onset, progression, and termination of each stage are largely unknown but environmental conditions and endogenous rhythms play a major role. For organisms that live for 1 year or less, including many invertebrates, particularly insects, passage through the stages occurs only once in a lifetime. Thus, the variations observed in their morphology and physiology may be attributed to genetic differences or phenotypic plasticity.

The following are some examples of various forms of migration across invertebrates and vertebrates that illustrate both phenotypic plasticity as well as flexibility within a phenotype. Hormonal bases, where known, of these migratory patterns are also discussed.

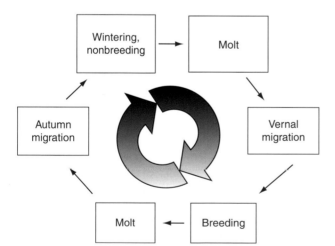

Figure 1 Life history stages of the annual cycle of a seasonal migrant. Peripheral arrows indicate the clockwise direction of the cycle. Central arrows illustrate annual change in photoperiod with increasing light coloration representing lengthening photoperiods of spring and summer and the darkening portions signifying short day lengths of autumn and winter. Points of contact of medial arrows represent winter and summer solstices.

Phenotypic Plasticity

Intergenerational Migration: Monarch Butterflies

Multivoltine species such as the Monarch butterfly (*Danaus plexippus*) produce several generations annually and together complete a migratory route that is an intergenerational roundtrip. This complex system accommodates the acquisition of a preferred food source and an overwintering site that provides constant temperature for diapause. One population migrates as far north as the Great Lakes region and the northeastern corridor of the United States during spring and summer months and then returns to overwinter in the highly localized Oyamel fir forests (*Abies religiosa*) of the transvolcanic mountains of Central Mexico (**Figures 2–4**). Phenotypic plasticity of longevity is apparent in this species with the adult population that migrates from the northern extent of the range south in autumn and overwinters in reproductive

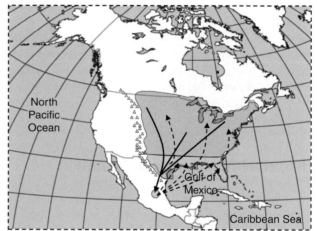

Figure 2 Round trip intergenerational route of the eastern population of Monarch butterflies. Multiple generations of summer breeding adults move north and east to breed throughout the summer range (pink-filled area). In autumn, the most northern populations commence a southward migration to the overwintering site (green-filled circle) in the Oyamel fir forests of the transverse neovolcanic belt of Mexico. Here, adults congregate en masse in reproductive diapause in the fir trees. The forest conditions and large number of individuals contribute to maintaining optimal temperatures for diapause. In March, the adults emerge from this state of repose and migrate to the gulf coast, identified as the spring range (gray-filled area), where they breed and oviposit on the southern milkweed plants (*Ascelpias*) and then succumb. Larvae hatch, feed on the milkweed, and later metamorphose into the summer, breeding adults to migrate into the breeding range. Successive generations move north throughout the summer months relying on the phenology of the northern milkweed. In autumn, the last breeding population migrates south and west to the Oyamel forests. Outward spring routes are indicated by hatched arrows and return autumn routes by solid black arrows. Portions of Rocky Mountain Range are indicated by small triangles. Modified from Brower LP (1996) Monarch butterfly orientation: Missing pieces of a magnificent puzzle. *Journal of Experimental Biology* 199: 93–103.

Figure 3 Swarms of Monarch butterflies moving south in autumn enroute to the Oyamel fir forests. Photograph by Karina Pais, National Public Radio.

Figure 4 Monarch butterflies spend the winter months in reproductive diapause in the Oyamel fir forests of Central Mexico. Photograph by John Edwards.

diapause for a period of 7 months before they breed and die. In contrast, the summer breeding adults migrate north in shorter distances from locations where they hatch to reach the successive flowering crops of milkweed plants (*Asclepias* sp.). Once they have fed, they breed and succumb. The lifespan of the spring adults is reduced to 2 months.

Such phenotypic plasticity is attributed to a distinction of the migratory syndrome within this population and regulated by the endocrine system. Onset of the breeding stage is regulated by juvenile hormone (JH), an acyclic sesquiterpene, produced by the neuroendocrine gland, the corpora allata. During the ontogenetic stages, JH regulates the development of immature characteristics. However, in the adult stage, the elevated hormone levels influence gonadal development and breeding behavior. For the Monarch Butterfly, JH-directed breeding is followed in quick succession with death. Synthesis of the active forms of juvenile hormone (JH I, II) is suppressed under reduced photoperiod and low environmental temperatures experienced by the adult Monarchs during autumn and winter months. In spring, increased photoperiod and elevated temperatures release the inhibition of JH I, II and promote migration, breeding, and eventual death. Studies of gene expression of Monarchs have noted that a suite of 40 genes with differential expressions appear to influence the behavior and physiology of these two states. The results link key behavioral traits with gene expression profiles in the brain that differentiate migratory from summer butterflies and thus show that seasonal changes in genomic function help define the migratory state. It is thought that the Monarch's locate the fir forests by relying on environmental cues that are dependent upon a genetic vector system integrated into an endogenous program.

One-Way Migrations: Desert Locusts

In contrast to other species, the migratory routes of most insects are not round-trip but one-way migrations. The adults do not necessarily return to locations where they were hatched. Though speculative, one explanation for this phenomenon is that migratory insects are 'hedging their bets' by depositing offspring in a wide variety of locations, which may prove productive for the next generation. Another possibility is that if food availability is unpredictable, then a strategy of nomadism relying on cues in the environment that may indicate spatial opportunism could be a decisive factor. A prime example of this is the desert locust, *S. gregaria*, a migratory species that shows extreme phenotypic plasticity. The ontogenetic life cycle is represented by a hemi-metabolous metamorphosis typical of the more advanced insects. This process consists of the egg, the multiple nymphal stages, and the adult. The nymphs (instars) or immatures resemble the adult in form and eating habits but differ in size and genitalia, and lack wings (**Figure 5**). Phenotypic plasticity is apparent among the adults and induced by the environmental variable – population density. When density is low, the solitary phenotype is prominent. Adults of this phenotype breed but are rarely observed in groups and are intolerant of close contact. If the density increases, the morphology, physiology, and behavior of individuals switch in short order (2 h) to the gregarious phenotype. In this form, individuals no longer show mutual repulsion. Rather they form massive swarms of up to 10^{11} individuals

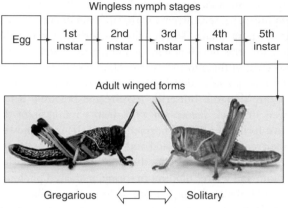

Wingless nymph stages

Egg → 1st instar → 2nd instar → 3rd instar → 4th instar → 5th instar

Adult winged forms

Gregarious ⟸ ⟹ Solitary

Figure 5   Ontogenetic life history stages of the migratory desert locust (*Schistocerca gregaria*). The fertilized egg hatches into four successive instars each of increasing size and development. All are wingless and terrestrial. The fifth instar has vestigial wings. The final instar molts into one of two fully winged adult phenotypes depending upon the environmental conditions. In low densities, the solitary form appears and is rarely observed. Yet, under crowded conditions, the gregarious form is apparent, forming huge swarms and migrates to locations where food is available, but which can be demolished in short order. Photograph by Tom Fayle.

that migrate en masse and land in agricultural areas where they deplete crops with devastating speed. Upon landing, adults may breed multiple times and at various locations throughout the broad geographical range of this species (**Figure 6**). Following this, the adults succumb.

Recent studies have identified a hormonal factor associated with the trigger for phenotypic transition from the solitary to gregarious phenotype. The levels of the neuropeptide, serotonin (5-hydroxytryptamine, 5-HT), increased in the thoracic ganglia in solitary adults following exposure to crowded conditions and acquisition of gregarious behavior. A highly conserved indolamine, 5-HT has been associated with neuronal plasticity in vertebrates, but the effect on the large-scale changes of population dynamics and on the onset of mass migrations, is recent.

Round Trip Routes

Ontogenetic migrations

Semelparous fish

Migratory movements are not restricted to the adult life history stage but may occur during ontogeny. Such migrations, however, transpire only once and do not cycle on an annual basis. A common example of this includes semelparous species that breed once and die and include diadromous fish that migrate between fresh and seawater. The most prominent organisms in this group are anadromous species – two Agnathan genera of lamprey (*Petromyzon* and *Lampetra*), teleost fish such as Pacific Salmonids (*Oncorhynchus* sp.) and Eels (*Aguilla* sp.), and a catadromous

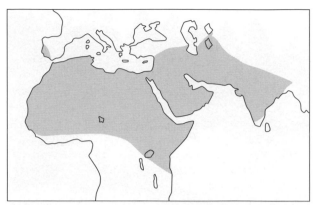

Figure 6   Map depicting portions of Africa, Southern Europe, Middle East, Southern Asia. Shaded areas indicate the geographical distribution of *Schistocerca gregaria*.

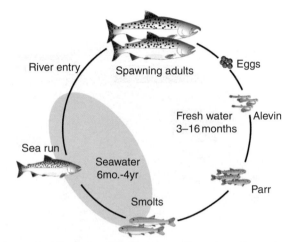

River entry — Spawning adults — Eggs
Fresh water 3–16 months — Alevin
Sea run
Seawater 6mo.-4yr — Parr
Smolts

Figure 7   Life cycle of Pacific salmon (*Oncorhrynchus* sp.) drawn by Kathleen Neely, NOAA, Seattle, WA.

teleost. Although there is variation among these species, the general life cycle of the Pacific salmon provides an illustrative example involving the following: at hatching, alevins emerge from the gravel of fresh water streams with yolk sacs attached and develop into fry, followed by the parr stage that can be identified by vertical line markings and the exhibition of rheotaxis (**Figure 7**). In preparation for seawater entry, the parr metamorphose into the smolts (called smoltification) with dramatic changes in physiology and behavior to accommodate saline conditions. Increases in thyroid hormones are thought to initiate the metamorphosis with prolactin and cortisol aiding in osmoregulatory changes as salinity changes. At this point, smolts enter seawater (sea run form) and travel into the Pacific Gyre where they spend differential amounts of time depending upon the species to complete growth. At the climax of this stage, sexual maturation is initiated, the hypothalamic pituitary gonad axis is activated and spawning migration begins as fish prepare for entry into fresh water navigating back into natal streams and rivers to spawn. Navigation to the natal waters has been attributed to imprinting of the chemical composition of the natal streams experienced by smolts during first migration to open water and guided by the memory of the olfactory cues accumulated earlier in life. Again, prolactin and cortisol are thought to regulate osmoregulatory changes as fish enter fresh water. Navigation back to natal streams is considered to be influenced by cortisol, which may prime or activate regions in the teleost brain, namely, hippocampus and other olfactory regions to recall memory and help to guide during the return trip. Increasing levels of sex steroids as gonads mature control changes in the morphology and behavior, leading to territoriality and reproductive behavior. Castration tends to prolong life, but only for a few months, suggesting that death postspawning is, indeed, programmed. The connection between reproductive hormones and programmed cell death in semelparous vertebrates is reminiscent of the life histories of the insect systems described earlier and presents a fruitful avenue for further comparative investigations.

Evolutionary explanations for Salmonid semelparity revolve around a combination of factors that include distance and arduousness of the return trip, which leave adults spent after the production of a multitude of gametes. Also, streams in which the young hatch are largely nutrient-poor, and the decaying carcasses of moribund adults can deliver dissolved elements, including N_2 and P, which serve to enrich the nursery conditions for the young fish.

African black Oystercatcher: a ploy to avoid adults?

In birds, another example of an ontogenetic migration is found among juvenile African Black Oystercatchers (*Haematopus moquini*). In this species, adults are sedentary and remain year round on the breeding grounds on the southern coasts of Namibia and South Africa. Following the postfledgling molt, young birds migrate away from the breeding grounds to specific locations along the coast called nurseries. After a period of years, the birds return to the natal territories to breed, assume a sedentary lifestyle, and never again migrate. Nothing is known of the endocrine mechanisms involved in the juvenile movement. Although speculative, the explanation for the ontogenetic migration may be that availability of specific foods, reduced competition, or low predation pressure at the nursery sites could play roles for the appearance and maintenance of this migratory pattern.

White Stork: delayed migration and breeding

White storks (*Ciconia cionia*) present an intriguing condition of delayed maturity in that the birds do not normally breed until they reach 3 or 4 years of age. This is a migratory species with populations that breed extensively from the northwestern tip of Africa north through Spain and extending eastward into Europe and beyond. Following breeding, birds return in autumn via western and eastern routes to overwinter in the south of the Sahara in eastern and southern Africa. At the end of breeding, young birds migrate with adults in autumn to the overwintering sites. However, in spring a few young birds return with the adults to breed. Most second-year birds remain south of the Sahara. Third and most fourth-year birds migrate north in spring for increasing distances but most neither complete the full trip nor breed successfully until their fifth and sixth years. This ontogenetic pattern of failure to complete the spring migratory route and delayed breeding suggests a tight linkage between both functions. The cycles of molt, body mass, fat, and reproductive hormones appeared similar between the juvenile and adult birds indicating that these regulatory mechanisms are not involved. Possibly, the capacity, to complete the spring migratory route and to breed require full maturation of the behavioral and physiological systems. Storks are diurnal migrants and rely on rising thermals to support soaring during migration which reduces the requirement of powered flight and energetic costs. Thermals are created during daylight hours as heat rises off the surface of the land or slopes in vertical columns of warm air rising through the atmosphere. Thus, storks avoid crossing open bodies of water, including the Mediterranean, effectively lengthening the flight distance for the outward and return journeys. Diurnal flight also leaves little time for feeding, replenishing fuel stores, and rest on a daily basis at stopover sites. Adult birds may be more competent in acquiring and storing fuel for flight at take off or locating novel food resources during the brief period of stopover. Juveniles could lack the 'know-how' to maintain themselves during spring migration. The delayed pattern of migration and breeding, therefore, may represent the time required to gain the experience necessary to successfully complete spring migration and arrive on the breeding grounds in condition to breed. This example suggests that both environmental conditions and migration strategies play a role in molding ontogenetic migration.

Seasonal Roundrip Migrations

Spiny lobster

The Caribbean spiny lobster (*Panulirus argus*) is found in the tropical and subtropical waters of the Atlantic and Caribbean Oceans extending into the Gulf of Mexico. It is nocturnally active, spending daylight hours in crevices and holes within the coral reefs. At night, animals move away from these protective sites to feed on a wide variety of marine invertebrates as well as scavenge for detritus along the ocean floor. In early autumn, juvenile and young adults are found feeding over a wide area in the Caribbean at depths of 3–10 m, a region of minimal cover with few rock and coral outcrops, patches of sea grass, and large colonies of sponges. However, by the end of autumn, these locations are plagued with seasonal disturbances associated with the southeastern hurricane season. Water temperature in the shallow areas drops as the ocean swells and wave action stirs up the sandy bottom, the turbidity increases, hampering feeding by lobsters. Autumn storms are thought to instigate mass movements of lobsters to deeper water where there is little effect of storms, and water temperatures are elevated above those of the shallows.

During migration, individual lobsters line up in a single file called *queuing*, with each resting its long antennae over the carapace of the individual in front to increase laminar flow and reduce drag (**Figure 8**). During the outward migration, the queue moves generally in a southerly direction and covers 30–50 km to reach reefs at depths from 10 to 30 m. The movement is recorded to occur during the night as well as by day, and animals have been shown to navigate by an internal magnetic compass using the earth's geomagnetic field. On the return migration, the spring movements appear to be less synchronized as animals reappear in the shallow areas in smaller numbers. At this time, breeding and molt ensue during the spring and summer months (**Figure 9**). Unlike other migratory systems, the mass migrations of spiny lobsters are not associated with reproduction directly but are rather thought to indicate an avoidance of cold water that impairs development, spawning, and survival of adults, particularly during molt. The endocrinology of this system is not well understood and deserves further attention.

Migratory song birds

The life cycles for many iteroparous migratory songbirds are variable and complex, and by far the most well studied. Some birds migrate between the breeding and overwintering sites solely. Others may migrate to special locations after breeding to molt, and when completed, will continue on to an overwintering location. Migratory patterns also vary within a species. For example, some members of populations, usually adults, will show no migratory activity and remain on the breeding territories year round. While other members, usually juveniles, migrate from the breeding grounds in autumn and overwinter in a location, but return the following spring to breed. In this case, the migratory pattern of the population is described as partial migratory.

A differential migratory pattern refers to the different wintering sites utilized by the members of a population. Here, males may overwinter at locations in closer proximity to the wintering grounds, whereas females (juvenile and adults) may migrate further away. In both partial and differential migratory patterns, competition over limited resources is usually the selective force influencing the mode of migration. These diverse patterns of migration represent distinct selective pressures on individuals within a population and illustrate the diversity of adaptations to seasonal variations and availability of resources.

For all taxa, the migratory life history stages are probably best known from the studies of passerine birds that have been the focus of scientific investigations for well over 100 years. The descriptions that follow are based on the migratory bird literature with a particular emphasis drawn from a long-distant migrant, the White-crowned Sparrow (*Zonotrichia leucophrys gambelii*) (**Figure 10**). Both the vernal and autumn migratory life history stages are composed of three phases (**Figure 11**). The first is the developmental phase in which all the cellular and molecular aspects of migration are set in place. The trigger for initiation of the spring developmental phase for many species is the vernal increase in day length, but for others, endogenous rhythms are most prominent, particularly for species that overwinter on or near the equator where changes in photoperiod are negligible. The trigger for the autumn phase, however, is poorly understood but is thought to be related to photorefractoriness that occurs at the termination of the breeding stage. The developmental phase for migration involves morphological changes that include muscle and liver hypertrophy,

Manuel Mola. Copyright © 2003

Figure 8   Migratory queuing of spiny lobsters (*Panulirus argus*). Photography by Manuel Mola, Copyright 2003.

and the physiological changes in the expression of enzymes that direct the synthesis and deposition of fuel for flight, namely, lipid and protein as well as the catabolism necessary for utilizing the stored fuels once flight

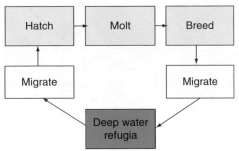

Figure 9 Life history stages of spiny lobster (*Panulirus argus*). Throughout late spring and summer months, juveniles and adults are found in the shallow waters (light blue-filled boxes) off the southeastern coast of the USA. In early autumn, individuals migrate in large numbers in lines or queues to deeper waters (dark blue-filled box) to avoid storm conditions in the shallow waters. By early spring, individuals return to the shallows to hatch young, molt, and breed.

Figure 10 White-crowned sparrow (*Zonotrichia leucophrys gambelii*). Photo by John C Wingfield.

begins. Also, the oxygen-carrying capacity of the blood improves with an increase in red blood cells as measured by hematocrit. This process entails the synthesis of the protein growth factor erythropoietin by the liver and kidney that is thought to be regulated by gonadal androgen and thyroid hormones. Behavioral changes include hyperphagia or an increase in appetite, and changes in social behavior, all of which culminate in the mature capability phase when migration actually begins.

During this phase, all the aspects that were developed in the preceding phase are now expressed, namely, cycles of fueling and flight. Fueling occurs both at the outset and throughout the migratory period at specific locations called stopovers. For birds and other species that do not feed while migrating, stopovers are a critical factor for refueling and rest. If prohibited, owing to the restriction of stopover sites because of ecological disturbance and reduction in resource availability, the success of migration is greatly jeopardized. Fueling is achieved by hyperphagia, lipogenesis, fat deposition, increased length of the gut and size of the digestive organs as well as flight muscle hypertrophy. The vernal increase in day length acts in conjunction with testosterone to organize the hypothalamic feeding centers in the brain to regulate feeding. NYP, central prolactin, and possibly, CCK, appear to play important roles, but the specific mechanisms in relation to migration are not well understood. The hormonal mechanisms regulating the digestive organ size and flight muscle are not well known.

At the end of this refueling substage, feeding ceases and birds rest prior to departure. In captivity, this state is described as the quiescent phase, which serves as a transition between the anabolic and catabolic stages, and may be required for an orderly transition. The plasma glucocorticoid, corticosterone, increases prior to departure, which is thought to be involved in the transition and/or preparation for flight. Factors that influence the timing of actual take-off may involve atmospheric conditions, lunar

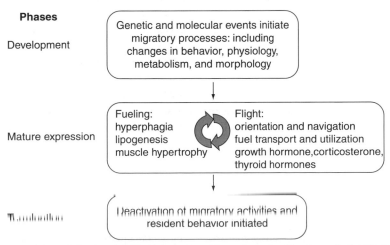

Figure 11 The three phases of the adult migration life history stage drawn from the literature of migratory passerine birds.

phase, or social inducement from flock members, but much remains unknown. After flight ensues, the levels of corticosterone rise further in conjunction with other mediators of lipid and protein utilization, all of which serve to meet the energetic demands of flight. Concomitantly, the digestive organs that were built up during the fueling phase are reduced in size and this is as a weight-saving strategy. At stopovers, the cycle begins again, with the anabolic functions taking over to replace lost fuel. How quickly and effectively birds regain their lost mass, fat, and muscle enroute is not known, but such information is critical to understanding the true costs of migration.

In addition to the physiological parameters, the behaviors expressed during migration change radically. Some species are nocturnal migrants in that they migrate during the night. Others are diurnal or day migrants and fly during the daylight hours. The distinction here is that the movement is oriented and the distances covered are much increased over daily activities observed in the previous or subsequent life history stages. To orientate and navigate to a specific location, migrants rely on a variety of exogenous cues. Many species utilize the earth's magnetic field, and solar and celestial cues; some even rely on olfaction. How birds know the route is an interesting point. Some species fly with adults and learn the routing. In other cases, young birds must rely on 'genetic knowledge' at least for their first autumn flight. Studies have shown that the route and ability to correctly perceive cues for orientation are 'learned' over time and resemble at some level the systems apparent in semelparous fish during spawning migration. Upon reaching the destination (overwinter or breeding sites), all the behavioral and physiological attributes of the migratory life stage are terminated allowing progression to the next stage, be it breeding or overwintering.

Irruptive and Nomadic Migrants

The Red-crossbill (*Loxia curvirostria*) is a classic example of an irruptive and nomadic migrant (**Figure 12**). This species relies on coniferous cone seeds that nourish both the adults and the young. Cone crops mature at regular intervals but in irregular quantities; thus, the amount of seed and how long it is available are unpredictable in space and time. Studies have demonstrated that fat deposition and migratory movements display a seasonal component that is linked to photoperiod and endogenous cycles. For example, in spring, birds show regular seasonal fattening and migratory movements. This is a time when seeds likely develop and are possibly easy to find with nomadic movements. Birds locating a sufficient supply will stop and commence breeding.

By midsummer, breeding is interrupted as birds complete their prebasic molt. In most years, locating mature seeds in the autumn is highly unpredictable. In response

Figure 12   Red crossbill (*Loxia curvirostria*). Photo by Eric Bjorkman.

to this seasonal uncertainty, birds show a second peak of fattening as an insurance policy in case food is not located. Should local cone crops be sufficient, birds may remain in the area and breed again. If, however, the seed crop at this site is poor, birds will respond with irruptive movements and commence searching for more productive coniferous stands. Should a rich cone crop be found, a second round of breeding can occur, but this time, during the winter months. The irruptive movements have been associated with elevations of plasma corticosterone. Experimental studies of captive birds have revealed that reducing or eliminating food results in increased locomotor activity as well as plasma corticosterone. It is suggested that this endocrine link plays a role in the initiation of the irruptive movements.

Conclusion

Considering migration as a life history stage within the context of the annual cycle provides a heuristic model for determining mechanisms. Compartmentalizing each life history stage allows of an investigation of the environmental factors that affect the onset, expression, and termination of each phase. Migration is an organism's solution to solving the problem of variable environments. This solution may be an expression of multiple phenotypes or phenotypic flexibility within the year, each matching the conditional requirements of ecology present at a given location. Or it may come in the version of multiple phenotypes at a particular time to match the conditions at that point as seen in insects (phenotypic plasticity). Thinking broadly, such ideas harken back to the earlier misconceptions of migration presented at the outset of this article. Certainly, migration is not a transformation of form from an aerial aerobic passerine to one capable of withstanding the conditions of winter by assuming a subaquatic (benthic) and anaerobic form. But the concept of alteration of form is, nevertheless, compelling.

Migrants prepare, express and terminate characteristic traits that allow for movement from local to distant habitats for breeding or survival. In doing so, the phenotype that is expressed, whether it is plasticity or flexibility, then matches the demands of the current local conditions. Thus, studies of migration that consider the concepts of alteration, coupled with movement, are key and propel future studies and advancements in the topic of migration.

See also: Bat Migration; Bats: Orientation, Navigation and Homing; Bird Migration; Collective Intelligence; Fish Migration; Food Intake: Behavioral Endocrinology; Insect Migration; Irruptive Migration; Magnetic Orientation in Migratory Songbirds.

Further Reading

Berthold P, Gwinner E, and Sonnenschein E (2003) *Avian Migration.* Berlin: Springer.

Birkhead T (2008) *The Wisdom of Birds, an Illustrated History of Ornithology.* USA: Bloomsbury.

Dingle H (1996) *Migration: The Biology of Life on the Move.* Oxford: Oxford University Press.

Dingle H and Drake VA (2007) What is migration? *Bioscience* 57: 113–121.

Greenberg R and Marra P (2005) *Birds of Two Worlds, the Ecology and Evolution of Migration.* Baltimore, MD: Johns Hopkins University Press.

Holland RA, Wikelski M, and Wilcove DS (2006) How and why do insects migrate? *Science* 313: 794–796.

Landys M, Ramenofsky M, and Wingfield JC (2006) Actions of glucocorticoids at a seasonal baseline as compared to stress-related levels in the regulation of periodic life processes. *General and Compared Endocrinology* 148: 132–149.

Newton I (2008) *The Migration Ecology of Birds.* New York, NY: Academic Press.

Quinn TP (2004) *The Behavior and Ecology of Pacific Salmon and Trout.* Seattle, WA: University of Washington Press.

Ramenofsky M and Wingfield JC (2007) Regulation of migration. *Bioscience* 57(2): 135–143.

Ramenofsky M, Agatsuma R, and Ramfar T (2008) Environmental conditions affect behavior of captive, migratory White-crowned Sparrows. *Condor* 110(4): 658–671.

Wilcove DS (2008) *No Way Home.* Washington, DC: Island Press/Shearwater Books.

Wingfield JC and Sapolsky RM (2003) Reproduction and resistance to stress: When and how. *Journal of Neuroendocrinology* 15: 711–724.

Wingfield JC, Schwabl H, and Mattocks PW, Jr (1990) Endocrine mechanisms of migration. In: Gwinner E (ed.) *Bird Migration,* pp. 232–256. Berlin: Springer-Verlag.

Relevant Websites

http://www.movebank.org/community – Community for Remote Animal Monitoring

Betta Splendens

C. V. Tillberg, Linfield College, McMinnville, OR, USA

Description and Natural History

Male *Betta splendens* have a body size of about 3 cm, with long flowing fins of an additional centimeter or more; adult males may exceed 6 cm in total length. Females are smaller, and have shorter fins. Coloration in domestic males ranges from deep red to deep blue, including purple, turquoise, green, orange, and even white. Some individuals may be combinations of these colors. Females are generally less colorful than males; the main color tends to be gray to brown, but with a slight hue of the colors found in males. Females also have a vertical striping pattern that darkens when they are sexually receptive. The dorsal, caudal, anal, and ventral fins of a male are very long, frequently as long as the body of the male fish itself. When not aroused, the male holds these fins in a relaxed, flaccid position. Upon arousal, the male erects these fins, forming the colorful, characteristic display of this species.

B. splendens are native to shallow freshwater ponds, including rice paddies, in Southeast Asia. Their labyrinth organ allows them to obtain oxygen by gulping air at the surface, which helps these fish tolerate low dissolved oxygen levels. Various strains of *Betta*s have been kept for ornamental and fighting purposes for centuries, creating a legacy of human selection for certain traits such as color, fin length, and temperament. Wild-type bettas tend to be less colorful, have shorter fins, and are less aggressive.

The behaviors of most interest to scientists and hobbyists alike center around the species' mating behaviors. Males establish and vigorously defend territories against other males, as well as court females, with a series of stereotypical behaviors. These behaviors include fin erection, flaring of the gill opercula and branchiostegal membrane, frontal and lateral displays, tail beating, exaggerated swimming motions, and, if allowed, chasing and biting of the other fish. These encounters may result in injury or death of one combatant. Also of interest is the parental care provided by males in a bubble nest they construct within their territory. The study of these behaviors has proved to be fertile territory for developing insights into animal communication, territoriality, sexual selection, and parental care. *B. splendens* also possess some basic cognitive abilities such as learning and memory. Thus, many researchers utilize the performance of stereotypical agonistic behaviors as a means to gain deeper insights about the organization of the neurological system of a nontetrapod model system. Finally, the ease with which the social and chemical environment of the fish is manipulated allows for investigation of the hormonal regulation of agonistic and sexual behaviors.

The Display

The agonistic and courtship displays of *B. splendens* include both morphological and behavioral visual cues. Morphological characters include color, body size, and fin length, while behavioral cues include posture, erection of fins and gill opercula, and swimming motions. Behaviorists have focused on these various aspects of the display to answer general questions about intra- and intersexual communication.

Male and female *B. splendens* distinguish sex on the basis of color pattern and fin length. When presented models of varying color pattern and fin length, males respond aggressively to unpatterned bodies and long fins. This combination of characters mimics a rival breeding male. Males also display courtship behaviors toward female-mimicking models with short fins and breeding pattern coloration. Females also appear to use these cues, as well as the behaviors of the fish they encounter, to distinguish among displaying females (a potential rival), courting males (a potential mate), and other fish.

Dorsal, caudal, anal, and ventral fins of domestic male bettas are considerably longer than wild-type males, and reflect a history of selection for the character. The trait appears to be under genetic and hormonal control. A dominant allele is responsible for long fins in the domesticated line; wild-type males are homozygous recessive for this allele. The manifestation of the long-fin morphology depends on the presence of sufficient androgen hormones. The levels of different hormones, and therefore fin morphology, in males are affected by the social environment in which the male develops. Males reared in isolation develop longer fins than communally reared males. Steroid levels reflect this morphological difference; long-finned isolated males have high levels of 11-oxysteroids and low levels of 5-β steroids, a steroid common in juvenile bettas. In contrast, communal, short-finned males had low 11-oxysteroid and high 5-β steroid levels. Further investigation of the communal males reveals a relationship between position in the dominance hierarchy and (i) fin length, (ii) 11-oxysteroid levels, and (iii) 5-β steroid levels. Dominant individuals have long fins, higher 11-oxysteroid, and lower 5-β steroid. Thus, fin length and sexual maturation are arrested in subordinate males when they are reared in a communal situation.

Body size and behavior interact in complex ways to affect the receptivity of females toward males. Domestic female bettas prefer larger males, but not necessarily the males with the most vigorous display. In fact, females that experience being around male bettas for a period of time will subsequently learn to avoid males and their harassing behavior. Thus, it might be that more vigorously displaying males are less attractive to females if the females associate display intensity with higher levels of harassment. Research on wild-type *B. splendens* found that these female bettas show no size preference on the basis of male size. Rather, larger wild betta males win more male–male contests. This might indicate that the importance of size and behavior in inter- and intrasexual selection varies between domestic and wild populations of *B. splendens*.

Honest Signaling

The charismatic morphology and behavior of *B. splendens* have made them useful model animals for investigating hypotheses arising from the theory of honest signaling. This theory proposes that accuracy of communicatory signals is enforced by the cost of sending the signal. If a signal is expensive or dangerous to send, then only strong individuals are capable of sending the signal correctly; it is too costly for less fit individuals to perform, which prevents cheating. As such, behaviorists have sought to test this theory by investigating the costs associated with various signals, the accuracy of the signal, and the fitness of the senders.

One of the earliest tests of honest signaling in bettas measured the overall display vigor and resource-holding power of males. Males with the more vigorous display were the victor of agonistic interactions in 11 of 12 trials: a result that supports the hypothesis that display vigor is an honest signal of fighting ability.

If display vigor is an honest signal, then one would expect this display to be costly. Measurements of the metabolic costs of aggressive encounters reveal an overall reduction in the muscular amino acid content, followed by a reduction of glycogen and an increase in free glucose. This is a clear demonstration that the overall agonistic encounter is metabolically expensive. But what of the various components of the agonistic display? Recent research on the costs associated with the erection of the opercula and branchiostegal membranes quantify oxygen consumption of displaying males and their ability to maintain the display in low oxygen water. During an agonistic encounter, males consume more oxygen. Males perform less opercula erection in hypoxic water. Erection of the opercula and branchiostegal membrane make oxygen exchange in the gills less efficient; therefore, this behavior fulfills the predictions of honest signaling as it is an expensive behavior to perform.

Cheaters should be rare in systems with honest communication, because of the high cost of dishonesty. Experimental tests of this hypothesis were performed by creating experimental conditions that induced males to cheat. Prior research (discussed below) indicated that isolated males briefly shown another betta will interact hyperaggressively. Thus, these isolated males were induced to cheat in the agonistic interaction by escalating to overt aggression, such as biting, much sooner than would be warranted in a normal male–male interaction. However, these isolated, hyperaggressive males were not of higher quality than their opponents; as such, they exhausted themselves at the beginning of the interaction and were usually defeated. Noncheating males went through the stereotypical series of agonistic displays and lost fewer fights than the cheaters. This demonstrates that ritualized agonistic behaviors can be stable in a population if dishonesty is costly.

Audience

Despite the territorial and aggressive nature of *B. splendens*, an important aspect of their behavior seems to be the presence of an audience during agonistic interactions. The audience effect changes the behavior of both the interactants and the observer. There are important social cues transmitted and obtained depending on the outcome of the interaction, as well as the sex and status of the observer. These studies have broadened our understanding of communication from considering it just as an interaction between the sender and the receiver to recognizing the importance of the social context in which communication occurs.

Observers of agonistic interactions gain information about the relative strength of the two interacting males. How this information is used by male observers has differed among some studies. Observer males hesitate longer before engaging with a male they have seen win an encounter than a male they have seen lose; this same male makes no distinction in attack latency between winners and losers of interactions it did *not* observe. Clearly, the outcome of the male–male interaction affects the way the observer interacts with the males it observed. However, once the observer engages with the winning male, the observer escalates the attack more rapidly, suggesting the observer has identified the winning male as a stronger opponent.

A separate investigation of how observation affects male behavior presented male observers with apparent male–male contests. The demonstrating fish actually interacted with separate, unseen fish, but to the observer male it appeared that the two males were interacting with each other. This may have created a more ambiguous distinction between winners and losers of these interactions, as it was possible that the demonstrating fish could have both won the interactions in which they were actually involved. The observing males responded more rapidly to the demonstrator male that had displayed more vigorously, that is, the apparent winner of the interaction as the observer saw it. This shorter latency to attack the winner is in contrast to the findings reported earlier, but both are in agreement that once initiated, the observer escalates the attack more rapidly.

In a similar study manipulating the appearance of competitive interactions, researchers allowed subject males to observe two demonstrator males in adjacent tanks. In this case, there were mirrors directly adjacent to, or 5 cm away from, the tanks between the two demonstrating males. To the observer, it appeared that these males were displaying to each other, when actually they were displaying to their own mirror image. The more distant mirror yielded lower levels of aggressive behavior in the demonstrator than the mirror immediately next to the tank. Therefore, the observer witnessed apparent interactions in which one male clearly behaved more aggressively than the other male. In these experiments, observer males behaved more aggressively toward the more aggressive male the apparent opponent of which had been artificially induced to behave less aggressively.

The effect of witnessing an aggressive encounter also affects how males interact with other males they did not observe. Bystanders of an aggressive encounter are motivated to behave more aggressively toward unfamiliar, nonbystander males. Naïve males are significantly less aggressive than males that have themselves witnessed aggression, suggesting that aggressive behavior in male bettas may be 'primed' merely by witnessing these kinds of interactions.

Females may also use the information they attain by witnessing an agonistic encounter between males. Female bettas allowed to observe agonistic encounters spend more time near the winning male.

Observing an agonistic encounter can provide important information about the relative quality of the interacting males for both male and female observers. However, the presence of an audience is not neutral in its influence on the combatants; the interacting males themselves are also affected by the presence and sex of an audience.

The mere presence or absence of other bettas affects aggression levels in males. Social isolation causes males to be less ready to perform aggressive displays initially. However, once isolated males are primed by observing a releasing stimulus, their levels of aggression increase steadily and exceed that of the nonisolated males. One model of aggression proposes that neural circuits responsible for agonistic behavior become miscalibrated in socially isolated fish because these circuits are never stimulated. Thus, an isolated fish lacks in readiness to respond to a challenger, but once it begins to respond,

it does so with more vigor than would be typical of a nonisolated betta.

Recent research demonstrates the effect of an audience as a primer of aggression in male bettas. Males are more aggressive if an audience is present prior to an agonistic encounter than if the audience and rival male are presented simultaneously. This difference holds even if the audience is viewed before the encounter, and then removed. This result suggests that at least a portion of the audience effect on interacting males is due to the priming of agonistic behavior by the presence of another fish. In the absence of this primer, aggression levels are lower; when present, aggression levels are higher.

The sex of the audience also influences how males interact with each other. Males behave differently in front of a female audience than in front of a male or no audience. When a female observer is present, males perform more displays such as gill erection and tail beats and fewer bites. The moderation of injurious fighting behavior in front of females might result in a lower likelihood of driving away the prospective female, while higher levels of aggression, including injurious behaviors such as biting, in front of a male audience may demonstrate a willingness to fight to an observer male.

Audience and nesting status interact to affect male motivation to fight. Without an audience, males are less aggressive when both own a nest than when one or neither owns a nest. When one or neither male owns a nest, male–male interactions are more vigorous in the presence of male observers than in the presence of female observers. Hormonal profiles of these interacting fish reflect their behavioral status. Regardless of the nest status, the presence of a female audience results in lower levels of 11-ketotestosterone. However, in the presence of a male or no audience, 11-ketotestosterone levels are low when both males have nests, but high when neither males have nests. These complementary studies demonstrate the importance of multiple interacting environmental cues modulating hormonal and behavioral responses.

A final use of information about one's audience is demonstrated by males that lose a fight in the presence of a female. When presented with the option to court either a female that watched them lose, or a naïve female, defeated males display significantly more courtship behavior toward the naïve female. The victorious males spend equal amounts of time displaying both to the female that observed them win as to the naïve female. This suggests that the defeated male is making an optimal decision in which female it should court; its chances of successfully courting the observer female are low compared to the naïve female. This remarkable behavior could be accomplished through a simple behavioral program, such as 'if lose, court unfamiliar female.' Alternatively, these results might suggest the ability of male bettas to understand the information available to, and motivations of, another individual – the observing female. The abilities to attribute mental states to one's self and to others are components of a 'theory of mind,' a cognitive capacity normally reserved for more 'advanced' vertebrates.

When kept in tanks with other individuals of the same sex, males and females both form dominance hierarchies. Dominant individuals get more food, display more frequently and for longer periods, and attack others more frequently than subordinate individuals. Furthermore, fish of similar rank interact agonistically with each other more frequently than with fish of very different rank in the hierarchy. It should be noted that these experiments were performed in situations of very high fish abundances; four fish were placed in less than 7 l of water, creating conditions of intense crowding with no possibility of escape. It is not clear whether the formation of a dominance hierarchy is a function of lab crowding or whether this level of social organization ever has the opportunity to arise in the wild.

Despite their cantankerous nature, bettas often choose to associate with other bettas rather than to be alone. Males and females given the opportunity to be with (i) another male versus alone, (ii) another female versus alone, (iii) three females versus alone, (iv) another female versus three females, and (v) a single male versus three females regularly choose the more social situation. The exceptions are that females choose solitude over a single male, and males do not distinguish between a single male and three females.

The habituation to neighbors by males results in a diminution of their fighting ability. Even though habituated males continue to display to other males by erecting fins and opercula, they reduce their biting attacks. Hence, when paired with a nonhabituated male, this reduction in the willingness to bite results in the defeat of the habituated males.

Domestic strains of *B. splendens* exhibit hormonal and behavioral differences from the wild strain. Comparisons of the aggressiveness of domestic and wild bettas against (i) a rival, (ii) a mirror, and (iii) a video of a displaying male demonstrate that bettas are significantly more aggressive in male–male interactions, attesting to the effect of centuries of selective breeding for this trait. Interestingly, there is no difference in aggression level between domestic and wild males when faced with a mirror or a video. Communicatory cues present in the actual male–male encounters may be critical in modulating aggression levels; these cues are absent in the mirror and video trials.

Behavioral and hormonal responses to the stresses of a novel environment and confinement differ between domestic and wild strains, as well. Both have increased opercular rate and cortisol levels during confinement, but only the wild strain respond in this way to a novel environment. Domestic fish are more placid during confinement, while wild fish are very active. These differences in response to stressful situations demonstrate that domestic fish are less anxious than wild bettas.

Parental Care

Upon mating, male bettas gather the fertilized eggs and deposit them in their bubble nest floating at the surface in their territory. The nest size varies among males and is positively correlated with the male body size. Females do not distinguish among males on the basis of the size of their nests, nor does the nest size correlate with male fighting ability. However, female preference for larger males should result in their also choosing males with larger nests. Male nest-holders vary in their aggressiveness at different times of the breeding cycle, with a peak in aggression after hatching of the eggs. During this time, males are most aggressive toward other males, moderately aggressive toward unfamiliar females, and least aggressive toward the egg-laying female.

Cognitive Abilities

B. splendens have been useful model organisms for investigating aspects of learning, memory, perception, and neurophysiology in the fish brain. The history of successful and failed agonistic interactions affects how bettas behave, suggesting some capacity for learning and memory. For example, winning males behave more aggressively in subsequent interactions, and losing males behave more submissively. The discovery that presentation with an agonistic stimulus, such as a mirror, could serve as a reinforcer has allowed for further investigation of the learning and memory abilities of bettas. Learning behaviors in response to agonistic stimuli is not limited to male bettas. Females also learn to perform a behavior, such as swimming through a ring, reinforced by exposure to a mirror. Male bettas learn best if given immediate reinforcement, less well with a 10 s delay in reinforcement, and poorly with a 25 s delay in reinforcement. A variable reinforcement schedule does not result in learning of the response, nor does it cause the fish to retain the behavior it learns. These are important differences in the effect of reinforcement schedule on bettas compared to tetrapod model organisms in which variable reinforcement elicits strong retention of the response.

Research on negative reinforcement of natural behaviors, such as agonistic displays, in bettas has yielded interesting results. Intermediate severities of punishment result in higher levels of performance of agonistic displays relative to unpunished trials, while severe punishment diminishes the performance of the agonistic behavior. This demonstrates that behaviors motivated by aversive stimuli, such as the presence of a rival, is enhanced by moderate punishment, such as a mildly painful shock.

The aggression bioassay has also been used to explore aspects of betta visual perception and processing. For example, this method has shown that texture is an important component of the eliciting stimulus. Models with a scale pattern elicit a vigorous agonistic response, while models lacking this texture do not. Another experiment using the aggression bioassay demonstrates brain lateralization in bettas by showing that individual males reliably display threats or courtship in postures that favor either their left or right eye. Finally, bettas are susceptible to subliminal stimuli. A brief, 1 s exposure to a mirror or to an image of a fish in the aggressive frontal display did not prompt a response in the exposed fish. However, these exposed fish were hyperaggressive in subsequent agonistic interactions. Susceptibility to subliminal stimuli may be an ancient cognitive property.

In bettas, the visual center of the brain (the optic tectum) is attuned to certain visual stimuli. Motion of the rival male or model is necessary to elicit activity. Furthermore, activity in this region of the brain spikes during specific portions of the observed display; when the rival male turns to face the observer, the optic tectum is most active. This motion and position give the most prominent view of the rival's erect opercula, and may be the height of the agonistic display.

B. splendens may also be a useful model organism in the study of the neurophysiology of aggressive behavior. Measurements of the effect of serotonin, its receptor agonists and antagonists, and selective serotonin reuptake inhibitors (SSRIs) on betta behavior show that administration of serotonin and its receptor agonist results in reduced aggression. Increased receptor antagonist does not increase aggression. Treatment with SSRI injections reduces brain serotonin levels, but does not alter fish behavior. However, treatment with Prozac, a SSRI drug, dissolved in the water did affect male behavior. Exposure to as little as $3\,\mu g\,ml^{-1}$ dissolved in the water for 3 h significantly reduced male aggression.

Sex determination in *B. splendens* is the result of both genetic and hormonal influences. Genetic females that have been ovarectomized may undergo a sex role reversal, including the elongation of fins, male-typical courtship and territorial behavior, and production of viable sperm. These so-called 'neomales' sire both female and male offspring, suggesting that, as in other teleosts, sex determination is not purely due to an XX/XY homo-/heterogamous system. The assumption of male-typical morphology and behavior is likely due to androgens produced by the newly developed testes in the ovarectomized females. Treatment of females with testosterone results in a reduction in aggression toward females and an increase in aggression toward males. If kept in a social situation, these treated females maintain their new hormonally induced social role after cessation of treatment. However, isolated treated females revert to female-typical behaviors. Thus, sexual identity and behavior are the result of a complex interaction between genes, endocrinology, and social environment.

See also: Aggression and Territoriality; Agonistic Signals; Communication and Hormones; Honest Signaling; Hormones and Behavior: Basic Concepts; Mate Choice in Males and Females; Mating Signals; Memory, Learning, Hormones and Behavior; Parental Behavior and Hormones in Non-Mammalian Vertebrates; Vision: Vertebrates; Visual Signals.

Further Reading

Abrahams MV, Robb TL, and Hare JF (2005) Effect of hypoxia on opercular displays: Evidence for an honest signal? *Animal Behaviour* 70: 427–432.

Bando T (1991) Visual-perception of texture in aggressive-behavior of *Betta splendens*. *Journal of Comparative Physiology a-Sensory Neural and Behavioral Physiology* 169: 51–58.

Cantalupo C, Bisazza A, and Vallortigara G (1996) Lateralization of displays during aggressive and courtship behaviour in the Siamese fighting fish (*Betta splendens*). *Physiology and Behavior* 60: 249–252.

Clotfelter ED, Curren LJ, and Murphy CE (2006) Mate choice and spawning success in the fighting fish *Betta splendens*: The importance of body size, display behaviour and nest size. *Ethology* 112: 1170–1178.

Clotfelter ED, O'Hare EP, McNitt MM, Carpenter RE, and Summers CH (2007) Serotonin decreases aggression via 5-HT1A receptors in the fighting fish *Betta splendens*. *Pharmacology Biochemistry and Behavior* 87: 222–231.

Dzieweczynski TL, Eklund AC, and Rowland WJ (2006) Male 11-ketotestosterone levels change as a result of being watched in Siamese fighting fish, *Betta splendens*. *General and Comparative Endocrinology* 147: 184–189.

Elcoro M, Da Silva SP, and Lattal KA (2008) Visual reinforcement in the female Siamese fighting fish, *Betta splendens*. *Journal of the Experimental Analysis of Behavior* 90: 53–60.

Haller J (1991) Biochemical cost of a fight in fed and fasted *Betta splendens*. *Physiology and Behavior* 49: 79–82.

Halperin JRP, Giri T, Elliott J, and Dunham DW (1998) Consequences of hyper-aggressiveness in Siamese fighting fish: Cheaters seldom prospered. *Animal Behaviour* 55: 87–96.

Herb BM, Biron SA, and Kidd MR (2003) Courtship by subordinate male siamese fighting fish, *betta splendens*: Their response to eavesdropping and naive females. *Behaviour* 140: 71–78.

Jaroensutasinee M and Jaroensutasinee K (2003) Type of intruder and reproductive phase influence male territorial defence in wild-caught Siamese fighting fish. *Behavioural Processes* 64: 23–29.

Lynn SE, Egar JM, Walker BG, Sperry TS, and Ramenofsky M (2007) Fish on Prozac: A simple, noninvasive physiology laboratory investigating the mechanisms of aggressive behavior in *Betta splendens*. *Advances in Physiology Education* 31: 358–363.

Matos RJ, Peake TM, and McGregor PK (2003) Timing of presentation of an audience: Aggressive priming and audience effects in male displays of Siamese fighting fish (*Betta splendens*). *Behavioural Processes* 63: 53–61.

McDonald CG, Paul DH, and Hawryshyn CW (2004) Visual sensation of an ethological stimulus, the agonistic display of *Betta splendens*, revealed using multi-unit recordings from optic tectum. *Environmental Biology of Fishes* 70: 285–291.

Verbeek P, Iwamoto T, and Murakami N (2008) Variable stress-responsiveness in wild type and domesticated fighting fish. *Physiology and Behavior* 93: 83–88.

Beyond Fever: Comparative Perspectives on Sickness Behavior

B. L. Hart, University of California, Davis, CA, USA

Introduction

Owners of domestic animals, as well as veterinarians and wildlife biologists, are aware of behavioral changes that occur when animals become sick and have a fever. These signs, which include depression, inactivity, sleepiness, anorexia, increased threshold for thirst, and reduction of grooming, are often the first indications that an animal is sick. Humans are no different from animals in showing these signs as markers of a febrile illness. Admittedly, animals are nonverbal, while people may say to themselves or others, 'I feel sick and I am depressed.'

While in many illnesses there are organ-specific signs, such as nasal discharge for respiratory illness or diarrhea from intestinal illness, the suite of behavioral signs that accompany fever are nonspecific. These nonspecific signs are seen in vertebrates in general, including mammals, birds, and reptiles. They are seen in diseases that are fatal as well as those that are rarely fatal, and in a wide range of diseases caused by viruses, bacteria, and multicellular parasites.

The nonspecific behavioral changes associated with sickness have been around for as long as animals and humans have harbored sickness and disease, which is to say, throughout evolutionary history. When physicians and veterinarians, or their forerunners, started thinking and learning more about the causes and treatment of illnesses, there was naturally a tendency to view the behavior of sick animals and people as the result of debilitation and reduced ability to obtain food and water. Until quite recently, in the scientific disciplines of animal behavior, and indeed in human behavior, the behavior of sick individuals was outside the field of inquiry. Such behavior was not considered 'normal' or adaptive; it was not recognized as being a response that could be the result of natural selection. The breakthroughs of Kluger and others, in the late 1970s, in understanding the pathophysiology of the fever response, and the physiological linkage between fever and depression, inactivity, sleepiness, and anorexia, worked out by Hart in the 1980s, revealed that the nonspecific behavioral signs are adaptive and normal for an animal in a state of illness. In contrast to earlier ideas, the absence of these signs of sickness behavior during an illness could be viewed as abnormal.

Experimental studies of the detailed behavioral changes associated with sickness have typically involved laboratory rodents in which the bacterial endotoxin, lipopolysaccharide (LPS), can be used to induce transient fever and sickness behavior without actually making the animals sick. The widespread generality of the sickness behavior syndrome, however, is seen in studies of animals much different from rodents. For example, goats given LPS show anorexia along with a reduction in grooming.

The pervasiveness of sickness behavior throughout human history is poignant in any number of literary works from a century or two ago. A befitting verse, 'Fragment,' was penned in the mid-1800s by the famous poet of his time, Thomas Hood, after he suffered for years from sickness-inducing maladies:

I am sick of gruel, and the dietetics,
I am sick of pills, and sicker of emetics,
I am sick of pulses, tardiness or quickness,
I am sick of blood, its thinness or its thickness,
In short, within a word, I'm sick of sickness!

The purpose of this review is, first, to document the role of sickness behavior in facilitating the fever reaction in combating active infections, especially for animals living in nature, and then discuss the recently explored ramifications of the sickness behavior phenomenon in animal and human behavior. Hart introduced the concept of the adaptive functions of the behavior of sick animals, linking the signs of sickness behavior to the survival value of the fever response. Since then, the concept of sickness behavior has been explored, apart from its link to fever, in a number of contexts and species as an interesting phenomenon on its own with widespread implications for a more comprehensive understanding of animal and human behavior.

Adaptive Value of Sickness Behavior and Relation to Fever

At the acute onset of an illness, the signs of sickness – inactivity, sleepiness, depression, reduction in appetite, and increased threshold for thirst – comprise a rather highly organized behavioral strategy critical to the survival of an animal or a human in nature. In the age of modern medicine, where our view of disease is shaped by an orientation on the importance of antibiotics and supportive therapy, and protection through immunization, it is easy to overlook that animals in nature have been exposed to, and survived, infections with pathogenic organisms through millions of years of evolutionary history. When a wild animal becomes sick with a pathogen, it may be at a life or death juncture. The behavior associated with being sick can be viewed as facilitating the infection-fighting febrile response, putting virtually all the animal's available resources into overcoming the invading pathogen.

Fever, especially coupled with a reduction in blood levels of iron, has the effect of inhibiting the growth of many bacterial and viral pathogens in addition to activating the immunological system. But, fever is very costly, metabolically, and for the response to have persisted over evolutionary time there are, by necessity, potential benefits. It is the cost of the febrile response that has led to the selection of a cluster of behavioral patterns that induce the animal to stay in one place, possibly lie down and curl up, and conserve its body resources. These programmed behaviors potentiate the fever response by reducing energy expenditure and conserving body heat.

When febrile temperatures reach a certain point, they can be associated with tissue damage to the heart, liver, or brain. Thus, one might wonder how this response could be selected over evolutionary time. From the standpoint of a wild animal, a high fever that produces tissue damage, even in vital organs, may still prove adaptive if the fever is effective at combating an otherwise lethal infectious disease. Of course, in the presence of veterinarians or physicians, we do not have to let things go this far.

The fever response is the result of the body temperature increasing because the body thermoregulatory setpoint is raised, much like the increase in the temperature of a house when the thermostat is set higher. People and animals with an elevated thermoregulatory setpoint feel chilly at temperatures that previously felt normal and feel comfortable only when their body temperature is elevated. The body puts resources into conserving and producing heat until the new thermoregulatory setpoint is reached.

The thermoregulatory set-point is elevated through the action of endogenous pyrogens (EPs), which are released from fixed tissue macrophages, blood monocytes, and lymphocytes upon exposure to bacteria and bacterial toxins such as LPS. Nowadays, EPs are also commonly referred to as inflammatory cytokine proteins or cytokines. One of the EPs, interleukin-1, is involved not only in causing fever but also in inducing other fever-related responses such as lowering blood concentrations of iron and increasing the excretion of sodium, which reduces the thirst response.

The metabolic costs of fever are met by reducing heat loss and increasing heat production through increased metabolism. Heat loss reduction is achieved behaviorally by curling up, huddling, and seeking warm places and physiologically by shifting blood flow from the skin to the interior parts of the body. Piloerection, to increase insulating properties of the fur coat, is also helpful in reducing heat loss. Heat production is gained through muscle contraction (shivering and nonshivering), nonmuscular physiological changes, and biochemical activities.

The increase in metabolism needed per 1 °C of fever in humans is estimated at 13%. However, the extent to which an increase in metabolism is needed to produce each 1 °C of fever among various species of animals is difficult to say. Obviously, with differences in body size (resulting in different surface-to-body-mass ratios), hair coat density, and habitat (e.g., temperate climates vs. arctic areas), the energetic cost of fever production will differ among species. The more efficient the heat conservation mechanisms are, the less energy will be spent on maintaining fever by costly metabolic means. When the body temperature is raised by shivering, the metabolic rate is increased two to three times over that of the resting level. It is the high metabolic cost of the fever response, needed to inhibit pathogen growth, and the threat of higher costs if shivering must be activated, that can explain the adaptiveness of sickness behavior typically associated with fever.

Of the signs of sickness behavior, the occurrence of anorexia seems particularly paradoxical. Febrile animals need calories to fuel the needs of an elevated body temperature and to reduce the demand for muscle breakdown for caloric needs.

However, looking at the situation of the animal in the wild, one can see that it may take considerable effort to forage or hunt. Both anorexia and a reduced thirst relate to the notion that an animal that does not feel hungry or thirsty has little motivation to move about in search of food and water. If an animal stays in one spot, it engages in much less muscular activity and thus can save on body energy reserves needed for the increased metabolic costs of fever. Natural selection tells us that conserving energy, by staying put and reducing heat loss, is more beneficial than continuing to forage or hunt. Also, by not consuming food the animal reduces the chance of raising blood concentrations of iron from foodstuffs containing iron. Reflecting the inhibitory effect on pathogens of reduced iron concentration in the blood, the animal's body is sequestering away iron into the liver and spleen, and the influx of new iron into the bloodstream would be counterproductive.

The terms used to describe the demeanor of a sick animal depend upon the background of the person making the description, and probably on how the observer would expect to feel as a sick animal. The term depression is often used to describe sick animals although this is not easily measured. Increased sleepiness, or a tendency to sleep during normal periods of wakefulness, is easily measured and corresponds with clinical observations and subjective reports of depression from people during febrile illnesses. There is experimental work showing that interleukin-1 induces excessive sleep in animals and humans. Increased sleepiness is complementary to anorexia in inducing the individual to remain in a heat conservation mode.

Although enhancement of sleep is undoubtedly important in recovery from the effect of an infectious disease, the signs of sleepiness in acute sick animals are particularly evident early in the disease process before tissue damage is brought about by the pathogen and before the body energy resources are depleted. Supporting these clinical observations, experimental studies with injections of interleukin-1 or LPS reveal an onset of excessive sleep within 1 h, which is when the rise in body temperature is evident. The curled-up position often seen in sick rodents and carnivores that are sleeping is more important for heat conservation in these smaller animals than in larger animals since the ratio of body surface area to body mass is proportionally greater in small mammals.

Animals that have been sick for several days often have a scruffy, dirty, and oily-looking hair coat, which is undoubtedly due to a marked reduction in grooming. Increased sleep in sick animals may account for part of the reduction in grooming, but there are reasons why an animal would actively inhibit grooming under the demands of a fever reaction, while letting the insulating benefit of conditioning the hair coat slip. For one thing, the body movement involved in grooming would result in greater heat loss from more exposure of skin surface and energy expenditure for the muscular activity involved in grooming.

For another thing, with oral grooming there is a loss of water through saliva used in licking. In rodents, oral grooming typically includes a sequence of grooming patterns starting with licking the paws and forelegs, interspersed with face washing, continuing onto other parts of the body. Small felids, such as domestic cats, also engage in extensive head-to-posterior grooming. With such extensive licking from head to tail, it is not surprising that calculations done with rats have determined that the water lost through grooming is comparable to the amount of water lost through urine. Rats deprived of water for 24 h can show a 50% decrease in oral grooming. Thus, for a febrile animal, with restricted access to water and an increased threshold to thirst, reduced oral grooming will conserve water. It appears as though the reduction in grooming at the time of sickness is a result of the effects of interleukin-1 or other EPs, independent of increased sleepiness and reduction in activity.

A dirty hair coat may not be the only cost or sign of reduction in grooming. Species as diverse as rodents, felids, and ungulates in which grooming is inhibited, show an increase in ectoparasites, such as fleas, lice, and ticks. Evidently, the immediate gains in water and energy conservation by cessation of grooming exceed the costs of a delayed, but increasing, parasite load. During the recovery phase, the parasite load can be reduced by renewed grooming activity.

As mentioned, the fever response, accompanied by anorexia, inactivity, reduction of grooming, and depression, puts much of an animal's resources into recovering from the disease. The general perspective is that for an animal unfortunate enough to be infected by microorganisms associated with a deadly disease, taking on the syndrome of sickness behavior represents the best strategy for surviving and later reproducing. Recent research, however, has revealed exceptions to the principle that an animal always goes into a full-on sickness behavior mode when confronted with an illness-inducing pathogen. These exceptions are explored in the section that follows.

To recapitulate this section, the behavior associated with being sick can be viewed as representing an all-out effort to overcome the infectious disease, putting the animal's resources into fending off the invading pathogens. The behavioral patterns that we commonly see in animals and people that are acutely ill, including complete or partial anorexia, sleepiness, depression, lack of interest in drinking water, and reduction of grooming activity, can be viewed as potentiating the energy-intensive fever response by conserving body heat and reducing energy expenditure through all behavioral means possible.

Figure 1 summarizes the stages of sickness that an animal typically goes through when it is infected with a pathogen. In this illustration, a lioness becomes infected by eating the leg of an antelope that was contaminated with a pathogen by being dragged around. The first stage is when infection occurs and the animal shows no visible signs of sickness as the pathogen multiplies in the body (**Figure 1(a)**). During this stage, macrophages and monocytes respond to the pathogen, and start to secrete inflammatory cytokines that bring on the second stage of sickness.

In the sickness stage (**Figure 1(b)**), the cytokines cause a resetting of the body thermostat, or thermoregulatory setpoint, causing fever. The cytokines also bring on a lowering of blood iron level, a loss or reduction of appetite, referred to as anorexia, an increase in sleepiness, and a reduction of activity. Here you have the classic image of sickness behavior, and in this pride of lionesses, the sick one naturally feels depressed and sick and sleeps almost continuously, conserving body temperature even while the rest of the pride is up and ready to go after the evening meal. The increase in body temperature associated with the fever, along with a lower level of blood iron, inhibits the growth of the pathogen, allowing the immune system lead time to get activated. The cytokines even give an extra kick-start to the immune system. The activated immune system starts killing off the pathogen while the sick lioness sleeps away.

The animal has now moved into the third stage, recovery (**Figure 1(c)**). The immune system is now fully in control removing the pathogen from the body. The revved-up cytokine activity stops, and the animal quickly feels better because the cytokines are no longer causing depression, sleepiness, and loss of appetite. The previously sick lioness is now quite hungry and is ready to join the others in the hunt.

Although the fever response and accompanying behavioral changes tend to be similar across different diseases,

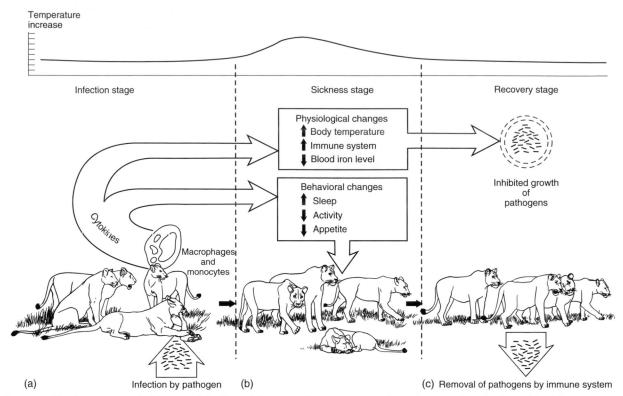

Figure 1   The three stages of sickness – infection, sickness, and recovery – are shown in a member of a lion pride. A lioness gets infected with a pathogen as she chews away at an antelope leg that has been dragged around, resulting in contamination of the leg by a local pathogen (a). Loss of appetite and a drive to sleep almost continuously characterize the sick-looking, infected lioness in the sickness stage (b). As the immune system gets rid of the pathogen, and the cytokines are no longer secreted, in the recovery stage, the lioness regains her appetite and is ready to join the prided in hunting for the evening meal (c). See the text for a full description of the events indicated in the three stages.

there is some variability. One disease may be characterized by a dramatic and acute onset of fever along with a sudden and complete anorexia and depression. In other diseases, anorexia and depression may come about much more slowly or be only partially manifest. Some diseases are characterized by bouts of febrile changes interspersed with more normal temperatures. The variability in behavior and febrile patterns may be due to differences in the disease-causing organisms. Differences in the immune systems among animals may also account for differences in their behavior when they are acutely ill. In addition, of course, many diseases are accompanied by gastric distress, vomiting, diarrhea, and/or coughing that are reflected in the animal's behavior while it is sick.

Adaptive Modification of Sickness Behavior

The ubiquitous nature and rather low threshold for the occurrence of the essential elements of sickness behavior undoubtedly stem from the adaptive value of the behavioral mode for animals and humans living in an environment

where life-threatening diseases can be around any corner. This said, it is not surprising to find instances in which the sickness behavioral mode is typically attenuated even though the animal has a febrile illness, reflecting fitness reasons that supersede the benefits of disease fighting.

One situation where there is suppression or dampening of sickness behavior in association with a febrile illness is in prey species where animals are found in herds. A predator may quickly notice an individual showing some aspects of sickness behavior and target that individual; a full manifestation of sickness-induced inactivity or lying down, may bring on immediate fatal consequences. As a countermeasure, some prey species, such as antelope, may inhibit, or cover up, the manifestations of anorexia and inactivity of sickness behavior.

For animals that collect food for future consumption, so-called food hoarding, one might expect that both hoarding and foraging for food for current use would be suppressed in animals with a febrile illness. However, appetite suppression for a few days while an animal fights a transient illness is different from suppression of food collection for the future that could cost the animal dearly over a long winter where the animal relies on stored food.

Thus, in a typical hoarder, such as the Siberian hamster, injections of LPS do not reduce food hoarding behavior even though immediate food intake is suppressed.

Seasonal effects can also alter the illness-induced suppression of eating. The long days of summer allow an animal some flexibility in missing a few feeding days while combating an illness; body condition can easily be regained once the animal is recuperating. In the short days of winter, however, foraging time is restricted and the quality or availability of food items may be much poorer than in summer. Here, a complete appetite suppression associated with a febrile illness may be too costly. Along this line, injection of LPS into Siberian hamsters housed under short-day, winter-like conditions does not suppress appetite in the manner seen in long-days typical of summer.

A modification of the sickness behavior accompaniment to a febrile illness may be evident, not surprisingly, in sexual responses. For females, devotion of resources to bearing and rearing young is paramount for lifetime inclusive fitness. But for males, fertilizing as many females as possible is the ticket to lifetime fitness. Studies using the universal sickness behavior inducer, LPS, have revealed the following picture. In nonpregnant, nonlactating female rats, a febrile illness brings on the full-blown suite of sickness behaviors; these not only help her survive the illness, but also take her out of circulation as a breeding partner, thus avoiding the added burden of pregnancy, while waiting for a better time to get pregnant. However, sexually active males with a febrile illness, but with breeding opportunities, do not show the activity-reducing aspects of sickness behavior but maintain their sex drive in the presence of receptive females; waiting for a better day may prove too late.

Once females have a litter of young, the picture changes with regard to the induction of sickness behavior in association with a febrile illness because of the importance of maternal care to the survival of dependent offspring. Using LPS to induce a transient infection-like fever in mother rats, it was shown that aggressive defense of the nest against intruders was not altered even though there was a loss of body weight and hair coat condition, reflecting cessation of eating and grooming.

Finally, there is one classical situation for animals in nature where the sickness behavior phenomenon is brought into play without a febrile illness. Such a situation is with accidental isolation of young animals in a novel environment. Typically the young animals initially go into an active phase where they try to escape; this phase is then followed by a passive phase, usually referred to as 'despair,' in which one sees the full suite of behaviors characteristic of sickness, including depression, anorexia, inactivity, and sleepiness. The activation of sickness behavior in this situation is adaptive in that it results in the isolated young laying low and conserving body resources until the young animal is found by the mother.

The Unwelcome Occurrence of Sickness Behavior in Humans with Chronic Illness

The close parallels between the sickness behavior that accompanies febrile illnesses in animals and humans were discussed in the first section; in humans, the suite of responses that comprise sickness behavior are manifested as depression, anorexia, listlessness, fatigue, malaise, sleepiness, and loss of interest in social interactions. Ancient humans, of course, got sick from time to time with infectious pathogens and the fever response, with the associated sickness behavior, served them well in recovery. The role of inflammatory cytokines in coordinating the fever response and elements of sickness behavior evolved long before chronic diseases, such as cancer, cardiovascular deterioration, and multiple sclerosis, were important players in human health. Dying from an infectious disease was the important variable in the evolution of sickness behavior.

However, nowadays, inflammatory cytokines are activated during chronic disease syndromes, but instead of facilitating recovery, the cytokine activation is counterproductive. Research over the past couple of decades points to numerous instances where inflammation is harmful; heart disease, Alzheimer's disease, and arthritis are examples. In this section, we touch upon some examples directly linking activation of the sickness behavior syndrome to some human health concerns.

The best way to set the stage of this section is to recall that while fever generally does not occur without sickness behavior, the sickness behavior syndrome can be activated without a febrile illness. This is evident in humans with some chronic disease syndromes in which the activation of depression goes beyond the debilitating aspects of the disease. One such nonfebrile, chronic disease in humans that activates the general sickness behavior syndrome, particularly depression, is in multiple sclerosis. The depression can exacerbate the disease, making life worse for the patient beyond nerve-damage-related effects. Physicians acknowledge that if the sickness behavior–depression syndrome could be controlled, then the disease outcome could be improved.

Another set of nonfebrile chronic diseases of concern is cancers of various sorts. Here, as in multiple sclerosis, activation of sickness behavior with loss of appetite and depression makes life and disease outcomes worse for the patient. Especially noteworthy in dealing with cancer is that several chemotherapeutic drugs activate sickness-related inflammatory cytokines, leading to depression, fatigue, anorexia, changes in sleep patterns, and/or loss of interest in social activities. Understanding the mechanisms by which the drugs activate the cytokines, as well as how this could be controlled, could be very important to patients being treated and could perhaps reduce the necessity of sometimes having to reduce or terminate chemotherapy.

Recent developments in the understanding of clinical depression have revealed that some individuals predisposed to depression have unusually high levels of the cytokines related to sickness behavior. This finding has implications, which are being explored by psychiatrists. One area involves looking at the way that different antidepressants used for treating clinical depression may act on the depression-inducing cytokines. It appears as though some, but not all, antidepressant drugs may alleviate depression by virtue of inhibiting the cytokines that cause depression. Finding the right antidepressant medication that inhibits the inflammatory cytokines can be a key to successful treatment.

Another area of research involves understanding the relationship between the immune system and depression because activation of the immune system increases the production of the inflammatory cytokines that induce sickness behavior. Treatment of individuals with intrinsically high levels of such cytokines and that may be predisposed to depression, with immune-boosting drugs, may precipitate a dangerous bout of clinical depression. Thus, physicians may have to evaluate the level of cytokines in such patients before administering an immune-boosting drug.

In conclusion, it is apparent that what started out in the mid-1980s as an insight into an unappreciated aspect of the behavior of animals when they get sick is now being recognized as a pervasive influence in many areas of animal behavior that are still under investigation. Understanding the relevance of the depression component of sickness behavior to human clinical medicine is now emerging as an important goal. Clearly, the syndrome of sickness behavior will be increasingly studied in research laboratories ranging from those dedicated to animal behavior to human clinical medicine.

Acknowledgments

Preparation of this chapter was supported in part by allocation #03-65-F from the Center for Companion Animal Health, University of California, Davis.

See also: Avoidance of Parasites; Conservation, Behavior, Parasites and Invasive Species; Ectoparasite Behavior; Evolution of Parasite-Induced Behavioral Alterations; Intermediate Host Behavior; Parasite-Induced Behavioral Change: Mechanisms; Parasite-Modified Vector Behavior; Parasites and Sexual Selection; Propagule Behavior and Parasite Transmission; Reproductive Behavior and Parasites: Invertebrates; Reproductive Behavior and Parasites: Vertebrates; Self-Medication: Passive Prevention and Active Treatment; Social Behavior and Parasites.

Further Reading

Hart BL (1985) Animal behavior and the fever response: Theoretical considerations. *Journal of the American Veterinary Medical Association* 187: 998–1001.

Hart BL (1988) Biological basis of the behavior of sick animals. *Neuroscience and Biobehavioral Reviews* 12: 123–137.

Hood T (1844) Fragment. In: *The Works of Thomas Hood* Vol. X, p. 51. London: E. Moxin, Son, & Co.

Kluger MJ (1978) The evolution and adaptive value of fever. *American Scientist* 66: 38–43.

Maxem A (2008) Sick and down. *Science News* 174 (available online).

Raison CL, Capuron L, and Miller AH (2006) Cytokines sing the blues: Inflammation and the pathogenesis of depression. *Trends in Immunology* 27: 24–31.

Tizard I (2008) Sickness behavior, its mechanisms and significance. *Animal Health Research Reviews* 9: 87–99.

Bird Migration

S. A. Gauthreaux Jr, Clemson University, Clemson, SC, USA

Introduction

Since the beginning of recorded history, the migration of birds has attracted the attention and intrigued the imagination of humans. Not surprisingly, the subject has garnered considerable interest from biologists who have sought answers to 'how' and 'why,' and their studies have produced a staggering amount of data and published findings on all aspects of bird migration. Many of the studies have been summarized in books devoted to the subject, and several of the more recent volumes can be found in the suggested readings at the end of this article. Because of the volume of literature on bird migration, even the most comprehensive reviews have been able to provide little more than a sketchy overview of the subject. This article reviews some of the findings about the behavioral aspects of bird migration: the flight directions and routes, the seasonal and daily timing of movements, the altitude of migratory flights, and the influence of weather on the density of migration.

Migration is a behavioral response to adversity that takes migrants from geographical areas of low suitability with respect to food availability, habitat, and climatic conditions to locations of higher suitability. Migration in birds is a continuum from *partial* migration where only some individuals of a population migrate to *complete* migration where all populations leave the breeding range of the species and move in some cases considerable distances to occupy a nonbreeding range of the species. Because of the annual climatic cycle, the suitability of temperate latitudes shows the greatest seasonal variability, and many migratory birds exploit the abundance of resources during the warmer months and breed and vacate these latitudes during the colder months when resources decline and disappear. Some species move relatively short distances and remain at temperate latitudes during the nonbreeding season (short-distance migrants), while others move to tropical latitudes and beyond (long-distance migrants). For western Europe, the proportion of breeding species that vacate and migrate farther south increases 1.3% on average for every degree of latitude. In eastern North America, the mean increase is 1.4% per degree of latitude. For the same latitude, about 17% more breeding species vacate eastern North America than Western Europe, because winters are colder in eastern North America. In the tropics, similar movement and breeding patterns occur in response to seasonal precipitation patterns. According to Peter Berthold, over 50 billion birds show

some form of seasonal migration between breeding and nonbreeding areas.

Migratory Routes

The longitudinal separation of species and populations of migratory land birds that characterize the breeding distribution often persists during the migration phase and the nonbreeding season. At a continent-wide scale, the reasons for these migration patterns are varied and likely relate to the locations of breeding and nonbreeding areas, major topographical features, availability of suitable resources on the migration route, peculiarities of life history, and prevailing directions of winds during the migration seasons. Among North America wood warblers (Parulidae) that migrate to the Neotropics, 40 species occur east of the Rocky Mountains, and 15 species are found west of these mountains. The western species winter almost entirely within a narrow strip of western Mexico from Sonora south through Guatemala. The eastern species generally winter in geographically separate areas of the Bahamas, West Indies, eastern Mexico, Central America, and northern South America. This pattern is found in other species as well. Western Tanagers (*Piranga ludoviciana*) breed west of the Great Plains and migrate to southern Mexico and the Pacific slope of Central America, and Scarlet Tanagers (*Piranga olivacea*) breed in the eastern United States and migrate to northwestern South America (**Figure 1**).

There is in general considerably more bird migration in the eastern two-thirds of the United States than in the western one-third. One reason for this pattern is that more migrants (species and individuals) breed in the East, but another contribution comes from species with breeding ranges that extend toward the northwest into Canada that migrate through the eastern United States. Although the breeding range of the Rose-breasted Grosbeak (*Pheucticus ludovicianus*) extends from the Atlantic Coast to the Northwest Territories of Canada, the migration of this species is primarily through the eastern half of the United States (**Figure 2**). The Swainson's Thrush (*Catharus ustulatus*) of North America is another species that shares this pattern. The two subspecies that breed in Pacific coastal woodlands west of the British Columbian Coastal Mountains, the Cascades, and the Sierra Nevada migrate southward along the coast to nonbreeding areas from Mexico to Costa Rica, but the two continental subspecies that breed east of these mountains migrate eastward before turning

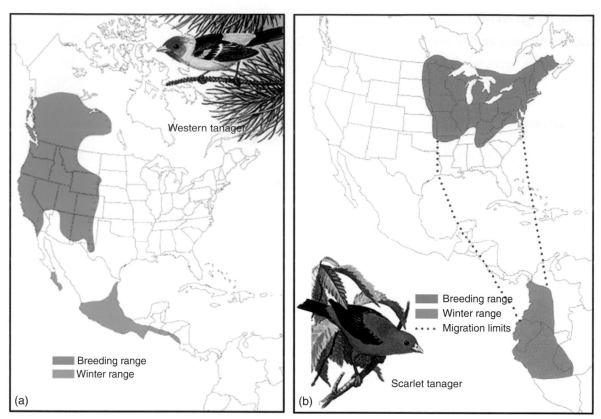

Figure 1   Breeding and nonbreeding ranges of two Neotropical migrants. (a) Western Tanager. (b) Scarlet Tanager (dotted red line delimits boundaries of migration route). Reproduced from Lincoln FC, Peterson SR, and Zimmerman JL (1998) *Migration of Birds*. Washington, DC: US Department of the Interior, US Fish and Wildlife Service. Circular 16. Jamestown, ND: Northern Prairie Wildlife Research Center Online. http://www.npwrc.usgs.gov/resource/birds/migratio/index.htm (Version 02APR2002).

southward to winter from Panama to Bolivia. Work with molecular genetic markers suggests that the breeding range of these subspecies expanded rapidly following glacial retreat. The east-then-south fall migration route of the continental subspecies is in keeping with birds returning to nonbreeding areas by 'backtracking' along their ancestral route of breeding range expansion. Approximately 33 species of land bird migrants in North America conform to this pattern. Thus, even though a number of land bird migrants breed considerably farther west and north of the eastern forests of the United States, they migrate through the eastern states.

Migration routes, timing, and duration obviously differ between different species, and in some cases, different populations within a species (e.g., the Swainson's Thrush). Individuals within a population of a species may differ with respect to these attributes not only because of age and sex differences (differential migration) but even within a group of individuals of the same sex and age. Although field observations and bird banding (ringing) and recovery of banded birds have produced a wealth of knowledge about migration routes of birds, the use of modern technology (satellite telemetry and geolocators)

has produced detailed, high-quality information about the route and timing of migrating individuals. Because small birds cannot carry satellite transmitters, most of the information gathered with satellite transmitter technology to date has come from larger birds.

Satellite telemetry has been used to study the fall migration patterns of the Osprey (*Pandion haliaetus*) in North America and Europe. For North America, the routes of migration differ among populations from the northwestern, central, and eastern United States (**Figure 3**). Birds from the northwestern population migrate mostly through California and winter mainly in Mexico, with some birds moving to El Salvador and Honduras. Birds from the central population follow three different routes to reach their nonbreeding areas in Mexico and locations south to Bolivia: (1) through the central United States and then along the eastern coast of Mexico, (2) along the Mississippi River Valley and then across the Gulf of Mexico and (3) through the southeastern United States and then across the Caribbean. Eastern populations migrate along the Atlantic Coast through Florida and across the Caribbean, and eastern birds wintered from Florida to Brazil. Departure dates differ among populations and females departed

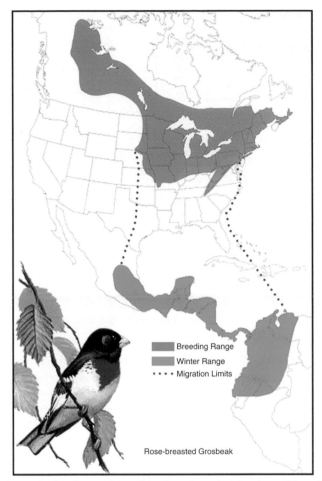

Figure 2 The breeding distribution and migration corridor of the Rose-breasted Grosbeak (*Pheucticus ludovicianus*). Although the breeding range of this species extends into northwestern Canada, its migration through much of the United States is confined to the eastern half of the country. Reproduced from Lincoln FC, Peterson SR, and Zimmerman JL (1998) *Migration of Birds*. Washington, DC: US Department of the Interior, US Fish and Wildlife Service. Circular 16. Jamestown, ND: Northern Prairie Wildlife Research Center Online. http://www.npwrc.usgs.gov/resource/birds/migratio/index.htm (Version 02APR2002).

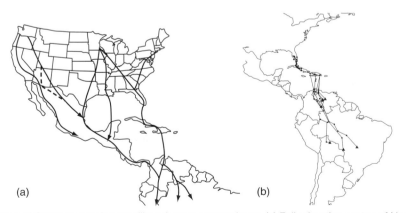

(a)   (b)

Figure 3 Migration routes of Ospreys carrying satellite telemetry transmitters. (a) Fall migration routes of North American nesting Ospreys. Dashed lines indicate movements of only one bird. Reproduced from Martell MS, Henny CJ, Nye PE, and Solensky MJ (2001) Fall migration routes, timing, and wintering sites of North American Ospreys as determined by satellite telemetry. *Condor* 103: 715–724. (b) Migratory routes of Florida breeding Ospreys. The circles indicate locations during migration, and triangles indicate nonbreeding areas. Reproduced with permission from Martell MS, McMillian MA, Solensky MJ, and Mealey BK (2004) Partial migration and wintering use of Florida by Ospreys. *The Journal of Raptor Research* 38: 55–61.

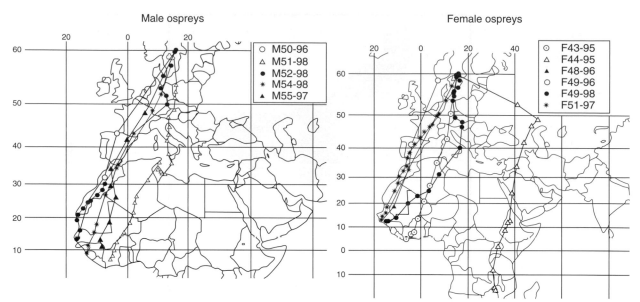

Figure 4 Maps of the migration routes of Ospreys tracked by satellite from the breeding site in Sweden to nonbreeding areas in Africa. (a) Five males. (b) Six females. Reproduced with permission from Hake et al. (2001).

from breeding areas before males. Females from the same population travel greater distances than males and have nonbreeding areas south of male nonbreeding areas (differential migration).

Ospreys equipped with satellite transmitters from several European countries have been tracked on autumn migration. Females start migration 2–3 weeks before males and young. Birds from the United Kingdom cross the English Channel pass through France and Spain into northern Africa and then some follow the coastline of western Africa and others cross the Sahara Desert. Most of the United Kingdom birds have nonbreeding areas in western Africa (e.g., Senegal). Birds tracked from Sweden moved from breeding areas through Spain, Italy, and the central Mediterranean, and a single female followed a more easterly route through the Red Sea (**Figure 4**). Most of the birds moved to nonbreeding areas in West Africa south of the Sahara (Gambia to the Ivory Coast), but one juvenile went to Cameroon and the female that followed the easterly route went to Mozambique – a 10 000-km journey. The average distance traveled between Sweden and nonbreeding areas was 6742 km (average of 174 km day^{-1} with stopovers included). When actively migrating, birds averaged 257 km day^{-1} with males moving fastest. Adults followed very straight routes toward a mean direction of 185–209° with angular deviation of only 6–33°. Young showed more variation in departure direction and orientation. Although most of the flights of Swedish birds occurred during the daylight hours (08.00–17.00 h) when the birds could exploit atmospheric thermals to minimize their cost of transport, there is evidence from other studies that some Ospreys migrate at night particularly over large bodies of water.

Robert E. Gill Jr., a biologist with the US Geological Survey, and colleagues used satellite telemetry to track the autumn migration of nine bar-tailed godwits (*Limosa lapponica baueri*) from breeding areas in western Alaska to nonbreeding areas in New Zealand. The birds departed Alaska between 30 August and 7 October on initial tracks strongly oriented in a southerly direction (193°) and crossed the entire Pacific Ocean within a 1800-km wide corridor. Tracking distances from Alaska to the first known landfall for seven females with surgically implanted PTTs ranged from 8117 to 11 680 km (10 153 ± 1043 km standard deviation), and for the two males with externally attached PTTs the distances ranged from 7008 to 7390 km. Duration of the flights for the females with implants ranged from 6.0 to 9.4 days (7.8 ± 1.3 days standard deviation) and for the two males with externally attached PTTs the flight durations were 5.0 and 6.6 days. Track speeds for the nine godwits on southward flights averaged 16.7 ± 0.6 m s^{-1} with a range of 8.7–25.5 m s^{-1}. These nonstop flights establish a new record for flight performance in migratory birds and have important implications for navigation and the physiological mechanisms that regulate migratory flight.

Because small birds cannot carry satellite transmitters, another technology has been used to monitor the migration of smaller birds. Engineers of the British Antarctic Survey developed a very small and low-weight (1.3–1.5 g) solar geolocator that records the change in light levels (time and intensity of sunlight) at different latitudes and longitudes and enables researchers to record the flight paths of smaller migrating birds. Because the device uses very little power and the recorded data are compressed, the device can record data for many years; however, the bird must be recaptured to download the recorded data. Scott Shaffer of the University of California at Santa Cruz and colleagues used miniature solar geolocators to track the migrations of Sooty Shearwaters (*Puffinus griseus*) and found that the birds follow a figure 'eight' pattern over the Pacific Ocean and cover as much as $64\,037 \pm 9779$ km on their annual migratory circuit of 262 ± 23 days. Birds traveled at rates as high as 910 ± 186 km day^{-1}, and each shearwater had a prolonged stopover in one of three areas off Japan, Alaska, or California before returning to New Zealand via a narrow pathway through the central Pacific Ocean.

Bridget Stutchbury of York University in Toronto and colleagues used solar geolocators to track the migration of 14 Wood Thrushes (*Hylocichla mustelina*) and 20 Purple Martins (*Progne subis*) from their breeding areas in Pennsylvania in the fall of 2007 (**Figure 5**). The following summer, the geolocators were retrieved from five Wood Thrushes and two Purple Martins, and the data were downloaded to examine migration routes and nonbreeding areas with an accuracy of ± 300 km. Both species showed rapid long-distance movements and periods of stopover. The two Purple Martins covered 2500 km to

reach the Yucatan Peninsula in 5 days (500 km day^{-1}), stayed in this area for 3–4 weeks, and then continued their migration to Brazil. Four Wood Thrushes stayed 3–4 weeks in the southeastern United States and then flew across the Gulf of Mexico to Yucatan Peninsula where two of the birds remained for 2–4 weeks before continuing migration to Honduras and Nicaragua. The overall rate of the return migration to the breeding areas in spring was 2–6 times faster than the southward migration in fall, and a Wood Thrush that flew around the Gulf of Mexico instead of across the Gulf took significantly longer to complete its spring migration (29 days *vs.* 13–15 days).

During discussions of migratory routes, the topic of migratory flyways often surfaces. Boere and Stroud have examined the flyway concept and suggest that although it is useful for water birds, the concept simplifies information about differences between-species and within-species with respect to migratory routes, and they recommend 'caution in applying the flyway concept to other migratory birds, given that ringing recoveries of passerines indicate widespread broad front migration across continental land-masses.'

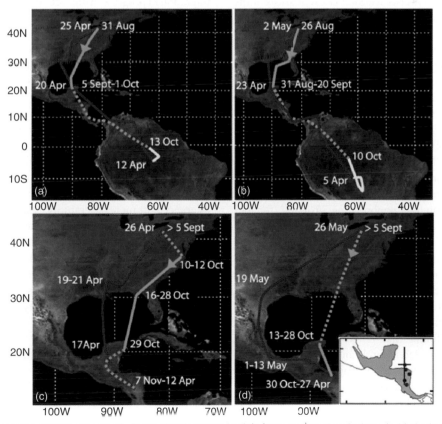

Figure 5 Migration tracks of birds breeding in Pennsylvania based on data from geolocation devices. (a, b) single purple martins; (c, d) single wood thrushes. Fall migration (blue), movement in nonbreeding area (yellow), and spring migration (red). Dotted lines indicate latitude could not be determined. Inset arrows locations of nonbreeding territories within species nonbreeding range (shaded) with the standard deviation for individual. Reproduced with permission from Stutchbury BJM, Tarof SA, Done T, et al. (2009) Tracking long-distance songbird migration by using geolocators. *Science* 323: 896.

Seasonal Timing

Much of what we know about the seasonal timing of bird migration comes from the work of field observers and bird banders, and their findings have been summarized in seasonal occurrence charts in state checklist and regional and national bird books. Migration schedules appear to be either 'hard-wired' in species that are considered calendar migrants or 'facultative' and flexible in species that are considered weather migrants. Calendar migrants begin migration irrespective of local conditions, whereas weather migrants may delay fall migration as long as local conditions (e.g., food availability, temperatures) remain favorable in the breeding areas. In spring, variation in weather conditions can greatly affect the timing of spring migration in some species. In an analysis of the variation in the timing of spring arrivals among 50 different species in comparison with the mean 40-year arrival dates, it was found that in late, cold springs migrants arrived later than in early, warm springs. Cold springs often have prolonged periods of cold, adverse winds from the north, while warm springs have prolonged periods of warm southerly winds. Thus, the advancement or retardation of the seasonal timing of migration can vary with year-to-year changes in continental wind patterns. Mathematical analysis of the timing of spring and fall migration has shown that those species that move north early return to southern latitudes late (e.g., waterfowl, sparrows). Breeding birds occupy their summer habitat as soon as it is habitable and depart as soon as they have finished breeding. The standard deviation of the timing of a species' migration is less in spring than in fall; hence, the birds are better synchronized in spring. During fall migration, some species show an almost bimodal timing with young and adults traveling at somewhat different times (differential migration). In spring, males of most species arrive before the females, and adults precede young. In waterfowl where pair bonds form on the nonbreeding grounds, females and males migrate in pairs and females show philopatry and site fidelity.

A number of factors must be considered when examining the seasonal timing of migration. The more important of these are vegetation development in the spring, food availability, and climatic factors in spring and fall. In a 18 year study of spring arrivals of migrants in Montana, W. Weydemeyer found that ranges in dates of arrival were greatest during late March and April and least in late May and June. In Norway, T. Slagsvold found that for the country as a whole there was a 6-day delay in bird arrival for each 10-day delay in vegetation development. Thus, the arrival of migrants at higher latitudes and elevations was faster than the development of vegetation. Earlier-arriving species varied considerably in arrival date at a particular locality from year to year, but late-arriving species had much less variation in arrival time.

In southern Michigan over a 7-year period, B.C. Pinkowski and R.A. Bajorek examined the spring arrival dates of 29 common or conspicuous migrants and summer resident species. They concluded that granivorous, omnivorous, and aquatic species tend to arrive earlier than strictly insectivorous species, and that earlier-arriving species have a greater variance in arrival time than the later-arriving species.

Daily Timing of Migration

The majority of small birds, including most passerines, migrate at night, and most waterfowl and shorebirds migrate both at night and during the day. Raptors, several woodpeckers, swallows, several corvids (crows, jays, and magpies), bluebirds, and blackbirds migrate during daylight hours. The determination of whether a species migrates at night or during the day has come from laboratory studies of *Zugunruhe* – migratory restlessness in caged birds; from data gathered when migrating birds collide with transmission and communication towers, buildings, or power lines or when migrants are attracted to, and killed at, lighthouses and ceilometers; and from direct visual studies of daytime migration in progress. Although some waterfowl and waders flock at night and during the day when migrating, the majority of songbirds migrate individually during the hours of darkness. When individually migrating nocturnal migrants are forced to fly during daylight hours (as when they are crossing the Gulf of Mexico), they form species-specific flocks just before dawn.

According to data gathered by surveillance radars at several localities in the United States and Canada, considerably more birds migrate at night than during the day. A number of studies have shown the temporal pattern of nocturnal migration. The initiation of nocturnal migration occurs about 30–45 min after sunset; the number of migrants aloft increases rapidly, peaking around 23.00 h in the spring and between 21.00 and 23.00 h in the fall. Thereafter, the number of migrants aloft decreases steadily until dawn, indicating that many migrants are landing at night. Daytime migration is initiated near dawn (sometimes earlier), peaks around 10.00 h, and declines to minimal density shortly after noon.

Directions of Bird Migration

Although considerable attention has been directed to laboratory studies of direction finding in migratory birds, there is an increasing emphasis on field studies of migratory orientation using direct visual means, radar, and satellite telemetry. Radar can provide detailed information on the direction of migratory movements when conditions for direct visual studies are poor and sample a

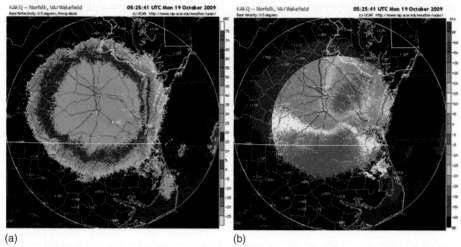

(a)   (b)

Figure 6 Displays of bird migration on Doppler weather radar located at Wakefield, Virginia, on 19 October 2009 at 05.25 UTC. (a) Relative reflectivity (dBZ) indicates the density of birds in the atmosphere. Note greater density at lower altitudes and reduced density as the 1° radar beam tilted 0.5° samples higher and higher altitudes. The edge of the pattern is where the radar beam is completely above the layer of migrating birds. Also note the return from some rainfall to the SE and E sectors of the display. (b) Radial velocity (measured in nautical miles per hour or knots) is the proportion of the velocity of birds directed toward the radar. Birds flying perpendicular to the radar beam have zero radial velocity (gray in image), while birds flying directly toward or away from the radar have maximum radial velocity (true ground speed). Negative velocities are for targets approaching the radar (cool colors) and positive velocities are for targets moving away from the radar (warm colors). Ambiguous velocities result from range folding (purple). Velocity corrected for wind direction and speed can be used to discriminate the types of targets (e.g., insects, birds).

fairly large geographical area. In the United States and in Europe, Doppler weather radars are operated at weather stations and many medium-sized and large airports. These radars readily detect migrating birds and in addition to estimating the density of birds aloft, information on their velocity and direction of movement can be obtained from the radial velocity of the targets (**Figure 6**). Smaller marine radars and tracking radars can be used to study bird movements in the atmosphere within 12–14 km of the radar, and these units are particularly useful for gathering detailed information on the flight direction, speed, and altitude of individual birds and flocks.

Although there are many geographical gaps in the coverage, a continental pattern of bird migration in North America is beginning to emerge. The axis of migration for most passerines is northeast to southwest in the eastern two-third of the United States, but in central southern Canada the axis of passerine migration is northwest to southeast. Wind direction exerts a strong influence on the direction and timing of migration, and the routes birds follow appear to be determined, at least in part, by the prevailing wind patterns in North America during spring and fall. Many species of shorebirds that migrate northward from nonbreeding grounds in Central and South America cross the Gulf of Mexico and arrive on the upper Texas coast and the coast of southwestern Louisiana. They continue their migration through the grasslands of the central United States to reach their breeding grounds in the Arctic. In the fall, the migratory routes southward are shifted considerably eastward as the

birds make their way toward their nonbreeding grounds. It is thought that the prevailing winds aloft from the east at lower latitudes bias the northward flights toward the west and the prevailing winds aloft from the west at higher latitudes bias fall flights toward the east.

In northwestern South Carolina, in spring, the prevailing winds blow to the northeast, and the average distribution of the directions of nocturnal migration in spring on calm nights (when wind directions are not an influencing factor) is toward the northeast (29.5°). Thus, in spring, the preferred direction of nocturnal migrants closely matches the prevailing wind direction. In fall, the winds in the same area usually blow toward the southwest, and the average direction of nocturnal migration on calm nights is toward the southwest (231.5°). These data were gathered using moon watching and ceilometer watching, but data gathered from radar show similar patterns. Wind direction in relation to the normal direction of migration can also influence the altitude of migration as well as the number of migrants aloft.

Altitude of Migration

Radar has provided the best data on the altitude of bird migration, and radar studies have shown that the altitude of bird migration varies considerably depending on the winds aloft. Most migration occurs below 1.5 km above ground level (agl) and density decreases up to 3 km or more. Radar studies in North America and Europe show that migration occurs at higher altitudes in spring than in fall, and except for certain geographical areas (e.g., the northern coast of the Gulf of Mexico) migration at night is at higher altitudes than migration during the day. In general, the larger the bird species and the faster its airspeed, the higher it flies during migration for minimum cost of transport. Bar-headed Geese have been seen and heard crossing some of the highest peaks of the Himalayas from India to central Asia at an altitude estimated to be 8840 m.

The distribution of nocturnal migrants in the airspace typically is strongly skewed to the lower altitudes. In a radar study of nocturnal migration in southeastern Louisiana, 70% of the migrants flew between 247 and 1127 m agl, and within this zone ~75% flew between 241 and 492 m agl. An analysis of weather radar measurements of the altitude of nocturnal migration on 70 spring nights and 35 fall nights at weather stations in New Orleans and Lake Charles, Louisiana; Athens, Georgia; and Charleston, South Carolina, showed that for 73% of 79 altitude measurements in the spring and 56% of 39 measurements in the fall, altitudes of peak densities of migrants were ~300 m agl. On some occasions, the altitude at which most birds were migrating was considerably higher than the usual 300 m, and on those occasions the peak densities were associated with favorable winds at those altitudes. With the exception of some shipboard navigation radar, most radar cannot detect birds very close to the ground, and consequently the minimum altitude of nocturnal migration displayed on radar cannot be measured accurately. Studies using direct visual means to detect migrating birds as they pass through a narrow vertical beam of light or the field of a vertically directed thermal imaging camera suggest that a considerable number of birds fly within 100 m of the ground at night. This is particularly so within an hour after the initiation of nocturnal migration and at the time birds are landing during the night. On some misty, cloudy nights, tremendous numbers of call notes from migrants aloft can be heard, and on many of these occasions the distance of the call notes overhead indicates the birds are flying within a few meters above the trees. The altitude of migration changes throughout the night. Usually, the maximum mean altitude of migration is reached about 2 h after initiation and thereafter slowly declines as birds begin to terminate their nightly migration.

Daytime migration usually occurs at altitudes below 300 m, and quite often flocks of daytime migrants can be seen moving just above tree level. This, however, is not always the case. When migrants are arriving on the northern coast of the Gulf of Mexico in spring during daylight hours after a trans-Gulf flight, they are usually at altitudes above 1500 m. A few flocks of arriving trans-Gulf migrants may reach altitudes of 3500–4500 m. When the migrants encounter *powerful* cold fronts and strong headwinds before they make their landfall, they will often fly within a few meters of the water's surface. On these occasions

when the flights are delayed and most of the migrants arrive at night, many will strike wires, towers, and other man-made structures, particularly if the structures have continuously illuminated lights and the weather is poor (low ceiling and misty). Migrants flying during the day will fly lower when there is poor visibility, dense cloud cover, and drizzle. When the migrants encounter weak, shallow cold fronts and headwinds as they cross the Gulf of Mexico in spring, they often gain altitude and fly in favorable winds above the cold front.

Weather Influences on the Density of Migration

Once calibrated, radar can be used to measure the quantity of migration, and it is possible to study the weather factors responsible for the night-to-night variation in the quantity of migration. It is generally accepted that in spring more migration occurs behind a warm front on the west side of a high pressure system and before a cold front and low pressure system. In fall, very large migrations occur just after cold fronts on the east side of high pressure systems. But what weather factors or combination of weather factors influence the density of migration? A number of studies have attempted to answer this question. Because weather factors interact in complex ways and show considerable covariation, multivariate statistical analyses (e.g., factor analysis, discriminant function analysis) must be used. The weather factors that have been shown to significantly influence the quantity of migration include: wind, rain, temperature, barometric pressure, and cloud cover. Of all the weather factors listed, rain, wind, and temperature are clearly the most consistently important factors. Rain usually prevents the onset of a migratory flight or terminates a migratory flight in progress if the migrants can land. Both wind and temperature are significantly interrelated. Thus, the largest spring migrations occur with winds from the south and southwest, which bring warming temperatures to a region, and the densest fall migrations occur with winds from the northwest and north, which usually bring colder temperatures to an area. Another point regarding the influence of weather on the quantity of bird migration should be mentioned. The amount of night-to-night variation in the quantity of migration explained by weather is 50–60% on average. The remaining variation is undoubtedly attributable to the actual number of migrants in an area physiologically ready to migrate. If weather conditions are excellent for migration and few migrants are physiologically ready to migrate, then the migratory movement will be a minor one. The weather conditions most often associated with migrants colliding with man-made objects (poor visibility, low ceiling, mist, and drizzle) are not those conducive to very large migratory movements. Why, then, do numbers of migrants occasionally collide with the guy lines of tall towers, tall buildings, and other obstructions during migration? The answer to this question is straightforward. When birds initiate a migration with favorable weather conditions, they sometimes move into areas where the weather has deteriorated (e.g., a stalled frontal system with low ceilings, mist, and drizzle), and these conditions cause the birds to fly at low altitudes where collisions occur. Occasionally, collisions happen under ideal weather conditions for migration, particularly when communication towers reach altitudes where birds typically migrate at night.

Future Research

Future discoveries related to tracking bird migration will come with advancement in technology. Perhaps the greatest need at present is the ability to track small birds over great distances with telemetry devices. The ideal device would weigh a small percentage of the bird's weight (preferably < 20%) and provide information on geographical location, altitude of flight, and some physiological measurements (e.g., body temperature, heart rate). In the meantime, work with color-banded birds, geolocators, stable isotopes, and genetic markers must continue and must be expanded to include additional species. Our understanding of the influences of competition, predation, and weather conditions on migratory behavior is its infancy, and if we are to measure responses of migratory birds to manipulation of these variables, we need to be able to monitor the migratory behavior of individuals over great distances and for months at a time. Only then will we start to understand the mechanisms that regulate these incredible seasonal journeys.

See also: Behavioral Endocrinology of Migration; Magnetic Orientation in Migratory Songbirds; Migratory Connectivity; Pigeon Homing as a Model Case of Goal-Oriented Navigation.

Further Reading

Alerstam T (1990) *Bird Migration.* Cambridge, UK: Cambridge University Press.

Berthold P (2001) *Bird Migration: A General Survey,* 2nd edn. Oxford: Oxford University Press.

Berthold P, Gwinner E, and Sonnenschein E (eds.) (2003) *Avian Migration.* Berlin: Springer-Verlag.

Boere GC and Stroud DA (2006) The flyway concept: What it is and what it isn't. In: Boere GC, Galbraith CA, and Stroud DA (eds.) *Waterbirds Around the World,* pp. 40–47. Edinburgh, UK: The Stationery Office.

Dingle H (1996) *Migration. The Biology of Life on the Move.* New York, NY: Oxford University Press.

Gill RE Jr, Tibbitts TL, Douglas DC, et al. (2009) Extreme endurance flights by landbirds crossing the Pacific Ocean: Ecological corridor rather than barrier? *Proceedings of the Royal Society B* 276: 447–457.

Greenberg R and Marra PP (eds.) (2005) *Birds of Two Worlds: The Ecology and Evolution of Migration.* Baltimore, MD: Johns Hopkins University Press.

Hake M, Kjellén N, and Alerstam T (2001) Satellite tracking of Swedish Ospreys *Pandion haliaetus*: Autumn migration routes and orientation. *Journal of Avian Biology* 32: 47–56.

Kerlinger P (1989) *Flight Strategies of Migrating Hawks.* Chicago, IL: University of Chicago Press.

Lincoln FC, Peterson SR, and Zimmerman JL (1998) *Migration of Birds.* Washington, DC: US Department of the Interior, US Fish and Wildlife Service. Circular 16. Jamestown, ND: Northern Prairie Wildlife Research Center Online. http://www.npwrc.usgs.gov/resource/birds/migratio/index.htm.

Martell MS, Henny CJ, Nye PE, and Solensky MJ (2001) Fall migration routes, timing, and wintering sites of North American Ospreys as determined by satellite telemetry. *Condor* 103: 715–724.

Martell MS, McMillian MA, Solensky MJ, and Mealey BK (2004) Partial migration and wintering use of Florida by Ospreys. *The Journal of Raptor Research* 38: 55–61.

Newton I (2008) *The Migration Ecology of Birds.* London: Elsevier.

Ruegg KC and Smith TB (2002) Not as the crow flies: A historical explanation for circuitous migration in Swainson's Thrush (*Catharus ustulatus*). *Proceedings of the Royal Society of London, Series B* 269: 1375–1381.

Shaffer SA, Tremblay Y, Weimerskirch H, et al. (2006) Migratory shearwaters integrate oceanic resources across the Pacific Ocean in an endless summer. *Proceedings of the National Academy of Sciences of the United States of America* 103: 12799–12802.

Stutchbury BJM, Tarof SA, Done T, et al. (2009) Tracking long-distance songbird migration by using geolocators. *Science* 323: 896.

Body Size and Sexual Dimorphism

R. M. Cox, Dartmouth College, Hanover, NH, USA

Introduction and Historical Perspective

Sexual dimorphism refers to any systematic difference in form between males and females of a species. By definition, males and females differ in the size of their gametes: males make many small sperm, whereas females make fewer and larger eggs. This fundamental difference in gamete size, or anisogamy, predisposes the sexes to different reproductive organs, which are termed primary sexual characters. However, sexual dimorphism typically refers to traits that are not directly associated with reproduction, termed secondary sexual characters. Although sexual dimorphism explicitly refers to morphology, such morphological differences are typically associated with sexual differences in behavior, physiology, and life history.

Sexual dimorphism holds a special place in the history of evolutionary biology. While developing his theory of sexual selection, Charles Darwin amassed a vast catalogue of sexual dimorphisms. He described dimorphisms ranging from the enlarged horns and mandibles of male beetles to the extravagant plumage and song of male birds and concluded with a detailed discussion of secondary sexual characteristics in humans. Darwin argued that these dimorphisms evolve because extravagant ornaments and songs are preferred by the opposite sex (intersexual selection), whereas weaponry and large body size confer an advantage in competition with members of the same sex (intrasexual selection). In either scenario, the key outcome is that individuals with a particular trait achieve greater mating success. Sexual selection has since been defined as selection arising through variation in mating success.

Darwin viewed sexual selection as distinct from natural selection in that it could favor the evolution of traits that reduce survival. This principle is illustrated by one of Darwin's favorite examples: the peacock's tail. Male peacocks (*Pavo cristatus*) have an elaborate train of ornamental feathers that they display when courting, but female peahens lack these showy ornaments. Because the male's extravagant tail requires a substantial energetic investment and greatly impedes his flight and movement, it should be disfavored by natural selection. However, this cost is offset by the increased mating success that he acquires due to female preference for the largest and most extravagant ornaments. This example emphasizes that sexual dimorphisms reflect the combined forces of selection on viability, fecundity, and mating success equilibrating at different optima in each sex.

Mating System and Reproductive Roles

Many of Darwin's observations were interpreted in light of the prevailing view that males compete vigorously and violently for females, whereas females are coy and selective in choosing mates. While this is often true, subsequent research has shown that these conventional gender roles and their associated selection pressures are reversed in many species. When this occurs, sexual selection can lead to the evolution of sexual dimorphism in which females are larger, brighter or more aggressive than males. Some biologists have concluded that this amazing and underappreciated natural diversity in gender and sexuality exposes major flaws in Darwin's theory of sexual selection. While many details of sexual selection theory are indeed contentious, modern evolutionary biology recognizes that sexual selection does not simply imply aggressive males and choosy females. Sexual selection can act on both sexes and its strength depends upon many interrelated aspects of behavior, ecology, and mating system.

One of the most important determinants of sexual selection is the ratio of sexually active males to fertilizable females, or operational sex ratio (OSR). When the OSR is large, competition for females is intense and the opportunity for sexual selection increases because many males fail to mate. This occurs because the strength of sexual selection is closely tied to variance in mating success; if all males have the same number of mates, sexual selection cannot occur, but if some males have many mates while others have few or none, sexual selection can be strong. The opportunity for sexual selection is also typically greater in polygynous mating systems, where some males mate with more than one female, than in monogamous systems, where each individual has a single partner. Studies of mammals, birds, and amphibians show that the degree of sexual dimorphism increases with the degree of polygyny across species. Sexual dimorphism can be particularly extreme under lek polygyny, where males compete on display arenas (leks) that are visited by females, leading to a combination of intra- and intersexual selection. Familiar examples include the large body size of male elephant seals, the antlers of fallow bucks, the extravagant ornaments and displays of male grouse, and the elongated eyestalks of male stalk-eyed flies.

When polyandry occurs and females have multiple mates but males do not, sexual selection can drive the evolution of dimorphism in which females are larger, showier, and more aggressive than males. American jacanas (*Jacana spinosa*) exhibit harem polyandry in which females aggressively defend territories containing several males. As predicted, female jacanas are larger and more brightly colored than males. Jacanas also exhibit another interesting phenomenon: sex role reversal. Anisogamy predisposes females to a greater initial investment in reproduction, which usually translates into a greater degree of subsequent effort raising offspring. However, male jacanas invest more heavily in parental care than females because they incubate the eggs. Consequently, a female's reproductive success is not constrained by her capacity for parental investment, and she increases her reproductive success by obtaining more mates. A similar situation occurs in the spotted sandpiper (*Actitis macularia*). Male sandpipers provide most of the parental care, which results in a female-biased OSR and causes female reproductive success to increase with the number of males that she is able to monopolize. This situation results in strong sexual selection, and females exhibit aggressive combat behavior similar to males of polygynous species. Sex role reversal and associated patterns of sexual dimorphism are also observed in pipefishes and seahorses, where males brood embryos in enlarged pouches and resemble pregnant females. Courtship roles are also reversed in giant waterbugs (Belostomatidae), where females actively court their mates and attempt to mate with multiple partners. Males decide which females to accept as mates, and parental investment is also reversed: males carry fertilized eggs on their backs for several weeks, during which time they aerate the eggs and protect them from predators.

Coloration, Ornaments, and Weapons

Sexual differences in coloration are referred to as sexual dichromatisms. In many cases, males are more brightly colored than females, which are often drab or cryptically colored to match their habitat. Familiar examples include the showy plumage of male birds, the colored badges displayed by male lizards, and the bright spots of male guppies. Males of many species also possess elaborate ornaments that are absent or reduced in females, such as the enlarged tails of male birds and the elongated fins of male fishes. Sexual dimorphism in coloration and ornamentation often evolves by female choice. Experiments on peacocks (*Pavo cristatus*), long-tailed widowbirds (*Euplectes progne*), and barn swallows (*Hirundo rustica*) show that females prefer not to mate with males whose tails have been shortened or rendered unattractive by removing ornamental feathers. However, artificially lengthening a male's ornament renders him more attractive and greatly increases his reproductive success. Studies of guppies (*Poecilia reticulata*) illustrate how natural

selection balances sexual selection for bright coloration. Male guppies are highly polymorphic in the size and number of colored spots on their bodies, and females prefer males with many bright spots, which stand out during courtship displays. However, bright spots also render males visible to predators. In the absence of predators, male guppies evolve greater numbers of spots due to female preference, but in the presence of predators, the number of spots decreases over subsequent generations due to natural selection.

Hypotheses concerning the benefits that females obtain from mate choice comprise one of the most contentious areas of evolutionary biology. Most studies distinguish between 'direct' benefits, such as nutrients or care provided by the male or territory quality to which the female gains access, and 'indirect' benefits, such as 'good genes' that are inherited by the female's offspring. For sexual dimorphism to evolve via preference for direct benefits, some aspect of the male phenotype must correlate with the benefit. For example, male bullfrogs (*Rana catesbeiana*) are larger than females, and females mate preferentially with the largest males. The large size allows males to defend optimal territories with low embryo mortality, so females that mate with large males receive a direct benefit in terms of improved offspring survival.

Models for indirect genetic benefits are based on the idea that coloration, ornaments, and weapons serve as indicators of male quality, and that quality is heritable. One subset of these indicator models proposes that elaborate traits act as 'handicaps' that incur a fitness cost. Bright coloration and elaborate ornaments are energetically expensive to develop and express, and may also render their bearers subject to increased predation and parasitism. Because only the best males can afford such a handicap, female preference for bright coloration or extravagant ornaments translates into preference for high-quality males. This provides an indirect genetic benefit to their offspring if some component of male quality is heritable. Other models for the evolution of extravagant traits propose that female preference evolves because it initially leads to an indirect benefit when the preferred trait increases the fitness of the offspring. This causes the genes for female preference and those for the male trait to become evolutionarily coupled, ultimately resulting in a 'runaway' process in which the evolution of the male trait exceeds what is favored by natural selection and is driven solely by correlated female preference.

Although coloration and ornamentation often evolve in response to mate choice, these structures can have a variety of functions, and it can be difficult to establish the relative contributions of intersexual, intrasexual, and natural selection. Male *Anolis* lizards have bright dewlaps that are absent or greatly reduced in females of most species. The dewlap is broadcast to conspecific and heterospecific males and females, and its function may reflect a combination of courtship display, male–male competition, and species recognition. By contrast, sexual dimorphisms in weaponry are usually more readily attributable to intrasexual selection. For example, male red deer (*Cervus elaphus*) possess large antlers that are used in violent male–male contests. Mating success is highly variable and related to fighting ability, creating strong sexual selection on this trait. Although fighting success is determined primarily by body size, it is also related to antler mass, and removal of antlers reduces fighting ability and dominance rank.

Body Size

Sex differences in body size are referred to as sexual size dimorphism (SSD), and are perhaps the most ubiquitous and well-studied form of sexual dimorphism. As a very broad generalization, females are larger than males in the majority of animals. Darwin attributed this to the fact that the largest females typically produce the most offspring, and modern evolutionary biologists recognize fecundity selection for large size as a primary cause of SSD. The prevalence of female-biased SSD stands in contrast to the common misconception that males are typically the larger sex, which is perhaps based on our familiarity with birds and mammals. Even within these two groups, females are larger than males in many lineages, including bats, rabbits and hares, raptors, and owls. Moreover, the magnitude of SSD often varies considerably within lineages and even within species. Among reptiles, for example, females are larger than males in most turtles and snakes, although males are the larger sex in tortoises, mud turtles, and vipers. By contrast, males are larger than females in crocodilians and the majority of lizards, although most lizard families include species with female-biased SSD. The Australian carpet python (*Morelia spilota*) displays an amazing degree of variation within a single species. Males from northeastern populations exceed females by 30% in body mass, but females from southwestern populations are more than ten times as massive as males.

In some lineages, 'giant' females are an order of magnitude larger than their 'dwarf' male counterparts. Examples include orb-weaving spiders, deep-sea anglerfishes, and the blanket octopus (*Tremoctopus violaceous*), which currently holds the record for extreme SSD. Females of this species exceed males by more than four orders of magnitude in body mass and two orders of magnitude in length. In many cases of extreme female-biased SSD, males live as internal or external parasites upon the larger female, existing as structurally reduced sperm donors. This phenomenon is particularly common in aquatic organisms, including marine worms, barnacles, and deep-sea anglerfishes. Dwarf males are usually associated with a skewed OSR in which females accumulate multiple mates, and with low population densities that place a

premium on the male's ability to locate and then remain with a female, rather than aggressively competing for multiple mates. Thus, while the large size of females presumably reflects selection for increased fecundity, the extremely small size of males cannot be fully explained without reference to the mating system and the resultant effects of sexual selection on male size.

Extreme male-biased SSD also occurs, although it never reaches the same magnitude as extreme female-biased SSD. Elephant seals (*Mirounga*) are a classic example of extreme male-biased SSD, with bull males exceeding females by two to seven times in body mass. The largest males have an advantage in combat, which involves bellowing, chasing, and slashing with enlarged teeth. Successful combatants can potentially monopolize an entire harem of 30–100 females. Strong intrasexual selection has clearly favored the evolution of large male size, but why have females remained small? One possibility is that, upon attaining reproductive maturity, females devote their energy to the production of energetically expensive offspring, rather than growth. The record for male-biased SSD belongs to a shell-spawning cichlid fish (*Lamprologus callipterus*) in which males are 12 times heavier than females. This extreme dimorphism reflects a combination of sexual selection for large male size, which allows the male to transport shells and defend his territory and its harem of females against other males, and natural selection placing a constraint on female size, which has to be small enough to permit entry into shells for spawning. However, SSD is usually subtler than these extremes. For example, adult human males are, on average, 7% taller than females, although there is substantial overlap between the sexes.

Disentangling the selective pressures that shape the evolution of SSD can be complicated because body size influences a diverse array of biological processes. Larger body size may be advantageous in competition for mates, production of eggs, and avoidance of predators, but it may also impede climbing, require greater skeletal support, or necessitate a greater absolute energy budget. Moreover, the relative strength of these various selective pressures can fluctuate as environmental conditions change. Male Galapagos marine iguanas are substantially larger than females because sexual selection favors the largest males, who dominate smaller rivals in contests for access to females. However, in years when El Niño storms cause their food resources to crash, the largest males suffer the heaviest mortality because they cannot meet the greater absolute energy demands of their large bodies. Despite this complexity, the general consensus across animal lineages is that SSD reflects a balance between fecundity selection, which favors large female size, and sexual selection, which tends to favor large size when males compete aggressively for mates and small size when males engage in 'scramble' competition for a limited number of widely dispersed females.

Ecological Dimorphisms

In addition to sexual selection and fecundity selection, males and females may also experience differences in natural selection that lead to sex-specific adaptation to different facets of their shared environment. One particularly striking example is the purple-throated carib (*Eulampis jugularis*), a Caribbean hummingbird in which males are the larger sex and possess a relatively short, straight bill, whereas females are small with a relatively long, curved bill. Although SSD in this species likely reflects sexual selection for large male size, the larger bill size of females is opposite the pattern predicted from sexual selection. However, bill size and shape correspond to the distinct nectar resources utilized by each sex. Males feed predominantly on the large, straight flowers of *Heliconia caribaea*, whereas females are most frequently observed feeding on the smaller, curved flowers of *H. bihai*. Moreover, each sex maximizes its feeding efficiently on the species of flower that matches its bill morphology. But why do the sexes partition their nectar resources in such dramatic fashion?

Most ecological hypotheses are based on the idea that morphological specialization allows males and females to lower intersexual competition for limited resources. For example, female sea kraits (*Laticauda colubrina*) are substantially larger than males and feed primarily on large conger eels, whereas male kraits prey upon smaller moray eels. However, males and females are similarly sized in populations where conger eels are absent, suggesting that the distribution of prey resources influences the degree of SSD. One major difficulty in establishing ecological causation is that divergence in resource use can be both a cause and a consequence of sexual dimorphism. Do female sea kraits evolve into a large size so that the species can utilize multiple prey resources, or is the ability to consume larger prey an indirect consequence of large female size resulting from other factors, such as fecundity selection? Moreover, many species exhibit differences in morphology without differences in resource use, or differences in morphology opposite those predicted by resource partitioning.

Sex-Limited Polymorphism and Alternative Reproductive Tactics

Although sexual dimorphism implies a simple dichotomy, more than one type of male or female often occurs within a single species. For example, all male *Papilio memnon* butterflies look similar, but females exhibit a variety of color patterns, each mimicking a different, noxious butterfly species. Other examples of female-limited color polymorphisms include lizards (*Anolis*), spiders (*Theridion*), and damselflies and dragonflies (Odonata). In many cases, sex-limited polymorphisms are associated with differences in behavior and reproductive strategies, termed alternative reproductive tactics (ARTs). Male ruffs (*Philomachus pugnax*) occur in two genetically determined forms: 'independent' males defend territories and are large with dark ornamental collars, whereas 'satellite' males do not defend territories and are small with white collars. Other examples of male ARTs include: 'fighting' fig wasps (*Idarnes*) with no wings and large mandibles versus 'dispersing' males with wings and small mandibles; 'guarding' dung beetles (*Onthophagus*) with large size and long horns versus 'sneaking' males with small size and short horns; and 'bourgeois' salmon with large size and aggressive behavior versus 'parasitic' males with small size and inconspicuous, sneaking behavior. The prevalence of ARTs reveals that, behaviorally and morphologically, there is often more than one way to be a biologically successful male.

In some systems, three ARTs may be maintained if the reproductive success of each tactic decreases as it becomes more common in the population. This is referred to as negative frequency-dependent selection, and it has been particularly well studied in the side-blotched lizard (*Uta stansburiana*). In this species, 'territorial' orange-throated males are successful against 'mate-guarding' blue-throated males, but vulnerable to yellow-throated 'sneaker' males. The relative frequencies of each morph cycle through time because each mating strategy is most successful when rare and least successful when common. Female side-blotched lizards also exhibit alternative reproductive tactics, with orange-throated females producing many small eggs and yellow-throated females producing fewer, larger eggs. As in males, these ARTs are frequency-dependent: the large progeny of yellow-throated females have a fitness advantage when rare, but this advantage disappears when they become common and selection shifts to favor progeny quantity over quality. These examples illustrate that variation within sexes is often as dramatic as that observed between sexes.

Proximate Mechanisms

Studies that attempt to address the factors driving the evolution of sexual dimorphism are often referred to as studies of ultimate causation. By contrast, studies that attempt to uncover the physiological, genetic, and developmental mechanisms by which males and females express different phenotypes are referred to as studies of proximate causation. In other words, studies of ultimate causation ask why males and females differ, whereas studies of proximate causation ask how they express these differences. A major area of current research concerns the integration of proximate and ultimate explanations for sexual dimorphism.

The first step in establishing proximate causation is to describe the ontogeny of sexual dimorphism, which refers to the developmental timing of sexual divergence with respect to the life cycle of the organism. Some dimorphisms develop early in embryology, others develop in juveniles, and still others are apparent only upon sexual maturation. Even within a species, the ontogeny of sexual dimorphism may differ dramatically for various traits. For example, male and female brown anole lizards (*Anolis sagrei*) differ in color pattern immediately upon hatching, implying an embryological origin of sexual dichromatism. Body size does not differ at hatching, but juvenile males grow more quickly than females and SSD is evident well before sexual maturation. Other dimorphisms, such as the nuchal crests and brightly colored dewlaps of adult males, develop later in ontogeny around sexual maturation. Developmental pathways may also differ among related species that share a common pattern of adult sexual dimorphism. In primates, adult males are typically much larger than females, but the growth trajectories that produce male-biased SSD differ considerably, even among closely related species. In the pygmy chimpanzee (*Pan paniscus*), males and females grow at the same rate, but males grow for a longer duration, whereas in the common chimpanzee (*Pan troglodytes*), males and females differ primarily in the rate of growth. Patterns of growth in other primates reflect a mixture of these two scenarios, as do patterns across animals in general. While developmental patterns establish the schedule of sexual differentiation, the next step is to identify genetic and physiological mechanisms that orchestrate sex-specific development.

There are two basic genetic mechanisms by which sexually dimorphic phenotypes can be produced from a shared genome: sex linkage on sex chromosomes, and sex-limited expression of autosomal loci. Because sex chromosomes exhibit sex-dependent dosage, their potential for regulating sexual dimorphism is obvious. Indeed, X-linked genes contribute to a substantial proportion of phenotypic variation in sexually selected traits across various taxa, and X-linked quantitative trait loci (QTLs) have been shown to regulate sexual dimorphism in polygenic traits such as body size. Although the Y chromosome typically contains less than 1% of the entire genome, Y-linked genes do contribute to sexual differentiation and to male fitness variation. Similar patterns are predicted in species with WW/WZ sex determination. Because of their role in sex determination, sex chromosomes are the ultimate basis for all sexual differences in species with genetic sex determination.

Although sex chromosomes provide a straightforward mechanism for the expression of sexual dimorphism, the majority of the genome resides on autosomes. Thus, it is likely that most sexually dimorphic traits are influenced by autosomal genes that are present in both sexes. This is particularly true for complex, polygenic traits such as body size. Sexual dimorphism in such traits is expected to arise from sex-limited expression of shared autosomal loci. Indeed, recent microarray studies reveal that sex-biased gene expression is widespread throughout the genome and across taxa. Sex-biased gene expression can occur via 'direct' genetic mechanisms, such as the sex-limitation of an autosomal locus via epistatic interactions with a sex-linked modifier locus, or via 'indirect' mechanisms, such as hormonal regulation (below). In addition to sex linkage and sex-limited expression, recent studies suggest that sex-biased inheritance can also be achieved by genomic imprinting, or by a form of cryptic mate choice in which females differentially produce sons and daughters with sperm from different sires. The details of these mechanisms are poorly understood, and future studies are likely to uncover novel genetic mechanisms for the expression of sexual dimorphism.

Sex steroids (androgens, estrogens, and progestins) are excellent candidates for the regulation of sexual dimorphism in vertebrates because they are produced and secreted in sex-specific fashion by the gonads. Moreover, sex steroids can coordinate the simultaneous expression of sex differences in morphology, behavior, and reproductive biology. For example, androgens influence the expression of bright coloration, territoriality, and courtship behaviors in breeding males of numerous birds, reptiles, and fishes. Likewise, estrogens and progestins coordinate the expression of bright nuptial coloration and associated rejection displays that females use to avoid male courtship once they become gravid. In many species, the female phenotype is the developmental default, such that bright male coloration, weaponry, and ornamentation develop only upon activation by androgens. However, the extravagant male phenotype is actually the default in many organisms, including the classic example of the peafowl, where the long tail and bright plumage of the male peacock develop unless inhibited by estrogens. Simple changes in the regulatory roles of hormones may facilitate the rapid evolution of highly divergent patterns of sexual dimorphism. For example, testosterone stimulates male growth in reptile species with male-biased SSD, but actually inhibits male growth in species with female-biased SSD. Other hormones, such as juvenile hormone, ecdysteroids, and insulin, are important regulators on sexual dimorphism and sex-limited polymorphisms in insects and other invertebrates.

Intralocus Sexual Conflict

Sexual dimorphism ultimately evolves because natural and sexual selection favor different fitness optima in each sex. However, because the sexes share a common genome and mechanisms for sex-limited expression are imperfect, genes that are favored by selection on males are often expressed in females, where they are detrimental. This creates a genomic tug-of-war referred to as intralocus sexual conflict. For example, male ground crickets (*Allonemobius socius*) with high reproductive success produce high-fitness sons and low-fitness daughters, whereas males with low reproductive success produce high-fitness daughters and low-fitness sons. Similar patterns have been documented in red deer (*Cervus elaphus*) and fruit flies (*Drosophila melanogaster*). Intralocus sexual conflict occurs whenever genetic correlations between the sexes prevent males and females from reaching their respective phenotypic optima. This conflict is gradually resolved as mechanisms for sex-limited gene expression evolve to facilitate the expression of sexual dimorphism. Recent studies comparing the strength of natural and sexual selection in males and females suggest that sexually dimorphic traits are frequently the source of ongoing sexual conflict in a variety of taxa. Sexual conflict theory has already provided a powerful conceptual framework for behavioral ecologists interested in male–female mating interactions, and it promises to lend similar insight into the studies of sexual dimorphism.

See also: Aggression and Territoriality; Bateman's Principles: Original Experiment and Modern Data For and Against; Cryptic Female Choice; Darwin and Animal Behavior; Female Sexual Behavior and Hormones in Non-Mammalian Vertebrates; Flexible Mate Choice; Mate Choice in Males and Females; Pair-Bonding, Mating Systems and Hormones; Reproductive Success; Sex Allocation, Sex Ratios and Reproduction; Sex and Social Evolution; Sex Change in Reef Fishes: Behavior and Physiology; Sexual Behavior and Hormones in Male Mammals; Sexual Selection and Speciation; Social Selection, Sexual Selection, and Sexual Conflict; Sperm Competition.

Further Reading

Andersson MB (1994) *Sexual Selection*. Princeton, NJ: Princeton University Press.

Arnqvist G and Rowe L (2005) *Sexual Conflict*. Princeton, NJ: Princeton University Press.

Badyaev AV (2002) Growing apart: An ontogenetic perspective on the evolution of sexual size dimorphism. *Trends in Ecology and Evolution* 17: 369–378.

Calsbeek R and Bonneaud C (2008) Postcopulatory fertilization bias as a form of cryptic sexual selection. *Evolution* 62: 1137–1148.

Cox RM and Calsbeek R (2009) Sexually antagonistic selection, sexual dimorphism, and the resolution of intralocus sexual conflict. *The American Naturalist* 173: 176–187.

Darwin CR (1871) *The Descent of Man, and Selection in Relation to Sex*. London: J. Murray.

Day T and Bonduriansky R (2004) Intralocus sexual conflict can drive the evolution of genomic imprinting. *Genetics* 167: 1537–1546.

Ellegren H and Parsch J (2007) The evolution of sex-biased genes and sex-biased gene expression. *Nature Reviews: Genetics* 8: 689–698.

Emlen ST and Oring LW (1977) Ecology, sexual selection, and the evolution of mating systems. *Science* 197: 215–223.

Fairbairn DJ, Szekely T, and Blanckenhorn WU (eds.) (2007) *Sex, Size and Gender Roles: Evolutionary Studies of Sexual Size Dimorphism*. Oxford: Oxford University Press.

Hamilton WD and Zuk M (1982) Heritable true fitness and bright birds: A role for parasites? *Science* 218: 384–387.

Oliviera R, Taborski M, and Brockman HJ (eds.) (2007) *Alternative Reproductive Tactics*. Cambridge: Cambridge University Press.

Roughgarden J (2004) *Evolution's Rainbow: Diversity, Gender and Sexuality in Nature and People*. Los Angeles, CA: University of California Press.

Temeles FJ, Pan IL, Brennan JL, and Horwitt JN (2000) Evidence for ecological causation of sexual dimorphism in a hummingbird. *Science* 289: 441–443.

Zahavi A and Zahavi A (1997) *The Handicap Principle: A Missing Piece of Darwin's Puzzle*. Oxford: Oxford University Press.

Boobies

H. Drummond, Universidad Nacional Autónoma de México, México

Introduction

Diverse aspects of sulid biology have been studied in recent years, including social behavior, sex ratio evolution, demography, life history evolution, dispersal, incubation, reproductive endocrinology, phylogeography, and speciation. This review is limited to the social behavior that has been most thoroughly explored: the mating system of the blue-footed booby and lethal sibling competition in the broods of three species of booby. Boobies are long-lived pelecaniform birds that nest on tropical oceanic islands and feed mostly by repeatedly plunge diving for small fish in the course of lengthy foraging trips. Blue-footed (*Sula nebouxii*), Nazca (*S. granti*), and brown boobies (*S. leucogaster*) periodically face severe food shortage when fish stocks are decimated by warm-water events of El Niño Southern Oscillation (ENSO). All three species nest colonially on the ground and show some degree of reversed sexual size dimorphism, and the blue-footed and Nazca boobies are highly philopatric.

Mating System of the Blue-Footed Booby

Blue-footed boobies are socially monogamous, partner fidelity in successive seasons is common, and social partners share all the duties of parental care roughly equally. Yet sexual conflict is rife: both members of a pair are liable to switch partners within or between seasons or have extra-pair relationships with colony neighbors, females sometimes dump eggs in the nests of their extra-pair partners, and the risks posed by infidelity and dumping are addressed by countermeasures such as vigilance, aggression, and infanticide. The following account is based mostly on studies by Hugh Drummond, Roxana Torres, Alberto Velando, and their collaborators on Isla Isabel, off the Pacific coast of Mexico.

Pair Formation and Copulation

Annually, blue-foot males stake out nesting territories of a few square meters on the ground and defend them against neighbors by fencing, jabbing, wingflailing, and pecking. Typically, females walk through the colony assessing territorial males, who may respond with parading, spectacular skypointing displays (**Figure 1**), and even aerial displays, until each female strikes up a relationship of mutual displaying with a particular male and joins with

him in territory occupation and defense. During the following days or weeks, each of the two partners progressively increases its attendance in the territory to about 8 h per daylight period; then the female lays an egg in one of the potential sites over which they have previously negotiated. Between pair formation and laying of the first egg, the two partners court reciprocally and copulate on the territory, peaking at 1.7 copulations per day in the 5-day presumed fertile period just before laying. If further eggs are to be laid, copulation continues until the clutch is completed 5–10 days later. Copulations are frequently followed by courtship, but females sometimes expel presumed ejaculate from their cloacas after the male steps down.

Assessing Mates

Bluefoots may assess the quality and condition of partners and potential mates on the basis of voice, morphology, or behavior, but one trait is particularly salient during courtship, that is, displaying foot color, which varies from turquoise to deep blue and differs between the two sexes (**Figure 2**). The highly vascularized webs reveal the male booby's condition: they brighten during the period of courtship, darken with age and after a day of unsuccessful fishing, and dull down during incubation and brood care. When her partner's feet are artificially darkened before she lays, a female displays to him less intensely and copulates with him less frequently; and if they are darkened after she lays her first egg, she makes her second egg smaller and lighter (and less viable). A fostering experiment showed that chicks cared for by males with naturally dull feet grow slowly, implying that such males are in poor condition and feed chicks less. Females have duller feet than males during courtship, but their webs reveal condition too: females in good nutritional condition have brighter feet and lay larger clutches, and they are courted more intensely by males.

Extra-Pair Behavior

In dense neighborhoods, territories are often contiguous and boobies interact extensively with their neighbors, repelling territorial intrusions and engaging in extra-pair liaisons. Before they start incubating, all males court extra females and roughly half of females establish an extra-pair sexual relationship with one or more male neighbors (**Figure 3**). As in within-pair relationships, extra-pair mates interchange courtship displays and copulate repeatedly

Figure 1 The female blue-footed booby skypoints to her male partner during an exchange of courtship displays on the pair's territory. Photo: Pablo Cervantes.

Figure 2 After a storm, the webs of the male (left) blue-footed booby are even more green than usual and those of his female partner even deeper blue than usual. Web color reflects a booby's sex, age, nutrition, health, breeding stage, breeding history, and capacity for caring for offspring. Photo: Hugh Drummond.

Figure 3 (a) A paired blue-foot male (left) pauses at the border of the territory of his extra-pair partner because her male partner (squatting behind grass) is threatening him with a yes-headshake display. (b) Later, the male (left) visits his extra-pair partner in her territory after departure of her male partner. Two extra-pair copulations followed. Photo: Hugh Drummond.

(in the territory of one of them), and copulations by females increase in frequency as laying approaches, peaking at 0.6 copulations per day in the presumed fertile period. Unfaithful females copulate roughly three times as often with their partners as with their extra-pair mates and their last two copulations before laying are usually with their partners, but new molecular data are showing that extra-pair males sometimes achieve fertilization.

Males and especially females tend to conduct their extra-pair liaisons on or near their territories. In the absence of their partners, males copulated with extra mates at twice the normal rate and females at six times the normal rate, but boobies forage at sea daily and every mated female and male is alone (unmonitored) during one-quarter of its time on territory. Both sexes double their rate of intrapair courtship after their partners court

with other individuals and sometimes attack those individuals, but there is no evidence that these behaviors inhibit infidelity. Mere presence may be key to monitoring and dissuading infidelity, and unfaithful boobies seem to evade monitoring: unfaithful females modify their daily pattern of presence on territory (although their partners match the modification), and both sexes tend to put a few extra meters between themselves and their partners before performing extra-pair activities.

For females at least, it seems that there is a penalty for creating suspicion that offspring might have been sired by extra-pair mates: if we briefly truncate a male's opportunity to monitor his partner's faithfulness by experimentally kidnapping him for 11 h in her presumed fertile period, there is an even chance that a few days later he will destroy the first egg she lays rather than incubate it. That this infanticide functions as a guard against investing parental care in a chick sired by another male is indicated by the observation that the male does not destroy the egg if kidnapped prior to the fertile period.

Quasiparasitism

Both paired and unpaired females sometimes lay single eggs in the empty nest sites of other boobies when laying by the resident female is imminent. All four cases of such dumping that were directly observed occurred when the resident male and female were simultaneously absent from their territory or distracted by a territorial defense battle, and in every case the dumper had copulated with the resident male a few days before. After laying, dumpers incubate their egg, defend it briefly against the returning residents, then yield and depart permanently, leaving the residents to take over incubation (which they do spontaneously after driving away an intruder).

Certainly, dumpers deploy a subtle strategy to parasitize the parental care of other boobies, but are male residents complicit (i.e., is this quasiparasitism) or are they dupes? Active complicity in dumping seems to be absent; all four dumpers apparently waited for the male to be absent or distracted before intruding and laying, and the only paired male that was present during dumping promptly destroyed the dumped egg. Males that returned to their territories after dumping showed only halfhearted nest defense and failed to drive the dumper away, possibly because they had some expectation of being the father of the dumped egg (molecular data have confirmed that this can happen). Unfortunately, we do not know whether paired males are more aggressive to dumpers with which they have not copulated (if such exist) than to those with which they have copulated. Maybe dumpers manipulate paired males into partially tolerating their dumping by courting and copulating with them, and more effective countermeasures have not evolved because manipulated males are often the fathers of dumped offspring. Other direct observations showed that fertilized unpaired females sometimes dump in the territories of unpaired males, which may enable them to instantly acquire a partner and territory.

Sibling Competition

Facultative Siblicide in the Blue-Footed Booby

Bluefoots lay one, two or, less commonly, three eggs at approximately 5-day intervals in a shallow scrape, where they incubate them immediately. Chicks hatch at 4.0-day intervals, resulting in substantial asymmetries in size and maturity that largely dictate the development of their dominance relationships. Hatching failure is commonplace, as is the death of the first-hatched, and especially the second- and third-hatched chicks during the first few weeks of life, but survival of complete broods is by no means unusual and 20% of three-chick broods survive intact through age 70 days. If parental food provision falls short before fledging, one or two nestlings are sacrificed,

a system of resource tracking known as facultative brood reduction. Hence, second and third eggs of blue-foots can function either to insure against failure of the first egg/chick or to furnish an extra fledgling when parents can satisfy the demands of their brood.

In many species of birds, facultative brood reduction occurs through begging competition, differential allocation of food, and selective starvation. In the blue-foot, it is mediated by sibling aggression and death can occur through enforced starvation or by expulsion from the nest followed by heat stress, starvation, or attacks by adult colony neighbors.

Dominance–subordination

Dominance–subordination is a learned relationship that is constructed in every dyad of blue-foot broodmates after hatching and maintained throughout the 3–4 month nestling period or until one chick (usually the subordinate) dies. This relationship secures feeding priority and superior growth for the dominant broodmate during the first few weeks and, more importantly, assures the survivor slot in the event of brood reduction. Surprisingly, subordinates catch up with dominants in growth by the time they fledge, implying that dominants eventually renounce or lose their feeding priority after the first few weeks.

In the typical two-chick brood, the elder chick starts pecking and biting the younger chick around the cranium, eyes, and nape as soon as it is able to do so (at about 8–10 days; **Figure 4**), then increases its daily attacking over the first few weeks of life. The younger chick is

Figure 4   As soon as its developing motor coordination permits, the first-hatched blue-footed booby chick (here about 10 day old) begins to peck at the head of its younger broodmate, to train it into submission. Photo: Hugh Drummond.

Figure 5 At age 17 days, the second-hatched blue-footed booby chick readily adopts a submissive bill-down-and-face-away posture when threatened by its dominant broodmate. Photo: Hugh Drummond.

Figure 6 Training effects trump size difference. In a dense blue-footed booby neighborhood, a dominant chick was accidentally adopted into a neighbor's brood of two older and larger chicks. Here the adoptee bites the resident subordinate chick, which signals submission despite being larger and more mature than its assailant. Photo: Hugh Drummond.

also initially inclined to attack, but is overwhelmed by its stronger rival and progressively learns to respond to attacks with a submissive bill-down-and-face-away posture that presumably inhibits attacking. The more pecks and bites it receives, the more submissive it becomes, until by age 17 days when it is absorbing five attacks per hour, it is submitting to 87% of them (**Figure 5**). Development of distinct agonistic personalities by the two broodmates was demonstrated by briefly pairing similar-sized unfamiliar chicks from different broods: two dominants fought fiercely, two subordinates stood there cheeping harmlessly and when a dominant and subordinate were paired each one assumed its accustomed role. Furthermore, when subordinates were fostered into nests with (slightly smaller) singletons (solitary chicks with no broodmate experience), they submitted to the singletons consistently over 10 days or longer, whereas dominants fostered with (slightly larger) singletons fought them fiercely and successfully on the first 2 days but by day 6 were equally likely to be losing. These experiments (involving no physical harm or effects on growth or survival) show that subordinates are trained losers and dominants are trained winners; they also imply that trained losing is a stronger learning effect.

In principle, the agonistic relationship between broodmates could be manipulated by parents to thwart or facilitate siblicide, modify its timing, or reduce the cost of broodmate strife. No blatant interference in agonism has been observed, but mothers could exercise subtle control through the relative size or composition of first and second eggs. Second eggs (laid near the end of a long ENSO event) did not differ from first eggs in total mass or in concentration of testosterone, androstenedione, or 5α-dihydrotestosterone (corticosterone was not measured), but did contain 10% less yolk. Possibly blue-foot mothers prepare their second chicks to thrive in a subordinate role

not by adjusting hormone titers but by providing them with less yolk than their siblings and hatching them roughly 4 days later. However, reduction of the size and androgen content of second eggs laid after sudden artificial darkening of a male partner's feet varied with ecological circumstances, hinting at facultative adjustment of egg characteristics by laying females.

After roughly the first 3–4 weeks, the dominance–subordination relationship of blue-foot broodmates is well established, aggression of dominant chicks declines to a modest two attacks per hour, and the two broodmates cohabit in relative peace. Experimental pairings showed that subordinates continually monitor the relative size and aggressive demeanor of their broodmates and will go on the offensive if they detect vulnerability, sometimes even overthrowing the dominant chick. But a rebellious subordinate can be beaten into submission by a substantially smaller dominant, so the attempt is seldom made. Likewise, in broods where the first-hatched chick is male and the second-hatched chick is female (the larger sex), and where consequently the subordinate chick outgrows its dominant broodmate at age 37 days, training effects prevail and dominance is stable through fledging (**Figure 6**).

Food deprivation and aggression

Ordinarily, dominants use restrained aggression to secure a moderate feeding advantage, but if parental provisioning declines they increase their attacking to the point where broodmates starve or depart in search of adoption in a neighbor's nest. Under experimental food restriction (with no long-term effects), young (<6 weeks old) dominant chicks suffered weight loss and tripled their rate of attacking, and their feeding advantage increased from 37% to 57% more attempted parental feeds than their broodmates received. After resumption of normal feeding, aggression and food allocation returned toward baseline values, implying a reversible mechanism.

Aggressiveness of dominant chicks is greater when a short hatching interval makes broodmates similar in size, but is not directly influenced by the number of brood-mates. Parents never appear to suppress or encourage broodmate aggression, but they could certainly regulate it by controlling hatch intervals, food ingestion, and even egg composition, and David Anderson has suggested that they suppress siblicide by making deep nests that thwart expulsion of the subordinate chick.

Dominance hierarchies

Broods of three chicks usually form a linear dominance hierarchy, in which ranks are initially assigned by age-related differences in maturity and size, then reinforced and maintained by trained winning and losing (which may be two largely independent axes of learning). It is quite possible that chicks do not recognize other chicks as indi-viduals and an experiment suggested that intermediate-rank chicks are not trained to an intermediate level of aggressiveness. Rather, the generally stable dominance rela-tionship between second and third chicks during the first few weeks may depend more on ongoing responsiveness to relative size.

Dominance and androgens

At age 11–20 days, the level of circulating corticosterone in subordinate bluefoots is twice as high as in dominants or singletons, implying that extra corticosterone is secreted in response to aggressive subordination or to the food depri-vation that accompanies it. Two days of experimental food deprivation, with consequent weight loss, induced increase in corticosterone not only in subordinates but also in dominants and singletons, implying that food deprivation may be the main cause of elevated baseline corticosterone in subordinates. However, in 4-h pairings of singletons with either a dominant chick or a subordinate chick, singletons and dominants (only) showed an increase in circulating corticosterone, implying that corticosterone can increase with the stress of a novel challenge (both singletons and dominants are unused to receiving aggression). But does corticosterone facilitate submissiveness? Corticosterone implants in subordinate chicks led to increased activity/ wakefulness in their broodmates, but there was little sign that implanted subordinates submitted to attacks more readily.

The challenge hypothesis predicts that dominant chicks should show more circulating testosterone than subordinates and singletons during establishment of dominance–subordination, and an increase in testosterone when experimentally challenged by a nonsubmissive nest-mate or provoked by food deprivation into increasing

attacks on their broodmates, but tests of these effects have so far drawn a blank.

Long-term effects of dominance and subordination

Subordinate chicks in two-chick broods grow up under severe disadvantages: daily impacts on the head and face (although this violence is increasingly substituted by threats as they age), psychological subordination during 3–4 months, early starvation (later followed by compensa-tory growth), and elevated circulating corticosterone. Such disadvantages in the infancy of a vertebrate are widely thought to engender deficiencies and poor perfor-mance in adult life, although confirmation of this has come largely from experimental studies where it is not always certain whether the challenges posed to infants are within the range experienced in nature.

Comparisons of 1167 fledglings from two-chick broods observed for up to 10 years disclosed surprisingly few differences between first-hatched and second-hatched birds (assumed to be mostly dominant and subordinate nestlings, respectively). Even more surprisingly, where there were differences these tended to favor subordinates. Although in some cohorts dominant fledglings were more likely to recruit into the breeding population than subor-dinate fledglings, across the whole sample recruiting rates were 41% and 37%, respectively, a statistically nonsignif-icant difference. In reduced broods, the longer a fledgling cohabited with its broodmate (before the latter's death), the less likely was it that the fledgling would recruit, indicating that growing up with a broodmate is costly; but this effect did not differ between dominant and sub-ordinate fledglings, implying that it is no more costly to suffer the oppression of a dominant broodmate than to be oppressing a subordinate broodmate. Nor did dominant and subordinate fledglings differ in their age, date, brood size, or nest success at first production or in their summed brood sizes or total nest success over the first 10 years of life. Contrary to expectation, in 81 pairs of broodmate fledglings that both recruited, subordinates tended to outperform dominants, for example hatching more chicks over the first 5 years of life.

Obligate Siblicide in the Nazca Booby and Brown Booby

Nazca and brown boobies are similar in ecology and populations of both species produce clutches of one or two eggs in roughly similar numbers, but only one chick is ever raised all the way to fledging. At Punta Cevallos on Isla Española in the Galapagos Islands (the focal Nazca booby population of this review, studied by David Anderson), two eggs are laid by higher-quality females and females given food supplements, so it is likely that food limitation constrains clutch size. Second eggs hatch

5.4 and 5.1 days after first eggs in Nazca and brown boobies, respectively, resulting in larger asymmetries in size and maturity between hatchling broodmates than occur in blue-foots. In Nazca boobies, longer hatching intervals accelerate siblicide and evidence suggests the hatching interval of this species is calibrated to ensure a prompt death of the second chick. Second eggs of Nazca boobies and probably brown boobies offer only insurance benefit. In the former, 99.95% of two-chick broods were reduced to one-chick broods, and of second chicks that survived, 70% substituted for first eggs that failed to hatch and 30% for first chicks that died. Experimental manipulation of clutch size showed that Nazca boobies are more likely to produce a fledgling from a two-egg clutch than from a one-egg clutch.

Parents of these two species are widely assumed to be unable to adequately feed two chicks through fledging and independence, and yet the estimated foraging capacity of the Nazca booby is sufficient for this task. Furthermore, when Nazca boobies on Isla Española were prevented from killing their siblings some pairs of broodmates survived a few weeks, and on Isla San Pedro Martir in the Sea of Cortez (the focal population of this review), 7% of two-egg brown booby clutches gave rise to two broodmates that grew up together for several weeks. However, the pairs of Nazca booby broodmates were undersized (wing chords 10–23% shorter and bodymass 12–17% lighter than single chicks), and in neither population was it shown that two broodmates can survive through independence and in good enough condition to be viable, or that parents' inclusive fitnesses are not compromised by caring for two chicks during 4 months or more. In theory, natural selection could favor first chicks eliminating their siblings despite parents being able to raise the whole brood, provided elimination of their competitor for parental investment improved their own inclusive fitness, even if elimination prejudiced their parents' inclusive fitness.

Dominance without subordination

In the obligate brood reduction of both species, the first chick attacks its broodmate at the earliest opportunity and sustains its pecking, biting, and pushing until the broodmate tumbles over the rim of the grass and twig nest to its eventual death. While in the nest, second brown booby chicks receive less than a quarter of the food their broodmates receive at the same age, and nearly all second chicks of Nazca and brown boobies die within 10 days of hatching.

Behavior of the second chick is hard to observe because the chick is short lived, normally concealed under a parent, and often overwhelmed by attacks when visible. But when 7-day-old brown booby second chicks were fostered for 11 days into blue-foot nests where the single 5.5-day older resident blue-foot chick was expected to dominate but not kill them, they did not respond

to attacks with submissive postures or show submission. Although they temporarily crouched and hid when attacked, these second chicks did not learn a subordinate role and were frequently aggressive. Some attacked their larger nestmate relentlessly, even managing to intimidate and expel it from the nest. Similarly, in those rare brown booby broods on San Pedro Martir where second chicks survived alongside their broodmates to age 30–60 days, they continued to attack at extremely high rates, fighting their natural broodmates tooth and nail on a daily basis.

This aggression–aggression relationship of the brown booby contrasts with the learned aggression–submission of the blue-footed booby. According to the Desperado Sibling hypothesis, despite the heavy odds against it the second brown booby chick goes on the offensive rather than acquiescing in subordination because it is doomed, and the only route to survival is by overthrowing and replacing its broodmate. And the urgent, uncompromising aggressiveness of the first brown booby chick is an evolved response to the threat from the second chick, which must be eliminated before the threat grows. More broadly, the agonistic strategy of younger broodmates in avian species with broodmate aggression is hypothesized to be a function of the cost of subordination: subordination is resisted to the extent that it reduces the probability of surviving. And the strategy of elder broodmates is an evolutionary response to the younger broodmates' strategy: the more they are likely to resist, the more they must be beaten into submission.

Parents of these two boobies build a shallow nest cup that may facilitate siblicidal expulsions, but in the brown booby at least, just as important in successful siblicide is the first chick's expulsion pushing (**Figure 7**), a motor pattern involving seizing and rushing forward that is

Figure 7   The brown booby first chick kills its sibling by expelling it over the nest rim. Outside the nest and unable to climb back in, the sibling will soon succumb to heat stress, predation, or starvation. Photo: Andrés de la Fuente.

apparently absent from the blue-foot's behavioral repertoire. Aggression is not expected to be conditional in obligate brood reducers, but an experiment showed that the less food brown booby first chicks ingest, the greater their rate of expulsion pushes. David Anderson inferred that Nazca boobies show high hatchling testosterone levels (compared to blue-foots), increase in circulating testosterone of both broodmates during fights, and long-term pathological effects of early androgen exposure on adult behavior.

See also: Aggression and Territoriality; Conflict Resolution; Dominance Relationships, Dominance Hierarchies and Rankings; Flexible Mate Choice; Food Intake: Behavioral Endocrinology; Infanticide; Mate Choice in Males and Females; Monogamy and Extra-Pair Parentage; Parent–Offspring Signaling; Social Selection, Sexual Selection, and Sexual Conflict; Spotted Hyenas.

Further Reading

Drummond H (2001) The control and function of agonism in avian broodmates. In: Slater PJB, Rosenblatt JS, Snowdon CT, and Roper TJ (eds.) *Advances in the Study of Behavior,* vol. 30, pp. 261–301. New York: Academic Press.

Drummond H (2002) Begging versus aggression in avian broodmate competition. In: Wright J and Leonard ML (eds.) *The Evolution of Begging: Competition, Cooperation and Communication,* pp. 337–360. Dordrecht, The Netherlands: Kluwer Academic Publishers.

Drummond H (2006) Dominance in vertebrate broods and litters. *Quarterly Review of Biology* 81: 3–32.

Drummond H, Torres R, and Krishnan VV (2003) Buffered development: Resilience after aggressive subordination in infancy. *American Naturalist* 161: 794–807.

Gonzalez Voyer A, Székely T, and Drummond H (2007) Why do some siblings attack each other? Comparative analysis of aggression in avian broods. *Evolution* 61: 1946–1955.

Humphries CA, Arevalo VD, Fischer KN, and Anderson DJ (2006) Contributions of marginal offspring to reproductive success of Nazca booby (*Sula granti*) parents: Tests of multiple hypotheses. *Oecología* 147: 379–390.

Mock DW and Forbes LS (1992) Parent–offspring conflict: A case of arrested development? *Trends in Ecology and Evolution* 7: 409–413.

Muller MS, Brennecke JF, Porter ET, Ottinger MA, and Anderson DJ (2008) Perinatal androgens and adult behavior vary with nestling social system in siblicidal boobies. *Plos One* 3: e2460.

Nelson JB (2005) *Pelicans, Cormorants, and their Relatives the Pelecaniformes.* New York: Oxford University Press Inc.

Nuñez de la Mora A, Drummond H, and Wingfield JC (1996) Hormonal correlates of dominance and starvation-induced aggression in chicks of the blue-footed booby. *Ethology* 102: 748–761.

Osorio-Beristain M and Drummond H (1998) Non-aggressive mate-guarding by the blue-footed booby: A balance of female and male control. *Behavioral Ecology and Sociobiology* 43: 307–315.

Osorio-Beristain M, Perez Staples D, and Drummond H (2006) Does booby egg dumping amount to quasi-parasitism? *Ethology* 112: 625–630.

Torres R and Velando A (2003) A dynamic trait affects continuous pair assessment in the blue-footed booby, *Sula nebouxii. Behavioral Ecology and Sociobiology* 55: 65–72.

Townsend HM and Anderson DJ (2007) Production of insurance eggs in Nazca boobies: Costs, benefits, and variable parental quality. *Behavioral Ecology* 18: 841–848.

Velando A, Beamonte-Barrientos R, and Torres R (2006) Pigment-based skin colour in the blue-footed booby: An honest signal of current condition used by females to adjust reproductive investment. *Oecologia* 149: 535–542.

Bowerbirds

C. B. Frith, Private Independent Ornithologist, Malanda, QLD, Australia

Introduction

Many ornithologists would concede that the behaviorally complex bowerbirds are at least as, if not more, interesting as any group of living birds. Indeed several eminent early ornithologists expressed the view that, in view of their behavior, bowerbirds should be set aside from all other birds as a distinct avian Class!

Bowerbirds are stout, typically heavy-billed, and strong-footed songbirds that range from the size of a Common Starling *Sturnus vulgaris* to that of a small slim Jackdaw *Corvus monedula*. To Europeans, many bowerbirds look generally not unlike a Common Starling or Eurasian Golden Oriole, *Oriolus oriolus*, or to Americans a Common Starling or a Gray Jay, *Perisoreus canadensis*, in general shape and proportions. The bowerbirds include both sexually monochromatic and strikingly sexually dichromatic species. Three socially monogamous and territorial species are known as catbirds and the remaining 17 species are confirmed or presumed to reproduce polygynously and apparently without defending territory other than their immediate bower or nest sites.

The promiscuous males of the polygynous bowerbird species are remarkable for their bower-building habits. They not only construct architecturally elaborate structures of sticks and other vegetable matter but also decorate them with numerous and various, often colorful, natural and (where available) human-made objects. In some species, parts of the bower structures are 'painted' with natural pigments prepared by the males and sometimes applied with a 'tool' of vegetable matter held in the bill to aid control and application of their 'paint.' In the light of these extraordinary abilities, and associated complex courtship behavior and vocalizations, bowerbirds have long been associated with high intelligence and artistic or esthetic senses beyond those of other birds (Diamond, 1982, 1987; Frith and Frith, 2008).

So structurally complex and 'artistically' decorated can some bowers be that some early European explorers reaching remote bowerbird habitats could not believe their local informants when told that the bowers were built by small birds. They therefore assumed that the attractive structures were built and decorated by human parents for the entertainment of their children.

The people of New Guinea (particularly the men) greatly admire bower-owning male bowerbirds for their industry in bower building, acquiring and retaining bower decorations, and their apparent artistry in displaying them. Papuan men perceive the birds' accumulated and defended bower decorations to be just like the wealth (some of which typically involve the feathers of birds of paradise, and rarely, those of bowerbirds) they must themselves accumulate to pay a 'bride-price' in order to acquire wives (for the men are polygamists). Past generations of Papuans living among bowerbirds doubtless had appreciated the biological significance of their bower structures, their decorations, and the associated behavior of their owning males thousands of years before science did.

Other than typically loud and harsh territorial (cat-like calls in the case of the catbirds) and bower advertisement calls, the vocalization of polygynous bowerbirds includes remarkably impressive vocal avian mimicry, mimicry of other animals, of mechanical sounds, and even of the human conversational voice.

The bowerbird literature is inundated with terms expressing a fundamentally erroneous interpretation of bowers and the behavior of birds at them. Thus, ornithologists commonly referred to bowers as 'playhouses' (and even as nests) and bird behavior at them as 'play.' Australian academic A. J. (Jock) Marshall, founding Professor of Zoology at Monash University in Melbourne, took exception to this widespread anthropomorphism and undertook bowerbird studies that sought to show that their bower building and associated behavior were exclusively dictated by the physiology of reproduction. In this, he was successful and he summarized his work in an important book about the bowerbirds (Marshall, 1954). His work and much subsequent, more sophisticated, research have failed to entirely quash the widespread misconception that bowers are some kind of nest or that the birds' activities at them represent some kind of avian recreational activity.

Origins, Relationships, and Taxonomy

Bowerbirds were long and widely considered most closely related to the birds of paradise (Family Paradisaeidae). Indeed several authorities placed both groups in the single family Paradisaeidae. Recent anatomical, behavioral, and genetic studies clearly demonstrate, however, that the two groups are quite distinct. Bowerbirds are placed within the Parvorder Corvida, which is an ancient lineage of Australo-Papuan passerine, or perching, birds derived from Gondwanan ancestors. The bowerbirds' radiation is thought to have occurred during the past 60 My, the group having diverged from the Australian lyrebirds (Family Menuridae) and scrub-birds (Family Atrichornithidae) some 45 Ma. Molecular studies time the separation of bowerbirds (superfamily Menuroidea) from the other corvines (superfamily Corvoidea, including the birds of paradise) at some 28 Ma. They also suggest that major lineages within the bowerbirds originated about 24 Ma. Several molecular studies indicate the close relationship between the bowerbirds, lyrebirds, and the Australian scrub-birds, these together forming a basal (or primitive) group of Australasian songbirds (Frith and Frith, 2004 and references therein).

The only known bowerbird fossils (from sites in Victoria, Australia; two from the Holocene and one the Pleistocene) are of the Satin Bowerbird *Ptilonorhynchus violaceus*. These are from locations remote from present-day wet forests and thus living Satin Bowerbirds, this attesting to more extensive subtropical rainforests in Australia's geological past. The Australian gray, *Chlamydera*, bowerbirds adapted to the steadily drying Australian continent from rainforest-dwelling ancestors of *Sericulus–Ptilonorhynchus* stock, and this heritage can be discerned in some of their plumage and courtship display traits (Frith and Frith, 2004).

As contemporary biomolecular studies have resulted in a taxonomic classification radically different in the number of bowerbird genera and constituent species (Kusmierski et al., 1993; Christidis and Boles, 2008) from traditional usage (reviewed in Frith and Frith, 2004), it is desirable and pertinent to clarify this situation here. The bowerbird genera and species had, except those suggested by one author in a regional avifaunal species list lacking discussion or justification, long remained stable. Almost all authors consistently acknowledged three major lineages within the family as follows: (1) *Ailuroedus*, (2) *Scenopoeetes–Amblyornis–Archboldia–Prionodura*, and (3) *Sericulus–Ptilonorhynchus–Chlamydera*. The traditional and the suggested new classification is shown in **Table 1**.

This new taxonomy suggests that the traditional eight bowerbird genera be reduced to five. It also shows a different location of the Fawn-breasted Bowerbird within the enlarged genus *Ptilonorhynchus* and a change of spelling of two specific names (to *maculatus* and *guttatus*). If this new taxonomy is accepted, the members of the genus *Sericulus* will probably also need to be combined within the enlarged *Ptilonorhynchus* of the new classification, because the external morphology of Lauterbach's Bowerbird is clearly intermediate in character between those of the traditional *Chlamydera* and *Sericulus* species. Thus, those that were traditionally treated as species of *Ptilonorhynchus* and *Chlamydera* (to now form only *Ptilonorhynchus* of the new classification) will all have to become species of *Sericulus* because *Sericulus* was

Table 1

Traditional classification	New classification
Monogamous species without a court or bower	
Genus *Ailuroedus*	**Genus *Ailuroedus*** (catbirds)
White-eared Catbird	White-eared Catbird
A. buccoides	*A. buccoides*
Black-eared Catbird	Black-eared Catbird
A. melanotis **(Figure 1)**	*A. melanotis*
Green Catbird	Green Catbird
A. crassirostris	*A. crassirostris*
Polygynous species with a court or bower	
Genus *Scenopoeetes*	**Genus *Scenopoeetes*** (court clearer)
Tooth-billed Bowerbird	Tooth-billed Bowerbird
S. dentirostris **(Figure 2)**	*S. dentirostris*
Genus *Amblyornis*	**Genus *Amblyornis*** (maypole bower builders)
Macgregor's Bowerbird	Macgregor's Bowerbird
A. macgregoriae **(Figure 3)**	*A. macgregoriae*
Streaked Bowerbird	Streaked Bowerbird
A. subalaris	*A. subalaris*
Vogelkop Bowerbird	Vogelkop Bowerbird
A. inornatus	*A. inornatus*
Yellow-fronted Bowerbird	Yellow-fronted Bowerbird
A. flavifrons	*A. flavifrons*
Genus *Archboldia*	Archbold's Bowerbird
Archbold's Bowerbird	*A. papuensis*
A. papuensis **(Figure 4)**	Golden Bowerbird
Genus *Prionodura*	*A. newtoniana*
Golden Bowerbird	
P. newtoniana **(Figure 5)**	**Genus *Sericulus*** (avenue bower builders)
Genus *Sericulus*	
Masked Bowerbird *S. aureus*	Masked Bowerbird *S. aureus*
Flame Bowerbird *S. ardens*	Flame Bowerbird *S. ardens*
Adelbert Bowerbird *S. bakeri*	Adelbert Bowerbird *S. bakeri*
Regent Bowerbird	Regent Bowerbird
S. chrysocephalus **(Figure 6)**	*S. chrysocephalus*
Genus *Ptilonorhynchus*	**Genus *Ptilonorhynchus*** (avenue bower builders)
Satin Bowerbird	
P. violaceus **(Figure 7)**	Satin Bowerbird *P. violaceus*
Genus *Chlamydera*	Lauterbach's Bowerbird
Lauterbach's Bowerbird	*P. lauterbachi*
C. lauterbachi	Spotted Bowerbird
Fawn-breasted Bowerbird	*P. maculatus*
C. cerviniventris	Western Bowerbird *P. guttatus*
Spotted Bowerbird	Great Bowerbird *P. nuchalis*
C. maculata	Fawn-breasted Bowerbird
Western Bowerbird	*P. cerviniventris*
C. guttata	
Great Bowerbird	
C. nuchalis **(Figure 8)**	

the first of these genera to be named and therefore must have priority under the rules of zoological nomenclature.

While hybridization between members of different bowerbird genera (intergeneric) is unknown, hybridization does occur, extremely rarely, within genera (intrageneric), being confirmed between two congeneric species of Amblyornis of *Sericulus*, and of *Chlamydera* (Frith, 2006).

Figure 1 Black-eared Catbird. One of the three monogamous bowerbird species, that are known as catbirds. Copyright Clifford Frith.

Figure 2 The Tooth-billed Bowerbird. The only court-clearing bowerbird. Seen here is an adult male calling over his leaf-decorated court. Copyright Clifford Frith.

Figure 3 Macgregor's Bowerbird. One of the mapole bower builders. An adult male is here perched at the edge of his bower mat. Copyright Clifford Frith.

Figure 4 Archbold's Bowerbird. Builder of a dispersed maypole bower that is mostly a mat of fern fronds. Here an adult male adds a decoration. Copyright Clifford Frith.

For continuity and convenience, the traditional classification, the presently familiar one and that found in all standard ornithological works of reference, is used herein.

Morphology and Diet

Bowerbird species vary in size from 21 to 38 cm in body length and 70–230 g in weight. Males are larger than

Figure 5 Golden Bowerbird. One of the maypole bower builders. Here an adult male attends to his bower decorations. Copyright Clifford Frith.

Figure 7 Satin Bowerbird. This avenue bower building species is the best studied bowerbird. An adult male reviews his decorations. Copyright Clifford Frith.

Figure 6 Regent Bowerbird. Here an adult male displays to an immature male in his avenue bower. Copyright Clifford Frith.

Figure 8 Great Bowerbird. This bowerbird, and some other members of its genus, add heavy stone pebbles to their bower decorations. In addition to representing 'rare' items on bowers distant from water, their significant weight may also tell females something of the owning male's strength and vigour. Copyright Clifford Frith.

females in all species except the Golden Bowerbird and the 'silky' bowerbirds of the genus *Sericulus*, in which males are smaller. Bowerbird family members include approximately 50 or so different plumages, given that most species wear a juvenile one followed by an adult male or female one, with some species also wearing a subadult male plumage for at least one year. Plumages of juvenile and immature individuals are similar to those of their respective adult females. Males of the polygynous species may wear female-like immature feathering for 6–7 years, before attaining their first adult plumage and presumably entering the breeding population. Females may, however, first breed when 2 years of age. Adult bowerbirds enjoy a high average expectancy of life, individuals of some species being recorded attaining an age of 20–30 years.

The three socially monogamous *Ailuroedus* catbird species are sexually monochromatic, while the polygynous genera are typically sexually dichromatic to some degree. Catbird sexes are alike, being green overall with white spotting on their breast, wings, tail, and about the head and throat. Sexes of the polygynous Tooth-billed Bowerbird are identically olive brown above and heavily streaked brown on whitish under parts. Adult males of other polygynous species of forest habitats are brightly pigmented gold or orange and black in Masked, Flame, Adelbert, Regent, and Archbold's Bowerbirds; glossy blue-black in the Satin Bowerbird; canary yellow and golden-olive in the Golden; or overall olive-brown with an orange or yellow crest in most 'maypole' bower building *Amblyornis* bowerbirds. The 'avenue' bower building *Chlamydera* bowerbirds of more open habitats are gray or brownish with only a small pinkish nape crests in three species. Females of the promiscuous species are drably plumaged in browns, olives, or grays and often with ventral spotting or barring.

Bowerbirds have an exceptional 11–14 secondary wing feathers (including the tertials), whereas typical songbirds have 9–10. An enlarged lachrymal (a part of the cranium near the orbit) is also a bowerbird trait that is otherwise found only in the Australian lyrebirds (Family Menuridae) within songbirds.

Bowerbirds have stout and powerful bills typical of a generalist avian omnivore. Exceptions to this are a longer and finer bill in Regent Bowerbirds as an adaptation to flower nectar feeding and a falcon-like toothed bill in Tooth-billed Bowerbirds as an adaptation to leaf eating. Complex structures on the Tooth-billed Bowerbird's inner mandible surfaces are used to masticate their unusual songbird diet. Most bowerbird species are predominantly frugivorous, but flowers, nectar, arthropods (mostly insects), and other animals, including smaller vertebrates, are eaten. Figs form a major component of catbirds' diets and they store, or cache, fruits about their territories, as do males of some polygynous species about their bower sites. Animal foods are important to the nestlings of the polygynous species. Some avenue bower building species form winter flocks (that may damage commercial fruit crops) and flocking Satin Bowerbirds will 'graze' on the ground on herbs and grasses (Frith and Frith, 2004).

Distribution and Habitats

Bowerbirds as a group inhabit tropical, temperate, and montane rainforests, riverine and savanna woodlands, rocky gorges, grassland, and arid zones of only New Guinea and Australia. Ten species occupy only New Guinea, eight only Australia, and two species are common to both places. Most species inhabit wet forests, up to 4000 m above sea level in the case of Archbold's Bowerbird. Several are highly localized (e.g., the Adelbert Bowerbird being confined to the Adelbert Mountains of Papua New Guinea, and the Golden and Tooth-billed Bowerbirds to rainforests above 900 m of the Atherton Region

of north Queensland, Australia), while others, notably New Guinea's Flame and Australia's Spotted and Great Bowerbirds, are distributed extensively.

Types and Functions of Bowers

The three catbird species do not clear a court or build a bower of any kind, their courtship involving a little more than a male chasing his mate through the trees of their traditional territory.

The Tooth-billed Bowerbird builds no bower but meticulously clears a forest floor court of debris before laying large leaves upon it, their paler undersides uppermost, as contrasting decoration.

The species of *Amblyornis*, *Archboldia*, and *Prionodura* all build a bower of the 'maypole' type, involving stacking or hanging sticks or other vegetable matter about or from a vertical sapling trunk or horizontal branches and laying out a mat of mosses or ferns beneath their structure. These bower mats are decorated with various colorful items that include fungus, tree resin, insect wings, fruits, flowers, snail shells, and more.

Males of the Satin Bowerbird (*Ptilonorhynchus*), the silky bowerbirds (*Sericulus* spp.), and the gray bowerbirds (*Chlamydera* spp.) all build 'avenue' bowers. The basic nature of the, traditionally sited, avenue bowers consists of two parallel walls of sticks erected vertically upon and into a foundation 'platform' of sticks laid haphazardly upon the ground. Simple avenue bowers consist of no more than this, being smaller and frailer in the *Sericulus* species, of medium size and strength in *Ptilonorhynchus*, and larger and stronger in the *Chlamydera* species. In the Fawn-breasted Bowerbird, the avenue bower is, however, elaborated upon by being built atop a bulky and tall foundation platform and in Lauterbach's Bowerbird, it is elaborated further by an additional wall being added at right angles to, and across, the entrances of the basic avenue.

American ornithologist E. Thomas Gilliard (1969) made the observation that adult males of the two latter *Chlamydera* species, which build the more complex and elaborate bowers, lack the colorful crest worn by other members of their genus that build simpler bowers. He noted this same correlation between bower complexity and adult male plumage within the *Amblyornis* bowerbirds. Adult males of each *Amblyornis* species differ in the extent of their colorful crest relative to the complexity of their bower structure; Macgregor's and Yellow-fronted Bowerbirds wear a large orange-yellow crest and build a simple maypole bower, Streaked Bowerbirds wear a smaller and duller crest but build a more complex and roofed maypole bower, and Vogelkop Bowerbirds have no crest but build a most complex roofed maypole bower.

Gilliard concluded that as a species evolves a more complex and elaborate bower, its secondary sexual characters of adult male plumage are transferred to bower architecture and colorful bower decoration. Sexual selection, through female choice, has apparently caused this transfer of colorful visual sexual signals from the plumage of adult males to their bowers and decorations. Females selectively mate with males owning superior bowers and in doing so, enhance bower architecture because males with inferior bowers fail to reproduce. The disadvantage (cost) of bright male plumage, in attracting predators to the bird and its bower (where males spend much time), would enhance the loss of such plumage (through natural selection) as the transfer of secondary sexual characters to the bower progressed. This hypothesis, known as the Transferal Effect (Gilliard, 1969), remained robust until a phylogenetic analysis of the bowerbirds indicated a lack of support for it (Kusmierski et al., 1993).

A number of hypotheses have recently been advanced to explain the function of bower structures. These include: the rape hypothesis, which suggests that females gain protection from forced copulations by a courting (or any other) male; the threat reduction hypothesis, which suggests that the bower provides a screen that females interpose between themselves and courting males in order to reduce the supposed threatening nature of males' sexual advances; and the hide-the-female hypothesis, which suggests that bowers may function to conceal the female from rival males while the bower owning male is attempting to court her (see Frith and Frith, 2004, and references therein). Future research will suggest other ideas about the function(s) of bower structures and it is quite possible that no one idea will provide an adequate answer to all bower types, if for even a single bower type.

Mating Systems and Nesting

Socially monogamous catbird pairs defend an all-purpose territory within which they nest and forage, pairs remaining together over several seasons if not for life.

Bower structures have nothing whatsoever to do with nests or nesting but are critical to the reproductive success of males of the polygynous bowerbirds. Males build bowers as a traditional focal point to which they attract multiple females for courting and mating each season. Thus, bowers function by representing symbols of each individual male's fitness 'symbols that are external to the males' body and that can, thus, be viewed and assessed even in the absence of the male owner. Such assessment is the concern of the females, and by comparing the bower sizes, materials, construction quality, numbers, types, and quality of bower decorations, females can obtain a measure of relative male fitness. To this, choosy females add an assessment of the health and vigor of males, their

plumage, and the intensity and quality of their courtship display movements, posturing, and vocalizations at their bowers, before soliciting the one of their choice.

If a female satisfies herself that the bower and its decorations, the male, and the frequency and intensity of his displays are acceptable, then she will solicit her chosen mate to be fertilized by him, typically within or at his bower. Once fertilized, females have no more to do with males and their bowers for that particular breeding season, unless the loss of a clutch or brood (i.e., to a predator) dictates the need to nest again in that season.

Older males, with greater experience, skills, and survival, are those typically selected by females and are thus more successful in obtaining matings. Courts and bowers are sited on topography exhibiting one or more required environmental features. No evidence of typical lekking by adult male bowerbirds exists except for the courts of Tooth-billed Bowerbirds, which appear to be unevenly dispersed through habitat to form denser aggregations. Because the males forming these aggregations are not in visual (only vocal) contact, they form 'exploded leks' as opposed to true ones (in which males are in visual contact).

Bower sites are occupied for decades, with individual males exhibiting long-term fidelity to them, with some of the Satin Bowerbird being occupied for 50 years or more. Because Golden Bowerbird bowers remain intact from one year to the next, it is conceivable that the same basic structure is actively in use for many decades. Female-plumaged immature male bowerbirds spend up to 5 or 6 years visiting rudimentary, or practice, courts or bowers of their own construction as well as bowers of adult males. Here, they acquire and learn the skills required for bower building, decorating, and displaying. Initially, they build inferior bower structures, which improve with experience.

Socially monogamous catbird pairs defend an all-purpose territory year round. Only female catbirds build nest, incubate the clutch, and brood the nestlings, but their mates feed them during these activities and do share in provisioning the nestlings to their independence. Offspring have a long period of post-nestling parent dependency. Females of the polygynous bowerbirds nest build, incubate their clutch, and raise their offspring entirely alone and unaided by the father or by any other conspecifics. Their nests are bulky open cup or bowl-shaped structure built upon a foundation of twigs, leaves, and vine tendrils into a tree fork, vine, or mistletoe tangle. Only the Golden Bowerbird nests in a tree crevice. The clutch consists of one to two, rarely three, eggs that are plain off-white to buff or blotched and vermiculated with pigmented scrawling lines, predominantly about their large end. Unlike those of most songbirds, bowerbird eggs of a clutch are laid on alternate days. The incubation period lasts about 21–27 days, and the nestling period approximately 17–30 days.

The Study of Bowerbird Behavior

Ever since they were first made known to the scientific world, people have studied the complex behavior of bowerbirds. Since the invaluable 1969 summary of Thomas Gilliard's life work, such studies have, however, increased exponentially. The majority of this recent work has involved those species more easily accessible: the Black-eared and Green Catbirds, Tooth-billed, Macgregor's, Archbold's, Golden, Regent, Satin, Spotted, and Great Bowerbirds and in terms of more experimental approaches the three latter species. Most long-term qualitative experimental studies have involved the Satin Bowerbird and the use of large numbers of automatically controlled video cameras at numerous bowers. These and other studies provide valuable insights into various aspects of bowerbird behavior and sexual selection. These include relationships between bower quality and mating success; the high skew among males in mating success, paternity; bower orientation; the function of bower paint; sexual competition through bower destruction and decoration theft by rival males; the cost of male displays; male mating success relative to their parasite loads and plumage condition; the function and significance of courtship vocalizations, with particular reference to vocal avian mimicry; the cost of bower display to males; mate-searching tactics by females; visitation patterns to the bowers of adult males by immature males; how males adjust their displays to the response of females; various other aspects and details of female choice including male avian vocal mimicry and other vocalizations; and more (for reviews see Borgia, 1995; Patricelli et al., 2004; Frith and Frith, 2004, 2009; Reynolds et al., 2007 and references therein).

Bowerbirds are 'Brainy'

A recent study demonstrated that bowerbirds do in fact have a brain that is relatively large compared with that of ecologically similar songbirds of their body size and zoogeographical distribution. The same author showed that the brain of bower-building species is larger than that of the non-bower-building species (Madden, 2001). A further study demonstrated that it is the temporalis fossae, or rear brain cavity, that is the enlarged part of the brain in the bower-building bowerbirds (Day et al., 2005). However, a larger brain does not necessarily dictate any level of corresponding higher intelligence, but may merely relate to more acute visual or other kinds of instinctive perceptions or abilities.

Some contemporary research has literally been looking at bowerbirds' bowers and their decoration in a different light (Endler and Day, 2006 and references therein). Bowerbirds, like all birds, perceive color in a very different way compared to human beings in that they are far more sensitive to the ultraviolet. Thus, interpretations of the colors of bower structures, their decoration, and of the plumages of the birds themselves are potentially fraught with dangers, as recent studies of such traits demonstrate. For example, male bowerbirds preferentially select certain colors for their UV-reflective qualities rather than for the colors as perceived by people. This opens up an entirely new field of investigation into the function and significance of both the plumage and external (bowers and bower decorations) secondary sexual characters of bowerbirds. In this field of study, as in the numerous others that their extraordinary behavior presents, a great deal of deeply interesting research remains to be performed on all bowerbird species (for a review see Frith and Frith, 2004 and references therein).

See also: Mate Choice in Males and Females; Mating Signals; Social Selection, Sexual Selection, and Sexual Conflict; Visual Signals.

Further Reading

Borgia G (1995) Why do bowerbirds build bowers? *American Scientist* 83: 542–547.

Christidis L and Boles WE (2008) *Systematics and Taxonomy of Australian Birds*. Melbourne: CSIRO Publishing.

Day LB, Westcott DA, and Olster DH (2005) Evolution of bower complexity and cerebellum size in bowerbirds. *Brain, Behaviour and Evolution* 66: 62–72.

Diamond JM (1982) Evolution of bowerbirds, animal origins of the aesthetic sense. *Nature* 297: 99–102.

Diamond JM (1987) Bower building and decoration by the bowerbird *Amblyornis inornatus*. *Ethology* 74: 177–204.

Endler JA and Day LB (2006) Ornament color selection, visual contrast and the shape of color preference function in Great Bowerbirds *Chlamydera nuchalis*. *Animal Behavior* 72: 1405–1416.

Frith C (2006) A history and reassessment of the unique but missing specimen of Rawnsley's Bowerbird *Ptilonorhynchus rawnsleyi*, Diggles 1867, (Aves: Ptilonorhynchidae). *Historical Biology* 18: 53–64.

Frith CB and Frith DW (2004) *The Bowerbirds*. Ptilonorhynchidae, Oxford: Oxford University Press.

Frith CB and Frith DW (2008) *Bowerbirds: Nature, Art & History*. Malanda: Frith & Frith.

Frith CB and Frith DW (2009) *Family Ptilonorhynchidae (Bowerbirds)*. In: Hoyo J, Elliot A, and Christie DA (eds.) *Handbook of the Birds of the World. Vol. 14. Bush-shrikes to Old World Sparrows*, pp. 350–403. Lynx Edicions, Barcelona.

Gilliard ET (1969) *Birds of Paradise and Bower Birds*. London: Weidenfeld and Nicolson.

Kusmierski R, Borgia G, Crozier RH, and Chan BHY (1993) Molecular information on bowerbird phylogeny and the evolution of exaggerated male characteristics. *Journal of Evolutionary Biology* 6: 737–752.

Madden JR (2001) Sex, bowers and brains. *Proceedings of the Royal Society of London B* 268: 833–838.

Marshall AJ (1954) *Bower-Birds, Their Display and Breeding Cycles – A Preliminary Statement*. Oxford: Oxford University Press.

Patricelli GL, Uy JA, and Borgia G (2004) Female signals enhance the efficiency of mate assessment in satin bowerbirds (*Ptilonorhynchus violaceus*). *Behavioural Ecology* 15: 297–304.

Reynolds S, Dryer K, and Bollback J (2007) Behavioural paternity predicts genetic paternity in satin bowerbirds (*Ptilonorhynchus violaceus*) a species with a non-resource-based mating system. *The Auk* 124: 857–867

Relevant Websites

www.life.umd.edu/biology/borgialab/ – Bowerbirds and sexual selection.

Caching

A. Brodin, Lund University, Lund, Sweden

Introduction

Many of us have observed the behavior of marsh tits or black-capped chickadees at a bird feeder. A particular bird will return at 1-min intervals, each time grabbing a peanut or sunflower seed and disappearing quickly into neighboring trees. If we follow an individual bird more closely through binoculars, it soon becomes obvious that the bird is not eating. Instead, it caches the nuts in bark crevices or under lichens. This behavior makes it possible for this little bird to secure a supply of food that it can retrieve and eat later, when the feeder is empty.

Animals need energy continuously but most animals obtain their food in discrete bouts of eating. This means that the intermittent events when energy is acquired from the environment must be buffered over short or long time intervals. In a short time perspective, digestion of food in the gut and hormonal regulation of the glucose level in the blood may be sufficient. In a longer perspective, for example when a hungry animal cannot eat, stored energy must provide the buffering effect.

Energy storing can occur both in- and outside an animal's body. Inside the body, energy will primarily be stored as carbohydrates or fat. In many cases, however, internal energy stores are not sufficient. Especially animals with well-developed cognitive abilities will also store food externally to buffer the energy need, a phenomenon called 'food hoarding' or 'caching.'

A few words about the terminology: The term 'hoarding' is more specific than the general term 'storing'; clearly, body fat deposits are stored but not hoarded. Even more specific is the commonly used 'caching,' a term that should be reserved for cases when hoarded food is actively concealed in some way. Concealment could be achieved by burying or covering caches or by storing them in a way that make them cryptic. The behavior of snowy owls that pile lemmings around the nest or acorn woodpeckers that store acorns in open holes in tree trunks can thus be categorized as 'food hoarding,' but not in a strict sense as 'caching.' I will use 'food storing' and 'hoarding' as alternatives for storing of food in general but 'caching' only for cases when animals conceal their hoarded food.

Little is known about possible differences in nutritional quality between hoarded and fresh food. Henceforward, I will consider hoarded food mainly from an energy perspective and not deal with its nutritional composition.

Food Hoarding in the Animal World

Food storing is especially well known in mammals and birds but also occurs in some invertebrates. To humans, the most familiar case of invertebrate food storing is honey production in bees. In summer and autumn, honeybees will gather nectar from flowers. They will transform the nectar into a form that is more durable for long-term storing, that is honey. The honey will then provide the colony with energy over the winter, unless, of course, a beekeeper replaces the honey by sugar water.

Among mammals, most predators will store the portion of their kill that remains after they have become satiated. Also, small predators such as shrews and moles may hoard surplus food. In its underground burrows, a single European mole may store 1–2 kg of earthworms. The mole keeps the worms in a half-alive 'zombie' state that maintains their freshness. To achieve this, the mole decapitates the worm and pushes its end into the earth wall, where it must stay until the mole eats it. Large-scale hoarders among mammals include rodents such as squirrels, mice, and beavers. In autumn, beavers will fill their pond with submerged parts of trees and bushes. In winter, they can use this supply.

Some bird species are widely known for their food-hoarding habits. The *Corvidae* (crows, jays, magpies, and their allies) is a passerine family that contains many food-storing species. Among these, nutcrackers and jays are especially well known. Some of these may store many thousands of seeds in scattered over a large area. They can remember these locations for at least a year if necessary. Nutcrackers may be the most highly specialized hoarders in the animal world. Below the beak, they have a special pouch that can be filled with up to 20 hazelnuts or the equal volume of pine nuts. When a load of nuts has been collected, the nutcracker may fly many kilometers (or miles) to reach especially suitable storing areas, for example, their breeding territory or areas where snow cover will be thin. Thin snow cover facilitates cache retrieval in winter. Most of the winter food for a nutcracker consists of stored nuts.

Nutcrackers have an almost 100% accuracy in nut recovery, even when they have to make excavations through deep snow months after storing. There are other corvids that are almost as specialized hoarders as nutcrackers, for example, pinyon, European, gray and Siberian jays, and the European rook.

Also, some species of nuthatches and titmice are food hoarders of this magnitude. There are a number of food hoarders in the family *Paridae* (chickadees and titmice), for example willow and marsh tits, and their American close relatives the black-capped and boreal chickadees. A single individual of one of these species may store 50–100 000 seeds and nuts in an autumn. The pygmy owl is an important predator of small northern birds such as chickadees and titmice. Even though these small owls are not scatter hoarders like corvids or parids, they deserve to be mentioned here for another reason. In winter, they hunt during the day, ambushing small birds that fly by them. A willow tit may store a large supply of seeds as winter food, whereas a pygmy owl may fill up cavities and nest boxes with willow tits!

The acorn woodpecker that occurs in Western North America is famous for its large and conspicuous larders. Acorns are wedged into small holes into one or a few adjacent tree trunks or telephone poles. These surfaces may be pierced by tens of thousands such holes. These granaries are not inconspicuous or covered. Instead, they are maintained and defended against acorn pilferers by a family group of woodpeckers. When acorn woodpeckers happen to make a hole in some man-made structure such as the wall of a box, they may continue to drop acorns in the empty space behind the wall even if they cannot retrieve the acorns.

Food Storing and Climate

Some types of food storing seem to occur irrespective of climate. As mentioned earlier, large predators may hide the remnants of a kill and return to the carcass later. Large-scale scatter hoarding, on the other hand, is a strategy that occurs in species that inhabits boreal or cool temperate regions. It can primarily be seen in autumn, thus being a strategy to increase food availability in winter. Typical examples are specialist hoarders such as jays, titmice, and squirrels in the northern forests but also

desert-dwelling rodents such as kangaroo rats may rely heavily on food hoarding.

The smaller an animal is, the less heat-producing body mass it will have compared to its body surface. This means that small birds that spend the winter in the boreal coniferous forests face a considerable energetic challenge. Species that weigh only 10–12 g, such as boreal chickadees, willow tits, and Siberian tits, are all highly specialized food hoarders that will store a large proportion of their winter food in the autumn. At high latitudes, low temperatures in combination with short days make their winter residency a remarkable feat. Undoubtedly, it is their massive food-storing effort in the autumn that makes it possible.

History of Food-Hoarding Studies

Early Studies

Food hoarding is a conspicuous behavior that has been noted by human observers for a long time. A food-hoarding study that was amazingly modern in its design was published already in 1720. In 'Angenehme Landlust' (agreeable country pleasures), Baron Johan Adam von Pernau describes how to make marsh tits store food in captivity. The marsh tit is a European close relative to the chickadees in America. The baron tells us that someone that searches for some kind of mind in animals should bring a marsh tit indoors, put a tree in the room, and let the bird get used to the environment. Just like in modern hoarding experiment, the baron then starved the bird for half a day to increase its motivation. Then, he provided it with hemp nuts and allowed it to store these. When the bird later was allowed to recover its stores, the baron concluded that the purposeful retrieval behavior showed that the bird considered its caching locations in such a way that its behavior cannot be explained by instinct.

In the 1940s and 1950s, ornithologists made long-term field studies of the hoarding behavior in several bird species. In Sweden, PO Swanberg described the high recovery success of nutcrackers. When nutcrackers make recovery excavations through snow, they usually leave cracked nutshells behind if they have been successful. Throughout the winter, the average success rate was almost 90%. Considering that some caches may have rotted, others pilfered, etc., the memory accuracy in the nutcrackers must have been almost 100%.

In Norway, Svein Haftorn described the hoarding behavior in three European parids that inhabit the coniferous forest, the willow tit, the crested tit, and the coal tit. Among other things, he showed that seeds that only were available in the autumn constitute an important part of the winter diet. He could show this by examining stomach contents of birds that he shot in late winter.

In the 1960s, a Soviet scientist, Krushinskaya, allowed Eurasian nutcrackers to store in an aviary and then performed hippocampus surgery (unanesthesized, with a spoon!) in the hoarding birds. Birds with hippocampus lesions stored normally but could not relocate the caches afterward. Similar studies (using anesthesia!) have been made afterward, both on nutcrackers and titmice. These confirm that nutcrackers use visual memory based on landmarks during retrieval and that hippocampal damage will impair retrieval success.

Modern Studies

The era of modern studies started with a game theoretical study by Andersson and Krebs, published in 1978. They showed that a costly and time-consuming behavior such as large-scale food hoarding has to pay for the individual hoarder if it should evolve. This means that there must be some mechanism that ensures that a hoarder has higher probability to relocate caches than scroungers. For species that scatter many caches widely, memory is a mechanism that evidently would provide such an advantage.

The same year, 1978, Stapanian and Smith published a study that spurred a number of studies of how animals should distribute scatter-hoarded food to minimize cache loss. Their idea was that a scrounger that happens to find a cache will search nearby potential caching locations. To reduce cache loss, a hoarder should disperse caches widely enough to reduce cache loss, at what they labeled 'optimal cache density.' Stapanian and Smith combined predictions from a theoretical model with empirical data on fox squirrels hoarding walnuts.

In some cases, coevolution has been demonstrated between scatter hoarders and the nuts that they store. Most large-seeded pines, for example, seem to depend on nutcrackers for their dispersal. In the same way, the Eurasian jay is important for dispersal of English oak and the blue jay for dispersal of pin oak. It may seem detrimental for a tree to gets its nuts taken by scatter-hoarding nut eaters. However, not all nuts are retrieved; many are left intact in the ground and hence already adequately planted.

Two other influential studies have been important for modern studies of scatter hoarding, spatial memory, and the hippocampus. In 1989, Krebs and coworkers and Sherry and coworkers showed that passerine families that contained hoarders on average have larger hippocampi than families with nonhoarders. The hippocampus is a brain structure that is important for the storage of memories in the brain. After these two studies, a number of comparative studies of hippocampal volumes in the context of food hoarding have been published. The phylogenetical level in these has ranged from family and species comparisons down to comparisons of geographically separated populations within one species.

Food-Hoarding Strategies

Stored food is an investment that must be protected in some way. There are two distinct strategies for this. The collected food can be gathered in one or a few caches, a behavior called 'larder hoarding.' Alternatively, food items can be spread out widely with only one or a few items in each cache, a behavior called 'scatter hoarding.' Although both types of caches can be protected for example by burying them, there is a difference in how they are protected. Scattered caches are primarily protected by concealment, whereas larders in many cases must be protected by physical defense.

Larder Hoarding

A chipmunk or mouse that stores nuts in a burrow may be safe from avian scroungers such as jays, but the cache may not be safe from other chipmunks or mice. If a competitor tries to take over the cache, it must be defended or it will be lost. The chipmunks of eastern North America provide an illustrative example. Dominant older individuals can defend their burrows, and hoard in larders, although younger, subordinate individuals cannot defend burrows and they are scatter hoarders.

Most birds that store nuts are scatter hoarders that cache their food. A group of acorn woodpeckers, however, store thousands of acorns in a so-called acorn granary. Acorns are clearly visible in holes in the tree trunks. However, they are wedged tightly into the perfectly sized holes, making it hard for squirrels to extract them. The woodpeckers can easily crack the acorns with their bills. Of course, the woodpeckers still have to defend their granary actively, for example against jays and other woodpeckers.

Beavers will stick trunks and limbs of trees in the bottom of their pond so that the upper ends protrude through the water surface. When winter comes, the trunks will freeze into the ice cover and the beavers can eat from the parts below the ice. The exit from the beaver lodge is below the water surface, meaning that the beavers can stay in their safe hut or below the ice as long as they have stored food left.

Scatter Hoarding

The term 'scatter hoarding' was coined in 1962 by Desmond Morris, who studied hoarding in the green acouchi, a South American rodent. He provided captive animals with trays that were filled with sand as hoarding substrate. He found that the acouchis distributed caches evenly between the trays in order to optimize dispersion.

Scatter hoarding occurs in many mammals, but the most striking examples are found among birds. Both the Eurasian and Clark's nutcrackers have been estimated to make 8000–10 000 caches in one autumn. Each cache will contain a 'pouch load' of pine or hazel nuts, making the total number of nuts more than 100 000. In the White Sea region in Northern Russia, individual Siberian and willow tits may cache up to 150 000 food items during the autumn. Tits place each item in a separate position. The tits cannot guard such a huge number of caches scattered over a large area, so they conceal them. These species (and their American close relatives the chickadees) will frequently store in trees. Caches are wedged into bark crevices or under loose fragments of bark or pushed in under pieces of lichen. The Eurasian nuthatch is even more elaborate when it caches. It excavates a small hole, for example in a tree trunk and then pushes a nut into this hole. After storing, it may push fragments of bark into the hole and cover the nut completely.

Nutcrackers and jays will cache acorns and nuts in the ground on the forest floor. First, they will make an excavation, then fill this with nuts, and, finally cover it with moss, grass, or leaves. The fact that some of these nuts will be left or forgotten is shown by that single oaks or hazels are growing in the coniferous forest, far from other trees of the same species. The rook is a crow-sized colonially nesting corvid that occurs in towns and in farmland in Europe. In the autumn, it will store hazelnuts, walnuts, and acorns until trees carrying these in the surroundings are emptied. The nuts are stored in substrates such as lawns where the rooks push them under the grass surface with their strong bills. In winter, the birds will spend much time managing their investments. They spend considerable time retrieving caches seeds, not consuming them, but moving them a short distance before they are stored in a new position.

Benefits of Food Hoarding

The benefit of food hoarding can be summarized in one sentence: it evens out the food supply. However, this benefit is valid for all types of energy storing. This makes it necessary to be more precise to identify the specific benefits of food hoarding. There are at least five more or less separate benefits:

1. Hoarding may increase the proportion of an ephemeral food source that a forager can secure. The difference in behavior between crows and gulls at a pile of fish parts can serve as an illustration of this. Gulls will typically scream and fight over the fish parts, and this conspicuous and loud behavior will attract more gulls. The scene may develop into something of a feeding frenzy; every time a gull succeeds in grabbing a large piece, it will be harassed by other gulls trying to steal the food. Most individual gulls seem to be rather unsuccessful. Crows, on the other hand, will calmly wait for the right opportunities and methodically cache as much fish guts they can, returning to eat later when the gulls have left.

2. Hoarding may increase the food supply during long periods of food scarcity. The large-scale food hoarding observed in autumn in many animals is a way of increasing the food supply during the winter. Bird species that are critically dependent on specific types of stored food will irrupt in irregular migration during years when their preferred crops fails in autumn.

3. The insurance effect. The mere existence of stored food may be beneficial even in cases when the food is not consumed. A hoarder that has access to stored supplies can afford to carry smaller internal fat deposits reserves than a nonhoarder. The stores make the hoarder less susceptible to rare periods with extremely bad foraging conditions. Imagine a morning in a northern forest when an overnight blizzard has covered many foraging substrates that normally are available to small birds. A nonhoarder that wakes up with small body fat reserves such a day may die, whereas a hoarder can change foraging strategy and start retrieving caches.

4. Food hoarding may decrease predation risk. The beavers mentioned earlier suffer very small predation risk as long as they can stay in the pond and eat pieces of trees that have been stored under the ice. The same is true for a rodent that keeps its winter food in its burrow and thereby can avoid exposure to owls.

5. Food storing may increase breeding success. The northwestern crow in North America eats animals such as clams that become available when tide is retreating. During breeding, the male will store such animals during retreating tide and then provide it later to the incubating female. The even supply of food makes it possible for her to spend more time on the nest.

Retrieval of Scatter-Hoarded Caches

Food that is stored in a central larder is easy to retrieve when needed. Food that is stored in thousands of scattered, cryptic caches, on the other hand, may be difficult to relocate. The obvious retrieval mechanism for a scatter hoarder is a precise memory for its caching locations. To remember thousands of caching locations, however, may be costly as it requires an advanced brain.

It is well known that many scatter-hoarding bird species excel in spatial memory capacity. The discovery by Swanberg that Eurasian nutcrackers seem to remember caching locations made in autumn throughout the winter has been verified experimentally on its close American relative, Clark's nutcracker. Birds that stored in captivity showed a slight decrease in retrieval success a year after storing, but not before that. Specialist hoarders not only remember caching positions accurately for a long time, but they also remember many such positions. Estimates indicate that a single nutcracker creates between 7000 and 10 000 caches in a year.

Parids (titmice and chickadees) do not seem to possess such a long-term memory. Many experiments have demonstrated an accurate caching memory in members of this family, but the retention intervals before retrieval have been short. Two species, the black-capped chickadee and the willow tit, have been subject to memory decay experiments. Captive birds were allowed to store and retrieve after various retention intervals in indoor aviaries. Both species relocated caches accurately shortly after storing, but retrieval success was down to chance levels in about a month. In the field, parids have been demonstrated to retrieve caches successfully months after storing. If they do not possess a cache memory of this duration, they must use other mechanisms, for example individual preferences for different types of hoarding substrates, individually separated hoarding niches, etc. This can be thought of as when people hide an extra front door key. Some will put it under a flower pot, others below the door mat, yet others behind a loose front board, etc.

Theoretical Models

Various phenomena in food-hoarding animals have been investigated with theoretical models. Many of these have been made in the framework of stochastic dynamic programming. Topics that have been modeled in this way include optimal daily fat gain trajectories in hoarders versus non-hoarders, effects of daily variation in predation risk, the significance of short-term hoarding versus long-term hoarding, effects of variations in food predictability, etc.

In evolutionary game theory, various behavioral strategies are compared, and the optimal solution is usually an evolutionary stable strategy, ESS. An ESS is a strategy that cannot be outcompeted by an alternative strategy when it dominates a population. Besides Andersson and Krebs (who compared hoarding and nonhoarding strategies), at least two game theoretical models have been published on food-hoarding behavior, one investigating if there are conditions when hoarding can evolve without a recovery advantage for hoarders, the second investigating how dominance rank within a group should affect optimal hoarding investment. The first one showed that for species that live in stable groups at least a mixed ESS (hoarding/nonhoarding) can evolve also without recovery advantage. The second shows that dominant individuals should store less than subordinate ones, but that optimal behavior for dominants depends on how predictable winter conditions are.

Other models have shown that caches in many cases should be considered as an alternative to body fat deposits. The reason is that, especially for small birds, it might be costly to carry much body fat. For example, a heavy bird could be easier to capture for a predator. If we compare a hoarding small bird species with a nonhoarding relative of the same size, the optimal level of body fat will differ since the hoarder can keep some of its energy reserves outside the body. However, there are always strategic differences between these types of supplies; the fact that small birds are scatter hoarders becomes important here. A starving animal may not have time to search for food. In order for caches to be an equal alternative to body fat, they must be reliably accessible. This means that they must be easy to retrieve (i.e., exact locations remembered or they must be stored in a larder) and predictably safe (i.e., not exposed to high pilfering risk).

Some Current Questions

Memory research on food-hoarding birds has broadened into studies of other types of cognitive abilities is food-hoarding birds. For example, it has been demonstrated that scrub jays, a food-hoarding American jay, possess an episodic memory of similar type as humans.

The comparative approach to the correlation between cognition (e.g., memory) and brain specialization (e.g., in the hippocampus) has been criticized by Bolhuis and others. The critics argue that the whole approach is flawed and that we cannot draw conclusions about function from morphology. This has spurred an intensive debate known as the 'neuroecology debate.' The most convincing comparative study this far is a comparison of black-capped chickadees across a climate gradient, from Alaska through Canada down to Central USA. Populations sampled from five locations showed a significant correlation; both hippocampal volume and hoarding propensity increased toward the north.

An enlargement of the hippocampus in food hoarders could depend on selection for increased memory capacity over evolutionary time. There could also be a direct physiological effect, during periods of intensive caching (and hence memorization) the hippocampus could grow, for example due to increased incorporation of baby neurons, increased vascularization, etc. In that case, training the brain would be comparable to training skeletal muscles. However, induced enlargement of the hippocampus in adult individuals has been questioned and proved hard to verify. Some current research is therefore attempting to measure hippocampal volume in living birds with MR (magnetic resonance) imaging techniques.

The coevolution between scatter hoarders and the nuts they store is another field in current research. As some trees seem to be more or less dependent on dispersal by corvids, knowledge about these systems may be important for future forestry practices.

See also: Ecology of Fear; Economic Escape; Empirical Studies of Predator and Prey Behavior; Internal Energy Storage; Risk Allocation in Anti-Predator Behavior; Spatial Memory; Trade-Offs in Anti-Predator Behavior; Vigilance and Models of Behavior.

Further Reading

Adams-Hunt MM and Jacobs LF (2007) Cognition for foraging. In: Stephens DW, Ydenberg RC, and Brown JL (eds.) *Foraging*, pp. 105–138, Chapter 4. Chicago, IL: Chicago University Press.

Brodin A (2007) Theoretical models of adaptive energy management in small wintering birds. *Philosophical Transactions of the Royal Society of London Series B: Biological Sciences* 362: 1857–1871.

Brodin A and Bolhuis JJ (2008) Memory and brain in food-storing birds: Space oddities or adaptive specializations? *Ethology* 14: 1–13.

Brodin A and Clark CW (2007) Resource storage and expenditure. In: Stephens DW, Ydenberg RC, and Brown JL (eds.) *Foraging*, pp. 221–269, Chapter 7. Chicago, IL: Chicago University Press.

Pravosudov VV and Smulders TV (in press) The integrative biology of food hoarding; a theme issue on scatter hoarding behaviour in animals. *Philosophical Transactions of the Royal Society of London Series B: Biological Sciences.*

Sherry DF and Mitchell JB (2007) Neuroethology of foraging. In: Stephens DW, Ydenberg RC, and Brown JL (eds.) *Foraging*, pp. 61–102, Chapter 3. Chicago, IL: Chicago University Press.

Vander Wall SB (1990) *Food Hoarding in Animals.* Chicago, IL: University of Chicago Press.

Caste Determination in Arthropods

A. G. Dolezal, Arizona State University, Tempe, AZ, USA

Introduction

Caste is one of the defining features in eusocial insects. At the most basic level, this system is characterized by morphology associated with the reproductive division of labor between sexually reproductive individuals and non-reproductive individuals who care for the reproductive's offspring. In the more derived groups of social insects, developmentally discrete castes fulfill these different behavioral roles. Reproductive division of labor is one of the defining characteristics of eusociality; eusocial insects include termites (Isoptera), ants (Hymenoptera), some bees (Hymenoptera), and some wasps (Hymenoptera).

In Hymenoptera caste divides individuals into one of the three categories: reproductive females (queens), reproductive males, and functionally sterile females (workers). Queens typically disperse, mate, find new colonies, and subsequently provide the reproductive function of the colony. Males disperse and mate with virgin queens; although there are some exceptions, males seldom perform any other activities and die shortly after mating. Workers perform all other tasks, including caring for young, nest building/maintenance, guarding, and foraging. All females are diploid and derived from fertilized eggs, while males are haploid and derived from unfertilized eggs. Female eggs are usually totipotent, that is, they have the potential to develop into either queens or workers. Developmentally, females are highly plastic and their physiology and morphology are influenced by various mechanisms, which result in caste differentiation. Developmental differentiation is primarily governed by nutritional stimuli, but other environmental factors such as climate and social interactions, as well as genetic effects (in some taxonomic groups), are important as well.

Termites (Isoptera) live in eusocial colonies with a similar system of reproductive division of labor to that of Hymenoptera. As with ants, bees, and wasps, a specialized sexual caste performs all reproductive functions for the colony, and a worker caste performs all other nest and foraging tasks. However, both termite sexes are derived from diploid, fertilized eggs, and thus both males and females are capable of developing into either workers or reproductives. Termite colonies are established by a winged queen and a winged king, which are known as primary reproductives. After colony founding, they shed their wings and begin producing workers. In many termite species, colonies produce secondary reproductives; these sexually capable individuals do not have

wings and are capable of superceding sick, injured, or absent parental primary reproductives. The worker caste is wingless and sterile. In some termite species in the families Kalotermititdae and Hodotermitidiae, a true worker caste has been replaced by preadult instars (referred to as nymphs or sometimes larvae). These preadult workers are not constrained from developing into reproductives later. Termites are hemimetabalous insects; unlike the holometabalous Hymenotpera, they do not undergo a pupal stage. Nymphs, or larvae, are physically similar to adults, and their metamorphosis is gradual. Thus, termite larvae function as efficient members of the colony workforce even before maturity. While less well-studied than Hymenoptera, termite caste determination appears to be affected by similar mechanisms, such as social interactions and endocrine cues.

In addition to the dichotomy between reproductive and nonreproductive individuals, many ant and termite species produce a polymorphic worker caste (e.g., workers vs. soldiers). In some ant species, extreme morphological differences exist among functionally sterile workers. In one pattern, exemplified by the red imported fire ant, *Solenopsis invicta*, variation is continuous, with workers ranging evenly throughout the entire gamut of body size. In others, like many *Pheidole* species, worker polymorphism is bimodal (or, in some cases, trimodal), and workers develop into small minor workers or large major workers. In termites, many species have distinct worker and soldier castes. Soldiers are generally larger and have more robust heads, usually with structures evolved specifically for defending the colony.

In both ants and termites, strong allometric scaling, in which the head or other stuctures grow disproportionately large relative to the rest of the body, is also observed. In these cases of allometry, major workers (or soldiers) increase not only in total size, but some characters (e.g., the head) grow nonlinearly, relative to the rest of the body. Most of the developmental determinants of reproductive caste are also key factors controlling worker caste polymorphisms.

Haplodiploid Sex Determination in Hymenoptera

In addition to differentiation between reproductive and nonreproductive castes, Hymenoptera have a unique system of sex determination. All Hymenopteran species are haplodipoid; males derive from unfertilized eggs, and are thus haploid. Females are derived from fertilized, diploid, eggs. When a reproductive female lays a male-destined egg, no sperm is released for fertilization; however, when a female-destined egg is laid, sperm is released and the egg is fertilized prior to oviposition (**Figure 1**). Eusocial Hymenoptera colonies are female-dominated, and male tasks are usually limited only to dispersal and mating. Thus, it is important that the queens regulate the sex of their eggs, or the colony will be burdened with useless males at inopportune times of the colony life cycle.

In honeybees (*Apis mellifera*), the decision to fertilize an egg is based on the cell type in which the egg is to be laid; in the comb, workers create different cell types for males, workers, and queens. Therefore, the workers produce appropriate numbers of each cell type and the queen must be able to differentiate among them and then lay a fertilized or nonfertilized egg as necessary. In other Hymenoptera, the mechanism underlying the queen's decision is not completely clear.

Physical Features of Caste

In most eusocial Hymenoptera, caste differentiation is based primarily on environmental factors experienced during egg and larval development. Morphological differences

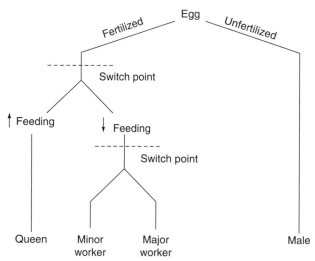

Figure 1 Model of differentiation possibilities of a totipotent Hymenoptera egg. If the egg is left unfertilized, it will develop into a haploid male. However, if fertilized, the egg can become a queen or a worker. The switch point between these caste fates is predominantly due to differential feeding regimens, with queen-destined larvae receiving more, and better quality, food. In some species, another switch point exists in which differentiation between worker morphological castes occurs (e.g., minor (small) vs. major (large, a.k.a. soldier)). In addition to a relationship with feeding regimen, hormonal changes also affect these switch points (e.g., **Figure 2**).

related to caste-specific tasks arise due to developmental differences triggered during growth and development. In general, reproductive females are larger, have larger ovaries with more ovarioles, and possess a spermatheca (an organ used for long-term storage of sperm). Workers are smaller, have smaller ovaries with fewer ovarioles, lack spermathecae (in more highly derived species), and have features more useful in nonreproductive tasks (e.g., the corbicula, or 'pollen basket' leg structure of honeybee workers).

Termite primary reproductives are generally larger, possess autotomous wings, and have larger and more well-developed eyes and a more heavily sclerotized body than nonreproductives. After shedding their wings, primary reproductive queens enlarge their ovaries, causing abdominal swelling. Workers usually lack compound eyes, are more lightly sclerotized, and are smaller. These developmental differences can result from a combination of several environmental factors, such as nutrition, the social environment of the colony, and climatological effects, as well as genetic factors.

Nutritional and Endocrine Controls of Caste Determination

With a few important exceptions, female eggs of eusocial Hymenoptera are not predestined for a specific caste fate and are genetically and morphologically indistinguishable from each other. After oviposition, however, environmental stimuli come into play to determine caste-specific developmental trajectory. In species across the spectrum of eusocial evolution, the predominant environmental factor in caste determination is nutrition. Specifically, a nutritional switch during development is the point of divergence between the reproductive and nonreproductive castes. The developmental timing of this switch is highly variable between species; generally, species with an earlier switch exhibit larger differences between the castes, since more developmental time is allotted towards caste-specific growth.

Nutritional determination of caste is best known from studies of the honeybee. As pointed out above, castes are reared in specific cell types in the comb. The cell type is important for determining the treatment of the larvae by workers; the nutrition delivered to the larvae determines the caste fate of any given individual. Honeybee larvae are fed with a mixture of mandibular gland secretions, hypopharyngeal gland secretions, pollen (a protein source), and sugar (from honey, a carbohydrate source); both queen- and worker-destined larvae are fed with all of these substances, but quantity and ratio differ. Queen-destined larvae are fed in high quantity with what is commonly referred to as royal jelly, a mixture of these four components that contains a higher concentration of mandibular gland secretion and sugar than what is provided to worker-destined larvae. Thus, queen-destined larvae are provided with a larger amount and higher quality food source, and the nutritional switch is flipped to push their development towards the reproductive phenotype (**Figure 1**). While ant, wasp, and other eusocial bee larvae vary in diet, this same basic pattern of nutritional caste determination applies in these groups.

The underlying causes of these trajectory shifts are likely due to changes in the larval endocrine system in response to nutritional stimulation during key periods. As larvae pass a critical size due to increased food intake, physiological changes that influence development are triggered, and the larvae proceed down the queen-specific developmental pathway. For example, juvenile hormone (JH), a sesquiterpenoid hormone known to govern developmental functions across insect taxa, is upregulated in response to increased nutrition. As food intake increases, so does body size and JH level; thus, larvae reach a critical threshold and the developmental pathway is changed to the larger form (e.g., from worker to queen, or from minor to major worker; see **Figure 2**).

In addition to JH, the IGF-1 IIS pathways (insulin/insulin-like growth factor-1-like signaling) have been implicated in controlling queen–worker differentiation. In the honeybee, several IIS genes are upregulated in queen-destined larvae compared to worker larvae. Also in honeybees, knockdown of the expression of *target of rapamycin* (*tor*), a central component in growth regulation response to nutritional stimuli, results in prevention of queen development, even in the presence of queen-specific nutritional cues. This emerging research focus

on the molecular level of metabolic pathways should shed light onto how the nutritionally mediated orchestration of caste determination is operating at the endocrine, genetic, and molecular levels.

This focus on the endocrine and physiological relationships in social hymenoptera has led to new ideas relating to the evolution of eusociality in insects. Work by West-Eberhard, and later Amdam and Page, has led to the presentation of what is known as the reproductive ground plan hypothesis, a framework that explains how the physiological mechanisms regulating reproductive behaviors in solitary ancestors may have been co-opted for the regulation of nonreproductive behavior in workers of extant social insects. In line with this hypothesis is the idea that, for eusocial societies to operate, a finely tuned relationship between both social environment and developmental responsiveness must evolve.

In many primitively eusocial bees and wasps, all females are potentially queens. Despite the variation in their size, all have complete reproductive systems and can lay both fertilized and unfertilized eggs after mating. In these systems, caste is determined less by larval development and more by behavioral dominance among adults. Some individuals may be larger, and can thus more capably dominate nestmates, but this is the point at which any developmental caste-differentiation ends. Since increased size usually denotes queen behavior/placement in the dominance hierarchy, larval nutrition and growth are also keys to the development of different caste roles. Recent research by Hunt has shown that, at least in the paper wasp *Polistes metricus*, some developing larvae facultatively go through diapause, resulting in increased storage protein synthesis and sequestration, and are thus more likely to express the queen phenotype as adults. In addition, the mechanisms of *Polistes* diapause have been proposed as an antecedent to sociality; similar to the predictions of the reproductive ground plan hypothesis, this work explains how the dichotomy between *Polistes* workers and queens may be based on the co-option of the bivoltine wasp life history seen in solitary species, which is split between early season nondiapausing individuals and late season diapausing individuals, for regulation of the social systems seen in *Polistes* wasps.

Termite caste differentiation is fundamentally different than that of Hymenoptera because the hemimetabolous developmental cycle of termites allows for larvae to be active members of the colony work force. However, like Hymenoptera, hormones are important regulators of caste differentiation. When larvae molt into their next instar, variation in endocrine activity can alter gene expression and produce caste-specific phenotypes. This system is poorly understood because studies on endocrine influences in termite caste determination are rare. The available evidence suggests that immature individuals that molt with low JH titers are able to express the

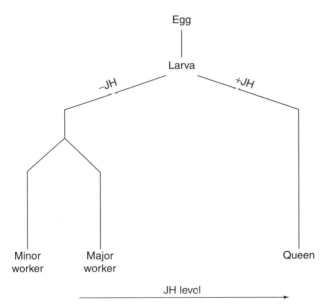

Figure 2 General diagram of developmental control points where juvenile hormone (JH) is known or suspected to play a role in caste determination. Those in larval or early pupal development are probably associated with nutritional switches.

reproductive phenotype, but those with higher JH titers have constrained gonad development and retain a worker phenotype. For example, in *Zootermopsis angusticolis*, winged queens require an increase in ecdysteroid levels and a decrease in JH levels to begin actual reproductive activity. While most studies have focused on JH, the underlying controls of caste fate in termites are most likely a complicated network of processes driven by JH, ecydsteroids, and other factors. Future research is needed, however, to elucidate the exact mechanisms behind these developmental changes. While this is true of all insect groups, termites are particularly understudied, especially compared to eusocial Hymenoptera.

Climate Effects on Caste

In many ant species, such as some *Formica* and *Myrmica*, larvae are more likely to develop into queens if they are exposed to winter or winter-like chilling effects. This is most likely an adaptation to prevent production of sexual castes during inopportune times of the year. If larvae that have not overwintered are less likely to develop into queens, the colony will be less likely to produce unnecessary virgin queens during nonmating seasons when the investment would be wasted.

Social and Maternal Influences on Caste

Since caste determination is usually based on nutritional and endocrine factors, the ultimate determinant is the treatment of the developing larvae. As larvae are completely dependent on workers for care, worker behavior is a key factor influencing caste determination. In many species of bees, wasps, and ants, worker behavior is modulated in some way by the mature queen. The queen's influence can change worker treatment of developing larvae and prevent the necessary dietary stimuli during key periods of larval development. In eusocial insects with relatively small colonies, queens can affect worker behavior with direct physical domination or intimidation, and thus influence the rearing/fate of developing larvae. More often, however, colonies are too large for a queen (or queens) to physically interact in such a manner. In these cases, queens instead release pheromonal signals that prevent workers from treating developing larvae appropriately to produce sexuals, or by inhibiting larval development directly.

For example, honeybees are noted for their use of queen pheromone to affect caste determination. When the queen produces this pheromone, workers will not build queen cells; without these larger, queen-specific structures, the expression of queen-rearing behavior cannot or will not occur. Queen pheromones, which are produced in the queen's mandibular gland, are primarily made up of 9-keto-(E)-2-decenoic acid (9ODA) and 9-hydroxy-(E)-2-decenoic acid (9HDA), both of which are highly specific in their effects on biological activity. The production of these compounds varies over a queen's lifetime, depending on her age, mating status, as well as the seasonality of the colony. For example, older queens will change their production to allow the workers to raise virgin queens and prepare for colony fission.

Thus, in honeybees, the queen influences caste determination even before egg-laying, let alone larval development by using pheromonal secretions to prevent the worker behaviors necessary for queen-destined larval development.

A similar method of worker control exists in vespine wasps, although the effects are somewhat reversed. Again, queen development is contingent on the production of appropriate cells by workers. Whereas queen pheromones prevent this from occurring in honeybees, Ikan's work showed that these cells are only produced when the wasp queen produces pheromonal secretions.

In some ant species, such as *Plagiolepis pygmaea*, *Tetraponera anthracina*, and *Myrmica rubra*, pheromones have other caste determining effects on larvae. Pheromones released by the queen can inhibit the development of queen-destined larvae via inhibition at a critical developmental window during the larval cycle. In species that require overwintering for queen production, pheromones can prevent larval hibernation from occurring. By preventing this key stage, the queen effectively removes the potential for queen development. Bumble bees, *Bombus terrestris*, exhibit a similar system in which exposure of developing larvae to a queen pheromone also precludes any chance of those larvae developing into queens themselves. Only when this pheromone is no longer produced can queen-destined larvae be reared.

In addition to pheromonally mediated inhibition of sexual development, queens can also influence the caste fate of their offspring during the egg stage. In some ant species, such as *Formica polyctena*, queens can bias otherwise totipotent eggs towards queen development by giving them a nutritional boost before oviposition. These eggs are larger and have better nutritional content than worker-destined eggs, which appear to lower the further nutritional stimulation necessary to switch larvae to queen development. Another method of maternal influence on egg fate is the sequestering of variable hormone quantities in eggs. In *Pogonomyrex* harvester ants, for example, queen-destined eggs have significantly lower levels of ecdysteroid hormones. This is in accordance with previous research, which has shown that JH application to a queen or her eggs results in significantly more queens, because JH depresses ecdysteroid level. In any case, larvae in species in which this occurs are not capable of as much caste plasticity, since their caste fate is determined in the egg stage.

Termites are poorly understood when compared to the eusocial Hymenoptera, but mechanisms of social control of caste determination appear to exist in termite colonies. The social interactions influencing reproductive status are often complex due to the presence or potential presence of both primary and secondary reproductives in termite societies. In some species, production of primary reproductives is thought to be regulated by worker manipulation of the developing larvae. In several species, workers destroy the wing buds of developing primary reproductives, removing their ability to disperse. Evidence in more primitive termite species, however, has shown that, like in many Hymenoptera, reproductive development may be mediated by inhibitory stimuli produced by current adult reproductives, though the mechanism of this inhibition is still unclear.

The production of secondary reproductives in established termite colonies can vary widely between species. In species that lack a true worker caste, in which nonreproductive tasks are performed only by immature individuals (pseudergates), secondary reproductives seem to derive from either older larvae or immature primary reproductives. The switch of an individual to a secondary reproductive role may result from changes in developmental timing, which produces an incomplete primary reproductive (i.e., wingless). In termite species with a true worker caste (that is, a caste of individuals that, even as adults, perform worker tasks), secondary reproductives can be derived from reproductively inactive primary reproductive offspring, or by the subsequent development of larvae or workers into reproductive morphology. The control over the sexual maturation of these individuals appears to belong to the functioning reproductives present in the colony, but, again, the mechanisms are unknown. In some species, primary reproductives physically dominate potential secondary reproductives, changing their caste fate before maturation. In other cases, functioning reproductives may inhibit sexual development via pheromone action on developing larvae, but there is no evidence to support this. Additionally, functional reproductives may also influence worker treatment of larvae, thus indirectly influencing caste determination. For example, workers of some species are known to remove extra secondary reproductives from the nest.

Genetic Effects

While there is extensive evidence for the environmental basis of caste determination across many taxa, findings in some taxonomic groups show that genetic factors can have significant effects on adult caste. Heritable variation in caste is seen in some ant, bee, and termite species.

Perhaps the best known case is the genetic caste determination seen in some *Pogonomyrmex* harvester ants.

Research on *Pogonomyrmex* exhibiting genetic caste determination revealed two distinct genetic lineages within individual colonies; queens are homozygous and belong to one of the two possible lineages, while workers are heterozygotes derived from cross-lineage mating. When the queen produces fertilized eggs, the lineage of the male whose sperm is used determines the caste of the newly-laid egg; one lineage of male results in queen development, while the other results in worker development. While the mechanism is still unclear, incompatibilities between the lineages that prevent queen production from heterozygous matings are proposed to exist. This system is particularly interesting because, for a queen to found a successful colony, she must mate with males from both lineages in order to produce both female castes.

Another example of genetic differences between castes is seen in the little fire ant, *Wasmannia auropunctata*. In this species, workers are produced normally, from fertilized eggs; queens, however, are produced from thelytokous parthenogenesis in which meiotic cells are fused in order to parthenogenetically produce unfertilized diploid eggs. Thus, diploid eggs are produced solely from queen gametes. This life history strategy has caused an almost complete separation of queen and male gene pools.

In some species of ants, such as *Dinoponera australis*, a true queen caste has been completely lost. Similarly, in the Cape bee, *Apis mellifera capensis*, colonies can reproductively operate in the absence of a queen. Without a queen present, some workers must reproduce. In many ant species with this trait, workers retain a spermotheca, and can thus successfully mate and reproduce normally. In the Cape bee, however, a reproductive worker produces males via unfertilized eggs and females via thelytokous parthogenesis, similarly to *W. auropunctata* queens.

In termites, genetic caste determination has been identified in the species *Reticulitermes speratus*; a single sex-linked locus with two alleles produces sex-specific traits. These alleles play a key role in the caste fate of the offspring.

The existence of the genetically influenced systems represents a puzzling evolutionary question. What is the benefit of reduced flexibility in caste production? The loss of totipotency results in a reduction in the ability of the colony to more fluidly adapt to quickly changing scenarios by changing/controlling caste ratios. For example, in the *Pogonomyrmex* genetic caste determination systems, queens are produced at proportions in line with predicted genotypic ratios independent of the age and colony size. In this scenario, colonies are unable to stop queen egg production, even at a very young colony age, when queen production is unnecessary and unwanted. While it has been shown that nascent colonies of *Pogonomyrmex* do not actually raise queen-lineage eggs to adulthood, the laying of these eggs at all still represents a potential waste of resources during a key time in the colony life cycle, and how this strategy is evolutionarily stable is puzzling.

Polymorphic Worker Castes

Among the eusocial Hymenoptera, only ants, and to a lesser extent, bumble bees, produce workers with morphologically complex worker castes. While relatively few ant genera produce polymorphic workers, some of the most ecologically successful often produce highly polymorphic work forces (e.g., *Pheidole* and *Camponotus*). Species with a broad spectrum of worker morphs tend to have a very early developmental switch for queen–worker differentiation, providing a maximum developmental time frame for intra-worker caste differences to develop (**Figure 3**).

The extreme variation seen between workers in ant species with a polymorphic worker caste has been explained by the reprogramming of the developmental pathways present in all eusocial Hymenoptera, specifically a revision of critical size and growth rules. While the framework is reprogrammed, development is still influenced by many of the same environmental stimuli that affect queen–worker differentiation, primarily larval nutrition.

The critical size of a larva is simply the size threshold it must reach to begin the developmental pathway into the next largest morph. Variation in the critical size needed for this transition (due to variable reaction norms of larvae) can give rise to polymorphic worker castes with continuous variation. Additionally, more strictly controlled thresholds exist, which can result in the production of more discrete worker size classes. Growth rules correspond to the allocation of energy to the growth and development of different body structures, which results in differences in the growth rate and final size of body characters, producing allometries. Both critical size threshold and the patterning of growth rules can be influenced during sensitive windows during development, and thus produce wide variations in worker morphology and size, even in colonies of highly related individuals. The location in developmental time where larvae actually make the switch to the larger caste-type is highly variable between species. Generally, the later the switch occurs during larval development, the more flexible the system; in species where the switch point occurs during the final instar, decisions about caste ratio can be made even at the figurative last minute of larval development (**Figure 1**).

Many species of termites also produce a soldier caste in addition to normal workers. In several species, increased JH levels in circulating hemolymph promote expression of the presoldier or soldier phenotype. This increase seems to coincide with the initiation of molting, caused by peaking levels of ecdysteroid hormones. These changes have been proposed to promote mandibular development and a more robust cuticle; recent evidence has also shown that these hormonal changes during development induce shifts in gene expression patterns to produce the soldier phenotype.

In addition to environmental factors like presence of predators or competitors and food availability, soldier production may also be limited by the same types of social influences that affect reproductive caste development. Soldiers, possibly through secretory products from the head, are able to use pheromones to inhibit the differentiation of developing larvae into new soldiers. In addition, there is some evidence that functioning reproductives are able to stimulate soldier production. The ability of the colony to regulate soldier number in these ways is most likely an adaptation to regulate caste ratio to maximize colony efficiency.

Conclusion

As noted, one of the defining features of social insect biology is the variation in caste between highly related, if not effectively identical, individuals within a colony. Undoubtedly, the determination of caste during development is affected by many factors, both within and between taxonomic groups. However, research into this field has led to many advances in our understanding of the various roles played by nutrition, hormones, climate, social organization, and genetics. While many questions remain unanswered as to the complexities underlying caste determination in insects, new tools in these fields of research continue to emerge and should prove invaluable in providing future information about the underlying factors influencing caste.

See also: Pheidole: Sociobiology of a Highly Diverse Genus; Termites: Social Evolution.

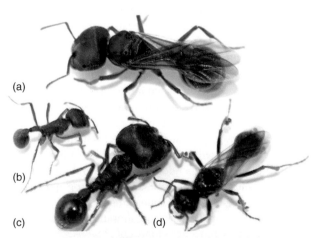

Figure 3 Example of different castes developed from totipotent eggs of the harvester ant *Pogonomyrmex badius*. (a) Queen (unmated, with wings still attached); (b) minor worker; (c) major worker; (d) male. Photo courtesy of Adrian A. Smith.

Further Reading

Brent CS (2009) Control of Termite Caste Differentiation. In: Gadau J and Fewell J (eds.) *Organization of Insect Societies.* Cambridge MA: Harvard University Press.

Hölldobler B and Wilson EO (1990) *The Ants.* Cambridge MA: Belknap Press of Harvard University Press.

Hunt JH and Amdam GV (2005) Bivoltinism as an antecedent to eusociality in the paper wasp genus *Polistes*. *Science* 308: 264–267.

Smith CR, Anderson KE, Tillberg CV, Gadau J, and Suarez AV (2008) Caste determination in a polymorphic social insect:

Nutritional, social, and genetic factors. *American Naturalist* 172: 497–507.

Smith CR, Toth AL, Suarez AV, and Robinson GE (2008) Genetic and genomic analyses of the division of labour in insect societies. *Nature Reviews Genetics* 9: 735–748.

Wheeler DE (1986) Developmental and physiological determinants of caste in social Hymenoptera: evolutionary implications. *The American Naturalist* 128: 13–34.

Wheeler DE (1991) The developmental basis of worker caste polymorphism in ants. *American Naturalist* 138: 1218–1238.

Winston ML (1987) *The Biology of the Honey Bee.* Cambridge MA: Harvard University Press.

Caste in Social Insects: Genetic Influences Over Caste Determination

N. Lo, Australian Museum, Sydney, NSW, Australia; University of Sydney, Sydney, NSW, Australia
M. Beekman and B. P. Oldroyd, University of Sydney, Sydney, NSW, Australia

Introduction

Eusocial insects dominate many ecosystems. Key to their success is the specialization of colony members into royal and worker castes. Royals are specialist reproducers, while workers are typically sterile or subfertile. The question of what determines whether a young social insect will become a royal or a worker has interested biologists for over a century. In the early to mid-twentieth century, there was disagreement on the relative importance of genotype in caste determination. Some proposed that each social insect larva had an equal capacity to develop into either the royal or worker caste, and that environmental factors alone decided their developmental fate. Others believed that an individual's genotype predisposed it to developing into a particular caste. Early experimental evidence supported the first hypothesis. A classic example of 'environmental caste determination' (ECD) was the development of honeybee larvae into queens when fed relatively high amounts of royal jelly, and into workers when fed a diet with low amounts of royal jelly. A variety of studies on caste determination in ants and termites also demonstrated the importance of environmental factors.

These results were in agreement with theoretical predictions that caste determination should be environmentally, rather than genetically, determined. Any allele or gene predisposing an individual to become a worker would be expected to be lost from the population, because of the sterile or subfertile nature of its bearer. Although a few early examples of 'genetically influenced caste determination' (GCD) were reported, by the end of the twentieth century it was widely believed that ECD was the rule in social insects.

Since the turn of the millennium, there has been a sharp increase in the number of demonstrations of GCD. Genotype has been shown to have a strong influence not only on the worker versus royal dichotomy, but also on the development of different forms of workers (e.g., large workers, small workers, and soldiers). This article presents an overview of the various examples of GCD, focusing on the royal versus worker dichotomy, and discusses the factors that may have led to their evolution. The term 'royal' is used only in reference to the winged, dispersing caste that initiates new colonies either alone, with a sexual partner, or with a swarm of workers. In some ant and termite species, a subset of individuals that are morphologically similar to workers (e.g., wingless) may become reproductive; these are termed 'ergatogynes' or 'ergatoids' respectively.

GCD, as defined here, includes a spectrum of caste determination mechanisms, ranging from cases where genotype has a slight, but significant influence on caste, through to cases where an individual's caste is essentially hard-wired on the basis of its genotype. Environmental stimuli thus play an important role in many, if not most, cases of GCD. In ECD, genotype has no significant effect on the probability of becoming one caste or the other.

Empirical evidence for GCD has primarily come from two methods. The first method is comparison of an individual's caste with its genetic profile – usually obtained using microsatellite markers. The advent and widespread use of this relatively straightforward method over the last two decades have enabled most of the examples of GCD to be discovered. A second method for detecting GCD is crossing of royals and ergatogynes/ergatoids and examination of the caste of offspring.

The examples presented here come primarily from the hymenopteran social insects ants and bees. As shown in **Figure 1**, these insects have a haplodiploid sex determination system. Unfertilized eggs are haploid and become male, while fertilized eggs are diploid and become female. Queens may be monoandrous (mating with only one male) or polyandrous (mating with two or more males).

Stingless Bees of the Genus *Melipona*

The first evidence for a genetically based mechanism of caste determination came from the Meliponini: a tribe of primarily monoandrous, stingless bees from tropical regions. Several species have been domesticated for honey production. In the 1940s, Warwick Kerr noticed that in the genus *Melipona*, queens and workers are reared in identical cells, are fed identical food, weigh the same after emergence, and that queen pupae are scattered randomly through the brood comb. This suggested that caste determination in *Melipona* was not determined by differential feeding or other environmental means.

Kerr examined the brood of several *Melipona marginata* and *M. quadrifasciata* colonies and noticed that the ratio of queens to workers in the brood was ~1:3 and 1:7 respectively. He realized that these ratios could be explained by the segregation of two or three caste-determining genes, respectively. Kerr proposed that queens are heterozygous at all caste-determining loci, whereas workers are homozygous at one or more caste-determining loci as shown in **Figure 2**. Notice how all *M. marginata* queens are AaBb and will therefore produce four kinds of gametes at equal frequency from a simple Mendelian segregation: 1 AB: 1 Ab: 1 aB: 1 ab. As with all other Hymenoptera, the Meliponini are haplo-diploid: the females are diploid, but the males are haploid and arise from unfertilized eggs. Therefore, males are produced by the doubly heterozygous *M. marginata* queens in this exact same genetic ratio. This means that there are four kinds of males in the population. For each of the four kinds of males, 1/4 of their diploid female offspring will be queens because they can only mate with doubly heterozygous queens (**Figure 2**). Similarly, for *M. quadrifasciata*, and six other species, 1/8 of female progeny will be queens. Thus, the proportion of potential queens in the population is kept constant because of genic balance at the caste-determining loci. The proportions of each allele are kept equal in the population: any rare allele is at a strong selective advantage because it is likely to be found in heterozygotes and thus passed on by queens, while any common allele is at a disadvantage as it is more likely to be present in homozygous workers and thus not passed on.

What happens to all the extra queens? *Melipona* colonies are founded by swarms that contain large numbers of workers (as opposed to independent colony founding by queens) and there is only one single-mated queen per colony. Thus, even with the three-locus system, where 'only' 1/8 of female brood becomes a queen, an excess of queens is produced. Queens emerge continuously, and whenever there is a laying queen in the colony, they are killed by the workers. Nonetheless, *Melipona* colonies have at least some capacity to reduce the number of queens produced when they are not required, for the number of queens produced declines in unfavorable seasons when food resources are scarce. Thus, there is a strong environmental component to the caste determination mechanism.

Melipona workers have five ventral nerve ganglia, whereas most queens have four, presumably because the two distal ganglia become fused at the pupal stage. Intriguingly, a small proportion of workers show the four ganglia typical of queens, and these were interpreted by Kerr as being genetic queens that had failed to develop phenotypically because of inadequate nutrition. The quantity of food provided to the brood cells by nurse workers seems to be the determinant of whether a genetic queen can develop the full queen phenotype.

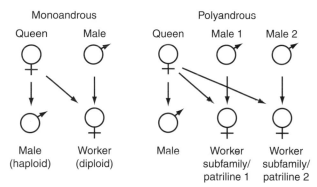

Figure 1 Sex determination and mating systems in ants and bees.

	Queen gametes			
Four kinds of male gametes	AB	Ab	aB	ab
AB	AABB	AABb	AaBB	AaBb
Ab	AABb	Aabb	AaBb	Aabb
aB	AaBB	AaBb	aaBB	aaBb
ab	AaBb	aaBb	aaBb	aabb

Figure 2 Model for caste determination in *Melipona marginata*. Individuals that have at least one of two genes homozygous develop into workers, while those with both genes heterozygous (shaded) develop into queens.

Melipona is the only genus of social bees that has equal-sized cells for workers and queens, and the only bees for which GCD has been proposed. Perhaps the reduced ability of workers to influence the caste of larvae developing in sealed cells is the reason why the GCD mechanism has evolved in these bees. It should be noted that Kerr's model has been questioned by some researchers, who suggest that a more complex model may be required.

The Slave-Making Ant *Harpagoxenus sublaevis*

In the slave-making ant *Harpagoxenus sublaevis*, colony initiation involves a queen finding a host colony, killing off all adults, and capturing the brood. Newly hatched workers then tend the eggs that she has laid. The queens' workers are able to raid other colonies for new slave brood.

Two forms of reproductives are known in *H. sublaevis*: winged queens, which are relatively rare within populations and wingless 'ergatogynes,' the most common form. Workers are morphologically very similar to ergatogynes, though slightly smaller and anatomically less complex. Following a series of crosses, Alfred Buschinger and coworkers proposed that the wing polymorphism is under the control of a single locus with two alleles: e and E. Queens are always e/e, while ergatogynes are either E/e or E/E. Workers can be of any genotype; however, e/e females have a significantly higher chance of becoming queens than E/e individuals have of becoming ergatogynes. E/E individuals have the lowest chances of becoming reproductives (i.e., ergatogynes). Thus, E is a dominant allele for winglessness and appears to slow development. Genotype thus determines whether a larva will become winged or wingless and whether it can become a queen. As for the case of *Melipona* stingless bees, environmental factors still play an important role in mediating phenotype (i.e., worker or reproductive). However, unlike *Melipona*, the alleles in the population are not balanced, and thus different caste ratios occur depending on the different types of crosses.

It has been speculated that the primitive condition of caste determination in the ancestors of *H. sublaevis* was environmental (i.e., fixation of an 'e'-like allele in populations). Mutation of e to E then occurred at some stage during the transition from the free-living, claustral colony-founding form to the parasitic, slave-making form. *H. sublaevis* initiates colonies via parasitism, whereby the founding queen invades an established colony of a related species. Selection may, therefore, have favored mating by individuals without wings, because of energetic benefits.

The GCD mechanism found in *H. sublaevis* has also been found in the ants *Leptothorax* sp. and *Myrmecina graminicola*, both of which are nonparasitic. The evolution of the equivalent version of the E allele in these cases may have been favored by the presence of patchy habitats, which favor colony foundation in close, rather than distant, sites and therefore do not require founding females to have wings. It remains to be seen whether a GCD mechanism underlies other cases of the queen/ergatogyne dichotomy, which is fairly common in ants.

Harvester Ants of the Genus *Pogonomyrmex*

Two *Pogonomyrmex* species, *P. barbatus* and *P. rugosus*, exist sympatrically in southwestern North America. Previous hybridization events between these two 'parental' species have resulted in three 'lineages' (H1, H2, and J1) with chimaeric genomes. There is little or no gene flow between lineages (or between lineages and the parental species). The three lineages, along with a fourth lineage (J2, which is either a divergent lineage of *P. barbatus*, or another hybrid lineage), have developed a GCD mechanism that differs greatly from the mechanism present in the parental species *P. barbatus* and *P. rugosus*. GCD in these lineages was discovered via microsatellite, allozyme, and mitochondrial genotyping studies.

Like the parental *P. barbatus* and *P. rugosus*, members of all four lineages are polyandrous. When a female and male of the H1 lineage mate, their offspring are able to develop only into queens (**Figure 3**). However, when an H1 female mates with an H2 male, their offspring develop into workers. Thus, an H1 female must mate with males of both H1 and H2 in order to produce both workers and queens. Similarly, matings

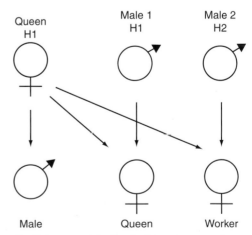

Figure 3 Colony reproduction in *Pogonomyrmex* lineages. The only way that a harvester ant queen (H1) can produce a fully viable colony with both workers and queens is by mating with males of both H1 and H2 lineages. Because she can only produce H1 males, she is reliant on male production by H2 queens for her worker production.

between H2 males and females result in queens, while female H2/male H1 matings result in workers. The same system is found in J1/J2 lineages. Each lineage effectively parasitizes the males of the other lineage to make workers and are therefore known as 'dependent lineages.' Matings between the H and J lineages and between the H, J, and parental lineages lead to inviable offspring.

It has been speculated that this GCD mechanism evolved as a result of queen–daughter conflict. To maximize their inclusive fitness, daughters are selected to avoid becoming workers and to enhance their chance of becoming queens. On the other hand, queens are selected to coerce their daughters into becoming workers rather than queens, thus enabling their young colonies to develop more rapidly. Indeed, in parental *P. rugosus* as well as dependent lineages, a maternal effect in incipient colony queens results only in workers being produced; overwintering is necessary before the queen caste can develop. The H and J lineages may represent cases in which the queen is no longer able to stop members of her own lineage from developing into queens. However, she is still able to influence hybrid offspring of the other lineage into becoming workers. An outcome of this is that intralineage offspring, which are able to become only queens, either die or do not develop properly during colony foundation.

The caste determination system of one of the parent species, *P. rugosus*, has recently been shown to be under genetic influence; however, it is very different from that present in the dependent lineages. In *P. rugosus*, the probability of becoming one caste or the other depends on the compatibility between the parental genomes. Thus, the offspring of a female who has mated with several different males have significantly different probabilities of becoming a queen. However, if the same males mate with a different female, these probabilities change. It is the combination of the two genomes that influences caste. Environmental factors play a much larger role in this system than in the dependent lineages: caste totipotency is retained in the former, but has been lost in the latter (at least in the case of intralineage matings).

A somewhat similar system to that found in the *Pogonomyrmex* dependent lineages has been discovered in a hybrid system involving the fire ants *Solenopsis geminata* and *S. xyloni*. Workers were hybrids of these two species, while almost all the queens were of pure *S. xyloni* ancestry. *S. xyloni* females have become developmentally constrained to become queens and are unable to become workers, even if there are no hybrid workers in the colony. On the other hand, hybrids can still become queens, but this happens only rarely, in the absence of pure *S. xyloni* offspring. In contrast to the *Pogonomyrmex* lineages, reciprocal hybridization is rare or does not occur: offspring of *S. geminata* queens are invariably fathered by other

S. geminata males, rather than *S. xyloni* males. Another difference is that *Solenopsis* colonies contain multiple, singly mated queens, rather than single, multiply mated queens. To produce both workers and queens, a colony must therefore contain both heterospecifically and conspecifically mated queens. This sets the stage for conflict between queens, since heterospecifically mated queens can gain direct fitness only via males.

The Japanese Termite *Reticulitermes speratus*

The majority (~80%) of termite species have an irreversibly wingless worker caste; larvae can follow one of two main developmental lines: the 'nymph' line (leading to alates) or the worker line (**Figure 4**). Since termites are hemimetabolous, two forms of immature 'neotenic' reproductives (one from each developmental line) can arise under certain conditions (e.g., in the absence of the queen or king). Laboratory crossing experiments of neotenic reproductives from the Japanese wood-feeding species *Reticulitermes speratus*, collected in Eastern Japan (Ibaraki), have shown an influence of genotype on caste determination in this species. Following four different crosses of nymphoids and ergatoids, eggs from each cross were reared under identical conditions (in the absence of the king and queen), and the caste and sex of offspring were examined. The same experiments were performed on offspring from parthenogenetic nymphoids and ergatoids. Parthenogenesis occurs via thelytokous automixis, resulting in highly homozygous offspring.

The four neotenic crosses and two parthenogenetic treatments resulted in very different offspring types; this was neatly explained by the presence of an X-linked locus named *worker* (*wk*) with two alleles A and B (**Figure 5**; unlike ants and bees, termites are diplo-diploid, and usually have an XY sex-determination system).

In the model, females inheriting two copies of A (i.e., wk^{AA}) become nymphs (i.e., future queens), while those inheriting one copy of A and B (wk^{AB}) become workers. Males inheriting a single copy of A ($wk^{A}Y$) become workers, while those inheriting a single copy of B ($wk^{B}Y$) become nymphs (i.e., future kings). The genotype wk^{BB} is lethal. Environmental factors still play an important role in the GCD mechanism. When reproductives are present during development, individuals with a worker genotype remain workers; however, 25% of females with a nymph genotype are modified into workers (male nymph genotypes have not been tested). Environmental stimuli, most likely pheromones, also control the development of nymphoids, ergatoids, and soldiers from their precursor castes.

The selective advantage of the GCD system may be enhanced inhibition of selfish reproduction by offspring.

As shown in **Figure 5**, offspring produced in developing king–queen *R. speratus* colonies have a genetic bias toward the worker, rather than the nymph, pathway. Workers have a relatively low propensity to develop into 'selfish' neotenics, requiring two molts and ~30 days, compared with one molt and ~12 days for nymphs. The evolution of genetically biased workers with a reduced capacity to become reproductive may have facilitated increased foraging behavior, permitting increased colony size and ecological dominance.

The predictions of the model in **Figure 5** have yet to be examined in field colonies from Ibaraki; a subsequent study of colonies in Central Japan (Kyoto), which employed microsatellite genotyping of colony members, showed that the model does not hold for colonies of this population. These colonies were almost always headed by a single king and numerous nymphoid neotenics, the latter of which parthenogenetically produced by the founding queen. No ergatoids were found, and both workers and nymphs in these colonies were shown to have

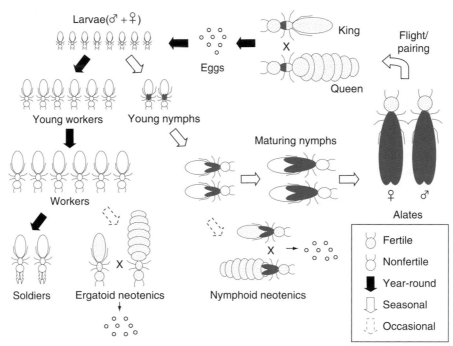

Figure 4   Generally accepted life cycle of termites with a worker caste. Reprinted Lo N, Hayashi Y, and Kitade O (2009) Should environmental caste determination be assumed for termites? *American Naturalist* 173: 848–853, with permission from The University of Chicago Press.

	wk^B Y ♂ Nymphoid		wk^A Y ♂ Ergatoid		Parthenogenesis
wk^{AA} ♀ Nymphoid	wk^{AB} ♀ Worker	wk^A Y ♂ Worker	wk^{AA} ♀ Nymph	wk^A Y ♂ Worker	wk^{AA} ♀ Nymph
wk^{AB} ♀ Ergatoid	wk^{AB} ♀ Worker	wk^A Y ♂ Worker	wk^{AA} ♀ Nymph	wk^A Y ♂ Worker	wk^{AA} ♀ Nymph
	wk^{BB} ♀ Lethal	wk^B Y ♂ Nymph	wk^{AB} ♀ Worker	wk^B Y ♂ Nymph	wk^{BB} ♀ Lethal

Figure 5   One-locus-two-allele model of caste determination in *Reticulitermes speratus*. Genotype matches phenotype under the experimental rearing conditions used (i.e., an absence of reproductive individuals). See text for details. Reprinted Hayashi Y, Lo N, Miyata H, and Kitade O (2007) Sex-linked genetic influence on caste determination in a termite. *Science* 318: 985–987, with permission from the American Association for the Advancement of Science.

alleles from both the king and the nymphoid queens. These results are not congruent with the predictions of the model shown in **Figure 5**, since inbreeding is not required to produce nymphs. The breeding biology of *R. speratus* in different parts of Japan is known to be variable. Colonies from some areas have been shown to contain ergatoid reproductives, which differs from the situation in populations from Kyoto. The caste determination systems Kyoto and Ibaraki may, therefore, have somewhat different caste determination mechanisms. Alternatively, the system proposed in **Figure 5** may be strongly overridden under field conditions.

The Fungus-Growing Ant *Acromyrmex echinatior*

The leaf-cutting and fungus-growing species *Acromyrmex echinatior* is found between Mexico and Panama. Queens of *A. echinatior* mate with multiple males, and, as a result, their offspring (queens or workers) consist of a number of full-sister lineages (patrilines; **Figure 1**). Ants of different patrilines are half-sisters. Two different-sized workers exist: small (SW), which specialize in tending the fungus garden, and large (LW), which forage outside the nest for plant material to feed the fungus with. It has been shown that one fifth of patrilines cheat on their nestmates by having a much higher chance of becoming a queen than a worker. These royal patrilines are shown to be rare both within individual colonies and in the population as a whole.

'Normal' patrilines are represented roughly equally among small and large workers and queens, whereas 'royal' patrilines show a bias toward queens and a bias toward either large workers (royal-LW patrilines) or small workers (royal-SW patrilines). As larvae destined to become LW also need to be fed more, the slightly higher than expected contribution of royal-LW patrilines toward queens is most likely a result of some larvae of these patrilines being fed extra to allow them to become queens. These patrilines are equally abundant as 'normal' patrilines within colonies. The real cheats are the royal-SW patrilines, which hardly contribute to the production of workers, and, if they do, produce mostly SW. Hence, feeding does not explain the bias toward queens in royal-SW patrilines. Cues that lead a larva to develop into one caste or another may be more complex than feeding alone. Royal-SW patrilines were shown to have an increased fitness of almost 500% relative to normal patrilines, whereas the fitness of royal-LW was not much higher than that of normal patrilines.

Despite the enormous increase in fitness, royal-SW patrilines are rare, as evolutionary theory would predict for real cheats. Most likely, the frequency of cheating genotypes is constrained by colony-level selection: colonies that contain too many royal cheats suffer a reduction in reproductive output. In addition, the direct feeding of ant larvae by workers gives nurse workers some control over the caste fate of larvae by evolving the capacity to recognize potential cheats (although such recognition has not yet been demonstrated; the rarer the cheats are, the more error-prone such recognition will be).

The Parthenogenetic Ants *Cataglyphis cursor*, *Wasmannia auropunctata*, and *Vollenhovia emeryi*

The ants *Cataglyphis cursor*, *Wasmannia auropunctata*, and *Vollenhovia emeryi* each have an unusual breeding system, whereby queens are produced parthenogenetically and workers are produced by sexual reproduction (queens are multiply mated in each case). In the case of *C. cursor*, and probably *V. emeryi*, automictic parthenogenesis occurs, leading to increased homozygosity of offspring, but not clonally produced offspring. In *W. auropunctata*, no meiosis occurs, resulting in clonally produced queens. Additionally, in *W. auropunctata* and *V. emeryi*, males (haploid) are clonally produced, apparently via elimination of the maternal genome in eggs. In all three cases, parthenogenesis allows queens to take advantage of the twofold increase of asexual production when producing further queens, while sexual production of workers provides the benefits of a genetically diverse worker population.

Although the genotypes of workers and queens in these species are clearly differentiated by high versus low heterozygosity, they do not necessarily have GCD, as defined here. In *C. cursor*, the caste determination system appears to be environmental. Studies of larval development in laboratory colonies indicate that both parthenogenetically and sexually produced eggs are not biased toward one caste or the other. Rather, it is the physiological state of nurse workers that determines the caste of larvae. In the field, queens lay parthenogenetic eggs in the spring, which then become queens. At other times, eggs are fertilized and develop into workers. Thus, cold temperatures are important in the switch of development of larvae from workers into queens.

The HoneyBee *Apis mellifera*

In large measure, the honeybee (*Apis mellifera*) was responsible for the development of the paradigm that caste determination is environmentally determined – indeed the queen bee production industry depends on ECD for the conversion of worker larvae into queens for sale. But even in the honeybee, it now appears that not all female larvae are created equal. When a queen honeybee dies, the workers select 5–20 female larvae (out of the several thousand that are available) and lavishly feed them so that they develop into queens. All the rest develop into workers. This seems strong evidence for ECD, but recent pedigree analyses using genetic markers show that some patrilines are strongly over represented in the queen larvae. Some scientists now believe that there are genetic determinants that make these larvae more attractive for rearing as queens. So even in honeybees, there may be a bit of GCD after all.

Conclusions

Evolutionary theory predicts that all individuals will behave selfishly, provided the cost of expressing such behavior (e.g., colony level costs) does not outweigh the benefits in terms of inclusive fitness. Hence, when female brood are totipotent, every individual 'prefers' to develop into a queen and prefers that her sisters develop into workers. This is so because an individual's relatedness to its own offspring is higher than its relatedness to those of sisters. Because the expression of selfish behavior by a large number of individuals will result in a reduction in inclusive fitness of most individuals, the outcome of selection at multiple levels (individual- and colony-level) will be the suppression of selfish behavior to a sustainable, low level.

How can the expression of selfish behavior be suppressed? In the case of Hymenopteran social insects, it is generally argued that adult workers hold the greatest overall power for several reasons. Because workers are mobile and numerous, they have access to more and better information than larvae. This is especially important when timing of queen production plays a role. Workers are the main brood-rearers and therefore have more opportunities to manipulate developing females by controlling larval nutrition and rearing temperature. For example, such manipulation prevents female larvae of the ant *Hypoponera* from cannibalizing other larvae and developing as queens (this is possible as size differences between queens and workers are small), since workers actively keep brood of different sizes separated to reduce such cannibalism. Being physically stronger gives workers the ability to kill noncompliant female brood. In *Myrmica* ants, workers not only feed larvae a spartan diet but also physically harass the larvae by biting them so that they

metamorphose and develop into workers. Manipulation by workers also occurs in termites, although it is not known how frequently. In the termite *Pterotermes occidentis*, immature individuals are prevented from metamorphosing into alate reproductives by other colony members, which bite and destroy their wing buds. Lastly, workers act as a collective, sharing an interest as they are equally related to female brood or, if within-colony relatedness varies, are assumed to be incapable of discriminating among different classes of kin. Being united into a 'community of interest' is likely to tip the balance of power further toward the workers.

GCD may have evolved as a means of curtailing selfish behavior by totipotent female brood, in cases where the collective workers are less able to directly influence the caste fate of female brood. This may be relevant to the cases of *Melipona*, *Pogonomyrmex*, and *Reticulitermes*, but not to *Harpagoxenus laevis* and other ants with a similar caste determination mechanism. The understanding of the molecular genetic basis of the cases of GCD described here may provide additional clues as to the selective forces governing its evolution.

See also: Caste Determination in Arthropods; Cooperation and Sociality; Kin Selection and Relatedness; Social Insects: Behavioral Genetics.

Further Reading

Anderson KE, Linksvayer TA and Smith CR (2008) The causes and consequences of genetic caste determination in ants (Hymenoptera: Formicidae). *Myrmecological News* 11: 119–132.

Crozier RH (1979) Genetics of sociality. In: Hermann HR (ed.) *Social Insects*, pp. 223–286. New York, NY: Academic Press.

Keller L (2007) Uncovering the biodiversity of genetic and reproductive systems: Time for a more open approach. *The American Naturalist* 169: 1–8.

Kerr WE (1950) Evolution of the mechanism of caste determination in the genus *Melipona*. *Evolution* 4: 7–13.

Matsuura K, Vargo EL, Kawatsu K, et al. (2009) Queen succession through asexual reproduction in termites. *Science* 323: 1687.

Queller DC and Strassman JE (1998) Kin selection and social insects. *Bioscience* 48: 165–175.

Wheeler DE (1986) Developmental and physiological determinants of caste in social Hymenoptera – evolutionary implications. *The American Naturalist* 128: 13–34.

Winter U and Buschinger A (1986) Genetically mediated queen polymorphism and caste determination in the slave-making ant, *Harpagoxenus sublaevis* (Hymenoptera: Formicidae). *Entomologia Generalis* 11: 125–137.

Categories and Concepts: Language-Related Competences in Non-Linguistic Species

L. Huber, University of Vienna, Vienna, Austria

Introduction

The human mind thinks about its world in a limitless and often unexpected way. This process is anything but effortless or immediate, and frequently goes wrong. Language and – more generally – the ability to form symbolic representations enables us to split and lump objects and events together in countless numbers of ways. This ability, most commonly called categorization, has thus been viewed as the groundwork of cognition, as a process by which the brain assigns meaning to objects or events by treating them as equivalents. However, while language depends on categorization, categorization does not depend on language. Therefore, non-human animals are adequate models to investigate the 'middle' range of categorization mechanisms – lying between perceptual discrimination of elementary stimuli and the linguistic manipulation of classes of objects, events, or ideas by using symbolic representations and by attaching to them verbal names. It can be viewed as the ability to treat similar but non-identical things as somehow equivalent, by sorting them into their proper categories and by reacting to them in the same manner.

In general, it is useful to distinguish between categorization and recognition. Categorization (or classification) refers to the perception of an object as belonging to a general class, while individual recognition is the identification of different images as depicting the same object, such as a specific face, despite changes in the viewing conditions. The brain of many animals solves both cognitive tasks in a natural, effortless manner and with an efficiency that is difficult to reproduce in computational models and artificial systems. How is this accomplished? What brain mechanisms have been evolved in the service of these tasks? Cognitive psychologists, ethologists, and neurobiologists have investigated these questions for decades and have converged on the common understanding that many different mechanisms have evolved to make categorization an adaptive cognitive trait. At least three different levels have been identified at which the individual groups objects or events: the perceptual, the associative, and the conceptual level. Most recently, neuroscientists have disentangled some neuronal processes underlying object recognition and categorization, by identifying the neuronal pathways and time courses of the processing and coding of category-relevant information.

Function

Considering, on the one hand, the vast amount of information arriving at the perceptual systems of mobile organisms and, on the other hand, the few behavioral output patterns possible in non-human animals, categorization may thus be conceptualized as an adequate solution to this 'informational bottleneck.' The drastic information reduction is a fundamental principle of cognitive economics and therefore widely dispersed among species. The evolutionary pressures to minimize processing requirements in small information-processing systems such as non-human animals are self-evident. Nevertheless, the specific solutions found by a wide range of species to compress the amount of retained information vary at different levels, from the purely perceptual (selective attention) to the level of learning or representation and finally, to the level of reasoning.

A good starting point is to consider an animal's stimulus problem in nature. In order to behave appropriately, an animal needs a description of the stimuli in any environment that results in little variation of the predictors of significant consequences of behavior. Somehow the animal must select particular aspects of experience to use as a basis for judgments on lumping and splitting. A balance must then be struck between the two kinds of processes. Different occasions may thus dictate different descriptions and different dimensions of invariance. Thus, categories are neither purely perceptual nor merely functional, but both. Categorization problems are jointly determined by physical variation in stimuli and the consequences of behavior. The evidence suggests that natural selection has equipped animals with considerable adaptations for dealing with the categorization problem in this sense.

Whether grouping is difficult or easy depends on at least three factors: (1) the relations among the objects, (2) the consequences of actions in their presence, and (3) the perceptual capabilities of the individual. Generally, the more variable an object class is, the more difficult the grouping is. Furthermore, the contingencies may be conditional, depending not on any particular object, but for example, on conjunctions and disjunctions of particular configurations (spatial or temporal) of objects. Therefore, researchers are now studying not only object recognition, as it has been done in the lab for a long time by presenting very impoverished stimuli, but also scenes, which are composed of many objects.

Sources of Natural Variation

Conceptually, it is helpful to distinguish between recognition (object constancy) and categorization. When is it justifiable to speak about natural categories rather than variations in the appearance of objects? Unsurprisingly, the distinction between natural categories and natural objects is a matter of degree rather than of kind. Natural objects are likely varying as the result of (1) the different physical conditions under which they are perceived, (2) the physical changes occurring to the object over time, or (3) their generative history, which is determined by more general sets of information.

Natural object classes – in the great majority – comprise of organisms or their products. These vary considerably, but not in a purely random manner. Biological diversity is variation of life at all levels of biological organization, a product of genetic variability among the populations and the individuals of the same species. The most important sources of biological variation are mutation, recombination, noise in pattern formation, and genetic drift. Developmental and functional constraints have restricted and channeled natural variation, thereby producing hierarchically organized ('nested') fields of similarities. These divergent lines of variation offer opportunities for lumping and splitting ('carving nature at its joints') to both animals and taxonomists alike.

The Structure of Natural Classes

Prototypical natural classes are open-ended, variable, and polymorphous. As such, the borders between classes are not fixed but vary, as they depend on the dynamic interaction between an organism and its environment. From the perceiver's point of view, the same object may deserve several responses in different contexts or times. This is especially true if the stimuli are themselves reactive, like other animals. For example, the same group mate may be a friend or foe when encountered on different occasions. Furthermore, the functional meaning of a stimulus is seldom determined by the mere presence or absence of one or few of its features, but on more complex concatenations of stimulus descriptors. The vast majority of natural classes are polymorphous. This means that no single isolated feature is necessary or wholly sufficient, but each contributes to some degree to determining class membership. Of course, we can list the so-called diagnostic features belonging to a certain category, but it is seldom possible to identify one single ('essential') property that reliably distinguishes that category from all others. Instead, each feature correlates – not necessarily perfectly, but to some extent – with category membership and determines the latter only in combination with other features typical of the respective category.

Given the evolutionary history of the most important stimulus classes for certain species, it is conceivable that perceptual and behavioral contingencies became conjointly correlated with environmental dimensions of variance. By virtue of natural selection, visual classes and natural categories were rendered coextensive. An illustrative example of this kind of co-evolution is the human face. It provides information about the age, sex, emotion, race, and cognitive states of the individual. The human-encoding system contains evolved and acquired ('expertise') components to interpret these kinds of information quickly and accurately. Because the fit between perceiver and signaler is never complete, it needs to be adjusted ('tuned') during lifetime. The accomplishment of this modification depends on learning and memory. Animals are equipped with several neuronal mechanisms to learn about and to store the relevant information in order to solve the categorization problem. Humans' abilities in this respect are not unique, nor did they arise de novo; rather, they may have clear non-linguistic parallels and origins in other (non-human) animals.

Categorization Mechanisms

Animals may represent (encode) classes of stimuli in (at least) six rather distinct ways: (1) pictorially (imaginably) as arrays of features or elements defined in their own absolute values, thus being a matter of assessing similarity; (2) associatively as collections of objects or events signaling the same consequence or follow-up event; (3) functionally as collections of items with the same inherent function; (4) abstractly as relations between two or more objects or events; (5) analogically as relationship between two or more other relationships; and (6) symbolically as relations to other classes.

Perceptual Classes

In fact, there are various ways in which an animal may generate image representations to solve a categorization task. First, the ability to categorize may depend on remembering each instance as well as the category to which it belongs, as purported by exemplar models. Second, as claimed by feature models, categorization may be based on the learned attention to diagnostic features, that is, a necessary set of defining features that characterize members of the same category. Discrimination and generalization are the primary candidates for the underlying mechanisms. Finally, according to prototype models, categorization may be accomplished by the generation of a summary representation of a category that corresponds to the average, or central tendency, of all the exemplars that have been experienced.

Exemplar theory

According to the exemplar theory, intact stimuli are stored in memory and classification of a novel stimulus is determined by the degree of its similarity to stored exemplars. This means that the item or idiosyncratic information is used for classification decisions and the individual exemplars have to maintain their memorial integrity. More sophisticated versions acknowledge the importance of single stimulus aspects that can be weighted differently and assume that classification is based on retrieving only a subset of the stored exemplars, presumably the most similar ones. Moreover, some exemplar models predict sensitivity to context, category size, exemplar variability, and correlated features. Nevertheless, the critical assumption of even those more liberal models is that stored exemplars play the dominant role in categorization.

Learning about perceptual classes in nature by recognizing a huge amount of instances may sound like an impossible feat, but some animals have quite striking capacities to retain long-term picture–response associations. For instance, Clark's Nutcrackers (*Nucifraga columbiana*) cache pine seeds in up to 6000 locations during the autumn and retrieve up to 90% of them months later with a high degree of accuracy. They use visual landmarks and a spatial memory system, which is robust and of long duration, to recover their caches. Some have argued that such impressive memory capacities are exhibited only by some specialists and is restricted to the specific domain for which it has been selected (e.g., food-caching). Recent comparative research with baboons (*Papio papio*) and pigeons, however, revealed very large capacities in nonstoring species to learn and remember visual images. Pigeons can memorize about 1000 arbitrary picture–response associations before reaching the limit of their performance, and baboons had not reached their limit with 5000 pictures after more than 3 years of testing. If extrapolated from the retention curves, baboons may be able to remember up to 14 000 pictures and their consequence. In both species, and possibly in many more, their capacity represents a rich library of information and experience to draw on during their daily activities.

Despite such huge capacities of item memorization, animals may use feature rules under different conditions. Actually, it seems as if the brain would be able to store two types of information simultaneously, item- and category-specific, and uses predominantly the type that is functional in a given task. The first type of information is needed to discriminate between instances of the same class, the second to discriminate between instances of different classes. For instance, pigeons, it has been shown, are able to tell male human faces apart using item-specific information, but when required to discriminate between male and female faces, they use the sex-specific categorical information to solve the task.

Differential within-versus-between-class generalization is commonly considered a key feature of visual categorization in animals. If a category-specific information is available as a kind of relational information about common properties of a category, then classification learning is restricted to learning about each individual stimulus separately and distinctly. If it were available, learning to use it would significantly facilitate classification, both in terms of acquisition and transfer to novel instances. The classic procedure to distinguish between rote learning and learning about category-specific features is the 'pseudo-concept' task, which involves the arbitrary assignment of category (e.g., 'tree') and noncategory exemplars ('nontree') to positive and negative reinforcement, respectively. If, as was in fact shown in many cases, pseudo-concepts are much harder to learn than 'natural categories' (e.g., tree vs. nontree), the latter was not learned by rote, thereby supporting the feature theory of categorization.

Feature theory

Proponents of feature models emphasize learning about features characteristic of the category. Acquisition of item-specific knowledge in categorization tasks is not denied, but is thought to be rendered nonfunctional and successively overshadowed by category-specific information during learning. In many experiments, the tested species have demonstrated some flexibility in both feature creation and selection. This means that in the course of categorization training diagnostic features are progressively extracted and continuously adapted to the current demands of categorization ('feature learning' and 'acquired distinctiveness'). The classifying individual exhibits flexibility not only with respect to the feature rule, but also with respect to the features themselves. In contrast to static conceptions of similarity, features are considered dynamic (i.e., varying with the context and the task) and to be a product of the perceiver, rather than of the objects themselves. As such the categorization process is a kind of perceptual carpentry, in which differential reinforcement selects from the available stimulus dimensions those that most accurately differentiate positive and negative instances and furthermore changes their saliency as a function of experience.

Recent experiments have shown that animals are able to decode various features from different dimensions and levels of complexity. For instance, pigeons have proved able to discriminate several categories at the same time, to use global and local stimulus properties, and to recognize isolated stimulus components, but if necessary, to comprehend the configuration of a stimulus: they may rely on 'simple' physical dimensions (e.g., intensity or color) in one task, but respond to a polymorphous class rule (some 'higher' feature) in another. With sufficient experience, pigeons may also become sensitive to spatial relationships among features. Altogether, it has become increasingly evident that the animals may form representations including complex feature rules rather than simple feature lists. Feature rules refer to some conjunctive, disjunctive, or additive combinations of features. Polymorphous categories seem to be represented best by an additive combination of features plus a threshold criterion (e.g., the 'm-out-of-n' rule).

Whether individuals try to find the diagnostic features or memorize whole pictures depends on both the content of the category members and the structure of the categories. Critical factors of the former are whether the diagnostic features vary independently and whether they are localized in contrast to a more configural, correlative compound of these. What matters for category structure is the size of the category, that is, how many instances are encountered, the similarity among members of the same class in relation to the (dis)similarity between classes, and the concatenation of class descriptors. If stimulus classes are composed like variations of a common theme, with the variations unlikely to follow along identifiable feature dimensions, the categorization strategy employed by humans has been described in terms of the prototype theory.

Prototype theory

Prototypes are viewed as a kind of summary representation of a category that corresponds to the average, or central tendency, of all of the exemplars that have been experienced. New exemplars can be classified on the basis of their similarities to this 'best example.' Intuitively, the prototype theory bears some validity, because if we are asked to imagine a category, usually the most typical instance of this category comes to our mind. In the laboratory, the theory gains empirical support from the so-called prototype effect. It refers to the fact that the prototype, or exemplars that bear a close resemblance to it, are instantaneously classified, sometimes even more readily than those exemplars of the category that have already been experienced.

Whether animals can form prototypes is not clear at the moment, and evidence in its favor is often overshadowed by simpler accounts like, for instance, in terms of 'peak shift.' The best evidence to date for prototype formation comes from experiments in which polymorphous stimulus classes have been composed of many stimuli that share a 'point of departure' or origin, but that cannot be separated in terms of simple feature rules.

Associative and Functional Classes

Many psychologists have been inclined to take feature or prototype learning as a sufficient criterion for saying that an animal has acquired a 'concept.' Others have restricted this assignment to cases that lack a perceptual basis for categorization. In the simplest case, the instances of a category share a common function. Such a common function as signaling food can be purely arbitrary as in the pseudo-concept tasks, or inherent as it is for edible items.

However, arbitrary classes lack two important properties of true functional (or equivalent) classes. Learning cannot be transferred to novel items and the members are not (i.e., independent of training) tied together a priori so that, if the reward contingency of one item changes, the contingency of the remaining ones is changing, too.

Ties between items of a class can emerge through special kinds of training, the result called 'acquired equivalence.' As a consequence, a change in response tendencies to some members of the class generalizes spontaneously to other members of the class, or the members of the class become interchangeable. Natural examples are 'food,' 'tool,' or 'enemy.' If the members of a class are tied together by a common function, the class as such can be easily manipulated and related to other classes. This kind of mental manipulation was exactly what some theorists have seen as the essence of concepts, thereby proposing the possession of a unique mental structure that is active when and only when an instance of that concept is presented. Of course, the concept would be more than the sum of its component features or component instances. For example, olive baboons were trained to discriminate only two pictures of objects, one food and one nonfood item. When subsequently tested, they immediately transferred this discrimination to many novel items of both classes. Even more impressive, chimpanzees learned first to categorize real items in food and tools, and then rapidly transferred their categorization to novel objects and later to pictures and arbitrary symbols (lexigrams) of the respective categories. This demonstrates that the apes could easily switch between different instantiations of the functional classes (picture-object-symbol equivalence) rather than generalize from stored features or exemplars.

Can other nonprimate animals also construct categories on the basis of their functional, rather than perceptual, properties? The answer is a qualified yes. For instance, pigeons can form functional equivalence classes via associations with a common response key, or a common delay, a common reinforcement history, or a common quantity of reinforcement. A particularly sophisticated method to test animals on the ability to form functional equivalence classes is the many-to-one mapping procedure. For example, in an experiment with sea lions, two perceptually distinct stimuli (let's say samples A and B) have been tied together via a third stimulus, comparison C, with which they were conditionally associated in separate trials. Their membership in a common equivalence class could be shown by an independent test following an intermediate training in which one of the two sample stimuli (say A) has become associated with a new comparison stimulus D. Evidence that something new has tied together samples A and B came from the test in which the sea lions spontaneously grouped sample B with comparison D. The nature of this so-called 'emergent relation' is a common representation of the two sample stimuli.

Relational Classes

A further question is whether animals themselves can find the binding relation among the members of a class, rather than learning it associatively. One of the most investigated and interesting relational concepts is 'sameness,' the ability to figure out that the members of a class share the common property of identical components. Sophisticated training procedures are necessary to prevent the animal from learning the stimulus differences in absolute, rather than in relational, terms. For instance, pigeons found a perceptual clue to solve the task. When trained to peck at pictures containing an array of identical icons, and to withdraw from pictures with nonidentical icons, they used the entropy difference between these two picture sets as a discrimination cue. Remarkably, honeybees (*Apis mellifera*) have learned to discriminate visual images on the basis of sameness in a matching to sample task. After reaching criterion, they were tested with novel images and learned to choose the correct ones much faster. However, only the spontaneous mastery of a test with radically novel stimuli, as for instance, stimuli perceived in other sensory modalities (sounds or smells), would determine whether the animals had acquired a true understanding of sameness as classificatory principle, uncontaminated by perceptual cues that covary with the abstract concept.

A further, even more difficult stimulus relation is 'transitivity.' It is an emergent relation among stimuli that are compared subsequently. If, for instance, A is bigger than B and B is bigger than C, an inference can be made that A is bigger than C. Here again, sophisticated experimental research has been required to control for purely associative accounts as alternative to a true inferential relationship. In nature, transitive inference may be part of a cognitive toolkit of social animals that are required to track and assess relationships among group members. As it seems unlikely that an individual in a large social group could ever observe interactions between all possible combinations of group members (in monkey groups of 80 animals, more than 3000 dyadic and more than 82 000 triadic combinations), it would be advantageous to conclude that if A is dominant to B and B is dominant to C, then A would probably also dominate C, even if it has never seen A and C interacting. Earlier studies of social categorization in primates, involving mother–infant bond and dominance–subordination relation in vervet monkeys and macaques were indicative, but not conclusive in this respect. More recent studies with social corvids, however, have provided clear-cut evidence that non-human animals can infer the dominance status of group members by representing somehow the transitive logic.

Emergent relations may also explain the human competence to classify despite inconsistent or incomplete information. A straightforward way to deal with this problem is to use inferential reasoning, the ability to associate

a visible and an imagined event. 'Inference by exclusion,' sometimes called 'fast mapping,' is defined as the selection of the correct alternative by logically excluding other potential alternatives. Humans are known to learn by exclusion, which is particularly evident in vocabulary learning by children. The ability to map a newly heard word to an object that does not have a known lexical entry is already present at a very early age. In animals, the first step to determine whether an individual is able to draw such inferences usually consists in investigating whether it will choose an undefined (no associated category membership) stimulus over a defined (already associated) one. The second step involves the choosing of the previously undefined, but then by exclusion defined, stimulus over a further undefined one. Only if the stability of the novel association in the presence of unfamiliar rather than familiar alternatives is shown, reasoning by exclusion can be inferred. Several studies with non-human species have suggested that at least some animals may also be able to solve inferences by exclusion. So far, only chimpanzees, sea lions, and domestic dogs have shown conclusive evidence for this.

Analogical and Symbolic Representations

The most advanced form of relational classes is based on the relations between relations, which are called analogies. One task that has been used for non-human, that is, non linguistic animals, is analogical or relational matching-to-sample. The solution to this task requires the animal to attend to the higher-order relations between relations, because none of the items in the sample set is presented in either of the two choice sets. So, for instance, in relational identity matching, the sample consists of one out of many pictures containing a set of several identical icons ('same') and then the animal has to choose from two comparison pictures the one that also contains a set of several identical icons (identity relation). To date, only great apes and a gray parrot called Alex proved able to understand analogies, but only if having received explicit symbol training before.

Summary

Memorization, feature learning, and acquired equivalence based on a common history of association are sufficient to account for many instances of categorization. The underlying perceptual and associative processes seem to be widely distributed across the animal kingdom. There is evidence that some corvids and parrots, as well as dolphins, some monkeys, and apes, show excellent transfer to wholly novel stimuli after training on relational classes. Bees and pigeons can do as well, but they would require much more stimuli to show a convincing transfer performance. Monkeys and pigeons have so far failed to acquire symbolic referential meaning for categorical relations of identity. Finally, the learning and understanding of higher-order relations between relations seems to require explicit symbol training. Only a few chimpanzees and a gray parrot, all of them having a kind of language training (e.g., lexigrams, sign language, vocalizations) before, have been shown to be able in this respect so far. Even so, this ability does not match the respective abilities of humans, of course, who have extensive practice with relational concepts and have made them an integral part of language.

See also: Alex: A Study in Avian Cognition; Animal Arithmetic; Decision-Making and Learning: The Peak Shift Behavioral Response; Metacognition and Meta-memory in Non-Human Animals; Multimodal Signaling; Non-Elemental Learning in Invertebrates; Referential Signaling; Social Recognition.

Further Reading

Chater N and Heyes C (1994) Animal concepts: Content and discontent. *Mind & Language* 9: 210–246.

Fagot J (ed.) (2000) *Picture Perception in Animals.* East Sussex: Psychology Press.

Gentner D (1998) Analogy. In: Bechtel W and Graham G (eds.) *A Companion to Cognitive Science (Blackwell Companions to Philosophy; 13)*, pp. 107–113. Malden, MA: Blackwell.

Hampton JA (1999) Concepts. In: Wilson RA and Keil FC (eds.) *MIT Encyclopedia of the Cognitive Sciences*, pp. 176–179. Cambridge, MA: The MIT Press.

Harnad S (ed.) (1987) *Categorical Perception: The Groundwork of Cognition.* Cambridge: Cambridge University Press.

Herrnstein RJ (1985) Riddles of natural categorization. *Philosophical Transactions of the Royal Society* B308: 129–144.

Huber L and Aust U (2006) A modified feature theory as an account of pigeon visual categorization. In: Wasserman E and Zentall T (eds.) *Comparative Cognition. Experimental Explorations of Animal Intelligence*, pp. 325–342. New York, NY: Oxford University Press.

Lamberts K and Shanks D (1997) *Knowledge, Concepts, and Categories.* East Sussex: Psychology Press.

Lazareva OF and Wasserman E (2008) Categories and concepts in animals. In: Menzel R (ed.) *Learning Theory and Behavior of Learning and Memory: A Comprehensive Reference* vol. 1, pp. 197–226. Oxford: Elsevier.

Mackintosh NJ (2000) Abstraction and discrimination. In: Heyes C and Huber L (eds.) *The Evolution of Cognition*, pp. 123–141. Cambridge, MA: MIT Press.

Margolis E and Laurence S (eds.) (1999) *Concepts – Core Readings.* Cambridge, MA: MIT Press.

Medin DL and Smith EE (1984) Concepts and concept formation. *Annual Review of Psychology* 35: 113–138.

Premack D (1988) Minds with and without language. In: Weiskrantz L (ed.) *Thought Without Language*, pp. 46–65. New York, NY: Oxford University Press.

Thompson RKR and Oden DL (2000) Categorical perception and conceptual judgments by nonhuman primates: The paleological monkey and the analogical ape. *Cognitive Science* 24: 363–396.

Zentall TR, Wasserman EA, Lazareva OF, Thompson RRK, and Rattermann MJ (2008) Concept learning in animals. *Comparative Cognition & Behavior Reviews* 3: 13–45.

Chimpanzees

J. C. Mitani, University of Michigan, Ann Arbor, MI, USA

Introduction

Darwin's theories of sexual selection and natural selection furnish powerful frameworks to interpret two prominent aspects of animal behavior: competition and cooperation. Our current understanding of male–male competition derives largely from the sexual selection theory. Because females typically invest more in offspring than males do, the number of reproductively active males exceeds the number of reproductively active females at any given time. As a consequence, males compete vigorously between themselves to obtain matings with females. Cooperation involves behaviors that increase the fitness of others, and creates a paradox for natural selection, a process that Darwin envisioned as "... acts solely by and for the good of each." Cooperation between males is especially puzzling because the primary resource over which these individuals compete, namely fertilizable females, is neither easily divided nor shared. Here, evolutionary theory predicts, and empirical research shows, that individuals nevertheless cooperate to increase their own fitness or to obtain indirect fitness benefits.

Chimpanzees (*Pan troglodytes*) represent a model system to investigate competition and cooperation in animals. Recent observations from long-term field research reveal that chimpanzees display these twin facets of behavior in somewhat surprising ways, with females competing for food, for space, and to reproduce, and males cooperating with others in interactions within and between communities. In this contribution, I summarize our current understanding of female competition and male cooperation in chimpanzees. This knowledge draws on long-term observations made on wild chimpanzees, and I begin with a brief description of the major field studies of these animals. I then provide a biological context for my discussion of female competition and male cooperation in chimpanzees by outlining selected aspects of their ecology, social organization, and life history.

Chimpanzee Field Studies

Along with their sister species, the bonobos, chimpanzees are humankind's closest living relatives. The close evolutionary relationship between chimpanzees and humans furnished the impetus to study their behavior in the wild, and chimpanzees are now one of the most intensively studied of all animals. Jane Goodall's well-known fieldwork on chimpanzees at the Gombe National Park, Tanzania, now approaching its fiftieth year, is the longest continuous study of any animal in the wild. Shortly after Goodall began her pioneering work, Toshisada Nishida started an equally important and long-running field study in 1965 at the Mahale Mountains National Park, also in Tanzania. Research at Gombe and Mahale stimulated others to initiate field research on chimpanzees at other sites across Africa. Additional long-term studies include projects at Bossou, Guinea (1976–present), Taï National Park, Ivory Coast (1979–present), Kibale National Park, Uganda (Kanyawara: 1987–present; Ngogo: 1995–present), and Budongo Forest Reserve, Uganda (1992–present). As a result of this research, we now possess a considerable amount of data on the behavior of wild chimpanzees derived from over 200 person years of fieldwork. This body of research forms a solid basis for describing the natural history of chimpanzees, including aspects of their ecology, social organization, and life history.

Chimpanzee Ecology

Chimpanzees are found only in Africa, with three generally recognized subspecies occupying geographically circumscribed areas in the western (*Pan troglodytes verus*), central (*Pan troglodytes troglodytes*), and eastern (*Pan troglodytes schweinfurthii*) parts of the continent. Chimpanzees are an extremely wide-ranging and cosmopolitan species and live in a variety of habitats. Most chimpanzees can be found in primary tropical rainforest, with others living in secondary forests and open woodlands. Some populations range across desert-like, xeric conditions. Regardless of the kind of habitat they occupy, chimpanzees live at extremely low population densities that rarely exceed five animals per square kilometer.

Like many non-human primates in the wild, chimpanzees spend most of their waking hours searching for, handling, and ingesting food. Chimpanzees are highly eclectic and catholic feeders and eat a variety of foods. These include the vegetative parts of plants, leaves and bark, the reproductive parts of plants, fruits and flowers, and other animals, both invertebrates and vertebrates. Despite their extremely varied diet, chimpanzees show a clear preference for ripe fruit, eating significantly more ripe fruit than do other fruit-eating primates with whom they live sympatrically.

The consumption of invertebrate and vertebrate prey is one aspect of chimpanzees' feeding behavior that has generated considerable interest and research. Early field observations by Goodall and Nishida revealed how chimpanzees make and use tools to harvest and ingest hard-to-obtain invertebrate prey, such as termites and ants. Goodall was the first to document hunting and meat eating by wild chimpanzees. Subsequent research at Gombe and elsewhere indicates that chimpanzees consume over 40 species of mammals. Most of these are other primates, with the red colobus monkey (genus: *Procolobus*) being the favored prey of chimpanzees wherever these two species live together. Chimpanzee predation on red colobus monkeys varies over time, with seasonal 'binges' or 'crazes' occasionally resulting in rates of predation that exceed one kill per day.

As they search for food, chimpanzees range over very large areas between 7 and 38 km^2. Most of the variation in territory size is due to the differences in habitat quality, with chimpanzees in poor habitats moving over larger areas than chimpanzees that occupy more productive habitats. Wherever multiple communities occupy contiguous space, chimpanzees display territorial behavior. Interactions between the members of different communities are typically hostile and characteristically involve physical aggression, chasing, and vocal battles. During some particularly severe interactions, hostilities escalate, and an individual sometimes falls victim and dies. Adult male chimpanzees are the primary participants in territorial encounters, although younger males and adult and adolescent females also participate occasionally. Male chimpanzees cooperate in the context of territoriality (see below), a striking feature of their behavior that can only be interpreted and understood in terms of their unusual social organization.

Chimpanzee Social Organization

In most primate species, individuals are social. Chimpanzees are no exception. Instead of moving around in stable and spatially coherent groups, however, chimpanzees live in fluid communities that vary in size between 20 and 160 individuals. The fluidity of chimpanzee communities is created by its members who fission and fuse to form temporary subgroups or 'parties' (**Figure 1**). Parties constantly change in size and composition, with large parties typically forming during times of ripe fruit abundance. During periods of fruit scarcity, feeding competition escalates, and chimpanzees reduce the levels of competition by gathering in small parties or by moving alone. Male chimpanzees also form relatively large parties whenever females come into estrus. Because female chimpanzees reproduce very slowly (see below), they are in estrus rarely, making large gatherings around them infrequent events.

Although chimpanzee parties change over time, they do not form randomly. Seminal observations made by

Figure 1 A chimpanzee party. Within communities, individual chimpanzees fission and fuse to form temporary subgroups or 'parties' that vary in size and in composition.

Figure 2 Male chimpanzees are gregarious. Males affiliate in several contexts, for example, by grooming each other.

Nishida over 40 years ago revealed that male chimpanzees are more social than females are. Male chimpanzees frequently affiliate by grouping together in parties. In addition, they groom each other often and maintain close spatial proximity to one another (**Figure 2**). Male chimpanzees are also well known for cooperating in several contexts. Males cooperate by forming short-term coalitions and long-term alliances, sharing meat with each other, and via joint participation in territorial boundary patrols (see below). In contrast, female chimpanzees, especially those in East Africa, often move alone or with their offspring. Although female chimpanzees form long-lasting relationships with their offspring, they affiliate and cooperate significantly less often than do males. Sex differences in chimpanzee affiliation and cooperation are due in part to a pronounced sex bias in dispersal, a process that plays a central role in the life history of female chimpanzees.

Chimpanzee Life History

As noted earlier, chimpanzees live at very low population densities in the wild. These low densities are partly attributable to the extremely low reproductive rates of female chimpanzees, which give birth only once every 5–6 years on average (**Figure 3**). Infant mortality is high, reaching up to 30% during the first 2 years of life. Disease, predation, infanticide, and maternal death are four factors that contribute to this high rate of mortality. Males which survive pass through successive stages juvenility (5–7 years) and adolescence (8–15 years) before reaching adulthood and physical and social maturity at 16 years of age. Male chimpanzees remain in their natal communities throughout their lives. They form enduring relationships with their maternal siblings and other male allies, a factor that accounts for their general gregariousness and sociability.

If females survive infancy, they start to show their first sexual swellings, a physical sign of estrus, when they are about 10 years old. Females continue to reside in their natal groups for another year, dispersing to another community at an average age of about 11. Virtually all females emigrate at sites where multiple communities of chimpanzees exist. At some sites, only isolated communities of chimpanzees have been studied, for example, Gombe and Bossou. With dispersal options limited in these situations, females do not always emigrate. After moving to new communities, young females continue to display sexual swellings and to mate males. But they experience a period of adolescent sterility for

Figure 3 Female chimpanzee and infant. Female chimpanzees in the wild give birth rarely, only once every 5–6 years.

the next 2–3 years, after which they give birth for the first time when they are 13–14 years old. One topic that has been the subject of persistent debate concerns whether female chimpanzees undergo menopause, the cessation of reproductive function before death. A recent compilation of long-term data from six populations of free-ranging chimpanzees indicates that females continue to reproduce till late in life, a finding consistent with the hypothesis that they do not experience menopause. Instead, menopause appears to be a uniquely human trait. Like humans, however, chimpanzees live an extremely long time, with animals in the wild occasionally reaching the age of 50 years. In sum, female chimpanzees display extremely slow life histories, characterized by prolonged development, delayed reproduction, and low reproductive rates. These factors set the stage for competition in the contexts of feeding and the use of space.

Female Chimpanzee Competition

Early field observations emphasized competition between male chimpanzees. Males are extremely gregarious, engaging in noisy, highly charged, aggressive confrontations daily. Because of their relatively asocial nature, female chimpanzees interact with other conspecifics rarely, and social relationships between them have been difficult to decipher. Showing scant inclination to strive for status, female chimpanzees are also notoriously hard to rank. Dominance rank relationships between females can be discerned, but usually only after several years of observation. As a consequence, female chimpanzees are often depicted as comparatively shy, secretive, and noncompetitive. Data derived from long-term research conducted at several study sites are beginning to transform this image of female chimpanzees, revealing a heretofore, unsuspected competitive nature.

Because female parental investment typically exceeds male investment, female reproduction is limited primarily by their ability to convert environmental energy directly into offspring. Female chimpanzees are, therefore, expected to compete mainly for food. As specialized feeders on ripe fruit, chimpanzees face a limited and unpredictable food supply that fluctuates in space and in time. To deal with the challenge created by their foraging regime, female chimpanzees move alone or with their dependent offspring over relatively small 'core areas' within the larger communal territory. Females settle in different core areas and remain faithful to them over time, becoming intimately familiar with the locations of food in the process. A growing body of evidence now shows that instead of competing directly for food during face-to-face encounters with others, female chimpanzees compete for high quality core areas.

Core areas occupied by females differ because chimpanzees range over heterogeneous habitats. Recent field studies reveal that high ranking female chimpanzees inhabit more productive core areas than do low ranking individuals, with dominant females living in areas that are smaller and possess more preferred foods than do subordinate animals. At Gombe, the size of core areas changes over time, and during periods of fruit scarcity when competition for food is heightened, the difference in size between the core areas of high and low ranking females grows larger. Additional observations indicate that rates of female aggression increase when new females immigrate into communities and attempt to establish core areas, with the immigrants eventually settling in areas away from those occupied by high ranking, resident females.

The preceding observations suggest that high ranking females out-compete low ranking females, with the former forcing the latter to settle in poorer habitats and to range more widely to search for food. With increased access to food, high ranking females spend less time foraging and eat higher quality foods than do low ranking females. Dominant females reside in prime feeding habitat and are heavier and show less variation in weight across seasons than do subordinate females. Finally, foraging success affects the reproduction of high ranking females in important ways. High ranking females live longer and have shorter birth intervals than low ranking females. Dominant females also give birth to more surviving offspring and produce daughters that reach sexual maturity earlier than those born to subordinate mothers.

The prior findings indicate that female chimpanzees compete with each other, albeit in indirect and subtle ways. Recent observations indicate that female–female competition can escalate dramatically during periods of increased competition for space. This occurs as females immigrate into new communities and attempt to establish a core area. When this happens, new immigrants receive considerable aggression from resident females. At Gombe and the Budongo Forest Reserve, resident females have been observed to form coalitions to attack and kill the infants of immigrant females. Cooperative behavior between female chimpanzees, illustrated by these coalitionary attacks, is relatively rare. In contrast, male chimpanzee cooperation is varied and has few parallels in the animal kingdom.

Male Chimpanzee Cooperation

Like female chimpanzees, male chimpanzees compete with members of their own sex. The extremely long interbirth intervals of female chimpanzees create male biased operational sex ratios, with many more reproductively active males at any moment in time than reproductively active females. With so few females available, male chimpanzees compete intensely for mates and reproductive opportunities.

Male–male competition in chimpanzees occurs both within and between communities. Within communities, male chimpanzees compete vigorously for status. Male chimpanzees continuously strive to dominate others and form linear dominance hierarchies, with individuals who achieve high rank obtaining substantial fitness benefits. Behavioral observations indicate that high ranking males mate more often than do lower ranking individuals, and paternity studies utilizing genetic markers reveal that males who rise to alpha positions sire up to 30–50% of all infants born during their tenures at the top of the dominance hierarchy. Male chimpanzees also compete with members of other communities via group territorial behavior.

While competition is a striking feature of male chimpanzee life, male chimpanzees are also well known to cooperate as they attempt to dominate others inside and outside their communities. Coalitionary behavior, meat sharing, and territorial boundary patrolling are three prominent examples of male chimpanzee cooperative behavior.

Coalitions

In several species of non-human primates, individuals form coalitions. Coalitions involve situations where two or more individuals cooperate and join forces to direct aggression towards others. Coalitions are a particularly conspicuous feature of male chimpanzee behavior, with males forming coalitions quite frequently in over one quarter of all aggressive interactions in which they are involved. High ranking individuals participate more often than low ranking animals, and coalitions play an integral role in male chimpanzee status competition. Specifically, males achieve and maintain high dominance rank via coalitionary support.

Revolutionary coalitions in which two low ranking individuals cooperate and direct aggression towards higher ranking animals occur very rarely in primates. In these cases, high ranking animals who are targeted are likely to counterattack, making it costly for coalitionary partners to cooperate. Such coalitions nevertheless occur occasionally between male chimpanzees, even when the dominance hierarchy is stable. In contrast, male chimpanzees form revolutionary coalitions much more frequently during periods of rank instability as males vie for alpha status. Anecdotal observations from several field sites indicate that male chimpanzees typically achieve the alpha position only with coalitionary support provided by others.

Although rare, revolutionary coalitions involving male chimpanzees have received considerable research attention because of the dramatic effects they have on male dominance relations and ultimately male fitness. Most coalitions that take place between male chimpanzees, however, do not upend social relations, but instead reinforce the status quo. Conservative coalitions involve situations in which two high ranking individuals cooperate to attack a lower ranking animal. By acting together in such coalitions, male participants reinforce their positions in the established dominance hierarchy. Conservative coalitions occur often, up to three quarters of the time, when the male dominance hierarchy is relatively stable.

Why do male chimpanzees cooperate by forming coalitions? In some cases, individuals improve their own fitness via coalitionary behavior. This clearly applies to males who participate in conservative coalitions. In these situations, both males benefit by maintaining their positions in the dominance hierarchy. Males who form revolutionary coalitions cooperate in an attempt to achieve higher dominance status, and may, therefore, benefit directly as well. Additional observations indicate that alpha males occasionally trade mating opportunities for coalitionary support. In these cases, males permit their allies to mate in their presence in exchange for support that they require to maintain their alpha position. Further studies reveal that males reciprocally trade coalitionary support and exchange other goods and services, such as grooming and meat, for support. These data are consistent with the hypothesis that males obtain direct fitness benefits by participating in coalitions via reciprocity. Finally, male chimpanzees form coalitions much more frequently with their maternal half siblings than do males who are distantly related or unrelated. In these situations, individuals gain indirect fitness benefits by helping their close relatives.

Meat Sharing

Wild chimpanzees frequently hunt and eat meat. They also often share the meat they obtain in hunts with conspecifics (**Figure 4**). At first blush, meat sharing appears paradoxical. Meat, after all, is a scarce and valuable resource and is highly prized. In addition, meat is very difficult to obtain.

Figure 4 Chimpanzee meat sharing. Male chimpanzees frequently share meat they obtain in hunts with each other. Reproduced from Mitani J C (2009) Cooperation and competition in chimpanzees: Current understanding and future challenges. *Evolutionary Anthropology* 18: 215–227.

At some field sites, chimpanzees engage in prolonged 'hunting patrols,' during which they move slowly and apparent deliberate fashion across their territory in search of prey, typically red colobus monkeys (see above). Hunting patrols can cover long distances and last up to 5 h of a 12-h active day. Chimpanzees may hunt red colobus monkeys upon encounter, but hunting is potentially costly because male colobus monkeys frequently mob chimpanzee predators, sometimes wounding them severely in the process. Because hunts are energetically costly and involve risks for hunters, it is puzzling why male chimpanzees share their hard-earned spoils with others.

Three hypotheses have been proposed and investigated to explain meat sharing by wild chimpanzees. The 'sharing under pressure' hypothesis takes its lead from the observation that successful hunts generate considerable frenzy. Several chimpanzees typically gather around individuals who possess meat, with the former harassing and attempting to grab pieces from the latter. This hypothesis suggests that meat sharing represents a form of 'tolerated theft,' and that possessors share meat primarily to reduce the costs of harassment imposed by beggars. Some evidence is consistent with this proposal, which helps to explain why chimpanzees share in certain situations. The hypothesis nevertheless fails to explain why alpha males, who frequently possess meat and are not especially vulnerable to harassment, readily share with others. Furthermore, the sharing under pressure hypothesis does not account for voluntary transfers that are frequently made by others.

Two additional hypotheses suggest important roles for cooperation between and within the sexes. The 'meat-for-sex' hypothesis proposes that male chimpanzees share meat selectively with estrous females to obtain matings with them. Data derived from most study sites do not support this provocative hypothesis. First, male chimpanzees do not always share meat with estrous females who beg. Second, matings do not regularly ensue between estrous females and males who share with them. Third, males who share meat with females do not mate more often than males who fail to share. Despite these findings, a recent study of chimpanzees in the Taï National Park, Ivory Coast, indicates that female chimpanzees there mate frequently with males who share meat with them over a relatively long period of time. Additional research will be necessary to resolve these conflicting results.

A third hypothesis suggests that male chimpanzees use meat as a political tool to maintain social bonds with others. Nishida and colleagues used observations of one particularly powerful alpha male at Mahale to generate this 'male social bonding' hypothesis. This male shared meat with others, but he did so selectively, typically only with his coalitionary partners. Coalitionary support provided by these individuals helped the alpha male maintain his top spot for over 15 years.

Observations of male chimpanzees living at Ngogo, Kibale National Park, Uganda, are largely consistent with hypothesis. Male chimpanzees there share meat nonrandomly with each other. In addition, males share meat reciprocally. Finally and significantly, male chimpanzees at Ngogo exchange meat for coalitionary support. In sum, these data provide a compelling example of how and why male chimpanzees cooperate with others in their own community. Group territoriality furnishes another case of cooperation, but one that involves individuals living in different communities.

Group Territoriality

Some of the most startling aspects of chimpanzee behavior occur in the context of territoriality. Male chimpanzees are extremely aggressive towards members of other communities and occasionally launch lethal coalitionary attacks on them. Boundary patrol behavior is an integral part of chimpanzee territoriality. During patrols, several chimpanzees, typically male, gather and move in single-file fashion to the periphery of their territory (**Figure 5**). When they arrive at the boundary, the behavior of patrollers changes dramatically. Chimpanzees fall completely

Figure 5   Territorial boundary patrol. Male chimpanzees patrol their territories by moving in single file fashion, occasionally making deep incursions into neighboring territories. Reproduced from Mitani J C (2009) Cooperation and competition in chimpanzees: Current understanding and future challenges. *Evolutionary Anthropology* 18: 215–227.

silent. They scan the environment, paying close attention to any movement in the tree canopy, and sniff the ground, inspecting signs left behind by other chimpanzees, such as feces and fruit remains discarded after feeding. Patrolling chimpanzees sometimes continue further, making deep incursions into the territories of their neighbors. Taken together, these observations suggest that patrollers actively seek information about or contact with neighboring chimpanzees.

Considerable heterogeneity exists in patrol participation. While some males patrol frequently, others do so far less often. These data represent a classic cooperation problem. Why do some males fail to participate, and by opting out, allowed to reap the communal benefits of territorial defense without paying the costs? Additional observations suggest that this problem may be more illusory than real. At Ngogo, males who patrol frequently mate more often than do males who patrol less frequently. These observations support the hypothesis that some males may derive greater fitness benefits by patrolling than others. Males who patrol often may do so to protect themselves, females in their community with whom they mate, and most importantly, their present and future offspring.

During patrols and other times, chimpanzees occasionally encounter members of other communities. Chimpanzees will attack strange conspecifics in these situations, but only if they possess overwhelming numerical superiority over their adversaries. By eliminating rivals, these cooperative group attacks effectively reduce the strength of neighboring communities and permit chimpanzees to expand their territories. Territorial expansion, in turn, increases the food supply, which has concomitant fitness consequences. Long-term observations from Gombe indicate that the size of chimpanzee territories changes over time and that the reproductive performance of females increases during periods of territorial expansion.

Concluding Comments

Chimpanzees are extremely long-lived animals, and yield the secrets of their lives only very slowly. The preceding findings, derived from years of dedicated work by multiple researchers at several sites, stand testament to the value of long-term field research. Whether it will be possible to continue this research in the future, however, is unclear as chimpanzees in the wild are highly endangered. Habitat destruction, a thriving bush meat trade, and recurrent outbreaks of infectious disease threaten the populations of chimpanzees throughout the African continent. A catastrophic decline in the number of wild chimpanzees during the past 30 years portends a grim fate for them. Effective action is required now to ensure that future students of animal behavior are able to continue to study and learn from these remarkable creatures.

See also: Cooperation and Sociality; Darwin and Animal Behavior; Kin Selection and Relatedness; Social Selection, Sexual Selection, and Sexual Conflict.

Further Reading

Boesch C, Hohmann G, and Marchant L (eds.) (2002) *Behavioral Diversity in Chimpanzees and Bonobos.* Cambridge: Cambridge University Press.

Boesch C, Kohou G, Nene H, and Vigilant L (2006) Male competition and paternity in wild chimpanzees of the Taï forest. *American Journal of Physical Anthropology* 130: 103–115.

Duffy K, Wrangham R, and Silk J (2007) Male chimpanzees exchange political support for mating opportunities. *Current Biology* 17: R586–R587.

Goodall J (1986) *The Chimpanzees of Gombe.* Cambridge, MA: Belknap Press.

Kappeler P and van Schaik C (eds.) (2006) *Cooperation in Primates: Mechanisms and Evolution*, pp. 107–119. Heidelberg: Springer-Verlag.

Langergraber K, Mitani J, and Vigilant L (2007) The limited impact of kinship on cooperation in wild chimpanzees. *Proceedings of the National Academy of Sciences of the United States of America* 104: 7786–7790.

McGrew W, Marchant L, and Nishida T (eds.) (2006) *Great Ape Societies.* Cambridge: Cambridge University Press.

Mitani J, Watts D, and Muller M (2002) Recent developments in the study of wild chimpanzee behavior. *Evolutionary Anthropology* 11: 9–25.

Muller M and Mitani J (2005) Conflict and cooperation in wild chimpanzees. *Advances in the Study of Behavior* 35: 275–331.

Murray C, Mane S, and Pusey A (2007) Dominance rank influences female space use in wild chimpanzees, Pan troglodytes: towards an ideal despotic distribution. *Animal Behaviour* 74: 1795–1804.

Nishida T (ed.) (1990) *The Chimpanzees of the Mahale Mountains.* Tokyo: University of Tokyo Press.

Pusey A, Williams J, and Goodall J (1997) The influence of dominance rank on the reproductive success of female chimpanzees. *Science* 277: 828–831.

Townsend S, Slocombe K, Emery Thompson M, and Zuberbuhler K (2007) Female-led infanticide in wild chimpanzees. *Current Biology* 17: R355–R356.

West S, Griffin A, and Gardner A (2007) Evolutionary explanations for cooperation. *Current Biology* 17: R661–R672.

Wilson M and Wrangham R (2003) Intergroup relations in chimpanzees. *Annual Review of Anthropology* 32: 363–392.

Circadian and Circannual Rhythms and Hormones

L. Ruggiero and R. Silver, Barnard College and Columbia University, New York, NY, USA

Introduction and Definitions

What are Circadian Rhythms?

Circadian rhythms are daily cycles in behavior and physiology, driven by an internal time-keeping system. These rhythms are important because they allow organisms to coordinate their activities with regularly recurring events in the environment, such as lighting, the timing of food availability, the presence of predators, and or the yearly cycles of changing seasons. Circadian rhythms are different from externally driven rhythms in that they persist even in the absence of environmental cues. For example, mice kept in constant darkness display rhythmic locomotor behavior on the basis of the timing of their internal clocks. Circadian rhythms are also present in physiology as seen in melatonin secretion by the mammalian pineal gland. If an animal goes into a deep cave with no cues to time of day, the daily cycle of changes in melatonin persists, with peaks starting at about the time that was previously night. This persisting rhythm is said to be free-running and has the periodicity of the internal clock.

While circadian rhythms occur in the absence of external cues, they are synchronized to the local environment; the process of synchronization is called entrainment. In order for synchronization to occur, the internal timekeeper, or clock, must be set to local time. The pacemaker's signal for local time can be one of a number of environmental cues, called zeitgebers, and include light, food availability, and social cues. The output signal from this clock can synchronize other tissues of the brain and body. This process can be seen in the example of melatonin secretion. In mammals, light (input) is transmitted to a brain clock (the timekeeper), which sends information about the light–dark cycle to the pineal gland. In the presence of light in the environment, the secretion of melatonin (output) is suppressed and the rhythm of melatonin is shifted. While we have focused on pineal melatonin secretion in this example, circadian rhythms can be seen in almost any number of physiological and behavioral processes one cares to measure.

Circadian rhythms in behavior are observed by examining activity under constant conditions and in response to environmental cues. In order for entrainment to occur, the timing of physiological and behavioral rhythms must be shifted. Some animals have biological clocks that free-run with a period longer than 24 h. For example, the human endogenous clock tends to be slightly longer than 24 h. This is analogous to a watch that runs too slowly. It has to shift forward each day to stay entrained to local time.

Other animals, such as mice, have an internal clock with a period shorter than 24 h, like a watch that runs too fast. These clocks must shift back each day to stay entrained to local time. In both cases, the internal clocks must adjust to local time on a daily basis. The resetting of the clock by an external cue is known as phase-shifting and occurs when a cue alters the temporal relationship between the timing of the stimulus and the phase of the endogenous rhythm. The degree and direction of the shift depends on both the time at which the stimulus is presented and its strength. Examples of circadian rhythms of behavior in the absence and presence of external cues can be seen in **Figure 1**.

Criteria for Identification of Endogenous Rhythms

To evaluate whether any particular response is under circadian control, three criteria must be met. First, the response must persist under constant conditions, that is, without any external cues from the environment, such as the example given earlier, in which an animal in a cave continued rhythmic secretion of melatonin. Second, the phase of the response must be reset by exposure to an external stimulus, or zeitgeber. Third, it must exhibit temperature compensation such that the rhythm is stable over a range of temperatures. This is important as many chemical processes are faster at high temperatures and slower in the cold. The circadian timekeeping mechanism compensates for fluctuations in temperature and circadian rhythms are not changed by thermal fluctuations.

Internal clocks are important not only for synchronizing the organism to daily time, but also for annual cycles, thereby helping to anticipate seasonal changes. Circannual rhythms are biological activities that occur each year and are synchronized with the seasons. Such seasonal activities include the timing of mating in fall or spring, optimized so that birth coincides with the greatest availability of food, and hibernation in the winter, which occurs as a way to conserve energy. This seasonal clock, like the circadian clock, functions in constant conditions; however, it too needs to be reset by environmental cues. Similar to circadian clocks that run close to 24 h, circannual clocks run close to 12 months. For example, ground squirrels have endogenous circannual clocks that control reproductive processes. In the absence of external cues, this clock has a period of about 10.5 months. Because it is slightly shorter than 12 months, it relies on external cues, such as day length, for resetting.

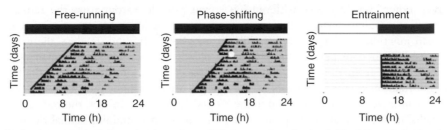

Figure 1 Hypothetical actograms depicting wheel running activity in a nocturnal animal. Black lines indicate wheel revolutions and signify periods of activity. In constant conditions (complete darkness), an animal displays free-running behavior based on its endogenous period. In this example, the animal's period is shorter than 24 h and so it begins its activity slightly earlier each day (left). When exposed to a pulse of light (white circle) an animal will shift its activity so that its next cycle of activity will begin based on the time of light exposure (center). An animal will entrain to a light/dark cycle so that it is active when the lights are off (black bar) and inactive when the lights are on (white bar) (right).

Evolution and phylogeny of circadian rhythms

Almost all living organisms, from single-celled bacteria to primates, display rhythmic cycles of biological activity with a period close to 24 h. This system is of great importance as it controls a number of biological functions including, bioluminescence in cyanobacteria, pupal eclosion in insects, and hormone production and secretion and the sleep/wake cycle in mammals. The simplest clock mechanisms are found in prokaryotic cyanobacteria and the fungus *Neurospora*. Clock machinery is found in the fruit fly, *Drosophila melanogaster*, and more complex components exist in vertebrates. Though differences in exist in complexity across species, all organisms have an internal pacemaker that detects and utilizes cues from the environment to synchronize physiological and behavioral outputs.

Circadian rhythms evolved to allow an organism to optimize its biological processes on the basis of predictable, recurring, daily, and seasonal events. For example, circadian rhythms in the earliest cells protected cellular processes, such as DNA replication, against the damaging effects of UV light during the day. In plants and some prokaryotes, an internal time-keeping system was needed to temporally separate the processes of photosynthesis, which relies on light from nitrogen fixation, which occurs during the night. In higher order animals, rhythms such as the sleep/wake cycle have likely evolved as way of conserving energy. Also, rhythms in organ systems, such as the digestive system, might have evolved to use and obtain nutrients most efficiently.

Mechanisms underlying rhythms in behavior and physiology

Rhythms in behavior and physiology are synchronized to daily and seasonal environmental changes. In some cases, animals rely solely on environmental cues to drive seasonal changes. For example, when days become longer, Syrian hamsters exhibit enlarged gonads and are reproductively competent. When the days shorten (predicting winter and decreased resources), the gonads regress and

reproductive hormone levels decrease. For this regression to occur, the Syrian hamster must first experience a period of long days. This suggests that the change in day length acts as a cue to the prevailing conditions for reproduction.

While Syrian hamsters rely on changes in the environment to predict the seasons, other animals rely on an endogenous clock. Temporal regulation of hormones also allows an organism to detect seasonal changes in the absence of environmental cues. While expressed as an endogenous rhythm, melatonin levels are synchronized with changes in day length; as the duration of night decreases, so does the duration of melatonin production. This mechanism enables seasonally breeding animals to anticipate seasonal changes and to time reproductive processes. By timing their breeding to in accordance with changes in day length, animals are able to ensure that offspring will be born at a time when the environment provides the best chances for survival. For example, shortening of days stimulates breeding in sheep, but inhibits breeding in Syrian hamsters. Sheep breed in the fall months, while hamsters breed in the spring, though both give birth in the spring during a time when the environment is favorable for their young.

The brain has a circadian clock and the eye provides entraining cues

The phenomena associated with daily and seasonal cycles suggest that there is an internal clock that is synchronized to the local environment by light. In mammals, this master clock is located in the suprachiasmatic nuclei (SCN) of the hypothalamus. The SCN is a bilateral structure that lies above the optic chiasm, comprising about 10 000 neurons in each side of the nucleus. The SCN is a heterogeneous structure made of different types of cells. Circadian rhythms in activity occur within individual cells in the dorsal or shell area of the SCN. Even if these cells are dispersed and thus disconnected from each other, they still show circadian rhythms of electrical activity. In order for the SCN tissue as a whole to produce a coordinated output, however, the individual cells need to be synchronized to each other. Synchrony is achieved by several mechanisms acting simultaneously, including synaptic connections, gap junctions, and perhaps diffusible signals. Cells in the ventral or core part of the nucleus receive photic input directly from the retina, and then communicate this information to the oscillators of the dorsal SCN. Once synchronized to each other and to the environment, the cells in the dorsal SCN produce a coordinated output signal to other brain regions (**Figure 2**).

The evidence that the SCN serves as the body's 'master clock' is very robust as it comes from many converging lines of evidence. Destruction of the SCN and blockade of its output by applying the sodium channel blocker tetrodotoxin eliminates all rhythms including drinking, locomotor activity, body temperature, and hormone secretion. The intrinsic physiological properties of SCN neurons are responsible for generating circadian oscillations because electrophysiological recordings from cultured SCN neurons and slice preparations show that oscillations in firing rates persist for several days in vitro. Finally, when SCN neurons are transplanted into the brains of animals whose SCN have been ablated, behavioral rhythms are restored with a period corresponding to that of the donor animal.

The free-running rhythm of the SCN is reset or entrained by signals from the environment and internal signals from the body. In mammals, the most potent entraining stimulus is light. Light enters the retina and the information is transmitted to the SCN, which is then synchronized to the day–night cycle. Information about light–dark cycles reaches the SCN directly by way of projections from the retina via the retinohypothalamic tract (RHT) and indirectly via the geniculohypothalamic tract (GHT). It was long thought that in mammals, the only light-sensitive cells in the body were those of the rhodopsin-containing rod and iodopsin-containing cone photoreceptors located in the outer retina. However, this notion was overthrown with the discovery of a novel opsin, termed melanopsin, a photopigment expressed in *Xenopus* oocytes and later found in mammalian retinal ganglion cells. Within the mammalian retina, ganglion cells receive light information from rod and cone driven pathways

and provide output from the retina to light-responsive parts of the brain. Melanopsin is found in 1–2% of retinal ganglion cells and renders these neurons intrinsically photosensitive. Melanopsin-containing retinal ganglion cells project directly to the SCN via the RHT, and in this way, they transmit light information to the brain along with rods and cones. In fact, the rods and cones synapse onto melanopsin-containing retinal ganglion cells and thereby exert their effects on the SCN through these cells. Though present in vertebrates and invertebrates, melanopsin is closely related to invertebrate opsins, suggesting that this photopigment was conserved throughout evolution.

Although the eye is the only light-sensitive organ in the body of mammals, nonmammalian vertebrates have additional sources of photic input. These include deep brain photoreceptors, which are located within the hypothalamus and the pineal gland, which lies between the forebrain and the cerebellum. The deep brain photoreceptors are composed of neurons that contact cerebrospinal fluid and transmit information about environmental lighting to the median eminence region of the hypothalamus and from there to the pituitary gland. These cells express specific opsin-like photopigments and are found within fish, amphibians, reptiles, and birds. They play important roles in circadian rhythmicity. Light penetrates the skull, reaches the brain, and enables entrainment. In some bird species, reception of light information by deep brain photoreceptors leads to testicular growth, implicating a role of these neurons in seasonal reproductive processes. In addition, illumination of the hypothalamus induces migratory behavior in birds, indicating that deep brain photoreceptors also plays a role in this aspect of circannual rhythms.

In nonmammalian vertebrates, the pineal gland provides an additional source of light sensitivity for photoentrainment. In mammals, the pineal does not respond to light directly, and information about the light: dark cycle reaches this gland only via the SCN. In nonmammalian vertebrates, the pineal itself is photosensitive and responds directly to light. In both mammalian and nonmammalian vertebrates, the pineal secretes melatonin during the night. The nightly secretion of melatonin occurs in proportion to the duration of darkness, while environmental light suppresses the secretion of pineal melatonin. In many nonmammalian species, including zebrafish, house sparrows, and chickens, the pineal organ has oscillator cells and is also capable of generating circadian rhythms and thereby supporting both photoreception and endogenous rhythm generation. By way of its role in tracking the duration of light and in producing melatonin, the pineal organ is important for regulating both daily and seasonal rhythms.

While light is the most salient, temperature is also an effective entraining cue. Small daily fluctuations in internal body temperature appear to directly entrain the mammalian SCN. Temperature sensitivity has also been demonstrated

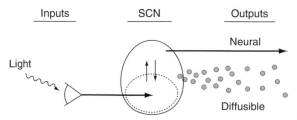

Figure 2 The suprachiasmatic nuclei (SCN). Cells in the ventral SCN receive light input from the retina and transmit it to cells of the dorsal SCN. These rhythmic cells are coordinated by input from the ventral cells and then produce output to other areas of the hypothalamus.

in isolated SCN in vitro; the cultured SCN shifts its rhythmicity following temperature pulses. This suggests that changes in hypothalamic temperature can entrain the master clock. Similarly, changes in body temperature can entrain peripheral oscillators. This is true of mice kept in different ambient temperatures show entrainment of clock genes in liver in response to changes in environmental temperature.

Molecular clocks

The discovery of molecular mechanisms underlying the intrinsic rhythmicity of individual cells has had a tremendous impact on our understanding of the circadian timing system and has enabled our exploration of the fundamental nature of these mechanisms. In 1971, Konopka and Benzer identified a gene in *Drosophila* that led to changes in the fly's endogenous rhythms. Mutations in this gene caused three phenotypes: a shortened period, a lengthened period, and arrhythmicity. This was a landmark finding in that it demonstrated a direct relationship between changes at the level of genes and behavior. This genetic basis for behavior led to a clearer understanding that the endogenous rhythms exhibited in the behavior of animals were actually gene-driven. The gene was termed *period* and was the first gene isolated that displayed a circadian phenotype. Additional so-called, 'clock genes' were later discovered and shown to be important not only in flies but also in mammals. Clock genes and their protein products comprise a transcription-translation loop with positive and negative feedback regulation. The rhythmic expression of clock genes is important for maintaining a functioning circadian clock, and disruption of these genes causes arrhythmicity (**Figure 3**).

Cellular clocks occur not only in the SCN but also in other brain areas and in peripheral tissues. Circadian clock gene expression in mammals has been detected in the liver, heart, muscle, kidney, pancreas, adipose tissue, and lung. The function of the SCN is to coordinate tissue-specific rhythms in the rest of the brain and the body, with each other and with external stimuli. When isolated from the SCN, individual peripheral cells drift out of phase with each other and the overall rhythm of the tissue as a whole dampens. When the intracellular processes within these cells are synchronized by the SCN, peripheral tissues are able to produce coherent rhythmic outputs.

Development of circadian rhythms

A number of rhythmic processes have been detected in utero, a time when regulation of nutrients and hormones may be important for normal growth and development. These processes include heart rate and breathing. While these rhythms are endogenous to the fetus, they are entrainable by maternal signals such as body temperature, metabolic pathways, uterine contractions, and hormones, including melatonin. The purpose of in utero entrainment is not clear; however, it probably occurs in order to prepare the fetus for its novel environmental conditions. Interestingly, even after birth, newborns are still entrained by signals from the mother. This has been demonstrated in rats in which lesioning the SCN of the mother causes desynchronization of rhythms among pups within the litter.

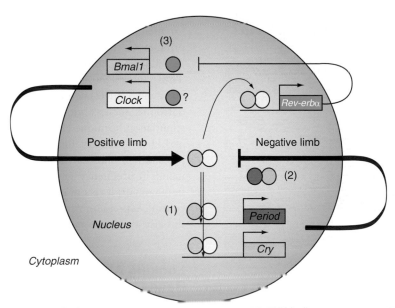

Figure 3 Diagram of the circadian clock mechanism. (1) The clock proteins, CLOCK (yellow) and BMAL1 (pink), drive the expression of clock genes (Per, Cry and Rev-erbα) in the nucleus. (2) PER (red) and CRY (light green) proteins in the nucleus inhibit CLOCK/BMAL1 action through negative feedback. They also down-regulate Rev-erbα (dark green). (3) When REV-ERBα protein is absent, Bmal1 (and possibly also Clock) genes are disinhibited and transcribed to produce new CLOCK/BMAL1 transcription factors that initiate a new cycle. Adapted from Albrecht U and Eichele G (2003) The mammalian circadian clock. *Current Opinion in Genetics & Development* 13: 271–277.

While entrainment and circadian rhythmicity are seen before birth, full maturation of the circadian system occurs after birth. For example, rhythms in within the SCN are detectable before birth. In children, a regularsleep-wake rhythm at about 3 months, though after this stage infants often have disrupted sleep because of hunger or teething discomfort (**Figure 4**). This time course of rhythm maturation is likely attributable to the development of connections from neurons of the SCN to target sites in the brain, and maturation of connections among SCN neurons themselves, thus, the networks that allow for synchronization of rhythmic cells are also not established until later in development.

The functions of circadian rhythms

The circadian timing system controls a vast number of physiological and behavioral processes including the

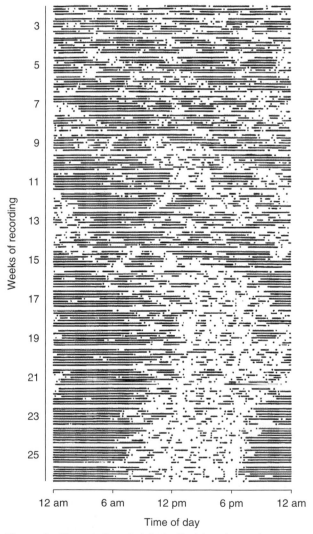

Figure 4 Sleep rhythms in infants. Dark bands are sleep periods of an infant over a 24-h day followed for several months. Adapted from Kleitman N and Engelmann T (1953) Sleep characteristics of infants. *Journal of Applied Physiology* 6: 269–282.

sleep–wake cycle, feeding, core body temperature, and hormone production. A key problem in understanding the function of the circadian system is the determination of how the rhythmicity is coordinated among various brain regions and how this affects other organ systems. Many breakthroughs in research occurred with the discovery of clock cells, in particular, that they were expressed throughout the entire organism. This led to the understanding that the SCN was not actually the driver of rhythms, but that the rhythms were generated within each cell so that it could display rhythmicity on its own. It then became clear that the role of the master clock in the SCN was to set the phase of these peripheral clocks, thereby synchronizing them.

Peripheral tissues can also be entrained without affecting the SCN phase. The brain clock provides timing information to the rest of the body to activate certain behaviors, such as when to eat. The nutrients provided by food stimulate the production of hormones and enzymes that act to set the phase of oscillators in peripheral tissues such as the liver. This phase setting, however, does not impact the phase of the SCN but only acts at the level of peripheral tissues (**Figure 5**).

Outputs of the SCN project to other parts of the hypothalamus, where they synchronize oscillators found in extra-SCN brain sites thereby modulating the timing of synthesis and/or secretion of neurotransmitters and neurohormones. The SCN communicates by direct synaptic connections with neurosecretory cells, such as gonadotropin-secreting neurons, and indirectly via multisynaptic autonomic pathways to endocrine glands, such as the adrenal gland through hormonal release (**Figure 5**). In this way, the SCN exerts neural and hormonal control over the physiology of the body. This is important for many processes. The circadian control of hormone production is important for reproductive processes including ovulation, estrus, fertilization, and pregnancy. Thus, the SCN projection to gonadotropin-releasing hormone (GnRH) neurons stimulates production of gonadotropins, such as luteinizing hormone (LH) and follicle-stimulating hormone (FSH). A close temporal regulation of LH is necessary for ovulation. Also, after ovulation, prolactin production is needed at a particular time to maintain pregnancy and promote lactation. In addition, in mammals, levels of corticosterone are under circadian control, expression rises prior to waking. It is thought that this rise allows animals to prepare for activity onset.

Looking Ahead: Applications of Research in Circadian Rhythms

Alterations in circadian rhythms have profound effects on the health of an individual, as a number of disorders are associated with circadian dysfunction. Phase shifting is seen in individuals experiencing jet lag. In this case, one's

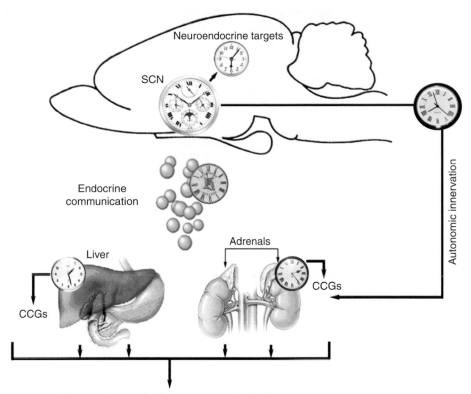

Temporally organized processes and behavior

Figure 5 SCN and peripheral tissues. The circadian system is controlled by neural and diffusible signals originating from the SCN. Peripheral tissues may be differentially regulated by SCN signals via local clocks allowing for more specific responsiveness based upon local needs and time of day.

rhythms must undergo phase adjustment to account for the change in local time. During the period of transition to the new conditions, the SCN and peripheral organs become desynchronized, and one experiences the fatigue, irritability, and insomnia of jet lag. Eventually, the various bodily rhythms entrain to the new phase, and return to appropriate synchrony with each other.

People who work night shifts also need to phase adjust, and the consequences appear to be more severe than for jet lag sufferers. Shift worker's schedules are not in phase with the light–dark cycle and they are therefore exposed to environmental light at the wrong time. When shift workers conform to the schedule of a typical day–night cycle on their days off, the rotation can lead to major disruptions of the circadian time. This leads to sleep deprivation and in more serious cases, cardiovascular dysfunction, altered metabolism, and mood disorders. Studies have shown that the incidence of on-the-job mistakes and industrial accidents is dramatically increased during night shifts, compared to day shifts. Because of the growing need in our society for around-the-clock work, understanding how to address the issues of circadian dysfunction is becoming increasingly important. Light and drug therapy have been used to treat circadian-based sleep disorders, seen in advanced and delayed sleep phase syndrome, jet lag, shift work and aging. Treatment protocols require careful

attention to circadian phase of light or drug application since these variables impact on the effectiveness of the treatment.

The disruption of circadian rhythms has widespread effects on the health of all organisms and therefore it is important for current and future research to address the potential impact of changing environmental conditions. This may be especially relevant in an age of global warming, where temperature conditions and resource availability are changing rapidly. While light is the most potent entraining stimulus in mammals, temperature affects clock gene expression and modulates the effects of photoperiod on the reproductive axis in birds. A better understanding of the effects of temperature on reproduction through effects on circadian and seasonal rhythms will be useful.

See also: Amphibia: Orientation and Migration; Bat Migration; Bats: Orientation, Navigation and Homing; Bird Migration; Fish Migration; Hibernation, Daily Torpor and Estivation in Mammals and Birds. Behavioral Aspects; Insect Migration; Insect Navigation; Irruptive Migration; Magnetic Compasses in Insects; Mammalian Female Sexual Behavior and Hormones; Maps and Compasses; Memory, Learning, Hormones and Behavior; Migratory Connectivity; Neural Control of Sexual Behavior; Parental Behavior and Hormones in Mammals;

Parental Behavior and Hormones in Non-Mammalian Vertebrates; Pigeon Homing as a Model Case of Goal-Oriented Navigation; Reproductive Skew, Cooperative Breeding, and Eusociality in Vertebrates: Hormones; Sea Turtles: Navigation and Orientation; Seasonality: Hormones and Behavior; Sleep and Hormones; Time: What Animals Know; Vertical Migration of Aquatic Animals.

Further Reading

Albrecht U and Eichele G (2003) The mammalian circadian clock. *Current Opinion in Genetics & Development* 13: 271–277.

Antlo M and Silver R (2005) Orchestrating time: Arrangements of the brain circadian clock. *Trends in Neurosciences* 28: 145–151.

Dunlap JC (1999) Molecular bases for circadian clocks. *Cell* 96: 271–290.

Dunlap JC, Loros JJ, and DeCoursey PJ (eds.) (2004) *Chronobiology Biological Timekeeping*. Sunderland, MA: Sinauer Associates.

Hastings MH, Herbert J, Martensz ND, and Roberts AC (1985) Annual reproductive rhythms in mammals: Mechanisms of light synchronization. *Annals of the New York Academy of Sciences* 453: 182–204.

Hazelerigg DG and Wagner GC (2006) Seasonal photoperiodism in vertebrates: From coincidence to amplitude. *Trends in Endocrinology and Metabolism* 3: 83–91.

Klein DH, Moore RY, and Reppert SM (eds.) (1991) *Suprachiasmatic Nucleus the Mind's Clock*. New York, NY: Oxford University Press.

Kleitmann N and Engelmann T (1953) Sleep characteristics of infants. *Journal of Applied Physiology* 6: 269–282.

Koukkari WL and Sothern RB (2006) *Introducing Biological Rhythms: A Primer on the Temporal Organization of Life, with Implications for Health, Society, Reproduction, and the Natural Environment.* New York, NY: Springer.

Kriegsfeld LJ and Silver R (2006) The regulation of neuroendocrine function: Timing is everything. *Hormones and Behavior* 49: 557–574.

Paul MJ, Zucker I, and Scwartz WJ (2008) Tracking the seasons: The internal calendars of vertebrates. *Philosophical Transactions of the Royal Society B* 363: 341–361.

Seron-Ferre M, Valenzuela GJ, and Torres-Farfan C (2007) Circadian clocks during embryonic and fetal development. *Birth Defects Research* 81: 204–214.

Yan L, Karatsoreos I, Lesauter J, et al. (2007) Exploring spatiotemporal organization of SCN circuits. *Cold Spring Harbor Symposia on Quantitative Biology* 72: 527–541.

Relevant Websites

http://www.srbr.org – This is the web site for the society for research in Biological Rhythms. Here one can find information about meetings, web sites for laboratories, and news in the field.

Cockroaches

C. A. Nalepa, North Carolina State University, Raleigh, NC, USA

Although the title of this article suggests that cockroaches are a homogenous grouping, the term 'cockroaches' applies to a sizeable group of morphologically and behaviorally diverse insects. Current species estimates are in the range of 4000–5000, with at least that many yet to be described. Cockroaches can resemble beetles, wasps, flies, pillbugs, and limpets. Some are hairy, several snorkel, some whistle, many are devoted parents, and males of several species emit light. Body sizes range from mosquito-sized species living in the nests of social insects, to mouse-sized burrowers weighing more than 30 g. Diversity is highest in warm, humid tropical rain-forests, but cockroaches are found on all continents and in nearly all habitat types where insects occur, including both arid and aquatic environments. Although their taxonomy is in flux on several levels, cockroaches are typically placed in the Order Dictyoptera along with termites and mantids, and divided into six families. Most cockroach species fall into just three of these families: the Blattidae, Blattellidae, and Blaberidae. Recent evidence indicates that technically, termites are cockroaches, as they are phylogenetically nested within cockroaches as a subgroup closely related to the cockroach genus *Cryptocercus*.

The great diversity of cockroaches makes them appealing for comparative studies, but with some exceptions, most of our knowledge of their behavior is based on just a slender fraction (~1%) of the group: the domiciliary pests and a handful of other species cultured in laboratories and museums. Although it is wise to be wary of generalizing any particular behavior to the entire group, there are excellent reasons why these cockroach species have become important subjects for studies of neurobiology, endocrinology, and behavior. They are easy to keep, as they can be fed conveniently on rodent or dog chow. They withstand and even thrive in crowded situations, and tolerate a wide range of temperature and humidity conditions relative to other tropical insects. Wounds heal quickly, and they have an astounding ability to survive without food or water. They are long lived and can be quite large (as exemplified by members of the genera *Periplaneta, Blaberus, Rhyparobia, Gromphadorhina*), facilitating both behavioral observations and surgical experimentation. Their size also allows for externally fitted mechanical and electronic devices, and internally implanted electrodes and canulae.

The cockroach nervous system is more decentralized than in many insects, with multiple nerve centers strung along the 'spinal cord' responsible for regulating vital functions. This is the reason they can live for some time without a head, and allows for their rapid escape response. When receptors on abdominal appendages (cerci) are stimulated by air displacement, the message travels straight to the legs, bypassing the cephalic brain. Individual behavior and its neurological basis have been extensively studied in domestic species. Cockroaches display associative learning (linking a visual cue with an odor) and have a mammalian-like place memory. Both stress and sleep deprivation can kill them. Cockroaches are apt models for age-related behavioral changes and have been used extensively in locomotory studies. Cockroach speed, stability, balance, righting behavior, climbing ability, and spatial orientation have been studied for their use in a variety of applications, including the development of blattoid walking robots. The first of these robots was developed in 1955, and they have since become sophisticated and miniaturized. Those used in a recent study of aggregation behavior were about the size of a matchbox.

Naturally, a fundamental reason for the extensive study of just a few key species has been the development of strategies to improve the management and control of structural pests. This applied research has resulted in a vast literature, with entire books written on the German (*Blattella germanica*) and American (*Periplaneta americana*) cockroaches alone. These species are not typical of their undomesticated brethren and are so intimately associated with humans that wild populations are no longer easily found. Consequently, the natural context of extensively studied behaviors is unknown. The most commonly utilized species typically have been maintained in culture for decades and are therefore apt to be lacking the variation expressed in free-living populations. Group dynamics, locomotor ability, growth rates, levels of endosymbiotic bacteria, and fecundity are known to differ between laboratory and wild strains, and crowded rearing conditions and the inability to emigrate can result in artificially elevated levels of density-dependent behaviors such as cannibalism and aggression. Even in cases where natural populations of cultured species exist, most are as yet unstudied in the field (e.g., *Gromphadorhina*). Obtaining field data on a group that is largely cryptic and nocturnal is a challenge, but the behavior of even diurnal, brightly colored species (largely Australian – **Figure 1**) is unknown.

Here I concentrate on two broad areas of interactions for which a relatively wide range of cockroach species have been investigated: mating behavior and social

Figure 1 The diurnal Australian cockroach *Polyzosteria mitchelli* digging a hole for her ootheca, which is fully formed and attached to her abdominal tip. The stereotyped behavioral sequence involved in oothecal deposition has been used as a taxonomic character. Photo by E. Nielsen, courtesy of David Rentz.

structure. There is a wealth of information on these subjects, and if the literature is traced, it becomes obvious that one man laid the foundation for these studies during the middle of the last century. Louis M. Roth was employed as a civilian entomologist for the United States Army when he was charged with the mission of investigating noninsecticidal methods for controlling insects of military importance. Fortunately, he had the foresight to recognize that successful control measures require a solid basis in the biology and behavior of the target insect, and initiated studies of courtship and mating behavior of three cockroach species during the late 1940s. This early work was published in 1952, but not before the adjective 'sexual' was removed from the description of cockroach behavior in the title; the Army Quartermaster General deemed it an unacceptable subject for association with a government agency. Even that proved insufficient, however, and Roth, along with co-author E.R. Willis, used their home addresses instead of the Army's to get the 64 page paper published. It was the first in a long series of publications that laid the groundwork for the modern study of cockroach behavior. Roth eventually published more than 200 papers on cockroaches, several of which were weighty monographs. He also trained, encouraged, worked with, or directly influenced a succession of scientists who not only went on to distinguished careers of their own but also spawned subsequent generations of biologists happy to ignore the stigma of working on the group.

That initial paper by Roth and Willis presented a detailed analysis of sexual behavior in three domestic pests and included a description of female American cockroaches (*P. americana*) producing a chemical that functions to attract males and instigate their mating behavior. It was the first report of a cockroach pheromone, several years

before the term was even coined. They discovered that when a water vial from a cage filled with virgin females was transferred to a cage containing males, the males responded by enthusiastically trying to mate with both the water vial and each other. That serendipitous discovery kicked off a storm of research on cockroach chemical ecology that continues to this day. During the 1960s and 1970s, there was a dramatic competition among laboratories to establish the physical source and chemical identification of the female sex pheromone of *P. americana*. Twists and turns included mistaken identifications, blind alleys, and reversals. In one study, researchers collected the fecal material from 100 000 virgin females and performed gut dissections on an additional 32 000 in an attempt to isolate the pheromone. It was finally characterized in the late 1970s, and its components synthesized for use in behavioral studies. It proved to be an incredibly stable compound. Refrigerated pheromones used for demonstration in classes at Cornell University in 1966 could still stimulate male courtship behavior in 1999. Continuing research has demonstrated that pheromones permeate every aspect of cockroach interactions. These include long-range, short-range, and contact chemicals that function in mate finding and courtship, trail following, kin-recognition, mother–offspring interactions, and alarm, aggregation, and oviposition behaviors. Cockroaches also produce an array of repellants, irritants, and sticky secretions as defense against natural enemies. Despite substantial research progress on the topic, examined cockroaches have a large number of distinct glands and glandular cells of unknown functional significance scattered over the body.

Mating Behavior

With few exceptions, the generalities of mating behavior are remarkably uniform among studied cockroaches. The typical sequence consists of mate finding, mate contact, a courtship that includes the male display of tergal glands, female attraction to these glands, and copulation. Volatile sex pheromones mediate the initial orientation and have been demonstrated in 16 cockroach species in three families. The pheromones are commonly female generated, but in some species, the initial roles are reversed, with males luring females. On the basis of limited available data, the general pattern appears to be that in cockroaches in which the male or both sexes can fly, it is the female that releases a long-range pheromone. Males release sex pheromones in species in which both sexes are earthbound. Once a potential mate is nearby, contact pheromones and short-range volatiles mediate sexual and species recognition and coordinate courtship. Antennal contact with the female instigates a male display in which he turns away and presents to her glands on the dorsal surface of his abdomen. Females respond by approaching the male, straddling his back, and feeding on the gland secretion. Tergal glands thus serve to maneuver the female into the proper position for mating and arrest her movement so that the male has an opportunity to engage her genitalia.

In some cockroach species, courtship is highly stereotyped, with an internally programmed, unidirectional sequence of acts; in others, male–female interaction is more flexible. Variations often take the form of behaviors that produce airborne or substrate-borne vibrations, particularly when males are courting reluctant females. These include rocking, shaking, waggling, trembling, vibrating, pushing, bumping, wing pumping, wing fluttering, 'pivot-trembling,' push-ups, anterior–posterior jerking, tapping, stridulation, and chirping. Some male cockroaches whistle at females with surprisingly complex, almost birdlike sounds that vibrate through the air and ground. Male Madagascar hissing cockroaches, *Gromphadorhina portentosa*, produce two different types of hisses during courtship (and a third during copulation). These courtship behaviors typically alternate with tergal display, until the female eventually responds or the male departs. Courtship can last well over an hour in some species, although females may react quickly to attractive males.

Mate selection in cockroaches involves a complex interaction between male competition and female choice. Fights among male conspecifics have been documented in most cockroach subfamilies and involve a rich repertoire of behaviors, with 19 individual acts documented in 13 species. The presence of a receptive female typically instigates battles, with males displaying sexually to the female and aggressively to each other. Fights usually involve butting, kicking, and biting, but vary in intensity. In the German cockroach (*B. germanica*) and other Blattellidae, fights are mildly aggressive, last but a few seconds, and never result in dismemberment. In the American cockroach, *P. americana*, bouts can last up to 3 min, and the longer the fight lasts, the more escalated and dangerous the behavior. Injuries occur but are rarely mortal; most fights end in truce or retreat. In some cockroaches, such as *Nauphoeta cinerea*, males form dominance hierarchies based primarily on the results of combat, and a few species are known to guard territories and food resources. Fighting in many species simply brings about 'spacing' whereby individuals are no longer in contact with each other. Interactions are more ritualized in some species than others, with distinct dominant and subordinate postures.

Male cockroach social interactions are difficult to classify because they are complex, species-specific, plastic, often density dependent, and may differ depending on the experimental design or research facility in which the studies are conducted. Laboratory studies of male–male social behaviors usually involve a no-choice face-off of two individuals or the interactions of a small, manageable group, while in nature most of these species are gregarious, living in sometimes extraordinarily large groups in caves, hollow trees, or in human structures. Field studies have been conducted on two species of *Blaberus* in the caves of Trinidad. Researchers found that male *B. colloseus* establish dominance hierarchies when at high densities, but hold territories when at lower densities, often trying to 'herd' females into their territory. Coveted spaces, in this species as well as other territorial cockroaches, may contain food or water, or are merely elevated sites. Female sex pheromones rise on air currents in natural habitats, so higher perches may be the best place to survey the olfactory environment. The stability of the pecking order varies in cockroaches that form dominance hierarchies and familiarity with members of the group can structure interactions.

The African species *N. cinerea* exhibits relatively stable linear hierarchies, but rank switching is common. An alpha male patrols his territory, lunging at subordinates and sending them cowering into corners of the cage. Male social rank in this species is recognized by both males and females via a combination of three chemical compounds that have individual, additive, and contrasting effects on status. The pheromonal blend that marks a high-status male, however, is not always the one preferred by females. Alpha males can nonetheless circumvent female choice by physically excluding rival males.

Cockroach females overtly choose mates on the basis of territory quality, dominance rank, and a variety of courtship acts and pheromonal cues that signal some aspect of male quality. Female hissing cockroaches prefer the longer, lower frequency courtship hisses produced by large males. *Nauphoeta cinerea* females favor experienced males, but are disinclined to accept males on which the chemical traces of a recent female consort can be detected. With age, females lower their standards and require less courtship from suitors than do their younger counterparts. The tergal feeding phase of courtship common to most studied cockroaches is a potential mechanism for female evaluation of potential mates. The tergal secretion of male German cockroaches is a complex mixture known to vary both individually and daily, and includes maltose, a potent phagostimulant for domestic cockroaches. Because these secretions exploit a female's underlying motivation to feed, they can be classified as sensory traps: they mimic stimuli that females have evolved, under natural selection, for use in other contexts. Because it smells like food, virgin, mated, and gravid females, as well as males and last instar nymphs are attracted to the tergal secretion. The sensory trap hypothesis thus provides one explanation for the frequency with which males respond to the courtship displays of other males (pseudofemale behavior).

The morphology and behavior of female cockroaches suggests sophisticated control over copulation, sperm storage, and sperm use. Males cannot force females to mate because female cooperation is required for the initial engagement of genitalia. Once engaged, females can break the connection by pushing or kicking at the male with their hind legs. Females may also accept a male for copulation but reject him as a father. They have sophisticated sperm storage organs, and so have the option of cryptic mate choice via biased sperm use when fertilizing eggs. There is evidence of sperm competition in three studied cockroach species; it occurs primarily after the female's first reproductive cycle.

Social Behavior

It is difficult to conceive of any group of animals that are as diversely social as cockroaches. Although individual taxa are typically described as solitary, gregarious, or subsocial, cockroach social heterogeneity is not so easily catalogued. Cockroaches that live in family groups are a rather straightforward category, and domestic pests (**Figure 2**) and a number of cave-dwelling species are without a doubt highly gregarious, forming groups that may include both kin and nonkin. Few species are convincingly described as solitary, although one category of loners may be those adapted to deep caves. For a variety of reasons, many others elude straightforward classification. For one thing, studies of cockroach social behavior are still in their early descriptive phase. With perhaps a score of exceptions, our concept of cockroach social organization is based largely on anecdotal evidence and brief observations during collection expeditions for museums. The data we have indicate considerable spatial and temporal variation in social structure, influenced by, among other factors, the age and sex of the insects, environmental factors, physiological state, population density, and harborage characteristics. Many cockroaches are nocturnal and cryptic; consequently, even those that live in laboratories can be full of surprises. Parental feeding behavior was only recently observed in hissing cockroaches, even though they are commonly kept in homes as pets, in laboratories for experiments, and in museums for educational purposes.

Even closely related species can vary widely in social proclivities. The German cockroach is strongly gregarious (**Figure 3**) and has been the test subject of the vast majority of studies of cockroach aggregation behavior; some congeneric species, however, have been described as solitary. One promising group for comparative study is the wood-feeding Panesthiinae cockroaches, which exhibit a variety of social structures (e.g., biparental care, maternal care, gregarious behavior), making them useful models for examining the basis of social variation in a closely related group. Kin recognition is well developed in some species. German cockroaches not only discriminate the odor of their own population, but also recognize siblings and modify their behavior accordingly. Neonates of several cockroach species orient to their mother in choice tests. There is even a report of hissing cockroaches in culture learning to discriminate between their human handlers.

Figure 3 An aggregation of German cockroaches in laboratory culture. A female at the bottom is forming an egg case. Photo by Colette Rivault and Mathieu Lihoreau.

Figure 2 German cockroaches infesting an urban structure. Photo by Colette Rivault and Mathieu Lihoreau.

Cockroaches that aggregate in sheltered areas during their period of inactivity are well studied because of the strong economic incentive to do so. Aggregations of domestic pests have been characterized in terms of group size and demographics, as well as preferred shelter characteristics, activity rhythms, site fidelity, and the sensory cues used in homing to harborage. Most species will orient to the faecal and cuticular residues and odors of conspecific, and often heterospecific, cockroaches. While these odors are probably not true aggregation pheromones, they are one cue that helps mediate the formation of groups. Not unexpectedly, age, sex, and reproductive status of individuals strongly influence their behavior. The proposed benefits of gregarious behavior include defense, favorable microclimate, proximity of potential mates, and both food and mutualistic gut microbes for neonates. The chief costs seem to be disease transmission and cannibalism. As in most studied animals that form groups, the behavioral rules governing aggregation behavior in cockroaches conform to a simple model of self-organization: individuals modulate their behavior in relation to social or environmental cues in their immediate vicinity. Researchers have successfully reproduced cockroach aggregation dynamics by using microrobots, and demonstrated that these robots can collectively 'agree' on an aggregation site in choice tests.

Cockroaches may exhibit the greatest range of parental care specializations among insects. Egg care is universal, but varies depending on reproductive mode. Oviparous females encase their eggs in a protective, hardened outer shell (ootheca), select and prepare the egg deposition site (**Figure 1**), and then conceal and in some cases, defend the egg case. A few species carry the ootheca attached to the tip of their abdomen until hatch. Ovoviviparous and viviparous females retract the egg case into their body, where it is incubated in a type of uterus (brood sac) throughout gestation. After hatch, the simplest parental interaction is brooding. Neonates shelter briefly beneath the mother's body until their cuticle hardens, then disperse. In species that exhibit more elaborate care, the female constructs a nest in which the offspring develop, or carries them either on her back beneath the wings (**Figure 4**) or clinging to the underside of her body. Some mothers can roll up (conglobulate), enclosing the ventrally attached nymphs. Specializations of the juveniles include appendages that aid in clinging to the female, and mouthparts that facilitate unique feeding habits.

Parental care is typically maternal, although biparental family groups are found in two genera of wood-feeding cockroaches. Exclusive paternal care is unknown. Parental care is significantly associated with egg retention, as internal incubation ensures the temporal and spatial proximity of mother and neonates at hatch. The one known parental oviparous genus rears its young in a nest: a condition that similarly sets the stage for more complex interactions.

Figure 4 A female cockroach in the subfamily Epilamprinae, possibly in the genus *Thorax*, carrying young on her back. The Indian species *Thorax porcellana* carries offspring under its domed forewings for their first two instars. Young nymphs are dependent on maternally provided food, reported to be either a glandular secretion or hemolymph. Photograph by Natasha Mhatre, all rights reserved.

A unique aspect of cockroach parental care is the wide range of mechanisms by which embryos and neonates are nourished. In the one known viviparous species, the brood sac acts as both uterus and mammary gland; it oozes a type of milk that is orally imbibed by developing embryos. In several taxa, mothers progressively provision neonates on bodily fluids. These fluids originate from either end of the digestive system, are expelled from the brood sac after the young emerge, or are secreted from specialized integumentary glands on the external surface of the body. Even in species in which direct feeding on maternal fluids is unknown, the ingestion of maternal feces or the feces of others in the social group (coprophagy) is thought to be important for establishing the microbial hindgut fauna of neonates.

Despite being distantly related, both genera of biparental wood-feeding cockroaches (*Cryptocercus*, *Salganea*) have offspring with altricial development. The neonates are dependent for food, and have a thin, fragile exoskeleton and reduced or absent eyes. These cockroaches provide insight into the behavioral and ecological conditions likely experienced by termite ancestors and suggest factors important to the evolution of eusociality in that group. Altricial development in the termite ancestor, for example, would have been a necessary precondition for alloparental care, just as it is in birds. Termites morphologically and behaviorally resemble young altricial cockroaches, suggesting a heterochronic origin from their cockroach ancestors. The complex system of caste polyphenism in termites is undoubtedly rooted in the physiology of cockroach social behavior. In cockroaches, tactile stimulation and short-range and contact pheromones not only mediate social interactions and serve as behavioral releasers, but also regulate physiological processes.

In young cockroaches, the social environment modifies behavior, synchronizes molting, and alters both developmental rates and body size. In adult cockroaches, social interactions have a complex influence on reproductive success, manifested in oviposition rate, hatching success, calling behavior, ejaculate quality, and the production of and reaction to sex pheromone. The phenomenon is best studied in *N. cinerea*, where male pheromones influence female longevity, the number and sex ratio of her offspring, and their rate of embryonic development. Such profound and fundamental physiological consequences of social interaction in cockroaches suggest that these are at the core of the complex caste system exhibited by their termite relatives.

See also: Dominance Relationships, Dominance Hierarchies and Rankings; Invertebrates: The Inside Story of Post-Insemination, Pre-Fertilization Reproductive Interactions; Orthopteran Behavioral Genetics; Predator Evasion; Social Evolution in 'Other' Insects and Arachnids; Subsociality and the Evolution of Eusociality.

Further Reading

Bell WJ, Roth LM, and Nalepa CA (2007) *Cockroaches: Ecology, Behavior, and Natural History.* Baltimore, MD: The Johns Hopkins University Press.
Gemeno C and Schal C (2004) Sex pheromones of cockroaches. In: Cardé RT and Millar JG (eds.) *Advances in Insect Chemical Ecology,* pp. 179–247. Cambridge: Cambridge University Press.
Huber I, Masler EP, and Rao BR (eds.) (1990) *Cockroaches as Models of Neurobiology: Applications in Biomedical Research,* 2 vols. Boca Raton, FL: CRC Press.
Jeanson R, Rivault C, Deneubourg J-L, et al. (2005) Self-organized aggregation in cockroaches. *Animal Behaviour* 69: 169–180.
Moore AJ, Gowaty PA, Wallin WG, and Moore PJ (2000) Sexual conflict and the evolution of female mate choice and male social dominance. *Proceedings of the Royal Society of London B* 268: 517–523.
Nalepa CA and Bell WJ (1997) Post-ovulation parental investment and parental care in cockroaches. In: Choe JC and Crespi BJ (eds.) *The Evolution of Social Behavior in Insects and Arachnids,* pp. 26–51. Cambridge: Cambridge University Press.
Roth LM and Willis ER (1952) A study of cockroach behavior. *American Midland Naturalist* 47: 66–129.
Schal C, Gauthier J-Y, and Bell WJ (1984) Behavioural ecology of cockroaches. *Biological Reviews* 59: 209–254.

Co-Evolution of Predators and Prey

E. D. Brodie III and A. Wilkinson, University of Virginia, Charlottesville, VA, USA

Introduction

From deadly toxins to stinging spines, from inconceivably accurate camouflage to wild warning colors, the elaborate antipredator mechanisms that animals employ in their own defense beg evolutionary explanations for why such adaptations should go so far in their protective abilities. The answer lies in a dynamic process in which evolutionary pressures are continually ratcheted forward, where selection itself evolves and becomes more intense with time, and the level of defense that is sufficient now will be left behind just a generation or two down the road.

Predator–prey co-evolution is an inherently reciprocal process in which defensive adaptation by prey results in stronger selection on predators to exploit those defenses. Evolutionary advances by predators, in turn, represent stronger selection for defense in prey. In this way, adaptive responses by one species merely result in the strengthening of selection for further responses in that same species. This seems an intuitive process, and it is tempting to assume that many of the elaborate defense or predatory mechanisms that we observe in nature have arisen from such an evolutionary feedback loop, but that perspective too quickly draws conclusions about process only from patterns of outcomes. The defining characteristic of co-evolution is reciprocity in adaptation.

The Phenotypic Interface of Co-evolution

Although co-evolution is often characterized as occurring between populations or species of predators and prey, the ecological interactions that ultimately drive coevolutionary change occur at the level of traits. The phenotypic interface of co-evolution is the set of traits in both players that mediate the outcome of interspecific interactions (**Figure 1**). For predators and prey, the interface might be centered on the amount of a defensive chemical in prey and the resistance to that chemical in predators. Each trait in the phenotypic interface acts as both a target and an agent of selection, thus each trait simultaneously evolves and drives the evolution of the opposing trait.

Traits found at the phenotypic interface are rarely simple, discrete traits but rather 'performance' phenotypes comprised of multiple underlying components that together determine the success of individuals involved in an interaction. Thus, while selective pressure may be acting on performance directly, it is the evolution of these underlying components that ultimately determines the manner in which participating populations will evolve. Fleeing speed may be the performance phenotype that determines escape for a prey, but speed is determined by muscle mass and coordination, enzyme activities, metabolic rate, and many other traits. Components involved in organismal performance may be the targets of selection on multiple fronts (e.g., sprint speed may be a function in both antipredator and foraging contexts), further complicating the process of co-evolution.

The ecological interactions that drive reciprocal selection are only one of the requisites for co-evolution. The traits mediating interactions must also have the potential to respond evolutionarily to that selection. This potential is captured in genetic complementarity, or heritability of the traits on both sides of the phenotypic interface. Heritability is a measure of the genetic variation underlying a trait – some variation in traits may be caused by environmental factors, and this variation is generally unavailable for evolutionary response. In order for reciprocal selection to result in adaptation by both species in an interaction, the traits at the interface of co-evolution must be heritable.

Coevolutionary Arms Races

Perhaps the most popular and intuitive metaphor for coevolutionary interactions between natural enemies is that of an 'arms race' (**Figure 2**). Reflecting on the Cold War process in which countries escalated and reciprocated with military might, in 1979 Dawkins and Krebs described the evolutionary interaction between predators and prey as an arms race. Just as each new submarine is better designed to sink enemy ships and new ships are better able to avoid destruction, predators and prey are caught in a counter escalating arms race of defense and exploitation. In a metaphorical race between foxes and hares, foxes may evolve, over time, greater running speeds to catch more hares, forcing hares to evolve greater speeds themselves.

Despite the appeal of the arms race metaphor, the directional evolutionary escalation of phenotypic traits implied in arms races is, in fact, uncommon in natural predator–prey systems. In a 1986 paper, John N. Thompson discussed many of the constraints to which arms races are subject. Because a one-to-one relationship rarely exists between phenotypes and underlying alleles, evolution of improved attack or defensive ability may simply cause the opposite member of the interaction to adjust its phenotypic or

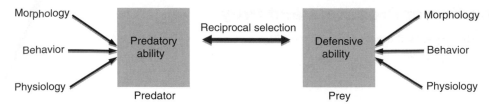

Figure 1 The phenotypic interface of co-evolution. The reciprocal selection that drives co-evolution takes place between traits in at least two species. These traits at the interface of co-evolution determine the outcome of interactions between individuals, and may in turn be comprised of other component traits.

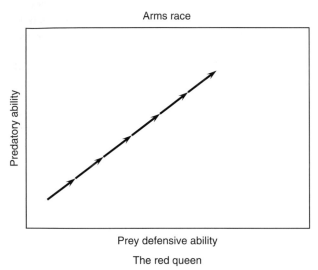

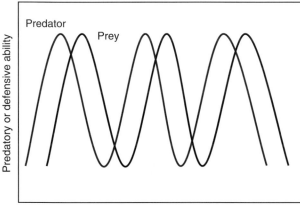

Figure 2 Coevolutionary dynamics. Co-evolution can follow a variety of dynamics, including the 'arms race' and 'red queen' processes. In an arms race, escalation by one species is followed by counter escalation in another, with the evolution of ever increasing abilities in both players. In a Red Queen, advances by one species are met with advances by the other, but frequency dependent processes cause the best fit type to change. This in turn creates cycles of genotype or phenotype frequencies over time.

behavioral response without having an effect on the underlying genes (i.e., counter escalations may be plastic rather than genetic, and therefore temporary in an evolutionary sense). Additionally, arms races are unable, once begun, to continue indefinitely. Limited genetic variation, as well as physiological and environmental constraints, places a limit on the level of exaggeration that can be obtained while asymmetric rates of evolution can result in one species evolving phenotypic traits beyond that which can be matched by the other member of the race.

The Life-Dinner Principle and Asymmetrical Selection

The relative rarity of coevolutionary arms races is often cited as being due, in part, to asymmetric selective pressures experienced by predators and prey. Dawkins and Krebs characterized this inequity as 'the life-dinner principle,' which claims that prey species experience greater selective pressure to win the race as a result of the relative severity of the alternative outcomes: a hare that loses the race has lost its life and all future reproduction, while a fox that loses the race has merely lost a meal, a relatively meager cost that potentially allows the predator to reproduce in future. The asymmetry is expected to result in greater evolutionary rates in the prey, thus allowing it to 'outrun' the predator and escape from the arms race, or the predator may simply fail to enter the race altogether.

This caricature, however, is a far too simplistic view of predator–prey interactions that, most conspicuously, underestimates the selective pressure experienced by predators. While the loss of a single meal may not directly result in the death of a predator, the loss of multiple meals certainly will. The rarity of predator–prey arms races is perhaps better explained in terms of the consistency with which predators and prey experience the consequences of a given interaction. Prey that fail to escape will immediately and consistently experience the consequences of the interaction, whereas predators are sometimes able to avoid many of the selective consequences that are possible within an interaction. Even if a predator 'only' loses a meal, the consequences of that lost meal may vary depending on the physical condition of the predator as well as the life history of that predator: generalist predators may be better able to cope with a lost meal because of the availability of alternative prey items, whereas specialist predators rely more closely on a specific prey species and so may view the lost meal as a rather more drastic consequence.

A predator that fails to capture a prey also often fails to physically interact with that prey, thus allowing itself to avoid any damage that may be inflicted directly by prey defenses, such as that caused by claws and toxins. In 1999, Brodie and Brodie explained how this difference in the predictability of fitness consequences might lead to generally weaker selection on predators than on prey.

In cases where predators and prey are forced to interact directly, many prey defenses, such as with spines and toxins, physically damage or even kill predators. These defenses can exert strong, consistent selective pressure on the predator, thus evening the playing field and allowing coevolutionary arms races to take place. One such example of a dangerous prey is the toxic newt, *Taricha granulosa*, found along the western coast of North America, from central California to southeastern Alaska. *T. granulosa*, commonly known as the rough-skinned newt, defends itself from predation by secreting, from glands in the skin, one of the most potent neurotoxins known: tetrodotoxin (TTX). TTX binds to voltage-gated sodium channels in muscles and nerves, blocking action potentials and resulting in the complete cessation of electrical current in blocked tissues. TTX poisoning generally results in the death of the potential predator via respiratory failure. However, some species of garter snakes, including *Thamnophis sirtalis*, have evolved resistance to TTX, in part due to mutations in the sodium channel genes that reduce the binding affinity of TTX. *T. sirtalis* are able to consume newts, though most do not fully escape the negative effects of TTX – even if the snake is not killed outright, TTX can act to inhibit locomotor performance in the predator. Depending on the toxicity of the newt and resistance of the snake in question, a snake may suffer only limited effects or be completely immobilized for several hours, rendering it unable to escape from predators or thermoregulate properly. In this manner, a selective advantage is gained by the most resistant snakes capable of escaping the more drastic effects of TTX poisoning. On the other side, newts often survive unsuccessful attacks from insufficiently resistant predators, either by crawling from the mouth of the dead or immobilized snake unharmed, or as a result of a behavioral response in the predator in which, for example, garter snakes actively reject too-toxic prey up to an hour after initiation of the process of ingestion.

Chase-Away Selection

Predator–prey co-evolution is not restricted to simple one-on-one interactions. Some of the most interesting dynamics occur when the strength of reciprocal selection is between two or more prey species eaten by a common predator. Many dangerous or unpalatable prey have evolved an additional level of defense against predators in the form of warning signals, often aposematic coloration, meant to deter predators prior to attack. Such is the case for many butterfly species that sequester toxic compounds from their food plants, brightly colored salamanders and frogs that secrete nasty compounds from their skin, and a range of other organisms that are chemically defended. The utility of warning signals depends largely on predator psychology, with predators learning to associate adverse effects with the visual or other sensory signals displayed by a prey item prior to capture. Unpalatable prey, therefore, only reap the benefits of a warning signal if the predator has previously attempted to consume a prey item displaying the same signal and found the result undesirable.

This avoidance of a particular signal by predators sets up a selection scenario on color patterns that drives reciprocal selection between prey. Undefended prey gain an advantage by resembling dangerous prey, so that predators confuse their identity and avoid both prey types. In such a mimicry system, usually labeled Batesian mimicry, one prey species is defended and others not. Predators learn to avoid the common pattern by experiencing the defended prey, but learn to eat the pattern when they encounter an undefended prey. This inequity generates to a tense dynamic because defended prey gain the biggest advantage when they are unique, but undefended prey adapt to resemble the defended species. In evolutionary terms, this leads to 'chase-away' co-evolution – predation drives the evolution of more precise mimics, resulting in directional selection on the defended prey away from the mimetic phenotype. Mimics effectively 'chase' models toward new phenotypes.

Other forms of mimicry involve multiple well-defended species of prey. In such Müllerian systems, all of the similar prey can help 'teach' individual predators to avoid a pattern. The most famous of these systems involve the South American *Heliconius* butterflies, which as larvae feed on *Passiflora* plants and sequester toxins leaving themselves well defended against predators as day-flying adults. Many species of *Heliconius* that inhabit an area share a common red, black, and yellow pattern. The dynamic here is not a chase-away, but rather a mutualistic convergence on a single pattern. Because a predator learns from encounters with any of the defended prey, individuals with the most common pattern will always have an advantage regardless of species. This force should generate stabilizing selection that pushes all the species to a common warning signal. Despite this prediction, some Müllerian systems still maintain a large degree of variation in color pattern. Understanding the forces that maintain this polymorphism is one of the major directions in current research on the mimicry process.

The Red Queen

When first introduced to the scientific community by Van Valen in 1973, the 'Red Queen hypothesis' was presented as a way to explain loglinear extinction rates observed in broad taxonomic groups. Van Valen proposed that the effective environment experienced by a group of species was comprised primarily of other organisms and that as one member of the ecosystem evolved, this caused a change in the effective environment to which all members of the ecosystem must adapt or face extinction. Across macroevolutionary time periods, therefore, continuous evolution in response to an ever-changing biotic environment was necessary to avoid extinction, even with a constant abiotic environment, and past adaptations did not improve the ability of a given taxon to adapt to future environmental changes. Van Valen likened this to the world in Lewis Carroll's book, *Through the Looking Glass*, wherein the Red Queen informs Alice that in that world '... it takes all the running you can do to keep in the same place.'

This metaphor has led the Red Queen hypothesis to be viewed as a large-scale manifestation of the concept of co-evolution, with evolutionary change in one species or group of species driving the evolution of other groups. However, a second, more specific, Red Queen has emerged that has come to refer directly to coevolutionary or population dynamics that proceed in a cyclical manner at the population level. Unlike arms race dynamics, cyclical Red Queen dynamics can be stable for long periods of time, with regular switches in the direction and intensity of selection, resulting less in an escalation of traits than a state of constant change (**Figure 2**).

Frequency-Dependent Selection

One mechanism by which Red Queen dynamics are maintained in natural populations is negative frequency-dependent selection, where selection acting against the most common phenotypes lends an advantage to the rare types in a population. Their relative fitness advantage causes rare types to increase in frequency and eventually become more common. At this point, they lose their advantage and begin to decrease in frequency again. Meanwhile the disadvantaged common morph became rare and gained a relative advantage, only to increase in frequency again. Thus, evolutionary trajectories of a given morph continually change direction and the composition of the population continues to fluctuate.

A beautiful example of the Red Queen in action involves feeding polymorphisms in the scale-eating cichlids of African Rift Valley Lakes, particularly the seven species found in Lake Tanganyika. These predatory fish approach potential prey from behind, attempting to bite scales from the flanks of the prey. Scale-eating cichlids appear specialized for this gruesome task thanks to an asymmetrical jaw joint that causes the mouth to open to either the left or the right side. First documented in *Perissodus eccentricus*, further examination revealed that all seven species of scale biters in Lake Tanganyika demonstrated a similar 'handedness' to their jaw openings.

The asymmetrical opening of the mouth maximizes the area of the prey flank that is in contact with predator teeth in a single bite, thus increasing the efficiency of each attack. This advantage, however, is realized only when the predator approaches its prey from a single direction: predators whose mouth opens to the right (dextral) must always attack from the left flank while predators with a mouth opening to the left (sinistral) must approach from the right. The two phenotypes exist simultaneously within a single population, but in unequal proportions that fluctuate over time. Indeed, examination of different age groups within a single population shows that the relative proportion of righty to lefty predators in larger, older fish is opposite that found in the younger age group.

Michio Hori's studies of *P. microlepis* in Lake Tanganyika demonstrated a clear rare morph advantage and the cycles of changing frequencies across generations that should result from negative frequency-dependent selection. The more common mouthed morph has limited hunting success, as prey vigilance appears to be directed toward the side that is most often attacked. This bias leaves the opposite flank less defended, providing a relative advantage to the rare mouthed morph. Increased feeding success by rare morphs results in increased reproductive success such that the next generation of predators will be composed primarily of the previously rare phenotype. Increased vigilance toward this new threat reverses the advantage, thus completing a single phase of the cycle.

A similar dynamic is also seen in the New Zealand freshwater snail, *Potamopyrgus antipodarum*, and its parasite, trematodes of the genus *Microphallus*. Although not strictly a predator–prey interaction, many parallels exist between host–parasite interactions and those among predators and prey and the study of one can provide valuable insights into the other. This is particularly true for parasites such as *Microphallus* that castrates both sexes of the snail host, effectively eliminating all future reproduction in a manner that is equivalent – at least, from an evolutionary point of view – to death and consumption by a predator.

P. antipodarum exists as either asexually reproducing female lineages or sexually reproducing lineages with males and females, sexual lineages being more common in populations sympatric with trematodes. Curt Lively's investigations of *Microphallus* and its host have demonstrated cycling genotypes as predicted by the Red Queen. The most common asexual lineage in a lake is disproportionally parasitized, causing rare lineages to spread. Sexual lineages gain an advantage over asexual ones because they can spin off variable genotypes of offspring that cannot be so easily tracked by parasite genotypes. Red Queen dynamics in this interaction are, once again, maintained through negative frequency-dependent selection. Trematodes locally adapt to the predominant phenotype of the host population, evolving increased virulence and ability to infect the most common genotype.

Rare phenotypes eventually become more common, driving selection on trematodes to target their genotype and relaxing selection to infect the formerly common type.

The Geographic Mosaic of Co-evolution

One of the major efforts in current studies of co-evolution is the investigation of the role of geography in dynamics. For many decades, co-evolution was discussed as though it played out at the level of species. In the 1990s, evolutionary ecologists took a lead from population geneticists, and began to recognize that co-evolution takes place at the scale of populations rather than range-wide distributions of species. The ecology at each locale varies and is expected to lead to spatial mosaics in selection. In terms of co-evolution, we should recognize 'hotspots' where strong reciprocal selection takes place, and 'coldspots,' where selection is weak, absent, or not reciprocal. Thus, a given pair of species may be engaged in a coevolutionary dynamic in one spatial location but not in another.

This geographic mosaic of co-evolution should lead to spatial variation in the traits found within a given species. Processes of gene flow, local extinction, and recolonization are all expected to interact with selection to generate a complicated geographic mix of variable traits and mismatches between the ecological abilities of interacting species.

Role of Local Ecology and Behavior in Generating Mosaics

The impact of local ecology on the spatial variation of reciprocal selection has been clearly illuminated by Craig Benkman's work on crossbills (*Loxia* spp.) that prey upon the seeds of various species of conifer. Crossbills have evolved a specialized beak morphology in which the upper and lower mandibles overlap in a manner that allows them to spread apart the scales of pine cones and thus reach seeds more easily (**Figure 3**). In response to crossbill predation, multiple conifers in both North America and Europe have evolved cone traits that protect against seed predation, including thicker scales with a high degree of overlap.

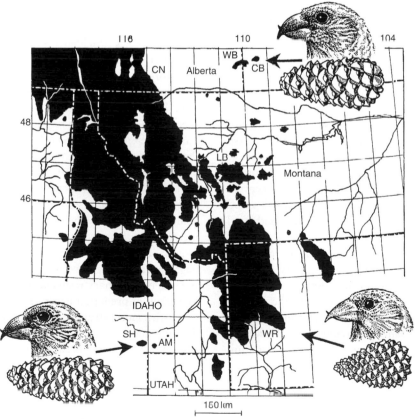

Figure 3   Crossbill pine-cone co-evolution. The distribution of lodgepole pine (black) and representative crossbills and cones in western North America, drawn to relative scale. Red squirrels (*Tamiasciurus hudsonicus*) are found throughout the range of lodgepole pine except in some isolated mountains, including the South Hills (SH), and Albion Mountains (AM). Red squirrels were absent from the West Block (WB) and Centre Block (CB) of the Cypress Hills until being introduced in 1950. Reprinted from Benkman CW, Holimon WC, and Smith JW (2001) The influence of a competitor on the geographic mosaic of co-evolution between crossbill and lodgepole pine. *Evolution* 55(2): 282–294.

Crossbills, in turn, respond to these traits by evolving larger, more decurved bills capable of sustaining the additional pressure needed to spread the scales. The coevolutionary interaction varies with locality and the specific species involved, however, and has resulted in the diversification of a wide variety of morphological and vocal 'types' of crossbill, each adapted to a local species of conifer.

Benkman has shown many ecological factors to play an important role in the local outcomes of the predator–prey interaction between crossbills and conifers. In Red crossbills (*Loxia curvirostra*) and the Rocky Mountain lodgepole pine (*Pinus contorta latifolia*), co-evolution between these two species is ameliorated by the presence of a competing seed predator, the red squirrel (*Tamiasciurus hudsonicus*). Squirrels are the dominant predator, outcompeting crossbills for seeds and thus claiming the role of primary selective agent on cone traits. Crossbills still appear to evolve beak traits specialized to match the mean cone phenotype of the local population, yet these specialized bills no longer appear to have a selective effect on cone morphology. In other areas, the strength of selection exerted on cone traits appears to increase with the size of the forest in which the interaction is taking place, possibly because of the abundance of resident crossbills that can be supported by large forested regions, in contrast to spatially smaller populations in which crossbills may be either less common or less persistent. Finally, studies examining the relationship between *L. curvirostra* and the European black pine (*Pinus nigra*) have shown that, because black pine do not hold seeds in the cone during the early summer months, coexistence and reciprocal selection between the two species can occur only when alternative seed sources are available during those months, thus allowing the crossbills to persist in the area.

Geographic mosaics can sometimes lead to a greater understanding of the evolutionary dynamics of a coevolutionary system. The garter snake–newt arms race described earlier shows dramatic geographic variation in both predator and prey phenotypes across western North America (**Figure 4**). A first glance at the geographic pattern of snake resistance and newt toxicity suggests a roughly matched level of exploitation and defense in the two species – snakes are non resistant outside the range of toxic newts, and the most toxic newts and most resistant snakes are seen in the same geographic areas. However, a closer look at apparent hotspots reveals a dramatic mismatch between the functional abilities of the two species. Indeed, nearly half the localities examined had predator and prey traits sufficiently different so that reciprocal selection is not expected to occur in the current populations. In every case of a mismatch, the TTX resistance levels of local predators were so extreme that any snake would be capable of consuming any sympatric newt with little or no adverse effects. This pattern appears to result from predators escaping from the arms race by evolving

extreme levels of resistance through mutations of large effect. By investigating many localities, we are able to examine coevolutionary dynamics at various stages in the arms race – we see some localities with populations of both species exhibiting ancestral ('prearms race') levels of phenotypes, others with exaggerated traits engaged in counterescalation via reciprocal selection, and still others where, at least temporarily, one population has escaped the evolutionary race.

One temptation emerging from geographic comparisons is to classify all examples of adaptive radiation as the result of coevolutionary interaction. Geographic variation in the phenotype of one species is not, by itself, sufficient to demonstrate a coevolutionary arms race. An elegant example of adaptive radiation across a geographic range as a result of unidirectional local selection is the soapberry bug (*Jadera haematoloma*) of North America. The soapberry bug is a seed predator that uses its needle-like beak to penetrate the fruit of multiple host species, pierce the seed coat, then liquefy and suck up the contents. Different populations of bugs vary considerably in beak length, with beak lengths matching the typical fruit size of the local host. In the last century, the introduction of the invasive 'round-podded' and 'flat-podded' golden rain trees (*Koelreuteria paniculata* and *K. elegans*, respectively) from east Asia, as well as the heartseed vine (*Cardiospermum halicacabum*), has provided a rare opportunity to chronicle the effect of local interactions on the phenotype of both predators and prey. As predicted, fruit size and shape of each host taxon appear to have driven the evolution of beak lengths locally adapted to the predominant host species of a given area. In regions where the native balloon vine (*Cardiospermum corindum*) has been replaced by the flat-podded golden rain tree, beak lengths of the soapberry bug have shortened to accommodate the smaller fruit. Similarly, in regions where the native soapberry tree (*Sapindus saponaria*) is displaced by the round-podded golden rain tree and heartseed vine, beak lengths have increased in length. The interaction, however, appears largely one-sided at this point in time: while soapberry bugs appear to have evolved in response to host phenotype, an evolutionary response of fruit size in the rain tree as a result of predation is yet to be seen.

Furthermore, strength of selection in a given location does not necessarily depend on the ecological intensity of the given interaction: predators may kill most of the prey of a population but if this does not result in differential survival or reproduction of the prey due to some defensive phenotype, the prey will be unable to evolve. This is apparent in the geographically varying coevolutionary interaction between Japanese camellia (*Camellia japonica*) and its obligate seed predator, the camellia weevil (*Circulio camelliae*) (**Figure 5**). Japanese camellia – an evergreen, broadleaf tree common throughout Japan – protects its seeds by enclosing them in fruit with an unusually thick,

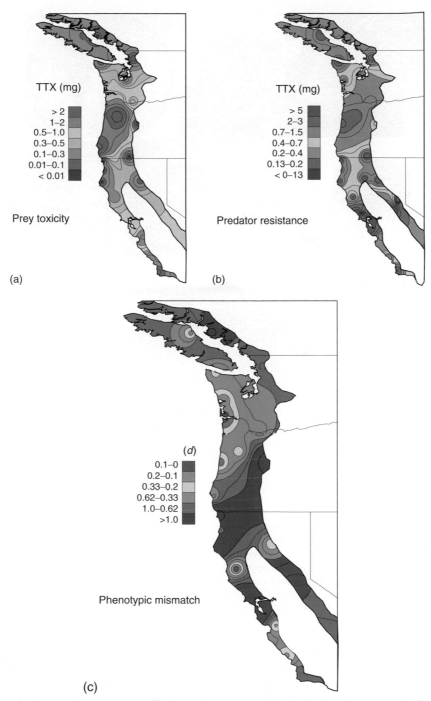

(a)

Prey toxicity

TTX (mg)

> 2
1–2
0.5–1.0
0.3–0.5
0.1–0.3
0.01–0.1
< 0.01

(b)

Predator resistance

TTX (mg)

> 5
2–3
0.7–1.5
0.4–0.7
0.2–0.4
0.13–0.2
< 0–13

(c)

Phenotypic mismatch

(d)

0.1–0
0.2–0.1
0.33–0.2
0.62–0.33
1.0–0.62
>1.0

Figure 4 Hotspots and coldspots in an arms race. The interpolated geographic distribution of prey toxicity, (b) predator toxicity, and (c) the degree of phenotypic matching of these traits across the geographic range of newts of the genus *Taricha* and the garter snake *Thamnophis sirtalis*. Note that the overall patterns of relative phenotypic exaggeration (yellow, orange, and red areas in (a) and (b)), are generally similar for newts and snakes throughout their range of sympatry. However, areas of exaggerated phenotypes do not necessarily have the strongest reciprocal selection as shown by an analysis of functional matching in (c). Reprinted from Hanifin CF, Brodie ED Jr, and Brodie ED III (2008) Phenotypic mismatches reveal escape from arms-race co-evolution. *PLoS Biology* 6: 471–482.

woody pericarp. The female camellia weevil is able to overcome this defense by using her long rostrum to bore a hole in the woody pericarp. Upon reaching a seed, she withdraws her rostrum and turns around to lay an egg in the seed. Each larva consumes a single seed upon

hatching, begins development in the fruit, and then burrows out to spend its first winter in the soil. The success of a given boring attempt is determined both by the thickness of the pericarp and by the length of the female's rostrum; thus selective pressure favoring thick pericarps drives

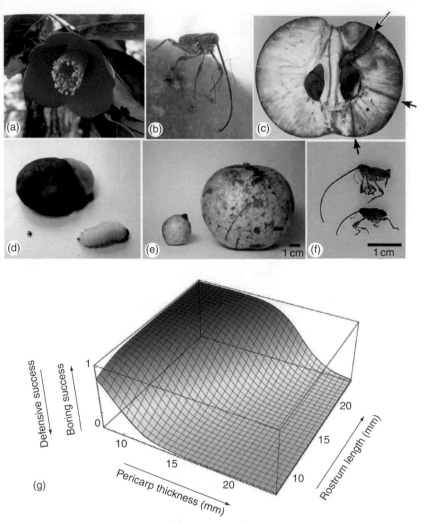

Figure 5 Arms race between weevils and *Camellia* fruits. The phenotypic interface of the interaction between Japenese camellia and camellia weevils is illustrated by (a) the flower of the Japenese camellia, (b) a female weevil drilling into the pericarp of a camellia fruit with her rostrum, (c) a cross section of fruit showing the holes drilled by weevils (solid arrows) and an exit hole of a weevil larva (dashed arrow), (d) a weevil larva and infested seed, (e) geographic variation in camellia pericarp thickness, and (f) geographic variation in weevil rostrum length. The three-dimensional selection surface (g) of the interaction shows the probability of boring success for various combinations of camellia and weevil phenotypes. Reprinted from Toju H and Sota T (2005) Imbalance of predator and prey armament: Geographic clines in phenotypic interface and natural selection. *The American Naturalist* 167(1): 105–117.

continued exaggeration of the plant defensive trait which in turn favors increased rostrum length in the predator.

The strength of reciprocal selection between these two species varies along a latitudinal gradient. Direct examination of the phenotypic interface of co-evolution at the population level has shown that, although weevils experience elevated boring success in northern populations and are capable of causing the death of more than half the seeds in a given population, selection on camellia to produce thicker fruits is relatively weak and rostrum length and pericarp thickness remain modest. This phenomenon indicates an imbalance in armament between predator and prey in favor of the predatory weevil, where each camellia, regardless of the thickness of its fruit in relation to other members of the population, remains vulnerable to weevil attack. In southern populations, however, camellia and weevil become more evenly matched in the arms race, the strength of directional selection on pericarp thickness increases (despite reduced ecological intensity of the interaction), and both predator and prey phenotypes are exaggerated.

Current Limitations and Future Directions in the Study of Co-evolution

Perhaps the predominant current limitation to understanding coevolutionary dynamics is the tendency of researchers to focus on one half of the equation. For example, we cite the garter snake–newt system is as an example of co-evolution, but research has focused primarily on the evolution of predatory adaptations (TTX resistance) in garter snakes, leaving the evolution of newt toxicity something of an enigma. Argument continues regarding the production of TTX, and the genetic basis of toxicity has not yet been adequately demonstrated. Although evolution of lateral asymmetry in scale-eating cichlids seems to be driven by differential feeding success between rare and common phenotypes, some argument remains as to whether the escape response of the prey is purely a learned behavior or the result of selection on analogous lateral asymmetry in the prey. The logistical difficulty in studying two species at an equivalent level of rigor is immense, but before coevolutionary interactions can be fully understood and conclusively demonstrated, significant steps must be taken toward balancing the body of research available for each species involved.

Perhaps because of their complexity, coevolutionary studies have tended to emphasize an observational approach to data acquisition in the past. Recent moves toward more experimental studies promise new insight into dynamics and have been particularly useful in revealing ecological factors important in driving reciprocal selection and geographic mosaics. Artificial laboratory systems using bacteria and phage predators are well suited to such studies as they churn out multiple generations everyday. However, such systems lack a certain focus on phenotypes that students of animal behavior seem to emphasize.

Finally, the incorporation of modern genetic and genomic methodologies promises to chart more of the genotype–phenotype map that ultimately drives the dynamics. Whether traits at the phenotypic interface of co-evolution are determined by one or many loci influences the likely coevolutionary dynamics at the population level. The extent to which co-evolution takes place between populations, traits, or genes will eventually be revealed when we have a fuller picture of the genetic basis of the traits that mediate ecological interactions.

See also: Defensive Chemicals; Defensive Coloration; Defensive Morphology; Parasitoids; Predator Evasion; Predator's Perspective on Predator–Prey Interactions; Specialization.

Further Reading

Abrams PA (2000) The evolution of predator–prey interactions: Theory and evidence. *Annual Review of Ecology and Systematics* 31: 79–105.

Benkman CW, Holimon WC, and Smith JW (2001) The influence of a competitor on the geographic mosaic of coevolution between crossbill and lodgepole pine. *Evolution* 55(2): 282–294.

Brodie ED, III and Brodie ED, Jr (1999) Predator–prey arms races. *Bioscience* 49: 557–568.

Brodie ED, III and Ridenhour BJ (2004) Reciprocal selection at the phenotypic interface of coevolution. *Integrative and Comparative Biology* 43: 408–418.

Dawkins R and Krebs JR (1979) Arms races between and within species. *Proceedings of the Royal Society of London B: Biological Sciences* 205: 489–511.

Hanifin CF, Brodie ED, Jr, and Brodie ED, III (2008) Phenotypic mismatches reveal escape from arms-race coevolution. *PLoS Biology* 6: 471–482.

Hori M (1993) Frequency-dependent natural selection in the handedness of scale-eating cichlids. *Science, New Series* 260 (5105): 216–219.

Lively CM (1996) Host-parasite coevolution and sex. *Bioscience* 46: 107–114.

Mallet J and Joron M (1999) Evolution of diversity in warning color and mimicry: Polymorphisms, shifting balance, and speciation. *Annual Review of Ecology and Systematics* 30: 201–233.

Thompson JN (1986) Constraints on arms races in coevolution. *Trends in Ecology & Evolution* 1(4): 105–107.

Thompson JN (2005) *The Geographic Mosaic of Coevolution.* Chicago, IL: University of Chicago Press.

Toju H and Sota T (2005) Imbalance of predator and prey armament: Geographic clines in phenotypic interface and natural selection. *The American Naturalist* 167(1): 105–117.

Cognitive Development in Chimpanzees

T. Matsuzawa, Kyoto University, Kyoto, Japan

Introduction

The genetic distance between humans and chimpanzees is extremely small; our DNA sequences differ by only about 1.2%. It should be possible to explore the evolutionary origins of the human mind through comparisons with the minds of chimpanzees. Shared traits of the two species must have had their origins in a common ancestor that lived around 5–7 Ma. In contrast, traits unique to us must have emerged during the evolution of modern humans from this last common ancestor.

Based on data from a total of 534 births recorded in the wild, we have learned many details of the life history of chimpanzees. Chimpanzees live for about 50 years and give birth at intervals of around 5 years. Mortality rate in the first 5 years of chimpanzee life is about 30%, while the average life span is about 15 years. Chimpanzee life history can be divided into the following five stages: individuals less than 5 years (0–4 years) are classified as 'infants,' those between 5 and 8 years are 'juveniles,' 9- to 12-year-olds are 'subadults,' those older than 12 years are 'adults,' and the 'elderly' are those aged over 36 years.

When considering human evolution, we must bear in mind the different routes through which evolution generates similarities and differences. Comparisons across a wide range of species test for convergence due to independently evolved adaptations to the environment. This kind of convergence is called 'analogy in evolution.' In contrast, closely related species are far more likely to share characteristics through having inherited them from a shared ancestor. Features similar by descent are referred to as 'homologies.' Both analogy and homology are important in understanding the evolutionary origins of human nature.

While understanding cognitive development in chimpanzees is in itself important, the study of chimpanzees has an additional unique benefit that sets it apart from studies of other non-human animals; it has the potential to tell us about evolutionary history of human cognitive development. There are two questions that may be answered through the study of chimpanzee cognitive development: first, 'what are the common cognitive traits shared by humans and chimpanzees?' and second, 'what traits are uniquely human?'

Chimpanzees Raise Offspring One at a Time While Humans Raise Multiple Children Concurrently

Humans often have brothers and sisters 2–3 years older or younger than themselves; in fact, the age gap is sometimes as small as 1 year. However, this is not the case for chimpanzees, whose reproduction is characterized by long interbirth intervals. Chimpanzee siblings are at least 5 years apart. Chimpanzee mothers are in a way similar to human single working mothers who take care of a single child at a time. Raising a child requires a lot of effort and investment, and only when the child has reached a certain level of independence can the mother afford to have her next infant. In sum, the chimpanzee way of rearing offspring is characterized by raising a single infant at a time.

Imagine that humans were to adopt the same way of rearing children as chimpanzees. Human infants develop more slowly than chimpanzees. Five years may not be long enough for them to reach independence; 8 years or so are needed. Now, suppose that a human female gives birth at 8-year intervals starting from the age of 18: she will have children when she is 18, 26, 34, and 42 years old. By the age of 50, she may not be able to have any more children. This means that if human females followed a more chimpanzee-like system they would only be able to produce a maximum of four offspring in their lifetime. If infant mortality rates were as high as they are in wild chimpanzees, this reproductive strategy would be unable to maintain human populations in the long run.

In the course of hominization – the process of human evolution – various changes occurred in reproductive and child-rearing strategies. While physiological mechanisms such as gestation are not under voluntary control and thus difficult to adjust, changing the way children were reared would have provided an alternative, more immediate solution. Humans evolved to wean earlier. They invented special foods that could be given to immature children. They gave birth after progressively shorter interbirth intervals. In these ways, humans were able to produce larger numbers of offspring, but faced the challenges of raising multiple children at one time. This new reproductive strategy required help from partners, grandparents, and other kin and nonkin members of the community.

Among the 220 extant primate species, humans have a comparatively strong tendency to maintain pair bonds. Females need the help of male partners to be able to raise multiple children concurrently. To keep the bond, human females conceal their time of ovulation. In contrast, female chimpanzees advertise ovulation with a swollen bottom that attracts the attention of the males. This human way of collaborative breeding may provide a driving force behind the division of labor between the sexes. In addition to the partner, additional help from kin is needed. The human way of rearing children may have brought about another solution; a longer postreproductive life span, particularly in females. Grandmothers no longer produce children themselves, but instead invest time and energy into their grandchildren's generation.

Stable Supine Posture

Many textbooks of anthropology point to upright posture and bipedal locomotion as the primary evolutionary forces behind human cognitive ability. The suggested scenario is as follows: the transition from quadrupedal to bipedal locomotion freed up the hands. The hands in turn began to manipulate objects, use tools, and even make tools to create other tools

However, there is another important but neglected issue related to human posture: the stability of infants in the supine posture (lying on one's back on the ground or other surface). Chimpanzee and other non-human primate infants cannot be stable when they are laid on their backs – they move the limbs in an attempt to grasp and cling to something. They struggle to turn over, in what is referred to as 'the righting reflex.' Only human infants are exceptional among primates in this respect; they can assume a stable supine posture and will lie quietly on their back.

Non-human primates have four limbs with which to climb trees and are well adapted to arboreal life. Primate infants cling actively to their mothers; in return they receive frequent embraces (**Figure 1**). The mother–infant relationship in primates, especially simians, is clearly characterized by this clinging–embracing. However, human mothers and their infants are often physically separated from each other. Human infants cannot cling to their mothers. When you touch the palm of a newborn, you will see the hands immediately curl around your finger in a grasping reflex. However, the strength of the grip is not sufficient to support the infant's weight: human babies are not able to cling to their mothers by themselves.

It must be also noted that non-human primate mothers receive no help from others during birth. In the final stages of delivery, the infant's face is directed toward the ventral surface of the mother. This helps the mother to retrieve the infant. The same does not apply to human births; human infants' heads are so large that they have to gradually twist

Figure 1 Primate infants cling actively to their mothers; in return they receive frequent embraces.

as they emerge through the birth canal until they finally face the mother's dorsal side. Human mothers cannot retrieve the infant by themselves. From birth onward, humans need the help of others to deliver and raise their offspring. Furthermore, just like many other mammals – cats, dogs, horses, and so forth – human mothers and infants are often physically separated from the birth; this is a trait unique among primates.

Given the human reproductive strategy of raising multiple children at the same time, it is highly adaptive to have infants who can quietly assume a stable supine posture. Human infants have an exceptionally high body fat percentage: they are wrapped in about 20% fat! This is in clear contrast with chimpanzee infants who have only about 4–5% fat. The large fat deposits of human infants are likely an adaptation for the cold nights of the open savanna, newly occupied niche for human ancestors, which left their forest habitats.

Stable supine posture brought with it three important innovations for human interaction. First, the supine posture of infants facilitated face-to-face communication with the mother and other members of the community. Second, it enhanced vocal communication. As the infant is physically separated from the mother, it needs to cry to attract her attention. Only human infants cry in the night, chimpanzee infants never do so since they are never separated from the mother. In return, human mothers chat to their infants from a distance. Thus, vocal exchange compensates for the lack of physical contact. Third, the hands are completely freed from having to support the body.

Human infants actually start manipulating objects much earlier than chimpanzees and other non-human primates. Human mothers often give rattles and other objects to their infants. Infants in turn pass these objects from one hand to the other (initially via the mouth) from a very early age. This behavior also facilitates manual–tactile–visual–auditory coordination. The early onset of object manipulation in humans is likely a precursor of tool use. In conclusion, from a developmental perspective, stable supine posture rather than bipedal upright posture enabled humans to become by far the most versatile and proficient tool users in nature.

Innate Mother–Infant Interactions

Primate infants are born with a set of innate reflexes: clinging, grasping, rooting, suckling, etc. Human newborns smile spontaneously with their eyes closed. Limited to the neonatal period, this behavior is referred to as 'neonatal smiling.' A recent discovery has confirmed that chimpanzee infants also show neonatal smiling. In addition, spontaneous smiling, in a rudimentary form, also appears to be shared with macaque infants. However, the spontaneous smiles of macaque infants are asymmetrical, similar to those produced during the very early phase of human neonatal smiling. In the case of humans, such initially asymmetrical smiles quickly develop into symmetrical ones.

The incidence of spontaneous smiles gradually decreases in both humans and chimpanzees until they finally disappear approximately 2 months after birth. Instead, infants begin to smile with their eyes open, when looking at faces or face-like stimuli. This is referred to as 'social smiling.' Generally restricted to social contexts and face-to-face situations, social smiles receive considerable attention and feedback from the infant's mother, father, grandparents, and other members of the community.

Just like human infants, chimpanzee infants exhibit neonatal facial imitation. When they see someone open their mouth in a face-to-face situation, infants also open theirs. When they see someone stick out their tongue, they stick out their tongue too. This kind of facial–gestural imitation disappears at about the age of 4 months in chimpanzees. Around the same time, the infants' ability to stand on four limbs first emerges, and they begin to take their initial steps away from the mother. They no longer imitate facial gestures, but instead show social smiling when looking at the faces of other individuals.

Eye-to-eye contact is another important characteristic of early mother–infant relationships. Mothers look into the eyes of their infants, while the infants look back into their mothers'. Mutual gaze is a truly unique feature shared by humans and chimpanzees, but not by, for example, macaque monkeys. In general, direct gaze is a sign of aggression or threat in both primates and nonprimate mammals. However, in the case of humans and chimpanzees, looking into one another's eyes often signals affection. Macaque mothers also pay attention to their newborn infants; however, it is unusual to see an exchange involving mutual gaze and social smiling between them. In contrast, chimpanzee mothers often lie on their back and use their feet to hold up their infants, while at the same time tickling them in the neck to elicit social smiling. At present, it appears that only humans, chimpanzees, and gorillas (and probably also orangutans) engage in this kind of face-to-face interaction (see **Figure 2**).

In sum, early mother–infant interactions in chimpanzees resemble those of humans in many respects. Such interactions are based on the innate behavior of newborns, such as neonatal smiling and neonatal facial imitation, and provide the basis for various forms of social exchange including social smiling and mutual gaze both in chimpanzees and in humans.

Synchronization: Coaction, Joint Attention, and Food Sharing

Chimpanzee infants almost always cling to the mother in the first three months of life. By around four months of age, the infants can assume the quadrupedal standing position and take their first steps away from the mother. At the same time, they also begin to interact with other members of the community. When 1-year-old, infants will often receive

Figure 2 Chimpanzee mothers often lie on their back and use their feet to hold up their infants, while at the same time tickling them in the neck to elicit social smiling in this kind of face-to-face interaction.

embraces from older siblings. Such allomothering or aunt-ing behavior is often seen both in the wild and in captivity.

Year-old chimpanzee infants have a strong tendency to synchronize their movement with others. For example, one infant may walk by supporting himself on both arms simultaneously and swinging forward (as if on clutches, see **Figure 3**); he will be followed by another infant moving in the same way. We may observe that one infant is staring intently into a bush – and then we'll see his peer looking just as intently at the same spot (**Figure 4**). This kind of joint attention and gaze following are obvious components of 1-year-old chimpanzees' behavior.

Figure 3   An infant walks by supporting himself on both arms simultaneously and swinging forward (as if on clutches), and he is followed by another infant moving in the same way.

Figure 4   An infant is staring intently into a bush – and his peer looking just as intently at the same spot. This kind of joint attention and gaze following are obvious components of 1-year-old chimpanzees' behavior.

The major difference between humans and chimpan-zees at this stage may lie in the establishment of triadic mother–infant–object relationships, in other words, social referencing. Suppose that a human infant encounters a new toy. She may look up at the mother *before* touching it. The mother may nod or smile, and only then will the infant actually start manipulating the object. While play-ing with the toy, the infant may often smile when showing it to the mother. The mother may smile back at her child and give social praise. In sum, human infants often manip-ulate objects within a social context. In contrast, chimpan-zee infants are much more likely to concentrate only on manipulating the object and seldom look up at the mother. The chimpanzee mother does not appear to care much about what her infant is doing as long as it isn't interfering with what she is doing! In sum, there are only dyadic relationships in chimpanzees' object manipulation.

Adult chimpanzees seldom actively share food. How-ever, food sharing from mother to infant often occurs. Careful observation tells us that it is usually the infant who makes the first move, requesting food that the mother is in the process of eating. The mother then responds by allowing the infant to take a bite out of the food she is still holding. Perhaps the most important difference between the two species concerns the bidirectionality of food shar-ing. In chimpanzees, food sharing is almost always one-sided, from the mother to the infant, and not the reverse. However, in humans, infants also have a strong tendency to give things to the mother. Let us imagine a situation where a plate of strawberries is placed on the dinner table. Even 2-year-old toddlers may not only eat the strawber-ries but also put a piece directly into the mother's mouth or try to hand one to a grandparent. Sharing food is a bidirectional affair in humans; we have a strong intrinsic motivation to share things among members of our community.

Imitative Processes

The next cognitive stage is imitation. Both humans and chimpanzees have a strong intrinsic motivation to copy others' behavior. There are various different imitative processes, including local enhancement, goal emulation, and true imitation. Irrespective of the precise mechanism, humans and chimpanzees often mimic behavior. The English verb 'ape' means to 'imitate.' In fact, monkeys do not ape, or at most ape seldomly.

Let us think about the game 'Do this!' (sometimes called 'Simon says'). Imagine that you touch your head and say 'Do this!' to your child. She will imitate the behavior well. This kind of interaction does not occur in macaque monkeys. You may, to some extent, see it in chimpanzees; however, even in their case, the imitation of actions is accompanied by some important constraints.

Actions that involve objects, such as putting a ball into a container, are easily imitated. Actions toward one's self, such as brushing one's hair with a comb, seem to be relatively difficult. But the most difficult actions to imitate are simple ones, without referencing objects. For example, it is very hard to teach chimpanzees to imitate actions such as waving, rotating one's arms, or tilting one's head. Generalized imitation – copying a new action immediately, whatever it may be, appears to be a uniquely human ability.

Suppose that you were to pick up a mobile phone and speak into it, and then place it on the floor in front of a chimpanzee. It is quite likely that the chimpanzee will pick up the phone and put it against her ear. In this case, the imitative behavior incorporates the very same object that the model used. There is a further, more complex level of imitation called 'pretence.' Young female chimpanzees in the wild have been seen using a dead hyrax (a raccoon-like animal) or a log (about 10 cm in diameter and 50 cm long) as a doll. These females carry the item around with them on their shoulders, or by pressing them between their thigh and belly as they walk, and often try to groom it, much like a mother chimpanzee might groom her baby. A yet more complex form of pretence incorporates no real object. A 3-year-old infant chimpanzee growing up in captivity was in the habit of gathering together wooden blocks scattered on the floor by dragging them along to the corner of the room using his fingers as rakes. One day, he performed the usual movements – but without any blocks being present! At the same time, his facial expression indicated that he was enjoying himself; he made a play face. It was clear that he was pretending to drag imaginary wooden blocks around the room.

Wild chimpanzees are known to have cultural traditions unique to each community. For example, fishing for termites using sticks or stalks is a well-known tool-using behavior among wild chimpanzees at Gombe, Tanzania. However, in several other communities, chimpanzees do not use tools to extract termites from their mounds, even though they catch them by hand and consume them when the insects emerge from the nest. Members of another community, in Bossou, Guinea, use a pair of stones to crack open nuts. Most other communities do not show this behavior despite them having both nuts and stones readily available in the environment. Thus, such intercommunity differences cannot be due solely to ecological constraints. The behaviors represent cultural traditions; social learning plays an indispensable role in passing on knowledge and skills from one generation to the next.

'Education by master apprenticeship' is a phrase coined to describe how chimpanzees acquire new behaviors through observational learning (**Figure 5**). Observational learning is characterized by the following four aspects. First, it is based on the long-term affectionate bond between mother and infant. Second, the mother takes on

Figure 5   'Education by master apprenticeship' describes how chimpanzees acquire new behaviors through observational learning.

the role of the 'model' who demonstrates specific behaviors in the correct context. Third, the infant has a strong motivation to copy the model's behavior. Fourth, the mother is highly tolerant toward the infant; she never scolds or neglects the infant, and allows him to observe her actions closely. The chimpanzee way of education clearly highlights features unique to human education. The latter is of course characterized by active teaching, including molding and verbal instruction, as well as a suite of other, more subtle behaviors such as watching, nodding, smiling, and social praise. Human infants are highly sensitive to any form of social encouragement from the mother.

In this stage of imitation, infants try to copy the actions of others and thereby acquire behaviors previously absent from their repertoire. As a result, infants are able to have the same experiences as others; expanding their behavioral repertoire through observing others' behavior and imitating their actions inevitably results in them experiencing the corresponding mental states. This may in turn provide the basis for understanding the minds of other animals in their social group.

Understanding the Minds of Others

Suppose that an infant has already experienced the mental state associated with a given behavior. Then, suppose that the infant witnesses that particular behavior performed by another individual. Such observation may evoke the corresponding mental state within the observer. This process is called 'empathy.' Moreover, the infant can consciously imagine or infer the other's mental state although no contagion of mental states has occurred (sympathy). This is called a 'theory of mind'; the understanding of someone else's mind.

Imagine the following scene. A mother and 2½-old female infant are moving through the trees. They come to a gap in the canopy. The infant whimpers because the distance to the next branch is too large for her. On hearing the infant's voice, the mother looks back and extends her hand to help the infant across (see **Figure 6**). At other times, she might pause as she is crossing from one tree to the next, in order to let the infant utilize her as a bridge!

There is a darker side to the same kind of intelligence: deception. The following episode observed in the wild provides an example. A mother chimpanzee arrived at a nut-cracking site to find that all the available stone tools were already being utilized by others who had reached the site before her. The mother approached her 9-year-old son who was already cracking nuts with stones. She groomed her son for a while, then stopped and assumed the quadrupedal standing position. This is a posture that signals the request to be groomed. The son stopped nut-cracking and dutifully began to groom back his mother. At this point, the mother suddenly snatched away his stones, and began to crack nuts herself, leaving him without his tools.

The understanding of other's mind can take many forms. The home range of the chimpanzees at Bossou borders a small human village. Adult males are the only chimpanzees brave enough to enter the village, and when they do so they often climb papaya trees to pluck their large sweet fruits. In some cases, the males in fact steal two papayas: one for themselves and one as a 'gift' for a reproductively receptive female.

In sum, the chimpanzees can have an understanding of each other's mind. Based on this, they help others but they also deceive others. The major difference between humans and chimpanzees in terms of altruism must be the reciprocal altruism. The origin of reciprocity is a topic to be answered in further studies.

Memory and Symbolization

Laboratory studies of chimpanzee intelligence have explored areas including perception, memory, and thinking. One of the most important findings in recent years concerns chimpanzees' extraordinary working memory for Arabic numerals. Young chimpanzees are extremely fast and accurate at memorizing the order and locations of numerals presented on a touch screen monitor. The best performer can memorize nine numerals in 0.67 s (see **Figure 7**). No human adult can solve the task so quickly and with such high accuracy.

What are the advantages of this kind of photographic memory in the wild? Suppose that a chimpanzee arrives at a huge fig tree and finds others already there. It will be important for her to quickly assess where the ripe red fruits are. It will also be important to bear in mind where in tree the highest-ranking male is, where the second highest is, and so forth, in order to avoid approaching them too closely and being prevented by them from eating the fruit. Another example might involve intercommunity encounters. Males patrol the boundary of their territory. Whenever they come across males of a neighboring community, they must be able to quickly assess the size of the potentially hostile band of neighbors. If the encounter turns aggressive, it will be vital to know how many individuals there are on both sides and where they are located.

In contrast to their extraordinary memory capability, chimpanzees are much less adept at the representation and use of symbols. Although they can readily be trained to match the color red to a corresponding symbol meaning 'red' or the color green to the symbol 'green' (a so-called matching-to-sample task), if the task is suddenly reversed and the symbols have to be matched to the colors, chimpanzees experience great difficulties in solving the task. Chimpanzees learn the concept 'If the color is

Figure 6 A mother and 2½-year-old female infant are moving through the trees. They come to a gap in the canopy. The infant whimpers because the distance to the next branch is too large for her. On hearing the infant's voice, the mother looks back and extends her hand to help the infant across.

Figure 7 Young chimpanzees are extremely fast and accurate at memorizing the order and locations of numerals presented on a touch screen monitor. The best performer can memorize nine numerals in 0.67 s.

red, then pick the "red" symbol' – but this experience does not automatically generalize to a symmetrical relationship between symbol and referent. In contrast, human children have no trouble with the spontaneous transfer. The two stimuli – the color and the symbol – become equivalent in the minds of humans.

What is the advantage of this kind of representation? One plausible answer concerns communication. A photographic memory is useful for memorizing things and grasping their details quickly. However, if you convert scenes into labels, you can carry the information with you and transmit it to other members of your community. Imagine that you encounter a creature with short brown hair, four legs ending in hooves, and a white stripe on its forehead. Your immediate memory of the creature may help you recognize and hunt it in the future. However, if you can convert the memory into a label ('horse'), you can share your experience with others in your community. You can then work together to capture the animal by hunting in a group.

Evolutionary Scenario of Cognitive Evolution and Cognitive Development

A trade-off may exist between photographic memory and symbolic representation. At some point in human evolution, brain capacity reached a limit. In order to accumulate new functions, old functions needed to be lost. Because of such trade-offs, humans may have lost much of their ability for olfactory processing and developed instead highly sensitive visual, auditory, and crossmodal functions. A similar scenario may be applied to the trade-off between memory and symbol use. The common ancestor of humans and chimpanzees may have had an extraordinary memory capacity, but this was subsequently lost in humans and replaced by newly acquired symbolic capabilities.

Similar trade-offs can occur during ontogenetic development. Young children may have better working memory than adults. According to neuroscientists, different parts of the brain mature at different rates. The association cortex responsible for symbolic representation may develop more slowly than that dealing

with perceptual processing and working memory. As they lose their photographic memory, humans instead acquire more advanced linguistic skills.

Finally, the same scenario may provide an explanation for individual differences in intelligence. In both humans and chimpanzees, brain size triples between birth and adulthood. This means that learning is essential for survival. We accumulate many knowledge and skills that allow us to adapt to the environment not through genetic channels but through learning. The same applies to humans and chimpanzees. Each individual accumulates different skills and knowledge based on the experience. In sum, all individuals are different – but in many ways they are also equal.

See also: Distributed Cognition; Empathetic Behavior; Mental Time Travel: Can Animals Recall the Past and Plan for the Future?; Problem-Solving in Tool-Using and Non-Tool-Using Animals.

Further Reading

Biro D, Inoue-Nakamura N, Tonooka R, Yamakoshi G, Sousa C, and Matsuzawa T (2003) Cultural innovation and transmission of tool use in wild chimpanzees: Evidence from field experiments. *Animal Cognition* 6: 213–223.

Emery-Thompson M, Jones J, Pusey A, et al. (2007) Aging and fertility patterns in wild chimpanzees provide insights into the evolution of menopause. *Current Biology* 17: 1–7.

Hawkes K, O'Connell JF, Blurton Jones NG, Alvarez H, and Charnov EL (1998) Grandmothering, menopause, and the evolution of human life histories. *Proceedings of National Academy of Sciences United States of America* 95: 1336–1339.

Hockings K, Humle T, Anderson J, et al. (2007) Chimpanzees share forbidden fruit. *PLos ONE* 9: 1–4.

Inoue S and Matsuzawa T (2007) Working memory of numerals in chimpanzees. *Current Biology* 17: R1004–R1005.

Matsuzawa T (2007) Comparative cognitive development. *Developmental Science* 10: 97–103.

Matsuzawa T (2009) Symbolic representation of number in chimpanzees. *Current Opinion in Neurobiology* 19: 1–7.

Matsuzawa T, Tomonaga M, and Tanaka M (eds.) (2006) *Cognitive Development in Chimpanzees.* Tokyo: Springer.

Premack D and Woodruff G (1978) Does the chimpanzee have a theory of mind? *Behavioural and Brain Sciences* 4: 515–526.

Takeshita H, Myowa-Yamakoshi M, and Hirata S (2009) The supine position of postnatal human infants: Implications for the development of cognitive intelligence. *Interaction Studies* 10: 252–269.

Collective Intelligence

S. C. Pratt, Arizona State University, Tempe, AZ, USA

Introduction

A collective intelligence is a group of agents that together act as a single cognitive unit. The iconic example is a swarm of honeybees cooperating to make decisions, build complex nest structures, allocate labor, and solve a host of other complex problems. Its defining characteristic is coordination without central control. Intelligence does not belong to a single knowledgeable leader, but instead is distributed across the entire group. Adaptive collective behavior emerges from interactions among a large number of individuals, each applying appropriate decisions rules to strictly local information. Although especially well-described in the social insects, collective intelligence is found in many systems, from the complex behavior of bacterial communities to the coordinated motion of fish schools and bird flocks. These examples have in turn inspired the development of artificial approaches to collective robotics and decentralized computational algorithms. This article reviews the major characteristics that allow collective intelligence to emerge from individual behavior, using both illustrative examples from well-studied cases and models that reveal the basic principles.

Self-Organization and Positive Feedback

A revealing example of collective intelligence is the forging of consensus decisions by colonies of the ant *Lasius niger.* These ants exploit sugary food using chemical trails that recruit nest-mates to rich sources. If a colony is presented with two artificial feeders filled with sugar water of different concentrations, it will soon develop a busy trail to the better one, largely ignoring the other (**Figure 1(a)**). This collective decision can be explained by a very simple model that does not require any ant to visit both feeders and determine which is better. Instead, the colony's choice emerges from the dynamics of recruitment behavior at each site. An ant that finds a feeder deposits a trail to it with a probability that depends on its richness. The trail summons nest-mates who reinforce it based on their own assessment, making it still more attractive to further recruits. Although recruitment happens at both feeders, quality-dependent reinforcement means faster growth for the trail to the better one. The difference is amplified as the stronger trail outcompetes the weaker one in attracting and retaining foragers. Eventually, the weaker trail dies out altogether, starved

of the reinforcement needed to overcome evaporation of the volatile trail pheromone.

This process illustrates several principles of collective intelligence. First, group-level order self-organizes from a large number of purely local interactions. Each ant applies appropriate decision rules to limited information about a single feeder or trail, and none has a synoptic picture of the whole problem. Second, control of foraging is highly decentralized, with no leader or hierarchy to guide the group's behavior. All ants are essentially identical in their behavioral rules and in their capacity to affect the behavior of others. Third, and most important, coordination depends on positive feedback. Small initial differences in trail strength are strongly amplified as each ant's reinforcement makes further reinforcement more likely. In this way, many small actions grow into a major group accomplishment.

Similar positive feedback underlies a broad range of complex collective behavior. Ants use pheromone trails to choose not only the better of two feeders, but also the shorter of two routes to the same feeder. Honeybees use another method of recruitment the dance language to allocate foragers among food sources or to choose a new home. For ants of the genus *Temnothorax*, tandem runs and social transports provide the recruitment needed for collective nest site selection. Positive feedback can also emerge without explicit signaling. It is sufficient for one animal to imitate the actions of another, thus becoming a model for still more imitators. In this way, a group of cockroaches can settle on a common aggregation site using only simple rules that make joining an aggregation more likely (and leaving it less likely) as its size increases.

Nonlinearity and Consensus

Although positive feedback is central to group coordination, mathematical models suggest that it is not enough for the clear decision-making shown by *L. niger* colonies. These models predict that consensus on a single option will happen only when feedback is highly nonlinear. That is, doubling a pheromone trail's strength must lead to more than a doubling of the rate at which it attracts recruits. Only then will differences between options be amplified sufficiently to eliminate all but a single trail. If the growth in trail attractiveness is linear, the best site will still be favored, but weaker recruitment will persist at lesser sites.

The consensus-building power of nonlinear responses shows very clearly when a colony of *L. niger* is presented with two identical feeders. The ants randomly choose one feeder and exploit it heavily while largely ignoring the other one (**Figure 1(b)**). The key to this consensus is amplification of random variation. If one feeder happens to be found first and thus gains a small advantage in number of visitors, nonlinear positive feedback will rapidly amplify this difference, allowing the early leader to monopolize the colony's foragers.

Nonlinear responses are called for whenever groups value consensus. Cockroaches, for example, may benefit from better predator defense and environmental homeostasis when they form a single large aggregation (**Figure 2(a)**). For *L. niger*, the benefit may be better defense of the honeydew-secreting homopterans that are a common natural sugar source. In other cases consensus is not ideal, and groups may do better with linear responses that produce split decisions. The honeybee's waggle dance is rather linear: if a bee doubles her dance effort, she

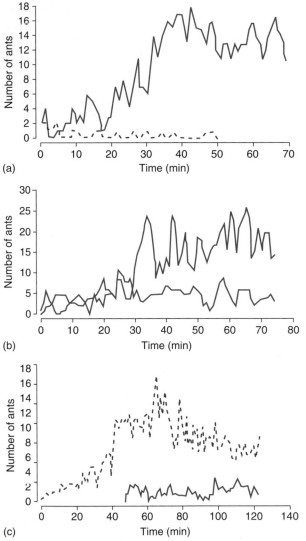

(a)

(b)

(c)

Figure 1 Collective decision-making by colonies of *Lasius niger* ants. (a) Change over time in the number of workers visiting two feeders, one with 1 M sucrose solution (solid line) and the other with 0.1 M solution (dashed line). (b) When presented with two identical 1-M feeders, the ants randomly choose one, largely ignoring the other. (c) If the 1-M feeder is presented after a trail is already established to a weaker feeder, the ants cannot switch their efforts to the better source. Adapted from Camazine S, Deneubourg JL, Franks NR, Sneyd J, Theraulaz G, and Bonabeau E (2001) *Self-Organization in Biological Systems*. Princeton, NJ: Princeton University Press.

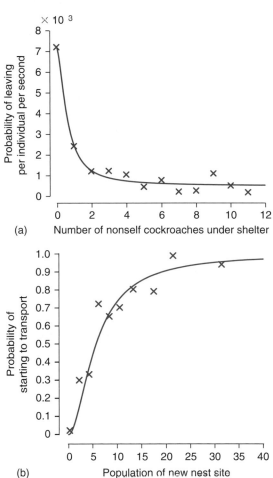

(a)

(b)

Figure 2 Nonlinear responses underlying collective behavior. (a) A cockroach's probability of leaving a shelter declines sharply as the number of roaches there increases. This nonlinear response helps a group of roaches reach consensus on one of two available shelters. Reproduced from Ame JM (2006) Collegial decision making based on social amplification leads to optimal group formation. *Proceedings of the National Academy of Sciences USA* 103: 5835–5840. (b) Nest site scouts of *Temnothorax curvispinosus* use a quorum rule when deciding whether to fully commit to a candidate site as their colony's new home. Crosses show the proportion of ants deciding to transport, rather than lead a tandem run, as a function of the population of the site being recruited to. Line shows a nonlinear function fit to these data. Reproduced from Pratt SC (2005) Behavioral mechanisms of collective nest-site choice by the ant *Temnothorax curvispinosus*. *Insectes Sociaux* 52: 383–392.

approximately doubles the number of recruits that she summons. Accordingly, a colony presented with two feeders exploits each at a level roughly proportional to its quality. This may allow colonies to respond more quickly if relative quality changes, a likely event for the ephemeral nectar flows on which bees depend.

On the other hand, consensus is critical for nest site selection by social insects, lest part of the sterile work force become separated from the reproductive queen. Interestingly, the two best-studied cases rely on linear forms of recruitment: honeybees use waggle dances and *Temnothorax* ants use tandem runs, in which recruits are led singly to a candidate site. Both groups are quite adept at reaching consensus on the best of several candidates, but how do they do so with these linear responses? The key appears to be re-introduction of nonlinearity in the form of a quorum rule (**Figure 2(b)**). Emigrations begin with a deliberative phase characterized by slow recruitment of scouts to multiple candidate sites. This gives way to rapid movement of the bulk of the colony to the first site whose population reaches a threshold. Models show that this nonlinear change in recruitment effectiveness increases the colony's likelihood of unanimously moving into the best nest, rather than splitting among several.

The Wisdom of Crowds

So far we have considered sociality as a constraint on intelligence: the group must reach a common solution despite its members' limited knowledge and influence. Theoretically, group living can offer a cognitive advantage, allowing many poor decision-makers to achieve greater accuracy than a well-informed individual. The basic insight was had by the Marquis de Condorcet in the eighteenth century. He described a jury of *n* members deciding between two options, each individual having probability *p* of making the correct decision. If each votes independently, with the group selection going to the option getting a majority of votes, then the probability that the jury's decision is correct rises with jury size, provided that $p > 0.5$. In other words, if everyone meets the rather low standard of exceeding a chance probability of being correct, the group as a whole can approach a 100% chance of making the right choice.

This 'wisdom of crowds' has many applications in human society, from democratic voting systems and jury trials, to prediction markets and internet search engines (Surowiecki, 2004). Similar advantages have been posited for animal groups, but few studies have been made. The best evidence is from experiments on different size groups of stickleback fish making movement decisions. Larger groups were better at choosing to follow the more attractive of two leaders.

Social enhancement of decision-making poses something of a paradox: it requires that group members influence each other, but also that each choice be independent. If individuals simply copy one another, then their mistakes become correlated rather than cancelling each other out. In humans, this is the problem of 'groupthink.' On the other hand, if everyone relies only on his own knowledge, then no one gains the benefit of others' wisdom. The solution lies in finding the proper balance of personal and social information. Foraging ants, for example, receive social information in the form of recruitment signals that bring them to options that others have found valuable. Once there, however, each one makes her own independent assessment before herself recruiting.

A simple model of collective choice suggests that striking the right balance is aided by the nonlinear responses described above. Consider a group in which each member chooses an option with a probability that depends both on its intrinsic quality and on the number of other group members that have already selected it. Inclusion of a social influence improves performance compared to purely independent decision-making, but this effect is much greater when the response to others is highly nonlinear. That is, accuracy increases if individuals follow the example of others only when their number exceeds a threshold. In this way, the group gains the advantage of pooled opinions without being misled by individual errors.

There are costs as well as benefits to the integrating power of nonlinearity. In a small proportion of cases, nearly all group members choose the wrong option, due to chance amplification of a few early mistakes. Experiments also indicate that very nonlinear recruitment systems restrict decision-making flexibility. When a colony of *L. niger* is given a high quality feeder after first developing a trail to a mediocre feeder, it is unable to switch its foraging to the better target (**Figure 1(c)**). The attractive power of the established trail is simply too great for a nascent trail to overcome. Honeybees faced with the same challenge can nimbly shift their foraging effort due to their more linear recruitment response.

Collective Motion

Some of the more spectacular examples of collective intelligence are seen in the acrobatic motions of fish schools and bird flocks, in which thousands of individuals execute rapid and near-simultaneous turns. The collective structures they form — parallel streams, spinning balls, toruses — may contribute to foraging efficiency or predator avoidance, or they may simply ensure that the group remains cohesive as it moves. Much of this coordinated behavior can be reproduced by self-propelled particle (SPP) models. In these models each animal chooses its direction and speed of motion based on two sources of information: (1) its own desired heading, perhaps guided by direct knowledge of the location of a food source, predator, or migration destination, and (2) the position and headings of its neighbors within the group. Cohesion is maintained by a policy of attraction to more distant group members, while collisions are avoided by turning away from neighbors who get too close. Common direction depends on alignment to others within a certain radius. Models of this type can account not only for cohesive movement, but also for the rapid transmission of changes in direction. The imitative behavior in these models plays the same role as recruitment in social foraging: it creates a positive feedback cascade that quickly spreads new information through the group.

Similar mechanisms may explain how a group can find its way to a destination known to only a few of its members. A honeybee swarm flies unerringly to its new home even though only 5% of its several thousand members know the location. Scouts appear to guide the ignorant majority by flying through the swarm at high velocity in the direction of the target site. An SPP model shows how these streakers could plausibly guide the swarm, if uninformed bees follow simple rules for avoidance at close distances and attraction and alignment at longer distances. A more general model of this type was developed by Couzin and colleagues. It shows that guidance is possible even when knowledgeable individuals fly at the same velocity as others and balance their directed flight with the same kind of social information used by uninformed bees. This model does not assume that group members can tell who is informed, but nonetheless predicts effective navigation toward the goal.

Decision-Making on the Move

What if the group contains two kinds of knowledgeable individuals, each with a different preferred heading? Models suggest that the outcome depends on the number favoring each direction, with even a small majority able to win over the whole group. If numbers are similar and desired headings are not too different, the group is expected to compromise on an intermediate direction. Above a critical difference in headings, the group either splits in two or reaches consensus on one of the two preferred directions (**Figure 3(a)**). In case of consensus, the choice of heading is random, unless there is a difference in how much each group weighs its preferred direction relative to social information. Even a small advantage in motivation will increase a subgroup's power to win over the whole group. This result suggests that leadership by a small number of individuals can have major influences on the behavior of otherwise self-organized groups.

Some support for these models is found in the observed behavior of homing pigeons traveling either alone or in pairs from a common release site (**Figure 3(b)**). After multiple solo flights, each pigeon develops an idiosyncratic route. When traveling together, pairs show three possible outcomes: they separate and follow their individual routes, they compromise on an average route, or they both adopt the preferred route of one bird. Pairs that stay together show the distance dependency predicted in the model: compromise at short distances and selection of one bird's preference at larger ones. Leadership was also evident, with some birds consistently more likely than others to prevail, perhaps on the basis of their higher social status.

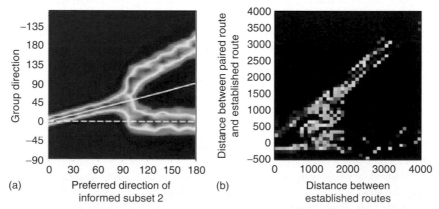

Figure 3 Guidance and decision-making in collective motion. (a) Results of a model showing that a small number of informed individuals can guide a large group. The model assumes two subsets of five informed group members with different desired headings, plus 90 uninformed members. When the difference in headings is small, the whole group adopts the average heading. Above a critical difference, the group reaches consensus on one heading or the other, with equal probability. Reproduced from Couzin ID, Krause J, Franks NR, and Levin SA (2005) Effective leadership and decision-making in animal groups on the move. *Nature* 433: 513–516. (b) Pairs of homing pigeons show a similar switch from compromise to leadership. When the difference between their preferred routes is small, the birds take an intermediate route. When the difference surpasses a threshold, the chosen route is nearly identical to that of one bird, and very different from that of the other. Reproduced from Biro D, Sumpter DJT, Meade J, and Guilford T (2006) From compromise to leadership in pigeon homing. *Current Biology* 16: 2123–2128. In each graph, color indicates the probability p of each group direction as a function of the difference in subset headings, ranging from blue ($p = 0.0$) to red ($p = 1.0$).

Pairs of pigeons also took shorter paths home than did solo birds, providing some evidence for the 'many wrongs' hypothesis. This idea holds that many individuals, each with a noisy estimate of the heading to a common destination, can average out their independent errors to find a much more precise group heading. In addition to improving navigation, models suggest that this kind of process can help animals follow subtle gradients toward food concentrations or better physical environments.

Phase Transitions

The switch from compromise to consensus in collective motion is one example of a phase transition, a central feature of nonlinear collective systems. Phase transitions are dramatic changes in behavior in response to continuous variation of a single key feature. Another example is seen in the trail-laying ant *Monomorium pharaonis*. Below a critical number of foragers, a colony cannot sustain a trail; above this number the positive feedback of trail reinforcement is strong enough to overcome evaporative losses. Thus, a group switches suddenly from solitary to group foraging as its size increases.

Similar sensitivity to group size is seen in the collective movements of gregarious locusts. Massive flying swarms of adult locusts are preceded by the assembly of younger insects into ordered marching bands. An SPP model shows how these bands can self-organize from each individual's tendency to align with nearby members, combined with a competing tendency to maintain its current heading. As their density increases, the locusts undergo a phase transition from random movement to ordered motion at a common direction and speed. For intermediate densities, direction frequently changes, but at higher densities it remains constant for long periods. Very similar phase transitions are in fact observed in groups of walking locusts confined to a ring-shaped arena.

Phase transitions may have significant impacts on a group's ability to match group behavior to changing circumstances. Near a phase transition, nonlinear systems have multiple stable states and can shift relatively easily from one to another, as demonstrated by the locusts' directional changes. This implies that adaptive selection of the best behavior may be easiest near the transition. For example, a single individual that detects a predator can more easily lead the group away from danger if it is near a phase transition.

Hysteresis and Group Memory

The existence of multiple stable states makes collective behavior dependent on a group's recent history. This characteristic of nonlinear systems is known as hysteresis.

Medium-sized groups of the ant *M. pharaonis*, for example, can forage either solitarily or with trails, depending on how the colony reached its current size. If reduced from a size at which a trail formation is easy, then the medium-sized group can maintain a trail already formed. If expanded from a size at which trail formation is impossible, the medium-sized group will not be able to build one from scratch. Thus, two otherwise identical groups can show very different behavior, based on their different histories.

Hysteresis creates a kind of group memory. Fish schools, for example, can form a remarkable variety of collective structures, ranging from disordered swarms, to parallel revolution about a central point, to cohesive directed motion. Mathematical models show that transitions from one structure to another can be achieved by simply changing the spatial range over which individual fish attempt to align themselves with their neighbors. For any given range, however, more than one structure may be stable. When fish in a disordered swarm increase their zone of alignment to a moderate value, they begin to swim in an orderly torus. When fish in a mobile, directed group decrease their zone of alignment to the same moderate value, they remain in their directed structure. Thus, the group's 'memory' of its former state determines its behavior. It is important to note that this is not an individual memory. Each fish follows precisely the same rules in the two conditions, and the difference in group behavior is an emergent property of the whole group.

Comparison to Neural Systems

The idea of collective intelligence is born from a fundamental analogy between societies and brains. Similar principles of feedback, nonlinearity, and multistability apply in both cases. Although the mechanistic details are vastly different, the structural similarity is sometimes very striking. For example, the house-hunting algorithms of ants and bees have great similarity to models of decision-making in the primate brain. Both assume competing streams of noisy evidence for different options borne either by sensory neurons or the recruitment behavior of scouts. This evidence accumulates, either as activity within a particular neural center or as a population of scouts advertising and visiting a nest site. A decision is made when the activity or population for one option surpasses a threshold, marking it as the chosen option.

Marshall and his colleagues made this loose analogy more rigorous by expressing both systems in the same modeling framework. Their results suggest at least one important functional difference: the neural system can achieve a statistically optimal tradeoff between decision speed and accuracy, because it includes mutually inhibitory connections between competing centers. No such connections are currently known for the ants or bees. However, the honeybee stop signal is an inhibitory behavior that conceivably plays such a role, leaving open the possibility that bees too can achieve statistical optimality.

Comparison to neural models underscores an important common feature of brains and societies: the role of forgetting. Neural decision systems depend on a steady loss of activity in the absence of new external input. This improves sensitivity to changes in the strength of a stimulus. In the same way, a honeybee scout eventually stops dancing for a candidate nest site that she has found, no matter how good it is. This helps the colony to avoid stalemates in which bees obstinately advertise more than one site. It also allows the colony to switch its attention from an early mediocre discovery to a better site found later.

Rationality

One might expect that an intelligent decision-maker would also be a rational one, but this is not always the case. When faced with certain kinds of challenging decision problem, animals and humans are likely to make errors that can prevent them from consistently maximizing their fitness. For example, decision-makers will change their preference between two options if a third, less attractive distracter option is also presented. These errors often occur when options vary in multiple attributes, such that no option is clearly superior in all attributes. This makes determining which is best a computationally challenging task. Individuals can deal with this by using simple rules of thumb based on local comparisons among options. Thus, if it is hard to say whether A or B is better, A may be chosen over B if it more clearly exceeds a distracter C than does B. Such rules may work well most of the time, but fail for particularly challenging cases.

Rationality has only begun to be addressed for collective decision-makers, but early work by Pratt and his colleagues suggests that collectives may be less prone to this kind of comparative error. They presented *Temnothorax* ant colonies with a choice that required them to trade off two prized features of nest sites – entrance size and light level. Colonies did not show the irrational changes in preference commonly seen when individual animals face a similar choice. An intriguing possibility is that the ants' highly distributed decision-making filters out irrational errors. Few individual ants know of all the options under consideration by the colony, and thus do not have the opportunity to make the comparisons that bring about irrationality. Thus, an apparent constraint – the relative ignorance of individual ants – may help the colony as a whole to perform better.

Applications

In recent years, swarm intelligence has proven a fertile source of inspiration for the design of artificial systems. In computer science, 'ant algorithms' provide an effective means of solving the hardest kind of optimization problems, where the total number of possible solutions is far too great for all to be tested. The basic idea, inspired by ant foraging, is to let distinct computational agents ('ants') sample the solution space, score the quality of each sample, and 'recruit' other agents to test variations of promising leads. This general approach has been used to design telecommunication networks and to schedule complex transportation routes. In robotics, ongoing research aims to design swarms of robots that can inspect dangerous and inaccessible places, efficiently monitor large areas, or build structures in remote and dangerous locations.

Engineers are attracted to several advantages of natural collectives. They are highly robust, working well if individual members are lost or if communication channels are broken. Cognitive sophistication is a feature of the whole group, not each member, so individual agents can be simple and cheap. Collectives work well at different population sizes without requiring wholly different control algorithms. They are also effective in the variable environments typical of real-world problems. In fact, randomness is an important component of natural collective intelligence, as when ant trails amplify random variation to select a single food source. Finally, swarms do not require unwieldy central control networks that can be extremely difficult to design and manage for large and complex systems.

In designing artificial systems, engineers often stray far from the original biological inspiration, as their goal is to solve a problem, not to mimic a natural system. Nonetheless, natural models are still crucial to the process, if only as a proof that solutions to certain difficult problems are attainable. In addition, the work of engineers and computer scientists can enhance the study of natural collectives, by providing useful analytical tools and concepts. Indeed, much of the work described in this article is a kind of reverse engineering, looking for the hidden mechanisms that explain complex collective behavior. Future discoveries will depend on the exchange of insights between engineers and biologists about both natural and artificial collective intelligence.

See also: Communication Networks; Consensus Decisions; Decision-Making: Foraging; Distributed Cognition; Group Movement; Honeybees; Insect Social Learning; Nest Site Choice in Social Insects; Rational Choice Behavior: Definitions and Evidence; Social Information Use.

Further Reading

Bonabeau E, Dorigo M, and Theraulaz G (1999) *Swarm Intellligence: From Natural to Artificial Systems*. New York: Oxford University Press.

Camazine S, Deneubourg JL, Franks NR, Sneyd J, Theraulaz G, and Bonabeau E (2001) *Self-Organization in Biological Systems*. Princeton, NJ: Princeton University Press.

Conradt L and List C (2009) Theme issue: Group decision making in humans and animals. *Philosophical Transactions of the Royal Society B* 364: 719–852.

Detrain C and Deneubourg JL (2008) Collective decision-making and foraging patterns in ants and honeybees. *Advances in Insect Physiology* 35: 123–173.

Edwards, SC and Pratt, SC (2009) Rationality in collective decision-making by ants. *Proceedings of the Royal Society of London B Biological Sciences* 276: 3655–3661.

Hölldobler B and Wilson EO (2008) *The Superorganism*. New York: Norton.

Marshall JAR, Bogacz R, Dornhaus A, Planqué R, Kovacs T, Franks NR (2009) On optimal decision-making in brains and social insect colonies. *Journal of the Royal Society Interface* 6: 1065–1074.

Seeley TD (1995) *The Wisdom of the Hive*. Cambridge, MA: Belknap Press of Harvard University Press.

Sumpter DJT (2009) *Collective Animal Behavior*. Princeton, NJ: Princeton University Press.

Surowiecki J (2004) *The Wisdom of Crowds*. New York: Doubleday.

Colony Founding in Social Insects

J. C. Choe, Ewha Womans University, Seoul, Korea

Introduction

From our anthropocentric perspective, we tend to think that the world is full of social or group-living animals, like human societies. On the contrary, the majority of species, at least in the animal kingdom, are solitary in their lifestyle. And yet, undeniably, some of the most ecologically successful animals spend much or all of their lives in organized social groups called colonies. According to the definition given by E. O. Wilson in his 1971 book *The Insect Societies*, a colony is 'a group of individuals, other than a single mated pair, which constructs nests or rears offspring in a cooperative manner.' Insect colonies, in particular, have long fascinated biologists because the members of a colony, however distinct they are as individuals, often act like a single organism, and are hence called superorganisms. Such a society exhibits features of organization and function analogous to the physiological properties of a multicellular organism.

In this article, I will limit my discussion to social insects because their colonies are most intricately organized and thus most rigorously investigated of all animal species with an exception of *Homo sapiens*. The term 'social' has a very broad and not easily delineated boundary. Here, I use the categorization scheme for insect social systems (**Table 1**) modified from the one proposed in the book by Choe and Crespi, *The Evolution of Social Behavior in Insects and Arachnids*. According to the Crespi–Choe categorization, eusocial systems exhibit qualities such as parental or biparental care, shared breeding site, cooperation in brood care, alloparental care, and irreversible caste formation. The following taxa are considered eusocial: essentially all species of ants (Formicidae) and termites (Isoptera); other hymenopteran families/subfamilies of wasps and bees – Stenogastrinae, Sphecidae, Vespidae, Apidae, Anthophoridae, and Halictidae contain eusocial species; and thrips (Thysanoptera) and aphids (Aphididae) with specialized 'resource defenders,' and ambrosia beetles (Scolytinae and Platypodinae). Such a broad definition of eusociality permits more extensive comparative tests for the origin and evolution of sociality. Research findings for insects falling within this broad definition of eusociality are the subjects of this review.

A colony of organisms has a lifecycle much like the cycle an individual organism goes through. George Oster and Edward O. Wilson conveniently divided the colony cycle into three stages – *founding*, *ergonomic*, and *reproductive*. The founding stage is a critically important phase in

that the risk of mortality is the highest in the life of a colony. High mortality during this phase of life history has led to adaptations for colony founding that are diverse as for the reproduction patterns of individual organisms. The process of colony founding provides a unique window of opportunity to test a variety of models for the origin and maintenance of eusociality, because individual reproductives may theoretically choose among various reproductive options.

Honeybees (*Apis* spp.) and some species of ants, such as Argentine ants (*Iridomyrmex humilis*), pharaoh ants (*Monomorium pharaonis*), and army ants (*Eciton* spp.), produce new colonies by the breakaway of a group of colony members from a mature colony. This process resembles vegetative propagation in some plants and various modes of asexual reproduction in single-celled organisms and invertebrate animals. In the majority of social insects, however, new colonies are founded independently, without the help of workers. Reproductives of social insects have the options of either initiating a new colony from scratch or taking over an existing young colony, which could be either a conspecific or another species. Unlike *coups d'état* in human society, usurpers of social insects work alone and thus do not derive any somatic investment from the mother colony. No 'helper' members of the original colony accompany reproductives.

In their highly influential 1977 paper, Bert Hölldobler and E. O. Wilson drew up a comprehensive diagram illustrating the possible routes of colony foundation and maturation in social insects. **Figure 1** presents a modification of their scheme that concentrates on the founding stage and includes colony usurpation as a legitimate founding mode. In this classification scheme, all modes of colony founding can be sorted into one of two categories that depend upon whether reproductives initiate colonies by themselves or are accompanied by workers. *Independent colony founding* involves the initiation of a new colony by reproductives without the aid of workers, while colony founding by budding and fission are grouped as *dependent colony founding*. Independent colony founding is further divided into two modes – usurping existing colonies or creating anew.

Independent Colony Founding

Among all the modes of colony founding, independent founding requires the least amount of investment by the

Table 1 Types of insect social systems

Type of society	Brood care	Shared breeding site	Cooperative brood care	Alloparental brood care	Castes
Subsocial	+				
Colonial	−	+			
Communal	+	+	+		
Cooperative breeding	+	+	+	+	
Eusocial	+	+	+	+	+

This categorization scheme modifies the one proposed by Crespi and Choe (1997). Here, brood care includes both parental and biparental care. Shared breeding sites are those that involve multiple females. Whether they cooperate in brood care or not is an important parameter in the evolution of sociality. Alloparental brood care refers to the presence of behaviorally distinct groups, with individuals specializing to be reproductives or helpers to those who reproduce. Following the definition given by Crespi and Yanega (1995), castes refer to 'groups of individuals that become irreversibly behaviorally distinct at some point prior to reproductive maturity.' This scheme allows the broadest possible taxonomic scale of comparative tests for explanation of the genetic, phenotypic, and ecological causes of variation in social systems.

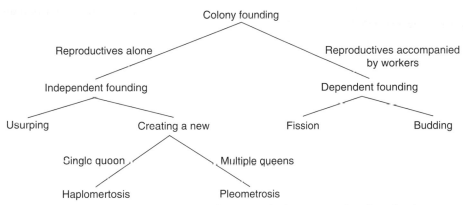

Figure 1 Modes of colony founding in social insects. Depending upon whether or not colony founding is accompanied by workers, it can be divided into dependent and independent founding. Dependent founding occurs in the form of either fission or budding. Independent founding has two types as well. The majority of social insect queens found a new colony singly (haplometrosis) or in group (pleometrosis). A sizeable minority usurp already established colonies of the same or different species.

mother colony. Inseminated females initiate new colonies alone or in a group. Unlike in budding and fission, the mother colony makes little investment as no workers accompany founding queens. This means that the colony of independently founding social insects passes through a solitary phase, however short it may be. This is curiously analogous to the reproductive cycle of multicellular organisms when they go through a single-celled haploid phase of sperm or egg. Under some ecological conditions, natural selection should favor independent founding, because independent reproductives can cover greater distances or wider areas than the ones escorted by workers. Independent founding is observed in a majority of ants and termites, halictine sweat bees, bumble bees (*Bombus*), hornets and yellow jackets (Vespinae), paper wasps (*Polistes, Parapolybia, Mischocyttarus,* and *Ropalidia*) as well as social aphids, thrips, and beetles.

Independently founding queens of social insects have the option of joining with other queens as a cofoundress in addition to starting one of their own. Although the majority of newly initiated colonies are founded by single queens (*haplometrosis*), a significant minority are founded by multiple queens (*pleometrosis*). Foundress associations during the colony founding stage are observed in a number of social insect groups, including ants, bees, wasps, termites, aphids, and thrips. Compared to the early recognition of worker sterility or queen–worker conflicts as a critically important topic in the evolution of insect sociality, students of social insects had not paid much attention to queen–queen conflicts until the mid-1980s, with a possible exception of wasp biologists, most notably, Mary Jane West-Eberhard. For the past two decades, however, facultative pleometrosis has been investigated with greater rigor, because these systems provide opportunities to analyze the relative contributions of genetic and ecological factors to the evolution of cooperation.

As far as the kin structure of colonies is concerned, it is interesting that cofoundresses are typically unrelated in ants and some bees. An exception to this 'rule' is *Lasius pallitarsis*, in which solitary and kin-pair queens produce workers, whereas nonkin pairs of queens do not. In eusocial wasps and bees, however, cofounding queens are generally close relatives. Transactional skew theory views foundress associations as a form of social contract that guarantees mutual benefits to both the dominant and subordinate foundresses. Skew models predict more or less evenly shared reproduction among genetically unrelated cofoundresses, because a greater incentive to help is required of nonrelatives. Indeed, low reproductive skew is reported among cofoundresses of ants and some bees. In contrast, reproductive outputs are highly skewed in foundress associations of eusocial wasps and many bees, in which per capita brood production generally declines as the number of cofoundresses increases. Like the counter-example found in ants, exceptions to the rule have also been discovered in paper wasps (*Polistes* spp.), tropical hover wasps (*Parischnogaster mellyi*), and social thrips *Dunatothrips aneurae*. The philopatric nature of foundresses naturally facilitates cofounding by nestmates from the previous year. In these insects, however, DNA microsatellite analyses revealed that their social organizations are much more diverse than previously assumed. Reproduction is often shared between related and nonrelated foundresses in some nests.

Evolution of Claustral Founding

Independent foundresses may be claustral, in which they confine themselves within the nests until the first brood of workers ecloses. Alternatively, they may be nonclaustral or semiclaustral, so that they forage outside of the nests while raising the first brood. In bees and wasps in which reproductives retain their wings throughout their lives and nests are essentially open all the time, the basal mode of colony establishment is independent, nonclaustral founding. Founding queens forage outside the nest to bring food to rear the brood. The dominant of a cofoundress association forces subordinate foundresses to forage, in which case she becomes functionally claustral but still does not exhaust her bodily resources like ant and termite queens. Foundresses of social gall aphids and thrips are also basically claustral, but have no need to forage outside or drain off their bodily energy reserve because their nests contain feeding sites. Founding females remain within the gall (or other types of domiciles) feeding on plant cell contents and producing offspring that develop to adulthood inside. They are so-called 'fortress defenders' in the terminology developed by David Queller and Joan Strassmann in their 1998 paper, exhibiting a form of resource-based sociality. When sites for shelter and food coincide, selective pressure is strong for the defense of expansible and food-rich nest sites. Indeed, this favored the evolution of morphologically specialized defenders or soldiers in both gall aphids and thrips.

Claustral colony founding is most common among ants. The ability to metabolize wing muscles is a preadaptation to the evolution of claustral founding. Claustral founding is also characterized by a suite of other physiological and behavioral adaptations, such as wing dropping after nuptial flight, the ability to pack away large amount of body fat, and ingluvial feeding, that is, feeding larvae on metabolized nutritional reserves. Mortality risk during foraging for founding queens is quite obviously the selective pressure that has led the evolution of claustral founding. The so-called 'higher' ants in the subfamilies Formicinae and Myrmicinae typically found new colonies claustrally, whereas semiclaustral founding is rather prevalent in the morphologically and socially more primitive ants of Paraponinae, Ponerinae, Myrmeciinae, and Nothomyrmecinae. This suggests that claustral colony founding is a derived state, while semiclaustral founding is an ancestral character in the evolution of colony-founding modes.

Contrary to the previous belief, however, recent studies revealed that semiclaustral colony founding occur at least in 28 species of 12 genera of myrmicine and 2 genera of formicine ants. Semiclaustral founding is obligate for foundresses of fungus-growing ants, *Atta* and *Acromyrmex*, because their fungus gardens would not survive without a constant supply of fresh leaves. Because they lack sufficient stored proteins for claustral founding, foundresses of *Messor andrei* also appear to be obligately semiclaustral. At least one species of harvester ants, *Pogonomyrmex californicus*, founds new colonies semiclaustrally in the laboratory setting. For most semiclaustral ants, their colony foundation habit seems facultative. Environmental variability may allow foundresses to adopt semiclaustral founding as a bet-hedging strategy.

Although claustral queens may benefit from much reduced predation, they must rely on the histolysis of their no longer needed wing muscles and stored fat bodies to supply the energy resources required to rear the first brood of workers. Thus, claustral foundresses can be called *capital breeders*, whereas nonclaustral or semiclaustral foundresses act like *income breeders*. Depending on the size of the capital, an obligately claustral foundress suffers a burden of trade-off between the number and size of brood. Despite a higher risk of mortality, a semiclaustral foundress may be able to practice a strategy to maximize the number and/or size of offspring within the capacity of a given environment. A colony founded by a capital breeder begins as a society of closed economy and then switches to an open economy once the workers start provisioning resources from the outside. A claustral breeder must budget her capital, that is, bodily resources, so that they are not exhausted before producing a sufficient worker force. Time and efficiency are the names of the claustral game.

Evolution of Foundress Associations

Among ants, pleometrosis occurs mostly in the 'higher' subfamilies, namely, Formicinae, Myrmicinae, and

Dolichoderinae, with possible occasional exceptions in Ponerinae and Nothomyrmecinae. Ants of the three subfamilies have evolved remarkably similar sets of modifications and variations on pleometrotic colony development. Cofoundress associations have also been observed among wasps of eusocial Polistinae, Stegogastrinae, and Sphecidae. All observed cases of pleometrosis among bees come from Halictini, Augochlorini, Ceratinini, and Allodapini. Among termites, soil-nesting species are more frequently pleometrotic than wood-dwelling species.

Cofounding has its own costs. Per capita reproductive output is likely to decrease, competition over resources increases, and close proximity can make the transmission of parasites and pathogens easy. Despite all these potential costs, however, pleometrosis is widespread, occurring in nearly all eusocial insect groups. This means that there must be sufficient benefits for cofounding counterbalancing the costs. A list, not necessarily mutually exclusive, of the advantages of pleometrosis suggested thus far is:

1. Production of the first worker cohort much larger and/or faster

In a number of species, colonies founded by multiple queens produce greater worker forces in the first brood, and in some cases, do so in less time. Producing a larger worker force in a given time or more quickly accrues a host of benefits in terms of colony defense, brood raiding, and foraging success. Pleometrosis appears to be an adaptation for intercolonial competition. Better success in defense, brood raiding, and foraging give multiple-queen colonies a competitive edge among incipient colonies. Colonies initiated by large numbers of queens better resist usurpation attempts. Larger or more rapidly formed worker forces can eliminate neighboring incipient colonies, but the importance of brood raiding in the colony-founding stage is debatable. The majority of brood-raiding observations are from laboratory studies and its prevalence in the field awaits further research. Brood raiding occurs frequently and is an important ecological parameter in *Solenopsis invicta*, an invasive species with an extremely high density of incipient colonies. On the other hand, a carefully planned field study on *Messor pergandei* turned up no evidence of brood raiding occurring in the field. Instead, increased foraging success by larger worker forces appears to clearly enhance colony survival and growth. *Cecropia*-nesting *Azteca* colonies with multiple queens produce more workers in less time than those with solitary queens, and those workers which chew their way out of the internode first can monopolize the supply of glycogen-rich Mullerian bodies secreted by *Cecropia*. The outcome of this scramble competition is the starvation of all other incipient colonies inhabiting the same *Cecropia* sapling. When many incipient colonies are clumped in an area, early brood production is

particularly important, given the probability of intercolonial brood raids and/or intense competition for limited food resources.

2. Increased survival of foundresses

To determine whether cofoundresses on average have higher survivorship than solitary foundresses, one has to analyze queen mortality due to intracolonial competition and intercolonial competition separately. Although enhanced survivorship may be a major benefit for foundress associations, most studies of pleometrosis have measured only mortality from intracolonial competition; the relationship between the number of cofoundresses and mortality is not always significant. Once intercolonial mortality caused by interference or scramble competition is added to the analysis, however, it becomes obvious why pleometrosis is favored over haplometrosis. Even though an individual foundress' probability of becoming the ultimate survivor of a foundress association decreases as the number of foundresses increases, the selection still favors pleometrotic associations because of the decisive advantages of larger worker forces during competition among incipient colonies.

3. Earlier maturation to the reproductive stage

Colonies with larger and more rapidly produced worker forces tend to survive better and produce reproductive offspring earlier than the ones with small or slowly produced worker forces. Enhanced production of workers may increase an individual foundress' reproductive output. Although a haplometrotic queen does not suffer from intracolonial competition, the pressures from intercolonial competition may be so great that joining other queens in colony founding may be the only way a queen can survive. Despite the heavy odds against surviving to become the ultimate reproducing queen in a pleometrotic colony, a queen may have a better overall chance of reaching the reproductive stage if she begins a colony with other foundresses than if she founds alone.

4. Better protection from parasites and predators

In the case of nonclaustral and semiclaustral founding, unattended nests and broods experience higher incidences of attacks from natural enemies. In *Polites* and *Mischocyttarus* wasps, foundress associations greatly reduce the rate of nest destruction by conspecifics and predators. Colonies started by solitary foundresses tend to leave their nests unattended for longer periods of time while foundresses are out foraging. Frequent switching and continuous guarding by communally nesting sweat bees effectively protect their nests from the attacks of kleptoparasites. The social thrips *D. anuerae* also has better protection against kleptoparasites than solitary thrips. Although multiple foundresses confined to close proximity may facilitate easier transmission of parasites and pathogens, cofoundresses can groom each other, which may reduce the mortality caused by fungal attacks.

5. Reduction of costs of nest construction and maintenance

Joint colony founding can also yield more immediate and direct benefit in terms of construction, defense, and repair of nests. Cofounding queens of several ant species cooperate in nest excavation and those of wasps and bees cooperate in building up the nest during the founding stage. In *S. invicta*, queens attempting to usurp other colonies are less successful when attacking a three-queen colony than a one-queen colony regardless of how many workers are present. Cooperating wasp cofoundresses also defend their nests from usurpers better and repair damaged nest more rapidly.

6. Assured fitness returns

Raghavendra Gadagkar argues that the subordinate cofoundresses can have the advantage of assured fitness returns by joining a foundress association. A foundress has no fitness gain if she chooses to start a colony alone but dies before her offspring reach the age of independence, but she is assured some fitness returns if she joins a pleometrotic colony and even assumes the role of helpers. Even if she cares for some larvae during their early stages and dies long before they grow to become independent, some other helpers are likely to care for the same larvae and bring them to independence. This model is most applicable to a life cycle of the sort exhibited by polistine wasps in which larvae are progressively provisioned and pupae need continuous protection by adults.

Cofoundress associations are dynamic systems of conflict and cooperation. The propensity for cooperation varies, and diverse ecological conditions may favor cooperative colony founding. Individual foundresses in a pleometrotic colony must contribute enough to enable their colony to outcompete neighboring incipient colonies, but at the same time, they must also carefully budget their energy reserves in order to outcompete fellow cofoundresses for the eventual ownership of the colony. They cooperate to survive during the early phase of the founding stage, but when the first workers eclose, they begin fighting fiercely, typically to the death. It is unclear whether workers actively choose the surviving queen and kill off other queens or if they passively eliminate already injured and less competitive queens. Cofounding queens of *Lasius niger* differ with respect to their oviposition rates and workers preferentially feed more fecund queens. Therefore, selection may favor queens that balance their energy budgets well and maintain good body condition until worker emergence.

Primary polygyny or extended coexistence of pleometrotic queens to the reproductive stage is extremely rare.

The transition from closed to open economic system provides a major explanation for the transition from pleometrosis to secondary monogyny. In claustral multiple queen associations, all available energy to rear the first brood comes from the stored bodily reserves of cofoundresses and there is a good reason why they must be cooperative. When the first workers start bringing food from outside, however, the colony's economy switches to an open system and cooperation among foundresses begins to collapse. This explanation is not complete because in most *Polistes* wasps, which are clearly nonclaustral, cofounding females stay together well into the reproductive stage, but a single female usually lays most or all of the reproductive-destined eggs. Primary polygyny has also been observed in semiclaustral *Acromyrmex* ants.

Unisexual founding in termites

Unisexual colony founding by female reproductives is the norm in the hymenopteran social insects but is an exception in termites. Colony founding in termites typically involves both sexes. In the case of pleometrotic founding, discussions often center around the adaptive significance of polygamous mating (mating between a single male with multiple females) in comparison to the usual monogamous mating. In the subterranean termite, *Reticulitermes speratus*, female reproductives which fail to mate sometimes found colonies alone or in female–female pairs. Foundresses without males reproduce through thelytokous automictic parthenogenesis. If a male is introduced to a female pair in the laboratory, only one female survives. If a partner male is absent, two females, but never more than two, found a colony in which they produce the first brood of workers in a cooperative manner. Recently, Kenji Matsuura and his colleagues discovered that secondary neotenic queens of *R. speratus* are produced almost entirely parthenogenetically by the founding primary queens, while workers and winged reproductives are produced by normal sexual reproduction.

Mixed-species pleometrosis in *Azteca* ants

Among all the observations of colony founding in social insects, the most enigmatic is cooperative founding by queens of two different species of *Azteca* ants. Dan Perlman conducted an extensive field study on the colony-founding processes of *Cecropia*-nesting *Azteca* ants in Monteverde, Costa Rica. Several species of *Cecropia* are myrmecophytes, obligately inhabited by *Azteca* ants. In Monteverde, *Cecropia* saplings (mostly *C. obtusifolia*) are occupied by two sympatric *Azteca* species, *A. constructor* and *A. xanthacroa*. *Azteca* colonies are founded in three distinct ways: by single queens, by single-species groups of queens, and by mixed-species groups of queens. Pleometrotic colonies of both single species and mixed species outperform haplometrotic colonies by producing larger first worker cohorts, and by eventually taking control of the entire tree through the monopolization of the main food source, Müllerian bodies grown by *Cecropia*. Field observations revealed that mixed-species foundress associations are nearly as cooperative and successful as their single-species counterparts throughout the founding stage. This means that mutualism is as important a selective factor as inclusive fitness for the evolution of pleometrosis.

Soon after landing on a *Cecropia* sapling, a newly inseminated queen explores the young upper internodes of the plant for a while, sheds her wings, and begins to chew into a thinner, unvascularized area in the wall of one of the uppermost internodes. *Cecropia*-nesting *Azteca* queens appear to settle in the first tree upon which they land, but once the tree is chosen, they appear to examine and choose among internodes. It is reasonable to presume that the queens can discriminate a recently occupied internode from an unoccupied one based on the presence of the loosely plugged entry hole. On the other hand, it may be rather difficult for late-arriving queens to estimate the number of foundresses that have already joined the association before entering the internode.

A queen can take as long as 2 h to push herself through the hole into the plant. She then plugs the hole with parenchymal tissue scraped from the inner wall of the internode. A late-arriving queen has options: she could choose to make a hole to enter an unoccupied internode or she could join other queens in already-excavated internodes. If she chooses the latter option, it takes only 5–12 min, thus saving energy and reducing exposure time to predation risk. A mature *Cecropia* tree is invariably occupied by a single colony with a single queen. Selection to produce a large worker force as quickly as possible to outcompete neighboring colonies in the same tree leads not only to foundress associations but also to the evolution of this remarkable case of interspecific cooperation in colony founding. Mixed-species colonies are quite common. One out of five queens is engaged in mixed-species foundress associations. All cofounding queens contribute to the production of workers whether they belong to single-species or mixed-species colonies. Like other pleometrotic species, *Azteca* queens start fighting against one another regardless of species identity, once the workers begin foraging and stockpiling Müllerian bodies. Colony founding in *Azteca* ants offers an ultimate testing arena for investigating the relative importance of genetic and ecological factors in the evolution of sociality and further studies will surely turn up many more exciting new discoveries.

Colony Founding by Usurpation

Queens that failed in attempts to found new colonies independently or that are subordinates in multiple-queen colonies can employ the alternative founding strategies of usurping established colonies or adopting orphaned colonies. Attempting to start a new colony when the season is well underway is hardly a viable option. Taking over nest, brood, and/or workers produced by other queens is clearly parasitism and therefore colony foundation by usurpation or adoption is called social parasitism. Such parasitism occurs within species as well as between species and the usurping queens taking possession of a colony of a different species are called 'inquilines.'

In wasps, usurpers typically kill eggs and early-instar larvae of the colony they have taken over. The usurper lays eggs while letting later-instar host larvae and pupae complete development so that they can be forced to provide care for the brood of the usurper. The success rate of usurpation is sufficiently high that selection could favor this as a primary founding strategy. Usurpers can save the energy required for founding a new colony and instead sit-and-wait to take advantage of another queen's investment. Special adaptations of more extreme social parasites include enlarged heads and mandibles of usurping queens as well as evolutionary loss of the worker caste. Social parasites exhibit a wide spectrum of integration with their host colonies. For instance, parasitic queens of the 'ultimate' inquiline ant *Teleutomyrmex schneideri* harmoniously coexist with the host ants of *Tetramorium* by allowing the host queen to continue to lay worker-destined eggs while the parasite produces reproductive offspring.

Dependent Colony Founding by Fission and Budding

At the opposite side of the colony-founding continuum, relatively large colonies of some species propagate vegetatively by forming a swarm that consists of reproductives and a sizeable worker force. This mode of colony founding has been called swarm founding, or more specifically, hesmosis in ant literature and sociotomy in termite literature. But the term 'swarm' is also used to refer the mating swarm of reproductives during nuptial flight and this dual usage creates occasional confusion with the colony-founding swarms. Dependent colony founding may be more accurately described as either fission or budding.

Colony fission occurs when a monogynous colony first produces a clutch of reproductives and then divides into two daughter colonies of roughly equal sizes. When daughter colonies are led by new queens, their worker forces are more or less even. When one of the daughter colonies retains the old queen, however, its sister colony with a new queen is often smaller in size. Colony fission is the norm of founding in honeybees, stingless bees, army ants, and some wasps. It appears generally the case that fission is the mode of colony founding adopted by species that achieve a very large colony size, but not all large-colony species reproduce by dependent founding. Fungus-growing ants of the genus *Atta*, *Vespula* wasps, and 'higher' termites of the Termitidae all attain colony sizes of thousands or millions of workers, yet invariably reproduce by independent founding.

When colony fission is present, it is the exclusive mode of colony founding for the species; facultative fission is unknown. Colony budding is a much less stereotyped process that often occurs concurrently with more standard independent colony founding. Colony budding involves the departure of already inseminated queens from polygynous colonies with a relatively small number of workers. Unlike fission, in which the daughter colonies disperse far enough to become independent, budding colonies remain in close proximity with one another and the mother colony. They often interchange workers among themselves, yielding a polydomous colony structure, in which a single large colony has multiple nests.

Dependent founding has evolved under the ecological conditions in which independent founding is not an option because colonies below a certain size are not viable. Colony fission and budding have clear selective advantages over independent founding, because the accompanying workers can construct the nest quickly and immediately begin colony-level performances via task specialization. Indeed, queens of species practicing dependent founding store much less body fat than independent founding queens.

Unanswered Questions and Future Studies

Considerable research has focused on foundress associations. Theoretical models assume that foundresses have the ability to assess costs and benefits of joining others. Data on the process of cofoundress formation and nestmate selection under field conditions are still needed. Provided that workers have little influence on the accession decision among cofoundresses, it is important to know how individual foundresses balance their energy budgets between individual and colony-level investment. In this both competitive and conflicting game of egg laying, she must lay as many eggs as possible if she hopes her colony to outcompete the neighboring colonies. At the same time, however, she should not exhaust herself before the physical combat against other cofoundresses. We must measure relative contributions of individual foundresses and observe whether foundresses can induce others to contribute more or prevent others from contributing too much.

For both pleometrosis and haplometrosis, long-term field studies are needed. Few studies of colony founding have followed queens long enough to estimate lifetime reproductive success. Surviving queens in pleometrotic colonies may be able to reproduce earlier and/or for a longer period of time than haplometrotic queens. Colony founding provides an excellent arena for testing various hypotheses for the evolution of sociality, and the approach of long-term ecological research will give us a much more complete picture.

See also: Ant, Bee and Wasp Social Evolution; Division of Labor; Kin Selection and Relatedness; Queen–Queen Conflict in Eusocial Insect Colonies; Reproductive Skew; Termites: Social Evolution.

Further Reading

Bernasconi G and Strassmann JE (1999) Cooperation among unrelated individuals: The ant foundress case. *Trends in Ecology & Evolution* 14: 477–482.

Bono JM and Crespi BJ (2008) Cofoundress relatedness and group productivity in colonies of social *Dunatothrips* (Insecta: Thysanoptera) on Australian *Acacia*. *Behavioral Ecology and Sociobiology* 62: 1489–1498.

Bourke AFG and Franks NR (1995) *Social Evolution in Ants.* Princeton, NJ: Princeton University Press.

Brown MJF and Bonhoeffer S (2003) On the evolution of claustral colony founding in ants. *Evolutionary Ecology Research* 5: 305–313.

Choe JC and Crespi BJ (eds.) (1997) *The Evolution of Social Behavior in Insects and Arachnids.* Cambridge: Cambridge University Press.

Crespi BJ and Choe JC (1997) Introduction. In: Choe JC and Crespi BJ (eds.) *The Evolution of Social Behavior in Insects and Arachnids*, pp. 1–7. Cambridge: Cambridge University Press.

Crespi BJ and Yanega D (1995) The definition fo eusociality. *Behavioral Ecology* 6: 109–115.

Gadagkar R (1990) Evolution of eusociality: The advantage of assured fitness returns. *Philosophical Transactions of the Royal Society of London B* 329: 17–25.

Hölldobler B and Wilson EO (1977) The number of queens: An important trait in ant evolution. *Naturwissenschaften* 64: 8–15.

Hölldobler B and Wilson EO (1990) *The Ants.* Cambridge, MA: Harvard University Press.

Hunt JH (2007) *The Evolution of Social Wasps.* New York, NY: Oxford University Press.

Keller L (ed.) (1993) *Queen Number and Sociality in Insects.* New York, NY: Oxford University Press.

Matsuura K, Vargo EL, Kawatsu K, et al. (2009) Queen succession through asexual reproduction in termites. *Science* 323: 1687.

Oster GF and Wilson EO (1978) *Caste and Ecology in the Social Insects.* Princeton, NJ: Princeton University Press.

Perlman DL (1992) *Colony Founding Among Azteca Ants.* PhD Dissertation, Harvard University.

Queller DC and Strassmann JE (1998) Kin selection and social insects. *Bioscience* 48: 165–175.

Wilson EO (1971) *The Insect Societies.* Cambridge, MA: Belknap Press of Harvard University Press.

Communication and Hormones

G. T. Smith, Indiana University, Bloomington, IN, USA

Introduction

Communication signals and hormones function in analogous ways at different levels of biological organization. Communication behaviors transfer information from one individual to another and coordinate the behavior and physiology of interacting individuals. The signaler produces a communication signal, which is detected by a receiver and changes the physiology and/or behavior of that receiver. Similarly, endocrine hormones coordinate the physiology of interacting cells, tissues, and organs within an individual. An endocrine tissue or gland integrates information about the internal milieu and external stimuli (either directly or via neural input and/or hormonal signals from other glands) and produces a hormonal signal that is received by other tissues and influences their physiology.

Both hormonal signals and communication signals are diverse (**Figure 1**). Hormones are derived from a wide range of biologically active molecules (e.g., protein and peptide hormones, steroids from cholesterol, prostaglandins from fatty acids, and biogenic amines and thyroid hormones from amino acids). Similarly, communication signals have evolved to exploit nearly every sensory modality. Although visual and acoustic signals are the most apparent ones, animals also use olfactory signals (pheromones); gustatory signals (e.g., contact chemosensory 'trails' in insects); tactile, vibrational, and seismic signals (e.g., substrate-borne signals in elephants, scorpions, spiders, and mole rats); and electrical signals (e.g., the electric organ discharges (EODs) of weakly electric fishes).

This article explores factors that influence the way hormones and behavioral signals effectively convey information, reviews the manner in which hormones regulate the production and reception of communication signals, and discusses how communication signals modulate hormones to affect receivers' physiology and behavior. Although hormones and communication signals regulate many aspects of social and organismal biology (e.g., feeding and energy balance, territoriality and aggression, and predator avoidance), this article focuses largely on hormones and communication signals used in reproductive contexts because the links between reproductive hormones and courtship signals are particularly strong and have been well-studied in a variety of model systems.

Parallels in the Design of Hormonal and Behavioral Communication Signals

Endocrine systems and communication behavior must overcome similar challenges to convey information efficiently. These challenges have shaped the evolution of signal properties depending on the type of information that they convey and how they convey it, and when and where the signals are produced.

Environmental Factors

The environment through which signals are transmitted influences the effectiveness of both hormones and communication behavior. There is a rich literature on how the properties of communication signals have evolved to maximize information transmission in particular environments. For example, the active space and optimal time course of olfactory signals (pheromones) are influenced by volatility, diffusion, currents in the medium (air or water), and chemical degradation. Broadcast attractant pheromones are often volatile compounds that can be transmitted over long distances, whereas territorial scent marks are often deposited on substrates and are less volatile and more persistent.

Visual signals are line-of-sight and are not useful for transmission in environments with obstructions. The effectiveness of visual signals is also influenced by the intensity and spectrum of the incident and the background lighting. Many studies have investigated how the light environment influences the evolution of vision and visual signals. For example, changes in light spectra with depth and water quality influence color signals in cichlids and killifish, and the evolution of plumage and ornament coloration is influenced by background coloration in bower birds.

The transmission of acoustic signals depends on the acoustic absorbance of the medium, the ambient noise, and the acoustic clutter created by objects and boundaries. The acoustic adaptation hypothesis suggests that acoustic signal properties (particularly those of long-distance signals) evolve to maximize transmission through the environment without degradation. Dense vegetation may attenuate high-frequency sounds more than mid- to low frequency (1.5–2.5 kHz) sounds, for example, whereas winds in open

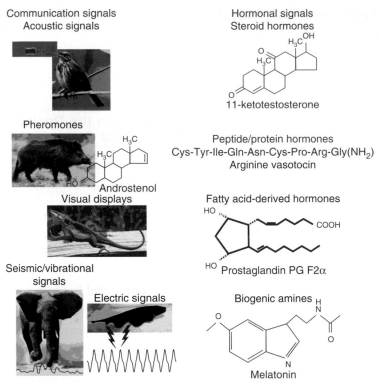

Figure 1 Diversity of communication and hormonal signals. (Left) Animals employ signals in every sensory modality to convey information. (Right) Similarly, hormones are derived from a wide range of organic compounds.

habitats make short-duration, high-frequency notes more effective signals because their temporal properties are less distorted. Accordingly, songbirds in forests tend to have songs with more low-frequency components, whereas songbirds in open habitats tend to have songs with short, high-pitched notes. Songbirds may also sing in particular locations (e.g., elevated perches) to influence song transmission through the environment. Similarly, the electrical communication signals of electric fishes are influenced by electrical noise sources (e.g., lightning and human-generated electrical noise) and by the electrical conductivity of the water in which the fish live. For example, three closely related species in the genus *Brachyhypopomus* that live in waters of different electrical conductivities have evolved different types of electric organs that 'impedance match' their environment and maximize the signal power.

Just as the ability of communication behaviors to transmit information is influenced by environmental factors, the efficacy of hormones to convey signals to particular tissues is affected by the transmission medium. For example, the transport of lipophilic hormones, such as steroids, in the aqueous medium of plasma is facilitated by the presence of binding globulins. The abundance and binding kinetics of these proteins regulate the availability of 'free' hormones to activate receptors in target tissues. Just as physical barriers obstruct the transmission of visual signals, physiological barriers, such as the blood–brain barrier (BBB), restrict the transmission of certain

hormonal signals into the brain. Small lipophilic hormones or hormones with active transport mechanisms can cross the BBB more readily than large protein hormones. Just as behavioral communication signals are subject to environmental degradation, local expression of enzymes that degrade or convert hormones in particular tissues influences the transmission of hormonal signals. For example, tissue-specific expression of aromatase, which converts testosterone to estradiol, or 5β-reductase, which converts testosterone to the inactive metabolite 5β-dihydrotestosterone, can shunt the potential for testosterone to act on androgen receptors in that tissue. Finally, just as environmental noise can mask communication signals, 'noise' created by exogenous environmental compounds that mimic or block hormone action (e.g., naturally occurring or man-made endocrine disruptors) can mask hormonal signals.

Intended and Unintended Receivers

Communication behaviors have evolved to transmit information between individuals, but these signals are also frequently intercepted by eavesdroppers. Signals that males produce to attract mates or defend territories, for example, may also make males more conspicuous to predators. In some cases, eavesdropping has led to the selection for 'private channels' for communication signals; that is, signals that are more easily detected by intended receivers than by eavesdroppers. Ultraviolet coloration in the tails of northern swordtails (*Xiphophorus*), for example, is relatively inconspicuous to their main predators, which have poor ultraviolet vision. Similarly, to avoid detection of their electrical communication signals by predatory catfish, weakly electric fish have evolved high-frequency electrical signals that can be detected by their own tuberous electroreceptors but not by the low-frequency-sensitive electroreceptors of catfish. Interestingly, some signals used by males to court females may reintroduce low-frequency components that make males more conspicuous both to predators and to females.

Hormonal signals must also overcome the problem of unintended reception. For example, mineralocorticoid receptors (MRs), which mediate the actions of aldosterone (e.g., resorption of water and sodium in the kidneys), are similar to glucocorticoid receptors (GRs) and can be activated by cortisol as well as aldosterone. This could create a problem because aldosterone concentrations are typically much lower than those of cortisol. The MRs in the kidney are 'protected' from cortisol activation by the expression of an enzyme (11β-hydroxysteroid dehydrogenase) that degrades cortisol. Another example of 'hormonal eavesdropping' occurs during pregnancy in mammals. Maternal steroid hormones, such as estrogens, have the potential to cross the placenta and activate fetal estrogen receptors. This unintended signal transmission is prevented by the expression of α-fetoprotein, which binds to maternal estrogens and intercepts them before they can reach the fetal receptors.

Coevolution of Signals and Receivers

The evolution of both hormonal and behavioral signals is intertwined with the evolution of the receivers of those signals. Signalers are likely to be selected to produce signals that are conspicuous to intended receivers. Male courtship signals, for example, may evolve to exploit preexisting biases in female sensory systems that initially are unrelated to detecting or selecting mates. Sensory biases have been suggested to explain the evolution of male traits and female preferences for complex vocalizations in Túngara frogs, visually conspicuous display ornaments in swordtails (*Xiphophorus* spp.) and sailfin mollies (*Poecilia latipinna*), and appendage-waving courtship displays of water mites (*Neumania papillator*). If signals are beneficial to receivers (e.g., if courtship signals accurately convey information about species identity or mate quality), their sensory systems may be selected to detect and discriminate signal properties.

Just as communication signals coevolve with the sensory systems of receivers, hormones and their receptors also influence each other's evolution. Analogous to the sensory exploitation hypothesis for the evolution of animal communication signals, work in Joe Thornton's

laboratory showed that novel hormones may evolve by exploiting the preexisting promiscuous binding properties of hormone receptors. Using elegant phylogenetic and molecular techniques to reconstruct the evolution of steroid receptors and their ligands, Thornton's group demonstrated that ancient GRs exhibited an affinity for the mineralocorticoid aldosterone long before the enzymatic pathways to produce aldosterone evolved. An additional copy of the ancestral GR gene was created in a genome duplication event, and the preexisting binding properties of the GR were exploited to allow the evolution of the tetrapod mineralocorticoid receptor (MR) from one of these copies. The evolution of aldosterone as a ligand for the MR then created subsequent selection pressure for increased cortisol specificity of the GR.

Hormonal Regulation of Signal Production

Communication signals may convey the individual, species, or sexual identity of a signaler as well as its social status, motivation, or physiological state. Hormone levels often differ between sexes and coordinate an animal's behavioral and physiological responses to changes in internal state or the social and physical environment. Consequently, it is not surprising that hormones have evolved as signals that can powerfully modulate communication behavior.

Hormones Affect Signals Across Multiple Time Courses

Hormones can regulate the production of communication signals over time courses ranging from seconds to lifetimes. Sex differences in hormone levels during development can have lifelong, organizational effects that underlie sex differences in reproductive communication signals. For example, experiments in Darcy Kelley's laboratory have shown that the development of sex differences in the calling behavior of the African clawed frog *Xenopus laevis* is regulated in part by the organizational effects of androgens on laryngeal muscles and brainstem vocal motor nuclei. Some organizational effects of androgens on the vocal development of male anuran vocal behavior are also gated by thyroid hormones and prolactin, which are key regulators of metamorphosis. Work in Michael Moore's laboratory revealed that the organizational effects of hormones also affect the development of throat patch (dewlap) coloration in male tree lizards that use alternative reproductive strategies. Males extend their colorful dewlaps in agonistic and courtship displays. High levels of testosterone (T) and progesterone (P) during development increase the probability that males will develop into territorial individuals with orange/blue dewlaps, whereas individuals with lower levels of T and P are more likely to develop into satellite/nomad males with orange dewlaps. Similarly, the expression of normal levels of flank marking (a behavior that deposits pheromones) by adult male hamsters depends on their exposure to testosterone during adolescence.

In adulthood, the activational effects of gonadal hormones may mediate sex differences and seasonal plasticity in communication signals. These effects are often complementary with the organizational effects of hormones. For example, seasonal plasticity in the production of vocal courtship signals by male fish, frogs, song birds, and mice is often regulated by gonadal androgens (testosterone, 5α-dihydrotestosterone (5α-DHT), and/or 11-ketotestosterone (11-KT)). Castration reduces signal production, and treatment with androgens restores male vocal production and can masculinize female vocal production. Similarly, dewlap extension displays in *Anolis* lizards, the production of male sex pheromones by goats and newts, and flank marking behavior in male hamsters, are reduced by adult castration and restored by testosterone treatment. The plasticity of female reproductive signals across seasons or the phases of estrus/menstrual cycles are also influenced by the activational effects of gonadal steroid hormones. The production of sex pheromones by female garter snakes, scent-marking behavior in female hamsters, and sexually proceptive ear wiggling behavior in female rats, are activated by estrogens. The enlargement of the sexual perineal skin around the time of ovulation in females of many primate species is activated by estrogens and inhibited by progesterone. The activational regulation of reproductive/courtship signals by gonadal steroids coordinates signal production with reproductive physiology.

Steroid hormones can also have rapid effects on communication behavior, allowing hormones to modulate signal production within minutes in response to changing social or environmental conditions. For example, playbacks of male vocalizations rapidly (within 20 min) elevate the levels of 11-KT in male gulf toadfish (*Opsanus beta*); 11-KT then rapidly facilitates male vocalizations.

Peptide hormones can also have rapid effects on the production of communication signals. For example, the related peptides arginine vasotocin (AVT) in fish, amphibians, reptiles, and birds, and arginine vasopressin (AVP) in most mammals have widespread and robust effects on social behavior, including the communication of signals. AVT increases the production of advertisement calls in many species of frogs, facilitates singing in some songbird species, and increases courtship displays by male bluehead wrasses (*Thalassoma bifasciatum*). Similarly, in male hamsters, AVP acts on the anterior hypothalamus to facilitate a behavior (flank marking) that deposits pheromones on the substrate. The effects of AVT/AVP on signals often interact with those of gonadal steroids and can be sex-specific and/or vary across seasons. In breeding bullfrogs

(*Rana catesbeiana*), for example, AVT enhances the production of release calls in males but suppresses release calling in females. AVT has little effect on release calls outside of the breeding season. In the electric fish *Apteronotus leptorhynchus*, AVT enhances the production of high-frequency, 'courtship' chirps (an electric communication signal) in males, but has little effect in females (**Figure 2**). In male canaries, AVT agonists increase song duration during the breeding season, but decrease song duration in the fall/winter. In sonic midshipmen fish (*Porichthys notatus*), different but related peptides influence vocal production in the two sexes and in different 'types' of males. AVT inhibits vocal motor circuits in territorial males, but has little effect in females or 'sneaker' males, whereas isotocin reduces vocal motor bursting in females and sneaker males, but not in territorial males.

In some cases, hormones or their metabolites can act directly as chemical communication signals between individuals. For example, many species of fishes release gonadal steroids and/or prostaglandin hormones or their metabolites into the water. These compounds can be detected by the olfactory systems of conspecifics, affect their physiology and behavior, and thus, act as 'hormonal pheromones.' Similarly, in pigs, the steroids 5α-androstenol and 5α-androstenone ('boar taint') are produced in large quantities in the testes of boars, are released in saliva, and robustly affect the reproductive physiology and behavior of sows.

Hormones Affect External Signaling Structures

Hormones influence signal production at different anatomical and physiological levels. For example, hormones promote the development of specialized coloration or external structures that are used as display ornaments. In many fish, male-specific color patterns used in courtship and/or aggressive displays depend on 11-KT (e.g., red ventral coloration in sticklebacks (*Gasterosteus aculeatus*), bright coloration in terminal phase male parrotfish (*Sparisoma viride*) and bluehead wrasses). Similarly, in some lizard species, dewlaps are larger and more brightly colored in males, and testosterone (and in some cases, progesterone) during development, masculinizes both the size and color of dewlaps. In lower vertebrates, changes in skin color following agonistic interactions (e.g., skin darkening in salmonids and eyespot darkening in *Anolis* lizards) are regulated in part by melanocyte stimulating hormone (α-MSH) and catecholamines. Although many external ornaments used by male birds in aggressive and courtship displays (e.g., combs and wattles in male fowl and beak color in zebra finches) depend on androgens, the hormonal control of sexually dimorphic plumage color in birds varies considerably across taxa. In some species, such as fairy wrens (*Malurus melanocephalus*) and house finches (*Carpodacus mexicanus*), more colorful male

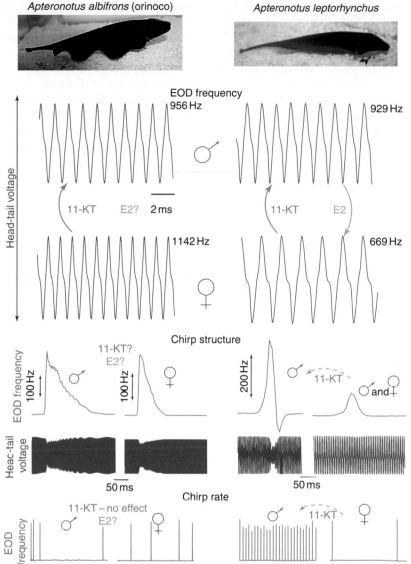

Figure 2 Sex differences and hormone effects on the electric communication signals of black ghost (*A. albifrons*) and brown ghost (*A. leptorhynchus*) knifefishes. (Top) Representative EOD traces (voltage difference between the head and tail of the fish over time) of males (top) and females (bottom) in each species. Each upward and downward deflection of the trace represents a single discharge of the electric organ. EOD frequency (EODf, the number of discharges per second) is sexually dimorphic in both black ghosts (left) and brown ghosts (right), but in opposite directions. In both species, androgens (11-ketotestosterone, 11-KT) masculinize EODf, but they do so by increasing EODf in brown ghosts, but decreasing it in black ghosts. Estrogens (estradiol, E2) feminize EODf in brown ghosts; the effect of estrogens on EODf in black ghosts is not known. (Middle) Sex differences in chirp structure. Chirps are transient changes in EODf or amplitude and are produced during aggression or courtship. The EOD voltage trace (red) shows the decrease in amplitude during some chirps, and the green trace shows the change in EODf. In black ghosts (left), male chirps are longer in duration than those of females. In brown ghosts (right), males produce 'high frequency' chirps that are only very rarely produced by females. 11-KT partly masculinizes chirp structure in brown ghosts. (Bottom) The chirp rate is sexually monomorphic and unaffected by androgens in black ghosts (left). In brown ghosts (right), males chirp much more often than females, and the chirp rate is increased by androgens. Based on data from Schaefer JE and Zakon HH (1996) Opposing actions of androgen and estrogen on *in vitro* firing frequency of neuronal oscillators in the electromotor system. *Journal of Neuroscience* 16(8): 2860–2868; Dunlap KD, Thomas P, and Zakon HH (1998) Diversity of sexual dimorphism in electrocommunication signals and its androgen regulation in a genus of electric fish, *Apteronotus. Journal of Comparative Physiology A* 183: 77–86; Kolodziejski JA, Nelson BS, and Smith GT (2005) Sex and species differences in neuromodulatory input to a premotor nucleus: A comparative study of substance P and communication behavior in weakly electric fish. *Journal of Neurobiology* 62: 299–315. reviewed in Zakon HH and Smith GT (2009) Weakly electric fish: Behavior neurobiology, and neuroendocrinology. In: Pfaff DW, Arnold AP, Etgen AM, Fahrbach SE and Rubin RT (eds.) *Hormones, Brain, and Behavior*, pp. 611–638. San Diego: Academic Press.

plumage is androgen-dependent. In other species, including ducks and peacocks, bright male plumage is a default condition, and the less colorful female plumage is estrogen-dependent. In other species, sexually dimorphic plumage coloration may be regulated by other hormones (e.g., luteinizing hormone) or by nonhormonal genetic mechanisms. In primates, the coloration and size of 'sex skin' can be hormone sensitive. Red sex skin in many female primates advertises fertility and is estrogen-sensitive. Similarly, red, but not blue, sex and facial skin in males is abolished by castration and restored by testosterone treatment. Some sexually dimorphic hair ornaments in male primates are also androgen-sensitive; for example, the 'cape' of male hamadryas baboons (*Papio hamadryas*) depends on testosterone.

Hormones Affect Muscles and Other Peripheral Signal Producing Tissues

Many communication signals are produced by specialized structures with dedicated muscles. Hormones can affect the anatomy and physiology of these muscles to facilitate seasonal and/or sex differences in signal production. The muscles in the vocal organs of fish, frogs, and songbirds are androgen-sensitive during development and/or adulthood. In plainfin midshipmen fishes, parental males that produce sustained courtship vocalizations have larger vocal muscles with more fibers and mitochondria-filled sarcoplasm than those of females or 'sneaker' males. Andrew Bass' laboratory found that androgen implants increased both muscle fiber size and the area of sarcoplasm. In African clawed frogs (*X. laevis*), several sexually dimorphic aspects of laryngeal muscle physiology are hormonally regulated. Males have larger larynges with greater cartilage and muscle mass (organizational effect of testosterone), the laryngeal muscles having more fast-twitch fibers that enable males to produce rapid trills (organizational effect of testosterone); female larynges have relatively strong neuromuscular junctions that allow strong responses to motor commands, but that prevent repetitive contraction–relaxation cycles at the rates needed to produce rapid trills (activational effect of estrogens). Similarly, in many songbird species with sexually dimorphic song, the vocal organ (syrinx) is larger in males than in females; and testosterone increases both the size of syringeal muscles and the expression of acetylcholine receptors. Muscles that control sexually dimorphic visual displays can also be androgen–sensitive. The muscles that control dewlap extension contain more fast-twitch fibers in breeding male green anole lizards (*Anolis carolinesnsis*) than in nonbreeding males, and this seasonal change in muscle physiology is controlled by testosterone.

Weakly electric fishes have electric organs that produce sexually dimorphic electrical communication signals (**Figure 2**). Most electric organs are developmentally and evolutionarily derived from muscles, and sex differences and short-term changes in the waveform of EODs arise from the effects of hormones on the anatomy and/or physiology of electrocytes in the electric organ. Andrew Bass found that in an African electric fish, androgens increased the surface area of the anterior face of electrocytes, and he hypothesized that the resulting change in passive membrane properties contributed to longer duration EOD pulses in males. Harold Zakon's laboratory found that androgens and estrogens change the expression of sodium and potassium channel subunits in the electrocytes of the South American electric fish *Sternopygus macrurus*. Differences in the kinetics of the currents carried by these channels result in slower electrocyte action potentials and longer EODs in males and faster electrocyte action potentials and shorter EODs in females. Lynne McAnelly and Michael Markham in the laboratories of Zakon and Philip Stoddard have also shown that the amplitude of the EOD changes with a circadian rhythm driven by melanocortin hormones (ACTH, α-MSH). EOD amplitude is higher when the fish are active at night, and the nocturnal increase in EOD amplitude can be reproduced with ACTH treatment. Melanocortins act via a protein kinase A/cAMP-dependent pathway to increase the amplitude of sodium currents, which in turn increases the current generated by the electric organ and the EOD amplitude. Similarly, Stoddard's laboratory has shown that circadian and socially mediated changes in the amplitude and waveform of the EOD in *Brachyhypopomus pinnicaudatus* are mediated by the effects of androgens and melanocortins on electrocyte excitability. Males produce EOD pulses of greater amplitude and longer duration than those of females. The duration and amplitude of the EOD in males is increased at night when the fish are active. Social interactions also increase EOD amplitude and duration. Melanocortins, serotonin, corticotrophin-releasing hormone, and thyroid-stimulating hormone mediate the social and circadian modulations of the EOD in this species. Melanocortins enhance the EOD by affecting the kinetics and relative timing of action potentials on the rostral and caudal sides of electrocytes within the electric organ.

Many pheromones act as territorial or courtship signals, and the scent glands that produce pheromones are often sensitive to gonadal steroids and other hormones. Testosterone stimulates the production of pheromones by femoral glands in male lizards. Testosterone and prolactin increase the production of sodefrin, a female-attracting pheromone, by the abdominal glands of male newts, and AVT stimulates the release of this pheromone in the water. Testosterone also causes hypertrophy of sebaceous glands and the production of estrus-stimulating (i.e., primer or 'male effect') pheromones by male goats.

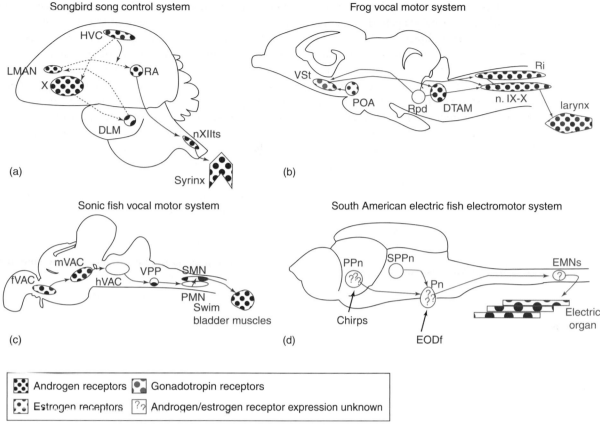

Figure 3 Schematic sagittal diagrams of brain circuits that control sexually dimorphic communication signals. The expression of hormone receptors is indicated by colored dots as shown in the legend. (a) The song control system of song birds consists of two subcircuits: (1) a direct motor pathway (solid red arrows) that controls song production and includes projections from HVC in the nidopallium to RA in the archipallium to the tracheosyringeal portion of the hypoglossal nucleus (nXIIts), which innervates the vocal organ (syrinx); and (2) an anterior forebrain loop (dashed red arrows) that is necessary for song learning and consists of an indirect pathway from HVC to RA via Area X in the basal ganglia, the medial dorsolateral thalamus (DLM), and the lateral magnocellular nucleus of the anterior nidopalium (LMAN). (b) The central pattern generator for vocalization in frogs consists of three brainstem regions: the dorsal tegmental area of the medulla (DTAM), the inferior reticular formation (Ri), and the nucleus of the ninth and tenth cranial nerves (n. IX–X), which innervates the larynx. DTAM receives descending input from the ventral striatum (VSt) and preoptic area (POA). The entire vocal motor system receives robust serotonergic inputs from one of the raphe nuclei (Rpd). (c) The circuit that controls vocalization in sonic fishes consists of networks of sensorimotor brain nuclei in the forebrain, midbrain, and hindbrain (vocal acoustic centers, fVAC, mVAC, hVAC) that project to brainstem vocal motor nuclei via the ventral prepacemaker nucleus (VPP). Pacemaker neurons (PMN), which generate the rhythm for vocalizations, are adjacent to the sonic motor nucleus (SMN), which innervates the sonic muscles on the swim bladder. (d) The electromotor system controls the timing of the electric organ discharge in gymnotiform (South American) weakly electric fish. EOD frequency is controlled by neurons in the medullary pacemaker nucleus (Pn). The Pn projects to spinal electromotor neurons (EMNs), whose axons form a nerve-derived electric organ (in apteronotid fish) or innervate a muscle-derived electric organ (in all other electric fish). Chirping is controlled by an excitatory projection to the Pn from the thalamic prepacemaker nucleus (PPn). The sublemniscal prepacemaker nucleus (SPPn) in the midbrain controls the jamming avoidance response. Although androgens and/or estrogens have dramatic effects on the physiology of EMNs, the Pn and the PPn, the expression of androgen and estrogen receptors has not been demonstrated. Based on Schlinger and Brenowitz (2009), Yamaguchi and Kelley (2003), Yang and Kelley (2009), Bass and Remage-Healey (2008), Zakon and Smith (2009).

Hormones Affect Brain Circuits That Control Signal Production

Hormones also act centrally on the brain and spinal cord to regulate communication signals. In many cases, the production of communication signals is controlled by dedicated brain circuits, which has allowed these circuits to be used to understand how hormones act on the brain to influence behavior. One of the best studied vertebrate brain circuits controlling communication is the song control system of songbirds (**Figure 3(a)**). This network of brain nuclei is subdivided into a descending motor pathway and an anterior forebrain 'loop.' The motor pathway includes direct projections from the telencephalic nucleus HVC to the premotor nucleus RA in the archipallium, which in turn projects to the tracheosyringeal motor nucleus (nXIIts) in the medulla. Motor neurons in nXIIts innervate the muscles of the vocal organ (syrinx). This pathway controls song production, and lesions of brain nuclei in this pathway disrupt normal song. In the anterior forebrain pathway, HVC projects indirectly to RA via brain nuclei in the basal ganglia (Area X), thalamus (DLM), and nidopallium (LMAN). This pathway is not necessary for adult song production, but is important for song learning.

In most temperate zone songbird species, males sing more often and sing more complex songs than females do. Several song control nuclei, including HVC, RA, Area X, and nXIIts, are much larger (in some cases, 5–6 times as large) in males than in females. These are some of the most dramatic sex differences in the vertebrate brain, and their discovery by Fernando Nottebohm in the mid-1970s fueled substantial interest in understanding how sex differences in brain structure are related to sex differences in behavior. Most of the song control regions contain receptors for androgens, and neurons in HVC also express estrogen receptors. Work in Barney Schlinger's laboratory has also revealed that the songbird brain expresses high levels of enzymes, such as aromatase and reductases, that can metabolize testosterone. In zebra finches, estrogen treatment during early development masculinizes the size of the song control nuclei and singing behavior. Sex differences in gonadal steroids during early development, however, may not explain sex differences in the song control system. Several lines of evidence suggest that intrinsic, genetically mediated sex differences in the brain contribute to the development of sexually dimorphic song control nuclei in zebra finches. For example, while in Art Arnold's laboratory, Juli Wade showed that genetic females induced to grow testes instead of ovaries had feminized song control nuclei, which suggests that testicular hormones are insufficient to masculinize these brain areas. Furthermore, Agate and colleagues found that a gynandromorphic zebra finch that was genetically male on the right side of its body and genetically female on the left side, had more masculine song control nuclei on the right than on the left, despite the fact that the two sides of the brain were presumably exposed to the same hormones from the gonads. These results suggest that the genetic phenotype of the brain, possibly in concert with local steroid metabolism, may cause song control nuclei to grow larger in males.

In many songbird species, testosterone has activational effects on the song control nuclei in adulthood, which contributes to sex differences and results in seasonal plasticity in these brain circuits. In some species (e.g., spotted towhees, *Pipilo maculatus*), HVC and RA more than double in size from the fall to the spring as testosterone concentrations rise in males. These dramatic changes in the brain have been hypothesized to underlie seasonal changes in song rate, song stereotypy, or song learning. The androgen-induced hypertrophy of RA in the spring reflects larger, more widely spaced neurons with longer dendrites and more synaptic connections. In contrast,

HVC grows in response to higher levels of testosterone in the spring because new neurons are added to this brain area. Fernando Nottebohm's discovery of neurogenesis in the HVC of canaries was one of the first demonstrations that parts of the vertebrate forebrain retain the ability to reconfigure circuits by adding neurons throughout adult life. The physiology of neurons in the song control system also changes seasonally and in response to hormones. In the laboratories of David Perkel and Eliot Brenowitz, John Meitzen found that several active properties of neurons in RA change seasonally in response to photoperiod and testosterone in Gambel's white-crowned sparrows. These changes make the neurons more excitable during the breeding season, when birds are singing more often and more stereotypically.

Brain circuits that produce vocalizations of frogs and sonic fish as well as those that produce electric communication signals in weakly electric fish have also emerged as strong models for understanding how hormones affect brain circuits that control sexually dimorphic communication (**Figure 3**). In plainfin midshipmen, the vocal muscles are innervated by motor neurons in the sonic motor nucleus in the medulla, and vocalization is controlled by a network of brain nuclei in the medial brainstem. These brain nuclei also receive descending inputs from forebrain and midbrain sensorimotor nuclei (**Figure 3(c)**). Like the avian song control nuclei, many of these areas contain receptors for androgens or estrogens. Vocalizations differ both between males and females and between two types of males that use alternative reproductive strategies. Parental males defend nests and attract females by producing loud, long hums; whereas the vocalizations of females and 'sneaker' males consist primarily of short duration 'grunts.'

Andy Bass and colleagues have studied the hormonal and neural regulation of acoustic signaling in sonic fish. Long-term treatment with testosterone increases the frequency of vocalizations as well as the size of motor neurons and the expression of aromatase in the medullary sonic motor nucleus. Steroid hormones also have rapid, nongenomic effects on the sonic motor system. Luke Remage-Healey in Bass' laboratory studied fictive vocalizations produced by in vitro preparations of the brainstem of midshipmen and oyster toadfish and found rapid effects of androgens, estrogens, and/or glucocorticoids on the vocal central pattern generator. In parental male midshipmen, but not in females or 'sneaker' males, estradiol and the nonaromatizable androgen 11-KT, but not testosterone, robustly increased the duration of fictive calls within 5 min. In females and 'sneaker' males, testosterone and estradiol, but not 11-KT, rapidly increased fictive call duration. Cortisol enhanced fictive calling in parental males, but reduced fictive call duration in females and sneaker males. The differences between sexes and male types in the rapid effects of androgens are consistent with endogenous hormone concentrations: parental males have higher plasma levels of 11-KT, whereas females and 'sneaker' males have elevated levels of testosterone. James Goodson in Bass' laboratory found similar sex- and type-specific effects of peptide hormones on fictive vocalization. AVT applied directly to the anterior hypothalamus–preoptic area (AH-POA) inhibited fictive vocal responses in parental males, but not females or 'sneaker' males; whereas applying isotocin to the AH-POA inhibited fictive vocalization in females and 'sneaker' males, but not in parental males.

Sexually dimorphic vocalizations in African-clawed frogs (*X. laevis*) are also controlled by a hormone-sensitive brain circuit (**Figure 3(b)**). The laryngeal muscles of frogs are innervated by motor neurons in the hindbrain motor nucleus of the ninth and tenth cranial nerves (n. IX–X). This nucleus receives a robust descending projection from the dorsal tegmental nucleus of the anterior medulla (DTAM), which is likely to be homologous to the parabrachial nucleus that controls vocalization in mammals. The pattern for sexually dimorphic calling (fast trills in males, slower raps and ticks in females) is thought to be generated by DTAM and interneurons in n. IX–X. The ventral striatum and preoptic area in the forebrain also regulate vocalizations via descending inputs to DTAM. Serotonin is a powerful neuromodulator of this circuit. The ventral striatum, DTAM, and n. IX–X all receive projections from the serotonergic dorsal raphe, and serotonin induces fictive vocalization in in vitro brain preparations. Hormone actions on the anuran vocal circuit and their role in the sexual differentiation of calling have been well studied in the laboratories of Darcy Kelley and Ayako Yamaguchi. The larynx, as well as neurons in both DTAM and n. IX–X express androgen receptors, and androgens have robust effects on the development, anatomy, and physiology of vocal neurons. Males have more and larger n. IX–X motoneurons with longer dendrites than those of females, and the developmental effects of testicular hormones partly regulate those sex differences. The physiology of n. IX–X motoneurons is also sexually dimorphic and possibly regulated by androgens. The presence of more substantial low-voltage-activated potassium currents and hyperpolarization-activated cation currents (I_h) in male n. IX–X motoneurons allow them to fire rapidly with high temporal fidelity to drive the laryngeal muscles when males produce trills. In vitro studies of isolated brain preparations in Yamaguchi's laboratory also revealed that the physiology of the brainstem central pattern generator for vocalization is androgen-sensitive. The male DTAM – n. IX–X network can produce two different vocal motor rhythms in vitro: one similar to the trills males produce to attract females and a second one similar to the slower release calls used by females to reject mating attempts. In contrast, brain preparations from females only produce the release call pattern. Brains from females treated with testosterone are masculinized; they can produce both types of vocal rhythms.

In electric fish that produce continuous, wave-type EODs, EOD frequency is often sexually dimorphic and serves as a signal of species identity, sex, and social status (**Figure 2**). Sex differences and individual variation in EOD frequency arise largely from the activational effects of gonadal steroids on brain circuits that control the EOD (**Figure 3**). In some cases, the direction of sex differences varies across species, and these reversals in sexual dimorphism are paralleled by reversals in the effects of hormones on neurophysiology and behavior. For example, in brown ghost knifefish (*A. leptorhynchus*), male EODs are significantly *higher* in frequency than those of females and androgens *increase* EOD frequency by increasing the firing rates of pacemaker neurons. Kent Dunlap and Harold Zakon found that in the closely related black ghost knifefish (*A. albifrons*), male EODs are significantly *lower* in frequency than those of females, and that androgens *decrease* EOD frequency. Joe Schaefer in Zakon's lab found that the effects of androgens and estrogens on EOD frequency are caused by their effects on the firing rates of neurons in the pacemaker nucleus and electromotor neurons in the spinal cord. Thus, species differences in sexual dimorphism of communication behaviors have evolved through reversals in the effects of hormones on the physiology of the neurons that control those behaviors.

Electric fish also modulate their EODs during social interactions to produce motivational signals known as 'chirps.' Sex differences in chirping have been most extensively studied in brown ghost knifefish (*A. leptorhynchus*). In this species, males chirp much more than females and produce 'bigger' chirps with more frequency modulation than those of females. Chirps are controlled by the thalamic prepacemaker nucleus (PPn). PPn projection neurons excite neurons in the brainstem pacemaker nucleus via non-NMDA glutamatergic synapses. This excitation causes a chirp by briefly increasing the firing rate of the pacemaker neurons, and thereby transiently increasing EOD frequency. Unlike the song control nuclei of songbirds, relatively few sex differences have been found in the size or structure of the PPn, despite robust sex differences in chirping. Instead, sex differences in chirping are associated with sex differences in the expression of the neuromodulators in the PPn. The PPn in males contains significantly more substance P and less serotonin than the PPn in females. These sex differences parallel the effects of substance P and serotonin on chirping. Leonard Maler's laboratory found that injecting substance P into the PPn induced chirping and caused fish to produce bigger chirps. Similarly, serotonin suppresses chirping. Furthermore, treating females with androgens both masculinizes their chirps and increases the expression of substance P in their PPn. Thus, androgens regulate the sexually dimorphic expression of neuromodulators in the PPn, which in turn regulate sex differences in chirping behavior.

Hormonal Regulation of Signal Reception

Appropriate responses to communication signals depend on the physiological and motivational state of the receiver. Because hormones coordinate physiological and behavioral responses to internal and external cues, they are logical modulators of both the ability of receivers to detect and discriminate among communication signals and the behavioral responses of receivers. Although in some instances, hormones may simply affect the responses of receivers through general effects on arousal and motivation (e.g., increases or decreases in sexual receptivity that modify responses to courtship signals), hormones can also directly affect the ability of sensory systems to perceive signals.

Hormones can influence the perception and response of pheromones by receivers. Cardwell and colleagues in Norm Stacey's laboratory found that in males of the cyprinid fish, *Puntius schwanenfeldi*, peripheral olfactory responses to the pheromone 15-keto-prostaglandin F2α (15K-PGF2α), which is released by females after ovulation, were androgen-dependent; electro-olfactorigram responses to 15K-PGF2α were significantly enhanced in juvenile males treated with androgens compared to blank-implanted control males. Androgen-regulated sensitivity to this pheromone allows males to be maximally sensitive to it during the breeding season. Similarly, in female newts (*Cynops pyrrhogaster*), estrogens and prolactin enhance the responses of olfactory receptors in the vomeronasal epithelium to the male proteinaceous pheromone sodefrin.

Joseph Sisneros found that gonadal steroid hormones mediate seasonal changes in auditory tuning in plainfin midshipmen. During the spawning season, high levels of estradiol and testosterone in females changes the tuning of the auditory organ (sacculus) so that the auditory afferents are better able to phase lock to higher frequency sounds found in the harmonics of the courtship calls of parental males. This change in tuning may make it easier for females to locate and discriminate the calls of potential mates when they are ready to spawn. Hormones can also regulate the physiology of central auditory neurons. Kathleen Lynch and Walt Wilczynski used in situ hybridization for the immediate early gene *egr-1* to show that treating female Túngara frogs (*Physalaemus pustulosus*) with gonadotropins enhanced the activation of neurons in the auditory midbrain by male courtship calls.

Hormones also mediate changes in the ability of electric fish to detect electrocommunication signals. The tuberous electroreceptors of electric fish that produce wave-type EODs are most sensitive to the frequency of the fishes' own EOD. This tuning is important because electric fish electrolocate by detecting distortions in their EOD and because electric fish perceive the EODs of other fish by detecting the amplitude and phase modulations created in their own EOD by the EODs of other fish.

Because EOD frequency changes in response to androgens and/or estrogens, the tuning of electroreceptors must also change in response to androgens and estrogens if they are to remain maximally sensitive to the fishes' own EOD. Harold Zakon and Harlan Meyer found that electroreceptor tuning is indeed steroid-sensitive in the electric fish *S. macrurus*. Androgens act in a coordinated manner on the pacemaker nucleus in the brain to change EOD frequency, on the electric organ to change EOD waveform, and on the tuberous electroreceptors to change their tuning.

Effects of Communication Signals on Hormones

The relationship between hormones and communication signals is reciprocal. Hormones influence the production and reception of communication signals in the many ways described above. In turn, communication signals can also have profound effects on hormone levels, and hormones often mediate the effects of communication signals on the physiology and behavior of receivers.

Early evidence that communication signals affect hormones includes studies in the 1950s and 1960s showing that social housing affects reproductive physiology in laboratory rodents. Van der Lee and Boot found that housing female mice together but without males prolonged and/or arrested their estrus cycles. The 'Lee-Boot effect' is mediated by pheromones in the females' urine. Similarly, Whitten found that pheromones in adult male urine accelerated and synchronized the estrus cycles of co-housed females. Adult male urine also contains pheromones that can accelerate the onset of puberty in juvenile female mice (the Vandenburgh effect). Pheromones also affect reproductive hormones in other mammals. 'Male effect' pheromones from rams and male goats increase the frequency of luteinizing hormone pulses in females, which synchronizes ovulation.

Pheromonal signals similarly affect hormones in fish and amphibians. For example, Sorensen's laboratory characterized the actions of a female goldfish preovulatory pheromone, which contains the progestin 17,20β-dihydroxy-4-pregnen-3-one (17,20β-P), and a postovulatory pheromone, which contains a prostaglandin. Both of these pheromones stimulate gonadotrophin II (GTH II) release in males. 17,20β-P also stimulates ovulation in conspecific females.

Acoustic communication signals can also have profound and rapid effects on hormones. For example, John Wingfield's laboratory examined the hormonal and behavioral responses of territorial male song sparrows (*Melospiza melodia*) to a simulated territorial intrusion (STI), in which a decoy of a territorial intruder and a song playback were presented on a male's territory. The testosterone levels of such males were significantly elevated in as little as 10 min after the STI. The signal responsible for this effect was multimodal; song alone or the visual stimulus of the decoy alone was less effective than the combined stimuli. The quality of male song also affects hormone levels in female songbirds. Rupert Marshall and colleagues found that female canaries exposed to songs with rapid trill rates ('attractive' syllables) had elevated levels of androgens and estrogens compared to when they were exposed to songs without attractive syllables.

Acoustic signals also have rapid effects in fishes. Luke Remage-Healey and Andrew Bass presented male oyster toadfishes with synthetic playbacks of 'boatwhistles,' a call produced by males to defend nest sites and attract females. The boatwhistle playbacks increased the males' own call production and significantly increased plasma 11-KT concentrations within 20 min of the playback.

In electric fish, glucocorticoids, brain plasticity, and electric communication signals interact reciprocally. Kent Dunlap and colleagues found that housing male brown ghost knifefish in pairs significantly elevated the cortisol levels and increased the production of EOD modulations (chirps) that function as agonistic and courtship signals. Implanting isolated males with cortisol also increased chirping. Subsequent studies by Dunlap and his students found that cortisol treatment or interaction with conspecifics dramatically increased cell division and the presence of radial glia in the PPn, which controls chirping. Presenting isolated fish with synthetic EODs mimicked the effects of social housing on cell division and the number of radial glia in the PPn. Thus, exposure to conspecific electric communication signals even in the absence of physical interactions triggers cortisol release, which in turn initiates long-term structural changes in brain regions that control the production of communication signals.

Conclusions and Future Directions

The examples in this article illustrate many commonalities in the hormonal control of communication behavior across vertebrates. These commonalities have allowed well-studied models of communication behavior (e.g., vocalizations in songbirds, frogs, and sonic fishes; electric communication signals in electric fish; and pheromones in fish, newts, and mammals) to provide general insights into hormone–brain–behavior relationships. The links between hormones, brain circuits, and communication behaviors have evolved in response to species differences in social organization. The Challenge Hypothesis, for example, has been broadly tested and explains interspecific variation in the relationship between androgens and agonistic communication signals (and other aggressive behaviors) as a function of mating systems. Because many communication signals are controlled by dedicated motor circuits in the brain, neural mechanisms underlying variation in the

hormonal control of communication signals can also be studied. Interspecific variation in the sexually dimorphic expression of androgen receptors in the song control regions of songbirds, for example, is correlated with sex differences in singing behavior. An exciting area for future studies lies in understanding links between sensory systems, hormones, and motor systems. Although communication signals profoundly affect hormone levels, we are only beginning to understand the neural mechanisms by which sensory systems discriminate among communication signals and transduce them into changes in hormone production and release. Studies using tools such as immediate early genes to map brain activity during exposure and responses to communication signals are likely to be particularly useful in addressing this question.

See also: Acoustic Signals; Active Electroreception: Vertebrates; Aggression and Territoriality; Agonistic Signals; Anthropogenic Noise: Impacts on Animals; Body Size and Sexual Dimorphism; Electrical Signals; Endocrinology and Behavior: Methods; Evolution and Phylogeny of Communication; Hearing: Vertebrates; Hormones and Behavior: Basic Concepts; Mate Choice in Males and Females; Mating Signals; Multimodal Signaling; Nervous System: Evolution in Relation to Behavior; Neural Control of Sexual Behavior; Neurobiology, Endocrinology and Behavior; Neuroethology: Methods; Neuroethology: What is it?; Olfactory Signals; Parasites and Sexual Selection; Playbacks in Behavioral Experiments; Seasonality: Hormones and Behavior; Signal Parasites; Smell: Vertebrates; Sound Production: Vertebrates; Swordtails and Platyfishes; Threespine Stickleback; Túngara Frog: A Model for Sexual Selection and Communication; Vibration Perception: Vertebrates; Vibrational Communication; Vision: Vertebrates; Visual Signals; Vocal–Acoustic Communication in Fishes: Neuroethology; White-Crowned Sparrow.

Further Reading

Arch VS and Narins PM (2009) Sexual hearing: The influence of sex hormones on acoustic communication in frogs. *Hearing Research* 252: 15–20.

Bass AH and Grober MS (2009) Reproductive plasticity in fish: Evolutionary lability in the patterning of neuroendocrine and behavioral traits underlying divergent sexual phenotypes. In: Pfaff D, Arnold A, Etgen A, Fahrbach S, and Rubin R (eds.) *Hormones, Brain, and Behavior,* 2nd edn., pp. 580–609. San Diego, CA: Academic Press.

Bass AH and Remage-Healey L (2008) Central pattern generators for social vocalization: Androgen-dependent neurophysiological mechanisms. *Hormones and Behavior* 53: 659–672.

Bass AH and Zakon HH (2005) Sonic and electric fish: At the crossroads of neuroethology and behavioral neuroendocrinology. *Hormones and Behavior* 48: 360–372.

Bridgham JT, Carroll SM, and Thornton JW (2006) Evolution of hormone-receptor complexity by molecular exploitation. *Science* 312: 97–101.

Cummings ME, Rosenthal GG, and Ryan MJ (2003) A private ultraviolet channel in visual communication. *Proceedings of the Royal Society of London, Series B* 270: 897–904.

Dixson AF (1998) *Primate Sexuality.* London: Oxford University Press.

Endler JA and Basolo AL (1998) Sensory ecology, receiver biases, and sexual selection. *Trends in Ecology & Evolution* 13: 415–420.

Endler JA, Westcott DA, Madden JR, and Robson T (2005) Animal visual systems and the evolution of color patterns: Sensory processing illuminates signal evolution. *Evolution* 59: 1795–1818.

Godwin J and Crews D (2002) Hormones, brain and behavior in reptiles. In: Pfaff D, Arnold A, Etgen A, Fahrbach S, and Rubin R (eds.) *Hormones, Brain, and Behavior,* 1st edn., vol. 2, pp. 545–585. San Diego, CA: Academic Press.

Hill PSM (2008) *Vibrational Communication in Animals.* Cambridge: Harvard University Press.

Hopkins CD (1999) Design features for electric communication. *The Journal of Experimental Biology* 202: 1217–1228.

Houck LD (2009) Pheromone communication in amphibians and reptiles. *Annual Review of Physiology* 71: 161–176.

Kimball RT and Ligon JD (1999) Evolution of avian plumage dichromatism from a proximate perspective. *The American Naturalist* 154: 182–193.

Marler P and Slabbekoorn H (2004) *Nature's Music: The Science of Birdsong.* New York, NY: Academic Press.

Nelson R (2000) *An Introduction to Behavioral Endocrinology.* Sunderland, MA: Sinauer.

Remage-Healey L and Bass AH (2006) A rapid neuromodulatory role for steroid hormones in the control of reproductive behavior. *Brain Research* 1126: 27–35.

Rosenthal GG and Ryan MJ (2000) Visual and acoustic communication in non-human animals: A comparison. *Journal of Biosciences* 25: 285–290.

Ryan MJ and Kime NM (2003) Selection on long-distance acoustic signals. In: Megala-Simmons A, Popper A, and Fay R (eds.) *Acoustic Communication,* pp. 225–274. New York, NY: Springer-Verlag.

Schlinger BA and Brenowitz EA (2009) Neural and hormonal control of birdsong. In: Pfaff D, Arnold A, Etgen A, Fahrbach S, and Rubin R (eds.) *Hormones, Brain, and Behavior,* 2nd edn., vol. 2, pp. 897–941. San Diego, CA: Academic Press.

Stacey NE and Sorensen PW (2002) Hormonal pheromones in fish. In: Pfaff D, Arnold A, Etgen A, Fahrbach S, and Rubin R (eds.) *Hormones, Brain, and Behavior,* 1st edn., vol. 2, pp. 375–434. New York, NY: Academic Press.

Stoddard PK (1999) Predation enhances complexity in the evolution of electric fish signals. *Nature* 400: 254–256.

Tokarz RR, McMann S, Smith LC, and John-Alder H (2002) Effects of testosterone treatment and season on the frequency of dewlap extensions during male-male interactions in the lizard *Anolis sagrei. Hormones and Behavior* 41: 70–79.

Wilczynski W, Lynch KS, and O'Bryant EL (2005) Current research in amphibians: Studies integrating endocrinology, behavior, and neurobiology. *Hormones and Behavior* 48: 440–450.

Wingfield JC, Hegner RE, Dufty AM, Jr, and Ball GF (1990) The "challenge hypothesis": Theoretical implications for patterns of testosterone secretion, mating systems, and breeding strategies. *The American Naturalist* 136: 829–846.

Wyatt TD (2003) *Pheromones and Animal Behaviour,* New York, NY: Cambridge University Press.

Yamaguchi A and Zornik E (2008) Sexually differentiated central pattern generators in *Xenopus laevis. Trends in Neurosciences* 31: 296–302.

Yamaguchi A and Kelley DB (2003) Hormonal mechanisms in acoustic communication. In: Megala-Simmons A, Popper A, and Fay R (eds.) *Acoustic Communication,* pp. 275–323. New York, NY: Springer-Verlag.

Yang E-J and Kelley DB (2009) Hormones and the regulation of vocal patterns in amphibians: *Xenopus laevis* vocalizations as a model system. In: Pfaff DW, Arnold AP, Etgen AM, Fahrbach SE, and Rubin RT (eds.) *Hormones, Brain, and Behavior,* pp. 693–706. San Diego: Academic Press.

Zakon HH and Smith GT (2009) Weakly electric fish: Behavior, neurobiology, and neuroendocrinology. In: Pfaff DW, Arnold AP, Etgen AM, Fahrbach SE, and Rubin RT (eds.) *Hormones, Brain and Behavior,* 2nd edn., vol. 1, pp. 611–638. San Diego, CA: Academic Press.

Communication Networks

M. D. Greenfield, Université François Rabelais de Tours, Tours, France

Introduction

The notion of animal communication networks arises from observations that more than two individuals often interact in the transmission and reception of signals. Although the classical approach to animal communication is built on the paradigm of a signaler who broadcasts his message (the signal) along a channel to a specific receiver, it is clear that several variations on the two-party theme may occur: the simplest variation ('broadcast networks') occurs when a signaler broadcasts an advertisement that offers information on his/her behavioral and physiological state to all individuals that happen to be within the receiving range. A second variation ('eavesdropping networks') occurs when a clear dialog exists between a given pair of individuals, but the signals of each, intended for the other member of the pair, are perceived by extra-pair individuals. This situation has often been termed 'eavesdropping,' but it should be noted that such interceptions by extra-pair individuals may or may not reduce the fitness of the primary signaler or receiver. A third category ('interaction networks') of networks occurs where individuals maintain dialogs with two or more neighbors. These several interactions may allow multiple opportunities for eavesdropping. Consider an assemblage of three neighbors: They interact directly with one another in pairwise dialogs, and each is also privy to information that the other two individuals exchange; that is, an individual both eavesdrops on its neighbors and communicates directly with them.

On the basis of these three observed variations, an animal communication network may be defined as any system of information exchange that includes three or more conspecific individuals. Our current interest in networks stems from the recognition that various processes occur when multiple individuals communicate that do not arise when communication is confined to pairwise exchanges.

One may note that the various networks described earlier also occur among individuals belonging to different species. For example, predators or parasites may eavesdrop on the sexual communications of their prey or host species and thereby locate them. Or individuals in one prey species may obtain information on the presence of predators by perceiving telltale changes in the communications among a second prey species; both of these phenomena might be categorized as interspecific eavesdropping. But many biologists restrict animal communication to information exchanges among conspecific individuals, and this restriction is extended to the treatment of networks presented here. First, each of these variations is discussed, followed by the presentation of a theoretical approach with which we can understand the evolutionary implications of communication networks. This article ends with an in-depth treatment of a special category of networks, choruses.

Broadcast Networks

Broadcast networks are commonly observed for sexual advertisement signals, territorial announcements, and certain types of messages emitted in animal societies, such as alarm calls. These signals may all be classified as 'public information' in that they are readily available to all receivers in the vicinity. The signals broadcast in these networks are not directed toward any specific receivers, and the signaler sends them without having specific information on the locations or even the presence of receivers. Nonetheless, signalers may broadcast at certain times and places where intended receivers are more likely to be present: males (or females) may advertise at sites where members of the opposite sex are expected to arrive and during periods of the day or season when they are normally most active and responsive. In social species, individuals may be more apt to emit alarm signals, which indicate danger such as the approach of predators, when genetic relatives are expected to be present than when many neighbors are unrelated individuals. Directionality of these broadcasts is generally minimum, and intensity may be maximum, features that would increase the number of potential receivers. In the case of alarm signals, the broadcasts may also bear features that make their own localization difficult for receivers, predators, and conspecifics alike. However, conspecific individuals need to detect only the alarm signal in order to benefit, information on the precise location of the signaler being unnecessary.

Eavesdropping Networks and Audience Effects

Eavesdropping networks are observed for diverse signals in which a clear dialog exists within a pair of individuals that exchange 'private information.' For example, a given male individual may perceive the courtship signals exchanged between a neighboring male and a female. Thus, the given male can surreptitiously learn of the presence of a receptive female, and he may subsequently arrive at the courting pair and intervene, possibly with success. Such potential events represent one source of selection pressure that favors minimum intensity in courtship signals and other private communications. These messages need to be sufficiently intense that the receiver can accurately evaluate the signaler, but often no more. In other situations, the signaler is not necessarily harmed by information revealed inadvertently to an eavesdropper. The aggressive signals exchanged between two individuals settling a dispute over territory ownership may be witnessed by a third, which then evaluates the motivation and physical abilities of the signalers. This third individual may then behave in accordance with its own motivation and perceived ability: behavior that could entail deferring to one or both signalers and moving away. The latter example shows how eavesdropping networks can bear some of the features characteristic of broadcast networks. Purportedly private information exchanged within a specific dialog may sometimes also serve as public information that is available – and indeed intended – for individuals outside of that dialog. In such cases, we might expect selection to favor signal intensities that exceed minimum levels because of the value of facilitating, as opposed to preventing, eavesdropping by third parties.

The distinctions between public and private information are blurred when a signaler engaged in a dialog could benefit from information revealed to third parties. When this happens, animals should sometimes adjust their signals when extra-pair individuals are present. These adjustments do occur, and they are termed 'audience effects': animals participating in a dialog modify their communications when 'bystanders' observe. Well-known cases of audience effects are found in various vertebrate species and entail raising the intensity or making other adjustments to signals in male–male interactions when conspecific males, or females, are present. In both the cases, the modification may serve to improve the signaler's status in future interactions with these bystanders: a male signals to other males that his competitive ability is potentially high, and he may thereby forgo the costs of overt fighting in a future aggressive encounter. Likewise, he signals to attending females that he is a potential mate of superior quality. While audience effects may be mediated via neural and hormonal pathways, their adaptive explanation could be economic: signal modifications are likely to be costly, and it may be worthwhile for a male to expend the necessary energy only when a long-term benefit, that is, an increase in overall status that transcends the immediate interaction with a single adversary, is expected.

Interaction Networks

Interaction networks are observed where animals occupy relatively stable positions in space or in a social unit and may exhibit some degree of individual recognition. Here, individuals regularly shift back and forth between the roles of active signaler or receiver and passive eavesdropper. By assuming these dual roles, an individual may be afforded the opportunity to make precise adjustments in its social interactions with members of the local population. For example, an individual perceiving the messages exchanged in a pairwise encounter between two of its neighbors may then adjust its own communications with these neighbors when future occasions arise: in the role of an eavesdropper, it may infer the relative hierarchy of its neighbors or their membership in social coalitions and then enter into its own communications with them in accordance with such information as well as its self perceptions and needs. Or, having learned the unique characteristics of the signals of its neighbors holding adjoining territories, it may then recognize a new arrival and respond aggressively, the so-called 'dear enemy effect': an individual's regular neighbors are territorial rivals that it has adjusted to out of necessity, but novel signals indicate the presence of an intruder from the outside that could upset a delicate balance of power. Here, overt aggression may be worthwhile.

Connectedness and Network Theory

A general property of animal communication networks is that messages are often transmitted and received indirectly via individuals that occupy intermediate positions between the initiating signaler and the final recipients. This property is most apparent in some broadcast networks, where a message that originates from one signaler may be transmitted to x receivers, each of which in turn transmits it to x new receivers. The advantage of this chain-reaction process for communicating to a maximum number of individuals is clear: following n cycles of such transmission, x^n individuals have been informed. Moreover, a message may be transmitted effectively across distances that would not be possible via direct transmission between one signaler and one or several attending receivers. In addition, the speed of transmission to the multitude of receivers is greatly enhanced. The dissemination of queen pheromone to a very high percentage of workers in a honeybee (*Apis mellifera*) colony on a daily basis is accomplished in this way. The queen may have direct contact with only a handful of workers, each of which receives a minute quantity of pheromone, but each of these workers in the queen's retinue transfers a portion of the chemical message it has received to a secondary cadre. Owing to extreme sensitivity to queen pheromone, four to five repetitions of this process serve to inform most workers and preserve the social structure of the colony. Alarm messages in many social insect colonies are also transmitted in this fashion.

The queen pheromone example serves to highlight some general properties of animal communication networks that are of interest to mathematical modelers. In the language and formulations of theoreticians, each animal is a 'node' that is connected via 'edges' to other nodes when regular communication occurs. The aggregate of all of these connections forms a global network the properties of which may reveal some key features about the individual components (animals) and the way in which they interact. Do most nodes (individual animals) have an average number of connections to other nodes, or are some nodes connected to many, many nodes, while the majority of nodes have relatively few connections? In the latter case, nodes with many connections, termed 'hubs,' can play a critical role in a network in that their removal may have a major impact on reducing or altering the incidence of subsequent interactions. Additionally, nodes with many connections, also termed high-'degree' nodes in situations where there is a continuous gradation in the number of connections among nodes, may tend to be connected to other high-degree nodes, while low-degree nodes are typically connected to other low-degree nodes. In other cases, most nodes may have an average number of connections, but groupings within the global network can exist wherein most nodes are interconnected but remain relatively unconnected to nodes belonging to other groupings.

Scientists have found applications of network theory in informatics and the World Wide Web, transportation systems, sociology, cell biology, epidemiology, and animal social behavior. In the context of animal communication networks, computer analyses using the methods of network theory have the potential to reveal the nature of social groupings. That is, the general patterns of who communicates with whom, the numbers of individuals each animal regularly communicates with, their locations and behavior, and their memberships in social coalitions can be determined.

Choruses

Choruses constitute a special category of animal communication networks wherein many individuals advertise sexual signals at the same time and exhibit mutual influences on their broadcasts. Viewed from our categorization of communication networks, choruses represent broadcast networks in which the various broadcasters themselves interact with one another and thereby establish collective displays that are temporally structured. While we normally associate choruses with acoustic displays, the basic phenomenon may occur in any signaling modality. The general explanation for choruses is that they represent the outcome of competition between neighboring broadcasters, but in some cases they may reflect cooperation.

Animal choruses exhibit different levels of temporal precision. The crudest level is observed in collective displays where individuals at a locality simply broadcast at the same hour of the day or night. Here, the individual broadcasters may be stimulated to begin and to continue signaling by perceiving the signals of neighbors. Via a chain-reaction process, such chorusing can spread among a great many individuals broadcasting over a wide area. However, the chorus may also arise because of common responses to an environmental cue, for example, photoperiod, in which case neighboring broadcasters do not necessarily display any mutual perception of one another. Various explanations are proposed for these choruses, particularly the dawn and dusk choruses observed in many songbirds, but a general one is that they represent a response to competition among individual broadcasters. In the typical case of males broadcasting sexual advertisements to females, a male must at least match the signals of his neighbors in order to influence and attract potential mates. Thus, if a male's neighbor signals, he is compelled to follow suit, energy permitting. Most of these choruses are collective displays of male acoustic signals, but some examples of visual, vibratory, and even chemical signals are known. And in cases of sex-role reversal where males exhibit some degree of mate discrimination and females assume a competitive role, female chorusing may also occur.

The next level of chorusing entails the repetition of 'bouts' of collective signaling many times during a daily activity period. The collective singing bouts observed in many acoustic insects and anurans are representative: 'bout leaders,' males that may have relatively more energy or motivation, initiate calling and are then joined by most of the other male signalers in the local population. Most participants maintain a high rate of calling for several minutes, after which the amount of calling gradually decreases to zero. Following an interval of silence of variable length, the cycle repeats. In some cases, the same individual male serves as bout leader over and over again.

In species where signal repetition is controlled by a central (nervous system) pattern generator, a regular, 'free-running' signaling rhythm results (e.g., a 100-ms signal emitted every 800 ms). This occurs in many arthropods and anurans, where a higher level of chorusing precision may occur. This precision is generally characterized by neighboring individuals that maintain specific phase relationships between their signal rhythms. Such phase relationships give rise to the collective synchrony or alternation that is often observed for acoustic, vibratory, and visual signals. The latter include both bioluminescent signals, as in the flashes of fireflies and certain marine crustaceans (ostracods), and signals relying on reflected light, as in the claw-waving displays of fiddler crabs. Rhythmic synchrony and alternation are not observed in chemical signaling, an absence that probably results from the inability to control the precise timing of the pheromone channel: chemical signals travel by diffusion and convection and are therefore relatively slow and rather subject to vagaries of the environment. Moreover, they fade out over a lengthy interval.

Interaction Mechanisms

Observers of synchronous and alternating choruses are led to ask two sorts of questions: What are the physiological mechanisms that lead to such interactions, and why do animals bother to interact in this way? Studies of the mechanisms responsible for synchrony and alternation have generally relied on controlled 'playback experiments': A given individual in the field or laboratory is isolated from its neighbor(s), which are supplanted by the presentation of precisely timed synthetic stimuli. These playback stimuli are typically adjusted incrementally in successive tests across a range of values for signal repetition rate, carrier frequency (pitch), and signal length and amplitude; the signaling of the test individual and the playback are then simultaneously recorded in each test. Thus, scientists have been able to discern some of the rules that regulate interactions among rhythmic signalers, those species in which signals are broadcast repetitively with a regular rhythm during a major portion of the hours of daily activity.

While chorusing assumes diverse forms and various mechanisms are apparently responsible for these displays, analyses of playback experiments also indicate some general patterns that account for many of the interactions observed. In many species that signal rhythmically, when an isolated individual perceives a single signal – either from a neighbor or a synthetic playback – it lengthens its concurrent intersignal interval but then returns to its regular, free-running rhythm following that interval (**Figure 1(a)** and **1(b)**). This response is called a 'phase delay' mechanism, because the signaler modifies only its phase relative to the neighbor's signal, not its actual rhythm. If a train of repeated signals are now presented as a playback stimulus, the signaler may then 'entrain' to that stimulus by broadcasting his signals at a given delay after each playback signal (**Figure 1(c)** and **1(d)**); that is, the signaler maintains a specific phase relationship with the stimulus. Depending on the nature of the signaler's phase-delay response and the rhythm of the playback stimulus, different signal interactions are possible. If the signaler's phase delay is relatively brief and the stimulus rhythm is equivalent to the signaler's free-running rhythm or slightly slower, the signaler and the stimulus will alternate (**Figure 1(c)**). On the other hand, if the signaler's phase delay is relatively long – that is, the length by which its concurrent intersignal interval increases in response to a stimulus approaches the mean length of his free-running interval – and the rhythms of the stimulus and signaler are similar, the two will synchronize (**Figure 1(d)**).

We now return to the natural interactions between two signalers and apply the rules determined from the playback experiments mentioned earlier. Replacing the playback stimulus with a neighbor, we observe that in species where signalers exhibit relatively brief phase delays, two males that have similar free-running rhythms will generally alternate. Moreover, the mutual responses of the two males will ensure an approximately 180° phase relationship between their signaling rhythms, which are slightly longer than during the free-running state (**Figure 1(e)**). This genre of signal interaction is most commonly observed in acoustic insects and anurans. Analogously, in species where signalers exhibit relatively lengthy phase delays, two males that have similar free-running rhythms will generally synchronize; that is, a 0° phase relationship will exist between their signaling rhythms (**Figure 1(f)**). Live signalers, however, are not perfect metronomes, and the likelihood that two males have constant and identical free-running rhythms is small. These two sources of stochastic variation will cause any resulting synchrony to be an imperfect one: typically, one male leads the other by a small phase angle, although the roles of leading and following usually pass back and forth between the males on successive signaling cycles (**Figure 1(g)**). Additionally, one male may occasionally drop out of the interaction for one or several cycles of signaling, only to return in approximate synchrony on the subsequent cycle. This imperfection would arise when a male happens to slow his relative free-running rhythm relative to the other individual, or when he happens to extend his phase delay beyond the normal length (**Figure 1(g)**). Synchrony as described above is found in various acoustic insects and also in some fireflies and fiddler crabs.

In other species, playback experiments have revealed strikingly different mechanisms in which signal interactions entail not only phase delays but also changes in the actual rhythms. Here, an individual signaler which perceives a rhythmic stimulus will exhibit a minor phase delay in its concurrent intersignal interval, but it will also extend its response over future signal cycles. If the stimulus rhythm is faster than its own, it will gradually accelerate its signal rhythm until the two rhythms are more or less 0° in phase. Should the stimulus rhythm be slower, it will gradually slow its own rhythm appropriately. And when live signalers interact, faster signalers get slower, slower signalers get faster, and the several signalers eventually approach a common rhythm (**Figure 1(h)**). This type of synchrony tends to be very precise: phase angles separating the signalers at any given cycle are

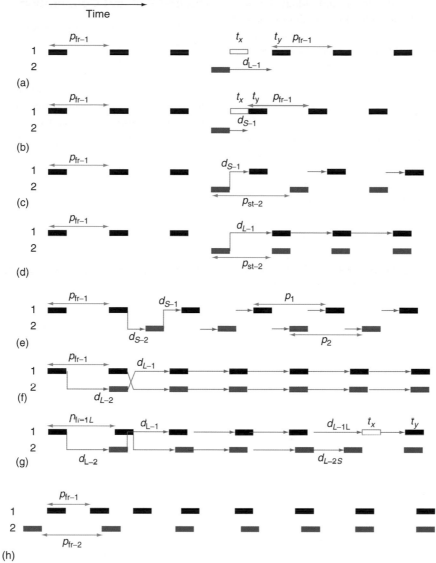

Figure 1 Phase relationships and signal interactions in choruses. In parts (a–d), the black horizontal bars (trace 1) indicate the timing of signals produced by an individual animal, and the red horizontal bars (trace 2) indicate stimuli, as presented in a playback experiment. In parts (e–h), the black and red bars indicate the timing of signals produced by two interacting individuals, represented by traces 1 and 2, respectively. For all parts, p_{fr} represents an individual's free-running signal period (green double-ended arrows), and open blue bars indicate the expected timing of signals, based on the free-running period, that were delayed for particular intervals (blue arrows) following a stimulus or a signal of the neighboring individual. (a) Presentation of a single stimulus results in the animal delaying its subsequent signal for a relatively lengthy interval d_{L-1} ($=t_y - t_x$), measured from the end of the stimulus; the animal returns to its free-running signal period immediately after that single modified cycle. (b) Same as (a) save that the animal delays its subsequent signal for a relatively short interval d_{S-1} ($=t_y - t_x$), measured from the end of the stimulus. (c) Animal entrains to a rhythmic stimulus, whose period (p_{st-2}) is slightly longer than the animal's free-running period, by broadcasting each signal after a short delay (d_{S-1}) timed from the end of the last stimulus. The resulting interaction is alternation, with the animal's rhythm lagging the stimulus by a phase angle = 166°. (d) Same as (c) save that the animal's period equals the stimulus period and it broadcasts each signal after a lengthy delay (d_{L-1}) timed from the end of the last stimulus. Here, the resulting interaction is synchrony; that is, the phase angle separating the animal's rhythm and the stimulus = 0°. (e) Individuals 1 and 2 have identical free-running periods and respond to each other by delaying their subsequent signal for a relatively short interval d_{S-1} ($=d_{S-2}$), measured from the end of the last signal broadcast by the other one. The resulting interaction is perfect alternation, with periods slightly longer (p_1, p_2) than the free-running period and the phase angle separating the animals' rhythms = 180° (compare with part (e), above). (f) Same as (e) save that individuals 1 and 2 respond to each other by delaying their subsequent signal for a relatively lengthy interval d_{L-1} ($=d_{L-2}$), measured from the end of the last signal broadcast by the other one. The resulting interaction is synchrony (phase angle separating the animals' rhythms = 0°). (g) Same as (f) save that stochastic variation occurs in free-running periods (see p_{fr-1L} of individual 1) and in delay intervals (see d_{L-1L} and d_{L-2S} of individuals 1 and 2, respectively). Thus, the animals' rhythms are not in perfect synchrony, and one animal (individual 1) drops out of the chorus for one cycle (compare with part (f), above). (h) Two animals initially have different free-running rhythms. Each makes slight phase delays in response to the other's signals; moreover, the faster one slows, while the slower one accelerates. Eventually, they achieve identical rhythms which are in perfect synchrony (phase angle = 0°).

extremely small, the rhythms of the individual signalers remain rather constant, and individuals seldom drop out of the collective display. It is best known in certain fireflies of southeast Asia and the Malay Archipelago, but variants also occur among some acoustic insects.

Selective Attention

Playback experiments and observations of pairs of interacting signalers offer some insight into the mechanisms that regulate choruses. However, as with communication networks in general, chorusing cannot be fully understood until we begin considering displays of three or more signalers, which invariably occur in the field. The complication generated by extending beyond pairwise interactions is most obvious for alternation: Individual A may alternate with individual B, but what occurs when individual C arrives and signals? In some cases, a three-way accommodation is possible where each individual signals with a 120° phase delay relative to a given neighbor, but the general solution to the problem of multiple neighbors is 'selective attention.'

Analyses of recordings made of actual choruses in the field have generally shown that a given signaler does not interact with every participant in the chorus. That is, groupings of interactive neighbors occur within the global communication network. Conceivably, these groupings might be based on any of various physical criteria: one interacts with individuals the signal intensities of which, as perceived, exceed a certain fixed threshold value, one interacts with a certain fixed number of individuals, or one interacts with the individual the signal intensity of which, as perceived, is maximum and all other individuals the signal intensities of which are within x dB of this most intense neighbor. Here, we note that the last criteria, a sliding threshold on the basis of the intensities of one's neighbors, would reflect basic physiological function. Animals usually exhibit sensory adaptation to the stimuli to which they are regularly exposed, and the range of neighbor signal intensities that elicit a male's attention is expected to slide up and down relative to the level of the background chorus that its neighbors generate. When neighbor intensities are high, the attention window rises, and when intensities are low, it descends accordingly. In addition to these basic rules of thumb, interactions in some choruses may be regulated by combinations of rules, or by rules that do not simply focus on the physical criteria of neighbors' signals. For example, a male might pay attention to male neighbors in the chorus on the basis of his history of interacting with them or their current behavior or signaling locations. He might even base his selectiveness on arbitrary rules; for example, interact with the two males signaling on the left, or on the right, and ignore the rest.

Competition and Cooperation

Signaling males in choruses generally compete with one another to attract local females, and we might suppose that the interactions which generate synchrony and alternation reflect signal competition at some level. This proposition is more obvious for alternation. Assuming that signals that are broadcast without interference from other signals will be more attractive to females, we may expect selection for phase-delay mechanisms that generate alternation. Thus, a male avoids having his signals masked by those of his neighbors. Moreover, it may be critical for a male to eavesdrop on neighboring signalers. This monitoring is not easily done during one's own signals because of the general reduction in perceptual sensitivity at this time. Alternation would serve to permit effective eavesdropping on nearby males and their broadcast networks. And when a male is surrounded by multiple neighbors, he may deal with any potential confusion by applying his phase-delay mechanism only to his nearest, most intense neighbors – generally his most important competitors – and behaving as if other members of the chorus were inactive. This type of selective attention is sometimes seen in frog choruses, where a male may insert his calls during brief 'windows' of relative silence that occur in the background cacophony.

A tidy explanation of chorusing as a response to competition seems less plausible for synchrony, a phenomenon the very name of which conveys the notion of cooperation. But in many cases, synchrony may also arise from signal competition between rival males. This possibility was indicated by analyses of female choice, which revealed that females in various species exhibit a striking orientation toward and preference for male signals that lead their neighbors' signals by brief time delays (10–400 ms). Such female preferences, which may result from general 'forward masking' or 'precedence effects' in receiver psychophysics, can exert considerable selection pressure on the mechanisms with which males adjust rhythmic signals: Since following signals are rather ineffective in attracting females, phase-delay mechanisms will be strongly favored when they result in a reduction in the number of signals that a male broadcasts immediately after a neighbor and also improve his chance of broadcasting a leading signal during the subsequent cycle. Imperfect synchrony then occurs when two males each sing with comparable rhythms and have relatively lengthy phase delays. One may then ask, given the disadvantage of signaling in the following role, why males ever produce any following signals and thereby generate synchrony. Following signals are produced, and produced quite regularly, because a 'motor delay,' lasting from 25 to 200 ms in different species, elapses between the instant when a signal is triggered by the central pattern generator and when it is actually broadcast. Thus, if a male perceives a neighbor's signal but his subsequent signal has already been triggered, he is compelled to broadcast his (following) signal despite its ineffectiveness (see **Figure 1(g)**).

Other cases of synchronous chorusing, particularly those where synchrony is very precise and generated by adjustments to free-running rhythms, may reflect cooperation among signalers. Cooperative signaling may be influenced by various ecological and behavioral factors. Signaling males often attract natural enemies (interspecific eavesdroppers) and conspecific females, but the attraction of enemies might be reduced when all males in a chorus broadcast synchronously. Here, the arrival of multiple signals from all directions at precisely the same time may challenge the ability of an enemy to localize any one signaler.

Second, females may need to perceive a particular signal rhythm or specific signal features of conspecific males before responding, and these characters may be obscured if males do not synchronize with their neighbors. Thus, a male must broadcast in synchrony, because to do otherwise would be spiteful: he might reduce the attractiveness of his neighbors to females, but he would also reduce his own appeal. Such cooperation has been found in a North American katydid species (*Neoconocephalus nebrascensis*) in which males produce 1 s chirps every 2 s, and neighboring males synchronize precisely. Female *N. nebrascensis* must hear the 1 s intervals between a male's chirps before responding to him, a preference feature that demands male synchrony. When local males alternate, females do not perceive the interchirp intervals from any male and respond to no one.

Third, owing to resource distribution and habitat structure, signaling males may be clustered in space, in which case the various clusters might compete with one another for attracting females. When such intergroup competition occurs, males might improve the attractiveness of their cluster by synchronizing with their neighbors and thereby maximizing the cluster's signal intensity. Of course, once females arrive at the cluster, the several males would compete with each other for mating opportunities, and we may imagine that additional signal interactions are generated in this intracluster context. That is, some choruses may reflect the combined influences of competition and cooperation.

Emergent Properties and Feedback Loops

The explanation of imperfect synchrony as an outcome of competition suggests that many choruses are best described as 'emergent phenomena.' According to this proposition, selection has favored the simple mechanisms that regulate signal interactions between individuals, but the resulting collective display is an event that just represents the summation of these interactions. Taken in sequence, we note that perceptual biases toward leading signals in receivers – which may represent mechanisms that improve the basic ability to localize the sources of stimuli – favor the evolution of certain adjustments in signal timing. These adjustments, which allow a male to reduce his incidence of following signals and to improve his chance of broadcasting leading ones in the subsequent cycle, happen to generate synchrony in some species and alternation in others. These outcomes depend on the basic features of signaling, particularly the adjustments made during signal interactions, but the specific format of chorusing that emerges is not under selection. That is, female receivers do not prefer any specific type of chorus, and males do not benefit from generating one.

Although a certain percentage of animal choruses may simply represent byproducts of the basic signal interactions between chorus participants, they present signaling environments that can have major influences on both signalers in the chorus and receivers attending to it. For example, in the case of alternation, an environment is created wherein signals are being broadcast nearly continuously. Because few gaps in the background noise appear, selective attention toward only a subset of neighboring signalers would be strongly favored by selection. By comparison, one may predict that such selection pressure would be weaker in species that happen to generate synchronous choruses: a signaler does not suffer if he pays attention to all of his neighbors. Similarly, chorusing interactions – either alternation or synchrony – may allow females to discern differences among male signalers, for example, signal rhythm, that are otherwise not readily perceived. Thus, once chorusing emerges, selection pressure maintaining female preference for leading signals may be strengthened, as females exhibiting this preference would now have the opportunity to benefit in a tangible way: the ability to choose higher quality mates more dependably. These examples demonstrate the various ways in which 'feedback loops' may connect the global signaling environment of an animal communication network and the basic interactions from which these networks are assembled.

See also: Acoustic Communication in Insects: Neuroethology; Acoustic Signals; Collective Intelligence; Mating Signals; Multimodal Signaling; Neuroethology: Methods; Playbacks in Behavioral Experiments; Signal Parasites; Vibrational Communication; Visual Signals.

Further Reading

Burt JM and Vehrencamp SL (2005) Dawn chorus as an interactive communication network. In: McGregor P (ed.) *Animal Communication Networks*, pp. 320–343. Cambridge: Cambridge Univsity Press.
Greenfield MD (2005) Mechanisms and evolution of communal sexual displays in arthropods and anurans. *Advances in the Study of Behavior* 35: 1–62.

Lim H-K and Greenfield MD (2007) Female pheromonal chorusing in an arctiid moth (*Utetheisa ornatrix*). *Behavioral Ecology* 18: 165–173.

McGregor P (2005) *Animal Communication Networks*, pp. 657. Cambridge: Cambridge University Press.

Newman M (2008) The physics of networks. *Physics Today* 61(11): 33–38.

Schwartz JJ, Freeberg TM, and Simmons AM (2008) Acoustic interaction of animal groups: Signaling in noisy and social contexts. *Journal of Comparative Psychology* 122(3 special issue): 231–333.

Strogatz S (2003) *Sync: The Emerging Science of Spontaneous Order*, pp. 338. New York: Hyperion Press.

Communication: An Overview

J. Lucas, Purdue University, West Lafayette, IN, USA
T. M. Freeberg, University of Tennessee, Knoxville, TN, USA

Introduction

Communication draws individuals together and maintains social bonds. It can also be a repulsive agent that pushes individuals away. Plants tell insects when pollen and nectar are available from their flowers. Prey tell predators that they are not worth stalking, either because they are too fast or too toxic. Bacteria tell each other when they need to move as a group. Males tell females that they are great mates. Females tell males whether or not they are ready to mate. Offspring tell parents how hungry they are. Signals can encode information about individual or group identity, dominance class, physiological state, and anatomical state. In short, communication is everywhere and is used by individuals of all animal species (if not by all species), and it is extremely important in the lives of organisms.

The simplest way to think about communication is that it is the sharing of information between a sender and a receiver through the form of a signal that must be constructed in such a way that the signal retains its information after propagating through the environment. Upon reception of the signal, the receiver will likely change its behavior as a result of the information it receives, and on average, this change of behavior should be advantageous to both the signaler and the receiver. The logic behind the last statement is simple: if the signal caused a change in behavior of the receiver that was detrimental to the sender, then the signal should not be sent to begin with. Similarly, if the signal-induced change in behavior is detrimental to the receiver, then the receiver should respond differently in a way that is ultimately advantageous to itself.

You will see this simple snapshot of communication repeated in every article in this section of the encyclopedia, because it is a robust way to think about animal communication. Nonetheless, organisms use a dizzying variety of ways to communicate with each other, and the relative value of that sharing of information is both subtle and complex. Indeed, these features of communication make this field both vibrant and ever expanding. The *Animal Communication* section is divided into four parts: *Signaling Mode, Signal Types, Adaptation and Signal Evolution*, and *Communication Across Species.*

Signaling Mode

There are seven essays on signaling mode. Signals are physical entities that are constructed or otherwise produced in some physical medium. Olfactory signals are chemicals detected by the olfactory sensory epithelium of the receiver. Relatively long-distance auditory signals are pressure gradients in air. Visual signals are derived from reflected or refracted light, or in rare cases from light generated by the signaler itself. These different categories of signals are called signaling modes. Six of our articles cover the five primary signaling modes used by animals: olfactory signals by Ginzel, visual signals by Fernandez-Juricic, auditory signals by Simmons and by Suthers, vibrational signals by Cocroft, and electrical signals by Stoddard. These articles focus on signals that are produced in a single mode. Animal behaviorists have recently begun to realize that many signals are multimodal, including, for example, both visual and auditory components, although this idea was introduced at least as early as Darwin in his seminal 1872 work, '*The Expression of the Emotions in Man and Animals.*' Multimodal signals are discussed by Uetz.

Signaling mode is important because different modes impose different advantages and constraints on both the sender and the receiver. Acoustic pressure gradients, for example, cannot be generated by animals that are too small. Given that most of the animals on this planet are small, it comes as no surprise that most animals are silent. Nonetheless, acoustic signals are used by many animals, because they propagate well over distances and can even wrap around objects (like small trees). Similarly, electrical signals are primarily used in freshwater, because salt water is too conductive to maintain a stable voltage that can be used to communicate with conspecifics. Note that constraints on signal use are relevant to the three sequential parts of communication: production, propagation, and reception. A visual signal, for example, is only useful if it can propagate through the environment (and unlike sound, light cannot wrap around trees), and if the information content in the signal can be detected and decoded by the receiver.

Signal Types

The next part of the *Animal Communication* section focuses on the variety of different messages that signals can carry. The enormous diversity in physical properties of signals parallels the enormous diversity of information those signals carry. For example, Gerhardt shows how signals carry information about mating interactions both between sexes (where individuals signal their willingness to mate and their quality, among other factors) and within sexes (where individuals use signals to jockey for access to mates). Some animals share information about food, although as Snowdon discusses, there are many reasons why these food calls might be given, and it can be difficult to determine the exact functional basis for the use of the calls. Horn and Leonard cover parent–offspring signals. As with any type of signal, there is a diversity of alternative reasons why parents and offspring should share information, ranging from honest information transfer to manipulation.

Mating, food, and parent–offspring signals generally represent amicable interactions between individuals. When there are serious conflicts of interest, interactions can become anything but amicable. Nonetheless, Gabor shows how communication can be critical in agonistic interactions particularly under circumstances where agonistic signals are given in lieu of the escalation of a contest over some resource. Both the winner and the loser can benefit if the information they exchange provides a relatively unequivocal consensus of the ultimate outcome of a contest – this notion relates to the key idea of ritualization of signals developed in the early years of Ethology as a field of study.

Breed shows that signals sometimes encode identity along with fundamental information about mating, food, or some other functional question. As the name implies, the signature whistle is a special type of vocalization used by dolphins to convey identity. Sayigh and Janik have studied these signature whistles extensively. The development of our understanding of the function of these signature whistles gives us a broader view of how scientists have grown to understand animal communication. This is so because it took some time to realize that dolphins used vocal signals and then to understand the context of those signals. This general process has been repeated many times and will continue to be repeated as we understand the enormous diversity of animal communication.

The presence of predators elicits alarm signals in a wide range of animals. One intriguing aspects of the predator alarm signals discussed by Slobodchikoff is that many arboreal species, such as birds, tend to produce alarm calls that are surprisingly similar across species. Similarly, terrestrial species share alarm call properties, but the alarm calls of arboreal species are very different from those of terrestrial species. The difference, in part, reflects the optimal escape response. Birds are less likely to be depredated if they are not located by the predator and if they freeze when a predatory hawk is located. Terrestrial animals, such as ground squirrels, are often relatively close to a refuge. Alarm calls in these animals are easier for their intended audience to locate, giving conspecifics information about where the predator is located and giving the predator the information that it has been spotted and therefore is less likely to be successful.

This part ends with a discussion of syntactically complex signals. The syntactical complexity of human language is unequivocally unique and truly extraordinary. Forty years ago, the notion that non-human communication had any properties of language was unthinkable, except by scientists such as Peter Marler and Charles Hockett, whose own theories and research bridged the fields of study of communication and language. On closer analysis, we are perhaps not completely different from other animals. Bird song, and in some cases nonsong 'contact' calls, can have a complex syntactical structure (i.e., vocalizations are governed in part by vocal element ordering rules) that makes some avian communication signals word like. As Bregman and Gentner discuss, non-human primates and perhaps whales may go one step further in generating longer patterns of sounds that convey meaningful information. The exciting aspect of this field of study is that we have just started to explore the true structural and functional complexity of animal communication.

Adaptation and Signal Evolution

The third part of the animal communication section covers an evolutionary perspective of animal communication. Communication involves the encoding of information in a signal, then the decoding of that information by the receiver. Thus, communication is fundamentally about information. Hailman discusses how we can quantify the information content in a signal and how signals vary in the types of information they encode.

We can think of the signaling process as the production and reception of signals encoding information about two nonexclusive entities: the referent of the signal (if there is one) and the motivational state of the sender and receiver. Signals may encode explicit information about some external referent, for example, a type of predator or food as discussed by Evans and Clark, or they could encode information about some internal physiological state of the sender. Owings also suggests that physiological state may influence how either the sender or the receiver focuses on the specific aspects of the environment.

One implication of everything we have talked about to this point is that the information encoded in a signal is correct, or 'honest.' In the late 1970s, Richard Dawkins and John Krebs questioned this assumption and asked why animals should encode honest information when there are circumstances where dishonesty seems valuable to the sender. We now know that dishonest signals abound. Batesian mimics, for example, are nontoxic animals that dishonestly mimic signals of noxious prey. Of course, if signals are always dishonest, then the receiver should simply ignore them. Hurd considers this important issue and shows that there are a number of circumstances where honest signaling is to be expected.

Up until the 1980s, there was a general notion that communication functions primarily between a single sender and a single receiver. A male advertises to a female. A contestant advertises to its opponent. In the 1980s, Patrick McGregor and Thor Dabelsteen suggested that communication may often involve signals transferred between more than two individuals. As discussed by Greenfield, these 'communication networks' can include information broadcast specifically between multiple individuals (as in a frog chorus), or they can individuals intercepting or eavesdropping on a private exchange of information between two other individuals. Examples of these include parasitoid *Corethrella* flies being attracted to the mate-attraction calls of male túngara frogs and a female chickadee eavesdropping on a contest between two males and using information derived from this contest to decide on a mate.

The final two articles in this section cover how signals change over generations. Current research indicates that signals can change in one of two main ways: culturally through information transferred from one generation to the next and genetically through inherited traits passed on from parents to offspring. These two modes of inheritance are by no means mutually exclusive. Cultural inheritance, covered by Freeberg, can affect a host of traits, including mating tendencies, which in turn affect factors such as gene flow. Ord suggests that we can think of the inheritance of traits more broadly in the sense that traits show a 'phylogenetic signal' representative of traits and shared phylogenies, and 'ecological determinism,' which is the degree to which ecological factors drive evolutionary change. Ord discusses techniques we can use to quantify these two mechanisms.

Communication Across Species

The final part of the *Animal Communication* section generally covers communication across species. Think about that snapshot of communication we described between a sender and a receiver. One would immediately think about communication between conspecifics partly because most communication does indeed occur between conspecifics, but not all communication is between conspecifics. Aposematic signals might come to mind, in which noxious prey signal their condition to predators. However, more recent work by Krams and colleagues on mixed species bird flocks suggests that communication between species can be very complex. Krams also discusses eavesdropping, a situation where signals intended for conspecifics are nonetheless used as sources of information by heterospecifics. Eavesdropping occurs when individuals detect predator alarm calls, and it also occurs when predators use mating signals as a means of locating potential prey, as suggested earlier regarding interception and eavesdropping in the discussion of communication

networks. Alcock expands on these ideas in a general discussion of signal parasites. Predators using mating signals to help locate prey would certainly be characterized as parasites of a communication system. Some parasites go further in producing a deceptive signal to their own advantage. For example, predatory fireflies mimic the light signal of their hosts and bolas spiders mimic the olfactory mating signal of their moth prey.

We end the *Animal Communication* section with a discussion of noise. This final article is not about communication across species per se but about one species (humans) impacting on the communication systems of many other species. As Slabbekoorn points out, the world is becoming a noisier place, largely because of human activity. This is true of both terrestrial and aquatic habitats. We need to know how this noise impacts on organisms. This is particularly true, given the critical role that communication plays in the lives of organisms.

See also: Acoustic Signals; Active Electroreception: Vertebrates; Agonistic Signals; Alarm Calls in Birds and Mammals; Anthropogenic Noise: Impacts on Animals; Communication Networks; Cultural Inheritance of Signals; Dolphin Signature Whistles; Electrical Signals; Evolution and Phylogeny of Communication; Food Signals; Honest Signaling; Information Content and Signals; Interspecific Communication; Motivation and Signals; Olfactory Signals; Parent–Offspring Signaling; Referential Signaling; Signal Parasites; Social Recognition; Syntactically Complex Vocal Systems; Vibrational Communication; Visual Signals; Vocal Learning.

Further Reading

Bradbury JW and Vehrencamp SL (1998) *Principles of Animal Communication.* Sunderland: Sinauer Associates, Inc.

Call J and Tomasello M (eds.) (2007) *The Gestural Communication of Apes and Monkeys.* New York: Lawrence Erlbaum Associate.

Espmark Y, Amundsen T, and Rosengvist G (eds.) (2000) *Animal Signals: Signaling and Signal Design in Animal Communication.* Norway: Tapir Academic Press.

Gerhardt HC and Huber F (2002) *Acoustic Communication in Insects and Anurans: Common Problems and Diverse Solutions.* Chicago: University of Chicago Press.

Greenfield MD (2002) *Signalers and Receivers: Mechanisms and Evolution of Arthropod Communication.* Oxford, UK: Oxford University Press.

Hailman JP (2008) *Coding and Redundancy: Man-Made and Animal-Evolved Signals.* Harvard University Press.

Hauser MD (1997) *The Evolution of Communication.* MIT Press.

Hauser MD and Konishi M (1999) *The Design of Animal Communication.* MIT Press.

Maynard Smith J and Harper D (2003) *Animal Signals.* New York: Oxford University Press.

McGregor PK (ed.) (2005) *Animal Communication Networks.* Cambridge: Cambridge University Press.

Owings DH and Morton FS (1998) *Animal Vocal Communication: A New Approach.* Cambridge, UK: Cambridge University Press.

Searcy WA and Nowicki S (2005) *Evolution of Animal Communication: Reliability and Deception in Signaling Systems.* Princeton, NJ: University Press.

Comparative Animal Behavior – 1920–1973

G. M. Burghardt, University of Tennessee, Knoxville, TN, USA

Introduction

The field of animal behavior has undergone a number of important transitions since it became an area of scholarly interest in the late nineteenth century. A number of writers asked important questions about the mechanisms underlying behavior, the role of environment and instinct, the development of behavior, and the origins of behavior even before Darwin, and much of this fascinating history is recounted in books and chapters by authors such as Hess, Burghardt, Boakes, and Burkhardt, among other sources. The major foundations of the conceptual apparatus of both American comparative psychology and European zoologically focused ethology were laid at this time, enriched by the methodological advances in animal physiology and animal learning. But it was really in the decades after WWI that the two fields, along with primatology, developed, came together, splintered, and reconverged.

I have chosen to limit this article to the years from the end of WWI to the award of the Nobel Prize in Physiology and Medicine to the two main founding fathers of ethology, Niko Tinbergen and Konrad Lorenz, along with Karl von Frisch in 1973. This was a highpoint of organismal behavior studies, before the impact of sociobiology, selfish gene theory, game theory, memes, evolutionary psychology, molecular genetics, the cognitive revolution, mathematical modeling, and modern neuroscience enriched, broadened, and to a large extent, refragmented the area of animal behavior, although along differing fracture points. In fact, the publication of *Sociobiology: The New Synthesis* by E. O. Wilson in 1975 built on the foundations of ethology while charting new directions with more explicit evolutionary theory. This moved animal behavior, with its origins in zoology, farther away from psychology, at least initially, than did ethology. Other articles in this volume are relevant to this story and should be read in conjunction. Drickamer covers the history of animal behavior up to 1900. Dewsbury gives a brief overview of what he terms 'animal psychology' during the entire twentieth century. In effect, this includes all the major developments involving behaviorism, ethology, pre-ethological zoological natural history, physiology, comparative psychology, behavior development and the nature–nurture controversy, animal cognition, cognitive ethology, primatology, behavioral endocrinology, behavioral ecology, and sociobiology. Taborsky covers the origins and history of European ethology. This leaves, in retrospect, very little to cover that is not an overlap of the other chapters. I have decided therefore,

that my most useful function will be to go over the most important topics and provide additional detail on some of them, and introduce others that have not found any mention at all. I also provide some additional references that are important to understanding the field of animal behavior. But there are others as well. Although secondary sources such as those cited in these entries are useful, it is often more enlightening to read the original texts. Now that more and more journal papers from the early decades of the twentieth century and even older ones have become available online, easier access to this material should, I hope, result in something of a renaissance of student engagement with the actual data and thinking of both early and recent contributors. This is something I set out to do in the study of animal play recently (see Chapter 2 in Burghardt, 2005). A number of collections of seminal papers are also available and some of these are listed in the references as well.

It is also important to note that the field of animal behavior, even during the period in question, was an international one. Unfortunately, there are many buried seminal findings in literatures that are not that represented in the field at large. English, German, French, Italian, Spanish, Russian, and Japanese contributions are rather familiar, at least via reviews and textbooks, but there are undoubtedly scholarly and empirical studies, as well as field observations, in several additional languages that are worth knowing.

Methodology

Although students of animal behavior in the nineteenth century frequently performed informal experiments, often quite elegant, the field was mostly descriptive and the descriptions themselves were often both anecdotal (single interesting stories) and anthropomorphic (labeling and interpreting behavior uncritically from the perspective of human psychology). Both anecdotes and anthropomorphism are problems when a species' ecology, normal behavior, physiology, sensory abilities, and brain function are not taken into account. In defense of the nineteenth century scientists, however, it should be noted that even experts had little knowledge then of any of these factors. Zoologists and psychologists did widely recognize the need for both careful description and experimentation by the first decade of the twentieth century. Experiments were needed to control variables to determine what stimuli animals could perceive (discriminate) as well as if and how rapidly they could learn, for example. Both classical and instrumental conditioning were developed by 1910, as was intelligence testing by Binet. By this time, the basics of statistical analysis of behavioral data had been developed. Francis Galton, Charles Darwin's cousin, was a pioneer in this field during the 1880s, followed by Pearson and others. The concept of correlation was devised during this time, for example. Formal hypothesis testing of the results of experiments using tests of significance really took off with the work of Ronald Fisher after WWI, but a change in attitude had already occurred among those studying behavior that had far-ranging future impact.

The early studies of animal behavior lacked clear documentation. Photography, and later movies, developed in the late nineteenth century and these, too, began having important consequences for evaluating claims. Audio recording and color photography came later, along with sonographic analysis of sounds. In the early days of bird song research, for example, researchers had to characterize such songs, using musical instruments and notation. Our ears are also not nearly as acute as are those of birds in terms of apprehending the nuances of rapidly changing notes and sequences. Sounds outside our hearing abilities were rarely considered. Early experiments by Spallanzani in the 1790s on bats using sounds we could not hear to avoid flying into objects in the dark were dismissed, since there was no way of measuring them. It was not until 1941 that Griffin and Galambos confirmed these results. Now ultrasonic, and even infrasonic, perception and communication has been documented in numerous species other than bats. Olfactory and other chemical cues underlying behavior also had to wait a long time before they could be objectively characterized and compared. We still have much to learn about communication via the chemical senses. However, during the time period covered by this review, our ability grew rapidly. One can compare the increase in scientific sophistication between the experimental demonstrations using insects by Henri Fabre before WWI with the pheromonal studies of ants by E. O. Wilson and chemistry colleagues during the 1960s.

The development of apparatus for testing physiological and sensory processes was rapid in the late nineteenth century, and by the beginning of the twentieth century scientists such as Thorndike, with the puzzle box, and Small, with the maze, were trying out instruments that could be used for studying a wide range of animals. This instrumentation phase of animal behavior studies grew rapidly during the 1920s, and reading the literature from this period reveals many ingenious testing methods. This process continues to the present time. However, the early apparatus was not automated, as the electronic revolution would not really take off until after WWII. By this time, however, many of the more imaginative methods, such as the delayed reaction box, were shown to have flaws in terms of comparing the cognitive processes of animals.

Thus, the work on comparative learning was restricted to a few species and a few instrumental and conditioning tasks. Often, the promise of such apparatus for comparing the abilities of animals was shown to not work as well as hoped. Nonetheless, these methods, especially for testing discriminations, made it much more likely that people could repeat, validate, and extend the work of others. In short, during the period under discussion, issues of apparatus, experimental design, and statistical analysis became increasingly prominent. In fact, much of the initial controversy resulting from European ethology, while ostensibly about theoretical issues, was really on methodological grounds. Ethologists were much more interested in studying the normal, species-characteristic behavior of animals under natural conditions and thus saw experiments primarily as a way of unraveling natural behavior and were less likely to use complex apparatus, sophisticated experimental designs, and rigorous controls, let alone statistical analysis. By the 1970s, this had changed and rigorous methods were being employed to study both naturalistic behavior such as foraging, courtship, and predator recognition and learning, discrimination, and problems solving.

Other methods were also employed. Selective deprivation from stimuli during development, such as birds hearing their normal songs or snakes having normal prey, led to the 'deprivation experiment' and much controversy on what the results meant. Although performed and interpreted before 1920, it was really when the controversy over instinct arose in the 1940s and 1950s that the nature of genes and environment in shaping behavior could begin to be resolved. At this time, genetic markers began to be employed to evaluate individual differences in behavior, as in the classic work of Margaret Bastock in the 1950s. Thus the idea, put forth by John B. Watson and Z. Y. Kuo in the 1920s, that environmental processes, not evolution and genes, were responsible for virtually all aspects of behavior, including innate responses and individual differences, could be firmly rejected. The understanding that behavior is a complex product of evolutionary history, genetic inheritance, and experiential history has continued to grow as mechanisms of gene expression are uncovered. On the other hand, methods such as observing and manipulating embryos in duck eggs, pioneered by Kuo, led to many innovative experiments by workers such as Gilbert Gottlieb, who cautioned against ignoring nonobvious sensory influences on behavior.

During the 1960s, detailed film analysis of the movements of animals, especially during courtship, allowed for more fine grained and objective sequences of behavior. As more aspects of behavior could be quantified, methods of reducing such behavior were developed including sequence analysis and sociometrics. Finally, the ability to use characters of all kinds, anatomical, physiological, and behavioral, to trace the phylogeny of behavior became more widespread, especially through the work of Heinroth on behavioral homologies in the 1920s, Lorenz's studies of waterfowl in the 1940s, and the development of cladistics and computer programs in the 1960s and 1970s. These are now much refined and are now able to process quickly behavioral data on hundreds of characters in hundreds of individuals or species and relate them to molecular genetic data as well. But the roots of many of these methods, and many others, are in this period.

Comparative Psychology

The essay by Dewsbury describes the major threads of animal psychology, animal learning, and comparative psychology as well as the impact of such recent developments as sociobiology and game theory. He described the roles of behaviorism and learning theory throughout the period 1920–1973 and beyond, but did not touch too deeply into ethology and comparative behavior studies. Except for the burgeoning field of primate studies in both laboratory and field, there was little comparative in traditional animal psychology into the 1960s. Jane Goodall's dissertation work on the chimps of Gombe in East Africa in the 1960 reawakened serious interest in non-human primate field studies, although Japanese scientists and American anthropologists were doing some pioneering work, building on the earlier studies of the psychologist C. R. Carpenter, the ethologist Adrian Kortlandt, and others. Comparisons of non-human primates and humans became the topic of best-selling books as shown by *The Naked Ape* by Desmond Morris and *The Territorial Imperative* by Robert Ardrey, both controversial best sellers in the 1960s.

Most comparative and animal psychologists continued studying rather simple conditioning and schedules of reinforcement in rats and pigeons, hoping that such work would uncover principles so general and useful that the natural behavior of animals, especially in the field, was not really necessary or even useful. Comparisons across species, if made, were generally with people and much of the field was explicitly anthropocentric. One exception to the general neglect of comparing animal species was the work of Bitterman. In the 1960s, he attempted to compare what he considered representative vertebrates on standard tasks, especially visual and spatial reversal learning and probability matching. He was using some of the methods being developed to go beyond the standard learning paradigms and standard species. He used goldfish as a representative fish, painted turtles as a representative reptile, pigeons as representative birds, rats as representative 'lower' mammals, and so on. He claimed to find discontinuities in the performances of these species on these tasks and concluded that this represented levels of cognitive performance in the respective taxonomic groups (e.g., fish, reptiles, birds) from where they came. The experiments themselves were well controlled. Bitterman attempted to

account for motivational, sensory, and motoric differences more elegantly than even some current researchers. Yet the project ultimately failed because of not taking into consideration the diversity within each of the classes of vertebrates in terms of behavior and learning prowess. Fish are now compared with primates in terms of some cognitive processes, and some birds, such as ravens, crows, and parrots outshine most primates in terms of problem solving, tool making and use, etc. What happened is that in the 1970s the cognitive revolution in psychology took hold and more diverse methods were developed to tap into myriad aspects of task performance, problem solving, social learning, implicit and explicit memory, self-recognition, retrospective memory, and other less 'molecular' processes. Starting in the 1990s, a field of comparative cognition arose that is now using a far more rich set of methods and problems and revisiting the problem of making explicit cognitive comparisons across closely related species such as the various apes, dogs, and wolves, and different species of jays in the western United States. Rarely even aware of Bitterman, these workers are revamping his program in ways that may be more successful. Controversy abounds, however, as in Bitterman's work 50 years ago! Still, the goal of developing a phylogeny of animal cognition, based not on a general intelligence divorced from the natural context, but one in which cognitive processes are considered more modular and specific to the social, foraging, and other behavioral contexts in which animals live and evolved, seems to be in the making. Comparing the papers in the *Journal of Comparative Psychology* from 1960s and the 2000s is quite revealing. Nevertheless, comparative psychologists still work on other problems including sensory processes, communication, parental care and behavioral development, reproduction and hormones, and other problems.

Comparative Ethology

Similar to comparative psychology, comparative ethology was an essential precursor to the modern field of comparative cognition, since ethological ideas of the need to consider the natural behavior of animals, along with primatology, became viewed as necessary to any successful comparative cognitive psychology. Indeed, such understanding of natural behavior is essential to physiological, perceptual, motivational, and other problems as well. Ethology, like comparative psychology, has roots in the writings of Charles Darwin, who wrote an important chapter on 'instinct' in the origin of species. Darwin used the comparative method to speculate on the origins and radiation of a number of behaviors including hive making in honeybees, slave making in ants, castes in social insects, and breeds of dogs and pigeons that had such remarkable differences in behavior. This use of the comparative approach also developed from an evolutionary framework but with the emphasis not on learning and laboratory experiments so much as it relied on insights from natural history and simple manipulations in the field. Clearly, to bring behavior into the evolutionary umbrella, Darwin needed to show that behavior was itself a product of natural selection.

One of the most important legacies of ethology was the emphasis on careful descriptions and classifications of behavior. These are often called 'ethograms,' and in many paper today, you will see a listing of behavior measures recorded with objective criteria for recognizing them. Ethology also developed a number of key concepts which have had varying periods of waxing and waning, but through the 1950s the main ideas being studied were key or sign stimuli, fixed action patterns, innate releasing mechanisms, action-specific energy, motivational conflicts (e.g., displacement behavior, redirected behavior), species specificity of behavior, value of behavior as a taxonomic tool, and specialized learning processes such as imprinting, among others. For ethologists, the evolutionary history of animals and their ecological requirements were essential for a complete understanding of behavior. Probably the most influential paper from the latter part of the period we are covering, and still frequently cited today, is the paper outlining the four major aims of ethological analysis by Tinbergen (1963). Here, he laid out the four main aims of ethological study which include the study of *causal mechanisms* (including sensory and neural processes, motivation, hormones, genetics, social contexts, etc.), *developmental processes* (including sensory and nervous system maturation, experience, etc.), *adaptive function* (What is the behavior 'good' for?, survival value, reproductive consequences), and *evolution* (processes and patterns underlying behavioral diversity within and across time and taxa). To these can be added the study of the personal and subjective experiences of animals, including people. While it is never possible to know exactly what even other people are feeling, we can make informed inferences that guide our responses to an injured, sick, grieving, or playful person, and this also applies to inferences to other species. These must be done carefully, using a *critical*, not naïve or sentimental, anthropomorphism to develop viable hypotheses.

Regardless, the ethological concepts, even in their initial and rather crude form, opened the way for far more rigorous studies of behavioral evolution than was possible in Darwin's time. The discovery that behavior of a species in a given context is often similar across individuals led to the comparative analysis of either the movement patterns themselves, such as the courtship rituals of birds, the head-bob displays of lizards, and the foraging tactics of fish, or sounds made in defense or communication. Movie and audio recording made it possible in quantifying large samples of data. It is also possible to compare the products of behavior such as spider webs, bird nests, and termite mounds. Similarly, when it was recognized that animals, when responding to objects in nature, including rivals, mates, prey, and enemies, often just cue in on a discreet visual, chemical, or auditory stimulus from the 'whole,' the way was open to analyzing these cues in a comparative and evolutionary fashion. It was also discovered that animals are often beset with conflicting drives or motivations, such as brooding eggs or getting food, and attacking an interloper or fleeing. It turns out that the behaviors resulting from such conflicts can be incorporated into displays. Also, behavior can shift from one context to another evolutionarily so that courtship behavior may contain elements of parental care, fighting, or even predatory behavior. Indeed, these systems may be linked so that female spiders that are more aggressive in predatory contexts are more likely to eat their mates than less aggressive ones. The process of transferring behaviors across contexts is often termed 'ritualization.' We see many comparable phenomena in cultural contexts in people. Many of the classic examples in game theory derived from observations of contests in animals. In short, while many of the classical ethological concepts are not commonly used today, the ideas behind them have been expanded upon and deployed in much recent work in behavioral ecology, sociobiology, neuroscience, and even in wildlife management and captive animal welfare.

An interesting phenomenon that intrigued the early ethologists, especially Konrad Lorenz, was imprinting. This is the attachment of a newly hatched precocial bird, such as a duckling or chick, to a parental figure. Within a fairly narrow window of time, the young bird would treat as its 'mother' and follow almost any object or animal, including people, that it encountered instead of its natural parent. This phenomenon involved learning, instinct, and social bonding and drew the attention of comparative psychologists. During the 1950s to 1970s, there were hundreds of experimental studies on this topic. It was a phenomenon that could be readily studied experimentally. Eckhard Hess was one of the first to develop a circular apparatus for imprinting birds on any type of moving object, and this apparatus was pictured in virtually every introductory psychology text for decades. The phenomenon could be extended to parents' bonding with their offspring, food and mate choice, and even habitat preferences. Strangely, work on imprinting has become almost extinct! However, as issues of captive rearing of endangered species, reintroductions, and other conservation issues arise for a host of reasons, including climate (and habitat) change, this research area may see a renaissance. But in some respects, it has a legacy in attachment theory, captive breeding, gene expression, speciation, and other fields.

The Legacy for Today

Few students today were even alive in 1973 when the Nobel Prize went to the ethologists. Yet it was an exciting time, a vindication of a field that began in 'mere natural history.' Today, many students and even some established researchers are unaware of or uninterested in the historical aspects of their fields of study, except, perhaps, a specific narrow research finding from the 'dark ages' that one should cite. This is a mistake. Without going into details, it is continually proving true that old findings are being rediscovered, often with new names. But also older data are providing new riches, as those developing phylogenies of behavior need to dig through the old literature to find details that are needed for their analyses. In any event, the period from 1920 to 1973 saw the establishment of several areas of animal behavior research as important scientific fields.

See also: Animal Behavior: The Seventeenth to the Twentieth Centuries; Behavioral Ecology and Sociobiology; Ethology in Europe; Future of Animal Behavior: Predicting Trends; Integration of Proximate and Ultimate Causes; Konrad Lorenz; Neurobiology, Endocrinology and Behavior; Niko Tinbergen; Psychology of Animals.

Further Reading

Burghardt GM (1973) Instinct and innate behavior: Toward an ethological psychology. In: Nevin JA and Reynolds GS (eds.) *The Study of Behavior: Learning, Motivation, Emotion, and Instinct* pp. 322–400. Glenview, IL: Scott, Foresman.

Burghardt GM (ed.) (1985) *Foundations of Comparative Ethology.* New York: Van Nostrand Reinhold.

Burghardt GM (1997) Amending Tinbergen: A fifth aim for ethology. In: Mitchell RW, Thompson NS, and Miles HL (eds.) *Anthropomorphism, Anecdotes, and Animals*, pp. 254–276. Albany, NY: SUNY Press.

Burghardt GM (2005) *The Genesis of Animal Play: Testing the Limits.* Cambridge, MA: MIT Press.

Burkhardt RW Jr (2005) *Patterns of Behavior: Konrad Lorenz, Niko Tinbergen, and the Founding of Ethology.* Chicago: University of Chicago Press.

Dewsbury DA (ed.) (1984) *Foundations of Comparative Psychology.* New York: Van Nostrand Reinhold.

Hess EH (1962) Ethology: An approach to the complete analysis of behavior. In *New Directions in Psychology*, pp. 157–266. New York: Holt, Rinehart and Winston.

Houck LD and Drickamer LC (eds.) (1996) *Foundations of Animal Behavior: Classic Papers with Commentary.* Chicago: University of Chicago Press.

Klopfer PH and Hailman JP (eds.) (1972) *Control and Development of Behavior: An Historical Sample from the Pens of Ethologists.* Reading, MA: Addison Wesley.

Klopfer PH and Hailman JP (eds.) (1972) *Function and Evolution of Behavior: An Historical Sample from the Pens of Ethologists.* Reading, MA: Addison Wesley.

Tinbergen N (1963) On aims and methods of ethology. *Zeitschrift für Tierpsychologie* 20: 410–433.

Compensation in Reproduction

P. A. Gowaty, University of California, Los Angeles, CA, USA; Smithsonian Tropical Research Institute, USA

Introduction

There are at least three concepts associated with the term 'reproductive compensation.' (1) Compensation may be an attempt to make up fitness losses from the possession of deleterious genes that are lethal. (2) Compensation may be an attempt to make up fitness losses from the possession of deleterious gene combinations associated with lethal effects, such as Rh factor, and may explain why parents sometimes have additional children after losing a child. (3) The generalized reproductive compensation hypothesis (RCH) described in a series of studies by P. A. Gowaty focuses on what constrained parents – those mating and reproducing under ecological and social constraints – may do to enhance the competitiveness of their surviving offspring. The generality of the third version of reproductive compensation comes from three sources. First is the recognition that ecological and social circumstances under which individuals make decisions about with whom to mate result in differences between parents in the survival probabilities of their offspring. Thus, competitive forces play out against ecological and social constraints acting on parents and prospective parents. Second, because pathogens in the parental generation are likely to be different than the pathogens of the offspring generation, parents face a challenge in the production of offspring phenotypes that will work against the newly evolved pathogens the offspring are most likely to face. Thus, the RCH addresses a general challenge encountered by all or most parents. Third, the RCH does not depend on the possession of particular genes, but instead emphasizes that individual flexibility is induced by challenges to individual reproductive success and survival. The other two hypotheses, unlike the generalized version, are specific to parents with deleterious genes or to deleterious gene combinations. The remainder of this article is about the generalized reproductive compensation hypothesis.

The conceptual antecedents of the RCH are in a series of reviews (see Further Reading) including those about the ecological and social constraints under which males and females live, the origins of monogamy and associated extra-pair paternity, and life-history variation that predicts individual's vulnerabilities to other's control. For example, when male–male combat reduces the access of some males to some females, males who lose the contests may be constrained to mate with individuals that they do not individually prefer. Thus, compared to males who win contests, the males who lose may have offspring of lower viability. It is these fathers, then, who would be under selection to compensate. Similarly, when ecological or social barriers limit females' abilities to freely express their preferences for mates, these females are constrained to mate with males they do not individually prefer, and selection will favor compensation by constrained females.

The Generalized Hypothesis of Reproductive Compensation

The RCH says that parents and prospective parents, mated to partners they do not prefer, flexibly and adaptively increase fecundity and/or allocations of resources to their offspring compared to parents and prospective parents mated to partners they do prefer. This version of the reproductive compensation hypothesis focuses on responses of parents able to adaptively and flexibly adjust their reproductive and parental behavior. The RCH posits that differences in the numbers of offspring surviving to reproductive age between parents breeding with partners they prefer and those breeding with partners they do not prefer is the selective differential that favors the evolution of compensation. Predicted mechanisms of compensation include (1) higher fecundity by females; (2) higher numbers of sperm ejaculated by males; (3) more parental resources provided to offspring during their development (e.g., larger eggs, longer bouts of nursing, higher ratios of immune components in mother's milk, etc.); and (4) parental effects accelerating the age of reproductive maturity of their surviving offspring (e.g., puberty acceleration in mammals). The RCH also predicts that parents who may have been in conflict over mating may nevertheless collaborate in their efforts to bring relatively uncompetitive offspring to maturity and entry into their breeding cohort.

The Assumptions of the RCH

The assumptions of the RCH include the following: (1) Pathogens and parasites, ubiquitous threats to the health of parents and offspring, evolve faster than their hosts. (2) Prospective breeders prefer potential mates with whom they would produce offspring with immune systems capable of successfully combating the pathogens of the offspring generation; that is, mate preferences predict offspring viability. (3) Common impediments to breeding with one's best (healthiest) offspring include barriers to dispersal and/or social control by others. (4) All individuals have the ability to compensate. (5) Variation between individuals in compensation is due to environmental factors such as access to food resources and/or developmental factors such as energetic resources amassed during developmental life stages. (6) The selection differentials favoring compensation are the differences in productivity (the number of offspring surviving to reproductive age) between unconstrained and constrained parents, but also between constrained parents that do and do not compensate.

The Red Queen's challenge to parents

The Red Queen told Alice, 'You have to run as fast as you can to stay in the same place.' Recently evolutionary biologists have referred to the arms races between pathogens and parasites as the Red Queen, because pathogens and parasites evolve more rapidly than their hosts, a fact that provides a strong advantage to pathogens and parasites relative to their hosts. The differential in evolvability between pathogens and their hosts is a selection pressure that favored the evolution in hosts of remarkable immune defenses. The reason hosts are not invariably killed by every disease organism to which they are exposed is because host immune systems allow them to sometimes defeat their pathogens. When this happens, hosts are 'staying in the same place.' 'The Red Queen's challenge to parents' is a figurative way to describe the constraints of Mendelian genetics on parents' abilities to produce offspring with genetically influenced immune defenses likely to work against pathogens in the offspring generation. **Figure 1** shows the Red Queens' challenge to parents in a graphical model.

Individual mate preferences predict offspring viability

The RCH assumes that individual mate preferences predict offspring viability; variation in offspring viability is the most important basis for mate preferences. This assumption does not exclude other criteria allowing mate preferences to work, and it does not mean that other grounds for preferences do not exist or do not matter. This assumption is not specific about the information or signals that mediate mate preferences, that is, it is not necessarily linked to the possession of indicator traits as is the classic idea of Hamilton and Zuk. Experimental studies in flies, mice, ducks, and other species produced results consistent with this assumption.

Constraints on mating with one's best partner for offspring viability are common

This assumption emphasizes that individuals are not always able to mate with partners that are uniquely best for them or better for them in terms of offspring viability. This is an important assumption because constraints are common and have predictable fitness effects associated

with reproductive decisions (who to find acceptable or unacceptable as a mate, who to mate, who to coerce, who to resist, which offspring to allocate to, etc.), yet many experiments testing mate preferences fail to control or eliminate the effects of some subtle or even dramatic social constraints such as intrasexual interactions or intersexual coercion. It is important to keep in mind that constraints can be ecological as well as social. Examples of social constraints include mechanisms of sexual selection and sexual coercion, biases in genetic substructuring of populations, and social mechanisms of dispersal limitation. Examples of ecological constraints include habitat limitations, the presence of predators, disease, ecological barriers, or ecological mechanisms of dispersal limitation. Field studies on American pronghorn demonstrate that females are sometimes constrained, and laboratory studies designed to evaluate the effects of constraints on the free expression of mating preferences demonstrated that constrained mate preferences affect offspring viability.

Given appropriate ecological or intrinsic resources, all individuals can compensate

Models of the evolution of compensation, defined narrowly as fixation of alleles at loci affecting the ability of individuals to compensate, show that compensation can evolve very rapidly. Loci that should be involved in the upregulation of compensatory mechanisms include those affecting (1) assessment of fitness (of competitors and offspring), (2) sensitivity to cues (such as mating with preferred or nonpreferred partners) indicating the need or potential need to compensate, and (3) responsiveness (such as upregulation of parental effects enhancing the likelihood that offspring survive to reproductive age). Even when costs are high, models of compensation show that compensation readily evolves. Whenever there are costs to compensation, compensating parents should have lower survival probabilities compared to parents without the resources necessary to compensate and compared to parents with no need to compensate.

Variation in compensatory ability

Intrinsic and extrinsic variation among individuals affects individual ability to compensate. Thus, constrained females with similar intrinsic resources or similar metabolic efficiencies with access to more resources will be able to compensate better than constrained females without such resources. Experimental studies on mallards, *Anas platyrhynchos*, support this assumption.

Some Predictions of the RCH

Compensatory mechanisms that constrained breeders may use to enhance the likelihood that their offspring survive to reproductive age can be premating, prezygotic, or postzygotic.

Increased fecundity increases the variation in offspring phenotypes

The most interesting of the RCH predictions is that constrained female breeders increase the variation in the phenotypes of their offspring either by increasing fecundity or by currently unknown mechanisms. Increasing mean fecundity in most cases will automatically increase the variance in traits among the offspring. A graphical model showing how increasing fecundity may work to increase variation in offspring phenotypes is in **Figure 1(b)**. **Figure 1(c)** shows how increasing variation in offspring phenotypes by some unknown mechanisms can also favor constrained parents' production of offspring with phenotypic variation similar to those produced by unconstrained parents. Elementary genetics shows that breeding individuals will necessarily produce offspring genotypes that influence offspring phenotypes that should work well against the pathogens in the parents' generations. Because parasites rapidly evolve new offensives to meet host defenses, the Red Queen's challenge to parents is to produce offspring whose phenotypes will thwart the pathogens and parasites of the offspring generation. Just as buying several lottery tickets with the same number will not increase a gambler's chances of winning the lottery, investing in offspring with phenotypes like their own would not increase parents' odds of producing offspring to fight against the pathogens of the offspring generation – unless, of course, the pathogens did not evolve between the parent and the offspring generations. The likelihood of stasis in the genomes of pathogen and parasite populations during the relatively long intervals between host generations seems low.

Increased ejaculate size provides more variable haplotypes among which females may choose

While constrained females may increase the numbers of eggs committed to a particular bout of reproduction, the numbers of eggs they lay, or the numbers of offspring born, constrained males could increase the numbers of sperm that they ejaculate. This would favor the production of offspring with variable phenotypes even if the genetic variation among a given female's eggs were very low. Indeed, collaboration between reproductive partners is expected whenever one or either partner is constrained relative to other same-sex individuals.

Extra-pair paternity

In socially monogamous populations such as some beetles, most birds, a few species of mammals including humans, either or both sexes may seek or accept extra-pair fertilizations. The RCH predicts that constrained parents of both sexes will adjust behavior, physiology, and parental effects to increase the likelihood that some of their offspring survive to reproductive age. If true, indicators of lifelong

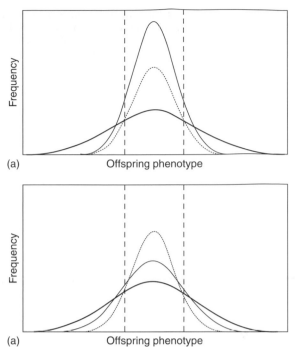

(a)　　　　　Offspring phenotype

(a)　　　　　Offspring phenotype

Figure 1 Models of the Red Queen's challenge to parents and reproductive compensation by (a) increasing fecundity (the number of offspring born or eggs laid) and (b) increasing variance in offspring phenotypes without increasing fecundity. Assuming that pathogens evolve more rapidly than their hosts, that mate preferences predict viability of offspring, and that ecological and social constraints keep some individuals from mating with their preferred partners, the wide, dark curve represents offspring phenotypes produced by unconstrained parents. The peaked areas of the curves in the center of the graphs between the vertical, dotted lines represent the offspring phenotypes parents are most likely to produce, that is, offspring phenotypes likely to resist the pathogens and parasites of the parental generation. The areas under the curves in the tails of the distributions represent those phenotypes likely to resist the novel pathogens and parasites in the offspring generation. The dotted curve represents the offspring phenotypes produced when parents are constrained to reproduction with partners they do not prefer. The narrow, solid line represents the offspring phenotypes produced when constrained parents attempt to compensate in (a) by increasing the number of eggs laid or offspring born and in (b) by increasing the variance in offspring phenotypes they produce. When constrained parents increase fecundity as in (a), the area under the curve in the tails of the distribution would be increased, enhancing the probability that some of their offspring are able to resist offspring generation pathogens. When constrained parents increase the variance in offspring phenotypes without increasing fecundity, perhaps through mating with more than one partner, the area under the curve in the tails of the distribution would be increased, enhancing the probability that some of their offspring are able to resist offspring generation pathogens. Reproduced from Gowaty PA (2008) Reproductive compensation. *Journal of Evolutionary Biology* 21: 1189–1200, with permission from JEB.

offspring health, such as relative mass or size at the end of the period of parental care or the likelihood of survival to reproductive age, will be greater in extra-pair offspring than within-pair offspring.

Table 1   Similar and contrasting predictions about *number of eggs laid or offspring born* or other traits such as increased egg size from the differential allocation hypothesis and the reproductive compensation hypothesis, when the differences in methods for evaluating consensus attractiveness and individual preferences are taken into account. Individuals mated with partners they individually prefer are nonconstrained (i.e., I-NC), and those mated with partners they individually do not prefer are constrained (i.e., I-C). An individual mated with consensus attractive partners is indicated by C-A; those mated with consensus unattractive partners by C-UA

Choosers' partners	Differential allocation hypothesis	Reproductive compensation hypothesis
C-A and I-NC	+	−
C-A and I-C	+	+
C-UA and I-NC	−	−
C-UA and I-C	−	+

Reproduced from Gowaty PA (2008) Reproductive compensation. *Journal of Evolutionary Biology* 21: 1189–1200, with permission from JEB.

Under RCH parental effects should enhance the likelihood of offspring survival

Constrained parents may increase the size of the eggs they lay, the intensity of fanning their clutches, the intensity or duration of incubating eggs, or brooding their young. Each of these activities may enhance the health of less competitive offspring. Constrained parents may also increase the size of the eggs they lay or the mass of the offspring they give birth too. The RCH also predicts that constrained parents allocate more nutritional or immunological elements to their zygotes. The important point is that these parental effects, under the RCH, are predicted to enhance the likelihood that their otherwise less competitive offspring survive at least to reproductive age.

Comparison of the RCH with the Differential Allocation Hypothesis

The differential allocation hypothesis (DAH) says that the relative attractiveness of individuals with biparental care influences the opportunities for each member of the pair to trade-off parental effort for mating effort. If a female is relatively more attractive than her partner, the DAH says she should allocate less parental effort to her current offspring, while her relatively less attractive partner should allocate more. The original version was egalitarian in that it used a single rule applied similarly to both sexes of parent to predict the different allocation levels of each of the parents to the care of their offspring. More recent interpretations limited the DAH to consideration of what mothers should do, not just in species with biparental care, but in species without paternal care of offspring. This revised version predicts that females allocate more parental effort to the offspring of more attractive males.

Superficially, the DAH and the RCH appear to make alternative predictions. However, caution is required (**Table 1**) primarily because important assumptions of the two hypotheses differ. The DAH assumes that sometimes fancy traits signal and mediate mate preferences, while the RCH assumes that individuals can assess offspring viability from information rather than specific signals. A problem thus commonly arises in attempts to compare the two hypotheses because the implications of the preferences are not taken into account when doing experiments (see **Table 1**). Second, under RCH, the key predictions are that compensatory mechanisms enhance the disease-fighting abilities of offspring so that offspring viability is enhanced, while under DAH, enhanced allocations are associated with normal mechanisms of parental care. Therefore, it is sometimes logical to expect both types of allocations, depending on how preferences are mediated. Third, the RCH applies even in species in which there is no postzygotic parental care, whereas the DAH does not apply in species without parental care. Fourth, the RCH stresses that both parents are under selection to compensate, even when only one of them is constrained to mating with a partner it does not individually prefer; thus, the RCH predicts parental cooperation, while the DAH does not.

In theoretical evaluations of the relative force of RC versus DA, it is crucial to keep in mind that if mate preferences do not result in enhanced offspring viability, the RC hypothesis is not applicable. It is also important to the RCH that constraints on the free expression of mating preferences matter: one would not expect unconstrained females to need or attempt compensation to enhance the competitiveness of their offspring. Theory designed to test if compensation can evolve begins by comparing the fitness of constrained parents, when some compensate and others do not.

See also: Differential Allocation; Flexible Mate Choice; Monogamy and Extra-Pair Parentage; Parasites and Sexual Selection; Social Selection, Sexual Selection, and Sexual Conflict.

Further Reading

Bluhm CK and Gowaty PA (2004) Reproductive compensation for offspring viability deficits by female mallards *Anas platyrhynchos*. *Animal Behaviour* 68: 985–992.

Burley NT (1988) The differential allocation hypothesis: An experimental test. *American Naturalist* 132: 611–628.

Byers JA and Waits L (2006) Good genes sexual selection in nature. *Proceedings of the National Academy of Sciences USA* 103: 16343–16345.

Gowaty PA (1996) Battles of the sexes and origins of monogamy. In: Black JL (ed.) *Partnerships in Birds. Oxford Series in Ecology and Evolution*, pp. 21–52. Oxford: Oxford University Press.

Gowaty PA (1999) Extra-pair paternity and paternal care: Differential fitness among males via male exploitation of variation among females. In: Adams N and Slotow R (eds.) Proceedings of the 22 International Ornithological Congress, Durban, University of Nata, pp. 2639–2656. Johannesburg: BirdLife South Africa.

Gowaty PA (2003) Power asymmetries between the sexes, mate preferences, and components of fitness. In: Travis C (ed.) *Women, Evolution, and Power*, pp. 61–86. Boston, MA: MIT Press.

Gowaty PA, Anderson WW, Bluhm CK, Drickamer LC, Kim YK, and Moore A (2007) The hypothesis of reproductive compensation and its assumptions about mate preferences and offspring viability. *Proceedings of the National Academy of Science* 104(38): 15023–15027.

Gowaty PA (2008) Reproductive compensation. *Journal of Evolutionary Biology* 21: 1189–1200.

Navara JK, Hill GE, and Mendonca MT (2006) Yolk androgen deposition as a compensatory strategy. *Behavioral Ecology and Sociobiology* 60: 392–398.

Navara KJ, Siefferman LM, Hill GE, and Mendonca MT (2006) Yolk androgens vary inversely to maternal androgens in eastern bluebirds: An experimental study. *Functional Ecology* 20: 449–456.

Sheldon BC (2000) Differential allocation: Tests, mechanisms, and implications. *Trends in Ecology and Evolution* 15: 397–402.

Conflict Resolution

O. N. Fraser, University of Vienna, Vienna, Austria
F. Aureli, Liverpool John Moores University, Liverpool, UK

Introduction

Conflict resolution is integral to the maintenance of group cohesion and the benefits associated with group living. While group living affords many benefits, such as reduced risk of predation, group defense of resources and increased access to mates, it also entails costs in terms of intra-group competition for access to limited resources. Conflicts of interest can arise between competitors when only one can gain possession of a critical resource, such as food or mates. In addition to competing for the same resource, group members may face difficulties when pursuing different objectives or have different motivations. For example, conflicts of interest may arise between potential mating partners as a result of differing interests for males and females, or between parents and offspring over weaning or scheduling of activities. Making decisions may also be a source of conflict of interest, from deciding on the direction of travel, to changing the group activity or the performance of behaviors requiring mutual consent such as grooming or play.

Conflicts of interest occur frequently in all group-living animals, but their consequences have the potential to compromise the benefits associated with group living. If conflicts of interest are not managed, they may escalate into aggressive conflict, which may be costly for all participants through risk of injury, energetic costs, physiological costs, such as increased stress levels, and potential damage to the relationship between opponents, resulting in the loss of the benefits afforded by the relationship. Where possible, therefore, conflicts of interest should be prevented from escalating into aggressive conflict by implementing (aggressive) conflict avoidance strategies, such as the use of appeasing and submissive behaviors. By selectively increasing behaviors that function to reduce tension and promote tolerance among partners during periods of tension or when high levels of competition are likely, the likelihood of conflicts of interest escalating may be reduced in the first place. Levels of grooming in chimpanzees (*Pan troglodytes*) and levels of play and sociosexual behavior in bonobos (*Pan paniscus*), for example, have been shown to increase prior to scheduled feeding times, when levels of tension are likely to be high. Spider monkeys (*Ateles geoffroyi*), whose society is characterized by a high degree of fission–fusion dynamics, face increased risks of aggressive conflict when two subgroups of individuals join each other after a period of separation. Postfusion embraces between members of fusing subgroups have been shown to reduce the likelihood of aggressive conflict.

Aggressive escalation is, nevertheless, a common occurrence in the lives of many group-living animals where the importance of the source of conflict of interest outweighs the risks of aggressive conflict. One of the reasons why aggressive conflict is possible without disrupting the benefits of group living is that post (aggressive) conflict mechanisms may enable the damage caused by aggressive conflict to be repaired.

Reconciliation

Prior to the systematic study of postconflict behavior in primates, the traditional view was that aggression caused dispersal in all animals and thus, a decreased probability of contact was predicted following aggressive conflict between opponents. Although some conflicts might end in dispersal, this hypothesis was questioned when de Waal and van Roosmalen showed for the first time that chimpanzees sought out their former opponents after a conflict and were actually more likely to engage in affiliative behavior immediately following the conflict than during subsequent interactions. The first postconflict affiliative interaction between former opponents was labeled 'reconciliation.' Although the term reconciliation implies a proven function of relationship repair, it was used as a heuristic term, from which predictions could be generated. Thus, demonstrating the occurrence of reconciliation is not the same as demonstrating a relationship–repair function, although the latter is implied by the term reconciliation.

Aureli et al. proposed a predictive framework within which the occurrence of reconciliation across species is determined according to the potential loss of benefits resulting from aggressive conflict and thus the need for relationship repair. Thus, reconciliation is possible in any species in which there are individualized relationships and intra-group aggression occurs, provided that aggression has the potential to disrupt relationships. If relationships are of sufficient value, the benefits of relationship repair should outweigh the risks of renewed attack, thus making reconciliation worthwhile. Accordingly, since the first study on reconciliation in chimpanzees, reconciliation has been demonstrated in over 30 primate species in which the behavior has been investigated with only a few exceptions, such as the red-bellied tamarins (*Sanguinus labiatus*), whose highly secure and cooperative relationships may preclude disruption by conflict and thus the need for reconciliation. Black lemurs (*Eulemur macaco*) present another exception to the occurrence of reconciliation as relationships between adults are either so valuable that aggressive conflict is extremely rare, and if it does occur is unlikely to disrupt the relationship, or are so hostile that the relationship affords no benefits to either opponent. If the relationship is of such little value, no benefits are lost in aggressive conflict and there is nothing to repair, so reconciliation is not necessary in either case.

Reconciliation is not specific to primates, indeed the predictive framework may be applicable to all gregarious animals. Although few studies have systematically investigated postconflict behavior in nonprimate species, reconciliation has been demonstrated in wolves (*Canis lupis*), domestic dogs (*Canis familiaris*), spotted hyenas (*Crocuta crocuta*), dolphins (*Tursiops truncates*), and domestic goats (*Capra hircus*). Interspecific reconciliation has also been observed between highly valuable partners such as cleaner wrasse *Labroides dimidiatus* and their client reef fish. Further anecdotal evidence is available for reconciliation in feral sheep (*Ovis aries*), dwarf mongooses (*Helegale undulate*), lions (*Panthera leo*), and mouflons (*Ovis ammon*).

The term reconciliation implies a relationship repair function, and this function has been demonstrated in all studies that have tested for it by showing that while unreconciled conflicts lead to a reduced tolerance around resources compared with baseline levels, following reconciliation, baseline levels of tolerance are restored. Reconciliation may reduce the costs of aggressive conflict in a number of ways. The original recipient of aggression is more likely to receive further aggression from both the original aggressor and from other group members. Following aggressive conflicts, recipients of aggression have also been shown to exhibit increased levels of self-directed behaviors, a behavioral indicator of anxiety in primates. This effect may be due to the uncertainty about further aggression or about the status of relationships that may have been damaged by the preceding conflict. Interestingly, some studies have also reported an increase in postconflict levels of self-directed behaviors in aggressors, suggesting that the degeneration of a valuable relationship through aggressive conflict is detrimental to both opponents. Reconciliation has been shown to reduce the likelihood of further aggression and reduce levels of self-directed behavior to baseline levels. The stress-alleviating effect of reconciliation has also been confirmed in human children using hormonal analyses.

Despite the advantages of reconciliation in terms of relationship repair and stress alleviation, not all aggressive conflicts are reconciled, as reconciliation also entails risks of renewed aggression. Thus, reconciliation is only likely if the benefits of reconciliation outweigh the risks.

The Valuable Relationship Hypothesis dictates that dyads with more valuable relationships are more likely to reconcile, as the benefits of relationship repair are likely to outweigh the costs of the risk of renewed aggression. In support of this, kin are usually more likely to reconcile than nonkin, friends more than nonfriends and partners who share food and support each other are likely to reconcile more than partners who do not. Furthermore, reconciliation is not expected, and was not found, in species in which aggressive conflict is absent between valuable partners, such as in rooks (*Corvus frugilegus*). The likelihood of reconciliation, however, is also based on the risks of renewed aggression, and thus the occurrence of reconciliation may be predicted on the basis of relationship quality characteristics other than just relationship value. Relationship quality can be viewed as consisting of three separate components: relationship value, compatibility, and security. The value of a relationship is a measure of the fitness benefits afforded by that relationship, whereas the compatibility of a relationship refers to the tolerance and general tenor of social interactions within a dyad. The predictability of a relationship, or the consistency of interactions over time, is known as relationship security. Thus, compatible partners are also likely to reconcile because the costs of renewed aggression are low. Partners with a valuable but also highly secure relationship, however, may not reconcile often, as the highly secure relationships may not be jeopardized by aggressive conflict, and thus reconciliation may be obsolete.

Measurement of Reconciliation

The standard procedure for investigating postconflict behavior is known as the PC-MC method, which allows postconflict behavior to be compared with baseline behavior. Following the cease of aggression, one of the conflict participants is followed for a given period of time (3–30 min, most often 10), known as the postconflict observation (PC). The same individual is then followed for the same period of time during a control period (matched control observation (MC)) to obtain baseline data, controlling for variables such as group activity and distance from the opponent. Thus, latency to first affiliative contact between opponents can be compared in PCs and MCs (see **Figure 1**). PC–MC pairs in which affiliation occurs first or only in the PC are known as attracted pairs. PC–MC pairs in which affiliation occurs first or only in the MC are known as dispersed. A measure of the overall conciliatory tendency, correcting for baseline levels of affiliation (corrected conciliatory tendency, or CCT) can be calculated as follows. (attracted pairs – dispersed pairs)/all PC–MC pairs. Reconciliation is usually demonstrated in a study group by comparing the proportion of attracted and dispersed pairs at the

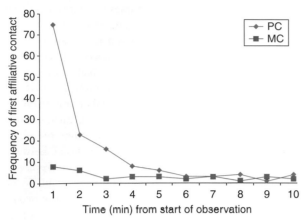

Figure 1 A hypothetical example of the latency to first affiliative contact between opponents in postconflict observations (PC) and matched control observations (MC).

individual level. CCT values can then be used to compare conciliatory patterns, such as for kin and nonkin. The function of reconciliation can also be investigated by comparing measures of anxiety or tolerance in PCs and MCs. Similar methods are used to investigate the occurrence, rates and functions of other postconflict interactions.

Bystander Affiliation

In addition to postconflict interactions between opponents, there is evidence that conflict participants are likely to engage in postconflict affiliative behavior with bystanders who were uninvolved in the previous conflict. The function of such interactions, however, are likely to vary according to the role the conflict participant played in the aggressive conflict, the initiator of the interaction and the quality of the relationships of those involved.

Consolation

When a bystander who shares a valuable relationship with the recipient of aggression offers them postconflict affiliation, the interaction is likely to serve a calming, or stress-reduction, function and is thus labeled consolation. Levels of self-directed behavior in chimpanzee recipients of aggression have been shown to be lower following postconflict affiliation from a bystander, supporting the consoling function. That consolation is provided by valuable partners has been shown in chimpanzees and is further supported by evidence that only mating partners offer postconflict bystander affiliation in rooks. Consolation occurs mainly when reconciliation does not, and thus may function as a stress-reduction alternative to reconciliation when risks of renewed aggression upon reconciliation are too high. Consolation, however, is unlikely to repair the opponents' relationship, as reconciliation does,

and thus reconciliation, where possible, is likely to be preferable to consolation. The mechanism driving consolation is unknown, but the fact that it is provided by valuable partners and serves a stress-reduction function supports the hypothesis that consolation is an expression of empathy, as valuable partners are expected to be more responsive to each other's distress. This, however, does not apply to affiliation initiated or solicited by the recipient of aggression, as, while both forms of consolation may serve similar stress-reduction functions, solicited consolation does not require the same cognitive capacity.

Appeasement

The recipient of aggression is not the only conflict participant to experience elevated levels of postconflict stress. Aggressive conflict may also leave the aggressor stressed, and thus postconflict affiliation from a bystander to the aggressor may also serve a stress-reduction function. This interaction is termed appeasement, as although it may calm the aggressor down, the primary motivation is likely to be to reduce the likelihood of the bystander becoming the target of further aggression from the aggressor. Thus, unlike consolation, the relationship between the bystander and the aggressor is unlikely to be valuable.

Self-Protection

Bystanders may also initiate postconflict affiliation with recipients of aggression in order to protect themselves from becoming targets of redirected aggression (see 'Aggressive Postconflict Interactions'). In chimpanzees, when no evidence for a consolation function was found, bystanders who affiliated with the recipient of aggression were those who were most likely to become targets of renewed aggression. Recipients of aggression did not target bystanders who had affiliated with them, but were no less likely to redirect aggression towards others after receiving affiliation. Thus, self-protective bystander affiliation does not appear to function by calming the recipient of aggression, and thus eliminating their need for redirected aggression. Behaviors used for self-protective postconflict affiliation from a bystander to the recipient of aggression are also likely to differ from those used for consolation, reflecting the differences in the bystander-recipient relationships. Bystander affiliation with a self-protection function is likely to be expressed with brief behaviors that minimize the risk to the bystander, whereas consolation offered by a valuable partner is more likely to involve extensive contact behavior, which is potentially more risky.

Relationship Repair

In some cases bystander affiliation may repair the relationship between opponents. Bystanders who have a valuable relationship with the aggressor, such as its kin, may be able to act as a proxy for the aggressor, and thus may affiliate with the recipient of aggression, reducing postconflict stress and repairing the opponents' relationship as if it were reconciliation. Such interactions are likely to take place when the risk of renewed aggression between the opponents is high, and thus the recipient of aggression is unlikely to allow the aggressor to get close enough for reconciliation to take place. Thus a relationship repair function for bystander affiliation is likely when the bystander has a valuable relationship with the aggressor, but the relationship between the opponents is of low compatibility, leading to a high risk of renewed aggression.

Support

Bystander affiliation may also serve to signal approval to and support for the aggressor, or to reconfirm or improve relationships between the bystander and the conflict participant. Furthermore, it may function to signal the strength of the alliance between the partners to other group members. Although such interactions may be directed from the bystander to either the aggressor or the recipient of aggression, they are more likely to be effective if directed at the aggressor as the bystanders are likely to gain more from showing their support for the winner of a contest. This form of bystander affiliation has also been suggested to reinforce the existing hierarchy when the recipient of postconflict affiliation is a high-ranking aggressor. There is some evidence for bystander affiliation to have a support function. Long-tail macaque (*Macaca fascicularis*) aggressors were more likely to initiate further aggression following bystander affiliation, presumably because the aggressors were encouraged by signals of support from bystanders.

Quadratic Affiliation

Aggressive conflict may have consequences not only for those involved but for the group as a whole. Dyadic conflicts frequently escalate to include other individuals, and a change in the relationship between conflict participants may affect other relationships in the group, thus merely observing an aggressive conflict may lead to an increase in tension. It is possible, therefore that bystanders affiliate with conflict participants in order to alleviate their own distress. It is also likely that affiliation with other bystanders might have the same effect, particularly where stress-reducing behaviors, such as grooming, are used. As aggressive conflict between two individuals may thus influence the behavior of two others, an affiliative postconflict interaction between two bystanders is the result of the behavior of four individuals, and so is labeled quadratic postconflict affiliation. Quadratic postconflict

affiliation in hamadryas baboons (*Papio hamadryas hamadryas*) has been shown to reduce levels of self-directed behavior, indicating that postconflict affiliation between bystanders has a stress-alleviating effect.

Aggressive Postconflict Interactions

Not all postconflict interactions are friendly. Targets of aggression, in particular, may redirect aggression towards other group members following the initial conflict. Sometimes such attacks are directed towards the aggressor's kin and thus may function to damage the aggressor by attacking their more vulnerable kin. Redirected aggression may also function to reverse the negative effects of losing a conflict, whereas aggression from the winner of the conflict to a third-party may reinforce the winning experience. Redirection of aggression from the victim of a conflict is likely to offer a safe outlet for venting frustration, particularly when the victim is not able to retaliate against the aggressor, and thus allows the victim to regain a measure of control over the situation and alleviate postconflict stress. Thus, redirected aggression is more common in species with strict dominance hierarchies, where bidirectional aggression is uncommon.

Following an aggressive conflict, bystanders may take advantage of the opportunity to attack the victim weakened by the previous conflict and, potentially, the losing experience. Thus, redirected aggression may offer the victim some measure of protection from further aggression as long-tailed and rhesus macaque (*Macaca mulatta*) victims were comparably less likely to receive aggression from bystanders when they redirected aggression than when they did not. Victims may also redirect aggression to bystanders as positive display to both their previous assailant and other bystanders of their postconflict condition despite the previous loss. Such a display may reduce the likelihood of the victim becoming the target of further aggression, but may also demonstrate their potential value to the aggressor in future conflicts and thus make reconciliation more likely. Hence, the target of redirected aggression may be largely irrelevant (provided they are of inferior competitive ability), as the act is performed primarily as a signal to other group members.

Conclusions

Different conflict resolution strategies play an important role in the maintenance of the benefits of group living and the management of conflicts of interest once they escalate to aggressive conflict (**Table 1**). Through postconflict interactions, valuable relationships can be repaired and reinforced, damage can be limited and emotional stress associated with aggressive conflict experienced by conflict

Table 1 Functions of postconflict interactions

Postconflict interaction	Participants and direction of interaction	Function	Function demonstrated?
Reconciliation	Between opponents	Relationship repair, stress reduction	Yes
Appeasement	From bystander to aggressor	Stress reduction, aggression avoidance	No
Consolation	From bystander to victim	Stress reduction	Yes
Solicited consolation	From victim to bystander	Stress reduction	No
Self-protection	From bystander to victim or aggressor	Aggression avoidance	Yes
Support	From bystander to aggressor	Reinforce valuable relationship	No
Relationship repair	Bystander (valuable partner of aggressor) to victim	Repair opponent relationship	Yes
Redirected aggression	From victim to bystander	Stress reduction, aggression avoidance	Yes
Quadratic affiliation	Between bystanders	Stress reduction	Yes

participants and bystanders can be alleviated, thus maintaining group cohesion, social relationship and associated benefits. The choice of postconflict interaction after aggressive conflict appears to be strongly dependent on the quality of the relationships among the individuals involved. Whereas reconciliation may offer the optimal solution for conflict participants in terms of reducing the costs of aggressive conflict, other postconflict interactions, such as the various forms of bystander affiliation, or redirected aggression, may offer alternative solutions for the opponents when reconciliation does not occur. Bystander affiliation may also offer bystanders the opportunity to limit the negative consequences aggressive conflict may have on themselves and rest of the group and may provide bystanders with the opportunity to reinforce relationships and strengthen alliances.

See also: Empathetic Behavior; Punishment.

Further Reading

Arnold K and Aureli F (2007) Postconflict reconciliation. In: Campbell CJ, Fuentes A, MacKinnon KC, Panger M, and Bearder SK (eds.) *Primates in Perspective*, pp. 592–608. Oxford: Oxford University Press.

Aureli F, Cords M, and van Schaik CP (2002) Conflict resolution following aggression in gregarious animals: A predictive framework. *Animal Behaviour* 64: 325–343.

Aureli F and de Waal FBM (2000) *Natural Conflict Resolution*. Berkeley: University of California Press.

de Waal FBM (2000) Primates – a natural heritage of conflict resolution. *Science* 289: 586–590.

de Waal FBM and van Roosmalen A (1979) Reconciliation and consolation among chimpanzees. *Behavioral Ecology and Sociobiology* 5: 55–66.

Fraser ON, Stahl D, and Aureli F (2008) Stress reduction through consolation in chimpanzees. *Proceedings of the National Academy of Sciences of the United States of America* 105: 8557–8562.

Judge PG and Mullen SH (2005) Quadratic postconflict affiliation among bystanders in a hamadryas baboon group. *Animal Behaviour* 69: 1345–1355.

Kazem AJN and Aureli F (2005) Redirection of aggression: Multiparty signalling within a network? In: McGregor PK (ed.) *Animal Communication Networks*, pp. 191–218. Cambridge: Cambridge University Press.

Koski SE, de Vries H, van den Tweel SW, and Sterck EHM (2007) What to do after a fight? The determinants and inter-dependency of post-conflict interactions in chimpanzees. *Behaviour* 144: 529–555.

Seed AM, Clayton NS, and Emery NJ (2007) Post-conflict third-party affiliation by rooks, *Corvus frugilegus. Current Biology* 17: 152–158.

Watts DP (2006) Conflict resolution in chimpanzees and the valuable-relationships hypothesis. *International Journal of Primatology* 27: 1337–1364.

Wittig RM and Boesch C (2003) 'Decision-making' in conflicts of wild chimpanzees (*Pan troglodytes*): An extension of the Relational Model. *Behavioral Ecology and Sociobiology* 54: 491–504.

Consensus Decisions

L. Conradt and T. J. Roper, University of Sussex, Brighton, UK
C. List, London School of Economics, London, UK

Introduction

Animals routinely face decisions that are crucial to their survival and fitness: they have to decide when and where to rest or forage, which individual to mate with, where to live, when to reproduce, and so on. Social animals have to make many such decisions not as individuals acting alone but collectively, as a group. A large proportion of collective decisions even require that all the members of a group reach a consensus. Consider a group of carnivores deciding where to move after a resting period, a shoal of fish deciding when to leave a foraging patch, or a colony of ants choosing a new nest site: unless all the members decide on the same action, some will be left behind and will be deprived of the advantages of group living, at least for the time being.

In animals, a consensus decision is defined as a decision in which members of a cohesive group choose, collectively, a single action from a set of mutually exclusive options, and that choice is binding in some way for all the members. In this context, consensus does not imply that all the group members necessarily share the same interest, or even like the decision outcome, but only that all members comply with the collective decision outcome to maintain group cohesion. Nor does the fact that the decision is collective mean that all the group members necessarily have the same influence on the decision outcome (see more on this below). Typical examples of consensus decisions are choosing between different movement destinations, nest sites, migration routes, or cooperative strategies, or the timing of group activities (e.g., foraging or resting).

Aggregation Rules

When animals make decisions, they typically have a choice between two or more options. In order to make decisions collectively (and achieve a group consensus), the preferred choices of individual group members have to be aggregated in such a manner that the group 'agrees' on one option. That is, an aggregation rule is required. Formally, an aggregation rule is defined as a function that assigns to each combination of individual inputs (e.g., choices or 'votes') a resulting collective output (e.g., a decision outcome). The classic example of an aggregation rule is majority voting between two options, under which the group selects the option that receives more votes than the other. However, a dictatorial decision rule, under which the group always adopts the choice of a single preordained individual, the 'dictator,' is also an aggregation rule. Humans often use aggregation rules that are based on majority voting, but in which only the choices of particular group members count (e.g., children are usually not allowed to vote in national elections).

In animals, empirical aggregation rules range from dictatorial ones to majority voting. For example, Andrew King and co-authors reported that in wild chacma baboons (*Papio ursinus*), the dominant male chooses the group's foraging patch, even if the choice is against the foraging interests of the majority of other group members. On the other hand, in wild red deer (*Cervus elaphus*), the majority of deer determine a herd's departure time from their resting site. Gerald Kerth and colleagues reported that in Bechstein's bats (*Myotis bechsteinii*), it can also be the majority of bats that decide when to change roosting sites. However, in most observed cases in non-human animals, the decision is made by several particular group members, not by all the members. This is very well documented by Thomas Seeley and colleagues' detailed work on honeybee swarms: new nest sites are chosen only by a few hundred informed scouts within the swarm. A group's aggregation rule is important, since it greatly influences the costs and benefits of the group's decision outcome to individual members and to the group as a whole (see also below).

Communication: Global and Local

While the implementation of complex aggregation rules in humans is obvious and familiar to all of us (e.g., national elections, parliamentary decisions), it is more difficult to see how animals could implement aggregation rules and thereby make decisions collectively. At first sight, their lack of a sophisticated language and their limited cognitive abilities seem to prohibit the necessary negotiations and voting that underlie many complex aggregation rules. How, then, can animals decide collectively in meaningful ways?

To address this question, it is helpful to distinguish between two types of animal groups: small groups and large groups. These groups should not be distinguished by the actual number of group members, but by the manner in which members can communicate with each other.

Small groups are defined as groups in which all the members can, at least in principle, communicate with all the others. In such groups, global communication is a possibility. Typical examples are groups of carnivores, primates, or some ungulate and bird species. In contrast, large groups are defined as groups in which global communication is no longer possible. Instead, group members can, at most, communicate with their neighbors (local communication). Typical examples are swarms of insects, large shoals of fish, large flocks of birds, or large herds of ungulates.

In small groups with global communication, group decisions could, at least in principle, be reached by general negotiations among all the members and explicit voting, or by central orders or coercion. Voting has been reported in several mammal and bird species, and dictatorial or coerced decisions in others. Empirical examples of voting behaviors include the use of specific body postures, ritualized movements, and specific vocalizations. In order to implement majority voting, animals do not need to be able to count explicitly but do need to assess relative numerousness (e.g., are more group members standing or sitting?). A recent review by David Sumpter and Stephen Pratt indicates that quorum responses, whereby an individual's probability of exhibiting a behavior is a sharply nonlinear function of the number of others already performing this behavior, could also be a plausible mechanism. In simpler terms, we speak of quorum responses when individuals are much more likely to perform a behavior if they find a threshold number of other individuals (the quorum) already performing this behavior than if they do not.

In large groups with only local communication, individuals are assumed to follow their own local behavioral rules, based on local information and local communication but resulting in a global group behavior that is not centrally orchestrated but self-organized. A good example of such self-organization is given by the movements of large flocks of starlings (*Sturnus vulgaris*), studied in great detail by Michelle Ballerini, Irene Giardina, Charlotte Hemelrijk, and their colleagues. Each flock member continuously tries to avoid collision with direct neighbors or obstacles but, at the same time, continuously tries to maintain cohesion with its neighbors. The overall result is the fascinatingly synchronized and well-coordinated flock movement that we observe in nature.

At first glance, self-organization seems to prohibit decisions made either by general negotiation and voting or by central orders or coercion. However, theoretical models suggest that aggregation rules in self-organizing groups can arise as intrinsic consequences of local behaviors of group members. They can range from majority rules (if all the group members adopt the same local behaviors, as illustrated by the work of Iain Couzin and colleagues in 2005) to aggregation rules in which only

certain members influence the decision outcome (by adopting more independent local behaviors than do other members). Empirical observations by Jens Krause and colleagues on roach (*Rutilus rutilus*) and by Herbert Prins on African buffalo (*Bufallo bufallo*) support the model predictions. Quorum responses also play an important role in large groups and have been described in honeybees (*Apis mellifera*) by Thomas Seeley and colleagues, in cockroaches (*Blattella germanica*) by Jean-Louis Deneubourg and colleagues, and in ants (*Leptothorax albipennis*) by Nigel Franks, Stephen Pratt, and colleagues. As in small groups, quorum responses are plausible mechanisms for decision aggregation.

Main Factors Influencing Consensus Decisions

The two most important factors influencing consensus decisions in social animals are (1) information and (2) interests. We address these in turn. Additionally, several side constraints can also play important roles, most notably time constraints. For more details on side constraints, we recommend the Further Reading section.

Information

In order to make advantageous decisions, decision makers require environmental information (e.g., about the quality of a foraging patch, the presence of predators, the best traveling routes, etc.). However, individuals typically have only incomplete and noisy information about the state of the environment. Groups of animals making decisions have the potential advantage, relative to solitary decision makers, that the private information of all their members taken together is likely to be more complete and more accurate than that of a single animal. This is because some group members might know about good foraging patches, some about good traveling routes, some about predators, and so on. Additionally, any false private information that one member might hold could be corrected by more accurate private information from others. Hans Wallraff suggested already in the 1970s that homing or migrating flocks of birds show better orientation than individuals would do on their own, and recent empirical work by Dora Biro lends further support to this notion. Stephen Reebs and Andrew Ward have made similar observations in fishes, and David Lusseau in bottlenose dolphins (*Tursiops* sp.).

In order to use the private information of all group members in a sensible way, the information has to be aggregated. The way in which the information is aggregated across group members can greatly influence the decision pay-offs or accuracy. For example, suppose a group has to make a choice between two options. Each member has some independent private information about which option is better, and that information is correct with probability p (where $0.5 < p < 1$). Condorcet's classic jury theorem shows that in this case, it is more likely that majority voting will yield a correct group decision outcome than does a dictatorial aggregation rule. And the accuracy of the majority decision will increase with the number of group members. However, if group members differ in the probabilities that their information is correct, or if different potential decision errors would result in different costs, other aggregation rules distinct from majority voting can result in more effective pooling of the available information.

Interests

The pay-offs of a decision outcome for a group of individuals obviously depend on whether the outcome is consistent with the members' interests. Group members can share the same interests during a decision. For example, when a swarm of honeybees is deciding on a new nest site, it is advantageous to all bees to choose the best site. There can be differences in information between bee scouts about which site is best, but there are no conflicts of interest: the nest site that offers the best survival and reproduction prospects for the swarm is the best nest site for all the bees. However, in many groups, members differ in sex, age, size, genetic relatedness, and physiological status and consequently have different requirements. This means that decision outcomes that are good for some members might be bad for others. For example, in many ungulates, females with young are more vulnerable to predation, and therefore prefer safer but lower quality foraging sites, than do males. Kathrin Ruckstuhl and Peter Neuhaus reported that conflicts of interest during collective ungulate movement decisions not seldom are so large that groups fail to reach consensus and split.

If there are conflicts of interest within a group, the way in which different individuals' preferences are aggregated can make a great difference to the group's overall pay-offs, and also to the individual pay-offs received by each group member. Often, aggregation rules that assign decision weight to a greater number of group members yield higher overall group pay-offs, and therefore, in human decision making are often considered as desirable and fair. However, since pay-offs are not necessarily higher for each individual group member and might even be lower for some members, the question of whether such fair and inclusive aggregation rules are likely to evolve in animals is complex. Work by Sean Rands, Iva Dostalkova and Marek Spinka and colleagues suggests that for pairs of individuals, aggregation rules that take the preferences of both partners into account are likely to evolve. However, in larger groups the evolution of aggregation rules can be very complex, and fair and inclusive aggregation rules are

not always guaranteed, but at least some skew in the influence of individual group members is often likely to evolve (see Further Reading).

Concluding Comments

The study of consensus decisions in animals is still relatively young, with the exception of studies of social insects. However, the topic has recently started to attract wider attention, and the literature is now expanding rapidly. The only review to date is still: Conradt L and Roper TJ (2005) Consensus decision making in animals. *Trends in Ecology and Evolution* 20: 449–456 (doi:10.1016/j.tree.2005.05.008). However, a themed issue (Group decision making in humans and animals) has just been published in Philosophical Transactions of the Royal Society London B, 364 (2009, eds L. Conradt & C. List; doi:10.1098/rstb.2008.0256). This issue consists of 11 contributions by natural and social scientists, and aims to introduce long-standing social science concepts on group decision making into the newly-emerging, relevant fields within the natural sciences.

Some classical theoretical papers are those by Iain Couzin et al. (2005) Effective leadership and decision-making in animal groups on the move. *Nature* 433: 513–516 (doi:10.1038/nature03236); and Conradt and Roper (2003) Group decision-making in animals. *Nature* 421: 155–158 (doi:10.1038/nature01294). A short note introducing Condorcet's jury theorem into this field is List (2004) Democracy in animals groups: a political science perspective. *Trends in Ecology and Evolution* 19: 168–169 (doi:10.1016/j.tree.2004.02.004). As a brief starting selection of mainly empirical papers, we recommend the publications in the Further Reading.

See also: Collective Intelligence; Decision-Making: Foraging; Group Living; Group Movement; Nest Site Choice in Social Insects; Rational Choice Behavior: Definitions and Evidence; Social Information Use.

Further Reading

Ame J-M, Halloy J, Rivault C, Detrain C, and Deneubourg JL (2006) Collegial decision making based on social amplification leads to optimal group formation. *Proceedings of the National Academy of Sciences of the United States of America* 103: 5835–5840.

Ballerini M, Cabibbo N, Candelier R, et al. (2008) Empirical investigation of starling flocks: A benchmark study in collective animal behaviour. *Animal Behaviour* 76: 201–215.

Biro D, Sumpter DJT, Meade J, and Guilford T (2006) From compromise to leadership in pigeon homing. *Current Biology* 16: 2123–2128.

Franks NR, Pratt SC, Mallon EB, Britton NF, and Sumpter DJT (2002) Information flow, opinion polling and collective intelligence in house hunting social insects. *Philosophical Transactions of the Royal Society of London Series B-Biological Sciences* 357: 1567–1583.

Kerth G, Ebert C, and Schmidtke C (2006) Group decision making in fission–fusion societies: Evidence from two-field experiments in Bechstein's bats. *Proceedings of the Royal Society B* 273: 2785–2790.

King AJ, Douglas CMS, Huchard E, Isaac NJB, and Cowlishaw G (2008) Dominance and affiliation mediate despotism in a social primate. *Current Biology* 18: 1833–1838.

Lusseau D (2007) Evidence for social role in a dolphin social network. *Evolutionary Ecology* 21: 357–366.

Meunier H, Leca JB, Deneubourg JL, and Petit O (2006) Group movement decisions in capuchin monkeys: The utility of an experimental study and a mathematical model to explore the relationship between individual and collective behaviours. *Behaviour* 143: 1511–1527.

Seeley TD and Buhrman SC (1999) Group decision making in swarms of honey bees. *Behavioral Ecology and Sociobiology* 45: 19–31.

Sueur C and Petit O (2008) Shared or unshared consensus decision in macaques? *Behavioural Processes* 78: 84–92.

Wallraff HG (1978) Social interrelations involved in migratory orientation of birds – possible contribution of field studies. *Oikos* 30: 401–404.

Conservation and Animal Behavior

R. Swaisgood, San Diego Zoo's Institute for Conservation Research, Escondido, CA, USA

Introduction

Conservation is the end result of a human value system that seeks to maintain the diversity of life forms on earth and ensure the ecological integrity of our natural heritage. People may be motivated to conserve nature by utilitarian values, such as recognition of the important services that a functioning ecosystem provides to humanity, or by a deep and abiding philosophy that other forms of life have intrinsic value as well. The biophilia hypothesis – championed by the founding father of biodiversity and sociobiology, E.O. Wilson – posits that humans are predisposed to an emotional attachment to nature that motivates them toward environmental stewardship. It may be argued that animal behaviorists, who spend long hours observing animals, have a more-than-average dose of biophilia, making their late arrival on the conservation stage surprising. Although the modern academic discipline of conservation biology has its roots in wildlife management that goes back generations, it was not founded until the 1980s. Conservation behavior – as the application of behavioral research to conservation is sometimes called – traces its formal, academic beginnings to the waning moments of the last millennium. As an emerging discipline, conservation behavior is still seeking to define itself and find its niche in both conservation biology and behavioral science.

Conservation behaviorists, a small group of individuals by any definition, are attempting to reinvent the way behavioral research is applied to conservation efforts. At the end of just over a decade of concerted effort, much of the heady promise of this nascent field has yet to come to fruition. A series of influential books and papers have, in recent years, addressed the many implications that behavior – particularly behavioral ecology – has for conservation. That the behavior of animals influences conservation is a truism, but the challenge for behavioral scientists has been to move from the implication phase of conservation behavior to more active applications to solve the real-world conservation problems. The discipline has experienced growing pains, but has real potential, some of which is just beginning to be realized.

Several subdisciplines within animal behavior have contributed to the emergence of conservation behavior, but to date most proponents have been rooted in behavioral ecology. Many of the topics in behavioral ecological research appear promising for conservation application – including mating strategies, mate choice, dispersal and habitat selection, behavioral responses to habitat fragmentation and anthropogenic disturbance, and the many behavioral facets of reintroduction programs. The strong theoretical framework afforded by behavioral ecology provides the basis for a hypothetico-deductive approach to conservation science. We are learning that a more integrated approach across the four levels of explanation in animal behavior – causation, development, adaptive utility, and evolutionary history – and across larger ecological scales – population, community, ecosystem, landscape – holds the most promise for the successful application of behavioral research to conservation.

Taking Stock of the Problem: What Are Some of the Challenges Facing Animal Conservation?

Today's natural world faces an onslaught of anthropogenic processes that threaten the functional integrity of ecosystems, leading to escalating rates of species loss seldom seen during the history of the planet. Habitat degradation and outright destruction are the single largest culprits, and the resulting fragmentation of habitat supporting wildlife has multiplicative rather than additive effects on loss of biodiversity. Widespread urbanization and intensive agricultural practices eliminate most of these lost species, whereas other human activities degrade much of the remaining natural areas. Chemical pollutants may affect survival and reproduction, and thus population recruitment rates. Noise pollution may likewise disturb mating and parenting behavior or lead to chronic stress, with its attendant consequences for immunocompetence, fertility, and allocation of energetic resources away from other demands of survival. A single aircraft overflight has been known to cause the immediate loss of most chicks in a white pelican nesting colony. Sadly, even our love for nature in the form of ecotourism may be harming animals. Ever-increasing numbers of nature enthusiasts converging on whales and dolphins may tip the energetic balance, causing chronic sublethal effects. A simple walk on the beach disturbs shorebirds, diverting them away from foraging, mating, or parenting behavior. Multiplied by thousands of beachgoers, the cumulative effects on survival and reproduction can mean population decline. Pets, acting as predators along edges of natural communities, can have reverberating effects on community dynamics deeper in the reserve. Studies have shown that even on-leash dogs can reduce diversity along a surprisingly wide trail margin.

Humans also often aid transport and colonization of invasive nonnative species. Those species with the right suite of behavioral and life-history characteristics sometimes get a foothold in their new environment and – without the ecological controls in their place of origin – undergo rapid population expansion. Over-run with exotics, the results for native competitors or prey species can be devastating. Take the case of the mountain yellow-legged frog (*Rana muscosa*) in California (**Figure 1**). The introduction of brown trout and the larger and more aggressive bullfrog have contributed to the decline of this endangered species. This high-mountain frog has little coevolutionary history with fish predators and its tadpoles are vulnerable to predation by nonnative trout. Populations have recovered rapidly when trout have been experimentally removed.

Overlaid on top of these long-known threats to animal populations is the unpredictable impact of anthropogenic climate change, which will exacerbate other impacts. Climate change may precipitate range shifts in many species, but fragmented habitat may prevent migration, calling upon even greater human intervention (such as widespread translocations of animals to help track shifting habitats).

Wildlife management is likely to become increasingly intensive in the future, as protected areas become more zoo-like, managed parks. Smaller, isolated populations will call upon the skills and resources of conservation biologists more than ever. Conservation behaviorists will undoubtedly play a larger role.

Just monitoring animal populations so that we know what we have will require considerable resources, though greatly aided by emerging technologies. Because behavior affects detectability, behaviorists will need to play an increasing role in survey and monitoring efforts. In managed areas, we will need to obtain a better understanding

of carrying capacity – the population size that can be sustained in a given patch of habitat. The ecological factor(s) limiting an animal's carrying capacity may be food, water, shelter, refuges from predators, or breeding sites. Modifying the factors limiting the carrying capacity can be one way to alter a population's size, for better or worse.

Some species have been rescued from the wild and placed in conservation breeding programs in zoos and other facilities. These small populations need to be managed for genetic diversity to avoid the deleterious effects of inbreeding and to preserve as much of the species evolutionary potential as possible. Managers combat both random genetic drift and, more importantly, artificial selection that causes domestication-like effects on captive or even wild populations living under increasingly artificial conditions. The long-term objective of conservation breeding and other small population management programs is to sustain a secure genetic reservoir until conditions in the wild can be improved. When suitable habitat is found or created, populations can be reestablished through reintroduction and translocation programs. Reintroduction of key species can serve to restore ecological integrity of natural areas, especially if the reintroduced species are ecosystem engineers. For example, fossorial rodents such as ground squirrels create burrows used by a variety of wildlife species. Experimental removal of kangaroo rats precipitates invasion of nonnative grasses (**Figure 2**). Removal of top predators, such as mountain lions, can lead to increased numbers of smaller predators (mesopredator release) that decimate prey species, such as songbirds.

Ecological restoration through reintroduction is one of the most promising areas for behavioral research contributions. Animal relocations – whether from captive-bred or wild-caught animals – require intensive behavioral research and management to prepare animals for release

Figure 1   One of the estimated 122 surviving adult mountain yellow-legged frogs in Southern California, decimated by invasive species and chytrid fungus. Photo by Ron Swaisgood.

Figure 2   The endangered Stephens' kangaroo rat of Southern California. A translocation program led by the San Diego Zoo is trying to re-establish this keystone species in suitable habitat. Photo by Ron Swaisgood.

to a novel environment and, in many cases, monitor and manage their postrelease behavior. Most reintroductions to date have taken a fairly simplistic approach to post-release management, relying primarily on soft-release practices such as acclimation pens and short-term provision of food and water resources. Behaviorists can redefine the meaning of 'soft' in these programs, replacing these practices with more ecologically relevant ones guided by a strong theoretical framework. That said, the application of conservation behavior is not limited to triage and is as diverse as the imagination and creativity of its proponents.

Tackling the Problem: How Can Behavioral Research Contribute to Conservation?

Animals on the Move: Space Use, Dispersal, and Habitat Settlement

Animal movement patterns figure critically in many aspects of conservation management. Movement patterns determine size and shape requirements for reserves and the degree of connectivity needed for adequate migration between protected areas to sustain genetically viable metapopulations across the landscape. Animals with large home ranges more often range outside the boundaries of protected areas, where they are no longer protected. In fact, wild dog home ranges are so large relative to most reserves that only those groups living well within the core of the reserve are not exposed to the risks outside reserve boundaries. The longest animal movements are seen in dispersal – generally the once-in-a-lifetime movement away from the natal home to settle in new habitat. Reserves need to be designed to accommodate these dispersal distances, which can be many times larger than the typical home range for the species. Dispersal is important for discovering and occupying new habitat and for inbreeding avoidance. Conservation planners need better data on spatial movements, including how far an animal can move, through what types of habitat it moves, and how this is influenced by topography and human-altered landscapes.

The last decade has witnessed a revolution in spatial ecology, brought on by emerging technologies such as GPS satellite telemetry and remote infrared-triggered camera and video traps. These technologies have proved especially useful in linking fine-scale movements of individual animals (of interest to behavioral ecologists) with larger landscape-level patterns (of interest to population and landscape ecologists).

In the fragmented habitats that many wildlife species occupy, animals must cross gaps to maintain connectivity among animal populations. Forest specialists, in particular, may be reluctant to cross a gap caused by clear cutting or a road. It is rarely physical ability that limits gap

crossing, but instead the animal's perception of risk from predators or stress-mediated avoidance of an unfamiliar landscape. Common sense might inform whether an animal can pass through a habitat, but experimentation is required to determine under which motivational circumstances they will actually do so. In this experimental approach, playbacks of conspecific vocalizations have been broadcast to lure individuals into or across particular landscape features, or animals have been experimentally relocated across gaps that vary in size and quality. These bioassays can be used to determine the permeability of different habitat types over varying spatial and temporal scales.

Behavioral constraints on dispersal may compromise conservation efforts. A common misconception is that if suitable habitat is present, then dispersers will find and occupy it, but such build-it-and-they-will-come approaches do not always work. Seminal research by Judy Stamps in the 1980s demonstrated that dispersers often prefer to settle near conspecifics, even among less social, territorial species. Such conspecific attraction in habitat settlement may serve to guide dispersers to suitable habitat without long, costly direct sampling of habitat quality. Research has further shown that dispersal is often the most risk-prone life history stage, exposing animals to the risks of predation and starvation in unfamiliar habitat. These risks select for more conservative dispersal strategies and the use of cues correlated with habitat suitability. Using this theoretical framework, conservation behaviorists have used bird song playbacks to recruit black-capped vireos to new areas, model decoys to attract terns to new colonies, white wash (mimicking droppings) to attract vultures, and rhino dung to encourage settlement in translocated black rhinos. This tool is proving particularly powerful in reintroduction and translocation programs, because, in fact, these conservation actions force a dispersal-like event upon animals whether or not they are prepared for dispersal. It is no surprise that most mortalities occur soon after release. Behavioral management of dispersal during this postrelease period could prove critical in determining the success or failure of the program.

Consider the example of caribou translocated from open prairies to mountain forests, which suffer higher mortality than those translocated from more similar habitat types. The difference lies in their behavioral response to the environment. In winter, mountain caribou move to the north slopes to forage upon lichens that grow prevalently on the cooler, moister slopes. In winter, prairie caribou forage on open tundra, digging under the snow to expose hidden lichens and when translocated, maladaptively move to the more open south slopes in an attempt to use their foraging strategy in the new locale. In another example, red squirrels translocated from Corsican pine forests traversed through Scotch pine in search of Corsican pine to settle. Conversely, squirrels captured in Scotch pine forests reject Corsican pine in favor of Scotch pine habitats. Such behavior can be explained by natal habitat preference induction (NHPI). More formally, NHPI occurs when dispersers prefer new habitats that contain stimuli comparable to those in their natal habitat. NHPI may help explain why relocated animals are so often prone to travel so long distances, at their peril – they are searching for someplace like home. Understanding this phenomenon enables the implementation of several new approaches to reintroduction programs, including efforts to match habitat type at capture and release site or to manipulate conspicuous cues in the postrelease habitat to match those from the animal's place of origin.

Foraging Ecology

Research on foraging ecology can provide key insights into animal–environment interactions important for conservation. NHPI provides but one example. Long-term studies across seasons may identify key limiting resources that determine the size of the population that can be sustained in an area (carrying capacity); such studies can be crucial for proactive conservation management to ensure those resources remain available, or can suggest ways to enhance these resources to support populations at risk. Conservation behaviorists may also identify key roles that animals under study provide for the ecosystem, such as seed dispersal, population control for smaller predators (by keystone predators), or interspecific competition for resources.

Security Areas

Security is another critical need for animals. Nesting and denning sites can be limiting resources. For example, endangered red-cockaded woodpeckers must excavate nests in live pine trees, a task that can take more than a year. This limiting resource shapes many aspects of the species' behavioral ecology, including the phenomenon of cooperative breeding. Because dispersal to new areas is limited by the availability of nesting cavities, young woodpeckers remain with their parents and help raise subsequent offspring, banking on future access to nest sites nearby. Understanding this system led to a novel conservation management action. Because of the high number of reproductively mature birds without access to nesting opportunities, construction of artificial cavities led to a rapid increase in the breeding population and was key to rapid recovery following loss of many cavities in a hurricane. New research is now suggesting that giant pandas may also be limited by access to quality den sites. After centuries of logging, few trees of sufficient girth to contain a panda-sized cavity remain, and panda females may have few quality dens to select from, limiting their ability to provide proper care for cubs.

Too Much of a Good Thing: Mating Strategies, Reproductive Skew, and Effective Population Size

Understanding mating patterns has several important consequences for conservation. Of those, reproductive skew may be most important. If some males have mating advantages over other males – either by direct male–male competition or female mate preferences – they will sire disproportionately more offspring. In small populations, this can further exacerbate loss of genetic diversity, and much larger populations will be required to maintain sustainable populations over the long term. A key concept is that of effective population size (N_e), which is the number of breeding animals in an ideal, randomly mating, population that would lose genetic diversity at the same rate as the actual population. The greater the reproductive skew, the smaller the N_e and the greater the concern for the long-term viability of the population. With the loss of genetic variability, species lose some of their evolutionary potential, their ability to adapt to changing environments, and may suffer consequences of inbreeding depression.

To maintain long-term population viability, there is a point where small populations require management to ensure that reproduction is distributed more equitably. Experimental manipulation of behavioral mechanisms related to mate choice is one course of action being pursued by conservation behaviorists. For example, Fisher and colleagues were able to control female preferences of threatened pygmy lorises by manipulating the level of odor familiarity with potential mates (**Figure 4**). These researchers reasoned that females should prefer more familiar-smelling males because males capable of monopolizing an area with their odors are more competitive than males that fail to do so. This demonstrates the importance of understanding the theoretical framework provided by behavioral ecology. If male–male competition is responsible for reproductive skew, then temporary removal of the most successful male can increase N_e, as has been done with threatened Cuban rock iguanas.

Relationships Matter: The Role of Social Behavior in Conservation

Several social processes have important ramifications for conservation. The Allee effect – the proposal that reproduction is maximized when animals are optimally aggregated – suggests that human-mediated alterations to animal population density may influence population growth and stability. Numerous studies have shown the benefits of sociality, even for relatively solitary species, suggesting one consequence of a reduction in animal densities is the breakdown of social processes that confer fitness advantages – ranging from predator avoidance to opportunities for mate selection. Indeed, these social benefits may drive the phenomenon of conspecific attraction, discussed earlier. The (mostly unproven) role for conservation behaviorists is to determine optimal social density. An example from captive breeding programs is instructive: mongoose lemurs housed in pairs near other conspecifics had 500% higher rates of reproduction than isolated pairs. Courtship behavior in flamingos is greatly enhanced by allowing them to view themselves in mirrors, effectively increasing the perceived animal density.

For highly social animals, we need to be mindful of the significant time and energy that animals invest in building and maintaining social relationships, and the advantages that these relationships confer. In translocation programs, for example, animals are often captured and relocated to another area, without due consideration given to social relationships. Debra Shier found that black-tailed prairie dogs had 500% higher survival rates if they were released in familiar groups than if they were released randomly with regard capture location.

We must also keep in mind that solitary animals are not asocial and that even territorial species invest in relationships with neighboring animals. In reintroduction programs, animals experience many challenges related to being placed in novel environments, and the added costs of renegotiating stable relationships may mean the difference between death and survival. This may explain why the normally solitary black rhinoceros experiences a surge in conspecific aggression following translocation. A team of researchers led by Wayne Linklater addressed this problem and have found, for example, that this aggression can be mitigated by releasing rhinos in larger reserves at lower social densities, which helps them to avoid direct encounters and gradually build social familiarity (**Figure 3**).

Scared to Extinction: Stress, Disturbance, and Antipredator Behavior

Human activities can disturb wild animals, altering their behavior in ways detrimental to their survival and reproduction. Behavioral researchers studying the stress response system are helping to understand human impacts and suggest new ways to mitigate against these impacts. Stress is the animal's behavioral and physiological response to biological challenge, and, when chronic, can negatively impact health and reproduction. There are numerous measures of stress, many of them problematic if observed in isolation, but immunoassay of concentrations of glucocorticoid hormones is most common. Researchers have documented a number of anthropogenic activities that activate the stress response system: noisy snowmobiles impact wolves and elk, radiocollars may affect wild dogs, noisy

Figure 3 Dr. Wayne Linklater spreads black rhinoceros dung in research designed to guide postrelease dispersal behavior in a large-scale translocation program in southern Africa. Photo by Ron Swaisgood.

Figure 4 Researchers manipulated mate preferences for a conservation breeding program for threatened pygmy lorises. Photo by Heidi Fisher.

crowds disturb captive giant pandas, and approaching hikers affect bighorn sheep, even if they do not run away. Human activities may also affect behavioral decisions that have energetic consequences. For example, sperm whales spend less time surfacing to replenish oxygen supplies when helicopters fly overhead and foraging grizzly bears consume fewer calories when hikers are nearby.

More recently, researchers have taken a closer look at human disturbance, using a more predictive approach from the theoretical framework provided by economic models of antipredator behavior. If animals perceive humans and their activities as potential predators, then much of their response can be predicted from our understanding of antipredator behavior. A growing literature indicates that human disturbance, like predation threat, alters animals' time budgets, with energetic and lost-opportunity costs. These indirect sublethal costs can have cumulative effects worse than actual predation. Scott Creel and colleagues found that temporal and spatial proximity to wolves dramatically affected foraging patterns, causing elk to avoid the best forage in open areas in favor of suboptimal forage near cover. The consequences of this decision-making process were substantial, leading to poor body condition, lower reproduction rates, and high calf mortality. The economic principles of antipredator behavior have proved effective at predicting many of the dynamics of animal responses to human disturbance, such as the effects of tourist group size, speed of approach, and distance to cover. Disturbance can also affect mate choice and parental investment in offspring in predictable ways.

But Wait, There's More: Other Applications of Behavioral Research to Conservation

Creative conservation behaviorists are exploring many other avenues to apply their craft in the service of conservation efforts. Several researchers have assisted in the many behavioral facets of reintroduction programs, most prominently, in the application of ontogenetic processes to train prey organisms in the antipredator behavior of their species prior to release. Debra Shier and Don Owings, for example, have shown that captive-born black-tailed prairie dogs can be trained to recognize and deal with predators more effectively than their untrained counterparts. When these investigators used wild-caught prairie dogs to demonstrate appropriate antipredator behavior to captive-reared conspecifics, these prairie dogs were more likely to survive when released to the wild than individuals that were trained with predators without such a conspecific demonstrator.

Behavioral plasticity is another key concept in conservation behavior. Understanding and predicting behavioral response to a changing world will inform many conservation decisions. For example, behavioral elasticity preadapts invasive species to colonize novel habitats, and lack of elasticity may spell the doom of native species that cannot adapt to the invader.

Natural populations comprise a number of different behavioral types – suites of correlated behavioral traits, analogous to temperament or coping styles. Aggressive animals may be predictably more active across many situations than less aggressive animals. While this aggressiveness may confer advantages in intraspecific competition, such an advantage may be offset by increased susceptibility to predation. In some cases, even immune function is influenced by behavioral type. The sum total of behavioral types for a population is called a 'behavioral syndrome.' A reduction in the range of types within a syndrome can cause conservation problems if the members comprising a population can no longer adapt and respond to temporally and spatially variable environmental factors. Conservation breeding programs, in particular, may not produce the full mix of behavioral types for release, especially if the captive environment does not contain the environmental features needed for the development of some behavioral types. Conservation behaviorists need to understand these suites of correlated traits so that they can ensure that release groups, of captive or wild origin, contain those individuals that perform different important roles, such as warning against predators or locating and exploiting novel foods. To do this, we must guard against differential reproduction among different types and provide enriched environments to shape ontogenetic processes for a plurality of personalities. There is also evidence that some populations contain dispersal phenotypes, individuals that are behaviorally and physiologically prepared for the risks and rigors of dispersal – clearly important for any relocation program. Thus, to maximize population persistence, it takes all types.

To move forward, conservation behavior needs to continue to become more integrated across disciplines and levels of analysis. Behavioral ecologists bring much to the table, such as the strong theoretical framework that allows hypotheses and predictions to be formulated. Psychologists and applied ethologists bring a rich tradition of applied science focused on proximate mechanisms. Because proximate mechanisms can be manipulated in the service of conservation, whereas adaptive value cannot, more focus on proximate mechanisms will help move this discipline from the implications to the applications phase of research.

See also: Anthropogenic Noise: Impacts on Animals; Conservation and Anti-Predator Behavior; Conservation and Behavior: Introduction; Habitat Imprinting; Habitat Selection; Learning and Conservation; Mate Choice in Males and Females; Mating Interference Due to Introduction of Exotic Species; Mating Signals; Ontogenetic Effects of Captive Breeding.

Further Reading

Anthony, LL and Blumstein, DT (2000). Integrating behaviour into wildlife conservation: The multiple ways that behaviour can reduce Ne. *Biological Conservation* 95: 303–315.
Bélisle, M (2005). Measuring landscape connectivity: the challenge of behavioral landscape ecology. *Ecology* 86: 1988–1995.

Blumstein, DT and Fernández-Juricic, E (2004). The emergence of conservation behavior. *Conservation Biology* 18: 1175–1177.

Buchholz, R (2007). Behavioural biology: An effective and relevant conservation tool. *Trends in Ecology & Evolution* 22: 401.

Caro, T (2007). Behavior and conservation: A bridge too far? *Trends in Ecology & Evolution* 22: 394–400.

Frid, A and Dill, LM (2002). Human-caused disturbance stimuli as a form of predation risk. *Conservation Ecology* 6: 11.

Gosling, LM and Sutherland, WJ (2000). *Behaviour and Conservation*. Cambridge: Cambridge University Press.

Griffin, AS, Blumstein, DT, and Evans, CS (2000). Training captive-bred or translocated animals to avoid predators. *Conservation Biology* 14: 1317–1326.

Holway, DA and Suarez, AV (1999). Animal behavior: An essential component of invasion biology. *Trends in Ecology & Evolution* 14: 328–330.

Linklater, WL (2004). Wanted for conservation research: behavioral ecologists with broader perspective. *BioScience* 54: 352–360.

McDougall, PT, Réale, D, Sol, D, and Reader, SM (2006). Wildlife conservation and animal temperament: Causes and consequences of evolutionary change for captive, reintroduced, and wild populations. *Animal Conservation* 9: 39–48.

Stamps, JA (1988). Conspecific attraction and aggregation in territorial species. *American Naturalist* 131: 329–347.

Stamps, JA and Swaisgood, RR (2007). Someplace like home: Experience, habitat selection and conservation biology. *Applied Animal Behaviour Science* 102: 392–409.

Swaisgood, RR (in press). The conservation-welfare nexus in reintroduction programs: A role for sensory ecology. *Animal Welfare*.

Swaisgood, RR (2007). Current status and future directions of applied behavioral research for animal welfare and conservation. *Applied Animal Behaviour Science* 102: 139–162.

Swaisgood, RR and Schulte, BA (in press). Applying knowledge of mammalian social organization, mating systems and communication to management. In: Kleiman, DG, Thompson, KV, Baer, CK, (eds.) *Wild Mammals in Captivity*, 2nd edn. University of Chicago Press.

Ward, MP and Schlossberg, S (2004). Conspecific attraction and the conservation of territorial songbirds. *Conservation Biology* 18: 519–525.

Relevant Websites

http://www.animalbehavior.org/ABSConservation – Animal Behaviour Society Conservation.

http://www.iucnredlist.org – IUCN Redlist online.

http://www.sandiegozoo.org/conservation/ – San Diego Zoo's Conservation Research.

http://www.zsl.org – Zoological Society of London.

Conservation and Anti-Predator Behavior

A. Frid, Vancouver Aquarium, Vancouver, BC, Canada
M. R. Heithaus, Florida International University, Miami, FL, USA

Introduction and Definitions

Deciding how much to invest in antipredator behavior, rather than in foraging or other activities, is a fundamental problem faced by most animals. Behaviors that lower an animal's probability of death by predation – vigilance, hiding, avoidance of risky sites, and others – carry the cost of reduced access to resources for growth and reproduction. Ecologists have approached this problem, using evolutionary principles. They have developed predation risk theory, often expressed mathematically, which assumes that prey maximize fitness (e.g., lifetime reproductive success) by making behavioral decisions that optimize trade-offs between predator avoidance and resource acquisition. Predation risk theory can be used to estimate the 'risk effects' of predators on prey: the lost foraging opportunities and lower levels of growth and reproduction experienced by prey investing in antipredator behavior (also known as nonlethal or nonconsumptive effects). Importantly, risk effects can be major components of trophic cascades: the indirect effects of top predators on the population processes of plants and animal species at lower trophic levels, as mediated by the density and foraging behavior of intermediate consumers (i.e., herbivores and mesopredators, collectively termed mesoconsumers). Thus, risk effects can potentially affect community structure and ecosystem function.

This article surveys how human influences can affect antipredator behavior, potentially altering the risk effects and mortality rates experienced by prey and the indirect effects of predators on ecosystems. For example, human consumption of space and resources (e.g., fishing, forestry, agriculture, urbanization, etc.) and climate change may create resource shortages for mesoconsumers, thereby limiting their scope for antipredator behavior and indirectly increasing the rates at which they are killed by non-human predators. Nonlethal human activities, such as ecotourism, can cause animals to experience energetic and reproductive costs that resemble the costs of predator avoidance. Notoriously, humans have eliminated top predators in many systems, relaxing the need for mesoconsumers to invoke antipredator behavior, potentially disrupting trophic cascades. As illustrated in the following sections, predation risk theory can provide tools for predicting, detecting, and potentially mitigating these and other conservation problems.

State-Dependent Risk-Taking: Why Human-Caused Resource Declines Can Increase Predation Rates

During the 1980s, Marc Mangel, Colin Clark, John McNamara, and Alasdair Houston pioneered models of state-dependent behavior (also known as dynamic state variable models), which, among their many capabilities, predict the effect of residual reproductive value on behavioral decisions that maximize fitness. State-dependent risk-taking, a subset of this theory, predicts that resource declines and associated losses of body condition should increase risk-taking and predation rates for individuals attempting to avoid imminent starvation or other net fitness losses. Bradley Anholt and Earl Werner provided early empirical support for this prediction by experimentally exposing bullfrog tadpoles (*Rana catesbeiana*) to predation risk from larval dragonflies (*Tramea laceratea*) under contrasting levels of food abundance. Tadpoles experiencing low food levels moved, on average, 1.5 times more frequently and at higher speeds than tadpoles experiencing high food levels. Higher movement rates under food scarcity, which reflect greater foraging effort, also increased exposure to predators and caused a 60% rise in mean predation rates, despite predator densities remaining constant (**Figure 1**). This and later experiments by Anholt and Werner provided ground-breaking evidence that, consistent with the theory on state-dependent behavior, dichotomous views about resource-driven ('bottom-up') versus consumer-driven ('top-down') effects on population regulation are simplistic. Rather, these mesocosm experiments suggested that synergisms between resources and predators are fundamental to population and community processes, and adaptive variation in prey behavior is inherent to these synergisms. Subsequent field studies suggest that these synergisms scale up to large vertebrates using vast landscapes. Data on sea turtles, ungulates, and other large vertebrates suggest that individuals in poor body condition spend more time in habitats with better food quality and higher predation risk and consequently, may suffer higher predation rates than individuals in better condition.

State-dependent risk-taking potentially has profound conservation implications because humans influence the global distribution and abundance of resources used by animals. For example, models of state-dependent behavior predict that overfishing can force harbor seals

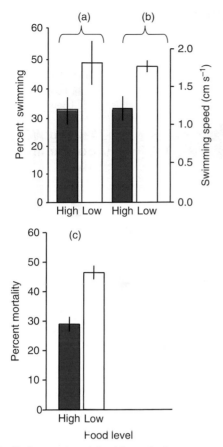

Figure 1 Early experimental evidence for how resource declines may indirectly increase predation rates via behavioral mechanisms of state-dependent risk-taking. (a) Mean percentage (±1SE, $N = 4$ for each treatment) of observations where bullfrog tadpoles were active in the presence of caged predators (larval dragonfly) at high and low food levels. (b) The mean speed of movement (±1SE, $N = 4$ for each treatment) of active bullfrog tadpoles in the presence of caged predators at high and low food levels. (c) Mean percentage (±1SE, $N = 20$ for each treatment) predation mortality of bullfrog tadpoles in the presence of uncaged predators at high and low food levels. Data from Anholt BR and Werner EE (1995) Interactions between food availability and predation mortality mediated by adaptive behavior. *Ecology* 76: 2230–2234. Redrawn with permission from Brad Anholt.

(*Phoca vitulina*) and Steller sea lions (*Eumetopias jubatus*) in western Alaska to increase foraging effort at the cost of increased exposure to Pacific sleeper sharks (*Somniosus pacificus*) and other predators. Through these behavioral mechanisms, overfishing may indirectly increase predation rates, potentially contributing to declining trends for some pinniped populations (**Figure 2**).

Nonlethal Interactions Between Wildlife and Humans

Humans interact with animals in many ways that are nonlethal yet potentially damaging. For instance, people on foot or on motorized vehicles often approach wildlife to the point of altering the animals' activity and eliciting antipredator behavior like fleeing. These responses by wildlife to humans are referred to as 'disturbance' and their context includes ecotourism (e.g., wildlife viewing, photography), resource extraction (e.g., machinery use, blasting, helicopter access to remote sites), and the nonlethal component of hunting (i.e., the search for quarry). Animal responses to disturbance stimuli may carry the same cost of predator avoidance – lost opportunities for feeding, mating, parental care, or other fitness-enhancing activities – and often include the energetic costs of locomotion while fleeing.

Theoretically, decisions by animals encountering humans should follow the same principles as antipredator behavior: optimization of trade-offs between access to resources (or net energy gain) and avoidance of perceived danger. Consistent with this hypothesis, factors that influence perceived risk of predation also affect animal responses to disturbance stimuli. For instance, when approached directly by a helicopter, Dall's sheep (*Ovis dalli dalli*) farther from steep rocky slopes (a refuge from predation), flee sooner than sheep closer to these slopes (**Figure 3**).

Antipredator behavior and responses to disturbance are analogous even when disturbance stimuli derive from modern technologies that prey had not encountered previously (e.g., motorized vehicles). This occurs because prey have evolved antipredator responses to generalized threatening stimuli, such as rapidly approaching objects that cross a threshold of perceived safety. Early support for this hypothesis was provided by Larry Dill's experiments in which zebra danios (*Brachydanio rerio*, a small fish) were exposed to real predators, a predator-shaped model, and a 'cinematographic' predator (a film of a black dot increasing in size, simulating an approaching object). In all cases, zebra danios fled when the angle subtended by the predator at the prey's eye reached a threshold rate of change (loom rate). The threshold loom rates depended on the size and speed of the approaching 'predator.' Thus, danios appeared to decide when to flee by relating the loom rate to a margin of safety, regardless of whether the predator was real, a model, or a film. Such generalized responses, however, are not mutually exclusive with predator-specific responses.

Importantly, lack of overt response to disturbance stimuli does not necessarily imply a lack of impact. According to models of state-dependent behavior, an individual's scope for antipredator behavior is influenced by the availability of its resources, its current body condition, and other factors affecting residual reproductive value. Thus, animals lacking alternative sites with adequate resources or struggling to maintain adequate body condition may be unable to afford to abandon their resource patch and flee from disturbance stimuli. (The same principle applies to the decision to abandon or care for dependent offspring.) People encountering wildlife often misinterpret lack of

Figure 2   Predictions from a model of state-dependent behavior on the relationship between fisheries and antipredator behavior (arrows show causal links). (a) Pacific sleeper shark caught as bycatch in a trawl net targeting walleye pollock (*Theragra chalcogramma*) in the Gulf of Alaska. On the one hand, the removal of top predators like sharks may relax predation risk and alter the behavior of mesoconsumers, such as (b) harbor seals, thereby disrupting indirect effects of top predators on species at lower trophic levels. The model predicts that shark removals will (c) greatly increase the proportion of dives to deep strata by seals, where both walleye pollock – the seals' most predictable resource – and sharks are most abundant. Consequently, shark removals should (d) greatly decrease and increase, respectively, rates of seal-inflicted mortality on (e) Pacific herring (*Clupea palassi*) in shallow and mid-depth strata and (f) pollock in deep strata. (Note that Panels c and d represent only seals in good body conditions, which have greater scope for antipredator behavior than seals in poor body condition.) On the other hand, fisheries depleting resources while top predators are still present may increase state-dependent risk-taking and predation rates for mesoconsumers. When herring are not sufficiently abundant to compensate for pollock declines and sharks are present at a constant density, overfishing of pollock should increase for seals the rates of (g) deep diving and, consequently, of (h) shark-inflicted mortality. Risk-taking and predation rates are exacerbated if seals are in poor body condition (g, h). This modeling approach can explore net effects of (a) concurrent removals of resources and top predators. Data are the outcome of computer experiments (*N* = 1000 forward iterations per treatment) simulating a 3-week winter period with the model and protocols described in Frid A, Baker G G, and Dill L M (2008) Do shark declines create fear-released systems? *Oikos* 117: 191–201. Photo credits: (a) Elliott Lee Hazen; (b) Alejandro Frid, (e, f) R.E. Thorne. In (b), the seal was captured for research purposes and is about to be released carrying recording devices that measure diving behavior.

fleeing as a neutral or even benign interaction. Yet beneath this superficial appearance, animals may experience decreased foraging efficiency due to increased vigilance, disrupted cycles of rest and digestion (which are particularly important for ruminants), and physiological responses to stress.

Similar to chronic risk effects from non-human predators, it is theoretically plausible for chronic disturbance to decrease long-term rates of energy intake and reproductive success. These risk effects can potentially lead to increased predation rates via mechanisms of state-dependent risk-taking, eventually causing population declines (**Figure 4**). Similarly, it is theoretically plausible

for chronic disturbance stimuli to indirectly influence plant community structure by causing herbivores to underutilize plant resources in areas perceived to be dangerous while increasing use of plants in areas that are perceived as safer.

Should animals not recognize that nonlethal stimuli do not warrant the costs of antipredator behavior? Animals rarely have perfect information and, theoretically, should maximize fitness by overestimating rather than underestimating risk. Overestimation costs, such as lost feeding opportunities, are lower than underestimation costs, which are death and loss of all future fitness. Thus, habituation to disturbance stimuli is only partial for

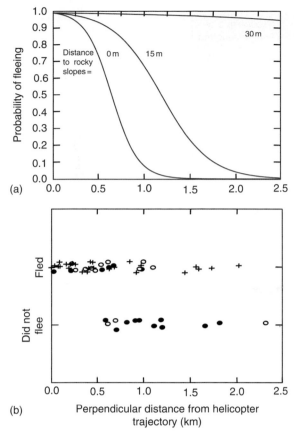

(a)

(b) Perpendicular distance from helicopter trajectory (km)

Figure 3 (a) Estimated probabilities of Dall's sheep fleeing during helicopter overflights as a function of two independent variables: the perpendicular distance (km) between the animal and the helicopter's trajectory (x-axis) and the animal's distance (m) to rocky slopes (a refuge from predation). Curves were generated from empirical data with logistic regression. (b) Scatterplot corroborating estimates of fleeing probabilities; dark circles represent sheep on rocky slopes, open circles represent sheep 5–20 m from rocky slopes (median = 20 m), and crosses represent sheep 25–1200 m from rocky slopes (median = 100 m). Points are jittered so that overlapping data can be read (i.e., there is no y-axis variability within response type). Note that the perpendicular distance from the helicopter's trajectory is a geometric correlate of angle of approach, with smaller distances implying more direct approaches by the helicopter. Both graphs suggest that sheep respond to the multiplicative (rather than additive) effects of the two independent variables. Modified and reprinted with permission from Frid A (2003) Dall's sheep responses to overflights by helicopter and fixed-wing aircraft. *Biological Conservation* 110: 387–399.

nondomesticated animals. Even corvids and squirrels in urban parks – archetypal examples of habituation – maintain levels of response to disturbance stimuli that are consistent with principles of antipredator behavior.

Importantly, when habituation levels for prey surpass those for predators, human infrastructure and its associated disturbance stimuli effectively provide prey with antipredator shields. For instance, Joel Berger found that moose (*Alces alces*) in the Yellowstone Ecosystem, USA,

generally avoid the vicinity of roads except during parturition. This choice of parturition sites may reduce calf losses to predation by grizzly bears (*Ursus arctos*), which avoid roads more than moose do.

Relationships Between Risk Effects and the Structure and Function of Ecosystems

Increasing evidence suggests that top predators can influence the structure and function of ecosystems via a combination of direct predation and risk effects on their prey. For example, the 1926 extirpation of wolves (*Canis lupus*) from Yellowstone National Park led to population increases and unrestrained browsing by elk (*Cervus elaphus*) released from the lethal and risk effects of wolves. The combined population and behavioral changes by elk lowered recruitment of woody riparian vegetation and upland deciduous trees. Since the 1995 reintroduction of wolves, plant recruitment has improved partly because direct predation by wolves has reduced elk numbers and consequently, the overall browsing pressure. Antipredator responses by elk, however, have influenced spatial variation in the strength of the indirect effects of wolves on vegetation. This trophic cascade was strongest in riskier areas where elk foraged least, sites with abundant logs, particularly in riparian zones, where poor visibility and obstacles hindering escape may increase the probability of death by predation, given an encounter with wolves (**Figure 5**).

Trophic cascades mediated by antipredator behavior can modify physical habitat structures. For example, in Zion National Park, unrestrained foraging by mule deer (*Odocoileus hemionus*) experiencing reduced predation risk from cougars (*Puma concolor*) appears to have eroded stream banks and reduced abundances of flora and fauna associated with riparian zones. Similarly, in Yellowstone National Park, the overgrazing of riparian trees and shrubs by elk released from wolf predation risk may have influenced the density and dam-building activities of beavers (*Castor canadensis*), which require woody plants to construct dams. These correlational studies suggest that, as mediated by ungulate herbivory and damming by beavers, the loss of large carnivores has the potential to indirectly alter hydrological processes that in turn influence the structures and composition of aquatic and terrestrial communities. More generally, there is evidence that herbivores across a wide range of body sizes and ecosystems (e.g., from grasshoppers to marine turtles) can influence recruitment and community structure of primary producers, but risk and lethal effects of predators limit this influence.

It is plausible for risk effects also to influence the dynamics of nutrient flow across ecosystems. For example,

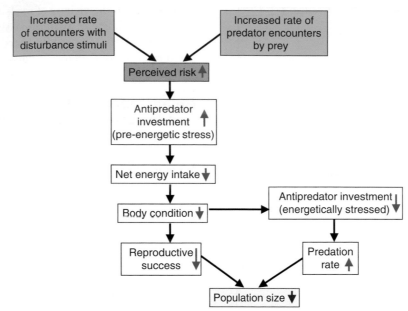

Figure 4 Conceptual model outlining the behavioral mechanisms by which increased rates of human-caused disturbance or predator encounters by prey could cause population size to decline. Downward-facing arrows inside boxes indicate a negative response and upward-facing arrows indicate a positive response. Modified and reprinted with permission from Frid A and Dill LM (2002) Human-caused disturbance stimuli as a form of predation risk. *Conservation Ecology* 6: http://www.consecol.org/Journal/vol6/iss1/art11/print.pdf.

a recent study by Joseph Bump and colleagues indicates that moose on Isle Royale, Lake Superior, spend much of their time foraging on aquatic macrophytes and transfer considerable amounts of aquatic nitrogen to terrestrial communities. Moose habitat use and, consequently, the spatial pattern and magnitude of their nitrogen deposition on land are influenced by wolf predation risk. The loss of wolves could, therefore, alter the nutrient dynamics of communities surrounding wetlands. Similarly, marine fishes that forage in seagrass ecosystems often shelter from predators among mangrove prop roots or in reefs, thereby transporting seagrass-derived nutrients and energy into reef and mangrove habitats. Declines of predators in sea grass system, therefore, can potentially alter the nutrient subsidies into these habitats.

Mescocosm experiments by Oswald Schmitz provide evidence that, for invertebrate predator–prey interactions in grasslands, the hunting mode of different predator species (spiders) indirectly influences ecosystem function, as mediated by predator-specific antipredator responses of herbivores (grasshoppers). Sit-and-wait ambush spiders leave persistent, point-source cues of predation risk. Consequently, grasshoppers can respond with chronic habitat shifts and their foraging pressure shifts from prefered plants associated with greater risk to less preferred plants that can be accessed more safely. As mediated by these risk effects on grasshoppers, the indirect effect of sit-and-wait spiders on ecosystem function is greater plant species diversity but lower primary productivity and nitrogen mineralization. In contrast, widely roaming spiders that

hunt actively do not leave predictable cues of predation risk. Therefore, grasshoppers are unable to respond with chronic habitat shifts and roaming spiders have stronger lethal than risk effects on grasshoppers. As mediated by lower densities of grasshoppers, the indirect effect of roaming spiders on ecosystem function is opposite to that of sit-and-wait spiders: lower plant diversity but greater productivity and nitrogen mineralization. The similar nature of risk effects across diverse taxa and ecosystems suggests that these processes scale up and that conserving species diversity within guilds of large predators might be important to the maintenance of many ecosystem functions.

While the plight of top predators in terrestrial sysems has long been recognized, only recently has it become apparent that marine ecosystems are experiencing catastrophic losses of large predators through target and bycatch fisheries. These losses may create risk-released systems mirroring those on land, where unrestricted grazing by herbivores or increased predation by mesoconsumers affects the foundations of food webs. For example, studies in Shark Bay, Western Australia, reveal that taxa ranging from herbivorous turtles and dugongs to piscivorous dolphins and seabirds modify their foraging locations and behaviors to minimize risk from tiger sharks (*Galeocerdo cuvier*) and suggest that these antipredator behaviors may influence the structure of seagrass communities. Changes in seagrass communities in regions where tiger sharks have been overfished further suggest that releasing marine herbivores from predation risk may alter benthic

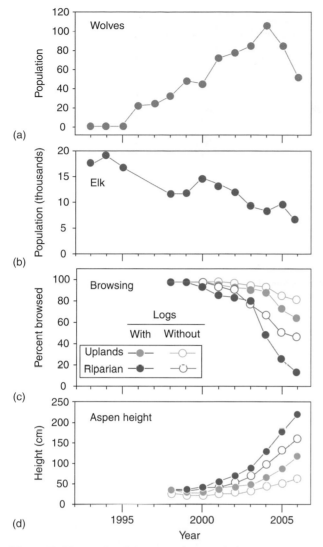

Figure 5 Time series data suggesting trends that are consistent with the hypothesis that risk effects and direct predation from wolves indirectly affect aspen heights, as mediated by the density and antipredator behavior of elk. (a) Wolf populations, (b) elk populations, (c) percentage of aspen leaders browsed, and (d) mean aspen heights in Yellowstone's northern range (early springtime heights after winter browsing but before summer growth). Direct predation by an expanding wolf population decreased elk numbers and contributed to decreasing browsing pressure on aspen. Risk effects, however, influenced the spatial variation in the strength of this trophic cascade, as illustrated by the stronger decrease in elk browsing and greater increase in aspen growth in areas with downed logs, particularly in riparian zones, where poor visibility and obstacles hindering escape may increase the probability of death by predation, given an encounter with wolves. Caption adapted and figure reprinted (with permission) from Ripple W and Beschta R (2007) Restoring Yellowstone's aspen with wolves. *Biological Conservation* 138: 514–519

communities. Losses of other large marine predators may also disrupt trophic cascades that are mediated by the antipredator behavior of mesopredators. Theoretical predictions suggest that the removal of Pacific sleeper sharks from northeastern Pacific ecosystems could shift pinniped predation from fishes in safer shallow waters to profitable fish species in deeper waters that pinnipeds might otherwise avoid to reduce predation risk (**Figure 2**). Just as in terrestrial systems, maintaining or restoring viable populations of large marine predators is likely important to marine conservation.

Anthropogenic Climate Change and Risk Effects

Global climate change caused by human activities is altering the resources, geographic distributions, phenology, and physiological and behavioral performance of many species at rates that exceed the range of natural historical variability. Therefore, it can potentially create novel ecological circumstances that attenuate or amplify risk effects or that indirectly increase predation rates through state-dependent risk-taking (**Figure 6**).

Climate-Related Mechanisms That Can Attenuate Risk Effects

It is plausible that climate change indirectly attenuates risk effects in some systems by altering environmental conditions that facilitate hunting success by predators. During winter on Isle Royale, the pack size and hunting success of wolves on moose increases with snow depth. Historically, the North Atlantic Oscillation has driven snow depth variation in the area. Under anthropogenic climate change, however, the frequency of winters with shallow snow packs could increase and moose could potentially experience more frequent winters of relaxed predation risk. Under this scenario, climate change would weaken the indirect effects of wolves on woody vegetation, as mediated by moose density and browsing behavior.

Climate change might also diminish the hunting effectiveness of some predators through physiological mechanisms. Pursuit-diving seabirds and pinnipeds are endothermic (warm-blooded) and consequently, their burst speed while hunting is unaffected by water temperature. In contrast, fish, their primary prey, are ectothermic (cold-blooded), and rising ocean temperatures are predicted to increase their burst speeds and escape ability, potentially reducing the risk effects of seabirds and pinnipeds on fish.

Another mechanism potentially attenuating risk effects is interspecific variation in climate-driven phenological responses. Christiaan Both and colleagues have analyzed responses to earlier spring warming by organisms at four trophic levels: deciduous oak trees (*Quercus robur*), caterpillars of the winter moth (*Operophtera brumata*) that feed on oak buds, several species of passerine songbirds that prey on caterpillars, and sparrowhawks (*Accipiter nisus*)

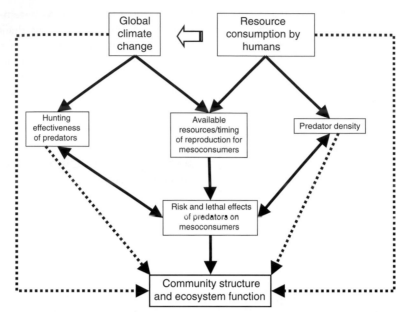

Figure 6   Simplified pathways illustrating hypothetical indirect effects of anthropogenic resource consumption and climate change on community structure and ecosystem function, as mediated by mesoconsumer antipredator behavior. Solid and dotted lines represent, respectively, direct and indirect links. Double pointed arrows indicate two-way relationships. To emphasize potentially dominant mechanisms and maintain visual clarity, some plausible relationships are not shown (e.g., climate change may influence the local density of top predators and predation risk may alter the timing of reproduction). Whether human influences are predicted to attenuate or amplify risk effects depends on ecological context (see text).

that prey on songbirds. Their findings indicate that over the last two decades, oak budburst has been occurring progressively earlier and moths are reproducing earlier, apparently tracking shifts in plant phenology. Similarly, songbird reproduction is occurring earlier, apparently tracking moth reproduction. Sparrowhawks reproduction, however, has not shifted to match the earlier peak abundance of songbird nestlings, which would enhance the ability of sparrowhawk parents to provision their nests. As underscored by this study, climate change may indirectly decouple some predator–prey relationships, with mesoconsumers (e.g., songbirds) experiencing a seasonal lowering of risk effects from their predators (e.g., sparrowhawks).

Climate-Related Mechanisms That Can Amplify Risk Effects

Range shifts and expansions in response to climate change will result in mesoconsumers encountering novel predators, and animals that were top predators previously may become mesopredators. For example, warming ocean temperatures are predicted to facilitate the expansion to higher latitudes of ectothermic sharks that are currently restricted to lower latitudes. These range expansions could amplify risk effects for endothermic marine mammals and other prey that currently live under lower predation risk from sharks at higher latitudes.

Climate Change May Indirectly Alter Mortality Rate Through State-Dependent Behavior

A general prediction derived from predation risk theory is that climate change, which affects resource availability, can indirectly alter rates of mortality inflicted by predators (or analogous agents: see below) through behavioral mechanisms of state-dependent risk-taking. Data for diverse taxa – including amphibians, polar bears (*Ursus maritimus*), and humans – are consistent with this prediction.

Efts (terrestrial juveniles) of the eastern red-spotted newt (*Notophthalmus viridescens*) compromise predator avoidance when dry conditions force them to invest more on behaviors that relieve desiccation stress. Live efts emit chemical cues signaling conspecific attraction, which facilitates huddling to reduce water losses. Recently killed efts, however, appear to emit a mixture of these attractive chemical cues and of repulsive chemical cues signaling predation risk. Experimental data indicate that efts experiencing moist conditions avoid chemical cues emitted by recently killed individuals, while these same cues attract efts stressed by desiccation. Drying trends associated with global climate change, therefore, can potentially exacerbate predator-inflicted mortality rates for amphibians via these behavioral mechanisms.

Notably, climate-induced resource losses may force top predators to invoke foraging modes that increase the risk of human-caused mortality, creating a situation resembling that of reduced antipredator behavior by energetically stressed mesoconsumers. Under normal conditions of sea ice, polar bears can meet energetic needs by hunting for seals at breathing holes on the ice pack and therefore, can avoid human settlements. In western Hudson Bay, however, earlier break-up of sea ice during spring has increased the period in which polar bears fast on land unable to access seals. Consequently, nutritionally stressed polar bears increasingly search for human-related food sources at settlements or camps, where humans kill them in self-protection. Sea ice break up is predicted to keep shifting to progressively earlier dates, raising concern for polar bear conservation and the management of human–bear interactions.

Humans are not exempt from their own form of state-dependent risk-taking. Historically, climate-driven resource shortages have influenced the decision by hungry societies to initiate wars that might never have occurred had human populations been well fed (**Figure 7**). Resource shortages induced by climate change can, consequently, exacerbate the potential for war or other human conflicts. Arguably, this is the biggest conservation issue of all, and – while not ignoring the complexities of human cultures and modern social institutions – principles of state-dependent risk-taking could contribute theoretical tools for anticipating and perhaps reducing conflicts between humans over scarce resources.

The converse of these examples is plausible when climate change is predicted to increase resource availability. For instance, Nicolas Lecomte and colleagues have shown that snow geese (*Chen caerulescens atlantica*) in the Arctic must leave their nests to drink. Drier conditions increase the dispersion of water sources and the duration of water acquisition trips, thereby amplifying nest losses to predation. Climate models, however, predict increased precipitation for the Arctic over the next two decades, which can potentially enhance nest guarding and reproductive success by Arctic-nesting snow geese.

Managing Irrevocable Changes to the Biosphere

The conservation challenges ahead are daunting. Most pressingly, climate change has already begun and will continue to reshape the biosphere. Encouragingly, a wide range of studies suggest that the basic mechanisms of risk effects and state-dependent risk-taking have great generality. Thus, predation risk theory might have important applications to ecosystem conservation and the management of irrevocable changes to the biosphere. Some examples are as follows

Conservation Implications of State-Dependent Risk-Taking

The conservation message of studies on state-dependent risk-taking is that human-caused resource declines should not be viewed solely as bottom-up impacts (e.g., nutritional stress), as is often the case. Instead, predictions and mitigation efforts should consider how the combined effects on resource availability of human consumption, nonlethal disturbance, and global climate change might limit the scope for antipredator behavior and potentially increase rates of predation for many species. Conversely, human-caused resource subsidies may indirectly decrease predation rates for some mesoconsumers, potentially altering some predator–prey interactions.

The framework of state-dependent risk-taking can also predict the indirect influence of human-caused disturbance stimuli on mortality rates inflicted by predators. Disturbance can functionally lower resource availability through increased vigilance or distributional shifts. Thus, wildlife managers might be able to predict scenarios in which chronic disturbance can increase energetic stress, thereby raising risk-taking while foraging and increasing predation rates indirectly (**Figure 4**).

Managing Risk Effects of Disturbance Stimuli on Wildlife

Predation risk theory can provide a rationale for managing disturbance stimuli without over-regulating humans. For instance, routes for motorized vehicles in remote areas (e.g., helicopters used for resource extraction) may be restricted to distances from known wildlife concentrations (e.g., raptor nests, ungulate birthing sites) that optimize the conflicting objectives of reducing disturbance to wildlife while avoiding excessive detours. Similarly, predation risk theory can be used to design wildlife viewing areas such that setback distances between people and animals optimize viewing opportunities and prevention of disturbance, rather than merely stressing animals that lack alternative habitats. Another concern is that hunting regulations generally consider only the lethal component of hunting, yet hunters disturb many more animals than they kill. Thus, predation risk theory could help develop hunting regulations that account for disturbance impacts on targeted game. Further, when top predators are endangered but game species are not (e.g., Florida panther (*Felis concolor coryi*) and white-tailed deer (*Odocoileus virginianus*), respectively), regulations might also account for the reduced hunting success that natural predators might experience because prey pursued by human hunters become more alert and difficult to capture.

Significantly, the use by prey of disturbance stimuli as safe zones that top predators avoid could diminish the effectiveness of national parks for protecting biodiversity. Although national parks usually have the dual mandate of facilitating recreation and conservation, access for recreation (e.g., roads, permanent campsites) might promote safe zones for prey that disrupt predator–prey behavioral interactions inherent to many aspects of biodiversity.

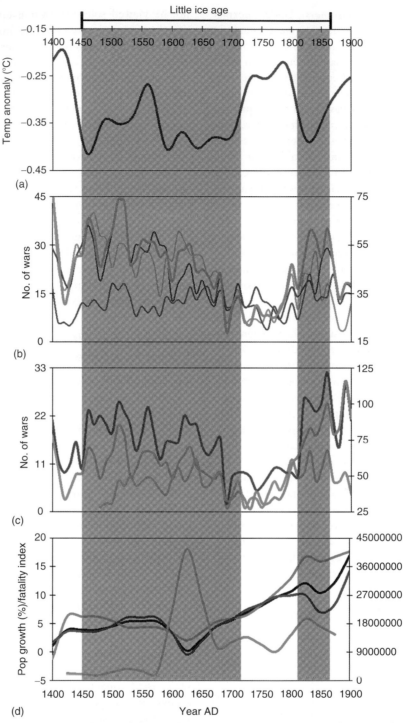

Figure 7 State-dependent risk-taking by humans? Time series data on paleo-temperature variation, war frequency, and population growth rate, AD 1400–1900, suggesting that climatic stress indirectly influences the willingness by humans to initiate conflicts and, consequently, the rates of war-inflicted mortality (a prediction stemming from theory on state-dependent risk-taking). The mediating factors between climatic stress and war (not shown here) are drops in agricultural production and rising food prices (see **Figure 2** of the original source). (a) Temperature anomaly (°C) in the Northern Hemisphere that is smoothed by 40-year Butterworth low pass filter. (b) Number of wars in the Northern Hemisphere (bright green), Asia (pink), Europe (turquoise), and the arid areas in the Northern Hemisphere (orange). (c) Number of wars worldwide (colors represent estimates by different authors: see original source). (d) Twenty-year population growth rate in Europe (turquoise), Asia (pink), and the Northern Hemisphere (blue) and the Northern Hemisphere 50-year fatality index (bright green). Cold phases are shaded as gray stripes. All war time series are in 10-year units. The bright green curves correspond to the right y axis. Caption adapted and figure reprinted (with permission) from Zhang DD, Brecke P, Lee HF, He YQ, and Zhang J (2007) Global climate change, war, and population decline in recent human history. *Proceedings of the National Academy of Science* 104: 19214–19219.

Predation risk theory could be applied to the design of parks so that these safe zones for prey are predicted and mitigated.

Restoring and Conserving Risk Effects

A growing body of evidence suggests that the risk and direct predation effects of predators on mesoconsumers across a wide range of body sizes and ecosystems can affect ecosystem function indirectly. If the influence of risk effects on trophic cascades and related processes is as widespread as these studies suggest, then conserving ecologically meaningful densities of top predators might be essential to many aspects of community dynamics. As experiments by Oswald Schmitz suggest, conserving predator diversity and its range of hunting modes, rather than merely conserving predator abundance, may be required to maintain some ecosystem functions (e.g., plant diversity, primary productivity, and nitrogen mineralization in grasslands).

Where extinct already, restoring populations of upper-level native predators may potentially reverse many aspects of ecosystem degradation via the reestablishment of risk and lethal effects; wolves reintroduced to Yellowstone provide a compelling case. In terrestrial systems where predator reintroductions are impossible or where human-dominated landscapes no longer support the habitat requirements of large carnivores, hunting can potentially become a management tool for reducing some ecological consequences of the loss of direct predation and risk effects on mesoconsumer populations. This potential, however, has yet to be met because hunting regulations and access logistics (e.g., the distribution of roads or navigable waterways) typically limit human hunting behavior to spatiotemporal distributions and patterns of prey selectivity that differ from those of non-human carnivores. Predation risk theory, combined with demographic analyses, could be used to optimize hunting regulations so that human hunters mimic some risk and lethal effects of non-human carnivores more closely without threatening human safety and prey populations themselves. Carnivore restoration and hunting as management tools, however, are difficult to reconcile with some values of human society, which could preclude some applications of predation risk theory.

A related concern is that predator reintroductions could cause high predation rates on prey that may have lost their antipredator skills after living under relaxed predation pressure for several generations. Data for some vertebrate taxa (e.g., moose, passerine birds, marsupials), however, suggest that predator-naïve prey can learn to recognize and avoid novel predators within one generation. Reintroduced predators, therefore, may have their highest predation rates on naïve prey early in their geographic expansion, but their kill rates may diminish as prey become savvier. Although the ability to learn antipredator skills does not negate concern that novel predators could drive extinct small populations of naïve prey, this issue generally applies to exotic predators invading small islands rather than to planned reintroductions of native carnivores. A related problem is that efforts to restore native mesoconsumers (e.g., endangered marsupials in Australia) through translocations may fail unless captive-raised mesoconsumers are trained to recognize and avoid predators before being released.

Marine top predators and their risk effects have been largely overlooked until recently. There is growing recognition that marine top predators like sharks, classic villains in popular culture, may have important indirect effects on marine communities, as mediated by the antipredator behavior of their prey. Sharks and other marine top predators have been declining almost worldwide because of target and bycatch fisheries. Their conservation and restoration may be essential to many aspects of marine ecosystem function, and predation risk theory can be used to support arguments for modifying fishery quotas and establishing marine reserves accordingly.

Although major losses of ecological integrity related to human resource consumption, climate change, and related processes will be inevitable, there is cautious optimism that predation risk theory can help predict and potentially reduce some of the damage. The maintenance of antipredator behavior over large ecological scales could well be a litmus test for our ability to conserve many levels of biodiversity.

Acknowledgments

We thank the following people for reviewing earlier drafts and helping us improve the manuscript: Larry Dill, Diana Raper, Anne Salomon, Brooke Sargent, and Ted Stankowich.

See also: Anthropogenic Noise: Implications for Conservation; Conservation and Behavior: Introduction; Ecology of Fear.

Further Reading

Berger J (2007) Fear, human shields and the redistribution of prey and predators in protected areas. *Biology Letters* 3: 620–623.

Both C, van Asch M, Bijlsma R, van den Burg A, and Visser M (2009) Climate change and unequal phenological changes across four trophic levels: Constraints or adaptations? *Journal of Animal Ecology* 78: 73–83.

Cairns D, Gaston A, and Huettman F (2008) Endothermy, ectothermy and the global structure of marine vertebrate communities. *Marine Ecology Progress Series* 356: 239–250.

Caro TM (ed.) (1998) *Behavioral Ecology and Conservation Biology.* New York, NY: Oxford University Press.

Clark CW and Mangel M (2000) *Dynamic State Variable Models in Ecology.* New York, NY: Oxford University Press.

Creel S and Christianson D (2008) Relationships between direct predation and risk effects. *Trends in Ecology & Evolution* 23: 194–201.

Festa-Bianchet M and Apollonio M (2003) *Animal Behavior and Wildlife Conservation.* Washington, DC: Island Press.

Frid A, Baker GG, and Dill LM (2008) Do shark declines create fear-released systems? *Oikos* 117: 191–201.

Frid A and Dill LM (2002) Human-caused disturbance stimuli as a form of predation risk. *Conservation Ecology* 6. http://www.consecol.org/Journal/vol6/iss1/art11/print.pdf.

Heithaus MR, Frid A, Wirsing AJ, and Worm B (2008) Predicting ecological consequences of marine top predator declines. *Trends in Ecology & Evolution* 23: 202–210.

Knight RL and Gutzwiller KJ (eds.) (1995) *Wildlife and Recreationists: Coexistence Through Management and Research.* Washington, DC: Island Press.

Lecomte N, Gauthier G, and Giroux J (2009) A link between water availability and nesting success mediated by predator–prey interactions in the Arctic. *Ecology* 90: 465–475.

Ray JC, Redford KH, Steneck RS, and Berger J (eds.) (2005) *Large Carnivores and the Conservation of Biodiversity.* Washington, DC: Island Press.

Ripple W and Beschta R (2004) Wolves and the ecology of fear: Can predation risk structure ecosystems? *Bioscience* 54: 755–766.

Rohr JR and Madison DM (2003) Dryness increases predation risk in efts: Support for an amphibian decline hypothesis. *Oecologia* 135: 657–664.

Schmitz OJ (2008) Effects of predator hunting mode on grassland ecosystem function. *Science* 319: 952–954.

Schmitz OJ, Grabowski JH, Peckarsky BA, Preisser EL, Trussell GC, and Vonesh JR (2008) From individuals to ecosystem function: Toward an integration of evolutionary and ecosystem ecology. *Ecology* 89: 2436–2445.

Schmitz OJ, Post E, Burns C, and Johnston K (2003) Ecosystem responses to global climate change: Moving beyond color mapping. *Bioscience* 53: 1199–1205.

Stirling I and Parkinson CL (2006) Possible effects of climate warming on selected populations of polar bears (*Ursus maritimus*) in the Canadian Arctic. *Arctic* 59: 261–275.

Conservation and Behavior: Introduction

L. Angeloni and K. R. Crooks, Colorado State University, Fort Collins, CO, USA
D. T. Blumstein, University of California, Los Angeles, CA, USA

Introduction

Separately, the disciplines of animal behavior and conservation biology are well-established, thriving fields of scientific inquiry, each with its own history and approach. Animal behavior is a chiefly theoretical discipline, while conservation biology is more of an applied science, established in response to the ongoing biodiversity crisis. Recent attempts have been made to apply principles of animal behavior to conservation problems to create a formally integrated discipline, commonly called *conservation behavior.*

There are several reasons why the union of these fields seems intuitive and profitable. First, behavioral ecologists focus on understanding how the state of an individual influences behavior. State-dependent behaviors include those that are plastic or flexible rather than fixed, whose expression can depend on environmental influences, such as temperature, resources, social cues, or maternal condition. Because the environment can influence the experience or condition of an individual, which can consequently influence its behavior, these state-dependent behaviors are of particular conservation relevance; in effect, they have the potential to be impacted by anthropogenic factors that alter the environment, such as disturbance and global climate change. There is growing evidence that anthropogenic disturbances are disruptive to behavioral processes. Disturbance may increase stress levels, thereby affecting behavior and physiology, and may also disrupt normal foraging, movement, communication, and mating patterns, among other behaviors.

Second, behavior is expected to have conservation relevance because it can affect demographic processes, such as survival and reproduction, and hence fitness. Quantification of fitness consequences is a major research area in behavioral ecology and is essential to conservation behavior because it is these fitness links that allow the development of individual-based models to better understand anthropogenic impacts. In fact, a shortcoming of most studies that investigate the effects of disturbance on behavior is that they fail to formalize the link to demography and fitness.

Third, many forms of behavioral information have been identified as useful conservation tools. For example, captive breeding and reintroduction programs have clearly demonstrated the importance of managing natural behavior (e.g., mating behavior, social behavior, and experience with predators and prey) to increase captive breeding and reintroduction success.

Fourth, behavioral diversity is a potential currency to consider, along with species and genetic diversity, when setting future conservation priorities. Behavior plays a key role in animal survival and evolution and its variation may allow species to respond to mounting changes to the environment. Conservation plans that incorporate the diversity of interesting behavioral traits also have the added benefit of harnessing public interest and support.

Several barriers have impeded the full integration of these disciplines, and these hurdles are being crossed only gradually. In this article, the history and current status of the field of conservation behavior are reviewed, barriers to the integration of behavior and conservation as well as tools to overcome them are described, and some of the conservation behavior topics that are covered in the other articles of this section are highlighted.

History

Although the formal integration of behavior and conservation is a recent effort, wildlife biologists have been using basic behavioral data to inform management decisions since the inception of the discipline of wildlife management. For the last century, wildlife biologists have been documenting the basic natural history of animals, including descriptions of fundamental behavior such as movement patterns, habitat selection, sociality, mating systems, and foraging. Indeed, the early history of wildlife biology was often dominated by naturalistic observations of the behavior of animals in the wild, employing traditional tools such as visual observations, binoculars, spotting scopes, and trapping. Ironically, the advent of new technologies, such as radiotelemetry, trip cameras, and remote-sensed satellite and GIS data, has allowed wildlife researchers to spend less time in the field directly observing wildlife, including their behavior. These new approaches, combined with an increasing focus on population-level processes such as estimating sizes and dynamics of populations, may have contributed to the decline of the importance of animal behavior in the field of wildlife biology in recent decades. For example, to become a certified wildlife biologist through The Wildlife Society, a course in animal behavior is not required but rather an elective; such is also the case for most undergraduate wildlife biology programs.

Conservation biology was formalized in the 1980s with the founding of the Society for Conservation Biology. The mission of this new 'crisis discipline' was to conserve the earth's biological diversity in the face of mounting anthropogenic impacts on the natural world. By necessity, conservation biology developed as a multidisciplinary science, merging fields of biological and social sciences to confront the growing biodiversity crisis. To accomplish their mission, conservation biologists have drawn heavily from the expertise in social sciences such as economics, political science, and sociology, in applied natural resource fields such as wildlife biology and forestry, and in theoretical natural sciences such as ecology and genetics. However, theoretical animal behavior was absent from almost all the early conceptual models of the interdisciplinary nature of the field. This disconnect remains today, with animal behavior typically receiving relatively little attention even in the most recent conservation biology texts.

In response to the perceived lack of integration between the behavior and conservation fields, the discipline of conservation behavior emerged in the mid-1990s when a number of symposia and subsequent publications highlighted the potential linkages between behavior and conservation and advocated greater overlap between the disciplines (see Further Reading). There was a great deal of promise for this new approach as scientists looked to a future where behaviorists and conservationists would work closely together, sharing knowledge and finding new ways for behavior to inform conservation practices. Since that time, a large literature has developed on the potential for the field of conservation behavior, on the recent work that has helped to initiate this field, and on the barriers that have slowed its development.

Several analyses have tried to address whether the new discourse on conservation behavior that began in the mid-1990s actually led to greater integration of the fields. Investigations of the scientific literature, which analyzed keywords, cross-citation rates, and the focus of articles in primary behavior and conservation journals, have found relatively little integration of research between the two disciplines over the subsequent decade. Where there is research overlap, it was found to be largely descriptive literature published outside of primary behavior journals.

The progress of this initially promising field of conservation behavior has been interpreted with different perspectives by two prominent contributors to the current discourse. Tim Caro has criticized behavioral biologists for not making greater advances in integrating with and contributing to conservation biology. He conceded that descriptive behavioral data have helped solve conservation problems, but argued that the primary theoretical advances in behavior 'have proved rather irrelevant in helping to solve the biodiversity crisis.' By contrast, Richard Buchholz has focused on the continuing development of this young field, arguing that 'the growing pains of conservation behavior are not symptoms of dysfunction, but rather positive signs of a thriving adolescence.' Only the future will tell whether conservation behavior matures beyond this adolescence and becomes more relevant to the biodiversity crisis.

Integration: Barriers and Tools

Given the potential for behavior to inform conservation, why has there been limited integration to date? An underlying cause is the historical and institutional separation of the fields of conservation biology and animal behavior. Typically, animal behaviorists and conservation biologists are housed in different departments, belong to different scientific societies, attend separate meetings, and apply for funds from different sources. They have been trained to ask very different research questions, with behaviorists focused on theory and conservationists focused on applied questions. As such, scientists studying animal behavior may feel that they have little to contribute to conservation biology or that the applied nature of the subject makes it less intellectually challenging and objective. In turn, those studying conservation biology may feel that animal behavior, particularly questions grounded in theory, has little relevance to their work. Ultimately, conservation biologists and animal behaviorists publish in journals that often have little overlap in topics, authors, readership, or scientific literature. Indeed, many behavior journals discourage articles with an applied focus, and conservation journals reject theoretical behavior papers without a conservation focus.

In addition, the two disciplines are focused on different biological scales that can be challenging to link. Behavioral ecologists address evolutionary questions at the level of the individual, whereas conservation biologists typically focus on processes occurring at the population, community, or landscape scales. When linking behavior and conservation, a primary challenge is, therefore, to relate behavior to fitness, and then relate fitness to the persistence, and hence conservation, of populations. The rapid development of computationally intensive individual-based modeling has the potential to help forge this link. Individual-based models rely on a fundamental understanding of factors that influence individual decisions. Such individual decisions can be thought of as mechanisms by which animals acquire fitness. Viewed this way, individual-based modeling may have an important role to play in developing population viability models.

Because conservation biology is a crisis discipline, critics argue that there may not be sufficient time to develop the necessary behavioral knowledge that can inform management. While acknowledging this, it is important to note that many conservation solutions will involve a long-term process, not a short-term intervention. Thus, by designing such programs to collect behavioral information along the way, behavioral knowledge and behaviorally inspired management strategies may be developed. Such adaptive management is an important part of current conservation biology.

The strategies for integrating behavior with conservation that were initially suggested in the 1990s apply even today. Behavioral biologists should collaborate with wildlife and conservation biologists and serve on management teams, providing expertise and advice on conservation issues. They should consult with conservation practitioners to determine how to modify their research programs to ask questions that are relevant to conservation. This may involve conducting research on rare species of conservation concern or species in disturbed environments, with reduced sample sizes as a potential consequence. The results of their research should then be publicized widely in conservation and behavior journals, as well as other formats – workshops, reports, popular writing, public speaking engagements – to make them accessible to conservation practitioners and the general public. There are a number of examples of successful integration of behavior and conservation that may serve as a model for those who are looking to do it in the future. In the following section, we highlight some of the behavioral topics that are beginning to be integrated with conservation issues by briefly reviewing the other articles within this section.

Topics at the Interface

The entries in this section emphasize that behavior and conservation will intersect in important ways when behavior is state-dependent and therefore potentially influenced by environmental disturbance, when behavioral change results in fitness or demographic change, and when behavioral knowledge can be used to provide useful tools in conservation programs. For example, the article by Henrik Brumm outlines how anthropogenic disturbance, specifically noise, can disrupt normal animal behavior. Noise can induce stress in animals, which can affect any number of behaviors (e.g., reproductive or antipredator behavior) that depend on being in good physiological state, but it can also interfere with animal communication by masking intentional acoustic signals as well as cues used to find prey and avoid predators. Anthropogenic noise has been shown to affect animal behavior in both terrestrial and aquatic systems for diverse groups including insects, fish, frogs, birds, bats, and whales. When it induces stress, when it masks sounds that are used to find prey and avoid predators, and when it interferes with courtship calls that are used to find mates or begging calls that are used to provision young, noise can impact individual fitness which may translate to demographic consequences at the population level. Some of these negative effects of noise can be mitigated behaviorally, for example by adjusting the communication signal so that it can be heard, by increasing visual attentiveness to

compensate for lost auditory awareness, or by displacing to a quieter habitat. However, these forms of behavioral compensation may also have negative fitness consequences themselves. In the end, further research is needed to formally assess the effects of anthropogenic noise on reproductive success and population viability.

Ulrika Candolin further develops the idea that human disturbance can negatively impact signaling between animals, in particular sexual signaling by males and the ability of females to assess male ornaments. Noise is not the only form of disturbance that can interfere with signaling between potential mates. Visual displays can also be muddled, particularly for aquatic animals like fishes, as nutrient pollution can lead to an increase in primary productivity resulting in eutrophication and turbidity. Chemical pollution and acidification of aquatic habitats also interfere with chemical cues that fish use to find appropriate mates. If these male ornaments advertise their quality to potential mates, then disruption of these signals may result in the choice of a lower-quality mate, potentially causing reduced fitness of the female or her offspring. Worse yet, anthropogenic disturbance may interfere with species recognition cues, causing individuals to hybridize across species boundaries, not only reducing fitness but also potentially influencing biodiversity. In some species, such as the threespine stickleback, males mitigate these effects of disturbance by displaying more vigorously, but this too may come at a cost of time, energy, and fitness. The conservation relevance of disrupted sexual signaling will depend on discovering its relative importance in determining fitness and population demography.

There are additional ways that anthropogenic effects can cause mating interference, as outlined by Alejandra Valero. Humans have introduced exotic species worldwide both intentionally and accidentally, which can negatively impact populations of native species. In addition to ecological interactions between native and exotic species (e.g., competition or predation), there can also be behavioral interactions that interfere with the reproduction of the native species. The signals of exotic species may mask the acoustic mating signals of the native species or may even jam their chemical pheromone receptors. Males may expend energy on courtship with females, and on rivalry with males, of the other species. Mating interference can reduce the ability of native females to choose the highest quality mate, can reduce the rate of successful copulation for the native species, and can result in hybridization across species boundaries. These behavioral effects, seen in a number of taxa, including insects, fishes, and lizards, provide another reason to prevent the introduction of exotic species and to control or eradicate those that are already introduced.

Humans impact not only the behavior of animals in the wild, but also those in captivity. Jennifer L. Kelley and

Constantino Macías García explore how behavior can be inadvertently altered in species of conservation concern when they develop as part of a captive breeding and reintroduction program. Only a small number of reintroduction programs have successfully established self-sustaining populations in the wild, in part because captive animals have not had the opportunity to develop normal foraging, antipredator, and social behavior. High densities and limited space in captivity can cause stress and aggression and can impede the development of territoriality, seasonal migratory behavior, exploratory behavior, and social behavior. The regular provisioning of food reduces the ability to develop foraging behavior and the lack of predators reduces the ability to develop antipredator behavior. Reduced exposure to parents and the opposite sex and a lack of mate choice may hinder development of appropriate reproductive behavior. Abnormal repetitive behavior sometimes develops as a result of frustration, fear, or discomfort, and it may be a sign of stress during early development. Fortunately, the negative behavioral effects of captive breeding can often be reversed once they are discovered, through greater exposure to conspecifics and predators, through environmental enrichment (including rearing in seminatural environments), and by acclimatizing individuals to natural conditions prior to release. These methods have enhanced the captive breeding and reintroduction programs for a number of birds and mammals.

Andrea S. Griffin further explores the way that experience and learning can impact captive populations that are later reintroduced to the wild, as well as wild populations that experience human-modified environments. Although additional postrelease monitoring is needed to establish the link between learning in captivity and fitness in the wild, there is much evidence that learning is important for the development of natural behavior needed to survive and reproduce in the wild. Postrelease survival of animals, such as houbara bustards and black-tailed prairie dogs, improves when captive individuals are trained to avoid predators, for example by experiencing dangerous stimuli and watching the alarm responses of others. A number of birds and rodents have been shown to learn about food aversions and food preferences, as well as the timing and location of food sources. Some species (e.g., some birds, mammals and fishes) also learn to identify appropriate mates and develop mating preferences from their social experiences. Thus, detailed knowledge about the behavior of the particular species in nature, although sometimes difficult to obtain for an endangered species, can provide useful tools for captive management. Outside of captivity, learning can also have conservation implications for wild populations that live in human-modified environments. Species that are thought to have greater capacity for learning, including larger brained birds and mammals, seem to have greater flexibility in their behavior. Behavioral flexibility also seems to increase survival in

harsh, modified, or novel environments, and large-brained birds and mammals are more likely to become established when introduced to new environments than species with smaller brains. Thus, knowledge about behavioral flexibility may help conservation biologists predict which species are most likely to be successful invaders, which are most likely to adjust to habitat modification and urbanization, and which to target for protection because of their vulnerability. However, while large-brained species may be more flexible in novel environments, they may also have life history characteristics such as delayed maturation and slow reproduction that make them more vulnerable to extinction and more difficult to rear in captivity. These counteracting forces must be evaluated before determining whether the ability to learn is a help or hindrance in our increasingly modified world.

Finally, Elisabet V. Wehncke makes the important point that the behavior of a particular animal species affects not only its own conservation status, but also the conservation status of other species that it interacts with, including that of plants. Animal behavior can affect seed dispersal by determining which seeds are dispersed, where seeds are deposited, and whether they survive after dispersal. In particular, the way that mammals, birds, reptiles, and insects are attracted to, prefer, handle, and process fruit determines which seeds will be dispersed. The social organization, including group size and degree of territoriality, and the movement patterns of the animal determine the pattern of seed deposition. How dispersers handle and deposit seeds will also affect the likelihood of subsequent seed mortality from desiccation, predation, damage, or competition. Understanding the link between animal behavior and its complicated effects on seed dispersal and plant demography will be important in predicting how habitat fragmentation, climate change, invasive species, and the loss of seed dispersers will affect plant populations. In turn, alterations in vegetative communities have the capacity to ripple throughout ecosystems, impacting animal populations and important processes such as nutrient cycling, hydrology, and succession.

Together, the entries within this section highlight several possible steps in the future development of the discipline of conservation behavior. Links could be identified between anthropogenic impacts and their effects on behavior. Further connections could be made between these modified behaviors and their effects on fitness and population demography, not only for the animal species in question, but also for other interacting species. In the end, this would allow us to develop individual-based models to predict how anthropogenic impacts might affect population persistence, to develop appropriate behavioral tools for conservation programs, to determine whether behavioral diversity is itself an important currency for conservation, and ultimately, to evaluate how animal behavior might be used to inform conservation biology.

See also: Anthropogenic Noise: Implications for Conservation; Learning and Conservation; Male Ornaments and Habitat Deterioration; Mating Interference Due to Introduction of Exotic Species; Ontogenetic Effects of Captive Breeding; Seed Dispersal and Conservation.

Further Reading

Angeloni L, Schlaepfer MA, Lawler JJ, and Crooks KR (2008) A reassessment of the interface between conservation and behaviour. *Animal Behaviour* 75: 731–737.

Blumstein DT and Fernández-Juricic E (2004) The emergence of conservation behavior. *Conservation Biology* 18: 1175–1177.

Blumstein DT and Fernández-Juricic E. (2010) A Primer on Conservation Behavior. Sunderland, MA: Sinauer Associates.

Buchholz R (2007) Behavioural biology: An effective and relevant conservation tool. *Trends in Ecology & Evolution* 22: 401–407.

Caro T (1998) *Behavioral Ecology and Conservation Biology.* Oxford: Oxford University Press.

Caro T (2007) Behavior and conservation: A bridge too far? *Trends in Ecology & Evolution* 22: 394–400.

Clemmons JR and Buchholz R (1997) *Behavioral Approaches to Conservation in the Wild.* Cambridge: Cambridge University Press.

Curio E (1996) Conservation needs ethology. *Trends in Ecology & Evolution* 11: 260–263.

Festa-Bianchet M and Apollonio M (2003) *Animal Behavior and Wildlife Conservation.* Washington, DC: Island Press.

Gosling LM and Sutherland WJ (2000) *Behaviour and Conservation.* Cambridge: Cambridge University Press.

Linklater WL (2004) Wanted for conservation research: Behavioural ecologists with a broader perspective. *Bioscience* 54: 352–360.

Soule ME (1985) What is conservation biology? *Bioscience* 35: 727–734.

Sutherland WJ (1998) The importance of behavioural studies in conservation biology. *Animal Behaviour* 56: 801–809.

Conservation Behavior and Endocrinology

L. S. Hayward, University of Washington, Seattle, WA, USA
D. S. Busch, Northwest Fisheries Science Center, National Marine Fisheries Service, Seattle, WA, USA

Introduction

Conservation biology requires a multidisciplinary approach. While some conservationists focus on communities and ecosystems, many of the best-funded conservation efforts focus on single species. Indeed, much of our strongest conservation legislation, including the US Endangered Species Act, is designed to promote the protection and recovery of specific species. Behavioral and endocrine data can serve as powerful diagnostic tools to focus efforts on single-species conservation.

Quantification of the behavioral or endocrine patterns of a species does not limit studies to organismal-level inquiry. Many critical ecological processes like pollination and seed dispersal are driven by patterns of behavior of a small set of species. An understanding of how environmental change impacts the behavior and endocrinology of a keystone species could shed light on community composition and the balance of the food web. For example, a change in browsing behavior by ungulates after reintroduction of wolves (*Canis lupus*) into the northern Yellowstone ecosystem had large effects on riparian vegetation structure and the ecosystem processes this vegetation structure affects. The change in ungulate behavior may have been potentiated by glucocorticoids, hormones often released in response to predators.

For those concerned with the conservation of a species, the most salient variables are usually survival and reproductive success. Unfortunately, direct measures of survival or reproductive success are often time-consuming and expensive to obtain, particularly for long-lived species. Additionally, it may be difficult to interpret the effect of a particular environmental factor or management strategy on survival and reproductive success because these measures represent the cumulative effects of multiple variables operating over extended time periods. Total experimental or statistical control over sufficient variables may be challenging, given the typical constraints on study duration and sample size. In contrast, both behavior and hormonal responses can be temporally tied to a particular disturbance, making them preferable as response variables, especially when conducting controlled experimental manipulations of the environment.

Although animal behaviorists were historically somewhat peripheral to the development of conservation biology, behavior studies can have important conservation applications. When managing for a threatened or endangered species, success or failure of a management action can hinge on understanding the behavior of that species. For example, in the late 1980s and early 1990s, a major California sea otter (*Enhydra lutris nereis*) translocation effort failed because of a lack of knowledge about otter homing behavior: many of the otters returned to their capture location. Meanwhile, behavioral data finding that sea otters travel further offshore than originally suspected led to legislation limiting set-net fishing in the otter's range, affording protection that ultimately contributed to population rebounds. Behavioral measures can also quantify disturbance impacts. Mexican spotted owl (*Strix occidentalis lucida*) behavior has been used to assess response to noise from helicopters and human recreation.

Endocrine data can similarly inform conservation decisions. For example, endocrine profiles indicate that historical hunting of African elephants (*Loxodonta africana*) continues to affect their physiology 20 years after the ban on the ivory trade was implemented. Females living in social groups altered by prior hunting activity and/or in close proximity to areas with high poaching risk have higher concentrations of glucocorticoid metabolites in their feces, indicating potential chronic stress. In addition, females living in social groups disrupted by hunting also have lower reproductive output, as measured by births and fecal reproductive hormone titers signaling pregnancy. These results indicate that the physiological effects of poaching can persist long after hunting is curtailed, making populations more susceptible to additional stressors.

While both useful in their own rights, hormone and behavior measures complement each other well. Simultaneously quantifying complementary response variables can be critical for an accurate assessment of environmental impacts because evaluating either alone may be problematic. Behavior often shows high individual variability in natural systems. High variation creates noise in datasets and necessitates large sample sizes to provide the power to detect a statistically significant signal. When dealing with threatened or endangered species, large sample sizes may not be feasible. Hormone titers, on the other hand, tend to be regulated within a narrow physiological range, resulting in less variation among the members of a population. Combining behavioral and hormonal data can be important because significant physiological changes sometimes occur in response to a stressor in the absence of a behavioral response. For example, a fear response may cause temporary immobility. This behavior can be almost indistinguishable from relaxed daytime roosting and sleeping in species like the northern spotted owl (*Strix occidentalis caurina*). Behavioral data can provide the social context needed to properly interpret endocrine data. For example, glucocorticoids are often elevated during breeding. Such an elevation may complicate interpretation of response to an environmental variable if a proper social context is not established.

Methodological Considerations Associated with Hormone Measures

Common Hormone Measures

Hormone titers are usually measured in blood or feces and, less commonly, in urine or feathers. By measuring hormones in blood, researchers can get an accurate endocrine profile of an individual at a discrete time point. However, the act of collecting a blood sample will inevitably alter the hormone profile of the research subject. Because steroid hormones are typically released in response to a cascade of other hormones and must be synthesized de novo immediately prior to release (they are membrane-soluble and cannot be stored in vesicles), titers do not change instantaneously. This delay allows a small window of time for researchers to collect blood with hormone profiles that represent a precapture baseline. The window is rarely longer than 3 min and many researchers prefer to collect blood within 1 min of capture, if possible. Typically, hormones can be measured in tiny volumes of blood and most animals recover quickly from sampling.

Researchers often take advantage of the dynamic nature of plasma hormone concentrations by collecting a series of blood samples to provide a profile of response to a given agent. For example, those studying stress biology typically use a series of samples collected postcapture to assess responsiveness of the hypothalamic–pituitary–adrenal (HPA) axis (the 'stress response'; **Figure 1**). The HPA axis is the system responsible for secreting adrenal steroids (**Figure 2**). A similar challenge involves collecting blood at regular intervals after administering an injection of a releaser hormone. In the case of the HPA axis, an injection of adrenocorticotropic hormone (ACTH) tests the responsiveness of the adrenal gland to the pituitary peptide that stimulates glucocorticoid synthesis and release. This chemical challenge can be critical for establishing whether responsiveness to a specific stressor is reduced (e.g., habituation to tourists) or whether the entire system has been downregulated – a situation that could be disadvantageous in the face of further stressors.

Measuring hormone metabolites in feces can achieve an integrated portrait of steroid secretion over time. The duration of the time interval varies with rates of both metabolism and digestion. The primary advantage of fecal hormone measures is that they provide a less invasive

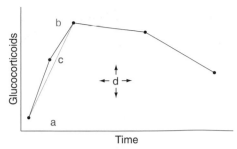

Figure 1 A generalized profile of a standard stress series in which plasma glucocorticoids (GCs) are measured at set intervals (represented by solid circles) in an individual after or during a challenge to the hypothalamo–pituitary–adrenal axis. Common challenges include capture and restraint or injection with adrenocorticotropic hormone. Features of the stress series valuable for comparisons include (a) the baseline level of glucocorticoid (measured within 3 min of capture and prior to injection); (b) the maximum level measured; (c) the rate of increase from baseline to max; and (d) the integrated area under the curve.

alternative to taking blood samples. Capture is not required, and, in some cases, free-ranging animals need not come into contact with researchers. Employing dogs (*Canis lupus familiaris*) trained to locate the feces of a focal species by scent is one effective way to obtain a large number of samples in a relatively short period of time.

Another advantage to fecal hormone measures is that the integrated hormone profile provided by the sample can be a more sensitive and accurate way to assess condition. For example, it may take a week to determine that a female baboon is pregnant from measures of estrogen and progestins in her plasma, but only 4 days from fecal measures of the same hormones. This discrepancy is due to the pulsatile secretion of the relevant reproductive hormones. Hormone pulses may be missed in blood draws, but accumulate enough to be detected in feces. Drawbacks to fecal steroid measures are that metabolites may vary with factors like sex and diet and may pose methodological challenges for assays. Extensive validations are required for each new species in which fecal hormone measures are employed.

Interpretation of Endocrine Data

The concentration of hormones circulating in the plasma or present in fecal matter can be measured with high precision. However, the fact that measurement of hormone titers is precise does not equate to the interpretation of hormone titers or fluctuations being equally precise. Hormone titers vary markedly among species. They are modulated on seasonal and daily cycles to promote and support various life history stages and in response to predictable changes in the environment. In addition, hormone titers can change rapidly, necessitating careful sampling procedures to accurately portray the

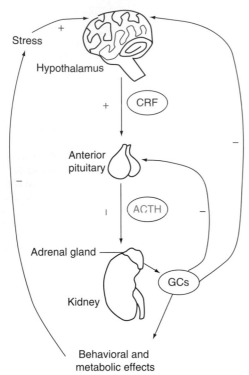

Figure 2 The hypothalamo–pituitary–adrenal (HPA) axis translates metabolic or psychological stress experienced by an individual into increased levels of glucocorticoids (GCs), cortisol or corticosterone depending on species. Secretion of GCs from the adrenal glands (or interrenal tissue in fish and amphibians) is stimulated by adrenocorticotropic hormone (ACTH). The release of ACTH from the anterior pituitary is stimulated by corticotropin releasing factor (CRF) from the hypothalamus. In the context of stress physiology, circulating GCs alter behavior and metabolism in a way that enhances the ability to cope with stressors. Other effects of GCs include suppressing the production of CRF and ACTH through negative feedback. At any point along the HPA axis, receptor sensitivity may be modulated to increase or decrease physiological responsiveness to stress. Conservation biologists must be careful to interpret levels of plasma or fecal GCs in the proper context before forming conclusions about the degree of stress an animal is facing. Low GC titers do not always signal adequate coping, as they sometimes signify a breakdown in coping mechanisms. Similarly, high GCs are not always detrimental to fitness, as increased energetic demands associated with breeding may elevate GCs. GC levels also vary by season, sex, social status and other factors.

range of baseline hormone titers in a population. Finally, due to temporal and individual variation in receptor types and densities, enzyme concentrations, and binding protein concentrations, the response of an individual to the same concentration of a given hormone can differ among species and individuals and in the same individual over time. To deal with these challenges in collecting and interpreting endocrine data, endocrinologists typically design studies as experiments (manipulative experiments or natural experiments), allowing them to isolate the hormone change to treatment alone. Hormone measures collected

without a control group are difficult to interpret for the myriad reasons why absolute hormone concentration change. When a control group is not available, it is important to at least control statistically for factors such as sex, life history stage, and season.

The link between hormones and behavior is undeniable. However, it is a misconception that hormones can cause certain behaviors. Instead, under the right circumstances, hormone concentrations potentiate specific behaviors, making them more likely to be expressed. Understanding this subtle difference in the link between hormones and behavior is crucial for those who simultaneously measure hormones and behavior, particularly in the context of conservation and when informing management decisions.

An Introduction to the Hormones Most Relevant to Conservation

Glucocorticoids

Glucocorticoids are the hormones most commonly employed in studies of conservation relevance. They are steroid hormones produced by the adrenal or, in many fish and amphibians, interrenal glands in response to signals from the pituitary and hypothalamus. The primary glucocorticoid in a species depends on its taxonomy: amphibians, lizards, birds, and some mammals (e.g., rodents) primarily produce corticosterone while most fish and other mammals (e.g., humans) produce cortisol. Glucocorticoids play two important roles in the body, depending on circulating hormone titers. Their primary role is basic energy regulation. At low to moderate levels, glucocorticoids influence feeding behavior and act on tissues to maintain adequate circulating levels of glucose and free fatty acids. At high levels, glucocorticoids orchestrate the physiological and behavioral changes associated with the emergency response and 'stress.' When experienced over short-term exposure (minutes to hours), high glucocorticoid concentrations can promote the behavioral and physiological changes needed to cope with unexpected challenges and can suppress behaviors that would otherwise distract the individual from survival. The effects of high glucocorticoid concentrations change with the duration of exposure. With long-term exposure (days to weeks), high glucocorticoid concentrations can be detrimental to health and fitness via altering both physiology (e.g., suppressing the immune system, growth, and metamorphosis) and behavior (e.g., inhibiting reproductive behavior).

Conservation biologists use measures of glucocorticoids to assess population health for two primary reasons. First, levels of glucocorticoids vary with a range of environmental factors such as food abundance, predator density, and anthropogenic disturbance. Second, in some species, glucocorticoids have been shown to predict survival and/or reproductive success, sometimes with excellent accuracy. For these reasons, glucocorticoids have the potential to make valuable response variables in studies of disturbance impacts.

Reproductive Steroids

Nearly all vertebrates have evolved sensitivity to environmental cues that helps them time breeding to maximize the likelihood of successfully rearing offspring and minimize risks to survival. Predictable cues that change little on an annual basis, like day length, are often used to prompt reproductive development well in advance of the time when young emerge or first establish independence. Fine-tuning of reproductive timing occurs in response to more annually variable factors such as temperature, plant growth, food abundance, body condition, and/or social cues. Signs of unfavorable conditions may signal to an animal to postpone reproduction or abandon the current effort and wait until the situation improves to try again.

The unconscious 'decisions' that an animal makes about when to breed and how much to invest in parental care are mediated by perception of the environment and then neural transduction into a range of peptide and steroid hormone secretions including gonadotropin releasing hormone (GnRH), lutenizing hormone (LH), follicle-stimulating hormone (FSH), kisspeptin, testosterone, estrogen, progesterone, and corticosterone or cortisol. These systems can have profound implications for conservation because some are more flexible than others, effecting how easily an individual may be able to cope with global change. Signals from the environment that conditions are favorable for breeding result in the release of GnRH from the hypothalamus, which in turn releases LH and FSH from the pituitary (**Figure 3**). LH and FSH coordinate the maturation of gonads and gametes. Once mature, gonads release sex steroids like estrogen and testosterone that mediate expression of secondary sex characters and sexual behaviors, like courtship and territorial defense. For species with dependent young, levels of sex steroids often fall and levels of progesterone rise later in breeding to promote investment in parental care. When attempting to assess reproductive status, steroids like testosterone and estrogen can serve as valuable response variables and can provide a direct index of reproductive condition. LH and FSH may also be useful measures, but are often more challenging to work with given the relative instability of the peptide hormones.

The dual objectives of survival and reproduction often impose energetic trade-offs on individuals that are mediated through hormones. At any stage in breeding, unfavorable conditions may interrupt sexual maturation or breeding efforts. Interruptions are often reflected in sex steroid titers. For example, reproductive dysfunction in captive animals can be diagnosed with sex steroid profiles.

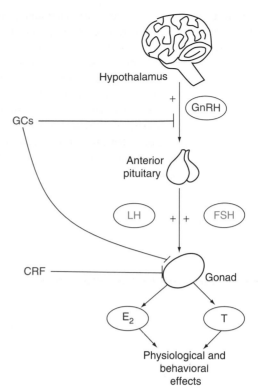

Figure 3 The hypothalamo–pituitary–gonadal (HPG) axis controls the release of sex steroids from ovaries and testes. Gonadotropin releasing hormone (GnRH) from the hypothalamus stimulates the release of luteinizing hormone (LH) and follicle-stimulating hormone (FSH) from the anterior pituitary. LH and FSH stimulate the gonads to produce estrogen, testosterone, and other sex steroids. Reproduction can be suppressed by the inhibitory effects of glucocorticoids (GCs) on sensitivity of the anterior pituitary to GnRH and of the gonads to LH and FSH. Corticotropin releasing factor also decreases the sensitivity of the gonads to LH and FSH. These inhibitory effects are modulated by a host of neurotransmitters.

Interruptions are usually mediated by increases in glucocorticoids, which suppress reproductive steroids. By combining measures of sex steroids with glucocorticoids, one can more accurately determine whether environmental stressors are severe enough to suppress reproduction. Environmental stressors can include anything from low food abundance to prolonged storm to disturbance by humans. For example, HPA and HPG systems can be used to determine whether population decline is a result of stress or failure of a modified environment to provide the correct cues for timing of breeding.

Another example of a trade-off between survival or investment in self and reproduction is the immunosuppressive effects of testosterone. The relationship between testosterone and immune function remains an area of active research. Artificially increasing testosterone may decrease immune activity in birds, for example. Field studies also show that testosterone and immune function inversely co-vary, and inducing immune activity can decrease testosterone. From an evolutionary standpoint, it might be adaptive for testosterone to suppress immune function. An animal with a short window in which to establish a territory and attract a mate (behaviors dependent on increased testosterone) may not be able to afford the time or available energy to launch the behavioral and physiological defenses associated with responding to infection. Furthermore, some speculate that increased disease pressure may be one reason that birds in the tropics or in colonies have generally lower peak testosterone levels than territorial birds breeding at higher latitudes.

Thyroid Hormones

Thyroid hormones are measured less commonly, but potentially provide a valuable index of nutritional condition, particularly when more integrated measures are taken from fecal metabolites. Thyroid hormones are derivatives of tyrosine produced by the thyroid gland. They act on almost every cell of the body to modulate metabolism. While thyroxine (T4) is the most prevalent thyroid hormone in the body, triiodothyronine (T3) is more biologically active and relevant to nutritional state. As food becomes scarce, T3 levels tend to drop, lowering the metabolic rate to help conserve reserves. Conversely, high T3 indicates ample nutrition. Thyroid hormone can be particularly valuable when testing hypotheses about the causes of decline in a sensitive species. For example, in a recent unpublished example, a population of killer whales (*Orcinus orca*) in the Pacific Northwest showed significantly lower average fecal T3 levels in 2008 compared to 2007. In 2008, populations of their preferred prey (Chinook salmon, *Oncorhynchus tshawytscha*) crashed on the west coast and the killer whale population suffered an approximate 8% decline. This result lends support to the hypothesis that food scarcity is a significant contributing factor to decline in this endangered population.

Organizational Versus Activational Hormone Effects

Hormone actions are broadly grouped into two categories: organizational and activational. At specific critical stages of development, hormone actions influence neural circuitry and tissue differentiation in a permanent way, having what is known as organizational effects. A classic example is the organizational effect of testosterone on mammalian males, without which the development of external genitalia would not occur. In contrast to organizational effects, activational effects of hormones are transitory or reversible. The role that testosterone plays in potentiating song in birds is an example of its activational effect. From a conservation or management perspective, it is important to recognize that the effects of exposure to contaminants that mimic hormones or disturbances that alter endogenous hormones will be very different depending on the window of development in which the exposure occurs.

Temperature-Dependent Sex Determination

The organizational effects of hormones are sometimes temperature dependent, with many reptiles exhibiting temperature-dependent sex determination. Such sensitivity may pose a conservation concern as temperatures increase globally and humans alter the landscapes of traditional nesting grounds. Conservationists have learned the hard way that understanding the basic physiology of a species is critical to captive breeding. In the 1970s, thousands of sea turtle eggs were taken from the wild and incubated artificially at temperatures that were later realized to produce only males. By releasing thousands of surplus males into the population, researchers increased intraspecific competition but not the reproductive capacity of the population, likely having the opposite effect on the population demography of what they had intended. Today researchers note a surplus of female hatchlings in many sea turtle populations, a consequence of increasing temperatures related to global change and, in some parts of the world, also to a reduction in shade due to deforestation.

Maternal Effects

A mother's circulating hormones have important organizational effects in many species of vertebrate from mammals to fish. For egg-laying vertebrates, yolk is a source of maternal steroids, and yolk hormone titers positively correlate with maternal titers. In several species, elevated maternal glucocorticoids translate into reduced fitness for offspring: young exposed to high maternal glucocorticoids during gestation are incapable of achieving maximal reproductive output as adults. It has been posited that this transgenerational effect of stress is one factor responsible for the crash seen in hare (*Lepus americanus*) populations after periods of high predator density. Similarly, female birds with elevated glucocorticoids hatch young that grow more slowly. In contrast to the negative influences of maternal glucocorticoids, mothers that deposit more androgens in their eggs often have offspring that grow faster and show other traits associated with high fitness.

These maternal effects emphasize that the consequences of disturbance may be intergenerational and may persist even after the source of disturbance has been removed from the environment. When assessing the impacts of an environmental factor on a population, it can be important to consider maternal effects. Similarly, the potential for maternal effects should be evaluated in captive breeding programs when suboptimal conditions experienced by the mother during yolk formation or gestation may decrease the reproductive potential of her offspring.

Reproduction

As environmental cues are altered by processes like urbanization, development of agricultural lands, pollution, and global climate change, reproduction may be impaired or postponed indefinitely in an animal that has evolved to 'expect' that conditions are likely to improve. In some cases, altered environments expose animals to negative stimuli, or stressors, which increase glucocorticoids and suppress reproductive steroids, disallowing or effectively aborting reproductive efforts. In other cases, altered environments simply fail to provide the stimulating cues necessary for full initiation of reproduction. In either case, hormone measures can help identify which individuals are experiencing stress or lacking stimulatory cues and can also help identify which environmental factors are responsible for the reproductive failure.

Captive Breeding

One of the earliest applications of conservation endocrinology and its intersection with behavior was supporting successful breeding of species in captivity. Captive breeding programs are integral to the conservation of many species whose numbers have dropped precariously in the wild. However, captive breeding efforts may be complicated by the fact that animals are sensitive to the environmental cues needed to initiate breeding. Fecal hormone measures can help identify periods of fertility and increase chances of fertilization. Fecal steroid measures can also be used to compare the hormone profiles of individuals with varying degrees of reproductive output. For example, in captive, endangered great hornbills (*Buceros bicornis*), females that lay eggs have different estrogen and corticosterone levels than females that do not lay eggs. Such results can provide a noninvasive method for assessing whether environmental enrichment is satisfactory to initiate breeding and help managers target activities to improve breeding success. Or, when a pair fails to mate, hormone measures may be used to determine whether one or both members have matured to full reproductive capacity. In more extreme cases, infertility relating to hormone disorders may be treated with the therapeutic application of exogenous hormones. Conversely, endocrine therapies have also been used successfully to suppress reproduction in invasive or pest species whose numbers have expanded in altered environments.

Because endocrine systems have been relatively well conserved throughout evolution, it is often possible to use knowledge from a common species to inform management of a closely related endangered species. For example, hormone samples collected from multiple breeding individuals of the common black-headed ibis (*Threskiornis melanocephalus*) were used to create a generalized endocrine profile to compare with isolated samples from the highly endangered Japanese crested ibis (*Nipponia nippon*). This approach was used to promote captive breeding of the crested ibis after all eight known individuals were taken into captivity. In addition, the promising results of initial gonadotropin application in promoting breeding of a another related ibis species (*Eudocimus ruber*) suggest that the approach may be successful for the crested ibis as well. Toward that end, genes for gonadotropins and FSH have been cloned in the crested ibis. This molecular technique makes possible the production of sufficient quantity of hormone for therapeutic purposes.

Coping with Change and Uncertainty

There is little dispute in the scientific community that humans are changing the environment in profound ways. Behavioral and endocrine data can be used to assess the effects of these changes on free-living animals and possibly to predict responses to future change. Here we explore the impacts of global climate change and anthropogenic use and modification of habitat on behavior and endocrinology.

Impacts of Tourism and Recreation

The nature of animals' response to human activities is variable and can change with repeated exposure. In some instances, human activities displace individuals from preferred habitat and change feeding behaviors and movement rates. For example, many animals flee from vehicles and some avoid areas with vehicle activity. While exposure to vehicles correlates with elevated fecal glucocorticoids in animals ranging from grouse (*Tetrao tetrix*) to elk (*Cervus canadensis*) and wolves (*C. lupus*), it does not in other animals such as the grizzly bear (*Ursus arctos horriblis*). Repeated exposure to an inescapable, noxious stimulus can induce habituation. Habituation results in an animal attenuating its response to or no longer responding to the noxious stimulus – in this case, exposure to humans. While behavioral habituation has been well documented, only a handful of studies have simultaneously documented behavioral and endocrine habituation.

Tourism

Research to date indicates that behavioral and physiological habituation often do not parallel each other. For example, an animal can seem calm in response to a human visitor, but simultaneously experience physiological change. Early studies of this disconnect between behavior and physiology focused on heart rate. For instance, some incubating sea birds do not respond behaviorally to human presence but do increase heart rate. Research on heart rate in wild animals is enjoying new popularity given recent technological advances.

Over the past decade, researchers have documented how exposure to tourists affects behavior, hormone levels, and the endocrine response to stress. Naive animals typically respond to tourists with both a behavioral reaction and an increase in glucocorticoids. With continued, predictable exposure to tourists, adults typically stop responding behaviorally or hormonally to tourists, indicating habituation. However, physiological habituation to tourist exposure may not always be complete and may not occur at all life stages. For example, while Magellanic penguins (*Spheniscus magellanicus*) can behaviorally habituate to tourists, tourist-visited hatchlings and adults have a different glucocorticoid response to a standardized stressor than unvisited birds. Furthermore, the nature of this HPA-axis change differs between life stages: the stress response in tourist-visited birds is increased in hatchlings and decreased in adults. Magellanic penguins are not a unique case; the stress response is also decreased with chronic exposure to tourists in adult marine iguana, *Amblyrhynchus cristatus*, and increased in juvenile hoatzin, *Opisthocomus hoazin*. While the effect of altered HPA-axis sensitivity to stressors is unclear, these data indicate that behavioral and physiological habitation can be dissociated.

Hunting

Ecotourism aims to disturb the observed animals as little as possible, but hunting, another common recreational activity, does not. Whether animals are able to habituate to hunting activities can be assessed by behavioral and endocrine data collected in tandem. For example, endocrinology can give insight into (1) whether the behavioral changes that result from hunting have negative health impacts or (2) whether physiological changes in response to hunting occur without alterations in behavior. To exemplify these points, we consider two studies on how chase by domestic dogs affects hunted prey. Data on cortisol and a suite of other physiological measures indicate that prolonged escape behavior (average of about 3 h) in red deer (*Cervus elaphus*) causes great physiological and psychological stress. These data indicate that there is indeed a physical and, likely, psychological cost to the deer of engaging in this unusual behavior. Exemplification of the second point: when cougars (*Felis concolor*) are chased repeatedly by hunting dogs, they display similar patterns of escape behavior as cougars chased just once. While the behavior of these animals is not changed, repeated chases cause cougars to downregulate their stress axis, potentially impairing their ability to respond appropriately to other stressors.

Beyond the direct impacts of hunting on the individual, hunting can also influence gregarious animals by removing members of their social group. Changes in group size, composition, and social hierarchy can cause changes in hormone levels and behavior. For example, glucocorticoid levels are related to group size in a variety of organisms from ring-tailed lemurs (*Lemur catta*) to cliff swallows (*Petrochelidon pyrrhonota*). In African elephants, glucocorticoids are impacted by the relatedness of individuals in the group and the age of the group's matriarch. If the matriarch is removed by poaching, other members of the disrupted family group show higher glucocorticoids and reduced reproduced success more than a decade later. Finally, an individual's social rank can impact both glucocorticoid and reproductive hormone titers.

Habitat Modification

Changes in habitat quality can induce changes in both behavior and hormone titers. For example, most studies find a positive relationship between habitat disturbance (e.g., urbanization, forest fragmentation, forestry activities) and glucocorticoids. Increases in glucocorticoids in response to habitat disturbance can be due to alterations in physical features (e.g., wind, solar radiation, predator pressure) and/or biological features of the environment (e.g., food availability and quality, predator abundance). However, it should be noted that some studies fail to find a relationship between glucocorticoid titers and the degree of habitat disturbance or, in fact, find a negative relationship between glucocorticoids and disturbance. The variety of possible results indicates that the response to habitat quality can depend on the limiting factors for the focal species, can be context dependent within and among species, and can be influenced by genetics and/or ontogeny. For example, the nature of the endocrine response to habitat change may be influenced by each individual's coping style (proactive or reactive) and the proportion of each coping style in the population: individuals that adopt increased tameness often dampen their glucocorticoid stress response.

Crowding

In some cases, habitat disturbance is great enough to make an area uninhabitable for a given species. As animals invade areas with suitable habitat, crowding can occur. Crowding typically induces agonistic behavior in both males and females, which can increase glucocorticoid titers and hormones related to aggression and inhibit reproductive physiology and behavior in adults. Crowding can also influence maturation in juveniles, increasing metamorphosis-inducing hormones in amphibians and decreasing sexual maturation in some mammals. The loss of habitat can have special consequences for territorial animals. Territory loss affects not only the individuals who need to find a new territory, but also territory owners who are challenged by the influx of new individuals. Engaging in territorial challenges can increase both glucocorticoids and testosterone. Alternatively, loss of habitat can reduce the expression of territoriality, which then modifies other behaviors, such as food caching.

Predator density

The impact of predators on their prey has been well-studied by behaviorists and endocrinologists. Data from these studies are applicable to conservation questions because invasive species often modify the predator–prey balance of ecosystems and the removal of top predators from ecosystems via hunting and habitat fragmentation can increase the abundance of meso-predators. Direct chase is the most obvious way that predators interact with their prey, but the sight or sound of attacks on others, the sight of the predator itself, and the presence of predator signs can also precipitate changes in physiology and behavior. The presence of predators can alter where and on what animals feed and the amount of time spent being vigilant, engaging in hiding and escape behaviors, giving warning calls, and mobbing. The presence and abundance of predators positively correlates with baseline and stress-induced glucocorticoids and fecal glucocorticoid metabolites. However, this response can depend on the type of predator (predator of self vs. of young) and the animal's ability to engage in escape behavior – namely, whether the animal perceives the predator as a threat and how serious the threat is considered.

The few studies that have simultaneously measured the behavioral and endocrine response to predators have all confirmed that behavioral modifications typically co-occur with elevated glucocorticoids. However, as with exposure to tourists, the endocrine response to predators can be dissociated from the behavioral response. For example, mobbing can occur without an increase in glucocorticoids, and some animals will increase glucocorticoids in response to a predator call without changing their behavior. In certain cases, elevated glucocorticoid titers released in response to predators are high enough to inhibit the production and release of reproductive hormones, such as testosterone, and can inhibit or delay reproductive behavior and/or activity.

Pollution

Over the past few decades, accumulating evidence from field and laboratory studies has associated exposure to environmental contamination with altered reproductive physiology in wildlife and humans. Specific contaminants known to affect endocrine systems are commonly referred to as endocrine disruptors. These compounds include pesticides, plasticizers (phtalates), flame retardants (polychlorinated biphenyls (PCBs) and polybrominated

diphenyl ethers (PBDEs)), industrial pollutants (heavy metals, dioxin), pharmaceutical agents (ethylestradiol from birth control pills), and surfactants (octylphenol and nonylphenol). Their effects range from altering the morphology of the gonads and sex organs to increasing the incidence of intersex individuals. They are also credited with reducing gamete production and fertility and decreasing hatching success and juvenile survivorship. Many of the effects of endocrine disruptors are likely organizational and therefore irreversible. In some cases, effects on methylation and other processes that effect gene expression may be passed on to subsequent generations. For all these reasons, endocrine disruptors may have significant population level impacts on reproductive success even at sublethal doses.

Endocrine disruptors may either stimulate or repress the effects of an endogenous steroid. When acting as an agonist, hormone mimics bind a receptor and initiate the secondary messenger pathways responsible for hormonal action at the cellular level. When acting as an antagonist, an endocrine disruptor sits inert on the receptor and blocks binding of the appropriate hormone.

It is the nature of hormones to work at miniscule concentrations on the order of parts per million. During critical developmental stages, animals are extremely sensitive to any compounds that bind hormone receptors. While steroids like estrogen may circulate in the body at concentrations much higher than those of environmental contaminants, they are often bound to binding globulins that regulate their interaction with receptors. In contrast, hormone mimics like DDT are not bound by globulins and are therefore free to bind receptors whenever they enter the bloodstream. For this reason, endocrine disruptors may have biologically significant effects at concentrations much lower than those of circulating endogenous hormones. On the other hand, many endocrine disrupting compounds (EDCs) bind with much lower affinity to receptors than hormones. For these, much higher concentrations of EDCs may be needed.

Toxicity studies can be notoriously difficult to conduct in the field, but are important to conservation. In natural systems, interactions among environmental contaminants, endocrine systems, and behavior may have direct and indirect population-level impacts impossible to predict from laboratory studies alone. Feeding is one example of a behavior that may alter exposure to contaminants in a natural system. In some cases, animals may preferentially forage in areas of heavy compound accumulation or eat at trophic levels that increase exposure through bioaccumulation. Alternatively, feeding behavior itself may be altered by toxin exposure. For example, animals may avoid foraging in contaminated areas or may need to forage more actively to cope with metabolic demands imposed by exposure to toxins. In either case, individuals may be at higher risk of malnutrition or predation and may have fewer resources to devote to parental care as a result of sublethal contamination.

Climate change

Climate change will alter habitat quality, but will also increase the severity of extreme weather events and disease incidence. Unusually high temperatures and unpredictable and unexpected stormy weather, including low temperatures and high precipitation and wind, can cause increased baseline glucocorticoids levels and an elevated stress response. The extent to which poor weather conditions elevate glucocorticoid levels may depend on the life-history stage of the individual. For example, the correlation between weather patterns and glucocorticoids in arctic-breeding songbirds is high during molt and can be less strong or absent during breeding. The endocrine response to severe weather events can also cause significant and severe declines in reproductive hormones, such as testosterone. Via affects on reproductive hormones, such events can potentiate decreases in aggressive behaviors, deactivation of territoriality, suspension of social hierarchies, and formation of groups or flocks. Other prolonged climatic perturbations such as drought can cause similar effects.

Animals have two behavioral strategies for dealing with severe weather: 'take it' or 'leave it.' Glucocorticoid titers can facilitate behaviors that promote either strategy. In the 'take it' strategy, glucocorticoid titers intensify the search for food and, when food is available, indicate that resources are sufficient to ride out the weather event in a refuge from the storm. This 'decision' may also be dependent on internal energy stores. When food is not available during a weather event, glucocorticoid titers prompt abandonment of the territory or locale and escape to areas where, potentially, conditions are better. Such irruptive migrations have been observed in many highly mobile taxa. The return to territories after severe weather events can also be linked to glucocorticoid titers.

Disease

As temperatures increase and in some locales, conditions become wetter, pathogens and their vectors will invade new areas and may increase in abundance, affecting disease prevalence and transmission. Diseases can cause a wide range of behavioral changes in animals, ranging from lethargy to mania. They can influence feeding behavior (e.g., diet choice and quantity of food consumed), reproductive behaviors, and attractiveness. The causal direction of the relationship between infection and alterations of hormone titers is not fully understood. For example, given that high glucocorticoid or testosterone titers may be immunosuppressive, high levels of these hormones could increase the likelihood of infection. In some cases, infections can also influence hormone titers. Pathogen infection often results in an increase in glucocorticoid titers, in which case behavioral changes may include those induced by the pathogen (e.g., feeding behavior), potentiated by glucocorticoids themselves (e.g., suppression of territoriality), and related to the decline in reproductive hormones caused by high glucocorticoid titers (e.g., fewer courtship behaviors).

Conclusion

In any conservation effort, time and resources are at a premium. While long-term demographic patterns are critical to an accurate evaluation of population health, most conservation researchers do not have the luxury of tracking survival and reproductive success over an adequate time span and collecting enough samples to evaluate the efficacy of specific management strategies. Even in systems where excellent demographic data have been collected, it is often impossible to determine the relative impacts of the disparate pressures. For this reason, behavioral and endocrine response variables can be invaluable as proxies for measures more directly related to lifetime fitness. Changes in behavior and hormone profiles can be directly linked to environmental factors in controlled experimental manipulations or in natural correlative experiments. Potentially, these measures will allow impacts to be assessed and management strategies implemented or modified *before* significant population declines occur.

Experience has shown that basic biology is fundamental to successful management and conservation efforts. At the same time, the current rate of global change makes conservation a pressing issue for any biologist. We hope that those with a background in behavior or endocrinology will conduct studies that may inform conservation efforts. We also hope that those involved with management will incorporate behavior and hormone data in their adaptive conservation efforts. At this stage, the theoretical frameworks guiding environmental endocrinology, specifically stress physiology, and the interaction between hormones and behavior in free-living animals are still being actively developed. The better we understand how changes in individual- or population-level behavioral patterns and endocrine profiles translate into changes in population ecology and demography, the better we will be able to assess the effects of environmental change and channel our limited resources to protect biodiversity.

See also: Conservation and Animal Behavior; Conservation and Behavior: Introduction; Endocrinology and Behavior: Methods; Field Techniques in Hormones and Behavior; Hormones and Behavior: Basic Concepts; Vertebrate Endocrine Disruption.

Further Reading

Arnold AP and Breedlove SM (1985) Organizational and activational effects of sex steroids on brain and behavior: A reanalysis. *Hormones and Behavior* 19: 469–498.

Boonstra R, Hik D, Singleton GR, and Tinnikov A (1998) The impact of predator-induced stress on the snowshoe hare cycle. *Ecological Monographs* 68: 371–394.

Clinchy M, Zanette L, Boonstra R, Wingfield JC, and Smith JNM (2004) Balancing food and predator pressure induces chronic stress in songbirds. *Proceedings of the Royal Society of London Series B-Biological Sciences* 271: 2473–2479.

Creel S, Fox JE, Hardy A, Sands J, Garrott B, and Peterson RO (2002) Snowmobile activity and glucocorticoid stress responses in wolves and elk. *Conservation Biology* 16: 809–814.

Gobush KS, Mutayoba BM, and Wasser SK (2008) Long-term impacts of poaching on relatedness, stress physiology, and reproductive output of adult female African elephants. *Conservation Biology* 22: 1590–1599.

Gray LE, Wilson V, Stoker T, et al. (2005) The effects of endocrine-disrupting chemicals on the reproduction and development of wildlife. In: Naz RK (ed.) *Endocrine Disruptors: Effects on Male and Female Reproductive Systems*, pp. 313–343. Boca Raton, FL: CRC Press.

Romero LM (2004) Physiological stress in ecology: Lessons from biomedical research. *Trends in Ecology & Evolution* 19: 249–255.

Romero LM and Wikelski M (2001) Corticosterone levels predict survival probabilities of Galapagos marine iguanas during El Nino events. *Proceedings of the National Academy of Sciences of the United States of America* 98: 7366–7370.

Suorsa P, Huhta E, Nikula A, et al. (2003) Forest management is associated with physiological stress in an old-growth forest passerine. *Proceedings of the Royal Society Biological Sciences Series B* 270: 963–969.

Touma C and Palme R (2005) Measuring fecal glucocorticoid metabolites in mammals and birds: The importance of validation. In: Goymann W and Jenni-Eiermann S (eds.) *Bird Hormones and Bird Migrations: Analyzing Hormone in Droppings and Egg Yolks and Assessing Adaptations in Long-Distance Migration*, pp. 54–74. New York, NY: Wiley.

Walker BG, Boersma PD, and Wingfield JC (2006) Habituation of adult magellanic penguins to human visitation as expressed through behavior and corticosterone secretion. *Conservation Biology* 20: 146–154.

Wasser SK, Davenport B, Ramage ER, et al. (2004) Scat detection dogs in wildlife research and management: Application to grizzly and black bears in the Yellowhead Ecosystem, Alberta, Canada. *Canadian Journal of Zoology* 82: 475–492.

Wikelski M and Cooke SJ (2006) Conservation physiology. *Trends in Ecology & Evolution* 21: 38–46.

Wingfield JC (2005) Flexibility in annual cycles of birds: Implications for endocrine control mechanisms. *Journal of Ornithology* 146: 291–304.

Wingfield JC and Ramenofsky M (1999) Hormones and the behavioral ecology of stress. In: Balm PHM (ed.) *Stress Physiology in Animals*, pp. 1–51. Sheffield: Sheffield Academic Press.

Relevant Websites

http://conservationbiology.net/ – Center for Conservation Biology, University of Washington.

http://www.si.edu/ofg/Staffhp/monforts.htm – Stevan L. Monfort, Smithsonian Institution.

http://ivabs.massey.ac.nz/resprog/cerg/research.asp – Conservation Endocrinology Research Group, University of Massey.

http://ase.tufts.edu/biology/faculty/romero/ – L. Michael Romero, Tufts University.

http://www.montana.edu/wwwbi/staff/creel/creel.html – Scott Creel, Montana State University.

http://post.queensu.ca/~bonierf/index.html – Frances Bonier, Queens University.

http://www.zoology.ufl.edu/ljg/ – Louis J. Guillette, University of Florida.

http://www.faculty.fairfield.edu/biology/sigxi/Faculty/Walker.html – Brian Walker, Fairfield University.

Conservation, Behavior, Parasites and Invasive Species

S. Bevins, Colorado State University, Fort Collins, USA

Introduction

The intersection between invasion biology, parasitology, and animal behavior is, at its most basic, an interaction among two species: an invasive parasite may adapt to a native host, or an invasive host could become infected with a parasite. The reality, as is often the case with animal behavior, is typically more complex, involving multiple host species and/or multiple parasite species. Examination of interactions among multiple species – and the behaviors that guide them – requires a community-wide approach. The mechanisms influencing a biological community, including parasitism, predation, and competition, must be taken into account as well.

A seminal experiment that first examined the dynamics of a parasite's effect on species interactions was completed by Thomas Park in 1948. Over a period of 4 years, 211 populations of flour beetles where sustained in a laboratory. These populations consisted of two species, *Tribolium confusum* or *T. castaneum*, which were either maintained alone or in mixed species colonies. Mixed-species colonies displayed marked competitive interactions, with one species always going extinct during the experiment. Park also added a naturally occurring protozoan parasite to some mixed-species populations and the results were striking: *T. castaneum* often went extinct when the parasite was present, but *T. confusum* died out when the parasite was absent. These were the first results to demonstrate the existence of parasite-mediated competition, leading to a flurry of research on parasites and their ability to regulate host populations, thus influencing the structure of biological communities.

There are now several examples of host and/or parasite behavior influencing competitive interactions among invasive and native species. Unlike the naturally occurring species in Park's experiments, an invasive species is generally defined as a nonnative species that spreads rapidly once established, with the potential to cause economic or environmental harm. Many species are introduced to new regions, but they are not necessarily invasive. The following case studies all involve species that have become invasive.

Parasite-Altered Behaviors and Invasion

For instance, take the thorny-headed acanthocephalan parasite, *Pomphorhynchus laevis* (the literal translation of the genus name is blister-snout, referring to the bulb-like front end, which is covered with spiny projections). This parasite lives in freshwater environments and the larval forms infect small aquatic crustaceans called amphipods. These infected amphipods must then be consumed by a fish in order for the adult form of the parasite to mature. Amphipods infected with the acanthocephalan exhibit behaviors that differ from their uninfected counterparts. For example, infected amphipods will often move toward light – or at least will not shy away from it like uninfected individuals – and this increases their chances of being preyed upon by fish. Other altered behaviors in *P. laevis*-infected amphipods include changes in antipredator behavior, and more conspicuous coloration. These changes have been shown to benefit the parasite, which requires a fish host to complete its life-cycle, but represents the ultimate cost for the amphipod. This particular acanthocephalan parasite is somewhat of a generalist, and it can infect several species of amphipod; however, the altered host behaviors that come along with the parasite are not necessarily consistent across host species.

In France, Bauer and colleagues examined a nonnative amphipod that has invaded freshwater habitats already supporting a native amphipod. These native amphipods exhibit classic behavioral alterations when infected with the acanthocephalan parasite, but the new invasive amphipod does not. The parasite was found in both species of amphipods, but only the native amphipod species had an altered response to light when infected with the parasite. This suggests that when infected, the native amphipod species is under more intense predation pressure than the invasive species – decreasing the competitive interactions between the two – which could ultimately aid the invader's ability to establish.

This system in an example of a parasite influencing the interactions between a native and an invasive species by altering host behavior, and there are many other documented examples that fall into this category. Calvo-Ugarteburu and McQuaid found that a mussel species native to South Africa (*Perna perna*) hosts two trematode parasites, one that alters behaviors and thereby reduces growth and one that castrates the host. A new invasive mussel is not infected by either parasite. This sets up a situation similar to that of the amphipods (seen earlier), with the parasites reducing the fitness of the native, but not the invasive species.

Parasite-altered host behaviors can also be extremely precise. In the case of the amphiphod, *Polymorphus minutus*, a laboratory infection study with an acanthocephalan parasite revealed that both infected and uninfected individuals remained benthic, presumably away from the surface of water, where they are subject to high predation pressure from the acanthocephalan's final avian host. Another amphipod species recently invaded the same area where *P. minutus* is found, but the invasive amphipod is larger and can prey upon smaller crustaceans, like *P. minutus*. Laboratory experiments revealed that when there is no predatory amphipod around, both infected and uninfected *P. minitus* spend a majority of their time at the bottom of a water column, but the addition of the predatory, invasive amphipod resulted in infected amphipods (not their uninfected conspecifics) dispersing to surface areas of a water column. This precise behavioral manipulation would allow the parasite to avoid predation by an unsuitable final host – an invasive predatory amphipod – while exposing its host to predation by the appropriate final host.

In another amphipod example, MacNeil et al. found a species of native amphipods that hosted a microsporidian parasite. This made the native more likely to be preyed upon by a larger, invasive amphipod that is resistant to infection. In addition, Georgiev and others examined native Mediterranean brine shrimp that are hosts to cestode parasites which alter coloration, as well as behaviors such as positive phototaxis, increased swimming time, and augmented surfacing behavior; all of these are behaviors that potentially increase the chances of cestode transmission to its final avian host. These parasites are not found in a new, rapidly invading brine shrimp species. Such parasites may also be examples of the 'enemy release hypothesis,' the idea that introduced species become invasive because they are no longer regulated by the natural parasites and predators of their native range.

This list of parasite-modified behaviors influencing invasions continues to grow as researchers delve deeper into species interactions. Invasions facilitated by parasites are relatively rare compared to the sheer number of invasions not known to be facilitated by parasites; however, these examples illustrate that parasite-modified behaviors can influence the invasion process and profoundly contribute to community structure, species establishment, and species extinction.

Impact on Species Conservation

Infections that translate to a fitness cost for the host can reduce population numbers. The introduction of a novel species can 'enhance' an existing parasite–host relationship, as was the case with *Erythroneura variablis*, a nonnative leafhopper insect that invaded California in 1980. The region's native leafhopper was parasitized by *Anagrus epos*, a wasp parasitoid that infected leafhopper eggs. The egg-laying behavior of the native grape leafhopper made it more prone to parasitism by the parasitoid

when compared with the invasive species. In this case, it is the intrinsic host behavior that when combined with parasitism gives the invasive species an advantage over its potential competitor. Settle and Wilson claimed a two-part advantage for the invasive leafhopper: (1) the parasitoid lowered native leaf-hopper population numbers allowing the invasive to establish and (2) the native suffered from an increased parasitoid population brought about by greater overall host numbers, as it was a multihost pathogen. This was one of the first documented instances of the indirect effects that parasites can have during an invasion.

The leafhopper had presumably existed with the parasitoid before the arrival of a nonnative leafhopper. After that introduction, the native leafhopper's population declined precipitously, because the large, combined leafhopper population provided by the invasive species allowed the parasitoid numbers to increase drastically, which in turn intensified parasitism pressure on the native species. This is known as apparent competition, in which it appears that one species' decline is due to the arrival of a new species, when in fact that decline is linked to a shared parasite or predator; Hudson and Greenman have reviewed that phenomenon. All of these indirect effects are based on the native leafhopper's egg-laying behavior, which make it more prone to parasitization than the invasive leafhopper. The decline of species because of invasive parasites, or invasive hosts that alter existing parasite/host systems, is a very real conservation problem.

Another example of invasion, parasites, and conservation biology, is the Puerto Rican parrot (*Amazona vittata*), which suffered high nestling mortality because of muscid botfly infections. These infections became a problem only after the population expansion of another bird, the pearly-eyed thrasher (*Margarops fuscatus*). Thrasher numbers are thought to have multiplied because of an increase in disturbed habitat related to a dramatic upswing in road-building. The thrashers appear to prefer disturbed habitats. This is another example of apparent competition, but one that is compounded by Puerto Rican parrot numbers having plummeted only to 13 animals in the 1970s because of habitat loss. Species of concern that already suffer from low population numbers and reduced genetic diversity may be more prone to ill effects. To that end, the invasion of parasites now plays a central role in conservation biology.

The Hawaiian archipelago, for instance, was home to a vast number of endemic avifauna, but no mosquitoes. *Culex quienquefasciatus*, the southern house mosquito, arrived with European sailors in 1826. The 1900s saw potentially hundreds of nonnative bird species purposely released in Hawaii, and some of the birds likely harbored avian malaria and avian pox. Migratory birds probably carried the parasites to the islands before human settlers arrived; however, there was no mosquito vector to transmit the malarial parasite. In addition, the parasite prevalence in migratory birds was most likely low, the migration to Hawaii being an arduous journey from even the closest landmass. The combination of introduced parasites, introduced vectors, and long-isolated avian populations lacking evolutionary experience with these pathogens proved catastrophic. More than 50% of Hawaii's endemic birds are now extinct.

The survival of some remaining species, however, may be attributable to altered behaviors. In the 1960s, Warner observed nonnative birds in Hawaii sleeping with their face and bill tucked in, and their legs pulled up into their feathers, while native Hawaiian birds slept with their faces and legs exposed. These behaviors left the native birds exposed to mosquito bites throughout the night, but observations from the late 1970s by van Riper and colleagues found native birds sleeping with exposed areas tucked. Presumably, the sleeping behaviors changed in response to the strong selection pressures exerted from parasites transmitted by mosquito bite. Daily altitudinal migrations also appear to have shifted, allowing some bird species to avoid peak mosquito biting times by periodically moving to higher altitudes that are mosquito-free.

Rabies is another disease that has been introduced to many areas of the world and also classically alters the behaviors of infected hosts. The rabies virus is transmitted by infectious saliva, initially infecting the peripheral nervous system of the host. Movement of the virus from the peripheral to the central nervous system is accompanied by a suite of behavioral changes, including increased aggression and biting behavior, as well as an abundance of saliva, the result of an impaired swallowing reflex. All of these behaviors appear to increase viral transmission opportunities, but it should be noted that not all organisms demonstrate classic behavioral changes, nor is the virus always fatal. Bats for instance, often survive rabies infection; however, rabies is known to aerosolize in highly aggregated bat roosts, offering the virus an alternative transmission route and potentially circumventing the need for transmission via aggressive host encounters.

Rabies records data back several hundred years in some parts of Africa, but its appearance in sub-Saharan Africa is thought to be relatively recent, suggesting an introduction with domestic dogs in the late 1800s. The impact on wildlife, especially native carnivores, has been severe. Rabies has led to the local extinction of African wild dogs from Serengeti National Park, and Ethiopian wolves – a species with fewer than 600 individuals remaining – suffer up to 70% mortality during rabies outbreaks. Rabies has been documented in many other native African species as well, but domestic dogs are believed to be the primary reservoir in this case.

Rinderpest is another virus accidentally introduced into Africa and it devastated native ungulate populations.

In, 'The Rise of our East African Empire,' Frederick John Dealtry Legard wrote, 'never before in the memory of man, or by the voice of tradition, have the cattle died in such numbers; never before has the wild game suffered.' The grouping behavior present in many large herds of African ungulates offered an easy target for a pathogen that is rapidly and directly transmitted in dense groups. In fact, it has been suggested that the Sahara acted as a natural barrier because the low ungulate densities would not have allowed the pathogen to cross over from Europe and the Middle East. Group sizes are often larger in open habitats, like those seen in some areas of Sub-Saharan Africa. While grouping behavior may reduce encounters with biting flies, it would make naive species living in African savannahs more readily exposed to introduced directly transmitted microorganisms. Wildlife populations rebounded from rinderpest after a massive campaign to vaccinate the reservoir host, cattle, and these rebounds showed just how profound the effects of this virus were. Wildebeest and other ungulate populations increased up to sevenfold, followed by a dramatic increase in predator numbers as well. Such dramatic population changes had system-level effects: nonungulate grazers such as zebra now had to compete for forage; fire-cycles were altered because foraging pressure changed the type and amount of plant material available, allowing some tree species regenerate. Parasites can truly have ecosystem-level effects.

Future Questions

Our understanding of the role animal behavior plays in parasite-involved invasions is still in its infancy, and many questions remain. While many named hypotheses have been suggested to explain invasion processes, it is often hard to disentangle the mechanisms driving an invasion. Is it a case of enemy release, or apparent competition? Is the invasive species in question simply a more aggressive competitor? Has it tapped into a previously limiting resource? Answering these questions is further complicated by invasion being a many faceted process, involving colonization, establishment, and spread. Behavior and/or parasitism may play a role only in one part of an invasion.

Robust, long-term studies are required if we are to decode the contribution that parasites and their behavioral effects have on the outcome of some invasions. Of course, we need experimental verification that the altered host behaviors actually do increase parasite transmission. Long-term studies are also needed to determine whether and how parasites evolve to infect new invasive hosts, or vice-versa. Invasive hosts may eventually pick up new parasites, but on what time scale, and what happens to those new associations over time? This is a new era, and there is no shortage of questions.

See also: Disease Transmission and Networks; Evolution: Fundamentals; Intermediate Host Behavior; Propagule Behavior and Parasite Transmission; Social Behavior and Parasites; Wintering Strategies: Moult and Behavior.

Further Reading

Bauer A, Trouve S, Gregoire A, Bollache L, and Cezilly F (2000) Differential influence of *Pomphorhynchus laevis* (Acanthocephala) on the behaviour of native and invader gammarid species. *International Journal for Parasitology* 30: 1453–1457.

Calvo-Ugarteburu G and McQuaid CD (1998) Parasitism and invasive species: Effects of digenetic trematodes on mussels. *Marine Ecology-Progress Series* 169: 149–163.

Elton CS (2000) *The Ecology of Invasions by Animals and Plants.* Chicago, IL: University of Chicago Press.

Holway DA and Suarez AV (1999) Animal behavior: An essential component of invasion biology. *Trends in Ecology and Evolution* 14: 328–330.

Hudson P and Greenman J (1998) Competition mediated by parasites: Biological and theoretical progress. *Trends in Ecology and Evolution* 13: 387–390.

MacNeil C, Dick JTA, Hatcher MJ, Terry RS, Smith JE, and Dunn AM (2003) Parasite-mediated predation between native and invasive amphipods. *Proceedings of the Royal Society of London Series B-Biological Sciences* 270: 1309–1314.

Medoc V, Bollache L, and Belsel JN (2006) Host manipulation of a freshwater crustacean (*Gammarus roeseli*) by an acanthocephalan parasite (*Polymorphus minutus*) in a biological invasion context. *International Journal for Parasitology* 36: 1351–1358.

Prenter J, MacNeil C, Dick JTA, and Dunn AM (2004) Roles of parasites in animal invasions. *Trends in Ecology and Evolution* 19: 385–390.

Torchin ME and Mitchell CE (2004) Parasites, pathogens, and invasions by plants and animals. *Frontiers in Ecology and the Environment* 2: 183–190.

Cooperation and Sociality

T. N. Sherratt, Carleton University, Ottawa, ON, Canada
D. M. Wilkinson, Liverpool John Moores University, Liverpool, UK

Introduction

One of the fathers of modern sociobiology, E.O. Wilson, defined a society as 'a group of individuals belonging to the same species and organized in a cooperative manner.' By this definition, all societies consist of aggregations of individuals, but not all aggregations are societies. Male mosquitoes, for instance, may form swarms but they lack the requisite level of cooperation to be considered social. Bird flocks and wolf packs, on the other hand, are generally considered societies because their members not only form groups, but also cooperate – for example, through alerting one another to the presence of predators, or through hunting in packs. Sometimes, the cooperative acts within societies come at a significant cost to the cooperators themselves, and such behaviors are termed 'altruistic.' Worker honeybees, for example, forego their own reproduction to help their queen reproduce and will even die in her defense.

Potential benefits of social living include a reduction in the rate of predation, improved foraging efficiency, improved defense, and improved care of offspring. For example, due to increased vigilance, the success rate of goshawk attacks on pigeons tends to decrease with increasing numbers of pigeons in a flock. Likewise by huddling together, emperor penguins help save energy and maintain a constant body temperature, thereby ensuring the successful incubation of their eggs. While it is often easy to see the benefits of group living, it is harder to understand why individuals do not free-ride on those benefits while giving nothing in return. Therefore, to understand how societies function and persist, we must understand the stability of the cooperative relationships that help define them. In his last presidential address to the Royal Society of London in November 2005, Robert May argued: 'The most important unanswered question in evolutionary biology, and more generally in the social sciences, is how cooperative behavior evolved and can be maintained.' Here, we review some of the principal solutions to understanding the evolution of cooperation – and hence societies – that have emerged over the past 50 years. Many of the solutions we discuss have also been applied to understanding the origin and stability of other forms of cooperation – including cooperation between cells in multicellular organisms and examples of cooperation between members of different species.

Kin Selection

While a form of a gene can spread because it enhances the carriers' own survival and reproduction, it can also spread in a population because it assists relatives who tend to share alleles identical by descent, even if this occasionally comes at a cost to the bearer's own reproductive success. Thus, cooperative behavior can frequently spread and be maintained because it favors the survival and reproduction of close relatives. While J.B.S. Haldane, R.A. Fisher, and others toyed with the idea in the 1930s, it was Bill Hamilton (in 1964) who recognized its full importance and who developed a formal quantitative theory, in which he introduced the logic of 'inclusive fitness.' This general body of theory is now known under the heading 'kin selection,' although strictly speaking this refers to the narrower subset of conditions in which individuals assist other individuals that share the same copy of a gene through their close genealogical relatedness. Even if natural selection favors nonaltruists over altruists within groups of individuals likely to share the altruism trait, the proportion of altruists may nevertheless increase if they do better overall than alternative groups without altruists.

Perhaps the easiest way to understand the logic of kin selection is through Hamilton's rule, which states that a form of a gene that causes an individual to perform an altruistic act will tend to spread so long as $rb - c > 0$, where b (broadly speaking) is the fitness benefit to the recipient from the altruistic act, c is the fitness cost to the altruist, and r is a measure of relatedness. The rule is shorthand for a full population genetics model, and, strictly speaking, it is only correct if we define the terms in very particular ways. For example, r formally measures how genetically similar two individuals are when compared to two random ones in the population with which the altruist will compete for entry into the next generation (indeed, r can be negative). Through its simplicity, Hamilton's rule serves to highlight the composite minimum conditions for kin-based cooperation, which are both ecological (mediated through b and c) and genetical (mediated through r).

Acts of kin-based cooperation are not confined to complex animals and can, for example, be seen in the social amoeba ('slime mold') *Dictyostelium discoideum*, in which solitary cells start to aggregate under harsh conditions to produce a 'slug.' Some cells eventually produce the spores that can colonize new areas, but only at the expense of a minority of other cells that collect together to hold the spore-producing cells aloft. The cells involved in producing the colony are frequently (but not always) genetically identical, so this extreme altruism can potentially be understood in terms of kin selection; rather like individual cells in a multicellular organism, here helping others reproduce is tantamount to helping yourself.

The colonies maintained by wasps, ants, and bees represent the classical examples of highly cooperative social groups. Here, sterile masses of individuals work by gathering food, cleaning the nest, and repelling predators, all in the service of a small reproductive minority. In a series of seminal papers, Hamilton proposed that their unusual genetics (specifically, their haplodiploid sex determination, in which fertilized eggs develop into females while unfertilized eggs become males) might help to explain the prevalence of cooperation in these groups. One implication of the genetics is that females are more related to their full sisters (relatedness 0.75) than they would be to their own offspring (0.5). Given this asymmetry, it is easy to see how sisters might be selected to forego their own reproduction if such behavior can help queens produce more sisters.

Unfortunately, while haplodiploidy may well *help* to explain the evolution and maintenance of cooperation in the highly social ('eusocial') insects, it cannot provide the complete solution. Once one factors in the 0.25 relatedness of sisters to brothers, the overall relatedness among siblings in haplodiploids is not especially high. Multiple queens and multiple matings with different males further act to reduce the average relatedness of offspring within a colony. While the manipulation of sex ratio and the ability of females to preferentially favor their sisters may each play a role in enhancing relatedness between the donor and the recipient of cooperation, other factors that also help tip the balance include punishment of workers that attempt to lay their own eggs. Thus, even if high relatedness does not provide the whole explanation, then it may help level the playing field considerably, making cooperative behaviors more likely.

Another good example of kin selection is seen in the formation of particular types of microbial mat. Laboratory populations of the bacterium *Pseudomonas fluorescens* rapidly diversify when maintained in unshaken broths, and a particular form – known as the 'wrinkly spreader' (produced by single mutations with large effects) – tends to build up at the liquid–air interface, creating a surface scum. The ability to live at the boundary layer allows the wrinkly spreader bacteria to avoid the oxygen-deprived conditions deeper in the water column, but it comes at some cost. To form and maintain the mat, the wrinkly spreaders have to make a cellulose polymer (glue), a metabolic cost that is not borne by other forms of the bacterium. Despite this cost, the mat can initially develop by kin selection (individuals helping to bind to the surface share the same trait for making glue), but it undergoes periodic collapses as nonglue-making 'cheats' invade.

The stability of some forms of cooperation may depend on both kin selection and other more direct forms of return. Providing 'parental' support to young that are not your own is relatively common in vertebrates. For example, in populations of birds such as Florida scrub jays and long-tailed tits, and mammals such as meerkats (suricates) and brown hyenas, there are nonbreeders that help raise young produced by dominant breeders. Why do not nonbreeders go it alone, rather than help look after another's offspring? Kin selection may play an important role – indeed, groups in cooperatively breeding species are typically made up of extended families, so that subordinates often help their relatives. Moreover, studies have shown that helpers sometimes provide their closer kin with preferential care. However, direct fitness benefits may also be important in maintaining cooperation in this type of system and may be even more important than kinship itself. In some cases, helpers may be forced into helping behavior to avoid punishment, but by helping they may also sometimes increase their chance of inheriting the territory of the breeding pair (they are 'paying the rent'). Likewise, the increased survival chances from grouping together may sometimes outweigh the costs of helping. In meerkats, for example, the foraging success and survival of all group members increases with the size of the group (an example of the 'Allee effect'). It is these and other nonkin routes to cooperation that we now consider.

Reciprocal Altruism

Kin selection may help to explain many cases in which individuals incur costs that benefit others, but as we have already seen, some of these examples involve additional phenomena that further maintain cooperation. At an extreme, how do we explain examples of cooperation among nonrelatives? Included in these examples are food sharing among unrelated crows and impala that groom unrelated individuals within the herd. Perhaps by helping others, the donor might subsequently be helped by the receiver when its own need arises? Evolutionary biologist Robert Trivers presented just such an explanation in the early 1970s, referring to the phenomenon as 'reciprocal altruism.'

To understand how cooperation might be maintained under these circumstances, mathematical modelers have spent a great deal of time and effort elucidating the types of strategy that would do well in a simple game, known as the two-person iterated Prisoner's Dilemma. In each round ('iteration'), two players decide simultaneously whether to 'cooperate' (C) or 'defect' (D), just as two prisoners accused of a joint crime may decide to cooperate with each other by staying quiet under interrogation, or defect on their criminal partnership by talking to the police in exchange for a lighter sentence. In the Prisoner's

Dilemma (see **Table 1**), mutual cooperation ('CC') pays more to both players than mutual defection ('DD'), but defecting while your partner cooperates ('DC') pays the defector most of all (reflecting a 'temptation' to defect) and a sole cooperator least of all ('CD,' the 'suckers payoff'). Many researchers consider the Prisoner's Dilemma the key metaphor for understanding cooperation, primarily because it captures the temptation to defect, but also because it can reflect some of the damaging effects of pure self-interest. However, in a one-off game, the most rewarding strategy is always to defect because whether your partner cooperates or defects, your best option is to defect. So, how can cooperation ever evolve? One solution to the problem arises if the players have a chance of meeting again.

In an iterated two-player game (played repeatedly, where the number of rounds is not known in advance by the players), the set of potential strategies is enormous – especially if long memories of previous interactions are allowed. One such strategy called 'tit-for-tat' (TFT – cooperate on first move, thereafter follow the partner's previous move) has been widely recognized as a successful strategy in these iterated games. TFT is thought to be particularly effective because it is 'nice' (in that its starting move is to cooperate), retaliatory (in that it follows a defection from the partner with defection), and forgiving (in that it subsequently matches any cooperative act with cooperation). However, TFT is not without its weaknesses: for example, if mistakes are occasionally made, two tit-for-tatters can get stuck into indefinite rounds of defection. As an alternative, the win–stay, lose–shift strategy (WSLS – keep to your strategy if your previous exchange was high paying (DC or CC) but otherwise change) can correct occasional mistakes, although it is still open to drift in a population of cooperators. While such analyses help explain why certain types of cooperative behavior are more successful than others, they should be viewed as providing aids to thinking – not specific quantitative predictions about behavior in the real world.

Can direct reciprocation explain examples of cooperation that we see around us? There do appear to be some good examples of reciprocation maintaining altruism, but

not many. The classical example is reciprocal blood sharing in vampire bats, in which females regurgitate blood meals to roost mates who have failed to obtain food in their recent past. While nest mates are often related, there appears to be more structure to the interaction. In particular, experimentally starved bats who received blood, subsequently gave blood to the former donors more often than one would expect by chance. Likewise, in laboratory experiments, cotton-top tamarin monkeys gave more food to a trained conspecific who regularly offered them food in the past, compared to an individual who never gave them food.

Studies of grooming have also produced some clear cases of reciprocal altruism. For example, on the African savannah, impala frequently approach one another and begin grooming. Like the vampire bat example, the benefit, in this case removing parasites, may be high, but the costs of grooming in terms of time, fluids, and energy may be relatively low. Here, individuals deliver grooming in bouts ('parcels' of 6–12 licks), and the number of bouts received and delivered is remarkably well matched: in this case, defection involves simply walking away or doing nothing. While the relationship is based on reciprocation, it seems very likely that parceling up the cooperative acts in this way helps reduce the temptation to defect. Business deals often show a similar structure to avoid exploitation – half paid in advance and the other half paid when the job is complete.

Male red-winged blackbirds in North America also appear to cooperate, sometimes coming to the aid of neighboring males in defending their nests and territories from potential predators such as American crows. One possibility is that the helpers are in fact the true fathers of some offspring on the neighboring territory and are selected to help out of sheer self-interest, that is, simple parental care. Alternatively, or in addition, the helper may benefit directly by removing any potential predator from the neighborhood ('not in my back yard'), and any benefit to the neighbor is incidental (a by-product 'mutualism'). R. Olendorf and colleagues recently put these and other explanations for cooperation to the test and ruled out any kin-based explanations on the basis of genetic analyses. However, they also looked for evidence of reciprocity by examining patterns of nest defense against a stuffed crow and simulating cheating by making it appear that a neighbor was not helping with the defense (a 'defection'). As anticipated, male blackbirds tended to decrease their defense against a potential nest predator after their neighbor appeared to defect in the earlier trial, suggesting that reciprocation was having an important role in maintaining cooperation – "I'll help mob your predators, if you help mob mine."

Table 1   An example of the payoffs involved in a two-player Prisoner's Dilemma

		Player A decision	
	Payoff to player A	Cooperate	Defect
Player B decision	Cooperate	3 (R)	5 (T)
	Defect	0 (S)	1 (P)

Participants must simultaneously decide whether to cooperate or defect. The defining inequalities of the dilemma require that $T > R > P > S$ (and, for technical reasons, $R > [S + T]/2$), such that the temptation (T) to defect exceeds the reward (R) from cooperation, which in turn exceeds mutual punishment (P). The sucker's payoff (S) the least that can be expected.

Indirect Reciprocity

By its very nature, direct reciprocity requires repeated dealings among the same sets of individuals, so it cannot apply to cases of helping strangers we might never see again. However, what if others were looking on? Perhaps by helping others one might gain sufficient reputation as a 'nice' individual that strangers would be willing to help you when your own need arose. So, instead of 'You scratch my back, and I'll scratch yours,' one could consider another, seemingly even more vulnerable, guiding principle 'You scratch my back and I'll scratch someone else's.' This is called the 'indirect reciprocity' route to cooperation. Although it may at first seem strange, bear in mind that what matters is that acts of cooperation are returned, not who returns them.

Examples of the importance of maintaining an untarnished reputation are widespread in human societies. For example, eBay in part relies on reputation to maintain honest transactions when it provides scores of partner satisfaction. Being a good person or good company to deal with, does not in itself explain cooperation, but it begins to suggest a role for reputation in partner choice.

Building on earlier arguments by Richard Alexander, mathematical modelers have demonstrated the theoretical plausibility of cooperation via indirect reciprocity by showing that behavioral rules can evolve in which individuals are more prepared to help strangers if these strangers have a reputation for cooperating. Of course, since reputable individuals tend to provide assistance to similar reputable individuals, kin selection may also play a key role here.

Can indirect reciprocity explain cooperation in humans? After all, humans frequently help others who may never have an opportunity to reciprocate, and it is possible that such acts of kindness are recognized and rewarded by others. Staged laboratory games support this view. For example, in a recent experiment, human subjects were repeatedly given the opportunity to give money to others, having been informed that they would never knowingly meet the same person with reversed roles (all donations were anonymous). Despite the anonymity, the personal histories of giving and not-giving were displayed for participants to see at each interaction. As one might expect, the authors found that donations were significantly more frequent to those receivers who had been generous to others in earlier interactions.

There are far fewer examples of indirect reciprocity in non-humans and they mainly include examples of cooperation between species rather than within species. One recent example comes from work on cleaner fish mutualisms. The cleaner fish *Labroides dimidiatus* remove skin parasites from their fish clients, but there is an apparent temptation for them to take a little more at the expense of the client by feeding on their mucus. Clients are faced with the challenge of getting cleaners to feed against their preferences if they are to come away unscathed. Field observations indicate that client fish almost always invite a potential cleaner to draw closer and inspect them if they have had the opportunity to see that the cleaner's previous interaction ended without conflict. By contrast, clients invite particular cleaners far less frequently if they observed that the last interaction of the cleaner ended with conflict, such as being chased away. So, a good reputation is good for the cleaner's business, and it may be an important way in which clients avoid 'defections.'

Strong Reciprocity

Everyday observation suggests that there is a sense of 'fair play' in human societies. This has been backed up by more formal, if abstract, experiments. For example, in the well-known 'ultimatum' game in economics, two players A and B have to agree on how a monetary reward has to be shared. Player A (the proposer) has one chance to suggest how the money is to be shared (e.g., 60% to player A, 40% to player B), but player B (the responder) can accept or reject the proposed division. If the bid is rejected, then both receive nothing, but if the bid is accepted then the proposal is implemented. The logical optimum is to offer the responder an almost negligible amount (1 cent, say, because 1 cent is better than nothing). However, this logical response may be grossly inadequate, because a common result in this type of game is that responders tend to reject proposals if the offer is anywhere less than about 25% (even when the sum is quite considerable).

Other primates may also exhibit what we might think of as a sense of justice. In an intriguing study entitled 'Monkeys reject unequal pay,' Sarah Brosnan and Frans de Waal investigated what happens when capuchin monkeys previously trained to exchange a pebble for a piece of cucumber started to see others being rewarded with a more favored food (a grape) – the monkeys tended to go on strike, refusing to exchange a pebble for cucumber even though the alternative was no reward at all.

Perhaps punishment can play a role in maintaining fair play and hence cooperation? Some may not see it as cooperation at all, bordering more on enforced slavery than on acts of 'kindness.' Nevertheless, when we see apparent examples of altruism we need to ask whether the threat of punishment is helping to maintain it. In a series of staged repeated games among human volunteers, Ernst Fehr and Simon Gächter showed that students are prepared to take on costs in order to punish those who had earlier shirked their opportunity to contribute to a public good. According to Fehr and Gächter, this altruistic punishment is simply a consequence of a 'negative' emotional reaction to the sight of somebody free riding (although this proximate negative reaction may ultimately have evolved for other reasons). As one might expect, those that were punished for not contributing learned their lesson and cooperated more in subsequent rounds.

Moreover, those games that prevented altruistic punishment altogether saw a marked reduction in the mean amount of cooperation over time. The behavioral tendency to cooperate for the public good yet punish noncooperators has been called 'strong reciprocity.'

Strong reciprocity may explain many examples of human-based cooperation, but it is difficult to understand how altruistic punishment might evolve as a consequence of natural selection. After all, if altruistic punishment is costly, then an individual that free-rides and lets others do the policing would tend to leave more offspring. The temptation to sit back and let others punish defectors has been termed a 'second-order defection' or 'twofold tragedy.' So, if strong reciprocity can explain cooperation, perhaps it has only replaced the problem with another one further down the line – why should you be the one to punish? Kin selection may provide one potential solution to this question, but note that kinship can reduce the underlying incentives to defect in the first place. For example, in many eusocial Hymenoptera, worker-laid eggs are killed by the queen and other workers. In a comparative analysis, Tom Wenseleers and Francis Ratnieks found that fewer workers reproduced when the effectiveness of policing worker-laid eggs was higher, indicating that these sanctions were an effective deterrent. However, higher relatedness among colony workers led to less policing, not more, a result which is consistent with the view that less policing is needed when workers are highly related. So, self-restraint based on kin selection can achieve for free what expensive policing could bring about.

Escaping from Prison

All adaptive explanations of altruism involve taking the 'altruism out of altruism,' either by showing how the actions can benefit other individuals carrying the same traits, or by showing how the nature of the interaction is such that it is in the ultimate interests of the altruist to cooperate. However, this commonality should not be taken to mean that all cooperation can be related back to the two-player Prisoner's Dilemma (or *n*-player version of it which can give rise to the 'tragedy of the commons'). One example of cooperation which is almost certainly not represented by a Prisoner's Dilemma comes from recent work on a species of fiddler crab on the northern coastlines of Australia, where males aggressively defend their burrows from other wandering males (intruders). Patricia Backwell and Michael Jennions found that male fiddler crabs may sometimes leave their own territories to help neighbors defend their territories against these floating intruders. Why be a good Samaritan? It turns out that reciprocity cannot explain it because the ally that came to the neighbor's assistance was always bigger than the neighbor itself. Here, it may directly benefit a resident to help its neighbor to defend a territory, so that

it can avoid having to renegotiate the boundaries with a new and potentially stronger individual. In this way, there is no temptation to cheat – large allies are helping themselves, and it is only incidental that helping the neighbor keep its territory is part and parcel of maintaining status quo.

Our interpretation of cooperation gets tested further when we observe that some individuals may actually pay a cost to acquire or enhance the by-product benefits produced by another (a phenomenon known as 'pseudoreciprocity'). Many lycaenid caterpillars, for example, produce sugary secretions that are consumed by ants and in turn the ants protect these individuals from predation. The sugar may be viewed as an investment, yet the protection arises as a consequence of general territorial ant defense. Likewise, many flowers attract pollinators using nectar. One might wonder why flowers do not save themselves the trouble and produce less nectar. Flowers are essentially in a 'biological market,' however, governed by simple laws of supply and demand, such that any flower that offers less than conspecifics may experience reduced pollination. So, while kin selection may be at the heart of much intraspecific cooperation, sometimes cooperation can be maintained by a complex interplay of several types of interaction including direct self-interest, reciprocity, reputation, partner choice, and the threat of punishment.

Acknowledgment

This study is based in part on a longer review in Sherratt and Wilkinson (2009) which provides full references to many of the studies described here. We thank our editor Joan Herbers for her constructive comments and helpful advice.

See also: Kin Selection and Relatedness.

Further Reading

Axelrod R (1984) *The Evolution of Cooperation*. London: Basic Books.
Axelrod R and Hamilton WD (1981) The evolution of cooperation. *Science* 211: 1390–1391.
Connor RC (1995) Altruism among non-relatives-alternatives to the Prisoner's Dilemma. *Trends in Ecology and Evolution* 10: 84–86.
Doebeli M and Hauert C (2005) Models of cooperation based on the Prisoner's Dilemma and the Snowdrift game. *Ecology Letters* 8: 748–766.
Dugatkin LA (1997) *Cooperation Among Animals: An Evolutionary Perspective*. Oxford: Oxford University Press.
Fehr E and Gächter S (2002) Altruistic punishment in humans. *Nature* 415: 137–140.
Hamilton WD (1964) Genetical evolution of social behaviour I. *Journal of Theoretical Biology* 7: 1–16.
Krause J and Ruxton GD (2002) *Living in Groups*. Oxford: Oxford University Press.
Noë R and Hammerstein P (1995) Biological markets. *Trends in Ecology and Evolution* 10: 336–339.
Nowak MA (2006) Five rules for the evolution of cooperation. *Science* 314: 1560–1563.

Nowak MA and Sigmund K (2005) Evolution of indirect reciprocity. *Nature* 437: 1291–1298.

Sherratt TN and Wilkinson DM (2009) *Big Questions in Ecology and Evolution*. Oxford: Oxford University Press.

Trivers RL (1971) Evolution of reciprocal altruism. *Quarterly Review of Biology* 46: 35–57.

West SA, Griffin AS, and Gardner A (2007) Social semantics: Altruism, cooperation, mutualism, strong reciprocity and group selection. *Journal of Evolutionary Biology* 20: 415–432.

West SA, Griffin AS, Gardner A, and Diggle SP (2006) Social evolution theory for micro-organisms. *Nature Reviews Microbiology* 4: 597–607.

Cost–Benefit Analysis

R. R. Ha, University of Washington, Seattle, WA, USA

Introduction

Cost–benefit analysis as applied to animal behavior predicts that if a behavior is adaptive, the benefits of a behavior must exceed the costs of that behavior. Fundamentally, the benefits and costs relate to fitness, though currencies such as time and energy are often used as proxies of fitness. In most cases, the benefits are measured in terms of increased energy intake, survival, and reproduction. The costs are typically related to reductions in energy and time available for alternative behaviors (**Figure 1**).

The cost–benefit approach has been extended to model the optimization of the benefit-to-cost ratio, and this extension states that an individual should maximize the benefit of the behavior while simultaneously minimizing any costs associated with the behavior. In other words, the benefit of any particular behavior should be traded off against the costs associated with the behavior. If the costs are greater than the benefits, then natural selection would not favor the behavior, and if some individuals in the population were better at maximizing the benefit relative to the cost, then they would leave more copies of their genes to future generations compared to individuals with marginal benefit-to-cost ratios. This does not require that the animals consciously evaluate costs and benefits, but that animals that behave in ways to maximize benefits relative to costs are more successful in terms of fitness. Many theorists have tested whether animals actually use a simple rule of thumb to make 'optimal' decisions.

While the ultimate currency is genes in future generations, many optimal decisions can be analyzed based on short-term impacts on energy and time. Cost–benefit analysis and the optimality model have been applied to diverse topics in behavior such as foraging, parental investment, sibling rivalry, dispersal, and the evolution of cooperation. They have also been applied to a wide range of taxa, including insects, crustaceans, fish, amphibians, reptiles, birds, and mammals. Optimality models make assumptions about the currency of relevance to the behavioral choice, and constraints on behavioral decisions. Because specific predictions can be made from these models on the basis of maximizing the net benefit, this technique has widespread application in the adaptive study of behavior (**Figure 2**).

History

The fields of behavioral ecology and animal behavior were revolutionized by the development of theories of optimization beginning in the late 1960s and 1970s, and the field of behavioral ecology is particularly reliant on an economic approach. Niko Tinbergen was the first animal behaviorist to illustrate the value of analyzing behavioral decisions based on tradeoffs between benefits and costs. He applied this concept to the removal of broken eggshells from the nest by black-headed gulls (*Larus ridibundus*, Tinbergen, 1953). He suggested that this behavior benefits the parents by reducing the risk of predation of the chick due to the conspicuous egg shell.

In the 1970s and 1980s, the application of this approach to foraging, in particular 'optimal foraging theory' (OFT), exploded. Somewhat as a reaction to this work, there were heated debates about the value of this approach. Critics of optimality theory argued vehemently that it was not reasonable to think that animals should behave in an optimal fashion. Supporters of optimization theory countered that the point was not that all animals make optimal decisions all of the time, but that consideration of the tradeoffs between costs and benefits may help us to understand the ultimate causation of behavior.

The debate, while heated, was likely beneficial in pushing behavioral ecologists and animal behaviorists toward our current use of dynamic optimization models and in addressing other limitations of simple optimization models. I discuss these issues in more depth under *Limitations*.

Assumptions

The goal of cost–benefit and optimality models is to measure the impact of behavioral decisions on fitness. Since fitness is difficult to determine in short-term studies, it is typical of optimality models to measure proxies of ultimate reproductive success or fitness. In order to accomplish this, researchers must make assumptions about the immediate currency of relevance to long-term fitness and about the constraints on behavioral decisions. Thus, it is typically necessary to understand something about the natural history of the animal in order to arrive at likely parameters for adaptive decisions. Here I review the common currencies and constraints used in predicting adaptive decision-making.

Currency

Schoener developed optimal foraging models that commonly use the long-term rate of energy (food) intake as a proxy for fitness, referred to as the rate maximization hypothesis. Alternatively, foraging researchers might assume that animals attempt to minimize the time required to find food (the time minimization hypothesis of Mangel and Clark). An alternative approach is to consider the time between food patches given that foragers must determine when to leave a depleted patch in search of a dense patch (Charnov's Marginal Value Theorem).

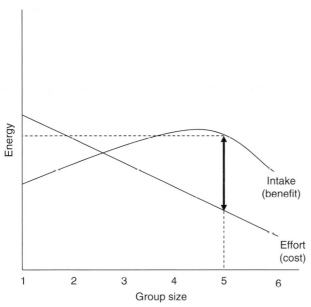

Figure 1 A simple cost–benefit analysis of group size. The energy costs (effort) versus the energy benefits (intake) for a hypothetical species at varying group sizes. The maximum difference between benefit and cost, where the benefit is greatest is at a group size of 5.

In modeling conflict situations, the costs associated with the behavior of competing individuals includes the likelihood of injury, the energy expended by the conflict, and increased risk of exposure to predation. The benefits of engaging in conflict includes access to resources such as mates, food, and territory, and a common approach is to model and measure these costs and benefits.

Further examples of common currencies used in these behavioral analyses include inbreeding avoidance, maximization of genetic benefits to offspring ('good genes,' or increased variation of the Major Histocompatibility Complex genes), maximization of body condition prior to migration or nesting, and avoidance of parasites.

Constraints

Making an optimal decision is frequently constrained by environmental factors, as well as morphological or physiological characteristics of the species. In the OFT literature, examples such as seasonal food fluctuations, competition, and food patchiness are commonly described. Likewise, animals may face constraints on optimal decision making based on cognitive (memory) or sensory limitations, incompatibility between two behaviors (foraging and vigilance), carrying load, and nutrient constraints. These constraints must be considered in realistic models and analyses of cost–benefit tradeoffs in order to produce valid results from the approach.

Limitations

Cost–benefit or optimality models are limited by the fact that the currencies and constraints invoked by the researcher may not match the animal's actual currency

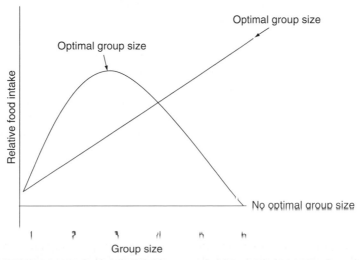

Figure 2 Three models for optimal group size. The relative energy intake (given costs and benefits) at varying group sizes for three hypothetical species. For species A (black line), the optimal group size is near 3. Species B (blue) shows that there is selection to favor higher group sizes (at least up to 6), while there is no effect of group size on species C (red).

and constraints. For example, foraging studies typically assume that animals attempt to maximize calories rather than a specific nutrient, but there are examples in which animals appear to be maximizing nutrients rather than calories (e.g., Belovsky's study of salt requirements in moose). In addition, it is likely that animals make dynamic decisions (decisions that change on a moment-to-moment basis) based on their current state, risk of predation, and ultimate fitness, but these models typically do not factor in all of these variables in a dynamic fashion. Thus, the models may not accurately predict the behavior of individual animals. For example, while conflict between competitors typically arises if the costs of fighting are less than the benefits of some desired resource, some studies have shown that conflict can arise in the absence of resources. This suggests that a future value (benefit) is overlooked in simple assessments of cost–benefit analysis; that is, the benefit of winning a current conflict is an influence on future conflict outcomes. Given competition between animals, it is clear that an individual animal may be prevented by other individuals from making an adaptive choice.

Another example of past–future influence is that of learning. Learning could be a factor in finding an appropriate breeding site, efficient foraging, and many other behavioral decisions. If learning influences decision-making, then clearly not all decisions can be immediately optimal. Many critics of optimality theory would argue that it is not reasonable to think that animals should behave in an optimal fashion. After all, they do not have all possible forms of information about the choice, nor is it likely that natural selection could create optimal decisions given the underlying genetic variability available upon which selection can act. In addition, it is likely that even if we can demonstrate optimality in some decisions, the lack of evidence may suggest that natural selection has not yet optimized the behavior, or has not kept up with a changing environment. Supporters of optimization theory would counter that the point is not that all animals make optimal decisions all of the time, but that a consideration of the tradeoffs between costs and benefits may help us to understand behavior.

Classic Examples

One of the classic examples of cost–benefit analyses is that of Zach's study of Northwestern crows (*Corvus caurinus*). These crows forage along the intertidal zone, and one high-value prey item are whelks, which they drop from the air onto rocks below. This serves to open the shell and expose the meat for consumption. By considering the calories available in whelks of different sizes, and the energy required to drop them at various heights, Zach was able to show that only the largest whelks were worth

the energy required to open them. In addition, he found that crows minimize the height at which they need to fly up and drop the whelk to break it open on the rocks. The rate-maximizing hypothesis would predict a higher flight than the crows actually average (5.2 m), but the currency in this example was minimizing the cost of the flight. Note that this minimum might also benefit the bird to find the broken whelk on the beach below.

Tinbergen's classic study of gulls discarding egg shells showed an interesting tradeoff between reducing overall predation and increasing cannibalism by neighboring gulls. He found that parents delay leaving the nest to discard of the shell when the chick is newly hatched, because of the risk of conspecific predation of chicks by neighboring gulls. This risk is significantly reduced once the chick's down has dried out and the chick is thus more difficult to swallow! Thus, parents discard the shell to reduce the overall risk of predation to their chick, but they wait ~1 h before they do so, which reduces the immediate risk of their neighbor eating their chick while they are off the nest.

Recent Examples

More recent work on optimality theory includes work on determining the neural mechanisms of cost–benefit analyses. For example, work by Gillette and colleagues on a predatory marine snail (*Pleurobranchaea californica*) showed a neural basis for the regulation of the tradeoffs of foraging and avoiding predators. Another example includes Hock and Huber's model evaluating winner and loser effects and their impact on dominance hierarchies.

Evolutionary Stable Strategies and Game Theory

The previous example touches on evolutionary stable strategies (ESS), and the tactic that would be adaptive for an individual to assume given the tactics of other individuals in the population. Hock and Huber suggest that securing future resources via a reduction of costs is more important for subordinates that currently lack those resources. This example is a case where the behavior of subordinates is dependent on the behavior of dominant individuals, and not just a simple immediate cost–benefit tradeoff.

A common definition of an ESS is that it is a strategy that, if adopted by all the members of a population, cannot be invaded by an alternative strategy. This is really just an extension of optimality theory in which there is the added component of frequency dependence. This extension of optimality theory suggests that the benefits and the costs to the individual are dependent on the strategy that other individuals in the population use. For example, if all the males in the population play the role of territorial nesting male, these conditions might favor a new strategy of a sneaker (nonpaternal investing) male. In this situation, even if the 'sneaker' male had fewer fertilizations than 'territorial' males, they would also get those fertilizations at a reduced cost of paternal care. More in-depth discussion of ESS and Game theory is given elsewhere in this volume.

Conclusion

Despite some controversy on the value of cost–benefit analysis in understanding animal behavior, analysis of individual decisions based on currencies such as energy and time has proved useful to behavioral biologists. Recent work in this area has included dynamic modeling and the neural basis for cost–benefit decision-making. It is likely that modern work will incorporate more sophisticated analyses of minute-to-minute decisions, as well as the impact of past and future consequences. These modern advances in the sophistication of cost–benefit analysis should continue to prove productive to our understanding of animal behavior.

See also: Caching; Digestion and Foraging; Foraging Modes; Game Theory; Habitat Selection; Hormones and Breeding Strategies, Sex Reversal, Brood Parasites, Parthenogenesis; Hunger and Satiety; Internal Energy Storage; Kleptoparasitism and Cannibalism; Optimal Foraging and Plant–Pollinator Co-Evolution; Optimal Foraging Theory: Introduction; Patch Exploitation; Wintering Strategies: Moult and Behavior.

Further Reading

Belovsky GE (1978) Diet optimization in a generalist herbivore: The moose. *Theoretical Population Biology* 14: 105–134.

Charnov EL (1976) Optimal foraging: The marginal value theorem. *Theoretical Population Biology* 9: 129–136.

Gillette R, Huang R-C, Hatcher N, and Moroz LL (2000) Cost-benefit analysis potential in feeding behavior of a predatory snail by integration of hunger, taste, and pain. *Proceedings of the National Academy of Sciences of the United States of America* 97: 3583–3590.

Goss-Custard JD (1981) Feeding behavior of redshank, *Tringa totanus*, and optimal foraging theory. In: Kamil AC and Sargent TD (eds.) *Foraging Behavior: Ecological, Ethological, and Psychological Approaches*, pp. 231–258. New York, NY: Garland Press.

Hock K and Huber R (2009) Models of winner and loser effects: A cost-benefit analysis. *Behaviour* 146: 69–87.

Hsu YY, Earley RL, and Wolf LL (2006) Modulation of aggressive behavior by fighting experience: Mechanisms and contest outcomes. *Biological Review* 81: 33–74.

Krebs JR and Davis NB (1993) *An Introduction to Behavioural Ecology*, 3rd edn. Oxford: Blackwell Science.

Krebs JR and Davies NB (eds.) (1997) *Behavioural Ecology: An Evolutionary Approach*, 4th edn. Oxford: Blackwell.

Mangel M and Clark CW (1986) Towards a unified foraging theory. *Ecology* 67: 1127–1138.

Maynard Smith J (1978) Optimization theory in evolution. *Annual Review of Ecology and Systematics* 9: 31–56.

Schoener TW (1971) Theory of feeding strategies. *Annual Review of Ecology and Systematics* 2: 369–404.

Tinbergen N (1953) *The Herring Gulls World.* New Naturalist Series. London: Collins.

Zach R (1979) Shell dropping: Decision making and optimal foraging in Northwestern crows. *Behaviour* 68: 106–117.

Relevant Websites

http://www.animalbehavior.org/ – Animal Behavior Society Web Site.
http://www.behavecol.com/pages/society/welcome.html – International Society for Behavioral Ecology.

Costs of Learning

E. C. Snell-Rood, Indiana University, Bloomington, IN, USA

Learning allows an individual to cope with environmental variation in both time and space. By learning, an individual can adjust its behavior to local conditions, and thus, across a range of environments, experience high performance, such as high energy gain, selection of quality mates, or avoidance of local predators. Given these benefits and the preponderance of environmental variation, we might expect learning to be under strong positive selection. Yet, learning abilities are not equally distributed across taxa, implying that the selection for learning varies among animals.

A consideration of the costs of learning may yield insights into why learning abilities vary so widely. These costs stem fundamentally from learning as a trial-and-error process. An increase in learning ability should correspond to an increase in the use of environmental 'information,' defined as data on the performance of a particular phenotype under certain environmental conditions (or with respect to specific resources). Any increase in the amount of information increases the chances that an individual will ultimately learn the behavior with the highest local performance. Thus, learning genotypes, relative to fixed or specialized genotypes, should (following learning) have high performance in a range of possible environments, but the costs associated with information may result in all genotypes having comparable fitness.

Phenotypic plasticity, the ability of a genotype to adaptively adjust its phenotype to local conditions, occurs not only in behavior (through learning) but also in morphology and physiology. The trial-and-error nature of learning sets many costs of learning apart from the costs of plasticity in other traits (e.g., morphology). For many other types of plasticity, information on the performance of alternative phenotypes accumulates over evolutionary time, and not developmental time, making the costs of learning much greater at the individual level.

This article reviews the importance of information as a cost in the evolution of learning. It (1) details how the acquisition, processing, and storage of information are costly at both the behavioral and neural levels; (2) outlines how these costs should result in various life-history trade-offs with learning; (3) reviews mechanisms by which the costs of information may be offset or reduced; (4) distinguishes global and induced costs; and (5) summarizes exciting areas of current and future research in the costs of learning.

Researchers have classified the costs of learning (and of plasticity in general) in many ways. A recent and useful classification system comes from Mery and Kawecki, who classify learning costs as 'operating' and 'production' costs.

Operating costs are the costs of the learning process itself, while production costs stem from the development of traits necessary for learning. Costs may also be classified from the perspective of information in development. This perspective focuses on the costs of information acquisition, processing, and storage. The costs of acquisition are primarily at the behavioral level, while the costs of processing and storage are mainly at the neural level.

Costs of Information Acquisition

The costs of information acquisition are manifest mostly at the behavioral level. The need to acquire information is often referred to as the 'cost of naïveté,' or the cost of being naive. A trial-and-error process means that an individual will necessarily make mistakes early in development, and the cost of being naïve stems from this suboptimal performance. These costs are measured relative to a specialist that could immediately perform well under local conditions. A good example occurs among bumblebee species that vary in the level of floral specialization. The cost of naïveté is revealed in the generalist species, which must take time and energy to learn to handle the flower on which the specialist can efficiently obtain nectar from the start.

The evidence for information acquisition costs comes from many systems. Generalist bird species are more likely to approach a wide range of novel objects than are related specialist species. Apparently, species that learn are more prone to actively explore their environment and such exploration is at least conceivably costly. There are various currencies in which the costs of exploration may be expressed. First, exploration takes time. Second, it requires energy. Exploration involves expression of a range of behaviors that may be metabolically expensive to perform. Third, exploration puts an individual at risk of mortality. They may be exposed to greater numbers or to a diversity of predators, and divided attention may make them less responsive to these predators.

The costs of information acquisition arise mainly at the behavioral level, but may also arise at the tissue level. Good learners may have to invest in extra tissue in sensory structures required to acquire relevant information. For instance, an increase in the number of photoreceptors may be required to acquire additional visual information. Increased sensitivity of sensory structures will, in turn, result in an increase in the amount of neural tissue that processes this information. For instance, the relative size of regions of the mammalian somatosensory cortex – which processes mechanosensory information – corresponds to the sensitivity of that tissue (e.g., hands vs. back).

Costs of Information Processing and Storage

Neural tissue is some of the most energetically expensive tissue in the body. The sodium–potassium pump alone is purported to account for ten percent of human total resting metabolism. Jerison's 1973 'principle of proper mass' originally suggested that an increase in cognitive or learning capacity would require an increase in neural investment. Theory about the design of neural networks suggests that at least part of this investment is likely to involve an increase in neuron number. This, in turn, suggests that the cost of learning at the level of information processing and storage may be reflected in the size of the brain or of brain regions involved in learning.

Hundreds of studies have tested for an association between cognition, including learning, and the size of the brain or brain regions. In particular, studies often focus on the association or 'learning' centers in the brain, such as the hippocampus in vertebrates and the mushroom bodies in invertebrates. The finding that cognitive ability correlates with brain region size generally assumes that size reflects neuron number. However, the size of brain regions may also vary because of synapse number, cell volume, or glial volume. Such variation may conceivably relate to learning ability – for instance, an increase in synapse density may increase the strength of an association – and may result in neural costs, but possibly to a lesser degree than neuron number.

Some of the best evidence for an association between learning and neural investment comes from studies of the insect mushroom body. Honeybees show an increase in mushroom body volume with age that coincides with learning the location of both the colony and foraging sites. Furthermore, generalist species of butterflies and beetles – for which learning is likely more important than specialists – show larger or more complex mushroom bodies. Recent work in *Pieris* butterflies suggests that families that are better able to learn to locate rare, red-colored hosts emerge with relatively larger mushroom bodies.

Studies in vertebrates also suggest a strong association between learning and neural investment. Several studies have found that species of birds that are more dependent on spatial learning for food caching have a relatively larger hippocampus. Some of these species have seasonal variation in hippocampus volume coincident with the need for spatial learning. There is also a correlation, at both the species and population level, between the relative volume of song-learning centers, and a bird's (learned) repertoire size. Finally, there is evidence that novel or innovative foraging behaviors – that are likely learned through trial-and-error learning – are more likely to arise in species of birds with relatively larger forebrains.

Neural investment comes with both production costs and operating costs. The development of larger neural investment is necessary for increases in both information processing and information storage. But the process of information processing and storage also comes with energetically expensive operating costs. For instance, forming long-term memories requires protein synthesis, which may explain why, in a recent study, Mery and Kawecki found that long-term memory formation may be costlier than short-term memory formation in *Drosophila*.

Information processing and storage is costly at the neural level because of a need for increased investment in and use of neural tissue. But it may also exact costs at the behavioral level, in particular, a necessity for sleep, which may cost time, energy, and reduced vigilance. Large amounts of information must be regularly consolidated such that only the most relevant information is actually stored in long-term memory. Sleep is thought to serve this function. Thus, there may be a correlation between learning and the costly need for sleep; some species (e.g., some aquatic mammals and birds) may mitigate this cost by sleeping with only one hemisphere at a time.

Direct Consequences of Learning Costs

The costs of information acquisition, processing, and storage may result in life-history trade-offs. The increased time and energy necessary early in life for sampling and increased neural investment may necessitate longer developmental periods and increased investment per offspring. Parental care may be necessary during learning periods as offspring suffer increased exposure to predators and reduced energy gain.

There is a good deal of comparative evidence linking neural investment, a proxy for learning ability, with life-history traits. Species of birds with relatively larger forebrains have relatively longer incubation, nestling, and fledgling periods, such that the overall parental investment per offspring is thus much higher in these species. There is even a suggestion in the anthropological literature that the evolution of human life history – in particular, delays in reproduction and high parental investment – has been driven by selection for learning and large brains.

The costs of learning may result in direct trade-offs with any number of traits. For instance, flies selected for increased learning ability show reduced larval competitive ability relative to control lines. Possibly, increased neural growth during this time may have created a trade-off with competitive ability. Finally, large neural investment may tradeoff with investment in other tissues, such as flight muscle mass in birds.

Global and Induced Costs of Learning

In studying the costs of learning, it is important to distinguish between global (constitutive) and induced (specific) costs of learning. Global costs are incurred by an individual regardless of the environment they experience. In contrast, induced costs are specific to an environment, and thus not necessarily suffered by every individual of a learning genotype. Global costs are more significant in the evolution of learning and are generally considered 'true costs of plasticity' because they can explain why good learners may fail to outcompete specialists. For example, a learning genotype that was innately biased to use environment A, but suffered only induced costs of learning in environment B, should outcompete specialists to environment A, and all organisms should evolve to be good learners. In contrast, the global costs of learning would be paid in both environment A and B, and both learners and specialists would persist.

There is abundant evidence that many of the costs of learning are induced or environment-specific. For instance, neural investment is generally very environment-dependent. The association between mushroom body volume and spatial learning has at least some experience-dependent component in insects and some vertebrates. Furthermore, it is well established that increased neural investment, and also learning ability, can be induced (in both vertebrates and invertebrates) by rearing individuals in more complex environments. In complex environments, learning should be more useful than in simpler, more predictable environments. Induced costs of learning can be found not only at the neural level but also at the behavioral level. Exploration of the environment often depends on the need for information.

Recent research in *Pieris* butterflies has suggested that naturally occurring variation in learning ability is correlated with variation in both global and induced costs. For example, families that are better able to learn to locate a rare (red-colored) host, emerge with larger mushroom bodies, regions of the brain involved in learning. However, specific learning experiences of individual butterflies also influence mushroom body volume: individuals that make more landings during learning and those with experience with the rare red host (relative to the common green host) have larger mushroom body volumes following learning. Thus, global costs may select against learning, but the evolution of innate biases and induced costs may facilitate the persistence of learning.

Offsetting the Costs of Learning

Animals should evolve mechanisms to minimize or offset the costs of learning. These mechanisms may increase the chance that learning will persist in a population, and may confound comparisons of learning costs between species. Changes in life history and development may offset learning costs. This is best illustrated by the idea that increased investment in learning may select for a longer life span. If the environment is constant for an individual's lifetime, a longer life span will allow an individual to reap the benefits of learning and offset the costs of learning. This prediction is supported by comparative evidence in birds. However, artificial selection experiments in *Drosophila* found that selection on learning ability results in a correlated decrease in the life span. These conflicting results may be reconciled if changes in learning first result in costs that translate into life-history trade-offs, but over time, species evolve mechanisms to reduce or offset these costs.

One of the simplest mechanisms to reduce the costs of learning is to reduce the need for learning across one's lifetime. By choosing familiar habitats and resources, individuals reduce the need to experience the costs of learning more than once. This may partly explain why habitat fidelity and the defense of familiar space are so common among animals.

The costs of learning may also be reduced through attention to particular types of information. Direct information, gained by individuals through direct interaction with the environment, is highly accurate, but quite expensive. In contrast, indirect information (an easily detected cue, such as photoperiod, that is correlated with an environmental state) and social information (information gained by observing an experienced individual) is much cheaper, but sometimes can be inaccurate, for instance, if an individual copies a conspecific while it is making a mistake. The low energetic and time cost of social information may explain why animals commonly attend to the learned behaviors or choices of conspecifics, even if conspecifics sometimes model suboptimal behavior.

The formation of innate biases appropriate to common environments may also offset the costs of learning. Organisms often encounter some environments and resources more commonly than others. The costs of learning can be significantly reduced by expressing innate behaviors that are favored under common conditions and allowing flexibility in these behaviors through learning under less common conditions. This idea is supported by recent work on *Pieris* butterflies, which show a strong innate tendency to search for common green hosts, but which can learn to locate very rare red hosts.

Questions for Future Research

Research on the costs of learning is an exciting and open field of study. Here are just a few of the areas that would benefit from further empirical and theoretical work.

1. While the costs of learning are often cited in reviews and models, and often detected through indirect measurements, we have very few instances where costs are directly quantified. One exception was the assessment of energetic requirements of mate sampling in pronghorns. It would be informative to measure the metabolic rates of individuals under different learning conditions, or from families with different learning abilities. Furthermore, while neural tissue is expected to be metabolically costly, it would be useful to quantify the differences in the energy budget of individuals that differ slightly in the size of brain regions involved in learning.

2. Many of the current associations between learning and neural investment result from gross measurements of neural investment, such as the mass or area of a brain region. Relatively little is known about whether this variation is underlain by variation in neuron number, synapse density, cell size, and/or glial density. Determination of neural mechanisms may have implications for just how costly neural investment really is: increases in glial density and cell size, for example, may be less costly than increases in neuron number.

3. It is well established that sleep is important for learning. But, we know little about the importance of sleep as a possible cost in the evolution of learning. Can reduced opportunities for sleep limit the evolution of learning in particular environments? For instance, might species that experience sleep deprivation in high predation environments suffer deficits in memory?

4. While there are many comparative observations that suggest that the costs of learning are linked to life-history trade-offs, there are few studies that make this link within species or suggest that this correlation is indeed causal. If life-history traits such as parental investment are manipulated, does learning ability respond in the predicted direction? Within species, to what extent are genetic or learned life-history traits associated with learning ability?

5. In general, little is known about the relative importance and prevalence of global versus induced costs of learning. For example, what proportion of interspecific variation in brain volume is constitutive versus developmental? Furthermore, little is known about the functional consequences of induced costs versus global costs: if an individual grows a large brain in a complex environment, does it result in an increase in learning ability as great as if it had emerged with such a large brain? Finally, to what extent do the induced costs of learning facilitate the maintenance of learning in conditions that might otherwise favor fixed behaviors?

6. Little is known about how the costs of learning select on development itself. Are these costs responsible for the evolution of innate biases and habitat selection, or are these mechanisms – which may reduce the costs of learning – a byproduct of selection on other traits?

In conclusion, by studying the costs of learning, the conditions under which learning may evolve can be more readily understood. For instance, if costs limit learning to species with certain life-history traits or those that live under particular predation pressures, an assessment of the learning costs may pave the way to understanding which species can learn innovative or novel behavior that allows survival in disturbed and changing environments. Furthermore, the costs of learning should apply to the development of any traits that develop through extensive interaction with the environment (e.g., acquired immunity). The study of the costs of learning is an exciting and active field in animal behavior, which may inform studies of conservation, phenotypic plasticity, and evolution.

See also: Cognitive Development in Chimpanzees; Development, Evolution and Behavior; Habitat Imprinting; Imitation: Cognitive Implications; Innovation in Animals; Memory, Learning, Hormones and Behavior; Optimal Foraging Theory: Introduction; Spatial Memory; Specialization.

Further Reading

Byers JA, Wiseman PA, Jones L, and Roffe TJ (2005) A large cost of female mate sampling in pronghorn. *American Naturalist* 166: 661–668.

Capaldi EA, Robinson GE, and Fahrbach SE (1999) Neuroethology of spatial learning: The birds and the bees. *Annual Review of Psychology* 50: 651–682.

Dall SRX, Giraldeau LA, Olsson O, McNamara JM, and Stephens DW (2005) Information and its use by animals in evolutionary ecology. *Trends in Ecology and Evolution* 20: 187–193.

DeWitt TJ, Sih A, and Wilson DS (1998) Costs and limits of phenotypic plasticity. *Trends in Ecology and Evolution* 13: 77–81.

Dukas R (1998) Evolutionary ecology of learning. In: Dukas R (ed.) *Cognitive Ecology: The Evolutionary Ecology of Information Processing and Decision Making*, pp. 129–174. Chicago, IL: University of Chicago Press.

Frank SA (1996) The design of natural and artificial adaptive systems. In: Rose MR and Lauder GV (eds.) *Adaptation*, pp. 451–505. New York: Academic Press.

Greenberg R (1983) The role of neophobia in determining the degree of foraging specialization in some migrant warblers. *American Naturalist* 122: 444–453.

Isler K and van Schaik C (2006) Costs of encephalization: The energy trade-off hypothesis tested on birds. *Journal of Human Evolution* 51: 228–243.

Iwaniuk AN and Nelson JE (2003) Developmental differences are correlated with relative brain size in birds: A comparative analysis. *Canadian Journal of Zoology* 81: 1913–1928.

Johnston TD (1982) Selective costs and benefits in the evolution of learning. *Advances in the Study of Behavior* 12: 65–106.

Laughlin SB, van Steveninck RRD, and Anderson JC (1998) The metabolic cost of neural information. *Nature Neuroscience* 1: 36–41.

Laverty TM and Plowright RC (1988) Flower handling by bumblebees: A comparison of specialists and generalists. *Animal Behaviour* 36: 733–740.

Lefebvre L and Sol D (2008) Brains, lifestyles and cognition: Are there general trends? *Brain Behavior and Evolution* 72: 135–144.

Mery F and Kawecki TJ (2005) A cost of long-term memory in *Drosophila. Science* 308: 1148.

Snell-Rood EC, Papaj DR, and Gronenberg W (2009) Brain size: A global or induced cost of learning? *Brain Behavior and Evolution* 73: 111–128.

Stamps J (1995) Motor learning and the value of familiar space. *American Naturalist* 146: 41–58.

van Praag H, Kempermann G, and Gage FH (2000) Neural consequences of environmental enrichment. *Nature Reviews Neuroscience* 1: 191–198.

Walker MP and Stickgold R (2006) Sleep, memory and plasticity. *Annual Review of Psychology* 57: 139–166.

Crabs and Their Visual World

J. Zeil and J. M. Hemmi, Australian National University, Canberra, ACT, Australia

Introduction

Crabs inhabit a large variety of habitats, from deep-sea vents, through the continental shelves, coral reefs, the intertidal zone, rain forests, and fresh water streams. For vision, each of these habitats offers different illumination conditions and different topographies of the visual world. We concentrate here on reviewing the visual–behavioral world of semiterrestrial crabs, because most of what we know about the uses of vision in crabs has been gathered about animals living in this particular niche. Crabs are found in all the different topographies of the intertidal zone: mudflats, sandy beaches, rocky shores, and mangrove forests. The visual problems that need to be solved and the behavioral repertoires of species inhabiting these various topographies are probably quite similar, as we will argue later, but the conditions for visual information processing and the cues offered by these different visual worlds are radically different. So what use is vision to crabs?

Orientation and Navigation

In one way or other, all semiterrestrial crabs are central place foragers, operating from places that offer protection. Even the Australasian soldier crabs (Mictyridae), which do not inhabit permanent burrows, return to the same slightly elevated areas of sand and mudflats to dig themselves in on the incoming tide. Therefore, one fundamental behavioral task that crabs face, and one that is greatly facilitated by vision, is to find their way to places of interest and back again to places of safety. Fiddler crabs, for instance, are known to use path integration to find their way back to their burrows. Path integration is a way of estimating the current location based on a previously known position and the distance and direction traveled since leaving that known position. This requires some form of compass information and a way of measuring distance traveled (odometry). Measuring the distance traveled can either be done by counting steps or by integrating optic flow. Fiddlers are likely to use leg-based odometry during path integration, because they underestimate the distance to the burrow when forced to run over slippery ground. Several studies indicate that some fiddler crab species, at least, can use the pattern of skylight polarization as a compass. It may be such an external compass reference, which allows foraging *Uca lactea annulipes* to counteract enforced turns. However, other species (*Uca pugilator*) appear to lack such an external reference. Other visual cues can serve the same purpose: the azimuthal position of the sun and the distant landmark panorama. The sandbubbler crab *Dotilla* appears to determine its straight foraging paths radiating from the burrow by relying on either the skylight polarization pattern or in its absence, the surrounding landmark panorama.

In fiddler crabs, information from the path integration system is used in unusual and interesting ways. During foraging excursions, it allows them to keep their longitudinal body axis aligned with the home vector, which plays an important role in enabling the crabs to monitor how close other crabs come to the invisible entrance of their burrow (see later). Path integration information helps male crabs of *Uca vomeris* to navigate back to the burrows of females they are currently courting and it makes crabs very good at detouring obstacles. As we will see later, crabs respond differently to bird predation, depending on how far away they are from their burrows, again information that is provided by the path integration system.

It is not clear at the moment to what degree crabs also use landmarks for navigation and homing. For rock crabs and crabs inhabiting mangroves, landmark guidance would seem the most useful navigational cue. Yet, we only know from the tree-climbing *Sesarma* that they are able to return to the same tree and on the tree to the same branches, day after day, indicating the use of visual memories. Ironically, the strongest evidence for the use of visual landmarks comes from species that inhabit relatively featureless sandy beaches and mudflats. Male ghost crabs, *Ocypode saratan*, advertise their mating burrows by sand pyramids they construct while excavating these burrows. Searching females, but also wandering males, are attracted by these pyramids and move from one to the next across the shore (**Figure 1**). Ghost crabs are guided by landmarks on their return to their burrows and male ghost crabs may be able to use their own pyramids as beacons when returning from foraging excursions, much like courting males in some fiddler crab species that build similar sand structures: their sand hoods serve as guideposts themselves, but also to attract mate-searching females. One prerequisite for the use of landmarks is the ability to recognize and remember visual patterns, and indeed, fiddler crabs do appear to discriminate between different visual shapes.

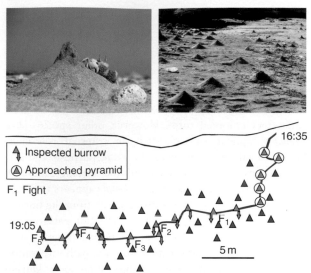

Figure 1 Sand pyramids as visual landmarks in ghost crabs (*Ocypode saratan*; Ocypodidae, Brachyura). Male ghost crabs excavate mating burrows and use the excavated sand to construct pyramids (top left photograph). These pyramids can reach high densities on the shore (top right photograph). The pyramids are approached by wandering males and females. The diagram below shows the path of a male through a stretch of beach and his responses to sand pyramids, associated burrows, and other males. Modified from Linsenmair E (1967) Konstruktion und Signalfunktion der Sandpyramide der Reiterkrabe *Ocypode saraten* Forsk (Decapoda, Brachyura, Ocypodidae). *Zeitschrift für Tierpsychologie* 24: 403–456. Photographs by Jochen Zeil.

Predator Avoidance

One of the reasons why it is so important for semiterrestrial crabs to be able to locate their burrow or rock crevices is the protection they offer against predators and, in the case of rock crabs, against breaking waves. Predator evasion is another critical, visually guided task all semiterrestrial crabs share, and its behavioral organization is likely to be similar across different habitats and species. However, in terms of visual information processing, the task differs widely, depending on the type of predator the crabs face and on the topography of their habitat. The simplest case is that of sand and mudflats. The geometry of these flat worlds offers some crucial, predictable, and invariant visual information that has shaped the way in which fiddler crabs, for instance, detect and identify danger from predators. In the flat, horizontal world of a mudflat, flying birds and everything that is larger than a crab itself will be seen in the dorsal visual field, above the visual horizon line. Most events seen by the dorsal eye, thus, signal potential danger. John Layne has shown this most convincingly in 1998 by monitoring inside an otherwise featureless cylinder crab responses to the appearance of a small object above or below the line of horizon. Crabs in this experimental situation respond with a startle running reflex to objects that appear above their

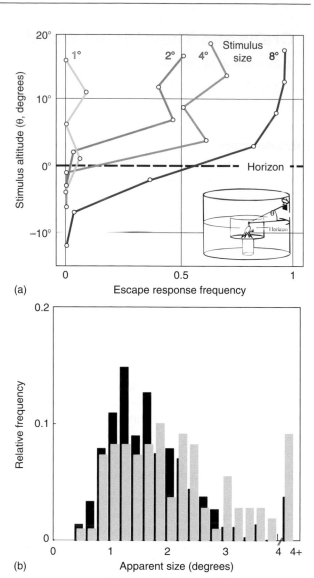

Figure 2 Predator avoidance in fiddler crabs. (a) Escape response frequencies of crabs confronted with horizontally moving black squares at different elevations (θ) in the visual field above and below the visual horizon. Redrawn from Layne JL, Land MF, and Zeil J (1997) Fiddler crabs use the visual horizon to distinguish predators from conspecifics: A review of the evidence. *Journal of the Marine Biological Association of the United Kingdom* 77: 43–54. (b) The apparent size of bird dummies at the moment fiddler crabs initiate their run toward the burrow in field experiments. Data replotted from Hemmi JM (2005) Predator avoidance in fiddler crabs: 2. The visual cues. *Animal Behaviour* 69: 615–625.

visual horizon, but rarely to objects moving below that line in the ventral visual field (**Figure 2(a)**). In their natural habitat, fiddler crabs respond to anything they see moving in their dorsal visual field by a distinct sequence of behaviors. On first detecting an object, they freeze. As the object comes closer, they run back to their burrow, where they stay at the entrance, before entering the burrow and disappearing out of sight, if the object continues to approach. They then stay a variable amount of time inside the burrow and upon surfacing, inspect the scene, before continuing with their activities. Each of these distinct stages of the response sequence is guided by visual cues that are increasingly correlated with risk. The crabs freeze and run home when the object barely subtends the visual field of one ommatidium (pixel) and has moved across at most a few ommatidia (**Figure 2(b)**). At this stage, then, a crab has little information on the size, the distance, or the direction of movement of that object. Close to the safety of its burrow, however, the crab can afford to collect more reliable information that is more directly related to risk, by letting the predator approach more closely. We do not know yet what visual cues the crabs are attending to, whether it is the apparent size of the predator, its elevation in the visual field, or maybe its time to contact. The main predators of crabs in these open mudflats are birds, like terns, which fly at a height of 2–4 m above the ground, scanning the mudflat for crabs without burrows. The birds have no chance to catch a crab that owns a burrow on the surface, despite the fact that the resolution of crab eyes is much worse than that of the hunting bird, because the crabs always detect the much larger bird when it is still 30 m or so away. However, in order to reach the safety of their burrow, the crabs respond to any fast movement in their dorsal visual field, whether it is a butterfly or a mangrove leave moving in the wind. The crabs, thus, face the problem of discriminating between relevant and irrelevant visual signals. They appear to solve it by learning. One of the striking features of fiddler crabs is their ability to habituate to repeated events that have proven harmless. Fiddler crabs are difficult to see because they respond very early to our approach and disappear below ground. However, if one sits still near a colony, the crabs quickly adjust. They reappear from their burrows and while they initially disappear again whenever they see the slightest movement, they quickly learn to ignore the novel feature in their environment. After a while, it even becomes difficult to chase them underground without actually approaching them. We believe that habituation is a fundamental part of the fiddler crab's antipredator strategy. By habituating, the animals reduce unnecessary costs by not repeatedly responding to harmless events. Habituation effects in the crab *Chasmagnathus* have been shown to last for more than a day. In addition, habituation is context specific and, therefore, an associative learning process. *Chasmagnathus* lives in a habitat rich of cord grass that constantly moves in the wind and, therefore, could be mistaken for potential danger. Long-term habituation is much weaker in the shore crab *Pachygrapsus*, a sympatric species that inhabits the more open areas of the same environment and, therefore, is not exposed to the same environmental motion noise.

Very little is known about predator avoidance in crabs living in mangrove forests and on rocky shores. It is likely that its organization is quite similar to that in fiddler crabs

and that rock and mangrove crabs also require a way of habituating to irrelevant image motion. However, because they live in a three-dimensionally complex environment, the visual topography of predation is much less well defined for these crabs. Depending on whether a crab sits on a thick or a thin branch, in the middle or at the edge of a boulder, or just emerges from a crack in a flat rock, danger is not only restricted to the dorsal visual field. Equally, a rock crab may need to habituate to some of the visual effects of wave motion and spray, but at the same time needs to be able to judge when a breaking wave is reaching its position on the rock. The rules of predator detection and of habituation to irrelevant image motion are, thus, bound to be different and more complicated in crabs inhabiting rocky shores, compared to mudflat dwellers.

Controlled laboratory studies on predator evasion in a number of different crab species (*Pachygrapsus*, *Heloecius*, and *Chasmagnathus*) have documented the typical animal response to approaching objects: running away in the opposite direction. The details of the visual cues triggering the onset of the escape response differ between experimental setup and species, but all studies clearly demonstrate that the crabs are sensitive to the direction of approach of an object. Under certain situations, crabs in nature respond just like those in the laboratory; they either run away from the dangerous stimulus, or confront it with their claws by striking a threat posture. For instance, crabs that have lost their burrow or refuge and find themselves in a part of the world without burrows and soldier crabs, which during their time of activity on the surface do not possess a burrow. However, in most cases, the safest option is not to run away from a predator, but to run toward a refuge. These observations clearly illustrate that predator avoidance behavior is context dependent. The choices made by animals do not simply depend on the visual signatures of a predator, but also on the behavioral and environmental context.

Foraging, Hunting, and Feeding

Both navigational and antipredator strategies differ depending on the crabs' mode of foraging. While many species such as most fiddler crabs and the sand-bubbler crabs *Scopimera* and *Dotilla* spend most of their topsoil-grazing life in the close vicinity of their burrows, other crabs like the fiddler crabs *U. pugilator* in America and *Uca signata* in Australia (personal observation) and, especially, soldier crabs (Mictyridae) and ghost crabs, go on long foraging excursions, covering tens to hundreds of meters. Often, these long distance foragers move in groups called 'herds' or 'droves,' which in the case of soldier crabs, often perform coordinated changes in direction and speed. As far as visual guidance is concerned, these herding crabs are in a different state compared to those operating close to their burrows. Their predator evasion strategy changes in two ways, because they have left the vicinity of their burrows. First, animals are now attracted to each other and coordinate their movements with their neighbors to form a 'selfish herd.' Second, when attacked, they run away from a predator and are attracted by objects that provide cover. Faced with persistent threat, soldier crabs will dig themselves into the ground, as do fiddler crabs when they find themselves on soft ground without a burrow (personal observation). The navigational guidance system also appears to change. Herding crabs are attracted to the water edge of tidal creeks or the open sea and appear to learn from their particular habitat in which direction to move from their burrow area high up in the intertidal zone to the feeding areas in the low intertidal. Crabs learn to use celestial compass information (both the sun and the pattern of skylight polarization) as a directional cue on these long-range foraging excursions. Herding crabs that are caught and tested in a circular arena with view to the sky or topped with a sheet of polarizer tend to move in directions that, in their local habitat, would bring them up the beach to their burrow area or down the beach to the water edge. There are some indications that they also incorporate the landmark panorama into this directional response. For pinpointing the burrow area of their colony or for locating their own burrow, they would need to be guided by close landmarks. Whether vision also plays a role in identifying good feeding sites, for instance, the moisture content of the soil or pools of water is unknown.

Compared to the topsoil-grazing fiddler and sand-bubbler crabs, ghost crabs are most versatile foragers. In addition to filter feeding, they scavenge along the deposit lines along the shore and are particularly attracted by vertebrate and invertebrate carcasses. The crabs wander quite some distance away from their burrows during scavenging, and when surprised by a predator, escape into the water, where they dig themselves in. Ghost crabs, however, also actively hunt. *Ocypode ceratophthalmus*, in Australia, for instance, go on extended foraging excursions, which often involve hunting, in particular for soldier crabs. In contrast to herding fiddler crabs which do not appear to return to their individual burrows, ghost crabs (*O. ceratophthalmus*) seem to be guided by landmarks around their burrows. Besides the long-range homing problem, the additional visual task these ghost crab hunters excel in is to detect, track, intercept, and catch moving prey. Fiddler crab males of the species *U. elegans* face a very similar task when attempting to lead a female to their burrow (see later). Males of this species wait near their burrow entrance for wandering females. Once they detect a female, they leave their burrow and run toward her. The males maneuver themselves into a position on the far side of the female and then guide the female back to their burrow, a seemingly challenging task that, however, can be performed with a very simple control system.

One interesting, but unresolved question in this context is how much 'hand-eye' coordination the crabs need in order to catch moving prey, to manipulate food, to fight, or to interact with other crabs. Quite generally, crabs are very dexterous invertebrates. They can use their claws as shovels, scissors, forceps, clamps, pliers, crutches, and both visual and acoustic signaling devices during feeding, grooming, cleaning, climbing, courtship, and fighting. To our knowledge, nothing is known about the visual control of this dexterity, although claw movements must take up a large part of the frontal visual field.

Territorial and Social Interactions

Apart from knowing their environment, being able to locate their refuge, to navigate between feeding places and their burrows, to hunt for prey and while doing all this to avoid predators, what else does a crab need to know to be a 'good crab'? Most semiterrestrial crabs live in dense populations, most conspicuously so the inhabitants of tropical and subtropical mudflats. So crabs need to be socially competent. We arguably know most about the social interactions, colony structure, and mating systems in fiddler crabs (Ocypodidae), partly because they are particularly easy to observe, compared, for instance, to rock or mangrove crabs (Grapsidae, Sesarmidae), and partly because their societies have a distinct spatial structure with highly conspicuous interactions, compared, for instance, with soldier crabs (Mictyridae).

The condition for a fiddler crab male to engage in social interactions and optimize his mating success is the possession of a burrow. Crabs that have lost their burrow wander across the mudflat and approach crabs on the surface. Resident crabs will retreat to their burrows on becoming aware of an approaching wanderer, who then will either contest the ownership of the burrow, or will start path integrating to its position when being chased away. A wanderer will then use this occupied burrow as a temporary refuge in case of danger, while continuing to search for an accessible burrow. At least two aspects of such an interaction are guided by vision: (1) wanderers are visually attracted by other crabs; (2) burrow owners are able to judge how close such a wanderer has come to its (invisible) burrow, a task that requires visual judgment of distance to the wanderer and information from the path integration system on the direction and distance of the burrow. This is no mean feat: depending on their size and eye height, fiddler crabs cannot see their burrow from more than about 10–20 cm away, because of perspective foreshortening and surface irregularities.

Once in possession of a burrow, the local neighborhood plays an all-important role for a crab. In one of the two fiddler crab mating systems, male fiddler crabs try to attract females to their burrow and repel other males. To do so, they first need to be able to distinguish them.

For *U. perplexa*, this seems to be possible at a distance of about 30 cm and for *U. pugilator* at about 10–15 cm. The presence or the absence of the male's large claw is likely to be important for this discrimination, but other cues such as the way the animals move might also play a role. Interestingly, females in this mating system have the status of crabs that have lost their burrows. They are vulnerable to predation and are attracted to burrow owners, a fact that must have driven the evolution of male signaling. In the other mating system, where males have to visit females at their burrow entrance to mate at the surface, the closer the resident females are, the less risky it is for males visiting them for mating. Indeed, *U. vomeris* males do know where the burrow of a female is whom they recently courted and can navigate back to that location, even when the female is not on the surface or when the burrow is covered by a sheet

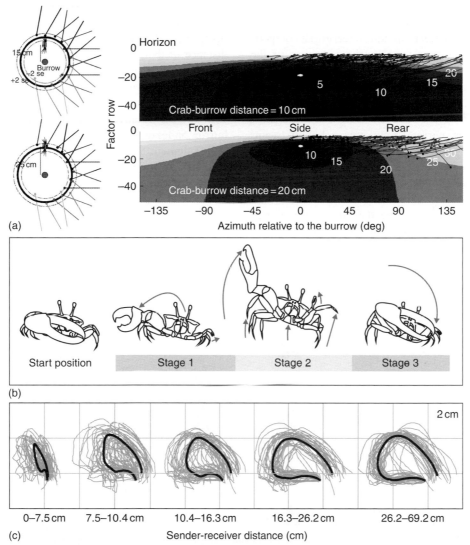

Figure 3 Distance judgments by fiddler crabs. (a) Burrow surveillance: foraging fiddler crabs respond to other crabs approaching their burrow in a burrow-centered fashion, whenever the other crab has approached a certain distance to the burrow, independent of their own distance to the intruder. Diagrams on the left show directions of dummy crab approaches (solid lines) and the position at the time of crab responses (large dots) for two different crab-burrow distances, as fitted by a statistical model. Diagrams on the right show how burrow-intruder relationships are seen in the visual field of a crab. Differently colored areas mark the visual field projections of circles of different radius around the burrow. Modified from Hemmi JM and Zeil J (2003) Burrow surveillance in fiddler crabs. I. Description of behaviour. *Journal of Experimental Biology* 206: 3935–3950; Hemmi JM and Zeil J (2003) Burrow surveillance in fiddler crabs. II. The sensory cues. *Journal of Experimental Biology* 206: 3951–3961. (b) The claw waving displays of male *U. perplexa*. (c) Males modify their display depending on the distance of receiver females (*x*-axis). The graph shows traces of claw tip paths (gray lines), normalized to claw size, and the mean shape of the display (black lines). When a female is far away, the males move their claws in a circular fashion that changes systematically as the female comes closer to a wave that contains mainly vertical movement components. Modified from How MJ, Hemmi JM, Zeil J, and Peters R (2008) Claw waving display changes with receiver distance in fiddler crabs, *Uca perplexa*. *Animal Behaviour* 75: 1015–1022.

of sandpaper. Male *U. capricornis* actually recognize their closest female visually by the female's individually distinct color pattern on the posterior carapace (**Figure 5**). Also, competition between neighboring males and females can be intense, so knowing your neighbors and the outcome of previous interactions are likely to be very important for a resident crab. In fact so much so, that some male crabs defend their male neighbors against larger intruders. In all these cases, crabs need to be able to remember the bearing and distance of neighbor burrows with respect to an external compass bearing that is in an allo- or geocentric reference system.

In both mating interactions and in territorial competition between males and females, the judgment of the absolute size of the other crab appears to be important. A number of species, for instance, mate in a size-assortative manner, females being more likely to mate with males of similar size. At what distance the crabs can make these discriminations is currently unknown, but it seems increasingly likely that they use the retinal elevation at which they see another crab as a distance cue to disambiguate apparent size and distance. The problem being that a small crab nearby can have the same apparent size as a large crab further away. In the flat world of mudflats, however, the further away something is on the substratum, the higher up in the visual field and closer to the horizon it appears. In mudflats, retinal elevation, thus, becomes a potent, panoramic, and monocular cue to distance, which has provided one of the selective pressures for the evolution of long vertical eyestalks in crab species inhabiting mud- and sandflats (see later).

The ability of fiddler crabs to assess where other crabs are is particularly striking in the case of burrow surveillance. As mentioned earlier, depending on their eye height, foraging burrow owners cannot see their burrow from more than 10–15 cm away. Yet, they are very responsive to other crabs coming too close to their (invisible) burrow and rush back to defend it. They can monitor this relationship between their burrow and an intruder, because (1) they have information on the direction and distance of their burrow from path integration; (2) they keep pointing with their longitudinal body axis in the direction of the burrow; (3) they are, thus, able to predict the retinal azimuth and elevation of the burrow, together with (4) the relative position of an intruder and the burrow, independent of the intruder's approach direction. This judgment heavily depends on the predictable visual geometry of a flat world, in particular, the relationship between distance on the surface and elevation in the visual field (**Figure 3(a)**).

In some fiddler crab species, absolute distance judgments are important in courtship: male *U. perplexa* modify their claw-waving display depending on the distance of a female receiver. When the female is far away, males

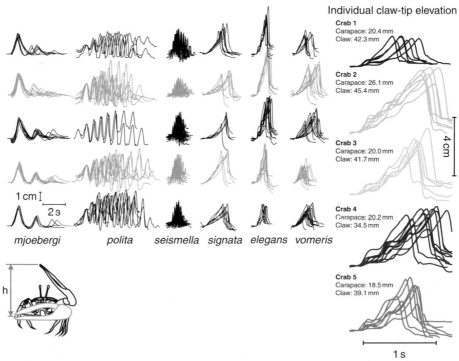

Figure 4 Claw-waving displays by fiddler crabs. (a) The vertical component of the male claw-waving displays in six Australian fiddler crab species. For each species, the displays of five individuals are shown. (b) Intra- and interindividual variation of the claw-waving display in *U. signata* in more detail. Modified from How MJ, Zeil J, and Hemmi JM (2009) Variability of a dynamic visual signal: The fiddler crab claw-waving display. *Journal of Comparative Physiology A* 195: 55–67.

engage a broadcast wave with large circular movements of the claw and, therefore, a large 'active space.' As the female comes closer, the male changes to simply moving the claw up and down, a waving pattern that reduces the horizontal extent of the wave and the apparent size of the wave envelope as seen by the female (**Figure 3(b)**). This distance-dependent tuning of signals most likely reflects different functions: first, a male needs to attract the attention of a female and needs a signal that is conspicuous from far away. Once the female gets closer, the information content of the male's signal increases beyond 'here is a male crab of species X,' to more subtle and robust indicators of male fitness, including his size, claw color, wave timing, and possibly seismic activity. Fiddler crabs show great diversity in claw-waving signals across the 97 recognized species. The signal structure is not just species specific, but shows intraspecific variation according to individual identity, geographic location, and fine-scale behavioral context (**Figure 4**). In some species, the males even synchronize their waving activity.

Semiterrestrial crabs, in general, employ an astonishing variety of species-specific signals, ranging from claw stridulation in ghost crabs, claw rapping in fiddler crabs, to static visual signals, like the sand pyramids of ghost crabs, hoods and towers in fiddler crabs, and dynamic signals like the threat postures in grapsid, ocypodid, mictyrid, and sesarmid crabs. Sexual dimorphism in claw size, shape, handedness, and color, is prevalent. In its extreme form, with the hugely enlarged claw of fiddler crab males, this sexual dimorphism goes hand in hand with species-specific color patterns on the main claw and species-specific movement-based signaling choreographies. Male claw color is used by *U. myobergi* females for species identification (at least against the sympatric *U. signata*) and male *U. capricornis* recognize resident female neighbors by their individually distinct carapace color patterns (**Figure 5**). Claw color in *Heloecius* varies systematically with size and sex and may, therefore, also contain socially significant information. At this stage, it remains unclear how these species-specific color patterns, especially on the enlarged claw of male fiddler crabs, interact as a signal with the specific claw-waving choreography and with polarization reflections that are known to be produced by the shiny and wet cuticle of crabs. Interestingly, fiddler crab carapace color patterns, but not claw colors (Zeil, unpublished results), change not only under the influence of endogenous rhythms but also on a much shorter time frame of minutes, under the influence of handling stress and predation threat.

Visual Systems

Semiterrestrial crabs have apposition compound eyes that are made up of many thousand ommatidia, each acting basically as one, albeit very sophisticated, pixel. In each

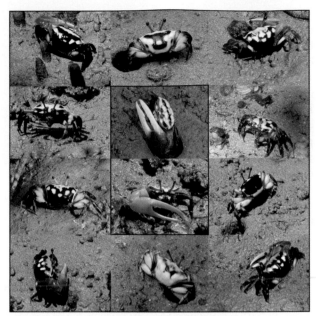

Figure 5 Claw and carapace colors in fiddler crabs. The hugely enlarged claw of males are colored in a species-specific pattern. The two images in the center of the panel show an *U. signata* male with a red and white claw on top and an *U. mjoebergi* male with a uniform yellow claw below. *U. mjoebergi* females distinguish between males of *U. signata* and their own males based on these claw color patterns. Certain species also have individually distinct color patterns on the posterior carapace and on the merus of their logo. This can be seen in the peripheral images of the panel that all show individuals of *U. capricornis*. Male *U. capricornis* recognize neighboring females by these color patterns. Modified from Detto T, Backwell PRY, Hemmi JM, and Zeil J (2006). Visually mediated species and neighbour recognition in fiddler crabs (*Uca mjoebergi* and *Uca capricornis*). *Proceedings of the Royal Society of London B* 273: 1661–1666. Photographs courtesy of Tanya Detto.

ommatidium, light is focused by an individual facet lens and a crystalline cone on the distal end of a thin, long light-guiding structure called the 'rhabdom.' The rhabdom contains light-sensitive pigments embedded in the membrane of microvilli that are formed by eight photoreceptor (retinula) cells. Retinula cell R8 forms the short, distal part of the rhabdom, in which microvilli directions are not uniform. The remaining length of the rhabdom is formed by the interdigitating microvilli packages of seven retinula cells. The microvilli of R1, R2, R5, R6 are vertically aligned, while those of R3, R4, R7 are horizontally aligned. Because light-sensitive pigments are elongated molecules that absorb light maximally when it is polarized parallel to the long axis of the molecule, and because these molecules tend to be aligned with the long axis of the microvilli, uniform microvilli directions in a rhabdom indicate high sensitivity to the plane of polarization of light. Each 'pixel', thus, contains eight parallel channels, four channels are most sensitive to vertically, polarized light, three to horizontally polarized light, and one (R8) is not polarization sensitive.

The distribution of spectral sensitivities in crab ommatidia is still somewhat of a mystery. Two crab opsins have been identified in *Hemigrapsus* (Grapsidae) and in fiddler crabs (Ocypodidae). In situ hybridization experiments in *U. vomeris* indicate that in most ommatidia, the two opsins are coexpressed in all photoreceptors, but not in R8. Electrophysiological measurements found evidence for either one or two spectral sensitivities, while microspectrophotometry on R1–7 has identified only one spectral sensitivity in four species of fiddler crabs. Recent intracellular electrophysiological recordings in *U. vomeris*, however, show in addition, evidence for an extra UV-sensitive photoreceptor. The situation is, in part, complicated by colorful screening pigments around individual rhabdoms that can have a strong influence on the spectral sensitivities of individual photoreceptor cells.

The outer shell of each ommatidium consists of a screen of dark absorbing pigment cells that prevent light reaching the rhabdom from any other direction, except through its own private lens. Because ommatidia are arranged perpendicularly to the curved surface of the compound eye, each points into a slightly different direction in space. The angular separation of the optical axes of neighboring ommatidia is called the 'interommatidial angle' $\Delta\phi$, which determines how densely the array of ommatidia samples the scene. Sampling density can vary across the eye, with areas of large local eye radius having the highest sampling density and areas of small local radius the lowest sampling density (resolution). Compared across the different species of semiterrestrial crabs, this regionalization of high resolution within the visual field is particularly interesting, because it reflects differences in the visual ecology between species and in the information content of the visual world they inhabit, as we will discuss further later.

Crabs carry their compound eyes on mobile eye stalks, which, however, are not used to make directed eye movements as we humans do, directing our gaze (and high-resolution fovea) to places of interest in the world, but to compensate for rotations of the body. Minimizing rotations of the visual system is of utmost importance for all animals, because visual information is computationally difficult to extract from the moving retinal image in the presence of rotations. These compensatory eye movements are controlled by both visual and mechanosensory information on body rotation, whereby the balance between the two inputs depends on the ecology of a species: in crabs that inhabit rocky shores, eye movements are largely driven by mechanosensory input from statocysts and legs, while in crabs living on mudflats, vision predominates. The reason being that vision in a flat world requires the visual horizon as a reference, as we have shown earlier, and in order to facilitate computation in a number of tasks, the visual system needs to be aligned with this important feature.

The overall organization of the visual system in semiterrestrial crabs, the length of their eye stalks, the distance between the two eyes, and the way in which spatial resolution varies across the visual field, is finely tuned to the topography of vision in different intertidal habitats. Crabs of the families Grapsidae and Sesarmidae that inhabit rocky shores and mangrove forests carry their eyes on short eyestalks, far away from each other at the lateral corners of their carapace (**Figure 6(a)**). The shape of their eyes is rather spherical, indicating that interommatidial angles are likely to be rather uniform across the eye. In contrast, crabs of the families Ocypodidae and Mictyridae that inhabit sand- and mudflats carry their eyes on long, vertically oriented eye stalks, which raise their eyes high above the carapace, but also bring them close together. The shape of their eyes is a vertically elongated oval, indicating that interommatidial angles ($\Delta\phi$) are much smaller in vertical, compared to horizontal directions because $\Delta\phi = A/R$, where A is the facet lens diameter and R is the local radius of the eye. Ocypodid crabs, thus, have a horizontally aligned, equatorial acute zone for vertical resolving power viewing the horizon.

These differences in eye design reflect the following differences in the conditions for spatial vision in spatially complex and flat-world habitats (**Figure 6(b)**). (1) The judgment of distance in complex habitats such as rocky shores and mangrove forests requires the binocular comparison of two images (binocular stereopsis). Stereopsis is the more accurate, the larger eye separation and the better horizontal resolution (the grapsid–sesarmid eye design). (2) In a flat world, the distance along the ground plane is encoded in retinal position, the further away something is, the closer it will be seen to the visual horizon line. Distance judgment can be performed monocularly and is, thus, independent of eye separation; it improves with resolution in vertical directions and with eye height above the ground (the ocypodid–mictyrid eye design). (3) In a flat world, everything that is larger than a crab itself will be seen above the crab's visual horizon, and the better its vertical resolution, the earlier and at greater distances it can make this discrimination (the ocypodid–mictyrid equatorial acute zone).

The equatorial acute zone for vertical resolution is not the only specialization in ocypodid compound eyes, related to specific tasks in a specific visual environment. There are two additional striking features of fiddler crab eyes indicating that they are not designed to optimally transmit all image information everywhere in the visual field. Instead, the dorsal eyes, for instance, are only optimized to detect small objects. The angular acceptance functions of receptors (their visual fields) in the dorsal eye are much narrower than the interommatidial angle, leading to large gaps between the optical axes of neighboring ommatidia (**Figure 6(c)**). The functional significance of such undersampling lies in the increased signal-to-noise ratio provided

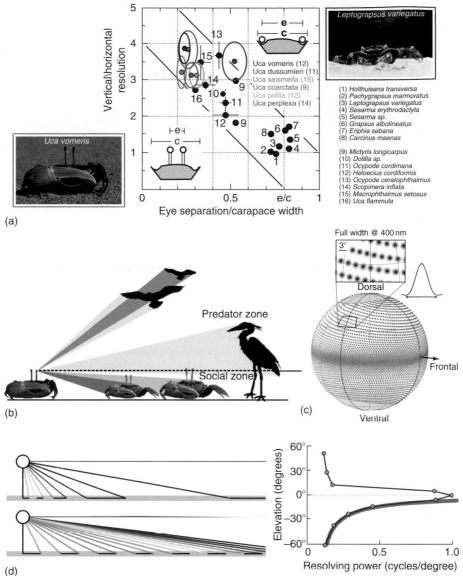

Figure 6 Visual system design in crabs. (a) The relationship between length of eye stalks, equatorial acute zone, and habitat complexity in semiterrestrial crabs. Species with high ratios of eye separation to carapace width (e/c) carry their eyes far apart close to the lateral corners of the carapace. At the eye equator, their eyes have a ratio of vertical to horizontal resolution close to one. These species tend to inhabit visually and three-dimensionally complex environments, such as rocky shores or mangrove forests. In contrast, species with long, vertical eye stalks (e/c = 1) have eyes with vertical resolution up to four times higher than horizontal resolution at the eye equator. These species tend to live in visually simple environments, such as mudflats. (b) The topography of vision in a flat world. Having eyes on long vertical stalks in a flat world has the consequence that visual information is conveniently distributed into a dorsal hemisphere viewing predators and a ventral hemisphere viewing the bodies of conspecifics. (a) and (b) modified from Zeil J and Hemmi JM (2006) The visual ecology of fiddler crabs. *Journal of Comparative Physiology A* 192: 1–25. (c) The distribution of resolving power in the fiddler crab compound eye. The viewing directions and angular acceptance functions of all ommatidia in a fiddler crab compound eye are mapped onto the unit sphere. Enlarged section shows heavy undersampling of visual space in the dorsal eye. Data and graphic courtesy of Jochen Smolka (Smolka J and Hemmi JM (2009) Topography of vision and behaviour. *Journal of Experimental Biology* 212: 3522–3532). (d) Ventral resolution gradient to counteract perspective foreshortening on flat ground. Diagram on the right shows a midsagittal vertical transect (red dotted line) through the resolution map shown in (c) that reveals a gradient of decreasing vertical resolution in the ventral (but also the dorsal) visual field. Data from Zeil J and Al-Mutairi M (1996) The variation of resolution and of ommatidial dimensions in the compound eyes of the fiddler crab *Uca lactea annulipes* (Ocypodidae, Brachyura, Decapoda). *Journal of Experimental Biology* 199: 1569–1577. This gradient can be modeled (thick green line) under the assumption that it serves to partially counteract perspective foreshortening. As a result, in the ventral visual field, equal stretches on the ground are seen by equal numbers of ommatidia (schematic diagrams on the left).

by narrow angular acceptance functions when viewing very small (distant) objects. The dorsal eye of fiddler crabs is, thus, an early warning system for approaching birds. Moreover, resolving power decreases in a systematic fashion away from the eye equator, both in the dorsal and the ventral eye (**Figure 6(d)**). The gradients are shaped in such a way as to minimize the effects of perspective foreshortening. In the ventral eye, this serves the purpose of efficiently imaging the flat ground and in the dorsal eye, birds flying parallel to the substratum move across approximately the same number of ommatidia, independent of how far away they are.

Outlook

Semiterrestrial crabs provide detailed examples of how the physical, the biological, and the social environment have shaped the optical and the neural design of visual systems. Most interestingly, their visual systems reflect the topographies of the different terrestrial habitats crabs have moved into since invading land about 40 Ma ago. In some cases, this has allowed us to understand the subtle selective pressures that have led to finely tuned visual systems, because we can identify and characterize visual tasks in some detail. However, in most cases, we know too little about the behavior of crabs and their natural visual information processing needs to be able to fully appreciate their visuo-motor competence. Behavioral polarization and color vision, claw-eye coordination, navigational and cognitive abilities, multimodal integration, learning and memory, photorecep-tor properties and neural information processing of color, polarization, motion, and shape, are just some of the funda-mental aspects of vision that wait to be fully explored – to the benefit of behavioral ecology, evolutionary biology, neuro-science, and robotics alike – in these dexterous, versatile, robust, diverse, and beautiful animals.

See also: Body Size and Sexual Dimorphism; Crustacean Social Evolution; Decision-Making: Foraging; Defensive Avoidance; Economic Escape; Empirical Studies of Predator and Prey Behavior; Evolution and Phylogeny of Communication; Games Played by Predators and Prey; Information Content and Signals; Insect Navigation; Life Histories and Predation Risk; Mate Choice in Males and Females; Mating Signals; Multimodal Signaling; Non-Elemental Learning in Invertebrates; Predator Avoidance: Mechanisms; Predator Evasion; Risk Allocation in Anti-Predator Behavior; Social Recognition; Vibrational Com-munication; Vigilance and Models of Behavior; Vision: Invertebrates; Vision: Vertebrates; Visual Signals.

Further Reading

Cannicci S, Ruwa RK, and Vannini M (1997) Homing experiments in the tree-climbing crab *Sesarma leptosoma* (Decapoda, Grapsidae). *Ethology* 103: 935–944.

Christy JH, Backwell PRY, and Schober U (2003) Interspecific attractiveness of structures built by courting male fiddler crabs: Experimental evidence of a sensory trap. *Behavioural Ecology and Sociobiology* 53: 84–91.

Crane J (1975) *Fiddler Crabs of the World (Ocypodidae: genus Uca)*. Princeton, NJ: Princeton University Press.

Detto T (2007) The fiddler crab *Uca mjoebergi* uses colour vision in mate choice. *Proceedings of the Royal Society of London B* 274: 2785–2790.

Hemmi JM and Zeil J (2005) Animals as prey: Perceptual limitations and behavioural options. *Marine Ecology Progress Series* 287: 274–278.

Herrnkind WF (1972) Orientation in shore-living arthropods, especially the sand fiddler crab. In: Winn HE and Olla BL (eds.) *Behavior of Marine Animals, Vol. 1: Invertebrates*, pp. 1–59. New York: Plenum Press.

Herrnkind WF (1983) Movement patterns and orientation. In: Vernberg FJ and Vernberg WB (eds.) *The Biology of Crustacea, Vol. 7: Behavior and Ecology*, pp. 41–105. New York: Academic Press.

Horch K, Salmon M, and Forward R (2002) Evidence for a two pigment visual system in the fiddler crab, *Uca thayeri*. *Journal of Comparative Physiology A* 188: 493–499.

Land MF (1997) Visual acuity in insects. *Annual Reviews of Entomology* 42: 147–177.

Land MF (1999) Motion and vision: Why animals move their eyes? *Journal of Comparative Physiology A* 185: 341–352.

Layne JE (1998) Retinal location is the key to identifying predators in fiddler crabs *Uca pugilator*. *Journal of Experimental Biology* 201: 2253–2261.

Nalbach H-O (1990) Multisensory control of eyestalk orientation in decapod crustaceans: An ecological approach. *Journal of Crustacean Biology* 10: 382–399.

Oliva D, Medan V, and Tomsic D (2007) Escape behavior and neuronal responses to looming stimuli in the crab *Chasmagnathus granulatus* (Decapoda: Grapsidae). *Journal of Experimental Biology* 210: 865–880.

Pope DS (2005) Waving in a crowd: Fiddler crabs signal in networks. In: McGregor PK (ed.) *Animal Communication Networks*, pp. 252–276. Cambridge: Cambridge University Press.

Rosenberg MS (2001) The systematics and taxonomy of fiddler crabs: A phylogeny of the genus *Uca*. *Journal of Crustacean Biology* 21: 839–869.

Salmon M and Hyatt GW (1983) Communication. In: Vernberg FJ and Vernberg WG (eds.) *The Biology of Crustacea, Vol. 7: Behavior and Ecology*, pp. 1–40. New York: Academic Press.

Shaw SR and Stowe S (1982) Photoreception. In: Atwood HL and Sandeman DC (eds.) *The Biology of Crustacea, Vol. 3, Neurobiology: Structure and Function*, pp. 291–367. New York: Academic Press.

Walls ML and Layne JE (2009) Direct evidence for distance measurement via flexible stride integration in the fiddler crab. *Current Biology* 19: 25–29.

Zeil J (2008) Orientation, Navigation and Search. In: Jørgensen SE and Fath BD (eds.) *Behavioral Ecology, Vol. 3 of Encyclopedia of Ecology*, pp. 2596–2608. Oxford: Elsevier.

Zeil J, Boeddeker N, and Hemmi JM (2009) Visually guided behaviour. In: Squire LR (ed.) *Encyclopedia of Neuroscience*, Vol. 10, pp. 369–380. Oxford: Academic Press.

Zeil J and Hemmi JM (2006) The visual ecology of fiddler crabs. *Journal of Comparative Physiology A* 192: 1–25.

Zeil J, Nalbach G, and Nalbach HO (1986) Eyes, eye stalks and the visual world of semi-terrestrial crabs. *Journal of Comparative Physiology A* 159: 801–811.

Crustacean Social Evolution

J. E. Duffy, Virginia Institute of Marine Science, Gloucester Point, VA, USA

Introduction

The Crustacea represent one of the most spectacular evolutionary radiations in the animal kingdom, whether measured by species richness or diversity in morphology or lifestyles. Its members range from microscopic mites of the plankton to fearsome giant crabs to sessile barnacles to amorphous parasites that are almost unrecognizable as animals. Crustaceans occupy most habitats on earth, from the deepest ocean trenches to mountaintops and deserts, and the dominance of the open ocean plankton by calanoid copepods makes them one of the most abundant metazoan groups on earth.

This ecological diversity suggests that the Crustacea should provide a wealth of interesting social and mating systems, and this is indeed true, as both classic and recent research has shown. Yet, despite their ubiquity and diversity, crustaceans have received surprisingly little attention from students of behavior compared with their younger siblings – the insects – or the vertebrates, no doubt due in large part to the aquatic habits of most crustacean species.

What are the ecological and behavioral consequences of the crustacean colonization of this range of habitats? What can they tell us about the generality of theory and the generalizations emerging from work on other, better studied taxa? Here, I highlight a few illustrative case studies of social systems in crustaceans, and discuss the broader implications of crustacean sociality for understanding some central issues in animal behavior and sociobiology.

A Primer in Crustacean Biology

Recent research in molecular systematics shows that the Crustacea is paraphyletic, with the insects (Hexapoda) nested within a pancrustacean clade that diverged in the Precambrian. Among the major branches in the crustacean family tree, the Malacostraca is the most diverse, both in morphology and in species, numbering tens of thousands. This group includes the large, ecologically and economically important crabs, shrimps, and lobsters familiar to the layperson. For all of these reasons, most of what is known about the social behavior of crustaceans comes from the Malacostraca.

Like their relatives, the insects, crustaceans share a basic segmented body plan divided into three regions: the head, thorax (pereon), and abdomen (pleon). The body is covered with a chitinous exoskeleton, which is shed periodically during growth. Each of the segments in the primitive ancestral crustacean body bore a pair of appendages, which have been modified during the evolution of the various crustacean groups into a wide range of structures used in feeding, locomotion, sensation, and communication. The bodies of most crustaceans are richly endowed with a wide variety of setae – stiff hair-like bristles of diverse form that are used for a wide range of functions. The two pairs of antennae, in particular, bear dense arrays of chemo- and mechanosensory setae, which are used in conjunction with directional currents of water generated by specialized appendages in the head region to distribute and collect chemical signals, and are important in social and mating interactions.

The mode of development strongly influences the potential for kin to interact, and thus the evolution of social systems in Crustacea. Most familiar decapods release microscopic larvae into the plankton, where they drift for some time – several months in some species – before settling to the bottom and transitioning to the adult lifestyle. In such species, populations are genetically well mixed and kin groups cannot form. In other species, however, eggs hatch directly into miniature versions of the adults in much the same way as eggs hatch into miniature adults (nymphs) in hemimetabolous insects such as grasshoppers and termites. This direct development is common to all peracarid crustaceans (isopods, amphipods, and their relatives) and is also found in some decapods. Crustaceans go through several molts as they grow, before reaching the adult stage.

Crustaceans display a wide range in reproductive biologies. While most species breed repeatedly during life and have separate sexes, brine shrimp and some *Daphnia* that inhabit temporary freshwater pools are cyclic parthenogens, and several shrimp are sequential or simultaneous hermaphrodites. Sex determination can be genetic, environmental, or involve some combination of the two.

Crustacean Mating Systems

The mating system is an important component of the social system in that it influences the size, composition, and kin structure of groups of interacting individuals. For example, establishment of monogamous relationships can lead to paternal care, and in some animals, avoidance of incest helps explain why adult helpers in social colonies do not breed. Crustaceans display a wide diversity of mating systems that are molded by the variance in mate availability in time and space, variation in female life history, and behavior. These range from situations involving fleeting encounters to various forms of mate guarding, monogamous pair formation, to harems. Here, I describe a few examples that provide insights into the evolution of more advanced social systems.

Precopulatory Mate Guarding and Its Consequences

A key trait influencing the mating system in many crustaceans is the limited time window of female receptivity, which results from the requirement that mating and ovulation take place immediately after a molt when her integument is soft. As a consequence, many crustaceans exhibit mate guarding, pair-bonding, and other behaviors that maximize a male's certainty of having access to a female when she is ready to mate. Precopulatory mate guarding (also called amplexus), in which the male carries the female for an extended period of time in anticipation of mating, is common in several groups of amphipods, including the familiar *Hyalella* species of North American lakes and *Gammarus* species of coastal marine waters, as well as many groups of isopods (**Figure 1**). Because of brief female receptivity, the operational sex ratio in such populations is highly male-biased, and this mate guarding allows the male to monopolize the female until she is receptive. In other species, including several crabs and lobsters, females can store sperm and so are not temporally restricted in mating time.

Mate guarding has been extensively studied in isopods and amphipods as a model system for understanding sexual selection and the resolution of intersexual conflict.

Males often do not feed while guarding so they incur a cost in exchange for the opportunity to mate. Females presumably also incur a cost in terms of reduced feeding, higher predation risk, and/or increased risk of being dislodged from the substratum. Indeed, experiments with the isopod *Idotea baltica*, conducted by Veijo Jormalainen and colleagues, showed that guarded females had lower glycogen (stored food) reserves and laid smaller eggs than females that had been mated but not guarded. Not surprisingly, female isopods often vigorously resist being guarded and the initiation of guarding tends to be a mutually aggressive affair. The proposed role of limited receptivity in selecting for mate guarding would seem to be proved by the exception to the rule: in terrestrial oniscoid isopods, females have extended receptivity and some can store sperm – using it for up to eight broods, reducing a male's ability to monopolize mating opportunities. Accordingly, these isopods lack prolonged guarding.

Sexual selection has molded the phenotypes of such mate-guarding species. Males are larger than females in several mate-guarding isopods, likely because larger male size is favored by both intrasexual selection, which favors larger size in competition among males, as well as intersexual selection generated by females resistant to guarding. Strong sexual dimorphism is also seen in some freshwater amphipods. Interestingly, among closely related species of the amphipod *Hyalella*, the dimorphism is reduced in species that inhabit lakes with fish, which impose strong size-selective predation on large individuals; in these populations exposed to predation, moreover, females show weaker preference for large males. Thus, phenotypic traits and behavioral preferences are molded by the trade-off between sexual selection for large male size and natural selection for reduced size to avoid predation.

Social Monogamy

Snapping shrimp (Alpheidae) are common and diverse animals in warm seas. Most live in confined spaces such as rock crevices, excavated burrows in sediment, or commensally within sessile invertebrates such as sponges, corals, or feather stars. Long-term heterosexual pairing, or 'social monogamy' is the norm among alpheids. Models predict that mate guarding can extend to long-term monogamous associations where male searching for mates is costly because of, for example, low population densities, male-biased operational sex ratios, or high predation risk outside the territory. All of these conditions are common among alpheids. As in the peracarids, pairing appears to have evolved partly as a male guarding response, as evidenced by the preference of males to associate with females close to sexual receptivity. But pairs of snapping shrimp also jointly defend a single territory, suggesting that other factors are also at play. Lauren Mathews conducted a series of experiments with

Figure 1 Precopulatory mate guarding in the estuarine isopod *Idotea baltica*. The larger male carries the smaller female for an extended period until she is receptive to mating. The initiation and duration of guarding often generates a struggle because of the conflicting interests of the male and female. Photo by Veijo Jormalainen, used with permission.

Alpheus angulatus testing the potential benefits of monogamy to the two partners. She showed that, in addition to its role in assuring males of mating opportunities, social monogamy is likely favored by benefits to both partners of sharing maintenance and defense of the joint territory. For example, females were less likely to be evicted from the territory by intruders when paired with a male than when unpaired, and males similarly were less frequently evicted when paired with a sexually receptive female. The tendency of both males and females to bring food back to the burrow may also have benefited their partners. Finally, paired females spent more time constructing the burrow than did paired males, possibly reflecting a division of labor in which males, with their larger snapping claw, took care of defense. As discussed below, the monogamous habit of these pair-living shrimp likely set the stage for the repeated evolution of multigenerational, cooperative societies in eusocial alpheids.

Sexual Selection and Alternative Male Mating Strategies

A more extreme case of mate monopolization occurs where males can assemble harems of females. This mating system is more common in situations in which female distribution is highly clumped, for example among habitat specialists, and in which males are capable of excluding other males from the habitat patch or group of females. An especially intriguing example from the Crustacea involves the isopod *Paracerceis sculpta*, which inhabits spaces within small intertidal sponges in the Gulf of California. Research by Stephen Shuster showed that large males may monopolize as many as 19 females in a given sponge. However, sexual selection driven by the intense competition among males for females has resulted in divergence of three alternative male mating morphs that coexist in the same populations. Alpha males are large and powerful and monopolize females by physically excluding other males. Beta males, in contrast, are similar to females in both morphology and behavior and gain access to sponges controlled by alpha males by mimicking females. Gamma males are very small and appear to mimic juveniles; although males attempt to exclude them, gammas can gain access to crowded sponges by slipping through male defenses unnoticed. Both beta and gamma males achieve some fertilizations in these highly competitive situations by subterfuge, providing an example of the 'sneaker' male morphs that co-occur with 'fighter' males in a range of animal taxa.

Larval Development, Parental Care, and Family Life

Social groups in most animals develop from nuclear or extended families. Thus, parental care and the concomitant

aggregation of kin in families are important prerequisites to more advanced social organization in many animals, including vertebrates, insects, and crustaceans. For example, one of the classical criteria of eusociality is cohabitation of multiple adult generations, which generally arises as offspring extend a long period of parental care and remain with their parents after maturity.

Parental care and associated social behaviors are only possible, however, when parents and offspring remain in spatial proximity where they can interact. In most decapods such as lobsters, crabs, and shrimp, planktonic larvae result in broad dispersal. In these species, families cannot form and thus kin selection cannot operate. Among 'direct-developing' crustaceans, such as amphipods, isopods, and a few decapods, the situation is different. In these species, extended parental care is relatively common (**Figure 2**). Care is typically provided only by the mother, initially in the form of carrying, grooming, and ventilation of embryos. But males also contribute in several species by building and defending burrows or other nest sites. In extreme cases, including the highly social bromeliad crab *Metopaulias depressus* and certain sponge-dwelling shrimp (see below), other individuals – generally older siblings – also provide some care in the form of nest defense or even food provisioning to young offspring.

A primary function of parental care in crustaceans as in most other animals is protection of the vulnerable young from predators and harsh environmental conditions. Active 'shepherding' by mothers of small juveniles faced with danger occurs in several species of crabs and caprellid amphipods (skeleton shrimp); in some cases, a mother picks up her young offspring and carries them away from predators, whereas in others, some (generally unknown) signal from the mother causes juveniles to aggregate or to enter her brood pouch. Mothers also

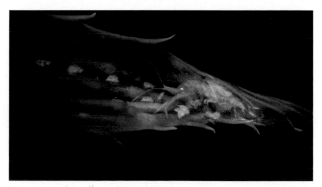

Figure 2 A mother of the Chilean marine amphipod *Peramphithoe femorata* with her young in their nest constructed on a frond of kelp. Ampithoid amphipods are common herbivores in coastal marine vegetation, where they build silken nests among algae and fouling material. Offspring of many amphipods remain with the mother for some time before dispersing, and in some species, are fed by the mother during this period. Photo by Iván Hinojosa, used with permission.

feed their offspring in several species of amphipods and, in desert isopods, even bring food back from extended foraging trips to provision offspring remaining at the nest, much as in birds, bees, and ants.

Not surprisingly, parental care tends to be better developed in habitats or situations where offspring face strong challenges from the biotic or abiotic environment. For example, several species of Australian semiterrestrial crayfish inhabit burrows in soil, sometimes far from open surface water (**Figure 3**). Burrows provide shelter from predators and harsh physical conditions, and are also a source of food in some species. Particularly in crayfish species that live far from surface water, the burrow may be complex and extend for >4 m into the ground. In these drier areas, burrows can only be dug during a limited time of year, and thus represent a valuable, self-contained resource. Juveniles often face harsh conditions and strong risk of predation outside the burrow, and the life history of the crayfishes has adapted accordingly. Semiterrestrial crayfish have no free-living larval stage as most decapods do; instead, juveniles cling to the mother's pleopods (abdominal appendages) after hatching and remain there for 2–3 molts before graduating into independent miniature versions of the adults. In *Procambarus alleni*, juveniles at this stage make short excursions outside the burrow but usually remain close to the mother, who helps them back into the brood area by raising her body and extending the abdomen. Females in some semiterrestrial crayfish also produce pheromones that attract the juveniles. Mothers in *Procambarus clarkii* also defend their juveniles, even those that are already foraging independently, against large males. Extensive cohabitation of mother and offspring reaches its most extreme manifestations in Tasmanian species of *Engaeus*, in which four generations – including mother and three year classes of juveniles – have been observed cohabiting in the same burrow. The prolonged associations between mothers and young, and the difficulty of establishing new territories outside the parental burrow, in these species recall the situations believed to foster the evolution of eusociality in insects, and in snapping shrimp as discussed below.

Kin Recognition and Kin Discrimination

The aggregation of genetic relatives – family members – provides opportunities for kin selection to mold cooperative behaviors. Maintaining cohesive kin groups is facilitated by the ability to recognize kin from nonkin. In most crustacean species, experiments suggest that parents are incapable of distinguishing their own offspring from unrelated juveniles. In these cases, family cohesion can be maintained by simple rules of context in which interactions occur. For example, mothers in many crustacean species accept juveniles found in the nest area but are very aggressive toward individuals approaching the nest from the outside.

At the other end of the spectrum, kin recognition is highly developed in certain desert isopods, which are the dominant herbivores and detritivores over wide areas of arid North Africa and Asia. In one such species, *Hemilepistus reaumuri*, parent–offspring groups share burrows, with both parents caring for the young for several months, and adults must make long excursions outside the burrow to forage (**Figure 4**). The burrow provides protection from the harsh environmental conditions of the desert and from predators. Because it represents a highly

Figure 3 The Tasmanian endemic semiterrestrial crayfish *Engaeus orramakunna*. This species lives in deep burrows that may house a mother with up to three successive cohorts of offspring all living together. Photo by Niall Doran, used with permission.

Figure 4 Two desert isopods, *Hemilepistus reaumuri*, at the entrance to their burrow. These animals live in family groups and have finely tuned kin recognition based on complex chemical mixtures that allow them to discriminate family members from intruders approaching the burrow after wide-ranging foraging trips. Photo by Karl Eduard Linsenmair, used with permission.

valuable shelter, competition and invasion are common threats, and recognition of kin is critical to maintaining group cohesion in the face of foraging traffic in and out of the burrow.

Research by Karl Eduard Linsenmair has demonstrated that kin recognition is remarkably finely tuned in *H. reaumuri*. Individuals in this species recognize one another using nonvolatile, polar compounds that are transferred by contact. Because the compounds can be transferred by touch, contact between unrelated individuals could easily lead to contamination of the family signal that would lead to attack upon return to the family burrow, where an attentive guard stands at the burrow entrance (**Figure 4**). Thus, individuals are scrupulous about avoiding contact with nonkin. A large series of experiments showed that the chemical 'badge' worn by each family is unique and genetically determined, and arises from regular close contact among family members in the burrow, which mixes the individual signals into a family-specific odor. This process is strikingly similar to the way in which common family odor is distributed among eusocial naked mole-rats within their familial burrows. Interestingly, attacks on newborn isopods and family members that have just molted are inhibited by another (undefined) chemical substance, allowing these individuals to acquire the family odor without harm. As a result of this finely tuned kin recognition system, isopod families are able to maintain their strict kin structure despite high population densities and frequent long foraging excursions to and from the burrow.

Individual recognition among crustaceans is not confined to kin but extends to unrelated individuals and even other species. Stomatopods (mantis shrimp) in the genus *Gonodactylus* are common inhabitants of tropical reefs, where they live in cavities in coral rock, along with various other fishes and invertebrates. Experiments by Roy Caldwell and colleagues have shown that these stomatopods can learn to identify other individual stomatopods based on chemical cues and that they use these cues, along with memory of the fighting ability of the individual, to determine how to approach a cavity that might be occupied. Interestingly, the stomatopods are also able to learn the odor of individual octopuses, which compete for the same cavities. The shrimp are much more hesitant and defensive when approaching a cavity occupied by a conspecific or an octopus that they have fought previously. These examples demonstrate that certain crustaceans are capable of quite finely tuned discrimination among individual animals, both conspecifics and other species.

Cooperative Breeding in Jamaican Bromeliad Crabs

About 4.5 Ma, a marine crab colonized the Caribbean island of Jamaica and moved up into the forests, radiating into at least ten endemic species of freshwater and terrestrial crabs. Among the most unique of this group is *Metopaulias depressus*, which lives exclusively in the small bodies of water that collect in leaf axils of bromeliad plants in the forested mountains (**Figure 5**). These small pools provide most everything the crabs need: water required to moisten the gills, molt, and reproduce; food in the form of plant matter, detritus, and small arthropods; and protection from predatory lizards and birds. Individual plants can live for several years and their leaf axils represent a reliable and stable water source that collects dew as well as rain and thus persist even through extended droughts. But because suitable bromeliads are scattered, in short supply, and surrounded by hostile habitat, finding and maintaining these nests presents challenges. As in many social insects, birds and mammals, these environmental challenges appear to have selected for a cooperatively breeding or even eusocial lifestyle in which delayed dispersal results in accumulation of large family groups that cooperate in raising the young. The story of the Jamaican bromeliad crab has been documented in an elegant series of studies by Rudolf Diesel.

Life History and Maternal Care

Bromeliad crabs breed once a year, during December and January, producing clutches of 20–100 eggs. When the eggs hatch, the larvae are released into the water in a leaf axil. Here, the larvae develop rapidly – within about 2 weeks – into small juvenile crabs. The young crabs then remain in the mother's territory for up to 3 months during which the mother provides extensive care for them, defending them against predatory spiders

Figure 5 A mother and young of the Jamaican bromeliad crab *Metopaulias depressus*. Mother crabs raise their larvae in pools of water that collect in the leaf axils of bromeliads and fastidiously manage the water chemistry by removing leaf litter and adding empty snail shells that raise the pH and concentration of calcium ions required by growing larvae. Older siblings also provide care in this cooperatively breeding species. Photo by Rudolf Diesel, used with permission.

and aquatic insect larvae, and provisioning them with food. But what is most remarkable about these crabs is the mother's extreme care in maintaining water quality in the leaf axils. By actively removing leaf litter and collecting and placing empty snail shells in the nursery pools, mothers more than doubled nighttime dissolved oxygen in the nursery pools, and raised pH and concentrations of calcium necessary for proper larval development. Indeed, mother crabs introduced more shells into nursery pools in which calcium concentrations had been experimentally reduced, confirming that they manage water quality actively and with a high degree of sophistication. Then, around the age of 3 months, the juvenile crabs begin to disperse from the nursery pool into other leaf axils on the same plant. They reach maturity after a year or more, and females live for up to 3 years.

Field studies have shown that the colony of crabs living on a single plant can consist of up to 84 individuals, but invariably harbors only a single breeding female. Generally, distinct annual cohorts of juvenile crabs are visible in a colony, and many colonies contain at least a few individuals of reproductive size that nonetheless do not breed. The size distributions of colony members suggest that most juveniles stay with the mother for at least a year.

In addition to maintaining good water quality and providing food for the larvae, experiments showed that mother crabs aggressively defended their nest against intruding crabs, even when the intruders were large, and sometimes even killed them. Mothers were also able to distinguish larger juveniles living in their own nests (presumably their offspring) from unfamiliar juveniles of the same size when both types of individuals were introduced experimentally into the nest; small juveniles were not attacked, regardless of whether they were familiar or not. Thus, Jamaican bromeliad crabs appear able to distinguish kin from nonkin.

Cooperative Brood Care and Social System

While a wide range of animals exhibit parental care of varying degrees of sophistication, what distinguishes cooperatively breeding or eusocial species is alloparental care, that is, care of young by individuals other than parents. Several lines of evidence confirm alloparental care in Jamaican bromeliad crabs. First, nonbreeding adult females from earlier cohorts that remained in the nest helped the mother defend the nest against unfamiliar intruders. Second, when the mother was removed, young ones in the nest survived and grew better in the presence than in the absence of nonbreeding adult siblings, presumably because the older individuals helped defend the nest and maintain good water quality.

Jamaican bromeliad crabs appear to be unique among crustaceans in the sophistication of brood care by both mothers and nonbreeding adult helpers, particularly in

comparison with other crabs, most of whom release larvae to face their fate in the plankton and provide no care afterwards. Indeed, Jamaican bromeliad crab colonies meet the criteria traditionally defining the most advanced social system, eusociality: overlapping adult generations, reproductive division of labor, and cooperative care of young.

What factors explain such advanced social organization in the bromeliad crab? As is true of many other social animals, both insects and vertebrates, the answer appears ultimately to involve ecological pressures that make independent reproduction difficult. In the case of bromeliad crabs, these pressures include the scattered nature of water-filled microhabitats, which are surrounded by unsuitable habitat, making dispersal dangerous. Moreover, because the bromeliad microhabitats are relatively rare, they are also in high demand and subject to invasion by competitors. Theory and data from other animals suggest that such ecological pressures favor delayed dispersal, which allows kin groups to form, and also provide an opportunity for the nonbreeding older offspring to help raise younger siblings, which provides inclusive fitness benefits. Moreover, field observations suggest that staying at home eventually pays off for some of the daughters either in inheriting the mother's territory when she dies, or colonizing an adjacent territory as the bromeliad sprouts new plants from the same rhizome. Such territory inheritance has similarly been suggested as a selective advantage to helping at the nest in eusocial termites.

Eusociality in Sponge-Dwelling Shrimp

Eusociality ('true sociality') is the most extreme manifestation of altruistic cooperation in the animal kingdom. Eusocial colonies historically have been defined on the basis of three characteristics: (1) presence of multiple adult generations living together, (2) reproductive division of labor, meaning that only a subset of colony members reproduce, and (3) cooperative care of young. This definition unites the familiar social bees, ants, wasps, and termites, which typically live in colonies headed by a single queen (and, in the case of termites, also a king) and containing many nonbreeding workers that cooperate in raising the queen's offspring, foraging for food, maintaining and defending the nest, and so on.

In 1996, social colonies were reported in the Caribbean coral-reef shrimp *Synalpheus regalis*, which consisted of a single breeding female – the queen – and tens to hundreds of other individuals, including many nonbreeding adults. Genetic analyses confirm that colonies of these eusocial shrimp consist of close relatives, and likely full siblings, the offspring of a single breeding pair, which evidently dominates reproduction for an extended period. Similar eusocial colonies have subsequently been discovered in several other species of *Synalpheus* (**Figure 6**). The colonies

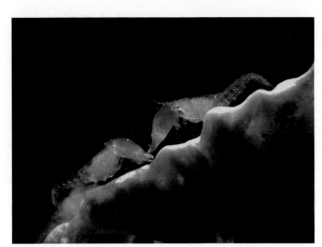

Figure 6 The Caribbean eusocial shrimp *Synalpheus regalis*. These shrimp occupy the internal canals of sponges on coral reefs. Several eusocial species, like this one, live in colonies of 10s to a few 100s of individuals with a single breeding female, the queen. Large nonbreeding individuals aggressively defend the colony against intruders. Photo by Emmett Duffy, used with permission.

consist of several generations living together, and the non-breeding colony members contribute to colony welfare by defending the nest, qualifying them as eusocial by the traditional definition. Eusocial colonies form only in certain species of *Synalpheus* that produce crawling offspring that typically remain in the same sponge where they were born, allowing kin groups to accumulate.

Eusociality poses a fundamental paradox for evolution by natural selection, as Darwin famously recognized: If adaptive evolution proceeds via differential survival and reproduction of individuals, how can a species arise in which most individuals never breed at all? As the only known case of eusociality in a marine animal, snapping shrimp have become valuable subjects for understanding general features of the evolution of advanced social organization in animals via comparisons with social insects and vertebrates. Why have a few species of sponge-dwelling shrimp, alone among marine animals, adopted this cooperative lifestyle? The search for an answer illuminates some key questions in understanding animal social life generally.

Natural History and Social Behavior in Sponge-Dwelling Shrimp

Shrimp in the genus *Synalpheus* are mostly symbiotic or parasitic, living their entire lives within the internal canals of living sponges and feeding on the tissues and secretions of their hosts. The common name snapping shrimp or pistol shrimp refers to the large claw carried on one side of the body, which produces a powerful jet of water and a loud snap when closed, and is used in aggressive interactions and fights. Unlike most alpheid shrimp, which

are aggressive toward all individuals other than their mate, eusocial *Synalpheus* species live in dense aggregations and are in nearly constant contact with other colony members.

The canals of host sponges provide a valuable resource in the combination of safe shelter and constant food, and shrimp populations fill nearly all suitable sponges on the reef, such that available habitat is 'saturated.' Because the host sponge combines food, living space, and a safe haven, there is a high premium on obtaining and defending it, and that necessity is clearly reflected in the aggression of resident shrimp against intruders, which sometimes ends in fights to death. Indeed, homeland defense appears to be the primary job of the nonbreeding helpers. Experiments with *S. regalis* reveal that, compared with juveniles or the queen, large helpers are more active, more aggressive, and more likely to be found near the periphery of a sponge, where intruders are a threat. In contrast, juveniles are sedentary and often congregate in groups to feed. Thus, shrimp show behavioral differentiation among classes of individuals reminiscent of the caste roles of certain social insects.

Social shrimp colonies also show coordinated activity. For example, in captive laboratory colonies, groups of shrimp have been observed cooperating to remove dead nestmates from the sponge. But the most striking example involves 'coordinated snapping,' during which a sentinel shrimp reacts to some disturbance by recruiting other colony members to snap in concert for several to tens of seconds. Experiments suggest that coordinated snapping in social shrimp is a specific and effective group warning signal to nest intruders, produced when individual defenders meet an unfamiliar shrimp and are unable to chase it away. The function of coordinated snapping as a specific warning to intruders is supported by its occurrence only after introductions of intruders, and its effectiveness at repelling them even after single snaps fail to do so. Coordinated snapping can also be considered an honest warning signal because the few intruders unable to flee in experiments were subsequently killed. Coordinated snapping in social shrimp thus represents a mass communication among colony members, a fundamental characteristic of highly social insects and vertebrates.

Genetics and Ecology in the Evolution of Shrimp Eusociality

Genetic relatedness between interacting individuals has occupied a central role in explaining the tension between conflict and cooperation since William Hamilton's seminal formulation of the concept of inclusive fitness (or kin selection). According to Hamilton, the evolution of behavioral interactions depends on both genetic relatedness among individuals and on the ecological factors that define the costs and benefits of their interactions.

In understanding the paradox of eusociality, in particular, kin selection has provided a key explanation and has stimulated four decades of highly productive research. Recently, it has been argued that kin selection is a consequence rather than a cause of eusociality, and that the ecological context driving competition and cooperation are the dominant pressures selecting for cooperation. Research on sponge-dwelling shrimp contributes to resolving this debate.

One powerful, albeit indirect, approach to evaluating evolutionary hypotheses is via phylogenetic comparative methods, which statistically separate the influence of recent common ancestry from that of ecological factors in shaping evolutionary change in a lineage. For example, comparative analyses among sponge-dwelling shrimp species in Belize controlled statistically for the close phylogenetic relatedness and the small body sizes of social shrimp, and supported the hypothesis that eusociality evolved as a result of both ecological benefits of group living and of close genetic kin structure. Eusocial shrimp species were more abundant and had broader host ranges than nonsocial sister species, supporting the basic hypothesis that cooperative groups have a leg up in ecologically challenging environments. But ecological advantages of eusocial colonies are not the whole story: eusociality arose only in species with nondispersing larvae, which form family groups subject to kin selection. Thus, superior ability to hold valuable resources favors eusociality in shrimp, but close genetic relatedness is nevertheless key to its origin, as in most social insects and vertebrates.

Adaptive Demography

In addition to the three classical criteria described above, eusociality is often recognized by the loss of totipotency, i.e., a transition to irreversible sterility or other form of specialization within a colony. In this sense, eusocial colonies are qualitatively different than other cooperatively breeding animal societies and evolution of sterility represents a threshold, which, once crossed, allows new evolutionary processes to act. Once workers are freed from selection for personal reproduction, their behavior, physiology, and body form can be molded by colony-level selection toward specialized phenotypes that benefit the colony as a whole, such as the soldiers, nurses, and other specialized castes that reach sometimes bizarre extremes in certain large-colony ant and termite species.

Among shrimp, the division of labor between reproduction and defense reaches its clearest manifestation in *Synalpheus filidigitus*, in which the queen's irreversible dependence on her colony is reflected in a physical metamorphosis. Queens of this species lack the typical large snapping claw, having replaced it with a second minor-form chela. This is strong indirect evidence for organized division of labor in the colony, since an alpheid lacking its fighting claw is helpless on its own. It also presents an interesting parallel with the advanced social insects, in which queens typically become nearly helpless egg-laying machines.

Colony-level selection may produce not only specialized individual phenotypes but also adaptive demography, that is, changes in the relative proportions of different types of colony members that benefit the colony by increasing its efficiency. Social shrimp also show trends suggestive of such adaptive demography. Growth allometry and body proportions of three eusocial shrimp species differed in several respects from that of their pair-forming relatives: allometry of fighting claw size among males and nonbreeding females was steeper, and queens had proportionally smaller fighting claws, in eusocial species. Shrimp are thus similar to other eusocial animals in the morphological differentiation between breeders and nonbreeders, and in the indication that some larger nonbreeders might contribute more to defense than others.

Eusocial shrimp species also tend to be smaller bodied than less social relatives, and this trend remains even after phylogenetic relationships are controlled for, as also reported for social wasps. This situation may result from selection for improved colony performance, that is, adaptive demography. Oster and Wilson argued that reduced body size could allow a colony to have a larger number of individuals and thus maintain more efficient operations, providing some redundancy, and maintaining a higher 'behavioral tempo' that enhances productivity. Whether this explains the patterns of smaller body size in social shrimp remains to be tested.

Conclusions and Comparisons with Other Animals

Evidence from crustaceans supports models based on study of insects and vertebrates that evolution of cooperative social systems is strongly influenced by ecological pressures and, in particular, the difficulty of obtaining and defending a 'basic necessary resource' in the parlance of Alexander, Crespi, and colleagues. For social snapping shrimp, this resource is the host sponge, which is in short supply and generally fiercely defended by competitors. For Jamaican bromeliad crabs, it is a host plant with a sufficient number and sizes of leaf axils to provide food, shelter, and a nursery for larvae. In desert isopods, and perhaps also Australian semiterrestrial crayfish, the resource is the burrow, which can only be built during a limited time after rain and is essential for survival under harsh conditions. In all of these cases, the aggregation of parent(s) with multiple cohorts of offspring creates kin groups that are presumably also essential to the evolution of cooperative behavior. Indeed, in sponge-dwelling shrimp, phylogenetically controlled comparisons confirm that eusocial groups evolved only in species with crawling larvae, which allow formation of close kin groups. Thus, these examples add to the list of examples from other taxa that show that advanced cooperative social life evolves in situations where cooperation leads to superior ability to hold valuable resources, and that cooperation is especially favored in kin groups, where helpers can receive inclusive fitness rewards for their efforts.

See also: Cooperation and Sociality; Group Living; Kin Selection and Relatedness; Mate Choice in Males and Females; Reproductive Skew; Social Recognition; Social Selection, Sexual Selection, and Sexual Conflict; Termites: Social Evolution.

Further Reading

Caldwell RL (1979) Cavity occupation and defensive behaviour in the stomatopod *Gonodactylus festai*: Evidence for chemically mediated individual recognition. *Animal Behavior* 27: 194–201.

Correa C and Thiel M (2003) Mating systems in caridean shrimp (Decapoda: Caridea) and their evolutionary consequences for sexual dimorphism and reproductive biology. *Revista Chilena de Historia Natural* 76: 187–203.

Crespi BJ (1994) Three conditions for the evolution of eusociality: Are they sufficient? *Insectes Sociaux* 41: 395–400.

Diesel R (1997) Maternal control of calcium concentration in the larval nursery of the bromeliad crab, *Metopaulias depressus* (Grapsidae). *Proceedings of the Royal Society of London, Series B* 264: 1403–1406.

Duffy JE (1996) Eusociality in a coral-reef shrimp. *Nature* 381: 512–514.

Duffy JE and Thiel M (eds.) (2007) *Evolutionary Ecology of Social and Sexual Systems: Crustaceans as Model Organisms.* Oxford: Oxford University Press.

Jormalainen V (1998) Precopulatory mate guarding in crustaceans: Male competitive strategy and intersexual conflict. *Quarterly Review of Biology* 73: 275–304.

Linsenmair KE (1987) Kin recognition in subsocial arthropods, in particular in the desert isopod *Hemilepistus reaumuri*. In: Fletcher D and Michener C (eds.) *Kin Recognition in Animals.* Chichester: John Wiley and Sons.

Shuster SM and Wade MJ (1991) Equal mating success among male reproductive strategies in a marine isopod. *Nature* 350: 608–610.

Thiel M (1999) Parental care behaviour in crustaceans – A comparative overview. *Crustacean Issues* 12: 211–226.

Thiel M and Baeza JA (2001) Factors affecting the social behaviour of crustaceans living symbiotically with other marine invertebrates: A modeling approach. *Symbiosis* 30: 163–190.

VanHook A and Patel NH (2008) Primer: Crustaceans. *Current Biology* 18: R547–R550.

Cryptic Female Choice

W. G. Eberhard, Smithsonian Tropical Research Institute; Universidad de Costa Rica, Ciudad Universitaria, Costa Rica

Charles Darwin distinguished two contexts in which sexual selection acts on males competing for access to females: Direct male–male battles and female choice. He apparently believed, perhaps because of cultural blinders to thinking about more intimate aspects of copulation, that sexual selection occurred only prior to copulation. He thus thought that a male's success in sexual competition could be measured in terms of his ability to obtain copulations. It is now clear that this view, which prevailed essentially unchallenged for about 100 years, is incomplete, and that males often also compete for access to the female's eggs after copulation has begun. This competition was originally called 'sperm competition' by Parker, but using Darwin's categories, sperm competition is now employed in a more restricted sense to refer to the postcopulatory equivalent of male–male battles; the postcopulatory equivalent of female choice is termed 'cryptic female choice' (CFC) ('postcopulatory' is generally used to refer to all events following the initiation of genital coupling). The word *cryptic* refers to the fact that any selection resulting from female choice among males that occurs after copulation has begun would be missed using the classic Darwinian criteria of success.

The term 'cryptic female choice' was first used in reference to female scorpionflies, which lay more eggs immediately after copulating with large males than after mating with small males (and thus presumably bias paternity in favor of large males). This idea, which is part of a general trend in evolutionary biology to realize that females are more active participants than was previously recognized, is discussed most extensively in two books by Eberhard. CFC has often been invoked to explain the rapid divergent evolution of traits such as male genitalia, as traits under sexual selection are known to tend to diverge rapidly.

In concrete terms, CFC can occur if a female's morphological, behavioral, or physiological traits (for instance, triggering of oviposition, ovulation, sperm transport or storage, resistance to further mating, inhibition of sperm dumping soon after copulation, etc. – see **Table 1**) consistently bias the chances that particular mates have of siring offspring when she copulates with more than one male. The result is selection favoring males with traits that increase the probability of certain postcopulatory female responses, as they are more likely to obtain fertilizations than others. Male traits associated with such female biases can be morphological (e.g., his genital morphology), behavioral (e.g., his courtship behavior during copulation),

or physiological (e.g., the chemical composition of his semen). There are more than 20 mechanisms by which postcopulatory female-imposed paternity biases can be produced (**Table 1**), and that have possible example species in which CFC may occur. Some female mechanisms involve the male's genitalia directly (relaxing barriers inside the female reproductive tract to allow the male to penetrate to optimum sites for sperm deposition); some involve manipulating sperm (e.g., discarding, digesting, or otherwise destroying the sperm of some males but not others); some involve her own gametes (e.g., modulation of ovulation, maturation of eggs, oviposition); and some involve postcopulatory investments of resources in particular offspring or resistance to the attempts of other males to mate with her. Still others involve female physiological processes such as hormonal changes that result in ovulation or maturation of eggs. The focus here is on behavioral traits of males and females.

Seen from another angle, CFC is the result of the difficulties that generally confront a male in his attempts to guarantee that the female's eggs will be fertilized by his sperm. He generally needs the female's help, because males almost never deposit their sperm directly onto the female's eggs in species with internal fertilization; it is also typical that the female, rather than the male or the motility of his sperm, is responsible for transporting the sperm on at least part of their subsequent journeys within her body. Similarly, copulation generally does not automatically result in transfer of sperm to the female, and insemination does not necessarily result in fertilization of all the female's available eggs. A male trait that increases the chances that the female will respond in a way that improves his likelihood of fertilizing her eggs can come under sexual selection by CFC. Thus, for instance, male traits that induce the female to permit the male's genitalia to reach that portion of her reproductive tract where his sperm will have the best chances of surviving and fertilizing eggs, or to refrain from ejecting his sperm from her body, could come under CFC.

Likelihood of Female-Imposed Postcopulatory Biases

Basic morphology suggests that CFC is probably more common than its better known precopulatory equivalent, classic Darwinian precopulatory female choice. Whereas precopulatory competition among males can occur with

Table 1 An undoubtedly incomplete list of possible mechanisms that are under at least partial female control which could bias paternity if the female mates with more than a single male, and thus exercise cryptic female choice

Remate?

Remate sooner rather than later?

Mature more eggs?

Ovulate?

Add more or better nutrients to eggs?

Oviposit more of already mature eggs more quickly?

Transport sperm to optimum sites for eventual fertilization?

Store sperm at different site from other sperm to allow selective use (in species with multiple sperm storage sites)?

Allow male genitalia to penetrate deeply enough to deposit sperm at optimum site for fertilization?

Interrupt copulation before male is entirely finished with sperm and semen transfer and courtship?

Flood reproductive tract with antibodies or other defenses against infections that might damage sperm?

Feed or otherwise nurture sperm received?

Kill sperm received?

Discard sperm from previous male?

Discard sperm from current male?

Abort zygotes from former males?

Resist abortion of zygotes from present male?

Allow male to deposit copulatory plug that impedes future insemination?

Produce copulatory plug that impedes future insemination?

Prepare uterus for implantation (mammals)?

Invest more heavily in rearing offspring prior to their birth?

Invest more heavily in rearing offspring following their birth?

Alter morphology following copulation that makes subsequent copulations difficult or impossible?

relatively little direct female influence, postcopulatory competition between males is generally played out within the female's own body. Even small changes in her reproductive morphology and physiology, such as the volume and chemical milieu of sites where sperm are stored and where they fertilize eggs, can have consequences for a male's chances of fertilization. The multiplicity of the CFC mechanisms is a result of this basic fact, and their large number and often largely independent controls increase the chances that one or more of these critical processes will come to be affected by males. The extreme power asymmetry between the tiny, delicate sperm, and the hulking, complex female, with her extensive array of morphological, behavioral, and physiological capabilities, also emphasizes the likely importance of female choice as opposed to sperm competition among postcopulatory selective processes. Males and their gametes are of course not completely powerless in determining whether or not fertilization will occur; but females are likely to influence the outcome. In an analogy with human sporting events, the female's body is the field on which males compete, and her behavior and physiology set the rules by which competitors must abide and which strategies will be effective. Even small changes in the female can tilt the competition in favor of males with particular traits.

Another reason to think that sexual selection by CFC is an important evolutionary force is that natural selection on many female reproductive traits is expected to easily lead to CFC. Take, for example, oviposition behavior. In most species, natural selection on females favors repression of oviposition until some stimulus associated with copulation signals that the female has sperm with which to fertilize her eggs. Natural selection on females will thus promote the ability to use cues associated with copulation, such as stimulation by the male or his seminal products, to disinhibit oviposition. Once females have evolved this ability to sense such stimuli and to trigger the processes associated with oviposition (e.g., change feeding behavior, search for oviposition sites, move her eggs from her ovaries to her oviduct), then improvements in male abilities to elicit these responses can come under sexual selection by CFC. If a female mates with more than one male, if her oviposition responses to males are not always complete (i.e., not all her mature eggs are always laid prior to mating with another male), and if some males elicit oviposition better than others, then (other things being equal) those males better able to elicit oviposition will outreproduce the others. Once such variant males appear in a population, selection can favor those females that accentuate this bias in fertilization even further. For example, females with higher thresholds for triggering oviposition would tend to lay eggs only after copulating with especially stimulating males, and would be favored because they would produce male offspring better able to stimulate females to oviposit in future generations. Changes in female thresholds, in turn, could set off a new round of evolution of male abilities to stimulate females. Another important consideration is that the polarity of the female responses expected to be favored by natural selection is consistently in the direction favorable to the male (*increase* chances of oviposition, ovulation, and sperm transport, etc.; *decrease* chances of remating, etc.), thus predisposing these female responses to be subject to further male accentuation.

The multitude of possible CFC mechanisms, the theoretical expectations that CFC can evolve rapidly, and the frequent finding that females mate with multiple males in nature suggest that it may also be widespread. Perhaps the most convincing indication that postcopulatory biases are of widespread importance involves the behavior of males during copulation. Male behavior was observed carefully during copulation in 131 arbitrarily chosen species of insects and spiders to determine whether males performed courtship during copulation. Using conservative criteria to define courtship behavior, copulatory courtship occurred in ?.00% of these species. Such behavior is paradoxical in the usual Darwinian interpretation that male courtship functions to induce the female to accept copulation: why should a male continue to court after he is already copulating? There are also reports of similar behavior in other groups, including nematodes, birds, scorpions, frogs, fish, reptiles, millipedes, mammals, molluscs, and crustaceans. If 80% is anywhere nearly representative, then female-imposed postcopulatory paternity biases are probably very common.

Nonetheless, the question of the general importance of CFC, like Darwin's idea of female choice before it, has been hotly debated. Because CFC was only recently carefully formulated and publicized, relatively few thorough experimental tests for its occurrence have been performed. Convincing rejections are intrinsically difficult to obtain, because there are so many different postcopulatory female processes that might be involved and that thus must be checked. In addition, it is harder to see what goes on inside a female than to observe precopulatory courtship. It can also be difficult to distinguish CFC from alternative explanations such as sperm competition and sexually antagonistic co-evolution.

Sexually Antagonistic Co-evolution: An Alternative Hypothesis

The benefit that a female is thought to derive from exercising CFC (i.e., from rejecting certain mates as sires) is improved quality in her sons. An alternative explanation for these rejections, which could also lead to rapid divergent evolution, is that the female thereby reduces the effects of male manipulations that damage her reproductive interests. For instance (to continue using the example of oviposition), a male ability to induce the female to oviposit more quickly following copulation could result in the female laying some eggs at suboptimal times or places. It is possible that male effects that originally evolved as means to win in competition with other males also incidentally result in a reduction in the female's reproductive output. Arnqvist and Rowe pointed out that rapid diversification could then result from sexually antagonistic co-evolution (SAC) of males and females, with each sex evolving new mechanisms to counteract recent advances by the other sex. Female evolution to reduce the number of offspring she loses due to this male effect could result in another round of male evolution to increase the ability to induce females to oviposit. Distinguishing SAC from CFC with direct observations is extremely difficult (and impossible in popular lab species such as *Drosophila* fruit flies and *Tribolium* flour beetles, in which natural habitats are unknown, and it is thus not possible to determine the natural reproductive payoffs of different behaviors). In addition, CFC and SAC are not mutually exclusive, and can act simultaneously or in sequence on the same traits.

There are two different versions of the SAC hypothesis. One emphasizes physical coercion by the male, and has been tested by looking for the predicted species-specific

mechanisms of physical or chemical male coercion of the female and species-specific resistance by the female. There is substantial evidence against such races, including the frequent lack of interspecific differences among females that correspond to the differences among males of the same species; a strong trend in allometric scaling of genitalia of insects and spiders that is opposite in direction to the trend predicted by SAC; the lack of the predicted correlation between coercive male mating attempts and rapid divergent evolution of male genitalia; and a general lack of female structures with mechanically appropriate designs for combating or repelling males. There are some female genital structures that mesh with species-specific male genitalia as predicted by SAC, but these are generally 'selectively cooperative' structures, such as pits, slots, or grooves that facilitate male coupling, rather than 'defensive' structures (such as erectible walls or poles that would prevent male coupling). The female structures are selective in that they facilitate coupling only with males that possess certain structures or forms (as expected under CFC).

A second SAC version emphasizes male stimuli which act as sensory traps. The male produces a stimulus that elicits a particular female response; this female response exists because previous natural selection in another context favored such a response to the same (or a similar) stimulus. An example would be the female oviposition responses to male stimuli during or following copulation that, as noted earlier, originally evolved to prevent the female from wasting eggs by ovipositing before she has copulated. By accentuating or elaborating the oviposition-eliciting stimulus, the male could obtain greater or more consistent female responses and thus win in competition with other males that copulated with the same female. But the female could lose offspring because of precipitous oviposition, and thus suffer net damage from the male manipulations.

The sensory trap version of SAC is less easy to distinguish from CFC, because it does not predict defensive morphological co-evolution in females. It seems less probable a priori, however, because it depends on the female not being able to evolve an effective defense against the male manipulation and thus eliminate the costs she suffers in reduced numbers of surviving offspring. Such a female defense would seem to be via easily evolved as a simple change in her stimulus response threshold, or a modification of her tendency to respond to the stimuli depending on the context in which she receives them. It also supposes that the inevitable benefit from a paternity bias that produces sons better able to stimulate females is consistently overbalanced by the male-produced damage, an empirically very difficult condition to demonstrate convincingly.

Of course, a priori arguments of this sort are less satisfying than conclusions based on data. The strongest empirical argument against this version of SAC is again from genitalia – the lack of the predicted correlation between coercive male mating attempts and rapid divergent evolution of male genitalia that was mentioned earlier. In a survey of many thousands of species of insects and spiders, the male genitalia showed no sign of a trend to diverge more rapidly in groups in which males control (or at least attempt to control) access to resources that are needed by females, and attempt to force or convince reluctant females which arrive to mate in order to gain access to the resources (i.e., in groups in which male reproductive interests are more likely to be in conflict with those of females).

Male Behavioral Traits Probably Under Cryptic Female Choice

Genitalic Morphology and Behavior

One of the most widespread trends in animal evolution is that the male genitalia of species with internal fertilization often evolve especially rapidly and divergently. The most probable explanation involves postcopulatory sexual selection, probably CFC. Data on genitalia are especially important and permit powerful tests because they are extremely abundant. This is because the taxonomists of many different groups discovered long ago that genitalia are useful in distinguishing closely related species, and detailed descriptions of genitalia are available in the extensive primary taxonomic literature on many groups.

Much of the diversity in genital morphology is probably intimately related to genital behavior. But genital behavior during copulation and the functional consequences of such behavior are neglected topics, long overdue for further research. Most studies of the functional morphology of genitalia are distressingly typological, often giving 'the' position of the male without taking into account the probability (given their often complex genital musculature) that the male structures move during copulation. Many surprising phenomena (such as the ability to sing to the female during copulation recently documented in a crane fly, and perforation of the female tract with long spines found in some beetles) probably remain to be discovered. Many details of genital behavior are normally hidden inside the female, but direct observations utilizing both copulating pairs and beheaded male insects (thus removing inhibition by the brain of posterior ganglia), and morphological studies of the genitalia of flash-frozen pairs and of muscle attachments and articulations, can give surprising amounts of information. Recent extension of these observations with X-ray imaging can give even more detailed ideas of genital function.

Probably the most complete studies of genital behavior to date with relation to CFC involve tsetse flies. One portion of the male's genitalia remains outside the female's body and delivers powerful squeezes to the tip

of her abdomen that show species-specific differences in frequency and duration. Earlier studies concluded that stimulation from some aspect of male copulation behavior (rather than chemical cues) induced the critical female responses of ovulation and rejection of future males. They were recently confirmed and extended by surgically removing and altering certain male genital structures, and blocking the female receptors contacted by these genital structures during copulation. These manipulations resulted in reduced sperm transport, reduced ovulation, and greater female acceptance of copulation with subsequent males. The morphology of these male structures varies between closely related species, while that of the portions of the female that they contact are uniform, favoring the cryptic female choice explanation over the sexually antagonistic co-evolution explanation for this case of rapid divergent genital evolution.

Nongenital Male Courtship During and Following Copulation

Male courtship behavior that involves structures other than his genitalia and occurs during or following copulation is common, but also poorly studied. 'Copulatory courtship' behavior patterns include waving, rubbing the female, licking, squeezing rhythmically, kicking, tapping, jerking, rocking, biting, feeding, vibrating, singing, and shaking. If these male behaviors function as courtship, the prediction is that they affect postintromission female responses that increase the male's chances of fertilizing her eggs. Very few studies have tested the prediction.

Tallamy and colleagues studied nongenital copulatory courtship in the cucumber beetle, *Diabrotica undecimpunctata*. The male waves his antennae rapidly over the female's head during the early stages of copulation, when the tip of his genitalia has penetrated to the inner portion of her vagina. If he waves them rapidly enough, the female relaxes the muscles surrounding this portion of her vagina, thus allowing the male to inflate a large membranous sac at the tip of his genitalia and deposit a spermatophore containing his sperm. If she does not relax these muscles, he is unable to inflate the sac and eventually withdraws his genitalia without having transferred a spermatophore. Some females mate with up to ten males before finally permitting a male to inflate his sac and transfer sperm. Females gain superior male offspring by screening males this way, as predicted by CFC. The sons of males which vibrate their antennae more rapidly also tend to vibrate their own antennae more rapidly when they copulate. Studies of three other insects have confirmed that nongenital copulatory courtship induces the female to favor the male's reproduction, by inducing the female to oviposit soon after copulation in a fly, to remain still rather than walking around during copulation in a flea, and to use the current male's sperm rather than that of previous males in a beetle (the female mechanism was not determined).

Again, there are indications, though less definitive than in the case of genitalia, that nongenital copulatory courtship is not the result of coercive coevolutionary arms races between males and females. Male copulatory courtship behavior is generally noncoercive, and inappropriately designed to force the female to continue copulation or to perform other responses leading to fertilization. Indeed, the sites where most possible female processes occur that could prevent fertilization are deep within the female's body, seemingly inaccessible to direct male manipulation via copulatory courtship.

Other Male Traits Possibly Under CFC

An apparently widespread trend for seminal products derived from male accessory glands to frequently affect female reproductive processes in insects and ticks suggests selection on male abilities to affect postcopulatory female reproductive processes via chemicals in their semen. Over 70 species have been studied, with the nearly uniform finding that male seminal products induce one or more of the following female responses: oviposit eggs that are already mature; ovulate or otherwise bring immature eggs to maturation; resist further mating; and (less frequently studied) transport his sperm. Such male products could evolve via CFC or SAC. Some studies in *Drosophila* suggest, though not conclusively, that the effects of seminal products may damage female reproductive interests (remaining doubts stem from the question of whether it is appropriate to draw conclusions regarding why given traits evolved based only on data obtained in fruit fly culture bottles).

CFC may affect the evolution of other nonbehavioral male traits, including sperm morphology, sperm proteins, and the egg molecules with which they interact, and CFC may also occur in plants, affecting both the properties of pollen tubes that influence their ability to grow down the style and find the ovules, and the ability of young zygotes to induce the mother to refrain from aborting them.

Summary

CFC and sperm competition extend the classic Darwinian context of sexual selection to include events that occur after copulation has begun. CFC has been demonstrated in a number of species, and there are reasons to expect that it can evolve readily. Several types of indirect evidence suggest that it may be a widespread and important evolutionary phenomenon, but there are as yet only a few direct demonstrations that it occurs. Further tests, preferably in a variety of different taxonomic groups, will be needed to determine the generality of its importance.

See also: Compensation in Reproduction; Invertebrates: The Inside Story of Post-Insemination, Pre-Fertilization Reproductive Interactions; Sexual Selection and Speciation; Social Selection, Sexual Selection, and Sexual Conflict.

Further Reading

Arnqvist G (2006) Sensory exploitation and sexual conflict. *Philosophical Transactions of the Royal Society B* 361: 375–386.

Arnqvist G and Rowe L (2002) Correlated evolution of male and female morphologies in water striders. *Evolution* 56: 936–947.

Arnqvist G and Rowe L (2005) *Sexual Conflict.* Princeton, NJ: Princeton University Press.

Birkhead T and Møller AP (1997) *Sperm Competition and Sexual Selection.* New York, NY: Academic Press.

Briceño RD and Eberhard WG (2009) Experimental modifications of male genitalia influence cryptic female choice mechanisms in two tsetse flies. *Journal of Evolutionary Biology* 22: 1516–1525.

Briceño RD, Eberhard WG, and Robinson A (2007) Copulation behavior of *Glossina pallidipes* (Diptera: Muscidae) outside and inside the female, and genitalic evolution. *Bulletin of Entomological Research* 97: 1–18.

Cordero C and Eberhard WG (2005) The interaction between antagonistic sexual selection and mate choice in the evolution of female responses to male traits. *Evolutionary Ecology* 19: 111–122.

Delph LF and Haven K (1997) Pollen competition in flowering plants. In: Birkhead T and Møller AP (eds.) *Sperm Competition and Sexual Selection*, pp. 149–174. New York, NY: Academic Press.

Eberhard WG (1985) *Sexual Selection and Animal Genitalia.* Cambridge, MA: Harvard University Press.

Eberhard WG (1996) *Female Control: Sexual Selection by Cryptic Female Choice.* Princeton, NJ: Princeton University Press.

Eberhard WG (in press) Rapid divergent evolution of genitalia: Theory and data updated. In: Leonard J and Cordoba A (eds.) *Evolution of Primary Sexual Characters in Animals*. Oxford, UK: Oxford University Press.

Marshall DC and Cooley JR (1997) Evolutionary perspectives on insect mating. In: Choe JC and Crespi BJ (eds.) *Mating Systems in Insects and Arachnids*, pp. 4–31. Cambridge: Cambridge University Press.

Tallamy DW, Powell RF, and McClafferty JA (2002) Male traits under cryptic female choice in the spotted cucumber beetle (Coleoptera. Chrysomelidae). *Behavioral Ecology* 13: 511–518.

Thornhill R (1983) Cryptic female choice and its implications in the scorpionfly *Harpobittacus nigriceps. American Naturalist* 122: 765–788.

Wiley RH and Posten J (1996) Perspective: Indirect mate choice, competition for mates, and coevolution of the sexes. *Evolution* 50: 1371–1381.

Cultural Inheritance of Signals

T. M. Freeberg, University of Tennessee, Knoxville, TN, USA

Introduction

Culture and cultural traditions are so fundamental to our own behavior as to be one of the key features that define us as human. Two people living a few city streets from one another can differ in the languages they speak, the clothing and body decoration they wear, the faith they practice or do not, and the various rituals they engage in as part of their lives, and these differences are often heavily influenced by the particular culture in which they are raised. Although the influence of culture and the social environment on human behavior has been appreciated for centuries, if not millennia, only in the past few decades has it really become apparent that the behavior patterns of many non-human animal species may also be impacted by the particular environment of social traditions in which they develop. Writers and politicians have noted that it takes a village to raise a child; it may be that it also takes a village-like setting to raise a greater spear-nosed bat, a bottlenose dolphin, an indigo bunting, or a pygmy marmoset.

The cultural transmission of behavior represents one of the many types of social influence on behavior, and is perhaps the one most likely to produce behavioral variants that persist for generations within animal populations. The cultural inheritance of communicative signaling can potentially impact on all the three main aspects of a communicative event between two individuals – (1) signal production: the specific signal variant that is produced by the signaler; (2) signal use: how, when, and toward which receiver that signal variant might be directed by the signaler, and (3) signal recognition: the perception of and locally adaptive response to the signal variant by receivers. Social learning and cultural transmission of behavior have long been assumed to be adaptive, but the functional significance of these learning processes is largely unknown – more research explicitly linking these developmental and evolutionary questions is needed. It is clear that these learning processes typically cause the behavior of the individual to become more like the behavior of the individual's group over time, and groups that are complex in their social structure may place greater demands for social and cultural transmission of behavior compared to groups that are relatively simple in their social structure. In this article, these points are expanded to attempt a summary of the quite new set of research programs on cultural inheritance of signals in the field of animal behavior.

Social Influences and Animal Cultural Traditions

The social environment of an individual can influence its communicative behavior in many different ways. In social facilitation, the presence of other individuals in the immediate social environment can lead to increased production of certain signals by a signaler, such as food calling by male jungle fowl in the presence of females. In social interference, the presence of other individuals in the immediate social environment can lead to decreased production of certain signals by a signaler, such as diminished grunting in the presence of high-ranking individuals in non-human primate species. Social learning is a set of processes by which the behavior of one individual develops through the experience of interacting with, or observing the behavior of, other individuals. For example, in some songbird species, aggressive interactions with adult males affect the development of songs in young males, whereas in other species affiliative interactions play an important role in song development. The presence of group members communicating in a specific area or near a specific stimulus may cause a naïve individual to attend to that area or stimulus, resulting in it learning something about communicative interaction in that area or near that stimulus through individual learning processes like classical conditioning or operant conditioning. Alternatively, the social environment of a young individual may be important only to signal learning of that individual through the set of acoustic stimuli generated, from which that young individual derives its signals. For example, in many songbird species, nestlings and fledglings of a territorial pair are often surrounded by a large number of territorial pairs that are also raising nestlings or fledglings, and the songs a young bird develops can often be predicted on the basis of songs of adults in their surrounding environment. In this case, it is possible that population-level differences in vocal signals may stem from the acoustic experiences of developing individuals alone, and not from the kinds of social interaction that are fundamental to social learning and cultural transmission (as defined in the following paragraph).

The cultural inheritance of signals refers specifically to the notion of communicative cultures (or communicative traditions) in animals. A communicative culture can be defined as a population- or group-specific system of signals, responses to those signals, and preferences for the individuals toward which those signals and responses might be directed, that is socially learned and transmitted across generations. There are four important components to this definition. First, a communicative culture is specific to a population or group. Behavioral variation that occurs only at the level of individuals or only at the level of species does not constitute the cultural inheritance of signals. Second, although signals are obviously important to a communicative culture, how those signals are used and the recognition of those signals and the ways in which receivers respond to those signals are also important (though these have been much less studied). Third, a communicative culture is socially learned, so it is important to determine the social mechanisms of this learning. For example, the communicative development of young individuals might be influenced by aggressive behavior of, or by affiliation with, older individuals, and by the nature and duration of the social relationship with older individuals. The fourth and final point relates to the multigenerational feature of cultural traditions. If a researcher conducts an experiment demonstrating that a set of young individuals has learned certain signals and responses characteristic of its communicative culture, can that set of experimental subjects then be used as a set of social models for an additional set of young individuals in the development of their communicative culture?

There have been two major approaches taken by researchers to try to document communicative cultures in non-human animals. The first major approach is a naturalistic observation study, in which a population or populations of a particular species are studied for many generations, or are behaviorally sampled at one time and then again many generations later. The researcher documents stability in the signals produced by individuals in each population over these long periods of time, and often tries to rule out other potential explanations for the behavioral stability, such as genetic differences or influences of the physical habitat. A strength of this approach is that the research is typically done in natural field settings, and so has obvious biological relevance. A weakness of this approach is that the researcher is often quite limited in the extent of causal inference that can be made about the specific role of social learning. The second major approach is an experimental study conducted in captive settings. The researcher controls the social environments of young and relatively naïve individuals, and documents the development of the communicative culture in those young individuals. A strength of this approach is that if the researcher finds that subjects raised in different social environments develop aspects of the communicative cultures of those different social environments (and has collected data pointing to the role of social interactions influencing that development), strong causal inferences can be made regarding social learning and cultural inheritance. A weakness of this approach is that if the experiment manipulates only certain, possibly less important,

characteristics of the communicative culture in the particular species, the work may be limited in its biological validity. There are a small number of research programs that have combined both approaches, however, to gain a deeper understanding of the social transmission of communicative traditions.

Social Traditional Influences on Signal Production, Use, and Recognition

In many species, there is considerable geographic variation in the communicative systems used by individuals: populations within species often differ from one another, locales within populations often differ from one another, and groups (flocks, pods, troops, etc.) within locales often differ from one another. For example, contact calls and social cohesion vocalizations differ among groups – even among groups in close proximity to one another – in a wide variety of species (black-capped chickadees, bottlenose dolphins, budgerigars, greater spear-nosed bats, killer whales, and pygmy marmosets, to name a few). In some of these species, when individuals possessing acoustically distinct contact calls or whistles are placed into the same groups, the acoustical properties of their vocalizations can change such that individuals converge on common acoustic themes. For example, in greater spear-nosed bats, a screech call is used in long-range signaling to facilitate group cohesion in foraging contexts. The screech calls of individuals from the same groups are acoustically similar to one another, but calls differ between groups. Boughman recorded individuals of different groups in captive settings, and then swapped individuals across groups, an experimental manipulation that models natural changes to group membership. Individuals placed into new groups modified their screech calls in ways that made them more acoustically similar to the screech calls of members of their new groups.

As these bat studies indicate, there is some evidence of acoustic modification of vocal signals in certain mammalian species, typically in contexts in which group membership or structure changes. In many mammalian species, however, the extent of acoustic modifiability seems relatively limited. For example, in non-human primates, cross-fostering studies involving different species often result in the cross-fostered individual producing signals that are acoustically indistinguishable from those of the adults of its own species (and not the cross-fostered host parents). On the other hand, there is more evidence of social experience playing a role in the development of vocal usage and in the development of vocal recognition processes. The major exception to limited vocal plasticity in mammals is cetaceans. There is evidence of processes like matrilineal inheritance of vocal patterns (with genetic inheritance ruled out) across generations of free-ranging cetacean species. There is also evidence of actual vocal learning under changing social contexts in smaller cetaceans housed in controlled captive settings.

Songbird species have long been studied by researchers interested in learning and communication, because there is considerable evidence that one of the key behavioral patterns used in courtship – male song – is strongly influenced by experiential background. Furthermore, evidence increasingly suggests that how (and how effectively) males use songs, as well as female preferences for songs, can be influenced by experiential background. For example, studies of captive groups of brown-headed cowbirds indicate that population-level differences in male song can be socially learned from adult social models by a first set of young experimental subjects. When those experimental subjects from the first set become adults, they can serve as adult social models and pass those population-level differences in song on to a second set of young experimental subjects. Female cowbirds that developed in those same experimental (and social experiential) backgrounds were found to prefer males, on the basis at least in part of those song differences, indicating that female preferences for song and female mating patterns could be socially transmitted across generations as well. Other studies indicate that not just mating patterns, but even mating systems (how and to whom cowbirds direct their communicative behavior, and whether individuals behave in polygynous or monogamous ways during the breeding season) can be impacted by the communicative culture in which young individuals develop.

Among the strongest research programs on communicative cultures are field-based studies with indigo buntings. The Paynes and their colleagues have documented male song repertoires, male mating success, and female mating decisions in specific populations for over 20 years. Local song traditions (locales within population) exist across generational time; certain song variants seen in some populations have been found to persist for several generations; and social learning of song seems to be the best explanation for the patterns of song use in these populations. Furthermore, the particular song types used by males during the breeding season correlate with mating success of those males – some song variations are strongly associated with successful mating and others are not. Captive studies that control the experiences of young males find that these males learn more vocal material from older males with which they can interact than from older males with which they cannot interact. Taken together, these various research programs are increasingly linking social learning processes to vocal signals, signal use, and preferences for those signals. Furthermore, these programs are beginning to link such signals and preferences to actual mating and reproductive success, which will bring us closer to understanding the biological significance of cultural inheritance of signals.

Biological Significance of Social Transmission of Signals

Researchers and theoreticians have often assumed that social learning and cultural inheritance are adaptive processes. There is clearly evidence that socially learned vocal signals, signal use, and signal recognition (or preference) can influence mating decisions. Thus, these processes of social transmission would appear to have obvious functional relevance. On the other hand, it is an open question whether variation for social learning ability within populations correlates with reproductive success. Indeed, there are a number of basic questions about the adaptiveness of the cultural inheritance of signals that we still cannot answer. For example, are communicative cultures as described here only found in a handful of avian and mammalian orders, or does the relative paucity of data on such cultures currently hide the fact that the cultural inheritance of signals is actually much more widespread? In species where communicative cultures have been demonstrated, is it the case that social learning processes are less costly than nonsocial learning processes in signal development, as is assumed by much theory and many models of the evolution of social learning? In many species, for example, there are populations that are resident in an area year-round and populations that are migratory. Simply being resident for a long time with a relatively stable number of neighbors could make possible extended periods of social interaction among young and older individuals. This could provide the context for cultural inheritance that might not be seen in migratory populations. Young individuals that migrate might be in groupings that are continually in flux and also might be in contact with a large number of individuals of behaviorally distinct populations, and the social learning of appropriate communicative behavior might be extremely costly. These patterns are illustrated in song repertoires of white-crowned sparrows. Sedentary populations of these sparrows in California show stable local dialects but migratory populations in eastern North America do not.

Unlike traits that are congenital or much more heavily influenced by genetic transmission, characters that are transmitted socially can spread rapidly through populations. The rapid spread of traits via social transmission can alter selection pressures and subsequent evolutionary change in those populations, as individuals modify or manipulate their social and/or physical environments with these new behavior patterns. This may be particularly true in the case of a communicative culture, when new signals, new ways of using signals, or new patterns of recognition or perception of signals emerge and are passed on to new generations. For example, in the aforementioned cases of indigo buntings and brown-headed cowbirds, patterns of mating and mate preferences can be influenced at least in part by the social traditional background of individuals. As such, communicative cultures can influence sexual selection processes operating on these populations.

It is thought that socially transmitted behavior is adaptive in populations in which the environment changes rapidly enough that strict genetic changes influencing those traits would be too slow for populations to track such changes effectively. On the other hand, it is thought that if environmental changes are too rapid, socially transmitted behavior (particularly the kinds of behavior that would be passed down from one generation to the next, as in the case of communicative cultures) would not be as effective a learning mechanism as individual trial-and-error learning. As was raised earlier, however, these arguments are still at the level of theory, and empirical data are needed to test the ideas.

There are a number of exciting research avenues that could be taken in developing a greater understanding of the biological significance of communicative cultures. The most profound results are likely to come from integrative methodological approaches to this very integrative question – making tighter experimental connections between communicative cultures as contexts for ontogeny, and communicative cultures as selection pressures operating on interacting individuals. We also need to understand more fully how environmental constraints influence cultural inheritance of signals. It is well known that different physical habitats can impact on signal transmission in different ways, regardless of the signaling channel. Thus, for example, populations of a particular species that reside in deep and undisturbed forested habitat are more likely to face different selection pressures for vocal signal design compared to populations of the same species that reside in more open and mixed habitat of urban environments – if so, the communicative cultures that might exist in these different populations would be constrained by these differences in habitat.

There are many unanswered questions in terms of the different learning strategies that young organisms might employ in the development of communicative cultures. Assuming a population in which young and naïve individuals develop signal production, use, and recognition at least in part through cultural inheritance, we need to begin to document how young individuals vary in the assessment strategies they use in their social learning. There are obvious costs to active assessment of social environments – for example, time, energy, opportunity, and predation costs. Different assessment strategies likely vary in potential costs and benefits. For example, should naïve individuals copy the communicative behavior of individuals that are reproductively successful, or should they avoid copying the behavior of unsuccessful individuals? Should naïve individuals play a conformist strategy and copy the most common communicative behavior patterns existing in the population, or should they base their decisions on other criteria such as their genetic similarity or social relationship with specific behavioral models? Do young and naïve individuals that are more socially interactive (i.e., extraverted in personality psychology terminology) exhibit richer patterns of communicative behavioral development than do individuals that are less socially interactive? The last two question touch on the notion that the complexity of social groups plays an important role in the social transmission of behavior, a subject to which we turn next.

Social Complexity and Cultural Inheritance

Over the past few decades, theory and empirical evidence increasingly point to the idea that species in which individuals live in large social groups may require greater cognitive ability and greater diversity in their communicative repertoires than species containing only solitary individuals or individuals living in smaller groups. This increased cognitive and communicative complexity is thought to be especially important in species that are long-lived, that reside in groups in which generations overlap, and in which individuals maintain long-term social relationships with others. In such groups, communicative systems are thought to be under selection pressure to increase in complexity, either to allow a greater number of distinct messages (such as detection of food or a predator) to be transmitted by group members or to allow signalers to convey a greater range of emotional and motivational signals to receivers in the group. Individuals may not be able to develop such complex communicative systems if only nonsocial learning processes of behavioral transmission exist, such as genetic inheritance or individual trial-and-error learning. Thus, in species in which individuals occur in large and complex social groups, cultural inheritance of signals may be necessary to make those complex systems of communication possible.

Almost all of the research on communicative cultures as described earlier, and on the links between social complexity and communicative complexity, has focused on vocal signals. Very little work has been carried out on visual displays, despite the fact that the same arguments made earlier for vocal signaling should hold for visual signaling and also despite the fact that we have known about the communicative importance of facial expressions and visual gestures at least since Darwin's *The Expression of the Emotions in Man and Animals*. Perhaps the main reason for this enormous difference in research effort and output is technological – for decades, it has been possible for researchers with small operating budgets to carry relatively lightweight audio recording equipment into field settings. The same had not been true for video recording equipment until very recently. Hopefully, studies of social learning and visual displays will become much more commonplace in the near future, so we can answer many of the questions raised earlier with visual signals. Nonetheless, some exciting findings are already emerging. Maestripieri, for example, analyzed gestures of three species of macaques, and found that the species with the simplest social organization (rhesus macaques) had the smallest visual signaling repertoire. This lends support to the notion that social complexity might drive signaling complexity. It would be interesting to know whether in each of these macaque species there is population-level variation in their repertoires of gestures and whether that variation may in part be due to social transmission. It is well known in humans that there are some gestures that are relatively universal, but others the meaning of which is very tightly linked to the particular culture of the signaler.

In psychology and anthropology, culture is often viewed as being a filter to possible human behavioral variation. It is not difficult to think about certain behavioral practices that your own culture views as perfectly acceptable and that another culture would prohibit or find abhorrent, and vice versa. A filter is a useful analogy for thinking about communicative cultures in non-human animals, in that a filter constrains the material passing through it and selects certain variants rather than others. Regarding the question of social complexity, we might expect populations or species in which individuals occur in complex groups to have more cultural filters affecting their behavioral development, compared to populations or species in which individuals occur alone or in simple groups. If so, species in which individuals occur in large and complex social groups should exhibit complex communicative cultures. As described earlier, researchers have noted in non-human primates (that tend to have complex social structures) that there is relatively little evidence of social learning in vocal production, but perhaps vocal use and vocal recognition are much more linked to the particular communicative culture in which young primates develop. Perhaps we need to focus more on visual rather than vocal signals in these species, if they rely considerably on expression and gesture to manage the behavior of others and to maintain social relationships.

The cultural inheritance of signals is a fairly new research area in animal behavior, but one that is quite exciting and may have large payoffs for our understanding of the developmental and evolutionary factors influencing systems of communication. Indeed, given the way communicative cultures link developmental and evolutionary questions, work in this area will likely elucidate the ways developmental processes can influence evolutionary processes operating on those populations, and in turn how evolutionary processes affect the way young and naïve individuals socially learn to communicate effectively with members of their group. Furthermore, increased knowledge about the cultural inheritance of signals in non-human animals will likely shed light on the evolution of culture, complex cognition, and communication and language in our own species.

See also: Acoustic Signals; Anthropogenic Noise: Impacts on Animals; Communication Networks; Dolphin Signature Whistles; Motivation and Signals; Visual Signals; Vocal Learning.

Further Reading

Avital E and Jablonka E (2000) *Animal Traditions: Behavioural Inheritance in Evolution*. Cambridge: Cambridge University Press.

Bonner JT (1980) *The Evolution of Culture in Animals*. Princeton, NJ: Princeton University Press.

Boughman JW (1998) Vocal learning by greater spear-nosed bats. *Proceedings of the Royal Society B* 265: 227–233.

de Waal FBM and Tyack PL (eds.) (2003) *Animal Social Complexity: Intelligence, Culture, and Individualized Societies*. Cambridge, MA: Harvard University Press.

Freeberg TM (2000) Culture and courtship in vertebrates: A review of social learning and transmission of courtship systems and mating patterns. *Behavioural Processes* 51: 177–192.

Freeberg TM and White DJ (2006) Social traditions and the maintenance and loss of geographic variation in mating patterns of Brown-headed cowbirds. *International Journal of Comparative Psychology* 19: 206–222.

Janik VM and Slater PJB (2003) Traditions in mammalian and avian vocal communication. In: Fragaszy DM and Perry S (eds.) *The Biology of Traditions: Models and Evidence*, pp. 213–235. Cambridge: Cambridge University Press.

Laland KN and Hoppitt W (2003) Do animals have culture? *Evolutionary Anthropology* 12: 150–159.

Maestripieri D (2005) Gestural communication in three species of macaques (*Macaca mulatta, M. nemestrina, M. arctoides*): Use of signals in relation to dominance and social context. *Gesture* 5: 57–73.

Payne RB (1996) Song traditions in indigo buntings: Origin, improvisation, dispersal, and extinction in cultural evolution. In: Kroodsma DE and Miller EH (eds.) *Ecology and Evolution of Acoustic Communication in Birds*, pp. 198–220. Ithaca, NY: Cornell University Press.

Rendell L and Whitehead H (2001) Culture in whales and dolphins. *Behavioral and Brain Sciences* 24: 309–324.

Snowdon CT and Hausberger M (eds.) (1997) *Social Influences on Vocal Development*. Cambridge: Cambridge University Press.

Whiten A and Byrne RW (eds.) (1997) *Machiavellian Intelligence II: Extensions and Evaluations*. Cambridge: Cambridge University Press.

Culture

S. Perry, University of California-Los Angeles, Los Angeles, CA, USA

Introduction

Over the past few decades, scientists have become increasingly interested in the question of whether animals exhibit 'culture.' The answer to this question depends, of course, on the way that culture is defined. The aspect of culture that makes the topic exciting to biologists is the fact that culture is created and maintained by social-learning mechanisms – that is, culture is a social (i.e., nongenetic) means by which traits can be inherited. Thus, most work on animal culture by biologists and psychologists has focused on the population dynamics of behavioral traits and the identification of social-learning mechanisms.

To most anthropologists, however, there is much more to culture than just social learning. Tylor, the father of cultural anthropology, defined culture as "the complex whole which includes knowledge, belief, art, law, morals, custom, and any other capabilities and habits acquired by man as a member of society" (Tylor, 1924, p. 1). There have been many modifications of this definition, most of which eliminate the possibility of culture in animals by definition. These definitions tend to emphasize not only the social inheritance of behavioral traits, but also group-specific sharing of ideas, beliefs, emotions, and rules (social norms or even explicit laws) for conducting social life. Many definitions also emphasize shared symbolic communication and public rituals in which social rules are symbolically reinforced and group identity is confirmed.

Such definitions, when applied to non-humans, are virtually impossible to test across species. How could you determine whether animals have particular beliefs, emotions, or a sense of group identity? Therefore, researchers of animal culture have largely ignored the topics of social norms, symbolic understandings, and group identity and have focused on the (somewhat) more tractable issue of documenting the extent to which social influence affects the transmission of particular behavioral traits. Those researchers who are uncomfortable using the label 'culture' for non-human animals use the less controversial term 'tradition,' which was defined by Fragaszy and Perry (2003, p. xxxi) as "a distinctive behaviour pattern shared by two or more individuals in a social unit, which persists over time and that new practitioners acquire in part through socially aided learning."

Experts in the field of animal culture hold a wide range of views regarding the issue of whether animal culture is essentially the same as human culture or different from it. This range is presented in the volume edited by Laland and Galef (2008), *The Question of Animal Culture*, and most of the ideas presented in this article have been expressed by one or more contributors to this edited volume, that is recommended as Further Reading. Researchers such as McGrew and de Waal tend to emphasize the continuities between human and non-human culture, noting that many of the differences emphasized by skeptics are quantitative, rather than qualitative, and that further research in this relatively new field is likely to reveal new information that closes, or at least narrows, the gap between humans and non-humans. Other researchers, such as Hill and Tomasello, are more convinced that there are profound differences in human and non-human forms of culture.

Do Non-human Animals Have Traditions?

Regarding the most crucial component of culture – social inheritance of traits – it seems clear that there are striking continuities between humans and non-humans. If culture is defined so that it is essentially synonymous with traditions, then it seems clear that 'culture' is a common feature of many animal societies. Social traditions have been documented in taxa as diverse as primates, rodents, bats, cetaceans, birds, fish, and insects (to name a few).

Although social learning has been implicated as the probable mechanism for creating and/or maintaining the patterning of traits often labeled cultural in these studies of animals, it is possible in many cases to quibble over the quality of the evidence. Similar quibbles could be raised for most claims of human cultural transmission as well, if these studies are held to the same methodological standards as studies of traditions in animals. Ironically, the scientific methodology for documenting traditions in animals is typically more rigorous than that used for documenting traditions in humans. However, even if we cannot confidently state the precise extent to which social influence affects trait distribution in humans and various other animal species, we can confidently state that social traditions are common features of many non-human animal societies.

Current methods for diagnosing traditions in the field and for counting elements in cultural repertoires are inadequate for making reliable cross-species comparisons between populations or species. Methodological differences make it particularly difficult to directly compare cultural repertoires and cultural dynamics between human and non-human populations. However, even in the absence of solid scientific evidence, it seems obvious that humans have larger cultural repertoires than any other species of animal. Further, comparisons among non-human taxa are premature; at this point in the development of the field, the size of the repertoire of traditional behaviors described for a species may tell us more about research effort than about actual species or population differences. In some of the better-studied species (e.g., chimpanzees, orangutans, macaques, and capuchins), it has been possible to document multiple putative traditions in each social group or population examined (9–24 in the case of the chimpanzees studied by Whiten and his colleagues), with each social group having a unique set of traditions. Clearly, non-human animals are not all 'one-trick ponies' when it comes to traditions. Humans and non-humans fall along a continuum with regard to cultural repertoire size, with humans being an extreme outlier. Finding out where, exactly, each species lies on this continuum is probably not a realistic research goal.

Do the Mechanisms of Social Learning Vary Between Humans and Non-humans?

Members of any species vary widely in the specific cultural traits that they exhibit. However, researchers tend to assume that all the members of a species share roughly the same cognitive abilities, that is, that there is within-species uniformity with regard to the types of social-learning mechanisms that all the members of any species can deploy. Species members may exhibit some adaptive and some maladaptive cultural behaviors. However, natural selection will act on phenotypes in a way that favors cognitive mechanisms that produce favorable outcomes most of the time. Thus, many researchers feel that cognitive mechanisms themselves, rather than any specific products of such mechanisms, should be the primary focus of researchers interested in explaining the evolution of cultural capacities. Consequently, researchers who are specifically interested in explaining the origins of the human form of culture have tended to focus on cognitive differences between humans and other apes.

A wide variety of social-learning mechanisms have been identified in both humans and non-humans. It is widely assumed that imitation (which I define as the precise copying of the motor patterns observed in another individual) and teaching (defined here as 'modified behavior by an experienced individual in the presence of a naïve individual, such that the naïve individual learns the behavior more quickly than it would otherwise and at some cost to the teacher') are the two means of social learning that can result in greatest fidelity in the social

transmission of motor skills. Many researchers have implied that imitation and teaching are the cognitive specializations that make human culture different from non-human culture. However, as more data have come in, it has become evident that an ability to imitate or teach does not provide a clean distinction between social learning in humans and non-humans. For example, there are many species (e.g., chimpanzees, orangutans, capuchins, pigeons, dolphins) that appear to be able to imitate under some circumstances, although they generally do not imitate as spontaneously, as frequently or as skillfully as do humans.

Recent experimental studies have compared the tendencies of chimpanzees and human children to copy the actions of models explicitly rather than to emulate (which I define as focusing primarily on the outcomes of a model's actions). Human children were more likely than chimpanzees to focus on the precise details of the techniques used to solve a problem, although both children and chimpanzees were prone to focusing on the outcomes of the actions they observed. Another putative difference between chimpanzees and human children is that humans seem to imitate for the sheer pleasure of imitating – that is, they engage in 'social imitation,' presumably because it helps them feel more like the people they are imitating. Researchers such as de Waal would argue that imitating because of a feeling of identification with others and a desire to 'fit in' is a feature not unique to humans. However, imitation of arbitrary traits is certainly more prevalent in humans than in non-humans.

Examples in non-humans of true teaching are few (though some reports are available for meerkats, ants, chimpanzees, and cetaceans). However, more may emerge as greater research effort is devoted to their discovery. It is worth noting that although teaching is a pervasive form of social transmission in developed societies where formal education is common, explicit teaching is actually rare in hunter-gatherer and horticultural societies. Nonetheless, members of hunter-gatherer societies manage to acquire an impressive array of culturally based skills.

Probably the prevalence of imitation and teaching (including both face-to-face teaching and teaching in the form of written instructions) in the human 'social-learning toolkit' is what enables humans to exhibit 'cumulative cultural evolution' (or the 'ratchet effect'). By faithfully copying others' innovations and then adding new twists to improve on current cultural variants, humans can accumulate technological improvements over many generations. Thus, new generations of humans do not have to 'reinvent the wheel.' They can build on the accomplishments of past generations. Although some fairly feeble examples have been proposed for 'ratcheting' in non-humans, it is clear that no other species comes close to humans in its ability to accumulate cultural modifications.

Does Cultural Content Vary Significantly Between Humans and Non-humans?

Most of the traditions that have been proposed for non-human animals relate to foraging strategies and tool use of various sorts. Vocal traditions (i.e., local dialects or group-specific calls) are common in some taxa of birds, but rare in mammals, with some notable exceptions among cetaceans, pinnipeds, and bats. Although there are a few noteworthy cases of gestural social conventions in the literature (e.g., capuchin bond-testing rituals, chimpanzee leaf-clipping displays, and grooming conventions in chimpanzees and macaques), cultural innovations that have a purely communicative function are rare in non-human primates and are probably rare in other animals as well, although there has been little research effort devoted to their discovery. In stark contrast, human cultures invariably include a rich mixture of social signals, normative rules, and linguistic traditions. Although human cultures also contain many technological components, the salient aspects of cultures tend to be social conventions.

Non-human communication is certainly not devoid of symbolic content (e.g., referential alarm calls have been documented in some primate species). However, it does seem clear that there is considerably more culture-specific symbolic content in human than in non-human communication systems. Also, core ideologies seem to link various traditions into clusters in human cultures, such that language family or geographic proximity is often a better predictor of what traditions will be present in a particular culture than ecological factors such as habitat type or subsistence strategy. Such symbolic linkage among traditions according to core ideas does not appear to be characteristic of ape cultures, though it is admittedly difficult to test whether apes share 'core ideas.'

Other features of human culture that appear to be lacking in non-human animals are social norms, laws, and moral codes. Many elements of human culture seem to have been designed to regulate the behavior of group members. Thus, human cultures typically contain rules about how to share or compete for resources and mates, rules governing the way in which labor is partitioned among group members, and rules governing the ways in which people of different genders and social status are allowed to interact. These rules and norms are often reinforced by public rituals in which all the group members participate, and these rituals are often highly emotional events that serve to promote a sense of group identity. Religious beliefs and stories may further reinforce norms by emphasizing the consequences of conforming or defecting.

It is interesting to note that in experiments with human children, subjects spontaneously protest against the actions of third parties who do not conform to arbitrary conventions in the way they carry out certain actions in experimental situations. Human children appear to be cognitively designed not only to copy behavior patterns that they witness, but also to spontaneously exhibit moralistic outrage toward deviant behavior and a desire to punish those who do not conform.

Although it seems clear that non-humans do not exhibit complex public rituals for the purpose of enforcing societal rules and establishing group identity (as such behavior would be highly conspicuous), it is difficult to determine whether non-human animals have social norms at all. Members of some monkey species have socially inherited ranks, with youngest daughters taking a place in the hierarchy just below the mother's rank and just above that of the next oldest sibling, and some researchers have suggested that this arrangement might qualify as a social norm or at least as a form of 'social culture.' Certainly, macaques (e.g., *Macaca mulatta*) learn their place in the social hierarchy from one another, and rapidly learn, either from direct experience or watching others, how to interact appropriately with dominant and subordinate individuals.

There are population-specific differences in aggressiveness in baboons (*Papio*) and tendencies to reconcile in macaques (*Macaca*) that may also qualify as social norms. However, knowing how these differences arise and are maintained is difficult. There is little or no evidence of punishment in response to violations of social norms in non-humans, and such evidence is critical to making the case that non-humans and humans have a similar cognitive underpinning for establishing how social behavior should be conducted.

Clearly more research is necessary before a final verdict can be reached regarding the presence or absence of social norms in non-humans. Such data will be hard to obtain because frequent behavioral observations over long periods are required to document the presence and maintenance of social norms. At present, it seems safest to assume that enforcement of social norms is, at best, far rarer in non-humans than in humans.

Assuming that humans are more prone to creating social norms, moral rules, and laws than are non-humans, what might explain this difference? One part of the answer may lie in the greater social complexity and propensity for cooperation with large numbers of conspecifics found in humans relative to non-humans. Unlike most animals, humans have multitiered societies, although hamadryas baboons (*Papio hamadryas hamadryas*) and bottle-nosed dolphins (*Tursiops truncatus*) have layered societies that approach the complexity of those of humans in that there is cooperation between higher-level social units. Humans are, however, probably unique in the extent to which there is cooperation even between individuals who barely know one another or have not even met, but who are either distant kin or are symbolically connected in some way. Such cooperation would probably be impossible without the aid of language, which enables people to bear messages to others.

Humans are also unique among vertebrates in the extent to which they have specialized division of labor. Although some gender- and age-specific partitioning of labor for the purpose of rearing offspring occurs among some monogamous species and cooperative breeders (e.g., many birds, social carnivores, and callitrichids (Callitrichidae)), human societies both exhibit a far wider variety of specialized labor and incorporate larger numbers of individuals into cooperative networks. The elaborate coordination of culture-specific ways of partitioning labor and cooperative roles would probably be impossible without language.

Another key element that probably differentiates humans from non-humans psychologically is the range of emotions experienced. Fessler has argued that shame, pride, and perhaps moral outrage are emotions experienced only by humans. It is difficult to test this idea directly, but if this view is correct, then it could explain why third party punishment is so common in humans relative to other species. Furthermore, the mechanisms of shame, pride, and punishment for deviation might explain why humans so easily establish arbitrary behavioral conventions such as particular dress codes or table manners.

Because human life has so many complex, culture-specific rules for how to behave properly, it is useful to an individual to prefer to establish cooperative relationships with others who share the same views and practices. Otherwise, coordination becomes difficult. McElreath argues that the need to reliably identify like-minded individuals sets the stage for the evolution of ethnic markers. Particular arbitrary traditional behaviors (e.g., ways of dressing or speaking) are symptomatic of a greater complex of rules for conducting social life. Members of the same cultural group have a strong sense of identity with their group and a sense that the ways of their own people are morally superior to those of other groups. Powerful emotions can be triggered by violations of a group's traditional behaviors.

Non-human animals may, in some cases, have a sense of group identity. For example, many species of animals exhibit striking levels of aggression toward members of other groups that is not a simple case of identifying some individuals as familiar and others as unfamiliar. For example, capuchin monkeys (*Cebus capucinus*) show sharp changes in the way in which they interact with former members of their group after they migrate to new groups. However, no case can be thought of in which witnessing the performance of particular traditional behaviors (e.g., a particular foraging technique or way of grooming) elicits a strong emotional response in individuals who exhibit a different way of accomplishing the same goal. Consequently, it seems unlikely that, in non-humans, particular behavioral traditions play a role as ethnic markers in establishing group identity. And why should they, if the function of ethnic markers is to facilitate identification of suitable partners for cooperative activities?

Because there is very little variation in the ways in which members of a non-human animal species conduct their daily lives, there should be fewer coordination costs in cooperating with members of different social groups than there are in humans. Also, for ethnic markers to serve a useful function, they must provide more information about cooperative potential than is available to individuals by direct observation or by means of reputation. In most animal societies that have complex cooperative relationships, there is so much face-to-face contact with potential allies that direct sources of information are likely to be far more reliable than information gleaned from 'ethnic markers.'

Conclusion

Traditions are prevalent in a wide taxonomic range of species, including primates, cetaceans, birds, rodents, fish, and even some insects. Thus, nongenetic inheritance of traits is likely to be an important feature of the behavioral biology of numerous species. Increased research effort regarding the precise mechanisms of social learning across taxa has blurred the prior distinction between humans and non-humans by revealing some instances of imitation and teaching in non-humans, although it still seems that humans imitate and teach with a far greater facility than other species.

Despite the important similarities between human and non-human animals, many researchers are uncomfortable using the term 'culture' to describe collections of traditions in non-humans because they feel that social inheritance is not the only important feature of culture in humans. Language enables humans to form and maintain large cooperative networks that can include individuals who have never met. Human cultures, unlike non-human 'cultures,' include much symbolic content and many social conventions that serve the purpose of regulating other group members' behavior.

In humans, there is considerable between-group variation in behavior, and it is easier to cooperate with individuals who share their own habits and social norms. These ethnic markers – seemingly arbitrary cultural elements that signal membership of particular ethnic groups – can serve as cues of group membership, and such emotionally salient traditions associated with group identity are an apparently unique feature of human culture.

See also: Apes: Social Learning; Imitation: Cognitive Implications; Punishment; Vocal Learning.

Further Reading

Box H and Gibson K (eds.) (1999) *Mammalian Social Learning: Comparative and Ecological Perspectives.* Cambridge, UK: Cambridge University Press.

de Waal FBM (2001) *The Ape and the Sushi-Master: Cultural Reflections of a Primatologist.* Cambridge, MA: Harvard University Press.

Fessler D (1999) Toward an understanding of the universality of second order emotions. In: Hinton A (ed.) *Biocultural Approaches to the Emotions*, pp. 75–116. New York: Cambridge University Press.

Fragaszy D and Perry S (eds.) (2003) *The Biology of Traditions: Models and Evidence.* Cambridge, UK: Cambridge University Press.

Heyes CM and Galef BG (eds.) (1996) *Social Learning in Animals: The Roots of Culture.* New York: Academic Press.

Laland KN and Galef BG (eds.) (2008) *The Question of Animal Culture.* Cambridge, MA: Harvard University Press.

McElreath R, Boyd R, and Richerson PJ (2003) Shared norms and the evolution of ethnic markers. *Current Anthropology* 44: 122–129.

Richerson P and Boyd R (2005) *Not by Genes Alone: How Culture Transformed Human Evolution.* Chicago: University of Chicago Press.

Tomasello M (1999) *The Cultural Origins of Human Cognition.* Cambridge, MA: Harvard University Press.

Tylor EB (1924) (originally 1871) *Primitive Culture*, 2 Vols, 7th edn. New York: Brentano's.

Whiten A, Goodall J, McGrew WC, et al. (1999) Cultures in chimpanzees. *Nature* 399: 682–685.

Whiten A, Horner V, and Marshall-Pescini S (2003) Cultural Panthropology. *Evolutionary Anthropology* 12: 92–105.

Zentall TR and Galef BG (eds.) (1988) *Social Learning: Psychological and Biological Perspectives.* Hillsdale, NJ: Erlbaum.

Dance Language

F. C. Dyer, Michigan State University, East Lansing, MI, USA

Introduction

The dance language of honeybees (genus *Apis*) is one of the most astonishing behavioral traits yet documented in the animal world. The dance is used by foraging bees to signal the location of food or other resources to their nest mates. What is most amazing about this is that the communication seems to involve a symbolic code – through movements of her body on the comb, the dancer encodes the direction and distance of a foraging site not currently in view, and the bees following the dance can use these instructions to fly to the same site even if the forager does not return. Outside human language, it is hard to find examples of animals communicating about objects, events, or locations that are remote from the context in which the information is exchanged. That an insect is doing this makes the behavior especially astonishing – all the more so to people who regard insects as simple automata.

Regardless of whether such preconceptions have any validity, the dance language raises deep questions about the physiology, development, and evolution of animal behavior; hence, it has long been a model system in ethology. The extensive scientific literature has been the focus of a number of comprehensive reviews. Karl von Frisch, who was responsible for decoding the dance and discovering many important elements of the underlying mechanisms, published a book in 1965 reviewing decades of work by himself and his students. This book is still a rich source of insights and an inspiring portrait of scientific discovery, even though more recent reviews have been published summarizing work done since the 1960s.

The major emphasis of this study – like most research on the dance language – focuses on the dance of the European honeybee *Apis mellifera*. However, there are a number of other species in the genus, all of which live in Asia, and all of which possess systems of dance communication. Comparisons of the dances of these other species will be useful for providing a broader picture of what dance communication is and how it evolved.

Dance Communication: History of Discovery

The dances of bees have been described by observers of nature as far back as Aristotle, but it was Karl von Frisch who first subjected the dance to systematic study, and who recognized what the dance is and generally how it works. As he described in his own writings, he recognized early on that the dance was somehow involved in allowing scout bees to recruit nest mates to food, but he initially misunderstood the nature of this communication.

Von Frisch observed two forms of the dance, which he called the 'round dance' and the 'waggle dance.' Both are performed on the vertically suspended comb inside the nest and can easily be seen in an observation hive. A round dancer circles repeatedly in place, occasionally reversing the direction of circling. A waggle dancer repeatedly runs in a straight line while waggling her body from side to side; upon completing each waggling run, the dancer loops back to perform another waggling run with her body aligned in the same orientation as the previous one. Von Frisch's initial interpretation was that round dances signaled the presence of nectar in the environment, and that waggle dances signaled the presence of pollen. Thus, at first, he missed the presence of spatial information in the dance.

Eventually, von Frisch realized that round dances were performed by bees that had found food a short distance from the hive, and that waggle dances were performed by bees that had flown farther, regardless of whether the resource provided nectar or pollen. Furthermore, by training bees to an artificial floor (scented sugar water) that was then moved to increasingly distant locations, von Frisch found that the length of the waggling run increased with flight distance. At the same time, von Frisch realized that the orientation of the waggling run correlated with the direction flown to the food. Specifically, the angle of the waggling run relative to the upward direction on the comb matched the angle the bee had flown relative to the current azimuth of the sun (**Figure 1**). An illustration can be seen in the accompanying video clip, which shows a

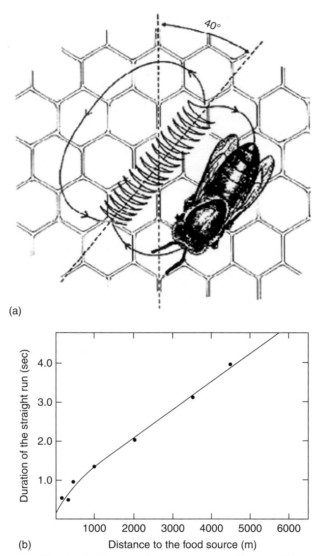

(a)

(b)

Figure 1 Waggle dance of honeybees. The dancing bee travels in a pattern roughly resembling a figure 8, with waggling runs alternating with return runs. The orientation of the waggling run relative to gravity (downward and to the left in this drawing) gives the direction of flight relative to the sun. The duration of the waggling run gives the flight distance, according to a population-specific code. Compare the form of this dance with the bee dancing in the video clip.

dancer indicating a food source opposite the direction of the sun.

These discoveries, first published in the middle of the 1940s, prompted a series of studies during the 1950s and 1960s that worked out much of the basic machinery of the dance. This work showed not only that the waggle dances contained spatial information, but also that followers seemed to be guided by this information – the distance and direction they flew to find food was correlated the information encoded in the dance. Other studies explored questions about the use celestial cues to determine the direction of flight, how bees measured the distance of flight, how they compensate for the effects of wind on the measurement of both direction and distance, the role of odors carried by the dancer in helping the recruits

pinpoint the location of the food, the use of dances to signal other resources such as water or nest sites, and the evolution of this remarkable communication system. Some of these questions will be addressed in more detail later. This body of work of over two decades had a remarkable impact on our understanding of animal behavior and helped justify the awarding of the Nobel Prize to Karl von Frisch (along with Tinbergen and Lorenz) in 1973.

Although von Frisch's work was acclaimed by the scientific community, a controversy arose in the late 1960s when an American biologist, Adrian Wenner, questioned whether the dance language worked in the way von Frisch claimed. Wenner acknowledged that the dance contained spatial information, but he challenged the evidence that the recruits used this information to find food. Wenner's alternative hypothesis was that recruits only obtained information about the odor of the food (from floral odorants adhering to the body of the dancer), then sought out the same source of odor in the field. The evidence for this 'Odor Search Hypothesis,' partly from Wenner's experiments and partly from von Frisch's own work, was the observations that the searching flights of recruits are indeed heavily influenced by odors in the environment. Indeed, odors can lead recruits to search in directions opposite what is signaled in the dance.

The difficulty that turned this observation into controversy is that both Wenner and von Frisch acknowledged an important role for environmental odors carried by the dancer in influencing the recruitment process. They simply disagreed on the nature of this role. Wenner claimed that odors are both necessary and sufficient for recruitment to a specific location, and that the spatial information in the dance is wholly unnecessary. Von Frisch claimed that odors normally supported dance communicating by guiding the final stage of the search, but were sometimes sufficient for recruitment when the food was close and the odor was very strong.

Wenner's claims were taken very seriously, and they spurred a series of very clever experiments designed to tease apart the roles of spatial dance information and environmental odors. Among the cleverest were Gould's 'misdirection' experiments, in which the visual system of the bees was manipulated so that the dancers and recruits used different references for interpreting the waggling angle. This created a situation in which the spatial information provided by the dances indicated locations removed from the source of any odors that would have been carried by the dancers. Provided that the environmental odors were not too strong, the recruits searched where the spatial signals directed them to go, thus vindicating von Frisch's characterization of the roles of spatial and olfactory information in the dance.

In the decades since, numerous other experiments have provided support for von Frisch's hypothesis. None has refuted it. The so-called dance language controversy is dead, but it is worth emphasizing how valuable this controversy was in leading to more rigorous tests of von Frisch's astonishing discoveries, and in highlighting the importance of odors in honeybee recruitment, which von Frisch tended to downplay. Indeed, the importance of the spatial signal itself may vary according to environmental circumstances. If experimentally excised from the dance (by turning the comb to the horizontal and depriving dancers and followers alike of the gravity reference they need to measure the alignment of their bodies), the colony will gain more mass (from nectar flowing into the hive) when dancers were oriented than when they were disoriented. However, this outcome occurs only under specific environmental conditions, namely, when floral resources were distributed in patches relatively far from the nest. By contrast, there is no difference in mass gain in oriented versus disoriented colonies when there was a superabundance of easily discovered food around the nest (when both groups gained mass) or little food at all (when both groups lost mass).

The Dance as a Window on the *Umwelt* of Honeybees

Understanding the dance language requires going beyond the act of communication to consider how bees acquire the navigational information they express in their dances, and how they integrate information about resource quality and colony need to modulate the dance signal. In fact, the importance of the dance language to biology is the power it gives to explore sensory, learning, and decision-making mechanisms that may be widespread in the insect world.

First, let us consider the flow of spatial information in the dance communication system. Dances communicate navigational information that the forager has acquired during her flight to the food and that will guide her subsequent flights. The dance then conveys this information to the bees that follow inside the nest. Then, they use it to guide the flight to the food (**Figure 2**).

The essential information encoded in the dance is the direction and distance traveled relative to the sun. The ability to use such vector information is widespread among animals, both vertebrate and invertebrate, but studies of honeybees have provided some of the deepest insights into how this works.

Using the sun for navigation offers certain advantages to diurnal animals, but it presents some major challenges. One challenge is that the sun's azimuth (its compass angle) moves during the day relative to fixed landmarks or feeding routes. Making this problem even more difficult, the azimuth shifts at a variable rate over the day – slow when the sun is rising and setting, and fast when it is high in the sky at mid-day (**Figure 3**). Also, this pattern of movement over time (the solar ephemeris) itself varies seasonally and with latitude. Many species are known to be able to take these variations into account in using the sun as a compass, but the clearest insights into this ability come from observations of the honeybee waggle dance, which provides a direct read-out of the angle of flight relative to the azimuth. And since the dance angle is expressed inside the nest out of view of the sun, it reveals information stored in memory about the current location of the azimuth. This allows us to measure the accuracy of this memory not only at the moment of the dance but also over time, as the bee compensates for the shift in the azimuth. Observing such compensation with the passage of time is possible because bees will continue to forage on completely overcast days when they cannot see any celestial cues. They find their way using landmarks, but when they return to the nest they base their dances on the remembered position of the solar

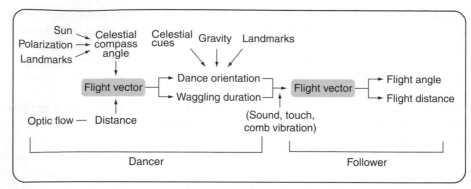

Figure 2 Information flow in the dance language, depicting the kinds of navigational information acquired by the forager, the encoding of this information in the waggle dance, the communication of this information to the follower bee(s) through the dancer's movements, and the decoding of this information by the follower for navigation to the food.

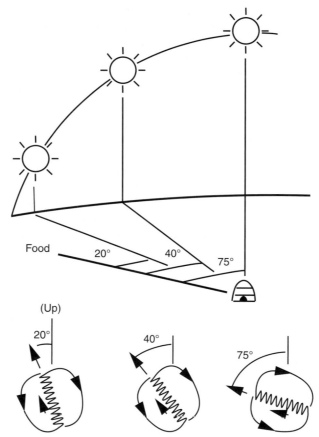

Figure 3 As the sun's azimuth (its projection to the horizon) shifts from left to right along the southern horizon (in the northern hemisphere), waggling runs performed to a fixed feeding place shift counterclockwise on the comb.

azimuth, compensated for time. This kind of accurate picture of the internally encoded solar ephemeris is possible only because of the dance language.

Another challenge that arises from depending on the sun as a navigational compass (and as a reference for communication) is that it sometimes disappears behind clouds. Obviously, the ability to learn the pattern of movement of the sun relative to landmarks provides a

way to communicate on cloudy days. Another solution to the problem was discovered early on by von Frisch, and again his discovery depended upon observations of dancing bees. When the sun is behind clouds but patches of blue sky are visible, honeybees can navigate using sun-linked patterns of polarized light that result from the sun's rays scattering in the atmosphere. Von Frisch discovered this when trying to understand why dancing bees with a view of blue sky were not aligned relative to gravity in the way predicted from their flight angle relative to sun. With the use of polarizing filters, he showed that the angle of polarization in the sky was the factor influencing their orientation. Thus, bees – and it turns out most arthropods – can extract useful navigational information from a quality of light to which we are completely blind. The study of this ability has proved to be an extraordinarily rich source of insights into the invertebrate visual system.

As for the communication of distance in the dance, von Frisch realized that it must be based on the operation of some kind of odometer during the flight to the food. His experiments suggested that the bee measured the consumption of energy during the flight. This hypothesis reigned as dogma for decades, but was overturned in the 1990s by evidence that energy consumption actually plays no part in odometry. Instead, the principal mechanism is the measurement of optic flow – the streaming of visual texture over the visual field as the bee flies.

To return to the analysis of information flow depicted in **Figure 2**, navigational information recorded by the forager is encoded in body movements that take place in a sensory context that is almost completely isolated from the context in which the information was recorded. Indeed, starting with von Frisch, it has been clear that the dance is controlled by sensory mechanisms that are partially distinct from those used to control the flight. If dancing bees can see celestial cues (sun and polarized light), then they can use these same cues to orient the waggle runs. In the darkness of the nest, however, gravity serves as a substitute for the sun, with body angle being measured with fields of sensory bristles between the head and thorax and between the thorax and abdomen. Until fairly recently, it was assumed that celestial cues and gravity were the only cues available for expressing directional information. However, the Asian honeybee, *Apis florea*, which nests in the open and is unable to use gravity as a directional reference, can orient to surrounding landmarks when celestial cues are obscured. So too can *A. mellifera*, provided that dancers are on a horizontal surface with no gravitational reference and that they have had an opportunity to learn the relationship between dance angles expressed relative to the sky and angles expressed relative to surrounding landmarks. This finding has implications for understanding the evolution of the dance language, as discussed later.

The expression of distance information in the duration of the waggling run (and the measurement of this signal by the follower bees) must be mediated by some kind of internal timer. What is striking is that the mapping from distance measurement to distance signal (and back again, for the follower bees) entails a large transformation of time scale. For example, in European honeybees, a flight of 1 km takes about 2 min and is then represented in a waggling duration of about 1 s. Nothing is known about how this rescaling of the signal takes place.

Now, we come to the question of how the signal passes from dancer to follower. Oddly, given how intensively this communication system has been studied, this is one of the most poorly understood steps in the process. Research has focused on several candidate sensory modalities, which are not mutually exclusive. These include the following: airborne sounds produced by the buzzing of the dancer's wings and detected by the antennae of the followers, tactile cues produced by the dancers body and again detected by the followers' antennae, vibrations of the substrate produced by the shaking of the comb by the dancer and detected by organs in the legs of the followers, and visual detection by the followers of the body movements of the dancer.

There is at least circumstantial evidence that all these could play a role in dance communication in one circumstance or another, although there are also situations in which each clearly plays no role whatsoever. For example, vision can play no role for dances that take place in the darkness of an enclosed nest, although it may be extremely useful in those species that nest in the open and dance in daylight. Sound may be important for cavity-nesting bees, but it is not used by *A. florea*, which nests in the open and has silent (but visually conspicuous) dances. Tactile cues may be useful for all species, but close observations suggest that direct antennal contacts between followers and dancers may be intermittent and hence unreliable. Finally, substrate vibration can play no role when dancers and followers are standing on the backs of other bees rather than on a common substrate; this is the case in the open-nesting species and in the reproductive swarms of all species. These patterns suggest that the modality of communication varies with context, depending upon either evolved species differences or facultative shifts in the stimulus conditions of the dance.

Regardless of the sensory modality by which the signal is detected by followers, an additional problem is how they decode the information as represented in the dance. This may be a relatively trivial problem in the case of the distance signal: whatever transformation occurs in the dancer to map the flight distance onto the waggling duration may simply operate in reverse for the followers. The problem could be more difficult in the case of the directional signal, however, given that dance followers may view the dancer from a variety of positions, and with their bodies in different alignments relative to the reference (e.g., gravity) to which the dancer's body is aligned. There is evidence, however, of a simple solution to this problem: followers standing behind the dancer, and aligned in the same direction as the waggling run, are more likely to get to the food than were followers viewing the dancer from other angles. Thus, provided she knows she is behind the dancer, determining the direction to the food may simply be a matter of a follower measuring her own alignment relative to gravity (and of course translating this into a flight angle relative to the sun).

Dance Communication in a Social Context

During the summer, a colony consists of a reproductive queen, tens of thousands of workers (of which about 25% may serve as foragers at any time), and a few thousand drones. Sustaining the growth of the colony, and its reproduction through colony fission, requires the acquisition of a large amount of resources. Estimates of the rate of nectar collection during the summer fall in the range of 2–3 kg day^{-1}, which is approximately equal to the total mass of the bees in the colony. This is possible in large part because of the dance language, which is an integral part of the social foraging strategy of the honeybee colony, and which allows for efficient allocation of foragers among shifting resources.

The advantage of the dance has always been assumed to be the precision of the spatial information that is communicated. However, this insight has been greatly extended over the past 25 years by research that has uncovered the rich away of social mechanisms that enhance the power of the dance by modulating its expression in different circumstances. The importance of these mechanisms arises because of the dynamic nature of both the food resources in the bees' environment and the demand of the colony for those resources. As ecological generalists, honeybees are ordinarily confronted with a diverse array of floral resources, which are commonly distributed in patches of various sizes and at various distances, and which also increase and decrease in availability over the day or over the season. Exploiting these resources efficiently requires the colony to marshal foragers at the appropriate places and times. This can be regarded as a decision facing the colony, in which dance communication plays a central role.

Research on the social mechanisms that lead to this outcome has yielded an important insight, which also applies to group-level decisions in many other social insect species: the foragers are allocated not through a centralized, hierarchical control process, but through a decentralized, distributed process in which no one worker bee ever has a synoptic view of the state of the colony or the options available to it (**Figure 4**). A pivotal role, however, is played by the returning forager, for it is she who decides whether to perform a dance on her return, and how many of repetitions of the waggling run to perform. This determines how many followers are contacted by her dance and how many recruits reach the food. The decision of how many waggling runs to produce is based on two sources of information. First, the dancer relies upon information about the intrinsic quality of the food, information that is available only to her. Information known to be used are the distance of the flower patch, the handling time in the patch, the sweetness of the nectar, and the presence of danger. Second, she relies upon information about the need of the colony for the food she is bringing back. This she determines by her reception upon arrival at the nest: the quicker she is greeted and unloaded, the more the colony must need what she is bringing, that is, the better her food is relative to other sources of nectar in the environment.

It might be assumed that some workers play the role of 'dispatchers' – monitoring the quality of food coming in and ordering foragers from better sites to be greeted more enthusiastically. Again, however, the control is decentralized; no forager is in a position to compare food quality from different sites let alone to direct the allocation of receiver bees. What determines the reception experienced by any particular forager is the overall influx of nectar into the nest. If a lot is coming in and most receiver bees are busy shuttling nectar into the storage areas of the colony, then the returning bees will have to wait longer to be unloaded. This influences the tendency to dance, but the foragers' assessment of the quality of the food also matters – as waiting time increases, only those bees returning from very high-quality resources will do dances. The combined effect is that the dances are performed to the best currently available sites.

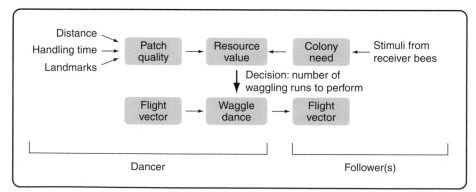

Figure 4 Information flow in the social modulation regulation of nectar gathering via the dance language. The dancer faces a decision of how many waggling runs to perform, which in turn influences the rate of recruitment to the food. This decision is based upon information about the intrinsic quality of the food and information about the needs of the colony.

Further tuning of the system is mediated by higher-order control systems, which are also partially influenced by actions of the returning foragers. For example, if bees returning from good resources are confronted with long unloading times, they may do 'tremble dances,' a series of shaking movements that cause bees engaged in other tasks in the colony to shift into the role of receiver bees. This increases the overall capacity of the colony to handle an influx of nectar. Furthermore, if nectar continues to flow over a longer time scale such that the comb becomes full, even an increase in the number of nectar-handling bees may not be enough to accommodate it. This situation stimulates comb building as a further extension of the colony's capacity.

Thus, the regulation of nectar foraging shows that the dance language is deeply integrated into the social organization of the colony. Social mechanisms serve to regulate the dance to suppress recruitment when food quality is low, to allocate recruits to the best resources, or to increase the capacity of the colony to handle a high influx of food. A further illustration of the importance of dances to the social life of bees is that it is used not only for nectar, but also other resources such as pollen (the colony's protein source), water (when the colony is heat stressed), and possibly propolis (plant resins used to seal up the nest cavity). Each of these roles of the dance is subject to its own mechanisms of social regulation, which undoubtedly interact with those used in regulation of nectar foraging.

Another important role for the dance is in 'house hunting,' the process colonies go through when moving to a new nest cavity during reproductive swarming or abandonment of the old nest. House hunting takes place when the swarm is clustered out in the open. It is one of the clearest examples of decision making on the social level, because it entails the selection of a single nest from an array of options. As in nectar foraging, the decision can be viewed as one facing the colony, but mediated by a set of individual decisions made by workers under a decentralized control system. Again, the scouts assess the inherent quality of the resource (a candidate nest site) and encode this assessment in the number of waggling runs performed, and then a set of social feedback mechanisms guarantee that the colony accumulates recruits at the best resource – with no central comparison of resource quality. Here, the control system diverges from that seen in the regulation of nectar foraging. There is no equivalent of the receiver bees to provide dancers a signal of the colony's demand for the resource. How, then, does the colony turn off recruitment to poorer nest sites and increase it to better ones? The answer is that each dancer has a tendency to cease dancing some time after she has begun; dancers to better nest sites drop out much more slowly, leading to a much more rapid build-up of recruits which also become dancers.

Evolution

As mentioned at the outset, few communication systems in the animal world rival the honeybee dance language in sophistication and flexibility. In addition to the question of how it works, any biologist would be interested in the question of where it came from. Attempts to understand the evolution of the dance language have focused on two questions: how it emerged from more primitive communicative behaviors, and how it has been optimized to enhance the efficiency of communication in different environments.

Both these questions have been addressed through comparisons of the communication systems of different bee species, including the members of the genus *Apis* and representatives of two other social bee taxa: stingless bees and bumble bees. All of the *Apis* species have symbolic dance communication akin to that which von Frisch described in *A. mellifera*. Stingless bees (Meliponinae) and bumble bees (Bombinae) have simpler forms of communication that may provide clues about the evolutionary precursors of the dance language of *Apis*.

What follows is a sketch of how the dance language could have originated and diversified, given what we know about the phylogenetic relationships among these bee groups and the distribution of communication behavior among them (**Figure 5**).

Landmarks used	Y	?	?	Y
Slope compensation	Y	N	N	N
Typical dance plane	0°–80°	90°	90°	90°
Gravity used	N	Y	Y	Y
Sky cues used	Y	Y	Y	Y
Nest exposure	open	open	cavity	cavity
	dwarf bees	rock bees	Eastern hive bees	western hive bees

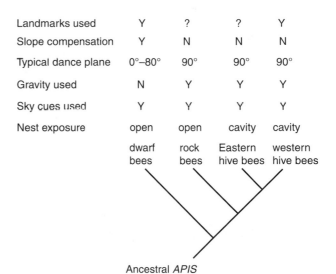

Ancestral *APIS*

Figure 5 Simplified phylogeny showing the distribution of traits associated with dance communication in the genus *Apis*. There are at least ten species in this genus, all but one of which lives in southern or eastern Asia. Most comparative studies of the dance have focused on four species, the dwarf bee *A. florea*, the rock bee *A. dorsata*, the eastern hive bee *A. cerana*, and the western hive bee *A. mellifera*. Each of these branches of the tree includes other species or geographical races, but the four branches shown capture essentially all of the known diversity in behavioral traits related to the dance language.

First, it is important to emphasize that all the bee subfamilies mentioned – honeybees, bumble bees, and stingless bees – have some kind of communication system in which returning foragers interact with their nest mates, and cause other bees to go out and look for food. In all cases, this involves vigorous body movements, sounds, and the sharing of food and environmental odors. The precise spatial communication system of *Apis* thus presumably arose from such a behavior. Long ago, Martin Lindauer, who was one of von Frisch's most accomplished students, suggested that dances signaling the direction and distance of food could have arisen in an ancestral species that nested on exposed combs like the modern *A. florea* or *A. dorsata*. By Lindauer's scenario, the excited body movements of returning foragers came to be oriented relative to the same celestial reference that was used to orient the flight, and acquired temporal features that correlated with flight distance.

According to Lindauer, the ancestral dance was also similar to that of *A. florea* in taking place on a horizontal extension of the nest and lacking the use of gravity, and instead using only celestial cues as an orientation reference. Later, the dance shifted to vertical flanks of a suspended, exposed comb, as in *A. dorsata*, and incorporated the use of gravity as a substitute for the sun, paving the way for a move into cavities as in the Asian and European hive bees. This evolutionary story can also accommodate other phylogenetic patterns not known to Lindauer. For example, the dances of *A. florea*, which never occur in darkness, are completely silent, but other species have both noisy dances and the need to dance in the dark. Sounds may therefore have been added to a dance that originally did not need them. Open-nesting species perform dances with exaggerated postures that may make a conspicuous visual target, but such postures are lacking (because they were presumably lost) in cavity-nesting species.

This hypothesis is largely consistent with modern phylogenetic analyses that support open nesting, and some other characteristics of *A. florea*'s dance, as the ancestral condition in the genus *Apis*. Other aspects of the story are more ambiguous, although this is perhaps not surprising when one considers how dramatically the dance languages of the *Apis* species, taken together, diverge from the communication systems of outgroup taxa that could be used to determine the direction of evolutionary changes as honeybees diverged from one another.

The second important evolutionary question about the dance concerns its adaptive modification for different ecological conditions. This has been intensively explored in the case of the distance code, which exhibits striking differences among honeybee populations and species. For example, both von Frisch and Lindauer made much of variation in the flight distance at which recognizable waggle dances rather than round dances are performed. More recently, it has been shown that the distinction between round dances and waggle dances is illusory; round dances actually contain directional information, and there is a continuum of dance forms between round dances and waggle dances. Nevertheless, some populations and species start doing well-oriented waggle dances at shorter flight distances than others. Furthermore, when one measures the duration of the waggling run as a function of flight distance, the relationship is steeper in some populations and species than in others.

This variation has a heritable basis, which suggests that it reflects evolutionary divergence among populations. This divergence need not have been driven by natural selection, of course, but there are intriguing ecological correlations that support a selective advantage of the shape of the distance code. The prevailing hypothesis is that these so-called distance dialects correlate with the distribution of food resources over the typical flight range in the environment where the dance communication takes place. The evidence is that populations with shorter flight ranges signal precise directional information at a shorter flight distance, and also have steeper dialect functions, which are thought to be more precise. The correlations are far from perfect, and in particular the parameters that determine the shape of the dialect function seem to be different in open-nesting than in cavity-nesting species.

The Dance as a Model System in Biology

Anyone who has watched bees dance for food, and is aware of the function of this behavior, cannot help but be amazed. Karl von Frisch's decoding of the dance language is certainly one of the great discoveries in modern biology. This is not only because of the inherent fascination that the dance holds for curious human observers. Even more important is the extent to which von Frisch's discovery laid the foundation for the study of deep questions about animal behavior. When we consider the role that the dance language has played in the study of vision, olfaction, audition, learning, circadian rhythms, decision making, social organization, and behavioral evolution, it is easy to see why von Frisch regarded the dance language as a 'magic well' of discovery. Furthermore, with advances in neuroscience, genomics, and evolutionary theory, it seems clear that the value of the dance as a model system will continue for many years to come.

See also: Honeybees; Information Content and Signals; Maps and Compasses.

Further Reading

Abbott KR and Dukas R (2009) Honeybees consider flower danger in their waggle dance. *Animal Behaviour* 78: 633–635.
Dyer FC (2002) The biology of the dance language. *Annual Review of Entomology* 47: 917–949.

Dyer FC and Dickinson JA (1996) Sun-compass learning in insects: Representation in a simple mind. *Current Directions in Psychological Science* 5: 67–72.

Gardner KE, Seeley TD, and Calderone NW (2008) Do honeybees have two discrete dances to advertise food sources? *Animal Behaviour* 75: 1291–1300.

Gould JL (1976) The dance-language controversy. *Quarterly Review of Biology* 51: 211–244.

Judd TM (1995) The waggle dance of the honey bee: Which bees following a dancer successfully acquire the information? *Journal of Insect Behavior* 8: 343–354.

Lindauer M (1956) Über die Verständigung bei indischen Bienen. *Zeitschrift für vergleichende Physiologie* 38: 521–557.

Michelsen A (1999) The dance language of honey bees: Recent findings and problems. In: Hauser M and Konishi M (eds.) *The Design of Animal Communication*, pp. 111–131. Cambridge, MA: MIT Press.

Michelsen A (2003) Signals and flexibility in the dance communication of honeybees. *Journal of Comparative Physiology A: Neuroethology, Sensory, Neural, and Behavioral Physiology* 189: 165–174.

Raffiudin R and Crozier RH (2007) Phylogenetic analysis of honey bee behavioral evolution. *Molecular Phylogenetics and Evolution* 43: 543–552.

Seeley TD (1995) *The Wisdom of the Hive*. Cambridge, MA: Harvard University Press.

Seeley TD and Buhrman SC (1999) Group decision making in swarms of honey bees. *Behavioral Ecology and Sociobiology* 45: 19–31.

Seeley TD and Visscher PK (2004) Group decision making in nest-site selection by honey bees. *Apidologie* 35: 101–116.

Sherman G and Visscher PK (2002) Honeybee colonies achieve fitness through dancing. *Nature* 419: 920–922.

von Frisch K (1967) *The Dance Language and Orientation of Bees*. Cambridge, MA: Harvard University Press.

Wehner R and Labhart T (2006) Polarization vision. In: Warrant EJ and Nilsson DE (eds.) *Invertebrate Vision*, pp. 291–348. Cambridge, UK: Cambridge University Press.

Wenner AM and Wells PH (1990) *Anatomy of a Controversy: The Question of "Language" Among Bees*. New York, NY: Columbia University Press.

Darwin and Animal Behavior

R. A. Boakes, University of Sydney, Sydney, NSW, Australia

For many thousands of years, at different times and in different parts of the world, humans have studied their fellow creatures in an attempt to obtain a better understanding of their behavior. Toward the end of the eighteenth century, an increasing amount of observational – and occasionally experimental – research on behavior took place in Western Europe. Nonetheless, the foundations of the contemporary science of behavior were mainly provided by the evolutionary theories and the ensuing debates of the nineteenth century. Of these, the key event was, of course, the publication in 1859 of Charles Darwin's 'On the Origin of Species by Means of Natural Selection, or the Preservation of Favoured Races in the Struggle for Life' (henceforth referred to as '*The Origin*').

From his youth, Darwin continued to maintain a keen interest in behavior. An early hobby was collecting beetles, and it is clear that he was intrigued as much by how they and other insects behaved as by their bodily structures. For decades, he maintained notebooks on behavior, read widely on the subject, and exchanged letters full of questions about the behavior of a wide variety of species, with correspondents throughout the world. Darwin's concern with behavior becomes evident in '*The Origin*' when he discusses what he saw as four major difficulties with his theory. The third of these was that of answering the question: "Can instincts be acquired and modified through natural selection?," and in Chapter 7, he gives his reasons for believing that behavior was as much subject to natural selection as a bodily characteristic. He starts by acknowledging that some forms of instinctive behavior may derive from habits acquired by a previous generation, as Lamarck had argued 50 years earlier. But the core argument of the chapter is that "it can clearly be shown that the most wonderful instincts with which we are acquainted, namely, those of the hive-bee and of many ants, could not possibly have been thus acquired." Spelt out with many examples, his simple but conclusive point is that in a number of insect species various innate behaviors are displayed only by sterile individuals. This means that "a working ant … could never have transmitted successively acquired modifications of structure or instinct to its progeny." He then proceeds to the "climax of the difficulty; namely, the fact that the neuters of several ants differ, not only from the fertile females and males, but from each other, sometimes to an almost incredible degree." Citing both his own measurements and data from others showing variation in the size and other characteristics of worker ants, Darwin concludes by explaining how natural selection operating on the parents could give rise to two or more kinds of neuter individuals. In so doing, he took the innate behavior of insects from being a key example of God's design to becoming important evidence for the power of natural selection (**Figure 1**).

The '*Origin of Species*' has justifiably been recognized as a magnificent book, and not just an extraordinarily important one. It is confident, passionate, and carefully constructed so as to convince the reader of two ideas: first, that no coherent account of the origin of species by special creation is possible; and, second, that natural selection is the primary process by which species evolve. As Darwin noted later, he deliberately played down issues that might divert attention from his two main arguments; some topics "would only add to the prejudice against my views." These included the importance or otherwise of Lamarckian inheritance and of sexual selection as secondary processes in evolution. He also postponed discussion of what would have been highly explosive in the predominantly religious society of mid-nineteenth century Britain, namely, that human beings were as much a product of natural selection as any other form of life. Famously, he simply notes just before the end of the book: "In the distant future I see open fields for far more important researches. Psychology will be based on a new foundation, that of the necessary acquirement of each mental power and capacity by gradation. Light will be thrown on the origin of man and his history." In 1859, arguing that other species had evolved was explosive enough.

Darwin's first aim was met within a remarkably short time. By the time the third edition of '*The Origin*' was published in 1861, he could write: "Until recently the great majority of naturalists believed that species were immutable productions, and had been separately created"; he then noted that this was no longer true. This rapid reversal was helped by the effective efforts of several of Darwin's scientific colleagues and friends, notably, Thomas Huxley, who relished the battle with orthodox and religious opinion. Huxley also boldly published the first book to contain a detailed argument for human evolution. His '*Evidence for Man's Place in Nature*' of 1864 started with a provocative and endlessly reproduced frontispiece in which a human skeleton heads a line containing skeletons of a gorilla, a chimpanzee, an orangutan, and a gibbon (**Figure 2**).

Alfred Wallace had developed the idea of natural selection independent of Darwin and, if Wallace's article describing natural selection sent from faraway Indonesia had not shocked Darwin into sudden urgency, '*The Origin*' would not have been published until much later than 1859 and probably in a less satisfactory form. In some ways, Wallace was more of a Darwinian than Darwin. He saw no need to accept any form of Lamarckian process to complement natural selection, and he argued that the idea of sexual selection was also unnecessary. On the other hand, having dismissed all other possible evolutionary processes except natural selection, he was unable to understand how human intellect and morality could have evolved. In 1869, Wallace appealed to supernatural intervention that had been applied to some human progenitor (**Figure 3**).

This time Darwin was shocked into publishing '*The Descent of Man and Selection in Relation to Sex*' of 1871 (hereafter referred to as '*The Descent*'). Darwin focused on three questions: "Whether man, like every other species, is descended from some pre-existing form"? What was "the manner of his development"? And what is "the value of the differences between so-called races of man"? Since Huxley and the German biologist, Ernst Haeckel, had already spelt out the evidence for evolution of the human body, Darwin concentrated on the human mind and on rebutting Wallace's claim that "natural selection could only have endowed the savage with a brain little superior to that of an ape."

Darwin's deep belief in human evolution went back to the day when, as a young biologist sailing on HMS *Beagle*, he landed on a beach in Terra del Fuego: "The astonishment which I felt on first seeing a party of Fuegians on a wild and broken shore will never be forgotten by me, for the reflection at once rushed into my mind – such were our ancestors." Nearly 40 years later, he took on the task of persuading his now large readership

Figure 1 A portrait of Charles Darwin around the time that he began to develop the theory of natural selection.

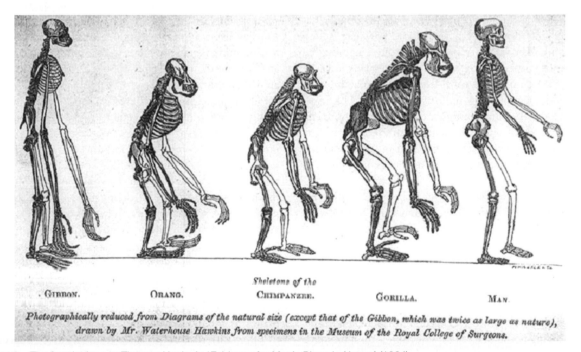

Skeletons of the

GIBBON.　ORANG.　CHIMPANZEE.　GORILLA.　MAN.

Photographically reduced from Diagrams of the natural size (except that of the Gibbon, which was twice as large as nature), drawn by Mr. Waterhouse Hawkins from specimens in the Museum of the Royal College of Surgeons.

Figure 2 The frontispiece to Thomas Huxley's '*Evidence for Man's Place in Nature*' (1864).

Figure 3 Alfred Wallace.

beyond that belief to the idea that the intellectual and moral sophistication of Europeans was not just related to the intellect and morals of Fuegians but had evolved from simple forms of life.

In arguing for mental evolution, Darwin aimed to undermine the view that animals were incapable of reasoning, did not display emotions, had no form of communication that in any way resembled human language, and never displayed behavior that could be described as 'moral.' In relation to reasoning, he cited various examples, mainly culled from his worldwide correspondence, of problem solving and tool use, mainly by apes. As for the emotional life of animals, he was unreservedly anthropomorphic: He had no doubt that 'elephants intentionally practice deceit' or that 'a dog carrying a basket for his master exhibits in a high degree self-complacency or pride.' In considering language, he pointed to examples of vocal communication in other species and vocal mimicry in birds like parrots. His argument for the evolution of morality took a similar approach, using examples of altruistic behavior in various species. Darwin concluded that "the difference in mind between man and the higher animals, great as it is, is one of degree and not of kind."

This first part of 'The Descent' lacked the confidence displayed in 'The Origin.' The evidence he put forward for his views was predominantly second hand, that is, gleaned from correspondence and reading rather than direct observation and experimentation. When in the second part of the book he describes his theories of sexual selection, it is as if with relief that he has reached safer ground. Here, he discusses ideas he had thought about for decades, based on an accumulation of detailed evidence. Having noted that sexual selection is most effective in polygamous species, in the final part of the book he united the two main – and to this point – apparently

unrelated themes. In human evolution, he suggests, sexual selection has played a dominant role both in the development of secondary sex characteristics – nakedness and male beards, for example – and intellectual ability. However, the latter is not spelt out. As for the third main question with which 'The Descent' started, that concerning the significance of racial differences, Darwin had no doubt that all humans were descended from a common ancestor, a view that directly contradicted the influential claim put forward by Louis Agassiz, the most important American biologist of that time.

We have seen that in 'The Origin' the behavior of insects was deployed as an argument against the adequacy of Lamarckian inheritance. In 'The Descent,' the behavior of vertebrates was used in the argument for human evolution, albeit with almost no appeal to natural selection but with a great deal to Lamarckian inheritance and some to sexual selection. Only a year after publishing 'The Descent,' Darwin published the third of his books in which the study of behavior was important. 'The Expression of the Emotions in Man and Animals' of 1872 (henceforth referred to as 'The Expression') has the same sense of excitement as 'The Origin,' with Darwin confident that his account of emotional expression within an evolutionary framework was far superior to its few predecessors.

'The Expression' was certainly superior in terms of its empirical base, in that for many years Darwin had been gathering a range of evidence on the topic. This evidence included the innovative use of photographs, ones of angry, fearful, sad, or happy children; of actors simulating such emotions; and even of inmates of an asylum for the insane. These were accompanied by prints – for example, of a snarling dog, a terrified cat, and of monkeys and chimpanzees displaying various moods – to illustrate the argument that human expressions were a product of evolution and that the same principles applied to both human and animal emotions.

These principles were based on the core idea that it is highly adaptive for individuals to signal their emotional states as clearly as possible: "With social animals, the power of inter-communication between members of the same community – and with other species between opposite sexes, as well as between the young and the old – is of the highest importance to them." The first of the three principles was based on the inheritance of "serviceable associated habits." In other words, some form of effective communicative behavior is first learned by a process of trial and error (although Darwin did not use this term), becomes an ingrained habit, and is then passed on via some genetic process so as to become instinctive in later generations. The second is the principle of antithesis: behavior expressing one emotional state – say, affection – is likely to be as different as possible from behavior expressing the opposite state – say, hostility. Remarkably, Darwin did not justify this principle in terms of more effective communication, as with hindsight we might expect from the author of '*The Origin.*' Instead, he appealed to 'the tendency to perform opposite movements under opposite sensations or emotions.' The third principle appealed to the 'constitution of the nervous system,' an unusual appeal in Darwin's works on behavior. He borrowed from a fellow evolutionist, Herbert Spencer, the idea that 'nervous energy' can overflow into 'less habitual' responses. For example, trembling is explained as the result of intense excitation of the autonomic system.

As in '*The Descent,*' there is almost no mention of natural selection in '*The Expression.*' Instead, the principle throughout is implied: individuals that can communicate better, using their species-specific behaviors, are likely to have more offspring. Darwin stressed the similarities between human and primate emotional expression, but found one example that he decided was uniquely human. Blushing, he argued, required self-consciousness, awareness that someone else might be looking at one's face; and thinking about one's face would automatically increase blood flow to this area (**Figure 4**).

Although Darwin made frequent reference to the acquisition of new behaviors that became habits, he does not seem to have had much interest in the processes by which such learning occurs. In contrast, this topic was of central concern to Herbert Spencer. In the 1860s and 1870s, Spencer was regarded by his peers, as well as by the general public, as important an evolutionary theorist as Darwin. Spencer had coined the term, 'survival of the fittest,' well before Darwin went public with the theory of natural selection. Nevertheless, Spencer maintained throughout his long and eccentric life that Lamarckian inheritance was the main driver of evolution and that natural selection was a secondary process – the reverse of Darwin's belief. As announced in 1855 in his first

edition of the '*Principles of Psychology,*' Spencer's main concern was with mental evolution: 'Mind can be understood only by showing how mind is evolved.' He believed that mental evolution is based on the transformation of reflexes into instincts and of instincts into intelligent behavior. In 1855, he proposed that the main driver of such transformations was what later would be known as Pavlovian conditioning. In 1871, in the second edition of his '*Principles of Psychology,*' he added a second learning process, based on the ideas of a contemporary psychologist and philosopher, Alexander Bain. The 'Spencer–Bain principle' stated that a response followed by some pleasant consequence will tend to be repeated.

Toward the end of the nineteenth century, Spencer's work was widely derided. His Lamarckianism, his psychology, his extreme laissez-faire politics, and his system of ethics were attacked from all sides. Yet his influence continued to be highly pervasive. In particular, the Spencer–Bain principle inspired the lively concern with trial-and-error learning that emerged in the 1890s.

Two years after publishing '*The Expression,*' the then 65-year-old Darwin invited to his home in the country a young physiologist, George Romanes. Darwin decided that Romanes was just the person to develop the ideas on mental evolution that Darwin had proposed in '*The Descent.*' Their admiration was mutual. Darwin became a revered father figure for Romanes who for the rest of his life vigorously defended every aspect of Darwin's theories, even those that after Darwin's death in 1882 began to look increasingly dubious, such as his theory of inheritance, 'pangenesis,' and his belief that instinctive behavior could evolve both as a result of natural selection and from inheritance of individually acquired habits. Romanes' aim in life became that of first accumulating systematic data on animal behavior and then using these to construct a detailed theory of mental evolution following the lines that Darwin had sketched (**Figure 5**).

Although as a neurophysiologist Romanes had proved to be a very able experimenter, the data he included in his first book, '*Animal Intelligence*' (1881), were predominantly anecdotal. By the standards of his time, he had reasonable criteria for judging whether to accept a report about some animal's remarkably intelligent behavior or indication that it had experienced a sophisticated emotion. However, the social status of the observer seems to have been as important a consideration as the thoroughness of the observation in assessing the reliability of some anecdote. Despite his self-appointment – and the general perception of him – as 'Darwin's heir,' Romanes' approach owed far more to Spencer. This is seen in his preoccupation with ranking different cognitive processes and emotions. For example, he considered the ability to operate mechanical appliances as indicative of a high level of intelligence and, since he had received many reports of cats operating latches so as to open doors, he ranked this

Figure 4 Herbert Spencer.

Figure 5   George Romanes.

Figure 6   Conwy Lloyd Morgan.

species' intelligence as being nearly as high as that of monkeys. Romanes seems not to have entertained the possibility that the relatively high number of reports concerning cats might reflect both the fact that this was one of the few species that a large number of humans observe daily and the fact that few other species have frequent opportunities to interact with mechanical devices.

For Romanes, all creatures that were capable of the most primitive form of learning – for example, including ones that displayed no more than what later became known as 'habituation' – possessed a mind, and this meant that even, say, a snail was to some limited degree conscious of the events impinging on its sensory organs. He believed that consciousness played an important role in instinctive behavior and stressed that instincts could be modified by experience. For example, although there was by then extensive evidence showing that in many species of birds their adult songs were influenced by early exposure to different sounds, for Romanes this was no reason against considering birdsong to be 'instinctive.' Within this framework, it was therefore quite appropriate to refer to the 'instincts of a gentleman.'

Romanes was a generous man. When he received an article sent from South Africa that was critical of his own work, he nevertheless appreciated its quality, supporting its publication and subsequently the career of its author, Conwy Lloyd Morgan. Prior to taking up a teaching position in South Africa, Morgan had studied under Huxley and absorbed his skeptical approach. Six years after returning to England on his appointment as a professor at what was to become Bristol University, Morgan published his important book, '*Animal Life and Intelligence*' (1890), followed

by his '*Introduction to Comparative Psychology*' (1895), the first book in English to bear such a title (**Figure 6**).

Morgan's influence on the study of behavior was substantial for three main reasons. The first was his insistence on the need for objective evidence based on careful experimentation or observation and the rejection of one-off anecdotal reports. Although he had become a close friend of Romanes and literary executor when Romanes died, Morgan had no hesitation in dismissing the kind of data on which Romanes had so often relied. Morgan developed many of his ideas from testing his dog, Tony. For example, he repeatedly threw a stick over a fence for Tony to retrieve and was impressed by how slowly the dog improved its ability to maneuver the stick through a gap in the fence. Just as Tony managed once to perform impressively, a passer-by stopped to watch for a few minutes: "Clever dog that, sir; he knows where the hitch do lie." Morgan noted that this was a characteristic – and in this case, entirely false – conclusion to draw from two minutes of chance observation.

Related to the need for careful and systematic observation of behavior was the need for careful interpretation of that behavior. To the extent that he is remembered today, Morgan is best known for his 'Canon.' This was essentially Occam's Razor, the scientific principle of parsimony, applied to behavior; where there are several possible explanations for why an animal behaved in a certain way, one should choose the simplest. What was new was that Morgan appealed to natural selection to justify its application to behavior. If a relatively simple process had evolved to the extent that an individual could respond appropriately in a particular context, then there would be no selective pressure to produce a more complex process capable of producing the same behavior. Morgan's most common demonstration of how his Canon should be deployed was in the analysis of what Romanes had seen as marks of high intelligence. Based partly on some informal experiments with chicks, Morgan argued that most of such examples could be better understood as the result of trial-and-error learning with accidental success. A key example for Morgan was that of an animal operating a latch to open a door or gate. From his study, Morgan had watched the regular attempts of his dog, Tony, to escape from the garden into the wide world beyond. The dog had repeatedly thrust its head through the fence railings here and there until once, apparently by chance, it inserted its head just below the gate latch and, on raising its head, the gate swung open. From then on, this appropriate action was performed with increasing rapidity to the point when a passing observer who had read Romanes might agree that it was an intelligent creature with some understanding of mechanical devices (**Figure 7**).

The third way in which Morgan made a lasting impact came from his rejection of Lamarckian accounts of the origins of instinctive behavior. This was partly stimulated by experimental work in the 1880s of the German biologist, August Weissman, whose failure to find any evidence for Lamarckian inheritance led him to propose the distinction between 'germ plasm' and 'body plasm' that laid the foundation for modern genetics. Morgan's final break with Lamarckian accounts of instinct came only in 1896 when he developed alternative ways of accounting for the kind of evidence that appeared to support the Lamarckians. The first was inspired by the work of a French writer, Gabriel Tarde, who in 1890 discussed the 'laws of imitation.' This led to the idea that social transmission could result in the rapid spread of some behavior that an individual animal had learned among a population of conspecifics and could support the continuation of that behavior over subsequent generations. When applied to humans, the difference between Fuegians and Europeans that Darwin had attributed to biological evolution was seen to lie in differences in cultural development. Morgan's second principle, 'organic selection,' was also proposed at the same time by at least two other theorists and ultimately one of the latter gained the credit for what became known as the 'Baldwin principle.' This supposes that, when an environmental change threatens the survival of an isolated group, those individuals who have the appropriate learning capacity to change their behavior in an adaptive way will have more descendants than those whose behavior is more resistant to change. Over the generations, the benefits of learning will buy sufficient time for adaptive innate behaviors to evolve by natural selection. The removal of Lamarckian processes meant that Morgan was now able to make a clear distinction between habit and instinct, as in his 1896 book of that name.

Early in the nineteenth century, biologists such as Darwin and Huxley endured long voyages in sailing ships that had not changed fundamentally in the four centuries since Portuguese mariners first left Western Europe to explore the globe. Later in the nineteenth century, steam ships were regularly plying the world's oceans. The expansion of the United States economy following the Civil War and the unparalleled development of the American university system meant that British evolutionists such as Huxley and Spencer could be paid to make the easy crossing of the Atlantic to give lecture tours. Morgan gave lectures on habit and instinct in Boston in 1896, and these very probably inspired a Ph.D. student at nearby Harvard who was looking for a new thesis topic. Edward Thorndike's subsequent experiments on trial-and-error learning represented the first quantitative studies of vertebrate behavior. His *Animal Intelligence* of 1898 provoked a generation of psychologists to undertake studies of what would much later be termed 'comparative cognition.' It also laid the groundwork for the behaviorist movement with its emphasis on learning theory that dominated American psychology until the 1960s. Ironically these developments occurred at a time when Darwin's theories were seen as outdated, so that the evolutionary framework in which studies of behavior had grown was disregarded.

See also: Animal Behavior: The Seventeenth to the Twentieth Centuries; Body Size and Sexual Dimorphism; Comparative Animal Behavior – 1920–1973; Evolution: Fundamentals; Imitation: Cognitive Implications; Motivation and Signals; Problem-Solving in Tool-Using and Non-Tool-Using Animals; Psychology of Animals; Sexual Selection and Speciation; Social Learning: Theory.

Figure 7 Morgan's dog opening the garden gate.

Further Reading

Boakes RA (1984; reprinted 2008) *From Darwin to Behaviorism.* Cambridge: Cambridge University Press.

Burkhardt RW (2005) *Patterns of Behavior: Konrad Lorenz, Niko Tinbergen and the Founding of Ethology.* Chicago: University of Chicago Press.

Darwin C (1859/2008) *On the Origin of Species: The Illustrated Edition.* New York: Sterling.

Griffiths PE (2008) History of ethology comes of age. *Biology and Philosophy* 23: 129–134.

Kruuk H (2003) *Niko's Nature: The Life of Niko Tinbergen and His Science of Animal Behavior.* Oxford: Oxford University Press.

Richards RJ (1987) *Darwin and the Emergence of Evolutionary Theories of Mind and Behavior.* Chicago: University of Chicago Press.

Deception: Competition by Misleading Behavior

R. W. Byrne, University of St. Andrews, St. Andrews, Fife, Scotland, UK

The natural world is full of wonderful examples of deception: a nightjar, safely camouflaged as dead leaves on the forest floor; a praying mantis, looking like flower petals, and waiting for a passing insect to land on the flower; an *Ophrys* orchid, emitting chemicals that mimic pheromones of a particular species of insect that its flower loosely resembles, pollinated by the insect's vain attempts at copulation; caterpillars, counter shaded to balance the light-dark effect of sunlight overhead and thus look flat, just like the leaves on and among which they live; hoverflies, dramatically patterned to look like stinging wasps, deceiving predators into leaving them alone. These examples can all be described by saying an animal or plant 'looks like ____ in order to ____,' which sounds teleological; but in fact this sort of deception is well understood to be the result of natural selection acting on body form. In none of these cases is there any indication that the deceivers have any understanding of how their deception works, or when it is appropriate: on the whole, their form is appropriate to their behavior, but the two are not linked. This is markedly shown when, for instance, a migrating nightjar chooses a temporary substrate on which its 'camouflage' makes it highly visible, perhaps a concrete block: it sits calmly, just as if it were invisible.

The issue of cognitive mechanism becomes more complicated when the deception is behavioral: at the very least, an animal that can use behavioral deception may have the option of doing it or not. Perhaps the simplest case is that of 'freezing,' where potential prey become immobile when they detect a predator. This trait functions in camouflage, since movement is much easier to detect than mere pattern, and is found in many animal species. As with deception in body form the cognitive mechanism may be simple: for instance, freezing may be triggered automatically when a predator is detected within a certain range (even closer and the reaction may switch to flight). Behavioral deception has also been reported in species lacking the elaborate nervous systems of birds and mammals. For instance, the *Photuris* firefly mimics the courtship flashing of other species, using the deception to lure individuals close enough to catch and eat. Stomatopod shrimps (*Gonodactylus* spp.), which use hammer like appendages to defend cavities as safe refuges, become soft and vulnerable when they molt their outer casing. Nevertheless, individuals in molt are actually more likely to use displays (normally signaling willingness to fight), apparently relying on bluff to deter competitors – whom they are in no position to fight. In all these examples, it is

likely that the deception is an evolved strategy the use of which is guided by relatively inflexible behavioral rules.

In some taxa of animal, however, 'tactical deception' has been described, in which behavior that normally functions in one (honest) way is seen to be used occasionally as a manipulative tactic, effective only if the audience is thereby led to misunderstand the situation. For instance, titmice (*Parus* spp.) have a call that is normally used when a sparrowhawk is sighted, and the reaction of hearers shows that the call functions as an alarm. But the same call is also used, at low frequency, in food competition: seeing competitors already exploiting a feeder, a tit may give the alarm call, scattering the competition, but then take the food itself. In neotropical mixed flocks, deceptive use of alarm calls may occur between species. In these permanent mixed-species assemblies, certain species of antshrike or shrike-tanager are typically flock leaders and act as sentinels, most often detecting the presence of hawks. The same species, however, sometimes give an alarm call when a bird of another species has just found a juicy arthropod, apparently distracting the competition by falsely suggesting a nearby predator. In these and other cases of false alarm calling by animals, interpretation is dependent on what the call normally means. If its message is equivalent to 'Danger,' then use in food competition is deceptive, whereas if its message is broader, equivalent to 'Go,' then no such interpretation is warranted. The issue of normal interpretation affects human communication in much the same way: anyone treating 'Have a good day!' or 'How do you do?' as meant literally is liable to attribute insincerity to the signaler, when these phrases are no more than ritual greetings. Unfortunately, it is often tricky to determine precisely what animal calls refer to, so it is difficult to be sure that 'false alarm calling' functions by means of deceit or not. What we conclude will vary according to whether a call refers to, say, a particular source of danger, a particular kind of escape strategy, or a general signal to move or leave.

In non-human primates, tactical deception takes a wide variety of forms. In gorilla groups, a single male normally restricts mating opportunities to himself; females, however, sometimes prefer to mate with other males in the group and sometimes use tactical deception to attain their aim. Female gorillas solicit the male of their choice; the couple remain behind when the group moves on and mate out of sight, suppressing their normal copulation calls (**Figure 1**). In baboon groups, mothers are normally solicitous of their juveniles' welfare and respond quickly to distress calls; however, while foraging mother and juvenile may be out

Figure 1 Tactical deception allows female gorillas a wider choice of mating partner than just the leading silverback. Here, a female solicits a young, nonleader silverback by standing at right angles to him and making 'head flagging' movements. The second panel shows the upshot of this interaction: the two remained behind when the group moved on, and mated, suppressing copulation calls. In fact, they were unlucky: note that their eyes are focused over the photographer's left shoulder, at the leading silverback who had just discovered them and who subsequently attacked and beat the female. (Photo credit: R.W. Byrne).

of sight and some juveniles take advantage of this to manipulate the situation. A juvenile approaches an unrelated adult that has found food, and then screams as if attacked. The mother apparently misinterprets the situation as one of danger and chases the other adult, while the juvenile simply appropriates the food. Tactical deception in primates, defined by the requirement that a benefit to the agent of the tactic is dependent on some audience misunderstanding the situation, shows a number of characteristic features:

1. The use of a particular tactic is not found in every member of the species or local population, but rather appears specific to one or a few individuals.
2. Although tactical use of an act is innovative compared to its normal use, the innovation is often a relatively small step from that normal use; learning from experience is thus quite feasible, even without insight on the part of the signaler into the (mental) mechanism of the deceit.

3. Tactical use is of low frequency compared to normal use, presumably because high-frequency use facilitates detection of the fraud.
4. Most often, the tactic functions by manipulating the attention of the audience, thus concealing information that would be of benefit. Much more rarely, the deception relies on commission, where the target audience must gain a false belief for the tactic to be effective.

These features raise a number of issues. How widespread among animals is the use of tactical deception for social manipulation, and what determines this distribution? How do individuals acquire tactics, and do they (ever) understand their mechanism of operation? In particular, is use of some tactics dependent on or aided by an understanding of false belief in others?

Distribution of Tactical Deception

Because of the low frequency of tactical deception, survey data are necessary to assess its taxonomic distribution. Despite appeals to the animal behavior community as a whole, few cases outside non-human primates and birds of the crow family have been reported. It remains unclear whether this is so because most animals are unable to use this sort of social manipulation, or whether most people who study other species do not recognize or record the cases that do occur. As a result, the understanding of what limits the use of tactical deception is almost entirely derived from data on non-human primates and corvids.

Deception has been reported widely across the primate order (**Figure 2,** upper panel), and the preponderance of records from cercopithecine monkeys and chimpanzees shows a strong bias of observer effort. Nevertheless, when correction is made for this bias, striking differences in rates remain. These turn out to correlate strongly with the species' neocortex volume, but not with the volume of the rest of the brain – or the species' typical group size, despite the fact that observers of larger groups have greater opportunities for seeing unusual behavior. It seems, then, that although use of deceptive tactics is possible for any primate, the frequency with which it is employed depends on the amount of cortical tissue available, most plausibly because very rapid learning is required. Consider the young baboon that screamed when it was not threatened and thereby manipulated his mother into driving off its competitor. It is plausible that his tactic was learnt from past experiences in which it was sufficiently tempted to approach a competitor, which did threaten him, causing him to scream in (honest) fear. If his mother were in hearing but out of sight, she might very well have come to his aid, with the unintended consequence of a food reward – reinforcing tactical use in the future. The learning requirements are stringent, however, the young baboon

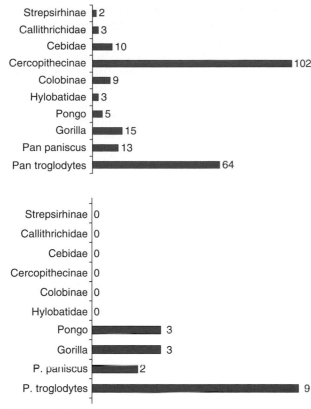

Figure 2 The taxonomic distribution of deception in primates. Tactical deception has been recorded in all major taxa: the numbers on the histogram give the actual numbers in the collation by Byrne and Whiten (1990). However, as the second panel shows, when these data are reduced to records giving convincing evidence of understanding something about other's mental states, the taxonomic spread is reduced to the great apes, alone. (Diagrams constructed by R.W. Byrne).

must learn to use the tactic only against competitors of lower rank than its mother and only when its mother is out of view but in hearing, from experiencing a very few learning situations.

Assuming that these principles apply to other kinds of animal, tactical deception should be expected among species with extensive social knowledge, which means those living in long-lasting social communities, and high neocortex volume, which in practice means large brains. Note that, since brain volume is an allometric function of body weight, few large-brained species are small bodied. Brain/body weight gives an index of the cost to species of 'affording' the brain, which is metabolically costly, and thus of recent selection on brain size. Over longer time periods, body size may be driven up by persisting demands for rapid learning.

Within mammals, obvious candidates are toothed whales, carnivores like wolves and hyaenas, and ungulates like pigs, horses, and elephants. Within birds, there are a number of families in which long-term sociality applies to groups larger than a monogamous pair, but only two where

some species are large-brained. These are the parrot and the crow families and although there are no field data of deception in either group, recent captive studies of corvid abilities have revealed several kinds of deception, considered in the next section. Avian brain anatomy is much less well understood than that of primates, but brains of corvids and parrots have a large investment in prefrontal areas, those generally associated with flexible learning skills.

Ontogeny of Tactical Deception

In non-human primates, use of deception varies between individuals under the same circumstances, suggesting a major role for the learning history rather than narrow genetic guidance, although it is clear that the underlying ability to learn rapidly in social circumstances is a prerequisite. Two very different research strategies have been applied to the issue.

With post hoc, observational records, researchers estimated the plausibility of an observation reflecting learning from past social circumstances, as sketched earlier. In the great majority of cases, learning from experience was considered quite reasonable, provided the animal was able to learn fast. However, in a number of cases involving monkeys and apes, the deceptive tactic required the animal to notice (i.e., represent mentally) the line of sight of a competitor and potential occlusion from its view. Intriguingly, the relatively few cases in which learning from experience seemed most improbable, because the necessary circumstances would seem outside the species' normal experience, were clumped in one taxon: the great apes (**Figure 2**, lower panel). This led to two controversial proposals: monkeys and apes must be able to represent and compute with the geometric visual perspective of others, and great apes must be able in some way to represent and compute with the mental states of other individuals. Both have now been confirmed by data from experiments in other areas of cognition that depend on similar attribution abilities, at least for the case of chimpanzees.

The experimental approach has relied on setting up circumstances, in which deceptive tactics should be profitable, in the hope that they will be displayed. For instance, one (subordinate) individual is given privileged information about food location, and then reunited with one or more regular social companions: the 'informed forager' paradigm. In chimpanzees, mangabey monkeys, and domestic pigs, it was found that dominants readily adapted to exploit the knowledge of the informed individual, resulting in the informed individual losing the food. In the primates tested, tactics of deception developed over several trials. Revealingly, the mangabey at first simply held back or wandered elsewhere, then – apparently noticing that the dominant was not near the food – it quickly retrieved it. The success of this maneuver resulted in its adoption as a regular tactic. Thus, the monkey may not have strived to create a false belief in the competitor, but simply learnt during the experiment a sequence of actions that worked to its advantage. The same was true for chimpanzees, but here more elaborate tactics were sometimes developed, including leading the competitor to an aversive object or a small food source before rushing to the large cache of food. In the case of domestic pigs, once the noninformed individual began to follow the other, its tracking was so close that the informed individual was never able to gain the food. Yet in this circumstance when no reinforcement learning can occur during the experiment, several exploited pigs apparently tried to deceive the dominant competitors. Variously, individuals were more likely to move to the food site when the competitor was further from the food, relatively further from the food than itself, moving away from the food or positioned out of line-of-sight from it. It is possible that these tactics might have been learnt during prior contests over different resources, provided that – like monkeys and apes – pigs can compute with the geometric properties of social situations. That is, pigs could learn these tactics if they could work out what was in view from another individual's perspective, and consequently categorize certain situations as 'unsafe' for them – even without understanding *why*, which might require knowledge of what others know. But the fact that the experimental pigs had spent almost all their short lives under controlled conditions makes that line of explanation speculative at best.

Consequently, no clear conclusion about the ontogeny of tactical deception in mammals is yet warranted. In most cases, learning from experience is a feasible explanation, even in animals unable to represent any mental states of others; although it is true that having such ability would make learning from experience more efficient, most researchers would consider that less plausible. In great apes and domestic pigs, the balance of plausibility tilts the other way, and accounting for the performance of deception (or attempted deception) in the absence of any likely reinforcement history is difficult unless these species are able to compute the effects of untried actions in advance – to plan in advance to change other's beliefs.

Extensive experimental research has also been carried out, using a rather similar paradigm, on several species of food-storing corvids. Many species of temperate-zone birds store food against future shortages, and in several corvid species, individuals also regularly pilfer others' food caches if they are able to see them made. In this case, just as in the informed forager work, it would pay individuals to deceive their competitors, and they might sometimes be able to do so better if they understood competitors' knowledge. Ravens cache preferentially when they are away from other ravens, and react with various tactics if they see others while they are caching: speeding up their caching process; covering the cache more thoroughly, even leaving the site to get better material for a covering;

recovering the food more quickly than usual; or sometimes refraining from caching it at all. Like non-human primates, ravens seem to understand the geometry of visual access, trying to hide from other ravens behind barriers when they cache. Like most primate tactical deception, raven deceptive tactics rely on withholding information; but in addition, they make false caches, carefully 'hiding' nonexistent food when competitors are watching. Pilfering ravens, too, use a range of tactics that seem to function in deceiving competitors: orienting or repositioning themselves to give good visual access while keeping some way away from the food-storing raven; delaying approaching an observed cache while the storer is still nearby, but rushing to exploit a cache if other potential pilferers appear; and, if the original storer is still around, searching at false, 'noncache' sites. Scrub-jays have similarly been found to use deceptive tactics in food storing. They prefer to cache when no competitor jay is in sight, and when they cannot avoid being seen, they cache as far from the observer as possible, in places that cannot be seen from the competitor's position, and if they cannot achieve that they choose dimly-lit areas for caches.

Although at present the ontogeny of corvid deceptive tactics is not fully understood, the logistical advantages of working with corvid species that can be hand-reared and grow to adulthood in a few years give distinct advantages, and already, some progress has been made. Scrub-jays react to others seeing them caching food, by recaching once they get the chance in private – but only if they have had the experience of *being* a pilferer themselves. Naïve jays, reared with no experience of pilfering other's caches, do not tactically re-cache their own, suggesting that the tactic is based on some understanding of the situation. In one clever experiment, a scrub jay was allowed to cache only in a certain area when observed by a particular competitor jay, and allowed to cache in another area when watched by another, different jay. Later, when given access to both areas while under observation, it re-cached or ate whichever food that particular observer had seen cached, but did not disturb other caches that the competitor had not seen made. Jays evidently remember who has seen what, and can somehow take the knowledge or ignorance of their competitors into account. Ravens have also been shown experimentally to be capable of taking into account whether their competitor is knowledgeable or ignorant, so it is entirely possible that all these corvid deceptive tactics are carried out with insight into the ignorance or false beliefs that they ensure in competitors.

The difficulties researchers face in attributing intent are not unique to animal work. Consider examples from everyday human life: a teenager who feels ill just before a much-dreaded cross-country run; or when we say 'You look great!' to a friend suffering from a serious long-term illness. In both cases, the agent is fully capable of mentally representing false beliefs, and they are apparently misrepresenting information to others. But are they deliberately falsifying, with full understanding of the falsity of the beliefs they are creating? Most people would suggest that some degree of self-deception is involved, and evolutionary theorists have long argued that self-deception should be expected in animal communication simply because it is harder to detect and therefore more effective deceit (a point often made in the Le Carré world of espionage and treachery). Moreover, even in cases where creation of false belief is deliberate and well understood, genetic predispositions and experiential learning are also liable to be important. When a kind friend gives an unwanted present, we most likely feign happiness, deliberately to create a false belief, in order to avoid giving pain or offence. But that tactic relies on voluntary control of facial expression, an option based on human genetic predispositions and not available to many other species; and we probably learned earlier in life the upsetting effects of too much honesty. Working out whether genetic constraints, access to learning opportunities, or the mental competence to represent social situations is the limiting factor in animal deception is likely to be a complex issue for future research.

See also: Conflict Resolution; Emotion and Social Cognition in Primates; Punishment; Social Cognition and Theory of Mind.

Further Reading

Bugnyar T (2007) An integrative approach to the study of 'theory-of-mind'-like abilities in ravens. *Japanese Journal of Animal Psychology* 57: 15–27.

Bugnyar T and Kotryschal K (2004) Leading a conspecific away from food in ravens (*Corvus corax*). *Animal Cognition* 7: 69–76.

Byrne RW (1996a) Machiavellian intelligence. *Evolutionary Anthropology* 5: 172–180.

Byrne RW (1996b) Relating brain size to intelligence in primates. In: Mellars PA and Gibson KR (eds.) *Modelling the Early Human Mind*, pp. 49–56. Macdonald Cambridge: Institute for Archaeological Research.

Byrne RW and Corp N (2004) Neocortex size predicts deception rate in primates. *Proceedings of the Royal Society of London B* 271: 1693–1699.

Byrne RW and Whiten A (1992) Cognitive evolution in primates: Evidence from tactical deception. *Man* 27: 609–627.

Byrne RW and Whiten A (1990) Tactical deception in primates: The 1990 database. *Primate Report* 27: 1–101.

Clayton NS, Dally JM, and Emery NJ (2007) Social cognition by food-caching corvids. The western scrub-jay as a natural psychologist. *Proceedings of the Royal Society: B* 362: 507–522.

Dally JM, Emery NJ, and Clayton NS (2006) Food-caching western scrub-jays keep track of who was watching when. *Science* 312: 1662–1665.

Emery NJ and Clayton NS (2001) Effects of experience and social context on prospective caching strategies in scrub jays. *Nature* 414: 443–446.

Halligan P, Bass C, and Oakley D (eds.) (2003) *Malingering and Illness Deception*. Oxford: Oxford University Press.

Mitchell RW and Thompson RS (eds.) (1986) *Deception: Perspectives on Human and Non-human Deceit*. New York: SUNY Press.

Decision-Making: Foraging

S. D. Healy and K. Morgan, University of St. Andrews, St. Andrews, Fife, Scotland, UK

Decision making underpins much of an animal's life, as the choices made can dramatically affect the fitness of both the animal making the choices and the fitness of that animal's mate and offspring. For example, a female zebra finch that chooses to mate with a male who helps her to feed their babies is likely to raise more of those babies than the female who chooses to pair with a male who is less helpful. Female rufous hummingbirds, on the other hand, do not get to make such a decision: male rufous hummingbirds never help with any of the offspring care (not even in building the nest). In its lifetime, an animal will have to make all manner of judgments, about such diverse issues such as mate choice, nest building, foraging, investment in offspring, and social interactions. The information on which an animal bases its decisions can also be diverse, and the decisions themselves can be made using a lot or a little information. As yet, we know little about the extent to which consciousness contributes to decision making in animals, but we do know that many decisions are made as a result of an animal's own experience. There are at least two reasons for investigating decision making in animals. Firstly, knowing what information an animal uses and in what contexts they use it can tell us a lot about how animals react to changing environments, which is increasingly interesting as we worry about climate change. Secondly, investigating decision making by watching an animal's behavior can help us to determine what is that is going on inside an animal's brain without being able to ask the questions verbally.

Despite the diversity of areas in which decision making affects fitness, most decision-making research has focused on foraging and mating. This emphasis occurs because we can readily understand the fitness benefits of choice in foraging and mating. In addition, both systems are amenable to experimentation. Mate choice and foraging have also shown us that a decision encompasses much more than simply what to eat or who to mate with as the animal making the decision may also take into account the quality of the individuals or items among which they can choose, spatial and temporal information as well as past experience. When choosing a potential mate, for example, animals may consider when and with whom to mate, where to nest, how many offspring to produce, the gender of those offspring, and so on. Furthermore, a choice that is suitable at one time and place might not be the best at another time and place, that is, how appropriate a decision is will be strongly dependent on the context in which it is made. To make the best decision, or even just a good

decision, can require access to a significant amount of information and the ability to process that information. It is not, then, terribly surprising that in attempts to cope with this information-processing dilemma, animals use 'short cuts' to reach a decision, such as copying the decisions made by other, often more experienced individuals. In mate choice experiments, for example, inexperienced females often seem to copy the choices of more experienced females.

Mate choice is a complex process of signaling by the advertising individual and signal interpretation by the choosy individual. Interpreting these signals depends on the other choices available (the context in which the choice is made), social influences, and the state of the individual making the choice (as well as the behavioral and morphological attributes of the potential mate and those of the decision maker). In spite of these complexities, the experimental study of mate-choice decisions has flourished, especially in the laboratory where investigators can often control the attributes of potential mates (e.g., an experiment might, for example, present a female with several males that differ in only one attribute). However, at least two significant problems occur with investigations into mate-choice decisions. The first is that it male traits often covary (e.g., big males have bright colors) so that investigators cannot isolate a single trait. For example, not only does body size vary, so do the size, shape, and color of sexually selected characteristics as does the behavior of the potential mates. In a typical laboratory mate-choice experiment, the experimenter presents a female (since females are usually the choosy sex) with two or more males. She can see, hear, and smell the males, but not make physical contact with them. Typically, the apparatus also prevents the males from seeing each other, and, sometimes, one-way screens are inserted between the compartments separating the female from the males so he cannot see her and respond to her (he might, e.g., sing more, or be more active, if the female comes close to him and thus encourage her to stay). These are concerns as the time the female spends in front of any of the male compartments is taken as a real indication of how much she would like to mate with that male.

The second problem with mate-choice studies is that experimenters have tended to concentrate on features of the animals that appear conspicuous to humans, classically, visual features such as red cheek patches on male zebra finches or red coloration in sticklebacks. Yet, we know that birds, for example, see colors in the ultraviolet that human cannot see. It follows that animals may use signals unavailable to our senses (e.g., UV, pheromones) when deciding with whom to mate.

Foraging

In spite of the appeal, then, of mate-choice decision making, most of the substance of what we know about decision making in animals comes from looking at foraging. Like choosing a mate, foraging can be a multifaceted problem. Where, when, and for how long to forage? What and how much to forage on and so on. Foraging is useful for addressing decision-making questions as the animal's choices have direct consequences, which can be learned by the animal and readily manipulated by the experimenter. These manipulations could be in amount, energetic consequence, or in the cost of obtaining the food. Other variables, such as the energy budget of the animal, the number of options available in a choice set, and the riskiness of the option, can then be manipulated to see whether they alter the choices made.

When to Stop Foraging?

The benefits to foraging are obvious. Foraging not only lowers the risk of death by starvation, it also provides the energy and material animals need to produce expensive signals such as antlers or colorful plumage, allowing animals to attract mates, the success of reproduction depends heavily on foraging success. However, foraging takes time away from other necessary behaviors. Additionally, the longer an animal forages, the longer it is exposed to predators, either because it is moving around and is, therefore, more likely to encounter predators or because it is less vigilant. Even for those animals for which predation is a relatively minor concern, such as top predators, increased time foraging may lead to decreased group or territorial defense and an increased likelihood of territory invasion. Animals would, ideally, minimize the time taken to find food while getting as much food as they can eat. As well as the time spent foraging, animals must consider how long to stay eating or looking for food in a particular patch before moving to another patch.

Foraging duration has important implications for parents, such as the songbird that has to leave defenseless chicks while collecting food with which to provision them. The length of time the parent spends foraging will have impacts for the offspring, the parent, and that parent's mate. An extreme example of this is a mother Emperor penguin that shortly after laying spends 9 weeks away foraging during which time the father shelters the egg and subsequently the young chick. Depending on how long the female is away, he may lose up to a third of his

body weight. In species with biparental care, conflicts of interest commonly arise over the level of care that each parent provides, especially in species that pair for a single breeding season. Parental care is costly, and the conflict of interest may mean that one, or both, parents may withhold some care.

Where to Forage?

For animals whose food sources replenish relatively rapidly (within a few hours) such as a territorial hummingbird foraging on nectar from refilling flowers, it is advantageous for the bird to remember information about the quality and quantity of resources in a patch within its territory as well as when it last visited that patch. However, for other animals, it might be more useful to remember where not to forage, as these areas will have been emptied of food (there are other reasons to avoid foraging in certain places, such as likelihood of the presence of predators). And most animals, at some point, need to decide whether to continue foraging in the current patch or whether to leave and search for a better patch.

For many animals, remembering good foraging locations is important, but for some, for example scatter-storing birds, which rely heavily on food they have hidden in many locations, such memories may reach into the hundreds or thousands of locations. Copying others may enable the decision of where to forage to be made more readily than by either remembering or relying on one's own searching abilities. One advantage to group living is possibly that animals may use the foraging success of others to decide the direction in which to search on the following day, that is, by following today's successful forager when they depart the group tomorrow morning.

When to Forage?

Deciding when to forage can be important for a number of reasons, which include the ability to exploit a renewable resource effectively, consuming a cached source before it spoils as well as avoiding foraging at times when predators are most active or effective. Being active only at night, for example, is one way for some prey animals to reduce their risk of predation. Many small nocturnal mammals use a circadian clock to maintain their nightly activity cycle, which is adjusted to keep their activities synchronized with changing patterns of light and dark. The entrainment of the clock allows the animal to track seasonal changes in day length.

The timing of foraging bouts has implications beyond predation. A territorial hummingbird, for example, should avoid returning to the flowers it has empty, and ideally should allow time for the nectar in a previously visited flower to replenish. To do this, the bird needs some mechanism that is sensitive to both the nectar secretion rates of flowers and the environmental frequency of floral visitor. Hummingbirds can accurately match the refill rate of flowers and can remember the difference in refill rate for different flowers. They also return sooner to flowers that are more likely to be emptied by competitors than to flowers from which competitors are excluded, demonstrating that their decisions about when to visit particular flowers involves the use of information about the flowers, themselves, and other animals.

How Much to Eat?

Obviously, an animal must eat enough to avoid starvation, and yet animals often both eat less and carry less fat than they can. This counterintuitive observation has led to the suggestion that there could be a trade-off between the benefits of fat storage and other costs such as predation risk (by being too heavy to outpace or outmaneuver a predator). One might, therefore, expect that decisions about body fat content might be responsive to the number of predators in an area although, currently there is evidence both for and against this possibility. For example, although in comparisons between closely related migratory and non-migratory bird species, some resident birds do not appear to carry as much fat as would be optimal, the addition of real or simulated predators does not always cause a drop in body weight as would be expected levels if the degree of fat stored depended on predation levels.

Factors other than predation risk can influence food-storage decisions. Hummingbirds, which hover at flowers to feed, rarely drink more at any one time than a third of the volume that their crop can hold. However, this is only the case for territorial hummingbirds with constant access to food. Nonterritorial birds in the same population drink more during each feeding bout because they need to take food when they can get it. This suggests a trade-off between the benefits of food acquisition and the costs of flying with a heavier crop. With a reliable food supply, territorial birds can afford to cut the flight costs but nonterritorial birds cannot.

Choice of Resource

Foraging animals often have to choose among foods that differ in many ways. Some foods provide lots of energy, others may provide more protein, and still others can be consumed or digested quickly. Additionally, a forager choosing among resources often faces a problem of incomplete information. For example, when a foraging bird encounters a conspicuous, brightly colored insect, this potential prey item could be tasty and nutritious or

noxious and unpalatable. The choice it makes as to whether to eat the new insect will depend on how similar that insect is to previous prey the bird has encountered. If the insect is very similar to prey the bird has learned are unpalatable, the bird is likely not to eat it. However, if the bird is really hungry, it may eat the new insect, even though the bird knows that the insect is likely to be unpleasant. As long as the level of unpalatability is such that the bird does not become very ill (or dies), the decision to eat the new insect would be sensible, in the circumstances. When the bird tries the new insect, however, it runs the risk that the insect is, indeed, lethal, and then the decision would seem to be a poor one. In situations of incomplete information, accuracy and relevance of memory for past experiences plus the animal's current state may lead to animals making costly decisions. Experimental support for this supposition comes from the greater consumption of noxious prey by starlings when they are hungry relative to when they are well fed.

Incomplete information is likely to underpin almost every decision an animal makes, even if they can remember very well what has gone before. This may be because their memory is not perfect or because animals often face situations that are similar to, but not the same, as a previous situation. They then have to decide whether the current choice is similar enough to one from the past that they choose a particular option. For example, nectarivores such as hummingbirds and bees cannot be sure what reward a new flower from a familiar species will provide. This is because flowers vary in the amount of nectar produced by different flowers on a single inflorescence, refill rates depend on temperature, humidity, and so on, and it is possible that the flower was recently emptied by a competitor. How animals respond to this variation inherent in many choices under varying conditions is referred to as 'risk sensitivity.'

When an animal faces a choice between one option that is variable in some way and between another that is constant, such that the mean rate of return is the same but the variance around that mean return differs, a preference for the constant option is considered to be 'risk averse' while a preference for the variable option is termed 'risk prone.' Typically, such decision making has been investigated in the laboratory in which the context in which the animal is allowed to choose between options can be readily manipulated. Aspects of the items can be manipulated (e.g., the size of a food item, the rate of delivery of food items) as can the animal's energetic state (by lowering its body weight through food restriction or increasing its energetic needs by lowering the temperature). In such experiments, animals tend to be risk prone when food varies in the rate of delivery but to be risk averse when food varies in amount. Although one would expect that animals on a positive energy budget would prefer the constant option (risk aversion) and when on a negative energy budget they would be risk prone (i.e., choose the variable option), there is little empirical evidence for this expectation.

Choice of the 'Best' Option

It is, perhaps, not surprising the animals can and will take a range of kinds of information into account when making decisions: the environment is rarely so stable that exactly the same decision is always the best one and the animal itself will sometimes be in a state more suited to one option than to another. It is still plausible that an animal might be able to determine its own state and that of the environment sufficiently accurately always to make the 'best' choice. However, in addition to the problem of incomplete information, decision making in animals has long been based on the assumption that animals assign a fixed value to each option and always choose the option with highest value. In humans, at least, this is not the case. We constantly make choices between options that can be altered by the presence of yet other options, even when those alternative options are clearly inferior. This is known as a 'violation of rational choice.' Rationality, in this context, describes the situation in which we assign fixed values to, for example, two options and then choose between them depending on their fixed values, without regard to the context in which the decision is being made. If we behaved rationally, the addition of an inferior item would not alter the choice between the first two items at all. However, this does not happen. The context in which the decision is made has a significant effect on the choice between two original items, such that it is necessary to know the context before an accurate prediction as to which item will be chosen (or preferred).

It now also appears that foraging bees and hummingbirds also make irrational choices when offered an inferior third option. Although at first glance it may appear that faulty decisions produce irrationality, it occurs because animals use rules of thumb (or heuristics) when making decisions. These rules of thumb work well in most situations and help animals to make faster decisions even without complete information. Although all of the work to date on 'context-dependent' choice in animals has addressed foraging decisions it is plausible that in many other decision-making situations animals choose irrationally.

See also: Caching; Internal Energy Storage.

Further Reading

Bateson M and Healy SD (2005) Comparative evaluation and its implications for mate choice. *Trends in Ecology & Evolution* 20: 659–664.

Hutchinson JMC and Gigerenzer G (2005) Simple heuristics and rules of thumb: Where psychologists and behavioural biologists might meet. *Behavioural Processes* 69: 97–124.

Ryan MJ, Akre KL, and Kirkpatrick M (2007) Mate choice. *Current Biology* 17: R313–R316.

Stephens DW, Brown JS, and Ydenberg RC (eds.) (2007) *Foraging*. Chicago, IL: The University of Chicago Press.

Decision-Making and Learning: The Peak Shift Behavioral Response

S. K. Lynn, Boston College, Chestnut Hill, MA, USA

Peak Shift Is a Directional Behavioral Bias

In a typical peak shift experiment, control subjects are trained to respond (e.g., button-press) to a positively reinforced stimulus (S+, for example a line of a particular orientation or a color of a particular hue). Treatment subjects are trained in a manner identical to control subjects with respect to S+ and also to withhold response to an unreinforced or punished stimulus (S–, a line or hue similar but not identical to S+). Both groups of subjects are then tested without reinforcement on a continuum of stimuli comprising various line orientations or hues, including the training stimuli. **Figure 1** depicts the common finding of such discrimination learning experiments: during the test, control subjects respond most frequently to the S+ stimulus. Treatment subjects, however, respond most frequently to a stimulus they have never encountered before.

The treatment subjects' expression of a preference for an unrewarded and novel stimulus over S+ is somewhat paradoxical. If all test stimuli are discriminable, why are the treatment subjects (S+/S– trained) not like the control subjects (S+ training only), responding most strongly to the stimulus they have learned is rewarding? The phenomenon that treatment subjects display in this type of experiment is known as peak shift. When the subjects' frequency of response is plotted as a function of stimulus value, data show a peaked response gradient. The stimulus receiving the maximum, or 'peak,' response by the treatment subjects is said to be 'shifted' relative to that of the control subjects.

An area shift is often noticeable, even in experiments that do not result in a significant peak shift. An area shift is characterized by an elevation of the rates of response to the novel stimuli on the side of S+ away from S–. Area shift often co-occurs with peak shift. In addition, a shift of the most strongly avoided stimulus is also produced, off of S– in a direction away from S+.

Peak shift is considered to be a general outcome of generalization accompanying discrimination learning. In a typical peak shift experiment, the stimuli are simple sensory perceptions, monotonically increasing in value on the stimulus domain. The experiment consists of a control group and a treatment group each undergoing training (discrimination) and testing (generalization) phases. Stimuli are presented one at a time, for instance, on a lighted button that the subject presses to indicate its response. The button is lit for a set length of time, and presses to the button while it is lit by a given color are the dependent variable. No reward is given during the testing phase. A variable intermittent reward schedule with all-or-none reward quantities may be given for correct responses during the training phase to minimize extinction during the testing phase. During training, incorrect responses to S+ and S– are followed by mild punishment (e.g., lights turn off and a delay is imposed prior to the next trial). Intertrial intervals last 2–3 s, during which the response button is not lit. The training phase lasts until the treatment group reliably responds to the stimuli (e.g., 80% correct). The number of S+ training trials may be balanced across both groups, and a variety of reinforcement schedules, stimulus dimensions, and species have been used. As a phenomenon of learning, the magnitude of shift is a function of stimulus presentation parameters such as S+ and S– similarity, reward value, and stimulus encounter rates.

Theoretical Accounts of Peak Shift

Peak shift might arise at any of several mechanistic levels of processing, including early peripheral sensory processes, well-learned associative mechanisms at the level of individual stimuli, or via a response to signal-borne risk occurring at the level of stimulus class. Peak shift is thought to arise from relatively uncomplicated mechanisms of learning. However, the 'right' explanation for a particular type of decision will shed light on larger issues of learning, preference establishment, and decision making.

Gradient Interaction Theories

Peak shift has been almost invariably attributed to Kenneth Spence's 1937 theory of overlapping gradients of excitation and inhibition. The summation of Spence's two gradients produces net excitatory and inhibitory behavioral impulses shifted from the training stimuli in the manner described earlier. On this and related accounts, differences in behavioral response strength arise from the level of behavioral excitation associated with each test stimulus. In terms of modern associative learning theory, bell-shaped gradients of positive associative strength (centered on S+) and negative associative strength

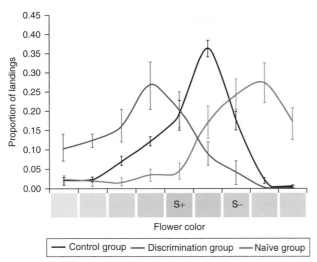

Figure 1 Typical results of a peak shift experiment. A control group was trained to approach the S+ stimulus. A discrimination group was additionally trained to avoid S−. At test, the full range of stimuli was presented. The discrimination group exhibited a shift in their preferred stimulus off of S+ in a direction away from S−. These data are from bumble bees trained to discriminate the colors shown while learning to forage on artificial flowers. In addition, a Naïve group received no color training prior to testing, and exhibited an innate preference for bluish flowers. Examination of the control group shows that their peak response was slightly shifted toward the innate preference rather than centered over S+ itself. Standard error of $n = 10$ bees per group is shown.

(centered on S−) overlap and interact additively, yielding a maximum net positive associative strength shifted off of S+ in a direction away from S−. Differences in behavioral response are due to differences in the association strength between the test stimuli and the reward or punishment associated with responding to those stimuli.

Neural network models of peak shift are a kind of associative account in which nodal weight strength drives behavioral response strength. Nodal weight strength can be mathematically identical to Rescorla–Wagner-based measures of association strength. Peak shift has also been modeled as an additive overlap of sensory receptor excitations, an account at the level of peripheral sensory organs.

All gradient-interaction accounts assume a model of stimulus generalization that involves a spreading of knowledge from training stimuli to similar novel stimuli. The spreading is usually a Gaussian (i.e., bell-shaped) function of the perceptual similarity among stimuli. The S+ and S− generalization gradients then interact additively during decision making, producing peak shift.

No satisfactory gradient-interaction explanation of peak shift has arisen that accounts for the various forms of peak shift and the variety of stimulus domains over which it can be induced. The various explanations fail to account for area shift, are not extendable to the full range of stimuli prone to peak shift, or cannot produce the variety of observed gradient shapes.

Adaptation-Level Theory

A peak shift can also be explained by a modification of Harry Helson's theory of sensory adaptation-level, developed in 1947. Over the course of exposure to simple stimuli, such as line orientation, a subject's perceptual system can become 'adapted' to the range of stimulus variation. The adaptation level is centered at the mean of the stimuli encountered. This account of peak shift posits that subjects represent the S+ and S− stimuli relative to the adaptation level rather than in more absolute terms. So, during training, subjects learn that S+ is located at, say, adaptation level +1 unit and that S− is at, say, adaptation level −1 unit. When, during testing, the adaptation level changes as new stimuli are encountered, responding to adaptation level +1 produces an apparent peak shift. The subjects, however, have not changed their response to the stimuli per se, and this peak shift is not driven by conventional learning parameters. Rather, the baseline against which stimulus differences are evaluated has changed. Because under the adaptation level account stimuli are represented relative to the range of stimulus variation, this kind of peak shift is known as a 'range effect.'

Although a scenario of changing adaptation levels does account for some peak shift results, researchers can control the forces that drive range effects with techniques such as probe-tests administered throughout training. Also, more complex stimuli possessing multiple components or dimensions (e.g., facial expressions, orientation of clock hands) are resistant to range effects. Range effects can, and must, be controlled in studies focusing on peak shift arising from discrimination learning.

The Signals Approach

An account of peak shift can also be derived from signal detection theory (SDT; **Box 1**). This approach postulates that during testing, subjects experience uncertainty about which response is appropriate to give to a particular stimulus. Under SDT, uncertainty of choice-making is due to perceptual similarity of S+ and S− stimulus classes and carries a risk of stimulus misclassification. Under a signals approach, peak shift arises from an attempt to optimize stimulus classification rather than as strictly determined by associative strengths. As a signal detection issue, peak shift can be characterized as an aversion to signal-borne risk associated with the uncertainty of stimulus classification.

The three parameters of SDT that govern choice (distribution, relative probability of occurrence, and payoffs – see **Box 1**) correspond to elements of a discrimination learning experiment. (1) The appearance of the S+ and S− stimuli constitutes signals. The signal distributions are interpreted as gradients of likelihood that, on the basis of perceptual similarity, a particular stimulus is from the

---

Box 1:   Signal Detection Theory and Extension to Nonthreshold-Based Decision Making

A signals approach to generalization and discrimination takes signal detection theory (SDT) as a descriptive, mechanistic model of decision making, as opposed to the theory's typical use as an analytical tool. SDT is a mathematical description of the trade-offs and risk inherent in the reception of signals (i.e., discerning one signal from another, or signal from noise). Classical SDT provides a functional description of how animals make choices among stimuli under conditions of uncertainty. Typically, that uncertainty is considered to arise from perceived variability in the appearance of stimuli. For example, variability in stimulus appearance could exist in the stimuli themselves or arise from noise in the sensory system. However, the uncertainty modeled by SDT may also arise from stimulus generalization, a process dependant on reinforcement history in addition to perception: inexperienced animals that are able to perceptually distinguish stimuli very well may yet be uncertain as to what response is appropriate to give to a particular stimulus. Like perceptual uncertainty, this response uncertainty can be described by signal detection theory.

Viewing generalization and discrimination as exercises in signal detection is straightforward. Consider a task in which subjects must approach yellowish stimuli and avoid bluish stimuli. In the overlapping greenish region, any given stimulus might be of one class (S+, deserving response) or another (S−, to be ignored). As a model, the signals approach posits that animals estimate information about stimulus encounters to determine to which stimuli it is on average most profitable to respond, such that the number of correct detections of S+ and correct rejections of S− are maximized while missed detections of S+ and false alarm responses to S− are minimized. Under SDT, response strength is determined by three signal parameters: (1) the stimulus distributions over a perceptual domain, (2) the relative probability of encountering stimuli of one class or another, and (3) the payoff (reward or punishment accrued) for responding to or ignoring S+ and S− stimuli.

These signal parameters can be combined in a utility function, the maximum of which locates the optimal placement for a response threshold on the stimulus domain:

$$U(x) = \alpha h P[CD] + \alpha m P[MD] + (1 - \alpha) a P[FA] + (1 - \alpha) j P[CR] \qquad [1]$$

where $U(x)$ is the estimated utility over stimulus domain x; $P[CD]$, the probability of correct detection, measured as the integral of the S+ distribution from threshold to negative infinity; $P[MD]$, the probability of missed detection, equal to $1 - P[CD]$; $P[FA]$, probability of false alarm, measured as the integral of the S− distribution from threshold to negative infinity; $P[CR]$, the probability of correct rejection, equal to $1 - P[FA]$; α, the relative probability of encountering an S+ signal, and $1 - \alpha$ equals the relative probability of encountering a signal from the S− distribution; h, the benefit of correct detection of S+; m, the cost of missed detection of S+; a, the cost of false alarm response to S−; and j, the benefit of correct rejection of S−. Costs may be negative or simply less positive than benefits, so long as $h > m$ and $j > a$.

In classical SDT, signal distributions over the continuous sensory domain are considered to be probability density functions (PDFs). The probabilities of correct detection, false alarm, missed detection, and correct rejection are calculated by integrating the respective PDFs from each possible threshold location to infinity (or by taking one minus that integral). This integration permits locating the optimal threshold placement (the maximum of eqn [1]). Threshold placement produces a stepped or sigmoid response gradient of dichotomous response strengths: equally strong response on one side of the threshold, and equally weak on the other. To apply SDT as a model of behavior to discrimination tasks in which subjects do not show a threshold-based response, such as peak shift, the assumption of integrated signal distributions can be changed. Substituting the integration of the PDF with the probability density, y_i, of a signal of a given value, x_i, allows signal detection theory to produce continuously variable response strengths.

One way to conceptualize this substitution biologically is to suppose that animals perceive signal variation discretely or construct discrete signal distributions rather than continuous probability density functions. Biologically, one may interpret this as an assumption that although signals may indeed fall along a continuous distribution objectively, animals perceive stimuli in flexible intervals of just-noticeable-difference, the dynamic width and placement of which is determined by contextual factors and the limits of their sensory acuity.

To reflect this substitution, eqn [1] can be modified to yield the expected utility of responding to a signal of $x = x_i$, a particular value, rather than a signal of $x \geq x_i$ as would result from PDF integration:

$$U(x) = [\alpha h f_{S+}(x_i) + (1-\alpha) a f_{S-}(x_i)] - [\alpha m f_{S+}(x_i) + (1 - \alpha) j f_{S-}(x_i)] \qquad [2]$$

where $U(x)$ is the utility of responding to a stimulus of a given value, x_i (correct detections and false alarms), less the utility of withholding response to that signal (missed detections and correct rejections); $f_{S+}(x_i)$, the relative frequency of a stimulus of value x_i from the S+ signal distribution; $f_{S-}(x_i)$, the relative frequency of a stimulus of value x_i from the S− signal distribution; and other variables are as for eqn [1].

Equation [2] produces a pulse-shaped gradient exhibiting peak shift. It is positive (utility > 0) for all stimulus values for which responding yields a net benefit. Like eqn [1], it provides a mechanism by which to make choices in the face of uncertain stimulus classification, not by reducing the uncertainty, but by allowing an animal to estimate to which signals it will be on average be most profitable to respond. Equation [2] can still produce threshold-based behavior, by solving for zero, the x-intercept, rather than maximizing the function.

S+ or S− stimulus class. (2) The relative probability of encountering an S+ or S− signal corresponds to the relative frequency of S+ and S− stimulus presentation during training. (3) The payoffs correspond to the value of reinforcement and punishment for responding to and ignoring stimulus presentations during training. The utility function (**Box 1**, eqn [2]) combines the signal

parameters learned during training to produce a pulse-shaped 'response gradient,' the maximum and minimum of which exhibit peak-shift (**Figure 2**). Peak shift can thus be interpreted as a signal detection issue. However, rather than using a threshold to dictate choice, as in other applications of signal detection theory, the maxima and minima of the pulse-shaped generalization gradient are used. The

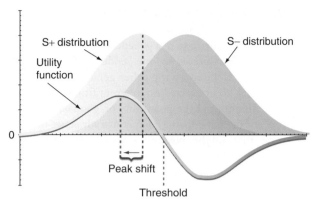

Figure 2 On the signals approach, S+ and S− signal distributions (shown here as yellow and blue bell-shaped gradients) represent the subject's estimate of the likelihood that a particular stimulus is from the S+ or S− stimulus class. Signal distributions may arise from actual signal variation, noise in the perceptual system, or cognitive generalization. Overlapping distributions produce uncertainty about which response (e.g., approach or avoid) is appropriate to give to any given stimulus. Behavioral response is dictated by a utility function (Box 1, eqn [2]) that integrates the signal distributions, the estimated probability of encountering an exemplar of either stimulus class, and the payoffs expected from correct and incorrect responses. The maximum and minimum of the utility function exhibit peak shift.

magnitude of the peak shift displacement is sensitive to variations in the three signal parameters.

Functionally, the peak shift experiment is a signal discrimination task in which animals are uncertain as to which response, approach or avoidance, is appropriate for any given test stimulus. Though the mathematics of the signals approach provides a functional account of peak shift, the approach can also be interpreted mechanistically, as an alternative to the gradient interaction accounts. According to this interpretation, peak shift is not a failure to perceptually discriminate S+ from similar signals, nor is it an artifact of overlapping gradients of excitation or associative strength. Rather, peak shift reflects an optimization of response (i.e., committing a number of unavoidable mistakes to achieve correct responses) in cases when subjects experience uncertainty about stimulus classification.

Decision Making at the Intersection of Comparative Psychology and Behavioral Ecology

Peak shift intrigues behavioral researchers for at least two reasons. First, its apparent universality makes peak shift a model system in which to study decision making. Second, when it occurs in situations in which decision makers evaluate stimuli linked to another organism's reproductive success, then peak shift has the potential to exert selective pressure on the evolution of morphology and communication.

Peak Shift Is a 'Model' Decision

Peak shift is taxonomically widespread: exhibited by birds; mammals, including humans; fish; and at least some arthropods. The phenomenon thus appears to reflect universal attributes of generalization, discrimination learning, and choice-making behavior. As such, peak shift is a 'model' type of decision making, suitable for comparative study at functional and mechanistic levels. Using peak shift as a tractable example of decision making, a variety of organisms can be studied, with strengths differentially well suited to phylogenetic, behavioral, neural, cellular, or molecular investigations.

In addition to being well suited to study at multiple levels, considerations of peak shift go beyond what is typically investigated in research on decision making. Many models of behavioral economics maximize utility: these models consider variability in (1) the costs and benefits of obtaining resources, and how those payoffs change with body state, and (2) the probability of encountering resources of some quality. Game theoretic approaches additionally account for the effect of others' responses on the decision maker's own behavior. However, these models overlook the fact that an animal's estimates of a resource's payoff and probability are based on sensory signals emitted by the resource. Outside of the laboratory, signals, such as color or tail length, vary. This variation may exist independently of any variation in the information encoded by the signals. For example, a signal that indicates a particular food quality (yellow skin on a banana signals ripeness) may vary even if the food quality itself does not (ten bananas of the same ripeness may not share the same yellow color). Typical utility optimization approaches account for variance in resource quality, not variance in the stimuli that signal that quality. Since real world signals are noisy, our understanding of choice behavior will be incomplete without accounting for signal variation and uncertainty. As a signal detection issue, peak shift experiments present an opportunity to investigate the role of this signal-borne risk in decision making and its interactions with those aspects of decision making more commonly investigated.

Significance of Peak Shift for Evolution of Signaling

Peak shift has also been recognized as a possible influence on the evolution of signaling systems. Theoretical development has explored the potential role of peak shift in the evolution of gender or species recognition characters, warning coloration, and sexually dimorphic exaggerated traits. Experiments have shown that peak shift can drive signal evolution in warning coloration, mimicry, and mate-selection systems. Cognitive underpinnings of behavior may thus have a role as a selective mechanism driving evolution, in addition to the better known interactions between genetic and environmental factors that form the basis of natural selection. Two examples are given in the following section.

Evolution of mimicry and crypsis

Bumblebees (*Bombus impatiens*) foraging for nectar exhibited peak shift when choosing the flowers to visit (and thus pollinate). In a laboratory study implementing a Batesian mimicry system, bees were trained to forage on artificial flowers (colored paper disks) under different signal parameter sets. During training, positions of 36 S+ and S− flowers, present simultaneously, were randomized in a 6×6 array on the floor of a flight cage. 'Baseline' bees received an arbitrary parameter set specifying the color and number of S+ and S− flower types, and the sugar-water reward for visiting the two flower types. Three groups of comparison bees each differed from baseline by manipulating one of the three signal parameters: increased variance of S− appearance (three S− flower colors used, whereas baseline used one), decreased relative abundance of S+ (28% of stimuli were S+ flowers, whereas baseline had 50%), and decreased reward for correct detection of S+ (33% sucrose concentration, whereas baseline used 50%). When tested on a range of nine colors (four exemplars each in random positions in the flight cage), the baseline bees exhibited peak shift relative to a control group that had received no S− training. Furthermore, as predicted by the signals approach, the comparison bees exhibited larger peak shift and area shift over and above that exhibited by the baseline bees, in accordance with the increased signal-borne risk of their training regimes. Simultaneous presentation of all test stimuli was used as a way to mitigate range effects. Also, range effects do not explain the greater shift produced by increased signal variance (which maintained the same adaptation level as the baseline condition). The results indicate that in natural situations of mimicry (two signals resembling one another) or crypsis (a signal being difficult to distinguish from noise), peak shift can influence the evolution of signaling traits.

Sexual selection

Male zebra finches (*Taeniopygia guttata*) exhibited peak shift when deciding which females to court. As nestlings, chicks imprinted on parental beak color, which was manipulated with paint. The nestling period thus corresponded to the training phase of a peak shift experiment. Male chicks learned that beak color could distinguish their father (the S− exemplar since courting another male will not lead to reproductive benefits) from their mother (the S+ exemplar, being a female). As adults, the males were tested by allowing them to court other zebra finches possessing a range of beak colors. The birds exhibited a preference to court birds with beak colors shifted

off that of their mothers in a direction away from that of their fathers. Models for sexual selection of exaggerated phenotypes typically require a genetic association between a sex-linked trait exhibited by one gender and a preference for that trait exhibited by the other gender. In this study, however, beak color carried no inherent fitness advantage, did not communicate the possibility of 'good genes' to the courters, and did not impart a competitive advantage to future offspring that might possess a particular beak color. The preference for the exaggerated trait was neither genetically predisposed nor based on a sensory bias, but was learned.

Where Does Peak Shift Fit in the Larger Space of Choice-Making?

Peak shift is characterized by uncertainty inherent to perceptual similarity of stimuli that vary on a continuum. A peak shift experiment and the phenomenon of the shift itself thus differ in several ways from topics more commonly treated as decision making, such as choosing among several discrete alternatives (e.g., diet choice), optimizing resource acquisition (e.g., foraging under time, energy, or predation constraints), and investment budgeting (e.g., parental care, life history pattern).

In peak shift, a 'hidden' preference for a novel stimulus is established over and above that for a known, rewarded stimulus (S+). From the perspective of models that do not account for generalization, the shift seems somewhat paradoxical. Additionally, preferences are being learned in the absence of the preferred stimuli and are shaped by the presence of unpreferred stimuli (S−). For example, changes in an unpreferred stimulus' encounter rate can influence choice, contrary to diet-choice models in which the abundance of less-preferred food items has no effect on intake of preferred items.

Many writers have highlighted features shared between peak shift and phenomena such as transitive inference, novelty seeking, extreme seeking, response to supernormal stimuli, artistic caricatures, esthetic preferences, and sensory bias. Though parallels can be seen, peak shift may not be responsible for any of these behavioral phenomena. For example, when driven by an aversion to signal-borne risk, peak shift does not usually produce a preference for stimuli that are extremely different from the training exemplars, or for novelty per se; the peak shifts are enough to only partially mitigate risk of mistakes. A unified mathematical description of choice making under risk and uncertainty, such as the signals approach, could help distinguish among these phenomena or make the parallels more mechanistically concrete.

See also: Decision-Making: Foraging; Rational Choice Behavior: Definitions and Evidence.

Further Reading

Blough DS (2001) Some contributions of signal detection theory to the analysis of stimulus control in animals. *Behavioural Processes* 54: 127–136.

Cheng K (2002) Generalization: Mechanistic and functional explanations. *Animal Cognition* 5: 33–40.

Green DM and Swets JA (1966) *Signal Detection Theory and Psychophysics*. New York: Wiley.

Lynn SK (2006) Cognition and evolution: Learning and the evolution of sex traits. *Current Biology* 16: R421–R423.

Lynn SK, Cnaani J, and Papaj D (2005) Peak shift discrimination learning as a mechanism of signal evolution. *Evolution* 59: 1300–1305.

Purtle RB (1973) Peak shift: A review. *Psychological Bulletin* 80: 408–421.

Sperling G (1984) A unified theory of attention and signal detection. In: Parasuraman R and Davies RR (eds.) *Varieties of Attention,* pp. 103–181. Orlando: Academic Press.

ten Cate C, Verzijden M, and Etman E (2006) Sexual imprinting can induce sexual preferences for exaggerated parental traits. *Current Biology* 16: 1128–1132.

Thomas DR, Mood K, Morrison S, and Wiertelak E (1991) Peak shift revisited: a test of alternative interpretations. *Journal of Experimental Psychology: Animal Behavior Processes* 17: 130–140.

Wiley RH (2006) Signal detection and animal communication. *Advances in the Study of Animal Behavior* 36: 217–247.

Relevant Websites

http://eebweb.arizona.edu/Animal_Behavior – Bumblebee Decisions.

Defensive Avoidance

W. J. Boeing, New Mexico State University, Las Cruces, NM, USA

Introduction

Predator–prey interactions are a major evolutionary driving force, mediating the behavior of both predator and prey. Ideally, in order to maximize its fitness, an organism would maximize the time spent foraging for food or finding a mating partner and reproducing. However, most animals have to take care not to become the meal while looking for one and have to evaluate the tradeoff between feeding and survival. Therefore, animals have evolved a vast array of behaviors to increase their chance of survival.

Avoiding an encounter with a predator by reducing spatial or temporal overlap is the single most effective way to guarantee survival and increase fitness and is thus a popular antipredator defense strategy. Predators can also be avoided by life-history (e.g., growing too large for a predator) or morphological adaptations. This article focuses on prey behaviors intended to prevent being detected by a predator or to elude them all together. The decision to flee once the predator has noticed the individual and starts pursuing it is covered elsewhere. Avoiding overlap with potential predators is achieved by adjusting feeding, mating, and breeding behavior and remaining in a secure place during high-risk times. Both the response time and time spent in hiding may vary greatly.

Instantaneous Responses

A prey animal that has not yet been detected but is aware of the imminent danger of a predator displays immediate defenses that are intended to minimize the risk of detection by the predator and reduce the probability of a dangerous encounter. An individual can perceive predation risk by visually recognizing a predator, by predator scent, or by acoustic cues either directly from the predator or via alarm calls from conspecifics. The animal then has the option to either duck down and freeze or move toward cover and stay in this safe location. Freezing or temporary immobility decreases conspicuousness of an individual and is used when an immediate burrow or cover is not in the vicinity, when the predator threat is not perceived as imminent or for protection of the young.

Visual cues can be provided in the form of the shadow of a predator (e.g., flying raptor overhead) or by directly seeing the predator. A large body of literature has focused on using predator silhouettes in laboratory settings.

Animals are able to recognize their potential predators and often enter and stay in their refuge after perceiving a visual predator cue. For example, small mammals like golden hamsters (*Mesocricetus auratus*) react to an owl silhouette by scurrying into their burrow and remaining there for some time before being active again. Animals are vigilant to increase the chance of detecting a predator before the predator detects them or to become aware of an approaching predator in time to escape. Vigilance increases when predation risk is higher (e.g., more predators, less cover, fewer alternative potential prey items). For examples teal (*Ana crecca*) normally forage by submerging the entire front end of their body underwater. However, with increased predation risk, they only put the beak below water and keep their eyes above water level, which decreases their foraging efficiency but provides them with a greater chance to see their predator first.

Olfactory cues are predominantly left in the environment by predators through urination and defecation. Rodents are best studied for their responses to predator smells. However, predator avoidance behaviors in response to olfactory cues have also been found in mammals, birds, and turtles. In laboratory experiments, rats commonly respond by freezing when exposed to the smell of cat urine. Dina Suendermann and colleagues demonstrated that gray mouse lemurs (*Microcebus murinus*) that were born and raised in captivity and had never been exposed to predators still responded to feces from potential predators.

In aquatic environments, prey use chemical substances (called kairomones) as indicators of predator presence and react accordingly within less than an hour. *Daphnia* (a type of zooplankton) lower their vertical position in the water column in response to fish kairomones, since deeper and darker water layers provide a refuge from visually hunting fish predators. In the presence of kairomones of *Chaoborus* invertebrate predators, *Daphnia* migrate upward. *Chaoborus* are found at greater depths to avoid fish predation themselves as well as harmful UV radiation. Thus, *Daphnia* have to be able to evaluate the tradeoff between predation by fish and *Chaoborus*, as well as other environmental factors such as food and UV radiation, to determine the most favorable vertical position.

Auditory cues can come directly from the predator or from conspecifics in the form of an alarm call. In a 2008 review by Daniel Blumstein and colleagues of studies examining the response of animals to auditory predator cues, out of 30 studies, 83% of organisms responded selectively to vocal sounds from their predators. They further found that yellow-bellied marmots (*Marmota flaviventris*) ceased foraging, increased vigilance, and often spent more time in their burrows after confronted with sounds of coyotes (*Canis latrans*), wolves (*Canis lupus*), or golden eagles (*Aquila chrysaetos*), differentiating the strength of their response to the respective predator. However, the strongest response was elicited by conspecific alarm calls. Immobility in the immediate environment of a predator has also been reported for insects, amphibians, reptiles, birds, and mammals. Notably female birds on their nests and young birds often respond with freezing as their best chance to remain undetected by their predator. This is especially effective when paired with camouflage.

Daily Responses

Certain daily rhythms help animals avoid high-risk habitats during high-risk times. When to be active or take refuge is one of the most important decisions animals make every day to avoid predators. Refuges are places (burrows, bushes, trees, open fields) or time periods that are safer from certain predators. Predator activity is not totally predictable on a daily, seasonal, or annual level. Therefore, most animals allocate predation risk on the basis of predator activity, the predator's ability to detect their prey, and their own ability to detect the predator, and they must adjust their behavior of when, where, and how long to be active accordingly. As such, juvenile grunts (Haemulidae) delay their off-reef migration to forage in open waters when attacks by models of predator lizardfish (*Synodus intermedius*) are frequent, and tadpoles and freshwater snails have been found to reduce their foraging activities when aquatic predators are present. Alternatively, instead of adjusting the time of when to feed, animals can choose certain, safer habitats for foraging. Safer habitats may offer food lower in quality or quantity. For example, small mammals typically prefer habitats that offer more cover and have a lower risk of being detected by a predator to habitats with less cover but that may have more food available. Similarly, larval tiger salamanders (*Ambystoma tigrinum*) forage in shallow waters among macrophytes in the absence of predacious beetles (*Dytiscus*) and move to less desired deeper foraging areas when beetles are present.

A circadian rhythm is an internal biological clock that can be entrained by external stimuli, the primary being photoperiod, and is found in most animals. It controls sleeping and foraging patterns of animals, and patterns of body temperature, brain activity, and hormone production are linked to this daily cycle. It is believed that circadian rhythms evolved very early in animals for cells to protect their DNA from UV radiation by replicating at night. Circadian rhythms can persist, even when the external stimulus (e.g., light) is removed or animals are kept under artificial, constant conditions in the laboratory.

Under laboratory conditions, fruit flies have been found to maintain their circadian rhythm for hundreds of generations. Most animals have some kind of resting phase and have specialized to be either diurnal or nocturnal. Animals that are active during the day often have pronounced visual abilities in addition to their keen senses of smell and hearing to detect predators. Because animals are most vulnerable during their resting phase (e.g., sleep), they need to find a secure place. By sleeping at night, they altogether avoid the entire suite of nocturnal predators. Resting places are chosen away from exposed areas: monkeys and birds often sleep up in trees; burrowing owls (*Athene cunicularia*) choose burrows that were previously dug by other animals like prairie dogs; many migrating ducks and sandhill cranes (*Grus canadensis*) may choose the middle of shallow playa lakes. On the other hand, advantages of being nocturnal include avoidance of diurnal predators and competitors as well as escaping the heat of the day, especially in desert environments where water loss can become critical. Many nocturnal foragers are less active during nights with bright moonlight, to lower their risk of being preyed upon. Nocturnal animals often have enlarged eyes (e.g., lemurs, owls) so that their eyes can capture more available light and they may still rely on their vision. Often their senses of hearing and olfaction are highly advanced to make up for reduced visual abilities, while some species have evolved unique ways to cope with low light levels at night (e.g., echolocation in bats). Many species of seabirds and sea turtles are diurnal but only visit their breeding sites at night to reduce the chance of predators detecting and looting their nests.

Daily migrations have been well documented in cervids (deer, caribou, elk, and moose). Cervids typically graze in open fields and meadows at dawn and dusk and then ruminate and digest in sheltered forest areas during daytime. For example, the elk (*Cervus elaphus*) in Yellowstone National Park exhibited this typical daily migration to graze in open fields and near riparian areas at dawn and dusk and stay in safer areas for the rest of the time. However, after the extirpation of wolves (*C. lupus*), the main predator of elk, in the 1920s, the elk population blossomed and elk changed their habitat selection pattern. Instead of spending most of the day in the safer, wooded areas, the elk stayed in riparian ecosystem, continuously grazing on willow (*Salix* spp.), aspen and cottonwood (*Populus* spp.) saplings, which had devastating effects on these tree populations. Fewer trees along the river banks in turn led to a degraded river ecosystem and soil erosion. With the reintroduction of the wolves in the mid 1990s, elk were forced to pick up their daily migrations again and head for cover during the daytime. With elk on the move again, the riparian ecosystems began to recover.

Diel vertical migration is one of the most widespread ecological phenomena on the planet. Microscopic zooplankton spend the night taking advantage of food sources and warmer temperatures in the upper, euphotic zone of lakes and oceans, but move down into the colder waters during the day, primarily to avoid fish predation. In some oceans, this means that these tiny organisms migrate over 100 m each day. This is an exceptional feat when considering the size of these organisms.

Seasonal Responses

Photoperiod is the main environmental parameter that many species use to make assessments about present climate conditions, food availability, or predator activity and is crucial for survival. Some circannual rhythms or seasonal changes in activity are directly aimed at predator avoidance. For example, herons (Ardeidae) forage during safer time periods of rainfall and dusk, and move from riskier mangrove habitat when overall predation is high to less profitable reefs and deforested inlets. Other seasonal behaviors like mating, breeding, or parental care require extra precautions against predators, while some annual rhythms including hibernation, estivation, and long-distance migrations are not primarily aimed at predator avoidance but are important adaptations to environmental factors (e.g., extreme cold or heat, or lack of water or food). Nevertheless, during dormant stages like hibernation or estivation, individuals are defenseless, extremely vulnerable to predation, and thus, have to find resting places that are habitable and provide protection from predators for long time periods.

Courtship and mating are behaviors during which predation risk drastically increases for most animals, and many predators have actually adapted their own behavior to take advantage of courtship and mating behaviors of their prey. Dragonflies conduct more attacks on mosquitoes (*Anopheles freeborni*) and are also more successful, when male mosquitoes are in swarms and when mating pairs leave the swarm. In return, behaviors of prey animals may be modified in the presence of predators. Andrew Sih and colleagues found that water striders (*Gerris remigis*) reduce their mating activity in the presence of predatory green sunfish. Male toads emit fewer calls when predation risk is high to reduce their chance of detection. Some animals (like some reptiles, octopus, squid, and some fishes) have the ability to quickly adjust their color patterns. They match their background and are cryptic but can display courtship colors for mates and warning colors for competitors. The males of the red-winged blackbird (*Agelaius phoeniceus*) have red and yellow shoulder patches that they can flash to attract females or to intimidate male competitors and are able to hide in the presence of potential predators.

Breeding and raising of the young is another key time during which an individual is not only more vulnerable to predation itself but also has to take care to protect its offspring. Selection of breeding sites is one of the most important decisions a parent can make. Next to old and sick individuals, newborns are by far the most vulnerable animals in a population. For example, bird nests that are built close to edges between habitats (where two ecosystems meet, e.g., forest and grassland) or in a small habitat patch are more likely to be looted by predators. If jays, magpies, and corvids are the main predators, the better strategy is to build the nest on the ground, while survival is higher in shrubs when foxes, badgers, and rodents dominate. Furthermore, when predators develop a search image for a certain nest type, predation can be density dependent and birds may need to look for nest sites that are far away from other, similar-looking nests. However, it has also been suggested that building several nests and only occupying one could serve as a predator defense. If a predator finds empty nests it might move to a different area to forage. Some birds have been found to build nests close to colonies of social insects that sting or bite like wasp nests or ant hills in order to discourage predators from coming too close. Other behaviors can limit nest detection too: the prairie warbler (*Dendroica discolor*) limits nest building activites to short periods for only part of the day and some bird species also limit the frequency at which they visit their nest for feeding, making nest detection less likely. Once the eggs are hatched, the parents remove the eggshells from the nest to reduce visibility. Terrestrial mammals often select underground or enclosed hidden spaces (also called lairs or dens) to bear and raise their young in a protected environment. Seals seek out haul-out sites on land or ice to temporarily leave the water between foraging to rest and for thermal regulation, to escape aquatic predators like sharks and orcas, or for mating and to safely raise their young. Haul-out sites are selected far from shore to avoid terrestrial predators.

Long-distance migrations are most common in birds (sandhill cranes, geese, burrowing owls, passerines, humming birds all migrate), but they also occur in mammals like caribou. Migrations are a way to circumvent hibernation, estivation, or diapause. Instead of entering a dormant phase, the animals inhabit two different ecosystems during different seasons. In the northern hemisphere, animals normally migrate south during the winter. In summer, they return to their more northern habitats to breed and take advantage of longer days to feed their young and thus higher reproductive rates. Migrations are again primarily aimed at escaping unfavorable conditions (temperature, lack of food) and not to avoid predation. In fact, predation can actually be higher during migration periods. The Eleonora's falcon (*Falco elenora*) for example has timed its breeding season with the migration of passerines, which it uses to feed its young. In response, some passerines have adapted nocturnal migration. However, another predator, the greater noctule bat (*Nyctalus lasiopterus*), takes advantage of this nocturnal migration and has become the main predator of these passerines.

Hibernation and estivation are prolonged states of torpor during cold and hot periods, respectively. Torpor is characterized by an animal's inactivity and strongly reduced metabolic rate. Hibernation is found in many mammals, especially rodents, and estivation occurs in crustaceans, snails, amphibians, reptiles, lungfishes, and rodents. The alpine marmot (*Marmota marmota*) hibernates for up to 9 months each year and seals its burrow with earth and their own feces to keep safe from predators and keep temperature and humidity fairly constant. Similarly, during estivation, crabs, toads, and the desert tortoise dig burrows and escape from the heat and remain safe from predators. These burrows may stay moist even during the summer. Lungfishes can burrow into the mud, surround themselves with a mucus cocoon, and survive the summer below a dried up lake safe from nonaquatic predators.

Life-History Responses

Some organisms have adapted to occupy habitats that are safe and have lower predation pressure to begin with during at least one of their life stages.

Diapause is a phase of dormancy in insects and zooplankton. Insects can undergo diapause during the summer or winter as eggs, larvae, pupae, or adults and may be used as a predator avoidance mechanism. Certain stimuli, such as shorter daylight, low or high temperature, lack of food, or high abundance of predators, are necessary to induce diapause. Production of 'duration eggs' (encapsulated eggs that can dry out or freeze and hatch once conditions are favorable again) occur in many zooplankton species (e.g., *Daphnia*, copepods, tadpole shrimps). They are typically found in temporary ponds and playa lakes but also occur in permanent lakes as a mechanism to guarantee survival of offspring in the face of intense periods of predation. Duration eggs may be ingested by fish predators; however, they are not digested and are still viable after passing through the intestine. Duration eggs are able to hatch, even decades after they were produced.

Embryonic diapause is a reproductive strategy found in mammals like some rodents, bears, and marsupials (e.g., kangaroos). Hereby, the embryo stays undeveloped in a state of dormancy and implantation in the uterus is delayed for up to a year. The purpose of embryonic diapause is to bear the offspring when its survival chances are optimal (e.g., food availability, mild environmental conditions, low predator activity).

Deep burial of the adult stages of clams occurs in the muddy intertidal zone. The soft-shelled clam *Mya arenaria* is most vulnerable to excavating red rock crabs (*Cancer productus* at lower beach elevations and thus digs itself deeper into the sediment. The maximum recorded burial depth is 25 cm. The clams use a siphon that extends to the sediment surface to filter small particles out of the water to feed.

Costs and Benefits of Defensive Avoidance

In the short-term, the benefit of avoiding a predator is obvious: increased survival and sparing the stress and energy expended when being pursued by a predator. However, there is a tradeoff, and prey organisms incur costs in the form of nonlethal predator effects (also termed indirect or nonconsumptive predator effects): Hiding places typically do not offer ideal food conditions and as such, reduce an individual's overall condition and biomass. To avoid starvation, an animal eventually has to leave its secure location. A recent review by Evan Preisser and colleagues suggests that these indirect predator effects have a similar or even higher impact on prey population density than direct consumption!

In the long-term, the cost of the reduced foraging due to predator avoidance is a decrease in number of offspring. However, theory predicts that the reduction in fitness is outweighed by the benefit of surviving for another day and having the option to reproduce at all; otherwise individuals will be out-competed by larger risk-takers.

To maximize their fitness in an ever changing environment, organisms have developed methods to assess predation risk and induce their defenses accordingly. For example, the amplitude of diel vertical migration in zooplankton is dependent on the density of predatory fish kairomone in the water, and migration may be abandoned in fishless lakes. Clams bury themselves deeper into the sediment at higher risk locations closer to the water. Elk in Yellowstone National Park ceased migration after the extirpation of the gray wolf.

Nevertheless, the dilemma that many organisms face in a multipredator environment is that minimizing risk from one predator may make them more vulnerable toward another predator. For example, birds that choose nest sites higher off the ground in shrubs to avoid rodents may experience higher predation by jays, magpies, and corvid predators. Small mammals that avoid open fields and forage under bushes to seek refuge from avian predators are more likely to encounter lie-and-wait snake predators. And zooplankton that migrate down in the water column during the day to escape visual predation by fish run a higher risk of being consumed by invertebrate predators that themselves have to stay in deeper water layers to hide from fish. Furthermore, predator avoidance might conflict with other environmental factors. Pinyon jays (*Gymnorhinus cyanocephalus*) are social birds that live in pinyon pine habitats of the foothills of western North America and build nests in trees (usually juniper, live oak, or pine). Nests that are more higher and farther away from the trunk are more visible from the air and more likely to be preyed upon by ravens and American crows. Nests that are more concealed and lower to the ground and closer to the tree trunk experience colder temperatures and are more often abandoned, presumably because the energetic cost to keep the eggs protected from cold temperatures are too severe.

In summary, organisms have to be vigilant and avoid multiple predators while simultaneously preventing starvation and, in case of birds and mammals, successfully rearing their vulnerable young. This delicate balance is one of evolution's driving forces that has created the dazzling diversity of species we find on our planet.

See also: Circadian and Circannual Rhythms and Hormones; Defensive Morphology; Economic Escape; Life Histories and Predation Risk; Optimal Foraging Theory: Introduction; Risk-Taking in Self-Defense; Vertical Migration of Aquatic Animals.

Further Reading

Barbosa P and Castellanos I (2005) *Ecology of Predator–Prey Interactions*, p. 394. New York, NY: Oxford University Press.

Blumstein DT, Cooley L, Winternitz J, and Daniel JC (2008) Do yellow-bellied marmots respond to predator vocalizations? *Behavioral Ecology and Sociobiology* 62: 457–468.

Caro TM (2005) *Antipredator Defenses in Birds and Mammals*, p. 591. Chicago, IL: The University of Chicago Press.

Chivers DP and Smith RJF (1998) Chemical alarm signalling in aquatic predator–prey systems: A review and prospectus. *Ecoscience* 5: 338–352.

Committee on Ungulate Management in Yellowstone National Park, National Research Council (2002) *Ecological Dynamics on Yellowstone's Northern Range*, p. 198. Washington, DC: National Academic Press.

Kerfoot WC and Sih A (1987) *Predation: Direct and Indirect Impacts on Aquatic Communities.* Hanover, NH: University Press of New England.

Lima SL (1998) Stress and decision making under the risk of predation: Recent developments from behavioral, reproductive, and ecological perspectives. *Advances in the Study of Behavior* 27: 215–290.

Lima SL and Dill LM (1990) Behavioral decisions made under the risk of predation: A review and prospectus. *Canadian Journal of Zoology* 68: 619–640.

Preisser EL, Bolnick DI, and Benard MF (2005) Scared to death? The effects of intimidation and consumption in predator–prey interactions. *Ecology* 86: 501–509.

Stephens DW and Krebs JR (1986) *Foraging Theory*, p. 247. Princeton, NJ: Princeton University Press.

Takahashi LK, Nakashima BR, Hong HC, and Watanabe K (2005) The smell of danger: A behavioral and neural analysis of predator odor-induced fear. *Neuroscience and Biobehavioral Reviews* 29: 1157–1167.

Tollrian R and Harvell CD (1999) *The Ecology and Evolution of Inducible Defenses*, p. 382. Princeton, NJ: Princeton University Press.

Defensive Chemicals

B. Clucas, University of Washington, Seattle, WA, USA; Humboldt University, Berlin, Germany

Introduction and Definitions

Chemical defense is perhaps one of the most widespread antipredator strategies among living organisms, from plants and bacteria to animals. Within the animal kingdom, defensive chemicals are found extensively in invertebrates (e.g., arthropods and molluscs, terrestrial and marine), but vertebrates also possess chemical defense strategies.

Defensive chemicals are substances utilized by prey to reduce predation risk. These chemicals include *noxious*, *odiferous*, indigestible, *toxic*, or *venomous* substances that repel, deter, injure/harm, distract, or prevent detection by predators. These substances can affect predator behavior by influencing the predators' olfactory, gustatory, or tactile sensory systems while they are searching, attacking, or consuming the prey. In addition, chemicals that when released warn conspecifics of presence of a predator can be considered a defensive chemical.

Chemical substances can be airborne, waterborne, or substrate bound. They can be released (e.g., sprayed) away from the animal creating an air- or waterborne substance, can be released externally and retained on the animal's *integument*, injected directly into another animal, or *sequestered* internally into the integument or internal organs. Depending on the medium they travel through or on (air, water, or substrate) and other physical characteristics (i.e., chemical composition, volatility), they can also have varying *active spaces* and time until dissipation. Defensive chemicals tend to have small active spaces and short duration due to the necessity of a targeted, fast acting effect on the predator's behavior. However, some chemical defense strategies, particularly waterborne chemicals, can have large active spaces, and certain defensive odorants can have a long duration.

Chemical Substance Acquisition

Proximate Methods of Acquisition

The chemical substances that animals use in defense against predators can be acquired in several ways. First, the substances can be *synthesized de novo* as in many insects. Second, chemicals can be sequestered internally by ingesting other organisms that contain the chemical (or a precursor for it). Finally, substances can be obtained externally by self-application. Animals can manually or orally spread a substance onto their integument or apply it by rolling or rubbing into the substance.

Evolutionary Origins and History of Acquisition

The evolutionary origins of chemical defense substances are likely as diverse as the various mechanisms (see later). Depending on whether the chemicals are synthesized de novo, sequestered, or self-applied, the history of acquisition could follow several evolutionary trajectories. Chemical substances that are synthesized de novo could have originally been substances or by-products of an unrelated metabolic process that eventually evolved into a defense chemical if individuals possessing them gained a survival benefit. These chemicals could still serve their original function, for instance if they are a by-product of digestion. Such substances could also have evolved into more complex defensive chemical compounds and have become coupled with the evolution of specialized releasing structures and behaviors.

Sequestration of defensive chemicals through diet is a common strategy, especially in insects feeding on toxic vegetation. Such acquisition most likely starts with the insects evolving a mechanism to digest the vegetation safely, followed by sequestration of the toxins for their own use. This toxicity gained through sequestration is also often coupled with the evolution of defensive coloration.

As beneficial as defensive chemicals are for prey, some are still costly, which may put constraints on the evolution of such antipredator strategies. Animals that synthesize their own defense substances, in particular, can face high metabolic production costs. In comparison, animals that sequester toxins from their diet may incur smaller costs. Similarly, animals that self-apply defensive chemicals externally do not have costs of production; however, the time taken to find the chemical source and apply the substances could have energetic and opportunity costs.

Mechanisms of Defense

Olfactory-Oriented Defense Mechanisms

Repellent

The most common form of chemical defense is repelling the predator with a chemical substance. These substances are characterized as odiferous, noxious, and/or unpleasant and, thus, usually affect the predator's olfactory organs.

Active repellents include spraying and releasing substances from exocrine glands or other orifices. Skunks (Mephitidae) are a classic example of an animal that sprays a self-made noxious substance to defend itself against predators.

The anal glands of these small carnivores contain sulfur-containing thiol chemical secretions that, with the help of specialized muscles, can be accurately sprayed out. Similarly, large and colorful birds called 'hoopoes' (*Upupa* spp.; also known as 'stink cocks') will spray an unpleasant secretion out of their preen gland when disturbed by predators to protect their eggs or nestlings. An odd example of a sprayed repellent is found in the Texas horned lizard (*Phrynosoma cornutum*), which will spray blood out of tear ducts by their eyes.

Many arthropods also spray chemicals as an antipredator behavior (Laurent et al., 2005). One of the best known are the aptly named bombardier beetles (comprising several subfamilies of Carabidae), which combines hydroquinone and hydrogen peroxide inside their bodies and spray out a hot, odiferous mixture when disturbed. Other examples of animals that spray repellents to repel their predators are several marine species of Cephalopoda and Gastropoda (e.g., squid, octopus, and sea hares). Gastropods, specifically sea hares (*Aplysia californica*), release colored ink through an ink gland (the color of the ink depends on the color of the seaweed they forage on). Cephalopods (i.e., squids and octopus), in contrast, spray ink that comprises melanin out of an ink sac. Inking can also have other effects on predators (see later and **Table 1**).

Defecation (releasing of feces) occurs in many animals during stressful situations, including encounters with predators. Although the function of such defecation is not clearly understood, it is possible that feces may repel predators by the offensive odor. It is also hypothesized that the releasing of feces lightens the animal allowing for more agile flight, and the subsequent repellent benefits might have been subsequently selected. Several species of birds are known to defecate upon being flushed, and several species defecate on their eggs or nestlings upon fleeing a nest (see **Table 1**). Some mammals are also known to defecate when in stressful situations (e.g., rodents). The nocturnal pen-tailed tree shrew (*Ptilocercus lowii*) will even roll onto its back and defecate and urinate when attacked during the day.

Animals can also use substances obtained from other organisms to repel predators. For instance, common wax-bills have been shown to utilize the feces of carnivores in their nests, which is thought to repel olfactory predators. In addition, several bird species are known to use shed snake skins in their nests, which possibly could affect predators relying on olfaction.

Olfactory camouflage or crypsis

Olfactory camouflage or crypsis is the simulation of the scent of uninteresting organisms or objects to avoid detection by predators or occurs when prey animals are rendered undetectable and unlocatable by means of olfaction. Studied cases of such camouflage are rare compared to those in the visual sense.

Graeme Ruxton reviewed several examples of olfactory camouflage in invertebrates. Caterpillars of the butterflies *Biston robustum* and *Mechanitis polymnia* have been shown to chemically match their own scent to the vegetation they feed and live on to avoid predatory ants. A limpet species (*Notoacmea palacea*) also appears to reduce predation risk by chemically matching the organisms they feed and live on. There is also a recent example of olfactory camouflage in a vertebrate species. Several ground squirrel species (*Spermophilus*) self-apply rattlesnake scent by chewing on snake-scented substances (e.g., shed skins or carcasses) and licking their bodies. Rattlesnakes exhibited less foraging behavior toward a mixture of squirrel and rattlesnake scent than squirrel scent alone, suggesting that they were unable to detect the prey's odor.

The specific placement of nests in several species of birds is suggested to provide olfactory crypsis for their eggs and nestlings. Birds also use an assortment of objects in their nest (e.g., shed snake skins, see earlier), and it is possible that some of these substances could render their nest undetectable by olfactory predators.

Animals can also reduce their own odor to prevent detection by predators. A nice example of this strategy is found in the red knot (*Calidris canutus*), a sandpiper species. These birds alter the chemical composition of their preen waxes (wax that is put into feathers while preening/cleaning) during the breeding season, and this has been shown to reduce their locatability by mammalian predators.

Olfactory mimicry

There are various examples in nature of one organism mimicking the scent of another organism. Olfactory mimicry is defined as the simulation of chemical characteristics of one organism (the model) by another organism (the mimic) that are perceived as the original organism by the predator (or 'dupe'). However, there are not many known examples of olfactory mimicry being used as an antipredator defense.

One possible example of olfactory mimicry is the carabid beetle *Anchomenus dorsalis* that aggregates (groups) with bombardier beetles (*Brachinus* spp.). Both species are aposematic and have similar color patterns, and both have unique chemical defenses. Nevertheless, *A. dorsalis* is known to 'rub' onto *Branchinus*, apparently obtaining the other species' chemicals on their integument. This self-application behavior is hypothesized to make *A. dorsalis* smell like *Brachinus*, potentially enhancing its signal of unpalatability to predators. Indeed, *A. dorsalis* does not display this rubbing behavior to nonaposematic carabid species. This case, therefore, may present an interesting example of Müllarian mimicry.

A tactic used by sea hares may also be considered as olfactory mimicry. These gastropods use the ink they eject when disturbed not only to repel their predators but to distract them as a *phagomimic*. That is, certain chemicals

Table 1 Examples of chemical defense strategies in animals

Defense mechanism	Animal species	Substance	Substance source	Employing mechanism
Repellent	Bombadier beetles (several subfamilies of Carabidae)	Hydroquinone and hydrogen peroxide	De novo	Spraying
Repellent	Stink bugs (Pentatomidae)	Cyanide	De novo, Metathoracic glands	Released from glands
Repellent, confusions (smoke screen), phagomimicry, alarm signal	Sea hare	Ink (mixture of glandular secretions)	Opaline gland (de novo) and ink gland (diet based for color)	Spraying
Repellent, confusions (smoke screen), phagomimicry (?), alarm signal	Cephalopods	Ink	Ink sac (not homologous with ink gland)	Spraying
Repellent	Gray partridge	Feces	De novo	Defecating upon being flushed
Repellent	Anatidae (water fowl)	Feces	De novo	Defecating on eggs and/or nestlings upon being flushed
Repellent	Common waxbills, Estrilda astrild (bird)	Feces	Carnivore feces	Applied to nest
Repellent	Skunks (Mephitidae spp.)	Anal gland secretions (thiols)	De novo (anal glands)	Spraying
Repellent	Rattlesnakes (Crotalus spp.)	Cloacal gland secretions	De novo	Spraying
Repellent	Hoopoes, Upupa spp. (birds)	UG secretions	De novo (uropygial gland)	Spraying
Reduce birds' smell	Red knots, Calidris cantus (birds)	UG secretions	De novo (uropygial gland)	Preened into feathers
Olfactory mimicry	Anchomorus dorsalis (carabid beetle)	Cuticular chemicals	Bombardier beetles	Rub on heterospecific beetles
Camouflage (chemical background matching)	Limpet (Notoacmea palacea)	Flavonoids	Surfgrass (diet based)	Sequestered in integument
Camouflage (chemical background matching)	Caterpillars (Biston robustum)	Cuticular chemicals	Sequestered via diet	Sequestered in integument
Olfactory camouflage	Ground squirrels (Spermophilus spp.) and chipmunks (Neotamias spp.)	Snake scent	Shed rattlesnake skins, dead rattlesnakes	Applied by licking substance into fur
Repellent, startle mechanism	Texas horned lizards (Phrynosoma cornutum)	Blood	De novo	Sprayed through tear duct
Alarm signal and secondary predator attractant	Ostariophysi fishes (e.g., minnows)	Damage-released alarm pheromones	De novo	Released during attack by predator into water
Alarm signal	Marine mud snail (Nassarius obsoletus)	Damage-released alarm pheromones	De novo	Released during attack by predator
Alarm signal	Echinoderms (e.g., black sea urchin, Diadema antillarum)	Damage-released alarm pheromones	De novo	Released during attack by predator
Alarm signal	Crayfish (Orconecte virilis and O. propinquus)	Damage-released alarm pheromones	De novo	Released during attack by predator
Alarm signal	Honeybees (e.g., Apis mellifera)	Alarm pheromones	De novo	Released with stinger

Continued

Table 1   Continued

Defense mechanism	Animal species	Substance	Substance source	Employing mechanism
Toxic, Repellent	Many insects (e.g., Monarch butterflies, *Danaus* spp.)	Alkaloids	Plants (sequestered via diet)	Releasing, spraying, on integument
Toxic	Melyrid beetles (Choresine)	Alkaloids (e.g., Batrachotoxins)	Plants (sequestered via diet)	Sequestered in integument
Toxic	'Poison frogs' (diverse genuses: *Melanophyrniscus, Pseudophryne, Dendrobates, Epipedobates, Phyllobates*)	Alkaloids (e.g., Batrachotoxins)	Arthropods via diet (e.g., ants and beetles)	Sequestered in integument
Toxic	*Pitohui* spp. and *Ifrita kowalei* (birds)	Alkaloids (e.g., Batrachotoxins)	Arthropods via diet (e.g., ants and beetles)	Sequestered in feathers and integument
Toxic	Red warbler (*Ergaticus rubber*)	Alkaloids	Unknown	Sequestered in feathers
Toxic	Newts (*Taricha* spp.)	Tetrodotoxin	De novo	Toxin in integument
Toxic	Pufferfish (Tetraodonitae)	Tetrodotoxin	Produced by bacteria obtained via diet	Toxin in internal organs and integument
Toxic	Nudibranchs (Nudibrachia)	Cnidocytes	Sequestered via diet	Cnidocytes sequestered in integument
Toxic, repellent	Gastropods	Terpenoids	Sequestered via diet	Sequestered in integument
Venomous	Sea urchins (Echinoidea)	Poisonous spines	De novo	Venom-coated spines
Venomous	Stingrays (Dasyatidae)	Poisonous barb	De novo	Venom injected by spearing with tail barb
Venomous	Cnidaria (e.g., sea anemones, corals, jellyfish)	Cnidocytes	De novo	Venomous cells
Venomous	Ants	Alkaloids	De novo	Venom released when biting or stinging
Venomous	Bees and wasp	Bee/wasp venom	De novo	Venom released when stinging
Venomous	Solenodons (*Atopogale cubana* and *Solenodon paradoxus*), shrews (*Neomys fodiens, N. anomalous*, and *Blarina brevicauda*)	Venom	De novo	Venom released when biting

in their ink induce feeding behavior in some predators toward the ink substance, which gives sea hares more time to escape while the predator is distracted.

Chemical alarm signals

Chemical alarm signals are substances released by prey in the presence of a predator that are triggered by attack or injury. These signals (sometimes called by the German name 'Schreckstoff') can reduce predation risk in several ways. First, chemical components released into the environment by a threatened or injured prey may recruit conspecifics to help mob the predator, functioning as a *pheromone*. This behavior is well studied in bees where alarm pheromones are released when bees sting a predator,

emanating both from the bee and the stinger left in the predator. This partly accounts for the quick defensive attacks on predators when even just one bee has stung it. There are many chemical compounds released with the stinger that could induce defensive behavior in bees; however, whether these compounds actually help localize the predator or simply increase searching behavior in still debatable.

A classic example of damage-released chemical alarm signals is in the Ostariophysan fishes. In particular, much work has been done by David Chivers and colleagues on how fathead minnows (*Pimephales promelas*) release chemicals upon being injured by a predator. Several functions are possible for these alarm signals. First, it is shown that the chemicals induce antipredator behavior in other minnows.

This conspecific alarm signaling is also found in other aquatic animals (e.g., gastropods, see **Table 1**). Second, damage-released chemicals from minnows have been demonstrated to attract additional predators, which benefits the attacked prey by increasing the probability that the other predators will distract is captor, and it will be dropped and can escape (Chivers and Smith, 1998), similar to a fear scream.

Contact-Based (Gustatory and Tactile) Chemical Defense

Unpalatability: Toxicity or distastefulness

Many prey animals are rejected, vomited out, cause a nauseating effect, or even kill predators upon being consumed due to toxic, noxious, or distasteful chemicals on or in their bodies. This unpalatability is often coupled with defensive coloration, which warns the predator of the prey's chemical defense. Well-known examples are the poison dart frogs in South and Central America (Dendrobatidae) and Monarch butterflies (*Danaus* spp.). These unpalatable amphibians and insects obtain their toxins through sequestration of alkaloids from their diet. Poisonous frogs have been shown to obtain alkaloids from the ant and beetle species they eat. Perhaps similarly, there are several species of birds (*Pithohui* spp., see **Table 1**) that have a comparable diet and also sequester batrachotoxins in their feathers. In addition, these 'toxic birds' are also very colorful, which potentially provides an aposematic signal to predators. Many marine organisms are toxic. Nudibranch species (Nudibranchia, sea slugs) can synthesize de novo chemicals, but other species sequester *cnidocytes* from their cnidarian prey into their skin to protect themselves. Similar to this sequestration of toxins, certain decorator crabs (Majidae) place brown alga (Dictyota) that contain defensive diterpene chemicals onto their carapace in order to reduce their palatability to predators.

One of the most toxic defensive chemicals found in animals is tetrodotoxin. This lethal toxin can be produced by bacteria that often are acquired through diet by the subsequently toxic animals. For example, pufferfish (Tetraodonitae) contain tetrodotoxin in some of their internal organs and on their integument rendering them unpalatable to most predators. Newts of the *Taricha* genus also possess tetrodotoxins; however, the source of the toxins is not bacterial in the rough-skinned newt (*T. granulosa*) and is thought to be self-produced. Interestingly, these highly toxic newts can be consumed by some predators in particular geographical areas. Certain garter snake species (*Thamnophis*) have evolved resistance to the tetrodotoxin and can eat the newts without succumbing to the toxin. This garter snake-newt predator–prey relationship is a classic example of co-evolution. Furthermore, it has been demonstrated that resistant garter snakes sequester the newts' toxins in their livers, which would make the snakes toxic to their own mammalian and avian predators.

Few mammals are toxic; however, slow lorises (*Nycticebus* spp.) produce a toxin, which is released from a gland on their arms, and these small primates will rub this toxic secretion onto their young as well to protect them from predators.

Venomous stings, spines, and bites

Prey animals are actively dangerous to their predators if they can inject venom with a stinger, spine, or by biting. Most notable are members of the Hymenoptera order (e.g., bees, wasps, and ants). Most bees and wasps have a needle-like organ on the end of their abdomen, a 'stinger,' through which they can release protein-rich venom after pricking a potential predator. Ants also have a stinger but inject an alkaloid-based venom. Other animals have analogous stinging weapons; for example, stingrays (Dasyatidae) have a venomous spine at the end of their tail, and sea urchins (Echinoidea) have venomous spines that protrude from their body. An example of a mammal with a defensive venomous bite is that of the aforementioned slow loris, which obtains a poisonous bite by licking the toxin produced on their arms before biting a predator.

Animals can use venomous bites or stings as both a defensive and an offensive weapon.

Several groups of reptiles have such venomous bites. Snake families that are venomous include Viperids (e.g., rattlesnakes, *Crotalus*) and Elapids (e.g., cobras, *Naja*), and lizards include the Helodermatids (e.g., beaded lizards, *Heloderma*). These reptiles use their venom when hunting to subdue prey and also in defense against predators. Rattlesnakes advertise their dangerous venom by rattling to warn predators. Cnidarians (e.g., jellyfish, coral, sea anemones) have stinging cells, or cnidocytes, which they use to capture their food, but also to protect themselves against predators. Several mammalian species have venomous bites, all in the order Soricomorpha (solenodons: *Atopogale cubana* and *Solenodon paradoxus*; shrews: *Neomys fodiens*, *N. anomalous*, and *Blarina brevicauda*). However, these venomous mammals appear to use their venom more to catch prey, rather than use it as a defensive chemical.

Distracting or startling substances

Some of the repelling chemicals mentioned earlier may not only deter predators by their olfactory properties but also function to distract the predator or startle it, giving the prey animal more time to escape. Defecation might serve such a purpose, and, as mentioned, several bird species are known to defecate upon being flushed by predators. Another potential example is inking by cephalopod and gastropods, which probably serves to startle and distract predators in addition to repelling.

See also: Defensive Coloration; Games Played by Predators and Prey; Honeybees; Olfactory Signals; Risk-Taking in Self-Defense.

Further Reading

Bartrom S and Boland W (2001) Chemistry and ecology of toxic birds. *Chembiochem* 2: 809–811.

Bradbury JW and Vehrencamp S (1998) *Principles of Animal Communication*. Sunderland, MD: Sinauer Associates.

Brandmayr ZT, Bonnaci T, Massolo A, and Brandmayr P (2006) What is going on between aposematic carabid beetles? The case of Anchomenus dorsalis (Pontoppidan 1763) and Brachinus sclopeta (Fabricius 1792) (Coleoptera Carabidae). *Ethology Ecology and Evolution* 18: 335–348.

Breed MD, Guzmán-Novoa E, and Hunt GJ (2004) Defensive behavior of honey bees: Organization, genetics, and comparison with other bees. *Annual Review of Entomology* 49: 271–298.

Chivers DP and Smith RJF (1998) Chemical alarm signaling in aquatic predator-prey systems: A review and prospectus. *Ecoscience* 5: 338–352.

Clucas B, Owings DH, and Rowe MP (2008) Donning your enemy's cloak: Ground squirrels exploit rattlesnake scent to reduce predation risk. *Proceedings of the Royal Society B* 275: 847–852.

Conover MR (2007) *Predator–Prey Dynamics: The Role of Olfaction*. Boca Raton, FL: CRC Press.

Daly JW (1998) Thirty years of discovering arthropod alkaloid in amphibian skin. *Journal of Natural Products* 61: 162–172.

Derby CD (2007) Escape by inking and secreting: Marine mollusks avoid predators through a rich array of chemicals and mechanisms. *Biological Bulletin* 213: 274–289.

Dettner K and Liepert C (1994) Chemical mimicry and camouflage. *Annual Review of Entomology* 39: 129–154.

Laurent P, Braekman J-C, and Daloze D (2005) Insect chemical defense. *Topics in Current Chemistry* 240: 167–229.

Ruxton GD (2009) Non-visual crypsis: A review of the empirical evidence for camouflage to senses other than vision. *Philosophical Transactions of the Royal Society, B* 364: 549–557.

Weldon P (2004) Defensive anointing: Extended chemical phenotype and unorthodox ecology. *Chemoecology* 14: 1–4.

Defensive Coloration

G. D. Ruxton, University of Glasgow, Glasgow, Scotland, UK

Introduction

The most commonly considered antipredatory use of coloration is avoidance of detection and/or recognition by the predator. This primary defense (called camouflage, crypsis, or masquerade) is covered in depth in another section. Here, we focus on the use of coloration as a secondary defense: that is, to influence the action of the predator subsequent to prey detection in ways that benefit the prey. First, we will discuss aposematism, which is the use of bright signals to warn would-be predators that a particular prey item is defended or otherwise unattractive to attack (more on these defenses are discussed elsewhere). The next section deals with the sharing of a single aposematic signal across several species (the phenomenon of mimicry). Finally, we cover signals that influence the point of attack on the body of the prey (deflective signals), and those that intimidate or confuse potential predators (deimatic signals).

Aposematism

Properties of Aposematic Signals

Given that attacks by predators are costly to the prey (in terms of survivorship, injury, or simply the time taken in repelling an attack), there are advantages to reducing the rate at which attacks occur and/or reducing their intensity. This holds even for prey with highly effective secondary defenses. Defended prey could achieve reduction in frequency of attack by crypsis, but an alternative is to provide predators with some indicator that defenses are present. Classical examples of such aposematic signals are the black-and-white stripes of a skunk and the yellow-and-black stripes common to many social wasps.

A common characteristic of aposematic signals is that they are conspicuous. In this context, conspicuousness describes a set of stimuli that attracts a predator's attention, thereby facilitating detection. Defended animals should benefit from making themselves distinctive so as to prevent predators from mistaking them for sympatric edible species. Since edible species are often cryptic, it seems inevitable that, in order to be distinctive, aposematic species adopt an appearance that is conspicuous.

Crypsis may bring benefits of low detection rates but may also impose opportunity costs on some prey species. Specifically, crypsis may require reduced movement and may restrict microhabitat selection. Conversely, prey may be more free to move around and to exploit environmental opportunities if they possess secondary defenses. Thus, aposematic displays may be more conspicuous because optimal conspicuousness may be higher for prey with secondary defenses than for those that lack them. A state of conspicuousness through simply not adopting hiding behaviors may be considered a prototype warning display. Once a prey is freed from the movement and microhabitat constraints of crypsis, it may be free to further heighten its level of conspicuousness for a variety of reasons unrelated to predation (such as mate attraction).

However, it appears that aposematic species are often much more conspicuous in appearance than would be required by the arguments mentioned here. Perhaps high levels of conspicuousness are needed to ensure signal honesty. Increased conspicuousness incurs the cost of increased attention from predators, and this may be too expensive to bear for prey that are not protected by secondary defenses. Conspicuousness may therefore be important because it confers some degree of signal reliability on an aposematic display, as only truly defended species could afford to draw attention to themselves (a more general treatment of signal honesty is given elsewhere).

Another explanation for the conspicuousness of all or part of a warning display is that the signal serves to draw a predator's attention to the presence of a 'visible' secondary defense. Some defensive traits may be manifest externally and be evaluated by predators without attacks taking place. Spines, claws, and inducible morphological defenses may be evaluated from a distance by predators, and aposematic 'amplifying' traits (such as the black-and-white contrasting coloration of porcupine quills) may help to draw a predator's attention to the defenses and aid in their evaluation. In such cases, the warning display contains both a manifestation of the secondary defense itself and some 'directing' or amplifying trait that draws attention to these defenses and discourage attack. Hence the defense is, to some extent, self-advertising. Warning displays that include the defense as part of the advertisement may be reliable, very hard to fake, and may provide accurate information regarding the quality being advertised.

If attacks on unprofitable prey are costly for both predators and prey, then we should expect a signaling system to evolve that matches the form of the warning display with cognitive capabilities of the predator such that prey avoidance is enhanced. At its most fundamental level, this means that if, for instance, relevant predators do not see in the ultraviolet, then selection will not favor potentially costly warning displays that function only in the ultraviolet. Thus, the general form of warning displays must lie within the operational boundaries of the perceptual systems of relevant predators. In addition, we should expect some match between signal and receiver that extends beyond the purely perceptual. There are at least four major components of predator psychology that may affect the way the predators and prey interact. These are (1) the capacity to show unlearnt wariness of prey items, (2) a capacity to learn to avoid defended prey, (3) memory retention, and (4) prey recognition. It is easy to see that a prey can benefit if it presents a predator with a warning display that (i) enhances unlearnt wariness, (ii) accelerates avoidance learning that occurs if wariness fades, (iii) reduces any tendency for predators to forget learned wariness, and (iv) maximizes accurate recognition so that the focal prey is not misidentified as belonging to a less-defended species. In turn, however, predators can themselves be subject to evolutionary change in each of these four psychological components. Hence, the evolution of the specific forms of warning signals may be a complex coevolutionary process between predators and prey.

Unlearnt wariness is likely to be a coevolved phenotype that prepares naïve predators for unprofitability in prey. Such unlearnt wariness may be effective as a defense against species with particularly potent, potentially lethal defenses (hence unlearnt wariness of snakes and spiders seems commonplace among birds and mammals). Notice unlearnt wariness may induce complete avoidance or simply more circumspect handling. Aposematism may enhance learning by simply accelerating the frequency of predator–prey encounters, or by causing cognitive changes that cause higher predator learning rates regardless of the rate of encounter. This second category of enhancement to learning is complex (including traits such as the novelty, distinctiveness, conspicuousness, and magnitude of a signal). It may, in part, represent an adaptive response in predators, such that they give more attention to reliable signals of unprofitability. Although much less studied, there may be important effects of warning displays on memory: slowing down forgetting and/or reinstating memories of the aversiveness of prey by memory jogging. Distinctiveness of aposematic signals may enhance their recognition by predators. After discriminations have been learnt, predators may heighten their avoidance response (and reduce mistakes) if characteristics of a warning signal are exaggerated.

Evolution of Aposematic Signals

When aposematic displays are common, it is clear that they can confer a selective advantage; predators rapidly learn to avoid them by killing a small proportion of the population, have frequent memory-jogging reminders, and make very few of the recognition errors that likely

beset defended cryptic animals. Hence, the per capita survival rate of individuals bearing the same aposematic signal is likely to be high when that signal is abundant. However, when rare, aposematism looks much less attractive: rare aposematic prey may be subject to very high per capita death rates because they lack the protection of crypsis enjoyed by nonaposematic individuals, and a large proportion (of a small number) of aposematic individuals are killed while predators learn to associate the signal with unpleasant outcomes of attack. These factors seem to conspire to make rare aposemes less fit than their cryptic counterparts.

However, there are several hypotheses for how the initial disadvantage of rarity might be overcome, explaining the evolution of aposematism in a population of originally cryptic individuals. The first hypothesis suggests that chance events may nullify the problems of rarity and conspicuousness. In particular, the temporary absence of predators and random genetic drift processes may be sufficient to explain how new conspicuous morphs could rise toward their critical density in a small locality. Whatever the stochastic mechanism, it seems possible that, in some populations at least, rare aposematic morphs could reach sufficiently high levels at a local level, so that the deterministic processes of selection would now strongly favor the aposematic prey. When an aposematic morph is at or near fixation in one locality, it may then destabilize crypsis in neighboring populations and initiate a narrow shifting cline that moves through the population, destabilizing and replacing crypsis as it moves. This stochastic/shifting balance theory is one of the most persuasive accounts of the evolution and spread of aposematic forms in the 'rare and conspicuousness' framework. However, it does require the assumption that there is 'something special' about conspicuousness even before the predator has coevolved with the signal bearers; even when they first appear, aposematic signals already speed-up learning (and/or reduce forgetting) and also reduce the costs of predator education.

A second solution to the problem of initial rarity is to consider the role that spatial and temporal aggregation of individuals may have played in the evolution of aposematism. The death of a few defended siblings would protect the rest of the group since predators would be unlikely to carry on attacking distasteful individuals in close proximity to those just found to be aversive. It is easy to see that defense alleles that are localized within kin groups have higher rates of survival, and thus reproduction, than competing alleles that do not confer defense and hence make all group members vulnerable to attack. Thus, the death of one or a few individuals could generate higher levels of protection for surviving aposematic kin than that conferred on nonaposematic individuals in the population. Note here that the benefits are not passed on to genetically related individuals, but to individuals of similar appearance that are close enough spatially to share the same predator individual. These may well be genetically related individuals but need not be. Once aposematism is established in this way, it may not require continuation of the aggregation of individuals for its maintenance.

As an alternative, there may be no problem of initial rarity if sexually selected traits could be used as warning signals by predators, becoming modified to serve the dual purpose of warning and sexual communication if a species acquired an effective secondary defense. In prey without secondary defenses, sexual selection may counteract natural selection and move a prey away from its original state of crypsis. Strong selection for the acquisition of secondary defenses may then act with the result that the animal becomes unprofitable and conspicuous. Similarly, several locust and grasshopper species show facultative aposematism: switching to a conspicuous appearance and changing their diet to develop toxic defenses only when local population density is high. The problem of initial rarity could be circumvented if the constitutive nature of many warning displays were derived from an ancestrally facultative state. Further, the initial hurdle of rarity may be reduced if aposematism frees mutants from the opportunity costs of crypsis discussed earlier.

Another hypothesis to explain the problem of the conspicuous mutant is the possibility that warning displays originated as amplifiers to visible secondary defenses, such as numerous sharp spines. So long as the benefits of predator deterrence are sufficiently large and the costs of reduced crypsis are small, explaining the evolution and exaggeration of such warning traits is straightforward. Once warning signals are common for visible secondary defenses and their association with unprofitability is well known, it is easy to see that they could be taken up by prey with invisible chemical or other defenses as a loose form of Müllerian mimicry (see later). The reluctance of predators to handle these mimics could mean that new mimetic mutants are not overly disadvantaged when rare and thus can spread to high density relatively easily.

Yet another hypothesis to explain the evolution of aposematism is that warning signals evolved gradually. In their initial stages they still provided high levels of crypsis when the viewer was at a distance, but served some beneficial aposematic effect from close range. If the 'first' warning signals were of this form, it is relatively easy to see that as they become common and recognizable their conspicuousness could gradually increase.

Mimicry

Müllerian Mimicry

In the last section, we discussed how aposematism is most effective when the signalers are at high density. A certain number of signalers will be attacked while a predator learns to associate the signal with undesirability in the putative prey. Since attacks are likely to be costly to the individual attacked, the larger the local prey population is, the less likely any given individual prey is to be selected for attack, and thus have to pay the cost. If two or more defended species shared the same signal (i.e., looked alike), then they could also share this cost of predator learning and so individuals of both species would benefit from the shared signal. Consider a predator that has to sample N prey of a given signal to learn to avoid such signalers in future. If two defended species have different signals, then individuals of each species must pay independent costs of predator education, whereas if both look alike and predators do not differentiate between the two species, then only N prey from across both populations will pay the price of educating predators. Thus, there should be selection for defended species in the same location to look alike, even if they are not closely related; this is the phenomenon of Müllerian mimicry. Examples of this have been reported in several insect types as well as in frogs.

It may be that the defended organisms of a given general type converge on a small number of distinct warning signals, but do not all converge on the same one. Each grouping with a particular signal is often called a 'Müllerian ring.' Examples include tropical butterflies and European bumble bees in which several distinct Müllerian mimicry rings appear to coexist in one place. Given that the proposed selective benefits of Müllerian mimicry center on reducing the burden of predator education, we should ask why do not all distasteful species evolve to have the same pattern. There are two general, nonmutually exclusive explanations. First, the different mimicry rings may contain members that are not completely overlapping in spatio-temporal distribution, so there is little or no selection pressure for phenotypes to converge. Second, the different mimicry rings may contain forms that are so distinct that any intermediate phenotypes are at a selective disadvantage. More empirical research evaluating these possibilities would be very welcome.

Batesian Mimicry

If the predator learns that a certain signal is associated with unattractive prey and thus avoids attacking individuals that carry that signal, then an undefended species that also carried this same signal would gain protection from predators. This is the phenomenon of Batesian mimicry. In this case, there is asymmetry in the relationship between the two species with the same signal: the defended (or otherwise unattractive) one is called the model, and its signal is copied by another undefended species, the mimic. If, while the predator is learning about the signal involved, it finds a substantial proportion of the signal-bearing individuals to be generally attractive as prey (i.e., to be mimics), then the predator will not learn to avoid bearers of this signal. Thus, we should expect Batesian mimics to be at low population density compared to their models and perhaps emerge later in a season, after the learned aversion by predators has been achieved. The more common the model is and the more unpleasant it is for the predator to attack it, the more effective the learned aversion will be and the more readily a population of mimics can be supported. The model likely pays a price for this mimicry. Even if the predator does eventually learn to avoid individuals bearing the signal, if during the learning period a small number of sampled prey individuals are actually palatable mimics, then the process of learning is likely to be slowed, and this may mean that a larger number of models experience the cost of being attacked.

If the model is disadvantaged by mimicry, why does not the model evolve as quickly away from the mimic as the mimic evolves toward it? One explanation is based on the relative success of rare mimic and rare model mutants. Any change in the mimic phenotype toward the model might provide a selective advantage (because there is an increased chance of being mistakenly misclassified as a model). In contrast, major mutants of the model species away from the mimic will not spread as rapidly because they are rare and not recognized as distasteful, and thus may face reduced fitness through higher predation risk. Even if models could readily evolve away from mimics, it is unlikely that models could ever 'shake off' mimicry completely since selection to avoid mimicry depends on the presence of a high mimetic burden in the first place. In essence, Batesian mimicry may be a race that cannot be won by models unless they adopt forms than mimics cannot readily evolve toward.

The mimicry need not be a perfect replica of the model in order to gain protection; it may just have to be similar enough to put doubt in the predator's mind. This phenomenon of imperfect mimicry can clearly be seen in hoverflies, which, although they have the distinctive colored stripe pattern of wasps, can often be readily distinguished from wasps by humans on the basis of differences in body proportions. Such imperfect mimicry may be possible when the model is particularly unpleasant for predators, making the predators much less likely to experiment with something that just might be a model.

Several Batesian mimicking species are polymorphic, with different morphs in different geographical regions mimicking different local models. This polymorphism may help to keep the local density of mimics of a particular model low in comparison to the population density of their model. Sometimes, Batesian mimicry may be limited to one sex. This dimorphism may stem from differential exposure to predators between the sexes and/or one sex having a greater need for coloration for other purposes. For example, male butterflies of such a species may have their appearance constrained by the need to use coloration to attract mates, whereas the appearance of females may be less constrained and can be mimetic of another species.

Deflection

Predators preferentially bias their initial strikes to certain parts of prey individuals' bodies. Our interest here is in whether prey species have evolved markings that manipulate this aspect of predatory behavior in a way that confers a fitness advantage to the prey. We would expect this advantage to be observed as an increased likelihood of escaping from an attack. There seems good evidence that brightly colored tails in reptiles can have a deflective effect, at least in laboratory experiments. No comparable evidence is available for any other prey type: although there are sufficiently tantalizing results for eyespots in fish and (particularly) in butterflies, and for tail coloration in weasels, to warrant further exploration. The current paucity of empirical evidence for such signals should not be taken as indicative of their likely scarcity in nature. Rather, it may reflect relative lack of scientific exploration of the phenomenon.

How Can Deflective Markings Evolve if They Make Prey Easier for Predators to Detect?

One potential drawback to brightly colored deflective signals is increased detection by predators. Cooper and Vitt explored this with a simple model, reproduced as follows.

Imagine that an individual with a cryptically colored tail is detected by a predator with probability P_d. After detection, the predator attacks and the probability of the prey escaping this attack is P_e. If these two probabilities are independent, then the prey individual's probability of being captured by a predator is

$$P_d(1 - P_e)$$

We assume that having a conspicuously colored tail increases the probability of detection by the predator by some value α, but also increases the probability of escaping by β.

Conspicuous tail coloration will be favored if the inconspicuous type is more likely to be captured by the predator, that is, if

$$P_d(1 - P_e) > (P_d + \alpha)(1 - P_e - \beta)$$

This rearranges to the condition

$$\beta > \frac{\alpha(1 - P_e)}{P_d + \alpha}$$

Hence, conspicuousness will be favored if β (the advantage that is gained from conspicuousness) is large or α (the increased probability of detection due to conspicuousness) is small compared to β and P_d (the probability of being detected in the absence of the conspicuous signal). Also unsurprisingly, a large probability of detection in the absence of conspicuous coloration (high P_d) favors selection of the conspicuous signal. More interesting is that a high probability of escape in the absence of conspicuous coloration (high P_e) favors the evolution of the conspicuous deflective coloration. This suggests that tail autotomy and, perhaps, associated behaviors that draw attention to the tail (increasing P_e) probably developed before the conspicuous coloration of these body parts in some species. More generally, this model demonstrates that deflective markings can still be selected for if they cause an increase in the rate at which their bearer is attacked. Specifically, they can still be favored by selection, provided their enhancement of probability of escape from an attack is sufficient to compensate for the potential cost of increased probability of attack.

Why Do Predators Allow Themselves to Be Deflected?

The empirical evidence provided earlier suggests that some prey may be able to reduce their risk of being captured in a predatory attack by inducing the predator to attack specific parts of their body. Why do predators allow such deflection to occur, if it costs them prey items? One might hypothesize that deflection occurs because of lack of familiarity with the prey type and predict that deflective markings would be relatively unsuccessful when used by prey species that the predator attacks frequently, compared to prey items that it attacks infrequently. Specialist predators should not be fooled by deflective markings, whereas generalist predators may have to accept such costs as a by-product of having evolved to be able to handle diverse prey types. The generalist predator may find deflective markings difficult to combat in one species encountered infrequently if similar visual cues are useful when attacking a different species that are encountered more frequently. This argument may provide a theoretical framework for consideration of why some styles of signal could be more effective at deflecting than others. It also raises the testable hypothesis that prey that use deflective signals will generally not be the main prey of predatory species that they successfully deflect. One might also expect to see predators habituating, such that their probability of being fooled by deflective marking declines with increased exposure. While this has been demonstrated repeatedly for startle signals (see next section), it has not been explored for deflective signals. However, predators of reptiles that shed their tails upon attack may not be under strong selection pressure to stop 'falling for this trick,' since they do end up with a substantial meal from the tail, particularly as tails are often used as fat stores.

Startle Signals

Edmunds (1974) defines startling signals (also referred to as 'frightening' or 'deimatic signals') as follows.

> Deimatic behaviour produces mutually incompatible tendencies in a predator: it stimulates an attacking predator to withdraw and move away. This results in a period of indecision on the part of the predator (even though it may eventually attack), and this gives the displaying animal an increased chance of escaping.

The classic example of a putative startling signal is the bright color patterns that some otherwise-cryptic moths and butterflies can be induced to display by disturbing them while at rest. They have brightly colored hind wings that they generally cover with cryptically colored forewings, but which can be voluntarily and suddenly revealed.

One key feature of startling signals is that they are induced by the proximity of a predator. It is generally postulated that the sudden appearance of a bright display or loud noise induces an element of fear or confusion in the predator, giving the prey individual an increased chance of fleeing before being attacked. Although often postulated, this possible mechanism has been subjected to little rigorous experimental testing. However, a recent series of laboratory experiments have suggested that some of the spot patterns of butterfly wings may function in this fashion.

Why Would Predators Be Startled?

It seems possible that the startled predator is misidentifying the prey item as something that could be a threat to it, rather than as something that is merely an unattractive food item (a more general treatment of deception is dealt elsewhere). Predators may be presented with conflicting selection pressures acting on their response to unexpected stimuli. A startle response may represent the best compromise between the costs of less efficient prey capture (because the delay in attacking allows some prey to escape) and the (potentially very high) cost of failing to respond rapidly to unexpected and imminent danger. However, one would expect predators to evolve mechanisms that allow them to habituate to startle signals from harmless prey. That is, one would expect predators to react to startle signals not by completely fleeing the scene, but by retreating to a safer distance from which they may be able to assess the potential threat. Fleeing would deprive it of the ability to learn about the source of the stimulus (except in as much as the other organism did not successfully pursue it). Thus the behavior of a predator toward a startle signal should not be fixed, but rather should be responsive to that individual's experiences subsequent to previous exposures to similar signals.

Such cognitive processes should be suitable to theoretical investigation, similar to those that have help to illuminate phenomena of aposematic signaling and mimicry.

See also: Alarm Calls in Birds and Mammals; Deception: Competition by Misleading Behavior; Defensive Chemicals; Hormones and Breeding Strategies, Sex Reversal, Brood Parasites, Parthenogenesis; Risk-Taking in Self-Defense; Vibrational Communication; Visual Signals.

Further Reading

Caro TM (2009) Contrasting coloration in terrestrial mammals. *Philosophical Transactions of the Royal Society B* 364: 537–548.

Cott HB (1940) *Adaptive Coloration in Animals*. London: Methuen.

Edmunds M (1974) *Defence in Animals: A Survey of Anti-Predator Defences*. Harlow: Longman.

Franks DW and Sherratt TN (2007) The evolution of multi-component mimicry. *Journal of Theoretical Biology* 244: 631–639.

Hill RI and Vaca JF (2004) Differential wing strength in Pierella butterflies (Nymphalidae, Satyrinae) supports the deflection hypothesis. *Biotropica* 36: 362–370.

Kawaguchi I and Saski A (2006) The wave speed of intergradation zone in two-species lattice Müllerian mimicry model. *Journal of Theoretical Biology* 243: 594–603.

Mappes J, Marples N, and Endler JA (2005) The complex business of survival by aposematism. *Trends in Ecology & Evolution* 20: 598–603.

Marples NM, Kelly DJ, and Thomas RJ (2005) Perspective: The evolution of warning signals is not paradoxical. *Evolution* 59: 933–940.

Ruxton GD, Sherratt TN, and Speed MP (2004) *Avoiding Attack: The Evolutionary Ecology of Crypsis, Warning Signals and Mimicry*. Oxford: Oxford University Press.

Sherratt TN (2008) The evolution of Müllerian mimicry. *Naturwissenshaften* 95: 681–695.

Stevens M (2005) The role of eyespots as anti-predator mechanisms, principally demonstrated in the Lepidoptera. *Biological Reviews* 80: 573–588.

Vallin A, Jakobsson S, Lind J, and Wiklund C (2005) Prey survival by predator intimidation: An experimental study of peacock butterfly defence against blue tits. *Proceedings of the Royal Society B* 272: 1203–1207.

Defensive Morphology

J. M. L. Richardson, University of Victoria, Victoria, BC, Canada
B. R. Anholt, University of Victoria, Victoria, BC, Canada; Bamfield Marine Sciences Centre, Bamfield, BC, Canada

Overview and Introduction to Types of Morphological Defenses

Defenses against predators can be divided into primary defenses, which reduce the chances of a predator encountering, detecting, or identifying the prey, and secondary defenses, which reduce the chances of an identified prey being approached, subjugated, or consumed. Primary defenses act in predator avoidance. Secondary defenses act in predator evasion. As we discuss different types of defense strategies, it is important to bear in mind that defenses may work at both levels, and rarely do prey have only one defense mechanism.

Morphological defenses are mechanical or physical properties of an organism that may help the individual to avoid predation. Often these defenses work in conjunction with a behavioral response and may be used in other contexts as well. We begin by outlining common types of morphological defenses and then consider some examples that illustrate the inherent interplay between morphological defenses and behavior.

Crypsis

Crypsis, or camouflage, can involve background matching, disruptive coloration that obscures recognizable body parts, or masquerading as an inedible object. A classic example of selection favoring camouflage to reduce detection by predators is that of the peppered moth, *Biston betularia*. As the industrial revolution in Europe led to a die-off of lichen on trees, leaving a darker background on which nocturnal moths rested during the day, Kettlewell showed that an initially rare dark form of the moth increased in frequency and that diurnal bird predators more readily detected and consumed pale moths on a dark tree. Rarity may also provide crypsis. Visual predators form search images when foraging, and the rare individual has an advantage if its form falls outside the predator's search image.

Disruptive coloration can decrease the chance of identification by predators. For example, many animals have a dark patch or stripe around their eye. The eye is a readily detected feature of an individual and thus, markings that obscure the eye can provide a substantial increase in camouflage. Similarly, patterning or projections from the body can help to disguise the body shape against the background. For moth-like artificial prey differing in color from their background, disruptive coloration around the edges of the wings (e.g., black areas that go to the edge on some part of the wing) decreases mortality by bird predators.

Some animals have body shapes, colors, and patterns that mimic inedible objects common in their environment, such as leaves or sticks. Both freshwater and saltwater neotropical fish species include some leaf mimics. Numerous examples exist of insects that have a body form that resembles twigs or leaves and that adopt body positions to further resemble twigs or leaves (e.g., praying mantids). Insects, such as caddisflies, build cases out of leaves, twigs, or, sand that provide both shelter and camouflage. Many predators rely on odor, sound, or vibrations to hunt, and prey that can smell, sound, or move in a way that matches that of something other than a prey item will presumably benefit from decreased detection and identification by a predator. Crypsis within sensory modalities other than sight are less well studied and could use more attention. More detailed discussion of crypsis is dealt with elsewhere in this volume.

Aposematism

Bright colors and pattern contrasts seen in many poisonous or venomous animals provide a morphological defense that works in conjunction with the chemical defense to prevent a predator from attacking. In animals living in an environment with a heterogeneous background, crypsis will be difficult to maintain if any movement is required. Both computer simulation studies and empirical observations support the hypothesis that for an active individual in a heterogeneous environment, selection leads to both unpalatability and aposematism. Use of toxicity and warning coloration as a defense requires that predators learn to associate the coloration with unpalatability, potentially leading to prey mortality by naïve predators.

This cost is minimized in coexisting prey that share predators by Müllerian mimicry of shared colors and patterns, to reduce the chance for any one individual of being killed by a naïve predator. This type of mimicry can lead to a striking variation in morphological patterning across a species range, with one species mimicking different coexisting congeners within its range, as seen in both *Heliconius* butterflies and in Asian green pit vipers.

Batesian mimicry, in which palatable species mimic the warning color patterns of unpalatable species, also occurs in groups such as hoverflies that mimic bees and wasps. Selection on such a trait is inherently frequency-dependent; if palatable mimics are too frequent, predators will kill many mimics prior to encountering an unpalatable individual and the benefit for the palatable mimic is lost. Further, the unpalatable species should experience selection to modify its warning coloration and/or pattern to allow predators to distinguish it from palatable mimics; frequency-dependent selection will, however, work against evolution of such differences at this stage as rare aposematic individuals are killed before the predator can learn to avoid them. More detailed discussion of aposematism, mimicry and toxicity is dealt with elsewhere in this volume.

Body Armor

Animals with body armor or a protective shell can respond to a predator attack by withdrawing into their shell, reducing the ability of predators to get at edible soft tissue. Evidence of the potential effectiveness of this strategy is seen among variants of a land snail found distributed among Japan and several other islands in the same region. Snails found on the same island as snail-eating snakes have a modified shell, with the shell extended into the aperture opening, changing both its shape and size. Predation trials with snail-eating snakes show that this narrowed aperture opening gives the snails a significantly increased chance of escaping a snake attack over snails of similar size with a rounded opening.

Body armor, such as the lateral plates seen on sticklebacks, can also protect an individual from injury during a predator attack. A related strategy, also used by sticklebacks as well as invertebrates such as dragonfly larvae, is to have sharp spines that will cause an attacking predator to let go prior to fatal injury occurring.

Weaponry

An animal under attack by a potential predator may go on the offensive if it has weaponry such as spines, horns, or large claws. These features may or may not have arisen through selection to avoid predation. For example, European clawed lobsters have one large crushing claw that is used in foraging and intraspecific male dominance competition, but this intact claw also significantly reduces predation. Lobsters that had lost the large claw through autotomy experienced 100% mortality when attacked by predators.

Porcupines provide an exemplary case of an animal with weapons evolved as a defense against predators. The spines of porcupines are designed such that when not erect they are flexible and not easily shed, but when erect the spines are readily released from the porcupine

with a minimal amount of force on the tips. This allows the porcupine to use its tail as a weapon, slapping it against a potential attacker to inflict injury.

Weaponry may also take the form of scent or poison glands. Threatened skunks release a noxious smell to fend off potential predators, even before the predator attacks. Many insects use muscular contractions to forcefully shoot a liquid containing quinone, acetic acid, or some other noxious compound toward an approaching predator. Millipedes, moths, and toads have glands that ooze out toxins at the moment of predator contact. More detailed discussion of attacks by prey on predators is dealt with elsewhere in this volume.

Body Shape and Size

Many species have modified body shapes or size when coexisting with predators, often in conjunction with other defense mechanisms. Changes in body shape in response to a predator have been observed in many taxa and are well studied in zooplankton, insects, anuran larvae, and fish. Changes in body size alone can also act to reduce predation risk. Physid snails with a larger body size have decreased mortality from crayfish, while *Daphnia* evolve smaller body sizes in the presence of size-selective fish predation. Changes in body shape that enhance escape success, such as a more streamlined body in fish and relatively elongated hindlimbs in lizards, may also reflect a response to selection by predators.

Eyespots

Eyespots refer to a circular color marking on the body of an animal; while the term is convenient because to humans they resemble eyes, little evidence exists as to whether predators interpret these patterns as eyes. Eyespots are most commonly seen in lepidopterans, but also occur in other insects and some fish. Eyespots are used in multiple ways as a defense. Moths, such as the hawkmoth, with cryptic forewings will move their forewings if attacked to reveal eyespots on the hindwings that are otherwise hidden when the moth is at rest. While the predator may be startled into thinking that it is now looking at its own predator's eyes, more likely the predator is simply overwhelmed by the presentation of a large amount of new visual information to process. The brief delay in the predator's attack while processing the new information can provide the potential prey with the split second it needs to get away and resettle cryptically somewhere else.

Alternatively, eyespots may act as deflection markers, deflecting a predator's potential attack away from the most vulnerable part of the body to a less vulnerable body part. A predator that aims its attack at an eyespot on the edge of a wing may provide the moth or butterfly with an opportunity to escape with only an injured wing.

Other body markings thought to act as a defensive mechanism by deflecting attack from vulnerable areas include the brightly colored tails of juvenile lizards, which are autotomized if grabbed by a predator, further increasing chance of escape. Many mollusc species similarly autotomize a tail, papillae, or siphon projections and most arthropods readily autotomize limbs.

In some fish families that are poisonous but typically remain cryptic in sandy substrate, a black eyespot on the dorsal fin appears to act as a warning signal to potential predators. Fish raise their dorsal fin to display a prominent black eyespot if disturbed.

A Further Characterization of Morphological Defenses

Morphological defenses can be further characterized as fixed or inducible. Fixed morphological defenses are evolved defenses that occur regardless of whether the predator is present in the habitat or not. Eyespots on moth wings, the presence of a shell on snails, and cryptic body coloration in moths, guppies, stick insects, etc. are all examples of fixed traits (also referred to as constitutive).

Inducible morphological defenses are those that arise in response to an environmental cue that is indicative of predator presence. Inducible defenses are more generally phenotypically plastic traits and can thus be further distinguished as either irreversibly or reversibly plastic. Irreversible plasticity is plasticity in a trait during development; presence of the appropriate cue at a particular point during development will cause a specific trait form that is then fixed for the remainder of that individual's lifespan. The majority of inducible morphological defenses will fall into this category, as morphology is not readily modified once development is complete. Reversible plasticity refers to an induced trait that can be modified multiple times during the lifespan of an individual. This includes nearly all behavioral traits – an animal can hide in a refuge while a predator is nearby, come out when the predator moves away, and then hide again when the predator is nearby. Some induced morphological defenses can be reversibly plastic, notably changes in coloration seen in insects, reptiles, squid, and fish. More information on developmental plasticity is discussed elsewhere in this volume.

Fixed Morphological Defenses

Fixed morphological defenses are likely to evolve when the benefits in the presence of predators exceeds the costs in the absence of predators. In addition, fixed defenses will evolve when developing the defense each time it is needed has too high a cost, no reliable cue for the presence of a predator is available, or the development of the defense cannot occur rapidly enough to be effective.

The evolution of morphological defenses and behavioral traits are inexorably intertwined. Cryptic individuals are only cryptic if the individual remains still on the appropriate habitat type, or travels in a manner aligned with its body pattern. Striped juvenile garter snakes that move in a straight line while fleeing have a good chance of escape, but a blotched individual fleeing in a straight line will not. Hidden eyespots are only advantageous if the moth moves its forewings to reveal the eyespots at the critical moment during a predator attack. Spines used as weaponry against potential predators must be locked into an erect position (a behavioral response) to be effective against potential predators. More generally, of course, all prey must recognize potential predators if they are to use any type of proactive defense. More information on trait evolution and selection are discussed elsewhere.

Inducible Morphological Defenses

The risk of mortality posed by predation varies in both space and time. Predators may move into or out of a habitat, or reproduction might occur in only some habitat patches. Predator life-cycles that lead to seasonal variation in abundance will also contribute to variability in predation risk. Antipredator defenses are, by definition, advantageous in the presence of predators because they reduce mortality rates, but the relative advantages will vary with risk of predation. Predation risk will rise when predators are more abundant, when alternative prey are rare, or when there are few refuges. In the absence of predators (i.e., 'relaxation' in selection), there is no benefit in having antipredator morphology, but there may well be costs. The defense may be expensive to produce or maintain, or it may reduce the efficiency of foraging by the prey leaving fewer resources to be allocated to other demands ranging from the immune system to attracting mates. In short, the cost might be anything that reduces other fitness enhancing activities of the prey.

If the costs of antipredator defenses are high enough, we expect that they will be expressed (induced) only when predation risk is high. This is effective only if the production of the defensive morphology happens fast enough to provide protection before the predators disappear again. Equally important is that the defense should not be induced when there is no risk. Whatever cue is used, it needs to be reliable. The presence of a predator and its associated chemical cues may not be reliable. When the predator is too small to be a threat, for example, inducing an antipredator morphology will incur the cost of inducing the morphology and of having the morphology without the attendant benefits.

A cue is information produced by the predator that is taken advantage of by the prey, but not provided intentionally by the predator. An ideal cue, from the perspective of the prey is one that the predator can do nothing about. The shadow cast by a hawk on a sunny day and recognized by a ground squirrel as evidence of a threat cannot be prevented by the hawk. Moving shadows are a reliable cue of the immediate threat of predation for ground squirrels. Similarly, the freshwater hypotrich ciliate *Euplotes* responds to proteins in the membrane of their predators by inducing a distinctive antipredator morphology with increased body width that thwarts gape-limited predators. These proteins function as predator recognition signals for the prey. Prey have evolved to use the same molecules to respond to the threat of predation because the cues are reliable.

Although theoretically necessary, costs associated with antipredator defenses when predation risk is absent may be so low that they are very difficult to measure. A simple calculation can show why this is so. Consider a prey that expresses a morphological defense that reduces the annual mortality rate from 80% of the individuals to 40%. In the absence of defense, the daily survival rate is $(1 -$ annual mortality rate) raised to the power $(1/365)$ to convert it from an annual to a daily rate, which gives us $(1 - 0.8)^{1/365} = 0.2^{1/365} = 0.9956$. In the presence of the defense, by similar logic, we have $(1 - 0.4)^{1/365} = 0.6^{1/365} = 0.9986$. Thus, the daily survival rate in the presence of the morphological defense has increased only by $(0.9986 - 0.9956) = 0.0030$ or 0.3%. If we try to measure the cost of the defense using a growth rate experiment, the difference in the daily growth rate between defended and undefended individuals has to be less than 0.3% for natural selection to favor the defensive morphology. In a 10 day growth experiment, setting the daily growth rate of the undefended individuals to 1, this is (undefended daily growth rate) raised to the power of 10 days – (defended daily growth rate) raised to the power of 10 days, which gives $1^{10} - 0.997^{10} = 1 - 0.97 = 0.03$. A 3% difference in the expected growth is so small that it may be difficult to detect. We need to be careful not to overstate the benefits of morphological defenses by measuring the benefits only when prey are attacked. We need to measure the frequency with which these benefits are realized in order to properly compare them to costs.

More information on antipredator trade-offs and mechanisms are discussed elsewhere.

Case Studies

Evolution of Fixed Defenses

Avoidance of dragonfly predators by Enallagma damselfly larvae

The damselfly genus *Enallagma* provides a natural replicated experiment of response to selection by a new predator type. The ancestral habitats of *Enallagma* larvae are lakes that have fish as top predators. Fish are highly visual predators, particularly keen at detecting movement, and

far faster swimmers than damselfly larvae. Damselfly larva in the presence of fish, minimize predation risk by reducing activity to minimize detection.

Lakes without fish have larvae of the dragonfly *Anax* as the top predator; these predators, not found in lakes with fish, do not have the same long-distance hunting strategies as fish and tend to perch on similar substrate to damselfly larvae. As a consequence, the strategy of remaining motionless that is effective against fish predators is deadly in the presences of *Anax* predators. In lab experiments, *Enallagma* from fish lakes allow an *Anax* predator to walk right up and consume them. Similarly, *Enallagma* from fishless lakes swim away from approaching predators and are thus consumed by fish, but often escape predation by *Anax*.

Morphology differs significantly between species found in the two habitat groups. *Enallagma* larvae have three caudal appendages, or lamellae, and species found with *Anax* predators (fishless lakes) have lamellae that are relatively larger and more round than those of species in fish lakes (**Figure 1**). Damselfly larvae swim using side-to-side undulation of their body and these morphological changes reflect selection for avoiding dragonfly predators through increased swimming ability. Individuals from a fishless lake species with experimentally reduced lamellae were less likely to escape a dragonfly predator, but the same reduction in lamellae size had no effect on the probability of predation for individuals from a fish lake species. This reveals the importance of considering morphological defenses in conjunction with behavioral defenses. Here, morphological changes in body structure reflect a response to selection by a predator, but the morphological defense is effective only when combined with a behavioral response, namely the propensity to swim away from a potential predator.

Gregariousness and aposematism in Lepidopteran larvae

Another example of an association between morphological and behavioral defenses is seen between the presence of aposematic coloration and gregarious behavior. This evolutionary relationship has been best demonstrated in lepidopteran larvae in which, across a range of species, aposematism in butterfly larvae is associated with gregariousness. Theoretical work suggests that while defenses and warning coloration can facilitate the evolution of gregariousness, gregarious behavior can also facilitate the evolution of warning coloration. In a large survey of over 800 species of tree-feeding lepidopterans, Tullberg and Hunter considered explicitly the presence of both warning coloration and defenses (physical or chemical) as potential precursors to evolution of gregariousness. Their results revealed that gregariousness evolved significantly more commonly in species that had either defenses or warning coloration.

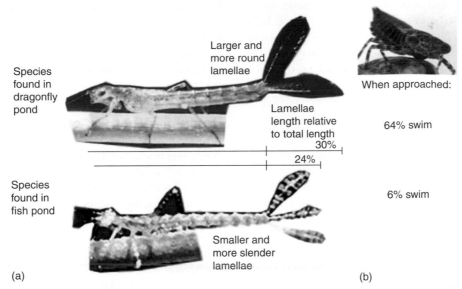

Larger and
more round
lamellae

Species
found in
dragonfly
pond

Lamellae
length relative
to total length
30%

24%

When approached:

64% swim

6% swim

Species
found in
fish pond

Smaller and
more slender
lamellae

(a)   (b)

Figure 1   Adaptive evolution of antipredator defense in damselfly larvae. (a) Larvae of species found in ponds with dragonfly larvae as top predators such as *Enallagma boreale* (top) have larger and more round lamellae leading to faster average swimming speeds than larvae found in ponds with fish as top predators such as *E. vespersum* (bottom) with smaller, more slender lamellae. Lamellae are transparent and thus reveal black background in images. (b) When approached by a dragonfly predator, most individuals swim away in species found in dragonfly ponds, while very few swim away (most remain still) in species found in fish ponds. Data from McPeek MA, Schrot AK, and Brown JM (1996) Adaptation to predators in a new community: Swimming performance and predator avoidance in damselflies. *Ecology* 77: 617–629.

Experimental work using naïve young chicks offered aposematic (red and black) and defended (secrete noxious compounds) bug larvae supports lower attack rate in aposematic prey with gregarious behavior. Work using wild-caught blue tits and novel prey (straws filled with suet) reveals that when prey distribution is clumped, attack rate on palatable prey is higher. Birds sampling an individual in a group of unpalatable prey not only dropped the prey, but also moved onto a different group of prey. Birds that attacked a palatable prey remained in the patch and took more prey before moving. Thus, regardless of warning coloration, if unpalatability or some other secondary defense has evolved, prey may benefit from gregarious behavior. More on group living as an antipredator behavior is discussed elsewhere.

Evolution of Induced Defenses

Daphnia

A century ago, German limnologists described seasonal variation in the morphology of freshwater crustaceans. As the season progressed, the cladoceran *Daphnia* developed enlarged helmets and tail spines (**Figure 2**). They gave it the name 'cyclomorphosis' to denote the annual cycle of changes in morphology. Explanations for this phenomenon focused on environmental causes such as temperature and turbulence for the next 50 years. The growing consensus is that these changes in morphology bear all the hallmarks of inducible morphological defenses against

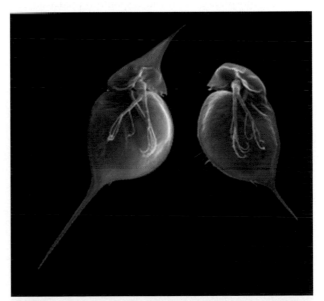

Figure 2   Individuals of the same clone of *Daphnia cullucata*. On the left, raised in the presence of predator cue, on the right raised in the absence. Photo: Ralph Tollrian.

predation. The defense can be induced by water that has contained predators. These infochemicals, or kairomones, are still unidentified. There is variation for how much individual clones can change their morphology. The distribution of clones with differing levels of inducibility is associated with variation in the risk of predation across the landscape. These defensive structures are made up of chitin and cannot be remodeled after they have been developed, so the level of defense lasts the lifetime of an individual. However, there is growing evidence that the level of expressed defense depends, in part, on the environment experienced by the mother. More on maternal effects is discussed elsewhere in this volume.

Euplotes

Remarkably, an ability to change morphology in response to predation risk is not limited to multicellular organisms. The hypotrich ciliate *Euplotes* can remodel its cytoskeleton within hours of encountering predatory ciliates or flatworms. The morphology shifts from a flattened ovoid to nearly round with extended lateral wings and a pronounced aboral ridge (**Figure 3**). This shift in morphology makes *Euplotes* too large to be consumed by their gape-limited predators. Induced morphs bear the cost of being less effective predators and having longer cell cycles. There is clonal variation in the ability to induce so it is likely that this variation should be distributed among habitats depending on variation in the risk of predation.

Hyla

A flexible morphology that can respond to predation risk can also be found in vertebrates. Larvae of the North American treefrogs *Hyla chrysoscelis* and *H. versicolor* raised in the presence of dragonfly larvae predators develop strikingly colored tails with a deeper shape and enhanced musculature (**Figure 4**). The change in body shape improves acceleration when disturbed and the tail coloration may direct predator attacks away from the more vulnerable

body. Larvae pay a cost in reduced efficiency in sustained swimming. Thus, when competition for food is high, there is a shift in morphology in the opposite direction to that expressed as an antipredator defense. More on foraging and predation risk is discussed elsewhere.

Usefulness of the Conceptual Framework

While this discussion has focused on organisms that can display behavior, the ideas developed in this context can usefully be applied to plants. Thorns can be induced by vertebrate grazing, and given that this is nonphotosynthetic tissue it must have some costs. Whether these costs are large enough to make induced rather than permanent defenses the better strategy is unknown. Plants respond to mechanical damage differently than they do to damage caused by insect feeding. It appears that there is a reliable cue in the feeding secretions of the insect that is used as a reliable cue to induce (upregulate) the production of antiherbivore defenses.

An inducible defense is a change in phenotype in response to an environmental cue. If the environmental cue provides no information about the future state of the environment, then there is no advantage to changing the phenotype. That lack of information can be because of the environment's being constant, so the cue provides no additional information, or because of the change in the environment being so rapid that no predictions can be based on the cue. Identical arguments can be made about

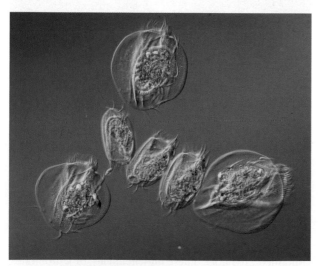

Figure 3 Members of the same clone of *Euplotes octocarinatus*. The three individuals in the center were grown in the absence of predators, the other three were grown in the presence of predatory flatworm extract. Photo: Juergen Kusch, University of Kaiserslauter.

Figure 4 Tadpoles of *Hyla versicolor* raised from the same clutch of eggs. The tadpole on the left was raised in the absence of predators and the one on the right was raised in the presence of a caged predator. Photo: B.R. Anholt.

the value of learning. If the environment is constant, there is no advantage to learning; if the world changes very rapidly too, learning is of little value. More on learning is discussed elsewhere in this volume.

Current Research Focus and Outstanding Questions

The synergistic relationship among antipredator traits has only recently been the focus of rigorous testing. While it has long been assumed that crypsis and inactivity are co-evolved traits, evidence that the association is adaptive has only recently been tested using stickleback predators and chironomid larvae. Specifically, Ioannou and Krause were able to show that each trait alone did not reduce predation risk and that only the combination of the two traits is effective. This type of mechanistic analysis of antipredator morphological and behavioral traits combined needs to be carried out in a variety of systems. Similarly, a recent simulation of how body patterns affect detection of a moving target reveals some general principles that can now be used to predict and test selection for body patterning in prey. Traditionally, work on antipredator traits have looked at traits in isolation, and while this made the questions more tractable, it is increasingly clear that results can be misleading. For example, the straight-line escape movement of striped snakes from a predator (as opposed to the reversals seen in unstriped snakes) may be taken as leading to decreased fitness without taking into account that a striped pattern hampers the ability of predators to estimate future position of the snake, making it unlikely that the attacking predator will successfully catch the snake.

Most prey experience multiple predators, and different predators may select for incompatible phenotypes. Understanding how prey experiencing different predators respond to these multiple selection pressures is another area that requires more research. How multiple predators and variable predator presence shape evolution in defense traits is an important area for future research. More on predator–prey evolution is discussed elsewhere in this volume.

Recent work suggests that the ability of prey to detect chemical cues from predators is more widespread and more sophisticated than previously recognized. Future work on isolating those compounds used as detection cues will enhance our ability to understand the role of cues in the evolution of inducible responses. The relationship between the timing of cues and induction also needs further elucidation. For example, does timing of cue presentation affect the degree of phenotypic response? Many of the animals showing inducible defenses have complex life histories that span different habitats – how do defenses induced in one life stage affect phenotype in a subsequent life stage? More on predation risk and life history is discussed elsewhere in this volume.

See also: Adaptive Landscapes and Optimality; Antipredator Benefits from Heterospecifics; Co-Evolution of Predators and Prey; Communication and Hormones; Costs of Learning; Defensive Avoidance; Defensive Chemicals; Defensive Coloration; Development, Evolution and Behavior; Developmental Plasticity; Evolution: Fundamentals; Future of Animal Behavior: Predicting Trends; Games Played by Predators and Prey; Group Living; Hormones and Breeding Strategies, Sex Reversal, Brood Parasites, Parthenogenesis; Levels of Selection; Life Histories and Predation Risk; Maternal Effects on Behavior; Microevolution and Macroevolution in Behavior; Phylogenetic Inference and the Evolution of Behavior; Predator Avoidance: Mechanisms; Risk Allocation in Anti-Predator Behavior; Risk-Taking in Self-Defense; Trade-Offs in Anti-Predator Behavior.

Further Reading

Caro T (2005) *Antipredator Defenses in Birds and Mammals.* Chicago, IL: University of Chicago Press.

Edmunds M (1974) *Defence in Animals: A Survey of Anti-Predator Defences.* New York, NY: Longman Group Ltd.

Gomez JM and Zamora R (2002) Thorns as induced mechanical defense in a long-lived shrub (*Hormathophylla spinosa*, Cruciferae). *Ecology* 83: 885–890.

Ioannou CC and Krause J (2009) Interactions between background matching and motion during visual detection can explain why cryptic animals keep still. *Biology Letters* 5: 191–193.

Kuhlmann HW and Heckmann K (1985) Interspecific morphogens regulating prey-predator relationships in protozoa. *Science* 227: 1347–1349.

McPeek MA (1990) Behavioral differences between *Enallagma* species (Odonata) influencing differential vulnerability to predators. *Ecology* 71: 1714–1726.

McPeek MA, Schrot AK, and Brown JM (1996) Adaptation to predators in a new community: Swimming performance and predator avoidance in damselflies. *Ecology* 77: 617–629.

Ruxton G, Sherratt T, and Speed M (2004) *Avoiding Attack: The Evolutionary Ecology of Crypsis, Warning Signals and Mimicry.* Oxford: Oxford University Press.

Tollrian R and Harvell CD (eds.) (1998) *The Ecology and Evolution of Inducible Defenses.* Princeton, NJ: Princeton University Press.

Tullberg BS and Hunter AF (1996) Evolution of larval gregariousness in relation to repellent defences and warning coloration in tree-feeding Macrolepidoptera: A phylogenetic analysis based on independent contrasts. *Biological Journal of the Linnean Society* 57: 253–276.

Tullberg BS, Leimar O, and Gamberale-Stille G (2000) Did aggregation favour the initial evolution of warning coloration? A novel world revisited. *Animal Behaviour* 59: 281–287.

van Buskirk J and McCollum GA (1999) Plasticity and selection explain variation in tadpole phenotype between ponds with different predator composition. *Oikos* 85: 31–39.

Development, Evolution and Behavior

A. L. Toth, Pennsylvania State University, University Park, PA, USA

Introduction

The goal of this article is to explore the relationships between evolution, development, and behavior. There are two main reasons why such considerations are important to the field of animal behavior – one is the important mechanistic links between behavior and development, and the second relates to conceptual insights that can be gained from the field of evolutionary developmental biology.

With respect to the first topic, it is abundantly clear that behavior and development are intimately linked. The brain develops both during embryological and adult stages, and this development can be a link between the environment and plasticity in individual behavioral responses. In addition, some forms of behavior develop and change over the lifetime of an individual, and thus can be considered developmental processes themselves. Second, studies of the molecular genetic basis of morphological development have progressed further than those of behavior. Molecular developmental biology paired with a comparative, evolutionary perspective has given rise to the field of 'evo-devo,' which can provide several lessons that can be applied to the study of behavior. These include the importance of conserved genes and changes in *gene regulation* in generating novel phenotypes and the utility of breaking down (behavioral or morphological) phenotypes into constituent parts, or '*modules.*'

In the sections that follow, both mechanistic and conceptual links between behavior and development are discussed. The first section explores the various ways in which behavior and development are interrelated and also how behavioral and morphological phenotypes differ but must be considered simultaneously. Then, the field of evolutionary developmental biology and some of the major tenets of evo-devo that can be applied to the study of behavior are described. Next, specific examples from across different animal taxa are reviewed that illustrate how considerations of the principles of evo-devo and the behavior-development relationship can advance our understanding of how behavior evolves. Finally, the article concludes by suggesting future directions for a more comprehensive integration of development and behavior that could lead to further insights into animal behavior.

Relationships Between Behavior and Development

In many ways, the study of behavior *is* the study of development. Like any other organ, the nervous system develops during both embryological and adult stages (though certainly to varying degrees) and its development is highly responsive to external and internal environmental factors. Interestingly, some of the same genes can affect both nervous system development during embryonic stages and neural function and behavior in adult animals. One example is the gene *fruitless* in the fruit fly *Drosophila melanogaster*. This gene is important for the development of male-specific neuronal projections into abdominal muscles used in mating, but also affects courtship behavior in mature animals. In fact, *fruitless* derives its name from a particular mutation that causes males to court other males.

In addition, there are many known instances of mechanistic links between certain forms of behavior and morphological or physiological development – this can be the result of *pleiotropic* effects of specific hormones or genes on both the nervous system and other organs. For example, in many vertebrates, testosterone is important in sex determination of the gonads and brain during embryological development, but can also have effects on male aggressive, territorial behavior during adulthood. In female insects, juvenile hormone levels during development influence ovary size and can also affect various forms of adult behavior, including egg-laying and foraging. A prominent example of the pleiotropic effects of insect hormones on both reproduction and behavior involves '*oogenesis*-flight syndrome,' in which ovarian development is associated with sedentary behavior, and a shut-down of ovarian development is associated with sustained flight. In two species of migratory locusts, *Locusta migratoria* and *Schistocerca gregaria*, this syndrome is exhibited in the extreme. Two entirely different locust morphs exist – larger solitary locusts that have narrow foraging ranges, cryptic coloration, and high ovarian development; and smaller gregarious locusts that fly hundreds of miles in search of food, are brightly colored, and have lower ovarian development. In both species, two hormones (juvenile hormone and corazonin) stimulate various aspects of the solitary phase including green coloration, reproductive physiology, and solitary-like behavior.

Although behavior and development are in many ways intimately linked, research in the fields of ethology and development have proceeded quite separately. Several reasons likely account for this separation. First, there is a general perception by biologists that behavioral phenotypes are farther removed from genes than developmental phenotypes; behavior is generated by neurons in real time at a rate that is much more rapid than even the fastest known changes in gene expression, whereas development proceeds gradually over hours, weeks, or even years.

In addition, whereas morphological phenotypes are stable or slowly changing, behavioral phenotypes are more unpredictable and fleeting, often making them harder to measure. For these reasons, in-depth studies of the genetic basis for behavior began later and have progressed more slowly than such studies of development.

Although some behaviors consist of stereotyped action patterns, many behaviors are more complex and require long-term maturation or a process of development to come to fruition. Such behavioral development may or may not require learning and/or restructuring of the nervous system. Notable examples of behavioral development include song learning and development by songbirds, the transition from hive work to foraging in honeybees, and juvenile play behavior (e.g., play fighting and play hunting) by a wide range of vertebrate animals, especially mammals. By recognizing behavioral phenotypes as developmental phenotypes themselves, it may be fruitful to apply a similar approach to the study of behavior as has been historically used to study development.

Basics of 'Evo-Devo'

During the early history of evolutionary thought, evolution and development were considered to be inseparable. Studies of embryology were used to infer evolutionary relationships among organisms, typified by Ernst Haeckel's famous insight that 'ontogeny recapitulates phylogeny.' Although such comparisons can be useful, subsequent studies in embryology showed that this view is an oversimplification. The adoption of a population-level focus on evolution with the rise of the modern synthesis of genetics and evolution led to a formal separation of the fields of evolution and development until fairly recently. With the application of molecular genetics to developmental biology, the fields of evolution and development were eventually reintegrated in the 'evo-devo synthesis,' and numerous important findings have already emerged from this relatively new hybrid field.

One of the major discoveries in developmental biology was the revelation that something so complex as multicellular development is genetically orchestrated via precise changes in the timing and location of gene expression. This insight stemmed from the elucidation of a hierarchical cascade of transcription factor genes that lead to the differentiation of segments during embryological development. This developmental cascade was first studied in the fruit fly D. melanogaster and the set of genes controlling early embryological development were elucidated in remarkable detail. This groundbreaking work by Lewis Nusslein-Volhard, and Wieschaus was awarded with the Nobel Prize in Physiology and Medicine in 1995. By using a comparative approach and studying the molecular genetics of development in other species, the pioneers of evo-devo made a startling discovery. It turns out that many of the same genes regulating early development in insects, specifically, the homeotic or Hox genes which determine the identity of segments, also control the development of segmentation in vertebrates, even in a similar anterior-to-posterior pattern (**Figure 1**).

Cross-species studies of the molecular basis for development have fueled the evo-devo synthesis, and in some cases, the findings have even caused biologists to rethink how new structures arise during evolution. A case in point involves the evolution of image-forming eyes in animals. The compound eyes of arthropods and the camera-like eyes of vertebrates differ hugely in their basic structure, and were long considered to be a classic example of convergent evolution. However, this interpretation was called into question with the discovery of the primary role of Pax6 genes, first identified as affecting vertebrate eye development and subsequently found in Drosophila. Further studies revealed another member of the Pax gene family to have a role in complex eye development in a jellyfish, suggesting Pax involvement in eye development may predate the evolution of the common ancestor of both insects and vertebrates. This finding suggested that vertebrate and insect eyes could have arisen from a proto-eye shared by a common ancestor. Yet another (less likely) possibility is that the Pax genes were co-opted to regulate eye development multiple times during animal evolution, and represent a remarkable example of convergent evolution on both genetic and phenotypic levels. Although these issues have not yet been completely resolved, the realization of these complexities of conservation and convergence would not have been possible without molecular genetic studies. A similar depth of study will be necessary to untangle these issues for behavior that are the apparent result of convergent evolution.

Evo-devo studies across a diversity of animals including butterflies, ants, and stickleback fish have provided additional insights into how evolution occurs. In particular, it has been fruitful to study convergent morphologies that have evolved repeatedly in several relatively closely related species. Studies of stickleback fish have shown that a convergent phenotype (the reduction of bony armor) can be attained via evolutionary changes in the same molecular pathways, but by altering different individual genes within the pathway. On the other hand, studies of winglessness in worker ants have shown that the genetic pathways that maintain a particular phenotype over evolutionary time can change. Although wing loss evolved only once early in ant evolutionary history, the network of wing development genes is interrupted at different points in different species of modern ants.

Insights to Be Gained from an Evo-Devo Approach to Behavior

As described earlier, comparative studies of the genetic basis for development have uncovered the deep extent of

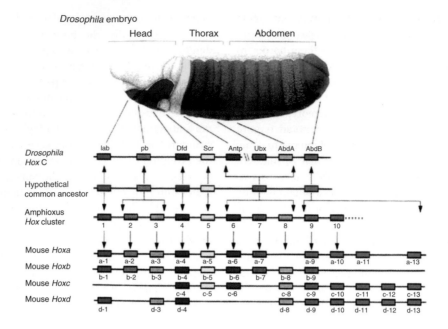

Figure 1   Diagram mapping *hox* genes to specific segments in both a *Drosophila* and mouse embryo. Adapted from Carroll SB (1995) Homeotic genes and the evolution of arthropods and chordates. *Nature* 376: 479–485.

conservation of gene function across organisms and led to a more careful consideration of the roles of conservation versus convergence in evolution. The field of evo-devo is also beginning to provide broadly generalizable principles about the evolutionary process itself, making it all the more important for behaviorists to consider an evo-devo approach. For example, some authors have suggested a shift in the focus of the levels of selection in evolution from the gene to the phenotype. In addition, evo-devo also forces one to consider the importance of nongenetic, or *epigenetic*, influences on phenotypes, including the external environment and social (e.g., maternal) effects. Some have suggested that epigenetic effects can lead to the evolution of novel phenotypes even before such changes are fixed by mutations in the gene sequence.

The field of evo-devo has led to several main insights, each of which can provide useful lessons for the study of animal behavior: (1) the importance of changes in gene regulation (in addition to changes in coding regions of genes) in generating morphological diversity, (2) the idea of a shared 'genetic toolkit' for development consisting of a core set of deeply conserved genes that are used repeatedly across taxa to generate diversity in form, and (3) the idea of modularity, that is, that morphology can be broken down into several distinct components that tend to be repeated in series and can be added, deleted, or shuffled, to create novel morphologies.

The Importance of Gene Regulation

Mutations in the coding regions of genes can disrupt the basic function of a protein, which can have severe if not lethal effects on the organism. Alterations in gene regulation involving the timing and location of the expression of genes, on the other hand, can result in more subtle changes in phenotype. Thus, regulatory changes have been proposed to be more likely targets for natural selection, which could result in more gradual evolutionary changes. There is a growing base of examples from evo-devo showing the importance of regulatory changes in generating morphological diversity. For example, changing the localization of transcripts of specific *Hox* genes can result in a variety of morphological novelties ranging from the patterns on butterfly wings to the shape and number of appendages. Studies of postdevelopmental changes in brain gene expression suggest that this may also be generalizable to behavior. Soon after the shocking discovery that humans and chimpanzees have 98% of their DNA sequences in common, biologists hypothesized that it is differences in gene regulation, rather than differences in coding sequence, that must explain the huge differences in behavior and intelligence between us and them. The application of global gene expression analysis to this question has indeed uncovered large-scale changes in brain gene regulation between chimps and humans. Another example of the importance of gene regulation comes from rodents. The localization of vasopressin receptors (*V1aR*) in a brain region (ventral striatum) in several species of voles makes all the difference between promiscuous, absentee fathers and monogamous, paternal males.

There are different levels at which changes in gene regulation can affect a phenotype – at the *transcriptional* or *translational* levels. In addition, a distinction has been made between two different types of transcriptional gene regulation: (1) *cis* regulatory change – a change in gene regulatory sequences that affects transcription of a given gene nearby, and (2) *trans* regulatory change – a change in one gene that regulates the expression of other genes that may be in a different part of the genome. Mounting evidence from developmental biology suggests that *cis* regulatory changes appear to be extremely important in the evolution of morphological changes across species. More detailed studies of the gene regulatory networks that affect variation in behavior both within and across species will be needed before an assessment of the importanc of *cis* versus *trans* regulation can be made for the field of behavior.

Genetic Toolkits for Behavior?

Studies of the genetic basis of development across a wide variety of taxa suggest that the existence of a 'genetic toolkit' for development, that is, a core set of genes or pathways that underlie morphological development and that are used repeatedly during evolution to generate diversity in body form. Prominent examples from development are homeotic (*Hox*) genes in segmentation and *Pax* genes in eye development across both vertebrate and invertebrate animals. Does a similar 'genetic toolkit' for behavior exist as well? Or do behavioral phenotypes rely more on new genes to generate behavioral novelty? Studies across both vertebrate and invertebrate animals suggest that such toolkits may indeed underlie several basic forms of behavior (aggression, reward, and sociality), as discussed below.

Many animals exhibit aggressive behavior, which can vary widely in the form of expression (from biting to stinging to highly ritualized aggressive displays) and in the context in which it is used (i.e., to defend a territory, to gain access to mates, to establish a position in a dominance hierarchy). Nonetheless, research on the mechanistic basis of aggressive behavior in both vertebrate and invertebrate animals suggests that these behaviors may share common molecular underpinnings. For example, low levels of the neuromodulator serotonin affect aggressive behavior in mice and have also been associated with impulsive aggression in humans. In lobsters, both extremely elevated and depressed levels of serotonin are associated with increased aggression. This is one example of the same molecule being associated with aggression across taxa in which opposite patterns of regulation can affect similar behaviors across species. Thus, the serotonin pathway may be evolutionarily labile, that is, easily changed during evolution to regulate behavioral differences, though the exact pattern of regulation may vary across taxa.

An important aspect of motivation is that the performance of some behaviors produce a self-reinforcing sensation of 'reward.' The reward system has long been known in mammals, typified by drug addictions in humans and mice that seek electric stimulation of the 'pleasure center' of the brain in preference to food. In vertebrates, the main *neurotransmitter* that has been associated with reward is dopamine. Dopamine is released in a certain brain region (nucleus accumbens) in response to eating and sexual activity. Elegant work on the molecular basis of pair bonding in voles has demonstrated a connection between expression of the vasopressin V1aR receptor in the ventral pallidum, but pair bonding can only occur when dopamine is actively present in the same brain region, suggesting that pair bonding has evolved to become a rewarding stimulus. Recent studies with insects suggest that invertebrates possess a reward system not so different from that of vertebrates. Research with crickets, flies, and honeybees suggest that dopamine instead mediates the learning of negative, aversive stimuli whereas a related neurochemical, octopamine, can affect learning and perception of rewarding stimuli such as food.

Eusociality, the complex form of social behavior that is defined by the presence of reproductive queens and workers that forgo their own reproduction to aid the reproduction of others has evolved multiple times in animals from termites to bees to naked mole rats. With striking convergent evolution of social form across such a wide variety of animal taxa, the study of the evolution of eusociality provides a good system to test for the existence of a 'genetic toolkit' underlying the evolution of complex social behavior. It has long been known that nutritional asymmetries among individuals within a social insect colony relate to differences in reproductive capacity and body form, and contribute to the development of different castes, including kings and queens, workers, and worker subcastes that specialize in particular colony tasks. Recent studies at the molecular level suggest that certain genes or pathways are associated with sociality across multiple taxa, many of which relate to nutritional and metabolic processes. For example, the storage protein Hexamerin is associated with caste differences in distantly related insects (termites and paper wasps). In addition, genome-wide studies of gene expression have repeatedly uncovered important differences in metabolic enzymes in numerous lineages (bees, wasps, and ants). Finally, differences in the regulation of deeply conserved genes that control feeding behavior (including the *foraging* gene and the insulin pathway) appear to regulate behavior across independently evolved lineages of ants, bees, and wasps. These studies suggest that eusociality, a complex form of social organization, evolved in part by changes in the regulation of deeply conserved genes regulating feeding and nutritional physiology. Further studies of the molecular genetic basis of eusociality in even more distantly related taxa, for example, mole rats, will provide a crucial test of this hypothesis, and will allow us to assess how broadly such a 'genetic toolkit' applies.

The Evolution of Behavioral Modules

Modules can be defined as distinct phenotypic units, developing more or less independently from each other, that make up part of a larger whole. It is intuitive that animal body plans are modular. Vertebrates have repeating series of vertebrae, and insect body plans are clearly organized into segments – just think of a caterpillar. Comparative anatomical studies and gene-level studies have shown that such modules can be reorganized to give rise to new body structures. Additional modules can be added, subtracted, or fused to form new structures. For example, in vertebrates, jaws evolved from modular series of gill arches in early fish, and skulls from fused elements of several vertebrae. In insects, repeated pairs of segmented appendages have evolved into specialized mouthparts and antennae, and the thorax has evolved from the fusion of three ancestral body segments.

Although somewhat less obvious than for morphology, some behaviors are also modular in structure. Many behaviors can be broken down into constituent parts, which often occur sequentially over time. This is true for both short-term sequences of behavior and behavioral phases that occur over the course of a lifetime. Breaking down complex behaviors into smaller component behaviors (or behavioral modules) can be a useful entree into detailed studies of the mechanisms underlying these behaviors. In the following paragraph, two examples of modular behaviors – one describing a set of behavioral modules expressed on a short time scale, and the other involving more long-term behavioral phases – are described. In each

case, modules appear to have been reorganized to generate novel forms of behavior during evolution.

Courtship behavior in *Drosophila* fruit flies is a complex affair. The general sequence consists of several stages (or behavioral modules), as follows: first, the male orients toward the female; then he taps her with his antennae; then he begins singing a courtship song by buzzing his wings; then he licks her genitals; then he mounts, and finally, if successful, he copulates. This is the general series of steps, but the sequence varies across species, with various elements that are either prolonged, shortened, or elaborated. The courtship song itself consists of modules of different forms of sound that vary widely across species. Elements of this courtship behavior vary across *Drosophila* spp., and there is evidence that in some cases, differences in courtship sequence, especially song, can act as species-isolating mechanisms. In the same way in which *Drosophila* courtship songs may help to isolate species, bird songs may do the same, facilitated by reorganizing different combinations of trills and whistles, which in some cases appear to be behavioral modules of song.

With regard to eusocial insects, a great mystery that has intrigued biologists since Darwin relates to the evolution of queens and workers. Given the fact that in most species, any female egg can become a worker or a queen, how can such extreme differences in morphology and behavior arise from the same genome? One hypothesis utilizes the idea of behavioral modules. If we imagine a solitary maternal insect, its behavior can be broken down into two distinct behavioral modules: (1) egg-laying and (2) maternal provisioning of brood with food collected during foraging. It has been suggested that an ancestral ovarian cycle consisting of these two basic modules of egg-laying and foraging/provisioning could be uncoupled. Instead of being separated in time as in solitary maternal wasps, the two behaviors could become separated into different individuals – queens that focus on egg-laying and workers that specialize in foraging/provisioning. Thus, worker behavior, which involves caring for siblings, may have evolved from maternal foraging/provisioning. Recent evidence at the molecular level supports the idea that worker behavior evolved from maternal behavior, similar patterns of brain gene expression underlie both maternal and worker behavior in primitively social *Polistes metricus* paper wasps.

Further expansions of an ancestral groundplan may have occurred among workers later in social insect evolution, in two contexts. First, colonies show a division of labor among nest workers and foragers; nest workers have higher reproductive capacity than foragers, and recent results suggest that the brain gene expression patterns of honeybee nest workers are indeed more queen-like than those of foragers. Second, we see a fine-tuned division of labor for foraging in honeybees; bees that forage for pollen have more well-developed ovaries and higher levels of expression of the egg-yolk protein Vitellogenin than bees that forage for nectar. Thus, these two ancestral modules of egg-laying and foraging may have been separated multiple times during social insect evolution to produce specialized individuals, giving rise to a division of labor.

The Co-evolution of Behavior and Development

Thus far, behavior and morphological development as separate phenotypic entities have been considered. However, in many cases, behavior and morphology coevolve. This may be due to similar selection pressures causing parallel evolution of the two or due to constraints imposed by pleiotropic effects of genes that affect behavior and morphology concurrently. As discussed earlier, there have been several studies of hormonal effects on both behavior and development suggesting the possibility of common mechanistic elements to the regulation of physiology, development, and behavior. However, to date, there have been few studies that have attempted to examine whether the same genes or pathways underlie both developmental and behavioral differences within and across species. This is an area ripe for study, and in the following section, two particularly promising models for addressing this question are summarized.

Three-spined stickleback fish (*Gasterosteus aculeatus*) have been important model systems for studying the evolution of development. These fish have evolved from marine to freshwater forms multiple times in several widely separated geographical areas. They thus provide a perfect system to examine the roles of conservation and convergence in phenotypic (both morphological and behavioral) evolution. Each time sticklebacks have invaded freshwater habitats, and this has been accompanied by a reduction in the presence of armored plates along the lateral side of the body as well as shortened pelvic spines, which are protection against predators. In many freshwater populations, sticklebacks have further diversified into distinct benthic (bottom dwelling) and limnetic (surface dwelling) forms, which show differences in jaw morphology that are related to differences in their feeding habits. These benthic and limnetic forms show consistently different patterns of foraging behavior, courtship, and aggressive behavior. It remains to be seen whether some of the same genes that regulate morphological differences are also used to regulate behavioral differences, or whether different toolkits are employed for each. If different toolkits exist, it is an intriguing question as to whether such toolkits coevolve *via* common regulatory elements that control numerous different pathways, or whether there are no such common regulatory elements to link pathways.

Horned scarab beetles are found worldwide, with striking variation in the presence/absence of horns and in their size and morphology. In some dung beetles, males take alternative forms: territorial, large-horned males, and nonterritorial small-horned males. The horns are used in combat between males for dung resources, and such contests help assure them possession of dung territories and access to females. Recent studies have begun to elucidate the molecular basis of horn development in dung beetles and suggests an important role for the insulin pathway in affecting energy allocation to horns (vs. other morphological features) resulting in *allometric* changes in horn size. There is also a correlation across species between horn size and behavior: beetle species that tunnel into dung have large horns, whereas those that roll dung on the surface do not. Given the role of various insulin pathway genes in regulating feeding and social behavior in insects, it will be intriguing to test whether the insulin pathway also affects tunneling and aggressive behavior in beetles.

Future Directions

Evo-devo has been extremely successful in elucidating several important principles about how morphological evolution can occur (as described in 'Basics of Evo-Devo'). Notably, the major insights from evo-devo have resulted from pairing molecular genetics data with comparative methods by studying a wide variety of species. The mechanistic basis of behavior, while traditionally believed to be harder to dissect than that of development, has nonetheless already hinted at similar findings to evo-devo – namely, that changes in the regulation of deeply conserved genes are likely to result in behavioral evolution and that a core set of genes, or 'genetic toolkit,' may be used repeatedly during the evolution of novel behaviors. The studies of the mechanisms responsible for the evolution of behavior have focused mainly on a handful of species (e.g., rodents, honeybees, and fruit flies). Reflecting on the history of evo-devo, it is clear that behavioral studies could also benefit greatly from a much expanded comparative analysis of behavior. This need may be fulfilled by a general broadening of the taxa considered for comparison to include distantly related species with similar patterns of behavior. Well-resolved phylogenies are needed in order to carefully choose species that are informative in a phylogenetic context (e.g., species in basal lineages or species within a branch of a phylogenetic tree that appear to have evolved similar behaviors independently).

One of the main obstacles to such studies has been the lack of gene sequence information and genetic resources for nonmodel genetic species. New technologies are quickly getting around this roadblock. For example, it is now possible to manipulate gene expression patterns in a number of model organisms through the use of pharmacological treatments or *RNA interference* (RNAi). In addition, next generation sequencing methods, which generate huge amounts of data at a fraction of the cost of traditional sequencing, are improving rapidly. Such methods are now being effectively used to generate large databases of expressed genes for a wide variety of ecologically and evolutionarily important species.

Such technological improvements can help pave the way for new and creative ways to study the molecular genetic basis of behavior in a wide variety of species. These advances, when coupled with an evo-devo perspective on behavior, promise to yield major insights into behavioral evolution in the near future.

See also: Caste in Social Insects: Genetic Influences Over Caste Determination; *Drosophila* Behavior Genetics; Evolution: Fundamentals; Genes and Genomic Searches; Honeybees; Integration of Proximate and Ultimate Causes; Play; Social Insects: Behavioral Genetics; Sociogenomics; Threespine Stickleback; Zebra Finches.

Further Reading

Abouheif E and Wray GA (2002) Evolution of the gene network underlying wing polyphenism in ants. *Science* 297: 249–252.

Barron AB and Robinson GE (2008) The utility of behavioral models and modules in molecular analyses of social behavior. *Genes, Brain, and Behavior* 7: 257–265.

Carroll SB (1995) Homeotic genes and the evolution of arthropods and chordates. *Nature* 376: 479–485.

Carroll SB (2008) Evo-devo and an expanding evolutionary synthesis: A genetic theory of morphological evolution. *Cell* 134: 25–36.

Carroll SB, Grenier J, and Weatherbee S (2004) *From DNA to Diversity: Molecular Genetics and the Evolution of Animal Design*, pp. 272. Malden, MA: Wiley-Blackwell.

Cresko WA, Amores A, Wilson C, et al. (2004) Parallel genetic basis for repeated evolution of armor loss in Alaskan threespine stickleback populations. *Proceedings of the National Academy of Sciences of the United States of America* 101: 6050–6055.

Emlen DJ, Lavine LC, and Ewen-Campen B (2007) On the origin and evolutionary diversification of beetle horns. *Proceedings of the National Academy of Sciences of the United States of America* 104(зupplcmcnt 1): 8661 8668.

Hudson ME (2008) Sequencing breakthroughs for genomic ecology and evolutionary biology. *Molecular Ecology Resources* 8: 3–17.

Kozmik Z (2005) *Pax* genes in eye development and evolution. *Current Opinion in Genetics & Development* 15: 430–438.

Love AC and Raff RA (2003) Knowing your ancestors: Themes in the history of evo-devo. *Evolution & Development* 5: 327–330.

Nusslein-Volhard C and Wieschaus E (1980) Mutations affecting segment number and polarity in *Drosophila*. *Nature* 287: 795–801.

Raff RA (2000) Evo-devo: The evolution of a new discipline. *Nature Reviews Genetics* 1: 74–79.

Robinson GE and Ben-Shahar Y (2002) Social behavior and comparative genomics: New genes or new gene regulation? *Genes, Brain, and Behavior* 1: 197–203.

Toth AL and Robinson GE (2007) Evo-devo and the evolution of social behavior. *Trends in Genetics* 23: 334–341.

West-Eberhard MJ (2003) *Developmental Plasticity and Evolution*, pp. 794. New York, NY: Oxford University Press.

Developmental Plasticity

A. R. Smith, Smithsonian Tropical Research Institute, Balboa, Ancon, Panamá

Introduction

Developmental plasticity is central to eusociality. The key feature of eusociality is reproductive division of labor. One or a few individuals reproduce, while the remaining members of a social group serve as workers. Such a distinction requires the ability to express the behavior and physiology required for reproduction by some individuals, and the expression of worker behavior without reproduction by others. Thus, developmental plasticity enables the expression of worker or reproductive alternative phenotypes. Some social insects are champions of developmental plasticity, producing queens and workers of such different morphology that a naïve observer would hardly guess they were of the same species. However, even without such extreme morphological differentiation between castes, the substantial behavioral differences between queen and worker castes in the smallest colonies highlight the central role of developmental plasticity in social insects.

In this article, I summarize developmental plasticity in reproductive division of labor in some representative social invertebrates, principally insects. I then discuss the role of developmental plasticity in the evolution of eusociality in insects, emphasizing the primitively eusocial species, because these species especially illuminate the evolution of division of labor. Developmental plasticity in ants and honeybees is covered by Adam Dolezal's study on caste, and in termites by Judith Korb's review on social evolution in termites. Throughout this article, I highlight relevant reviews as a gateway to more detailed study.

In its broadest interpretation, 'social behavior' includes any interaction between two conspecific animals. Here, my focus is on social organization based on reproductive division of labor, meaning that only one or a few individuals in a social group reproduce while the rest serve as workers assisting the reproductive individual(s). While the term 'eusocial' has been subject to constant argument and redefinition, the combination of reproductive division of labor and cooperative brood care encompasses the core of what makes the social insects so interesting to the animal behaviorist, regardless of which definition of eusociality one chooses. My focus is on the determination of the reproductive division of labor – the factors that influence the development of an individual into a reproductive queen or a nonreproductive worker.

Developmental plasticity, defined as 'the ability of an organism to react to an internal or external environmental input with a change in form, state, movement, or rate of activity,' is broad enough to include most of animal behavior (see West-Eberhard's (2003) book for a more detailed discussion of this definition). In the context of reproductive division of labor, developmental plasticity can be thought of as the ability of a single genotype to produce both reproductive and worker phenotypes. In general, social insect caste determination results from developmental plasticity rather than genotypic differences, although there are a few instances of genetic caste determination. By 'determination,' I mean the adoption of one of two alternative developmental pathways (such as reproductive or worker) following a decision point. Prior to this determination, the individual has the potential to develop along either pathway. The point during development at which such determination occurs, and to what extent it can be reversed, varies immensely across social insect taxa. Except for the termites and the shrimp, all the examples discussed below are from the insect order Hymenoptera (the bees, wasps, and ants). It must be kept in mind that hymenopteran societies are exclusively female (males disperse and mate, usually with little role in the life of a colony), so the discussion is focused on queens and their daughters – workers and gynes.

Developmental Plasticity and Reproductive Division of Labor

A major theme in social insect developmental plasticity is the interaction of nutrition, the social environment, and endocrine regulation of reproduction to generate two alternative discrete phenotypes: the queen and the worker. In small-colony insect societies, these variables are integrated through adult behavioral interactions and are typically reversible. In larger-colony, more derived (socially specialized) species, caste is typically determined by a nutritional switch during development. If nourishment is at a sufficient level when the switch is reached (nourishment which is typically under social control), endocrine triggers begin development into the queen phenotype. If nourishment is low, the immature develops along the worker phenotype. In these cases, caste is typically not reversible.

Adult Determination of Reproductive Caste

One way to induce worker behavior is for a reproductive female to inhibit the reproduction of her daughters who would otherwise be fully capable of reproducing on their own. This appears to be the case in the neotropical sweat bee (Halictidae) *Megalopta genalis*, which can nest either solitarily or in social groups of, typically, 2–3 females. Even two-bee nests have reproductive division of labor: one female rarely leaves the nest, lays eggs, and has enlarged ovaries, while the other forages and has slender ovaries. Yet, when queens were experimentally removed from these simplest of societies, workers enlarged their ovaries and reproduced at the same rate as naturally solitary-nesting reproductives in the same population. These results suggest that *Megalopta* females remain totipotent, that they are not inherently hopeless reproductives who have chosen to stay and help, and that caste is induced by social interactions between adults. Why these females rather than other offspring that left the nest chose to stay and work (or why queen dominance was directed toward them rather than the other offspring) remains an open question. The nature of these interactions in *Megalopta* is currently under investigation, but research in other small-colony halictid bees shows that aggression from the queen suppresses ovarian development.

In one study of the halictid *Lasioglossum zephyrum*, which has somewhat larger colonies than *Megalopta*, repeated nudging from a steel ball manipulated by a magnet simulating queen aggression was sufficient to inhibit ovarian development. Further studies of *L. zephyrum* showed that the bees form a dominance hierarchy with the queen at the top. Queens directed their aggression disproportionately at the worker directly below them in the hierarchy, and, when queens were experimentally removed, it was not the highest worker in the hierarchy, but the second highest who became the replacement reproductive, illustrating the cumulative effect of queen aggression.

Stenogastrine wasps (hover wasps), which recent phylogenetic studies show are not monophyletic with the social paper wasps discussed later, also have very small colonies (as small as two females in some species) and apparent adult totipotency. They also form dominance hierarchies, and these function as a reproductive queue: the next dominant worker assumes the queen position upon death of the current queen. As with the bee *Megalopta*, there is no evidence that wasps that would otherwise be poor reproductives, become workers. Stenogastrines and some other primitively social wasps such as *Mischocyttarus drewseni* (Polistinae) are atypical among social insects in that queen replacement is relatively frequent. However, even in other small-colony species in which natural queen replacement is less frequent, the ability of workers to develop into the queen caste upon experimental queen removal shows the importance of queen dominance behavior in suppressing worker aggression.

Polistes, and Other Independent-Founding Paper Wasps (Polistinae)

By far the most thoroughly studied social insect with adult determination of reproductive castes is the paper wasp genus *Polistes*. While *Polistes* has long been one of the most prominent examples in animal behavior of aggression-based dominance hierarchies socially regulating reproduction, it has long been clear that differential larval nutrition can influence their future caste. The biology of *Polistes* is relatively similar to the other independent-founding paper wasps (i.e., those that initiate a new nest with one or a few individuals, rather than a reproductive swarm) in the genera *Mischocyttarus*, *Belanogaster*, *Ropalidia*, and *Parapolybia*.

A *Polistes* nest is initiated by a single foundress or a few cofoundresses. In cofoundress groups, the wasps establish a dominance hierarchy through aggressive interactions. The dominant female becomes the queen and may monopolize reproduction through suppressing ovarian development of the subordinates by physically dominating them and by eating the eggs of other females and replacing them with her own. Dominant *Polistes* foundresses meet incoming subordinate foragers to receive building material, solid food, and regurgitated liquid food through trophallaxis. If the subordinate forager resists, the queen often bites her until she offers food. Thus, the dominant wasp has a twofold advantage in maintaining her status: she avoids the energetically expensive task of foraging and can direct the flow of food toward herself.

There is an endocrine component to dominance as well. Queens have elevated levels of the hormones, juvenile hormone (JH) and ecdysone, both of which increase aggression. JH is produced by the corpora allata (part of the insect brain) and ecdysone by the ovaries. In *P. dominulus*, corpora allata and ovary size correlate with the establishment of dominance: wasps with more active endocrine glands are more likely to be dominant, and experimental hormone treatment affects behavior. Dominant foundresses not only had increased hormone titers relative to subordinates, but relative to solitary foundresses as well, suggesting both social and reproductive influences on hormone titers. How JH and ecdysone interact to affect behavior, and the interaction and feedback of reproductive and social effects on hormone titers are both open questions being pursued by current investigation.

When the first *Polistes* offspring on a nest emerge as adults, they are typically dominated by the queen as described earlier for cofoundresses, and as a result, they also have small ovaries, lowered hormone titers, and develop into workers. However, both subordinate cofoundresses and workers can become replacement queens if the original queen dies or is experimentally removed while their ovaries are in a developing phase, thus removing the inhibition on subordinate reproduction. Studies of the independent-founding *Ropalidia marginata*, which has a similar life history as *Polistes*, suggest that the totipotency demonstrated by queen removal studies may not be complete. For example, when young females were isolated, fed ad libitum, and allowed to initiate nests alone, many did not, despite the complete lack of social competition. A later experiment showed that there apparently is a reproductive queue within colonies as to which female develops into a replacement queen that is not related to any observable dominance hierarchy. This illustrates that much about the social regulation of developmental plasticity still remains to be discovered, and that it may be different in different taxonomic groups of social insects.

Recently, James Hunt and colleagues, building on both their own and others' earlier studies, proposed that there are actually two developmental pathways for temperate zone *Polistes* based on larval nutrition. In one pathway, termed 'G1' (first generation), wasps emerge with few hexamarin storage proteins and need to feed in order to enlarge their ovaries and develop eggs. In the second pathway, termed 'G2' (second generation), wasps emerge with high levels of hexamarin storage proteins and more developed ovaries. The importance of hexamarin storage proteins, at least in temperate zone *Polistes*, results from the seasonal nature of the colony cycle: colonies are founded in the spring, G1 workers produced during the summer, and then, in late summer or autumn, a second generation of prereproductive gynes is produced. These gynes mate, and then go into diapause to survive the winter before emerging in spring to initiate their own colony. Because diapause is energetically expensive, gynes must be provisioned with extra nutrients – the hexamarin proteins. While gynes are typically produced at the end of the favorable nesting season, experiments manipulating larval nourishment show that the G1 or G2 developmental pathways are determined by levels of larval nutrition. G1 and G2 do not equate with worker and queen. G1 wasps may become replacement queens, although they are disadvantaged in competing for reproduction by their lack of nutrient stores and small ovaries. And G2 gynes may end up as subordinate, nonreproductive cofoundresses the subsequent spring. Thus, while *Polistes* females are totipotent, with adult social interactions inhibiting reproductive behavior to create workers, the G1 and G2 are apparently worker- and queen-biased developmental pathways that predispose females to the worker and queen phenotypes, respectively.

An obvious question raised by this work is, what about the tropics? Despite the lack of winter, most tropical environments have a wet and dry season, only one of which may be favorable. Thus, there may still be production of 'immediate worker' and 'future reproductive' phenotypes, although this remains to be tested. Also, in the tropical *Polistes* studied to date, colonies do not last forever, even though they often last through more than one wet and dry season, suggesting that even if a colony cycle is not imposed

by winter, there may still be a terminal period in some species during which the colony switches from provisioning for G1 females and switches to G2's. This possibility should be examined though to date there is no evidence for it in tropical *Polistes*. Some species of *Polistes*, and other tropical social wasps, such as *Ropaladia marginata* and the Stenogastrines, have apparently evolved division of labor without determinate nesting cycles. The applicability of the G1–G2 developmental pathways to nondiapause tropical *Polistes* (not to mention the little-studied basic biology of most tropical *Polistes* and other paper wasps) is an open question. Likewise, the extent to which gyne- and worker-biased developmental pathways are common to other small-colony, primitively social insects, and the seasonal reproductive characteristics of nonworker-containing species closer the threshold of sociality, remains to be studied.

Morphological Castes

In many primitively social insects in which caste results from aggressive interactions, body size plays an important role, possibly because it confers an advantage in aggressive interactions. For instance, *Megalopta* bee foragers tend to be smaller than dispersing reproductives and larger *Lasioglossum* bee foragers are more likely to become replacement queens. Nevertheless, in all these groups, size is only a weak correlate of caste, and is often less important than other factors, such as age or timing of emergence from diapause. In some other social insects, queens and workers fall into discrete morphological distributions. The Hymenoptera (the bees, ants, and wasps) are holometabolous insects, meaning that after larval growth, they pupate and then undergo a final molt into their adult body. Adults cannot grow or shrink, and cannot change the shape of their exoskeletons. Thus, discrete morphologies between castes indicate discrete preadult developmental pathways. The divergent pathways usually result from a nutritional switch based on differential larval provisioning. For instance, in one of the best-studied examples, honeybee (*Apis mellifera*) larvae fed with royal jelly (a substance produced by the workers) rather than the typical worker diet will experience a rise in JH titer during the fourth and fifth larval instars. Queen and worker developmental trajectories diverge after this point. However, artificial supplementation of food for worker-destined larvae, or experimental treatment of these larvae during the fourth and fifth instar with JH can cause the larvae to develop as queens. Thus, in honeybees, a nutritional switch triggers an endocrine response that separates the queen and the worker developmental pathways. The timing of this switch during development is variable across social insect groups, and can be as early as the egg stage in some ants (queen-destined eggs are supplemented with more nutrition than worker-destined ones). Earlier divergence between queen and worker trajectories permits more caste-specific development time, and thus the potential for more morphological differences between the castes.

Developmental Plasticity in Termites

Termites are members of the order Isoptera, which is hemimetabolous. This means that, in contrast to holometabolous insects such as the Hymenoptera, they can continue to molt throughout their adult life. In most groups of termites, 'workers' are undifferentiated immatures that do little to help rear offspring and can potentially develop into replacement reproductives, dispersing winged reproductives, soldiers (morphologically specialized nest defenders), or remain as undifferentiated workers. Both sexes can become workers. Termites have both morphological castes and a potential for adult caste determination. While all termite lineages exhibit extreme developmental plasticity through sequential molts, the details of caste and caste plasticity differ dramatically between groups. For example, in some termites, certain castes are limited to only one sex. A more detailed discussion on developmental plasticity in termites is discussed elsewhere in this Encyclopedia.

Developmental Plasticity in Social Shrimp

A social system remarkably similar to the termites in many respects has evolved in *Zuzalpheus* snapping shrimp (Crustecea), which live inside tropical marine sponges. Like termites, most nonreproductives are morphologically undifferentiated immatures. Also, like termites, because these shrimp live inside their food source, they do not require active provisioning of offspring, but do require the nest defense of morphologically specialized soldiers. Unlike any of the social insects, the queen continues to grow through successive molts. A review of the social snapping shrimp is discussed elsewhere in this Encyclopedia.

Developmental Plasticity and the Evolution of Reproductive Division of Labor

From one perspective, the transition from solitary living to social groups with division of labor represents the origin of a spectacularly successful new phenotype. Social insects are widespread, ecologically important, and speciose. However, from the perspective of developmental plasticity, the queen and the worker phenotypes are simply incomplete versions of the ancestral solitary bee or wasp: queens are reproductives who do not provision their young, and workers express parental care without ever reproducing. The challenge for understanding the evolution of reproductive division of

labor, then, is to understand how reproduction and provisioning were uncoupled into queen and worker phenotypes.

Mary Jane West-Eberhard proposed the 'ovarian groundplan hypothesis' to explain how reproduction and provisioning could be uncoupled. The hypothesis is based on the links between ovarian development, hormone expression, and competitive behavior in solitary wasps and bees when found in groups. A typical reproductive female solitary wasp develops eggs in her ovaries. As the egg nears maturation, she constructs a cell in which to lay the egg. When the egg is fully developed, she lays the egg in the cell. In at least some progressively provisioning species, her ovaries are then significantly smaller than before, due to the recently laid egg. The solitary progressively provisioning female wasp at that phase of her cycle forages for prey with which to provision her offspring. Thus, the wasp undergoes ovarian enlargement, with accompanying queen-like behavior (building a new cell and laying an egg), and ovarian diminishment, with accompanying worker-like behavior (foraging). The link between ovarian physiology and behavior is hormones – specifically JH. JH presumably increases with ovarian enlargement and decreases with ovarian diminishment, though this has not been studied in any solitary progressive provisioner. As discussed earlier, JH also increases aggressive behavior. Thus, a reproductive could dominate and withhold nourishment from her daughter such that the daughter could not respond to rising JH levels with ovarian development, leaving her 'socially castrated.' This subordinate would thus be 'stuck' in the foraging phase of the ovarian cycle, resulting in a two-wasp group with reproductive division of labor.

The behavioral sequence described by West-Eberhard is common among solitary wasps. JH expression is sensitive to nutritional and social influences. The links between ovarian development, hormone expression, and behavior are all supported circumstantially by data from other solitary insects (e.g., *Drosophila* and locusts) or social Hymenoptera (best studied in *Polistes* and honeybees), but have never been tested in a solitary bee or wasp. Many 'solitary' bees and wasps often cohabit a nest, either as groups of reproductives or as mother–daughter associations, and often establish dominant–subordinate relationships with queen–worker like behavioral patterns. However, the extent to which these then suppress ovarian development and associated hormone expression, and the circumstances that lead socially castrated females to stay on the nest rather than flee, remain to be tested.

Some of the strongest empirical support linking reproductive physiology to the expression of worker behavior comes from work done by Rob Page, Gro Amdan, and colleagues on artificially selected strains of honeybees (*Apis mellifera*) in the framework of the reproductive groundplan hypothesis. The different terminology (reproductive vs. ovarian) connotes a focus not just on ovarian and endocrine regulation, but reproductive physiology and genetics more broadly. The ovarian groundplan hypothesis originated from studies of honeybee lines subject to long-term artificial selection for or against the colony-level trait of pollen hoarding. Workers from the pollen hoarding line show a suite of traits suggesting a predisposition toward reproduction, while those from lines selected not to hoard pollen lack this predisposition. It should be noted that these studies are not of bees predisposed for queen or worker caste fate, but among workers. Workers typically do not reproduce in honeybee colonies except in the absence of the queen when they may lay unfertilized male eggs. These studies demonstrated a difference in foraging preference (high-pollen strain bees tend to forage for pollen, while low-pollen strain bees forage for nectar). This preference was linked not only to behavioral and sensory traits affecting foraging, but also reproductive development and physiology. Most notably, high-pollen females emerged with larger ovaries and higher titers of JH and vitellogenin, the egg yolk precursor protein. Given that solitary insects typically forage for carbohydrates (e.g., nectar) when not engaged in reproduction and forage for protein (e.g., pollen) when developing eggs, these results are consistent with the hypothesis of social insect reproductive castes evolving from solitary ancestral behavioral regulation.

Because honeybees are model organisms amenable to lab work and with a sequenced genome, ongoing research has elucidated the genetic and endocrine signaling and regulatory mechanisms of the reproductive and behavioral differences exhibited between these two strains of workers in spectacular detail. However, because honeybees are highly derived social species and the reproductive groundplan hypothesis addresses only differences between workers, the question of the origins of reproductive division of labor remains open. Future studies using the considerable genetic and physiological insights from honeybees to test hypotheses for the regulation of reproduction and behavior in solitary and primitively social bees and wasps will be especially useful in this regard.

One such study by Toth and colleagues used data from the sequenced honeybee genome and associated studies of gene expression, caste, and physiology to test gene expression patterns in *Polistes metricus*. Genes associated with foraging and reproduction, respectively, in honeybees were similarly expressed in *P. metricus*, suggesting that similar modifications of nutritional and reproductive physiology were involved in the independent evolutions of sociality in both groups. Moreover, independent *P. metricus* foundresses (who had to provision offspring as well as reproduce) showed similar patterns of gene expression as later workers (who provisioned without reproducing), supporting the hypothesis that worker behavior derives from ancestral reproductive maternal behavior, minus the reproduction.

Conclusion

Developmental plasticity permits social insects to express either reproductive queen or nonreproductive worker phenotypes depending on their environment and nourishment. In small-colony species, worker reproduction is inhibited by social competition, including overt aggression and queen control of nutrition. In larger-colony, more derived species, a developmental switch determined by larval nutrition (itself under social control) typically determines the expression of queen and worker phenotypes.

The evolution of queen and worker phenotypes likely resulted from subjecting the ancestral solitary reproductive physiology to social control in order to decouple reproduction and associated parental behaviors. Recent genetic and physiological studies have strongly supported the hypothesis that queen and worker phenotypes did not evolve de novo, but from selectively modifying the ancestral solitary reproductive system. Future studies comparing other species to the well-studied *Polistes* and honeybees will be crucial for testing current hypotheses for the evolution of division of labor. For instance, despite frequent invocation of the 'solitary ancestor,' endocrine control of reproduction and behavior has not been measured in detail in a solitary bee or wasp species. Understanding the physiological development of caste in these species, as well as facultatively social and other primitively social species, will greatly expand our knowledge of how developmental plasticity in the expression of reproductive castes evolved.

See also: Caste Determination in Arthropods; Crustacean Social Evolution; Termites: Social Evolution.

Further Reading

Bloch G, Wheeler DE, and Robinson GE (2002) Endocrine influences on the organization of insect societies. In: Pfaff D, Arnold A, Etgen A, Fahrbach S, Moss R, and Rubin R (eds.) *Hormones, Brain, and Behavior*, vol. 3, pp. 195–235. San Diego, CA: Academic Press.

Gadagkar R (2001) *Social Biology of Ropalidia marginata – Toward Understanding the Evolution of Eusociality*. Cambridge, MA: Harvard University Press.

Hunt JH (2007) *The Evolution of Social Wasps*. New York: Oxford University Press.

Korb J and Hartfelder K (2008) Life history and development – A framework for understanding developmental plasticity in lower termites. *Biological Reviews* 83: 295–313.

Michener CD (1990) Reproduction and castes in social halictine bees. In: Engels W (ed.) *Social Insects: An Evolutionary Approach to Castes and Reproduction*, pp. 77–121. Berlin: Springer-Verlag.

O'Donnell S (1998) Reproductive caste determination in eusocial wasps (Hymenoptera: Vespidae). *Annual Review of Entomology* 43: 323–346.

Page RE and Amdam GV (2007) The making of a social insect: Developmental architectures of social design. *BioEssays* 29: 334–343.

Roisin Y (2000) Diversity and evolution of caste patterns. In: Abe T, Bignell DE, and Higashi M (eds.) *Termites: Evolution, Sociality,*

Symbioses, Ecology, pp. 95–120. Dordrecht: Kluwer Academic Publishers.

Ross KG and Matthews RW (eds.) (1991) *The Social Biology of Wasps.* Ithaca, NY: Cornell University Press.

Schwarz MP, Richards MH, and Danforth BN (2007) Changing paradigms in insect social evolution: Insights from halictine and allodapine bees. *Annual Review of Entomology* 52: 127–150.

Smith AR, Kapheim KM, O'Donnell S, and Wcislo WT (2009) Social competition but not subfertility leads to a division of labour in the facultatively social sweat bee *Megalopta genails* (Hymenoptera: Halictidae). *Animal Behaviour* 78: 1043–1050.

Toth AL, Varala K, Newman TC, et al. (2007) Wasp gene expression supports an evolutionary link between maternal behavior and eusociality. *Science* 318: 441–444.

West-Eberhard MJ (2003) *Developmental Plasticity and Evolution.* New York: Oxford University Press.

West-Eberhard MJ (1996) Wasp societies as microcosms for the study of development and evolution. In: Turillazzi S and West-Eberhard MJ (eds.) *Natural History and Evolution of Paper-wasps*, pp. 290–317. New York: Oxford University Press.

Wheeler DE (1986) Developmental and physiological determinants of caste in social Hymenoptera: Evolutionary implications. *The American Naturalist* 128: 13–34.

Dictyostelium, the Social Amoeba

J. E. Strassmann, Rice University, Houston, TX, USA

Introduction

The social amoeba *Dictyostelium discoideum* is an odd model system for behavior: it lacks a nervous system, is not an animal, and is usually single-celled. On the other hand, it is hard to imagine an organism more ideally suited to advancing our understanding of social behavior. Its social life is fascinating, and tools developed by hundreds of cell and molecular biologists over the last few decades allow a gene-based approach to understanding its sociality. Studies of *Dictyostelium* provide a crucial independent test of social evolution theories, since these theories were developed with social insects and vertebrates in mind, not social amoebae.

D. discoideum is a eukaryote that lives most of its life as independent amoebae in the forest soil, eating bacteria, and dividing about around every 4 h when food is abundant. But when they run out of food, a much more intense social stage begins (**Figure 1**). The amoebae aggregate in thousands and form a multicellular motile organism. Ultimately, the multicellular slug organizes itself into a fruiting body in which about 25% of cells die to form a rigid cellulose walled stalk while the other cells form hardy spores at the top of the stalk, where they are more likely to be dispersed. The group of spores is called the sorus. This is one-stop sociality, with a single, magnificent altruistic act by some formerly independent cells that benefits the rest. It can be compared both to a major transition to multicellularity and to the altruism of social insect workers. In some ways, the social-insect comparison is apt because, unlike most multicellular organisms, which consist of clones of cells, dicty arrives at multicellularity by aggregation. Therefore, as in social insects, we might expect both altruism favored by kin selection and conflicts between the different genotypes in an aggregation. Given the genetic tools available for dicty, the potential for understanding the mechanisms of altruism and the control of conflict in this organism is great, making it a rich field for graduate students. This piece introduces the group, points out some of the most important molecular and genomic tools, summarizes what is known of its social behavior, and suggests promising future directions.

Background

Where Is *Dictyostelium* on the Tree of Life?

D. discoideum is the best-studied member of the Dictyostelia which is in the Amoebozoa, a kingdom that is sister to the node that is made up of animals and fungi. We will henceforth call *D. discoideum* by its vernacular name, dicty. Other members of the Dictyostelia are much less studied and have not acquired common names, and so will be referred to by their scientific names. Dicty occupies a fascinating place on the Tree of Life, with many cellular processes shared with fungi and animals, including humans.

There are about 80 named species in the Dictyostelia, but it is clear that this number will increase greatly as more wild-collected clones are sequenced. The named species are divided into four main groups, with genetic divergence between them as great as that between hydra and humans. Dicty is in Group 4, the dictyostelids, according to an excellent recent phylogeny from the Schaap and Baldauf groups (**Figure 2**). This group has the hardiest, most easily cultured species.

Polysphondylium is a name given to some Dictyostelids with branched fruiting bodies that are imbedded in the *Dictyostelium* phylogeny, and so should not really have a different genus name (**Figure 2**). It can be seen that *P. violaceum* is in or close to the dictyostelids, while *P. pallidum* is in Group 2, the heterostelids. Another genus embedded in *Dictyostelium* is *Acytostelium*, a small apparently monophyletic group of species characterized by a social stage that does not require the sacrifice of any cells in the heterostelids. Instead, it forms tiny stalks entirely from cellular secretions. It will be made clear later that social variation in the Dictyostelia greatly enhances their value as a model social group.

Where Dicty Lives

Dictyostelia live in the upper layers of soil where they are predatory on bacteria, eating them by engulfment. Dicty is particularly common in autumn when leaf litter is abundant. Some species are more widespread than others, with *D. mucoroides* and *P. violaceum* among the most ubiquitous. Dicty was first described by Kenneth Raper from a site just off the Blue Ridge Parkway near Mount Mitchell, NC, USA. It is abundant in forest soils of the Appalachians above about 1000 m elevation, but it also occurs generally in the eastern United States, with collections made from Houston, TX, to northern Minnesota and Massachusetts. Other samples assigned to this species have been collected as far South as Costa Rica. It has also been found along the eastern coast of Asia, including China and Japan, but not in Europe or Africa.

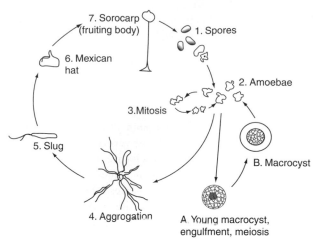

Figure 1 The life cycle of *Dictyostelium discoideum*. (1) A hardy spore which can last for years in the soil germinates, releasing a motile amoeba. (2) The haploid amoeba hunts bacteria, eats them, and grows. There are several kinds of social communication among amoebae, including quorum sensing. (3) After about 4 h of eating, the amoeba divides mitotically. This individual stage can last for months. (A) When food becomes scarce, under certain conditions, a sexual stage is initiated. Amoebae aggregate, and the first two individuals of opposite mating types fuse, forming a diploid individual. Thousands of others join the aggregate, and amazingly are consumed by the zygote, which becomes a giant cell. This giant cell undergoes meiosis and then divides many times. (B) A hardy macrocyst is formed of recombined spores. (4) Another pathway can also result from food scarcity: the asexual multicellular pathway which is initiated with aggregation. (5) A motile multicellular slug visible to the human eye is formed, and this slug migrates toward heat and light. (6) At a new location, if it has migrated, the slug reorganizes in a form called a Mexican hat. (7) The cells form a sorocarp or fruiting body in which about 20% of cells die to form a stalk which the other cells flow up and become hardy spores at the top.

Life Cycle

There are three important cycles in the life of dicty, asexual division, sexual aggregation and meiosis, and the social cycle (**Figure 1**). During the feeding stage of their life, dicty exists as independent amoebae that move through the soil by advancing pseudopods and engulfing any bacteria they encounter. The amoebae divide about every 4 h when bacteria are plentiful (**Figure 1**, steps 2 and 3). At this stage in their life, their existence is essentially solitary since they do not depend on others to eat, move, or divide. However, it is clear that communication among amoebae is maintained through small signaling proteins like CMF and PSF, which function as quorum sensing molecules and more. This communication is important because starvation may be near, and this is when either the social stage or the sexual stage begins. When an amoeba has stopped finding enough bacteria for food and senses that there are sufficient other amoebae nearby, a dramatic change transpires. When starvation occurs in dark, moist, warm conditions lacking in phosphorus, with calcium present, the sexual stage is initiated

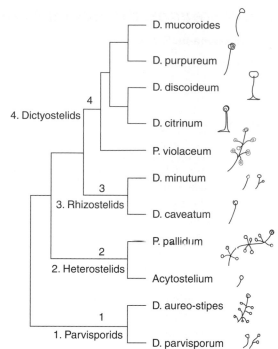

Figure 2 The Dictyostelidae is divided into four groups, each represented here by two to five species: Group 1, the parvisporids; Group 2, the heterostelids; Group 3, the rhizostelids; and Group 4, the dictyostelids (following Schaap et al., 2006). The group is as diverged as hydra to human, or all of animals, but does not show nearly that much morphological variation. The group is made up of three previously named genera, *Dictyostelium*, *Polyspondylium*, and *Acytostelium*. Only *Acytostelium* is monophyletic. Please see text for information on these species. Drawings are not to scale. The species indicated here are *D. parvisporum, D. aureo-stipes, Acytostelium leptosomum, Polysphondylium pallidum, D. caveatum, D. minutum, P. violaceum, D. citrinum, D. discoideum, D. purpureum,* and *D. mucoroides*. Schaap, P., T. Winckler, et al. (2006). "Molecular phylogeny and evolution of morphology in the social amoebas." Science 314: 661–663.

(**Figure 1**, steps A and B). Two cells of opposite mating types fuse, forming a diploid zygote. During the amoeba stage, the cells are haploid, so no change is necessary before fusion. The zygote is attractive to the thousands of nearby amoebae, which are engulfed and eaten by the zygote, which grows to an enormous size (for a dicty), forming a macrocyst (**Figure 3(d)**). The macrocyst then divides meiotically and then mitotically to form thousands of recombinant cells. Unfortunately for students of the system, laboratory conditions for hatching these recombinant cells have not been well worked out.

Under the multicellular system, there is little cell division and no recombination (**Figure 1**, steps 4–7). The starving amoebae begin to signal to each other with cAMP released to the environment. They not only release cAMP, but also move toward it. They elongate as they move, and a cAMP gradient is produced along their cells, so others move toward the end that is away from the

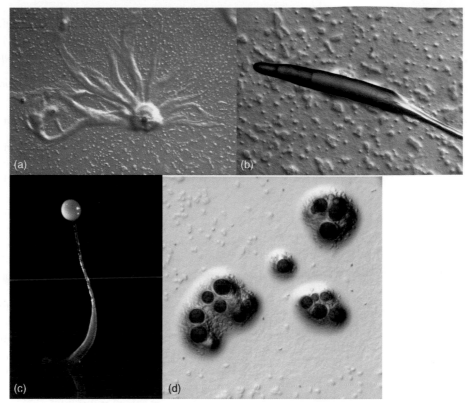

Figure 3 Multicellular stages of *Dictyostelium discoideum*. (a) Aggregation of formerly independent cells into a multicellular body. (b) Motile multicellular slug moving towards light. (c) Fruiting body consisting of a basal disc, a stalk, and a sorus, or spores. The basal disc and the stalk are formed of formerly living amoebae that have died to form this supporting structure. (d) Macrocysts, the sexual stage of *D. discoideum*. (Courtesy of Owen Gilbert).

highest concentration. As more and more starve, they concentrate in great streams of dicty cells, flowing toward a center in a process called aggregation (**Figure 3(a)**). After a few hours, this center concentrates into a mound, which then elongates slightly and begins to crawl around toward light and heat and away from ammonia (**Figure 3(b)**).

This translucent slug looks like a tiny worm, but differs from it in some important ways. As it crawls through a sheath largely made up of cellulose, it drops cells at the rear, and these cells can feed on any bacteria they discover, effectively recovering the solitary stage. The slug moves more quickly and farther than any individual amoeba could move: an important advantage to the social stage. Though the slug lacks a nervous system, there are differences among the constituent cells. Those at the front direct movement and ultimately become the stalk. There is a recently discovered class of cells called sentinel cells that sweep through the slug from front to back picking up toxins and bacteria, functioning simultaneously as liver, kidney, and innate immune system, before they are shed at the rear of the slug.

Slugs move farther and for a longer time when the environment lacks electrolytes, when it is very moist, and when there is either directional light or no light. When they cease moving, the cells of the slug concentrate into a tight form known as a Mexican hat. Then,

in a process called culmination, the cells that were at the front of the slug begin to form cellulose walls and to rise up out of the mass as a very slender but rigid stalk (**Figure 3(c)**). These cells die. The remaining three-quarters or so of the cells flow up this stalk, and at the top they form hardy spores. At this point, the spores, stalk, and basal disk comprise an erect structure called a fruiting body (**Figure 3(c)**).

Thus, some of the cells sacrifice their lives so that the others may rise up and sporulate a millimeter or so above the soil surface, or into a gap between soil particles. Others sacrifice themselves as sentinel cells picking up toxins and bacteria as they made their way through the slug. Still others were shed from the rear of the slug during their normal movement. If these do not encounter bacteria, or enough other shed cells to form a new, smaller fruiting body, then they also perish.

How Dictyostelids Are Obtained, Collected, and Cultured

Many studies can be performed using previously collected clones obtained from the stock center for the price of postage. This stock center is accessed through www.dictybase.org and preserves thousands of clones. Most of these are genetically modified versions of the type clone, NC4. Early modifications allowed for growth in a bacteria-free shaking medium; these axenically grown clones are referred to as Ax4 and related names. There are many clones that have one gene knocked out that are of interest to students of social genes. The stock center also has hundreds of unmodified clones collected from the wild. Many are dicty, but there are also quite a few other species represented in this stock center. There the clones are preserved in liquid nitrogen tanks, and shipped out on request to researchers.

It is a lot of fun to culture your own dicty or other Dictyostelids from the wild. This process involves placing a few drops of a dilute soil sample on a weakly nutritive agar plate that has been previously inoculated with a bacterial strain to provide food for the Dictyostelids. The isolation process is basically a race to visualize and isolate Dictyostelids from competing fungi, bacteria, and other living organisms. Detailed instructions for collecting soil samples and culturing and isolating Dictyostelids are at the www.dictybase.org and www.ruf.rice.edu/~evolve.

What Can Sociobiology Tell Us About Dicty?

What Are the Benefits to Grouping in Dicty?

The social stage in the dicty life cycle involves the clear cost of death for about a quarter of all cells, and so there should be a compensating advantage. This advantage cannot accrue to the dying cells, but there could be a kin-selection benefit to genetically identical clonemates that joined the same aggregation. We first discuss the advantages, then the disadvantages, to grouping with nonclonemates, and then the genetic relatedness within cooperating groups.

An early stage in aggregation is the slug, which can move tens of centimeters, through a protective cellulose sheath. This movement may bring the constituent cells to a new location where bacteria are more plentiful. Cells that are shed during movement can themselves take advantage of any food sources that are encountered. Clearly, movement is facilitated by the social stage compared to the movement of individual amoebae, and it occurs in the relative protection of the cellulose sheath. Slugs made up of larger numbers of amoebae move farther than those with fewer amoebae. Once the slug has finished moving, it forms a stalk of dead cells that the living cells migrate up. This stalk provides a benefit in lifting the spores above the substrate where they can sporulate and where dispersal is facilitated. Larger groups both make longer stalks and invest a slightly smaller proportion of individuals in the stalk. Nearly all species

except dicty form a stalk from the beginning of migration (instead of a free slug), which facilitates gap crossing in the three-dimensional soil matrix, but it means that cells die and are lost from the migrating group. This places a cost and a limit on the distance traveled.

What Are the Costs of Grouping with Nonrelatives in Dicty?

The advantages to grouping in dicty may not accrue equally to all genotypes if multiple genotypes are represented in a single fruiting body. In particular, clones that succeed in avoiding contributing to the dead stalk cells will be more represented in the next generation. Some clones may be able to avoid stalk contribution when paired with others. When two clones are mixed, one often predominates among the spores while avoiding contribution to the stalk cells. In a round robin tournament, where every clone is paired against the others, there is a dominance hierarchy in which some clones consistently dominate in spore contribution. This is interesting and puzzling, for if they are consistently dominant, we would not expect the losing forms to persist in nature, particularly in the same habitat. This puzzle can be solved if different clones dominate under different conditions, if there are tradeoffs in dominance, or if the environment is changeable enough that the system is not at equilibrium. This result that clones compete in fruiting bodies and do not pay the costs of stalk formation equally is interesting and important and sets the stage for future investigations.

If there is conflict within an aggregation regarding which becomes spore and which becomes stalk, we expect that it may also be expressed earlier, as the slug migrates. Since the front of the slug is the organizing center that directs movement and later becomes stalk, cells in a chimera of two or more clones may be less willing to join this altruistic region, and this hesitancy may slow slug movement. This is exactly what happened. For a given number of cells, chimeras moved less far than pure clones (**Figure 4**).

What Is Relatedness Within a Dicty Fruiting Body?

One of the challenges of working on a microorganism is that they are hard to see. Even though fruiting bodies of dicty measure 1–4 mm and so are visible without magnification, they are hard to find in forests. Naturally occurring fruiting bodies on deer feces were first seen near the main building at Mountain Lake Biological Station on 15 October 2000. However wild-collected fruiting bodies were not successfully genotyped until a few years later, and those on dung exhibited high genetic relatedness of

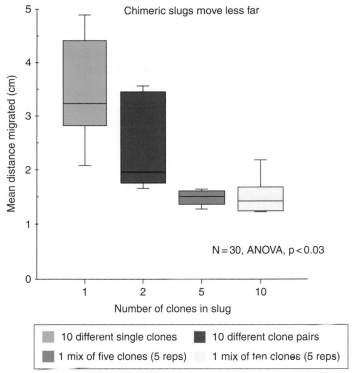

Figure 4   When equal numbers of cells are mixed, those made up of multiple clones produce slugs that travel less far toward light (Foster et al., 2002).

close to 0.90 within fruiting bodies. This is estimated from spores, not stalk, since it has not yet been possible to genotype dead stalks. Within 0.2 g soil samples, there can be as many as six genetically distinct clones represented, but we do not know at what frequency they form chimeric fruiting bodies. This is an area that could use more work.

Does Dicty Recognize Kin?

If dicty can recognize and exclude nonkin, then exploitation by nonrelatives will not be a big problem. Most experiments that take clones from a single location indicate that there is considerable mixing with nonkin. However, a study that investigated chimera formation of clones from a wide geographic range found that genetically distant nonkin clones mix less freely. More work in this area is needed.

Clearly, dicty has a social structure that is amenable to further study. It has a solitary and a social stage. In the social stage, it is clear who is benefiting and who is paying costs. Genetic diversity occurs at a scale where interactions are likely. Chimeric groups show costs compared to groups of pure clones, as would be expected with social conflict. The standard variables of sociobiology, costs and benefits, relatedness, and recognition, are all important in dicty and can be both manipulated and measured. An exciting frontier involves experimental evolution,

since dicty goes through its social stage in only a few days. Much can be learned from dicty. In the following section, we discuss how the availability of genetic approaches makes the system even more attractive.

What Can Dicty Tell Us About Sociobiology?

Does Cheating Have a Genetic Basis?

One of the advantages to a microbial system is that genes can be knocked out and the impact of their lack can be evaluated. In dicty, one way this is achieved is by a process known as REMI, restriction enzyme mediated integration. It is used to insert a labeled cassette conferring antibiotic resistance into the DNA at sites cut by the cointroduced restriction enzyme. Dosages are tweaked so each cell receives either no insertions or a single insertion. Then those lacking insertions are killed with an antibiotic. The pool of mutants can be selected.

One of the most interesting selections for students of social behavior involved favoring knockout mutants that increased the knockout's ability to become spore and not die in stalk. The process to attain these knockout mutants involved beginning with a pool of knockout mutants and allowing them to form fruiting bodies repeatedly, with each round beginning with spores from the previous round. Thus, knockout mutants that preferentially attain spore status will be overrepresented. The hundred or so genes identified using this process are a rich source of future study subjects. The molecular pathways involved in cheating are diverse and worthy of further study. Nevertheless, we know something about how some specific genes influence social competitiveness.

How Is Cheating Controlled in Nature?

Cheating in natural populations of dicty presents a number of problems. If cheating is common, then social cooperation itself can be threatened. This is so because a clone would not benefit from sociality if it became stalk while another clone became the fertile spores. If a gene conferred an advantage to its bearer in all environments, then it might increase in frequency until it was fixed in the population. Cheating can be controlled if the amoebae that aggregate together are highly related because then the benefits of cooperation would go to relatives and cheaters would cheat other cheaters.

A mutant that is a cheater but confers a cost on its bearer in terms of fruiting body success will spread only when it is the rarer partner in a fruiting body. A study of the gene *fbxA−*, also known as *chtA−*, showed that the knockout mutant was consistently a cheater, becoming overrepresented among the spores compared to the frequency in the original mixture. However, another impact is that chimeric mixtures produced fewer spores, and *fbxA−* by itself produced defective fruiting bodies with essentially no spores. The cost of this cheater mutant means that it can only thrive at low frequencies with respect to this locus, when it is in a minority in the fruiting body and can exploit other genotypes. In fact, the point where the advantage of cheating crosses the loss of spore production is at a frequency of only 0.25, much lower than that found in wild fruiting bodies. Thus, it is no surprise that in a search of morphologically defective mutants among wild-collected spores none were found in a sample of 3316 spores.

Pleiotropy is another way that cheating may be controlled. If a gene that favors a fair balance between spore and stalk also has some other essential function, then it could not easily be defeated. This is so because it would also lose the essential function. Such a gene is *dimA*. When this gene is knocked out, the bearer cannot respond to differentiation inducing factor, DIF, a hormone that is normally produced by the spores that will become spore and induces other cells to become stalk. The *dimA−* cells do not respond to DIF and in the slug stage appear to cheat by contributing less to the prestalk region. But by the time the fruiting body is formed, they are actually underrepresented among the mature spores, because wild-type cells have actively transdifferentiated from prestalk to prespore. The loss of an essential function, whose exact nature is still unclear, means that *dimA−* cells cannot

lose cooperation and become cheaters, without losing more fundamentally in other ways.

What Is the Evidence for a Green Beard Gene in Dicty?

Hamilton realized that if a single gene encoded (1) a recognizable signal, (2) recognition of the signal in others, and (3) altruistic behavior toward bearers of the gene, then altruism could evolve with respect to this gene, no matter what its implications were for the rest of the bearer's genome. Dawkins quickly picked up on this and called the trait a greenbeard gene, where the recognizable trait is a green beard, but he considered genes with such complex effects improbable. Haig wisely surmised that if there are greenbeard traits, a possible candidate would be an adhesion gene, since in this case the multiple functions might be unseparable. It seems that the dicty gene *csaA* functions in this way. It is a homophilic adhesion gene. When this gene was first successfully knocked out, the knockout appeared to function as well as the wild-type. But then the investigators realized that in a chimera with its parent, on agar, it was a cheater, contributing more than its fair share to the spores. This was so because the reduced adhesion caused it to slide to the back of the slug where prespore cells are found. But this reduction in adhesion had another effect. On agar, the *csaA* knockouts suffered no deficits in aggregation, but on soil, their reduced adhesion meant that they often failed to make it into the fruiting body. On soil, chimeric mixtures produced fewer knockout mutant spores. Thus, *csaA* is a greenbeard gene. The recognition and the action are the homophilic binding. The binding likewise ensures that the altruism is directed preferentially toward those that share the gene.

One might wonder whether variation in the *csaA* gene contributes to present day recognition among clones. Apparently it does not. There is little variation in the gene as seen in present populations. This may be something else expected from a greenbeard gene. It has become fixed in a form such that everyone has the recognized trait, the ability to recognize, and the altruistic behavior, so discrimination, stable or unstable, based on this locus, is no longer possible.

Clearly, this is only the beginning of a very interesting period of research as genes for social traits in dicty are discovered and characterized, leading to new insights into social behavior and evolution.

How Does Social Behavior Vary Across the Dictyostelia?

In this article, we have focused on dicty because it is by far the best-studied species, but there are other interesting species awaiting further work. The Dictyostelia are an ancient group with as much molecular diversity as is found in all animals and a divergence time of around 900 My (**Figure 2**). Compared to animals, Dictyostelia vary little in form; whether this is because of the greater levels of conflict in a social organism with physical cohesiveness like a multicellular organism but lacking a single cell bottleneck is an interesting question. All Dictyostelids have a group of hardy spores on top of a dead stalk. They differ in whether an aggregation center forms one or many fruiting bodies, in the number of spore groups there are, and in exactly where they are on the stalk. Species assigned to the polyphyletic genus *Polysphondylium* have tree-like fruiting bodies with both side and terminal balls of spores (**Figure 2**). Some species like *Dictyostelium polycephalum* have a group of spore balls at the end of a stalk. *Dictyostelium rosarium* has beads of spore balls running up a curved stalk.

Form does not differ only in the final social stage; there are differences along the way. Some species do not form a migratory slug, but culminate on the spot. Of those that do form a motile slug, most begin to form the stalk immediately, moving ahead at the end of the dead stalk cells. Dicty is one of only three species with cells that are not terminally differentiated before fruiting.

There is variation in the chemoattractant that first causes amoebae to aggregate. In dicty and its close relatives in Group 4, the dictyostelids, the chemoattractant is cAMP. In *Polysphondylium*, the chemoattractant is a dipeptide called glorin. In other species, it is folate.

Dicty and its relatives in the dictyostelids have lost the ability to form spores except for during the social process, but this is not true for members of the other three groups where many of the species have been found to form spores that are not as hardy as those from the social stage. These are called microcysts. In some ways, these species may be interesting to study, for the members have a nonsocial option for hard times. Does this solitary option make the social contract regarding fair contributions to stalk more enforceable?

A tantalizing glimpse of what else might be discovered in this novel social system comes from *D. caveatum*. A single clone was isolated by Kenneth Raper from a slurry of bat guano from Blanchard Cave, Arkansas. This clone is a predator on other *Dictyostelium* from all four groups. It aggregates right along with the others, and then delays progression through the multicellular stages so it can munch on the others. A 1% initial frequency of *D. caveatum* in a blend can result in nearly all *D. caveatum* spores. No doubt other social exploiters of novel ways lurk in the bacteria-rich corners of the planet.

There are collections of these other species in the stock center. Indeed, wild culturing techniques most often yield *D. giganteum*, various *D. mucoroides*, and *D. violaceum* all among the hardy Group 4 dictyostelids. There is a sequenced genome for *D. purpureum*, and genome sequences are on the way for five to ten additional species, including members of the dictyostelids, the rhizostelids, the heterostelids, and the parvisporids.

See also: Kin Recognition and Genetics; Kin Selection and Relatedness.

Further Reading

Bonner JT (1967) *The Cellular Slime Molds*. Princeton NJ: Princeton University Press.

Chen G, Zhuchenko O, and Kuspa A (2007) "Immune-like phagocyte activity in the social amoeba." *Science* 317: 678–681.

Crespi BJ (2001) "The evolution of social behavior in microorganisms." *Trends in Ecology and Evolution* 16: 178–183.

Ennis HL, Dao DN, Pukatzki SU, and Kessin RH (2000) "*Dictyostelium* amoebae lacking an F-box protein form spores rather than stalk in chimeras with wild type." *Proc. Natl. Acad. Sci. USA* 97: 3292–3297.

Fortunato Angelo, Queller David C, and Strassmann Joan E (2003) "A linear dominance hierarchy among clones in chimeras of the social amoeba, *Dictyostelium discoideum*." *Journal of Evolutionary Biology* 16: 438–445.

Foster KR, Fortunato A, Strassmann JE, and Queller DC (2002) The costs and benefits of being a chimera. *Proceedings of the Royal Society of London, Series B* 269: 2357–2362.

Gilbert OM, Foster KR, Mehdiabadi NJ, Strassmann JE, and Queller DC (2007) "High relatedness maintains multicellular cooperation in a social amoeba by controlling cheater mutants." *Proceedings of the National Academy of Sciences USA* 104: 8913–8917.

Kessin RH (2001) *Dictyostelium: evolution, cell biology, and the development of multicellularity*. Cambridge UK: Cambridge University Press.

Mehdiabadi NJ, Talley-Farnum T, Jack C, Platt TG, Shaulsky G, Queller DC, and Strassmann JE (2006) "Kin preference in a social microorganism." *Nature* 442: 881–882.

Raper KB (1984) *The Dictyostelids*. Princeton NJ: Princeton University Press.

Santorelli LA, Thompson CRL, Villegas E, Svetz J, Dinh C, Parikh A, Sucgang R, Kuspa A, Strassmann JE, Queller DC, and Shaulsky G (2008) "Facultative cheater mutants reveal the genetic complexity of cooperation in social amoebae." *Nature* 451: 1107–1110.

Schaap P, Winckler T, Nelson M, et al. (2006) Molecular phylogeny and evolution of morphology in the social amoebas. *Science* 314: 661–663.

Shaulsky G and Kessin R (2007) "The cold war of the social amoebae." *Current Biology* 17: R684–R692.

Strassmann JE, Zhu Y, and Queller DC (2000) "Altruism and social cheating in the social amoeba, *Dictyostelium discoideum*." *Nature* 408: 965–967.

Strassmann JE and Queller DC (2007) "Altruism among amoebas." *Natural History* 116: 24–29.

West SA, Griffin AS, Gardner A, and Diggle SP (2006) "Social evolution theory for microorganisms." *Nature Reviews Microbiology* 4.

Differential Allocation

N. T. Burley, University of California, Irvine, CA, USA

Overview

Introduction and Definitions

The Differential Allocation Hypothesis states that, among iteroparous, sexually reproducing species, natural selection favors individuals that allocate costly reproductive resources in direct proportion to the relative mating attractiveness of their sexual partners. 'Mating attractiveness' refers to the extent to which alternative phenotypes are preferred by members of one sex in a population. Attractiveness reflects relative mating quality, such that by securing attractive mates, individuals obtain direct and/or indirect fitness benefits. Direct benefits are those that impact the number of offspring produced. Indirect benefits are those that impact offspring fitness; principal among these are additive genetic benefits that enhance the viability, fecundity, and/or mating attractiveness of offspring ('offspring quality'). The basis for the expectation that an individual will commit less effort to a current reproductive attempt when mated to an unattractive partner than when mated to an attractive one is that the return on reproductive effort devoted to the offspring of an unattractive partner is lower. This expectation is contingent on the assumptions that future mating opportunities are likely and that they may involve sexual partners with different levels of attractiveness.

Differential allocation involves per capita adjustment of parental investment to individual offspring, adjustment in the amount of focused mating investment (effort spent to acquire a particular mate), and/or adjustments that influence the number of offspring produced in a reproductive bout; in any case, the future reproductive capacity of an individual practicing differential allocation will vary inversely with its current reproductive effort. When per capita parental investment is varied, differential allocation may constitute an adaptive parental effect.

While usually studied in the context of indirect fitness benefits, differential allocation can also influence direct benefits. A female might allocate more eggs than average to a male whose phenotype indicates high fertilization capacity (such that a smaller-than-average fraction of her eggs remains unfertilized), or she might lay larger eggs that produce larger hatchlings with higher survivorship. However, variation in egg size or egg number does not necessarily reflect differential allocation. In a species with biparental care, a female might provide more eggs to a male because she judges him to be a superior provider of parental care to young in ways that would lower her total cost of rearing young to independence; in this case, her allocation of eggs would not constitute differential allocation.

Generality and Significance

Depending upon the mating system, individuals of one or both sexes may benefit from optimization of their reproductive contributions to offspring of a given mate, such that differential allocation is a routine component of their reproductive strategy. Where both sexes make mate choice decisions, the relative attractiveness of mating partners should be central to allocation decisions. Thus, individuals that experience mate-getting difficulties due to low attractiveness may increase parental investment to obtain or maintain mates.

First suggested as a reproductive tactic applicable to species with biparental care in 1986, differential allocation began to receive widespread interest by investigators a decade later. Evidence now indicates that differential allocation occurs broadly among animal taxa, including those with uniparental care and those that lack postzygotic investment in offspring. Theoreticians have recently begun to develop quantitative models that explore the range of life historical and ecological conditions that favor this reproductive tactic. Implications of differential allocation for the evolution of sexually selected traits, mating system evolution, and sex allocation have been addressed at varying levels.

Historical Perspective

Origin of Hypothesis

Nancy Burley's investigations of mate choice in socially monogamous birds that display biparental care, which began in the 1970s, led her to propose the Differential Allocation Hypothesis. Burley sought to identify major mechanisms by which sexual selection might operate in such species. She had found for two socially monogamous species that both sexes participate in mate choice and, as a result, relative mating attractiveness influenced mate-getting ability of both sexes. Under ideal conditions, mate choice by both sexes should generate population-wide patterns of positive assortment for mating attractiveness. This pattern results from the greater access to the most attractive individuals of each sex to each other, leading individuals to pair with others whose relative attractiveness is similar to their own. However, because of various constraints, such as that organisms have finite time and other resources to devote to searching for mates, individuals might often need to settle for a mate of lower mating attractiveness than their own. Thus, Burley wondered how selection would favor the reproductive cooperation required to rear one or more broods when partners were not closely matched for mating attractiveness, since quality mismatches between mates should increase sexual conflict.

During the time Burley was at work on this problem, evolutionary biologists emphasized the possibility that disparity in parental effort between the sexes resulted from one sex having lost a major contest in the evolutionary 'battle of the sexes,' due to the tendency of the other sex to benefit from mate desertion or deceit. To understand how parental workloads might reflect the outcome of evolved tactics involving negotiation between the sexes, and how a high workload might benefit a care giver, Burley focused attention on the case in which failure to provide care by one parent is not an option because a single parent cannot successfully rear offspring. Under such circumstances, she reasoned, individuals might behave as if bargaining to achieve a favorable workload, and partners able to agree on a 'fair' division of labor would tend to reproduce successfully together and outperform those that failed to agree. Thus, an individual with lower mating attractiveness might be able to sustain a cooperative reproductive relationship with a mate of higher mating attractiveness by undertaking a greater-than-average share of parental investment typical for that sex. The benefit of this arrangement to the less attractive individual would be to increase offspring survivorship and/or quality, while the benefit to the more attractive individual would be to increase its lifetime fitness by reducing its current reproductive effort.

First Experimental Test

Testing this hypothesis was a challenge because many factors can influence parental care (e.g., an individual's caretaking abilities and prior breeding experience, as well as age and residual reproductive value). Moreover, an individual's parental ability may influence its mate-getting ability, thus complicating interpretation of observed patterns. The discovery that the color of plastic leg bands (regularly used to permit individual recognition of birds) influenced mating attractiveness of both sexes of zebra finches (*Taeniopygia guttata castanotis*) made it possible to control confounding variables, and Burley proceeded to test the Differential Allocation Hypothesis using captive breeding populations in which individuals of one sex were randomly banded with colors that were either attractive, unattractive, or of neutral attractiveness to members of the opposite sex. She predicted that parental expenditure of individuals of the noncolor-banded sex

would vary in direct proportion to the color-band attractiveness of their partners, and that expenditures by the color-banded sex would vary inversely with their own band attractiveness. (Previous research had validated key assumptions underlying this experimental design, namely that zebra finches respond to color-banded conspecifics as if band attractiveness were a heritable aspect of an individual's phenotype, and one that impacts offspring quality.) Parental expenditure was measured as time spent in parental care activities during observation sessions throughout the nesting phase.

Results of two experiments (males color-banded in one, females in the other) were consistent with predictions of the Differential Allocation Hypothesis. The amount of care provided by color-banded parents varied inversely with the attractiveness of their band color, and the parental expenditures of the noncolor-banded sex varied directly with their mate's band attractiveness. The experiment in which males were color-banded ran for 2 years. During this time, the parental expenditure of attractively banded males decreased, while that of unattractive males increased, suggesting that parental roles were subject to ongoing negotiation even among well-established pairs. The relative cost of parental expenditure was indexed by mortality rate. Unattractively banded birds had higher mortality rates than attractive ones of the same sex. This finding supports the assumption that a high parental workload imposes a long-term reproductive cost, such that 'parental expenditure' reflects 'parental investment.'

Extra-pair activities were not investigated in these experiments, but subsequent research showed that unattractively color-banded male zebra finches were at greater risk of losing paternity through their social mate's extra-pair copulations than were attractive males. Since low paternity confidence does not favor high parental investment by males (nor does high confidence favor low investment), the results strongly supported the conclusion that both sexes of zebra finches evaluate relative mating attractiveness and adjust their willingness to incur parental investment as predicted by the Differential Allocation Hypothesis.

Extensions and Experimental Approaches
Applicability to Other Mating Systems

In the following decades, researchers studying a wide range of taxa (including arthropods, amphibians, fish, mammals, and birds with precocial young as well as those – like zebra finches with highly altricial young) have reported that females engage in differential allocation; male response has been largely unstudied. Notably, many of these investigations have been performed on organisms that do not show biparental care, indicating that the hypothesis is applicable to a wide range of conditions. Unfortunately, for taxa in which most parental investment occurs at or before egg deposition (such as many arthropods), researchers often refer to the variable allocation of reproductive resources depending on mate quality as postcopulatory or 'cryptic' mate choice, without differentiating differential allocation from other phenomena (or mate choice from parental investment); thus, less is known about differential allocation in these species.

Major Experimental Designs

Two principal experimental approaches have been used to study differential allocation. One involves manipulation of male attractiveness and has mainly been undertaken using birds, for which it is often possible to perform phenotype manipulations during the interval between initial mate choice and onset of reproduction. This approach has the advantage of dissociating male phenotype from genotype, such that the interpretation of possible treatment differences in offspring performance variables is not confounded by paternal qualities normally linked to attractiveness. (For example, attractive males might provide material benefits to females that enhance increase female fecundity.) Where females are found to increase investment in offspring of males whose ornamental traits (traits that function in mate attraction) have been experimentally enhanced, alternative hypotheses become less plausible.

Manipulation of male ornamental traits has been used to study differential allocation in barn swallows (*Hirundo rustica*), a species in which females show mate preferences for males with long tails. When the tail length of recently mated males was experimentally manipulated by shortening or extending feathers, females responded by varying the rate at which they fed offspring; they provided more food to offspring of males with long tails. Since the foraging ability of males with elongated tails was impaired by the manipulation, this outcome might have been explained by female compensation for male inability to provide parental care. This interpretation, however, does not account for the findings that females mated to attractive, long-tailed males laid more clutches and that they reared a greater number of offspring than those mated to males with shorter tails. Thus, results of this experiment support the prediction that females practice differential allocation in response to the relative attractiveness of their mates' tails.

The second experimental approach, which has been undertaken on a wider taxonomic range, involves assignment of mates of naturally varying levels of attractiveness to females and observing reproductive consequences. A weakness of this approach is that assigned mates may be less acceptable to females than those which they have chosen; if so, females may fail to reproduce or reproduce less successfully than they would have done with a chosen partner, which undermines the methodological goal of achieving random mating through mate assignment. Also, in mating systems in which females typically experience active choice, an unnatural lack of choice (or very small range of choices) might influence female perception of future reproductive opportunities, causing those with current mates deemed 'minimally acceptable' to increase current reproduction in light of uncertain future opportunities. Thus, negative evidence for differential allocation using this approach should be interpreted cautiously.

Variables Investigated

Investigators have studied a range of response variables, including parental care, egg size, maternal contributions of specific substances to eggs, clutch size, offspring growth patterns, and overall reproductive success of individuals mated to partners of variable attractiveness. Where response involves only one variable, interpretation may be straightforward. Studies on birds with precocial offspring and on fish have shown that females produce larger eggs when mated to attractive males. Larger eggs typically contain more protein and lipids, which influence hatching success and hatchling size. The number and size of eggs commonly varies inversely across clutches, however; where this occurs, additional information is necessary to interpret whether female reproductive effort has been altered.

Maternal egg allocations of substances such as hormones, antioxidants (including carotenoids), and immune factors (including antibodies) may also vary. Recent studies indicate these allocations may represent maternal investment that influences offspring quality. For example, high egg androgen titers are associated with enhanced early development and growth of chicks in some species. Also, androgen deposition appears costly to mothers, suggesting that provisioning of high hormone levels constitutes parental investment.

To investigate whether variable maternal provisioning of androgens reflects differential allocation, those who study zebra finches and barn swallows have adopted the phenotype manipulation techniques discussed earlier. Diego Gil and colleagues pioneered the investigation of tactical maternal yolk androgen allocations using color-banded zebra finches. They found that females deposited greater amounts of androgens in eggs when mated to attractively color-banded males than when paired with unattractive males. A similar result was obtained in a field experiment on recently pair barn swallows, which involved manipulation of male tail length. In this study, the yolk androgen level of eggs was positively correlated with paternal tail length following the manipulation. Also, maternal androgen allocation was determined to be independent of egg sex and shown to correlate positively with offspring growth rate. A similar experiment on another population of barn swallows, however, found no effect of mate's tail length on a female's yolk androgen application.

Establishing Costs of Allocation

Studies reveal that differential maternal provisioning is practiced in a range of ways, but few beyond Burley's original study have investigated whether the observed allocations are costly to the individuals providing them. One exception is the study by Heinz-Ulrich Reyer and colleagues on water frogs (genus *Rana*) that form a European species complex; the complex includes two species (referred to as 'LL' and 'RR'), as well as the naturally occurring hybrid between them ('LR'). Tadpoles produced from matings between hybrid LR males and both LL and LR females typically die before they can metamorphose into adult frogs, and females of both genotypes prefer LL males. However, females usually have little choice of mating partner, because males have the ability to tightly clasp ('amplex') a female they encounter and hold on to her until she produces egg masses, which are then fertilized by the clasping male. Researchers compared clutch sizes produced by female frogs that were randomly assigned to be mated by males of genotype LL or LR. Both LL and LR females released more eggs (adjusted for female body size) when mated by LL males. In addition, those females that laid smaller clutches achieved higher postmating condition, via resorption of unspawned eggs, and were consequently able to produce larger clutch masses the following season. This study demonstrated the cost of reproduction underlying the rationale for strategic allocation of reproductive resources and indicated that females practice differential allocation even in a mating system that involves sexual coercion by noninvesting males.

Evolutionary Implications

Differential allocation has additional implications for sexual selection and mating system evolution, as well as sex allocation practices.

Ornament Evolution

There is growing recognition that differential allocation contributes to the evolution of ornamental traits. In the typical case involving female choice for ornamented males, females display differential allocation by increasing per capita parental investment in offspring of highly ornamented males (and by decreasing investment in offspring of males with poor ornaments). This response can amplify paternal genetic effects (alleles underlying the ornamental trait) on the attractiveness of adult sons, because the expression of ornamental traits is often condition-dependent. For example, imagine a species of bird in which females are brown and males have colorful plumage that includes red tails. Males vary in the brightness of their tails, and females prefer to mate with those that have the most intense red color. The intensity of an individual male's tail color increases with the amount of maternal care he receives as a nestling (his 'environment'). Tail color is also influenced by a male's genotype, such that males that have inherited favorable alleles from their father achieve redder tails than sons of males with different alleles. (Tail color may also be influenced by interaction effects reflecting the combination of particular alleles and environmental factors.) Thus, through differential allocation, males with alleles that confer the most intense red tails also receive the greatest care, and they develop the reddest tails. These males tend to have high reproductive success, which causes the alleles for intense red tail coloration to increase in frequency in the population. (Daughters may also benefit from having this parental combination.) Over evolutionary time, this enhanced response to selection for red tails may contribute to further evolution of tail color if, for example, a mutation creates a new allele ('vibrant') that increases tail redness even further. When males with vibrant tails first occur, they will be rare, and females will be willing to incur great parental effort to rear their offspring. Their maternal behavior will propel the spread of the vibrant allele faster than through female choice alone.

Mating System Evolution

Because an individual's mating attractiveness varies with the availability of potential mates and superior competitors, differential allocation may also impact mating system dynamics. When one sex is in short supply, individuals of that sex have enhanced mate-getting ability and, in species with biparental care, may be able to negotiate decreased parental workloads. For example, in an experiment using zebra finches and in which adult sex ratio was varied among breeding populations, the average parental expenditure of males varied directly with the proportion of adult males in the breeding population, as predicted by the Differential Allocation Hypothesis. To the extent that the population sex ratio is influenced by the relative parental investment of the two sexes, the occurrence of differential allocation will therefore tend to exert balancing selection on the adult sex ratio, as high investment by the common sex will cause moderation of the sex ratio. However, where factors other than parental investment influence population or operational sex ratios, one sex may remain less common; if so, individuals of that sex should be able to extract greater parental investment from their mates on a continual basis. This will lead to adjustments in the reproductive roles of the two sexes, which may in turn impact the evolutionary trajectory of the mating system. One study suggests that this dynamic has contributed, for example, to the tendency of male birds to provide relatively high levels of care, which in turn facilitated evolution of altricial young.

Sex Allocation

The Trivers–Willard Hypothesis is one of the most influential hypotheses in sex allocation theory. It states that, for species in which variance in reproductive success differs between the sexes, individuals of the high-investing sex (usually females) profit by varying offspring sex in relation to their own condition at the time of conception. Thus, females in good condition should tend to produce offspring of the sex with higher variance in reproductive success, because they are more likely to produce successful offspring of that sex than are mothers in poor condition.

Numerous studies on birds and mammals have provided inconsistent support for the Trivers–Willard Hypothesis. Knut Røed and colleagues perceived that the Differential Allocation Hypothesis might better predict sex allocation patterns in species in which male body size is important in intrasexual combat, as occurs in many polygynous mammals. Such species tend to show sexual size dimorphism, with sons receiving more maternal investment than daughters. When male body size is heritable, the authors reasoned, females could achieve indirect fitness benefits by allocating greater maternal investment and producing sons when mated to large males. They tested this hypothesis using reindeer (*Rangifer tarandus*) from experimental herds in which they systematically varied the body size (mass) of males present during the short mating season. Results were consistent with the occurrence of differential allocation and did not support the Trivers–Willard Hypothesis: average size of males in herds significantly predicted progeny sex ratio, and female condition correlated with calf mass at birth but not with offspring sex. Notably, mass of neonates was not influenced by paternal mass, and offspring sex was not predicted by paternal dominance status in the experimental herds. These last findings suggest that the observed patterns were not due to coercive tactics employed by males. Also, outside the brief mating season, females were housed with males of a wide range of sizes, so females experienced an appropriate context in which to evaluate the relative attractiveness of their mating partners.

Alternative Hypotheses and Future Directions

Numerous studies involving a broad taxonomic distribution support the hypothesis that female animals allocate a range of reproductive resources in direct proportion to the sexual attractiveness of their mates. Relatively few studies, however, have considered alternative hypotheses for the causes of this allocation or have investigated whether allocations are actually adaptive. One alternative possibility is that females are sometimes deceived or coerced into making high parental allocations by males. If this were true, a female's high parental expenditure in a coercive male's offspring should reduce, rather than increase, her fitness. To demonstrate that differential allocation benefits its practitioners will requires multigenerational experiments that explore the possibility that effects on offspring fitness result from material contributions provided by males, including contributions transmitted during mating. One taxon in which both negative and positive effects of seminal contents on female reproduction have been reported is the insects. This is a promising group for further study, because male seminal fluid/spermatophore contributions are often large compared to the body size of breeding females.

The investigation of material contributions to eggs or zygotes by both parents is an area in which much work is needed. It is important to establish the consequences of such contributions to the parents providing resources (does current contribution impact future reproduction?) and offspring. Thus, for example, variable deposition of yolk androgens might have evolved in birds as a maternal tactic to manipulate offspring begging rate and thereby influence paternal care levels. If studies were to obtain results consistent with this interpretation, additional research would be needed to determine which (if either) parental contributions – male care or female egg androgens – represent differential allocation.

Reproductive Compensation

Not all studies investigating differential allocation have found supportive evidence; more interestingly, a number of studies have reported results opposite to those predicted. Patty Gowaty developed the Reproductive Compensation Hypothesis, which states that when individuals are constrained to mate with nonpreferred partners, they may adaptively increase their per capita investment to compensate for fitness deficits their offspring would otherwise experience as the result of having an inferior parent. Gowaty hypothesized that this response is likely in mating systems typified by sexual coercion and, more generally, whenever an individual's choice of mates is substantially constrained by social and ecological circumstances. Thus, although both hypotheses assume that individuals are limited in their choice of mates, the Reproductive Compensation Hypothesis and the Differential Allocation Hypotheses make opposite predictions regarding the relationship between an individual's parental investment and its current mate's attractiveness. These hypotheses are not mutually exclusive, however. An individual might practice compensation when its current mate is of very low quality and its expectation of future reproductive opportunities involving better mates is very low;

yet, if social circumstances were altered, it might practice differential allocation when it expects that favorable future opportunities are more likely. Studies that investigate this possibility would be very useful.

To investigate the circumstances under which females might experience fitness benefits from increasing versus decreasing maternal investment in response the attractiveness of their mates, Edwin Harris and Tobias Uller developed a theoretical model using a state-based approach. They concluded that reproductive compensation is most likely to benefit females when the quality of available mates is low and foregoing reproduction is not an option. The inability to forego reproduction would likely occur due to sexual coercion.

Causes of Individual Variation

The modeling approach of Harris and Uller represents the first published mathematical treatment of the circumstances favorable to the occurrence of differential allocation. While these authors concluded that differential allocation is favored under a wide range of the life-history conditions they investigated, they also found that an individual female's circumstances (e.g., her age and the effect of increased investment on her future reproductive capacity) have large influences on the benefit that may be obtained from this reproductive tactic. This result may prove useful in explaining variation in results among studies (sometimes on the same species) in the tendency of experimental subjects to practice differential allocation. It also underscores the importance of viewing mating system components as dynamic (i.e., the relative parental workloads of males and females may be highly variable within and among populations) and investigating possible causes of the variation in tendency of individuals to practice differential allocation that is observed in experimental studies.

See also: Flexible Mate Choice; Mate Choice in Males and Females; Social Selection, Sexual Selection, and Sexual Conflict.

Further Reading

Burley N (1986) Sexual selection for aesthetic traits in species with biparental care. *American Naturalist* 127: 415–445.

Burley N (1988) The differential-allocation hypothesis: An experimental test. *American Naturalist* 132: 611–628.

Burley NT and Johnson K (2002) The evolution of avian parental care. *Philosophical Transactions of the Royal Society of London B* 357: 241–250.

Charnov EL (1982) *The Theory of Sex Allocation*. Princeton, NJ: Princeton University Press.

De Lope F and Møller AP (1993) Female reproductive effort depends on the degree of ornamentation of their mates. *Evolution* 47: 1152–1160.

Eberhard WG (1996) *Female Control: Sexual Selection by Cryptic Female Choice*. Princeton, NJ: Princeton University Press.

Gil D, Graves JA, Hazon N, et al. (1999) Mate attractiveness and differential testosterone investment in zebra finch eggs. *Science* 286: 126–128.

Gowaty PA (1996) Battles of the sexes and origins of monogamy. In: Black JL (ed.) *Partnerships in Birds*, pp. 21–52. Oxford: Oxford University Press.

Harris WE and Uller T (2009) Reproductive investment when mate quality varies: Differential allocation versus reproductive compensation. *Philosophical Transactions of the Royal Society B* 364: 1039–1048.

Mousseau TA and Fox CW (eds.) (1998) *Maternal Effects as Adaptations*. Oxford: Oxford University Press.

Reyer H-U, Frei G, and Som C (1999) Cryptic female choice: Frogs reduce clutch size when amplexed by undesired males. *Proceedings of the Royal Society of London B* 266: 2101–2108.

Røed KH, Holand Ø, Mysterud A, et al. (2007) Male phenotypic quality influences offspring sex ratio in a polygynous ungulate. *Proceedings of the Royal Society B* 274: 727–733.

Sheldon BC (2000) Differential allocation: Tests, mechanisms and implications. *Trends in Ecology and Evolution* 15: 397–402.

Stearns SC (1992) *The Evolution of Life Histories*. Oxford: Oxford University Press.

Wolf JB (1998) Evolutionary consequences of indirect genetic effects. *Trends in Ecology and Evolution* 13: 64–69.

Digestion and Foraging

C. J. Whelan, Illinois Natural History Survey, University of Illinois at Chicago, Chicago, IL, USA

Introduction

Foraging is central to life – all organisms must acquire nutrients and energy. But because foraging consists of external, or ecological processes, as well as internal, or physiological processes, the study of foraging has been pursued mostly independent of the study of processing and absorbing ingested food. Nonetheless, the ecological and physiological processes involved in foraging, digestion, and absorption must be coadapted. Foraging strategies obviously affect the ability of foragers to encounter, capture, handle, and ingest particular resources from those available. But, resources are not accepted and consumed randomly, often because of limits on the ability of the forager to process all encountered items. As we shall see, the ability to adequately process foods postingestion often strongly affects diet selectivity. Similarly, behavior can have large consequences for the processing of ingested foods. Foragers of most taxa have evolved abilities to adjust both behavior and physiology in ways that facilitate efficient acquisition of nutrients and energy.

Digestive Processing

Consider a squirrel eating a walnut, a highly preferred food. After encountering the walnut, the squirrel carries it to a safe place for consumption. The squirrel breaks through the hard exterior of the walnut shell, removes the edible interior, and chews and swallows the walnut kernel. The digestive system of the squirrel now goes to work, breaking down the kernel into its more basic constituents, including fatty and amino acids, for absorption and assimilation.

The important process of digesting and assimilating foods occurs principally in the gastrointestinal tract – the gut – with the help of associated organs. Digestive organs and enzymes break down foods (complex carbohydrates, fats, proteins) into assimilable constituents (simple sugars, fatty acids, amino acids, and small peptides), and active and passive mechanisms move those constituents from the interior of the intestines (the lumen) into the blood stream. Although all animal guts share many basic features, each appears adapted to the particular diet consumed. In the following sections, we first consider key aspects of digestion and absorption. We then examine the interaction of foraging behavior and gut processing in relation to ecology and life history.

Modulation of Gut Structure and Function

Virtually all foragers face an environment that changes over short and long time scales, and over small and large spatial scales. Moreover, energetic and nutrient demands of foragers vary over the annual cycle and with changing seasons. Consequently, resource consumption is not constant over time and space. To accommodate variable resource consumption, many species modulate (flexibly adjust) various components of their digestive processing machinery. Modulation may entail adjustment of a single property of the gut, such as gut volume, or several components, such as digestive enzymes, transport mechanisms, and throughput rate. Via modulation of gut structure and function, a forager can adjust gut properties and processes in response to changes in food availability much as different species have specialized guts suited to their particular food habits. Gut modulation confers upon the forager a marvelous degree of flexibility with respect to improving efficiency in the face of ever-changing food availabilities.

Constraints or Limits

Does an animal cease to forage because it is satiated, because it has exceeded the ability of its gut to process ingested foods, or because it must seek safe quarters from its own predators? Ecologists and physiologists tend to think about these factors quite distinctly. Both ecologists and physiologists consider feeding rates to be determined by intrinsic and extrinsic factors. But to ecologists, intrinsic factors include the forager's habitat preferences, search strategies, and its susceptibility to predation. Extrinsic factors are properties of the environment, such as abundance and distribution of resources and predators, and properties of the resource, such as detectability (crypsis) and defense mechanisms (behavioral, physical, and/or chemical). To physiologists, intrinsic factors are gut structure and function, including gut size, digestive enzymes, and mechanisms of absorption. Extrinsic factors are properties of the resource, including relative digestibility, nutrient content (especially N, P, and C), energy content, and chemical defenses. Both perspectives provide valuable insights, but they also lead to somewhat unproductive controversies, such as whether ecological or physiological processes constrain foraging. It is more useful to consider ecological and physiological mechanisms as coordinated, linked steps of a single process, in which either ecological or physiological steps can be rate-limiting. The interesting question is under what circumstances are the ecological or the physiological steps rate limiting.

Gut Capacity

Gut capacity is the ability of the gut to process food. Gut capacity is a function of gut size, digestive enzymes, and absorption mechanisms, along with throughput or retention time of food in the gut, and, in some species, rates of fermentation of microbial symbionts. Gut capacity is not a fixed property of foragers because each of the components that contribute to it can be adjusted to particular diets and circumstances.

Gut size

Guts obviously have finite size (nominal length, volume), and hence gut size influences ingestion. Gut length and volume affect the frequency of foraging bouts, the potential quantity of an ingested meal within a foraging bout, and the extent to which that meal can be processed. In general, greater gut size implies increased times between feeding bouts, increased meal size, and longer retention time within the gut, allowing more complete digestion and absorption. Nonetheless, most species do not maintain maximal gut size, instead flexibly adjusting size to load. The lesson is that the gut is an expensive organ to maintain, and typically, foragers maintain a gut that is big, but not too big.

Digestion

Digestion relies on the action of digestive enzymes, proteins that convert larger molecules into smaller constituent parts. Physical reduction of food particle size by chewing in the mouth or grinding action in a muscular gizzard or stomach aids enzymatic action. Enzymatic digestion occurs primarily in the small intestine, though some enzymes in some species are secreted in the mouth and stomach. Digestion of proteins and complex carbohydrates, but not lipids, involves both extracellular and membrane-bound enzymes.

Absorption

The small intestine uses a variety of fascinating absorption mechanisms, described with seemingly intimidating terminology, including passive and carrier-mediated mechanisms (see **Figure 1**). Passive mechanisms include transcellular diffusion, in which particles (mainly lipophilic compounds) move through the cells, and paracellular diffusion or solvent drag, in which particles (mainly water-soluble compounds, including sugars, amino acids, some vitamins) move between the cells to the circulatory fluids. Paracellular solvent drag has the advantage of an almost instantaneous fine-tuning of the match of absorption to digestive loads,

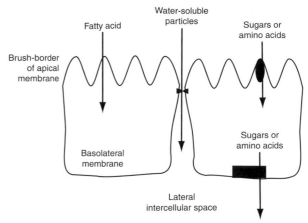

Figure 1 Schematic drawing of enterocytes of intestinal mucosa. Two enterocyte cells are joined together by tight junction. The brush-border apical membrane lines the lumen of the intestine. The basolateral membrane faces away from the lumen. Membrane-bound proteins (black oval in brush border of right enterocyte) transport some nutrients, like amino acids and sugars, across the apical membrane. Lipids and some other molecules are able to pass through the brush border by diffusion. Membrane-bound transporters aid movement of amino acids and sugars across the basolateral membrane to circulatory fluids. Paracellular solvent drag occurs when water drawn from the lumen flows through the tight junction, bringing some small molecules with it. Modified from Karasov WH and Martinez del Rio C (2007) *Physiological Ecology: How Animals Process Energy, Nutrients, and Toxins*. Princeton, NJ: Princeton University Press.

because transport is proportional to concentration at cell junctions, which is proportional to the rate of digestion (hydrolysis). A drawback is that it is nonselective and can lead to inadvertent uptake of toxins or secondary metabolites. Carrier-mediated transport across the (apical) brush-border and basolateral membranes involves carrier proteins. Carrier-mediated transport is either active (requires energy to transport the substance against an electrochemical concentration gradient) or facilitated (substance is transported down an electrochemical gradient). In both cases, saturation of the carrier molecules places an upper bound on transport, following classical enzyme kinetics.

Digestive symbioses

Symbiotic microflora inhabiting the guts of termites allow them to digest cellulose. Such symbioses are common. Virtually all foragers have symbioses with a complex microflora, including bacteria, fungi, and protozoa (exceptions may be limited to a number of marine Crustaceans). In humans, resident bacteria outnumber human cells by a factor of 10. The community of microflora residing within the gut has been likened to a metabolically vital organ. The gut is sterile at birth and subsequently undergoes colonization and, likely, continuous turnover of associated microflora. These microflora have been characterized as resident (autochthonous) if they colonize and achieve some stability within the gut, and commensal (allochthonous) if they are noncolonizing and require consumption for repopulation. The roles of the gut microflora are subjects of intense investigation, as they are involved with many aspects of host health, from development of the immune function to digestion and absorption of nutrients. Numerous factors, including diet, host health, and associations with conspecifics, affect the community of microflora.

Interaction of Behavior and Gut Processing

Foods are not all equal. For any given organism, some foods represent higher rewards and/or lower costs than other foods. For instance, foods differ with respect to the amount of energy (e) potentially assimilated per unit time spent securing and processing them (h). Foraging ecologists often assume that a forager prefers foods in decreasing order of the ratio of e/h. The classic diet model of optimal foraging theory predicts that a food will be included in the diet only when higher-ranked items fall below a specified threshold of abundance – an extremely abundant, but low-ranked food may be excluded by a consumer foraging optimally with respect to e/h.

Foods differ, however, beyond energy content – they also differ with respect to chemical (nutrient) composition and ease and extent of enzymatic degradation. Foods (or portions of them) resistant to degradation are referred to as being refractory to digestion. Many foods contain toxins. These additional food characteristics often lead to the inclusion (or exclusion) of a specific food item in the diet in apparent violation of expectation based on a simple ranking of e/h. Moreover, the need to include foods with particular nutrient rewards may vary with the life history or the annual cycle of a particular forager.

Food Quality

Nutritional relationships

Eating a 'balanced' diet, one that includes different types of food, from meats and dairy products, to fruits, vegetables, grains, and nuts, is important for human health. Such varied diets ensure consumption of the different sorts of nutrients needed for body maintenance. In humans, consumption of both rice and beans provides more useable protein than consuming only rice or only beans because the different sorts of amino acids found in those food groups. Meat, even if derived from different species (say rabbits and squirrels consumed by a hawk), are usually more or less equivalent with respect to energetic and nutritional reward. Such foods that are largely similar in chemical composition and nutritional reward may be substitutable with respect to physiological and fitness

consequences. Joint consumption of substitutable resources results in fitness equal to that predicted by consumption of the linearly weighted sum of the resources. Foods that differ in chemical composition and nutritional reward may be complementary with respect to physiological and fitness consequences. Joint consumption of complementary resources (like rice and beans) results in fitness greater than predicted from consumption of a linearly weighted sum of the two (or more) resources.

Most ecological models of foraging assume that foods are substitutable. When animals consume substitutable foods, intuition may suggest that there should be few consequences for digestive processing. However, if the foods differ in abundance temporally or spatially, theory demonstrates that modulation of gut function may occur if the foods differ slightly in ease of digestion or absorption. The result of modulation, even for substitutable foods, is that foods become somewhat antagonistic. This quasiantagonism will drive a switch from inclusion of one food type to the other, thus promoting specialization.

Many different foods are likely complementary resources. Mechanistically, complementarity can arise for various reasons relating to nutrition, toxicity, physical characteristics, or some combination of two or more factors. Complementarity promotes diet mixing, as the more of a single food type is consumed, the better off is the forager from consuming the alternative food type(s). Because, by definition, complementary resources differ with respect to chemical and/or physical composition, we may expect their consumption to promote a more generalist (jack of all trades) digestive processing machinery.

Although nutritional relationships among food resources have proved a useful conceptual framework for guiding ecological investigations, they have been largely ignored in studies of digestive processing. Hence, we know little about how digestive processing accommodates foods that differ in their nutritional relationships. Theoretical models demonstrate that when foraging on foods that differ in profitability, richness, and ease of digestion, adjustment or modulation of gut size and throughput rate leads to digestive-system specialization (**Figure 2**). Modulation of digestive physiology to a particular food type causes other food types to become antagonistic resources. This organizing framework may be a particularly rewarding pathway for further integrating ecological and physiological understanding of diet selection and resource exploitation systems.

Food bulk

Foraging ecologists and their models deal mostly in the currency of energy, with a primary concern being the energy assimilated relative to time expended handling and consuming the food, e/h. More recent models of foraging incorporate gut processing as a component of food handling time and effort. Some of these models identify another critical component of food – its bulk.

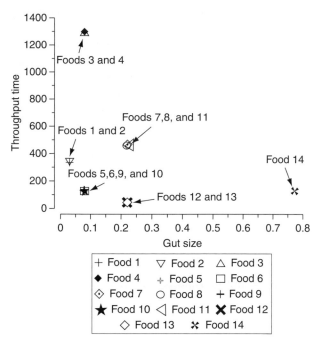

Figure 2 The effect of 14 different foods on optimal gut size and throughput time. Foods differ in energetic reward, bulk, ease of absorption, and external handling. Joint adjustment or modulation of gut size and throughput time results in six apparent digestive physiological syndromes. Increasing food richness (energy-to-bulk ratio) leads to smaller gut volumes and little change in throughput times. Higher absorption rates favor shorter throughput times with smaller effects on gut volume. See Orlando et al. (2009) for details.

With this new component, foods vary along two axes: *energy:handling* (*e/h*) and *energy:bulk* (*e/b*). Foraging ecologists traditionally refer to *e/h* as food reward. We now consider *e/b* to represent food richness. Once gut processing and food bulk are incorporated into models of harvest (ingestion), expectations of diet selectivity are greatly enriched. Instead of food preferences being ranked in descending order of food reward (*e/h*), foods preference now depends upon both food reward and food richness (*e/b*).

Rankings of foods by *e/h* need not be identical to rankings by *e/b* – a food with high reward (*e/h*) may concomitantly possess low richness, while a food with low reward may possess high richness. Preference partially depends upon previous consumption – as foods are consumed and the gut becomes 'bulked up,' the universe of acceptable foods declines. Because ingested foods 'compete' for processing within the gut, foods achieve a sort of complementarity, which in turn promotes partial selectivity. With gut processing, we thus expect that preference can change during a meal or foraging bout, and that in addition to complete selectivity for one resource or opportunism for all encountered resources, foragers may exhibit partial selectivity (take all high-ranked items encountered and some proportion of low-ranked items encountered).

Fluctuations in Resource Abundance

Foragers often face fluctuations in resource abundance. These fluctuations may occur over both long and short time frames. Many resources fluctuate in abundance seasonally, necessitating changes in diet selection. For instance, many songbirds switch between a diet dominated by invertebrates during the breeding season to one dominated by either fleshy fruits or seeds in the nonbreeding season. At temperate latitudes, invertebrates can reach great abundances during the summer growing season, whereas many seeds and fruits reach peak abundances in autumn. Species such as the American robin (*Turdus migratorius*) may forage almost exclusively on invertebrates during the growing season and exclusively on fruits and seeds during the nongrowing season. Invertebrates, such as insects, differ in chemical and physical characteristics from fleshy fruits and seeds. Insects are highly proteinaceous, with chitinous exoskeletons. Fruits, in contrast, tend to be bulky (due to fiber, seeds, or pits), rich in nonstructural carbohydrates or lipids (rarely both), and low in proteins.

Foragers, like robins, which switch from a diet of insects to one of fruits, modulate gut structure and/or function. The most common digestive adjustment with this particular diet switch is modulation of retention or throughput time. On fruits, throughput time is fast, whereas on insects, it is slow. Some species, like the yellow-rumped warbler (*Dendroica coronata*), modulate amino acid transporters in the intestine to match load (when fed diets high in protein), but show no such adjustment for glucose transporters. In contrast to these findings, other vertebrates, including rabbits, cats, mice, and chickens exhibit glucose transporter modulation in relation to load. The difference appears related to the ability of some bird species, in diverse taxa, to absorb glucose passively through solvent drag.

Fluctuations in resource abundance may also occur over short time frames, as when different resources are spatially heterogeneous, or when one (or more) resource undergoes a temporary pulse (emergence of periodical cicadas, Cicadidae: *Magicicada*). Diet switching caused by short-term fluctuations can result in low digestive efficiencies when the resources differ with respect to carbohydrate, lipid, or protein. Research suggests that adjustment of the gut following abrupt switches in resource use can take two to three days, or more. Immediately following the switch, efficiency is low, and it increases over several days as the gut becomes habituated to the new resource. Again, this is because gut modulation, as noted earlier, causes resources to assume a certain degree of antagonism to each other

Frequency of Feeding

Some foragers feed only intermittently – examples include some ambush predators, many snakes, turtles, and all crocodilians. The preeminent example of an intermittent feeder is the Burmese python (*Python molurus*) – these large constrictors tolerate long fasting periods punctuated by consumption of large prey that may equal if not exceed their own body mass. In intermittent foragers, the gut is small while fasting, and presumably, its maintenance costs low. Following consumption, the gut undergoes a remarkable transformation, increasing in size and gastric function, and upregulating intestinal digestive enzymes and nutrient absorbers (transporters). To a human observer, a python, following a meal, appears inert and quiescent. But while processing the meal, the metabolic rate of a python has been likened to that of a thoroughbred racehorse in a dead heat.

Reproduction

Reproduction is energetically demanding while requiring mobilization of particular nutrients, such as protein, calcium, and sulfur. The energetic and nutrient requirements may be met in some animals by utilizing endogenous nutrient reserves (so-called capital breeders), or, in other animals, from changes in diet quality and quantity (so-called income breeders). For income breeders, adjustment of gut structure and function accompanies the associated changes in diet quality and quantity. The intestinal mass of brown-headed cowbirds is about 10% larger during egg laying than prelaying, apparently to accommodate increase food intake rate needed to garner the energy and specific nutrients required for egg production. In mammals, peak energy demand occurs during lactation, although it is also heightened during fetal development. During lactation, many organs involved with food processing increase in size and function, including the liver, the pancreas, the absorptive surface of the intestines, and the length of the entire gut. These changes in organ size permit an increase in the capacity to absorb nutrients through upregulation of nutrient transporters.

Migration

Many animals undergo seasonal migrations between different areas. Here, we focus on bird migration, but presumably similar challenges and adaptations are also likely in other taxa. In birds, migration poses some perplexing challenges. On the one hand, migration is energetically costly and requires ample fat to fuel it. Thus, migrants markedly increase food intake, a process we call 'hyperphagia.' As we saw in previous sections, increases in food consumption typically induce increases in gut mass to permit efficient processing of the increased intake of food. On the other hand, an increase in gut mass makes flight more expensive to fuel and increases susceptibility to predation. A larger gut may also preempt internal space that could otherwise be used for fat stores and increased

muscle mass. How do migrants resolve these conflicting challenges?

One solution is to adopt a fueling strategy in which the forager takes a longer time to amass fuel deposits using a small gut. This strategy, used by red knots (*Calidris canutus piersmai*) migrating northward from NW Australia, trades off a small gut for a longer fueling period. Later in their migration, red knots grow a larger gut that enables faster refueling, necessary for timely arrival in their far northern breeding areas, where time available for breeding is extremely limited.

A possibly more common strategy is to reduce gut size, just before or during nonfeeding flights between staging areas. Reducing gut size could provide three benefits: (1) an ability to more easily accommodate stored fuel and muscle within a limited body cavity; (2) a reduction in the cost of maintaining an expensive organ; and (3) an increased agility to avoid predators. Both red knots and bar-tailed godwits (*Limosa lapponica*) decrease gizzard size just prior to migration. Both during migratory flights and while experimentally fasted in captivity, great knots (*Calidris tenuirostris*) decrease mass of digestive organs. Black-cap (*Sylvia atricapilla*) and garden (*Sylvia borin*) warblers also decrease digestive organ mass under migratory flight and fasting in captivity. For both the waders and the passerines, the reduced gut mass compromises their ability to process and assimilate food immediately upon refeeding. A return to normal food processing is quickly attained, although more so for the warblers than the waders.

Migrating animals often encounter different food resources as they move along their migratory routes – such as caterpillars in one place and berries in another. As discussed earlier, these transitions can reduce the efficiency of nutrient uptake while the digestive system adjusts to the new food type. Digestive adjustment may take several days or more, and this delay may affect the length of time a migrant needs to rebuild fuel stores for the next leg of its journey. The penalty of reduced food utilization that results from diet switching could potentially affect various characteristics of migration, such as stopover location, stopover length, flight distances, and resource selection. Speed of modulation may influence whether migration tends to be rapid and smooth or slow and jerky. Thus, such diet switching and gut modulation could ultimately be an important determinant of the time course of migration.

Conclusions

Foraging is a coordinated, whole organism process involving both external (ecological) and internal (physiological) processes. Depending upon circumstances, either ecological or physiological processes may be rate-limiting. When digestive physiology – the structure and function of the gut – is rate-limiting, foragers typically adjust the gut to increase capacity and/or efficiency. Many factors cause changes in diet quality or quantity, including short- and long-term fluctuations in resource abundances, seasonal changes in environmental conditions (temperature), and aspects of the life cycle, such as reproduction and migration. In most organisms, changes in diet corresponding to such changes in environmental conditions or demands of the life cycle are accompanied by concomitant modulation of the gut that permits high efficiency and processing capacity in response to those changes. A recent meta-analysis found strong, quantitative support for four hypotheses proposed to explain flexibility in gut structure and function as an adaptive accommodation to changes in diet. Changes in diet may arise inevitably due to changing environmental conditions outside the control of the forager, or due to purposeful changes in foraging behavior. Changes in diet (for whatever reason) induce modulation of gut structure and function. It is possible, at least in some species, that circannual rhythms trigger gut modulation in anticipation of changing demands and associated diet quality or quantity. We can thus view the flexible gut as a vital and integral component of an animal's foraging strategy.

See also: Behavioral Endocrinology of Migration; Caching; Circadian and Circannual Rhythms and Hormones; Cost-Benefit Analysis; Food Intake: Behavioral Endocrinology; Food Signals; Foraging Modes; Habitat Selection; Hibernation, Daily Torpor and Estivation in Mammals and Birds: Behavioral Aspects; Hormones and Breeding Strategies, Sex Reversal, Brood Parasites, Parthenogenesis; Hunger and Satiety; Kleptoparasitism and Cannibalism; Migratory Connectivity; Optimal Foraging and Plant–Pollinator Co-Evolution; Optimal Foraging Theory: Introduction; Patch Exploitation; Seasonality: Hormones and Behavior; Trade-Offs in Anti-Predator Behavior; Water and Salt Intake in Vertebrates: Endocrine and Behavioral Regulation; Wintering Strategies: Moult and Behavior.

Further Reading

Jeschke JM, Kopp M, and Tollrian R (2002) Predator functional responses: Discriminating between handling and digesting prey. *Ecological Monographs* 72: 95–112.

Karasov WH and Martínez del Rio C (2007) *Physiological Ecology: How Animals Process Energy, Nutrients, and Toxins.* Princeton, NJ: Princeton University Press.

McWilliams SR and Karasov WH (2001) Phenotypic flexibility in digestive system structure and function in migratory birds and its ecological significance. *Comparative Biochemistry and Physiology Part A* 128: 579–593.

McWilliams SR and Karasov WH (2005) Migration takes guts. Digestive physiology of migratory birds and its ecological significance. In: Marra P and Greenberg R (eds.) *Birds of Two Worlds*, pp. 67–78. Washington, DC: Smithsonian Institution Press.

Naya DE, Karasov WH, and Bozinovic F (2007) Phenotypic plasticity in laboratory mice and rats: A meta-analysis of current ideas on gut size flexibility. *Evolutionary Ecology Research* 9: 1363–1374.

Orlando PA, Brown JS, and Whelan CJ (2009) Co-adaptation of foraging behavior and gut processing as a mechanism of coexistence. *Evolutionary Ecology Research* 11: 541–560.

Piersma T and Lindstrom A (1997) Rapid reversible changes in organ size as a component of adaptive behaviour. *Trends in Ecology and Evolution* 12: 134–138.

Schmidt KA, Brown JS, and Morgan RA (1998) Plant defenses as complementary resources: A test with squirrels. *Oikos* 81: 130–142.

Speakman JR (2008) The physiological costs of reproduction in small mammals. *Philosophical Transactions of the Royal Society B* 363: 375–398.

Starck JM and Wang T (eds.) (2006) *Physiological and Ecological Adaptations to Feeding in Vertebrates.* Enfield, NH: Science Publishers, Inc.

Stephens DW, Brown JS, and Ydenberg R (eds.) (2007) *Foraging.* Chicago, IL: University of Chicago Press.

Whelan CJ and Brown JS (2005) Optimal foraging under gut constraints: Reconciling two schools of thought. *Oikos* 110: 481–496.

Whelan CJ, Brown JS, and Moll J (2007) The evolution of gut modulation and diet specialization as a consumer-resource game. In: Jørgensen S, Quincampoix M, and Vincent TL (eds.) *Advances in Dynamic Game Theory: Numerical Methods, Algorithms, and Applications to Ecology and Economics,* vol. 9, pp. 377–390. Annals of the International Society of Dynamic Games. Boston, MA: Birkhauser.

Whelan CJ, Brown JS, Schmidt KA, Steele BB, and Willson MF (2000) Linking consumer-resource theory with digestive physiology: Application to diet shifts. *Evolutionary Ecology Research* 2: 911–934.

Whelan CJ and Schmidt KA (2007) Foraging ecology: Processing and digestion. In: Stephens DW, Brown JS, and Ydenberg R (eds.) *Foraging*, pp. 140–172. Chicago, IL: University of Chicago Press.

Disease Transmission and Networks

D. Naug, Colorado State University, Fort Collins, CO, USA

Introduction

Animals living in large groups are particularly vulnerable to infectious diseases. The close proximity of individuals offers excellent transmission opportunities to a pathogen that is spread by direct contact between hosts. Many studies show a positive relationship between group size and parasitism in terms of prevalence (proportion of infected individuals in a group) and intensity (number of pathogens per individual). If the host population is homogeneous in exposure and susceptibility to a pathogen, the birth and death rates of the host and the contact rate between susceptible and infected individuals are sufficient to predict the infection dynamics. However, groups are rarely homogeneous and individuals differ among themselves in various respects such as age, sex, physiological state, behavior, and spatial location. This causes individuals to differ in their probability of becoming infected and transmitting the infection, making it more difficult to predict the trajectory of an infection.

The rate at which an infection spreads and whether it persists in the population depend on the magnitude of the key epidemiological parameter, R_0, or the mean number of infections caused by a single infected individual. In order to stop an epidemic outbreak, R_0 must be maintained below 1. According to the mass-action SIR (susceptible-infected-recovered) model, the most basic model of epidemic spread, $R_0 = \beta TS$, where β is the transmission coefficient that incorporates both infectiousness and contact rate of the infected individuals, T is the duration of infectiousness, and S is the available number of susceptible individuals. The simple SIR model has provided many important insights into the epidemiology of a wide range of pathogens but its fundamental assumption of homogeneous mixing among individuals is clearly unrealistic. Population-level estimates of R_0 can obscure the considerable variation in contact rate and infectiousness among individuals. Several studies have shown that typically, 80% of the transmission events are contributed by 20% of the host population: a trend that is referred to as the 80/20 rule. This was highlighted during the recent global epidemic of severe acute respiratory syndrome (SARS) when a few infected individuals were responsible for giving rise to an unusually large number of secondary cases. Whether or not infected individuals have contact rates that are disproportionately higher than the population average has important implications because public-health programs generally rely on the immunization of only a fraction of the hosts to protect the entire population.

Network Theory

The effects of host heterogeneity on the spread of infectious disease can be most simply modeled by dividing a population into subpopulations with different within-group and between-group transmission rates. A more explicit approach is to use models that incorporate the structure of the actual contact network in the population. Unlike the continually changing set of contacts in random mixing models, each individual is assigned a finite set of contacts to who they can transmit infection and from who they can be infected. Predictions from network models can be considerably different from those that use mean-based approaches. Although individuals may have the same number of contacts per unit time in both network and mass action models, the fixed contact structure in networks can lead to rapid, localized spread of an infection followed by a slowing down of the process as the number of susceptible individuals depletes locally. This makes disease extinctions more likely than outbreaks though the latter are more explosive if they occur.

The use of network models also has bearing on the evolution of the pathogens themselves. Given their high reproductive rates, pathogens are likely to undergo rapid selection to adapt to the available transmission routes between infected and susceptible individuals. Both theoretical and experimental results show that high transmission rates are selected in localized networks where there is intense competition for susceptible hosts while networks that are more global in their connectivity select for lower transmission rates due to lack of such competition. Localized contact structure also selects for a higher diversity in the pathogen population in contrast to a randomly mixed host population where cross-immunity to similar strains structures the pathogen population into discrete, nonoverlapping strains.

A transmission network is generally defined by a matrix X that describes the connections among all the individuals within a group. In its simplest form, the matrix is unweighted, with $x_{ij} = 1$, if there is one or more interactions that can transmit an infection and $x_{ij} = 0$ if there is none. The matrix is also generally undirected, meaning that infection can pass either way across an interaction or $x_{ij} = x_{ji}$. More detailed models can be constructed using weighted, directed networks. The structure of the transmission network can be characterized by a number of parameters that can be quantified from these matrices. The most commonly used ones are (1) degree, the number of connections an individual has; (2) density, the proportion of existing connections out of all possible ones; (3) path length, the average number of links that connect any two individuals; and (4) clustering, the density of the local neighborhood or cliquishness. Focal measures such as degree can identify high-risk individuals in the population and can be used to inform surveillance and infection control strategies. Network level measures such as average path length and clustering coefficient can make predictions about the spread of the infection in the population. Critical points that reflect order of magnitude shifts in network properties and the consequent propagation of an epidemic can be identified from phase transitions in network parameters.

Network models are difficult and time consuming to build because they require information about the connectivity between every pair of individuals in a group. In this effort, researchers have mainly relied on infection tracing that describes the actual connections through which the infection spreads or contact tracing that looks at all the potential connections from a source individual. Network models are also complex in terms of their statistical evaluation unlike differential equations based mass-mixing models. Moreover, as different diseases are transmitted via different transmission pathways, network models are disease specific and cannot be easily generalized. In the face of these difficulties, simulating networks with different structures (**Figure 1**) and studying the parameters that influence transmission dynamics has been an important and influential research paradigm.

Types of Networks

Random Networks

In these types of networks, each individual has a fixed number of random connections, resulting in a network with no clustering and short path lengths. The early growth rate of an infectious process and the final epidemic size are lower in these networks compared with the mass-action model, largely because of the quick depletion of the local environment of susceptible individuals around an infected individual.

Regular Networks

In these networks, individuals are connected only to their adjacent neighbors, leading to a homogeneous network with high clustering and long path lengths. This leads to an even stronger depletion of the local environment and thus the growth rate of the infection.

Small-World Networks

The transmission properties of small-world networks have generated a lot of interest and are important to understand because many biological networks including human social networks show small-world properties. They lie somewhere between regular and random networks, displaying high clustering but small path lengths due to the existence of a few long-range connections. Even though the transmission process is still largely localized, the few long-range links allow the infection to spread relatively quickly and more synchronously over the entire network. Small-world networks may or may not have a scale-free structure.

Scale-Free Networks

These networks are characterized by an extreme heterogeneity in connectivity, the number of contacts per individual being described by a power law distribution. A few highly connected individuals, called superspreaders in the epidemic context, have a disproportionately high influence on the transmission process. Networks of human sexual contacts have been shown to follow such a distribution and the transmission and maintenance of sexually transmitted diseases thus depends mainly on a few promiscuous individuals. In such networks, control measures

Figure 1 Four common types of networks, from left to right: random, regular, small-world and scale-free. Adapted from Watts DJ and Strogatz SH (1998) Collective dynamics of 'small-world' networks. *Nature* 393: 440–442 and Strogatz SH (2001) Exploring complex networks. *Nature* 410: 268–276, with permission from *Nature Journals*.

directed at random individuals are quite ineffective while targeted interventions work really well. By immunizing the superspreaders, the contact network becomes sparser by orders of magnitude and brings about a drastic reduction in the number of transmission events.

A Model Network for Experimental Epidemiology

Complex network models being hard to parameterize can lead to predictions that are no more reliable or maybe even worse than those from simpler frameworks. However, with the relative dearth of suitable experimental systems with sufficient social complexity, opportunistically obtained data about the course of natural epidemics in humans have been the only major recourse for testing network models. In this context, the honeybee colony can prove to be an ideal model system. Honeybees not only provide the setting of a crowded social group that is susceptible to a vast array of infectious diseases but they are also extremely amenable to a variety of experimental paradigms at both the individual and the social level. The long association of honeybees and their pathogens over evolutionary time provides a backdrop to test how the network of social interactions in the colony could serve as the central arena for host–pathogen dynamics. The pathogens can exploit the network to rapidly spread across the colony, while the host can use its structural properties as a mechanism to resist the spread. The recent finding that honeybees possess only one-third as many genes for immunity as other insects strongly suggests that the structure of social organization is an important mechanism that compensates for a lower physiological immunocompetence.

Interaction networks in a social insect colony could be organized according to one of the following designs: (1) work-chains with each individual performing all the required parts of a given task, (2) work-chains with each individual performing one and only one part of a task with one or more other individuals completing the rest, and (3) work-chains with each individual performing only one part of a task at any given time but performing all the parts equally frequently. The efficiency and the reliability of material and information flow are substantially incremented in each successive type of network, which is adaptive for ergonomic purposes. It is however less recognized that the same design features will also promote the transmission efficiency of pathogens, increasing the vulnerability of the colony to an infectious disease.

Food, information, as well as pathogens primarily enter a honeybee colony from the environment through the foragers. Nearest-neighbor based interactions drive the subsequent transfer process, spreading these across the

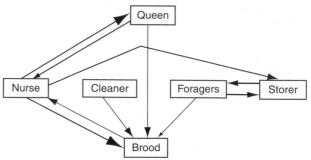

Figure 2 Idealized social network within a honeybee colony with arrow widths indicating interaction frequency.

colony. With the individuals spatially distributed within the colony according to their ages, this results in a centripetal flow from the oldest individuals at the outer edge of the colony to the youngest ones residing at the center. This flow pattern imparts some amount of protection to the most valuable youngest members from invading pathogens: a phenomenon that can be termed 'organizational immunity.' This social contact network in the colony is therefore highly structured and nonrandom, leading to a pool of individuals that is heterogeneous with respect to its probability of contacting, manifesting, and transmitting an infection, presenting an invading pathogen with the challenge of negotiating this complex landscape (**Figure 2**).

Superimposed on this general age-based interaction pattern, one also sees that only a minority of the individuals in the colony are the primary drivers of the majority of the transfer process. This gives the interaction network an appearance of a scale-free structure. In contrast to such heterogeneous connectivity observed in the large honeybee colonies, individuals are more uniformly connected to each other in social insect species with smaller colony sizes. Within-species comparisons suggest that colony size is a primary driver of network structure and complete mixing becomes more and more improbable with increasing number of individuals. There is considerable variation in network structure even among colonies of similar sizes and it has been shown that the density of contact network in the colony determines the spread of a contagious pathogen within it.

Small perturbations in the structure of social organization can bring about large changes in transmission dynamics. Many honeybee diseases, which remain in the background at a low level in the colony, can rapidly turn lethal and erupt into an epidemic under certain conditions generally referred to as 'stress.' Investigation of these so-called 'stress' conditions suggest that they translate into disruption of the normal social organization in the colony in the face of contingencies such as a nectar flow in the environment, high demand for a certain task, or a rapid increase in colony population size. The resulting higher activity level and more generalization of labor profiles can lead to higher contact rates or other changes in social network structure.

A disease can also bring about some restructuring of the social organization in the colony. Disease at an individual level is defined as a disruption of homeostatic mechanisms, leading to an alteration in the normal set point of an organism and its symptoms are the physiological mechanisms that restore it. This definition can be extended to an epidemic being a process that disrupts the social organization critical to the functioning of a group and its symptoms are mechanisms that, via collective action of its members, attempt to restore the social structure. It has been speculated that a disease symptom such as bees starting to forage at a younger age is an adaptive response on the part of the host that serves to reduce within-colony transmission of the disease by keeping infected bees outside. However, it is equally plausible that such a response can in fact increase transmission rates by contaminating the food they collect. Behavioral fever in response to an infection, which can inhibit the development of a pathogen, requires bees to cluster more tightly that can in turn increase the contact rate among them. Bees infected with a pathogen have also been shown to incur an energetic stress that increases their hunger level, leading them to be more eager solicitors but more reluctant donors of food. This could lead uninfected and infected bees to occupy different positions in the contact network in terms of sources and sinks in the transmission chain. It is important to note here that the structure of the social network in the colony is an emergent property that arises from individual behavior, which can be altered by simple pathophysiological mechanisms arising from a disease.

Areas for Future Research

For disease ecologists interested in using network theory, the development of network statistics remains a major research focus. A second area of rapidly developing interest is dynamic networks which account for the possibility that the structure of the contact networks might not remain constant over time, maybe partly as a consequence of the disease outbreak itself. More importantly, empirical research has lagged behind the pace of theoretical work made possible by increased computational power. Matching efforts to develop laboratory experimental systems are urgently needed to explore the interaction between network structure and disease dynamics. Integration of behavioral biology and physiology to the already existing framework of ecology, evolution, and mathematical modeling would also be critical to our understanding of the structural and functional properties of biological networks.

Research on the proximate basis underlying the behavioral interactions among individuals will give insights into the role of demographic and environmental factors on disease dynamics via their effects on social structure. It will also help answer the important questions of how the pathophysiology of a disease can alter the structure of the contact network and whether such symptomatic restructuring benefits the host or the pathogen. In social insect groups where the colony social network is considered to be primarily a product of ergonomic considerations, it is important to explore whether pathogens have played any selective role in its design. This addresses the broad issue of how any group of interconnected units, whether a bee colony or a computer cluster, deals with the challenge of shielding its network from attacks without seriously compromising its performance.

Conclusion

It is being increasingly recognized that excessive use of antimicrobials to treat diseases selects for resistant strains of pathogens that can no longer be eliminated by the same drugs. Intervention measures that have short-term epidemiological benefits but long-term evolutionary repercussions have led to the recent resurgence of many diseases and the heightened virulence of pathogen populations. This has led to the suggestion that understanding the natural dynamics of a disease from an evolutionary, ecological, and behavioral perspective might provide pointers to preventive and curative methods that are more sustainable. There are plenty of accounts concerning behavior and customs in humans that affect the transmission of infectious diseases. Agricultural practices such as the clearing of land and irrigation have brought increased contact between human populations and animal reservoirs of diseases such as schistosomiasis and malaria. Urbanization has brought about increased transmission of lyme disease, cholera, dengue, and leishmaniasis. Changes in sexual behavior have had a large influence on the spread of human immunodeficiency virus (HIV), human papillomavirus (HPV), chlamydia, gonorrhea, and other sexually transmitted diseases. With the current threat of these numerous emerging diseases, it has become extremely important to understand the dynamics of infectious processes in the context of crowded living conditions that characterize many animal groups and humans. An understanding of the behavioral processes that define the structure of a social group will help identify the transmission pathways used by pathogens to spread and suggest possible ways to manage the social structure as a counteractive measure to both prevent and control the spread of a likely epidemic.

See also: Consensus Decisions; Group Movement; Life Histories and Network Function; Nest Site Choice in Social Insects.

Further Reading

Keeling MJ and Eames KTD (2005) Networks and epidemic models. *Journal of the Royal Society Interface* 2: 295–307.

Naug D (2008) Structure of the social network and its influence on transmission dynamics in a honeybee colony. *Behavioral Ecology and Sociobiology* 62: 1719–1725.

Naug D and Camazine S (2002) The role of colony organization on pathogen transmission in social insects. *Journal of Theoretical Biology* 215: 427–439.

Newman MEJ (2003) The structure and function of complex networks. *SIAM Review* 45: 167–256.

Watts DJ and Strogatz SH (1998) Collective dynamics of 'small-world' networks. *Nature* 393: 440–442.

Disease, Behavior and Welfare

B. V. Beaver, Texas A&M University, College Station, TX, USA

In the world of veterinary medicine, we have come to appreciate the closeness of behavior and disease. The behavioral changes associated with illness are the main indicators that animal owners use to assess welfare. There is a long and complex interaction between behavior, welfare, and disease.

Behavior and biology cannot be separated, even at the most superficial level. Not only is the brain related to the performance of specific behaviors, behaviors will affect it and other body systems as well. These changes can be subtle or complex. As an example, stress affects the release of cortisol, one measure of the stress an animal may be experiencing. It can also release more neutrophils into the bloodstream to be ready for potential invasions by microorganisms. Depression can affect the adrenal glands' response to life in general, much less their response to stress, and it affects the ability of the thyroid gland to normalize the body's activity. Drugs used to treat various medical conditions frequently have secondary effects that impact quality-of-life issues. Excessive sleepiness from seizure medications may make their use undesirable because then the pet is not a good companion. Behaviors in the extreme, such as a phobia to thunderstorms, may make drugs ineffective. And now we must add the fact that supplements can also affect welfare. Antioxidants are popular and their actions can help remove free radicals in the brain to help hold off the behaviors associated with cognitive dysfunction. Foods eaten provide the precursors to the neurotransmitters, and those are related to brain function in sites that impact behavior. They can also be the basis for food allergies, which can come with a different set of behavioral changes.

Illness and Behavior

Some of the complexity of the interrelationships of illness and behavior can be seen in the body's response to an infection (**Figure 1**). When a microbe causes an infection, the body responds with phagocytizing cells going after the organism. These produce interleukin 1, which is associated with both fever and initiation of slow-wave sleep. At the same time, the organism can produce endotoxins, to which the body responds with leukocytic endogenous mediators. Fever and perhaps interleukin 1, can also produce these mediators. The resulting production helps the body sequester plasma iron, amino acids, and zinc, making them unavailable to the microorganisms. There is also an increase in the acute phase reactant glycoproteins for antibody production and leukocytosis in the form of white cells to help fight the infection.

Fever also directly results in the sequestering of plasma iron, the increase of glycoproteins and leukocytosis. In addition, it will increase body metabolism by approximately 25% for a 2 °C increase in body temperature. Pathogens do not grow as well at higher body temperatures, so the fever helps reduce their efficiency.

Slow-wave sleep helps in survival and is one of the obvious behavioral changes noticed by owners (**Figure 2**). The reduced movement results in energy conservation. This is important because the animal is not eating as much, and metabolism has increased, meaning the body is using its surplus energy for survival. Sleep reduces grooming behaviors, which in turn helps minimize body heat loss secondary to saliva evaporation, and it conserves energy because hunting is not done. Anorexia is another part of the slow-wave sleep changes. If the animal does not have to hunt food, it conserves energy as well as prevents the intake of iron, which is a good resource for many infectious organisms.

Thus, a reduction in activity, desire to eat, and grooming are behaviors that owners notice in the sick animals. But these are not restricted to domesticated animals; the responses occur in mammals of all types. And these are just the beginning of the interrelation of behavior, welfare, and disease.

The Brain and Behavior

Without going into depth about the interrelationship of the brain and behavior, because it is being covered elsewhere, it is important to note in this article that diseases that affect the brain also affect behavior. Conditions such as brain tumors, cerebral vascular accidents, and hydrocephalus can affect any part of the brain, with the resulting symptoms being quite variable. Diseases also affect the brain directly. As examples, the canine distemper virus can result in encephalitis, seizures, and focal changes on the retina of the eye. Feline panleukopenia is associated with cerebellar hypoplasia if kittens are infected in utero or before 4 weeks of age.

The question is the effect of these things on welfare. In the case of tumors, encephalitis, and seizures, few would doubt that an animal's welfare is negatively affected, at least as the disease progresses. The bigger questions could

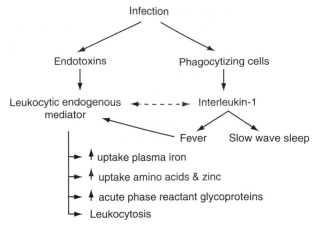

Figure 1 The body responses to infection.

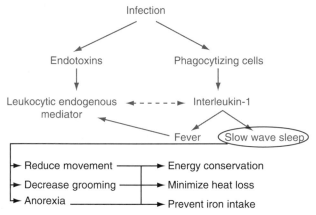

Figure 2 Responses associated with increased sleepiness.

be argued relative to hydrocephalis, cerebellar hypoplasia, and other similar conditions. In these cases, the animal does not seem to suffer significantly, or at least is not apparently aware that it is different from other animals. Humans make the distinction. And does the fact that the animal is affected make it more likely to be singled out as prey, reducing its welfare?

Behavior and Medical Conditions

There are a number of medical conditions that can be expressed as behavioral problems only, without the typical clinical signs associated with that same condition. Others of these problems are parallel to human conditions, and still others should be part of differential diagnoses considered in a patient with certain presenting signs. Meaning unclear please check if the sentence conveys the intended meaning. Certainly, these categories are artificial and do overlap. They are discussed here based on the primary way in which they present to a veterinarian.

Medical Problems that Can Be Seen as a Behavioral Problem Only

This category can be difficult to diagnose unless the veterinarian is aware of the behavioral expression of the condition. We certainly do not recognize all of the possibilities at this time, but our knowledge is growing. The following are some examples that we know about.

Hypothyroidism

Animals typically affected with hypothyroidism show a number of systemic signs, such as thinning skin, symmetric alopecia, and reduced muscle tone. However, some individuals do not show those signs and only show a behavioral change, such as aggression. Unfortunately, diagnosis is not always straightforward because there are a number of causes for aggression. Blood work is typically used to rule in or rule out this condition; however, interpretation can be difficult. A number of things can artificially elevate or suppress the thyroid values, so the veterinarian is faced with evaluating all information to make a determination.

Seizures

Seizures can have a number of different causes, and in that way, they parallel human conditions. However, seizures can have a behavior-only expression in animals. Aggression, tail chasing, air snapping, or star gazing may be the only manifestation of a seizure. Treatments, when appropriate, involve antiseizure medications. The problem occurs with animals showing aggression, especially when the bouts are widely spaced. It is difficult to know if the time without aggression is related to the medication controlling the seizure activity, or whether it is just a time without abnormal firing within the brain. Because aggressive bouts are potentially dangerous to those around the animal, treatment options become more difficult.

Hormonal abnormalities

Hormonal changes, even in neutered animals, can present as a behavioral problem. Aggression is the most common of the complaints brought forth by owners. In mares, a condition has been identified in which the mare shows stallion-like behavior and has an excessively good muscle tone for her conditioning program. Hormonal measurements show very high testosterone levels. There are a number of possible sources of the testosterone, since cholesterol, estrogen, and testosterone can be converted between each other, but glands that need additional evaluation include the ovaries, adrenal glands, or pituitary gland. One mare is known to have produced a filly foal that went on to exhibit similar signs.

Occasionally by female dogs will show an increase in aggression for a week or so, about every 6 months, even though ovaries and uterus have been removed. Because hormone cycling is tied into the cycling of the entire body systems, periodic expressions of behavior are certainly possible even though the ovaries are no longer part of the picture.

Behavioral Problems that Parallel Human Conditions

Researchers and clinicians are beginning to find a number of human conditions that have an animal model. This not only provides researchers with a valuable resource, it may provide the animal with better treatments and owners with a better understanding of what is wrong with their animal.

Interstitial cystitis

Interstitial cystitis is a condition in humans and cats in which the individual experiences painful urination and straining. The signs are typical of those associated with a bladder infection, but in these cases, there are erosions of the bladder wall. The problem in cats is difficult to diagnose because the urethra is very difficult to catheritize, and thus, it is difficult to use an endoscope to visualize the inside of the bladder. Surgical biopsies are seldom done. In both species, there seems to be a connection with stress, so treatment is centered around controlling or eliminating the stress in their lives.

Portosystemic shunts

Dogs and cats can have a shunting of blood such that some of it bypasses the liver. Because of this, the liver is not able to detoxify the blood or pull out those things that allow that organ to function normally. Affected animals will typically show the worst signs of their condition following a meal, returning to near normal after several hours. Their presenting signs can vary from neurologic disease (seizures, blindness, and head pressing) to behavioral signs such as poor learning ability, aggression, stereotypies, and hyper behavior. Most of the affected animals will also have a stunted body size because the nutrients cannot be properly processed.

Hydrocephalus

Miniaturization of dogs has been associated with the increased tendency for them to have hydrocephalus. There is also an apparent hereditary factor; there is a high prevalence in bull terrier dogs. Affected animals may present for medical problems such as seizures, but behavioral changes are even more common. Poor learning ability, inability to housetrain, aggression, and hyper behaviors are common. Stereotypies such as the circling shown by bull terriers, can become very extreme. The surgical correction of shunting spinal fluid from the brain into another part of the body is done in humans and severely affected animals.

Hyperactivity/hyperkinesis

There is an animal version of hyperactivity/hyperkinesis, and perhaps even the model to show that there are at least two separate versions. Affected animals are presented because they just cannot seem to settle down, they are hard to train, and habituate poorly. As with humans, there are often coexisting conditions as well. In veterinary medicine, we have found that the hyperkinetic dog will calm down with a stimulant, as usually happens with attention deficit hyperactivity disorder (ADHD) people. This is probably seen in other species as well. As an example, race horses have been medicated with a stimulant 'to get their mind right,' implying that the condition exists in them.

Hyperactive animals are presented with the same signs, but they will calm down to normal with mild doses of a tranquilizer instead of the stimulant. Unfortunately, the imaging modalities that can be used in humans are not available to use on unanesthetized animals, so for this condition, the animal would not make a good research model for humans.

Because the presenting signs are the same, diagnosis is made by response to therapy. Other conditions, such as hyperthyroidism, inadequate exercise for caloric intake, allergies/atopy, and drug reactions, can cause the same presenting signs. These complicate the diagnosis. Coexisting problems may also mask the primary problem and their treatment may be less than successful unless the 'hyper' problem is also treated. In fact, if the 'hyper' problem is successfully diagnosed and treated, the coexisting problems may actually disappear.

Feline hyperesthesia syndrome

A relatively recent connection between an animal condition and one in humans has been made for the feline hyperesthesia syndrome. Affected cats will be acting normally and suddenly their eyes will dilate and the skin will twitch along their backs. Often they vocalize and then may dash around the house. These are episodes of explosive arousal that may be spontaneous or triggered by tactile stimulation. Researchers have shown that in a few of these cats, there is spontaneous electrical activity in the epaxial muscles. Biopsies demonstrated vacuoles that are similar to those seen in human-inclusive body myositis/myopathy.

Narcolepsy

Several animal species are available as models for the study of narcolepsy. The most common sign is the abrupt onset of a sleep episode during a time or event where this would not be normal. Restless sleep at night is also common. A brain chemical, hypocretin, helps regulate staying awake and stabilizes rapid eye movement (REM) sleep. Affected individuals have low hypocretin levels, although it is not understood why this happens yet. The Doberman

Pincher dog model has been used most as a genetic model and has been important in establishing various treatments.

Stereotypic behaviors

Stereotypic behaviors are repetitive behaviors that are not functional. They are often rhythmic in manner and may start as a normal response to a specific situation. Over time, their expression becomes more rigid. Animals have shown a number of different stereotypic patterns. Some are oral in nature, such as tongue play, excessive grooming, cribbing/windsucking, rubbing of teeth, and licking. Locomotor patterns are seen as stall/kennel circling, walking back and forth with a head flip, tail chasing, weaving, digging, kicking, or pawing. Freezing in location, nose rubbing, head shaking/nodding, playing in water, tail swishing, self-rubbing, and self-mutilations are more generalized and thus may constitute a third category.

The development of stereotypies may be an initial response to stress or other and in this acute phase, the expression is less rigid than at later stages. It is believed that the expression of the behavior may release neuroendorphins, and thus have a stress-relieving effect. This pattern can usually be broken by addressing the stressors, tightening up the animal's schedule, and increasing exercise.

If allowed to continue, the acute phase gives way to the chronic phase, and some work suggests that this will occur after about 6 weeks. In the chronic phase, the pattern of behavior becomes more rigid and the brain apparently changes to make this behavior a default one. Regardless of treatments, the animal will revert to this pattern with any new stress, even if the owners can successfully stop the problem in the first place. Some recent work with equine cribbing/wind sucking (aerophagia) suggested that there may be medical conditions related to the development of this stereotypy. Affected horses showed the behaviors more often at a time when ingesta was reaching the cecum. Their fecal pH was lower than nonaffected horses. These individuals were also more likely to have gastric ulcers and show improvement with antacids. Baseline cortisol levels were higher, suggesting that they were more susceptible to stress, and beta endorphin levels were lower. Another study of psychogenic alopecia in cats found that 19 of the 21 cats had underlying skin disease. These findings indicate that stereotypic behaviors may have complicated etiologies and need complex treatment protocols.

Obsessive compulsive disorders

An obsession is a pervasive thought that is considered to be intrusive and senseless. The compulsion is the intentional behavior performed in a repetitive fashion in response to that obsession. Unfortunately, we are not able to ask animals about their thoughts, and so will never know for certain that the behaviors are the animal equivalents of the human obsessive compulsive disorders (OCDs). However, the determination of affected individuals to perform the specific behavior strongly suggests that OCD does exist.

The patterns shown by individuals with presumed OCD vary. They can range from flank sucking, acral lick dermatis, fly snapping, light chasing, cribbing/windsucking, circling, to tail chasing. In some cases, the behavior is a stereotypic one, but not always. Certainly not all OCDs are stereotypies and not all stereotypies are OCDs. This suggests that there may be preexisting brain differences that predispose individuals to developing OCDs, but at this time, we do not know.

Treatment protocols are only partially successful at eliminating the problems. In a study of Doberman Pinchers with acral lick dermitis, drug therapy helped only about 50% of the dogs, and then, only about half of those had reasonably good responses.

Self-mutilation syndrome

The self-mutilation syndrome is a self-directed behavior in which the animal bites itself, usually to the point of creating a wound. Stallions represent 70% of horses showing the behavior, with geldings and mares about equally divided in the remaining 30%. The majority of affected individuals bit both sides of their body, usually near the point of the shoulder, but other behaviors are also common, such as the 40% that rub, spin toss their head, or roll; 40% that buck; 39% that are hypersensitive to touch; and 32% that vocalize. The self-mutilation syndrome has been compared to Tourette's syndrome in humans. Affected people show a variety of tics, and many have uncontrolled vocalizations too. The etiology is a combination of genetic and environmental factors, which seems to hold true in affected horses as well. The condition has been seen in racing stallions that are injured but maintained on their regular performance rations. Since it is more common in Arabians, Quarter Horses, and Standardbreds, there is some question about the genetic implications. This, however, has not been proven yet.

Conditions with Medical and Behavioral Differential Diagnoses

There are several examples of conditions that may have both medical problems and behavioral problems that need to be considered before the actual diagnosis is reached. The following examples are primarily associated with domestic animals, but would not have to be limited to those species. It should be recognized, however, that some of the behavioral differentials would not necessarily occur in wild populations.

Housesoiling by urination

When a dog or cat urinates in the house instead of outdoors or in a litterbox (as in the case of most house cats), it can be a sign of a medical or behavioral problem. If not quickly and accurately dealt with, the owner might choose to get rid of the animal, although there are cases where owners have lived with the problem for years, and recognize it as a problem only when they are going to get a new carpet.

Differential diagnoses would include a number of medical conditions such as an infection of the urinary bladder, interstitial cystitis, pyelonephritis, polydipsia, metabolic/endocrine diseases, urinary incontinence, and neurologic disease of the spine or brain. These things must be worked up to ensure proper treatment.

Differential diagnoses must also include behavioral differentials such as incomplete/improper housetraining, poor litterbox management, and lack of or poor access to appropriate elimination areas. Dog owners are notorious for assuming that the dog is housetrained after a few weeks of working with them, and then expect the puppy to tell the owner when it needs to go out. They also will leave puppies alone longer than the bladder capacity will allow. These are not realistic expectations. Cat owners tend to have too few litterboxes, locate them too far away, and neglect to clean them often enough. When the litterbox is put far away from the daily activity of the kitten, it encourages accidents when the urge to urinate comes on suddenly. The location of the box may also be associated with undesirable things such as noisy clothes, dryers, or half doors that must be vaulted over. Since most owners clean the litterbox no more than once a week, the accumulation of urine and fecal matter and the associated odor can prove repulsive to a cat.

Housesoiling by defecation

A fairly common problem among housebound dogs and cats is defecation in the house. This can be the result of the sudden need to defecate, or it can be the buildup of feces that was not expelled when the animal was outside. As an example, a dog that normally defecates outdoors in the morning before the owner leaves for work might not do so if the weather is bad. Then the dog is indoors and the urge to defecate becomes so strong that the animal can no longer control that need. The differential diagnoses for this problem include a number of medical conditions such as intestinal parasites, colitis, inflammatory bowel disease, and several neurologic diseases. Animals with neoplasia, megacolon, and some metabolic/endocrine diseases will also show housesoiling. Musculoskeletal problems such as osteoarthritis or lumbosacral instability, and dietary problems that result in very dry feces can make defecating or posturing to defecate painful. The animal will avoid going until it is absolutely necessary, and so may not be in the right place. The same behavioral conditions mentioned with urination problems above also need to be considered.

A somewhat similar problem exists with some horses, where they defecate in their water buckets or manger instead of in another stall location. For them, the behavior seems more likely to be a behavioral choice rather than a medical condition.

Cognitive dysfunction

Age changes occur in the brain of older individuals of many species. With Alzeheimer's disease in humans, there is a deposition of beta amyloid plaques, cortical shrinkage with ventricular dilation, neurofibrillary tangles, and a decreased blood supply. Similar changes have been shown in dogs and cats, among other species; although, the neurofibrillary tangles do not occur. Clinical signs such as forgetfulness (going to the wrong door and housesoiling) and reduced social interactions can be similar to humans. In animals though, there are a number of geriatric-related conditions that must be considered as differential diagnoses that must either be ruled out or determined to coexist with cognitive dysfunction (**Figure 3**). The primary complaint that dog owners have is that the animal is now urinating or defecating in the house, while cat owners notice the increased vocalization.

Treatments for cognitive dysfunction are showing a lot of promise, and are certainly being looked at in human medicine as well. The response to antioxidants and drugs with that effect shows a lot of promise. Other drugs that increase dopamine levels or blood flow to the brain are being used as well.

Coprophagy

Coprophagy is the eating of feces. While the behavior is normal for the young of most species, probably to help them establish intestinal flora, and for the dams of young of some species, to keep the nest area clean, it can occur for less desirable reasons. In some cases, the behavior is associated with an animal kept in a barren environment, thus resembling a behavior of 'boredom.' Coprophagy has also been seen as the result of various medical conditions. These include an exocrine pancreatic insufficiency, hydrocephalus, and high parasite burdens.

Eating the feces of another species is commonly done by dogs. Cat feces is high in protein and the smell/taste

Cognitive Dysfunction

Differential diagnoses

    Reduced sensory perception
    Musculoskeletal pain
    Urinary tract disease
    Separation anxiety
    Environmental phobias
    Obsessive compulsive disorders
    Inadequate housetraining
    Urine marking
    Peripheral or central neurologic disease

Figure 3 Differential diagnoses for cognitive dysfunction.

seems to attract dogs. Horse feces has predigested vegetable matter and attracts some dogs in the same way that regular grass does. In this case, however, the predigestion by the horse's gastrointestinal system allows the vegetable matter to be processed in the dog since the cellulose bonds have already been broken.

In summary, it is not always possible to separate medical from behavioral etiologies, and in some animals, one may be complicated by the other. Certain traits tend to support medical problems over behavioral ones, such as conditions seen in the very young or very old. Conditions with an abrupt onset or a change in the character of the animal will also support medical problems. There are breed predispositions for both medical and behavioral conditions that create long lists to rule in or rule out for a diagnosis. Behavioral problems often have an identifiable trigger or associated event that suggests environmental cue responses rather than physiological ones. Other preexisting medical conditions are suggestive that the new problem could be associated either as a variation or as something else to add to the list of the animal's specific problems. Establishing a cause and/or treatment for a problem, so that that the animal can experience best welfare, can be a diagnostic challenge.

See also: Insect Flight and Walking: Neuroethological Basis.

Further Reading

Beaver BV (2003) *Feline Behavior: A Guide For Veterinarians*, 2nd edn. St. Louis: Saunders.
Beaver BV (2009) *Canine Behavior: Insights and Answers*, 2nd edn. St. Louis: Saunders.
Hart BL (1988) Biological basis of the behavior of sick animals. *Neuroscience & Biobehavioral Reviews* 12: 123–137.
Hart BL, Hart LA, and Bain MJ (2006) *Canine and Feline Behavior Therapy*, 2nd edn. Ames: Blackwell Publishing.

Distributed Cognition

L. Barrett, University of Lethbridge, Lethbridge, AB, Canada

Introduction

The notion that our brains alone make us the people we are is one that permeates Western thought and has infiltrated our popular culture. A quick Google search reveals that there are at least around 30 films that involve some form of brain or mind transference, where one person's mind or brain somehow ends up in another person's body. Clearly, we find this idea very appealing and, however implausible the actual mechanics are, we seem happy enough to buy into the notion that if a person's brain is moved to another body, that person would, to all intents and purposes, remain the same. This is a very Cartesian perspective: as a plot device, 'body swaps' or 'mind transfers' revolve around René Descartes' famous dictum 'I think, therefore I am' and the idea that we are made up of 'mind stuff' (or at least 'brain stuff') that is entirely distinct from our body stuff. Of course, bodies are needed to carry our brain-minds around, but for the most part, if it's a healthy functional body, there are assumed to be no adverse consequences of finding 'ourselves' in somebody else and, it's a neat way for the hero or heroine to learn some important life-lessons by wandering about, quite literally, in someone else's shoes.

Even if one disregards the notion that there is some kind of incorporeal 'mind stuff,' and adopts a purely materialist notion that our brains are our minds (although how this actually works is anybody's guess ...), we still have a strong sense that our cognitive abilities reside solely with our brains. Many of us are constantly reminded to 'use our heads, not our hearts' when it comes to decision-making, and when a man is accused of thinking with something other than his brain, it is rarely a compliment. Bodies are a necessary encumbrance, then, but have nothing to do with how we think about the world. This is probably because, as linguistic, brainy creatures, we tend to focus only on certain cognitive processes, such as logical, linguistically based rational problem solving, as the main job that our brains do for us. It is also the case that we assume that all these cognitive processes are, in the words of Andy Clark, securely bound by our 'skin and skull,' and happen only in our brains. But is this really accurate? If we adopt a broader perspective, we can see that perhaps cognition isn't all in the head, perhaps our bodies are more involved than we suppose, and perhaps some of our cognitive processes are distributed even more widely than that, reaching out into the environment itself. Perhaps we need to think again about the nature of thinking?

Embodied Cognition

Let's start with the idea that bodies are an integral part of cognitive systems. That is, let's consider the idea that cognition is embodied. If one thinks about it from an evolutionary perspective, the idea of embodied cognition makes perfect sense – after all, all animals possess bodies, and they all did so before they possessed anything remotely resembling a brain. Indeed, the term 'embodied cognition' is something of a misnomer because all cognition (outside of a computer science laboratory) is, by definition, embodied. Looked at from this perspective, it then becomes easier to see that brains must first have evolved as a means for animals to gain greater control over their physical actions in the world, enabling them to respond to unpredictable environmental changes in a more flexible, and so more effective, manner; a brain is for doing things, for behaving in intelligent ways, not for thinking intellectually about them. It seems that our own abilities to engage in intellectual, 'inactive' thought has led us to a very anthropocentric view of cognition, and so we often fail to appreciate that many of the things we take for granted, such as making a cup of tea, recognizing a familiar face in a crowd, or even walking across a room without falling over, are also feats of immense skill.

Another way to look at this is to use the distinction that Andy Clark makes between mind as a 'mirror of the world' versus mind as a 'controller of action in the world.' Our 'classical' view of cognition is the 'mirror' view, where the brain stores 'passive' inner descriptions (representations) of the external world, which it manipulates and processes to produce an output, which is fed to the motor system to produce action in the world. The sequence of events can be characterized as 'sense-plan-act,' with a clear separation between perception and action. In contrast, an embodied perspective highlights the fact that being a physical creature results in a high level of interconnectedness between different bodily systems – changes to one component will affect all the others. Consequently, we should never treat sensory and motor systems as separate, but as tightly coupled, and we should see the brain/mind as a 'controller' of action in the world. In this case, an animal's inner states are not passive 'pictures' of the external world, but are, instead, 'plans of action' for engaging with the external world.

Action-Oriented Representation

This idea has been most actively promoted by the MIT roboticist, Rodney Brooks, who rightly points out that, when we view the vast sweep of evolutionary history, it is immediately clear that most of the time has been spent perfecting the so-called 'simpler' sensorimotor mechanisms that enable survival in a dynamic world. The 'high-level' forms of cognition, such as planning, logical inference, and formal reasoning, which we tend to associate with cognitive processes, evolved very late in the day and did so very quickly. This implies that all the 'higher' cognitive faculties – those that we consider to be the most complex – must actually be quite simple to implement once the essential perceptual and motor processes that allow an organism to act in the world are in place. What is more, these perceptual and motor processes must, as a direct consequence, underpin the evolution and elaboration of these 'higher' functions, so that they are not free of bodily influence in the manner we tend to assume. In other words, an animal's knowledge of the world is fundamentally tied to its physical actions in it. Consequently, because an animal's representations of the world are linked to, and controlled by, its acting body, they should be heavily 'action-oriented' (Clark, 1997); that is, they should describe the world by depicting it in terms of the possible actions an animal can take.

An elegant example of action-oriented representation is given by the work of another roboticist, Maja Mataric. She designed a robot rat that had the ability to construct an internal 'map' to guide its movements in a cluttered environment. This map was made up of a combination of the robot's motion and its sensory readings as it moved around. As the robot encountered a wall, this landmark was not represented in the map as 'a solid vertical object' but instead was stored as a combination of actions such as 'moving straight, with short lateral distance readings heading south.' The map formed by the robot was simultaneously a map of the layout of the environment and an action plan. This means that, contrary to the 'classical' view of cognition, there was no need for the rat robot's sensory perceptions of the environment to be transformed (by some form of cognitive processing) into an action plan, because the perception of the environment was already specified in terms of the actions of the rat. Interestingly, if one takes a real rat and prevents it from moving its legs as it is carried around a novel environment, then there is no activation change in its hippocampus (the part of the brain associated with spatial mapping). This strongly suggests that, in real rats, motion is also crucial to the generation of an internal map, and that perhaps real rats also use similar kinds of action-oriented representations to construct their maps of the environment.

An embodied, 'action-oriented' approach has at least three consequences for how we think about cognition. First, it means that in some circumstances, there will be no need for an organism to possess any form of internal 'mirror-like' representations of the external world at all. If to sense something in the world is simultaneously to generate an action plan for what to do next, an animal can simply rely on what is in the world to guide what it should do. It can, in Brooks' words, 'use the world as its own best model.' This makes eminent sense evolutionarily, because building an internal representation of the world and then using that as the basis for action, and 'throwing the world away' is costly in terms of brain tissue and energy expenditure. Given that evolution is a thrifty process, tightly coupled perception–action mechanisms that do not require costly internal cognitive processing should be quite common across the animal kingdom. Second, if concepts of the world are grounded in the ways in which an animal acts in the world, then we must think twice when attributing human-like processes to other species. A creature with fins, wings, or flippers is unlikely to understand the world in the same way as large, hairless, bipedal apes, like ourselves.

The third consequence of an embodied approach is that we need to rethink our assumption about the link between the complexity of an animal's behavior and the level of internal cognitive complexity that such an animal possesses. When adopting the classical view, we are prone to assuming a direct one-to-one mapping between the complexity of behavior produced by an animal and the complexity of the proximate mechanism that produced it; an animal capable of complex behavior is assumed to be in possession of an equally complex cognitive architecture. An embodied approach shows us that this assumption is false and that there is no necessary relation between behavioral and cognitive complexity. Simple mechanisms can produce highly complex behavior as a result of the interaction between an organism's brain, body, and environment. Building a termite nest is an immensely complex behavior, for example, but an individual termite is not a psychologically complex animal. All a termite needs to know is how to make a ball of dirt, impregnate it with pheromone, and carry it around until it encounters other similarly impregnated balls of dirt, and drop it next to them.

Behavioral Complexity and Cognitive Complexity

It is easy to see this with termites, of course, but for other organisms, such as many species of birds, primates, and cetaceans, it can be more difficult, and this means that we run the risk of mistakenly attributing more complexity to the animals than is warranted, as well as over-estimating the cognitive requirements of a given task. An extremely powerful demonstration of this effect is provided by work didabots, which are small-wheeled robots.

When placed into an arena in which a series of obstacles have been placed (polystyrene blocks), the robots trundle around, pushing them together into clusters and apparently 'tidying up' the arena. On the face of it, one would immediately assume that the robots possessed some internal rule(s) for detecting objects and then pushing them together. In fact, the robots are equipped with two sensors on either side of their bodies that, when activated by an object within a certain distance, lead the robot to turn in the direction away from the object. In other words, the robots are programmed exclusively to avoid obstacles. Clustering behavior occurs because of the specific configuration of the sensors on the robots' 'bodies.' The sensors are placed at an angle on the front end of the robot. As the robots move forwards, the sensors can detect cubes off to the side, but not straight in front. This means that, although the robots turn away and avoid cubes on either side, a cube directly in front of them is pushed along, because the didabot cannot 'see' it (i.e., its sensors receive no stimulation from it). If the didabot then encounters another cube off to the side, triggering its sensor, it produces avoidance behavior, moving off to the left or right, and leaving the object it has just been pushing next to the object it has just avoided. In other words, the didabot clusters the two cubes, and over time, this simple self-organizing process produces an ever-larger cluster of cubes and a very tidy arena. Change the robots' bodies, however, and you change their behavior: moving one of the sensors directly to the front of the robot results in the complete absence of clustering behavior, because objects directly in front of the robot are now avoided in the same way as those off to the side, which means that no pushing behavior occurs.

Morphological Computation

The role of the body in reducing the demand for specific neural control of behavior (and so, reducing the demand for expensive brain tissue) has been termed 'morphological computation,' and some striking examples have emerged from the field of artificial life. Simply by adopting a particular spacing the facets (the ommatidia) in the compound eye of the fly, one can create an eye that compensates for motion parallax (the way in which objects to the side of an organism travel faster across the visual field than those at the front). Specifically, the ommatidia should be clustered together more densely toward the front of the eye, because this arrangement can automatically perform the 'morphological computation' that

would otherwise have to be performed in the fly's brain. Similarly, Puppy is a four-legged running robot, with a total of 12 joints (one at each hip and shoulder, one at each knee, and one at each ankle) with springs that connect the lower and upper parts of each leg. There are also pressure sensors on the feet that indicate when the foot is in contact with the ground. The control system of the robot is extremely simple – there are motors that simply move the shoulders and hips backwards and forwards in a rhythmic fashion. If you place Puppy on the ground, it will scrabble around for a bit, as it gains purchase on the surface, and then settle into a remarkably life-like running gait. This is due to a tightly coupled interaction between the control movements of its hip and shoulder joints, its anatomy (its overall shape and how the springs are attached) and the environment (the friction on its feet produced by the ground surface and, of course, the force of gravity). There needs to be no sensory feedback or central (brain-like) control of Puppy's movements because its artificial 'muscles' – the springs – perform the necessary morphological computation that helps keep Puppy on an even keel. The human knee joint does much the same thing. As you will have noticed, your knee has a remarkable freedom of movement compared to some other joints. This ability to wobble around a bit is what makes it easy for us to cope with uneven ground when we walk rapidly and smoothly; we do not need any specific sensory feedback to be sent to our brains, followed by the activation of motor neurons to activate specific muscles. Instead, our knees morphologically compute the necessary adjustments, allowing both greater speed and requiring less neural tissue.

If the examples from artificial animals seem somewhat removed from the world of real animals, work on the behavior of rat pups, real as well as robotic, should help to reveal the necessity of taking an animal's body and environment into account when trying to explain the complexity of behavior. Robot rats, built with a completely random control architecture (i.e., no internal 'rules' for how to behave with respect to other rat pups), were found to display patterns of behavior that were either intermediate between, or identical to that of, 7- and 10-day-old rat pups. The robotic rats showed the same kind of 'goal-directed' behavior as real pups, as they followed the walls of their arena ('thigomotaxis'), huddled together with other robot 'pups,' and borrowed into corners. In each case, the 'goal-directedness' of the robots' behavior resulted simply from the interaction of the geometry of their bodies and that of the arena in which they were placed. When a wall was contacted, the tapering nose of the robot meant that it slid along the wall, with its direction determined by its angle of approach. The options for any other kinds of movement (i.e., those that allowed it to move away from the wall) were constrained by this contact, resulting in a high probability of wall-following. If the robot encountered a corner, the effectiveness of any other movement at all became extremely limited, with the only option being a kind of backing-up maneuver. Even this option was limited, though, if other robot rats randomly encountered the robot in this position. As other robots pressed in from the sides, so behaviors like 'huddling' and 'corner burrowing' were seen, just like in real rat pups. This does not mean that real rat pups have only a random neural architecture, but it does mean that they need not be equipped with any dedicated sensorimotor routine or a specific kind of neural processor that produces thigmotaxic or huddling behavior. Rather, and as Brooks has long suggested, the bodies of the rat pups may be so tightly coupled to the environment, and so mutually constraining, that no cognitive control at all is required to produce the behaviors seen.

These examples, therefore, highlight beautifully that the proximate means by which a behavior is produced need bear no relation whatsoever to the form that behavior takes (who, e.g., would imagine that a rule for object avoidance would be a good way to produce object clustering?) and completely destroys any notion that there is, by necessity, a simple one-to-one mapping between the complexity of a proximate mechanism and the complexity of the behavior that it produces (as body-world coupling can be sufficient to rule out the need for any specific control process). If we take on board the lessons that the didabots, Puppy, and robot rats offer, then it is clear that a focus on the kinds of emergent, contingent 'mind' that brains, bodies, and environments can achieve in concert may prove more productive than persisting with the idea that 'intelligent' behavior is achieved solely by raw brain power.

The consort behavior of male baboons provides us with another neat illustration of what this more holistic approach entails. Male baboons socially and sexually monopolise adult females during their fertile periods, remaining in close proximity to them at all times. These close spatial relationships ('consortships') can last from a few hours to a week, depending on the specific population of baboons. Among East African populations, these consortships are often disrupted by aggression from other males, who then take over the consort male's position. There are various social tactics that males can employ to either avoid or facilitate a take-over. Anthropologists, Shirley Strum and her colleagues, were able to show that much of the behavior that is often held up as an example of 'Machiavellian Intelligence' (i.e., as sophisticated cognitive strategizing to achieve a specific goal) may actually be the result of how particular animals are either constrained or afforded certain courses of action by the environment. Their analysis focussed on one tactic that they called 'sleeping near the enemy.' This designation was based on the observation that, while older males were able to resist consort take-over attempts by younger and more aggressive males during the day, they were less able to do so at sleeping sites, where younger males were able to displace them and leave with the female in the morning. A change in topography, from the plains, where the animals foraged during the day, to the cliffs, where they slept at night, was the key factor leading to this difference.

Older socially experienced males could resist take-over on the plains by using social tactics to divert aggression, such as grabbing a younger animal and using it as a 'buffer' against attack by the male. Such tactics require a high degree of visual contact with others, a significant amount of behavioral coordination and, therefore, sufficient experience with other animals to deploy them successfully. On the sleeping cliffs, however, these tactics were constrained by topography. The height and narrowness of the cliffs resulted in changes in the mobility of males, their proximity to other animals, and a reduction in overall visibility. All of these factors served to reduce older males' ability to manipulate the situation socially while at the same time, favoring the more direct, aggressive tactics of younger males. Male behavior could, therefore, be more simply accounted for by recognizing that males were employing those behaviors in their repertoire that were afforded by the environment, and were prevented from using others due to the constraints that the environment imposed, rather than varying their tactics in a Machiavellian fashion to thwart and outwit their rivals.

Distributed Cognition

Acknowledging that both the body and the environment form an integral part of biological cognitive systems has further implications for theories of cognitive evolution. Specifically, it means that, just as animals can use morphological computation to reduce the strain on their brains, so we should expect animals to use the structure of the environment, and their ability to act in it, to bear some of their cognitive load, and save on expensive brain tissue. Recognition of the embodied nature of cognition, and the interaction between animals and the environment as part and parcel of cognitive processes, naturally gives rise to the idea that cognition is 'distributed.'

The basic idea behind a distributed approach to cognition is that actions in the world are not merely indicators of internal cognitive acts but actually cognitive acts in themselves. In this regard, David Kirsh has made a distinction between 'pragmatic acts' that move an individual closer to task completion in the external environment, and 'epistemic acts' that do not aid in the completion of the task itself, but place an individual in a better state in its cognitive environment so that the task becomes easier. Epistemic acts, then, are actions that help improve the speed, accuracy, or robustness of cognitive processes, rather than those that enable someone to make literal progress in a task. To give a human example, moving Scrabble tiles around makes it easier to see the potential words that can be formed, and can therefore, be considered as an 'epistemic act.' Jigsaw puzzles are also superb examples of how cognition is a distributed process. One simply cannot solve a jigsaw puzzle by thinking about it, and planning each successive move in advance. It just doesn't work – the task is too perceptually complex for a person to make any headway. Instead, solving a jigsaw puzzle is performed partly in the head and partly in the world, as one physically sorts, rotates, and moves the different pieces. One starts simply by joining complementary pieces and thereafter responds dynamically to the particular local patterns that appear. This close interleaving of physical and mental actions allows us to significantly reduce the complexity of the task and achieve our goal much more efficiently than using either physical or mental actions alone. Think of the way in which we use post-it notes, memory-sticks, notebooks, computer files, whiteboards, books, and journals to support our academic work, or the way in which we lay out all the ingredients we need for cooking so that what we need comes to hand at the moment we need it. As Andy Clark puts it, there is a true sense in which the real 'problem-solving machine' is not the brain alone, but the brain, the body, and the environmental structures used to augment, enhance, and support internal cognitive processes.

We can also look at other animals besides ourselves from this perspective. Kim Sterelny has argued that all animals can be considered to be 'epistemic engineers,' changing the world around them to change the nature of the informational environment. The contact calls that many bird and primate species produce to advertise their location simplify the task of keeping track of other individuals in the environment, for example. Similarly, the use of moss to reduce the conspicuousness of their nests is a means by which birds can engineer their environment to make the cognitive task of their predators that much more difficult. In a similar vein, the late James Gibson wrote extensively on how animals made use of the information structure of the environment to achieve their goals. His theory of 'ecological psychology' was aimed at illustrating how psychological phenomena were not to be found inside an animal's head alone but were produced by the interaction of an animal with its environment.

At present, we do not know the full extent to which animals epistemically engineer their environments, nor the degree to which they make use of epistemic acts in their problem-solving. This is simply because we have not investigated these ideas very deeply as yet. Incorporating an embodied, embedded approach into comparative psychology is a promising and exciting prospect for the twenty-first century, and should lead us away from the idea that intelligent behavior is the sole province of those creatures that possess large brains. Hopefully, there will also be wider acknowledgment and acceptance of the notion that cognition is not a property of the brain alone, but of the embodied, environmentally situated, fully integrated complex that makes up what we know more familiarly as an 'animal.'

See also: Cognitive Development in Chimpanzees; Morality and Evolution; Sentience; Social Cognition and Theory of Mind.

Further Reading

Barrett L, Henzi SP, and Rendall D (2007) Social brains, simple minds: Does social complexity really require cognitive complexity? *Philosophical Transactions of the Royal Society, Series B* 362: 561–575.

Brooks RA (1999) *Cambrian Intelligence: The Early History of the New A.I.* Cambridge, MA: MIT Press.

Clark A (1997) *Being There: Bringing Brain, Body and World Together Again.* Cambridge, MA: MIT Press.

Clark A (2008) *Supersizing the Mind: Embodiment, Action and Cognitive Extension.* Oxford: Oxford University Press.

Kirsh D (1996) Adapting the environment instead of oneself. *Adaptive Behaviour* 4: 415–452.

Maris M and te Boekhorst R (1996) Exploiting physical constraints: Heap formation through behavioural error in a group of robots. In: Asada M (ed.) *Proceedings of IROS'96: IEEE/RSJ International Conference on Intelligent Robots and Systems*, pp. 1655–1660. IEEE.

Mataric MJ (1990) Navigating with a rat brain: A neurobiologically-inspired model for robot spatial navigation. In: Meyer J-A and Wilson S (eds.) *From Animals to Animats: International Conference on Simulation of Adaptive Behaviour*, pp. 169–175. Cambridge, MA: MIT Press.

May CJ, Schank JC, Joshi S, Tran J, Taylor RJ, and Scott I (2006) Rat pups and random robots generate similar self-organized and intentional behaviour. *Complexity* 12: 53–66.

Pfeifer R and Bongard J (2007) *How the Body Shapes the Way We Think.* Cambridge, MA: MIT Press.

Pfeifer R and Scheier C (1999) *Understanding Intelligence.* Cambridge, MA: MIT Press.

Sterelny K (2004) Externalism, epistemic artefacts and the extended mind. In: Schantz R (ed.) *The Externalist Challenge*, pp. 239–254. Berlin: Walter de Gruyter.

Strum SC, Forster D, and Hutchins E (1997) Why Machiavellian intelligence may not be Machiavellian. In: Whiten A and Byrne RW (eds.) *Machiavellian Intelligence. II. Extensions and Evaluations*, pp. 50–87. Cambridge: Cambridge University Press.

Division of Labor

J. H. Fewell, Arizona State University, Tempe, AZ, USA

Introduction

Division of labor occurs when different individuals within a group specialize in the different tasks necessary for the maintenance and growth of a social group, from food gathering to production and care of its offspring. Division of labor is a fundamental attribute of sociality and is considered perhaps the key contributor to the success of eusocial insects – ants, bees, wasps, and termites. In these taxa, division of labor allows the colony to produce complex structures, such as the comb hives of honeybees and the elaborate domes of some termite nests, and highly organized and efficient systems for food gathering and storage. However, division of labor is not limited to the eusocial insects. It can be found within aggregations and communal insect societies. It has been reported in shrimp, caterpillars, dung beetles, and spiders, and has even been produced spontaneously in artificial associations of normally solitary insects. Parallel patterns of division of labor are seen outside of the invertebrates, particularly in the social mammals, although the specific mechanisms by which division of labor is produced may vary.

As defined here, division of labor is a statistical, rather than an absolute pattern of behavioral differentiation, and can be measured across different time scales. Included within this category are the queen–worker dimorphisms of many eusocial insects which are absolute across their lifetimes. Also included is preferential foraging for pollen versus nectar by individual bees, which may perform the task of foraging for only a few days. As illustrated by these two examples, division of labor is generally divided according to whether it involves differentiation for reproduction or for other behaviors central to colony function, generally referred to as tasks. Because of the different consequences and mechanisms for these two different sets of tasks – reproductive and nonreproductive – they are generally treated as separate categories in evolutionary and behavioral research.

Reproductive Division of Labor

Reproductive division of labor occurs when only one or a few individuals within the colony are responsible for producing the colony's offspring. The division of a colony into a few reproductive individuals and multiple sterile workers is a critical transition in social evolution, as it involves a shift from working to maximize one's own direct fitness to assisting another individual to reproduce. The division into reproductive and sterile helper castes, or workers, is thus one of the essential criteria for categorizing a social group as eusocial.

As defined by Wilson in his classic *The Insect Societies*, eusocial colonies are those in which there is (1) overlap of generations (both parents and adult offspring are present in the colony), (2) cooperative brood care (individuals help rear offspring that they did not produce), and (3) reproductive division of labor. This last category, reproductive division of labor, is considered the litmus test for eusociality. In the highly eusocial Hymenoptera, including most of the ants, and some wasps and bees (most notably the honeybee), there is generally only one reproductive female present in the colony, and she produces all the thousands to millions of eggs necessary for colonies to grow and reproduce. All other females are sterile workers that perform the other tasks necessary to keep the colony functioning (although workers can on occasion produce male-destined eggs). Males produced by the colony generally perform only the behavior of mating with, and fertilizing new queens. Because Hymenopteran males generally do not do work related to the maintenance of the colony, they are considered to fall outside of the reproductive caste system.

The rationale behind workers in the eusocial Hymenoptera being female lies in large part with their system of haplo-diploid sex determination. Female Hymenoptera are diploid (they have two sets of chromosomes), receiving half of their genome from their queen mother and half from their father. In contrast, males are produced from unfertilized eggs, usually laid by the queen. This unusual system of gene transfer results in high levels of relatedness among female workers. Because males have only one set of chromosomes to pass on to their daughters, all workers with the same father automatically have half their genome in common as a result. They additionally receive on average one-fourth of their genome in common from their mother, making them highly related to each other, and to any new reproductive females the queen may produce. From an evolutionary perspective, this means that, for females, becoming sterile comes at the cost of losing one's own reproductive output, but with the benefit of helping a close relative (up to three-fourths of their genome in common) produce many more offspring. In contrast, males have only the genomic information of their queen mother. Without the additional contribution of patrilinial DNA, they are on average only one-fourth related to their sisters; this reduces the evolutionary benefit of giving up reproduction to become a worker.

Although, workers in the eusocial Hymenoptera are universally female, this is not the case for the other major group of eusocial insects, the termites. Like the eusocial Hymenoptera, the higher termites (family Termitidae) show complex systems of division of labor, including reproductive castes. However, their reproductive and worker castes contain both males and females. Correspondingly, termites are diploid, and males and females each receive two sets of chromosomes, one from each parent. Although a diploid system cannot produce the high levels of relatedness seen for haplo-diploidy in the absence of other factors, levels of relatedness within termite colonies are often high. One possible reason for this may be high levels of inbreeding, or mating within a family group, increasing the probability that individuals within the colony have high proportions of their genomes in common. In the case of termites, inbreeding would produce high relatedness for both males and females.

Determination of Reproductive Caste

Most of the brood production in eusocial colonies is production of workers, which contribute to colony size but because they are sterile are not considered reproductive output. In many of the ant taxa and in the honeybees, colonies produce functional queens and males only during narrow windows of time. These new queens mate, and begin new colonies of their own. In a system in which some individuals reproduce and others remain sterile, the mechanisms by which a queen is chosen become important. A general principle in reproductive division of labor is that queen and worker castes should be determined by environment rather than by genetics. The argument behind this principle is simple: if one genetic variant produces sterility while another produces offspring, the variant for sterility would be quickly selected out of the population. This principle is upheld in the vast majority of cases. However, there are some rare and interesting exceptions. In some populations of harvester ants, *Pogonomyrmex*, queen versus worker castes are determined by genotype, so that females heterozygous for multiple markers become workers and homozygous females become queens. In these populations, new queens must mate multiply to ensure that they have sperm from males both genetically similar (to produce daughter queens) and different from them (to produce new workers).

In most systems of eusociality, however, the differentiation between workers and queens is a developmental process primarily associated with environment. As we move through stages of eusociality, from primitively to highly eusocial, there is an increase in the degree of physiological separation between queens and workers, associated with changes in the environmental factors affecting development and behavior. In primitively social colonies, queens and workers are differentiated by function rather than physiology. Multiple females besides the queen may be physically able to mate, and may even have developed ovaries. Determination of who becomes the queen is associated with dominance interactions, and is primarily a function of the social environment.

In more derived eusocial systems, queens and workers differentiate in size and development, such that queens become much larger and contain more fat stores than workers, allowing development of functional reproductive organs only in queens. Differentiation of queens and workers is driven via variation in nutritional quality and quantity during larval development. In the most highly eusocial insects, such as honeybees and most ants, the difference between worker and queen is still nutritionally based, but the different nutritional elements fed to the developing larva and their programmed physiological responses are highly integrated. Larvae destined to be new honeybee queens are fed a mixture of pollen, nectar, and royal jelly, a mixture dense in protein and fats, but also hormonally rich. The larva fed this compound develops into a queen that is not only much larger, but also has different physical apparatus, from reproductive organs to other neural and physical structures. Ant larvae that are fed differentially according to whether they are queen or worker destined also develop significantly different anatomical characters; workers are wingless, but queens emerge from pupation with wings used to fly for mating. After mating, queens chew off these wings before establishing their nests, and will remain wingless for the rest of their lives, generally spent underground.

Division of Labor and Worker Specialization

Although reproduction is the ultimate determinant of fitness, the work needed for a social insect colony to maintain itself and grow goes far beyond reproductive output, including care and feeding of the queen and brood, nest construction and repair, waste management, nest guarding and defense, and foraging. All these chores must be allocated appropriately, such that they are performed when needed and, also importantly, reduced in performance when need is low. The system by which colonies flexibly distribute workers across tasks is often termed 'task allocation,' but workers are not actually assigned tasks by some external supervisor. Instead, the colony is a distributed system, in which the organization of work emerges from the cumulative decisions made by individual workers across the colony based on local information received and given. The mechanisms producing division of labor in a colony are an integration of the different genotypes and developmental trajectories among individual workers that predispose them to specific tasks, linked with social communication of task availability and performance.

If division of labor is measured as the degree to which different members of a group specialize in different tasks, then the fundamental building block of division of labor is task specialization. As with division of labor itself, task specialization is a statistical concept. Individuals specialize when they preferentially perform one task over all other tasks available to them. There are multiple mechanisms at the genetic and developmental levels that contribute to variation in individual task performance and specialization. These include morphological castes, in which individuals vary in morphological or physical features associated with specialization on different task sets; age polyethism, in which individuals perform different tasks at different ages or developmental stages; genetic (or intrinsic) task specialization, in which individuals of the same age or morphological group differ in the tasks they preferentially perform because of genetic and/or developmental variation. The degree to which each of these mechanisms applies to a social group varies with level of sociality. Both age polyethism and morphological castes are associated with more advanced eusociality, while division of labor based on intrinsic variation shows up even in the noneusocial taxa.

Morphological Castes

Physical worker castes are a less common manifestation of individual task specialization, but are often what is immediately thought of when considering division of labor in the social insects. Most morphological variation associated with task performance occurs as variation in body size with smaller workers more likely to perform in-nest tasks, while larger individuals work as foragers or soldiers. It costs colonies more metabolically to produce and maintain a larger worker body size, but size does convey an advantage for some tasks. For example, larger workers in bumblebee colonies can regulate body temperature better in colder environments and can carry larger nectar loads, while harvesting ants with larger head widths often transport larger seeds.

Variation in worker size is most common in more highly eusocial and larger colonies. This pattern occurs both across species, and ontogenetically as individual colonies grow from a queen and a few workers to several thousand individuals. In ant species in which size polymorphism occurs, newly formed colonies generally have smaller workers and less size variation than larger and older colonies. An extreme example of size polymorphism is found in the leafcutter ants. Large colonies of *Atta cephaloides* can have worker sizes ranging from minims with headwidths of 1 mm or less to majors with head widths of 7 mm or more. Minims tend the brood and the

fungus gardens that feed them. Medium-sized workers perform both in-nest and foraging tasks, while the largest workers act as soldiers, clear trails, and carry the largest leaves back to the colony. Majors often carry passengers, tiny minims that ride on the head of the major or on the leaf she carries, to repel parasitic phorid flies that can lay eggs in the crevices of the major's head.

Some species of ants and termites also show specialization associated with morphological changes in specific body parts, such as the mandibles, head, or abdomen. Morphological differentiation that significantly changes the allometry of shape is generally absent in the flying social insects. This makes sense, as flight itself requires a delicate balance in terms of load distribution; small changes in shape or size of the head can have dramatic consequences for lift and drag. However, in some of the more derived ants and termites, we see worker castes with highly specialized body parts. As an extreme example, soldiers of the nasutitermitine termites have heads molded into a long tubular nasus that squirts a sticky repellent at invaders.

Size and morphological variation may also associate with changes in neural processing and visual acuity that would further contribute to differences in task performance. Tasks such as foraging require interaction with a much more complex and visually rich environment than in-nest tasks (especially as the nest environment is often dark). In bumblebees, larger foragers have stronger visual fields that potentially allow for better flower discrimination. In ants and bees, foraging is also associated with increased size in the mushroom bodies of the brain, which are associated with visual processing and memory.

Age Polyethism and Foraging for Work

A common mechanism of task specialization in eusocial colonies is age or temporal polyethism, in which workers perform different tasks as they age from newly emerged through to old age. In the general schedule of age polyethism, as often diagrammed for honeybees, newly emerged workers often perform the duty of cell cleaning. At ~4 days to 2 weeks they transition to other in-hive tasks, including feeding brood, which requires associated physiological changes in the mandibular and hypopharyngeal glands to produce substances that transform pollen into nutritional brood food. Workers of this age range may also begin to produce comb wax from honey, with concurrent development of the wax glands.

From these tasks, they may transition to food storage, and finally to foraging, which is generally performed by the oldest workers. Bees often begin foraging at ~3 weeks of age, but this timescale varies considerably with genetics and environment. In a colony stressed for food or in the midst of a resource flow, workers can begin foraging at 7 days or earlier. An individual worker is also unlikely to proceed along a preprogrammed or set schedule. Workers vary in the rate at which they switch from one task to another, such that some forage early and others may never forage. Some tasks, such as undertaking (removing dead bees and larvae) or guarding are performed only by a small subset of the colony. This variance illustrates that age polyethism is not a hard rule, but is instead tempered by genetic variation among workers in their task propensity and developmental schedule, and by the social dynamics of the colony itself.

Models suggest that, to some degree, the age polyethism schedule can be driven by the social environment, rather than purely from an intrinsic developmental schedule. In ants, shifts in task performance with age correspond to a general shift in location of activities from the nest center, where the brood is reared, to the periphery and outside the nest. The 'foraging for work' model was one of the first and most provocative explorations of whether social dynamics, in this case coupled with spatial parameters, can generate division of labor. In the model, workers with no current task move through the nest 'foraging for work' until they encounter a task; they then perform that task until it is no longer needed. Because workers emerge at the nest center, they essentially displace other workers searching for tasks toward the nest periphery. The oldest workers become foragers, with its associated high mortality. The model is important in illustrating how the iterative effects of local social interactions (self-organization) can generate what are generally considered intrinsically driven behavioral patterns.

Genetic Task Specialization

Even in groups where physical or age-based castes are absent, we still see patterns of task specialization and division of labor. A central mechanism driving these patterns is intrinsic variation among group members in their preference for different tasks, generally termed intrinsic or genetic task specialization. The link between genotype and individual task performance has been made for numerous taxa across the range of eusocial insects, including taxa, such as bumblebees and eusocial (Vespid) wasps, in which age polyethism is weak or absent. In each of these systems, workers are capable of performing different tasks when need or opportunity is high, but tend to preferentially perform different tasks based on matriline or patrilineal differences. Genotype has also been shown to play a role in individual task ontogeny, suggesting a link between age polyethism and genetic task specialization.

Colonies with strong genetic task specialization face the problem that individual specialization can potentially limit task flexibility. In some eusocial systems this problem is alleviated in part by polyandry, or multiple mating, by the queen. In a typical mating flight, a honeybee queen may mate with from 10 to 30 or more males. This extreme polyandry contributes to the genotypic diversity and related task propensities of the worker offspring she will then produce.

Response Thresholds and Division of Labor

The interactions between genetically based task specialization and social dynamics have been explored empirically and theoretically using a series of models collectively called response threshold models. The primary assumption of these models is that individuals within a group vary in their task thresholds, an internal set point at which they respond to an external stimulus to perform a given task. Thresholds can be produced genetically, and in some models, are lowered by experience so that successful performance of a task reinforces specialization. In the threshold models, as stimulus levels for a task rise, those group members with lower thresholds perform it first. When they do so, they reduce stimulus levels, thus reducing the probability that other group members will perform it. They become, by default, the specialists for that task. Because individual thresholds vary across tasks, different group members specialize in different tasks, so that some group members are more likely to forage, while others are more likely to tend brood or remove refuse.

The model can be expanded to consider how a social insect colony, such as a honeybee hive, responds to a dynamic environment. For example, bees regulate collection of pollen around colony-level set-points that are related to the amount of pollen stored in the colony, and the amount of brood currently consuming it. According to the threshold model, when need for pollen is low, only a narrow genotypic subset of workers should collect it. However, as need or opportunity is increased, the stimulus levels for pollen collection should increase, and the thresholds of a wider diversity of workers met. Tests of this model show an excellent fit to these predictions. When pollen is added to test hives with workers from diverse genetic sources, genotypes of marked pollen foragers are skewed toward one or a few genetic sources. When pollen is removed, the number of pollen foragers increases; these new foragers represent a more diverse and evenly distributed source genotypes.

The response threshold model provides a simple but powerful framework for understanding how division of labor can be generated via social dynamics. It is a good fit with studies linking variation in genotype with task preference in both ants and honeybees. However, the applicability of the model extends to social systems beyond the eusocial ants and bees.

There is evidence that division of labor can emerge spontaneously within any group in which individuals vary in thresholds, even in the historical absence of social evolution. Females of solitary ground nesting bees (Halictidae) that are forced together into artificial social groups show a division of labor where different individuals tend to specialize on nest excavation and guarding. In these bees, task differentiation is also a product of aggression and consequent spatial dynamics. The bees dig a narrow central tunnel, and movement from the bottom of the nest (where excavation occurs) to the top (for guarding) often requires that individuals pass each other.

In another case of forced sociality, harvester ant queens that normally found nests alone also show division of labor when placed into artificial associations. Both the harvester ant queens and ground-nesting bees can be compared with closely related species in which females actually form communal associations to cooperatively construct nests and rear brood. Interestingly, levels of division of labor within the forced associations are generally stronger than in the related species with evolved communal associations. These communal societies contain unrelated and fully reproductive females. When division of labor emerges, the variation in task performance among them has been associated with variation in survival costs, with some females taking over more risky or physically costly roles. Thus, as division of labor spontaneously emerges in these groups, there is the potential also for the emergence of 'cheating,' in that individuals gain fitness benefit from the costly work performed by others. This argument provides one hypothesis for the observation that levels of division of labor are often relatively low in communal groups.

Division of Labor as a Case of Self-Organization

The response threshold and foraging for work models provide examples of how explorations of complexity theory, and particularly self-organization, can contribute to our understanding of social organization. Self-organizational processes are those in which local interactions among individuals generate nonlinear effects at the global or group level. Self-organizing systems contain no central or external controller to dictate patterns of interaction; instead, the dynamics occur locally as individuals interact and as a result change each other's behaviors.

This is a good fit with social insect colonies. Even in colonies containing queens, the task behaviors of the colony members are primarily distributed, based on local interactions and cues. The behavior of each individual in the colony alters the behavior of those around her, either because she provides information that stimulates the performance of a task, or because she reduces the stimulus level by performing the task herself. These interactions generate a series of positive feedbacks, in which they help amplify individual differences in task performance and specialization. As we move from the elements of division of labor seen in incipient social groups through to the task allocation systems of eusocial colonies, selection shapes social dynamics along with the physiological and developmental attributes of the workers themselves. At its pinnacle, division of labor becomes the highly organized but flexible system that has contributed to the tremendous ecological success of the eusocial insects.

See also: Ant, Bee and Wasp Social Evolution; Kin Selection and Relatedness; Queen–Queen Conflict in Eusocial Insect Colonies; Termites: Social Evolution.

Further Reading

Beshers SN and Fewell JH (2001) Models of division of labor in social insects. *Annual Review of Entomology* 46: 413–440.

Bonabeau E, Theraulaz G, and Deneubourg JL (1996) Quantitative study of the fixed threshold model for the regulation of division of labour in social insect societies. *Proceedings of the Royal Society of London B* 263: 1565–1569.

Camazine S, Deneubourg JL, Franks NR, Sneyd J, Theraulaz G, and Bonabeau E (2001) *Self-Organization in Biological Systems.* Princeton, NJ: Princeton University Press.

Fewell JH and Page RE (1999) The emergence of division of labour in forced associations of normally solitary ant queens. *Evolutionary Ecology Research* 1: 537–548.

Gadau J and Fewell JH (2009) *Organization of Insect Societies.* Cambridge, MA: Harvard University Press.

Hölldobler B and Wilson EO (1990) *The Ants.* Cambridge, MA: Belknap Press of Harvard University.

Hölldobler B and Wilson EO (2008) *The Superorganism.* Berlin: Springer Verlag.

Oldroyd BP and Fewell JH (2007) Genetic diversity promotes homeostasis in social insect colonies. *Trends in Ecology & Evolution* 22: 408–413.

Page RE and Erber J (2002) Levels of behavioral organization and the evolution of division of labor. *Naturwissenschaften* 89: 91–106.

Robinson GE (1992) Regulation of division of labor in insect societies. *Annual Review of Entomology* 37: 637–665.

Seeley TD (1982) Adaptive significance of the age polyethism schedule in honeybee colonies. *Behavioral Ecology and Sociobiology* 11: 287–293.

Seeley TD (1995) *The Wisdom of the Hive.* Cambridge, MA: Harvard University Press.

Tofts C and Franks NR (1992) Doing the right thing: Ants, honeybees and naked mole rates. *Trends in Ecology & Evolution* 7: 346–349.

Wheeler DE (1986) Developmental and physiological determinants of caste in social Hymenoptera. *The American Naturalist* 128: 13–34.

Wilson EO (1971) *The Insect Societies.* Cambridge, MA: Harvard University Press.

Dolphin Signature Whistles

L. S. Sayigh, Woods Hole Oceanographic Institution, Woods Hole, MA, USA
V. M. Janik, University of St. Andrews, St. Andrews, Fife, Scotland, UK

Introduction

The term 'signature' has often been applied to animal vocalizations when an individually distinctive pattern was found in them. The vast majority of animals achieve this by means of voice cues, which are cues that result from individual variability in the shape and size of the vocal tract. Thus, they are determined by genetic and developmental factors and are independent of the call type that is produced. The term 'signature whistle,' however, refers to a much more complex signal. Signature whistles are individually distinctive acoustic signals of dolphins, which indicate the identity of the caller. Unlike recognition signals in most other animals, identity is encoded in a frequency modulation pattern that is learned or invented early in life. Dolphins produce different frequency modulation patterns in different call types and thus not all call types carry the identity information encoded in the signature whistle.

History of the Study of Dolphin Signature Whistles

Early research on dolphin communication, initiated by John Lilly and others, focused on finding language-like components in the vocal repertoire. However, this work was hindered by the fact that dolphins do not make any consistent external movement associated with their vocalizations, and thus it was impossible to associate specific sounds with specific individuals. (Although these sounds are commonly referred to as vocalizations, they are not produced in the larynx like the vocalizations of terrestrial mammals. Instead, they are produced by phonic lips near the blowhole, where air is passed between nasal sacs for sound production.) This problem was overcome by Melba and David Caldwell, who recorded captive dolphins that had been isolated for medical attention. Through this work, they found that isolated dolphins produced large numbers of stereotyped, individually distinctive whistles (generally on the order of 90% of all whistles in this context), which they called signature whistles. Although signature whistles have been found in several species of dolphins (including common dolphins, *Delphinus delphis*, Pacific white-sided dolphins, *Lagenorhynchus obliquidens*, and spotted dolphins, *Stenella plagiodon*), the majority of research has focused on the bottlenose dolphin (*Tursiops truncatus*, **Figure 1**).

The Caldwells' defined the signature whistle as the predominant frequency contour produced by a dolphin in isolation. The Caldwells' early descriptions of signature whistles still largely hold amidst numerous later studies. They found that the fundamental frequency of signature whistles ranged from 1 to 24 kHz (although more recent work has found that upper frequencies can extend to above 30 KHz), and typically lasted for about 1s. Many signature whistles were found to consist of repetitive elements, called loops, which could be connected together or separated by brief, stereotyped intervals of silence. Often these multilooped whistles contained a distinctive introductory and/or terminal loop, with varying numbers of central loops. Although dolphins may vary whistle parameters such as duration and absolute frequencies, the overall contour, or shape, of the whistles usually remains remarkably stable for periods of up to several decades. These contours can be distinguished from each other by means of visual representations, or spectrograms, which are plots of frequency versus time (**Figure 2**). The Caldwells hypothesized that these individually distinctive contours functioned in individual identification.

The obvious next question to address in the study of signature whistles was whether or not socially interactive dolphins produced them as well. This question was first examined by Peter Tyack, using light-emitting devices worn by dolphins on their melon (forehead). These devices enabled the vocalizing dolphin to be identified. In his study, Tyack found that socially interactive, captive dolphins not only produced signature whistles, but also imitated the signature whistles of their tank mates. Later studies showed that captive dolphins primarily produced signature whistles when they were out of visual contact even if their separations were voluntary. This supported the idea that they were used for group cohesion and individual identification. However, the question still remained whether dolphins in the wild produced signature whistles or whether they were unique to dolphins held in captivity. This question was resolved by recording members of a resident bottlenose dolphin community near Sarasota, Florida, USA, which has been the focus of a long-term (35+ years) research program, coordinated by Randall Wells. Approximately once per year since the mid-1980s, researchers have carried out brief capture–release events, during which dolphins were recorded with suction-cup hydrophones placed directly on the melon. This provides a rare opportunity to record whistles from known individuals in the wild. During brief capture–release events, the vast majority of dolphins

Figure 1 Photos of bottlenose dolphins (*Tursiops truncatus*) in Sarasota, Florida. Courtesy of Chicago Zoological Society Sarasota Dolphin Research Program.

produce one predominant whistle contour, as the Caldwells found for captive dolphins.

Wild dolphins in undisturbed conditions also produce signature whistles. Whistles recorded from known members of the resident Sarasota dolphin community during undisturbed conditions were compared with those recorded from the same individuals during brief capture–release events. This comparison showed that approximately 50% of whistles produced by free-swimming dolphins were signature or probable signature whistles. This percentage is lower than that observed for temporarily isolated dolphins, either in the wild or in captivity, but still indicates that signature whistles are an important component of the vocal repertoire of bottlenose dolphins. Whistle imitations have also been found both in captivity and in the wild and have been hypothesized to function in addressing other individuals, but this remains to be proved. When groups of bottlenose dolphins first meet at sea, they either ignore each other or exchange signature whistles and then join each other.

Nonsignature Whistles

What of the approximately 50% of whistles produced by free-swimming dolphins that are not signature whistles?

Little is known about the variety of whistle types that dolphins produce in addition to their signature whistles. Interestingly, studies of both captive and wild dolphins indicate that general voice cues that are present in non-signature whistles are not used by dolphins to recognize one another. Changes to gas-filled cavities that occur with changes in depth may have necessitated a mechanism for individual recognition different from that used by terrestrial mammals.

Signature Whistle Development

The development of the modulation pattern of signature whistles is strongly influenced by vocal production learning. Production learning is relatively rare among mammals and even fewer animals use it in the development of recognition calls. Besides dolphins, only some bat and parrot species as well as humans apply vocal production learning to individual recognition. In one study, captive dolphins were trained to copy novel sounds and even to associate such novel sounds with specific objects. Studies with both wild and captive dolphins have found evidence that young dolphins learn sounds from their acoustic

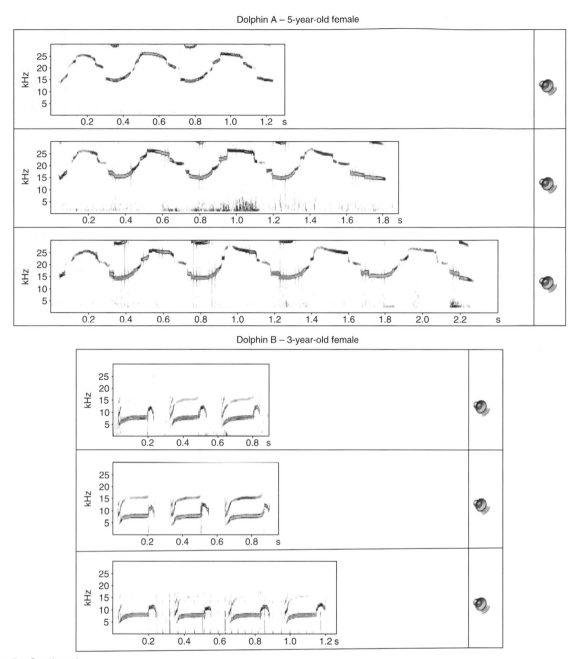

Figure 2 Continued

environments. One study of captive calves found that they were more likely than wild calves to incorporate a constant frequency component typical of the trainer's 'bridge' whistle, used to reinforce behaviors. Among the resident Sarasota, Florida, dolphins, about one-third of calves developed a signature whistle that was similar to that of their mother (with males being more likely than females to do so). Others appeared to learn their whistles from siblings or infrequent associates; more work is needed to determine what factors influence the choice of whistle contour in these calves.

Signature Whistle Functions

The Caldwells' hypothesis that signature whistles are used in individual recognition has been supported by several experiments. Dolphins in Sarasota, Florida, responded more strongly to playbacks of signature whistles of kin than to those of nonkin. Similar results were found when synthetic signature contours were played back, indicating that the contour alone provides sufficient information for individual recognition. This feature of signature whistles is quite rare among animals; as was mentioned previously,

Dolphin C – 16-year-old female

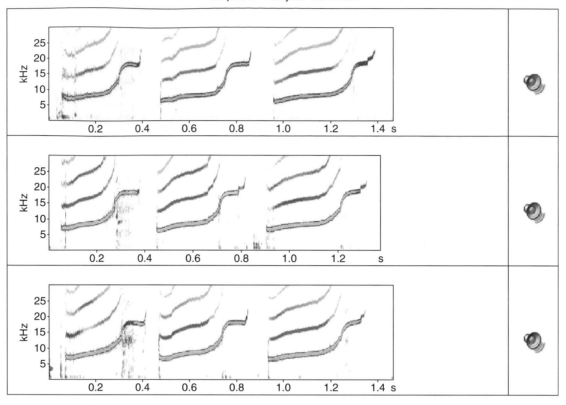

Dolphin D – 26-year-old female

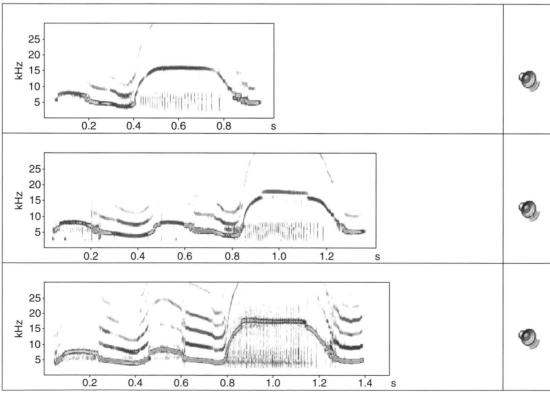

Figure 2 Continued

Dolphin E – 33-year-old female

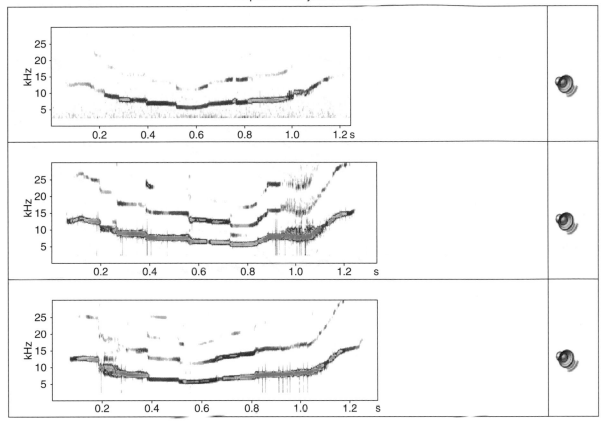

Dolphin F – 41-year-old male

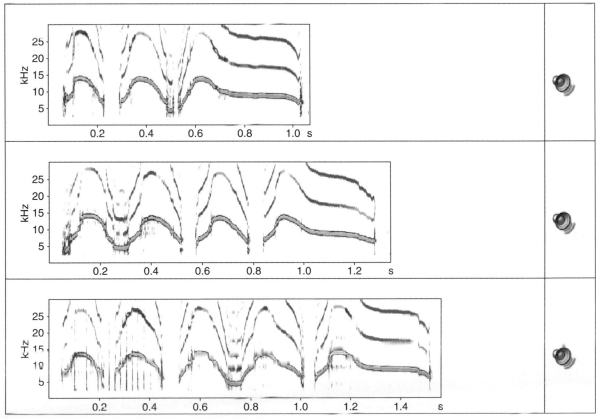

Figure 2 Continued

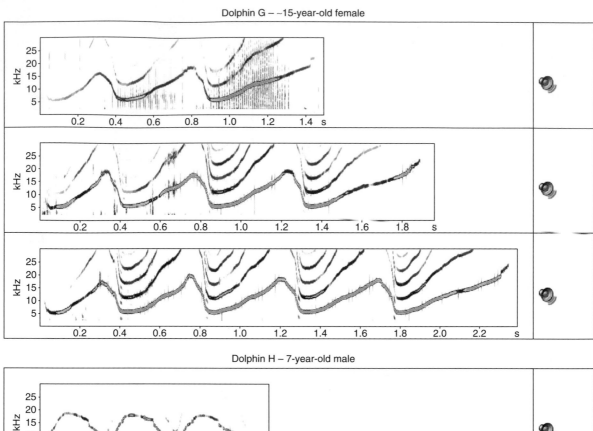

Figure 2 Continued

most animals use voice cues for individual recognition. Furthermore, free-swimming dolphins primarily produce signature whistles when they are out of visual contact with other group members, rather than when swimming in close association. This suggests that signature whistles also play

an important role in maintaining group cohesion. In Shark Bay, Australia, free-swimming calves that had separated from their mothers were more likely to whistle when initiating a reunion with their mothers than during other contexts. (The species of dolphins that occurs in Shark Bay,

Dolphin I – ~15-year-old female

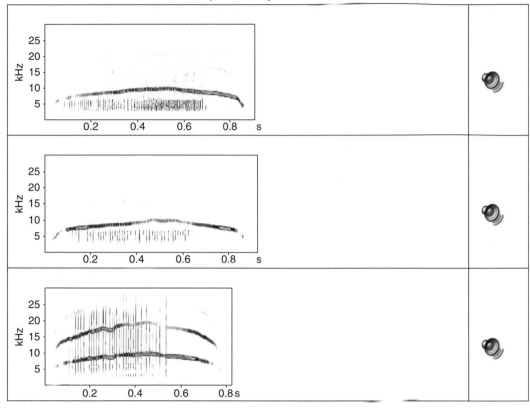

Dolphin J – 3-year-old male

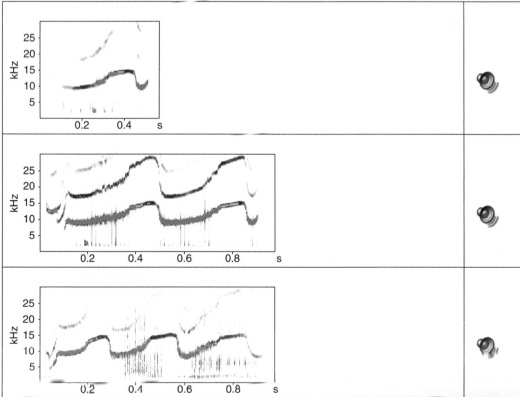

Figure 2   Continued

Australia, is *Tursiops aduncus*, not *Tursiops truncatus*.) During brief capture–release events, mother–calf pairs tend to exchange whistles back and forth while separated.

Signature whistle repetition rates appear to signal the level of stress, or level of arousal, of an animal. In Sarasota, individuals produce many more signature whistles during brief capture–release events than during undisturbed conditions. Rates of signature whistle production are also higher at the beginning than at the end of a capture–release session, and during an individual's first capture–release session than during later sessions. Dolphins also vary other whistle parameters, such as frequency, number of loops, and duration (**Figure 2**), and these parameter variations could provide additional information about the vocalizer. However, these variations appear to be idiosyncratic, and thus may require knowledge of the vocal patterns of each individual animal.

Signature Whistle Stability

Although dolphins may vary aspects of their signature whistles, the overall contour does remain strikingly stable both within and across recording sessions. In order to demonstrate this, we (along with Carter Esch and Randall Wells) asked naïve judges to sort spectrograms of 20 randomly selected whistles from each of 20 dolphins, although the judges had no knowledge of how many dolphins' whistles were present in the sample. The judges grouped together a mean of 18.9 out of 20 whistles for each dolphin, and included in these groups an average of only 0.5 of a whistle from the 380 whistles of other dolphins. This study demonstrated not only that dolphin signature whistles are highly stereotyped, but also that human observers are highly capable of classifying these contours. While signature whistles of females are remarkably stable throughout their lifetime, males sometimes change the contour of their signature when forming close alliances with other males. These alliances result in an almost permanent association between males and their signature whistles tend to become more alike over time. The benefits of such changes

are still unclear, but it has been reported from field sites in Sarasota and Australia.

Whistle Classification

Developing automated methods for whistle categorization remains one of the great challenges for researchers of dolphin communication. Although several researchers have developed computerized methods of classifying signature whistles, none has been as effective at grouping together externally validated categories of whistles (i.e., those known to have been produced by the same dolphin) as human observers. In some cases, only subtle features may differentiate one signature whistle from another, whereas most computerized techniques tend to discount such features. However, a recently developed program using neural network architecture enables researchers to identify signature whistles versus nonsignature whistles from recordings of free-swimming dolphins. This program will allow the study of signature whistles in areas where temporary capture–release programs are not feasible. By identifying signature whistles in specific areas over time, they can be used to monitor individual movements as well as population sizes using mark–recapture methods that have previously been applied only to photoidentification data.

Summary and Conclusion

In summary, bottlenose dolphins produce stereotyped, individually distinctive whistle contours called signature whistles, which function in individual recognition and in maintaining group cohesion. Dolphin signature whistles are qualitatively different from individually distinctive signals seen in other mammalian species: they are learned, individually distinctive labels that seem to function similarly to human names and are one of very few such signals in the literature to date.

Figure 2 Spectrograms and associated sound files of three signature whistles from each of 10 bottlenose dolphins, recorded during brief capture–release events in Sarasota, Florida. Age and sex for each individual are noted (data courtesy of the Chicago Zoological Society's Sarasota Dolphin Research Program). Several different signature whistle contour types are illustrated, including multiloop whistles with varying numbers of connected or disconnected loops (dolphins A, B, C, D, E, G, H, J), including examples of distinct introductory (dolphin J) and terminal loops (dolphins D, F, G, H), and whistles with no loop structure (dolphins E, I). Note the stability of the contour even with variation in frequency parameters (e.g., dolphins A, D, G) and duration (e.g., dolphins I and J). Spectrograms were made in Avisoft SASLAB Pro, using a 256 pt FFT, 50% overlap and FlatTop window. The color scheme ranged from light blue–dark blue–purple–red–yellow–light green–green, with green being the loudest portions of the signal. Recordings were made with different types of recording equipment; prior to 1989, most recordings were made on Sony or Marantz stereo cassette recorders, with upper frequency limits of 15–20 kHz; thus, harmonics are less noticeable in these recordings. Later recordings were made on hifi video cassette recorders, with frequency responses extending to above 30 kHz. High-pass filters, ranging from 500 Hz to 2.5 kHz, were used on some sound files to reduce extraneous noise.

See also: Acoustic Signals; Cultural Inheritance of Signals; Parent–Offspring Signaling; Referential Signaling; Social Recognition; Sound Production: Vertebrates; Vocal Learning.

Further Reading

Caldwell MC, Caldwell DK, and Tyack PL (1990) Review of the signature-whistle-hypothesis for the Atlantic bottlenose dolphin. In: Leatherwood S and Reeves RR (eds.) *The Bottlenose Dolphin*, pp. 199–234. San Diego: Academic Press.

Cook MLH, Sayigh LS, Blum JE, and Wells RS (2004) Signature-whistle production in undisturbed free-ranging bottlenose dolphins (*Tursiops truncatus*). *Proceedings of the Royal Society of London B* 271: 1043–1049.

Deecke VB and Janik VM (2006) Automated categorization of bioacoustic signals: Avoiding perceptual pitfalls. *Journal of the Acoustical Society of America* 119: 645–653.

Esch HC, Sayigh L, Blum J, and Wells R (2009) Whistles as potential indicators of stress in bottlenose dolphins (*Tursiops truncatus*). *Journal of Mammalogy* 90(3): 638–650.

Esch HC, Sayigh L, and Wells R (2009) Quantifying parameters of bottlenose dolphin signature whistles. *Marine Mammal Science* 25(4): 976–986.

Fripp D, Owen C, Quintana-Rizzo E, et al. (2005) Bottlenose dolphin (*Tursiops truncatus*) calves appear to model their signature whistles on the signature whistles of community members. *Animal Cognition* 8: 17–26.

Janik VM (2000) Whistle matching in wild bottlenose dolphins (*Tursiops truncatus*). *Science* 289: 1355–1357.

Janik VM, Dehnhardt G, and Todt D (1994) Signature whistle variations in a bottlenosed dolphin, *Tursiops truncatus*. *Behavioral Ecology and Sociobiology* 35(4): 243–248.

Janik VM, Sayigh LS, and Wells RS (2006) Signature whistle contour shape conveys identity information to bottlenose dolphins. *Proceedings of the National Academy of Sciences of the USA* 103: 8293–8297.

Janik VM and Slater PJB (1998) Context-specific use suggests that bottlenose dolphin signature whistles are cohesion calls. *Animal Behaviour* 56: 829–838.

Miksis JL, Tyack PL, and Buck JR (2002) Captive dolphins, *Tursiops truncatus*, develop signature whistles that match acoustic features of human-made model sounds. *Journal of the Acoustical Society of America* 112: 728–739.

Richards DG, Wolz JP, and Herman LM (1984) Vocal mimicry of computer-generated sounds and vocal labeling of objects by a bottlenosed dolphin, *Tursiops truncatus*. *Journal of Comparative Psychology* 98: 10–28.

Sayigh LS, Tyack PL, Wells RS, and Scott MD (1990) Signature whistles of free-ranging bottlenose dolphins, *Tursiops truncatus*: Mother–offspring comparisons. *Behavioral Ecology and Sociobiology* 26: 247–260.

Sayigh LS, Esch HC, Wells RS, and Janik VM (2007) Facts about signature whistles of bottlenose dolphins (*Tursiops truncatus*). *Animal Behaviour* 74: 1631–1642.

Sayigh LS, Tyack PL, Wells RS, Scott MD, and Irvine AB (1995) Sex differences in signature whistle production of free-ranging bottlenose dolphins, *Tursiops truncatus*. *Behavioral Ecology and Sociobiology* 36: 171–177.

Sayigh LS, Tyack PL, Wells RS, Solow AR, Scott MD, and Irvine AB (1999) Individual recognition in wild bottlenose dolphins: A field test using playback experiments. *Animal Behaviour* 57: 41–50.

Smolker R and Pepper JW (1999) Whistle convergence among allied male bottlenose dolphins (Delphinidae, *Tursiops* sp.). *Ethology* 105: 595–617.

Tyack P (1986) Whistle repertoires of two bottlenosed dolphins, *Tursiops truncatus*: Mimicry of signature whistles? *Behavioral Ecology and Sociobiology* 18: 251–257.

Tyack PL and Sayigh LS (1997) Vocal learning in cetaceans. In: Snowdon C and Hausberger M (eds.) *Social Influences on Vocal Development*, pp. 208–233. Cambridge: Cambridge University Press.

Watwood SL, Tyack P, and Wells R (2004) Whistle sharing in paired male bottlenose dolphins, *Tursiops truncatus*. *Behavioral Ecology and Sociobiology* 55(6): 531–543.

Domestic Dogs

B. Smuts, University of Michigan, Ann Arbor, MI, USA

Introduction

I live with a couple of wolves who go almost everywhere with me. I leave them alone with small children, and I trust them not to bite when I take food out of their mouths. I even share my bed with them. Although you may think I am reckless, I am no different from millions of other people who greatly value the companionship of *Canis lupus familiaris* – the subspecies of wolf that we call dogs.

Some scientists consider domestic dogs and wolves to be the same species because their genomes are virtually indistinguishable (e.g., their mitochondrial DNA has a maximum sequence divergence of 0.01), and they can mate and produce fertile offsprings. Other scientists argue that, despite genetic similarity, dogs and wolves should be considered separate species (*Canis familiaris*) because dogs are domesticated and wolves are wild. However we classify them, wolves and dogs are each other's closest living relatives, and it is helpful to keep their kinship in mind as we explore the evolution, behavior, and cognition of man's (and women's) best friend.

Dog Evolution

Wolves were the first wild animals that became domesticated, but it is not known exactly when, where, or how this happened. Dog skeletons can be distinguished from wolf skeletons in numerous ways: dogs have smaller skulls in relation to body size, more tightly packed teeth, and wider snouts. The earliest archaeological evidence of dogs dates to about 14 000 years ago. Domestication must have begun some time before this, since it took time for a dog-like morphology to evolve.

Recent comparisons of mitochondrial DNA from dogs and modern wolves suggest that the ancestors of domestic dogs most likely diverged from a population of East Asian wolves sometime between 15 000 and 135 000 years BPE. The large gap between the earliest fossils and the older estimates of divergence can be reconciled if, for thousands of years, dog skeletons changed very little and only recently evolved the features that allow archaeologists to distinguish them from wolves. Alternatively, advances in genetic techniques may shift the estimated divergence time to be more in accord with the fossil evidence.

In any case, it is clear that the earliest stages of domestication occurred when humans were hunter-gatherers.

In the late nineteenth century, Darwin's cousin Francis Galton envisioned such humans stealing wolf puppies from dens and rearing them as hunters and guard dogs. This scenario, popularized in the 1950s by Nobel Laureate ethologist, Konrad Loranz, dwindles in appeal when we consider scientists' experiences in raising wolves in captivity. If they are to coexist with humans, wolf puppies must be removed from dens very early (about 10 days after birth). They must be intensively nurtured, bottle fed, and kept away from other wolves for at least 4 months. Some adult, human-reared wolves remain fearful around unfamiliar humans, and some have bitten the people who reared them. These wolves require a lot of meat and abundant exercise to remain healthy. They ignore human commands and will run away unless confined to carefully constructed enclosures. In short, it seems unlikely that mobile hunter-gatherers could have maintained relationships with fearful, sometimes aggressive, disobedient, escape-prone adult wolves.

Lorenz's adoption scenario reflects the widespread assumption that humans deliberately domesticated wolves, but many researchers now think wolves took the first decisive step toward domestication when they began to feed on human leftovers. The notion of wolves as camp followers, subsisting at least in part on what humans discarded, is made more plausible by contemporary accounts of wild carnivores, including wolves, spotted hyenas, black bears, and red foxes feeding on human trash. If they ate human refuse then the wolves least wary of humans would get more food and if, over many generations, these wolves ceased associating with other wolves, successfully reproduced among themselves, and passed their feeding habits to their offspring, natural selection could produce wolves bold enough to begin interacting with humans. At some point, humans likely found the presence of these wolves useful; perhaps their warning barks to other wolves alerted people to approaching predators like lions or bears. Once people recognized such advantages, they might have intentionally left scraps of food behind for the wolves, sealing a mutualistic bargain that radically altered the lives of both species.

This scenario gains indirect support from experiments with captive silver foxes bred for reduced fear of humans. After just ten generations of selection, some of the foxes followed humans, played with them, and licked their faces, just like dogs. After 35 generations, novel morphological features began to appear, including floppy ears, curly tails, and spotted coats – all traits seen in some domestic dogs but never in wolves. Since selection was based purely on behavioral criteria, these dog-like physical traits appear to be evolutionary by-products. If something similar occurred in a population of wolves through natural selection operating over hundreds of generations, then humans did not invent dogs at all.

Archaeological evidence of dogs buried in graves with people suggests that by 14 000 years BPE or earlier, dogs played a role in human spiritual life. By the beginning of written history, some dogs were helping people in hunting and battle, while others were serving as companions. Still, we do not know what proportion of the total dog population found such favor. It is possible that throughout their history, many dogs lived as the majority of dogs do today, scavenging among people who barely tolerate them, always keeping a watchful eye out for the rare handout or angry rock-thrower.

Dog Behavior and Interactions with Humans

Ability to Understand Human Gestures

Whether they lived as pampered lap dogs or skulking scavengers, dog survival and reproduction must have been strongly influenced by their interactions with humans. In particular, the dogs most able to read human emotions, anticipate human actions, and understand human communication would have had an evolutionary advantage. This logical idea received little scientific attention until about 10 years ago, when researchers in the United States and Hungary independently began studying dog–human interactions under controlled, laboratory conditions. The experiments described in the following section used pet dogs living at home; the animals were never harmed during the observations.

Dogs' responses to human communicative gestures became an especially popular research topic. To investigate responses to human pointing gestures and other aspects of human body language, scientists used a two-choice task, in which one container is baited with food and the other is empty. The baited container is determined randomly, so that each one conceals the prize 50% of the time. Dogs quickly learned to go to a container, and in the presence of neutral humans, they chose the baited container about half the time, indicating that they could not smell their way to success. If a human pointed to one of the containers, however, dogs chose it significantly more than half the time, even if they had received no prior training with pointing (experimenters always pointed to the baited container).

This ability may seem unimpressive, but some very close non-human primates, including chimpanzees, typically failed to understand the human pointing gesture without intensive training. Dogs may have caught on quickly because they had been raised by humans, unlike most captive chimpanzees. However, when researchers compared home-reared puppies experiencing abundant human contact and kennel-reared puppies with little human contact, both groups followed the pointing gesture with equal aptitude, indicating that intensive human contact early in life was not a prerequisite for understanding this gesture.

These results supported the domestication hypothesis, which posits that dogs evolved specialized abilities to understand humans during domestication. As a further test of this idea, the two-choice task was given to captive wolves, which were expected to succeed less often than dogs. Even though previous experiments had suggested that rearing environment did not matter much for dogs, the researchers went out of their way to raise young wolf puppies and dog puppies in exactly the same way, with nearly constant human attention and nurturing. At 4 months of age, the wolf puppies performed no better than chance, but the dog puppies succeeded in following the human gesture. Another key test of the domestication hypothesis involved the silver foxes selected for reduced fear of humans. In the two-choice task with human pointing, the selected foxes performed more like dogs than wolves.

The domestication hypothesis has been widely cited by scientists and has also received much media attention. New findings, however, call it into question. First, the researchers who reared the wolf puppies discovered that when the wolves grew up (by the age of 2 years), they performed the two-choice task as well as dogs, without any training. Second, although dogs overall usually chose the baited container more often than chance, in all experiments many individual dogs failed the test, even after repeated trials. If domestication selected for dogs that could read human cues, why was their performance so variable? Third, nondomesticated species (fur seals, gray seals, and bottlenose dolphins) tested in captivity proved at least as successful as dogs in following the pointing gesture in the two-choice task. Finally, in 2008, a study comparing wolves and dogs reported slightly superior performance in the wolves. Clearly, much additional research will be necessary to clarify the factors affecting animals' ability to understand human pointing and other gestures.

Communication and Cognition

Along with (some) dogs' abilities to understand pointing, scientists have studied many other canine behaviors that appear to facilitate interactions with humans. For example, dogs can often learn how to do something simply by watching a human doing it. In one experiment, dogs that failed to solve a detour task to obtain a food reward succeeded after a human demonstrated the solution. There is even one report of a dog able to copy a variety of different human actions when told 'do as I do.'

Dogs seem to be keenly aware of where humans place their attention. Positive attention facilitates learning.

For example, dogs did better in the detour task if the human made eye contact with them while demonstrating the solution, and they were more likely to obey a simple command when issued by a human who was looking directly at them, compared with a human looking away from them or attending to another human. In addition to hindering responsiveness to commands, lack of human attention also reduces the probability of interaction. For instance, dogs were more likely to approach and beg from a human whose eyes they could see and were more prone to drop a ball at the feet of people facing them. Finally, even subtle changes in attention can encourage malfeasance. We are not surprised when a dog eats forbidden food as soon as we leave the room, but in test situations, pet dogs also tend to take the food when the human is still present if she closes her eyes, turns her back, or plays a computer game. This is an experiment easily replicated if you have a dog at home.

Dogs are skilled at reading human body language, but what about human spoken language? It seems as if they often know what we are saying, but a dog may run to get the ball when you say 'Where's the ball?' not because she understands that the word ball signifies a specific object, but rather because she associates that phrase and the intonation with which you say it with both positive emotions *and* an object she plays with often. You can test this possibility by asking a dog to respond to a familiar request using the same words but in a different voice and with different intonation. In this circumstance, many dogs fail to respond. The take home lesson is to train your dog using a variety of different voices.

Sometimes, however, dogs do understand that words refer to specific objects. In a controlled test situation a border collie, Rico, was asked to retrieve various objects (e.g., children's toys, different kinds of balls) whose names he had learned through day-to-day interactions in his family. His accuracy in this test demonstrated knowledge of the names of over 200 objects, giving him a vocabulary size similar to that of language-trained great apes or parrots. In addition, Rico could retrieve a novel item from a room with seven objects whose names he knew and one object he had never seen before, indicating that he used a process of exclusion to deduce which object he was supposed to get. In still another test, Rico was taught the names of eight novel items just one time, after which they were removed from his home. Four weeks later, his success in retrieval tests indicated that he remembered the names of half of the novel items.

Clearly, dogs do understand a considerable amount of what we say or do, but how good are they at communicating similar information to us? Many dogs make clear requests to humans through their body language and vocalizations (e.g., staring at the cookie jar or poking the leash with the muzzle). Scientists tested this ability with a simple experiment you can do at home. They asked a person well known to a dog to leave the room while another, in front of the dog, hid a treat somewhere the dog could not reach. When the familiar person reentered, they could find the treat based solely on the dog's behavior.

In the preceding experiment, the dog indicated the location of the treat by the natural behavior of moving toward it and looking at it, alternating with looking back at the human. But, could a dog learn to use abstract symbols to communicate what he wants, as we do with language? One experiment suggests they can. A pet dog, Sofia, was first taught to associate each of a variety of moveable stimuli, or signs, with specific objects and/or activities (e.g., half of a rubber ball signified a toy; a cup signified food). Sofia learned that when she wanted something, such as a toy, if she pressed the correct sign with her paw, a person would give her what she asked for. After this period of training, the signs were replaced with arbitrary symbols known as lexigrams, such as a circle or letter, located in the same place the signs had been. Eventually, the lexigrams were available only on a large keyboard, which prevented Sofia from remembering their meaning by their location. She spontaneously pressed specific keyboard lexigrams, apparently asking for what she wanted. People viewing tapes of Sofia's actions just before and after pressing a key nearly always guessed correctly which key Sofia had pressed, demonstrating that Sofia was pressing keys nonrandomly. Prior to this experiment, only great apes and dolphins had shown the capacity to communicate with humans using abstract symbols.

Emotions

The abilities described in the previous section reflect not only dogs' cognitive skills but also their emotional connections with humans. To objectively evaluate dogs' emotional bonds with humans, experimenters used a method called the strange situation test (SST), originally designed to assess a child's attachment to caregivers. In this test, the majority of children are less likely to interact with a stranger than with the caregiver, show distress when the caregiver briefly leaves the room and greet the caregiver enthusiastically when she returns. Most pet dogs showed very similar behaviors when tested with their human caregivers, but human-reared wolves did not – at least so far. Attachments can form quickly; shelter dogs become attached to an unfamiliar human after just three 10-min handling periods in 3 days. If, over evolutionary history, the human environment of dogs was as unstable as it often is today, then selection may well have favored dogs that could rapidly form new attachments.

Another experiment showed that shelter dogs were more likely to form a new attachment with a handler who massaged them compared to one who trained them. This might be because massage and other kinds of intimate touching trigger the production of oxytocin, a hormone that facilitates maternal bonding in mammals, including humans. When people stroke their dogs for about 10 min, blood oxytocin rises significantly in both humans (it nearly doubles) and dogs (it increases fivefold), and in another experiment, humans who frequently received gazes from their dog showed a significant increase in urinary oxytocin. In some mammals, exposure to oxytocin decreases social avoidance and increases the ability to read subtle indicators of emotion in another's face. Interestingly, the wolf puppies that failed to follow the human pointing gesture rarely looked at the demonstrator's face during the test, but the dog puppies and the older wolves that performed well did. Domestic dogs vary considerably in their willingness to maintain eye contact with a human who is looking at them, and it would be interesting to determine whether differences in how dogs respond to the pointing task reflect differences in their oxytocin levels.

Attachment to humans influences dog problem-solving. In one test, experimenters placed food on the other side of a fence and observed the responses of dogs that lived in the home with those of dogs that spent most of their time in the yard. The yard dogs figured out how to grab the food right away, but the home dogs hesitated to act and instead looked back at the human. Only after the human encouraged them to get the food did they take it. When the human-reared wolves that failed to show attachment behavior toward their caregivers confronted a difficult problem, they kept trying to solve it on their own, but dogs reared the same way gave up after a few attempts and turned to look at the human instead.

Attachment to and dependence on humans clearly benefit dogs, but in some situations, they can be disadvantages. One study showed that human actions during a problem-solving test interfered with dogs' abilities to reason by inference. Human behavior can also compromise a dog's own preferences. When confronted with two plates containing different numbers of small pieces of food, dogs tended to choose the plate with more pieces. However, in an identical situation, if the human caregiver behaved enthusiastically toward the food on the plate with fewer pieces, many dogs chose that plate instead.

Human–Dog Cooperation

Despite the fact that dogs and humans have worked together on tasks like hunting and sheep-herding for hundreds of years, few studies have examined exactly how they cooperate. One study of 34 blind humans walking with their guide dogs revealed interesting patterns of interaction. Within most of the human–dog pairs, dogs initiated more actions (range 40–80%) than the humans did, but the two individuals frequently traded the initiator role back and forth. Since most types of actions (stop, turn,

step down, etc.) could be initiated by either partner, the initiator role was not predetermined by the specific action taken but seemed to depend more on subtle cues or rhythms that remain to be studied.

Benefits to Humans of Associating with Dogs

Much of the research on human–dog interaction asks dogs to solve problems invented by humans in human-created settings like university laboratories. By its very nature, this research highlights the ways in which dogs depend on humans. To understand how much we depend on them, we need scientific studies of human–dog interactions during challenges more natural to dogs, such as tracking people or finding scat from endangered species.

Dogs are good for us. People who live with dogs have better health, including more rapid recovery from heart attacks. In addition, the presence of a dog reduces stress, facilitates social interaction, and enhances learning in a variety of subjects, including children in an elementary school classroom, disabled children, children with reading problems, people in retirement homes, and Alzheimer's patients. Numerous anecdotes show that dogs can often reach depressed or withdrawn people who fail to respond to other humans. As the benefits of contact with dogs become more widely recognized, they will become increasingly welcome in schools, offices, hospitals, nursing homes, and psychological and physical therapy settings. To avoid exploitation of dogs, it will be crucial to implement practices guaranteeing their safety and well-being, such as biscuit breaks and vacation time, as well as daily opportunities to exercise, play games with people, and interact with other dogs.

Dog–Dog Interactions

Social Behavior Among Companion Dogs

Play

When dogs that live with humans visit dog parks or other multidog settings, they tend to engage in frequent, vigorous play. Most studies of playing dogs focus on interactions within pairs, or dyads. During dyadic play, domestic dogs, like wolves, show many behaviors reminiscent of fighting and hunting, and like wolves they use various signals, including the play bow, to demonstrate playful intent. Although play can escalate to aggression, this does not occur often.

Some researchers have suggested that if a dog is rarely or never in the 'top dog' role during play, he will lose interest in playing with that partner, and that dogs, therefore, tend to switch roles back and forth to make play roughly symmetrical. Only two studies have quantified roles adopted during dyadic dog play and neither supported this hypothesis. Analysis of videotapes of over

50 pairs of adult dogs that interacted repeatedly showed that although role switching did occur, in most pairs one dog adopted the 'top dog' role significantly more often than the other. Many of these pairs, including ones in which one dog never achieved the top dog role, continued to play despite this inequity.

Being older than the partner and out-ranking the partner predicted adoption of the top-dog role, but because age and rank were positively correlated, researchers were unable to assess their independent effects. The younger member of playing pairs also self-handicapped more often (behavior that puts them in a vulnerable position, such as lying on the ground with belly up) and gave play signals more often. These results indicate that young dogs are so motivated to play that they keep the game going, even when they usually end up on the bottom.

Another study videotaped puppies playing with littermates during the first few months of life. Littermate pairs, like adult dogs, tended to adopt asymmetric roles and asymmetry increased over time. Young puppies initiated play with some littermates more often than others, and these preferences intensified over time, suggesting that puppies may figure out early which siblings they prefer to interact with. Stray dog littermates studied in India remained together for at least 4 months, so under some circumstances, sibling preferences could have adaptive significance.

Dominance, conflict, and reconciliation

Anyone who watches dogs interacting will notice that when two dogs come together, they often show opposite body language. Darwin clearly described how one dog will stand tall with tail up and ears forward while the other dog crouches slightly, flattens the ears and lowers the tail. In wolves, these ritualized behaviors reaffirm well-established status differences, which promote relaxed, friendly relations between wolves of different rank. Something similar seems to be going on in dogs, except their body language tends to be less exaggerated than that of wolves, and some dogs seem to care little about status. Remarkably, scientists have not yet conducted detailed studies on ritualized interactions in dogs.

Serious aggression is rare in well-socialized dogs, in part because of the ritualized communication described already, and few systematic data exist on dog–dog conflict. One study based on veterinarian records suggested that aggression involving dogs from different households most often involves two males, whereas aggression within households more often involves two females and also tends to be more severe.

When two dogs get into a conflict, they tend to reconcile through affiliative contact soon afterwards, as many other social mammals do, including wolves, spotted hyenas, and many primate species. In one study, dogs most familiar with each other were less likely to fight,

but when they did, they were more likely to reconcile. This finding and evidence of play partner preferences suggest that dogs form special relationships, or friendships, with specific partners.

Stray and Feral Dogs

As mentioned earlier, many dogs scavenge on the fringes of human society, and these dogs interact with other dogs at least as much as they interact with humans. Since there is little reason to consider this a recent phenomenon, dogs have undergone selection not only for living with humans but also for living with other dogs. Dogs are the ultimate two-species socialites.

Noncompanion dogs can be classified into two types: stray dogs that tolerate humans and feral dogs that avoid humans. Reactions to humans are largely determined by the presence or absence of human contact during the first few months of life. Feral dogs have not been socialized to humans and will avoid them throughout their lives. Stray dogs are often former companion dogs or their offspring, and strays sometimes join feral dogs. These facts led some scientists to claim that companion, stray, and feral dogs in a given area comprise a single population, and that feral dogs, therefore, cannot evolve adaptations specific to their way of life. These ideas remain untested.

Stray and feral dogs typically feed, at least in part, on human refuse. In most early studies of stray dogs, observers followed individuals, pairs and small groups of dogs as they searched for trash in urban or suburban neighborhoods. They observed stray females mating with more than one male and mothers alone with young puppies. On the basis of these observations, researchers described social relationships among stray dogs as ephemeral and involving little, if any, cooperation.

More recent studies in a wider variety of habitats have dramatically altered this picture. Stray and feral dogs in and near Italian villages also moved about as individuals or in small groups, but these dogs belonged to larger groups that defended shared territories against dogs from other groups. Like wolves, these dogs treated fellow group members and nonmembers very differently. Conflicts within groups were usually resolved through ritualized, noninjurious aggression, but nongroup members were attacked and even killed. These studies also reported that individual male dogs sometimes remained near particular mothers, chasing away intruders that got too close to the puppies.

Stray dogs in a town in West Bengal, India, formed small groups and defended territories much like the Italian dogs. Although females occasionally mated with more than one male, most mated with just one, and the mates of these monogamous females guarded them and their young puppies. One male was even seen regurgitating food to his puppies, as wolf fathers do. In another instance, two females reared their pups in the same den and nursed them communally. Although companion dogs and Italian stray dogs exhibited two mating periods per year, the Indian dogs came into heat in synchrony once a year, just like wild canines. Researchers also reported ritualized dominance and submission within groups and discerned clear within-group dominance hierarchies.

In an ongoing study of feral dogs subsisting on garbage near Rome, Italy, researchers observed additional wolf-like behaviors. These dogs live in territorial groups with 25–40 members, much larger than groups reported for stray dogs. Observers frequently witnessed competitive interactions, particularly in the context of feeding and mating, and in the most intensively studied group, they documented a strict linear hierarchy among adults. As in wolf packs with multiple litters, the highest-ranking female reared her pups in the core of the group's territory, and these pups received male protection. Low-ranking females gave birth closer to the periphery of the territory and received less male assistance.

Dingoes and New Guinea singing dogs are descendants of domestic dogs brought to Australia and New Guinea by seafaring people about 4000–5000 years ago. Although dingoes sometimes associated with Australia's native people, mostly they remained independent, forming small packs characterized by individual hunting of small game and cooperative hunting of larger game, territorial defense, and feeding of young by all pack adults, just like wolves. It is unclear whether feral dogs that survive mainly by hunting occur outside Australia and New Guinea.

Conclusion

In a best-selling modern book about dogs, Elizabeth Marshall-Thomas asked: 'What do dogs want?' The research described here suggests that dogs want to interact with humans, and that they may have evolved specific abilities to do so. It also indicates that dogs want to interact with other dogs, and that some dogs retain the ability to form adaptive social groups whose members cooperate. Studies of dog behavior have increased dramatically year by year for at least a decade. We can look forward to a lot more information about who dogs are and what they need to be happy and healthy. This knowledge will undoubtedly benefit their best friends as well.

See also: Empathetic Behavior; Social Cognition and Theory of Mind; Social Learning: Theory; Spotted Hyenas; Wolves.

Further Reading

Boitani, KL and Ciucci, P (1995). Comparative social ecology of feral dogs and wolves. *Ethology Ecology and Evolution* 7: 49–72.

Coppinger, RP and Coppinger, L (2001). *Dogs: A New Understanding of Canine Origin, Behavior and Evolution*. Chicago: University of Chicago Press.

Crockford, SJ (2000). *Dogs Through Time: An Archaeological Perspective. British Archaeological Reports International Series 889*. Oxford: Oxford University Press.

Hare, B, Brown, M, Williamson, C, and Tomasello, M (2002). The domestication of social cognition in dogs. *Science* 298: 1634–1636.

Jensen, P (ed.) (2007). *The Behavioural Biology of Dogs*. Trowbridge, UK: Cromwell Press.

Lindsay, SR (2000). *Handbook of Applied Dog Behavior and Training*. vol 1. *Adaptation and Learning*. Ames, IA: Iowa State University Press.

Marshall Thomas, E (1993). *The Hidden Life of Dogs*. New York: Houghton Mifflin.

Miklosi, A (2007). *Dog Behavior, Evolution, and Cognition*. Oxford: Oxford University Press.

Olmert, MD (2009). *Made for Each Other. The Biology of the Human–Animal Bond*. Cambridge, MA: A Merloyd Lawrence Book by Da Capo Press.

Savolainen, P, Zhang, Y, Luo, J, Lundeberg, J, and Leitner, T (2002). Genetic evidence for an East Asian origin of the domestic dog. *Science* 298: 1610–1613.

Serpell, J (ed.) (1995). *The Domestic Dog. Its Evolution, Behaviour and Interactions with People*. Cambridge: Cambridge University Press.

Smuts, BB (2001). Encounters with animal minds. *Journal of Consciousness Studies* 8: 293–309.

Trut, NT (1999). Early canid domestication: The farm-fox experiment. *American Scientist* 87: 160–169.

Vila, C, Savolainen, P, Maldonado, JE, et al. (1997). Multiple and ancient origins of the domestic dog. *Science* 276: 1687–1689.

Wells, DL (2007). Domestic dogs and human health: An overview. *British Journal of Health Psychology* 12: 145–156.

Dominance Relationships, Dominance Hierarchies and Rankings

I. S. Bernstein, University of Georgia, Athens, GA, USA

Dominance and Dominance Hierarchies

Investigators witnessing aggressive interactions between animals have noted that the interplay begins symmetrically; both participants engage in similar behavior during which one injures the other or is seen as threatening to injure the other. Such fights continue until the behavior of one changes to break off the interaction. The loser is inferred to have discovered that it cannot limit the punishment that it is receiving, which exceeds what it is willing to accept, by aggressive means. The loser stops initiating further aggression and runs away, or uses signals that are described as 'Submissive' because they seem to reduce further aggression by the opponent and bring the interaction to an end. Sometimes, however, investigators note the absence of the first symmetrical phase and try to explain why one of the two immediately moves to submissive behavior, or flees whenever the other shows any sign of aggression toward it. The explanation is often 'Dominance,' which assumes that one or both animals have learned from previous encounters and that the submitting individual anticipates the expected outcome, terminating its responses as soon as it perceives that the dominant is likely to initiate aggressive behavior toward it. Learning is invoked when there is no apparent physical change in the individuals that would account for the change in the pattern of interaction, and this change is attributed to prior experience.

This is not to say that occasionally the two will not follow their usual pattern or that the relationship is permanent. Deviation from the expected behavior provokes research into the specific factors responsible for the unexpected sequence and any permanent change in the relationship as a consequence of the unusual encounter. Rowell (1974) drew attention to the fact that it is the subordinate individual that shows the most change in behavior when a dominance relationship has been established, and if at any time the subordinate refuses to submit, the relationship may be challenged or even cease to exist. For this reason, she argued that dominance relationships should more appropriately be called 'subordinancy' relationships.

Of course, there are other possible explanations for the fact that agonistic responses may not begin with symmetrical displays of aggression. While dominance implies a learned relationship based on the outcomes of previous contests between the same two individuals, two often cited alternatives are 'Territoriality' and 'Trained Losers.' When one individual is described as a trained loser, a generalized learned response to avoid all aggressive encounters, regardless of the identity of the opponent, is postulated. Territoriality is invoked when the geographic location of the encounter correlates strongly with which of the two interactors shows aggression and which terminates the encounter as quickly as possible. When the directionality of agonistic signals is not influenced by geography and when the same individual submits reliably to some opponents but shows aggression toward others, we invoke the concept of a dominance relationship.

One complication in demonstrating that dominance is established through learning based on a previous history of agonistic encounters is that some dyads seem to establish dominance on their very first encounter without a period of obvious contest. This is likely to occur when there is a great disparity between the two, for example, one is fully adult and the other is immature. In such cases, we assume that socially sophisticated individuals do learn dominance relationships on the basis of a past history of agonistic interactions and that such learning can be generalized such that an immature individual that has lost numerous dyadic fights with adults will learn to yield to any individual having the same properties as the adults that it has lost to in previous dyadic encounters. This is, admittedly, hard to demonstrate empirically, although socially deprived immature animals often fail to yield to adults and may launch suicidal attacks against much more formidable opponents.

Alternative Measures, Causes, and Consequences

Witnessing a single encounter thus provides insufficient information to infer dominance. It requires many observations to support the hypothesis that a past history of losing to a particular individual is responsible for an individual submitting to that individual (but not all other individuals) at the first sign of aggressive behavior and that this is true regardless of the location of the encounter. In order to find more efficient means of identifying dominance relationships, many researchers have attempted operational definitions of dominance on the basis of the inferred consequences of a dominance relationship, and then measured these to infer dominance. The argument is that if one individual can aggress against another with no fear of retaliation, then such a relationship would allow the dominant individual to use aggression, or the threat of aggression, in competition for resources, or to coerce the behavior of the subordinate in any way advantageous to the dominant. A dominant individual could thus gain priority of access to incentives and control the behavior of the subordinate.

Operational measures of dominance might thus measure which individual obtains the most food, or drinks first, or gains access to preferred locations or partners. All of these are considered incentives because they presumably enhance an individual's genetic fitness, and evolution should favor individuals that used dominance to enhance their genetic fitness. Of course, this assumes that the dominant individual is at least as motivated as the subordinate to acquire the 'incentive' and not either sated or motivated to achieve a different goal at the moment.

Measures of priority of access to incentives usually correlate quite well with more laborious measures determining agonistic sequence outcomes but, even though most correlations are often significant, the degree of correlation varies and is seldom 1.0. A correlation of less than one indicates that priority of access is not synonymous with dominance. The causal relationship of variables is also unclear and in some cases may come about because both dominance relationships and other relationships are influenced by the same variables, for example, kinship influences grooming and kinship alliances influence dominance.

Although the argument that dominance must improve an individual's genetic fitness is compelling, dominant individuals do not always enjoy the highest genetic fitness. This is so, in part, because dominance status is not a lifelong attribute whereas genetic fitness is measured over a lifetime. Even a short-term measure such as breeding success or reproductive success may not correlate well with either dominance or genetic fitness as alternative strategies exist to maximize genetic fitness. For example, if attaining dominance significantly improves reproductive success but decreases longevity, then an alternative to achieving dominance would be to maintain a less costly dominance position and reproduce for a longer period of time.

Instead of looking for the consequences of dominance as a measure of dominance, one can ask why a particular individual is successful in its agonistic encounters with others and then look for physical or other attributes of dominant animals that can be used to predict dominance. Surely physical size and strength, weaponry, and aggressiveness must contribute to an individual's ability to prevail, even if not absolutely, at least relative to an opponent. If the correlations are very high, then it is a simple matter to measure the physical properties of individuals, compare the measures, and decide which will be dominant even if the two have never met each other. But especially in social species, primates, for instance, this type of comparison breaks down. Even fighting is social, not only in the sense of an interaction between the opponents, but also in the sense of the involvement of other group members, rather than just a dyad, in many contests. Social skills in forming alliances and using such alliances appropriately may thus contribute far more to dominance relationships than individual physical abilities or determination. Dominance predictions for a group on the basis of paired comparisons are notoriously unreliable as indicators of dominance relationships once the full group has been established. In fact, the dominance relationship between any dyad can be readily reversed by changing the social context in which agonistic encounters take place, much as location can change relationships in territorial forms of aggressive encounters. Measuring patterns of social alliances, the reliability of such alliances, and the skill of an individual to manage the context of a contest such that its own allies are present and ready to intervene whereas few of its opponent's allies are present or indicate a willingness to intervene, is challenging indeed. The contributions of all of these factors to dominance can be summarized as the 'power' or 'resource holding power' of an individual when discussing dominance. Such terminology (especially the latter term) implies that dominance is all about competition for resources and is the primary determinant of competitive outcomes. Not all competition, however, involves contests and scramble competition is, at times, far more significant than contests that can be influenced by dominance.

Efforts to measure dominance relationships would be much simpler if a single aggressive or submissive gesture was always associated with dominance or subordination. Efforts to find such 'formal' signals of dominance relationships have been found wanting. Although a subordinate macaque monkey will grimace to a dominant individual, grimaces also occur when an animal is in pain, frightened, or in a number of other social contexts. A monkey grimacing to a snake should not be regarded as indicating the 'dominance' of the snake to itself. Most primate expressions and communication signals can take on different meanings on the basis of the context in which they are embedded. A rhesus monkey male approaching a female prior to mounting, or at about the time of ejaculation, may grimace, but should not be considered as submitting to the female (or indicating pain). An estrous female presenting her hindquarters to a male may be communicating something very different than when one male presents its hindquarters to another or two juveniles interrupt play fighting to quickly present their hindquarters to one another.

Hierarchies

Investigators of agonistic encounters have long recognized the predictability of the outcomes of encounters between individuals and have described groups on the basis of which individuals reliably won encounters with other group members. In describing the network of dominance relationships in a group, it was clear that some individuals dominated most or all group members and some very few. Quantification was used to indicate the rank order of group members on the basis of the number of others that they could dominate. Ideally, one might expect a linear hierarchy but departures from strict linearity were recognized. Triangles could be identified wherein A dominated B, which in turn dominated C but, contrary to expectations of transitivity, C dominated A. Departures from linearity could be measured using the Landau or a similar index, or linearity could be forced on a hierarchy by ignoring dyadic analyses and focusing on total agonistic wins and losses and calculating some kind of ratio. Constructing hierarchies is particularly troublesome when not all individuals interact with all other individuals in a group, or with different frequencies. Hierarchies can be constructed, nevertheless, on the basis of the number of opponents defeated (corrected by the number lost to) or the total number of fights won regardless of the identity of the opponent (corrected by the number of fights lost). As with dyadic analyses, both methods suffer when there are missing data cells and unequal participation. Transforming data to percentages could be used to correct for unequal scores and assumptions of transitivity could be used to resolve the problems of missing cells in the matrix. Row and column totals used to calculate expected cell frequencies also have to be adjusted because agonistic matrices always have a zero cell diagonal (individuals do not ever win or lose fights with themselves).

After witnessing sequences in agonistic interactions, it is easy to understand that aggression might end when one of the participants runs away and is out of reach of the aggressor, but it is more difficult to understand how submissive signals work to mollify the aggressor so that it ceases further aggression against a still-present victim. It can be assumed that submissive signals must somehow serve to appease or placate the aggressor thus reducing the motivation to attack, but given the plethora of aggressive and submissive signals, it is unclear as to which submissive signal is a sufficient response to which aggressive signal. It is clear that there are no fixed sequences such that a particular submissive signal is elicited by a particular aggressive signal. Maxim (1978) attempted a quantitative approach to determine the numerical value of each aggressive and submissive signal. He collected data on complete sequences of agonistic behavior and reasoned that such sequences end when the value of the submissive signals is equal to the value of the aggressive signals. By running each sequence as a series of simultaneous equations, he attempted to assign values so that each sequence would achieve a zero sum. This logical and ambitious program seemed very promising, but somehow the animals did not seem to end sequences when the

equations balanced. Was this because of our failure to distinguish the intensity of different signals or to recognize all of the signals? Perhaps the failure was not in the observer but in the animals that might have failed to notice one or more signals? Or perhaps, the internal motivation of the aggressor was not always totally expressed in the signals produced, or some individuals were deceptive and signaled greater intensities than they actually experienced? Was it possible that the animals were incapable of calculating the elegant equations that the investigators were using and hence incapable of recognizing a balanced equation?

The same problem arose when investigators became dissatisfied with an ordinal rank system for dominance, since such scales preclude most parametric tests correlating dominance with quantitative outcomes. Efforts were therefore made to come up with a metric to assess dominance on an equal interval, if not ratio, scale (e.g., Boyd and Silk, 1983; Zumpe and Michael, 1986). Clearly, numerical approaches to assigning ranks on the basis of the ratio of wins and losses for each dyad were far superior to subjective assessments of rank obtained just by watching the animals. Perhaps then there was some mathematical way to assess rank differences on the basis of the ratio of wins and losses? If a ratio of wins to losses of <0.5 indicated that the rank order was incorrect, could wins minus losses divided by the number of encounters then be used to scale rank differences? A dyadic approach was less than satisfactory because if one added the distance between 1 and 2 and between 2 and 3 and summed those distances, they often failed to equal the distance between 1 and 3. Worse still, sometimes number three won against number one more than vice versa producing a rank triangle. If animals truly had a linear rank order, this rank order pertained to the group and not to the dyad and so it was reasoned that the total number of agonistic wins and losses should be used, regardless of the identity of the partner. Thus, one could use row and column totals of wins and losses and get a proportion ranging from 0 to 1 for each animal in the group. By correcting for the number of individuals in the group, one could calculate the mean distance that should separate all individuals in the hierarchy and then see whether individual pairs had larger or smaller differences in their ratios than would be predicted, thus indicating the distance between individuals of adjacent rank. If they had identical ratios, they could be viewed to be tied, but this was based on their rank in the 'group' rather than which of the two more often bested the other in agonistic encounters. Thus, if a particular dyad had a very high rate of agonistic encounters that was consistently unidirectional, then the loser might be placed low in the hierarchy regardless of the fact that it seldom, if ever, lost to any one other than the number one. A high frequency of losing was seen as an indication of low rank, but was it possible that high frequencies of interactions indicated close, and therefore unsettled or challenged dominance relations? When dominance relationships are very clear, the dominant seldom needs to assert dominance and the subordinate, in the absence of aggressive signals, shows few submissive signals. In dominance relationships that are not secure, the dominant may need to be more assertive and may perceive the slightest change in behavior as a challenge from the subordinate. (Note that there is some disagreement among investigators as to whether dominance is an outcome of agonistic encounters or whether some agonistic encounters are an outcome of individuals striving for dominance status, e.g., see Mason (1993).)

These group analyses had been designed not only to rank order individuals, but also to indicate the gradient of dominance in a group. Bayly et al. (2006) compared eight such methods and found that although each may indicate something about the relationships among the animals, the eight did not correlate well with one another. Other investigators had compared a wide range of techniques to measure dominance and found that although many correlate significantly with one another, the correlations were sometimes <0.2 and when the same measures were repeated, the correlations varied. It is clear that, although we can force equal interval scales onto dominance hierarchies, it can be only through our mathematical sophistication, and this may be beyond the means of our subjects, which use much simpler methods to manage their dominance relationships. Whereas we perceive the group as an entity and understand transitive relationships such that we would assume that if A wins over B and B wins over C, A should therefore win over C; for the animals, with less than perfect self-concepts, they only know which individuals win against them and which individuals they can defeat. For a group of 100 individuals, each individual must remember 99 relationships. If the individual were to recognize every dyadic combination in the group, it would have to remember 4950 relationships. I suggest that few of us can do that, let alone monkeys living in groups of that size. One must remember that the outcome of fights is not based on individual attributes. For each of the 99 relationships that an individual must remember, it must also remember which individuals come to its aid and which individuals come to the aid of each of its 99 potential opponents. Since fights are most often solved using social techniques, predicting the outcome and deciding whether to fight or flee may require recognizing the opponent and deciding whether the strength and number of the opponent's allies that seem ready to come to its aid is greater than the number and strength of the individuals indicating a willingness to come to your aid. This is a daunting task indeed. Watching fights in which you are not involved provides information assuring you that you are not a target and that the animals that you aid are not involved and also may provide information about which individuals side with the opponents, one or both of

which are sometimes your opponents. I doubt that they are also making transitive inferences about whether they should be able to best one of the other in future fights based on the outcome of the current fight, in which they are not involved.

What Do Individuals Know?

What does the individual subject know? Who it can win against, who comes to its aid, and perhaps who comes to the aid of its opponents. This is a formidable amount of information to remember even in a moderate-sized group. It is also why I wonder whether an animal can keep track of every other animal that it has received favors from or granted favors to and the time scale of each … Perhaps they only remember a short period of time in the past? Perhaps they only remember a general positive or negative relationship: Aid your friends (if it is not too costly) and attack your enemies whenever given the opportunity to do so. This is perhaps why aiding the aggressor is more common than defending the victim. It is less risky. Yes, it does make hierarchies conservative (bridging alliances and revolutionary alliances will be rare, but this is a functional outcome and not a motivation). In small groups of 3, there are 3 possible dyads but each individual is a member of only 2. In a group of 4, there are 3 possible opponents but 6 possible pairs. In larger groups of, say, 30, each individual is a member of 29 pairs but the number of dyads is 30 times 15 or 450. Group sizes may be limited by how many simultaneous relationships an individual can track. If self awareness is limited in animals, then transitive inference may be beyond the capacity of the subject when they are one of the things to be represented mentally. Note, however, that if individuals aid aggressors against victims and are more likely to aid an individual that they are dominant to (and is therefore unlikely to turn the attack against them), such aiding will foster transitive dominance relationships.

In summarizing dominance for whole groups, one may note different dominance 'styles.' For example, in some groups, aggression is almost invariably unidirectional. In others, there may be greater mutual exchange of aggressive signals. In some groups, dominant individuals are virtually never challenged, whereas in others, dominant individuals seem to be frequently resisted by subordinates. Attempts have been made to characterize such patterns as despotic and egalitarian and/or as tolerant or intolerant. Although these patterns may characterize a particular group for a period of time, it is also apparent that some styles are more common in groups of one species than another. Reconciliation, the tendency to engage in affiliative interactions more quickly following an agonistic encounter than if no such encounter had occurred, also correlates with dominance style. These concepts are understandable to investigators but there is still a lively debate as to which agonistic patterns co-correlate and whether there is a taxonomic predictor of such styles. Certainly, invoking these concepts in describing particular groups does not imply that the individuals in such groups have these concepts or are directly guided by such concepts.

See also: Body Size and Sexual Dimorphism; Chimpanzees; Conflict Resolution; Cooperation and Sociality; Forced or Aggressively Coerced Copulation; Social Selection, Sexual Selection, and Sexual Conflict.

Further Reading

Bayly KL, Evans CS, and Taylor A (2006) Measuring social structure – A comparison of eight dominance indicies. *Behavioural Processes* 73: 1–12.

Bernstein IS (1981) The baby and the bathwater. *Behavioral and Brain Sciences* 4: 419–457.

Bernstein IS and Gordon TP (1980) The social component of dominance relationships in rhesus monkeys (*Macaca mulatta*). *Animal Behaviour* 28: 1033–1039.

Boyd R and Silk JB (1983) A method of assigning cardinal indices of dominance rank. *Animal Behaviour* 31: 45–58.

Chapais B (1988) Experimental matrilineal inheritance of rank in female Japanese macaques. *Animal Behaviour* 36: 1025–1037.

Chapais B (1992) The role of alliances in social inheritance of rank among female primates. In: Harcourt AH and deWaal FBM (eds.) *Coalitions and Alliances in Humans and Other Animals*, pp. 29–59. Oxford: Oxford University Press.

Clark DL and Dillion JE (1973) Evaluation of the water incentive method of social dominance measurement in primates. *Folia Primatologia* 19: 293–311.

Drews C (1993) The concept and definition of dominance in animal behaviour. *Behaviour* 125: 283–313.

Farres AG and Haude RH (1976) Dominance testing in rhesus monkeys; comparison of competitive food getting, competitive avoidance and competitive drinking procedures. *Psychological Reports* 38: 127–134.

Maestripieri D (1999) Formal dominance: The emperor's new clothes? *Journal of Comparative Psychology* 113: 96–98.

Mason WA (1993) The nature of social conflict: A psychoethological perspective. In: Mason WA and Mendoza SP (eds.) *Primate Social Conflict*, pp. 13–47. Albany, NY: State University of New York.

Maxim PE (1978) Quantification of social behavior in pigtail monkeys. *Journal of Experimental Psychology Animal Behavior Processes* 4: 50–67.

Richards SM (1974) The concept of dominance and methods of assessment. *Animal Behaviour* 22: 914–930.

Rowell TE (1966) Hierarchy in the organization of a captive baboon group. *Animal Behaviour* 14: 430–443.

Rowell TE (1974). The concept of social dominance. *Behavioral Biology* 11: 131–154.

Rushen J (1984) Should cardinal dominance ranks be assigned. *Animal Behaviour* 32: 932–933.

Sade DS, Altmann M, Loy J, Hausfater G, and Breuggeman JA (1988) Sociometrics of *Macaca mulatta*: II. Decoupling centrality and dominance in rhesus monkey social networks. *American Journal of Physical Anthropology* 77: 409–425.

Zumpe D and Michael RP (1986) Dominance index: A simple measure of relative dominance status in primates. *American Journal of Primatology* 10: 291–300.

Drosophila Behavior Genetics

R. Yamada and E. A. McGraw, University of Queensland, Brisbane, QLD, Australia

Introduction

In the early 1910s, Thomas Hunt Morgan identified the first white-eyed fly in the 'Fly room' at Columbia University. Then, Morgan and his three students, Sturtevant, Bridges, and Muller reported a series of fundamental concepts in the chromosomal theory of heredity, including the sex-linked inheritance of white eyes, recombination and linkage between sex-linked genes, and the first chromosome maps based on linkage. These major scientific breakthroughs were discovered in *Drosophila*, and the field of modern genetics was founded.

Since then, *Drosophila melanogaster* has been one of the major model organisms in genetics. *D. melanogaster* has many benefits for genetic research. It is easy to rear in the laboratory, has a short generation time (10 days at 25 °C), produces large number of progeny (each female can lay over 100 eggs), and has a high tolerance of inbreeding. Moreover, the larval salivary gland contains giant polytene chromosomes that exhibit banding patterns useful in the identification of chromosomal rearrangements and deletions by visual inspection. Thousands of mutations with visible phenotypes served as genetic markers and provided essential tools for genetic analysis. In 1968, Lewis and Bacher described the efficient method of inducing mutation using ethyl methane sulphonate (EMS). EMS mutagenesis facilitates the most important tools of *D. melanogaster*, the power of forward genetic screens to dissect the genes that affect a specific phenotype. In 1980, Nusslein-Volhard and Wieschaus extended this approach to the first large-scale mutagenesis project that attempted to isolate most of genes involved in the embryonic development. Following the discovery of homeobox genes, Lewis, Nusslein-Volhard, and Wieschaus received the Nobel Prize in 1995. With advanced genetic screen techniques such as modifier screens and clonal screens together with genetic transformation techniques, it is possible to screen for almost any biological process, including complex behavior.

William E Castle was the first person to use *Drosophila* for a genetic study in the laboratory at Harvard University in 1901. Subsequently, Castle and his students began to study simple behaviors, including phototaxis (an organisms movement in response to light), geotaxis (response to gravity), and later mechanosensory and olfactory responses. This was followed by a series of studies using more extensive genetic strategies such as quantitative genetic analysis and selection experiments for behavioral

phenotypes in *Drosophila* and other organisms including mice and rats. One difficulty in studying the genetics of behavior is that the heritability of behavioral phenotypes is highly sensitive to the environment and genetic background. This is thought to be due to the involvement of multiple gene networks in complex behavior. The consistent conclusion from the early studies was that the genetic basis of behavior is complex and multigenic. Hence, behavior has often been considered a more complex set of phenotypes than either developmental or anatomical defects. Seymour Benzer was the first to report the successful isolation of a behavioral mutant with respect to phototaxis using genetic screens.

Classic Single Gene Mutant Studies of Behavior

The publication of Seymour Benzer's (1967) paper was a seminal moment in the history of behavioral genetics. Before his report, the idea that single genes control complex behavior was not accepted. Traditional approaches to solving the question 'how do genes influence behavior?' had been carried out by selective breeding for the behavioral trait of interest from natural populations. However, Benzer took a different approach, using mutagenesis and genetic manipulations, to quantify a series of behaviors, including phototaxis, circadian rhythms, learning and memory, courtship, etc. His strategy was straightforward: induce mutations by feeding EMS to male flies, screen their offspring for behavioral phenotypes, then use genetic crossing to isolate single-gene mutations responsible for these altered behaviors. Since then, hundreds of scientists have continued Benzer's experimental philosophy and referred to it as 'neurogenetics.'

Circadian Rhythm

Circadian rhythm mutants are one of the excellent examples of the original discoveries in neurogenetics. Circadian rhythms are cycles of behavior and physiology found in nearly all organisms that sets their internal clock time to an approximate 24 h cycle. Circadian rhythm allows organisms to adapt to external environmental factors such as the regular cycles of light and temperature that pervade the biosphere. This clock is conserved in some bacteria, protozoa, plant, and animals, reflecting four billion years of evolution of life on a rotating planet with an oscillating cycle of day and night. In animals, the rest period in the activity cycle is the best understood aspect of circadian behavior. Most of our current understanding of the molecular basis of circadian rhythm has come from studies in *Drosophila*.

In 1971, Konopka and Benzer performed a simple screen for the phenotype of altered eclosion rhythm and the locomotor rhythms. They identified *period (per)* mutants, the first clock mutants in any organisms. The allele *per⁰* is arrhythmic; *perˢ* has a short circadian period of 19 h; and *perˡ* exhibits a long circadian period of 29 h instead of the normal 24 h. With the cloning of *per* gene by the groups of Rosbash and Hall, and the group of Young in 1984, *per* transcripts and PER protein were subsequently both shown to oscillate in abundance with circadian rhythms, giving rise to the autoregulatory feedback model of how clock gene products might underlie the core mechanism of the biological clock. Importantly, more recent studies suggest that the molecular basis for the circadian clock is generally conserved between flies and mice. Detailed descriptions of these mechanisms are reviewed by Sehgal and Allada.

Courtship Behavior

The earliest descriptive studies of courtship behavior in *Drosophila* were conducted by Sturtevant in the mid 1910s. Bastock and Manning then characterized serial steps of stereotypical actions in males: orientation, tapping, wing vibration, licking, and copulation. This was followed by a series of studies revealing that males and females exchange sensory modalities in each step by visual, acoustic, pheromonal (pheromone: molecule emitted by one individual that alters the behavior or physiology of conspecifics), and tasting signals (**Figure 1**). In 1976, Hotta and Benzer used genetic mosaics (gynandromorphs: flies composed of male and female tissue) to roughly map the portions of the nervous system that control the courtship. Using mosaic analysis, in 1977, Hall refined the techniques and identified anatomical foci in the brain and the thoracic and abdominal nervous system that are required for sex-specific courtship behavior.

The gene *fruitless (fru)* is the best-studied gene involved in courtship behavior. The bisexual *fru* mutant was originally identified as a male-sterile variant in the 1960s. Further studies also revealed that *fru* was required for various aspects of male courtship. The *fru* gene encodes multiple forms of a transcription factor that are required not only for male-specific courtship behavior but also for viability in both sexes. One of the *fru* transcripts is male-specifically spliced and responsible for male-specific courtship behavior. Ectopic expression of the male-specific form of *fru* alters almost every feature of male courtship. A reduction in male-specific

Figure 1 Steps in courtship by *D. melanogaster*. The colored arrows represent the known sensory modalities by which flies communicate: (+) for stimulatory and (−) for inhibitory signals. Reprinted from Greenspan RJ and Ferveur JF (2000) Courtship in *Drosophila*. *Annual Review of Genetics* 34: 205–232, with permission from Annual Review of Genetics.

fru expression in the median bundle exhibits faster copulation, skipping early steps in the behavior such as orientation, tapping, wing vibration, etc. Females ectopically expressing male-specific *fru* show male courtship behavior. Recent studies of specific labeling of neurons that express male-specific *fru* revealed a precise map for core neural circuits involved in male courtship behavior. A set of genes that operate downstream of *fru* remains to be identified.

Male courtship represents an innate behavior, but is also modified by experience. Immature males, in their first day after eclosion, produce female-like pheromones that stimulate mature males to show active courtship toward them with a wing vibration that produces courtship song. Exposure to the song enhances the success in copulation once they are mature. Another example is that male flies tend to exhibit decreased courtship vigor in response to virgin females (receptive) if they have previously experienced rejection from mated females (unreceptive). While these experiments indicate that flies can learn and remember, further analyses are desirable to confirm the experience-dependent modifications in courtship.

Learning and Memory

The first learning mutant, *dunce (dnc)*, was isolated by Quinn, Harris, and Benzer in 1974, using a genetic screen for the Pavlovian olfactory paradigm. In this study, flies were trained to avoid an odor paired with electric shock as a negative reinforcer. After training, the relative avoidance of the shocked odor was scored as a learning index. Using this paradigm, a series of learning mutant flies were identified, initiating a growing list of genes known to be involved in olfactory memory formation.

The cloning of the *dnc* gene, encoding cAMP phosphodiesterase, and the *rutabaga (rut)* gene, encoding $Ca^{2+}/$calmodulin-responsive adenylyl cyclase, revealed the importance of cAMP signaling pathways in learning. In addition, both *dnc* and *rut* show preferential expression in a part of the brain called the mushroom bodies (MBs). Considerable evidence supports the importance of MBs as a major structure for olfactory memory and storage. The chemical ablation of MBs disrupted olfactory learning ability and learning mutants show anatomical defects in MBs. The rescue experiment revealed that expression of *rut* activity in the MB neurons of *rut* mutants is sufficient for olfactory memory formation. Using the *Shibire* transgene, a fine genetic tool to block synaptic transmission, it has been revealed that the α/β neuron in MBs are responsible for olfactory memory. While these accumulating evidences support a dominant role for MB neurons in olfactory learning, their relevance to overall memory is still uncertain.

Natural Variation of Behavior in Populations

Most of the mutations identified by genetic screens cause severe perturbation in the function of specific genes well beyond that observed in the natural variation found in wild populations. Genes responsible for a behavioral phenotype are difficult to isolate from natural variants because most behavior is likely to be regulated by multiple gene networks. When this is the case, behavior phenotypes tend to dissipate during the course of crosses for genetic mapping, resulting in a failure to isolate the relevant gene(s). One exception has come from the area of foraging behavior.

Searching for Food: Foraging Behavior

Individual flies in natural populations of *D. melanogaster* can be categorized according to the type of food-searching behavior they exhibit, as either rovers or sitters. Rovers search wider areas for food than sitters. The typical phenotypic frequencies of the two phenotypes in natural populations are 70% rover and 30% sitter. Sokolowski performed density-dependent selection experiments to reveal that rovers are dominant under crowded conditions and sitters under less crowded conditions. The phenotypic differences in behavior are attributed to variation in a single gene called *foraging (for)*. Gene *for* encodes a cGMP-dependent protein kinase (PKG). Rovers have 12% more PKG enzyme activity than sitters, suggesting that this small difference might be sufficient to make

variation in behavioral phenotype. These results give us an idea of how behavior is altered and selected in natural populations.

Latitudinal Clines in Clock Genes

The clock gene, period, involved in circadian rhythm (as described earlier), harbors DNA sequence variation in natural populations of *D. melanogaster*. The variation found around the coding region for a threonine-glycine (Thr-Gly) repeat ranges from 14 to 23 copies (**Figure 2(a)**). Recent studies revealed that the distribution of copy number in the Thr-Gly repeat correlates with latitude in both hemispheres. The most frequent alleles in the northern hemisphere are *(Thr-Gly)$_{17}$* and *(Thr-Gly)$_{20}$*. The *(Thr-Gly)$_{20}$* variant is more prevalent in the north and the *(Thr-Gly)$_{17}$* in the south. A similar latitudinal cline of the *(Thr-Gly)$_{20}$* variant was also found in Australia, further suggesting the existence of climatic selection (**Figure 2(b)**). Subsequent studies demonstrate that these variants show different circadian temperature compensations and abilities in maintaining a constant circadian period under different environmental temperatures. The *(Thr-Gly)$_{20}$* variants show a very consistent circadian period at different temperatures and exquisite temperature compensation, while the *(Thr-Gly)$_{17}$* variants show poor temperature compensation resulting in shorter circadian periods at lower temperatures. These results suggest that the *(Thr-Gly)$_{20}$* variant might be adapted to the colder and more thermally variable environments at higher latitudes. This thermal explanation for the latitudinal cline of Thr-Gly repeat has been supported by the observation in 'Evolution Canyon' on Mt. Carmel in Israel. The northern-facing slope of this canyon is colder than the southern-facing slope, and the frequencies of *(Thr-Gly)$_{20}$* and *(Thr-Gly)$_{17}$* are significantly different in a manner consistent with a thermal explanation.

A similar natural polymorphism is found in *timeless*, another clock gene involved in circadian rhythms that generates two different length TIM isoforms. The allele generating the longer TIM isoform is more common in the south, while that encoding the shorter isoform is more prevalent in the north in European natural populations of *D. melanogaster*. Although the functional relevance of these polymorphisms remains to be characterized, the studies of natural variation in clock genes provide us with a novel approach for behavioral research leading which can illuminate animal adaptation within an evolutionary and ecological context.

Selection for Aggressive Behavior

Aggression is a complex behavior that is heritable in natural populations of *Drosophila*. Sturtevant reported the first description of fly aggression in 1915. Subsequent

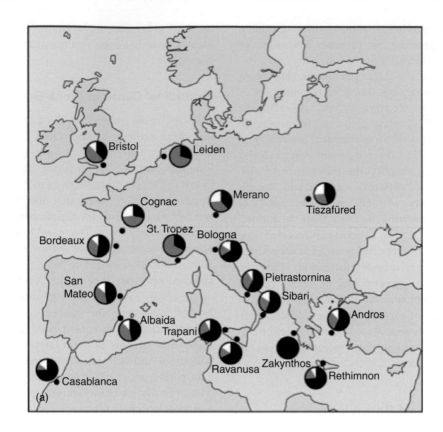

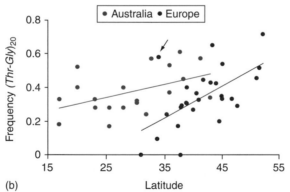

Figure 2   (a) Distribution of *per* alleles in Europe. Frequencies in the various regions for the alleles *(Thr-Gly)₂₀* (gray), *(Thr-Gly)₁₇* (black), and all other alleles (white). Cited from Costa et al. (1992) and reprinted from Greenspan RJ (2007) The world as we find it. In: *An Introduction to Nervous Systems*, pp. 123–139. New York, NY: Cold Spring Harbor Laboratory Press, with permission from Cold Spring Harbor Laboratory Press. Cited from Costa et al. 1992 and reprinted, with permission of the Cold Spring Harbor Laboratory Press, from Greenspan 2007 Proc Biol Sci. A latitudinal cline in a *Drosophila* clock gene. Costa R, Peixoto AA, Barbujani G, Kyriacou CP. 1992 Oct 22; 250(1327): 43–49. (b) Latitudinal cline of *per* alleles. *(Thr-Gly)₂₀* frequency in Australian (blue) and European (red) natural populations. Reprinted from Kyriacou CP, Pexoto AA, Sandrelli F, Costa R, and Tauber E (2008) Clines in clock genes: Fine-tuning circadian rhythms to the environment. *Trends in Genetics* 24: 124–132, with permission from Elsevier.

studies showed the detailed description of the ethological perspective on aggression. In 1988, Hoffmann revealed that enhanced aggression could be achieved by artificial selection experiments. Despite the importance of and ongoing interest in aggression, the gene(s) involved in aggression have never been identified. The complexity of aggressive behavior and the lability of the aggression phenotype make it difficult to perform a standard genetic screen. In 2002, Kravitz and colleagues characterized nine

distinct patterns of aggressive behavior, such as fencing, lunging, holding, and boxing, to facilitate a quantitative analysis of aggression. More recent work by Dierick and Greenspan isolated candidate genes involved in aggression, using microarray analysis of selected lines. They developed a population-based selection system to increase aggression in a laboratory strain of *D. melanogaster*, by picking the most aggressive males from a population cage that contained 120 males and 60 females with multiple territories. After 11 generations of selection, the lines showed more aggression than the control lines. After 21 generations of selection, microarray analysis was performed to characterize genes that were differentially expressed in selected and control lines, instead of taking a traditional genetic mapping approach. Consequently, 42 candidate genes for aggression were found. Subsequent mutant analysis then revealed that a mutation in a gene encoding cytochrome P450 6a20 (*Cyp6a20*) significantly altered aggressive behavior (reviewed by Robin et al., 2007). An independent study by Anderson's group also supports the involvement of *Cyp6a20* in aggression. They showed that social experience increased the expression of *Cyp6a20* to suppress aggression and that *Cyp6a20* is expressed in pheromone-sensing olfactory tissue. These findings revealed that *Cyp6a20* plays a common role mediating heritable and environmental influences on aggression. These studies also represent a more rapid approach for isolating behavioral genes from natural variation.

Courtship Song and Species Recognition

Species Recognition

Species recognition is a major factor in premating reproductive isolation between species. In *Drosophila*, species recognition depends on chemical sensing of pheromones and courtship songs produced by wing vibration. Artificial application of foreign pheromones on the fly body surface is sufficient to induce unusual courtship behavior between different *Drosophila* species. Dummy flies covered with female pheromones can attract males. The male courtship song is not essential for mating, because wingless males are able to copulate although they take longer to be successful. However, making *D. pallidosa* males wingless drastically enhanced interspecies mating with *D. ananassae* females, but reduced intraspecies mating, implying the importance of song for species recognition.

Song Rhythm

The interpulse intervals (IPI), the most important parameter for species recognition, are species-specific and average 35 ms in *D. melanogaster* and 50 ms in *D. simulans*. In 1980, Kyriacou and Hall noted another rhythmic component of song from *D. melanogaster*. The oscillation of IPI, known as the IPI cycle, is also species-specific with a period of around 1 min in *D. melanogaster*, 35–40 s in *D. simulans*, and 75 s in *D. yakuba*. These song rhythms are important for female receptivity, as females prefer their species-specific IPI and IPI cycle for mate choice. Surprisingly, Kyriacou and Hall found that the *period* mutations, which regulate the circadian rhythm described earlier, also altered the song rhythms, with the short-day mutant (*per^s*) reducing the IPI cycle to ~40 s, the long-day

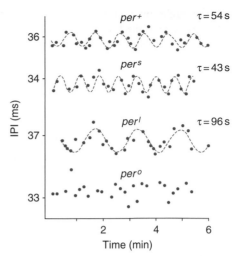

Figure 3 Song rhythms in *per* mutants. Rhythmic oscillation of interpulse interval (IPI) in the male courtship song in normal males (*per^+*) and per mutant males: short day (*per^s*), long day (*per^l*), and arrhythmic (*per^o*). τ: the period of the song rhythm. Cited from Kyriacou and Hall 1980 and reprinted, with permission of the Cold Spring Harbor Laboratory Press, from Greenspan 2007 Circadian rhythm mutations in *Drosophila melanogaster* affect short-term fluctuations in the male's courtship song PNAS November 1, 1980 vol. 77 no. 11 6729–6733.

mutant (*per^l*) extending it to ~80 s, and the *per^o* mutatn showing an arrhythmic phenotype in both phenotypes (**Figure 3**). Interestingly, circadian rhythm and courtship behavior are correlated and likely to be involved in the process of speciation.

Evolution of Song

Kyriacou and Hall mapped the species-specific song rhythms of *D. simulans* and *D. melanogaster* to the X chromosome, a finding coincident with the *per* gene's location on the same chromosome. Subsequent interspecific transformation experiments revealed striking evidence that a *D. melanogaster* male containing a *D. simulans period* transgene sang with the *simulans*-like short cycle, revealing a single gene control of an important species-specific parameter. In contrast, the mean IPI remained intact in the transformants, implying that those two important parameters of song rhythm, IPI and the song cycle, are regulated independently. Similar interspecific transformation experiments revealed that the species-specific mating rhythm is also controlled by *per*.

Different fly species also have distinctive mating rhythms (the pattern of mating with respect to time of day). *D. melanogaster* flies have a peak in their mating rhythm late in the day which is maintained at high levels in the night, while *D. pseudoobscura* flies have two different peaks, one around dusk and the other in the middle of the night (**Figure 4(a)** left). The *D. melanogaster* transformants carrying a *D. pseudoobscura per* transgene showed a *pseudoobscura*-like peak of mating preference during the middle of the night in contrast to the *D. melanogaster* pattern (**Figure 4(a)** right bottom). Subsequent studies revealed that if males and females of two types of transformants (carrying the *pseudoobscura per* transgene or the *melanogaster per* transgene) are mixed together, they prefer to mate with flies harboring the same type *per* transgene (**Figure 4(b)**). This assortative mating could be related to the differences of mating rhythm alone, since wingless

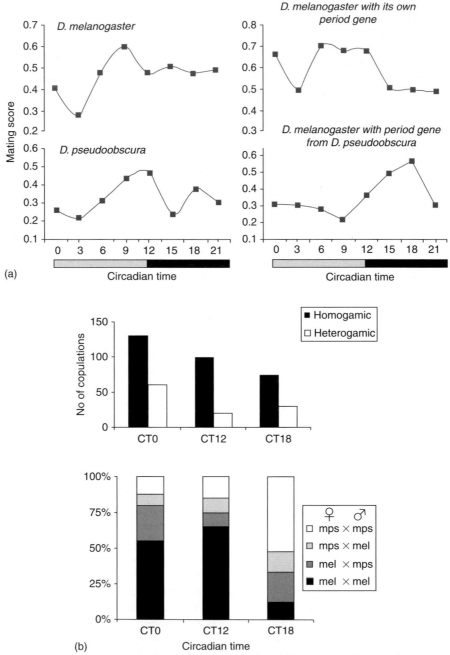

Figure 4 (a) Mating rhythms in *D. melanogaster* and *D. pseudoobscura* (left) and *per* transgenic *Drosophila* (right). Graphs show proportion of pairs mating at various time of day. Genetically engineered transgenic flies carrying the *D. pseudoobscura* period show a peak of mating during the night that is absent in normal *D. melanogaster* but present in normal *D. pseudoobscura*. Cited from Tauber et al. (2003) and reprinted from Greenspan RJ (2007) The world as we find it. In: *An Introduction to Nervous Systems*, pp. 123–139. New York, NY: Cold Spring Harbor Laboratory Press, with permission from Cold Spring Harbor Laboratory Press. (b) Assortative mating in transgenic flies carrying *D. melanogaster per* or *D. pseudoobsucura per*. (top) The number of homogamic mating (black bars, *mel* × *mel* and *mps* x *mps*) and heterogamic mating (white bars, *mel* × *mps*). *mel*: *D. melanogaster* carrying its own *per* gene, *mps*: *D. melanogaster* carrying *D. pseudoobscura per* gene. (bottom) Relative proportion of the different types of male/female pairings for the data shown in top graph. CT: Circadian time. Reprinted from Tauber E, Roe H, Costa R, Hennessy JM, and Kyriacou CP (2003) Temporal mating isolation driven by a behavioral gene in *Drosophila*. *Current Biology* 13: 140–145, with permission from Elsevier.

males were used in these experiments to avoid any effects of the song rhythm influenced by the *per* transgene (because *per* also alter song rhythm as described earlier).

These studies demonstrated the possible involvement of a single gene, in this case *per*, in the speciation process by influencing both species recognition and mate preference through male song and mating rhythm. However *per* is not likely involved in female song preference as *per* mutant females still prefer the wild-type song rhythm, indicating that the *per* influence on the male song and the female reception are not co-opted. Understanding the genetic basis of female preference is the next challenge for this field.

Bacterial Infection and Insect Behavior

Circadian Rhythm and Immunity

Circadian rhythm is important not only for behavior but also for more basic physiological processes such as immunity. The effects of a disrupted circadian rhythm on infection and disease in mammals are well documented, although the molecular mechanisms underlying these interactions are unknown. Recent studies by Schneider and colleagues have revealed the functional relationship between circadian rhythm and innate immunity in *D. melanogaster*. They found that infection by bacterial pathogens disrupted circadian rhythm, with sick flies moving constantly all day resulting in sleep deprivation. Further studies have shown that circadian mutants per^{01} and tim^{01} died significantly earlier than wild-type control flies when exposed to a lethal dose of pathogenic bacteria. Lee and Edery took a different approach to studying the impact of circadian regulation on immunity. They found that the survival rate of files that are infected with lethal pathogenic bacteria depends on the time of day when they are infected. Flies infected in the middle of the night showed better survival rates (about threefold greater) than flies infected during the day. Similar to Schneider's study, the per^{01} mutant showed higher mortality than the wild-type control in their experiments. These studies provided evidence of a novel interaction between bacteria and fly behavior as well as a new avenue for immunity research, which is applicable to medical strategy based chronobiology.

Influence of *Wolbachia* Symbiont on Behavior

A few studies report symbiont-based behavioral manipulations in *Drosophila*. For example, *Wolbachia* has been shown to increase the male mating rate. *Wolbachia* are maternally inherited intracellular bacteria that infect a broad range of invertebrate hosts. Current estimates suggest that the total number of infected arthropod species might be around 66%, and notably about 30% of flies in the Bloomington *Drosophila* stock center (one of the biggest centers in the world) are infected. *Wolbachia* commonly manipulate host reproduction in a variety of ways,

resulting in embryonic lethality, thereby favoring their own persistence and spreading into host populations. While the reproductive phenotype of *Wolbachia* has been studied extensively, little is known about its effects on host behavior, despite its presence in nervous tissues. de Crespigny and colleagues found that *Wolbachia* infected males show higher mating rates than uninfected control males in *D. melanogaster* and *D. simulans*. A recent study by McGraw and colleagues showed that *Wolbachia* infection influences olfactory cued locomotion in *Drosophila* in a species-specific manner. In *D. simulans*, the olfactory response was increased in response to infection, but it decreased in *D. melanogaster*. The influences of *Wolbachia* infection on behavior found in these studies are relatively moderate compared with the differences found in a number of mutant studies. However, because mating rate, locomotion, and olfaction are essential behaviors in nature, the subtle alteration of these behaviors by *Wolbachia* could have a significant impact on their fitness. Further studies are needed to examine the effects of *Wolbachia* in both the laboratory and field. In addition, the genetic and molecular basis of the interaction between *Wolbachia* and host insect remain to be identified. Almost all behavioral genetic studies in *Drosophila* do not mention the *Wolbachia* infection status of the flies studied. Future work in fly behavioral genetics should take into account the presence of the microbe and its possible role in insect behavior.

Conclusion

Neurogenetic research in *Drosophila* paved the way for the fruit fly becoming a model system in the study of complex behaviors such as circadian rhythm, courtship, learning and memory, foraging, aggression, etc. These studies revealed that behavioral genes are pleiotropic. For example, *period* influences circadian rhythm and courtship, *fruitless* is involved in both the development and functioning of the nervous system that regulates various aspects of male courtship, and *foraging* alters food searching, olfactory learning and memory, and epithelial fluid transport. In addition, most of the behavioral mutations turned out to be hypomorphic partial loss of function alleles. One simple explanation is that null mutations tend to be lethal, whereas milder mutations such as those that alter splicing patterns, expression levels, or enzymatic activity, often produce informative behavioral phenotypes. Such 'kinder,' milder mutations are identified through genetic screens of variants from natural population as well as from selection experiments with wild-type strains, instead of through gene disruption strategies such as gene knockout.

These principles from neurogenetics allow us to study complex behavior, using the tiny fly. More recently, van Swinderen demonstrated attention-like processes in *Drosophila* by measuring brain activity responding to visual stimuli. Shaw and colleagues characterized the behavioral sleep state in *Drosophila* and a subsequent series of studies elucidated striking similarities in features of sleep between human and fly, revealing its regulation by homeostasis and circadian rhythms, the pharmacological responses to drugs such as caffeine, methamphetamine, and antihistamines, and the influence from sex and age. As for sleep, general anesthetics induce immobility and increased arousal thresholds in flies, responses resembling human ones. In addition, with advances in technology, *Drosophila* behavior is now being studied from the diverse lenses of many biological disciplines such as genetics, molecular biology, biochemistry, cell biology, anatomy, and physiology. Consequently, the accumulating evidence and depth of understanding of process and mechanism mean that *Drosophila* has become a medically important model organism, with particular contributions made in the areas of insomnia, drug sensitivity, human neurodegenerative diseases, and even consciousness.

On the other hand, the advent of genome sequencing technology for any organism, together with the ability to test gene function with RNAi in which genetic analysis is not essential, creates the potential for most organisms to become behavioral genetic models. In the near future, it will be possible to study the molecular basis of far more intriguing behaviors than those of the fly, for example social behavior in ants and honeybees, swarming behavior in locusts, and behavioral regulation of host insects by symbionts or parasites. Nevertheless, *Drosophila* has provided us with the state-of-the art technology for behavioral genetic research and will continue to play a pivotal role in this field.

Acknowledgments

The authors thank two anonymous reviewers for their helpful comments on the article. E. A. McGraw and R. Yamada were supported by ARC Discovery Project #DP0557987.

See also: Honeybees; Locusts; *Tribolium*.

Further Reading

Benzer S (1967) Behavioral mutants of *Drosophila* isolated by countercurrent distribution. *Proceedings of the National Academy of Sciences of the United States of America* 58: 1112–1119.

Costa R, Peixoto AA, Barbujani G, and Kyriacou CP (1992) A latitudinal cline in a *Drosophila* clock gene. *Proceedings of the Royal Society B: Biological Sciences* 250: 43–49.

Costa R, Sandrelli F, and Kyriacou CP (2007) Evolution of behavioral genes. In: North G and Greenspan RJ (eds.) *Invertebrate Neurobiology*, pp. 617–646. New York, NY: Cold Spring Harbor Laboratory Press.

Davis RL (2005) Olfactory memory formation in *Drosophila*: From molecular to systems neuroscience. *Annual Review of Neuroscience* 28: 275–302.

Ferveur JF (2007) Elements of courtship behavior in *Drosophila*. In: North G and Greenspan RJ (eds.) *Invertebrate Neurobiology*, pp. 405–435. New York, NY: Cold Spring Harbor Laboratory Press.

Greenspan RJ (1997) A kinder, gentler genetic analysis of behavior: Dissection gives way to modulation. *Current Opinion in Neurobiology* 7: 805–811.

Greenspan RJ (2001) The flexible genome. *Nature Reviews Genetics* 2: 383–387.

Greenspan RJ (2007) The world as we find it. *An Introduction to Nervous Systems*, pp. 123–139. New York, NY: Cold Spring Harbor Laboratory Press.

Greenspan RJ (2008) Seymour Benzer (1921–2007). *Current Biology* 18: R106–R110.

Greenspan RJ and Ferveur JF (2000) Courtship in *Drosophila*. *Annual Review of Genetics* 34: 205–232.

Hall JC (1977) Portions of the central nervous system controlling reproductive behavior in *Drosophila melanogaster*. *Behavior Genetics* 7: 291–312.

Hotta Y and Benzer S (1976) Courtship in *Drosophila* mosaics: Sex-specific foci for sequential action patterns. *Proceedings of the National Academy of Sciences of the United States of America* 73: 4154–4158.

Konopka RJ and Benzer S (1971) Clock mutants of *Drosophila melanogaster*. *Proceedings of the National Academy of Sciences of the United States of America* 68: 2112–2116.

Kyriacou CP and Hall JC (1980) Circadian rhythm mutations in *Drosophila melanogaster* affect short-term fluctuations in the male's courtship song. *Proceedings of the National Academy of Sciences of the United States of America* 77: 6729–6733.

Kyriacou CP, Pexoto AA, Sandrelli F, Costa R, and Tauber E (2008) Clines in clock genes: Fine-tuning circadian rhythms to the environment. *Trends in Genetics* 24: 124–132.

Manoli DS and Baker BS (2004) Median bundle neurons coordinate behaviours during *Drosophila* male courtship. *Nature* 430: 564–569.

Mehren JE, Ejima A, and Griffith LC (2004) Unconventional sex: Fresh approaches to courtship learning. *Current Opinion in Neurobiology* 14: 745–750.

Nusslein-Volhard C and Wieschaus E (1980) Mutations affecting segment number and polarity in *Drosophila*. *Nature* 287: 795–801.

Peng Y, Nielsen JE, Cunningham JP, and McGraw EA (2008) Wolbachia infection alters olfactory-cued locomotion in *Drosophila* spp. *Applied and Environmental Microbiology* 74: 3943–3948.

Quinn WG, Harris WA, and Benzer S (1974) Conditioned behavior in *Drosophila melanogaster*. *Proceedings of the National Academy of Sciences of the United States of America* 71: 708–712.

Robin C, Daborn PJ, and Hoffmann AA (2007) Fighting fly genes. *Trends in Genetics* 23: 51–54.

Sehgal A and Allada R (2007) Circadian rhythms and sleep. In: North G and Greenspan RJ (eds.) *Invertebrate Neurobiology*, pp. 503–532. New York, NY: Cold Spring Harbor Laboratory Press.

Shaw PJ, Cirelli C, Greenspan RJ, and Tononi G (2000) Correlates of sleep and waking in *Drosophila melanogaster*. *Science* 287: 1834–1837.

Shirasu-Hiza MM, Dionne MS, Pham LN, Ayres JS, and Schneider D (2007) Interactions between circadian rhythm and immunity in *Drosophila melanogaster*. *Current Biology* 17: R353–R355.

Sokolowski MB (2001) *Drosophila*: Genetics meets behaviour. *Nature Reviews Genetics* 2: 879–890.

St Johnston D (2002) The art and design of genetics screens: *Drosophila melanogaster*. *Nature Reviews Genetics* 3: 176–188.

Sturtevant AH (1965) *History of Genetics*. New York, NY: Cold Spring Harbor Laboratory Press.

Tauber E, Roe H, Costa R, Hennessy JM, and Kyriacou CP (2003) Temporal mating isolation driven by a behavioral gene in *Drosophila*. *Current Biology* 13: 140–145.

van Swinderen B (2005) The remote roots of consciousness in fruit-fly selective attention? *BioEssays* 27: 321–330.

Werren JH and O'Neill SL (1997) The evolution of heritable symbionts. In: O'Neill SL, Hoffmann AA, and Werren JH (eds.) *Influential Passengers*, pp. 1–41. Oxford: Oxford University Press.

Ecology of Fear

J. S. Brown, University of Illinois at Chicago, Chicago, IL, USA

Introduction

A tourist's video from South Africa posted on YouTube records an amazing sequence of events. A group of lionesses attempt the foolhardy. They attack a group of African buffalo, separating out a calf and dragging it into the waterhole, presumably to drown it and to fend off any counterattack from the buffalo. The buffalo retreat, but a crocodile unexpectedly grabs the calf. The lionesses have no choice but to drag the hapless calf back out of the water. This emboldens the buffalo which return, drive off the lionesses, and rescue the apparently mauled but otherwise healthy calf. This choreography of buffalo, lions, and crocodile illustrates the nexus between antipredator behaviors and their ecological consequences for births and deaths. The mother buffalo almost lost her calf, the crocodile almost got a meal, and the lionesses in losing their meal deprived a crocodile and inadvertently saved the calf.

The ecology of predator–prey interactions involves how predators kill prey and prey feed predators. The behavioral ecology of predator and prey studies stealth and vigilance, and the topics of many of the preceding articles. The Ecology of Fear is the study of how the fear and antipredatory behaviors of prey and the stealth and hunting behaviors of predators influence predator–prey interactions, population dynamics, species coexistence, and evolutionary dynamics. Long before I applied this term to the consequences of predator–prey behaviors on the ecology of predator–prey systems, others had recognized the strong connections. To name but a few: Michael Rosenzweig noted how fear responses of prey could radically change the dynamics and stability properties of predator–prey interactions. Steve Lima points out how a single goshawk swooping over a flock of flamingos could directly kill one, but it could set off a stampede that might kill several more.

Oswald Schmitz showed how the nonlethal effects of predators could be as or more important than the lethal effects. He and his lab enclosed grasshoppers with spiders.

In some treatments, the spiders were potentially lethal, whereas in others, the spiders had been rendered harmless; their chelae glued together. Relative to control populations, the number of grasshoppers declined equally whether the spiders were lethal or simply scary. Even without direct mortality from spiders, the grasshoppers experienced reduced fitness as they abandoned perceived risky habitats and forewent other feeding opportunities for the sake of fear.

As the ecology of fear, we shall see how predation risk represents an additional activity cost for the prey. These fear responses then influence the population sizes and dynamical stability of predator–prey interactions. Fear responses also structure the spatial and the temporal landscapes of their prey, thus influencing and determining habitat suitability for both predator and prey. Fear responses can also create behavioral cascades up and down food chains. Fear responses may influence the length of food chains, and the coexistence of multiple prey species and predator species. Risk of injury from hunting creates fear responses from the predators. Finally, fear responses may be the unit of conservation, may be necessary for ecosystem health, and may provide behavioral indicators for the status of the prey and the whereabouts of their predators.

Predation Risk as an Activity Cost

Predators, as a direct effect, increase the mortality rate of prey by killing them. Similarly, the consumption of prey by a predator enhances her survivorship and fecundity. The lethality of predators directly affects the per capita growth rates of the prey and predators. But, these direct effects become modified as soon as the prey exhibit flexible fear responses. The prey do this by either allocating time away from risky times and places, or using vigilance behaviors to reduce risk while engaged in some risky activity such as foraging. In both cases, the prey may be forgoing other fitness-enhancing opportunities to avoid being killed by the predator. The fear responses reduce the direct effect of the predator on the prey. For this reason, the effects of fear on the lethal effects of predators have often been referred to as 'trait-mediated effects,' 'behavioral indirect effects,' 'nonlinear effects,' or even 'higher-order interactions.' Regardless, fear can best be thought of as the prey viewing predation risk as an activity cost. How big should this predation cost be?

As an adaptation shaped by natural selection, the prey's cost of predation should integrate three components: predation risk (μ), survivor's fitness (F), and the marginal fitness value of the activity ($\partial F/\partial e$). For a foraging animal, $\partial F/\partial e$ describes how much its fitness prospects increase with an incremental increase in energy gained. For a foraging animal, the predation cost of the activity would be: $\mu F/(\partial F/\partial e)$. In other words, an animal's cost of predation can double either by doubling the level of predation risk, doubling what the animal has to lose from being killed in terms of future reproduction and survivorship, or halving the marginal value of energy to enhancing the animal's fitness. This cost of predation forms the basis for the prey's fear responses – the greater the cost of predation, the greater will be the prey's fear responses. Interestingly, this means that hungry animals (high $\partial F/\partial e$) or animals in a poor state (low F), should be easier to catch. The foraging (or activity) cost of predation forms the basis for antipredatory behaviors that influence predator–prey interactions, population dynamics, and species coexistence.

The Paradox of Fierce Carnivores

Consider a textbook predator–prey model. In the state space of predator density versus prey density, the prey's isocline is hump shaped, and the predator's is a vertical line (**Figure 1**). The prey's (or predator's) isocline shows all combinations of prey and predator numbers such that the prey's (or predator's) growth rate is zero. The prey's isocline rises at low prey numbers as the prey experience safety in numbers, but then it reaches a peak and declines as the negative fitness consequences of competition among the prey outweighs benefits of reduced predation risk. The predator's isocline is independent of the number of predators and only depends on prey abundance. In its simplest version, the predators only interact with each other indirectly through the consumption of prey – they do not interfere with each other. Hence, the predators have a subsistence abundance of prey that they require to just get by.

Now, the paradox of such a predator–prey system. If the predators are very inefficient and require a high subsistence abundance of prey, then their isocline intersects the prey's to the right of the hump (see **Figure 1**). This yields a nice stable equilibrium, although there may be damped oscillations as the prey and predator dynamics

approach this equilibrium. But, if the environment was to degrade and support fewer prey, then the predator's face extinction should the prey population numbers become too low. More efficient predators do not face this problem. They only require a low abundance of prey for subsistence and their isocline intersects the prey's to the left of the hump. But, such an intersection produces nonequilibrium dynamics with permanent limit cycles or fluctuations that may result in the extinction of the prey, predator, or both. Inefficient predators may be extinction prone from extrinsic perturbations to the environment, while efficient predators may create intrinsic instabilities in the dynamics resulting in extinction. Prey fear responses may rescue the paradox and promote the stability of predator–prey system with fierce carnivores.

Fierceness is not really a quality of the predator. Squirrels prey on acorns, but acorns do not scream, or get up and run away – hence squirrels are not fierce predators. Fierceness emerges because the prey can perceive and respond to the predators – the capacity of the prey to fear and respond makes a predator fierce. Let's revisit the isoclines and imagine a fierce predator. When the predators are very scarce, the prey should have a low predation cost of foraging and should show little vigilance or fear responses. To a predator, they are easier to catch – such a predator may be quite efficient and require only a low density of prey to subsist. At low predator numbers, the predator's isocline starts as an efficient predator. But, as predator numbers increase, so does the prey's predation cost of foraging, and the prey become harder to catch. With more predators, the predators now have a higher subsistence level of prey. The predator's isocline is not vertical but has positive slope (**Figure 1**). Because of the prey's fear responses, the predators are extremely efficient on unwary prey when predators are scarce (resistant to extrinsic perturbations), and the predators are extremely inefficient on highly wary prey when predator's are abundant (resistant to intrinsic instabilities in population dynamics). Fear responses may be a critical component in predator–prey dynamics, and the stability and persistence of predator–prey systems with fierce carnivores. Mule deer and mountain lions, and wolves and elk in Yellowstone provide examples of such systems.

Landscapes of Fear, and Habitat Quality for Prey and Predators

In thinking about the wolves and elk of Yellowstone, John Laundre developed the concept of the landscape of fear as an important component of habitat suitability in addition to landscapes of productivity, vegetation cover, food availability and/or physical properties of the landscape. One can imagine a vegetation map, or a

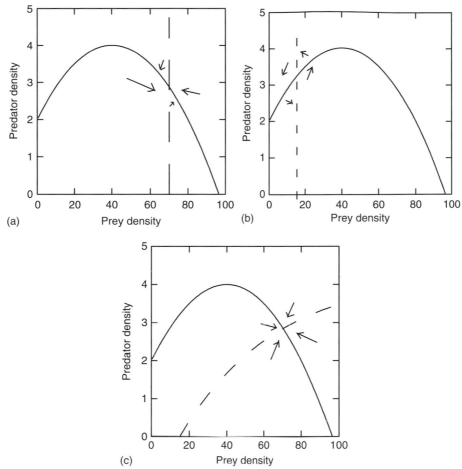

Figure 1 How prey fear responses can stabilize predator–prey interactions. The graphs show various configurations of predator–prey dynamics as shown using isoclines diagrams. The hump-shaped curve is the prey's isocline showing all combinations of prey and predator densities such that the prey have zero growth rate (prey growth rate is positive below this isocline and negative above). The predator isoclines (all combinations of prey and predator densities such that the predator has zero population growth rate) represent three different scenarios and outcomes. The vertical predator isocline in (a) shows an inefficient predator that needs a high density of prey to subsist. A drop in prey number below this line would result in the predator's extinction. The vertical predator isocline in (b) shows a highly efficient predator (low density of prey required to subsist). Because the isocline intersects the prey's in a region of positive slope, the equilibrium point is unstable yielding nonequilibrium dynamics that could result in the extinction of the predator. The predator isocline in (c) that curves gently upward shows a predator whose prey exhibit appropriate fear responses. At low predator densities, the prey can be fearless and hence easy to catch (the predator is highly efficient); at higher predator densities, the increasingly fearful prey become harder to catch (the predators become highly inefficient). The net effect of this positively sloped isocline is a stable equilibrium and a predator population that is well buffered from declines in prey abundances.

topographic map of elevation, or a mapping of the soil characteristics. The landscape of fear is a topographic map where the 'elevation' lines represent lines of equal predation cost (**Figure 2**). Creating this map from direct observations of risk would be nigh impossible. Typically, this mapping of fear is measured by setting out a grid of depletable food patches and measuring the foragers' giving-up densities. Spatial statistics can then extrapolate these into lines of equal giving-up density. Or, if the giving-up densities are converted into quitting harvest rates, the lines of equal foraging cost can be presented in units of energy per unit time. Either way, changes in the 'elevation' of this map represent changes in the predation cost of foraging.

The landscape of fear can be compared to other mappings of productivity, vegetation structure, and other physical features to assess how these features may relate to predation risk. A perspective on both the prey's and predator's habitat quality emerges by comparing the landscape of fear with the animal's overall level of feeding activity. We can simplify this into a 2×2 matrix where columns represent 'Low' versus 'High' activity levels, and rows represent 'Low' versus 'High' predation cost. Various combinations represent core habitats, refuge habitats, unsuitable habitats, and valuable but risky habitats. This last habitat provides the greatest opportunities for the predator – high prey activity under conditions of high risk. Such habitats may be core to the predators (**Table 1**).

Spatial variation in predation risk provides opportunities for habitat selection. If prey move freely between habitats and they do not interfere with each other, theory predicts an ideal free distribution where prey distribute themselves between habitats so that each habitat provides the same fitness reward (expected per capita growth rate to the individual). If habitats simply vary in productivity, then the distribution of individuals between habitats should match productivities (resource matching). If habitats also vary in the risk of predation, then individuals should distribute themselves in a manner that roughly matches the ratio of reward to risk ('μ/f rule' where μ is predation risk and f is feeding rate).

If the predators are also free to move so as to equalize their fitness opportunities, then habitat selection becomes a predator–prey foraging game. One outcome has the predators matching the resources available to the prey and the prey distributed so as to equalize the predators' opportunities. Additional distributions emerge when the predators interfere with each other (this promotes a more even distribution of predators), when the prey can respond with vigilance behaviors (this tends to even out the distribution of prey), and when the prey exhibit safety in numbers (this may cause prey to herd up in one habitat or the other). The interplay between resource opportunities, predation risk, and available fear responses strongly influences the distribution and abundance of prey, and subsequently that of the predators via foraging games that may involve three or more trophic levels.

Behavioral Trophic Cascades

Why is the world green? All those leaves, blades of grass, and small plants represent food for herbivorous insects, vertebrates, and many other taxa. Furthermore, over-browsing and overgrazing do occur. In fact, the efficient consumption of phytoplankton and periphyton produces the clean, clear oligotrophic lakes that we humans prefer over pond-scummy eutrophic ones. Three hypotheses include the idea of a 'green desert.' Herbivores may eat most of the nutritious plant material, leaving behind the unfavorable bits. Or, plants are highly defended with silicates to grind down herbivore mouth parts, secondary compounds meant to poison and deter, and/or structural defenses such as spines, or bark meant to harm or discourage the herbivores. Such defenses represent the plant's noncognitive fear responses to their own predation cost of nutrient foraging. Finally, predators may consume herbivores that otherwise would have eaten plants. This represents a classic trophic cascade where the predators indirectly benefit the plants by consuming their herbivorous predators – from the plant's perspective, 'the enemy of my enemy is my friend.'

All three of these hypotheses may have analogs in the ecology of fear. It has been suggested, but not empirically verified, that plants deter herbivory by producing tissue of lower nutritional value than otherwise would be optimal. Foragers have higher giving-up densities on foods of low quality or high handling times. Structural and chemical defenses raise the forager's costs or lower its harvest rate – both will raise giving-up densities.

Finally, predators can create a behavioral trophic cascade. The presence of the predators raises the herbivore's predation cost of foraging, and the herbivores in response

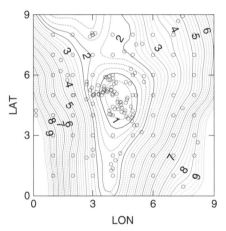

Figure 2 A landscape of fear for a colony of eight cape ground squirrels (*Xerus inauris*) measured by M. van der Merwe and I in Augrabies Falls National Park, South Africa. The lines of equal quitting harvest rate (converted from giving-up densities in experimental food patches) have units of kilojoules per minute. Differences between lines show how much the predation cost of foraging changes in space. The extrapolation of this map was made from an 8 × 8 grid of food patches (blue open circles) spaced at 10-m intervals (LON refers to 10-m increments from West to East; and LAT indicates 10-m increments from South to North). In this desert grassland, the squirrels' perception of safety increased with the proximity of burrows (shown as red open circles).

Table 1 How the prey's landscape of fear (as measured by giving-up densities, the amount of food left behind by prey in food patches), and the prey's activity patterns can indicate the Prey's Refuge, Unsuitable, and Core habitats, and the Predator's Core habitat

		Giving-up density	
		Low	*High*
Feeding activity	Low	*Refuge*: safe and poor in opportunities	*Unsuitable*: risky with low opportunities
	High	*Core*: safe and rich in opportunities	*Predator's Core*: risky and rich in opportunities

forage less intensively on the plants. Abramsky and colleagues showed how a behavioral cascade can be reversed very quickly. Desert spiny mice (*Acomys cahirinus*) consume just a few desert snails. But when numerous rock refugia were added to their home ranges, the spiny mice, now much more protected from their predators, consumed vast numbers of snails in a matter of days. Many cases of overgrazing or overharvesting may have less to do with too many herbivores, and much to do with herbivores that are not fearful.

Fear and Mechanisms of Species Coexistence

A mechanism of coexistence requires environmental variability and tradeoffs among the organisms exploiting this variability. If variability takes the form of habitat heterogeneity, then one species may be better at competing in one habitat while the other may have the competitive edge in another. Fear responses of prey to their predators likely create some coexistence mechanisms and promote biodiversity. The mechanism of coexistence generally involves some sort of spatial or temporal variability in predation risk, and the tradeoff among the different prey species is either quantitative (one prey is better at avoiding predators, while the other is better at garnering food) or qualitative (one prey feels safer under one set of conditions, while the other feels safer under another).

Differences in antipredator strategies between nighttime and daylight provide a stark example of qualitative tradeoffs. Daylight plays into the sense of vision for both prey and predators. Darkness favors hearing and smell. While many species may be active both day and night, most have a strong preference for either daylight, darkness, or as crepuscular animals, the twilight that separates them. While thermoregulatory and water regulation may play large roles in determining temporal patterns of activity, predation risk and the latitude for fear responses likely loom large for most organisms choosing their times of day or night.

Zooplankton exhibit a regular pattern of daily migration up and down in the water columns of temperate ponds and lakes – darker depths provide safety during the day from visually orienting fish, the upper water column provides food and relief from predatory invertebrates. Coral reefs experience an almost complete change from the 'day-shift' to the 'night-shift' community. The colorful reef fishes seem to gain a visual edge over their predators by day with barracudas patrolling just off the reef, and other predators attempting to wait in ambush among the corals. But, by night, they retreat completely to refugia among the coral; presumably, the reduced visibility prevents these reef fishes from distinguishing between a harmless competitor and a predator. Meanwhile, the sea

urchins have been hunkered down all day – to come out during the day invites attacks from trigger fish (they can literally blow the urchins onto their backs exposing the urchin's vulnerable underbellies) and the like. But, come night, the triggerfish retreat, and urchins can more safely patrol and graze throughout the reefs and even the adjacent sandflats. Coral reef fish of the night offer bizarre morphologies of spines, poisons, and behaviors of puffing up. The separation of communities by night and day is not just a matter of predation pruning each species into its temporal niche; rather, these organisms exhibit strong coadaptations of behaviors and morphology that allow them to avoid periods of danger and seek periods of relative safety. Fear responses strongly stabilize the coexistence and dynamics of these temporal communities, as each 'shift' avoids intruding on the others.

Space provides much the same opportunity for mechanisms of coexistence based on tradeoffs in food and safety. Among Heteromyid rodents, there are situations where kangaroo rats (*Dipodomys* sp.) seem to have the competitive edge in the risky, open microhabitats, while pocket mice (*Chaetodipus* and *Perognathus*) enjoy an edge in the safer bush microhabitat; although, as we shall see, snakes versus owls can wonderfully complicate this simpler scenario. Rosenzweig tested for these effects by augmenting cover on some plots in the desert and removing shrub cover from others. Via population dynamics, we might expect over weeks or months a slow steady shift in numbers as differential mortality and fecundity favored kangaroo rats on the coverless plots and favored pocket mice on the cover-augmented plots. Not so. Within a day or two, the kangaroo rats quickly abandoned the augmented plots, and the pocket mice quickly abandoned the open plots. Fear accelerated the dynamics and likely drove the pocket mice away, but another feedback was likely at work in cover plots. Less fearful pocket mice likely depressed seed resources (remember the behavioral cascade) to the point where kangaroo rats could not profitably feed; they needed to go elsewhere.

Fear as a driver of coexistence can operate in unexpected but marvelous ways. My favorites are fox squirrels and gray squirrels of the Midwest United States. Fox squirrels are orange in color, slightly larger, and have a temperament that seems to focus on 'managing' their predators, that they will mob or harass. Gray squirrels, on the other hand, are most of the time gray (with melanistic ones making up most of the rest), slightly smaller, and seem to have a temperament focused on managing other squirrels and competition for food. They are smaller, but they are the interference dominant over fox squirrels. Fox squirrels predominate on the riskier and less productive wood margins or savannas, while gray squirrels predominate in the safer deep woods. The temperament, body size, and even coat color (a bit of apomatism?) of fox squirrels may all be antipredator adaptations allowing a competitive edge in riskier habitats. Habitat heterogeneity in fear is necessary for their coexistence. But, the presence of squirrels alone is probably not responsible for the large predator community (hawks, coyotes, foxes) that exists in these habitats and creates this fear! Squirrels form a relatively small component of these predators' diets, and if there were only squirrels around, there probably would be many fewer predators. Rather, voles, chipmunks, cottontail rabbits, and white-footed mice likely feed the predators that create the mechanism of coexistence for the two squirrel species. In the absence of these more accessible prey, there would be many fewer predators, and gray squirrels would likely outcompete fox squirrels in this much safer world. We see a strong interaction of lethal and nonlethal effects in creating this coexistence.

Predator Facilitation

Coexistence among predators may be facilitated if predators can exploit the prey's predator-specific fear responses. Predator facilitation likely is widespread and may promote differences in predator-hunting tactics (sit-and-wait vs. active pursuit), as well the coexistence of diverse predator taxa. As a direct effect, horned vipers and barn owls would be labeled as 'competitors' as both seek to kill gerbils inhabiting the sand dunes of Middle Eastern deserts. But, predator facilitation is at work and snakes likely make it easier for owls to kill gerbils and vice versa. Appropriately, gerbils head for the cover of shrubs in response to owls. Snakes exploit this fear tactic by lying in ambush under shrubs – fear of owls may reinforce the snake's behavior. The presence of snakes under shrubs then drives gerbils into the waiting talons of the owls.

The lionesses hunting the buffalo calf may have been predator facilitation gone awry. The presence of crocodiles may make the approach of the buffalo to the waterhole slow and deliberate. This may increase the chances for the lions to conduct an ambush of their own. Furthermore, wariness for lions may increase the crocodile's chances of capturing an antelope. Split attention, predator-specific fear responses, and the overall penalty of multitasking may enhance both prey and predator diversity through predator facilitation.

What Are the Predators Afraid of?

The ecology of fear influences predators in at least two ways. First, the predators themselves may be at risk from each other or from other predator species. Second, the predators may be at risk of injury or even death from capturing their prey. Intraguild predation, or additionally, intraguild harassment recognizes that predators, while hunting for the same prey may harass, injure, or interfere with each other. A predator such as a leopard in Chitwan National Park, Nepal, or a red fox in the Midwestern United States loses twice when they face an even larger predator such as tigers or coyotes, respectively. Not only do the tigers and coyotes frighten and compete for the same prey as the leopard and fox, but the leopard and fox become less effective predators as they must vigilantly watch their backs while hunting. The reappearance of coyotes at the Morton Arboretum, Lisle, Illinois, was marked by the rapid disappearance of red foxes. It is unlikely that the coyotes killed many if any foxes, but the foxes now found their former habitat unprofitable in the face of these twin penalties. Intraguild predation adds a predation-like cost of foraging to mesopredators caught in the middle. Mesopredator release refers to the increase in numbers and impact of midsize carnivores when a larger predator has been extirpated from the system. Like overgrazing by herbivores, much of this release may stem from the mesopredator's ability to forage so much more efficiently once freed from its predation cost of foraging.

Large carnivores may actually be quite fragile. In the absence of any assistance from a social group, a large carnivore may face death if it suffers even a moderate injury that temporarily reduces its capacity to hunt. A swollen paw or a sprained muscle may render a cheetah incapable of capturing its prey – starvation may be imminent. This suggests that large carnivores may need to hold back in hopes of reducing the risk of injury. Perhaps they even forgo opportunities to make a risky kill. Interestingly, this scaling of boldness or derring-do allows predators to tradeoff risk of injury with likelihood of prey capture. A cautious predator risks little injury but has a concomitantly lower chance of successfully capturing its prey.

Derring-do and risky prey likely give predators additional degrees of freedom to manage their hunting tactics. A hungry or down-and-out-on-its-luck mountain lion can cope by being more daring, or by going after riskier prey such as porcupines. Lions and tigers in the extreme may turn to man eating. Similarly, when prey are abundant or the predators are in fine shape, such predators can be more cautious or stick with less risky prey. In going after the young buffalo, were the lionesses exploiting a unique opportunity or exhibiting a degree of desperation? Regardless, the ecology of fear, by permitting predators to ramp up their daring or seek riskier prey, likely stabilizes their population dynamics and forestalls starvation during lean periods.

Furthermore, predators may experience a tradeoff between their prey's catchability and risk of injury. Such is the case for the mountain lion facing deer (hard to catch, little risk of injury) and porcupines (easy to catch, yet high risk of injury). These prey present the predator with a tradeoff between catchability and risk. Such prey may provide a mechanism of coexistence for a predator that is more agile yet injury prone and one that is less agile but brawnier (e.g., cheetah vs. leopard). A tantalizing example may be the mountain lion (agile) and jaguar (brawny) of North and South America preying upon deer and peccaries, respectively. The ranges and diets of these two cat species overlap considerably, but not entirely. The traditional range of the mountain lion corresponds almost exactly to that of deer and guanacos and other South American llama-like antelope – these may be hard to catch but relatively risk-free prey. The original range of the jaguar overlaps almost exactly with that of the collared and white-lipped peccary – easier to catch but more likely to cause injury.

Fear, Behavioral Indicators, and Conservation

The games of fear and stealth that go on between predator and prey greatly enhance the behavioral sophistication exhibited in nature. With the elimination of predators, such fear behaviors may decline or become less intense. The prey may become behaviorally less sophisticated and more focused on the business of feeding – recall the behavioral cascades. In domestic animals, we see an evolved reduction or elimination of many fear responses. Goats and sheep likely outforage native antelopes simply because they are less fearful, less distracted from feeding, and less vigilant – we have bred them that way. Retaining the full suite of fear behaviors in prey may be a conservation goal in itself.

The absence of predators can reverse the behavioral cascade and result in herbivores being far more destructive on their own resources. The return of wolves to Yellowstone has become iconic. Frightened elk forage less intensively, particularly on riverine willows. The return of willows has returned meanders to the meandering streams, provided more food for moose, nesting and foraging sites for several bird species, and renewed opportunities for beavers. The initiation of these effects could be seen in the rapid and dramatic reappearance of fear behaviors by elk to wolves. Fear behaviors and their role in behavioral cascades may be critical to maintaining or restoring natural areas.

Fearless prey may provide valuable indicators of current habitat quality. Starving animals or animals with greatly reduced feeding prospects will perceive a high marginal value of food ($\partial F/\partial e$) and a reduced sense of future fitness opportunities (F). Both these changes in response to a degrading environment will lower the predation cost of foraging. Such animals will forage their resources more intensively and exhibit fewer fear responses to their predators (such animals will also be easier to catch!). Noting a drop in fear responses even in the face

of continued predation threats may provide managers with a valuable indicator of a prey population with a degrading habitat.

Fearless prey may also indicate the absence of predators. Mahesh Gurung and Som Ale used the vigilance behaviors, herd sizes, and habitat selection of blue sheep and Himalayan tahr, respectively, to determine the presence or absence of snow leopards from stretches of the Annapurna and Everest ranges of Nepal. They could also use more subtle patterns of fear responses to determine the actual whereabouts of snow leopards in valleys frequented by these cats. Fear behaviors can provide an indicator of the long-term presence or absence of predators, and indicators of the imminent presence of a predator.

See also: Defensive Avoidance; Empirical Studies of Predator and Prey Behavior; Games Played by Predators and Prey; Trade-Offs in Anti-Predator Behavior.

Further Reading

Abrams PA (2010) Implications of flexible foraging for interspecific interactions: Lessons from simple models. *Functional Ecology* 24: 7–17.

Altendorf KB, Laundre JW, López González CA, and Brown JS (2001) Assessing effects of predation risk on foraging behavior of mule deer. *Journal of Mammalogy* 82: 430–439.

Brown JS and Kotler BP (2004) Hazardous-duty pay and the foraging cost of predation. *Ecology Letters* 7: 999–1014.

Brown JS, Laundre JW, and Gurung M (1999) The ecology of fear: Optimal foraging, game theory, and trophic interactions. *Journal of Mammalogy* 80: 385–399.

Hugie DM and Dill LM (1994) Fish and game: A game theoretic approach to habitat selection by predators and prey. *Journal of Fish Biology* 45(supplement A): 151–169.

Laundre JW, Hernandez L, and Altendorf KB (2001) Wolves, elk, and bison: Reestablishing the 'landscape of fear' in Yellowstone National Park, USA. *Canadian Journal of Zoology* 79: 1401–1409.

Lima SL and Dill LM (1990) Behavioral decisions made under the risk of predation – a review and prospectus. *Canadian Journal of Zoology* 68: 619–640.

Manning AD, Gordon IJ, and Ripple WJ (2009) Restoring landscapes of fear with wolves in the Scottish Highlands. *Biological Conservation* 142: 2314–2321.

Ripple WJ and Beschta RL (2004) Wolves and the ecology of fear: Can predation risk structure ecosystems. *Bioscience* 54: 755–766.

Rosenzweig ML and MacArthur RH (1963) Graphical representation and stability of predator–prey interaction. *American Naturalist* 97: 209–223.

Schmitz OJ, Beckerman AP, and Obrien KM (1997) Behaviorally mediated trophic cascades: Effects of predation risk on food web interactions. *Ecology* 78: 1388–1399.

Stephens DW, Brown JS, and Ydenberg R (eds.) (2007) *Foraging: Behavior and Ecology.* Chicago, IL: University of Chicago Press.

Van der Merwe M and Brown JS (2008) Mapping the landscape of fear of the Cape ground squirrel (*Xerus inauris*). *Journal of Mammalogy* 89: 1162–1169.

Economic Escape

W. E. Cooper, Jr., Indiana University Purdue University Fort Wayne, Fort Wayne, IN, USA

Escape Theory

Flight Initiation Distance

In order to optimize lifetime reproductive success, prey must balance activities that enhance fitness, such as foraging and social behavior, against the need to avoid predation. Wariness in the presence of predation risk is only the first step in a series of adaptive decisions prey must make to avoid being captured. Prey must make escape decisions that trade the degree of risk they are willing to accept before fleeing against the loss of fitness that could be obtained by feeding, mating, or other behaviors if they do not flee. These considerations have led to several economic models, that is, models based on costs and benefits to fitness of escape behavior.

In 1986, Ydenberg and Dill published a model that predicts flight initiation distance, the distance separating predator and prey when the prey begins to flee. Flight initiation distance is also known as approach distance, flush distance, and flight distance, the last being misleading because it also has been used to mean the distance that an animal moves once it starts to flee. Economic models of escape apply to short-term encounters between predators and prey rather than to longer-term effects of predation risk when predators are not immediately present.

In the scenario of cost–benefit escape models, a prey detects an approaching predator. Before the costs of escaping, including loss of opportunities to perform other behaviors that increase fitness and energetic cost of fleeing, were considered, it was widely believed that a prey should flee as soon as it detects a predator. Ydenberg and Dill predicted that a prey, having detected a predator from a distance, should not flee immediately, but should monitor the predator's approach until the cost of staying (not fleeing) is equal to the cost of fleeing, and only then should it flee. The cost of staying is the expected loss of fitness that the prey incurs by not fleeing, and the cost of fleeing is the opportunity cost (benefit forgone by fleeing) plus escape costs (energetic cost, cost due to injury while fleeing). As the predator approaches, the cost of staying increases. The cost of fleeing decreases as the predator continues to approach because by not fleeing the prey has already accumulated some portion of the maximum benefit obtainable during the encounter.

As the distance between predator and prey decreases, the risk and the corresponding cost of staying curve rises, but the cost-of-fleeing curve falls (**Figure 1**). The predicted flight initiation distance occurs where the two curves intersect: The prey assesses both costs and flees when they are equal. If the costs were known in fitness units, the model could predict flight initiation distance precisely, but measuring costs is notoriously difficult. The model has been extremely useful in predicting relative flight initiation distances for larger and smaller costs. Suppose that the cost of fleeing curve is fixed. If two cost of staying curves do not intersect, the predicted flight initiation distance is longer for the higher one, for which risk is greater at all nonzero distances. Similarly, if the cost of staying curve is fixed, predicted flight initiation distance is shorter for the higher of two cost of fleeing curves.

The Ydenberg and Dill model guided research on escape behavior for over 20 years, and its predictions have been widely supported. Some modifications have been proposed. If a predator is not detected until it is closer than the predicted flight initiation distance, the prey should flee immediately. Beyond some distance, prey may be unable to detect predators or risk may not vary with distance (i.e., the predator does not pose enough threat to monitor). Between the predicted flight initiation distance and the minimum distance at which risk does not vary with distance, escape decisions are expected to be based on economic considerations.

Another modification proposes that prey that escapes to refuge incorporate a margin of safety into escape decisions, fleeing soon enough to reach refuge before a predator by some distance or time. This hypothesis was supported by data for woodchucks (*Marmota monax*). An Australian parrot (galah, *Cacatua roseicapilla*) fled at approximately half the distance at which it appeared to detect the predator. This may be a novel form of spatial assessment related to margin of safety. The margin of safety hypothesis is appealing because prey must begin to flee soon enough to escape, but presumably cannot assess variation in risk with distance precisely. Prey may use simple rules of thumb to select flight initiation distances that approximate the best available decisions.

The Ydenberg and Dill model has had great heuristic value, and its predictions have been widely supported. However, prey could make escape decisions that would lead to greater fitness than allowed by the model. The Ydenberg and Dill model does not lead to optimal escape decisions because it neither takes the prey's fitness when the encounter begins (initial fitness) into account nor discounts initial fitness in relation to predation risk. According to the asset protection principle, animals having assets such as high residual reproductive value should behave

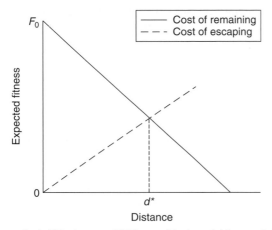

Figure 1 In Ydenberg and Dill's graphical model the predicted flight initiation distance is the distance between predator and prey at which cost of staying (expected loss of fitness) equals the cost of fleeing. d^* – predicted flight initiation distance, F_0 – initial fitness.

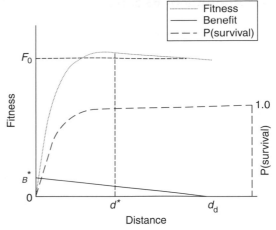

Figure 2 In Cooper and Frederick's optimality model the optimal flight initiation distance (d^*) is the distance at which expected fitness after the encounter is maximized. Benefits than may be gained by not fleeing reach a maximum (B^*) when the predator reaches the prey. Expected fitness is the sum of initial fitness (F_0) and B discounted by probability of survival (energetic costs are omitted in the figure).

conservatively to protect them. Thus, prey should select longer flight initiation distances to protect larger initial fitness. Another difficulty with the Ydenberg and Dill model is that the prey's fitness can never be greater after an encounter than before it, which is contradicted by sexual cannibalism and cases in which male external fertilizers can increase lifetime fitness by staying even if they are killed.

Two recent optimal escape models summarized by Cooper and Frederick address these problems, allowing prey to select the flight initiation distance that maximizes expected fitness when the encounter ends. In both optimality models, an encounter begins when the prey detects the predator. Prey often can select optimal flight initiation distances for which expected fitness is greater after than before the encounter. The difference between the two optimality models is that all fitness is lost upon death due to predation in one model, but fitness acquired during an encounter due to reproduction is retained after the prey is killed in the other.

Unlike the graphical Ydenberg and Dill model, the optimality models are mathematically explicit. Expected fitness is calculated based on the prey's initial fitness, benefits gained during the encounter, costs of escaping, and probability of survival (**Figure 2**). The optimality models feature a fixed initial fitness and curves relating benefits, escape costs, and predation risk to distance between predator and prey. The optimal flight initiation distance is the distance for which the sum of the fitness components (initial fitness, benefits, and escape cost) discounted by the probability of survival is greatest (**Figure 2**).

Although many qualitative predictions of the Ydenberg and Dill model and optimality models are identical, optimality models are preferable because they overcome the shortcomings of the former and provide a theoretical framework that permits a better understanding of the effects of factors that affect escape decisions. Autotomy, voluntary shedding of a body part to permit escape when contacted by a predator, has multiple effects. Loss of the entire tail by lizards precludes further autotomy until regeneration occurs, increasing the lethality of the predator when it reaches the prey. Loss of lipids stored in the tail and of social status after autotomy may reduce initial fitness in subsequent encounters. Loss of social status also reduces potential reproductive benefits. Running speed decreases following autotomy, increasing predation risk. Predictions that flight initiation distance should increase to compensate for lower escape speed after autotomy may be incorrect if the combined effect of decreased speed and increased lethality on predation risk is outweighed by decrease in initial fitness.

Two factors not incorporated into current models of flight initiation distance are alternative strategies available to cryptic prey and the influence of the time spent assessing risk. A model for cryptic prey identified two optimal escape strategies. Immediate escape upon detecting a predator is favored by a low cost of fleeing, delay in detection by the predator, and by a greater likelihood of escaping if the prey initiates escape rather than reacting to attack. Remaining motionless until detected is favored by effective crypsis and high escape cost. This model might be modified to optimize flight initiation distance on the basis of the probability of being detected. Escape decisions by Columbian black-tailed deer depend on the time spent evaluating threat, and statistical models based on assessment time were more strongly supported than those based on distance.

Other Aspects of Escape and Refuge Use

Although theory has focused on flight initiation distance, this variable reflects only the decision to begin an escape attempt, not decisions made during escape. Prey may hide or remain accessible during escape, alter speed and direction of flight, and decide when to stop fleeing. These topics have received little theoretical attention although early theory examined predator and prey trajectories. Two escape variables may be predictable using current theory. The logic underlying predictions about flight initiation distance appears to apply to the distance fled between flight initiation and termination and in refuging species, the probability of entering refuge. Prey should flee further when risk and initial fitness are greater and flee shorter distances when escape costs and opportunity costs of fleeing are greater. The probability of entering refuge should increase as risk and initial fitness increase and decrease as the opportunity cost of entering refuge and risk in refuge increase.

Prey that enter refuges must decide when to emerge. Factors that affect hiding time (the latency between entry and emergence from refuge, also called emergence time) are similar to those affecting flight initiation distance. Risk decays as the time spent in refuge increases, but the prey may lose valuable foraging and social opportunities while hiding. Ectothermic prey in refuges that are cooler than the outside environment may undergo progressive decline in body temperature as hiding time increases, making them less efficient runners and requiring basking to elevate body temperature upon emergence. Models of hiding time are isomorphic to Ydenberg and Dill's escape model and optimal escape models, but with the time since entering refuge substituted for distance, the cost of remaining in refuge for the cost of fleeing, and the cost of emerging for the cost of staying. Selection of hiding time by prey and how long to wait for prey to emerge by predators has been modeled using game theory, the main conclusion being that prey should wait longer than predators.

Empirical Findings

Flight Initiation Distance

Risk factors

Aspects of predators that affect risk include starting distance (the distance between predator and prey when escape begins), approach speed, directness of approach, and predator size, type, and number (**Figure 3**). Starting distance may affect perceived risk because continuing approach may indicate that the predator has detected or is likely to detect the prey. Starting distance and flight initiation distance have been found to be positively correlated in snakes, Australian and Tasmanian bird species,

and Columbian black-tailed deer. Among lizards, starting distance has little or no effect in ambush foragers, which wait motionless for prey to approach: flight initiation distance increased slightly during rapid approach in one species, but not during slow approach in four species. The lack of effect of starting distance during slow approach may be a consequence of lizards assessing risk as very low until a predator is very close, in which case the expected survival curve would not vary with starting distance. In actively foraging lizards, which move through the habitat searching for food, flight initiation distance increases markedly as starting distance increases.

Predator approach speed is a reliable cue to risk that strongly affects flight initiation distance in many prey. Flight initiation distance increases as approach speed increases in insects, fish, lizards, birds, and mammals. Directness of approach also affects flight initiation distance in diverse prey, but not as greatly as approach speed. A directly approaching predator is likely to have detected the prey or to do so soon. A predator approaching on a path that will bypass the prey becomes increasingly less likely to attack as minimum bypass distance increases. Flight initiation distance has been found to be longer during direct than indirect approaches in grasshoppers, several lizards and mammals, and one of two gull species. The reverse occurred in a fiddler crab because of its inability to assess distance during direct approach.

Flight initiation distance has been found to increase as predator size (or eye size) increases in four fish and a lizard. Variation in response to predators of different types and species indicates that prey respond to various predator traits.

When more predators pose greater threat, flight initiation distance is expected to increase with the number of predators. This effect has been found in a lizard, one of two birds, and a gazelle, but no effects was found in mayflies. Interpretation of these disparate findings requires knowledge of circumstances in which risk increases or is unaffected by the number of predators and, especially for invertebrates, of sensory and cognitive capacities of prey. Escape cost may also affect the relationship between flight initiation distance and the number of predators. It was hypothesized that the cost of fleeing may be greater for an omnivorous bird because of time-consuming search for animal prey than for a granivorous bird, which can resume feeding upon returning after escape.

Numerous aspects of prey and their environments that influence predation risk affect flight initiation distance, including amount of cover, distance to refuge, degree of crypsis, body armor, body condition, parasites, experience with predators, factors that affect escape ability (running speed, gravidity, tail loss, body temperature), and prey group size (**Figure 3**). Sparser cover is associated with longer flight initiation distance in a fish, numerous lizards and birds, and a few mammals. Exceptions may occur if

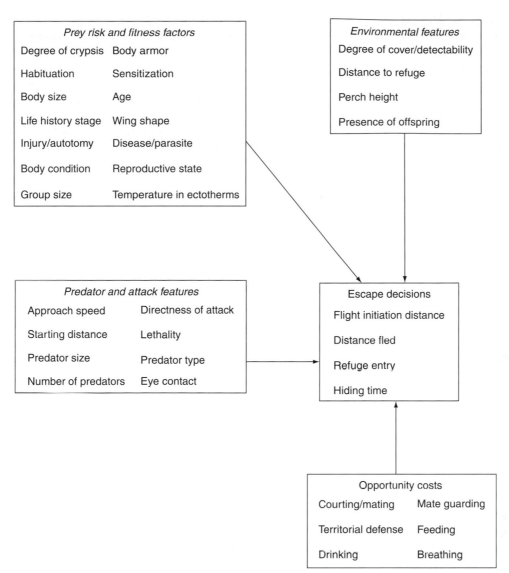

Figure 3 Many factors affect escape decisions and can be categorized into opportunity costs of escape or hiding and factors that affect predation risk and the prey's initial fitness, including intrinsic features of prey, characteristics of predators and their approach, and environmental features. Escape decisions that have been studied from an economic perspective are shown in order of occurrence in a predator–prey encounter.

predators are in cover or juveniles have different predator suites than adults (e.g., absence of effect of cover in juvenile *Psammodromus algirus* lizards).

Flight initiation distance increases as the distance to refuge increases in many species. The distance to refuge had the strongest, most universal effect in a recent review and has been found in crabs, fish, frogs, lizards, birds, and mammals. This effect can be very strong in prey that flee into or stop adjacent to refuge, occurring in a crab, seven of eight lizards, and woodchucks. The exception was a lizard that does not use the nearest refuge.

For prey that climb vertical objects to escape from terrestrial predators, risk decreases as perch height increases. Flight initiation distance decreases as perch height increases in numerous lizard species. When climbing is not important

for escape, perch height may have the opposite effect or be unrelated to flight initiation distance. *Anolis* lizards on fence posts had longer flight initiation distance as perch height increased because individuals higher on posts had to flee further down to escape. Northern water snakes, which escape into water, had longer flight initiation distances when on higher perches, presumably because of greater exposure on high perches. For birds, the advantage of perch height for escape is less clear. Flight initiation distance decreased as perch height increased in three of six raptors, but was unrelated to perch height in the others. Of 34 nonraptorial birds, flight initiation distance was unrelated to perch height in 27 species, increased as perch height increased in seven, and decreased in three. However, tall trees were absent in the study areas. More consistent relationships might be detected using a greater range of perch heights, but many birds may not assess risk as decreasing as perch height increases, perhaps because they are hunted by aerial predators or people.

The more prey rely on crypsis, the shorter the predicted flight initiation distance. Evidence for several species of homopterans, fish, lizards, *Craugastor* frogs, and mammals is consistent with this hypothesis, but is equivocal because it is based on the differences between pairs of species, sexes, or male color phases, or other factors are uncontrolled. Experimental and comparative studies are needed to assess the effect of crypsis.

Because body armor decreases predator lethality, armored fish and lizards have shorter flight initiation distances than unarmored species. With the exception of a study of cordylid lizards, studies have involved few species, indicating the need for confirmation by comparative studies.

Risk assessment is influenced by experience with predators, leading to habituation in the absence of attack and wariness following attack. Prey that are habituated to people have shorter flight initiation distances than prey that are infrequently exposed to people (28 of 32 lizards, birds, and mammals). A reversal occurred in a lizard eaten by local people. Human density did not affect flight initiation distance in three birds. Assessed risk is predicted to increase when a predator approaches more than once in succession. Seven of 11 species, including grasshoppers, lizards, and birds, behaved as expected, but in two birds and two lizards, flight initiation distance was greater during the first than the second approaches. Reversals may occur because of habituation if the first approaches are not very threatening. Additional evidence of learning risk by prey is that several birds and mammals have longer flight initiation distance during hunting season than other times and that flight initiation distance is longer in marmots and moose in cases where hunted than not hunted. Other effects of experience include greater flight initiation distance by lava lizards on islands where feral cats are present and by a fish, the zebra danio, following experience with a predator model.

Factors affecting escape ability include morphological traits affecting speed and maneuverability, loss of body parts, reproductive state, and body condition. In a study of 83 birds, flight initiation distance was greater in species with more pointed wings, as predicted by reduced lift and thrust associated with pointed tips. Yellow-bellied marmots permit closer approach on substrates that permit faster running. Some species of lizards that have lost tails suffer reduced sprint speed and consequently have shorter flight initiation distances due to greater risk.

While gravid, pregnant, or carrying clutches of unfertilized eggs, females may be slowed. Rather than increasing flight initiation distance, gravid lizards of some species change tactics, relying more on crypsis and/or staying closer to refuge. Flight initiation distance did not differ between gravid and nongravid female striped plateau lizards or between sexes in many crabs, fish, lizards, snakes, and birds, but was greater in males than females in several lizards and in females than in males in a lizard and a gazelle. Because sexes may differ in detectability, escape tactics, and vulnerability, predictions about sex difference in flight initiation distance must be based on detailed knowledge of each species.

Predicted flight initiation distance increases as the body condition increases, because prey in good condition have greater initial fitness and perhaps lower escape costs than prey in poorer condition. This prediction has been verified in Balearic lizards and turnstones (birds), and a comparative study of 133 bird species in which more heavily parasitized species had shorter flight initiation distances. Exceptions have been reported in woodpigeons and chinstrap penguin chicks (based on immune response), and may be consequences of reduced escape ability by prey in poor condition or the presence of food that is more valuable to prey in poorer condition.

Body size may affect flight initiation distance if the range of prey size is large enough to affect escape ability or predator lethality. Flight initiation distance in a crab, a fish, a snake, and some lizards was not affected by body size; decreased as body size increased in mayflies and brook sticklebacks; and increased as body size increased within species in a fish and several lizards and between species in birds. Flight initiation distance may decrease as body size increases because of decreased risk, but may increase as body size increases if prey become more suitable for larger predators as prey size increases.

Parental care affects flight initiation distance because parents have a large investment to protect. In several avian species, the day of incubation and presence of eggs versus nestlings did not affect flight initiation distance, but in alpine accentors, flight initiation distance was greater after eggs hatched than earlier. Mourning doves on nests sometimes do not flee and various birds, notably killdeer, permit closer approach by predators while engaging in parental distraction displays than at other times. Flight initiation distance was shorter during nesting than other activities in four bird species, did not differ between foraging and nesting in three species, and was shorter during foraging than nesting in tricolored herons. These differences may reflect variation in detectability and profitability of foraging. In mammals, flight initiation distance was longer for females associated with offspring than for other individuals in five ungulates, but shorter in the bobuck, a marsupial. A detailed knowledge of the ecology of each species, including consequences of alternative activities, is needed to predict effects of offspring presence. The difference between birds and ungulates may be a consequence of greater mobility of young ungulates than of eggs and nestlings.

Little is known about relationships between flight initiation distance and other aspects of natural history. However, in a study of 150 birds, flight initiation distance was greater in omnivores and carnivores than herbivores and in cooperatively than noncooperatively breeding species.

Because escape ability of ectotherms is impaired at low body temperatures, predicted flight initiation distance is longer at cooler temperatures. Some ectotherms maintain body temperatures in narrow ranges by behavioral thermoregulation, allowing efficient escape despite variation in environmental temperature. Flight initiation distance is predicted to increase as body temperature decreases, but not necessarily as air temperature decreases. However, when body temperature falls too low for escape by running, lizards switch tactics. Lizards with cooler body temperature had longer flight initiation distance than warmer ones in seven species, but flight initiation distance was unrelated to air temperature in several lizards, increased as air temperature increased in one species before completion of warming by basking, and decreased as air temperature increased in another species. Flight initiation distance was not affected by water or air temperatures in one snake species, but was greater in warmer than cooler individuals in two others, suggesting increasing reliance on escape as escape ability improved.

The relationship between group size and predation risk is multifaceted, and some facets may have opposing effects on flight initiation distance. Risk dilution among group members suggests that flight initiation distance may decrease as group size increases because of lowered risk that an individual will be the one targeted by the predator. Group defense leads to the same prediction. On the contrary, the many eyes hypothesis suggests that probability that a group member will detect a predator increases as group size increases. If risk is great enough to elicit immediate escape, the many eyes hypothesis predicts that flight initiation distance increases as group size increases. Not surprisingly, findings are mixed. Flight initiation distance was found to increase as group size decreased in fish, but to increase as group size increased in mammals and birds. Quadratic relationships between flight initiation distance and group size occurred in two species. In a water strider, flight initiation distance was maximal at intermediate group size, whereas in barred ground doves, it was minimal at intermediate group size. Juvenile chinstrap penguins permitted closer approach when near than far from a breeding subcolony, suggesting risk dilution near the group.

Opportunity costs

Opportunities for social activities, foraging, or other fitness-enhancing activities may be lost by fleeing or allow benefits to be gained by not fleeing (**Figure 3**). Flight initiation distance is reduced when fleeing entails loss of opportunities by males to guard or court females, defend territories, or fight with rivals in a cricket, a fish, several lizards, and a gazelle. Female striped plateau lizards have shorter flight initiation distances when interacting with males than when alone. In Columbian black-tailed deer approached by a person holding a simulated gun, males have shorter flight initiation distance than females during the breeding season, which might reflect sexual motivation, and male moose have shortened flight initiation distance during rut.

Giving up foraging opportunities may be costly to hungry individuals, but little research has addressed the effect of hunger on flight initiation distance. Flight initiation distance decreased in hungry vipers and in large, but not small mayflies. Presence of food is associated with shortened flight initiation distance in all species studied, including crayfish, fish, lizards, and a reindeer. Flight initiation distance decreases as the amount of food present increases in a mayfly, a crayfish, a fish, and a lizard. Food mass did not affect flight initiation distance in a hermit crab, but this exception is artifactual because more food was present than could be consumed during trials. In house sparrows, flight initiation distance was greater when seed density was higher, possibly because of satiation.

Distance Fled, Refuge Entry, and Hiding Time

Because distance fled is strongly affected by predator behavior and in prey that hide, by distance to refuge, these factors are controlled by stopping approach by an investigator or model immediately when escape begins and by discarding trials in which prey enter refuges. In lizards, distance fled is greater when predation risk is greater as indicated by greater predation pressure, faster approach, more direct approach, greater distance to refuge, sparser cover, tail loss by autotomy, and successive approach. In another lizard, distance fled was unrelated to the number of predators, but proximity of refuges might have obscured an effect. Distance fled is also shorter when food is present than absent and decreases as the amount of food increases. Findings for other taxa also indicate that distance fled increases with risk. Distance fled was greater in a hermit crab when predator odor was in the water, and distance jumped by *Craugastor* frogs increased with body size during fast, but not slow approaches. Distance fled increased with the approach speed in juvenile chinstrap penguins. Distance fled was not affected by the approach speed in Columbian black-tailed deer, but increased as flight initiation distance increased among individuals unhabituated to human presence. Distance fled by female bobucks was shorter when offspring were present.

The probability of entering refuge varies as predicted with risks and cost. Higher proportions of lizards entered refuge after a faster than a slower approach, after a more direct than an indirect approach, and after the second of two successive approaches. In striped plateau lizards, which suffer reduced running speed after autotomy, a higher proportion of autotomized than intact lizards entered refuges. The probability of entering refuge is much greater when the investigator pursues lizards rather than stopping as soon as they begin to flee.

Thermal cost of refuge use and risk of heat exhaustion while fleeing when thermally stressed affect the probability of entering refuge. In one lizard, refuge entry was more likely when refuge and air temperatures were equal than when air was warmer than the refuge temperature. In another lizard, refuges were used frequently at both high and low, but not at intermediate temperatures. Refuges were used at low temperatures when the thermal cost of entering refuge was low and escape ability was limited by inefficient running, and at high temperatures at which running to escape might lead to heat exhaustion. In a third lizard, refuge entry was more likely before completion of morning basking, when lizards and refuges were both still cool, than later when lizards were much warmer than the refuge.

As predicted by optimality theory, hiding time is affected by the risk of emerging, which decreases as hiding time increases, and by the cost of staying in refuge, which increases as hiding time increases. Hiding time varies as predicted for various factors affecting the cost of emerging. Hiding time is greater when a predator has approached than not approached in two fish, increases as approach speed increases in five lizards, increases interactively with food availability in a marmot, increases as directness of approach increases in four lizards and a fiddler crab, increases as proximity of the predator to the refuge increases in a crab and two lizards, increases with the duration of handling by a predator before refuge entry in a turtle, and increases with repeated approach in a crab and four lizards. Hiding time also increases following autotomy in striped plateau lizards, consistent with increased risk upon emergence because of lower running speed.

Costs of remaining in refuge include loss of social and foraging opportunities, decrease in body temperature leading to impairment of escape ability in ectotherms, risk from a predator in the refuge, and reduced oxygen supply in refuge. When a potential mate or a rival male is outside the refuge, hiding times by males are shorter than in their absence. Hiding time is shorter when food is present outside refuge in a polychaete worm, a fiddler crab, and two lizards, but not in a hermit crab. In a marmot, hiding time decreased when food was present after slow approach, but increased in shy individuals in the presence of food following rapid approach. Hiding time decreased as body condition declined in a lizard and as hunger increased in a barnacle, two fish, and a bird. Hiding time decreased as refuge temperature became progressively lower than body and outside air temperatures in three lizard species. Presence of a predator in the refuge shortens hiding time in a

lizard. In a turtle, hiding time withdrawn into the shell is shorter in water than on land, possibly due to greater safety in water. Reduced dissolved oxygen concentration in the water is associated with shorter time hiding with a closed shell in a clam.

Other Aspects of Escape

Because prey must cope with several risks and costs simultaneously during each predator–prey encounter, the ability to assess multiple simultaneous risks and costs is a necessity. In fact, prey can evaluate more than one factor at once. Flight initiation distance and hiding time are affected by each of two simultaneous risks and by simultaneous risks and costs. In 10 of 13 studies, effects were interactive, indicating that risk or cost increases more rapidly for one factor than the other.

Economic approaches have not been applied to some features of escape, in particular, escape speeds and angles and changes in them. Escape speed affects risk and the need to use refuges, and may be varied in relation to predator position and speed. The importance of escape speed is suggested by a positive relationship between flight initiation distance and maximum running speed in a wolf spider. Escape angles selected by fish and lizards often appear to be a compromise between fleeing directly away from the predator and the need to monitor the predator during escape. A lizard species selected escape angles that reduced the overall risk from predators approaching simultaneously from different directions. On the other hand, cockroaches escape by fleeing unpredictably in one of several preferred directions. In many cases, escape angles may be determined by refuge location and may change during escape, depending on the predator's pursuit path. Initial escape angle seems amenable to economic analysis, but models taking into account the movements of both pursuer and pursued are needed to predict changes in speed and direction during encounters.

Effects of different escape tactics have not been examined economically. When escaping from owls, voles flee in a start–stop manner, whereas spiny mice frequently change direction. Some prey use predator-specific escape tactics. When a herd of Thomson's gazelles is approached by a cheetah, which pursues briefly, few individuals perform pursuit-deterrent displays, and displaying ceases when pursuit begins. When approached by African wild dogs, which pursue for long distances, a higher fraction of the herd displays, and display continues during pursuit until the dogs select one gazelle. Cost–benefit models seem appropriate for predicting use of different escape strategies, but remain to be developed.

Changes in escape tactics alter the relationship between predator–prey distance and predation risk. Heavy gravid females in some species may compensate for reduced speed by staying closer to refuge and relying more on immobility to avoid detection. They may permit closer approach than when nongravid because distance to refuge and the probability of being detected are diminished. Tactical shifts may occur when risk is high, but can be reduced by alternative behaviors.

Summary

Economic models predicting flight initiation distance and hiding time are very successful. Apparent failures of predictions are explicable by differences in escape tactics or ecology that alter the relationships between risks or costs and antipredatory behaviors. Predictions of optimality theory are strongly supported for the probability of entering refuge. Predictions about distance fled have been verified, but less consistently. Most risk and cost factors have been examined for few species, and more information is needed to assess the effects of differences in ecological factors, but cost–benefit analysis has greatly improved our understanding of escape behavior.

See also: Defensive Coloration; Defensive Morphology; Ecology of Fear; Group Living; Predator Avoidance: Mechanisms; Risk-Taking in Self-Defense; Trade-Offs in Anti-Predator Behavior.

Further Reading

Blumstein DT (2003) Flight-initiation distance in birds is dependent on intruder starting distance. *Journal of Wildlife Management* 67: 852–857.

Broom M and Ruxton GD (2005) You can run – or you can hide: Optimal strategies for cryptic prey. *Behavioral Ecology* 16: 534–540.

Caro TM (2005) *Antipredator Defenses in Birds and Mammals.* Chicago, IL: University of Chicago Press.

Cooper WE, Jr (2009) Theory successfully predicts hiding time: New data for the lizard *Sceloporus virgatus* and a review. *Behavioral Ecology* 20: 585–592.

Cooper WE, Jr and Frederick WG (2007) Optimal flight initiation distance. *Journal of Theoretical Biology* 244: 59–67.

Cooper WE, Jr and Frederick WG (2007) Optimal time to emerge from refuge. *Biological Journal of the Linnaean Society* 91: 375–382.

Hugie DM (2003) The waiting game: A 'battle of waits' between predator and prey. *Behavioral Ecology* 14: 807–817.

Kramer DL and Bonenfant M (1997) Direction of predator approach and the decision to flee to a refuge. *Animal Behaviour* 54: 289–295.

Martín J and López P (1999) When to come out from a refuge: Risk-sensitive and state-dependent decisions in an alpine lizard. *Behavioral Ecology* 10: 487–492.

Martín J, López P, and Cooper WE, Jr (2003) Loss of mating opportunities influences refuge use in the Iberian rock lizard, *Lacerta monticola*. *Behavioral Ecology and Sociobiology* 54: 505–510.

Stankowich T (2008) Ungulate flight responses to human disturbance: A review and meta-analysis. *Biological Conservation* 141: 2159–2173.

Stankowich T and Blumstein DT (2005) Fear in animals: A meta-analysis and review of risk assessment. *Proceedings of the Royal Society of London, Series B, Biological Sciences* 272: 2627–2634.

Ydenberg RC and Dill LM (1986) The economics of fleeing from predators. *Advances in the Study of Behavior* 16: 229–249.

Ectoparasite Behavior

M. J. Klowden, University of Idaho, Moscow, ID, USA

Introduction

Arthropods that feed on vertebrate blood are particularly important because of their opportunity to infect their hosts with pathogens that can cause serious illness. As vectors of these pathogens, arthropods such as mosquitoes and ticks must first identify a host, feed on its blood, and use that rich resource to their own advantage. Finding this resource in a complex environment is a considerable challenge for an animal with a relatively unsophisticated nervous system, and the identification and location of these hosts is dependent upon stereotyped behavior patterns that are genetically programmed into the central nervous system and elicited when key stimuli are integrated and associated.

Ectoparasites undoubtedly evolved from free-living ancestors. A parasitic way of life can be evolutionarily approached by several routes, with just a few basic requirements for channeling an organism in this direction. First, two or more organisms must coexist and have ecological opportunities for contact. Second, the smaller of the two must have preadaptations for feeding on the larger, and third, the resulting association must benefit the parasite by increasing its reproductive potential. Protein is often rare in nature, and animals capable of utilizing this rich resource in the blood of vertebrate hosts would reap huge reproductive rewards. Examples are autogenous mosquito species that are able to mature their first batch of eggs by carrying over metabolic reserves that were acquired during the larval stage. However, a larger number of eggs can be produced if the female takes a blood meal and additionally utilizes the external protein source.

The fossil record is rather poor for arthropods in general because their body walls did not resist the pressure of their surroundings, and those that were parasitic and attached to their hosts were easily damaged during the process of carcass decomposition. Despite this limited record, it is thought that beginning about 145 Ma, arthropod hematophagy evolved independently at least six times following two major routes. Lice and fleas probably took the first route. In this scenario, the arthropod initially developed a prolonged intimate association with the host, perhaps by residing in animal burrows that maintained heat and humidity and a source of organic matter. While host feces might have been the first major source of nutrients, skin and hair may also have been encountered, ultimately selecting for arthropods that could utilize these resources more efficiently with their more primitive chewing mouthparts. Behavioral adaptations allowing the arthropod to visit the host directly were eventually followed by the morphological adaptations in mouthpart and digestive system structure and function that were suited to blood ingestion and utilization.

Along the second route toward hematophagy, followed by mosquitoes, bedbugs, and ticks, the arthropod presumably first had the preexisting morphological adaptations that allowed it to pierce the integument of other arthropods or feed on plants and then tend more toward synanthropy. Attracted to other insects aggregating around a vertebrate host or various host products, such contact could eventually lead to hematophagy on the vertebrate host itself. Some unusual insects that follow this path are noctuid moths (genus *Calyptra*) that have diverged from their plant-feeding relatives to feed on the blood of large vertebrates, feeding not only on eye exudates, but also penetrating skin.

There are also two strategies for removing the blood from a host while feeding. In one, the blood is removed surgically through slender stylets that inject pharmacologically active components in the saliva before feeding. This approach is found in the mosquitoes, blood-sucking bugs, and lice. In the other approach, scissors-like mouthparts lacerate the skin and cause the blood to flow to the surface where it then can be ingested. This mode of feeding is characteristic of ticks, horse flies, black flies, and stable flies.

Most insect ectoparasites that are hematophagous and feed on blood are not parasites in the strict sense, but rather micropredators that prey on host blood and are otherwise free-living. Unless they are present in truly large numbers and reduce the host's fitness by the removal of enough blood to cause anemia, insect ectoparasites generally do not have a measurable impact on vertebrate host fitness unless they are also vectors of pathogens. The true parasites, such as the lice that live their entire life cycles on a vertebrate host, and a few ticks that complete most of their life cycles on the same hosts, have no need for the more sophisticated host-finding behaviors that are the most essential and challenging behaviors for ectoparasites. Because many hosts also display defensive and grooming behaviors in response to the feeding attempts by their ectoparasites, the parasites have been forced to evolve additional behaviors to contend with these defensive activities.

Adaptations That Make Behaviors Possible

A suite of morphological and physiological adaptations is necessary in order to transition from a free-living to an ectoparasitic, hematophagous way of life. Among these are the mouthpart adaptations required to pierce skin and inject saliva; digestive enzymes to process the large amounts of protein; an excretory system that can deal with the brief presence of large amounts of nitrogen, water, and toxic heme; sensory receptors that can detect host odors and activate the associated behavioral subroutines that bring the parasite to the host; and the development of pharmacological components in the saliva itself that circumvent the host response to feeding. The morphology of the entire digestive tract differs substantially in hematophagous and free-living animals. In addition, those ectoparasites that act as vectors of disease agents have had to coevolve relationships with the agents, dealing with the infection of their own tissues and attempts of the agents to modify their behavior.

Behaviors for Host-Location

To parasitize a host, hematophagous arthropods must first identify and locate the host amid all other stimuli that are present: a considerable challenge in a complex environment replete with signals. This generally involves a series of linked behaviors that are sequentially expressed when varying degrees of host stimuli are perceived. Each behavior appears to be linked to the next by the orderly perception of stimuli that are related to host proximity. Although the ultimate aim is to feed on blood, the behaviors are activated not by the blood itself, but by stimuli that are associated with hosts that contain blood.

For many ectoparasitic arthropods, the first behaviors in host-finding are initiated before host stimuli are even at hand. The initial strategy is to survey the environment for potential host stimuli by simply becoming more active, thus allowing sensory receptors to sift through new territory. The locomotor path that the insect follows is relatively straight and its sensitivity to stimuli is high. This increased level of activity is known as appetitive searching behavior or ranging, and is unrelated to the actual detection of host stimuli. As the period of food deprivation increases, the intensity of these appetitive behaviors also increases, ultimately making the insect more likely to encounter host stimuli in the environment. There is a strong circadian component to this behavior, with activity rhythms often related to the time of day. Mosquitoes that feed during crepuscular periods range during the evening hours, while day-biting species range throughout the day. Ixodid, or hard ticks, feed on blood once during each of their three life stages, leaving the host to molt and then engage in another attempt to find a host. Unable to fly, their ranging behavior involves a different 'sit and wait' strategy consisting of questing, that is, their ascent to the tops of vegetation where they wait and wave their first pair of legs in the air to acquire stimuli from a passing host

to grab as it passes. Because questing makes them susceptible to water loss, this behavior alternates with quiescence in the lower, more humid, litter zone that allows them to rehydrate. Although primarily regulated by circadian rhythmicity, the questing behavior is lengthened when humidity is higher. Immature ticks tend to quest at lower levels as they search for smaller hosts such as rodents, while adult ticks that feed on larger vertebrates quest on vegetation at a higher level.

Once this ranging behavior provides clues that hosts are in the environment, ectoparasites use olfactory stimuli to pinpoint the host's location. Living things exude metabolic products that ectoparasites recognize and that are generally classified as kairomones, chemicals that benefit the receiver and not the emitter. Olfactory stimuli are perceived in mosquitoes by sensory receptors on their maxillary palps and antennae, and the responses of these receptors are filtered and integrated by the central nervous system before a particular behavior is expressed. Among the most important of the host odors is carbon dioxide, a product of protein, carbohydrate, and fat metabolism that is central to the activation and orientation behavior of all ectoparasitic arthropods. Ambient levels of carbon dioxide are between 0.04% and 0.1%, varying with the vegetative landscape and time of day. By comparison, exhaled human breath consists of 4–5% CO_2. Fluctuating levels of CO_2 that may be present in the environment give blood-feeding arthropods their first indication that a vertebrate host is nearby. Emanating as filaments from a plume of host odors, CO_2 in pulses as low as 0.05% activates the behavior of mosquitoes, causing an increase in flight speed and a change in flight track. The CO_2 receptors on the maxillary palps quickly become adapted to the new levels, and with CO_2 changing in concentration as the host becomes nearer, provide a good indication of the progress toward host location. Because of its attractive nature, CO_2 is commonly used to supplement the attractiveness of light traps that are set to assess mosquito densities.

While CO_2 is largely responsible for signaling that a live host is nearby, it is produced by all living things so it is not responsible for the discrimination in host preference displayed by many biting arthropods. For instance, when confronted with both a cow and a human, the malaria vector, *Anopheles gambiae*, will choose the human to feed upon even though the CO_2 levels are greater from the cow. There are a number of other host emanations that allow mosquitoes to discriminate between various hosts, and CO_2 has the additional function of increasing the sensitivity of antennal receptors that perceive these other host odors. Mediated by the central nervous system, CO_2 perception by receptors on the maxillary palps significantly boosts the responsiveness of antennal odor receptors that respond to minor skin emanations and brings the mosquito into the next phase of host-finding. In addition, starvation can lower the thresholds for responding; various other physiological states can raise them. All factors considered, an ox can be recognized by a tsetse at about 90 m away, by horse flies at about 30 m, and mosquitoes become activated by calves at about 20–80 m. Once the changes in CO_2 are detected, the insect transitions to its next suite of behaviors involving more specific attraction. With the exposure to more specific stimuli as a result of behavioral activation, other host odors become perceptible both as a result of closer proximity and increased sensitivity of other receptors. Host preference may also in part be related to mosquito resting behaviors and the degree of passive environmental contact between ectoparasite and host that occurs under natural conditions. The typical resting behaviors of both are instrumental in bringing the host and ectoparasite together.

The volatile skin odors that initiate these next behavioral phases of host-finding include lactic acid, acetone, 1-octen-3-ol, and several carboxylic fatty acids, produced in sweat and breath. L-Lactic acid is a major component of human sweat and used as an orienting cue by mosquitoes, but works synergistically with CO_2 and has little effect alone. Octenol also works best as a supplement to CO_2. The ratios of these minor components, not merely their presence, are probably most important for registering host preference. For example, the tsetse responds strongly to two of the components in cow urine when presented together, but its response is much reduced when each is presented singly. Individual components of human emanations presented separately are usually much less attractive than the gestalt. Combinations of host byproducts in certain ratios may allow the insect to discriminate between different hosts and specifically select its preferred host. This importance of ratios may parallel sex recognition in the many insects that use sex pheromones, where several components may be shared among different species but the blends of those components allow discrimination of one's own species in order to engage in mating. Similarly, for host recognition, particular hosts may produce blends of kairomones that are recognized by a nervous system that is genetically programmed to activate specific behaviors when a particular profile of chemicals is perceived.

Once the insect enters the host odor plume and has been activated, the difficult task of orientation begins, that is, tracing the plume back to the host. As the wind disperses the odor plume from the site of release, it creates and stirs discrete odor packets of varying concentration. Because the dilution by air currents largely uncouples the relationship between host proximity and concentration, it is believed that absolute concentrations are ignored and the composition of the packet is evaluated as simply being above or below a behavioral threshold concentration. When traveling within a packet containing host stimuli, the insect maintains a straight line of upwind flight, but upon leaving a packet, its behavior changes to a zig-zag

flight at right angles to the wind. This change in flight behavior presumably maximizes the chances of locating another odor packet as it diffuses from the plume. Some moths cease their upwind flight and fly in a zig-zag manner across the wind in a maneuver known as 'casting' in order to find a lost plume. Tsetse entering a plume tend to turn upwind significantly less often than when leaving a plume. Within a packet, the flying insect may engage in optomotor anemotaxis in which visual cues that flow from front to rear across the optical receptors help to maintain forward movement against the wind.

As the insect's orientation behavior allows it to close in on the host on the basis of olfactory information, visual stimuli begin to become more important in the final stages of host location. During this final approach, insects must reduce their flight velocity so as not to overshoot their targets. Distances at which visual information becomes of greater consequence depend on whether the outline of the animal is easily discernable contrasted against the foliar background, the time of day and light levels present, the size of the animal being hunted, and the resolution of the compound eye. Compared to the 2° resolution of the tsetse, mosquito resolution is generally poorer, ranging from 12° for the day-biting *Aedes aegypti* to more than 40° for the nocturnal *A. gambiae*. The visual pattern of the target animal is more attractive when it is relatively simple; complex patterns strongly interfere with visibility from a distance. It has been suggested that some zebras evolved their stripes as a response to tsetse, with their stripes breaking up the visual image of their body form and reducing their visibility against the background vegetation to minimize the ability of the flies to feed on them and transmit trypanosomes. With a solid target, the distance at which vision becomes important ranges between 5 and 20 m for mosquitoes and 50 and 180 m for tsetse.

Many biting flies show preferences for different parts of the body before they alight. As they get closer, some black flies and mosquitoes prefer to land on a host's extremities rather than the body core. Different species of horse flies show preferences for different areas on the backs of cows. Although three related species of *Anopheles* mosquitoes prefer different hosts, they all choose to feed on the legs and feet of those hosts. In contrast, another *Anopheles* species that feeds on humans tends to feed on the upper body. Ectoparasites that are fairly host specific may show preferences for specific parts of the body, while those that are generalists tend to land randomly. Since many blood-feeding insects land on clothes and insert their mouthparts through the fibers, the tactile stimuli from a host are probably less important for these behaviors. Ticks may wander about the host until they identify sites that provide them with their preferred tactile and olfactory stimuli.

Odor becomes essential again for landing behavior, along with moisture and temperature, although this landing phase of the behavioral continuum is probably the least understood. Some antennal receptors on the mosquito antenna are sensitive to temperature changes as small as 0.2 °C that could be sensed as close as 40 cm from a human extremity. Lice, the true parasites that live their entire life cycle on a vertebrate host, will leave the host when its body temperature falls after death. Although mosquitoes can detect changes in humidity with their antennal hygroreceptors, humidity gradients have not been shown to significantly influence landing behavior. In general, the same components used for attraction appear to be involved in landing behaviors.

In the Ixodid hard ticks, the larval, nymphal and adult stages each feed on blood once before falling off the host, with immatures then molting to the next instar and female adults laying eggs. The female dies after laying eggs, but the newly hatched larva and recently molted nymph and adult engage in questing behavior on vegetation, responding to carbon dioxide, various host odors including ammonia, lactic acid, butyric acid, changes in light intensity from the shadow of the host, and vibrations resulting from host movement. Once on the host, they may crawl over it for several hours before they finally attach and secrete cement at the feeding site that secures them for several days of blood ingestion.

Argasid soft ticks employ a completely different feeding strategy. Argasids typically remain within the nest or burrow of their host and have no need to quest. They often feed repeatedly on the same host for much shorter periods, ranging from minutes to an hour and do not secrete a cement during attachment. In between feeding, they molt to the next instar, and adults subsequently produce a batch of eggs after each blood meal. Some field species of soft ticks live on hosts in open areas and drop off according to a circadian rhythmicity that may also be synchronized with host behavior patterns. This positions them in areas where the hosts may normally be inactive and therefore more accessible for the next instar. These ticks engage in rapid crawling behavior toward a potential host in response to nearby host odors.

Behaviors for Feeding

Once on the host, the ectoparasite must begin to feed. The stimuli that initiate the probing of the mouthparts into the skin have not been well studied and appear to occur as the terminal behavioral event after alighting or site selection. Once the mouthparts are inserted, engorgement behaviors are dependent upon various components present in the blood that allow feeding to continue. A number of hematophagous ectoparasites are influenced by the cellular components of the blood, including adenine nucleotides such as ATP and ADP. Others, including many anopheline mosquitoes and sand flies, will feed on fluids that have the optimal ionic balance resembling

physiological saline and do not require ATP for ingestion. This information is extremely important for the design of artificial membrane feeders to maintain vectors in the laboratory in the absence of live hosts.

Given that the ectoparasite continues to feed on the host when driven by positive stimuli, at some point this behavior must end when nutritional requirements are fulfilled, lest the insect should burst from too much blood for its gut capacity. In many vectors, stretch receptors monitor abdominal distention and provide negative feedback to the central nervous system to terminate ingestion when the blood meal reaches a certain volume. Indeed, mosquitoes that have had their ventral nerve cords surgically interrupted so as to interfere with the signals from the abdomen to the brain will feed until they actually burst, yet continue to feed afterward. This occurs with the presence of the stimuli that favor feeding and the absence of negative stimuli that terminate it.

Other Behavior Associated with Feeding

Protein is often scarce in nature, and the ability to feed on blood partially satisfies the nutritional requirements of those hematophagous ectoparasites. Some exclusively hematophagous insects like the tsetse have symbiotic microorganisms that supply the remaining essential nutrients. A contributing factor in directing this evolutionary course has been the reproductive use of the blood by females to provide a source of protein for their egg development. Although some species also can produce an initial batch of eggs autogenously, or without a blood meal, blood ingestion significantly increases their reproductive capacity. This use of blood primarily for reproductive purposes is found in many vectors, where the female feeds on blood while the male feeds only on plant sugars.

Although the advantages of feeding on the nutritious blood of a larger host are many, there are also drawbacks. Hosts are usually not cooperative and discourage feeding attempts by displaying defensive behaviors, including swatting and pecking, that pose significant risks to the life of the insect. Many day-biting mosquitoes that feed when their hosts are active and defensive have evolved behavioral mechanisms to deal with this risk. Immediately after ingesting a large blood meal, abdominal distention causes a complete inhibition of host-seeking behavior that lasts as long as the distention remains. Small blood meals that fail to trigger the distention do not inhibit host-seeking behavior, and these mosquitoes will attempt to feed again. As the large blood meal is digested and excreted and abdominal distention reduced, host-seeking behavior returns briefly until the eggs develop and a second mechanism of inhibition takes over. This second inhibition continues until the mature eggs are laid and results from the release of a peptide. Thus, given that the blood meal is large enough for eggs to develop and then the eggs do indeed go on to develop, the female is prevented from taking the unnecessary risks of approaching a host when it would not be of any reproductive benefit. Consequently, rather than contend with host behavior that might end the female's life and prevent her from reproducing at all, the mosquito is programmed to refrain from attempts at feeding and minimize the risks from defensive hosts by turning off her host-seeking behavior until after she lays her eggs. Anopheline mosquitoes that generally bite at night when hosts are sleeping or otherwise less active have not had the degree of defensive host behavior to deal with and have not evolved such effective mechanisms as to inhibit host-seeking behavior during reproductive periods. These mosquitoes tend to continue seeking a host throughout their life cycle. These repeated blood meals are able to supplement reproductive capacity of mosquitoes in this group.

Mosquitoes and several other hematophagous vectors undergo holometabolous development, characterized by a radically different immature stage compared to adults. In contrast to the completely terrestrial adults, the larvae are aquatic and must develop in water. The terrestrial female is challenged by the requirement to lay her mature eggs near or on water, necessitating a behavior change toward the end of egg development so that the eggs will be laid near a medium that will allow them to survive. During the process of egg maturation, not only is host-seeking inhibited and the female less responsive to host stimuli, but also is another behavior, preoviposition, activated in response to oviposition site stimuli. During preoviposition, the female becomes more responsive to odors from potential oviposition sites and orients toward these sites rather than to hosts for feeding. This behavior results from the release of a preoviposition factor when mature eggs are present and allows the female to orient toward an aquatic site during an appropriate period of her life.

See also: Parasite-Modified Vector Behavior.

Further Reading

Balashov YS (1984) Interaction between blood-sucking arthropods and their hosts and its influence on vector potential. *Annual Review of Entomology* 29: 137–156.

Beelitz P and Gothe R (1991) Investigations on the host seeking and finding of *Argas (Persicargas) walkerae* (Ixodoidea: Argasidae). *Parasitology Research* 77: 622–628.

Belan I and Bull CM (1995) Host-seeking behaviour by Australian ticks (Acari: Ixodidae) with differing host specificities. *Experimental and Applied Acarology* 19: 221–332.

Cardé RT and Willis MA (2008) Navigational strategies used by insects to find distant, wind-borne sources of odor. *Journal of Chemical Ecology* 34: 854–866.

Carroll JF, Mills GD, Jr, and Schmidtmann ET (1998) Patterns of activity in host-seeking adult *Ixodes scapularis* (Acari: Ixodidae) and host-produced kairomones. *Journal of Medical Entomology* 35: 11–15.

Cooperband MF and Carde RT (2006) Orientation of *Culex* mosquitoes to carbon dioxide-baited traps: Flight manoeuvres and trapping efficiency. *Medical and Veterinary Entomology* 20: 11–26.

Grant AJ and O'Connell RJ (2007) Age-related changes in female mosquito carbon dioxide detection. *Journal of Medical Entomology* 44: 617–623.

Klowden MJ (1990) The endogenous regulation of mosquito reproductive behaviour. *Experientia* 46: 660–670.

Klowden MJ (1996) Endogenous factors regulating mosquito host-seeking behaviour. In: Bock GR and Cardew G (eds.) *Olfaction in Mosquito-Host Interactions Ciba Foundation Symposium 200*, pp. 212–223. London: Wiley & Sons.

Lane RS, Mun J, Peribanez MA, and Stubbs HA (2007) Host-seeking behavior of *Ixodes pacificus* (Acari: Ixodidae) nymphs in relation to environmental parameters in dense-woodland and woodland-grass habitats. *Journal of Vector Ecology* 32: 342–357.

Lane RS and Stubbs HA (1990) Host-seeking behavior of adult *Ixodes pacificus* (Acari: Ixodidae) as determined by flagging vegetation. *Journal of Medical Entomology* 27: 282–287.

Lu T, Qiu YT, Wang G, et al. (2007) Odor coding in the maxillary palp of the malaria vector mosquito *Anopheles gambiae*. *Current Biology* 17: 1533–1544.

Takken W and Knols BGJ (1999) Odor-mediated behavior of afrotropical malaria mosquitoes. *Annual Review of Entomology* 44: 131–157.

Electrical Signals

P. K. Stoddard, Florida International University, Miami, FL, USA

Introduction

Electric sense is the ancestral vertebrate condition, retained by most of the primitive fish orders, including the lobe-finned fishes from which tetrapods arose. Electric sense is derived in a few teleost fishes, some fully aquatic amphibians, the platypus, and the echidna. While all animals produce electricity through muscle and nerve action, just a limited assortment of fish have evolved specialized tissues the sole function of which is to generate electric fields in the water outside their bodies. These bioelectric fields serve different functions in different taxa. The stargazers, a group of bottom-dwelling marine perch, produce strong defensive electric pulses that deter sharks and rays, though the stargazers themselves are not electroreceptive. Torpedoes (Torpedinidae), the electric eel (*Electrophorus electricus*), and the electric catfish (*Malapterurus electricus*) generate electric discharges in excess of 100 V that immobilize small fish long enough to allow these predators to ingest the meal before it can recover and swim away. Weakly electric fish of three orders generate electric fields measuring only a fraction of a volt per centimeter, used for imaging nearby objects in dark and murky waters, and for communication. These weakly electric fish include the Mormyridae and Gymnarchidae of Africa (order Osteoglossiformes), the electric knifefishes of the Neotropics (order Gymnotiformes), and a few sinodontid catfish from Africa (order Siluriformes). Signals produced by weakly electric fish will constitute the main topic of this article.

Electrogenesis

Electric signals are the products of large arrays of excitable cells, electrocytes, that fire action potentials en masse to produce the Electric organ discharge (EOD). Electric organs are made up of many electrocytes arrayed in series, like batteries in a flashlight, summing their individual action potentials to produce a larger voltage. Groups of electrocytes in series can be arranged in parallel arrays to increase current (**Figure 1**). Mass flow of sodium ions into the electrocytes polarizes the skin of the fish, setting up electrostatic fields in the water outside the animal. Within the electric organ, the net sodium flux in the headward direction polarizes the fish's head positive relative to the tail (**Figure 2**). A bare wire in the water near the fish's head can be used to detect this positive potential. Displayed on an oscilloscope, this potential is seen as an upward (positive) voltage pulse. If the net flux within the electric organ is in the tailward direction, the tail becomes positive relative to the head, thus the head becomes negative relative to the tail and the oscilloscope displays a downward (negative) voltage pulse. The net flow of positive ions changes direction with successive action potentials from different membranes. Electric organs of different weakly electric fish species can generate one to five distinct phases in a single EOD (**Figure 3**). Depending on the species, sex, and reproductive condition, an EOD can last from 0.1 ms to as long as 10 ms.

Some electric fish produce sinusoidal EOD trains consisting of successive EODs with intervals equal to or less than the duration of the EOD pulse itself (**Figure 4**) Amplified and played through an audio speaker, the EOD wave train is heard as a tonal sound, the fundamental frequency of which is determined by the discharge rate. Discharge rate, in turn, is generated by pacemaker neurons in the medulla and conveyed to the electrocytes by spinal motoneurons. Pacemaker cells of these so called 'wave fish' are the most temporally stable bio-oscillators known. Other electric fish produce their EODs less frequently, with silent intervals considerably longer than the EOD duration. If the EOD trains of these 'pulse fish' are amplified and played through an audio speaker, they are heard as ticks or buzzes, depending on the rate. Most gymnotiforms keep their discharge going all the time, day and night. Mormyrids and the electric eel, the one high-voltage gymnotiform, produce highly irregular discharge trains, often remaining silent for brief periods when they are inactive.

No terrestrial organism generates electric signals, because air is such a poor conductor of electricity. Conversely, the high salinity of marine environments tends to short-circuit in the high-conductivity water. In the highly conductive saltwater of marine environments, only large fish can support the large number of cells in parallel needed to produce high-current electric signals necessary to spread over useful distances. In the ion-poor waters of the tropics, smaller fish can generate electric signals because little current is needed to sustain a detectable voltage (this property follows directly from Ohm's Law $V = Ir$). Thus, the relatively small weakly electric fish have radiated extensively in the tropical freshwater river systems of the New World and Africa.

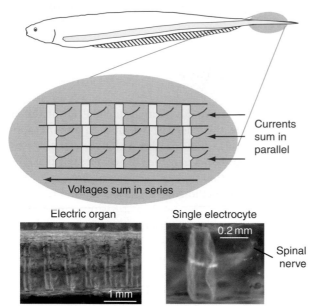

Currents
sum in
parallel

Voltages sum in series

Electric organ

1 mm

Single electrocyte

0.2 mm

Spinal
nerve

Figure 1 Electrocytes in series (rows) sum their voltages, like batteries in a flashlight. Multiple rows of electrocytes sum their currents. The electric organ and electrocyte shown are from the gymnotiform *Brachyhypopomus pinnicaudatus*. Photos courtesy of Michael Markham.

Embryology and Development

Electrocytes are derived embryonically from muscle cells, myocytes. As the electric organ develops, myocytes fuse and stop expressing the contractile proteins and sarcomeres that characterize skeletal muscle. Instead, they transform to produce a neuron-like physiology, packing their membranes with the voltage-gated ion channels and ion transporters needed for larger-scale ion flux (**Figure 5**). These cells are large enough to view with the naked eye and can be seen by shining a light through the fish's tail (**Figure 1**). Electrocytes can be nearly 1 mm across. In one gymnotiform family, the Apteronotidae, larval fish develop myogenic electric organs, which they replace over the course of few weeks with electric organs derived from the axons of spinal motoneurons. These neurogenic electric organs can sustain the higher discharge rates (500–2000 Hz) that typify this family's EOD. Ghost knifefish (*Apteronotus* spp.), common in the tropical fish pet trade, are the best-known members of this large family.

Electric organs are bilateral structures. In gymnotiforms the main organs run from just behind the gill to the tip of the pointed tail (**Figure 6**). In mormyrids, electrocytes are restricted to the caudal peduncle, the tissue between the trunk and the caudal fin.

Electroreception, Passive and Active

Electricity is detected by electroreceptors, specialized sensory cells embedded in the skin, typified by jelly-filled pores open to the surrounding water. Structurally and embryonically, electroreceptors resemble the sensory hair cells of the lateral line system, minus the mechanoreceptive hairs. Lampreys and the Condrostei (sharks, rays, sawfish, paddlefish, sturgeons) have electroreceptors, as do the sarcopterygian lobe-finned fishes, the coelacanths and lungfishes. Thus, our ancestors certainly possessed electroreceptors before they crawled from the sea. Electroreceptors were lost in the modern teleost lineage of bony fishes, probably because their large eyes enabled color vision, a more useful sense during daylight hours. Electroreception reevolved independently in select groups of nocturnal teleost fish (**Figure 7**). One such group was a lineage of Osteoglossiformes that includes two families of weakly electric fish, Mormyridae and Gymnarchidae, plus the Notopteridae of which some members are electroreceptive but not electrogenic. The second electroreceptive teleost lineage includes the sister orders Siluriformes (catfish) and the Gymnotiformes (electric knifefish). The 'ampullary' electroreceptors, whether primitive or derived, detect the extremely weak electric fields produced by muscle action and ventilation of small invertebrates, and thus are used for passive electrolocation of prey. Such electroreception is said to be 'passive' because it does not depend on the receiver to generate a signal. Passive electroreception is extremely sensitive. The ampullary system may detect electric fields as weak as 100 μV cm^{-1} in freshwater species, or just a few microvolts per centimeter in marine species (**Figure 8**). Ampullary electroreceptors detect electric fields in the spectral range of 1–100 Hz. Ampullary electroreceptors of marine species have best frequencies in the low end of this range, catfish in the middle, and weakly electric fish at the high end.

Mormyrids and gymnotiforms evolved new types of 'tuberous' electroreceptors specifically adapted to detect EODs. In contrast to ampullary electroreceptors, the tuberous receptors are tuned to the higher frequencies of the species-typical EOD. Mormyrids have two varieties of tuberous electroreceptors, the mormyromasts, used for active electroreception of nearby objects, and the Knollenorgans, used for detection of conspecific signals in communication contexts. Gymnotiforms have only a single morphological class of tuberous electroreceptor, used for both active electrolocation and communication. Some of the gymnotiform tuberous receptors are specialized for encoding temporal features of the EOD, and others for encoding EOD amplitude. In general, tuberous electroreceptors have higher sensory thresholds than ampullary electroreceptors. However, they can encode exquisitely small changes in the self-generated electric field, enabling active electrolocation.

During active electrolocation, objects in the water within half a body length of an electric fish distort the self-generated electric field in ways that can be detected by arrays of cutaneous tuberous electroreceptors (**Figure 9**).

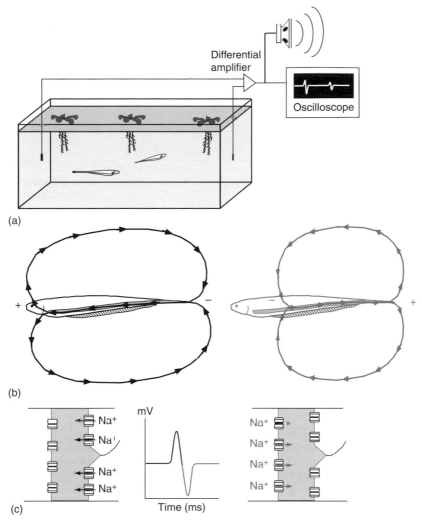

Figure 2 (a) Signals of electric fish can be recorded by placing two noncorrosive wires in the water and amplifying the voltage differences between them. Electric signals can be visualized on an oscilloscope or heard by playing the amplified signals through an audio speaker. (b, c) Gymnotiform pulse fish may produce biphasic EODs by generating two action potentials from each electrocyte. The innervated, posterior face fires an action potential (c, left), causing a headward sodium ion flux. These fluxes sum through the electric organ (b) producing a positive polarization of the head and a negative polarization of the tail. The anterior, noninnervated face of the electrocytes fires an action potential (c, right), causing a tailward flux of sodium ions. The tail becomes positive relative to the head. These two phases sum to produce a biphasic EOD (c, center).

Objects more resistive than the surrounding water produce an electric shadow on the skin, whereas objects more conductive than the surrounding water produce electric hotspots on the skin. Objects with capacitive properties (most living things) phase shift the EOD, distorting the waveform. Some electroreceptors respond to these phase shifts. Electric fish can identify the shape, distance, size, and material of a nearby object. They can even tell a small, proximate object from a large, distant object. By this means, electric fish can see in the darkness of night or in 'whitewater' rivers opaque with suspended minerals. Electric fish cannot resolve distant objects with active electrolocation, but they can 'see' through submerged mud, sand, or root tangles. With similar sensory abilities, we would readily view the contents of each other's pockets; we would know whether our friends' teeth

have dental fillings and whether those fillings were made of acrylic or silver amalgam.

Electric Communication – Signal Production

If one animal is signaling for its own purposes, active electrolocation for instance, another fish can listen in. If that fish is a predator, the signaler has an immediate problem. Indeed, hostile eavesdropping from electroreceptive catfish is believed to have led to the origin of more complex, high-frequency signals that escape detection by catfish equipped with ampullary electroreceptors. But gymnotiforms and mormyrids with high-frequency sensitive

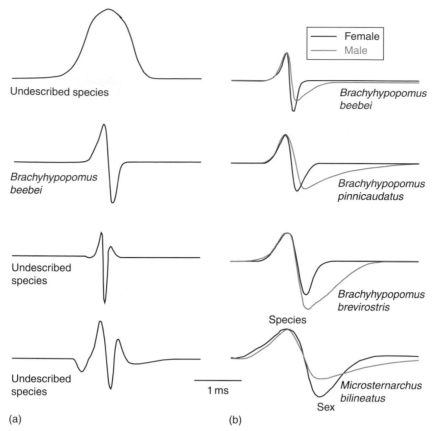

Figure 3 EOD waveforms can vary considerably, even within a single family. Shown here are EODs from the gymnotiform family Hypopomidae. (a) EOD of species in this family may have from one to five phases. Many species were undescribed at the time these recordings were made. (b) Even simple biphasic EODs may encode multiple properties. Here the first phase duration is distinctive of species, while the second phase duration is distinctive of the sex.

tuberous electroreceptors can readily detect EODs of other fish, opening the channel for electric communication.

Electric fish use their signals to communicate the standard conversational topics that interest most animals: identity, sex, aggression, and real estate. That said, most of what we know of electric communication derives from signal usage, rather than response to EODs. First, each taxon has characteristic EOD waveforms. Visually indistinguishable mormyrid species may be readily discriminated by striking differences in EOD waveform. In many but not all taxa, males and females have different waveforms. The most common sex difference for males is to discharge at lower discharge rates or to produce EOD waveforms with lower-frequency spectra (**Figure 3**), though this pattern is reversed in some species. In some taxa, males even alter their EODs to shift energy into the spectral band of the ampullary electroreceptor system, readily detectable by predators. Such signals may prove a male's quality to prospective mates, as long as he can survive. In those taxa in which signal intensities of sexually mature individuals have been measured without disturbing them, males appear to have stronger intensities than females, another signature of a sexually selected communication signal.

Electric fish alter their temporal discharge patterns in the presence of conspecifics (**Figure 10**). Some taxa generate characteristic variations in tempos or rhythms that serve as electric songs for courtship or intrasexual challenge. Changes in rate and tempo can be subtle, but when played through an audio speaker some of these compound signals are obvious to the human ear, often being described as chirps, accelerations, rasps, or silences. Absent such changes in discharge rate, it can be hard to know whether an electric fish is attempting to communicate, since they also discharge to electrically image their surroundings.

These compound signals are most commonly given by sexually mature males and appear most frequently during social encounters (**Figure 10**). The structure of the compound signals varies with the sex of the fish being encountered – some signals are typical of same-sex aggressive encounters, while others are given during courtship and spawning. Brief silences are used to punctuate communication sequences, contrasting sharply with rapid-fire EOD trains. But silences can have just the opposite function as well. Weakly electric fish of the genus *Gymnotus* are a preferred food of electroreceptive predators, including the electric eel (*E. electricus*). Upon

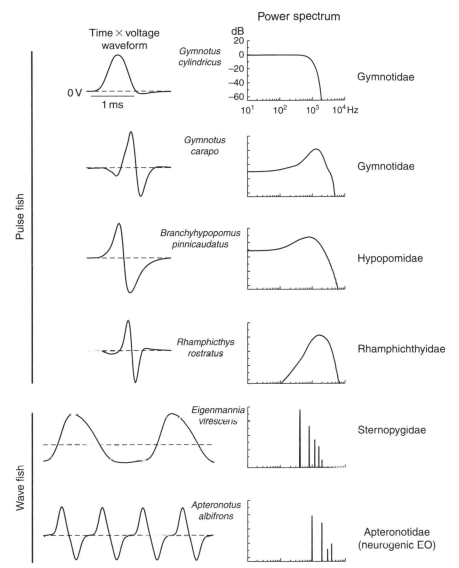

Figure 4 EODs may be delivered either as individual pulses with significant intervals between EODs, or one directly after another with small intervals. Fish with these two temporal patterns are called pulse fish and wave fish, respectively. Power spectra of pulse EODs are broad, regardless of discharge rate, whereas those of wave fish are narrow, with discharge rate setting the fundamental frequency. For comparison, white noise would generate a power spectrum with equal energy at all frequencies, and a pure tone would generate a power spectrum with energy at a single frequency. Shown here are typical EODs of the five gymnotiform families. Members of the Apteronotidae produce the highest frequency signals, accomplished with an electric organ (EO) comprised of motor nerve axons instead of muscle-derived electrocytes. Among the weakly electric osteoglossiforms (not shown), the mormyrids are pulse fish and the sole gymnarchid is a wave fish.

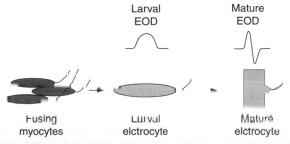

Figure 5 During larval development, muscle cells (myocytes, left) fuse to become electrocytes (center). In the process they stop expressing contractile proteins and become specialized for electrogenesis. In some species, electrocytes change morphology (right), deriving two active faces, which enables them to fire separate and opposing action potentials, making more complex waveforms possible.

detecting EODs of electric eels, or even crude mimics in the lab, *Gymnotus* may go silent for minutes at a time: an apparent strategy to avoid being detected and eaten. Individual females that are not ready to spawn, but find themselves confined to an aquarium with a persistently amorous male of their own species, will silence their EOD entirely for a few minutes, rendering themselves difficult for their suitor to locate. Viewed under infrared light, which fish cannot see, this scene is amusing to watch – the male visibly startles when he suddenly loses track of the female he has pursued relentlessly up to that point.

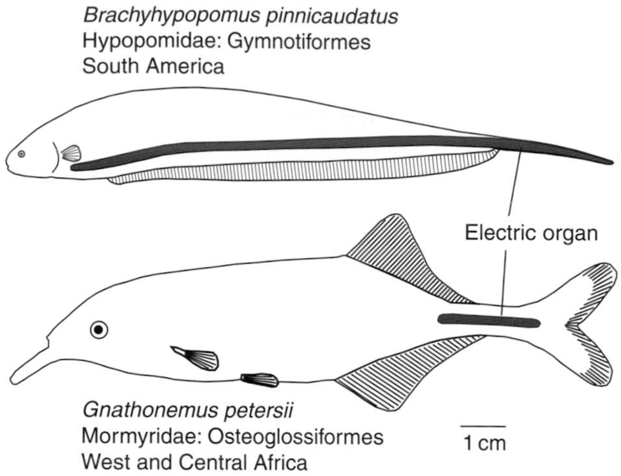

Electric Communication – Signal Detection and Identification

Mormyrids use their sensitive Knollenorgan electroreceptors to analyze EODs of other individuals. The Knollenorgans encode the time and polarity of the rises and falls of an EOD. Distinctive phase durations and intervals thus characterize the mormyrid EODs. Gymnotiforms have a special problem in that few can turn off their own signal. Imagine the problem of trying to listen to other individuals while talking nonstop. Their solution appears to be distortion analysis – gymnotiforms take advantage of their exquisitely sensitive distortion analysis circuitry to analyze each other's EOD waveforms as distortions of their own EOD waveforms. In fact, they should be nearly deaf to EODs of other individuals unless they themselves are signaling. One exception to this principle would be ampullary detection of the small number of fish that

Brachyhypopomus pinnicaudatus
Hypopomidae: Gymnotiformes
South America

Electric organ

Gnathonemus petersii
Mormyridae: Osteoglossiformes
West and Central Africa

1 cm

Figure 6 Gymnotiform and mormyrid electric fishes, the two largest groups of weakly electric fish, have different morphologies. In gymnotiforms, the electric organ runs the length of the body, whereas in mormyrids, electric organ is confined to the caudal peduncle.

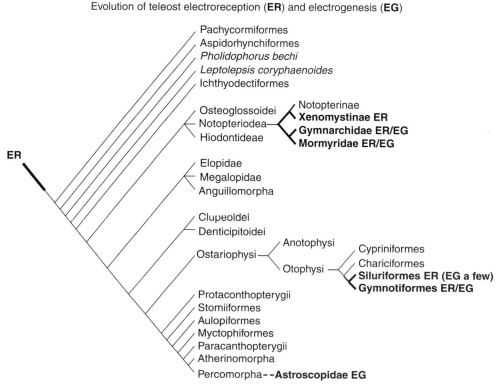

Evolution of teleost electroreception (**ER**) and electrogenesis (**EG**)

Figure 7 Electroreception was lost in the modern bony fish lineage Teleostei, then reemerged in two separate lineages. Electrogenesis evolved independently in the Gymnarchidae/Mormyridae lineage, the Gymnotiformes (electric knifefishes), and at least twice within the Siluriformes (catfish). Electrogenesis emerged in the stargazers (Astroscopidae), which are not electroreceptive.

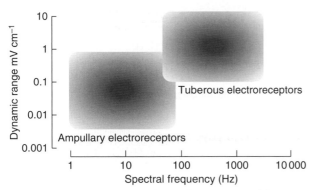

Figure 8 Approximate dynamic sensitivity ranges of the ampullary and tuberous electroreceptors. As a general rule, ampullary electroreceptors encode lower frequencies and are more sensitive than tuberous receptors.

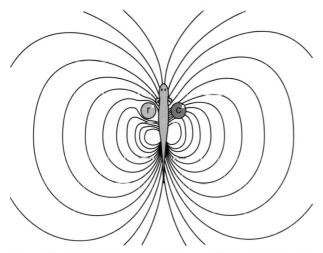

Figure 9 Electric fish are exquisitely sensitive to distortion of their self-generated electric fields. Objects more resistive than the surrounding water ('r' object on left) produce a 'shadow' in the electric field detected on the skin, measurable as a reduction in field intensity and a phase shift in the waveform. Objects more conductive than the surrounding water ('c' object on right) produce an electric 'hot spot' on the skin, measurable as a localized increase in electric field intensity. The brains of these fish process these localized electric field distortions to determine location, distance, size, shape, and material of nearby objects.

include low-frequency energy in their EODs. At the time of writing (late 2008), studies have just begun to emerge showing that, as one would expect, mormyrids actually do recognize species and sex by the EOD waveform. The same has yet to be shown rigorously for many gymnotiforms, though in one study, *Gymnotus carapo* discriminated individual conspecifics by their waveforms.

Electric Communication – Risky Signals

Risky behavior is a widely adopted mate-attraction strategy among sexually selected species. Male electric fish of

a few species produce EODs with energy in the spectral sensitivity ranges of both the tuberous and ampullary electroreceptors. These dual-spectrum signals are presumed to attract females. Lab studies have shown that these signals definitely attract the unwanted notice of predatory catfish and electric eels for which weakly electric fish constitute a significant fraction of their diet in the wild. In the majority of weakly electric fish taxa, the EOD has little or no energy in the range of the ampullary spectrum. Males of many such species nonetheless can briefly modify their signals to produce energy in the spectral range of ampullary electroreceptors. Extreme or prolonged depolarization of neurons in the EOD control center overdrives the electrocytes in the electric organ so the resulting EODs are distorted in ways that temporarily shift the spectrum.

Hormonal Regulation

Sex differences in EOD waveforms are under the regulation of steroid hormones. Over the course of days or weeks, androgens alter the discharge rate of pacemaker neurons, producing male-typical discharge rates. Estrogens exert the opposite effect, causing rate differentiation in the female-typical direction. In the wave fish, change in discharge rate is matched by a commensurate change in pulse duration to maintain an even duty cycle, the ratio of signal to silence.

Many electroreceptors are most sensitive to particular parts of the frequency spectrum. These tuning properties of the electroreceptors are also under the regulation of steroid hormones, so the receptors stay in tune with the electric signal to maintain sensory acuity. Interestingly, an increase in circulating androgen levels during the breeding season changes the tuning of ampullary electroreceptors in male stingrays to make their passive electrosensory system less sensitive to prey and more receptive to the bioelectricity produced by passing female conspecifics.

In some taxa, androgens alter the morphology of the electric organ, increasing both the size of the tail and the size of individual electrocytes. Within the electrocytes, androgens alter gene expression and RNA splicing of ion channel subunits to change the discharge waveform characteristics at the level of a single cell.

In about half the gymnotiform taxa, pituitary melanocortins, a family of peptide hormones, remodel the electrocytes in minutes to boost the signal power by as much as 300%, or to exaggerate masculine traits in the EOD waveforms. Melanocortin actions augment EOD waveforms during the night hours, and particularly during social encounters when the signal is used to communicate. At other times, these fish diminish their EOD waveforms, reducing energy expenditure and predation risk. These characters change on a circadian rhythm – fish kept under

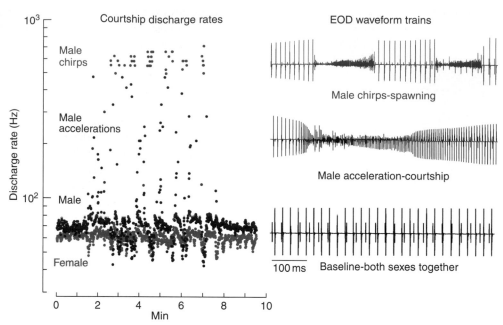

Figure 10 Electric fish can produce complex, compound signals by increasing the rate of discharge until individual EODs diminish in size. These signals are given during social encounters. Shown here are two types of complex signals given by the gymnotiform *Brachyhypopomus pinnicaudatus* during courtship and spawning. During the courtship phase, males run their snouts into the females and accelerate their discharges ('acceleration signals'). If the female proves interested, the males leads her to his chosen ovoposition site and lowers his basal discharge rate (lowest frequency dots) while giving high-frequency 'chirps.' The female deposits eggs, the male fertilizes them, and the cycle begins again. Shown here are six such courtship–spawning cycles between a captive breeding pair in an aquarium.

constant light and fed on a random schedule will continue to augment and diminish the EOD on a nearly 24 h rhythm for days or weeks.

Ecophysiology of Electric Signals

Each EOD gives the signaler a brief image of its surroundings, just as you would get by closing your eyes tightly and taking fast, blinking peeks. Successive EODs allow the fish to build up a more detailed image. In the rapidly changing environment of a turbulent stream or river, higher discharge rates allow the fish to track moving objects in the turbulent currents. Gymnotiform species that inhabit the fastest waters of riffles, rapids, and waterfalls may discharge as fast as 1000–2000 EODs per second. In contrast, species that dwell in ponds or sluggish rivers, where life is slower, may produce only 2–20 EODs per second. Wave fish select a fundamental frequency and stick to it for hours, days, or even months at a time. To change their discharge frequency by more than a few percent, they must also change the tuning of their electroreceptors to maintain the signal-sensory frequency match without losing sensory acuity during active electrolocation. These changes are possible, but require slow and careful hormonal coordination. Pulse fish, on the other hand, have broadly tuned electroreceptors and readily adjust their

discharge rates without incurring a sensory mismatch. Most pulse fish, gymnotiforms and mormyrids alike, discharge at low rates during the day when they are less active, and greatly increase their discharge rates at night, when they need to track moving objects.

Even during the day, electric fish continue to discharge. In the field, one commonly sees water snakes, herons, and tail-biting fish species hunting in the same habitats where electric fish hide during daylight hours. An ongoing discharge pattern maintains an early warning system. At the slightest disturbance in the water, even a puff of wind rustling the emergent vegetation in which they hide, the fish increases its discharge rate, thereby increasing the temporal and spatial acuity of its active-electrolocation sense.

Physical factors also affect electric fish ecology. For example, dissolved oxygen limits which electric fish can live where. Wavefish with the highest discharge rates are specialists of well-oxygenated moving waters. Most wavefish cannot survive long in low-oxygen conditions. One exception is *Gymnarchus niloticus*, the sole wave discharging species in Africa. *Gymnarchus* is an obligate air breather and is emancipated from dissolved oxygen restrictions. Another wavefish that survives low-oxygen water is *Eigenmannia virescens*, a Gymnotiform wave species with enlarged gills. Beyond those specialists, pulse fish can better withstand the daily anoxia that typifies floating meadows, and the episodic anoxic conditions that follow flushing of organic material into rivers during flood conditions. Some have speculated that pulse fish are specialists of low-oxygen waters, limiting their exposure to predatory fish.

Energetics

Producing a single EOD has no energetic cost because opening ion channels allows sodium and potassium cations to flow down their electrical and chemical gradients. The energetic cost of electric signaling comes from pumping these cations back to the side of the cell membrane where they started. Electrocytes thus use ATP to restore the ionic gradients typical of a polarized cell. For most weakly electric fish, EOD production appears to consume about 2–4% of their energy budgets. But males of some species greatly boost signal power around the time of courtship, so that electric signals may consume as much as 25% of their energy consumption.

Biophysical Advantages and Limitations of Communicating with Electric Fields

Electric signals have very different physical properties than light or acoustic signals. EODs are not electromagnetic waves like light, radio, or X-rays, but rather are electrostatic fields, such as you would get if you placed a battery in water: current flows out one end and in the other. Unlike electromagnetic waves, electrostatic fields do not propagate. Electric fish approximate electric dipoles, at least at a distance. The electric field intensity, measured in volts per unit distance ($mV\,cm^{-1}$), attenuates at the reciprocal of the cube of distance from the dipole center. This sharp attenuation curve contrasts with sound waves, which attenuate through spherical spread at the reciprocal of the square of distance. Thus, electric fields are useful for communication only within a few body lengths of the signaler. Active electrolocation is limited by coherence of the image distortion, and thus is useful at no more than half a body length.

Unlike sound, electric fields are not subject to harmonic distortion or reverberation. In the tropical rainy seasons, lightening constitutes the major source of background noise. If a thunderstorm is raging within 100 km, the water is filled with electrostatic noise. Electric fish can overcome the background noise through regular repetition of the EOD (wavefish have mastered this science) or by boosting the signal power, the strategy employed by mormyrid pulse fish with irregular EOD rates. An interesting hypothesis is that fish can produce irregular discharge rates to hide their signals in the background noise. Perhaps mormyrids use irregular discharge rates to make themselves less conspicuous to electrosensory catfish. Likewise the electric eel may discharge irregularly to elude detection as it moves in on gymnotiform prey.

Special Case of the Electric Eel

The electric eel (*E. electricus*) is unique among electric fish as the sole species capable of producing either a low-voltage or a high-voltage discharge. A mature electric eel can protect itself or stun fish prey by delivering a 1-s burst of EODs peaking at 600 V with a current of 2 A. The eel's low-voltage EOD, used for active electrolocation and communication, is a mere $10\,V\,cm^{-1}$, still 10–100 times stronger than a typical gymnotiform EOD. Electric eels reach sexual maturity at a length of 1 m, and large individuals have been reported in the upper Amazon approaching 3 m.

When the eel detects its small fish prey, it forms its body into the shape of the letter C with the prey situated in the gap. The high-voltage discharge immobilizes smaller fish, which immediately go belly-up. The eel reverts to its low-voltage EOD as it locates the immobile prey and sucks them down.

Electric eels deliver a high-voltage shock at the slightest provocation, a memorable experience for any adventurous zoologist seeking to experience one of the true marvels of the natural world. It seems unlikely that any large predator, such as a caiman or anaconda, would persist in attempts to eat an electric eel. Even a baby electric eel 10 cm long can deliver a painful shock. Like *G. niloticus* of Africa (also a big fish), *E. electricus* is an obligate air breather. As such it lurks near the bottom of shallow, sluggish waters during the day, rising to take a breath every few minutes. Local fishermen and unwary biologists have stepped on resting eels with unpleasant consequences that range from sharp pain to heart fibrillation and drowning.

Other Interests in Electric Signals

Electric fish are excellent animals in which to explore the integration of behavior, neural circuits, neurochemistry, and ion channels. Studies of electrosensory systems have revealed the neural mechanisms that underlie temporal hyperacuity, image processing, background cancelation, parallel processing, and network switching. Studies of electric organs have improved our understanding of ion channels, ion pumps, and evolutionary adaptation of these molecules.

See also: Active Electroreception; Vertebrates; Communication and Hormones; Electroreception in Vertebrates and Invertebrates; Neurobiology, Endocrinology and Behavior; Neuroethology: What is It?.

Further Reading

Bullock TH and Heiligenberg W (eds.) (1986) *Electroreception.* New York: Wiley.

Bullock TE, Hopkins CD, Popper AN, and Fay FR (eds.) (2005) *Electroreception.* Ithaca, New York: Cornell University Press.

Heiligenberg W (1991) *Neural Nets in Electric Fish*. Cambridge, MA: MIT Press.

Ladich F, Collin SP, Moller P, and Kapoor BG (eds.) (2006) *Communication in Fishes,* vol. 2. Enfield, NH: Science Publisher, Inc.

Moller P (1995) *Electric Fishes History and Behavior*. London: Chapman & Hall.

Stoddard PK (2002) Electric signals: Predation, sex, and environmental constraints. *Advances in the Study of Behaviour* 31: 201–242.

Stoddard PK and Markham MR (2008) Signal cloaking in electric fish. *Bioscience* 58: 415–442.

Electroreception in Vertebrates and Invertebrates

S. P. Collin, University of Western Australia, Crawley, WA, Australia

Introduction

Electroreception is an ancient sensory modality, having evolved more than 500 Ma, and has been lost and subsequently 'reevolved' a number of times in various vertebrate and invertebrate groups. The multiple and independent evolution of electroreception emphasizes the importance of this sense in a variety of animal behaviors from different habitats, including prey detection, orientation, navigation, and the use of bioelectric stimuli in social interactions. The detection of weak electric fields has probably occurred via two different mechanisms: the induction of electrical signals in specialized receptor cells and the interactions of magnetite crystals with the earth's geomagnetic field. This brief account predominantly concentrates on passive electroreception, where the weak electric fields produced by other organisms or anthropogenic sources are detected using specialized end organs. However, some mention will also be made of the use of geomagnetic cues in navigation and orientation and the subsequent induction of electric fields in some animals. Active electroreception and electrocommunication, which utilizes self-generated signals to navigate through electrosensory space, will be examined elsewhere in this volume.

The Evolution of Electroreception

In vertebrates, electroreceptive sensory structures can be broadly categorized into two distinct classes, ampullary and tuberous organs, based primarily upon the cellular morphology and shape of the receptor organs and secondarily on their respective frequency tuning characteristics. Ampullary receptors are broadly tuned to low-frequency fields ($<0.1-25$ Hz), while tuberous receptors are tuned to higher frequency fields from 50 Hz to over 2 kHz. Thought to be a primitive vertebrate character, the ability to detect weak electric fields has thus far been found in agnathans (lampreys but not hagfishes), chondrichthyans (sharks, skates, rays, and chimeras), cladistians (bichirs), chondrosteans (sturgeons and paddlefish), a small number of species within the osteoglossomorphs (knife fishes), and three orders of teleosts: mormyrids (African electric fishes), gymnotids (South American electric fishes), and siluriforms (catfishes). Within the aquatic realm, electroreception has also been found within the sister group of the Actinopterygii fishes, the Sarcopterygii, which comprises the dipnoan

lungfishes and the actinistian coelacanth. Magnetite crystals have been localized in the olfactory epithelium in teleosts.

In contrast to anuran amphibians (frogs and toads), urodele amphibians (axolotls, salamanders, and newts), and larval members of the Gymnophiona (caecilians) also possess electroreception. Amphibian ampullary organs are similar to ampullary organs in both teleost and non-teleost fishes. To date, there have been no studies showing passive electroreception in reptiles and birds although some birds are known to detect low-frequency fields via the induction of electrical signals in somatosensory receptors and are hypothesized to possess magnetite crystals.

The monotremes (egg-laying mammals that suckle their young) are one of the few groups of mammals to have evolved electroreception. Passive electroreception occurs in the platypus, *Ornithorhynchus anatinus*, the echidna, *Tachyglossus aculeatus*, and the long-nosed echidna, *Zaglossus bruijnii*, via cutaneous receptors that are distinctly different from those of fishes and amphibians. Therefore, the unique morphology of these specialized cutaneous receptors and their afferent pathways to the central nervous system may reflect this long evolutionary separation and support the independent evolution of this sensory modality for a third time. The star-nosed mole, *Condylura cristata*, also shows behavioral responses to electric fields, but this is yet to be confirmed anatomically and physiologically.

Only a small group of invertebrates have been found to possess the behavioral basis for electroreception. The freshwater crayfishes, *Cherax destructor* and *Procambarus clarkii*, respond to low-level electrical signals of biological relevance by changing their posture and/or activity patterns. However, the anatomical basis of the electrical sense in these crustaceans is not well known and may be shared with other sensory cells. The nematode, *Caenorhabditis elegans*, is known to orient to electric fields by stimulating their amphid sensory neurons, whereas electrostatic forces from frictionally charged surfaces alter insect locomotory behavior. The cockroach, *Periplaneta americana*, can detect the presence and strength of static electric fields using specialized cells within the sensory hair plate of the antennae.

Structure and Function of Passive Electroreceptors

Vertebrates

Passive electroreception is generally mediated by either ampullary receptors (agnathans, chondrichthyans, teleosts, nonteleosts, and amphibians) or mucous glands (monotreme mammals) (**Figure 1**). The receptors are superficial structures embedded within the dermis and connected to the surface by a pore or canal filled with a mucopolysaccharide gel (ampullary organ) or mucus (gland). At the base of the ampullary canal is an ovoid capsule containing aggregations of receptor cells, each of which supplies an afferent axon that projects to the medulla. The canal walls are formed by squamous epithelial cells joined by tight junctions, which create a high resistance

(**Figure 1(b)** and **1(c)**). The canals radiate in all directions from an ampullary cluster, providing a method of directionally sampling the electric field surrounding the animal. The length of the canal changes according to the position of the receptors over the head and, in some animals, to environmental factors such as osmoregulatory constraints and concomitant changes in skin resistance.

In some aquatic vertebrates, up to 400 ampullary tubes radiate from a single cluster. Freshwater fishes have short canals and low numbers of receptor cells in comparison to

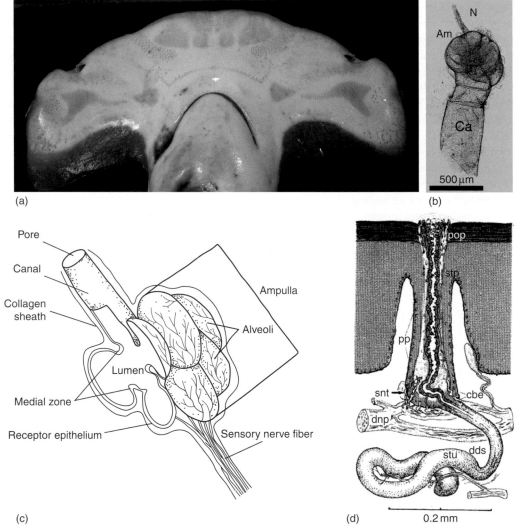

(a)

(b)

(c)

(d)   0.2 mm

Figure 1   (a) Ventral surface of the hammerhead shark, *Sphyrna lewini*, showing the large number of the pores of ampullae of Lorenzini, which are nonhomogeneously distributed. Photograph kindly provided by L. Litherland. (b) Ampullae (Am) of the sawfish, *Anoxypristis cuspidata*. Note the alveoli are positioned in a grape-like structure (multialveolate type). The nerve (N) extends from the sensory ampulla. Photomicrograph kindly provided by B. Wueringer. (c) Schematic representation of a single ampulla of Lorenzini of a rhinobatid shovelnose ray. From a somatic pore, the canal extends, widening proximally to an ampullary bulb. The ampulla is formed by several alveoli in a grape-like arrangement. The epithelium of adjacent alveoli and the canal is separated by the medial zone. A sensory nerve fiber extends from the proximal end of the ampulla. Reproduced from Wueringer BE and Tibbetts IR (2008) Comparison of the lateral line and ampullary systems of two species of shovelnose ray. *Reviews in Fish Biology and Fisheries* 18: 47–64, with kind permission from Springer Science + Business Media. (d) Diagram of an echidna sensory mucous gland: pop, pore portion of gland; stp, straight portion; dds, dermal duct segment; stu, secretory tubule. Sensory nerve terminals (snt) arise from a dermal nerve fiber plexus (dnp). The intraepidermal segment of the gland shows a club-shaped enlargement (cbe) of the papillary portion (pp). Here, the sensory nerve terminals penetrate the epidermal layer. Reproduced from Proske U, Gregory JE, and Iggo A (1998) Sensory receptors in monotremes. *Philosophical Transactions of the Royal Society: Biological Sciences* 353: 1187–1198, with permission from the Royal Society.

marine species, which possess long canals and numerous receptor cells. One afferent fiber typically innervates each organ or cluster of ampullary organs in teleosts in contrast to the situation in elasmobranchs, where up to thousands of receptor cells within a single alveolus are contacted by up to 15 afferent fibers. The sensory epithelium of an ampullary organ usually comprises receptor cells bearing a single kinocilium, which are surrounded by supportive cells that bear numerous stereocilia/microvilli (as found in cladistians and dipnoans), although in lampreys, teleosts, and chondrosteans, a kinocilium is lacking. There is a wide variation in the number and distribution of electroreceptor organs (**Figures 1(a)** and **2**), ranging from six tubules in the rostral organ of the coelacanth, *Latimeria* to 75 000 in the paddlefish, *Polyodon*.

The sensory cells of gymnophionan ampullary organs possess a kinocilium surrounded by microvilli much like the sensory cells of ampullary organs in *Polypterus*. In contrast, sensory cells of urodelian ampullary organs possess no kinocilium but have numerous microvilli. Electrophysiological data on adults of other gymnophionan species, that is, *Thyphloneetes* also show a sensitivity within the range of electroreceptors found in fishes. Unlike almost all anuran larvae, larval *Ichthyophis kohtuoensis* are predators, which feed on *Daphnia*, insect larvae, and tadpoles, where electroreception is used to detect and localize larger prey.

The structures in the skin responsible for the electric sense in monotremes (platypus and echidna) have been identified as sensory mucous glands with an expanded epidermal portion that is innervated by large-diameter nerve fibers (**Figure 1(d)**). The gland consists of a coiled tube (closed at its end) lined with secreting cells. The secreted mucus represents a low-resistance pathway from the skin surface down to the nerve endings in the deeper papillary portion of the gland. Afferent recordings have shown that in both platypuses and echidnas the receptors are excited by cathodal (negative) pulses and inhibited by anodal (positive) pulses. Estimates give a total of 40 000 mucous sensory glands in the upper and the lower bill of the platypus (**Figure 3**), whereas there are only about 3000 and 100 in the tip of the snout in the two species of echidnas, that is, *Zaglossus* and *Tachyglossus*, respectively. The platypus uses its electric sense to detect electromyographic activity from submerged prey in the water and for obstacle avoidance. The role of the electrosensory system

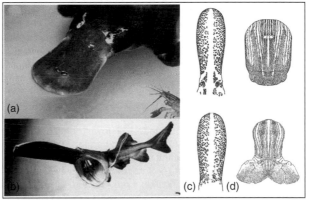

Figure 3 Comparison of platypus with paddlefish to show the similarity of the paddle-shaped bill organ and rostrum, respectively. (a) The bill organ of the platypus (and one of its prey, the yabbie, *Cherax* spp.). About 40 000 electroreceptors and 60 000 mechanoreceptors form a sensory array that enables the capture of small benthic invertebrates, as well as some vertebrates such as fishes, without any assistance from vision, hearing, or olfaction. Note the numerous pits formed by the opening of the mucous gland electroreceptors on the bill; these pits are barely visible where they are in sharper focus on the more proximal region of the bill, near the eye. (b) Young paddlefish attacking the dipole wires as if trying to capture plankton: its rostrum is innervated by about 60 000 ampullary electroreceptors that enable the capture, using the prey's electrical discharge, of free-swimming invertebrates such as *Daphnia*. (c) and (d) Electroreceptor arrays of dorsal (upper illustrations) and ventral (lower illustrations) surfaces of the rostral bill organ of the paddlefish (c) and platypus (d). Reproduced from Pettigrew JD and Wilkins L (2003) Paddlefish and platypus: Parallel evolution of passive electroreception in a rostral bill organ. In: Collin SP and Marshall NJ (eds) *Sensory Processing in Aquatic Environments*, pp. 420–433. New York: Springer, with kind permission from Springer Science + Business Media.

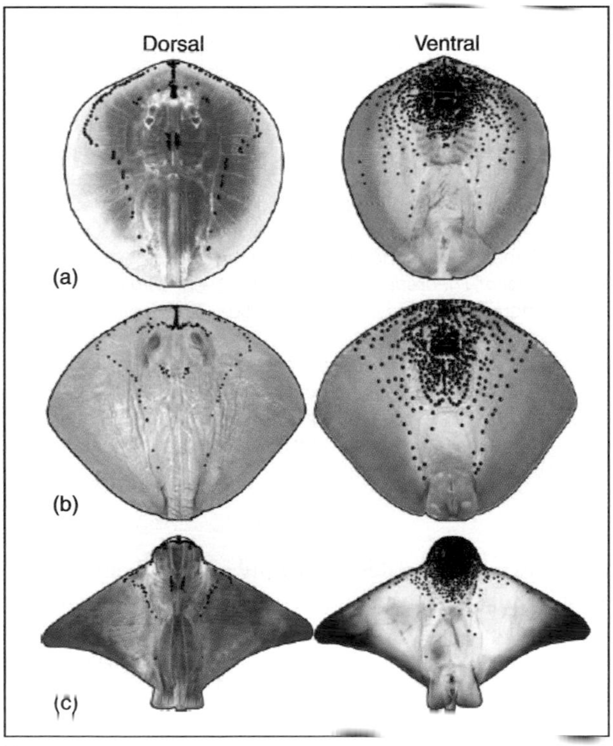

Figure 2 Dorsal and ventral electrosensory pore maps of one representative individual of three species of myliobatiformids. (a) *Urobatis halleri*. (b) *Pteroplatytrogon violacea*. (c) *Myliobatis californica*. Reproduced from Jordan LK (2008). Comparative morphology of stingray lateral line canal and electrosensory systems. *Journal of Morphology* 269: 1325–1339, with kind permission from Wiley.

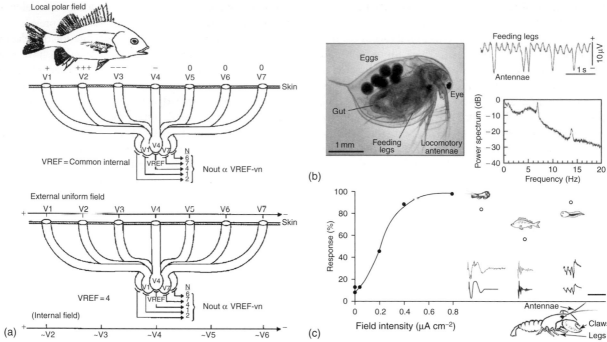

Figure 4 (a) Model for the encoding of extrinsic polar and uniform electric fields by the elasmobranch electrosensory system. Any living prey produces a weak polar electric field formed by the differential distribution of charges on (or in) the organism. This creates weak potentials (C, −) in the water that surrounds the body of the prey. When the prey approaches, the surface pores of the ampullary system sample the field potentials (Vn) across the surface of the skin. The potentials at each pore are conducted to their individual ampullae and stimulate sensory neurons (N). In this scenario, the voltage gradient from the prey does not influence the common internal reference potential (VREF) at the ampullary cluster (and basal surfaces of receptor cells), and the effective stimulus for all ampullae is represented by the voltage drop across the skin. Neural output for each individual ampulla (Nout) is proportional to the difference between VREF and the voltage at its associated surface pore (Vn). Reproduced from Tricas TC (2001) The neuroecology of the elasmobranch electrosensory world: Why peripheral morphology shapes behavior. *Environmental Biology of Fishes* 60: 77–92, with kind permission from Springer Science + Business Media. (b) A single *Daphnia*. Upper right panel shows the electric potential from a tethered *Daphnia* measured at a distance of 0.5 cm. The oscillations result from higher frequency feeding motions of the legs and lower frequency swimming motions of the antennae. The lower right panel shows the power spectrum of the signal from a single, tethered *Daphnia* showing the approximately 7-Hz feeding frequency as the sharper peak and the broader 5–6 Hz swimming frequency. Reprinted from Freund et al. (2002). Behavioral stochastic resonance: How the noise from a *Daphnia* swarm enhances individual prey capture by juvenile paddlefish. Freund JA, Schimansky-Geier L, Beisner B, et al. (2002) Behavioral stochastic resonance: How the noise from a Daphnia swarm enhances individual prey capture by juvenile paddlefish. *Journal of Theoretical Biology* 214: 71–83, with permission from Elsevier. (c) Response of the freshwater crayfish *Cherax destructor* to different voltages of a step function. Points are the total number of responses for all animals in each test (out of 40, 30, 40, 50 stimuli presentations for the curve, crayfish, tadpole, and fish signals, respectively). Responses to electrical signals from swimming animals are indicated by the animal icons. Waveforms shown beneath for *C. destructor* (left), *C. carpio* (middle), and *Rana* sp. (right): the top signal of each is the recorded swimming signal, the lower signal is the analog. Time scale bar: lower right is 1.5 s for the tadpole and 100 ms for the other two signals. Inset: lower right is *C. destructor*. The body and labeled appendages were monitored for movement changes. Reprinted from Patullo BW and Macmillan DL (2007) Crayfish respond to electrical fields. *Current Biology* 17: R84–R85, with permission from Elsevier.

in the echidna has not yet been established but is thought to be able to detect (1) the presence of weak electric fields in water, and (2) moving prey in moist soil.

Invertebrates

There have been no morphological analyses of cells, which preferentially respond to weak electric fields in invertebrates. However, in a study on the crayfish, *Procambarus clarkii*, chemo- and mechanoreceptive neurons concomitantly revealed responses to electric fields that initiated behavioral changes in activity (touches, grabs,

and tugs at electrodes) when subjected to an electric field. Similarly, the fine structure of electroreceptors in cockroaches has not been described. However, the antennae in cockroaches are known to move in response to static electrical fields and the imposed deflection is abolished after the antennae are removed at the head-scape joint. Therefore, it appears that specialized receptors for the detection of electric fields may not be present. However, the influences of electric fields on long antennae and the behavior mediated by scape hair plates underpin the avoidance evoked by cockroaches confronted with static electric fields.

The Role of Passive Electroreception in Behavior

In aquatic environments, ampullary organs are used to detect animate and inanimate electric fields by measuring minute changes in potential between the water at the skin surface and the basal surface of the receptor cells. Epidermal pores and the jelly-filled canals comprising each electroreceptor organ ensure that the potential within the ampullary lumen is the same as that at the surface. The hair cells of each receptor act as voltage detectors and release neurotransmitter onto the primary afferent neurons according to the difference between the basal and apical potentials. The primary afferent neurons encode stimulus amplitude and frequency information that is sent to the brain, where a sophisticated set of filter mechanisms are used for extracting the weak electrosensory signals from a much stronger background noise, predominantly created by the animal's own movements. As each ampulla functions independently, the distribution of the ampullary organs on the body surface provides information about the electric field's intensity, its spatial configuration, and possibly the direction of its source (**Figure 4(a)**). The behavioral relevance of this level of sensitivity was uncovered in a series of experiments involving both experimental and free-living elasmobranchs, which induced feeding responses toward either buried fish or a pair of buried electrodes, that could not otherwise be detected using other sensory modalities.

The high sensitivity of electroreceptors allows some species to localize prey by detecting the very faint potentials associated with the ionic leakage of the gills (modulated by ventilatory movements) of buried teleosts. Crustaceans produce higher bioelectric fields ($1000\,\mu V\,cm^{-1}$) (**Figure 4 (b)**). Ampullary organs respond to DC or low-frequency electric fields, for example, 1–8 Hz in elasmobranchs and 6–12 Hz in silurid teleosts. Sensitivity thresholds vary across taxa: $0.1\,\mu V\,cm^{-1}$ in lampreys (*Lampetra tridentata*), $5\,nV\,cm^{-1}$ in elasmobranchs, and $5\,nV\,cm^{-1}$ in marine teleosts and $1\,\mu V\,cm^{-1}$ in freshwater teleosts.

A theoretical mechanism in which sharks or rays process directional information in the immediate vicinity of an electrical field has been suggested. This approach algorithm predicts that as an animal enters a local electric field, that is, emanating from potential prey, it corrects its swimming course to keep the angle between its rostrocaudal body axis and the detected electric field constant. It is suggested that this behavior is independent of the angle of approach, the polarity of the prey's field, changes in the strength and the direction of the field, and the position of its source.

The paddlefish, *Polyodon spathula*, possesses a rostrum that is adorned with ampullary electroreceptors, which act as an antenna (**Figure 3(c)**). The rostrum is considered a sensory device with sufficient sensitivity to detect the electric fields of planktonic prey with a sensitivity threshold of $10\,\mu V\,cm^{-1}$, a considerably higher sensitivity than the sensitivity of individual electroreceptors. Paddlefish use this rostrum to laterally strike at planktonic prey using its electric sense passively without the use of visual, chemical, and hydrodynamic senses at distances of 8–9 cm. Higher concentrations of receptors along the edges of the rostrum and its saccade-like motion through the water may serve to enhance prey detection by increasing the width of the electrical scan field.

In the axolotl, *Ambystoma mexicanum*, gill beat frequency increases and feeding behavior is elicited in the presence of an electrical field. A pair of electrodes hidden in the substrate is excavated in response to a threshold voltage of $10\,\mu V\,cm^{-1}$ with more reliable responses elicited above a voltage of $25\,\mu V\,cm^{-1}$. Studies with other amphibians (larvae of *Salamandra salamandra* and adult aquatic *Triturus alpestris*) also behaviorally demonstrate the capacity for prey detection by electroreceptors in other species within the urodeles. Although the threshold sensitivity of single fibers of the anterior lateral line nerve to square pulses of 0.5 s is found to be $100\,\mu V\,cm^{-1}$ or less (not an extremely high sensitivity), for single afferent fibers this is quite adequate to justify the conclusion of electroreception.

Much like the rostrum of the paddlefish, the bill of the platypus also possesses electroreceptors arranged in parasagittal stripes alternate with stripes of mechanoreceptors (**Figure 3(a)** and **3(d)**). The role of the electroreceptors is to find submerged prey as the bill probes the substrate with characteristic side-to-side movements. This is an important sense for the enigmatic platypus since the eyes, ears, and nostrils are all closed in water. The bill acts somewhat like an antenna, where the electrical field decay across the bill can be used to locate crustaceans by temporally integrating volleys of signals. A maximal sensitivity of $50\,\mu V\,cm^{-1}$ has been recorded. Interestingly, behavioral thresholds are one to two orders of magnitude below the physiological thresholds recorded from the peripheral electroreceptors.

Like the platypus, the star-nose mole is semiaquatic and fossorial, and probes the substrate for live worms and crustaceans using lateral head movements. Its nose is adorned with highly tactile (short and long) rays, which, along with the head, are extremely mobile during prey localization. Although the structural basis for electroreception has not yet been confirmed, these mammals appear to respond behaviorally to electrical fields. However, further conclusive evidence is required in order to isolate the responses of electroreceptors and confirm this species as being the first example of a eutherian mammal in passive electroreception.

The behavioral experiments carried out in freshwater crayfish reveal that at least one group of invertebrates can respond to DC and low-frequency electric fields (**Figure 4(c)**). However, behavioral thresholds to electric fields vary between *C. destructor* ($3\text{--}7\,\mathrm{mV\,cm^{-1}}$) and *Procambarus clarkii* ($20\,\mathrm{mV\,cm^{-1}}$), which could be attributed to either the different behaviors analyzed in the two studies and/or electric field shape. Despite these differences, these behavioral thresholds are extremely high compared to typical fields of animate and inanimate objects found within the natural environment of these species.

Cockroaches exhibit a clear avoidance of electric fields at applied electrode voltages over $1\,\mathrm{kV}$, which is equivalent to a modeled electric field strength of $8\text{--}10\,\mathrm{kV\,m^{-1}}$. Behavioral experiments in both flies (*Drosophila*) and cockroaches (*Periplaneta*) show that electric fields can be detected in both species and that the information is used to evoke an escape response. With an 'avoidance threshold' to a static electric field strength of $8\text{--}10\,\mathrm{kV\,m^{-1}}$, cockroaches change their behavior in response to a variety of anthropogenic electric fields such as those produced by high-voltage wires and household appliances. Although there is some doubt about whether the electric sense is used in nature, it seems counter intuitive to assume that the ability to detect weak electric fields is not found throughout the animal kingdom and that with further research, more examples of electroreceptive invertebrates will be uncovered.

Spatial Distribution of Electroreceptors and Localization of Prey

Many large electroreceptive species show a wide variation in receptor number and distribution, reflecting differences in habitat and feeding strategy rather than being constrained by phylogeny (**Figure 2**). This is especially relevant for elasmobranchs, which possess a range of body forms; skates and rays are dorsally compressed, while the body form of sharks is more conical. The distribution of pores over the head of skates and rays is most sensitive to the horizontal component of an electric field, while the pores in sharks are able to provide sensitivity in three dimensions. The spatial differences in the density of ampullae have been examined in 40 species of skates. Species that feed predominantly on benthic invertebrates possess high densities of pores on their ventral surface, especially around the mouth, providing a greater resolution for locating, manipulating, and ingesting prey excavated from the substrate. On the other hand, skates that feed on more mobile fish prey possess low pore densities, where the need for increased resolution is not as critical and may be concomitantly mediated by the visual system. Hammerhead sharks (Sphrynidae), with their unique cephalofoil head morphology, constitute an intermediary between the flattened disk of skates and rays and the conical head shape of pelagic carcharinids. Sphyrnids possess a higher number of pores on the ventral surface of the head, while the sandbar shark *Carcharhinus plumbeus* possesses an even distribution over both the dorsal and the ventral surfaces of the head. Recent research in the lesser spotted catshark, *Scyliorhinus canicula*, reveals that the electroreceptors are derived from neural crest cells, and their distribution is, at least partly, determined by specific developmental signaling processes.

The omnihaline bull shark, *Carcharhinus leucas*, possesses an uneven distribution of ampullary pores over the dorsal and the ventral surfaces with a higher density of ampullary organs in the anteroventral areas of the head. This suggests a high dependence on electroreception in the capture of fish and other prey in the dark, murky waters characteristic of many coastal estuaries and riverways.

The sensory mucous glands of the platypus are also nonhomogeneous and are distributed over the inner and the outer surfaces of the bill of both the upper and the lower jaws, and on the front surface of the shield (**Figure 3(c)**). Each gland is supplied with up to 30 myelinated sensory axons. A total of $30\,000\text{--}40\,000$ mucous sensory glands are innervated by an estimated $380\,000\text{--}640\,000$ stem axons. The maximal sensitivity of $50\,\mathrm{\mu V\,cm^{-1}}$ mentioned earlier is attained approximately $80°$ from the rostral pole of the bill and $20°$ down, indicating that information from the edge of the bill is important to the platypus as it searches the substrate for prey. Sensitivity of the animal to a stimulus varies over two orders of magnitude from the preferred orientation. Given the large difference in the spatial distribution of mucous glands, the head saccade behavior of the platypus helps to locate the direction of a stimulus using reorientation of the head to present the most sensitive part of the bill to the stimulus. In contrast, the sensory mucous glands in the echidna are all crowded into the tip of the snout and comprise a small fraction of the population of axons in the trigeminal nerve. The number and distribution of electroreceptors in the two species of echidna studied thus far suggests that electroreception is not as important as in the platypus.

Irrespective of the location and spatial resolution of the ampullae/mucous glands, the electrosense is only a short-range sense for prey detection with the ability to detect both AC and DC bioelectric potentials at close range (at least for marine elasmobranchs) and is most likely augmented by visual, chemoreceptive, and mechanoreceptive input in localizing and successfully capturing the prey.

Passive Electroreception in Different Aquatic Environments

Several studies have revealed that the sensitivity of passive electroreception of low-frequency bioelectric fields emanating from prey may vary enormously in a range of different aquatic environments. Canal length varies significantly in the ampullary organs of saltwater and freshwater vertebrates and is thought to be an adaptation for adjusting sensitivity. The long canals of elasmobranchs enable low-frequency electric fields to be virtually unattenuated because of the high resistance of the canal walls and the low resistance of the jelly core. Marine ampullary receptors enhance sensitivity to voltage gradients since there is a relatively small transepidermal voltage difference due to a low skin resistance relative to salt water. In contrast, the skin resistance of freshwater ampullary receptors is very high, due to its role in osmoregulation. Since the skin has a much larger resistance than the internal tissues, a long canal is unnecessary, where a more 'superficial' organ will detect a large voltage drop. Despite the two methods of measuring potential differences, marine elasmobranchs are sensitive to voltage gradients of at least $5\,\mathrm{nV\,cm^{-1}}$, while freshwater rays, that is, *Potamotrygon circularis*, are sensitive to gradients between 50 and $100\,\mathrm{\mu V\,cm^{-1}}$. Similarly, some marine catfish are at least one order of magnitude more sensitive ($0.08\,\mathrm{\mu V\,cm^{-1}}$) than some freshwater teleosts ($1\,\mathrm{\mu V\,cm^{-1}}$).

The effects of habitat salinity also have influence on the structure of the ampullary organs in teleosts, which are reduced to 'microampullae' in freshwater plotosid catfish, that is, *Plotosus tandanus*. These microampullae consist of short canals ($50\,\mathrm{\mu m}$ in length) and contain low numbers of receptor cells (10–15) and appear different to the ampullary organs described for marine *Plotosus anguillaris*, which are characteristically defined as resembling the ampullae of Lorenzini in elasmobranchs. In contrast, the canals of the ampullary organs in *P. anguillaris* measure in centimeters and the ampullae include hundreds of receptor cells. Similarly, elasmobranchs such as the bull shark, *C. leucas*, which migrate between saltwater and freshwater rivers also possess differences in receptor types and therefore the ability to detect bioelectric fields. Although the average number of ampullary organs (2052) found in juvenile *C. leucas* is in agreement with the characteristically high numbers of organs in other carcharhinids, juvenile *C. leucas* collected from freshwater reaches of the Brisbane River in Australia possess ampullae of Lorenzini that differ morphologically from those previously described for marine elasmobranchs. These structural differences may allow the ampullae to function more effectively in freshwater habitats by shunting a portion of the electrical signal passing through the ampullae. A similar ecomorphological difference has also been described for the ampullae in the freshwater stingray, *Himantura signifer*. These 'miniampullae' are relatively simple with a short canal and a reduced sensory epithelial surface area and are presumably adapted for operating in a freshwater environment.

A recent study of electrosensitivity in the euryhaline Atlantic stingray, *Dasyatis sabina*, reveals that sensitivity to electric fields is significantly reduced in freshwater. In order to elicit a feeding response, stingrays tested in freshwater required an electric field 200–300 times greater than stingrays tested in brackish and saltwater environments and the maximum orientation distance is reduced by 35.2%, from 44.0 cm in brackish and saltwater environments to 28.5 cm in freshwater.

In addition to differences in canal length, receptor structure, and topographic arrangement in species that frequent both saltwater and freshwater, the associated changes in temperature and the ionic composition between the two environments may also have effects on sensitivity. Electrophysiological recordings show that the sensitivity of the apical microvilli in freshwater catfish, *Ameiurus nebulosus*, is reduced by 80% when exposed to a hyperosmotic solution. Discharge rates of ampullae of Lorenzini appear to be temperature sensitive over a wide physiological range, where sensitivity appears to increase as the temperature rises, providing a potential for the electric sense to be useful as a biomonitoring system for water pollution.

A comparison of 40 skate species, ranging from depths of 63–2058 m reveals that both the number of alveoli and the overall size of the ampullae increase significantly with depth. This trend suggests that species inhabiting deeper regions of the water column, where sunlight may fail to penetrate, possess higher numbers of receptor cells and may rely more heavily on electroreception in this relatively prey-depauperate environment. The depth-related modifications to the ampullae in the deep-sea skate, *Raja radiata*, are mostly restricted to the mouth region (along the superficial ophthalmic and mandibular clusters), emphasizing the importance of prey localization. An increase in the number of alveoli and ampulla size in this region would enhance both receptor sensitivity and the signal-to-noise ratio, thereby mediating the perception of a slightly weaker bioelectric field than its shallow water counterparts. The resultant reduction in the signal-to-noise ratio may also allow these deeper dwelling species to search for prey higher in the water column and hence cover a greater area per unit time. Interestingly, the region of the dorsal nucleus in the medulla that receives input from the superficial ophthalmic ampullae occupies a disproportionately large input (40%) than ampullae from other clusters.

Echidnas probe subterranean nests for prey and become particularly active after rain. They appear to probe the soil in an exploratory fashion with their nose and may dig furiously to expose a moving beetle or caterpillar. It is suggested that the movements of prey generate short-range electric fields, which are detected by the electroreceptors in the snout tip. However, since the number of mucous glands in echidnas is so low, electroreception cannot always provide a reliable method of locating food.

The Detection of Potential Predators

The selection pressures to evolve elaborate mechanisms for the detection of potential predators are always intense

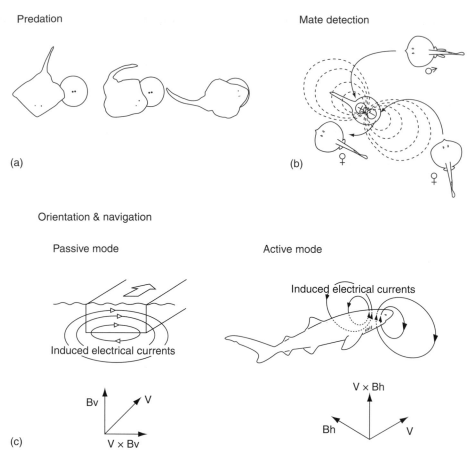

Figure 5 (a) Little skate, *Raja erinacea*, aroused by food odor, orienting to a DC dipole electrical field of 0.2 μV cm^{-1} at the perimeter of the test area (20 cm from electrodes buried in the sand). Images from video frames are in left to right order. (b) Mating stingrays, *Urolophus halleri*, in shallow water of the Sea of Cortez orient to a buried plastic model during playback of the low-frequency bioelectric field recorded from a female. Males approach, explore, and sometimes dig up buried models (as they do actual females) in an attempt to mate. Females also locate and approach the model and often bury next to it. (c) Diagrams illustrating motional electric fields available to elasmobranchs for navigation in the ocean. Large-scale fields associated with ocean streams may be used by the fish to set a heading with respect to the stream (passive mode), or the fish's own swimming movements within the earth's magnetic field create fields whose polarity and intensity depends on the fish's compass heading and velocity (active mode). Bh, Bv, earth's magnetic field horizontal, vertical vector; V, velocity vector. Reproduced from Bodznick D, Montgomery JC, and Tricas TC (2003). Electroreception: Extracting behaviorally important signals from noise. In: Collin SP and Marshall NJ (eds) *Sensory Processing in Aquatic Environments*, pp. 389–403. New York: Springer, with kind permission from Springer Science + Business Media.

especially early in development (**Figure 5**). However, only a few examples exist that illustrate the importance of electroreception in predator avoidance than in elasmobranchs. Elasmobranch embryos, pups, and juveniles are particularly susceptible to predation. Embryos of egg-laying elasmobranchs are naturally predated upon by teleosts, other elasmobranchs, and marine mammals. At this early stage in development, the stimulus invoking a predatory attack is induced by the embryo circulating water around the egg case alerting potential predators, which presumably use their mechanoreceptive lateral line to localize these stationary delicacies. Predation is potentially avoided by ceasing all ventilatory streaming, a freezing behavior elicited by the embryo's electrosense, which responds to sinusoidal electric fields between 0.5 and 1 Hz. This frequency band corresponds to the ventilatory pulses produced by large predators.

Pups and juvenile elasmobranchs also possess increased sensitivity (five times that of embryos) due to an increase in the ampullary canal length brought about by a twofold increase in disk size as in the skate, *Raja erinacea*, for example. This relationship between canal length and voltage sensitivity enhances the ability of these vulnerable stages to detect potential differences between the skin surface and the internal ampullary cluster in order to avoid predation. In the catfish, *Clarias gariepinus*, an increase in the number of ampullary receptors forms the basis of nearly a fourfold increase in sensitivity during the first four months of development, thereby providing a finer spatial resolution for the detection of an electrical stimulus in the form of a potential predator.

The Role of Bioelectric Stimuli in Social Behavior

The sensory specializations of peripheral electrosensory receptors enable prey detection by stimulation of individual ampulla by bioelectric stimuli at close range. Although few studies have concentrated on the role of electroreception on behavior, some studies in elasmobranchs have revealed that weak bioelectric fields can provide the stimulus for the localization of mates and conspecifics. By modulating the ionic potentials produced by the spiracles, mouth and gill slits during ventilation, buried female round stingrays (*Urolophus halleri*) can be located by actively searching males in the absence of any other sensory cues. Female rays also use this weak stimulus to locate buried consexuals (**Figure 5(a)** and **5(b)**). This ability is mediated in both skates (which can also encode the weak electric organ discharges produced by conspecifics during social and reproductive interactions) and stingrays, such as the Atlantic stingray, *D. sabina* (which do not possess electric organs for communication). The primary afferent neurons are most sensitive to stimuli that vary sinusoidally at the same frequency as the natural respiratory movements.

The Role of Electroreception in Migration and Geomagnetic Orientation

Behavioral studies have demonstrated that a range of animals, including representatives of all five vertebrate classes, are able to sense the earth's magnetic field and use it as an orientation cue while migrating, homing, or moving around their habitat. Relatively little is known, however, about the physiological mechanisms that underlie this sensory ability with numerous hypotheses proposed including processes mediated by electromagnetic transduction, a magnetite-based detection system, magnetic effects on melanophores, a physicochemical mechanism of optical pumping, and chemical reactions to provide a physiological magnetic compass. Yet despite these theoretical analyses, little direct neurobiological or anatomical evidence exists to support any of the proposed mechanisms. In no case yet have primary magnetoreceptors been identified with certainty.

Some animals are known to regularly migrate over short and long distances, where individuals return to favorable environmental conditions and/or a preferred habitat over a range of time frames. On a macroscale, white sharks, *Carcharodon carcharias*, are able to migrate up to 3800 km, showing a bimodal preference for depths of 0.5 and 300–500 m. On a microscale, some species migrate in and out of a region with changes in daily tidal flow or light–dark cycles, while others exhibit some sort of homing behavior, returning to particular regions on an annual or seasonal basis. These types of behaviors may be considered directed migrations, but true navigation or the ability of an animal to know its position in relation to some specific destination has not been demonstrated convincingly in elasmobranchs. However, a number of studies showing highly directional swimming suggest that these animals must be able to sense and react to a number of environmental factors, such as light irradiance, temperature, and directional landmarks such as ridges, valleys, and even the sun. Many of these orientational behaviors over long distances may not necessarily involve the electroreceptive sense (although it is often implicated) but may involve visual, olfactory, and/or lateral line input.

It is postulated that the electroreceptive system is the basis for geomagnetic orientation in elasmobranchs and possibly other animals. In the ocean, sharks and rays are exposed to electric fields resulting from two sources. (1) Electric fields produced by their own motion through the water in the presence of the earth's magnetic field, where the horizontal component of the animal interacting with the horizontal component of the magnetic field produces a vertical electromotive field and (2) electric fields associated with ocean streams and ionospheric circulation, where animals are thought to use the horizontal electric field produced by the interaction of the horizontal movement of the ocean stream with the vertical component of the earth's magnetic field. Type 1 fields may provide 'active electro-orientation,' where an animal could maintain a heading by avoiding any change in the electric field induced by a change in direction through the horizontal component of the earth's magnetic field. Type 2 fields may provide 'passive electro-orientation,' where an animal may ascertain its course, for example, its current speed and direction, by detecting the voltage induced by the motion of saltwater through the vertical component of the earth's magnetic field.

Ampullae of Lorenzini are considered unable to measure DC voltages and the induced voltage due to water flow in the ocean is not uniquely interpretable in terms of the speed and direction of flow at the point where the electrical measurement is made (**Figure 5(c)**). A recent theory suggests that the electric sense is used to determine a compass bearing as it swims by comparing the inputs of both the electroreceptors and the hair cells in the semicircular canals. According to this theory, the directional cue is the directional asymmetry of the change in induced electroreceptor voltage during turns.

Other studies suggest that the directional movement patterns tracked by some sharks are due to a type of 'topotaxis,' an orienting behavior to boundaries between different geomagnetic levels in sea-floor magnetization. The tracks made by some species of pelagic sharks actually match topographic features such as ridges and valleys, where at 100 m above the sea floor, geomagnetic intensity is known to vary by $1400\,\mathrm{nT\,m^{-1}}$ over a distance of 1 km. While sharks basking at the surface would swim in a straight line, while orienting to the earth's main magnetic field, individuals were shown to swim up and down in the water column would swim along more tortuous paths, orienting to local magnetic topography. Therefore, these sharks may possess a method of tracking according to geomagnetic intensity. Minerals such as oxides of iron and titanium in the earth's crust form patterns in association with seamounts and bands, providing the ability to detect geomagnetic gradients in sea-floor magnetization. A robust behavioral assay has recently been determined to reveal whether sharks detect magnetic fields and to measure their detection thresholds. Captive sharks are conditioned by pairing activation of an artificial magnetic field with presentation of food over a target. Conditioned sharks subsequently converge on the target when the artificial magnetic field is activated but no food reward is presented, thereby demonstrating that they are able to sense the altered magnetic field.

How could this gradient be detected? The electrosensory system is certainly a candidate where deposits of magnetic minerals or simply different bottom types may produce both local and regional electric fields that could facilitate orientational cues. However, an alternative mechanism may be mediated directly via a magnetite-based sensory system. Although single domain magnetite is not yet described in elasmobranchs, magnetite crystals have been identified in teleosts, where they lie within the olfactory epithelium. Using confocal microscopy, magnetite-containing cells in the nose of the trout are innervated by the ros V nerve, which is one branch of the fifth cranial nerve (the trigeminal). Electrophysiological recordings from this nerve have revealed units that respond to magnetic stimuli consisting of abrupt changes in field intensity.

Static geomagnetic fields have also been shown to affect the navigation abilities of birds, although magnetite crystals have again not been identified. Electrophysiological and behavioral data suggest the existence of two separate magnetoreceptor systems in birds. The first system, which is associated with the visual system, may provide directional (compass) information and is based on a series of chemical reactions. A retinal-based biochemical reaction takes place where the subsequent spin of electrons is affected by subtle changes in magnetic gradients. A subsequent reaction between a photoreceptor protein called 'cryptochrome' and oxygen produces a complicated entanglement of scattered electrons, which spin in different ways and are affected by changes in the earth's magnetic field. The second system is theoretically based on magnetite and is associated with branches of the trigeminal nerve. Therefore, birds might detect features of the earth's field in two different ways that can be used in assessing geographic position (map information).

See also: Active Electroreception: Vertebrates.

Further Reading

Bodznick D, Montgomery JC, and Tricas TC (2003) Electroreception: Extracting behaviorally important signals from noise. In: Collin SP and Marshall NJ (eds.) *Sensory Processing in Aquatic Environments*, pp. 389–403. New York: Springer-Verlag.

Brown BR (2002) Modeling an electrosensory landscape: Behavioral and morphological optimization in elasmobranch prey capture. *Journal of Experimental Biology* 205: 999–1007.

Bullock TH and Heiligenberg W (eds.) (1986) *Electroreception.* New York: John Wiley and Sons.

Kajiura SM (2001) Head morphology and electrosensory pore distribution of carcharanid and sphyrnid sharks. *Environmental Biology of Fishes* 61: 125–133.

Kalmijn AJ (1978) Electric and magnetic sensory world of sharks, skates and rays. In: Hodgson ES and Mathewson RF (eds.) *Sensory Biology of Sharks, Skates and Rays*, pp. 507–544. Washington, DC: Government Printing Office.

Kalmijn AJ (1982) Electric and magnetic field detection in elasmobranch fishes. *Science* 218: 916–918.

Klimley AP (1993) Highly directional swimming by scalloped hammerhead sharks, *Sphyrna lewini*, and subsurface irradiance, temperature, bathymetry, and geomagnetic field. *Marine Biology* 117: 1–22.

Lohmann KJ and Johnsen S (2000) The neurobiology of magnetoreception in vertebrate animals. *Trends in Neuroscience* 23: 153–159.

Newland PL, Hunt E, Sharkh SM, Hama N, Takahata M, and Jackson CW (2008) Static electric field detection and behavioural avoidance in cockroaches. *Journal of Experimental Biology* 211: 3682–3690.

Patullo BW and Macmillan DL (2007) Crayfish respond to electrical fields. *Current Biology* 17: R83–R84.

Pettigrew JD and Wilkens L (2003) Paddlefish and platypus: Parallel evolution of passive electroreception in a rostral bill organ. In: Collin SP and Marshall NJ (eds.) *Sensory Processing in Aquatic Environments*, pp. 420–433. New York: Springer-Verlag.

Proske U, Gregory JE, and Iggo A (1998) Sensory receptors in monotremes. *Philosophical Transactions of the Royal Society: Biological Sciences* 353: 1187–1198.

Raschi W (1986) A morphological analysis of the ampullae of Lorenzini in selected skates (Pisces, Rajoidei). *Journal of Morphology* 189: 225–247.

Solov'yov IA and Schulten K (2009) Magnetoreception through cryptochrome may involve superoxide. *Biophysical Journal* 96: 4804–4813.

Walker MM, Diebel CE, Haugh CV, Pankhurst PM, Montgomery JC, and Green CR (1997) Structure and function of the vertebrate magnetic sense. *Nature* 390: 371–376.

Zakon HH (1988) The electroreceptors: Diversity in structure and function. In: Atema JR, Fay R, Popper AN, and Tavolga WN (eds.) *Sensory Biology of Aquatic Animals*, pp. 813–850. New York: Springer-Verlag.

Emotion and Social Cognition in Primates

L. A. Parr, Yerkes National Primate Research Center, Atlanta, GA, USA

Introduction

Social cognition is a complex and multifaceted topic that can be broadly defined by what individuals know about their social environment. This knowledge can be defined and studied at many different levels, ranging from straightforward questions about social perception (how individuals recognize one another) to higher-level cognitive questions (what individuals know about minds of others – a concept called theory of mind). Human studies have focused on this latter view and as such, social cognition has become synonymous with theory of mind. Such an approach, however, is difficult for comparative studies because theory of mind cannot be directly measured using an observable behavioral analog. This article will focus on another critical, but often understudied, aspect of social cognition: how individuals visually communicate using facial expressions. It will review early ethological studies of primate facial expressions and their function in communicating motivational tendencies. It will describe how chimpanzees perceive and process facial expressions and will suggest similarities between some facial emotions in humans and chimpanzees. The goal is to provide an understanding of social cognition based on the interactions that occur between individuals in their social environment.

Social Cognition in Non-human Primates

The Psychological Approach

Most studies of social cognition in humans and other animals have taken a very high-level cognitive focus, asking specifically what individuals know about the intentions and beliefs of others – a concept referred to as theory of mind. This psychological perspective on social cognition implies that individuals are able to understand the mental states of others and then use this information to make predictions about future events. Humans do not achieve this skill until they are around 4 or 5 years old, well after the onset of basic language skills, as evidenced by the false-belief task. In this task, children have to answer a simple social puzzle by taking the perspective of the main character; for example, they are asked how this character will respond in a particular situation. Because solving the task requires this verbal instruction, the false-belief task has yet to be adapted for use with nonverbal organisms, such as non-human primates.

Instead, studies of social cognition in non-human primates have almost exclusively put animals in a situation where a dominant and subordinate individual compete for a food reward. In a typical setup, the subordinate chimpanzee will only attempt to access food when the dominant cannot, or has not, seen it. In these experiments, the measurable outcome is the proportion of trials in which the subordinate successfully takes the food, and the interpretation presumes that the subordinate knows what the dominant can and cannot see. Although the data from these studies are intriguing, their interpretation relies heavily on unobservable processes that lend chimpanzees the ability to mentally represent what others know, thus predicting how they will act, and then altering their own behavior accordingly. However, the ability to understand social behavior and respond flexibly to these complex and often unpredictable interactions, while being a hallmark of primate sociality, is not always best explained by high-level cognitive interpretations. Perhaps more importantly, by narrowly focusing on theory of mind as the standard for studying social cognition, researchers overlook other important social skills, such as visual communication, that are necessary for primates to navigate their social worlds. This is not to say that animal researchers should abandon their studies of theory of mind. On the contrary, it only suggests that this approach is not the only lens through which social cognition and social interactions should be studied in other species.

An Alternative, Behavioral Approach

The ability of non-human primates to navigate their complex social environment depends on a series of basic social skills. Individuals must, for example, be able to recognize each other and their relationship history: whether they are familiar or unfamiliar, relative or friend, alliance partner or rival, etc. These relationships are not only recognized with regard to oneself, but individuals must also be able to recognize and monitor the changing dynamics of each other's social relationships. To keep track of these relationships, individuals recognize specific behavioral patterns and social cues, how these cues are used, and their meaning within different social contexts. Importantly, individuals must effectively communicate their motivations, intentions, and emotions in ways that maximize their own benefits within the social community. Therefore, important building blocks for social cognition are a number of innate and acquired skills that visually orient individuals to meaningful social cues, such as faces, which enable the recognition of specific individuals, and facial expressions, which enable social communication and the expression of emotion.

While it is extremely unlikely that this type of social complexity could be explained by a series of innate or learned stimulus-response contingencies (e.g., cue x means y and only y), there is also little justification for imposing more high-level, cognitive explanations that invoke unobservable mental states. Moreover, some have argued that the attention focused on social cognition as a high-level, mentalistic ability, like theory of mind, only reflects our own inherent Cartesian biases about how the brain affects behavior. Just because primates live in social groups and have large brains does not confirm the presence of selection pressures to drive the evolution of advanced socio-cognitive strategies and counter-strategies, for example, the evolution of mentalizing, nor does it negate the fact that, in most cases, complex behavior can be explained by simpler rules. It has been argued, for example, that individuals (chimpanzees) can engage in seemingly complex social interactions by detecting regularities in ongoing behavior and then forming more abstract concepts, or general heuristics, from those behaviors. For example, witnessing conspecifics rapidly orienting, through both gaze and posture, to a specific location can lead to a general heuristic about the presence of something important at that location, not and presumption that those individuals know what is there. Ultimately, the behavioral approach allows social cognition to be studied in terms of observable, distributed events, like the use of facial expressions during social interactions, rather than unobservable, private ones. This type of behavioral flexibility is a defining feature of primate social complexity and is particularly relevant for emotional communication, as will be described later.

Facial Expressions in Primates

Early ethological studies in non-human primates catalogued facial expressions in terms of their appearance, antecedent conditions, and potential function in beautiful objective detail. Facial expressions were identified as important social displays comprised of elemental movements, controlled by the facial musculature and embellished by features of the head and face, such as the scalp, eyes, eyelids, ears, lips, jaw, and mouth corners. It was also noted that facial expressions were not random assortments of independent movements, but rather some movements occurred together more frequently than others. However, the detailed and objective focus of these early ethologists led them to concentrate on the individual movements as the most useful unit of analysis. In fact, it was only a few years later that human researchers began studying

universal facial expressions of emotion among different cultural groups and devised an intricately detailed and clever coding system (the facial action coding system (FACS)) to describe facial expressions according to the appearance changes produced by the underlying facial musculature. Thus, among humans and non-human primates, the muscular basis for facial expressions was seen as the basic unit of analysis providing the most accurate and objective descriptions of the behavior.

Next, in order to understand whether these expressions functioned as communicative displays, researchers focused on when specific patterns of facial movements/ expressions were likely to be produced. If expressions were produced in predictable social contexts, then they had the potential to serve important communicative functions and thus influence group members in meaningful ways. A particular facial configuration used just prior to an attack, for example, could be interpreted by group members as conveying an increased likelihood of attack, using a mechanism similar to the behavioral heuristics described earlier. Therefore, this expression provided important information about the future motivation of individuals who use the expression based on the general regularity of its consequences. This was an important distinction for the field of primate communication as previously researchers simply used their subjective intuition, based on their own expertise with a particular species, to determine the meaning of specific behaviors. These anthropocentric interpretations were fraught with difficulty not only because people form different opinions, but they failed to provide any objective or descriptive terminology for comparing these behaviors across situations and species. Moreover, humans cannot help but to interpret behavior using their own psychological constructs which, as mentioned earlier, do not necessarily translate to other species, no matter how similar the behaviors are in appearance. Humans, for example, automatically interpret facial expressions emotionally and have a very difficult time overcoming these projections even in other species, much to the profit of the advertising industry. Objectively identifying specific movement patterns, however, removes any need for anthropocentric or context-dependent interpretation. This is what led researchers studying cross-cultural emotions to develop the FACS, because an objective quantifiable comparative tool, similar facial expressions across cultural groups would automatically be interpreted using emotion labels, when whether they communicated the same emotion was the question of study.

The ability to compare similar facial configurations across species and social contexts has provided a much more detailed and accurate understanding of the homology of these displays. An example can be made using the primate bared-teeth display (see **Figure 1**). This expression was observed to be given by subordinate monkeys

Figure 1   An illustration of the bared-teeth display in a human, chimpanzee, and rhesus macaque.

after a fight, leading to its characterization as a submissive grin, or colloquially the fear grin. After more careful study in different species, the same expression appears to be used in numerous other contexts, such as during play or when approaching others in a friendly manner, and its meaning appears to have undergone considerable phylogenetic transformation. Among some species of macaques, such as the rhesus monkey, the bared-teeth display is used almost exclusively by recipients of aggression and appears to signal submission, as noted earlier. However, among macaque species with a more egalitarian social organization than the rhesus monkey, such as the Barbary macaque, the expression is also used during greetings and other affiliative contexts and appears to have an overall reassuring function. No longer limited to the context of submission, the bared-teeth display appears to convey something more like benign intent, or 'I mean no harm.' Among humans, the bared-teeth display has been proposed to be homologous to the human smile, also serving a reassuring and appeasing function and exhibiting considerable behavioral flexibility in its use. Thus, researchers can gain a better understanding of the meaning of facial displays by having a broad and detailed understanding of how, when, and in what species they are produced. Recently, the importance of multimodal cues, i.e., the combination of facial, gestural, and vocal cues, has been shown to provide more context dependent meaning and elaborate on an already sophisticated communication system.

After describing primate facial expressions in great objective detail, early ethologists went on to conduct structural analyses of chimpanzee behaviors, including facial expressions. These analyses sort behaviors into specific factors representing statistical regularities in their appearance, which can later be interpreted by the researcher. Behaviors grouped by the same factor must be associated with the same motivations and tendencies and thus communicate similar meaning. For chimpanzees, the factors that emerged were descriptively named affinitive, play,

aggressive, submissive, and excitement. The researcher can then identify the particular facial expressions that are the most strongly associated with each factor. The silent-bared teeth display was most strongly associated with the affinitive category, not the submissive category. More interestingly was that this analysis revealed functional differences associated with several subtypes of the bared-teeth display that contained subtle differences in lip retraction, some associated more strongly with the submissive factor, while others were grouped with the affinitive context. Using these procedures, the motivation behind these expressions can be described objectively using emotional terms and tied to a more general type of socioemotional context, rather than a specific situation.

Homology with Human Emotional Expressions

To understand the evolution of specific behaviors, comparative studies are needed and this requires a standardized system of description and measurement. The FACS represents the gold standard for measuring facial movement in humans, so it should be no surprise that researchers have adapted the FACS for use with other species, both chimpanzees (ChimpFACS) and rhesus macaques (MaqFACS), following similar guidelines used to develop FACS. This involved first, confirming the presence of similar facial muscles, or mimetic muscles, across different phylogenetic groups of primates. With very few contemporary studies in this area, researchers have confirmed that non-human primates share a very similar pattern of mimetic facial muscles, refuting earlier ideas that both the number and control of mimetic muscles had become elaborated throughout phylogeny, with humans showing the most advanced repertoire of facial movement and expressions. Next, the way in which the contraction of specific muscles functioned to change the appearance of the face was identified using intramuscular stimulation studies, and then these actions were labeled and assigned various criteria for coding. In total, 43 movements were described for the chimpanzee, 17 of which relate to specific facial muscles and the remaining are miscellaneous action descriptors, such as head and eye movements. Interestingly, some movements common in humans, such as the knitting of the brow caused by contraction of the corrugator and associated muscles, were never observed in the chimpanzee. So, differences were present but overall, remarkable similarity was demonstrated between chimpanzees and humans.

As a first use of the ChimpFACS, researchers set out to confirm the basic categories of chimpanzee facial expressions by examining their pattern of movements. Over 250 photographs of chimpanzee facial expressions were coded using ChimpFACS, for example, by noting the presence of a specific component movement. These were also labeled using descriptions published in existing ethograms of chimpanzee facial behavior. A discriminant functions analysis was then used to calculate the agreement between these two classification types. This revealed a 70% agreement between the a priori category labels and the common movement configurations coded using ChimpFACS. Furthermore, by selecting specific expression with the highest percentage of agreement (>90%), it was possible to identify the most prototypical movement combinations for each expression category. **Table 1** illustrates some of these prototypical expression categories.

Of interest for homology is whether any of these expressions resemble human emotional expressions, so the latter are also plotted along with their specific movement configurations. Many of the matching human expressions are not identical to those described by Paul Ekman as basic facial emotions (anger, fear, happiness, sadness, surprise, and disgust), but represent some variation of these. Sadness may be most similar as both young human infants and chimpanzees have pout faces used in very similar situations. Happiness, conveyed best by laughter, is thought to be homologous to the chimpanzee play face, but in humans, the upper teeth are fully exposed while the chimpanzee play face conspicuously covers the upper teeth. Note that chimpanzee play faces also have a distinct laughter-type vocalization. Smiling, which can also convey happiness, shares most in common with the bared-teeth face and based on the earlier discussion, at least some versions of this facial expression are best characterized as appeasing/reassuring in both chimpanzees and humans. Anger shares some features in common with the chimpanzee bulging-lip face, although chimpanzees do not show the brow knitting that is a characteristic of human anger. Anger can also be conveyed by screaming, as in strong protests or rage, and chimpanzee screams are largely expressions of protest as opposed to real terror or pain. Although these are only semiquantitative comparisons, they provide a clear visual representation of how similar many of the facial expressions are between

chimpanzees and humans, enabling informed assumptions about their underlying emotional content.

Communication and Cognition

Categorization of Facial Expressions

Also important for social cognition is how different species perceive and categorize social signals, like facial expressions. The signals must be easily distinguished from one another or they would fail to have any predictive function. Seemingly at odds with this prediction is the well-established finding that primate, particularly chimpanzee, facial expressions are highly graded, both in terms of their form, the exact movements involved in the display, and their overall intensity. Some researchers have even noted that chimpanzees have some expressions that appear to be blends, showing features in common with several different types and that these expressions can be used in many different contexts. Explaining how chimpanzees perceive and process these highly variable signals would be difficult based on cue-response contingency learning. By invoking the concept of behavioral/cognitive flexibility, however, regularities in the use of these specific movement patterns can come to communicate general tendencies or motivations that can then be parsed in to more detailed meaning by incorporating additional factors, such as actor, environment, and antecedent conditions. Thus, the concept of behavioral/cognitive flexibility nicely fits with existing data on emotional communication. At this stage, perceptual discrimination studies in chimpanzees have only utilized the most prototypical examples of facial expressions, like those shown in **Table 1**. However, it is very plausible that graded signals can provide meaning to group members mentioned earlier and beyond that related to motivational intensity. Small changes in expressions or blends between expressions may alter meaning in completely new and, as yet, unstudied ways.

Parr et al. (1998) were the first to examine facial expression discrimination in chimpanzees using a computerized, joystick-testing paradigm. According to this procedure, subjects first must orient to a sample stimulus, for example, a facial expression, by contacting it with a joystick-controlled cursor. Then, two comparison images appear on the screen, one showing another example of the same expression type as the sample and the other showing a neutral face (see **Figure 2**). In these experiments, all individuals were different, so subjects could not match based on individual identity. Of the five expression categories presented – bared-teeth display, pant hoot, scream, relaxed lip face, and the play face – subjects learned to discriminate all but the relaxed-lip and the neutral face. This was interesting as these two faces are emotionally-neutral, but the relaxed-lip contains the distinctive droopy lower lip.

In an effort to understand the role of unique and distinctive features in expression discrimination, such as mouth position, visibility of the teeth, staring eyes, etc.,

Table 1   A comparison of chimpanzee facial expression prototypes and similar emotional expression in humans

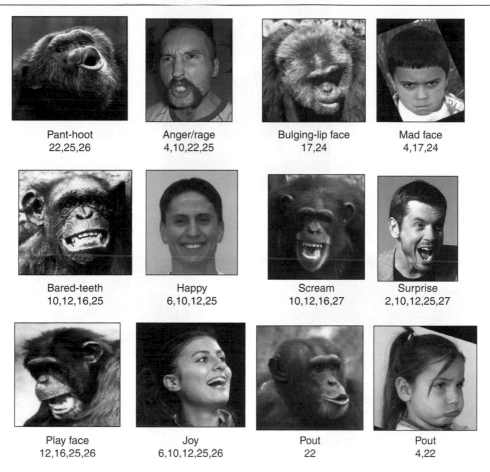

Pant-hoot 22,25,26	Anger/rage 4,10,22,25	Bulging-lip face 17,24	Mad face 4,17,24
Bared-teeth 10,12,16,25	Happy 6,10,12,25	Scream 10,12,16,27	Surprise 2,10,12,25,27
Play face 12,16,25,26	Joy 6,10,12,25,26	Pout 22	Pout 4,22

Numbers refer to action unit codes according to FACS and ChimpFACS (see text for details).

a follow-up experiment was performed in which each expression category was paired with every other, so every pair-wise combination of expression types was represented. Performance discriminating these dyads was then correlated with the number of features shared between the two expression types, such as mouth open, teeth visible, etc. If performance was based on the detection of specific features, like a droopy lip, then expression dyads that shared these features would result in poorer performance than expression dyads that had no overlap in features. This hypothesis was supported for only some expression types – the bared-teeth display, pant-hoot, and relaxed-lip face – but not others, such as the play face or scream. Thus, it was concluded that the features of some expressions are more salient than others.

After the development of the ChimpFACS, this experiment was replicated using a standardized set of chimpanzee facial expressions generated using a three-dimensional chimpanzee model (see **Figure 3**). In this way, all facial expressions of the same category contained identical movements and each was standardized to its peak intensity. The pair-wise matching task was then replicated and the pattern

of errors made by subjects was analyzed using a multidimensional scaling analysis. This creates a graph of how similar or different each expression is to each other. Those that cluster together are perceived as highly similar, while more dispersed expressions are perceived as more distinct. Then, the dimensions of the plot can be interpreted based on the features of each expression. One dimension appeared to reveal the degree of vertical mouth opening or closing and the other was the degree of horizontal mouth opening with lips either being retracted (as in the bared-teeth display) or puckered as in the pout.

In a slight modification of the task, subjects were required to match each expression prototype by selecting only one of its individual component movements, thus asking what individual movement, if any, was the most representative of each expression prototype. For example, the bared-teeth display is comprised of three movements, one raising the upper lip (action unit (AU) 10), one lowering the lower lip (AU16) and one retracting the lips back (AU12). These were combined into three trials, pairing the bared-teeth prototype with each combination of these three movements (see **Figure 4**). Subjects showed

Figure 2   An example of the matching-to-sample task used to study expression categorization in chimpanzees. The image on top is the sample, the one to match. The correct choice is the matching expression (lower left).

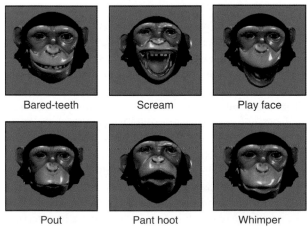

| Bared-teeth | Scream | Play face |
| Pout | Pant hoot | Whimper |

Figure 3   An example of standardized, prototypical chimpanzee facial expressions created using 3D animation software.

a clear preference for which movement was most representative of each prototype and, moreover, this single movement explained most of the errors found in the previous cluster plot. Most expressions were relatively easy to

discriminate and little confusion occurred, but when subjects did have trouble it was not because the two expressions in the pair shared the most number of features in common, as was the earlier hypothesis, it was that the two expressions shared their most salient movement in common. It is perhaps these salient movements that are involved in a general pattern recognition heuristic and provides chimpanzees with an ability to categorize facial expressions into basic groups, despite the degree of blending or grading, the meaning of which can be elaborated on by other factors, like context.

Conclusion: The Role of Emotion in Social Cognition

This article has argued that social cognition is best understood through detailed and objective measurements of observable behavior, such as facial expressions, with an appreciation for the dynamic environment in which these signals take place. By comparing how these emotional signals look and function in related species, we can

| Bared-teeth display | AU10 | AU12 | AU16 |

Figure 4 The Poser chimpanzee model, showing the bared-teeth display prototype and each individual component movement it contains. Reproduced from Parr LA, Waller BM, and Heintz M (2008) Facial expression categorization by chimpanzees using standardized stimuli. *Emotion* 8: 216–231, with permission from · · ·.

begin to understand their evolutionary origins, thus better understanding their overall function and meaning. What has been learned is that the primate social environment is extremely complex and most behavioral interactions do not lend themselves to interpretation based on learned or innate cue-response contingencies, nor do they warrant rich interpretations based on unobservable mental states. Instead, it is suggested that primates interpret their social interactions using more flexible behavioral heuristics.

Primates have the ability to recognize specific behavioral cues, such as faces and facial expressions, performed in a variety of social contexts, providing them with an understanding of the basic motivational states of others. This is then filtered through an assessment of other factors, such as social context and previous experience, giving more detailed information about the overall emotional meaning of the signal, and what the sender is likely to do. As a result, the cognitive and emotional state of others can be interpreted and maybe even predicted, not from mind reading, but from the perception of the others' behaviors as it occurs dynamically within a specific environment.

The dynamic nature of social communication creates an even more complicated scenario, as these interactions occur rapidly and can change course frequently, necessitating behavioral adjustments based on ongoing processing. Facial expressions do not appear in a simple sequence: behavior a, then expression b, leads to behavior c, interpreted as expression b functions to change behavior from a to c. Rather, facial expressions are embedded within dynamic social interactions that takes the form of multiple feedback loops, dependent on the reaction of the social partner and changing in ongoing needs. To understand the real essence of social cognition, one must study these interactions within the distributed cognitive system in which they occur, where behaviors do not occur in linear cause and effect sequences, but rather communication emerges within a structure of a dynamic and ever changing society.

See also: Cognitive Development in Chimpanzees; Distributed Cognition; Empathetic Behavior; Interspecific Communication; Social Cognition and Theory of Mind; Social Recognition; Visual Signals.

Further Reading

Andrew RJ (1963) The origin and evolution of the calls and facial expressions of the primates. *Behaviour* 20: 1–109.

Barrett L, Henzi P, and Rendall D (2007) Social brains, simple minds: Does social complexity really require cognitive complexity? *Philosophical Transactions of the Royal Society B* 362: 561–575.

Burrows AM (2008) The facial expression musculature in primates and its evolutionary significance. *BioEssays* 30: 212–225.

Dunbar RIM (1988) *Primate Social Systems*. London: Chapman & Hall.

Parr LA, Cohen M, and de Waal FBM (2005) The influence of social context on the use of blended and graded facial displays in chimpanzees (*Pan troglodytes*). *International Journal of Primatology* 26: 73–103.

Parr LA, Hopkins WD, and de Waal FBM (1998) The perception of facial expressions in chimpanzees (*Pan troglodytes*). *Evolution of Communication* 2: 1–23.

Parr LA, Waller BM, and Heintz M (2008) Facial expression categorization by chimpanzees using standardized stimuli. *Emotion* 8: 216–231.

Parr LA, Waller BM, Vick SJ, and Bard KA (2007) Classifying chimpanzee facial expressions using muscle action. *Emotion* 7: 172–181.

Povinelli DJ and Vonk J (2003) Chimpanzee minds: Suspiciously human? *Trends in Cognitive Sciences* 7: 157–160.

Preston S and de Waal FBM (2002) Empathy: Its ultimate and proximate bases. *Behavioral and Brain Sciences* 25: 1–72.

Preuschoft S (1992) 'Laughter' and 'smile' in Barbary macaques (*Macaca sylvanus*). *Ethology* 91: 220–236.

Tomasello M, Call J, and Hare B (2003) Chimpanzees understand psychological states-the question is which ones and to what extent. *Trends in Cognitive Science* 7: 153–156.

van Hooff JARAM (1967) The facial displays of the Catarrhine monkeys and apes. In: Morris D (ed.) *Primate Ethology*, pp. 7–68. Chicago, IL: Aldine.

van Hooff JARAM (1972) A comparative approach to the phylogeny of laughter and smiling. In: Hinde RA (ed.) *Nonverbal Communication*, pp. 209–241. Cambridge: Cambridge University Press.

Waller B and Dunbar RIM (2005) Differential behavioural effects of silent bared teeth display and relaxed open mouth display in chimpanzees (*Pan troglodytes*). *Ethology* 111: 129–142.

Relevant Websites

http://userwww.service.emory.edu/~lparr/index.html – Dr Lisa Parr's website.

www.chimpfacs.com – The ChimpFACS website.

http://www.do2learn.com/games/facialexpressions/face.htm – Do2Learn-Facial expressions.

Empathetic Behavior

F. B. M. de Waal, Emory University, Atlanta, GA, USA

Definitions of empathy commonly emphasize two aspects, which is the sharing of emotions and the adoption of another's perspective. The latter, cognitive aspect remains controversial for many species, but the first, emotional aspect is hard to deny, and was already recognized by Darwin in *The Descent of Man*. "... many animals certainly sympathize with each other's distress or danger."

Empathy allows the organism to quickly relate to the states of others, which is essential for the regulation of social interactions, coordinated activity, and cooperation toward shared goals. Even though perspective-taking is often critical, it is a secondary development. This is even true for our own species, as Hoffman noted: "humans must be equipped biologically to function effectively in many social situations without undue reliance on cognitive processes."

The selection pressure to evolve rapid emotional connectedness likely started in the context of parental care. Signaling their state through smiling and crying, human infants urge their caregiver to come into action, and equivalent mechanisms operate in other animals in which reproduction relies on feeding, cleaning, and warming the young. Offspring signals are not just responded to but induce an agitated state, suggestive of parental distress at the perception of offspring distress. Avian and mammalian parents alert to and affected by their offspring's signals must have out-reproduced those that remained indifferent.

Once the empathic capacity existed, it could be applied outside the rearing context and play a role in the wider network of social relationships. The fact that mammals retain distress vocalizations into adulthood hints at the continued survival value of care-inducing signals. For example, primates often lick and clean the wounds of conspecifics, which is so critical to healing that migrating adult male macaques have been observed to temporarily return to their native group, where they are more likely to receive this service.

One of the first experimental studies of animal empathy was Church's (1959) study entitled '*Emotional Reactions of Rats to the Pain of Others*.' Having trained rats to obtain food by pressing a lever, Church found that if a rat pressing the lever perceived that another rat in a neighboring cage receives a shock from an electrified cage floor, the first rat would interrupt its activity. Why should this rat not continue to acquire food? The larger issue is whether rats that stopped pressing the lever were concerned about their companions or just fearful that something aversive might also happen to themselves.

Church's work inspired a brief flurry of research during the 1960s that investigated concepts such as 'empathy,' 'sympathy,' and 'altruism' in animals. This included studies of monkeys, which showed a much more dramatic empathy response than rats. Monkeys will for many days refuse to pull a chain that delivers food to them if doing so delivers an electric shock to a companion. In order to avoid accusations of anthropomorphism, however, authors often placed the topic of their research in quotation marks, and their studies went largely ignored in ensuing years.

In the meantime, human empathy became a respectable research topic. In the 1970s began studies of empathy in young children, in the 1980s in human adults, and finally, in the 1990s, neuroimaging of humans watching others in pain, distress, or with a disgusted facial expression. Mirror neurons are commonly invoked to explain human empathy, but despite the fact that these neurons were discovered not in humans, but in macaques, animal empathy research has lagged.

Half a century after Church's study, however, there is a revival of interest in animal empathy and a basic mechanism common to humans and other animals has been proposed. Accordingly, seeing another in a given situation or display certain emotions reactivates neural representations of when the subject was itself in similar situations or experienced similar emotions, which in turn generates a bodily state resembling that of the object of attention. Thus, seeing another individual's pain may lead the observer to share the bodily and neural experience. These reactions are so automatic and instantaneous that it is not unusual for humans to shout 'ouch!' while watching a child scraping its knee.

The Perception–Action Mechanism (PAM) of Preston and de Waal manifests itself very early in human life, such as when newborns cry contagiously, and is increasingly suggested for other animals. Examples range from involuntary facial mimicry during play in orangutans, yawn contagion in both primates and canines, to heart-rate increases in geese while watching their mated partner in a fight. Physiological continuity between the ways humans and chimpanzees process emotional stimuli is suggested by the apes' drop in skin temperature while watching aversive images as well as human-like lateralized changes in brain temperature.

Emotional Contagion in Mice

Langford and colleagues put pairs of mice through a so-called 'writhing test.' In each trial, two mice were placed in two transparent Plexiglas tubes such that they were able to see one another. Either one or both mice were injected with diluted acetic acid, known to cause a mild stomach-ache. Mice respond to this treatment with characteristic writhing movements. The researchers found that an injected mouse would show more writhing if its partner was writhing, too, than it would if its partner had not been injected. Significantly, this applied only to mouse pairs that were cage mates.

Male (but not female) mice showed an interesting additional phenomenon while witnessing another male in pain: its own pain sensitivity actually dropped. This counter-empathic reaction occurred only in male pairs that did not know each other, which are probably also the pairs with the greatest degree of rivalry. Was that rivalry suppressing their reaction, or did they actually feel less empathy for a strange rival?

Finally, Langford and colleagues exposed pairs of mice to different sources of pain – the acetic acid as before and a radiant heat source. Mice observing a cage mate writhing because of the acid injection withdrew more quickly from the heat source. In other words, their reactions could not be attributed to mere motor imitation, but involved emotional contagion, because seeing a companion react to pain caused sensitization to pain in general.

Consolation Behavior

A well-studied primate response to others' distress is so-called *consolation* behavior, that is, friendly, reassuring contact by an uninvolved bystander toward a distressed party, such as the loser of a fight (**Figure 1**). Similar behavior in children is typically attributed to *sympathetic concern*. That consolation serves to alleviate distress is suggested by the finding that such contact is directed more at recipients of aggression than at aggressors, and more at recipients of intense than mild aggression (**Figure 1**).

Other studies have confirmed consolation behavior in chimpanzees, gorillas, and bonobos as well as canines and corvids. However, when de Waal and Aureli set out to apply the same observation protocol to detect consolation in monkeys, they failed to find any, as did others. The consolation gap between monkeys and the Hominidae (i.e., humans and apes) extends even to the one situation where one would most expect consolation to occur: macaque mothers fail to comfort their own offspring after the receipt of aggression. Content analysis of hundreds of reports confirms that reassurance of distressed parties is typical of apes yet uncommon in monkeys.

After initial failures to find effects of consolation among chimpanzees, a comprehensive analysis demonstrated that the rate of self-directed behavior (i.e., self-scratching and self-grooming) is elevated following aggressive conflicts, but drops significantly as soon as individuals receive

Figure 1 Consolation behavior is common in chimpanzees, and functions to reassure distressed parties. A juvenile puts an arm around a screaming adult male who has just been defeated in a fight with a rival. Photograph by Frans de Waal.

consolation. Self-directed behavior serves as an index of anxiety, hence these observations suggest that consolation effectively counters anxiety induced by agonistic conflict.

The early primate literature contains many striking qualitative accounts of consolation among apes. In 1925, Yerkes reported how his bonobo was so concerned about his sickly chimpanzee companion, Panzee, that the scientific establishment might not accept his claims: "If I were to tell of his altruistic and obviously sympathetic behavior towards Panzee I should be suspected of idealizing an ape." Ladygina-Kohts noticed similar empathic tendencies in her young chimpanzee, Joni, which she raised at the beginning of the previous century. Kohts, who analyzed Joni's behavior in the minutest detail, discovered that the only way to get it off the roof of her house after an escape (much better than holding out a reward) was to appeal to his sympathy: 'If I pretend to be crying, Joni immediately stops his plays or any other activities, quickly runs over to me . . . he tenderly takes my chin in his palm, lightly touches my face with his finger, as though trying to understand what is happening.'

Empathic Perspective-Taking

In 1996, I suggested that apart from emotional connectedness, apes have an appreciation of the other's situation. Psychologists usually speak of empathy only if it involves such understanding combined with the adoption of the other's point of view. Thus, one of the oldest and best-known definitions stresses the 'changing places in fancy' with the sufferer. The prevailing view of empathy as a cognitive affair dependent on imagination and mental state attribution explains the occasional skepticism about non-human empathy. But perspective-taking by itself is, of course, hardly empathy: it is so only in conjunction with emotional engagement.

Menze was the first to investigate whether chimpanzees understand what others know, setting the stage for studies of non-human theory-of-mind and perspective-taking. After several ups and downs in the evidence, current consensus seems to be that apes, but perhaps not monkeys, show perspective-taking both in their spontaneous social behavior and under controlled experimental conditions.

One manifestation of empathic perspective-taking is so-called 'targeted helping,' which is help fine-tuned to another's specific situation. For an individual to move from emotional sensitivity toward an explicit other-orientation requires a shift in perspective. The emotional state induced in oneself by the other now needs to be attributed to the other instead of the self. A heightened self-identity allows a subject to relate to the object's emotional state without losing sight of the actual source of this state. The required self-representation is hard to establish independently, but one common avenue is to gauge reactions to a mirror. The *coemergence hypothesis* predicts that mirror self-recognition (MSR) and advanced expressions of empathy appear together in both ontogeny and phylogeny.

Ontogenetically, there exists compelling evidence for the coemergence hypothesis. Bischof-Köhler found that the relation between MSR and the development of complex forms of empathy holds even after controlling for age. Gallup was the first to propose phylogenetic coemergence, a prediction empirically supported by the contrast between monkeys and apes, with compelling evidence for both MSR, consolation, and targeted helping only in the apes.

Apart from the great apes, the animals for which we have the most striking accounts of consolation and targeted helping are cetaceans and elephants, which is why the coemergence hypothesis predicts MSR in these taxa. The mark test, in which an individual uses a mirror to locate a mark on itself that it cannot see without a mirror, has now tentatively confirmed this prediction for dolphins and elephants. MSR is believed to be absent in all other mammals. The coemergence hypothesis may have its own neural correlate, that is, the presence of Von Economo Neurons, or VEN cells. These special neurons have been found only in the brains of Hominoids (not other primates), cetaceans, and elephants.

In the future, we may be able to address the self–other distinction more directly through neural investigation. In humans, the right inferior parietal cortex at the temporoparietal junction helps distinguish between self- and other-produced actions, and this distinction may be critical to full-blown empathy.

Altruism and Prosocial Behavior

The common claim that humans are the only truly altruistic species, since animals are driven by return-benefits, assumes that animals not only engage in reciprocal exchange, but do so with a full appreciation of how this will ultimately benefit themselves. However, return-benefits generally remain beyond the cognitive horizon of animals, that is, occur too distantly in time to be linked to the original act. Since animals cannot be motivated by future events that they cannot predict, intentionally selfish altruism involves highly speculative (and as yet unproven) assumptions.

When animals alert others to an outside threat, work together for immediate self-reward, or vocally attract others to discovered food, biologists may speak of altruism or cooperation, but such behavior is unlikely to be motivated by empathy with the beneficiary. One category, however, termed *directed altruism* (i.e., altruistic behavior aimed at others in need, pain, or distress), is traditionally explained as a product of empathy in humans, and may share the same proximate causation in animals.

Apart from evidence for consolation behavior (seen earlier), there exists a rich literature on primate support in aggressive contexts, costly cooperation, and food-sharing. Primates engage in the latter even when separated by bars, hence while protected from physical pressure to share. Emotional activation is often visible in the facial expressions and vocalizations of both altruists and beneficiaries. Empathy is a perfect candidate mechanism for directed altruism, since it provides a unitary explanation for a wide variety of situations in which assistance is dispensed according to need. Perhaps confusingly, the mechanism is relatively autonomous in both animals and humans. Thus, empathy often reaches beyond its original evolutionary context, such as when people send money to complete strangers, primates bestow care on unrelated orphans, or a bonobo saves an injured bird.

Experimentation on risky altruism is ethically problematic, yet there are increasingly experiments on low-cost altruism, also known as 'other-regarding preferences.' A typical paradigm is to offer one member of a pair the option to either secure food for itself or food for both itself and a companion. Recent experiments in my lab have shown well-developed prosocial tendencies in monkeys, such as marmosets and capuchins. In these studies, the role of reward could be ruled out in that the prosocial choices were made even if they delivered no extra rewards compared with selfish choices, and one study ruled out punishment in that the least vulnerable parties (i.e., dominants) turned out to be the most prosocial. These results are explained more easily by empathy with another's situation than cost/benefit calculations.

With regard to chimpanzees, the same tendencies have proved harder to establish until a series of experiments by

Warneken and colleagues yielded an outcome in line with the overwhelming observational evidence for spontaneous aiding behavior in this species.

Empathy as Evolved Proximate Mechanism

The PAM model predicts that the more similar or familiar subject and object, the more their neural representations will agree, and hence the more accurate their state-matching. Generally, the empathic response is amplified by similarity, familiarity, social closeness, and positive experience with the other (see Table 1 in Preston and de Waal's 2002 paper). In studies in humans, subjects empathize with a confederate's pleasure or distress if they perceive the relationship as cooperative, yet show a counter-empathic response – also known as *Schadenfreude* – if they perceive the relationship as competitive. These effects of previous experience have been confirmed by fMRI research: seeing the pain of a cooperative confederate activates pain-related brain areas, but seeing the pain of an unfair confederate activates reward-related brain areas, at least in men.

Relationship effects are also known for rodents, in which emotional contagion is measurable between cage mates, but not strangers. In monkeys, empathic responses to another's fear or pain are enhanced by familiarity between subject and object, and prosocial tendencies in capuchin monkeys vary with social closeness, being strongest between partners that spend most time together. This means that empathy and prosocial tendencies are biased the way evolutionary theory would predict. Empathy is (1) activated in relation to those with which one has a close or positive relationship and (2) suppressed, or even turned into its opposite, in relation to strangers and defectors. The latter, retaliatory aspect is well-documented in chimpanzees, which show both reciprocation of favors within positive relationships and a tendency to square accounts with those that have acted against them.

A common way in which mutually beneficial exchanges are achieved is through investment in long-term bonds to which both parties contribute. This reciprocity mechanism is commonplace in non-human primates, and has been suggested for human relations as well. Individual interests may be served by partnerships (e.g., marriages, friendships) that create a long-lasting communal 'fitness interdependence' that flows from mutual empathy. Within close human relationships, partners do not necessarily keep careful track of which did what for which, as also indicated for reciprocal exchange among chimpanzees.

Conclusion

Empathy covers all the ways in which one individual's emotional state affects another's, with simple mechanisms

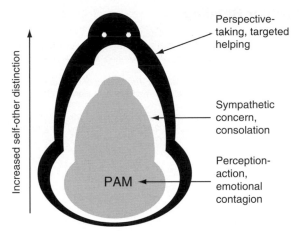

Figure 2 The Russian Doll model of empathy and imitation. Empathy induces a similar emotional state in the subject as the object, with at its core the perception-action mechanism (PAM). The doll's outer layers, such as sympathetic concern and perspective-taking (which includes understanding the reasons for another's emotions), build upon this hard-wired socio-affective basis. Even though the doll's outer layers depend on prefrontal functioning and an increasing self-other distinction, they remain fundamentally linked to its core.

at its core and more complex mechanisms and perspective-taking abilities as its outer layers. Because of this layered nature of the capacities involved, we speak of the Russian Doll Model, in which higher cognitive levels of empathy build upon a firm, hard-wired basis, such as the PAM (**Figure 2**). The claim is not that PAM by itself explains sympathetic concern or perspective-taking, but that it underpins these cognitively more advanced forms of empathy, and serves to motivate behavioral outcomes. Empathy may provide the main motivation that makes individuals that have exchanged benefits in the past to continue doing so in the future. Instead of the cognitively demanding assumption of cost–benefit calculations and learned expectations to explain such behavior, the assumption here is one of empathy-based altruism mediated by bonding and emotional sensitivity. It is summarized in the following conclusions:

1. Evolutionary parsimony assumes similar motivations to underlie directed altruism in humans and other animals.
2. Consistent with kin selection and reciprocal altruism theory, empathy is biased toward close, familiar individuals and previous cooperators, and biased against defectors.
3. Empathy, broadly defined, may characterize all mammals and birds.

See also: Conflict Resolution; Emotion and Social Cognition in Primates.

Further Reading

Bischof-köhler D (1988) Uber den Zusammenhang von Empathie und der Fäshigkeit sich im Spiegel zu erkennen. *Schw. Z. Psychol.* 47: 47–159.

de Waal FBM (1996) *Good Natured: The Origins of Right and Wrong in Humans and Other Animals*. Cambridge, MA: Harvard University Press.

de Waal FBM (2008) Putting the altruism back into altruism: The evolution of empathy. *Annual Review of Psychology* 59: 279–300.

Decety J and Jackson PL (2006) A social-neuroscience perspective on empathy. *Current Directions in Psychological Science* 15: 54–58.

Gallup GG (1982) Slef-awareness and the emergence of mind in primates. *American Journal of Primatology* 2: 237–248.

Langford DJ, Crager SE, Shehzad Z, et al. (2006) Social modulation of pain as evidence for empathy in mice. *Science* 312: 1967–1970.

Menzel EW (1974) A group of young chimpanzees in a one-acre field. In: Schrier AM and Stollnitz F (eds.) *Behavior of Non-Human Primates,* Vol. 5, pp. 83–153. New York: Academic Press.

Preston SD and de Waal FBM (2002) Empathy: Its ultimate and proximate bases. *Behavioral and Brain Sciences* 25: 1–72.

Empirical Studies of Predator and Prey Behavior

W. Cresswell, University of St. Andrews, St. Andrews, Scotland, UK

Introduction

Studying predator–prey behavior often requires observational studies combined with field experiments, rather than traditional laboratory experimental methods. This article describes the particular characteristics of predator–prey systems that lead to specific methods for their study.

There are some special characteristics of predator and prey behavior that need to be considered when designing empirical studies.

1. Optimum behavioral strategies of both predator and prey are dynamic, changing very quickly depending on conditions.
2. There are two broad classes of predator and prey behavior that require different approaches of study: *primary behaviors* concerned with encountering prey or avoiding encountering predators, and *secondary behaviors* concerned with increasing the probability of capture of prey or escape from predators that become important only if an attack occurs.
3. The fitness and ecological effects of antipredation behavior are usually indirect, not involving direct predation mortality, but rather arise through the nonlethal costs of carrying out avoidance or capture-reducing behaviors.
4. The spatial and temporal scale over which predator–prey interact is frequently large and involves both landscape features and alternative predators and prey.
5. Experiments involving manipulating predation rates or predator–prey occurrence are frequently unethical.

As a consequence of the first four characteristics of empirical studies, the measures of the consequences of any one behavior in isolation, or under one set of conditions, may not give a representative idea of how important the behavior is generally. A priori predictions based on these measures are then usually impossible, or at best complex and condition dependent. Both avoidance and capture-reducing behaviors need to be studied simultaneously to fully understand the fitness consequences of a behavior, and the fitness consequences of avoidance behaviors need to be measured using a *starvation–predation risk trade-off* approach. Overall a strictly experimental approach may be impossible, difficult or misleading.

A variety of specific empirical methods have been adopted to deal with the particular characteristics that arise when studying predator–prey behavior. The main approach is a long-term observational field study involving both predator and prey interacting in natural systems. Long-term studies need to employ technological solutions that allow individual predators and prey to be monitored remotely over varying temporal and spatial scales. Ideally, these capitalize on existing 'natural' experiments where predators and prey systems have been already altered by man or compare populations that are disturbed and undisturbed by predators, or contain innovative field experiments utilizing the starvation–predation risk trade-off approach, or models of predators or prey. The limitations of using an observational approach are then dealt with using statistical techniques to control for the effects of confounding variables and to suggest causation.

In short, the dynamic, spatial and temporal nature of predator–prey behavior means that we have to adopt field-based observational studies of the whole system, involving measurements of starvation–predation risk trade-offs, targeted field-experiments, and statistical methods to measure biologically significant effects and to infer causation.

Special Characteristics of Empirical Studies of Predator and Prey Behavior

Dynamic, Multiple and Interacting Predator–Prey Behaviors

If a predator is to catch prey, it must outwit its prey. If a prey is to avoid being caught by a predator, it must outwit its predator. As a consequence, predator behavior can be understood only in the context of prey behavior, and vice versa (**Box 1**).

A range of behaviors from either predator or prey can affect the probabilities of encounter/avoidance and capture/escape in a very short space of time. Both predator and prey continually have to reassess these probabilities and change their behavior as the other does the same. For example, when a hawk hunts small birds, it chooses areas where it expects to encounter vulnerable prey – such as prey in small groups, far from protective cover, and feeding

---

Box 1 Predation behavior equation

From the prey's point of view:

P(survive) = p(avoidance) + p(behaviors on attack lead to escape)

From the predator's point of view:

P(capture) = p(encounter) + p(behaviors on attack lead to capture)

---

intently rather than prioritizing vigilance. As it encounters such vulnerable prey, the prey immediately changes its behavior by forming larger groups, moving to cover, and prioritizing vigilance, so becoming less vulnerable. If alarm calls are given, then this change in vulnerability may even precede the hawk's arrival, causing a hawk to change its intended location even before encountering the prey, because the spatial arrangement of prey changes. But the sudden appearance of a single vulnerable prey might suddenly cause the predator to remain in the area and to attack, altering the vulnerability of all prey in the area. Because the nature of predator–prey behavior is interactive, studying only prey or predator behavior is likely to give results that are impossible to interpret correctly.

The range of behaviors available to predators and prey means that there may be certain behaviors suitable for particular circumstances or individuals, and these alternative strategies may give similar fitness. For example, small birds may vary vigilance levels, exposure times, feeding patches, distance to cover, mass and group size to minimize predation risk. A value of any single one of these variables or combination of these variables may give equal fitness for any given ecological condition. Therefore, measuring only one behavior may lead to no correlation between that behavior and fitness.

The two broad classes of predator–prey behavior (avoidance/encounter and escape/capture: **Box 1**) also lead to different levels of selection on the expression of behavior. Any variation in antipredation behavior should translate into some variation in survival, but general defenses, like predator avoidance, should have a greater effect on survival because they decrease the likelihood that an individual is attacked, whereas more specific defenses in response to attacks can only reduce risk of capture. This means that we can only understand the degree to which selection is operating on a behavior, consequences for fitness of that behavior, or the biological importance of that behavior by studying encounter or avoidance behaviors alongside capture or escape behaviors. Any experiment that simplifies a system (because that is what experiments have to do to make clear predictions and tests) may cut out a major structuring variable that is the primary point of selection. For example, an estimate of the importance of vigilance in the reduction in capture rate during a laboratory experiment may not accurately represent the level of selection on vigilance behavior if, in the natural system, the prey almost always avoids the predator.

The two classes of behavior also require a different empirical approach because behaviors such as avoidance involve tracking movements of predators and prey, as well as the availability and behavior of alternative predators and prey, over a large spatial and temporal scale. In contrast, measuring how capture probability varies involves measuring variables over very short time scales, because predator attacks may only last a few seconds and involve many interacting behaviors.

Nonlethal Effects

The fitness and ecological effects of antipredation behavior are usually indirect and do not involve direct predation mortality, but rather they arise through the nonlethal costs of carrying out avoidance or capture-reducing behaviors. This means that the fitness consequences of avoidance behaviors need to be measured using a starvation–predation risk trade-off approach, and it may be hard to identify and evaluate the relative importance of avoidance behaviors.

There may often be little direct evidence of predators consuming prey if selection acts to favor individuals that respond to predation risk (i.e., that then show nonlethal effects) primarily by avoiding predators, by reducing their activity in a predator's presence, or by carrying out behaviors that reduce the chance of being attacked by a predator (e.g., a lion will not attack alert gazelles). This means that we need to measure the fitness consequences of antipredation behaviors indirectly using a starvation–predation risk trade-off. An animal's use of resources, or inability to exploit resources fully, because of antipredation behaviors, should significantly reduce the time or level of resources available for other activities. Although reducing predation risk through behavioral and physiological means may increase survival in the short term, reducing predation risk will often result in a decrease in the time available to gather enough resources for survival in the long term, or for fecundity, reducing the overall fitness. In other words, any behavioral response to predation can, depending on circumstances, increase or decrease the risk of predation, increase or decrease the likelihood of starvation (or long-term survival), or increase or decrease fecundity. But any increase in the time spent invested in one factor is likely to result in a decrease in the time spent on the other factors, so that changes in predation risk, starvation risk, and fecundity will not be independent of each other. Therefore, the appropriate framework in which to measure the effects of predation risk is via a trade-off. The costs (or benefits) of diverse nonlethal traits (i.e., any antipredation behavior from an increase in vigilance to avoidance of a dangerous area) can then be measured by a common currency of changes in survival or reproductive output (or correlated proxies such as reduction in foraging intake or body condition).

The prevalence of nonlethal effects may then make it hard to identify and evaluate the relative importance of avoidance behaviors. If avoidance behavior is a common consequence of predation risk, then studying the effects of predation may be difficult: the absence of direct interactions between predator and potential prey may be because avoidance prevents interaction, or simply because the potential predator does not interact at all with the suggested prey. We can overcome this by examining systems that have been perturbed by changes in ecological conditions, for example, during severe weather periods, when animals may have to prioritize foraging over predator avoidance (see section 'Natural Experiments'). Alternatively, the conventional experimental approach may then be the only way to identify nonlethal effects. For example, if we wish to test the hypothesis that bats are nocturnal to avoid predation by day-flying hawks, then we either look for situations where nocturnal bats have been forced to forage diurnally, or experimentally create the conditions where this will occur.

Predators themselves also respond to other predators, through interference competition or through intraguild predation risk. Many predators are themselves prey, or prey on other predators, dependent on the availability of alternative prey, predator density, and distribution and ecological conditions that promote the overlap or segregation of predators. Consequently, predators are frequently constrained in their behaviors because they avoid intraguild predation risk, and therefore a knowledge of both top-down and bottom-up trophic effects is necessary to fully understand predator behavior.

This also means that there may be systematic biases in the predator–prey behavior literature because avoidance behaviors where predators and prey do not interact are hard to study directly (or less interesting to publish), whereas capture or escape behaviors can be more easily be related to predation risk and fitness.

Scale

Predators hunt over a large area because they require a large number of prey, or a small number of large prey that also range widely. Most predators are generalists, and there are frequently a whole guild of predators that can feed on the same prey. Consequently, preys have to respond to a variety of predators, and the predators themselves respond to other predators. Predation events are rare and the nonlethal effects of predation risk become obvious only with a good behavioral–ecological field study of a species and its predators. This means that long-term field studies are necessary, over a large enough area sufficient to encompass a reasonable sample size of predators and focal prey, as well as alternative prey. This also means that realistic experimental systems are difficult to construct, and instead, innovative field experiments are needed. Any observational system also requires innovative methods to measure all relevant and confounding variables, particularly when the spatial and temporal scale varies so widely.

Ethics

Much interesting predator–prey behavior is from behaviorally flexible and, therefore, intelligent species. Predation involves fear, suffering, wounding, and death. Therefore any experiment that manipulates predation risk, particularly in vertebrates, creates ethical problems in a modern society where unnecessary animal suffering is not sanctioned. This means that observational approaches may

actually be the only legal approach. The legal obligation to use the observational approach means that statistical techniques to control for confounding variables are vital. This also means that experimental techniques that can manipulate predator occurrence without causing death of prey are important when studying predator–prey behavior. Stress and the nonlethal effects of predation risk remain of course, although many scientists rationalize this type of stress as being a normal part of the experimental animal's life in any case. Some experiments are carried out without recourse to ethics however: behavioral experiments found routinely in the literature can involve staged predation events between spiders of different sizes or handicapped through leg removal, dragonfly larvae eating tadpoles, or fish eating other fish. Again scientists may rationalize this (rightly or wrongly) in terms of the 'normal' degree of predation that is present in a natural system, or that less 'intelligent' organisms require less ethical consideration.

Empirical Methods Specific to Studying Predator and Prey Behavior

Traditional laboratory experiments can tell us much about the causes and consequences of predator and prey behavior, but here we concentrate on why alternative approaches may be much more productive. As the specific characteristics of predator–prey systems outlined earlier suggest, a traditional, strictly experimental approach may be impossible or inefficient or misleading. Predator behavior depends on prey behavior and both depend on particular circumstances. This means that experimental systems can rarely capture details that may be crucial in eliciting a whole range of behaviors that determine encounter/avoidance and capture/escape probabilities. Many experiments may then give only very specific results and not necessarily give any indication of the biological importance (e.g., the relative importance) of the behavior being tested. This also means that a priori predictions in more realistic experiments are complicated: indeed many of the important findings in studies of predator–prey interactions have not been those expected by the experiment designers.

The solution is to employ a range of specific empirical methods. These invariably involve an intimate knowledge of the natural system, based on long-term and widespread field observation to determine biological significance. This then allows the careful targeting of field experiments or statistics to determine or infer causation.

Observational Studies

Arguably, a very important approach to studying predator–prey behavior is an observational field study involving both predator and prey interacting in natural systems. This means that what might be considered 'old-fashioned' skills of the naturalist – patient and systematic observation of animals in the field – comprise one of the most important components of studies of predation behavior. Long-term studies are needed to deal with the rarity of events and the low density of predators. Careful field skills are also needed to ensure that any behaviors observed are not profoundly affected by the predator and prey both potentially responding to the presence of the observer.

Methods involved in observational studies are centered on systematic and unbiased observations of predator and prey behavior. Focal sampling, time budgets, or mapping can measure hunting and avoidance behavior, as well as baseline anticapture behavior such as group size and vigilance. Data from observation of attacks are usually opportunistic because attacks are difficult to predict. This makes them particularly prone to bias. For example, it is easy to conclude that peregrine falcons *Falco peregrinus* chase their prey for long periods as a common hunting strategy because such hunts are much more likely to be noticed by an observer than the very short-duration surprise ambush hunts that are actually more common. Generally circumstances that allow the frequent and easy observation of hunting behavior may not be representative of how a predator and prey system operates: an owl hunting in daylight is observable and may yield many data on hunting behavior, but this may have no bearing on how it normally hunts or how prey respond to it during darkness. Such examples are obviously biased, but there are many circumstances where biases are very subtle. As a consequence, any unsystematic observation of hunting behavior, or anecdotal descriptions of antipredation behavior are rarely very useful in understanding predator or prey behavior. But, because of the rarity of any data on predators and prey interacting during attacks, conclusions based on these types of observations are common.

A key factor in any observational study (and indeed any study) is to ensure that sampling is done appropriately. A large number of observations from a single prey animal and a single prey predator could never tell us as much about the general population as sampling several prey animals and several predators. Studies of predators are particularly prone to an inappropriate level of sampling (i.e., pseudoreplication) because predators may range over large areas and are at low density, precluding a researcher from sampling more than a few animals. Marking of focal animals is therefore very important to determine the level of sampling. This often involves capture of prey and predators to attach tags or rings, but can also involve use of individual variation in appearance (e.g., whisker patterns in lions *Panthera leo*) as a means of individual recognition.

A major problem of observational studies is an effect of the observer. Observers have their own 'nonlethal' effects on prey and predator distribution via disturbance and avoidance, or increased vigilance in the prey. The net effect might well then be as misleading as an artificial lab experiment where many of the normal behaviors that might be used by predators or prey are constrained by the experimental conditions. Studying populations of animals that are habituated to humans but not other predators is a way around this problem.

Remote Monitoring of Predators and Prey

A potential solution to gather unbiased observational data is to continuously monitor animals, using a remote monitoring or automated recording system. There are a variety of technological solutions that allow individual predators and prey to be monitored remotely over varying temporal and spatial scales, such as geographical positioning system (GPS), radio, satellite, and passive integrated transponder (PIT) tags. The advantages of these systems are that representative data can be collected, because they do not depend on the animal being visible, and they allow observation of animals when they are less conspicuous. A key advantage is that they can remove observer effects, although the tags themselves may alter behavior. The disadvantages are that the animal must be caught and many animals need to be tagged to get representative data. Tags are expensive and difficult to deploy, and so many studies end up with many observations from a few animals or biased samples because tagged data are not collected systematically or from a representative sample of the population.

Remote cameras can be used to collect data from predators or prey. Camera traps put out in an unbiased way can be used to map predators and prey, identify individual variation in behavior, and even record very rare predation events when cameras record prey continuously for long periods. Some classes of predator and prey behavior occur very rapidly: a whole range of attack and escape behaviors in hawk attacks on small birds typically last only a few seconds. Analysis of slowed down video records is then needed.

Predator behavior can also be studied indirectly through studying prey remains. Many predators from octopuses (Octopoda) to song thrushes *Turdus philomelos* leave uneaten, undigestable or partly digested parts of prey as a record of what, where, and how frequently prey have been consumed. Many elusive predators such as big cats may leave conspicuous dung which allows studies of prey choice with respect to age, sex, and even identity when DNA is extracted.

Natural Experiments

Observational studies result in correlations but do not necessarily identify causation. Use of existing 'natural' experiments where predators and prey have been already altered by man, however, can allow reasonable causation to be established. Natural experiments are particularly relevant to studies of predator–prey behavior because predators have been removed from many areas of natural occurrence, or are reestablishing. Introductions of predators and prey species to new ecological circumstances, or where anthropogenic change alters existing ecological circumstances for predators or prey are common and can also be exploited. A classic example of this are studies of how small birds compensated for predation risk by managing their fat reserves in areas where hawks had been accidentally exterminated by man with areas where hawks were still present. Alternatively 'natural' experiments can be exploited on a smaller scale when particular environmental effects and constraints apply occasionally, in a similar way to changing conditions in a controlled experiment. An example of this might be comparing the characteristics of a prey group attacked with the nearest group that was not attacked. When the two groups occur in more or less identical conditions apart from the variable of interest (e.g., one group is larger than the other) and have just been encountered by a mobile predator, then it is reasonable to argue that this is more or less equivalent to a choice experiment.

Field Experiments

Field experiments, where conditions might be experimentally altered within the context of natural systems, are a good approach because they isolate single variables to identify causation, without restricting the full range of other behaviors or factors, so also indicating biological, rather than just statistical, significance. An example of a field experiment might be testing how changing perceived predation risk affects vigilance behavior by providing food at different distances from protective cover. A key approach is to observe, in natural systems, how prey behavior affects predator behavior and vice versa. This can be a short cut to determining the truly important behaviors in predator–prey behavioral interactions, and so focussing in on the most relevant experiments.

A major class of field predation experiments involve altering predation risk via the abundance of predators. This can be done through the use of models or exclosures. For example, even approximate models of hawks flown down lines toward feeding birds, tested against control shapes, will elicit major behavioral changes consistent with their responses to natural hawk attacks. Human experimenters are often used to simulate predation risk to assess variation in behavioral responses: an experimenter might, for example, approach a bird on a nest in a standard way to determine how investment in nest defense depends on the stage of breeding. Predators may be placed into natural systems such as predatory fish in ponds, but restricted behind transparent barriers so that perceived

predation risk can be manipulated without the ethical problems arising from direct predation. Some experiments on invertebrates may surgically alter predators, for example gluing mouthparts together to allow the introduction of predators without any lethal effects. Alarm calls, predator scents, predatory remains, and other indices of predation risk in natural systems are also used to experimentally manipulate perceived predation risk.

Predator abundance can be decreased by the use of exclosures that will prevent the lethal effects of predators, but may not necessarily change perceived predation risk: an animal may not recognize that it is in a situation of lowered predation risk during the experiment, or its perception may change during the experiment confounding the results. In all of these experiments, varying the predation variable (i.e., the model type) and/or using proper controls are essential to determine whether the animals are responding as if they perceive predation risk as actually varying. This also, of course, relies on a detailed knowledge of how the animals behave in natural predator encounters.

Predator behavior can also be experimentally manipulated by presentation of prey models. This is a frequent lab experimental technique to investigate optimal foraging and prey choice, but can also be used for field experiments. For example, stuffed prey models presented in pairs to hunting sparrowhawks *Accipiter nisus* in the field have been used to test whether hawk prey choice depends on the conspicuousness of prey and state of vigilance. Similarly, artificial nests with eggs from farmed birds such as quails *Coturnix coturnix* have been used widely to determine nest defense behavior and nest–predation behavior, although their effectiveness, as with all field experiments, depends on how closely they mimic the characteristics of the natural system.

Another major class of field experiments involves using the starvation–predation risk trade-off approach, where food patches are provided for animals under varying conditions that are thought to affect predation risk, such as distance from predator concealing cover. According to optimal foraging theory, and backed up by many empirical studies, animals will visit a risky patch only if it is profitable, and so will deplete safer patches to a higher degree. The technique measures what is called giving up density (GUD). The advantage of this technique is that varying dimensions of predation risk can be integrated into a single currency: how much foraging gain will an animal forgo to stay in a safe area. The disadvantage of the technique is that often different species visit different patches to different degrees, and predation risk may vary within a patch with time, so making any more than very general conclusions impossible. A variant on this type of field experiment is to provide animals with a choice of patches that vary in predation risk and profitability and to compare relative use. If rewards between patches are varied, then the point at which an animal will switch from a poor but safe patch to a profitable but risky patch indicates the relative importance of predation risk. Such techniques can also be used to identify and rank predation risk by identifying which variables result in the largest differential in patch profitability necessary to cause a feeding animal to shift between patches.

Statistical Techniques

The major disadvantage of the observational approach imposed on many aspects of the study of predator–prey behavior is that only correlations can be established, not definitive causation, and observed effects may be confounded by uncontrolled variation. We can, however, use statistical techniques to control for the effects of confounding variables and to suggest causation. Statistical modeling techniques such as generalized linear models can evaluate the effects of a range of potential predictor variables on a measure of behavior: potentially confounding variables measured secondarily in a study can also be effectively tested. For example, a model testing how fox hunting success depends on prey density might include a number of confounding variables such as time of year or time of day. Prey density might not turn out to make a difference to hunting success, but time of day might: a general modeling approach can allow many variables to be tested simultaneously. Exact causation can then be confirmed using a follow-up field experiment. This process can be much more efficient than carrying out a number of separate experiments that might only be able to test for the effects of one or two variables at the same time. Such a modeling approach requires large sample sizes, but because they can account for repeated sampling of the same individuals, they can test for the effects of changes in predictor variables within, as well as between individuals. This means that multiple samples from the same individual can be very informative, as long as a reasonable number of different individuals are also sampled. Other statistical techniques such as path analysis can allow directions of causality to be inferred.

Conclusion: Planning Empirical Studies of Predator and Prey Behavior

Studying predator–prey behavior presents a number of specific problems, but also some unique opportunities to increase our understanding of animal behavior. We need to consider predators and prey simultaneously in the context of natural systems, using predominantly an observational and field experimental approach, or realistically scaled and complicated lab experiments. We need to separate avoidance or encounter and escape or capture behaviors because each requires a different approach. We need to have sufficient time or a good system or a very technical approach. Overall, this means that observational studies are the most valid approach. These are of course useful only if they consider a range of variables so that potential confounding effects can be identified using statistical techniques.

Perhaps the three most important points to consider when planning any empirical study of predator and prey behavior are: first, prey behavior arises from the starvation–predation risk trade-off: therefore antipredation behavior must be measured in conjunction with foraging behavior to determine the true costs and benefits of a behavior. Second, any predator–prey pair that is studied exists within an interacting food web of many potential other predator–prey combinations: therefore behaviors can be fully understood only in the context that they are part of more complex systems. And third, and most importantly, prey behavior changes with predator behavior and vice versa, over very short time scales: therefore behaviors can be properly understood only in the context of dynamic systems that consider both predators and prey simultaneously. Overall, this makes empirical studies of antipredation behavior a continuing challenge for behavioral scientists.

See also: Experiment, Observation, and Modeling in the Lab and Field; Experimental Design: Basic Concepts; Games Played by Predators and Prey; Playbacks in Behavioral Experiments; Predator's Perspective on Predator–Prey Interactions; Robotics in the Study of Animal Behavior.

Further Reading

Ajie BC, Pintor LM, Watters J, Kerby J, Hammond JI, and Sih A (2007) A framework for determining the fitness consequences of antipredator behavior. *Behavioral Ecology* 18: 267–270.

Caro TM (2005) *Antipredator Defenses in Birds and Mammals*. Chicago, IL: University of Chicago Press.

Cresswell W (2008) Non-lethal effects of predation risk in birds. *Ibis* 150: 3–17.

Cresswell W and Quinn JL (2004) Faced with a choice, predators select the most vulnerable group: Implications for both predators and prey for monitoring relative vulnerability. *Oikos* 104: 71–76.

Lima SL (1993) Ecological and evolutionary perspectives on escape from predatory attack – A survey of North-American birds. *Wilson Bulletin* 105: 1–47.

Lima SL (1998) Stress and decision making under the risk of predation: Recent developments from behavioral, reproductive and ecological perspectives. *Advances in the Study of Behavior* 27: 215–290.

Lima SL (2002) Putting predators back into behavioral predator–prey interactions. *Trends in Ecology & Evolution* 17: 70–75.

Lima SL and Dill LM (1990) Behavioral decisions made under the risk of predation: A review and prospectus. *Canadian Journal of Zoology* 68: 619–640.

Lind J and Cresswell W (2005) Determining the fitness consequences of anti-predation behaviour. *Behavioral Ecology* 16: 945–956.

Minderman J, Lind J, and Cresswell W (2006) Behaviorally mediated indirect effects: Interference competition increases predation mortality in foraging redshanks. *Journal of Animal Ecology* 75: 713–723.

Roth TC, Lima SL, and Vetter WE (2006) Determinants of predation risk in small wintering birds: The hawk's perspective. *Behavioral Ecology and Sociobiology* 60: 195–204.

Endocrinology and Behavior: Methods

K. L. Ayres, University of Washington, Seattle, WA, USA

Endocrine Measures

Researchers have long been interested in the interplay between animal behavior and 'secretory blood-born products,' later termed hormones. The most basic definition of a hormone is an organic chemical messenger between cells. The endocrine system is one of the most ancient chemical messenger systems in living organisms and plays a large role in physiology and behavior. A secretive cell secretes hormones, which then travel through blood and/or tissue fluids and act on target cells. Target cells can be including different cells throughout the body or the cell that secreted the hormone (i.e., the secretive cell sends a signal to itself). Hormones help integrate, regulate, and coordinate physiology and behavior; therefore, they often have multiple, simultaneous functions.

Hormones can be broken down into three main types:

1. *Modified lipids and phospholipids*: derivatives of cholesterol and fatty acids such as steroid hormones (e.g., glucocorticoids, androgens, progestins, estrogens, etc.) and the prostaglandins.
2. *Peptides*: hormones that are a string of amino acids (i.e., a protein).
3. *Modified amino acids*: derivatives of tyrosine and tryptophan such as thyroid hormones and the catecholamines, respectively.

Hormone Receptors

Hormone receptors are proteins that bind hormones. Once bound, the hormone/receptor complex initiates a cascade of cellular effects resulting in some modification of physiology and/or behavior. Hormones usually require receptor binding to mediate a cellular response. Receptor binding and the associated cellular cascades amplify the hormone signal allowing hormones to act at very low concentrations, sometimes as low as parts per trillion! Scientists who study hormone receptors are interested in the cells and tissues that possess the receptors (i.e., the ovary, the testis, the stomach lining, specific parts of the brain, etc.) The location of the receptor shows where the hormone should be biologically active.

Binding Globulin

Binding globulins are plasma proteins with low solubility that bind hormones, such as the steroids and thyroid hormone, and help carry the protein through the blood. Measurements of the 'total' concentration of a hormone in blood include hormones that are bound to a binding globulin and hormones that are 'free' or unbound to a carrier protein. Free hormones are only those that are not bound to binding globulins.

Total Hormone Concentration

The total hormone concentration is the combined concentration of hormones bound to binding globulins as well as hormones that are 'free' or unbound.

Free Hormone Concentration

The free hormone concentration is the concentration of hormones that are circulating unbound or 'free' from binding globulins.

Hormone Matrix

A substance that contains hormones and/or hormone metabolites is called a hormone matrix. Hormones and hormone metabolites can be measured in a number of matrices such as blood, saliva, urine, fecal material, blubber, and hair. Each hormone matrix and hormone type has its own methodological pros and cons so the hormone and the matrix of interest will depend largely on the question being asked and the animal species that the researcher is studying.

Blood

Most hormones can be measured in blood, making this a very reliable matrix for analyzing almost any hormone of interest. If the study animal can be managed and bled, the blood hormone concentrations can be measured using RIA or ELISA. Blood is often a more invasive approach; however, some captive animals will allow blood to be drawn willingly if they are trained with positive reinforcement techniques. For example, captive dolphins at marine parks are trained to present their tail flukes so that blood may be drawn from the fluke. Blood hormones are secreted in pulses. Therefore, a researcher must be careful when working with circulating hormones, since results can vary drastically if the animal is sampled at the height of secretion versus a trough. This usually means that multiple individuals must be sampled multiple times to adequately access average hormone concentrations. Also, handling the animal may elicit a physiological stress response. This may confound results if the researcher is interested in endocrine measures that are affected by stress, which are common in endocrinology. For that reason, it is often important to obtain the blood sample within a certain amount of time before the adrenals can secrete glucocorticoids that may interfere with the study. On the other hand, if the physiological stress response is the focus of the study, than blood is useful for testing the magnitude of the physiological stress response, which is not easy in other hormone matrices.

Saliva

Saliva is considered a good matrix for measuring biologically active 'free hormones' that are not bound to binding globulins. The binding globulins are thought to be too large to pass into the saliva, and therefore saliva should only contain free hormones separate from the binding globulins. Saliva has become a very popular way for measuring hormones in humans as this approach is fairly noninvasive and relatively sanitary. The researcher and/or the subject require very little training to obtain a saliva sample. In fact, there are companies that now offer saliva hormone tests commercially, where you can send in your own saliva sample and they will analyze your hormone levels. Usually these commercial services test steroid and thyroid hormone concentrations. Some hormones that can be measured in saliva include estrogens, androgens, progestins, melatonin, and glucocorticoids.

Saliva may not be the best option for sampling non-human species, because obtaining a saliva sample may be somewhat invasive for an unwilling or sensitive animal, not to mention an animal with a bad bite! This means that some animals would still require capture and restraint to obtain a saliva sample.

Feces

Fecal hormones have become very popular for studies of sensitive and/or endangered species; elusive species; and in studies where handling may confound results (i.e., stress physiology studies). Fecal hormones are also compounded over time making them less sensitive to the issues of pulsatile secretion associated with blood sampling. However, while fecal sampling for steroid hormones and modified amino acids is becoming common, peptide hormones are difficult, if not impossible, to measure in feces, because they are broken down during digestion.

When conducting fecal hormone studies, it is important to thoroughly mix samples as hormones are usually unevenly distributed in the fecal material, causing hormone 'hot spots' and 'cold spots.' Samples should also be lyophilized, because increased food intake results in a higher concentration of water in the feces. Once freeze dried and extracted, hormone concentrations should be analyzed per gram dried fecal material for more accurate results. It is also important to consider the lag time associated with metabolism, excretion, and digestion of hormones. The digestive lag time is species specific and ranges around a few hours for small mammals and birds to around 12–48 h in larger mammals. Many fecal hormone issues have been addressed in Wasser et al. (1994, 2000). Some hormones that can be measured in fecal material include androgens, estrogens, progestins, glucocorticoids, and thyroid hormones.

Urine

Urine sampling is similar to fecal sampling, but is more common in captive animal studies. Captive animals can be trained using positive reinforcement techniques to urinate on cue as is done for captive dolphins in marine parks. The researcher can also line the bottom of enclosures to catch the urine whenever the animal urinates, making this approach very noninvasive with many of the same benefits as of fecal sampling. Unlike fecal sampling, the animal must usually be captive and/or present for the researcher to collect a urine sample.

Researchers usually analyze urinary hormones per ml creatinine to control for differences in urinary excretion rates. Some hormones that can be measured in urine include androgens, estrogens, progestins, glucocorticoids, aldosterone, catecholamine metabolites, and leutinizing hormone.

Hair

Recent studies have shown that some hormones can be extracted and measured from hair samples. For example, Gleixner and Meyer showed that you could extract and measure estradiol and testosterone in the hair of cattle. The extraction requires an agent to break the keratin bonds of the hair and an agent to facilitate the release of hormones from the hair into solution. They also found that different color hairs had different concentrations of hormones so hair color would need to be taken into account for hair hormone studies.

Blubber

Blubber sampling is popular for studies of marine mammals, which cannot easily be captured and/or restrained for drawing blood. The researcher uses a projectile dart to biopsy a small amount of skin and blubber from the animal, which can then be analyzed for hormones as well as some hormone receptors. This technique has become popular for assessing reproductive state in marine mammals, where it is both difficult and dangerous to capture and/or sedate the animal. Some hormones that can be measured in blubber include progesterone, estradiol, and testosterone.

Egg Yolk

Studies of hormones in egg yolk have been used to test nongenetic maternal effects on offspring development in egg-laying vertebrates. The environment experienced by the laying mother can affect what type and concentration of a particular hormone she deposits in the egg. This can have a significant effect on development and behavior after the chicks are hatched and reared. For example, Daisley and colleagues showed that higher levels of testosterone in the egg yolk of Japanese quails result in more 'bold' or 'proactive' behaviors after hatching in both sexes.

Hormone Extractions

Before a researcher can assay hormone concentrations in their samples, it is important to select and validate an appropriate extraction solvent and protocol. Appropriate extraction protocols depend on the chemical properties of the hormone of interest and the biological matrix from which one is extracting the hormone. For example, steroid hormones are fairly nonpolar and are often extracted in methanol or ethanol.

Measuring Hormone Concentrations

Gas Chromatography–Mass Spectrometry

GC–MS is a method of detecting and measuring the amounts of a chemical, using the separation of chemicals by chemical properties and mass. This technique is very accurate, but can be expensive and sometimes more specific than is needed for many hormone studies compared to immunoassays, which are cheaper and better for high-throughput analyses. However, GC–MS is often used to initially validate hormone measures in immunoassays.

High-Performance Liquid Chromatography

High-performance liquid chromatography or high-pressure liquid chromatography (HPLC) is a type of column chromatography used to separate, identify, and quantify hormone concentrations. A column is packed with chromatographic packing material called the stationary phase. The solvent and the extracted hormones are the mobile phase and are moved via a pump through the column to a detector. The detector records the retention times of the hormones in the sample. Retention time varies depending on the interactions between the stationary phase, the hormones, and the solvent used.

Immunoassays

Immunoassays have become very important for measuring hormone concentrations. Radioimmunoassay and enzyme-linked immunoassays tend to be the most utilized assays today and are very sensitive, which is important because hormones can be biologically active in concentrations as small as part per trillion!

Radioimmunoassay

RIA was developed in the 1960s and 1970s and allows incredibly sensitive measures of hormone concentrations. The assay relies on a hormone specific antibody, which competitively binds the unknown hormone of interest with a known concentration of radioactively tagged hormone. The amount of bound radioactive hormone is compared with a standard curve, which allows one to calculate the unknown hormone concentration.

Enzyme-Linked Immunoassay

ELISA works the same way as RIA except that instead of radioactively tagging the known hormone, the hormone is tagged with a color-changing agent that can be measured using a spectrometer. This method does not involve handling radiation, which can be a logistical advantage over RIA.

Hormone Validations

Hormone validations are a series of experiments that are conducted each time a new hormone, matrix, and/or species is going to be studied. These experiments test the presence of the hormone(s) in the matrix, the ability to extract the hormone(s) given a specific extraction protocol, and the ability to measure the hormone concentration(s) effectively given a specific quantification technique.

Recovery

A recovery experiment is a validation experiment that tests the ability to recover a known amount of 'tagged' hormone from a given matrix. The tag is usually a radioactive isotope. The matrix of interest is spiked with a known amount of tagged hormone. The researcher then conducts the desired extraction protocol and tests for the percent recovery of the tagged hormone spike to see whether the extraction protocol was effective and sufficient to recover most, if not all, of the spike.

Parallelism

Parallelism is a validation experiment conducted when beginning a study on a new species, matrix, extraction technique, antibody, and/or assay. Parallelism tests for the presence of the hormone of interest in the extract and the ability to detect a range of hormone concentrations in the assay. To test parallelism, the researcher makes a series of serial dilutions from the unknown extract. Each serial dilution is quantified in an immunoassay. The hormone concentrations of the serial dilutions are then compared with the known standards that were used in the assay. If the hormone is present and can be quantified over a range of concentrations, then the serial dilution concentrations should parallel the standard curve. If an extract shows nice parallelism, usually the dilution that yielded about 50% binding in the assay is used to test for accuracy. Fifty percent binding is the dilution that maximizes the number of samples that will fall on the standard curve when the researcher starts analyzing the actual samples for the study. However, if an unusually high- or low-concentration sample is used for the parallelism experiment, then the 50% binding will not be very useful to this effect. Therefore, if it is possible, parallelism should be run on a pool of different kinds of samples from the population of interest.

Accuracy

Accuracy is a validation experiment conducted when beginning a study on a new species, matrix, extraction technique, antibody, and/or assay. Accuracy tests the ability of an assay to accurately measure a known hormone concentration when spiked with the extract that yielded about 50% binding in the parallelism validation. The test is conducted by running a normal set of standards in the assay along side another set of standards spiked with the concentration that yielded 50% binding. A zero is also run which contains only the 'spike.' After the assay is run and the hormone concentrations are quantified, the 'spike' concentration is subtracted from each of the spiked standard concentrations and the results are graphed against the normal (un-spiked) standards. If the antibody in the assay measures the hormone concentration accurately with no interference, then concentrations of the normal standards graphed with the concentration of the spiked standards minus the spike should yield a line with a slope that is not significantly different from 1.

Challenge Experiments

Challenge experiments help validate that a physiological change in circulating hormone concentration is reflected in the matrix of interest. For example, one common challenge experiment is an ACTH (adrenocorticotrophic hormone) challenge. ACTH is the hormone released from the brain that tells the adrenals to release glucocorticoids. In an ACTH challenge, the animal is injected with ACTH and samples of the matrix of interest are collected before and after ACTH injection. Glucocorticoid concentrations are then assayed to see whether they have increased after the 'challenge' with ACTH. If the concentrations have increased after the challenge, then it shows that glucocorticoids in that matrix are an effective measure of adrenal activity during the physiological stress response. Another common challenge is a TSH (thyroid stimulating hormone) challenge, which tests whether or not measures of thyroid hormones in a certain matrix reflect thyroid function.

Experimental Manipulation

Many behavioral endocrinologists are interested in manipulating study organisms by introducing excess hormones/hormone activity or suppressing hormone concentrations/activity to see how these changes affect behavior. Experimental methods employed by behavioral endocrinologists can be generally described as ablation, replacement, and overexpression manipulations. R. Silver outlined the necessary criteria used in behavioral endocrinology to test whether a particular hormone affects a specific behavior or the behavior affects hormone concentrations. These criteria include the following:

1. A hormonally dependent behavior should disappear when the source of the hormone is removed or the hormone actions are blocked (ablation).
2. After the behavior stops or is reduced, restoration of the missing hormonal source or its hormone should restore the absent behavior (replacement).
3. Hormone concentrations and the behavior should covary. The behavior should be observed more often when hormone concentrations are higher and less often or never when hormone concentrations are lower.

Ablation Manipulations

Ablation manipulations remove the hormone of interest or block its actions.

Surgery

The oldest ablation manipulations in endocrinology are surgical organ/tissue removals. For example, the first recognized experiment in behavioral endocrinology was conducted by Berthold in 1849. In this study, Berthold removed the testes of or gonadectomized roosters. The gonadectomized roosters or 'capons' did not crow and exhibited more feminized morphology.

Antagonists

Antagonists or 'blockers' are used as a way to chemically 'knockout' or reduce hormone function. An antagonist binds the hormone receptor, but does not cause the down stream cellular effects that are characteristic of the hormone, resulting in less bioactivity by the hormone of interest.

Blocker

See section 'Antagonists.'

Genetic Knockouts

Genetic knockouts or partial knockouts can occur naturally by mutation or can be engineered to decrease hormone and/or hormone receptor production. These mutant phenotypes can be used to test the effects of a particular loss of hormone function on morphology and behavior.

Replacement Manipulations

Replacement manipulations restore the source of the hormone or the hormone itself that was lost in an ablation manipulation. In Berthold's rooster experiment, he transplanted testes from a normal rooster into the gonadectomized capons. This restored crowing and male sexual behavior, while also restoring the typical rooster phenotype.

Hormone Implants

Implants are commonly used in behavioral endocrinology to replace a lost hormone or to overexpress a hormone of interest. Most implants involve silastic tubing packed with the hormone. The tubing is implanted under the skin and releases the hormone steadily over time to test how increased hormone levels affect behavior or to restore a lost phenotype and/or behavior from an ablation manipulation/mutation.

Agonists

Agonists or 'mimics' are similar to antagonists in that they are chemicals that bind hormone receptors, but unlike antagonists they *do* activate the cellular cascades indicative of the natural hormone.

Mimic

See section 'Agonists.'

Genetic Mutations

Mutations that cause the overexpression of a given hormone occur naturally or can be engineered. These mutant phenotypes overexpress a given hormone and/or receptor and can be used to test the effects of hyper expression of a given hormone on morphology and behavior.

Conclusion

The interplay of hormones and behavior is a fascinating and rapidly evolving field. There are many methods from which to choose when conducting research in this field and new questions and methods are emerging every day. The most effective method for a study will depend largely on the research question and the focal species. The researcher must decide on the types of hormones that are relevant to understanding their question. They must decide on the best matrix for their hormone of interest and how they will accurately extract and measure the hormones from that matrix. Once these questions are answered, endocrine measures can be very powerful tools for understanding how hormones affect behavior and how behaviors influence physiology.

See also: Aggression and Territoriality; Behavioral Endocrinology of Migration; Communication and Hormones; Conservation Behavior and Endocrinology; Experimental Approaches to Hormones and Behavior: Invertebrates; Female Sexual Behavior and Hormones in Non-Mammalian Vertebrates; Field Techniques in Hormones and Behavior; Fight or Flight Responses; Food Intake: Behavioral Endocrinology; Hormones and Behavior: Basic Concepts; Hormones and Breeding Strategies, Sex Reversal, Brood Parasites, Parthenogenesis; Horses: Behavior and Welfare Assessment; Hunger and Satiety; Immune Systems and Sickness Behavior; Internal Energy Storage; Invertebrate Hormones and Behavior; Memory, Learning, Hormones and Behavior; Molt in Birds and Mammals: Hormones and Behavior; Pair-Bonding, Mating Systems and Hormones; Parental Behavior and Hormones in Non-Mammalian Vertebrates; Pets: Behavior and Welfare Assessment; Poultry: Behavior and Welfare Assessment; Reproductive Skew, Cooperative Breeding, and Eusociality in Vertebrates: Hormones; Seasonality: Hormones and Behavior; Sex Change in Reef Fishes: Behavior and Physiology; Sexual Behavior and Hormones in Male Mammals; Sleep and Hormones; Stress, Health and Social Behavior; Tadpole Behavior and Metamorphosis; Vertebrate Endocrine Disruption; Welfare of Animals: Behavior as a Basis for Decisions; Wintering Strategies.

Further Reading

Berthold AA (1849) Transplantation of testes. Translated by D. P. Quiring. *Bulletin of the History of Medicine* 1944, 16: 399–401.

Daisley JN, Bromundt V, Mostl E, and Kotrschal K (2005) Enhanced yolk testosterone influences behavioral phenotype independent of sex in Japanese quail chicks Coturnix japonica. *Hormones and Behavior* 47: 185–194.

Gleixner A and Meyer HHD (1997) Detection of estradiol and testosterone in hair of cattle by HPLC/EIA. *Fresenius Journal of Analytical Chemistry* 357: 1198–1201.

Kellar NM, Trego ML, Marks CI, and Dizon AE (2006) Determining pregnancy from blubber in three species of delphinids. *Marine Mammal Science* 22: 1–16.

Milspaugh JJ and Washburn BE (2004) Use of fecal glucocorticoid metabolite measures in conservation biology research: Considerations for application and interpretation. *General and Comparative Endocrinology* 138: 189–199.

Nelson R (2005) An Introduction to Behavioral Endocrinology. Sunderland, MA: Sinauer Associates, Inc.

Robeck TR, Schneyer AL, McBain JF, et al. (1993) Analysis of urinary immunoreactive steroid metabolites and gonadotropins for characterization of the estrous cycle, breeding period, and seasonal estrous activity of captive killer whales (Orcinus orca). *Zoo Biology* 12: 173–188.

Wasser SK, Hunt KE, Brown JL, et al. (2000) A generalized fecal glucocorticoid assay for use in a diverse array of nondomestic mammalian and avian species. *General and Comparative Endocrinology* 120(3): 260–275.

Wasser SK, Monfort SL, Southers J, and Wildt DE (1994) Excretion rates and metabolites of oestradiol and progesterone in baboon (*Papio cynocephalus*) faeces. *Journal of Reproduction and Fertility* 101: 213–220.

Ethograms, Activity Profiles and Energy Budgets

I. S. Bernstein, University of Georgia, Athens, GA, USA

Ethograms, Activity Profiles, and Energy Budgets

When we study behavior, we examine all the actions of an individual over time. Behavior may be a continuous stream of activity, but we attempt to reduce this continuum to a series of discrete actions that we can examine and quantify. We ask what all this activity is about, what the functions of the actions are and what stimuli in the environment or the organism are responsible for the behavior. Behavior is therefore divided into elements that we believe are meaningful and such a process is inherently subjective. Nonetheless, we can operationally describe the structure of the units selected and produce what amounts to a dictionary of behavior, an Ethogram. Ideally, an Ethogram should include all the actions typical of a species. Idiosyncratic behavior or behavior found in only a few individuals, or behavior that is unusual, is usually not included in an Ethogram; it is placed in a general or 'wastebasket' category.

The Ethogram is then used in an attempt to measure the frequency and duration of the units of behavior, recognizing some as brief acts, or events, and others as prolonged states. We correlate these measures with observable stimuli to see what immediately preceded the element and therefore might be responsible for the behavior, and what followed as an apparent consequence of the behavior and therefore might be the function of the behavior. Function does not, of course, require intention at the time of the action, as many acts may be reflexive, resulting from prior selection acting on the ancestors of the individuals so that genes contributing to acts with favorable consequences (with regard to the individual's genetic fitness) were selected for. Other behavior may come about through experiences during the lifetime of the individual, some of which may be learned. So we now have some idea about all four of Tinbergen's 'Why' questions regarding the structure under consideration: the proximal cause, the ontogenetic cause, the function, and the evolutionary cause (which may be quite distinct from the current function).

One problem with Ethograms is that they may be different in the case of each investigator, and it may be impossible to compare data collected on the same individuals by different investigators, each using their own Ethogram. Comparing across Ethograms is not as easy as deciding whether one was a lumper and the other a splitter. If that were the case, then all one would need to do is combine the splitter's categories into the larger units used by the lumper.

For example, one person defined severe aggression as biting or chasing, whereas another defined and scored biting and chasing separately and did not have a 'severe aggression' category. On the other hand, if one investigator defined severe aggression as biting or chasing and mild aggression as a slap or lunge, while another investigator defined severe aggression as biting and slapping and mild aggression as chasing, lunging, and a specific facial category, no ready comparison of their data is possible. You will note that the second included a facial expression, not scored by the other, in one category and combined the other categories differently from the first observer. There is no way to directly compare their quantitative data for severe and mild aggression. If we all used the same Ethogram of course, we would have no problem, but no universally agreed upon Ethogram currently exists.

Data Collection Methodology

Every investigator of behavior should begin with a defined Ethogram, for at least a subset of the animal's activity, for example, reproductive behavior and correlated responses. None of us can study all that an animal does. When we try to do so, we score only what seems important to us at the moment and may stop scoring some elements as irrelevant after a while and add others that we decide are important. Unsystematic data collection of this sort can only tell us that we did see something, but not necessarily how often or for how long. Jeanne Altmann's classic paper on data collection methods assumes that we begin with a discrete Ethogram (1974). She describes how we must focus on a single individual and record what happened, when and for how long, but for a predefined list of activities (Focal Data). Alternatively, we may try to watch all of the individuals in a group simultaneously and score some particularly noticeable action every time it occurs (All Occurrence Scans).

Yet another alternative is to focus on a single individual at a time (as in Focal Data) but only for a brief instant, at predetermined intervals, to indicate whether or not it was in one of several particular states (Instantaneous Scans). In designing such a system, we should select intervals far enough apart so that our samples for each individual are relatively independent of what it was doing on the previous scan. Determining the proper interval for collecting

Instantaneous Scan data requires some quantitative information on the duration of items in the Ethogram and whether they are random with regard to one another or occur in sequences in which the same elements are repeated. Pilot data are used to obtain such estimates.

For example, if one were to use Instantaneous Scans to determine what percentage of the day humans spent sleeping, it would hardly do to collect 1000 scans at 10 s intervals. The probability of a human being asleep 10 s after having been observed to be asleep is considerably higher than the random probability of seeing a human asleep. The 1000 scans collected in a 3 h period might falsely lead one to conclude that humans never sleep or that they spend 100% of their time asleep. Clearly, if humans sleep for an average of 7 h at a time, then instantaneous scans taken at intervals of less than the mean sleep duration will not yield independent data. The interval used between scans must thus be based on the mean duration (plus 1.96 standard deviations to insure 95% confidence of the independence of each entry) of the longest state in the Ethogram being measured, or instantaneous scans must be performed at different intervals, depending on the item being examined. Of course if the number of scans is very large, then the inflated N produced by too frequent scanning will not jeopardize conclusions that would have been reached on the basis of a much smaller sample at longer intervals inasmuch as the increased power of tests with larger Ns becomes negligible as Ns become very large.

Other data collection techniques include trying to keep a running log of whatever we see as it strikes us, just to find out what does occur (ad libitum data collection). Whereas ad libitum data collection may be necessary as pilot work to decide what seems worth studying more systematically, and as a prequel to systematic data collection, ad libitum data provide information only on the existence of acts and states. True Individual Focal Data, in contrast, will provide information on both the frequency and durations of all items in the Ethogram. Focal data collection is expensive in terms of time, for only a single individual can be followed at one time and in order to collect 1 h's worth of data on a group of 100 individuals, one would need 100 observers working simultaneously, or a single observer to spend 100 h collecting 1 h's worth of data on each individual. (If you are studying social interactions, 100 h of data actually provide 2 h of data on each subject as you have 1 h on the subject and 1 h on each possible partner that could interact with the subject.)

Instantaneous Scans are also done by observing one individual at a time, but since data collection is instantaneous, a group of 100 individuals can be done as fast as the observer can find each on a predetermined list. A predetermined list and time frame is necessary so that data collection is random for all individuals. If one scores individuals opportunistically, then if A is grooming B at the scan for A, and B is scored next because it is available, then the data for B are not independent of what was just scored for A. Instantaneous Scan data can be used to say what percentage of scans included the behavior and hence, what percentage of time the individual spends in that activity.

All Occurrence data will give us true frequencies, but if and only if all individuals under study are continuously visible and the behavior lasts long enough so that a scanning observer will be able to check all individuals in a time period shorter than the duration of the action or state.

Another method is called 'One–Zero' data collection in which, for predetermined periods of time, one enters whether or not a particular action occurred (but not how often) during that period of time. Such data yield neither frequency nor duration data but do allow one to calculate the probability of seeing an act at least once in the predetermined interval of time. Some may question the biological significance of such data, but sometimes the probability of an act occurring can be more significant than how often it will occur or how long it will last.

For example, when an individual must maintain proximity to another for a set period of time to achieve some goal, then the individual's concern may be the probability of being attacked in that period of time, rather than how many blows will be struck (frequency) or how long the attack will last (duration). Being able to predict how others will respond is the essence of social skill. The probability of various responses to one's behavior, by another, influences what course of action should be selected to achieve a goal at the least cost. The probability of favorable and unfavorable outcomes will certainly vary with the identity of the partner and partner identification is essential for the development of social skills, but it may also be important to predict the probability of a particular response on the basis of the duration of proximity with that partner. If the duration of the behavior an individual wishes to engage in while in proximity to a particular partner exceeds the typical duration that that individual will tolerate that individual's presence at that distance, then it might be wise to postpone or forego that behavior. One must assess not only the distance that must be maintained between oneself and a dominant individual that may wish to contest for a particular resource, but also the probability that the opponent will actually respond to one at a particular distance. One–Zero data are collected by investigators, but animals assess the probability of action by an opponent on the basis of what they are doing, at what distance, and for how long. Although One–Zero data provide no information on the frequency or duration of an opponent's response, sometimes it is the likelihood of a response that is key to a decision, rather than the frequency or duration of that response. Being attacked can be of more concern than the duration of the attack or the frequency of elements in that attack.

Activity Profiles

Having collected data with a specific Ethogram, an investigator then may try to categorize a group or species by the amount of time it spends in each major activity. These data are summarized by time of day and by season or weather conditions to detect diurnal patterns in behavior, if they exist, and to note seasonal and weather adjustments of behavior by an individual or a group. Differences in how members of specific age–sex classes allocate their time can be examined and questions can be answered about how food supply or rainfall influences the amount of time spent resting, feeding, traveling, etc. Sometimes nonintuitive results are found as when animals travel less far while searching for food during the dry season, when food is scarce and most difficult to find. This may occur because the animals cannot search for food in some areas during the dry season because they need water to drink at some measured intervals and the time required to travel from the food sources in these areas to the nearest water source exceeds the time interval that the animals can go without drinking.

Activity profiles can be used to compare groups under different ecological conditions, to compare the same group seasonally or diurnally, to compare different age–sex classes, and to compare groups of different species. The latter is especially useful when two species are sympatric, living in the same geographically defined area. Differences in activity profiles between the two species may indicate that, although the two live in the same place, they have different ecological niches, that is, they use the available resources differently. Such differences can allow two species to share the same home range while minimizing competition between them.

Energy Budgets

Investigators may also try to theoretically determine how animals must allocate energy for survival in an area. This results in an Energy Budget that considers what the various needs of an individual are for survival, how long it takes to satisfy each need, and how long the individual can go before it is once again necessary to respond to that need.

For example, an individual may need to eat every 6 h and therefore must never be more than 6 h away from food, and it must drink every 24 h and so must never be more than 24 h from water. At the same time, it may need to rest in a secure location every 12 h. If resting takes 4 h, eating takes 1 h each time, and drinking takes 5 min, one can find whether the distribution of resources in an area can accommodate all of an individual's needs after considering the travel time between the various resources.

For example, in this case if food and water are not in the same place, then food must always be within a 5 h walk from the nearest water (they must eat every 6 h and it takes 1 h to eat) and resting locations must always be within a 2 h walk to food (they rest for 4 h and must eat every 6 h). When resource distributions change, or when the time to obtain the resource changes (e.g., feeding on soft fruit may require little processing time, whereas searching for insects, or extracting embedded nuts may require far more processing time to obtain the same amount of energy), one can calculate the adjustments in time allocations that must be made in order to satisfy the individual's basic needs.

Of course, something like food is complex for food represents not only calories but also specific nutrients which vary by the type of food. Moreover, most foods contain some level of toxins and the individual can only detoxify a certain amount in a given period of time thus limiting the amount of that kind of food that can be eaten at one time. An Energy Budget must consider all of the various aspects of 'food' in considering how much time must be allocated to obtaining each required type and what maximum benefits can be obtained from one type before limits, such as the ability to neutralize toxins in the food, are reached. By calculating an Energy Budget and plotting the location of resources, one can assess potential habitats as either suitable or not suitable for a particular taxon. This may explain why animals are not present in some locations that, at first glance, seem like suitable habitat. Such considerations are very important in conservation planning and refine estimates of potential habitat on the basis of broad ecological classifications.

Social Factors

The study of social interactions adds complexity to the study in that at least two individuals must be considered simultaneously and each acts as a particular stimulus for the other and may change its stimulus value by changing its behavior (communication). Social resources are also harder to analyze in terms of how often they are needed, how long it takes to reestablish social relationships after separations, etc. Reproduction in most animals is necessarily social as is the rearing of young. Defense against attack, huddling for warmth, and even hunting for prey are social in socially living animals. Although hunting need not involve joint hunting or cooperation or sharing, socially living animals must coordinate their activities with other group members so that these activities do not disperse group members to the extent that they cannot readily reunite. In every case, the investigator must define the Ethogram of responses used to study the area of interest and produce quantitative data that can be used for analyses. Depending on the hypotheses under investigation, different aspects of behavior may be studied and we may complain that an experiment failed to collect data on some aspect of particular interest to us (but not necessarily to the original investigator).

No study can score everything and so each of us must carefully explain why we scored what we did and how we did it. A good experiment should be replicable, and to do that the methods, including the Ethogram, must be presented in detail. In our replications, we can collect additional data but the replication must use exactly the techniques used to collect the original data. In this way, Science builds on the past.

Further Reading

Altmann J (1974a) Observational study of behavior: Sampling methods. *Behaviour* 49: 227–265.

Altmann SA (1974b) Baboons, space, time and energy. *American Zoologist* 14: 221–248.

Ardener E (1984) Ethology and language. In: Harre R and Reynolds V (eds.) *The Meaning of Primate Signals*, pp. 111–115. Cambridge: Cambridge University Press.

Bernstein IS (1991) An empirical comparison of focal and ad libitum scoring with commentary on instantaneous scans, all occurrence and one–zero techniques. *Animal Behaviour* 42: 721–728.

Bobbitt RA, Jensen GD, and Gordon BN (1964) Behavioral elements (taxonomy) for observing mother–infant–peer interaction in *Macaca nemestrina*. *Primates* 5: 71–80.

Dohlinow P (1978) A behavior repertoire for the Indian langur monkey (*Presbytis entellus*). *Primates* 19: 449–472.

Hinde RA and Rowell TE (1962) Communication by postures and facial expressions in the rhesus monkey (*Macaca mulatta*). *Proceedings of the Zoological Society of London* 138: 1–21.

Reynolds V (1975) Problems of non-comparability of behaviour catalogues in single species of primates. In: Kondo S, Kawai M, and Ehara A (eds.) *Contemporary Primatology*, pp. 280–286. Basel: S. Karger.

Reynolds V (1976) The origins of a behavioral vocabulary: The case of rhesus monkey. *Journal for the Theory of Social Behaviour* 6: 105–142.

Van Hooff JARAM (1962) Facial expressions in higher primates. *Symposium of the Zoological Society of London* 8: 97–125.

Ethology in Europe

M. Taborsky, University of Bern, Hinterkappelen, Switzerland

The Very Beginning

Like for other aspects in biology, Charles Darwin can be safely viewed as the great grandfather of ethology. In his endeavor to understand the continuity between biological traits across different taxa, including humans, he realized that behavior was an important component of his evidence for biological evolution. He devoted a whole chapter of *On the Origin of Species* to the study of instinct, and in *The Expression of the Emotions in Man and the Animals*, Darwin observed: "With mankind some expressions, such as bristling of the hair under the influence of extreme terror, or the uncovering of the teeth under that of furious rage, can hardly be understood, except on the belief that man once existed in a much lower and animal-like condition."

Despite such precursors, the systematic study of animal behavior in Europe began with the work of Oskar Heinroth, director of Berlin Zoo and Aquarium from 1913 to 1945. The term 'ethology' was coined in America by the zoologist Morton Wheeler in 1902, and the eminent American biologist Charles Otis Whitman, and his student Wallace Craig, had already gained significant insights into the behavior of animals by their systematic and comparative studies of pigeons. In Europe, Heinroth published his *Beiträge zur Biologie, namentlich Ethologie und Psychologie der Anatiden* (contributions to biology, especially ethology and psychology of the anatids) in the *Verhandlungen des 5. Internationalen Ornithologen-Kongresses in Berlin* in 1910, which might be considered as the first substantial 'ethological' publication. Oskar Heinroth was particularly interested in the behavioral displays, social signals, and rituals among conspecifics. In good Darwinian tradition and using ducks, swans, and geese, he argued that species-specific instinctive behavior, such as morphology, could be used to determine the genetic affinities of different taxa, a view he shared with C.O. Whitman from Chicago. Together with his wife Magdalena, Heinroth raised numerous young of a great variety of bird species in 'Kaspar Hauser conditions,' that is, in isolation from other birds, with the ambitious aim to systematically study the instinctive behavior of all the bird species of central Europe. This massive piece of work was published from 1924 to 1934 in the four-volume classic of *Die Vögel Mitteleuropas* (The Birds of Central Europe).

Development of Concepts

Heinroth's work was influential to a number of naturalists, mostly ornithologists, who readily accepted his view that behavior is a species-specific trait suited for comparative studies in order to reconstruct phylogeny. Most prominent among his disciples was to become the Austrian zoologist Konrad Lorenz, a zealous observer of behavior with an unrivalled intuitive knowledge of animals. Lorenz was strongly influenced by the idea that instinctive behavior patterns are innate and invariable, rendering them suitable to reconstruct phylogenies just like any other bodily structure. In this view, instincts are equivalent to organs. Lorenz regarded the concept of instinctive behavior patterns not as a hypothesis to be tested but as a fundamental assumption. Like his much-admired mentor Heinroth, Lorenz used mainly birds as model systems, particularly because of the opportunity to raise them by hand and due to their similarity to humans with regard to sensory biology, conditions greatly facilitating systematic study. Lorenz' declared approach was to start with impartial observations devoid of any hypothesis. On the basis of these observations, he developed ideas and concepts, above all, the hypothesis of 'innate releasing mechanisms.' He also took up Heinroth's concept of imprinting (which the latter had never called by this term) and elaborated it, first in his seminal paper on *Der Kumpan in der Umwelt des Vogels* (The companion in the bird's world) that was published in German in 1935.

Interestingly, Lorenz and the other early ethologists in Europe focused on questions that had been of interest mainly to psychologists, despite their highly divergent approach. The emphasis was on internal states and environmental influences, but the ethologists' concern extended beyond the question of immediate causation of behavior. Comparative Ethology was, thus, intended to be an alternative to (North American) behaviorism – being mainly a psychologists' domain – and Pavlovian 'reflexology.' The European naturalists' aim was to teach psychologists the comparative and evolutionary approach, and the necessity to understand the behavior of whole animals in their natural environment. The preeminent advocate of this approach was Niko Tinbergen, a Dutch naturalist, who turned out to become, besides Lorenz, the second founding father of ethology. Together, they developed the concept of instinct based on a hierarchical organization of drives, assuming as key categories of behavioral causality action-specific energy and consummatory acts, together with innate releasing mechanisms. These concepts were strongly shaped by an imagination of energy flow, but without assuming a material counterpart. The most conspicuous concepts of the day were Lorenz' psycho-hydraulic model and Tinbergen's hierarchical organization of drives.

The concepts of instinct, developed first and foremost by Lorenz and Tinbergen, were important sources of testable hypotheses enabling young researchers to scrutinize resulting predictions by experimental, physiological, and cybernetic approaches. It is important, however, to consider these developments in a historical context. In the end, the original concepts proved to be too simplistic and partly mistaken, and considered in retrospect, they seem to have lead nowhere. If the concepts of the 'classical' period of ethology might now be viewed as scientific dead ends, at the time they were immensely important for the emergence and development of a behavioral science. Researchers like Gerard Baerends at Groningen, Robert Hinde at Oxford, and the American comparative psychologists Theodore Schneirla and Daniel Lehrman, were sparked by early ethological theory and took it to its first acid test. The intense discussion that ensued proved to be as important to the field as was the preceding development of the theory.

Lorenz and Tinbergen

Despite their common interest and general agreement, the roles of that important, odd couple, Lorenz and Tinbergen, diverged greatly. Lorenz was mainly concerned with the causality of behavior as a result of the interaction of external stimuli and innate, pre-programmed action patterns. He was interested in imprinting as a mechanism causing long-lasting effects of early experience and in the origin and function of ritualized social interactions. His major tool was the handraising of birds, mainly ducks and geese, which enabled him to observe his study organisms at close hand and largely undisturbed. Lorenz was no keen systematic experimenter but had a rather philosophical interest in the mechanisms underlying behavior at any level. In contrast, Tinbergen loved the systematic experimental approach, which he preferably used under field conditions. He formalized the concept of instinct and strived for closer connections between the causal analysis of behavior and a physiological approach. Like Lorenz, he was a keen naturalist and an exceptionally gifted observer, but in addition, he was also an excellent experimenter.

Over several decades, the rise of ethology in Europe was firmly in the hands of these two characters. Even if Karl von Frisch shared the 1973 Nobel prize of medicine and physiology with them for establishing ethology as a scientific discipline, von Frisch's role in this endeavor was negligible. He was an outstanding experimenter and his research on the dance language of bees is a masterpiece, but he was not actively involved in the conceptual and practical establishment of ethology. In contrast, Lorenz and Tinbergen, in their own ways, strived for constructing the foundations for a scientific study of behavior. While Lorenz was the hyperactive motor developing and disseminating concepts of behavioral regulation, outspoken, self-confident, sometimes dogmatic, but well connected among his colleagues, Tinbergen was more modest and careful in his attempt to gain attention for his ideas, and for the sake of ethology, always integrative in his actions and arguments, but nevertheless pushing hard towards the establishment of ethology as a biological science on an equal footing with, say morphology or physiology. Tinbergen realized that confining ethology to the study of causation would be too narrow a discipline. He wanted survival value and evolution incorporated in a systematic study of behavior, just as ontogeny and causation – the four 'Tinbergen questions' that today still serve as the undisputed guideline for the study of behavior.

Eventually, ethology gained momentum, and Lorenz and Tinbergen obtained prestigious positions with the Max-Planck Society and at the University of Oxford, respectively. Their continuous effort to establish this new discipline included the foundation of scientific journals, with *Ethology* being first in the field, cofounded by Lorenz under the name *Zeitschrift für Tierpsychologie* (*Journal for Animal Psychology*) in 1937. They organized international conferences that became increasingly popular, with the 1952 meeting at Lorenz' Max-Planck Institute in Buldern being the first of a series of 'International Ethological Conferences' (IECs) that still continue today, with biannual intervals. These activities spawned numerous disciples, and ethology groups emerged at various universities all over Europe. The most conspicuous centers of ethology in postwar Europe were the Max-Planck-Institute for Behavioural Physiology at Seewiesen, where Lorenz finally gained ground, and groups at the Universities of Oxford, Cambridge, Leiden, and Groningen. For several decades, Germany, Great Britain, and The Netherlands stood at the forefront of this new discipline. While the research directions pursued in German and British ethology diverged increasingly, ethology in the Netherlands kept its broad scope.

Regional Divergence

The policy of the German Max-Planck Society was to merge ethology with physiology, which defined the research activities at the Max-Planck-Institute for Behavioural Physiology at Seewiesen from the start (1958) right to its end (1999). This was an obvious combination, not only because, in essence, behavioral causation is a physiological problem, but also because of the strong position of physiology in biological science in Germany; its keen focus on sensory and neural physiology prepared the ground for the study of behavioral mechanisms at the level of organisms, organs, and cells. In Britain, on the other hand, where the tradition was strong in ecology and evolutionary thinking, the interest in the adaptive value of behavior gained momentum. While Tinbergen at Oxford strived to keep behavioral research in balance between the ultimate and proximate levels of explanation, his successors abandoned three of the four legs of ethology and capitalized entirely on the adaptive value of behavior – giving birth to the most conspicuous subdiscipline of ethology called behavioral ecology.

Behavioral Ecology and Beyond

The onset and expansion of behavioral ecology may be seen as the second success story of ethology, after its hype in postwar Europe that followed the birth of the discipline. It is also worth noting that behavioral ecology brought about the unification of behavioral research in Europe and America, despite the divergent history of this science on these two continents. The consequent application of evolutionary theory to the study of behavior revolutionized not only ethology, but the whole of organismic and integrative biology. William D. Hamilton of University College, London, achieved a breakthrough with his theoretical model of *The genetical evolution of social behaviour* published in 1964, where he concluded that "Species following the model should tend to evolve behavior such that each organism appears to be attempting to maximize its inclusive fitness." This caused a paradigm shift from the focus on the maximization of individual fitness to the inclusion of effects on kin, which helped to understand the evolution of altruistic behavior such as known from eusocial insects – a dodgy puzzle to Charles Darwin.

The other significant conceptual progress in the early days of behavioral ecology was its focus on the fitness effects of biological traits, which paved the way to the cost-and-benefit analysis of behavior. Role models of this development include John Krebs and Nicolas Davies from Oxford University, who developed elegant experimental tests of qualitative and quantitative predictions of foraging, aggressive, and reproductive behaviors. They spread the ideas and concepts of this new approach to the study of animal behavior by editing an influential series of books (*Behavioural Ecology: An Evolutionary Approach* I–IV, 1978–1997). John Maynard Smith from Sussex and Geoffrey Parker from Liverpool proposed a game theoretical approach to the study of behavioral decisions, which had been used already long before in economy and military strategy. At the Max-Planck-Institute in Seewiesen Lorenz' successor, Wolfgang Wickler, pursued a strict functional approach to behavior that had lasting effects on the development of ethology in Europe by the dispersion of his disciples. In the Netherlands, behavioral research centers at the Universities of Groningen, Leiden, and Utrecht thrived with a balanced approach to behavior that, remarkably, was not as narrowly confined to the study of its adaptive value as was the custom at many other centers of ethology. Countries in Scandinavia and Southern Europe developed their own strong research programs in behavioral ecology, which have contributed substantially to the progress in theoretical and empirical behavioral research in Europe.

Despite this positive development, the unfavorably narrow focus of behavioral ecology on the adaptive value of behavior proved to obstruct a comprehensive understanding. The study of the causation of behavior was largely left in the hands of specialists from other disciplines, such as neurophysiology, endocrinology, developmental biology, and experimental psychology. Despite the scientific progress in these fields, a synthetic view of the origin, causation, function, and evolution of behavior has been largely missing. It seems safe to conclude that in recent years, ethology in Europe has somewhat regained its balance. Behavioral development, causal mechanisms, adaptive value, and evolutionary patterns, all have their firm place in behavioral research and university curricula at literally hundreds of institutions distributed all over Europe. National and supranational behavioral societies thrive, small and large meetings on all aspects of behavioral research are popular, and the job market for young researchers in behavior shows promise for a bright future of this discipline. Ethology, or 'behavioural biology' as it is often called, seems to have come of age.

See also: Behavioral Ecology and Sociobiology; Comparative Animal Behavior – 1920–1973; Darwin and Animal Behavior; Development, Evolution and Behavior; Future of Animal Behavior: Predicting Trends; Integration of Proximate and Ultimate Causes; Konrad Lorenz; Neuroethology: What is it?; Niko Tinbergen; Psychology of Animals; William Donald Hamilton.

Further Reading

Bateson PPG and Klopfer PH (eds.) (1973–1995) *Perspectives in Ethology,* vols. 1–11. New York: Plenum Press.
Burkhardt RW (2005) *Patterns of Behavior. Konrad Lorenz, Niko Tinbergen, and the Founding of Ethology.* Chicago: University of Chicago Press.
Hamilton WD (1964) The genetical evolution of social behaviour. I and II. *Journal of Theoretical Biology* 7: 295–311.
Hinde RA (1966) *Animal Behaviour. A Synthesis of Ethology and Comparative Psychology.* New York: McGraw-Hill.
Kappeler P (ed.) (2010) *Animal Behaviour: Evolution and Mechanisms.* Berlin: Springer.
Krebs JR and Davies NB (eds.) (1978–1997) *Behavioural Ecology: An Evolutionary Approach,* vol. 1–4. Oxford: Blackwell Science Ltd.
Lorenz K (1950) The comparative method in studying innate behaviour patterns. *Symposia of the Society for Experimental Biology* 4: 221–268.
Maynard-Smith J (1982) *Evolution and the Theory of Games.* Cambridge: Cambridge University Press.
McFarland D (1985) *Animal Behaviour.* Essex: Addison Wesley Longman Ltd.
Tinbergen N (1963) On aims and methods of ethology. *Zeitschrift für Tierpsychologie* 20: 410–433.

Evolution and Phylogeny of Communication

T. J. Ord, University of New South Wales, Sydney, NSW, Australia

The Legacy of Evolutionary History

In broad terms, the way in which animals communicate with one another is dependent on three key factors: the function of a signal, the ability of receivers in the environment to detect a signal, and the historic interaction of these factors that may or may not be the same as they are today. For instance, many mating calls or displays given by males to entice females to mate with them are under selection to advertise each male's quality as a mate; for example, that he is in good condition or likely to be a good parent. All animal signals must also travel through the environment before reaching receivers, during which time the message of the signal can become degraded because of reverberation, masking caused by noise, or a number of other factors. Certain types of signals are more resistant to environmental degradation than others and some signal types are also better at stimulating the sensory system of receivers than others. Yet, how communication evolves in response to social and environmental selection pressures is dependent on an animal's evolutionary history. Put simply, the evolution of communication generally builds on what was present in evolutionary ancestors. Two species may produce a signal to attract mates and even face the same challenges of transmitting that signal through a noisy environment, but the type of signal each species evolves might be quite different because each species is descendent from different evolutionary ancestors.

Studying how the evolutionary history of a species has contributed to present-day forms of communication presents a difficult challenge for researchers. Behavior is an aspect of an animal's phenotype that rarely leaves a trace in the fossil record. How then is it possible to know what communication was like in evolutionary ancestors, let alone attempt to understand the selective pressures that might have acted on historic signals and shaped the subsequent direction of signal evolution? By comparing the similarities and differences in the form of communication used among closely related species, we can map this variation onto an evolutionary tree, known as a *phylogeny* (**Figure 1**), and extrapolate back in time to decipher what evolutionary ancestors might have been like and the types of selection pressures they probably faced.

Let us consider an example of mate choice in a group of freshwater fish species from Central America known as swordtails. These fishes got their name because males possess an elongated tail filament that looks like a long sword protruding from the tail fin. Researchers have shown that females prefer males with longer swords to males with shorter swords. The sword as it is used today advertises to females that a male is in top-notch condition because he can deal with the 'handicap' of having an exaggerated ornament that impedes swimming, yet he can still forage and avoid predators successfully.

Surprisingly, females in other closely related fishes in which males do not possess a sword also exhibit a preference for swords when researchers artificially attach swords to male tails. Obviously, the handicap principle cannot explain the presence of a female preference for tail filaments in species that lack this male ornament. With a clever series of mate choice experiments on different species that did and did not naturally possess swords and mapping findings onto a phylogeny, researchers were able to reconstruct the evolutionary history of both the sword and the female preference for the sword. This is because closely related species often share similar features inherited from a common ancestor. If females of two sister species both show a preference for male swords, it is likely that the common ancestor of these two species also had a similar preference. Using this comparative approach to extrapolate back in time, it was discovered that females evolved a preference for the sword *before* the sword evolved in males. How can females exert a preference for an ornament as specific as a sword when it does not actually exist in males? One explanation is that females did not initially fancy male swords per se, but large male size more generally. Body size in many animals is a good indicator of condition. A cheap way for males to tap into a female's preference for large size is to evolve an elongated tailfin to give the appearance of large size. As swords became increasingly longer over evolutionary time they subsequently became reliable indicators of condition in their own right because of the increased costs associated with impeded swimming performance. By combining information on phylogeny, the communication systems used by species today and the factors that influence signal production and reception, it is possible to use the comparative method to gain considerable insight into how communication has evolved.

In the rest of this section, I will elaborate on the general concepts of how phylogeny and the comparative method can be used to understand the direction and mode that evolution has taken and how phylogenetic approaches can be used to identify the adaptive processes that have shaped the design of animal signals. In the final section, I will provide a brief overview to some of the available methods

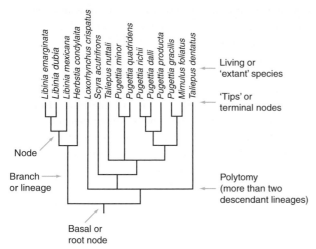

Figure 1 Common phylogenetic terms illustrated by a phylogeny of the decorator crabs. Courtesy of Kristin Hultgren.

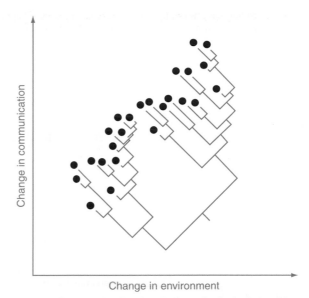

Figure 2 Scatter plot showing the hypothetical relationship between the form of communication used by different species and the environment they live in, combined with the underlying phylogenetic relationships among species. Disregarding phylogeny, a strong positive relationship appears to exist. However, in actual fact much of the variation in communication and the type of habitat species live in is inherited from evolutionary ancestors.

that comparative biologists can use to reconstruct ancestor states, estimate the extent to which communication is dependent on evolutionary history, and the ways in which researchers can test hypotheses for the adaptive significance of communication.

Phylogeny and the Trajectory of Signal Evolution

A lot can be learnt from simply documenting when and to whom animals produce signals, and how those signals subsequently affect the behavior of receivers. Experiments might then follow to test the function and/or selection pressures believed to act on signal production through direct manipulation. This might consist of calls being recorded and played back to animals to determine that some call types are preferred by females to others, or the experimenter might induce changes in physiology that affect the production of signals (e.g., manipulations of the general condition of the signaler through dietary supplements) to confirm that signals convey reliable information on the condition of the sender. If a factor has been a general influence on the evolution of communication, then we would also expect other species in the same situation to exhibit similar characteristics and this leads to the obvious comparison of communication systems across species.

In much the same way that you might look more similar to your brother or sister compared with somebody randomly picked out from a crowd, closely related species often share behaviors and have similar ecologies, because they retain those attributes from a common evolutionary ancestor. We therefore need to be a little careful in how we perform comparisons across species. If we do so without regard to the phylogenetic relationships of the species examined, we could erroneously conclude that an association between a signal characteristic and some other factor exists when in fact they occur together because both the

signal and putative 'causal' factor have been inherited together from a common ancestor. Statistically, the complication of treating data from closely related species as independent when in fact they are not is known as 'pseudo-replication' (**Figure 2**). The extent to which closely related species share similar phenotypes, including the form of communication they use, is measured by estimating *phylogenetic signal*. A high phylogenetic signal indicates that the evolutionary relationships between species predict phenotypic similarities between those species – species that share a recent common ancestor also share a particular trait – whereas low phylogenetic signal reflects that species' phenotypes are unrelated to their phylogenetic relationships – species that share a recent common ancestor do not share the same trait (**Figure 3**). A related term is *historical contingency*: the tendency for evolutionary elaborations or changes in descendant species to be modifications of historic phenotypes. Historical contingencies therefore relate to changes contingent on what has already evolved in the past.

The degree to which species diverge from evolutionary ancestors is dependent on a number of factors, but at the core of these factors is the rate of genetic mutation. If mutations are rare, the genetic basis of communication will change very little over long periods of evolutionary time, even in the face of strong selection for modification. However, it is difficult to determine the extent to which low mutation rates explain the retention of particular forms of communication over evolutionary time.

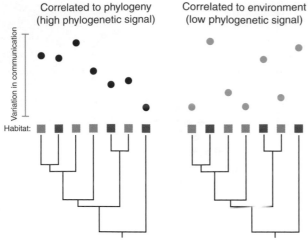

Figure 3 Animals may share communication systems through common ancestry or because they have converged on similar forms of communication when living in similar habitats through natural selection. Each dot represents the mean value of a species for some aspect of their communication, while the squares represent the type of habitat occupied. For example, the graph could illustrate the frequency of a mating call in frogs living in either open grasslands (orange) or closed tropical rainforests (chocolate).

There are statistical methods available for estimating the correlation between communication signals and phylogeny (i.e., the phylogenetic signal), but this can reflect a number of factors, not just low mutation rates. For example, if morphology influences the production of a communication signal (e.g., the shape of the vocal track influences the type of calls that an animal can produce) and if morphology exhibits little evolutionary change because it needs to maintain a functional capacity in another unrelated context (e.g., modifications to the vocal track might influence the ability of an animal to breathe or feed properly), then this will constrain evolutionary change in communication. By the same token, communication signals may change over evolutionary time not because of an adaptive response to selection, but because of random genetic change that culminates in arbitrary changes in signal design.

Ecological determinism is sometimes billed as the antithesis of historical contingency: it refers to conditions where the ecology of an animal primarily drives evolutionary change. Ecological determinism is likely to be an important factor affecting communication, because communication is a phenotypic trait that is often influenced by environmental conditions. Indeed, closely related species living in different habitats are expected to produce divergent forms of communication as species evolve signals suitable for communication in their respective environments. Conversely, distantly related species occupying similar habitats should converge on similar forms of communication (e.g., **Figure 3**). Such *convergent evolution* in which remarkably similar characteristics evolve independently in different species in

response to common selection pressures provides some of the most compelling evidence for adaptive evolution.

Both divergent and convergent evolution are expected to be associated with low phylogenetic signal, because communication in species that exhibit these forms of evolutionary change is dependent on the ecology and not the phylogeny of the species in question. However, this is not to say that high estimates of phylogenetic signal automatically exclude the possibility of adaptation. Ecological determinism can also lead to *stabilizing selection* in which forms of communication are conserved even after species split from evolutionary ancestors. For example, *niche conservatism* occurs when closely related species occupy similar environments, perhaps because they are already adapted to a certain habitat type. While the environment is still an important source of selection acting on communication, because closely related species live in very similar environments, they will also tend to produce similar forms of communication through selection. Communication that conveys honest information on an animal's condition can also exhibit high estimates of phylogenetic signal. There are only so many signal characteristics that can serve as quality indicators and once these evolve, they will tend to be retained with little modification over evolutionary time. Thus, communication may be very similar between species and exhibit high phylogenetic signal, yet still be under the influence of selection.

Mode, Pattern, and Rate of Signal Evolution

A question of special interest to evolutionary biologists concerns the mode by which evolution occurs. The classical perspective of Charles Darwin views evolution as small incremental changes accumulating over long evolutionary time scales. This mode of evolution predicts a series of 'intermediate' links between ancient species and those in existence today, but these are often lacking in the fossil record. Instead, new species seem to appear suddenly in the fossil record and live relatively unchanged for long periods of evolutionary time.

Some paleontologists have suggested that this pattern of sudden changes followed by stasis is a true representation of how evolution occurs. According to this nonincremental view, the formation of new species is accompanied by an intense period of selection that results in a rapid spurt of evolution, followed by long periods of relative stasis where species change very little (perhaps because of stabilizing selection). This process would lead to a more 'punctuated' rather than a 'gradual' mode of evolution (**Figure 4**). Behavioral ecologists might find this debate somewhat esoteric, but it can be quite relevant to the study of communication.

The use of signals during mating is a popular topic in the study of communication, especially how female mate choice might lead to *directional selection* on males for

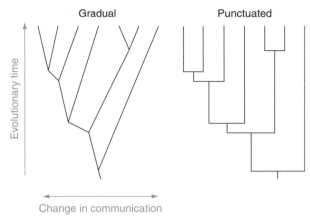

Figure 4 How the evolution of communication would appear under a gradual mode of evolution, where change accumulates over time, or under a punctuated mode of evolution, where change is concentrated at species events (represented by the nodes of the phylogeny, see **Figure 1**) followed by periods of stasis. Some forms of sexual selection on communication can be expected to produce a gradual accumulation of change over time, while communication important in species recognition should follow a more punctuated pattern.

increasing elaboration of sexual ornaments and other signals. This process should result in a gradual mode of evolution as modifications culminate over evolutionary time into increasingly more complex mating signals. To produce viable offspring, females must choose a male that is not only in good condition (for example) but also of the same species. Signal characteristics important in species recognition should be under considerable stabilizing selection and only subject to change during speciation when divergence in communication systems between populations is expected to be rapid, a process that should result in a punctuated mode of evolution.

The mode of evolution, whether it is gradual or punctuated, can also help to elucidate the function of different components making up the same communication signal. One of the main hypotheses explaining the evolution of elaborate multicomponent or multimodal signals (signals made up of components that use different sensory modalities, such as an acoustic signal – song – and a visual signal – colorful plumage or courtship dance) is the need to convey multiple messages. One component of a mating signal might communicate "I am of good condition" while another might communicate "I am of the right species," which will lead to predictable differences in the mode of evolution that each component has taken. We can use phylogenetic comparative methods to determine which component has evolved gradually – consistent with sexual selection – or which component exhibits rapid bursts of evolution followed by relative stasis – consistent with species recognition. Despite being a powerful approach for testing the functional significance underlying the

evolution of different signal components, few studies have documented whether different components used by animals in communication have evolved via different modes of evolution.

We can also infer the presence of potential adaptive functions driving the evolution of communication by studying other phylogenetic patterns. Of particular interest is whether evolutionary changes are skewed toward the tips or base of a phylogeny, and whether new signal components are added to or replace previously existing components (**Figure 5**). Communication critical to species recognition should result in evolutionary changes in signal components skewed toward the tips of the phylogeny, because the evolution of species-typical signals will tend to result in new signal characteristics replacing preexisting forms, essentially 'erasing' similarities in communication between sister taxa. Communication that advertises the condition of the signaler or some other quality indicator will tend to be conserved with little modification in descendant taxa. This will tend to result in new signal components evolving early or toward the base of the phylogeny. Novel signal components may subsequently evolve, but because characteristics conveying honest information are costly to maintain (otherwise they could be 'faked'), innovations will tend to replace previously existing signal components. Overall, this will result in low diversity in signal designs across species because signal components will tend to be retained from evolutionary ancestors, but will also result in more instances of evolutionary convergence between distantly related species as similar honest indicators evolve independently in different groups.

We can also expect the rate of evolution to differ depending on the type of selection acting on communication. Unless constrained by low mutation rates, signal characteristics subject to intense forms of selection should evolve extremely rapidly. Yet, even when subject to the same selection pressure, different components making up the same signal can exhibit very different rates of evolution. Multimodal signals are the product of different physiological and morphological attributes in the sender and rely on different sensory systems in receivers for detection. Elaborations of song are generally expected to evolve more freely than ornaments such as body coloration or elaborate plumage, because the latter are tied to morphological rather than behavioral adaptations. Vocalizations are also dynamic signals that can be turned 'on' and 'off,' but conspicuous ornaments are static and remain permanently 'on' unless animals can shed them during periods when they are not required (e.g., winter plumages in birds are sometimes drab compared with the bright, colorful breeding plumages in the summer months). Conspicuous forms of communication attract the attention of not only intended receivers such as mates, but also unintended receivers such as predators. If animals live in an environment with lots of predators, this opposing selective

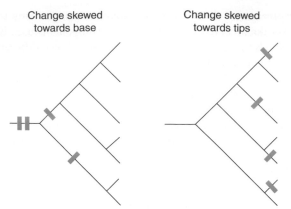

Change skewed towards base   **Change skewed towards tips**

Figure 5 Hypothetical distribution of character evolution where signal components evolved early in the history of a group and are concentrated at the base of the phylogeny, such as components that function as quality indicators, or where the evolution of new signal components appears skewed towards the tips of the phylogeny, such as those that might be important in species recognition.

pressure can limit evolutionary change in conspicuous morphology, while dynamic signals are more free to vary.

Correlated Evolution

I have outlined how communication that is important in species recognition or subject to different types of sexual selection or genetic/physiological/morphological constraints will leave telltale signatures in the mode, pattern, and rate of signal evolution. We can also explicitly test for associations between signal characteristics and the factors predicted to result in these evolutionary signatures. For example, the evolution of mating signals will be tightly linked to the intensity of sexual selection. For those species in which the sex ratio is heavily male-biased, females are a limited resource and males must compete intensely with one another for mating opportunities. One result of this increased competition is the evolution of more elaborate mating signals in males, which predicts a positive correlation across species between how elaborate male mating signals are in species and the degree to which sex ratios are male-biased. Or perhaps we have a hypothesis that certain habitats select for particular forms of communication (e.g., **Figure 3**). In both instances, we can incorporate phylogeny into our statistical analyses to partition out the potential confounding affect of shared evolutionary histories and measure how variation in communication across species can be accounted for by social or environmental factors.

Indeed, tests of correlated changes between communication and predicted social or environmental influences are the most common use of phylogenetic comparative methods by behavioral ecologists and there are many studies that provide examples of what can be learnt by

this approach. In my own work on the evolution of color signals in dragon lizards, I have used phylogenetic correlation analyses to show how the diversity of colorful morphologies found across species can be accounted for by the intensity of sexual selection males experience. Furthermore, the type of color signals species have evolved is heavily dependent on whether species live in habitats where lizards are more prone to predation by birds or where communication is more difficult because of visual obstructions and poor habitat light. Other examples are listed in the suggested reading.

Phylogeny, Cultural Inheritance, and Plasticity

Explaining the diversity in animal communication need not be limited to investigating differences and similarities in communication across species. Different populations of the same species can also vary in communication. It could even be argued that if we truly want to understand the processes leading to signal divergence between species, we really should be investigating differences in communication among populations within species, which represent the starting points of evolutionary divergence. A critical step in the evolution of new species is the formation of reproductive barriers between populations. One of the key factors believed to limit members of different populations from interbreeding with one another, and to allow speciation to subsequently occur, is divergence in communication systems, especially those important in mating. The only way to detect divergences in signals that occur prior to speciation is to investigate variation in communication systems at the population-level.

A cautionary note needs to be made here about the assumptions underlying how changes in signal characteristics are acquired when studying signal variation among populations compared with studying signal variation between species. The implicit assumption when studying differences in communication between species is that there is a genetic basis to these differences. This may not always be true between populations, especially when aspects of communication are learned or culturally inherited. A distinction is sometimes made between 'vertical' – meaning phylogenetic – versus 'horzitonal' – meaning cultural – transfer of signal characteristics (**Figure 6**). One solution is to measure the phylogenetic signal; if it is high, then signals are more likely to be genetically inherited and if it is low, signals could be culturally inherited. The difficulty with interpreting what phylogenetic signal actually reflects is applicable here. As is true for comparisons across species, error in the measurement of signal characteristics inflates the estimated variation among populations (or species) and can lead to false inferences of low phylogenetic signal. Conversely, culturally inherited signals may exhibit high phylogenetic signal because adjacent populations are more likely to share

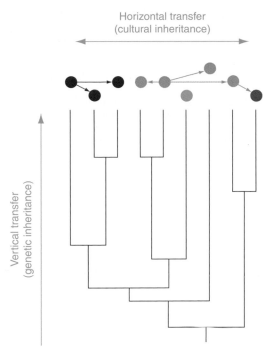

Figure 6 Communication transferred horizontally between populations through cultural inheritance, compared to vertical transfer via ancestry. In this example, the color of each dot might represent a song 'dialect' in different populations of a bird that is transferred through frequent contact or dispersal between populations. The connections between dots represent the social network or direction of dialect transfer.

similar culturally inherited signals because of their geographic proximity compared with more distant populations. Adjacent populations are also more likely to be genetically related and hence a phylogenetic analysis would indicate that communication is correlated with phylogeny.

'Plasticity' is another related issue particularly relevant to the study of the evolution of communication. Plasticity reflects the ability of animals to change their signals depending on the environmental or social conditions experienced at the time of communication. For example, birds in noisy habitats, such as those singing near highways, might produce songs that are much louder than birds singing in quieter areas. This difference could be genetic or plastic; birds in noisy habitats may be genetically predisposed to produce loud songs or alternatively birds in noisy habitats may have learnt to increase the volume of their songs. In the first instance, evolution may have occurred, while in the latter instance, it has not. The consequence of plasticity on the evolution of communication remains an open question and is relevant to investigations of signal variation at both the population and species level.

Neither cultural inheritance nor plasticity should necessarily exclude adopting a phylogenetic comparative

approach, but interpretations of the underlying mechanism used to explain signal variation should be done with caution. Consider an example where we wish to test the hypothesis that species produce communication signals ideally suited for transmission through the environment in which communication is typically conducted. When populations live in different types of habitat, environmentally induced divergence in communication is expected and should in turn promote reproductive isolation between populations. An obvious test is to confirm that differences in communication are correlated to differences in habitat, but we are not sure whether communication is culturally inherited or whether animals can learn to tailor signals to prevailing environmental conditions. Currently, the best approach is to apply a phylogenetic comparative method that explicitly measures the correlation between characteristics and phylogeny and identifies the remaining variance that is associated with the environment. This will control for any pseudo-replication resulting from phylogenetic relationships or factors that might mirror phylogenetic relationships (e.g., closely related, adjacent populations inheriting signal components culturally) and determine whether the environment facilitates divergence in communication. Unfortunately, it is difficult to assess whether communication divergence reflects genetic adaptation or plasticity. Obtaining the answer to this question requires intensive empirical study, such as detailed observations of how animals change the way they communicate according to environmental fluctuations and/or so called 'common garden' experiments in which individuals from one habitat type are transferred to another habitat type and assayed for behavioral change.

New methods are being developed that incorporate geographic proximity into comparative analyses, providing a more direct means to estimate how much of the variance in communication between populations is the result of genetic (phylogenetic/vertical transfer), geographic (cultural/horizontal transfer) or environmental factors. A phylogenetic comparative method that incorporates social network analysis to quantify more accurately the degree to which interactions between individuals from different populations influence the transfer of signal characteristics would be especially useful (e.g., **Figure 6**). For now, however, investigators will need to remain cautious with their interpretations of what the underlying mechanisms are that have lead to the observed correlations.

A Primer to Phylogenetic Comparative Methods

The number of programs available for applying phylogenetic comparative methods is daunting, so much so that it is difficult to know where to start or even what method is most appropriate for the question of interest.

Box 1

There are a number software packages available that combine a variety of methods together and can be run regardless of a user's platform (PC or Mac). All can be downloaded for free following a quick search online.

- *BayesTraits:* Can be used to test correlations between traits, estimate phylogenetic signal, the mode of evolution, and whether evolutionary change is concentrated towards the tips or base of the phylogeny.
- *COMPARE: Phylogenetic Comparative Methods:* Provides several programs to test correlations between traits using likelihood based methods or independent contrasts, estimates phylogenetic signal, and performs analyses for estimating the rate of evolutionary change in traits. Ancestor reconstructions can also be calculated using likelihood.
- *Mesquite: A Modular System for Evolutionary Analysis*: A particularly useful program for reconstructing ancestor states graphically, using parsimony and likelihood based methods. Options are also available for testing correlations using independent contrasts and other methods.

Most programs are free and available as downloads off the internet or by request from the program's author. I provide a list of some popular methods in **Box 1** and a brief outline of some commonly used techniques below. Readers interested in details of various techniques are referred to the suggested reading at the end of the section.

Ancestor State Reconstructions

There are several methods for reconstructing what communication systems may have been like in the past (**Figure 7**). The simplest methods are those based on parsimony, an algorithm that maps ancestor 'states' onto a phylogeny by favoring solutions requiring the least amount of evolutionary change. Parsimony approaches are often favored because of their straightforward computation, but they have also been criticized for lacking statistical rigor and for not presenting a realistic view of how evolution occurs. Other methods such as those based on least squares, maximum likelihood, and Bayesian techniques apply a probabilistic approach to finding a mathematical model of evolution that best fits the observed distribution of data across species on the phylogeny. These methods fit various scenarios of how evolution might have occurred – for example, gradual or punctuated, rapid or slow rate of evolutionary change – and calculate the probability that each explains present-day variation in traits. Once the model of best fit is identified, it is then used to assign ancestor states onto the phylogeny.

Phylogenetic Signal, Patterns, and Rates of Evolution

In fitting different models to the data, it is possible to use the parameters of the best fitting models to infer something about the correlation between characteristics and phylogeny or the mode of evolution a characteristic has likely followed. Methods that estimate phylogenetic signal do so by applying mathematical models that in effect transform the phylogeny, essentially stretching or shrinking phylogenetic branches, to simulate the evolution of a characteristic as if it were heavily dependent on phylogeny

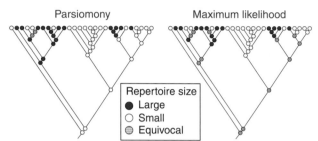

Figure 7 Alternative ancestor reconstructions of the repertoire size or number of distinct components making up communication in species. It is important to note that ancestor reconstructions will depend on both the method and the phylogeny used. In this example, differences are apparent at several nodes throughout the phylogeny. For instance, parsimony reconstructs a small repertoire in the root or most basal ancestor, while maximum likelihood assigns an equivocal state, meaning it is equally likely that this ancestor had a large or small repertoire.

or not at all. Similar transformations to the underlying phylogeny are used to estimate the likelihood that evolution has occurred via bursts during speciation followed by relative stasis, or through a more gradual mode of evolution. Regression slopes of estimated evolutionary change in a characteristic as a function of time since divergence from evolutionary ancestors can be used to estimate the rate of evolution: steeper slopes reflect more rapid rates of change compared with shallower slopes.

Correlation Tests of Adaptation

All taxa are related to each other in one way or another. Not incorporating phylogeny into statistical comparisons across species can subsequently lead to inflated rates of Type I statistical error (i.e., erroneously concluding that a significant effect exists when in fact it does not). Phylogenetic independent contrasts are the most commonly used comparative method for conducting correlation tests. It corrects for phylogenetic nonindependence by transforming species data into a set of differences or 'contrasts' between immediate relatives (**Figure 8**). In doing so, it assumes a null hypothesis that

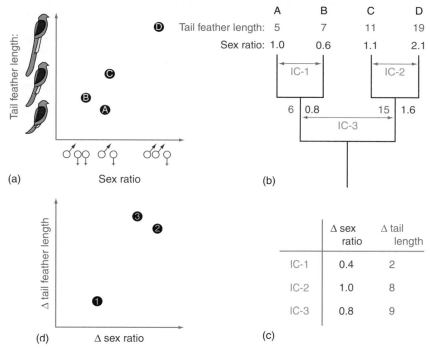

Figure 8 Hypothetical example of how a phylogenetic independent contrast analysis is conducted. Plotting tail length in males by sex ratio suggests increased competition for mates is correlated with more showy tail plumage. (a) To correct for potential bias resulting from shared ancestry, difference scores are calculated between species pairs and ancestor nodes on the phylogeny, (b) to transform the data into phylogenetically independent contrasts, (c). Following this conversion, a positive relationship remains between tail length and sex ratio, (d) suggesting that an evolutionary relationship might exist between the elaboration of tail ornaments and competition for mates.

variation among closely related species is explained by the phylogenetic relationships between those species. Newer methods use likelihood and Bayesian techniques to estimate the relationship between phylogeny and trait expression and control for this level of phylogenetic signal during correlation tests.

Annotated Bibliography

A detailed overview of phylogenetic comparative methods for behavioral ecologists can be found in Ord and Martins (2009). Nunn and Barton (2001) also provide a useful practical guide to several methods. There are a number of books available but most are dated in the methods they present. Nevertheless, they still provide a solid conceptual foundation to comparative biology. These texts are Harvey and Pagel (1991), *The Comparative Method in Evolutionary Biology,* and Martins (1996), *Phylogenies and the Comparative Method in Animal Behavior.* Examples of studies using comparative methods to study the evolution of communication include: acoustic signals in birds (Seddon, 2005) and frogs (Ryan and Rand, 1995); color signals in birds (Doucet et al., 2007), fish (Garcia and Ramirez, 2005), and chameleons (Stuart-Fox and Moussalli, 2008); the sword in swordtails (Basolo, 1990);

dynamic visual displays in lizards (Ord and Martins, 2006); vibration signals in insects (Henry and Wells, 2004); and electric signals in fish (Turner et al., 2007).

See also: Electrical Signals; Phylogenetic Inference and the Evolution of Behavior; Swordtails and Platyfishes; Túngara Frog: A Model for Sexual Selection and Communication; Visual Signals.

Further Reading

Basolo AL (1990) Female preference predates the evolution of the sword in swordtail fish. *Science* 250: 808–810.

Doucet SM, Mennill DJ, and Hill GE (2007) The evolution of signal design in manakin plumage ornaments. *American Naturalist* 169: S62–S80.

Garcia CM and Ramirez E (2005) Evidence that sensory traps can evolve into honest signals. *Nature* 434: 501–505.

Harvey PH and Pagel MD (1991) *The Comparative Method in Evolutionary Biology.* New York: Oxford University Press.

Henry C and Wells MLM (2004) Adaptation or random change? The evolutionary response of songs to substrate properties in lacewings (Neuroptera: Chrysopidae: Chrysoperla). *Animal Behaviour* 68: 879–895.

Martins EP (ed.) (1996) *Phylogenies and the Comparative Method in Animal Behaviour.* New York: Oxford University Press.

Nunn CL and Barton RA (2001) Comparative methods for studying primate adaptation and allometry. *Evolutionary Anthropology* 10: 81–98.

Ord TJ and Martins EP (2006) Tracing the origins of signal diversity in anole lizards: Phylogenetic approaches to inferring the evolution of complex behaviour. *Animal Behaviour* 71: 1411–1429.

Ord TJ and Martins EP (2010) The evolution behavior: Phylogeny and the origin of present-day diversity. In: Westneat DF and Fox CW (eds.) *Evolutionary Behavioral Ecology.* pp. 108–128. New York, NY: Oxford University Press.

Ryan MJ and Rand AS (1995) Female responses to ancestral advertisement calls in Tungara frogs. *Science* 269: 390–392.

Seddon N (2005) Ecological adaptation and species recognition drives vocal evolution in neotropical suboscine birds. *Evolution* 59: 200–215.

Stuart-Fox D and Moussalli A (2008) Selection for social signalling drives the evolution of chameleon colour change. *Public Library of Science, Biology* 6: e25.

Turner CR, Derylo M, de Santana CD, Alves-Gomes JA, and Smith GT (2007) Phylogenetic comparative analysis of electric communication signals in ghost knifefishes (Gymnotiformes: Apteronotidae). *Journal of Experimental Biology* 210: 4104–4122.

Evolution of Parasite-Induced Behavioral Alterations

F. Thomas, Génétique et Evolution des Maladies Infectieuses, Montpellier, France; Université de Montréal, Montréal, QC, Canada

T. Rigaud, Université de Bourgogne, Dijon, France

J. Brodeur, Université de Montréal, Montréal, QC, Canada

Host Manipulation Within an Evolutionary Context

The evolutionary origins of host manipulation are not known with certitude, but according to several authors, this strategy may have been consecutive to the establishment of complex parasitic life cycles – that is, life cycles that require more than one host for completion. For instance, any individual parasite able to modify the behavior of its host with a resulting increase in the success of its transmission to the next host would have been favored over its conspecifics by natural selection. Given enough genetic variation, what originally may have been an incidental side effect of infection could then have been shaped by selection into a refined manipulation mechanism. Within an evolutionary perspective, one must also keep in mind that behavioral changes are not always the product of natural selection in a given host–parasite interaction; the capacity to manipulate can also be inherited from an ancestor of the parasite. For this reason, the evolution of host manipulation must be envisaged within a phylogenetic context. When similar behavioral changes are induced by phylogenetically unrelated parasites experiencing comparable selective pressures, convergence is a reasonable explanation since similar manipulation of host behavior could have arisen independently in different parasite lineages. There is currently evidence of such convergence in different parasite taxa. In other systems, however, such as acanthocephalans in which all species are known to manipulate their hosts, it seems more parsimonious to suggest that this capacity has been inherited, with subsequent modifications, from a common ancestor. When manipulations do not evolve independently (i.e., they are ancestral legacies), their adaptive value can nonetheless be maintained since they still significantly increase the probability of successful transmission.

Evolutionary Processes Leading to Parasitic Manipulation

Compared with the large number of empirical studies that have described examples of host manipulation by parasites, there have been relatively few attempts to determine the processes through which this strategy can evolve in host–parasite systems. Three main evolutionary routes have been hypothesized to have led to host manipulation in a large range of systems: (1) the manipulation sensu stricto, (2) the mafia-like strategy, and (3) the exploitation of compensatory responses. Routes (2) and (3) correspond to scenarios where the host genotype is involved in the evolution of manipulation.

Manipulation Sensu Stricto

Under this scenario, parasite-induced host changes that are adaptive for the parasite transmission are considered as illustrations of the 'extended phenotype' concept proposed by Dawkins, in which genes in one organism (i.e., the parasite) have phenotypic effects on another organism (i.e., the host). The extended phenotype perspective is a decidedly parasite-centered view. Despite this limitation, it has been, and remains, useful to envisage the evolution of manipulation within a theoretical perspective. Poulin published a key paper on this aspect, in which a fundamental and reasonable assumption was to consider that the energy spent by parasites into host manipulation would not be available for other functions (such as growth, survival, fecundity, etc.). Some studies on trematodes and acanthocephalans indeed revealed that parasites act on host central nervous system, but precise mechanisms involved in these disruptions remain unclear.

Genes of the parasite are thus expressed in the host phenotype, but there is a cost for the parasite. Resulting compromises, or trade-offs, therefore suggest that natural selection would optimize, not maximize, the influence of parasites on host behavior. In other words, there should be optimal manipulative efforts (ME*). Several predictions can then be derived on how ME* should increase or conversely decrease in relationship with various factors.

One variable likely to influence ME* is the passive transmission rate of the parasite, that is the transmission rate without manipulation. It can be predicted that ME* should decrease with an increase in the passive transmission rate (**Figure 1**, case 1). For instance, when the intermediate host of a parasite in a common prey species for definitive predatory hosts, there would be only weak selective pressure on parasites to manipulate the host, since successful transmission can be achieved without

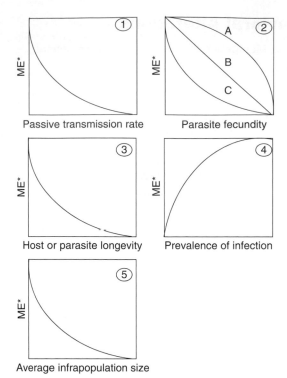

Figure 1 Predicted relationship between the optimal manipulative effort ME* and various ecological variables (passive transmission rate, parasite fecundity, host or parasite longevity, prevalence of infection, and average population size). For fecundity, the three curves illustrate the three most likely scenarios: (a) ME* is only reduced when fecundity is very high, (b) ME* gradually decreases as fecundity increases, and (c) small increases in fecundity lead to large decreases in ME*. Reproduced from Poulin R (1994) The evolution of parasite manipulation of host behaviour: A theoretical analysis. *Parasitology* 109: S109–S118.

manipulative effort. We thus expect ME* to be the highest in systems where passive transmission is negligible. Parasite fecundity is also expected to significantly influence ME*. Theory predicts that not all life-history traits can be maximized at the same time, and consequently, any investment in one trait is likely to occur at the expense of another trait. In such a context, we expect ME* to be smaller in parasite species with highly fecund adults than in less fecund species (**Figure 1**, case 2). Parasite longevity inside the host, as well as host longevity, may determine ME*. Transmission success is indeed ultimately influenced by both the instantaneous transmission rate and the life expectancy of the host–parasite association. A small increase into a manipulative effort can strongly influence transmission success if both the parasite and the host have a long life. In this context, we can predict that, unless increasing the generation time is counterselected in the parasite, there should be, in many cases, a negative relationship between ME* and host/parasite longevity (**Figure 1**, case 3). Another relevant variable for the

evolution of ME* is the prevalence of infection. For instance, when a parasite life cycle involves a prey species as intermediate host and a predator as definitive host, a small manipulative effort can be sufficient to attract the attention of predators and to favor transmission when prevalence of infection is low in the intermediate host population. However, when the prevalence increases, individual parasites must perform a higher ME* for their host to be preferentially captured by foraging predators. Therefore, everything else being equal, ME* should be positively related to prevalence of infection (**Figure 1**, case 4). Finally, when individual parasites (in gregarious species) can share manipulative costs, infrapopulation size can theoretically influence ME*. In such a case, it is expected that the manipulative effort of each parasite would decrease when the infrapopulation size increases (**Figure 1**, case 5). There is however no empirical demonstration that parasites can cooperate and share manipulative costs (Thomas et al., 2005).

Most studies consider the scenario of manipulation *sensu stricto* as the main process used by parasites to take the control of their host behavior; the majority of these studies also assume that costs are inevitably associated with manipulation. The latter assumption is undoubtedly reasonable (Poulin et al., 2005). However, despite the existence of suitable systems with which hypotheses about cost might be tested, at this stage, speculation has proven more attractive than data collection. This gap currently limits our understanding of the evolution of manipulative processes based on this scenario.

Interactive Scenarios of Manipulation

Because parasitic manipulation often dramatically reduces host fitness, it is frequently presented (e.g., as in the scenario of manipulation sensu stricto) as a game with evident winners (i.e., parasites) and losers (hosts). A different perspective has been suggested recently, one that argues that the ability of parasites to manipulate host behavior inevitably results from a long-term coevolutionary interaction which probably leads to complex and interactive mechanisms. These considerations are relevant in an evolutionary context since they mean that behavioral changes in infected hosts, even when they result in significant fitness benefits for the parasite, are not necessarily an illustration of the extended phenotype of the parasite alone. They can also be direct products of natural selection acting on the host genome as well. At the moment, very few studies have explored the degree to which parasite-manipulated behaviors could be a compromise between host and parasite strategies. Two main processes have been proposed (mafia-like strategy and exploitation of host compensatory responses); both rely on the idea that 'making the host do *something*' can be achieved when the *something* is better than nothing for the host.

Mafia-like strategy

This scenario, proposed by Zahavi, is undoubtedly the most extreme illustration of the possible interactive nature of the relationship between parasites and hosts in determining host behavior. In this scenario, parasites select for collaborative behavior in their hosts by imposing extra fitness costs in the absence of compliance. Such a parasite can adopt a facultative virulence commensurate to the rate of collaboration displayed by the host. When the host does not behave as expected, this parasite can increase its virulence, thereby making any noncollaborative behaviors a more expensive option for the host than collaborative ones. This 'mafia-like strategy' can, in theory, force the host to accept behaving in ways that benefit the parasite. Even 'suicidal' behavior (see **Box 1**) could be selected with this scenario, since, indeed, a reduction in survival is not synonymous with a reduction in fitness. From an evolutionary point of view, the key parameter to consider is net fitness and not survival. In this way, a host that cooperates with the parasite, even to the point of displaying suicidal (manipulated) behavior, could be favored if it had reduced fecundity compared to the complete castration experienced by an uncooperative host with a retaliatory parasite. Of course, these collaborative behaviors do not result from conscious choices or appreciation of the problem to be solved. Instead, selection is expected to produce, through time, population-specific phenotypic plasticity and to act on the pattern of condition-dependent expression of behavior, causing individuals to behave differently when infected. In particular systems, as with birds (see the example later), the infected host could develop behaviors that mitigate extra costs of parasitism; in other words, parasites could 'teach' the host that it is better to 'comply' than to resist.

The great spotted cuckoo (*Clamator glandarius*)–magpie (*Pica pica*) association is currently the best example of the mafia hypothesis (**Figure 2**). The host can raise at least part of its own young along with those of the cuckoo. Soler and colleagues found that ejector magpies suffered from considerably higher nest predation levels by cuckoos than did accepters. The interpretation was that the cuckoo retaliates and punishes noncompliant hosts. As a result, the frequency of 'accepting genes' from the avian mafia is more likely to increase in the host population than 'rejecting genes.' Because most host pairs that failed in their first reproductive attempt managed to lay a replacement clutch, cuckoos could in theory benefit from this second attempt if such hosts learned from their previous experience and therefore were more prone to accept the cuckoo egg during a subsequent breeding attempt. This prediction was supported in areas with a high density of cuckoos.

Recently, Hoover and Robinson provided experimental evidence for the mafia strategy with another brood parasite, the brown-headed cowbird (*Molothrus ater*). In manipulating ejection of cowbird eggs and cowbird access to nests of their warbler host, they showed that 56% of ejector nests were destroyed by cowbirds, compared with only 6% of accepter nests. This mafia behavior selects for collaborative hosts not only in evolutionary time by

---

Box 1   Examples of the Diversity of Parasite-Induced Behavioral Alterations

Numerous trophically transmitted parasites alter the behavior of their intermediate hosts in a way that increases their vulnerability to predatory definitive hosts (Figure A, see also the section written by Janice Moore). Parasites can also manipulate host habitat choice; arthropods harbouring mature nematomorphs or mermithids seek water and jump into it, thereby allowing the parasitic worm to reach the aquatic environment needed for its reproduction (Thomas et al., 2002, Figure B). Parasitic wasps can make their spider host weave a special cocoon-like structure to protect the wasp pupae against heavy rain (Eberhard, 2000, see also Brodeur and Vet, 1994), or can even cause the host to seek specific microhabitats where parasitoid mortality from natural enemies is reduced (Brodeur and McNeil, 1989). Viruses may stimulate superparasitism behavior in solitary parasitoids to achieve horizontal transmission (Varaldi et al., 2003). Some digeneans drive their molluskan intermediate hosts toward ideal sites for the release of cercariae (Curtis, 1987). 'Enslaver' fungi make their insect hosts die perched in a position that favors the dispersal of spores by the wind (Maitland, 1994). Vector-borne parasites can render their vertebrate hosts more attractive to vectors, and/or can manipulate the feeding behavior of vectors to enhance transmission (Hamilton and Hurd, 2002).

Figure A The amphipod *Gammarus insensibilis* manipulated by the trematode *Microphallus papillorobustus* (Photo P. Goetgheluck); infected gammarids display a positive phototactism, a negative geotactism and an aberrant evasive behavior. Indeed, instead of hiding under stones, they swim toward the surface and are more frequently eaten by aquatic birds (definitive hosts).

Figure B The Gordian worm (*Paragordius tricuspidatus*) exiting the body of a cricket (*Nemobius sylvestris*) (Photo P. Goetgheluck). Mature worms manipulate cricket behavior, causing them to jump into water. Once it has emerged, the parasite swims away to find a mate (Thomas et al., 2002).

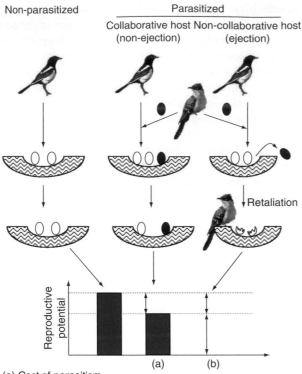

(a) Cost of parasitism
(b) Cost + extra-cost

Figure 2 Mafia behavior in the cuckoo *Clamator glandarius* parasitizing the magpie host *Pica pica* (inspired from Ponton F, Biron DG, Moore J, Moller AP, and Thomas F (2006b) Facultative virulence as a strategy to manipulate hosts. *Behavioural Processes* 72: 1–5.

decreasing the proportion of hosts that bear rejector genes, but also within an individual host's lifetime through a learning processes. Learning occurs in systems where individual females are likely to be parasitized repeatedly within or across breeding seasons.

Theoretically, retaliation can evolve even when hosts rear only the parasite's young (e.g., the case when nests are parasitized by *Cuculus canorus*). This is possible if, during the breeding season, nonejectors enjoy lower rates of parasitism in later clutches compared to ejectors, making nonejectors able to rear a clutch of their own following the rearing of a cuckoo nestling. This however implies that brood parasites have a good memory for the location and status of nests in their territory.

For this strategy of manipulation to evolve, several conditions must be met. For instance, both the host and the parasite must be able to adjust their life-history decisions in a state-dependent manner. Numerous lines of evidence suggest that host species are able to recognize environmental cues, including parasitic infection, and to adjust their life-history traits accordingly. There are recent suggestions that parasites are also able to perceive a large set of environmental variables and to respond in a state-dependent

manner. There is evidence that parasites can recognize many biochemical and physiological variables of their internal host environments that are of selective importance (e.g., sex and age of the host, presence/absence of other parasites, etc.). There are also good reasons to believe that parasites are able to perceive numerous cues of the external environment of their hosts such as, for instance, their host population density, the presence of predators, or sexual partner or competitors of the host. Thus, parasites could have some capacities to evaluate some of the phenotypes displayed by the host they are infecting (e.g., phototaxis, hyperactivity).

Mafia-like strategy of manipulation works from a theoretical point of view but concrete examples in the living world are still scarce; appropriate studies are lacking. Indeed, experiments must test infected hosts that exhibit noncompliance and study the fitness consequences of such noncompliance. Despite these difficulties, from an evolutionary perspective, it is important to demonstrate that infected hosts behaving in a way that benefits the parasite lack fitness compensation. In the absence of such a demonstration, the mafia scenario becomes a potential alternative to the manipulation sensu stricto hypothesis.

Exploitation of host compensatory responses

Recently, Lefèvre and colleagues proposed that certain parasites could affect fitness-related traits in their hosts (e.g., fecundity, survival, growth, competitiveness, etc.) in order to stimulate host compensatory responses, because these responses can match at least partially with parasites' transmission routes.

Most organisms are able to alter some life-history traits plastically, to deal with adverse environmental conditions and alleviate the fitness costs incurred. For instance, one way of responding to poor environmental conditions is to reproduce earlier in life and produce larger numbers of offspring; a phenomenon called 'fecundity compensation.' Analogous to other environmental factors, parasites have the potential to influence the evolution of plastic compensatory responses. When resistance is not possible by other means, selection is likely to favor hosts that react to parasite-induced fitness cost by adjusting their life-history traits. For example, parasitized hosts can respond to a fitness loss due to infection via mechanisms such as an enhanced courtship behavior, increased rate of egg laying, higher offspring number and/or size and/or stronger parental effort. Compensation is thus a widespread strategy among living organisms and parasitism can be a decisive environmental variable initiating compensatory responses in their hosts. In some cases, parasites can directly exploit the compensatory responses that have been selected in other ecological contexts, by mimicking the causes that induce them. In other cases, because of their significant effects on host fitness, parasites can themselves be the causal agents of the compensatory response displayed by the host.

Although no study has yet been specifically designed to test this recent hypothesis, there are in the literature several examples of manipulation sensu stricto that could, in fact, illustrate an exploitation of host compensatory responses. For instance, an increased predation risk is a change that can potentially be of interest for trophically transmitted parasites since, by definition, this type of parasite requires a predation event to complete its life cycle. Parasitized hosts are often more vulnerable to predators because they forage more to compensate for the negative effect of infection. Inducing energy depletion could then in certain cases be viewed as a cost induced by trophically transmitted parasites to trigger a compensatory response that favors the transmission to a predatory definitive host. Vector-borne parasites have been shown to manipulate several phenotypic traits of vectors in ways that favor parasite transmission. For instance, infected insect vectors usually display increased probing and feeding rate. Because increased biting is usually associated with mechanical interference, this could be interpreted as manipulation sensu stricto. However when infected vectors, which also frequently display a decreased fecundity, are free to bite more, they can in certain cases recover a normal level of fecundity (i.e., equal to uninfected conspecifics). In this view, the increased biting rate may represent a host compensatory response to parasite-induced fecundity reduction.

Other potential examples exist. For instance, parasites often reduce survival of their host, and infected hosts are expected to respond by increasing their reproductive effort. Parasites with direct transmission would benefit from decreasing the reproductive outlook of their host, because this should promote compensatory sexual activities that increase social interactions, and hence parasite transmission. The sexually transmitted ectoparasite, *Chrysomelobia labidomera*, reduces the survival of its leaf beetle host (*Labidomera clivicollis*). In response, infected males develop higher sexual motivation before dying. This behavioral modification clearly benefits the sexually transmitted parasite since enhanced inter- and intrasexual contact (i.e., copulation and competition) provide more opportunities for transmission.

It thus appears that parasites could achieve transmission by triggering host compensatory responses, when the latter fit, at least partially, with the transmission route. However, we are left with the same question as before: is this strategy common? Further studies are clearly needed to answer this question, even if it is sound from a theoretical point of view. This type of host manipulation seems parsimonious for several reasons when compared with the hypothesis of manipulation sensu stricto, in which the parasite must maintain a certain degree of manipulative effort with putative fitness costs. Indeed, if among the arsenal of compensatory responses displayed by the host, some are beneficial for transmission, selection is likely to favor parasites that exploit these responses, not only because this meets their objectives, but also because this requires no manipulative effort: the host is doing the job. The main task of the parasite is to induce a fitness-related cost to the host, something that, by definition, parasites normally do. Another good reason to believe that exploiting host compensatory responses is a likely scenario from an evolutionary perspective comes from the fact that it is also advantageous for the host: once infected, it is better for the host to behave in a way that alleviates the costs of infection, even when this also ultimately benefits the parasite (Dawkins' 'aligned desiderata'). Under these conditions, resistance is less likely to evolve than when there is no compensation for the host. Based on these considerations, we could logically predict that manipulation sensu stricto should have mainly evolved in systems where there is nothing suitable for the parasite transmission within the repertoire of host compensatory responses. As a possible example of such a situation, we suggest the case of the well-known textbook example involving the small liver fluke (*Dicrocoelium dendriticum*). It is indeed difficult to imagine what kind of compensatory responses could make the ant climb to the tip of grass blades.

Manipulation sensu stricto and exploitation of host compensatory responses appear to be different scenarios, but it is important to keep in mind that they are not mutually exclusive; instead, these two processes can occur along a continuum (**Figure 3**). At least, exploitation of compensatory response offers a good hypothesis for a starting point of the evolution of host manipulation. However, among the repertoire of compensatory responses that a parasite can induce in its host, all are not potentially useful for parasite transmission. Therefore, selection would favor parasite variants exploiting a response with transmission potential, or variants that induce specific costs to the hosts, for inducing the response of interest. As soon as such a process begins, because it is hard to imagine that such selection would have zero cost for the parasite, manipulation could evolve from an exploitation of compensatory response.

Origin of the Manipulation: Exaptation?

It has been hypothesized that certain parasites such as trematodes manipulate the behavior of their hosts by being encysted in the host brain, but this initially is favored because it allows the parasite to avoid an immune response from the host. The word 'exaptation' (defined in 1982 by Gould and Vrba as characters that evolved for other functions, or for no function at all, but which have been co-opted for a new use) could be proposed in such a context. However, whatever the term used, there is an obvious limitation when attempting to distinguish those changes that are true adaptations versus fortuitous payoffs: for most parasite-induced phenotypic changes, it will be impossible

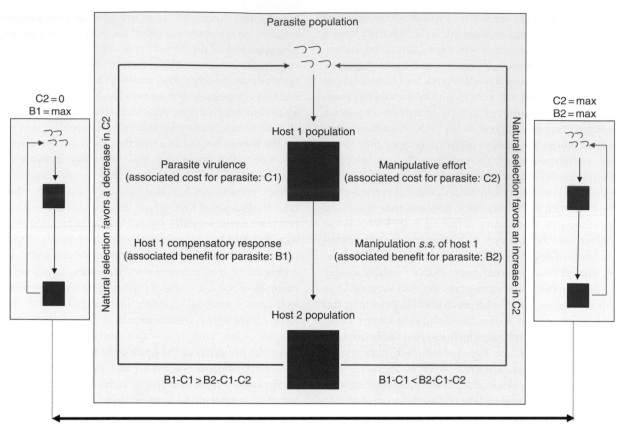

Figure 3 Schematic representation of Lefèvre et al. (2008) of the evolutionary dynamics of two manipulative strategies: (i) the manipulation *sensu stricto* and (ii) the exploitation of host compensatory responses. The two strategies can be represented by a continuum (arrow in bold) along which a parasite can both induce a host compensatory response (via its virulence, e.g., fecundity reduction) and invest energy in manipulation *sensu stricto* (i.e., a mixed strategy). Such a parasite has fitness costs associated with the manipulative effort of manipulation s*ensu stricto* **(C2)** and with the virulence incurred to the hosts **(C1)** and gains transmission benefits from the *sensu stricto* manipulation **(B2)** and from the host compensatory responses **(B1)**. When B1-C1 > B2-C1-C2, natural selection favors a decrease in manipulative effort (i.e., decrease in C2) and thus favors the strategy based on the exploitation of host compensatory response. Inversely, when B1-C1 < B2-C1-C2 natural selection favors a higher investment in the manipulation *sensu stricto* strategy. At the extreme left is shown the case in which the parasite induces a host compensatory response that matches totally the parasite's objectives. At the extreme right, no compensatory response exists in the host phenotypic repertoire; host behavior that benefits the parasite can be achieved only by manipulation *sensu stricto*. This simplistic view has the advantage of emphasizing the fact that when the compatibility between the type of host compensatory response and the parasite's objective is strong enough, exploitation of host compensatory response is the best strategy of manipulation.

to know the selective landscape in which they evolved. Thus, although it is always interesting to consider historical information, one must be aware that taxa vary greatly in the availability of such information. This is probably the reason that little attention has been devoted at identifying the role of exaptation in manipulative changes.

Parasitic Manipulation: An Ongoing (Co) Evolution?

Is host manipulation a fixed trait in many parasites, or is it a phenomenon still under selection? It is generally thought that parasites and their hosts are constantly coevolving. Parasites have to continually adapt to hosts because hosts are not their passive victims. Since most

parasites are harmful to them, the hosts may either tolerate parasites (i.e., reduce parasitism costs as discussed earlier) or resist parasites. When hosts resist, parasites are under pressure to evade this resistance. If a given parasite strain or variant can do so, it will be selected and will spread in the host population, because most of the host genotypes will be sensitive to this variant. But if a rare host variant (genotype) resistant to this spreading parasite variant is already present in the population or if it appears by mutation, this new resistant host variant will be able to fight the parasite, and co-evolution will continue. Hosts and their parasites should therefore be under antagonistic co-evolution that should at least in part be genetically determined, and genetic variation must be present in both parasite and host populations for this process to work. This evolutionary frequency-dependent arm race is often referred as the 'red queen' hypothesis, a metaphor taken from Lewis Carroll's book *Through the Looking Glass*. Just as Alice and the Red Queen had to run as fast as they could in order to stay in the same position, living organisms always have to evolve and adapt to other living organisms to maintain themselves. This hypothesis has been investigated theoretically for many years, but only recently received empirical and experimental supports. The evolutionary scenarios described earlier proposed processes that could be at the root of the evolution of behavioral manipulation. But should we consider parasites altering host behavior as definitive winners, or should we consider these parasites as 'typical' parasites coevolving with their hosts?

The arms race is based on the assumption that there is genetic variation in both the host's ability to resist the parasite and the parasite's ability to exploit its host. In the case of parasites altering host behavior, it has often been reported that individuals in a given parasite population do not modify host behavior with similar intensity. There are even hosts infected by a mature parasite that are not manipulated at all. However, to date, no investigation has asked if this variation has a genetic basis, either on host or on the parasite side. We only have an indication that, in one nonnatural case of parasites changing the host behavior, the host genotype matters: two laboratory inbred strains of mice are not equally sensitive to behavioral changes induced by *Toxocara canis*.

A consequence of the arm race is the occurrence of local adaptations. Species are always composed of more or less discrete populations, linked by various proportions of migrants. In cases of host–parasite associations, it is predicted that parasites should be more adapted to the host with which they coevolved (local host), than with hosts belonging to other populations. However, the pattern of local adaptation could be reverse: some hosts are adapted to resist their local parasites. The outcome of who is locally adapted to whom is determined by several factors, one of the most important being the relative migration rate of hosts and parasites between populations. If the parasite migrates more than the host, parasite local adaptation is predicted because high migrant parasite frequency increases the probability of occurrence of a parasite genotype adapted to local hosts. If hosts migrate more than parasites, the hosts would be locally adapted to resist parasites. To date, there is no study of local adaptation of parasites altering host behavior.

Finding either genetic variation in parasite's ability to manipulate host behavior (or host ability to resist manipulation), or a local adaptation pattern would provide helpful information about how this peculiar way of exploiting a host has evolved and continues to evolve. Such an approach could (and should) be combined with the trade-off approach described in section 'Mafia-like strategy,' because due to the arms race between hosts

and parasites, maintenance of variation in manipulation, could prevent the predicted 'optimal manipulation effort' to be reached in field populations.

Even without such evidence, there are data suggesting that manipulating host behavior is an evolving interaction rather than a static process. We have already noted that some parasite phyla such as the acanthocephalans are entirely composed of manipulative members. But all acanthocephalans do not alter their host behavior in the same way and do not change the same behaviors in all their hosts. A good example could be found in two sympatric acanthocephalans infecting the same intermediate host, *Gammarus pulex*. One, *Pomphorynchus laevis*, infects fish as final hosts; the other, *Polymorphus minutus*, infects aquatic birds as final hosts. These two parasites modify different intermediate host behaviors: *P. laevis* reverses phototaxis (gammarids are attracted by light when infected, making them more prone to be found in the river drift where fish are hunting), while *P. minutus* reverses geotaxis (gammarids are attracted by the water surface when infected, making them more prone to grab on floating material where waterfowl are feeding). However, these modifications are specific: *P. laevis* does not change geotaxis, *P. minutus* does not affect geotaxis, and the central nervous system is not disrupted in the same way by the two parasites. This clearly suggest that, even if these two acanthocephalans share a common manipulative ancestor, the manipulation evolved in an adaptive way for the two descendants in parallel with their adaptation to different ecological niches (fish and bird, respectively). We can conclude that enough genetic variation in behavioral manipulation existed in their common ancestor to allow evolution. There is no reason to suppose that this variation has been lost in contemporary species, especially if local adaptation should exist, because this process can maintain genetic variability in both hosts and parasites.

Concluding Remarks and Future Directions

As illustrated in this chapter, there are different evolutionary routes allowing parasites to make their hosts behave in ways that favor transmission. This research topic is however still in its infancy. The issues of manipulation sensu stricto versus interactive scenarios, as well as other questions about parasite-induced behavioral changes, have much to gain from attention to mechanisms. Indeed, elucidating the proximate mechanisms mediating changes in host behavior could considerably help our understanding of manipulative processes. Collaborative and multidisciplinary research approaches are necessary to show the physiological, the neurological, and ultimately the genetic basis of behavioral changes in parasitized organisms.

Understanding the evolution of host manipulation by parasites also requires considering manipulated hosts within their ecological context. We indeed need to have an accurate knowledge of the selective pressures really experienced by both the host and the parasite. For instance, most experiments do not take into account the fact that, in natural conditions, other predators unsuitable as hosts may also take advantage of the manipulation. This phenomenon is nonetheless critical to our understanding of the costs and the benefits of parasitic manipulation. In some cases, certain features of parasite-induced behavioral changes seem more relevant to limiting the risk of predation by the wrong (non-host) predator than to increasing transmission to appropriate hosts. Moreover, hosts in nature are usually infected by multiple phylogenetically unrelated parasites, which may have opposing interests in their use of the host. In certain cases, parasites have been shown to sabotage the manipulation exerted by other parasites, turning back infected hosts to a normal behavior. Manipulative parasites affect the structure of the parasite communities that exploit the same host and the responses are clearly of an evolutionary nature. More generally, infections involving conflicting parasites are likely to explain some of the variation observed in behavioral changes associated with infections by manipulating parasites.

Multidimensionality in host manipulation by parasites has received little attention so far. In most cases, manipulated hosts are not simply normal hosts with one aberrant trait (e.g., behavior); instead, they are deeply modified organisms with a range of modifications, some of which may favor parasites, and some of which may favor hosts. It is currently unknown whether multiple changes in host phenotype are related or independent; why and how the multidimensionality of host manipulation evolved are fascinating questions requiring collaboration among parasitologists and researchers from other disciplines, especially physiology, morphology, and developmental biology.

See also: Intermediate Host Behavior; Parasite-Induced Behavioral Change: Mechanisms.

Further Reading

Adamo SA (1999) Evidence for adaptive changes in egg laying in crickets exposed to bacteria and parasites. *Animal Behaviour* 57: 117–124.

Biron DG, Ponton F, Marché L, et al. (2006) 'Suicide' of crickets harbouring hairworm: a proteomics investigation. *Insect Molecular Biology* 15: 731–742.

Brodeur J and McNeil JN (1989) Seasonal microhabitat selection by an endoparasitoid through adaptive modification of host behaviour. *Science* 244: 226–228.

Brodeur J and Vet LEM (1994) Usurpation of host behaviour by a parasitic wasp. *Animal Behaviour* 48: 187–192.

Combes C (1998) *Parasitism, the Ecology and Evolution of Intimate Interactions*. London: The University of Chicago Press.

Curtis LA (1987) Vertical distribution of an estuarine snail altered by a parasite. *Science* 235: 1509–1511.

Dawkins R (1982) *The Extended Phenotype*. Oxford: Oxford University Press.

Eberhard WG (2000) Spider manipulation by a wasp larva. *Nature* 406: 255–256.

Franceschi N, Rigaud T, Moret Y, Hervant F, and Bollache L (2007) Behavioural and physiological effects of the trophically transmitted cestode parasite *Cyathocephalus truncatus* on its intermediate host, *Gammarus pulex*. *Parasitology* 134: 1839–1847.

Gandon S, Capowiez Y, Dubois Y, Michalakis Y, and Oliveri I (1996) Local adaptation and gene-for-gene coevolution in a metapopulation model. *Proceedings of the Royal Society of London B* 263: 1003–1009.

Haine ER, Boucansaud K, and Rigaud T (2005) Conflict between parasites with different transmission strategies infecting an amphipod host. *Proceedings of the Royal Society B* 272: 2505–2510.

Hamilton JGC and Hurd H (2002) Parasite manipulation of vector behaviour. In: Lewis EE, Campbell JF, and Sukhdeo MVK (eds.) *The Behavioural Ecology of Parasites*. London, UK: CABI Publishing.

Helluy S and Thomas F (2003) Effects of *Microphallus papillorobustus* (Platyhelminthes: Trematoda) on serotonergic immunoreactivity and neuronal architecture in the brain of *Gammarus insensibilis* (Crustacea: Amphipoda). *Proceedings of the Royal Society B* 270: 563–568.

Hurd H (2003) Manipulation of medically important insect vectors by their parasites. *Annual Review of Entomology* 48: 141–161.

Lagrue C, Kaldonski N, Perrot-Minnot MJ, Motreuil S, and Bollache L (2007) Modification of hosts' behavior by a parasite: Field evidence for adaptive manipulation. *Ecology* 88: 2839–2847.

Lefèvre T, Adamo S, Missé D, Biron D, and Thomas F (2009) Invasion of the body snatchers: The diversity and evolution of manipulative strategies in host–parasite interactions. *Advances in Parasitology* 68: 45–83.

Lefèvre T, Roche B, Poulin R, Hurd H, Renaud F, and Thomas F (2008) Exploitation of host compensatory responses: The 'must' of manipulation? *Trends in Parasitology* 24: 435–439.

Levri EP (1998) The influence of non-host predators on parasite-induced behavioural changes in a freshwater snail. *Oikos* 81: 531–537.

Maitland DP (1994) A parasitic fungus infecting yellow dungflies manipulates host perching behaviour. *Proceedings of Royal Society of London B* 258: 187–193.

Michalakis Y (2008) Parasitism and the evolution of life-history traits. In: Thomas F, Guégan JF, and Renaud F (eds.) *Ecology and Evolution of Parasitism*. Oxford: Oxford University Press.

Miura O, Kuris AM, Torchin ME, Hechinger RF, and Chiba S (2006) Parasites alter host phenotype and may create a new ecological niche for snail hosts. *Proceedings of Royal Society of London B* 273: 1323–1328.

Moore J (2002) *Parasites and the Behavior of Animals*. New York: Oxford University Press.

Moore J and Gotelli NJ (1990) A phylogenetic perspective on the evolution of altered host behaviours: A critical look at the manipulation hypothesis. In: Barnard CJ and Behnke JM (eds.) *Parasitism and Host Behaviour*, pp. 193–233. London: Taylor and Francis.

Ponton F, Biron DG, Moore J, Moller AP, and Thomas F (2006b) Facultative virulence as a strategy to manipulate hosts. *Behavioural Processes* 72: 1–5.

Ponton F, Lefèvre T, Lebarbenchon C, et al. (2006a) Do distantly parasites rely on the same proximate factors to alter the behaviour of their hosts? *Proceedings of Royal Society of London B* 273: 2869–2877.

Poulin R (1994) The evolution of parasite manipulation of host behaviour: A theoretical analysis. *Parasitology* 109: S109–S118.

Poulin R (2003) Information about transmission opportunities triggers a life history switch in a parasite. *Evolution* 57: 2899–2903.

Poulin R (2007) *Evolutionary Ecology of Parasites,* 2nd edn. Princeton, NJ: Princeton University Press.

Poulin R, Fredensborg BL, Hansen E, and Leung TLF (2005) The true cost of host manipulation by parasites. *Behavioural Processes* 68: 241–244.

Rigaud T and Haine ER (2005) Conflict between co-occurring parasites as a confounding factor in manipulation studies? *Behavioural Processes* 68: 259–262.

Soler M, Soler JJ, Martinez JG, and Møller AP (1995) Magpie host manipulation by great spotted cuckoos: Evidence for an avian mafia? *Evolution* 49: 770–775.

Thomas F, Adamo SA, and Moore J (2005) Parasitic manipulation: Where are we and where should we go? *Behavioural Processes* 68: 185–199.

Thomas F, Brown SP, Sukhdeo M, and Renaud F (2002) Understanding parasite strategies: A state-dependent approach? *Trends in Parasitology* 18: 387–390.

Thomas F and Poulin R (1998) Manipulation of a mollusc by a trophically transmitted parasite: Convergent evolution or phylogenetic inheritance? *Parasitology* 116: 431–436.

Varaldi J, Fouillet P, Ravallec M, Lopez-Ferber M, Boulétreau M, and Fleury F (2003) Infectious behaviour in a parasitoid. *Science* 302: 1930.

Wellnitz T (2005) Parasite-host conflicts: Winners and losers or negotiated settlements? *Behavioral Processes* 68: 245–246.

Yanoviak SP, Kaspari M, Dudley R, and Poinar G (2008) Parasite-induced fruit mimicry intra tropical canopy ant. *The American Naturalist* 171: 536–544.

Zahavi A (1979) Parasitism and nest predation in parasitic cuckoos. *The American Naturalist* 113: 157–159.

Evolution: Fundamentals

J. M. Herbers, Ohio State University, Columbus, OH, USA

Evolution: What Is It?

Evolution is any change in the genetic structure of a population from one generation to the next. Populations that experience genetic change over succeeding generations may become sufficiently different from the original state and from other similar populations to produce new species.

The concept of biological evolution implies mutability of species: species are not static entities but rather can shift from one generation to the next as a result of some evolutionary force. Many scientists accepted the concept of mutability prior to Darwin; indeed, Jean-Baptist Lamarcke proposed a cogent (albeit incorrect) mechanism to drive biological evolution. Charles Darwin (along with Alfred Russell Wallace) proposed the principal mechanism currently accepted, and called it natural selection. Darwin described a process whereby individuals with heritable variations that confer an advantage leave more offspring than others; as Darwin pointed out, this process is exactly analogous to the form of selection imposed on crops and domesticated animals by culling and selective breeding. We now know that natural selection is not the only process that can produce evolutionary change, and much contemporary research is devoted to understanding the relative roles of different evolutionary forces.

Evolutionists often distinguish between microevolution and macroevolution. The former refers to short-term changes within populations, whereas macroevolution refers to long-term changes that involve speciation and extinction events. Microevolution and macroevolution are the results of the same underlying processes, but measured over different time scales. Here, the fundamentals of microevolution are reviewed.

First, Some Terminology

Note that our definition of evolution involves changes in genetic structure of populations. Populations evolve as the genotypes of individuals within it are replaced by other genotypes. Individual organisms do not evolve. Individual organisms live or die; reproduce or fail to do so. The aggregate result summed across different individuals within a population can sometimes produce changes in the population genetic structure.

Genetic structure refers to DNA sequences as well as the various ways that combinations of alleles become packaged within individuals. Thus, a population's genetic structure includes the component of allele frequencies in the gene pool as well as genotype frequencies within the genotype pool, combinations of alleles at different genetic loci, and even chromosomal rearrangements. The simplest genetic structure is seen for haploid individuals, with allele frequencies identical to genotype frequencies. Few animal species have all-haploid individuals, but there are many for which the male is haploid and the female is diploid. Some prominent haplodiploid groups are hymenopterans (ants, bees, wasps, sawflies), mites, and thrips.

Phenotype refers to a trait, and most of the time we are interested in traits that reflect DNA sequences. Thus, RNA sequences are phenotypes as are amino acid sequences in proteins, morphological traits, chemical signatures, behaviors, and even extensions of the individual such as nest structure.

Markers are particular types of genes/alleles that provide information about other genes or phenotypes encoded by other genes. Geneticist have used markers for decades to map out chromosomes, and in the last 20 years, behaviorists have used markers like microsatellite DNA sequences, randomly amplified DNA sequences (RAPDs), and SNPs (single nucleotide polymorphisms) to gain insight into how complex behavioral traits evolve.

Evolutionary Forces: Natural Selection

Darwin's formulation of evolution by natural selection has attained the status of true theory in science: it is universally accepted as the predominant mode for adaptive evolution. Adaptive evolution occurs when a population improves its fitness from one generation to the next. Fitness itself is a somewhat slippery concept and can be defined for alleles, genotypic combinations, individuals, groups, and populations.

Fitness is defined as reproductive success, which itself has two components, survival and reproduction. That is, an individual's fitness depends on its probability of reaching adulthood (survivorship) as well as its probability of leaving offspring (fertility). Natural selection occurs when individuals vary in their fitness, that is, when there is differential reproductive success.

Fitness can be measured only in the context of a particular environment: it is not an absolute. A trait advantageous in one environment can be deadly in another. The classic story of industrial melanism in the

British Isles demonstrates this principle neatly. The peppered moth is a common insect of British forests that is active at night and that spends its days resting on tree trunks. It has two distinct morphs that correspond to different genotypes. In forests near coal-burning industries, dark-colored moths gain camouflage against soot-blackened trees. In forests away from pollution sources, those same moths stand out against their background and are easy prey. By contrast, the speckled morph is easily spotted by predators in polluted woods but camouflaged in unpolluted woods. Not surprisingly, the abundance of these two color morphs is a strong function of pollution: in unpolluted woods the black morph is rare and the speckled one common, whereas in polluted woodlands the reverse is true. Fitness is a function both of genotype and environment.

Species can arise when two populations derived from one ancestral population become genetically distinct over time. There are many mechanisms by which speciation can occur, and the prevailing mode is allopatric speciation: divergence as a result of populations occurring in separate environments with different selection pressures. A prime example is speciation observed on islands. Darwin examined diversity among Galapagos finches to infer their long-term evolutionary history of modification from their ancestors in mainland populations. Similarly, hundreds of speciation events have occurred among *Drosophila* flies, as the Hawaiian Islands were formed and individual flies migrated to new habitats; their descendants then formed new populations that adapted to local conditions and diverged from populations experiencing other environments.

Darwin, of course, knew nothing about genetics. Rather, he discussed traits of organisms, what we now call phenotypes. Phenotypes are the results of complex genetic and developmental processes; in turn the environment in which an animal lives can affect those processes. We therefore must be clear about the fundamental distinction between natural selection and evolution: they are not the same thing. Natural selection discriminates between phenotypes and evolution is a change in genotypes within populations. Natural selection is one mechanism that can cause evolutionary change, if there is a connection between genotype and phenotype. This is best illustrated in a figure adapted from Lewontin's book (**Box 1**).

Evolution occurs in the blue Genotype Space, whereas natural selection occurs in the red Phenotype Space. The two are related via Transformation Rules that dictate how changes occur. The transformation Rule τ1 dictates how phenotypes arise from genotypes of a cohort of fertilized eggs (zygotes). As those individuals mature, natural selection (τ2) can alter the phenotypic distribution within the population. The resultant new phenotypic distribution has a genotypic distribution that can be inferred from τ3. Adults mate and have offspring, which themselves have

a genotypic distribution dictated by τ4, principally the laws of Mendel.

Note that natural selection, which occurs in phenotype space, is only one of the four transformation rules that dictate evolutionary change. Note also that natural selection by itself need not produce evolutionary change, if there is no underlying shift in the genotypic distribution of the population. We therefore distinguish between natural selection (the difference between P1 and P1′) and a response to selection (the difference between G1 and G2).

This framework for evolution also allows us to think about how populations in different environments experience different transformation rules, especially in phenotype space. Those differences can cause two populations to move in different directions both genetically and phenotypically. Ultimately, that divergence in genotype space can produce separate species.

Evolutionary Forces in Addition to Natural Selection

So far we have described only one mechanism that produces evolutionary change, the mechanism proposed by Darwin and known as natural selection. Students of animal behavior are prone to interpret most of what they study as the product of natural selection, but it is important to keep selection as a working hypothesis rather than unverified assumption. Few would dispute that natural selection is a major force in the evolution of animal behavior, but many traits cannot be explained in light of natural selection alone. The following are the chief evolutionary forces.

1. *Mutation:* changes in genetic structure from one generation to the next as a result of random alterations in DNA sequences during replication. Mutations occur exclusively in genotype space and arise constantly in populations. Mutations in any given gene are rare, occurring roughly once in every million or so DNA replication events per gene. New mutations introduce variation into gene pools and are included in the transformation rule τ4 in **Box 1**. Most mutations are deleterious and selection weeds them out (unless protected as recessive alleles in heterozygous condition), but on occasion a mutation confers an advantageous phenotype to its bearer and can increase in frequency in the population. The genotypic variation we observe in natural populations ultimately derives from numerous mutation events. It is important to note that some entire classes of mutations are not subject to any force of natural selection because they produce no phenotypic change. The best such example is the third codon of DNA triplets: in the universal genetic code, the third codon is often 'silent' and thus mutations in that codon induce no change in the amino acid in the

---

Box 1   Schematic Diagram of the Evolutionary Process, adapted from Lewontin (1974)

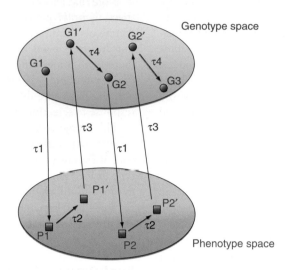

Evolution occurs within populations and involves processes operating on genotypes as well as phenotypes. In genotype space, shifts in genetic structure occur, whereas changes in trait distributions occur in phenotype space. While evolution is defined as a change in genetic structure between generations, evolutionary change is mediated by shifts in both genotype and phenotype space that occur via a set of transformation rules.

In this figure, the blue genotype space has five points. In generation 1, the newly formed fertilized eggs have a certain genetic structure G1 whereas those same individuals as adults may have a different genetic structure G1′. In the next generation, G2 represents the genetic structure of the newly fertilized eggs whereas G2′ represents the genetic structure of the next generation's adults. Similarly, in the red phenotype space the newly fertilized eggs have a distribution of phenotypes given by P1, and P1′ is the distribution of phenotypes of the resulting mature adults. P2 is the phenotypic distribution of the next generation of fertilized eggs, P2′ of the resultant adults, and so on.

The pathway of biological evolution moves within and between the spaces along a trajectory governed by what Lewontin termed 'transformation rules':

- τ1 represents the link between genotype and phenotype for fertilized eggs. It thus represents the set of processes from DNA transcription through translation and metabolism to give rise to the zygotes' phenotypes.
- τ2 represents processes driven by the environment that selects among phenotypes from zygote to adult stages. The chief force is natural selection.
- τ3 represents the rules governing how phenotypes reflect genotypes. Not all phenotypic change induces genotypic change, especially for quantitative traits.
- τ4 represents the rules by which parental genotypes are reflected in the genotypes of their offspring, chiefly Mendel's laws.

With this framework, we can formally define evolution as a change from G1 to G2 (or, alternatively from G1′ to G2′, if adults are the point of reference). Evolution by natural selection proceeds from genotype space to phenotype space and back.

---

corresponding protein. Because the amino acid does not change even though the DNA sequence changes, such mutations are 'silent' and cannot produce phenotypic change. Another example is a change in repeat numbers for microsatellite DNA loci, which typically has no impact on fitness.

2. *Genetic drift:* chance variations in gene frequencies that result from random sampling error. Genetic drift is the trickiest concept to grasp since it relies on an understanding of probabilities. Let us examine the transformation rules τ4 above from a different perspective. Consider the sex ratio of offspring within litters of puppies. In canines, just as in humans and most mammals, XX individuals develop to become females, whereas XY individuals become males. Mendel's laws

predict that in any given litter half the puppies are male and half are female. Yet we understand intuitively that for a litter of two puppies, some consist of two males, some litters contain two females, and some contain one male and one female. Furthermore, in litters of four puppies, some contain two females and two males but some litters contain three of one sex and one of the other, while some litters contain only males or only females. We accept this variation among litters as a consequence of random events governing which type of sperm fertilizes each ovum. Any deviation from expectation resulting from such events is called genetic drift.

Now let us expand from a single litter of puppies to three litters of 4 puppies each. The 12 total puppies across these litters are expected to produce 6 males and 6 females, but we would not be surprised to see 7 and 5, 8 and 4, or 9 and 3. In fact, the probability of having exactly 6 puppies of each sex in a pooled litter of 12 is only 22.5%. Thus in our example, genetic drift is more likely than an exact Mendelian sex ratio.

Clearly, the strength of genetic drift as an evolutionary force depends crucially upon the number of parents reproducing in a population. If a population is small (e.g., endangered species or captive populations in zoos), the effects of drift can produce major changes in genetic and thus phenotypic structure. In large breeding populations (such as mosquitos and Norway rats), the effects of genetic drift are minor.

Random sampling error can also occur when a small number of individuals leave a large breeding population to initiate a new population. For example, a few mice might cross a frozen lake to populate an offshore island, and those few mice almost surely do not represent the full range of genetic variation in their original population. The alleles those individuals carry constitute the gene pool for an entire population to be established; the resultant reduction of genetic variation in the new population reflects a form of genetic drift known as the Founder Effect.

3. Nonrandom mating refers to any pattern that deviates from random pairing between males and females. It has three principal syndromes:

(a) Assortative mating occurs when mates preferentially choose phenotypes like their own (positive assortative) or unlike their own (negative or disassortative mating). Humans mate assortatively for height, ethnicity, IQ, religion, and socioeconomic status. In the wild, assortative mating has been shown for snow geese and goby fish, among others. Disassortative mating is best encapsulated by the rare male effect, by which a male with an unusual phenotype captures an inordinate number of matings. While we have some good examples of these mating schemes, assortative mating is more limited in scope than other evolutionary forces.

(b) Inbreeding refers to preferential mating between relatives, and is widespread in the animal kingdom. Brother–sister mating is the rule for many insects, and father–daughter mating is not uncommon in highly structured social groups. Inbreeding usually reflects very low dispersal as juveniles mature to reproductive age.

Inbreeding can produce pathological conditions, because repeated mating between close relatives tends to increase the frequency of homozygosity within a population. Indeed, population geneticists typically infer inbreeding from a deficit of heterozygotes (fewer than expected from Hardy–Weinberg frequencies, to be covered in the following

section). Homozygosity by itself is not necessarily problematic, but inbreeding allows deleterious alleles that have been 'hidden' from natural selection in heterozygous condition to become homozygous. Thus, inbred populations can develop pathologies that derive from genetic homozygosity of rare deleterious alleles. The result is that the entire population can experience reduced fitness, a condition called inbreeding depression. Managers of captive populations (such as in zoos and preserves) go to great lengths to avoid inbreeding for this reason.

(c) Sexual selection is an extremely important evolutionary force in the animal world and is fully described in another section. Sexual selection is a form of natural selection whereby one sex exerts selection on the other. Females may choose males as a function of their phenotype, or vice versa. The same transformation rules given earlier for natural selection apply to sexual selection.

The Concept of Equilibrium

Many traits do not change from one generation to the next. For example, horseshoe crab morphology has changed very little over the past 200 My. Stasis of phenotypes implies evolutionary equilibrium. Equilibrium, however, does not mean that there are no evolutionary forces. We must distinguish between neutral equilibrium, when no forces are acting on a trait, and balanced equilibrium, when two forces act in equal and opposite directions.

The concept of neutral equilibrium is encapsulated in the Hardy–Weinberg law. It is easy to show mathematically that if there are no evolutionary forces (no selection, no mutation, large population size, and random mating), the allele frequencies and genotype frequencies are stable from one generation to the next and related to each other (see **Box 2**). The Hardy–Weinberg law rarely reflects natural situations, but it serves as the starting point for the large field of population genetics.

If a trait is affected by more than one evolutionary force, and if those forces act in opposition to each other, then evolution should ultimately come to represent a balance between opposing forces. For example, the peacock's tail is a marvel of adaptation that reflects two opposing selective forces: males have tails that serve to attract females (i.e., females exert sexual selection on males to have ever more elaborate tail plumage); those same tails, however, also slow males down and increase predation (predators exert natural selection to reduce tail elaboration). The tails we see in nature presumably reflect a state that is a balance of these two selective forces.

Population geneticists have defined conditions that produce balances between mutation and selection, between opposing selection forces, between mutation

Box 2 The Hardy–Weinberg Equilibrium

This algebra forms the starting point for population genetic theory. Here we treat the simplest possible genetic structure for any trait: a single gene controls behavior, and that gene has just two alleles (A_1 and A_2) in the population.

We first define allele frequencies as the proportions of the two alleles A_1 and A_2 in the gene pool. We will let p equal the proportion of A_1 and q the proportion of A_2. Note that $p + q = 1$ by definition, since there are only two alleles in the gene pool. (Do not confuse this lower-case p with the uppercase P in **Box 1**.)

With two alleles A_1 and A_2, there are exactly three genotypes segregating in the population: $A_1 A_1$, and $A_2 A_2$ (homozygotes) and $A_1 A_2$ heterozygotes. These three genotypes can vary in relative proportion within the population, and we assign values as follows:

D = the proportion of $A_1 A_1$ homozygotes
E = the proportion of $A_1 A_2$ heterozygotes
F = the proportion of $A_2 A_2$ homozygotes

Note that $D + E + F$ must sum to 1, since there are only three genotypes in the population. It is easy to calculate the allele frequencies if we know the genotype frequencies:

$p = D + 1/2E$ since the A_1 allele makes up 100% of the D genotype and half the E genotype
$q = F + 1/2E$ since the A_2 allele makes up 100% of the F genotype and half the E genotype.

It is obvious then that $p + q = D + E + F = 1$.

Let us take a simple example. Suppose our starting population has 50% $A_1 A_1$ homozygotes, 20% $A_1 A_2$ heterozygotes, and 30% $A_2 A_2$ homozygotes. Then by definition,

$D = 0.50$
$E = 0.20$
$F = 0.30$

and

$D + E + F = 1$

Also our definition above allows us to calculate that the allele frequencies are

$p = D + 1/2E = 0.60$
$q = F + 1/2E = 0.40$

So far all we have done is to define terms for the genetic structure of this population.

Now let us consider what happens when the population starts to mate and produce offspring. We assume there are no evolutionary forces operating at all: there is no selection, mating occurs at random, there is a large population size, etc.

Then we can set up a table to determine offspring genotypes from these parents:

		Paternal genotype		
		A_1A_1	A_1A_2	A_2A_2
Maternal genotype	A_1A_1	100% A_1A_1	50% A_1A_1 50% A_1A_2	100% A_1A_2
	A_1A_2	50% A_1A_1 50% A_1A_2	25% A_1A_1 50% A_1A_2 25% A_2A_2	50% A_1A_2 50% A_2A_2
	A_2A_2	100% A_1A_2	50% A_1A_2 50% $A_2 A_2$	100% A_2A_2

The frequencies of genotypes in the next generation are calculated from the frequencies of different kinds of pairing multiplied by the proportion of offspring from the pairing.

$A_1 A_1$ offspring arise as follows:

Parental cross	Frequency of that cross	Proportion $A_1 A_1$ offspring
$A_1 A_1 \times A_1 A_1$	D^2	100%
$A_1 A_1 \times A_1 A_2$	DE	50%
$A_1 A_1 \times A_2 A_2$	DF	0
$A_1 A_2 \times A_1 A_1$	DE	50%
$A_1 A_2 \times A_1 A_2$	E^2	25%
$A_1 A_2 \times A_2 A_2$	EF	0
$A_2 A_2 \times A_1 A_1$	DF	0
$A_2 A_2 \times A_1 A_2$	EF	0
$A_2 A_2 \times A_2 A_2$	F^2	0

Continued

Box 2 Continued

In the next generation, $A_1 A_1$ offspring will occur with frequency:

$$D^2 + DE + \frac{1}{4} E^2 = (D + 1/2E)^2 = p^2$$

Similarly, the proportion of $A_1 A_2$ offspring is $2pq$ and the proportion of $A_2 A_2$ offspring is q^2. Repeating this approach shows that these proportions remain stable for generations thereafter – the genetic structure is at equilibrium.

This, then, is the Hardy–Weinberg law: in the absence of any evolutionary forces, within one generation, a simple and straightforward relationship arises between allele frequencies and genotype frequencies:

$D = p^2$
$E = 2pq$
$F = q^2$

It turns out that these proportions also occur at equilibrium conditions for many balanced polymorphisms as well. Therefore, we can use genotype frequencies to make inferences about other aspects of the population's genetic structure.

One important and well-studied evolutionary equilibrium represents a balance between mutation producing deleterious alleles and selection removing them. One can easily show that if the mutation rate is one mutation in every million DNA replication events, the frequency of recessive lethal alleles in the gene pool is 0.001. That is, one in every thousand alleles in the gene pool is an allele that causes death in homozygous condition. Using the Hardy–Weinberg ratios given earlier, we can then estimate the proportion of individuals who are 'carriers' for that allele, having one copy in heterozygous condition as $2pq = 2 \times 0.999 \times 0.001 = 0.002$. We therefore infer that 1 of every 500 individuals in the population is a heterozygote carrying the lethal allele.

---

and drift, and so on. In each case, the resulting balanced polymorphism maintains genotypic diversity within populations. Furthermore, at equilibrium we see the relationship between alleles and genotypes predicted by the Hardy–Weinberg algebra. We can use those relationships to estimate the strength of selection, the rate of mutation, and other forces of evolutionary change.

Two Approaches to Studying Evolution

Given that evolution occurs partly in genotype space and partly in phenotype space, we have two general approaches for studying evolutionary change. The field of population genetics starts in genotype space and assumes relatively simple relationships between genotype and phenotype. Refer again to the diagram in **Box 1**. Population genetics examines the transformation laws τ2 and τ4 explicitly, but tends to ignore the τ1 and τ3 transformation laws. The Hardy–Weinberg equilibrium forms the foundation for this vast field, which focuses on how allele frequencies in populations change over time. Furthermore, the kinds of traits that this approach considers have fairly simple genetic bases and the typical one-to-one correspondence of genotype and phenotype allows for straightforward predictions about the trajectory of adaptive evolution.

By contrast, the field of quantitative genetics starts in phenotype space and tries to understand how variation in the τ2 transformation laws affects evolution in both phenotypic and genotypic space. Quantitative traits (sometimes called polygenic traits) are the norm in studies of behavior. Traits such as running speed, plumage coloration, or diet choice must be affected by many genes simultaneously and also by the environment in which the animal lives. For such traits, the correspondence between genotype and phenotype is fuzzy, and populations display continuous variation to produce a distribution of phenotypes. Modeling such situations with the algebraic machinery of population genetics is unwieldy, and thus a quite different approach has been developed for the study of quantitative traits.

Quantitative genetics typically examines the distribution of phenotypes and how that distribution changes from one generation to the next, thereby implying change in the underlying genetic structure of populations. The field of quantitative genetics focuses almost exclusively on selection as an evolutionary force, because it derives from the agricultural tradition of artificial selection, or selective breeding.

This approach examines the distribution of phenotypes when selection is applied (see **Box 3**). If there is no change in that distribution from one generation to the next (i.e., there has been no response to selection), then the trait under study must have no underlying genetic basis: the variation in phenotypes is caused by factors in the environment rather than genetic variation among individuals. A response to selection implies otherwise, that phenotypic variation arises at least in part from genetic variation in the population.

The contribution of genetic variation to phenotypic variation is encapsulated in the concept of heritability. Heritability, denoted by the symbol h^2, has a minimum possible value of 0; $h^2 = 0$ implies that the phenotypic distribution is caused entirely by environmental differences experienced by individuals. A group of animals

---

Box 3   The approach taken by the field of Quantitative Genetics

This field deals explicitly with phenotypes for complex traits that show continuous variation and for which there is a distribution across the population:

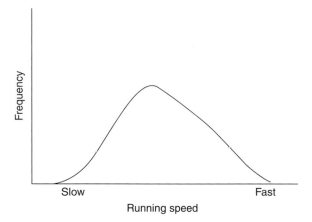

Typically, such traits are affected by many genes and also are affected by the environment of the bearer. For example, running speed in animals varies within a population. That variation reflects some genetic variance for running ability, but also reflects factors such as differential nutrition and disease. Trying to identify all the factors that cause a particular animal to run at a particular speed is a large task, and the field of quantitative genetics attempts instead to understand the relative importance of genetic variation versus other factors that give rise to the phenotypic distribution observed.

The following is the basic equation of quantitative genetics:

Phenotypic Variance = Genetic Variance + Environmental Variance + Gene-Environment Interaction

or

$$V_P = V_G + V_E + V_{G \times E}$$

In fact, each of the components of phenotypic variance can be further subdivided. For example, total genetic variance includes additive genetic variance (variance due strictly to allelic differences) as well as epistatic variance (variance resulting from interactions among genes) and dominance variance (variance resulting from dominance/recessiveness characteristics at different loci affecting the trait).

Fundamental to understanding evolutionary change in phenotypic distributions is heritability, indicated by the symbol h^2. Heritability is defined as the relative contribution of genetic variance to total phenotypic variance:

$$h^2 = \frac{V_G}{V_E}$$

Heritability ranges from 100% (complete correspondence between phenotype and genotype) to 0% (no relationship between genotype and phenotype). Traits under selection can only evolve if those traits are heritable:

$$R = h^2 S$$

where R is the response to selection (measured as a change in the distribution of phenotypes from one generation to the next) and S is the strength of selection. If selection is strong, the phenotypic distribution shifts more rapidly from one generation to the next than if selection is weak. Alternatively, strong selection on a trait with low heritability may produce only a weak response to selection.

Natural selection reduces genetic variation by eliminating alleles contributing to low fitness. That effect has the interesting consequence that as populations become adapted to their environments, there is less and less genetic variance upon which selection can act.

---

that are clones can be reared in different individual environments to produce variation that has a heritability of zero.

The maximum value of heritability is observed when the phenotypic distribution is generated exclusively by genotypic differences between individuals. It is theoretically possible to generate a heritability of 1 by rearing animals in a constant environment. To approximate that condition, many studies of evolution are carried out in tightly controlled conditions; another alternative is to reduce environmental variance contributing to phenotypic variance via a 'common garden' experimental design.

The fundamental equation of quantitative genetics is

$$R = h^2 S$$

where R is the response to selection (measured as the change in the phenotypic distribution of a trait, see **Box 4**), h^2 is heritability, and S is the selection pressure. Our fundamental equation has an important corollary that any response to an imposed selection pressure

---

Box 4   Types of selection modeled in evolution studies

Scientists distinguish three broad categories of selection in nature, which differ in the way they change distributions of phenotypes that have a genetic basis. In the diagrams below, phenotypic distributions are given for a trait such as running speed. In each column, we start with the same distribution. However, note that the shading patterns are different. The intensity of blue within each bar indicates the relative fitness for an organism with that trait.

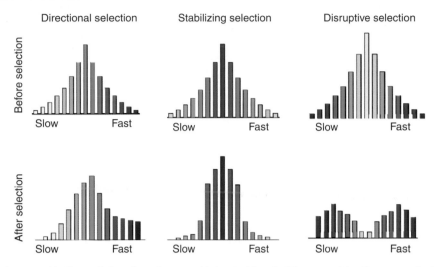

Suppose we are interested in the evolution of running speed in tropical lizards. We can envision three ways that selection acts upon running speed. In the first column, faster lizards have higher fitness than slower lizards; such a situation might be imposed by a very fast predator. Selection in this case is directional: it moves the distribution toward the high end. The response to selection (second row) shows how the distribution of running speeds changes as a result of the differential predation. The response would be measured as a change in the average running speed within the lizard population.

The second column illustrates the case if both fast and slow lizards have lower fitness than lizards that run at medium speed; this situation could result if fast lizards reach physiological exhaustion quickly and thus are preyed upon at the same rates as slow lizards. In this case, the two tails have lowered fitness: selection acts to stabilize the middle of the distribution. After selection, the resulting distribution has the same average value, and the response to stabilizing selection is evident as a reduction in the variance of running speed.

Finally, the third column supposes that fast and slow lizards have higher fitness than lizards with intermediate speed; this situation might result if predators fail to catch fast lizards and fail to notice slow lizards. Called disruptive selection, this form selects for the tails of a distribution and produces a phenotypic response to selection that is bimodal.

---

implies genetic variation for the trait under selection. Therefore, scientists use the equation to estimate heritability in the lab by exerting a known selection pressure upon a trait of interest and then measuring the response to selection.

Estimating heritability in nature is considerably trickier, because it is impossible to eliminate all environmental variation that contributes to phenotypic variation. The approach in field studies typically requires an a priori estimate of genetic variation in the target population, for example, from the knowledge of family structure. With that information, scientists can infer the relative contributions of genetic variation versus environmental variation in producing the total range of phenotypic variation.

In recent years, a blended approach has been employed to search for quantitative trait loci (QTLs). Using sophisticated genetic and statistical techniques, scientists have been able to not only estimate how many genes underlie complex traits, but also assess their relative contributions to

those traits. Identification, mapping, and characterization of such loci have been facilitated by the development of molecular markers. Together, these approaches can provide new insights into how complex traits have evolved.

See also: Bateman's Principles: Original Experiment and Modern Data For and Against; Caste in Social Insects: Genetic Influences Over Caste Determination; Compensation in Reproduction; Cryptic Female Choice; *Dictyostelium*, the Social Amoeba; Differential Allocation; *Drosophila* Behavior Genetics; Flexible Mate Choice; Forced or Aggressively Coerced Copulation; Genes and Genomic Searches; Helpers and Reproductive Behavior in Birds and Mammals; Infanticide; Invertebrates: The Inside Story of Post-Insemination, Pre-Fertilization Reproductive Interactions; Kin Recognition and Genetics; Marine Invertebrates: Genetics of Colony Recognition; Mate Choice in Males and Females; Microevolution and Macroevolution in Behavior; Monogamy and Extra-Pair Parentage; *Parmecium* Behavioral Genetics; Sex

Allocation, Sex Ratios and Reproduction; Sex Changing Organisms and Reproductive Behavior; Sexual Selection and Speciation; Social Insects: Behavioral Genetics; Social Selection, Sexual Selection, and Sexual Conflict; Sperm Competition; Unicolonial Ants: Loss of Colony Identity.

Further Reading

Futuyma DJ (2009) *Evolution*. Sunderland, MA: Sinauer Press.
Lewontin RL (1974) *The Genetic Basis of Evolutionary Change*. New York, NY: Columbia University Press.
Lynch M and Walsh B (1998) *Genetics and Analysis of Quantitative Traits*. Sunderland, MA: Sinauer Press.

Experiment, Observation, and Modeling in the Lab and Field

K. Yasukawa, Beloit College, Beloit, WI, USA

Introduction

For those of us who study animal behavior, the principal reward is finding out how and why animals do what they do. The first step is to ask a question about behavior, which might be broad and generally descriptive or narrow and more hypothetical. As a research program develops, it tends to generate more specific questions, which tend to be influenced by earlier observations, questions, and potential answers.

The behavior of red-winged blackbirds has been studied by hundreds of researchers. Early observations of the behavior of this species of bird show that males try to keep other males out of small portions of nesting habitat. In other words, males defend territories, and this leads to the initial question: How do male red-winged blackbirds defend their territories?

The next step is to do some background reading or make preliminary observations, and propose some possible answers, called working hypotheses.

Reading published descriptive studies and observing male red-winged blackbirds at a marsh in spring would show that males perch at prominent locations, scan the habitat constantly, sing, and show their red-and-yellow epaulets, which are the colorful wing patches after which the species is named. These observations yield at least two potential answers to our question:

Singing is used to defend the territory.
The red-and-yellow epaulets are used to defend the territory.

The most important aspect of these potential answers is that each predicts the results of studies that we can perform. A proposed explanation that makes testable predictions is a scientific hypothesis. We then perform those studies to see whether we get the predicted results. This process is called hypothesis testing, and having more than one possible answer means that we are using multiple working hypotheses.

To make it easier to talk about our hypotheses, we give them descriptive names: the song hypothesis and the epaulet hypothesis for territory defense. What predictions do they make? A productive approach is to use what philosophers of science call conditional scientific predictions, or an if–then construction.

If red-winged blackbird song is used to defend territory, then males should sing when they are on territory, but not when they are away from the territory, and males that are unable to sing should be unable to hold their territories.

If red-winged blackbird epaulets are used to defend territory, then males should show their epaulets when they are on territory, but not when they are away from the territory, and males that lack epaulets should be unable to hold their territories.

Once testable predictions have been stated, the next step is to choose a research design, including a statistical method, to test the predictions. The rest of this article is devoted to the methods by which hypotheses are tested in animal behavior.

Hypothesis Testing

The key aspect of hypothesis testing is whether the method is appropriate to test a specific prediction and whether the prediction, and therefore the hypothesis, can be rejected. Philosophers of science call it the hypothetico-deductive method of hypothesis falsification. (If you are interested in the philosophy of science, try searching the web for the names Karl Popper and Imre Lakatos.) According to this viewpoint, critical tests could produce results that are contrary to the hypothesis and its prediction, so any study with the potential to falsify a working hypothesis has the potential to add to our knowledge of animal behavior.

Have predictions of the song and epaulet hypotheses been tested? Observational studies demonstrate that male redwings do not show their epaulets while trespassing on other territories, or when they are establishing their territories, but once they establish ownership, they show their epaulets during encounters with other males. Experiments show that males whose epaulets are blackened with hair dye are more likely to lose their territories than males receiving a sham treatment. Both the observations and experiments support the predictions of the epaulet hypothesis.

Observations show that song is the most common and conspicuous vocalization male redwings give on their territories, but that trespassing males do not sing. Experiments show that males that are surgically prevented from singing have much more difficulty holding their territories than males that are given sham operations. These observational and experimental studies also support the predictions of the song hypothesis.

The study of animal behavior uses two major categories of hypothesis tests: empirical studies and modeling.

Empirical methods run the gamut from descriptive field research to controlled laboratory experiments. Modeling uses mathematical analysis of equations and computer simulation of systems that cannot be solved analytically.

Empirical Methods

Research in animal behavior that involves gathering data can be divided into experimental and descriptive work, but we are better served thinking about empirical research as a continuum, with purely descriptive field-work at one end, tightly controlled laboratory experiments at the other, and many other kinds in between. The critical dimension is not where the data are gathered (field or laboratory) or how they are gathered (observation or experiment), but the degree of control the researcher has over the conditions of the study, and therefore the implications of the results. No one kind of research is inherently better than another because each has advantages as well as limitations. A thorough understanding of animal behavior requires a combination of methods.

Internal and External Validity

The degree of control is important because it affects the validity of the results. There are two kinds of validity to consider. External validity is how well results of a study generalize to other situations or conditions. (External validity is similar to ecological validity, or how well the study resembles the real world.) In contrast, internal validity is the extent to which an effect can be attributed to a specific cause. The practices that enhance external validity also reduce internal validity, however, and the methods that produce high internal validity are effective because they remove the complications of the real world.

Field researchers want to experience many different conditions to see a full range of behavior, so they make no attempt to control conditions. Descriptive field studies thus have ecological and external validity, but the conditions that make such studies externally valid also prevent the researcher from identifying causation.

At the other extreme, a controlled laboratory experiment achieves the so-called rule of one variable – control and experimental groups differ only in the variable that is manipulated by the researcher – so that any difference between groups must occur because of the experimental variable; there are no other, confounding variables. Controlled laboratory experiments therefore have high internal validity, but the high degree of control necessary to achieve the rule of one variable makes it impossible to generalize the results to the real world, where everything varies.

Between these extremes, but closer to the descriptive field study, is a natural experiment in which the researcher takes advantage of some change in the environment and compares behavior before and after the event. Although the researcher does not manipulate the environment, there is a weak sense of control in that the researcher makes a comparison of before and after groups. For example, the eruption of Mount St. Helens and an outbreak of periodical cicadas have produced natural experiments in the study of male red-winged blackbird behavior.

Closer to the other extreme, in a field experiment or quasi-experiment the researcher controls some, but not all conditions. These experiments typically include a manipulation by the researcher, and compare control and experimental groups, but because not all possible factors are controlled, the rule of one variable is not fully achieved. Singing and silent loudspeakers have been used in the field in to assess the ability of song alone to defend an otherwise empty red-winged blackbird territory.

Studies of cache recovery by birds that store (cache) food items have used a variety of empirical methods. Field studies show that caching birds store food items and find them, apparently with great accuracy. Researchers hypothesized that caching birds have spatial memories that allow them to remember many (thousands in some cases) specific cache sites, and to return to them accurately, many months later. This memory hypothesis was subsequently tested in controlled laboratory experiments. These studies show that caching birds accurately return to cache sites even if the food item and all other suspected cues are removed. These experimental results prompted further research, including quasi-experiments in the field.

Preliminary Considerations

Empirical data can be used to address Niko Tinbergen's four central questions of animal behavior: (1) What causes the behavior to occur? (2) How does the behavior develop? (3) How does the behavior affect survival, mating ability, and reproductive success? (4) What is the evolutionary history of the behavior? Before behavior is measured, however, some fundamental questions must be answered.

What is the best level of analysis?
    Choose a level of analysis, from fine detail of individual movements to complex social interactions, that provides the right amount of detail – not too little to be worthy of note and not so much that it is overwhelming.
What is the right species?
    Choose a species that is appropriate for the topic. Among the things to consider are the availability and ease of observation, tolerance of human observers, appropriate life-history characteristics, social organization, and existence of suitable background information.

Where should I make my observations?

The full richness of behavior occurs in the field, but field studies present practical difficulties, including the need for permission to work in a particular place, the difficulty of making good observations, and the logistics of traveling to and from the field site. Behavior can also be studied at zoos or farms, but these venues present their own challenges and limitations, including restrictions on what can be done and when it can be done, and the number of animals that can be observed. Behavior can also be studied in the laboratory, where conditions can vary from somewhat naturalistic to very artificial.

When should I make my observations?

Behavior must be observed at the appropriate time of year and day, and scheduling observations can help to reduce bias in data collection, but circumstances may dictate that schedule (e.g., the study area or laboratory building is closed a night) even though a different schedule would be superior.

Once these four questions are answered, you need to consider observer effects, anthropomorphism, and ethics.

Observer Effects

An observer can have subtle or substantial effects on the behavior of animals. These effects can be mitigated by concealing yourself in a blind (hide) or by making a video recording of the behavior, but being restricted to a blind or using a video camera might make observation more difficult. An alternative is to spend time making the animals accustomed to your presence, but it is difficult to assess the effectiveness such habituation attempts.

Anthropomorphism

When observing and describing animal behavior, it is tempting to assume that the animals are just like humans, with human thought processes and emotions. We often hear people say, "My dog is feeling guilty," or "My cat is jealous." But animals are not just like us – they can differ dramatically from us in their sensory abilities, behavioral responses, and ability to learn. Using human emotions and intentions to explain the behavior of (non-human) animals can thus prevent us from understanding that behavior. On the other hand, viewing animals as machines is not productive either, and a bit of projection might lead to interesting hypotheses to test.

Ethics

Any study of animal behavior should balance the information to be gained against the harm to the animals. When examining the ethics of behavioral research, there are three important questions: Will the research increase scientific understanding? Will the research produce results beneficial to humans or to the animals themselves? How much suffering will the research inflict on the animals? The benefits measured by the first two questions must be weighed against the cost measured by the third. A valuable tool in determining this balance is the Guidelines for the Treatment of Animals in Research and Teaching, which can be found on the web site of the Animal Behavior Society.

Keeping your question or hypothesis in mind, you next need to make preliminary observations, identify the behavioral variables to measure, and choose suitable recording methods for making the measurements.

Principles of Animal Behavior Study Design

Good studies of animal behavior are designed according to three principles.

Replication Must Be Independent

Independence means that one observation or animal (a replicate) does not influence or affect another. Independence of replicates is important because we use the differences among the replicates to estimate the overall differences in whatever we are studying. For example, if you observe one individual many times, each observation is not independent of the others because the same animal is involved. Such an improper use of repeated observations is called pseudoreplication, and it leads to improper statistical analysis and interpretation of results. Attempts to avoid pseudoreplication can also, however, lead us astray. Suppose we want to know whether schooling fish respond differently to large and small predators. If we use a single school of 20 fish to observe reactions to large and small predators, then obviously each fish is not an independent replicate because each school member is affected by the other fish in the school, so we end up with only one replicate (the school). To avoid pseudoreplication and to generate a more useful sample size (number of replicates), we might observe each member of the school separately, thus producing 20 independent replicates. Unfortunately, although we have generated a statistically valid design, we have also produced a biologically meaningless (invalid) one because schooling fish do not encounter predators individually.

Variables Must Not Be Confounded

If we observe schools of fish responding to large and small predators, but do the large-predator observations in the morning and the small-predator observations in the afternoon, then we cannot attribute a difference to the size of the predator only. Fortunately for us, there are several sampling designs that address this issue. One is to randomize the order of observations using a randomization method such as flipping a coin or rolling dice, or using a table of random numbers or a computer's random number generator. Random sampling is also an important requirement for statistical testing, so we have another reason to use randomization procedures. Of course, if a sequence is truly random, then sometimes we will get a long series of the same group. A solution to this problem is to balance in combination with randomization. For example, randomly choose large or small predator first, observe the opposite next, and then repeat the process. This procedure produces balanced pairs of observations. By the way, if two or more observers are involved in the study, then confounding applies to observers as well as to subjects, so do not have one observer watch responses to large predators and another watch responses to small predators.

Known but Unwanted Sources of Variation Must Be Removed

A way to deal with known confounds is by blocking (or matching). Block designs allow us to compare like with like or matched observations. So, in our fish school example, we would observe each school's reaction to both large and small predators (but of course not always in the same order because order effects are another source of confounding). Block designs can be quite complex, so a full discussion of them is beyond us here.

Our next two steps in testing our hypotheses are to identify appropriate behavioral variables and then to choose suitable methods to record them.

Behavioral Variables

We must break the continuous stream of behavior into distinct categories to make useful measurements, and we need names for the categories. Behavior can be described by its structure (postures and movements) or by its consequences (effects). Structural descriptions are objective, but they can be needlessly detailed. Describing behavior by its consequences is simpler, but the presumed consequences can be wrong. Neutral or descriptive labels avoid this problem. For example, nestlings of many bird species call and are subsequently fed by their parents. This calling could be described in great detail as a series of movements, postures, and sounds, or it could be called begging, or it could be called cheeping, which sounds like the call itself (an onomatopoetic name), but does not attribute a consequence. Another form of description uses spatial relationships (where and with whom) rather than what an animal does. For example, the parent bird bringing food to its nestlings could be said to approach and depart the nest.

Observations of behavior can be divided into three types of measurements: latency, duration, and frequency. Latency is how long until a behavior occurs; duration is how long a behavior lasts. Frequency (rate) is how often a behavior occurs in a given period. When choosing among these types of measurement, it helps to consider the continuum from events to states of behavior. At one extreme, events are discrete behavior patterns (e.g., copulations) that can be counted to produce a frequency. At the opposite extreme are prolonged activities called states (e.g., resting) whose duration is measured.

Recording Methods

Rules for the systematic recording of behavior are critical to designing good studies, and the choices involve two distinct levels: sampling rules (which subjects to watch and when) and recording rules (how behavior is recorded). There are four kinds of sampling rules: ad libitum, focal animal, scan, and behavior sampling. Ad libitum sampling is simply noting whenever something of interest occurs. This method is simple, but it is biased in favor of the most conspicuous individuals and behavior. Focal animal sampling limits observations to specific individuals or groups for a specified period. The sequence of focal animals should be chosen according to the three principles of study design that we have already discussed. One problem with focal animal sampling, especially in the field, is that the focal animal may disappear during the specified sampling period. A method to reduce this problem is scan sampling, in which a group of individuals is scanned at specified intervals and the behavior of each individual at that instant is recorded. One last method is to focus on categories of behavior rather than on the individuals performing them. This method is called behavior sampling and it involves watching a group of animals and recording each occurrence of a particular behavior along with the individuals that perform it. Focal animal, scan and behavior sampling can be combined, for example by taking a scan or behavior sample each time all of the focal animal samples have been completed.

There are two kinds of recording rules: continuous recording and time sampling. The goal of continuous (all-occurrences) recording is an exact record of frequencies, start and stop times, and durations of behavior, but this method is difficult to implement. An alternative is time sampling, in which behavior is sampled periodically at a specified sample point at the end of a sample interval. Time sampling can be subdivided into instantaneous sampling and one–zero sampling. Instantaneous sampling is used for events, but it is not appropriate for rare behavior, which would be missed too often. The result is the proportion of sample points in which the behavior occurred. One–zero sampling also uses sample points and produces a proportion of periods in which the behavior occurred, but unlike instantaneous sampling, we note whether the behavior occurred at any time during the previous interval. The length of the sample interval makes a difference, so choose the shortest possible interval (and thus the most sample points).

Recording Medium

Our next consideration is the medium used to make recordings. Camera phones have made video recording easy, at least for short periods (think YouTube), but non-video alternatives include voice recorders (for detailed verbal descriptions), chart (event) recorders, automatic data recorders, and check sheets. Note that high-tech methods are not necessarily better than paper-and-pencil methods. Video and audio recordings have the advantage of instant replay, but the field of view of a video camera is limited and if you make a recording you then have to analyze it, which can be extremely time consuming. When I began studying red-winged blackbird behavior in 1973, I wanted to construct a time budget of male activities as well as to calculate frequencies of behavioral events. As a graduate student with limited resources, I used paper-and-pencil methods to do continuous (all-occurrences) recording by setting up a data sheet with 15 rows, each representing 1 min and consisting of 60 equally spaced dots, and a short-hand code for each behavior. I recorded time budget categories by noting when each started, and events by writing a letter code for each category. Armed with a clipboard, windup stopwatch, and lots of pencils (first rule of field work: always have more than one pencil), I spent my mornings observing and recording the behavior of male red-winged blackbirds.

These days there are commercially available methods of recording behavioral data that will run on desktop and laptop computers.

For observational studies, the final methodological question to answer is, How much data should I collect? Answering this question with statistical analyses is beyond the scope of an undergraduate project, but graduate students will want to learn how to perform a priori power analyses. A rule of thumb is, gather as much data as possible given the logistical constraints.

Once the methodological decisions have been made, it is finally time to observe behavior, and then to analyze the data, using appropriate statistical methods. Although proper statistical analysis is critical, this topic is also too large and complex for us to consider here.

Throughout these stages, it is important to remember that your purpose remains testing hypotheses to answer the four principal questions of animal behavior.

Experiments

Everything we have discussed so far applies to all empirical studies of animal behavior, but experiments have additional considerations. Experimental design is a huge and complex topic, but we need some understanding of basic principles to conduct even a simple experiment. A good starting point is a list of the desirable properties of experimental design: good estimation of treatment effects, good estimation of random variation, absence of bias, precision and accuracy, wide applicability, and simplicity in both execution and analysis.

Experimental Designs

The design of treatments is basic but crucial because it defines our hypothesis tests. Treatments can be broadly divided into unstructured (random differences) and structured (fixed differences) designs of which there are many. The design of layout, or how we assign treatments to experimental subjects, is a complementary consideration to treatment design. Five commonly used designs in studies of behavior are completely randomized one factor, randomized block, nested, Latin square, and completely randomized two factor (factorial) designs. Analysis of variance (ANOVA) can be used to analyze data from these designs.

In contrast to the previous designs, which use specific levels of each factor (e.g., low-, medium-, and high-hormone treatments), in gradients we attempt to assess behavioral response to a continuous range of treatments (e.g., hormone concentration). Analysis of gradients uses statistical testing such as correlation or regression because both the measurement variable and the treatment variable are numeric.

When two (or more) treatment variables are categorical and our measurement variable is a count (number of occurrences), enumeration methods such as goodness-of-fit tests and tests of independence are appropriate.

Once the experimental methods are set, it is time to do the experiment and analyze the results with the proper statistical test. It is also important to remember once again that your purpose remains testing hypotheses to answer the four principal questions of animal behavior.

Modeling

Modeling involves the behavior of a set of equations or computer simulation rather than of animals, but the purpose is still hypothesis testing. Models are common in animal behavior and in everyday life. Think about giving directions to your house. A map of the route would probably be quite elementary – a few lines representing the streets to take and maybe a few major landmarks. This simple map is a model of reality, but it is not meant to be the real. Like maps, models are simplified versions of reality.

Models can be deterministic or stochastic, and static or dynamic, but all are formal and mathematical.

A mathematical model uses equations to describe the essential aspects of behavior by presenting our understanding of that aspect of behavior in a testable form. The model's equations can be solved mathematically to examine how behavior might occur under very clearly described circumstances, which are called the model assumptions. In animal behavior, a commonly used method is game theory, although many other methods are used as well.

Game theory was introduced to animal behavior by John Maynard Smith, who first used it to analyze contests for an important resource such as food, territory, or mates. Maynard Smith tried to answer a question that had been puzzling animal behaviorists for many years: Why do animals use display (like disputing neighbors shaking their fists at each other) rather than more violent means to settle disputes? At one time the answer was "because fighting would produce injuries, which would be bad for the species." Explanations that rely on advantages to the species or other groups of individuals are called group selection hypotheses, but evolutionary analyses in the 1960s and 1970s showed that these hypotheses are usually inadequate. Maynard Smith's model was the now-classic game that compares the behavioral strategies dove and hawk.

A behavioral strategy is simply a fixed and predictable way of behaving in a contest. It does not imply that contesting animals make conscious decisions. The purpose of a game-theory model is to compare alternate strategies to see whether one is evolutionarily stable. An evolutionarily stable strategy (ESS) cannot be invaded by any other strategy. Maynard Smith was able to show that, contrary to the group selection hypothesis, a display-only ('dove') strategy could not resist invasion by a fight immediately ('hawk') strategy because a hawk will always defeat a dove. Perhaps surprisingly, the hawk strategy is also not stable against dove because dove does not pay the cost of injury. This game-theory model demonstrates that neither strategy is an ESS.

Hawk and dove are certainly not the only ways that an animal might behave in a contest, and other strategies have been studied. If you are interested in learning how to develop game theory models, try Gamebug, a teaching and learning resource.

In some cases, however, the relationships are too complex for mathematical (analytical) solutions, so a computer model can be used to simulate behavior. Like mathematical models, computer simulations attempt to model a particular behavioral system to gain insight into how the system works, but they require a computer (or even a network of computers) programmed to perform the tedious calculations and to display the results in a useful way. The first such simulation was of nuclear detonation for the Manhattan Project during WW II. Simulation was used because the scale of a detonation was far greater than blackboards and mathematical models could handle.

Simulations such as stochastic dynamic programs and genetic algorithms follow a specified procedure, but others are purpose-built to test hypotheses for particular circumstances or species. What all simulations share is a set of representative scenarios for which a complete enumeration of all possible states would be impossible. Like mathematical models, simulations start with assumptions and are typically run under different conditions to investigate changes in those assumptions or other conditions.

A study of bowerbirds provides an example of both game theory modeling and computer simulation. In many species of bowerbirds the males build amazing structures (bowers) and decorate them with artifacts. Females mate with males with the best bowers, so just a bit of thought suggests that a male bowerbird might do one of three things to be successful. He could spend time constructing and defending a bower against raiding by other males ('defender'), or he could split his time between defending his own bower and visiting other bowers to steal their decorations ('stealer'), or he could split his time between defending his own bower and visiting other bowers to destroy them ('destroyer'). Using measurements of the costs and benefits of these strategies in terms of access to females, the game-theory model shows that both destroyer and stealer are stable against defender under most circumstances. Simulations show that defender is stable if intruders have to travel long distances between bowers or if residents are able to repair damaged bowers quickly.

Regardless of the modeling method used, as with empirical methods, our purpose remains testing hypotheses to answer the four principal questions of animal behavior.

Conclusion

A very real risk in writing or reading an article like this is losing the forest for the trees – we tend to get caught up in the fine details and thereby lose sight of the big picture. For those of us who have dedicated our lives to the study of animal behavior, the big picture remains explaining how and why animals do what they do. Our use of the methods outlined in this article has produced a lot of valuable information, but perhaps the most important conclusion for you, the reader, is that much more remains poorly understood or completely unknown. If you find animal behavior fascinating, then you can use the methods described here and elsewhere to answer the most general question, "How and why do animals do what they do?"

See also: Endocrinology and Behavior: Methods; Ethograms, Activity Profiles and Energy Budgets; Experimental Design: Basic Concepts; Game Theory; Measurement Error and Reliability; Neuroethology: Methods; Niko Tinbergen; Playbacks in Behavioral

Experiments; Remote-Sensing of Behavior; Sequence Analysis and Transition Models; Spatial Orientation and Time: Methods.

Further Reading

Bart J, Fligner MA, and Notz WI (1998) *Sampling and Statistical Methods for Behavioral Ecologists.* Cambridge, UK: Cambridge University Press.

Dawkins MS (2007) *Observing Animal Behaviour: Design and Analysis of Quantitative Controls.* Oxford, UK: Oxford University Press.

Dugatkin LA and Reeve HK (1998) *Game Theory and Animal Behavior.* Oxford, UK: Oxford University Press.

Holland JH (1992) *Adaptation in Natural and Artificial Systems: An Introductory Analysis with Applications to Biology, Control, and Artificial Intelligence.* Cambridge, MA: MIT Press.

Hutchinson JMC and McNamara JM (2000) Ways to test stochastic dynamic programming models empirically. *Animal Behaviour* 59: 665–676.

Kamil AC (1988) Experimental design in ornithology. In: Johnston RF (ed.) *Current Ornithology* 5, pp. 313–346. New York, NY: Plenum Press.

Lehner PN (1996) *Handbook of Ethological Methods,* 2nd edn. Cambridge, UK: Cambridge University Press.

Mangel M and Clark CW (1988) *Dynamic Modeling in Behavioral Ecology.* Princeton, NJ: Princeton University Press.

Martin P and Bateson P (2007) *Measuring Behaviour: An Introductory Guide,* 3rd edn. Cambridge, UK: Cambridge University Press.

Ploger BJ and Yasukawa K (2003) *Exploring Animal Behavior in Laboratory and Field.* San Diego, CA: Academic Press.

Sokal RR and Rohlf FJ (1994) *Biometry: The Principles and Practices of Statistics in Biological Research,* 3rd edn. New York, NY: W. H. Freeman and Company.

Tillberg CV, Breed MD, and Hinners SJ (2007) *Field and Laboratory Exercises in Animal Behavior.* London, UK: Academic Press.

Whitlock MC and Schluter D (2009) *The Analysis of Biological Data.* Greenwood Village, CO: Roberts and Company.

Zar JH (2010) *Biostatistical Analysis,* 5th edn. Upper Saddle River, NJ: Prentice Hall.

Relevant Websites

http://www.animalbehavior.org/ – Animal Behavior Society.
http://www.jwatcher.ucla.edu/ – Jwatcher observation software.
http://www.noldus.com/animal-behavior-research/ – Noldus observer software.
http://hoylab.cornell.edu/gamebug/ – Gamebug.

Experimental Approaches to Hormones and Behavior: Invertebrates

S. E. Fahrbach, Wake Forest University, Winston-Salem, NC, USA

Introduction and Definitions

What Is an Invertebrate?

Invertebrates constitute an estimated 97% of all animal species on earth. Current taxonomies describe more than 30 invertebrate phyla, some of which have been evolving as separate lineages for hundreds of millions of years. What links these diverse phyla? Invertebrates are typically defined in dictionaries and encyclopedias as 'animals lacking a backbone or a notochord.' Modern reference sources typically note that 'invertebrate' is not a scientific term, although it is widely known and instantly understood by nonbiologists.

A scientific view of animals that reflects molecular evolutionary analyses divides the bilaterians (animals with a bilaterally symmetric body plan) into two great lineages (clades): deuterostomes and protostomes. These terms were originally developed on the basis of patterns observed during early embryonic development, including whether the blastopore (first opening) forms the mouth (protostomes) or anus (deuterostomes). Phylum Chordata, the phylum that is the taxonomic home of the vertebrates but which also includes some animals without a backbone, is one of a relatively small number of deuterostome phyla. The vast majority of bilaterians are protostomes. Estimates of the age of the last common ancestor of the deuterostomes and protostomes range from 600 to 1200 Ma. There is a lack of consensus on what this animal was like. Was it an animal similar in complexity to modern bilaterians, possibly with appendages? Or a microscopic flatworm? This is of interest to us in that understanding it will help us to identify the ancient and conserved elements of neuroendocrinology. This is important not only for our understanding of animal evolution, but also for the estimation of the value of invertebrate models to solve problems related to the health of humans and their predominantly vertebrate domestic animals.

Our knowledge of the behavioral neuroendocrinology of invertebrates is slight compared with the number of invertebrate species, typically estimated to be some tens of millions. Many reasons account for this gap in our knowledge. Often, invertebrates are minute in size; many that are macroscopic live in habitats difficult to study (such as the deep ocean); others are poorly suited to life in the laboratory. Another reason why we know so little is that relatively few biologists have the taxonomic training needed to study invertebrates. As a consequence, investigators study a handful of model organisms, make broad generalizations on the basis of extremely limited data, and rue the missed opportunities for a truly comparative analysis.

Hormonal Regulation of Behavior in Invertebrates: An Overview

Many invertebrates have an open circulatory system, which means that their tissues are bathed in a fluid that serves most of the functions of vertebrate blood. This fluid is commonly referred to as hemolymph in arthropods and as coelomic fluid in some other groups. Although open circulatory systems are low pressure and inefficient relative to the closed circulatory systems of vertebrates, a muscular heart is typically present. Invertebrate hormones can therefore be defined as chemical messengers present in the fluid that circulates through the tissues, and invertebrate endocrine cells as the sources of those chemical messengers. As in the case of vertebrates, autocrine and paracrine chemical messengers are also found in invertebrates.

Comparative studies have revealed that, like vertebrates, invertebrate animals rely heavily on circulating protein hormones, steroid hormones, and biogenic amines to coordinate behavior and physiology. Peptide hormones, which are small protein hormones (typically 30 amino acids in length or less) that regulate behavior are typically produced in the central nervous system and are often referred to as neuropeptides. If we compare the endocrine systems of vertebrates with those of invertebrates, we find that the signaling molecules themselves are often far less conserved than the receptors through which they signal. For example, a neuropeptide found in fruit flies and other insects (DH31, diuretic hormone) but not in vertebrates binds to a G-protein coupled receptor that is homologous to the mammalian corticotropin releasing hormone receptor. Another example of receptor conservation is found in the similarity of the ecdysone receptor, an important nuclear receptor for the steroid hormones of arthropods, to the liver X receptor alpha of mammals. Such observations have led to the general conclusion that extant invertebrates and vertebrates do not use the same hormones but that hormone receptors reflect evolutionary conservation of ancient signaling pathways. However, the surprising persistence of some signaling molecules across the vast reaches of animal evolution is now well-documented. Two excellent examples are the oxytocin family of peptides, found in animals as different as earthworms and humans, and the relaxin/insulin-like peptides and growth factors currently identified in chordates, arthropods, molluscs, and nematodes. Even more surprising is the growing body of evidence that the general function of a category of signaling molecule may be broadly conserved. For example, as is the case in vertebrates, the insulin-like peptides of insects play important roles in metabolism, the regulation of growth, and aging.

In vertebrates, much research on the relationship of hormones and behavior has focused on reproductive behavior. Many invertebrate taxa are also characterized by male and female individuals that display overt and/or covert sexual dimorphisms and that have characteristic sex-specific behaviors. A major difference between vertebrate and invertebrate taxa is that there is relatively little evidence that sex differences in invertebrates reflect developmental actions of hormones. Other important topics in vertebrate neuroendocrine research, such as the hormonal regulation of ingestive behaviors, the hormonal correlates of stress, and hormonal modulation of aggressive behavior, are also poorly represented in the invertebrate literature. A major point of convergence between vertebrate and invertebrate research is found in the study of hormonal modulation of life history transitions. Species-typical responses to a changing environment – whether such changes are predictable, as in the case of seasons, or unpredictable, as in the case of severe weather events – are often directly reflected in changes in individual patterns of hormone synthesis and secretion, which in turn produce coordinated changes in both physiology and behavior. Many of the best-known examples of such responses in invertebrates come from studies of insects, a circumstance that reflects the economic importance of many insect species, either because they are pests or pollinators. These examples include choices between migrating and settling, between entering and foregoing a form of developmental arrest called diapause, and in the special case of social insects that maintain a fixed nest, a choice between tending larval brood and foraging away from the nest. One example of particular importance is found in what is likely one of the most common animal life history transitions on our planet: the shedding of a cuticle by an arthropod. Experimental attempts to identify the mechanisms regulating this life transition have driven the development of almost all the key experimental approaches important for the study of hormones and behavior in invertebrates.

Arthropod Molting and Metamorphosis

Arthropod molting and metamorphosis can only be understood in terms of the hormonal coordination of ontogeny, nutritional status, and environmental cues. The terms molting and metamorphosis are often used interchangeably by nonbiologists, but biologists use them to refer to specific different aspects of the process. Molting is the shedding of the exoskeleton (cuticle) of the previous life stage. Without molting, an insect cannot grow. Once a new cuticle is fully formed, an insect engages in species-specific stereotyped patterns of movements (molting behaviors) that liberate it from the cuticle of the previous stage. The timing of a molt is critical for two important reasons. First, the molt will fail (and the insect will die) if it is attempted prior to complete deposition of a new cuticle. Second, the fresh cuticle of a newly molted insect is pale and soft because the chemical reactions (tanning) that result in hardening and darkening take hours to days to be completed. The insect is extremely vulnerable to predators before the new cuticle has tanned. A molt should therefore occur only when the insect is in a safe, sheltered location, or at the time of day when its predators are inactive. Molts are now known to be activated at specific times by complex coordinated actions of multiple neuropeptides on target neurons in the central nervous system, providing an important example of the modulatory role of peptides on neural circuits that control behavior. These neuropeptides are secreted when both the nutritional status of the insect and the time of day (or season) signal that a successful molt is possible.

The term metamorphosis refers to the development of winged, reproductively competent adults from feeding larval stages (**Figure 1**). It is not a behavior per se (although metamorphic changes in the nervous system can result in important changes in behavior, such as the

Figure 1 Several adult worker honeybees (*Apis mellifera*) on a wax honeycomb from a standard removable frame hive typically used for beekeeping and research. Maggot-like white larvae shaped like the letter 'C' fill the cells of the honeycomb. The honeybee is an excellent example of a holometabolous insect. Both the molts (shedding of the cuticle of the prior stage) and metamorphosis (development of stage appropriate tissues and structures) are controlled by hormones. Photograph by Professor Z. Huang, Michigan State University, East Lansing, MI, USA. Used with permission.

shift from crawling to flying), but is rather a sequential process of tissue-specific regulation of gene expression In insects designated as holometabolous, the process of metamorphosis involves a series of feeding larval stages, each separated by a molt, followed by a quiescent pupal stage that precedes the molt to the adult stage. The larvae of holometabolous insects are typically soft-bodied caterpillars or maggots. The morphological and behavioral changes that accompany the larval–pupal and pupal–adult transitions in such insects can be dramatic: for example, larval insects may possess very simple photoreceptors that permit the discrimination of light and dark, while the compound eyes of the adult provide the capacity for sophisticated color and form vision to the extent that some social insects are able to recognize specific conspecifics on the basis of their idiosyncratic patterns of facial markings. This life history strategy, which is also referred to as complete metamorphosis, is characteristic of many of the most familiar groups of insects, including moths, butterflies, flies, bees, and beetles. By contrast, insects designated as hemimetabolous progress through a series of feeding larval (also called nymphal) stages, each separated by a molt, during which they gradually acquire adult characteristics. In appearance, nymphs are typically miniature versions of the adults of their species. A final molt to the adult stage is associated with the attainment of reproductive maturity. A hemimetabolous life history strategy is often referred to as incomplete metamorphosis. Cockroaches, crickets, and grasshoppers provide familiar examples of insects that experience incomplete metamorphosis.

Because of their importance to the development of invertebrate neuroendocrinology, examples of key experimental approaches for the study of hormones and behavior in invertebrates will be drawn primarily from studies of metamorphosis and molting in insects. Studies of the hormonal regulation of tissue metamorphosis have led to the definition of the major categories of insect hormones; studies of the hormonal regulation of molting have revealed the existence of a complex hierarchy of peptide signals that ensures that performance of this behavior is restricted to appropriate times in the insect's life and that, once initiated, is carried through to completion. A brief account of techniques used to study hormonal regulation of polyphenisms in social insects follows the overview of major methods.

Experimental Approaches: Overview of Major Methods

Ablation

Ablation of known or presumed tissue sources of hormones, followed when possible by replacement, was historically an extremely important technique in invertebrate behavioral neuroendocrinology, although its use is limited to species large enough to withstand surgical manipulation. A related technique called ligation exploits the fact that insects use a system of tracheal tubes connected to the body surface at openings called spiracles to supply oxygen to their tissues. This allows the insect body to survive division into separate compartments, with each compartment supplied only with hormones produced within that compartment. Ligation is often accomplished by tying a stout silk thread as tightly as possible around the boundary between the head and the thorax or between the thorax and the abdomen. Carefully timed ligations of caterpillars were used in the classic early experiments of Kopeć to demonstrate that a factor produced by the brain of the gypsy moth is required for metamorphosis. In an astonishingly prescient study published in the *Biological Bulletin* in 1922 titled 'Studies on the necessity of the brain for the inception of insect metamorphosis,' Kopeć named this factor 'brain hormone.' Given that this publication preceded by several years the 1928 report of Ernst Scharrer describing what eventually came to be called 'neurosecretory neurons' in the fish hypothalamus, Kopeć can fairly be said to have been among the first to recognize that the brain is an important endocrine gland, although the significance of his results was not recognized until decades later. During the late 1930s and early 1940s, Fukuda used a double ligation technique to establish that, in addition to the brain, glands in the anterior thorax also produce a circulating factor ('molting hormone') required for molting and metamorphosis. In the 1930s, Fraenkel used abdomens of the relatively large larvae of calliphorid flies ligated into anterior and posterior abdominal compartments as a sensitive bioassay for the molting hormone. In these ligated larvae, the anterior compartment forms a normal, darkly tanned pupal cuticle, while the posterior compartment will remain white and soft, as is typical of larval cuticle, unless injected with a substance that contains molting hormone activity. This bioassay was subsequently used to purify (from a half ton of silkmoth pupae) sufficient molting hormone so that its chemical identify could be determined. We now know that hormone produced by the brain is the neuropeptide called prothoracicotropic hormone (PTTH); its target is a steroid-synthesizing gland in thorax called the prothoracic gland, which is the source of the molting hormone (**Figure 2**). The molting hormone obtained from the silkmoth pupae and the molting hormones of all insects were discovered to be steroids. They are members of a C27 group of steroids given the generic name 'ecdysteroids.' A potent ecdysteroid with bioactivity in many insects is 20-hydroxyecdysone. Ecdysteroids act as transcription factors by binding to nuclear receptors in target cells. The historic first evidence that any steroid hormone acts via regulation of transcription came from studies of ecdysteroid regulation of puffing in the polytene chromosomes found in the cells of the salivary glands of flies. A model proposed by Ashburner to account for the regulation of gene expression in the salivary glands is now

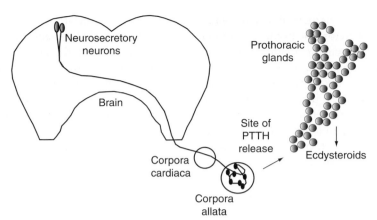

Figure 2 Schematic diagram of the prothoracicotropic hormone (PTTH) neuroendocrine system in the model insect *Manduca sexta*, commonly known as the tobacco hornworm moth. Two pairs of neurosecretory neurons in the brain (for clarity, only one pair is depicted in this figure) synthesize the neuropeptide PTTH, which is transported to varicose release sites on the surface of the corpora allata glands. The prothoracic glands respond to PTTH by synthesizing ecdysteroids, the steroid hormones of insects. Diagram is based on the immunolocalization of PTTH (O'Brien et al., 1988), a technique described in the *Chemical Neuroanatomy* section of this article.

accepted as a general mechanism for the action of ecdysteroids on all tissues that have been studied, including the nervous system.

Parabiosis (surgical union of the circulatory systems of two individuals) is a specialized technique for replacement of a hormone or putative hormone historically important in research on insect hormones. Parabiosis-based studies were used by Wigglesworth, who studied the blood-sucking bug *Rhodnius prolixus*, to establish that a third factor determines whether a molt is status quo (e.g., larval–larval or pupal–pupal) or metamorphic (e.g., larval–pupal). Wigglesworth named this factor juvenile hormone. Williams subsequently used the parabiosis technique to establish that the abdomen of male silkmoth pupae is an enriched source of juvenile hormone, which eventually permitted the identification of the sesquiterpene chemical identity of the juvenile hormones. Two small glands found at the base of the brain called the corpora allata are the major sites of juvenile hormone synthesis and release (**Figure 2**).

Detection and Measurement of Hormones in Fluids and Tissues

As is true in vertebrate endocrinology, the modern era of research in invertebrate hormones and behavior has its origins in the development of techniques for accurate measurement of low concentrations of hormones in fluid and tissue samples. The specific challenge posed by most invertebrates, however, is their small size and the correspondingly minute volumes of their hemolymph. The development of radioimmunoassay and other antibody-based assays and the use of high-performance liquid chromatography permitted the accurate measurement of hormones in small volume samples and freed researchers from dependence

on technically challenging, time-consuming bioassays that at best yielded semiquantitative results, but it has often proved necessary to pool samples to make essential measurements. It is also often impossible to obtain sequential samples of hemolymph from a single individual, which poses a particular difficulty for detecting transient pulses of hormone secretion.

Studies of insect molting and metamorphosis have advanced by the development of sensitive radioimmunoassays for ecdysteroids and juvenile hormones. Because these hormones are not proteins, they are conjugated to proteins such as keyhole limpet hemocyanin or bovine serum albumen to elicit an immune response in mammalian hosts to in order to generate the antibodies required for quantitative assays.

Advances in analytical chemical techniques for separating and identifying complex mixtures of proteins from samples as small as the contents of a single cell are now beginning to be applied to the analysis of invertebrate neuroendocrine cells. Capillary separations and innovative mass spectrometry techniques (e.g., matrix-assisted laser desorption/ionization time of flight mass spectrometry (MALDI-TOF)) have been used to perform the first global characterizations of peptides in the nervous system of the mollusc *Aplysia* and the insect *Apis mellifera* (the honeybee). These methods are particularly important for the study of invertebrates as the small size of many species, including the fruit fly *Drosophila melanogaster*, makes the application of traditional protein chemistry methods to the characterization of insect hormones a heroic endeavor.

Chemical Neuroanatomy

Techniques used on vertebrate nervous tissue typically also work well on invertebrate preparations. One example is the use of the aldehyde fuchsin method to stain peptidergic granules in neurosecretory neurons. The classic studies of Berta Scharrer (from the 1930s on) demonstrated, using cytological techniques, the likely existence of neurosecretory neurons in the central nervous system of many if not all invertebrates, including the mollusc *Aplysia* and cockroaches. The subsequent development of antibodies specific for insect peptides and their use with the same standard immunocytochemical techniques developed for the study of vertebrates has permitted the mapping of neurosecretory neurons in the brain as well as the identification of endocrine cells outside of the central nervous system, such as the Inka cells of the epitracheal glands. The use of fluorescent labels for antibodies in combination with laser scanning confocal microscopy has greatly advanced our knowledge of the distribution of peptides in the insect brain. Such studies, for example, confirmed the existence of lateral and medial groups of neurosecretory neurons in the protocerebrum of the insect brain. These neurosecretory neurons send their axons to a neurohemal release site outside of the brain called the corpus cardiacum. For example, the neurosecretory neurons that synthesize Kopeć's brain factor PTTH send their axons to the corpus cardiacum, from which it is released into the hemolymph in which it travels the short distance to the steroidogenic prothoracic glands. The paired corpora cardiaca glands of insects also contain intrinsic cells that release hormones synthesized at that site directly into the circulatory system. These antibody-based studies have therefore confirmed the idea championed by Berta Scharrer and others that the brain–cardiacum–allatum system of insects is analogous to the hypothalamic-hypophysial system of vertebrates.

Although the corpora allata and corpora cardiaca are separate glands in most insects, some dipterans, including the fruit fly *D. melanogaster*, have a fused ring gland that unites the corpora allata, corpora cardiaca, and prothoracic glands in a single structure.

Bioinformatics Analyses, Studies of Gene Expression, and Transgenics

The burst of genome sequencing projects that marked the start of the twenty-first century included the genomes of several invertebrates important for behavioral research, including the honeybee *A. mellifera*. One exciting by-product of these projects has been the annotation of genes encoding peptides and their cellular receptors, which are almost all members of the G-protein coupled receptor superfamily of transmembrane proteins. In several species (including the fruit fly, honeybee, mosquito, the red flour beetle, and the nematode *Caenorhabditis elegans*), attempts have been made to identify the complete set of genes that encode peptide precursor proteins and/or the complete set of peptide hormone receptors.

A Hydra Peptide Project has also been completed, extending our phylogenetic analyses to cnidarians. These projects have led to the discovery of many new hormones (the existence of many of which were predicted by earlier bioassay studies) and, through projects in which GPCRs are cloned and expressed, the matching of endogenous peptide ligands to their specific GPCRs. A project in which a bioinformatics analysis was combined with modern techniques of chemical analysis was used to predict the complete set of peptides in honeybees. These new data provide tools for both neuroendocrinologists and evolutionary biologists, as the former can test the function of the newly identified chemical signals and the latter can reconstruct their phylogeny.

The development of molecular biological tools for studying gene expression through measurements of RNA extracted from tissues (including Northern blotting, RNase protection assays, DNA microarrays, and quantitative RT-PCR) has allowed regulation of gene expression in the brain and other to be used as an endpoint in endocrine studies. Microarray studies have focused on changes that reflect endogenous patterns of hormone secretion, as in a 1999 study by White and colleagues of changes in gene expression during metamorphosis in the fruit fly, or responses to hormone treatment, as in Whitfield's 2006 study of the maturation of foraging behavior in honeybees. In situ hybridization, in which nucleic acid probes are used to identify cells in prepared tissue that contain specific mRNAs, can also be used to identify hormone sources in invertebrate tissues. For example, the two pairs of ventromedial neurosecretory neurons that express eclosion hormone, one of the insect neuropeptides involved in 'turning on' molting behavior, were identified in the brain of the tobacco hornworm (*Manduca sexta*), using the method of in situ hybridization.

Identification of the genes that encode peptide hormones and their receptors also permits, in certain species of invertebrates, gene knockouts through creation of transgenics and gene knockdowns by treatment with double-stranded (ds) RNA (also referred to as RNA interference, RNAi). Studies of transgenic flies using genetic ablations of specific populations of neurosecretory neurons by Clark and others have revealed a surprising complexity in the hormonal regulation of behaviors associated with molting. A study in which four of the neuropeptides associated with molting behavior (eclosion hormone, ecdysis-triggering hormone, crustacean cardioactive peptide, and bursicon) and their receptors were knocked down in the red flour beetle (*Tribolium castaneum*) using the RNAi method allowed Arakane and colleagues to ask if findings from studies of the fruit fly are generally applicable. The results were surprising in that the phenotypes of the mutant fruit flies did not successfully predict the phenotype of the beetles treated with RNAi. Studies using RNAi are likely to play an increasingly important role in the analysis of hormone–behavior relationships in invertebrates, as they permit the study of species for which mutants cannot be readily created in the laboratory (which at present for invertebrates includes almost all invertebrate species except the fruit fly *D. melanogaster* and the nematode *C. elegans*).

Hormonal Regulation of Morphological and Behavioral Polyphenisms in Social Insects

Social insects (bees, wasps, ants, and termites) present opportunities to study the neuroendocrine regulation of complex social behaviors in a context free from cultural and ethical constraints. Social insects are characterized by polyphenisms: the occurrence of multiple phenotypes in a population that are not due to different genotypes, but rather reflect the impact of the environment (broadly defined) on gene expression during development. A fascinating and well-known example of a polyphenism in honeybees is the ability of a fertilized female egg to develop into either a reproductively active queen or a sterile worker (**Figure 3**). This phenomenon, which is often referred to as caste determination, results from differential larval nutrition (the feeding of royal jelly to future queens), which in turn causes changes in the temporal patterning of secretion of ecdysteroids and juvenile hormones, which in turn results in what Evans and Wheeler have described as 'differential expression of entire suites of

genes involved with larval fate.' Studies using microarrays to study the differences in gene expression profiles in developing honeybee queens and workers have revealed that, among other changes, many genes related to metabolism are upregulated in future queens. Gene expression profiles have now been tracked throughout the development of reproductive and worker castes in several species, including ants and social wasps.

An example of a purely behavioral polyphenism in social insects is found in the age-based division of labor characteristic of honeybee workers: in a typical colony, younger workers perform tasks within the hive such as comb building and tending larval brood, while older workers forage outside the hive for pollen and nectar (**Figure 4**). The transition from hive worker to forager typically occurs when an adult bee is 3 weeks old. Foragers have significantly higher circulating levels of juvenile hormone than younger workers, and treatment of younger bees with synthetic juvenile hormone induces the precocious onset of foraging. It was therefore surprising when an ablation study (surgical removal of the corpora allata, the sole source of juvenile hormone in the worker bee, on the first day of adult life) revealed that workers were able to make the transition to foraging in the absence of juvenile hormone. A breakthrough in the understanding of this behavioral transition occurred with the recognition of an interaction between circulating levels of juvenile hormone and stores of the yolk protein precursor, vitellogenin.

Figure 3 View of the three adult members of a honeybee colony: the stocky male drones (white arrows), the female reproductive or queen (center), and the female workers (unmarked). In honeybees, as in most insects and unlike vertebrates, sex differences in morphology are not the result of gonadal hormone action on peripheral structures. The differences between the two female castes (queen and workers) are instead the result of nutritionally induced differences in the temporal profile of the secretion of ecdysteroids and juvenile hormone and reflect the feeding of the queen with royal jelly when she was a larva. Photography by Professor Z. Huang, Michigan State University, East Lansing, MI, USA. Used with permission.

Figure 4 A worker honeybee foraging at a flower for pollen and nectar. The honeybee colony is characterized by age polyethism, the division of labor according to worker age. In a colony with a typical age structure, the oldest workers will forage while the younger workers will maintain the physical structure of the hive, tend the queen, and feed the larvae. The transition from hive bee to forager typically occurs 3 weeks after the completion of metamorphosis and is associated with high titers of juvenile hormone. Individually number-tagged bees are often used in behavioral research. The tags are glued to the dorsal thorax on the first day of adult life, before the young bee is able to sting or fly. Photograph by Professor Z. Huang, Michigan State University, East Lansing, MI, USA. Used with permission.

Studies in which RNAi knockdown of vitellogenin in young bees led to extremely precocious onset of foraging suggest that this protein may function either directly or indirectly as a hormone that regulates behavior. The discovery of a role for vitellogenin in the regulation of the transition to foraging in honeybees may eventually lead to an understanding of the regulation of the behavior of sterile workers in terms of the reproductive physiology of their fecund ancestors.

See also: Caste Determination in Arthropods; Caste in Social Insects: Genetic Influences Over Caste Determination; Developmental Plasticity; Division of Labor; Hormones and Behavior: Basic Concepts; Social Insects: Behavioral Genetics.

Further Reading

Amdam GV, Nilsen KA, Norberg K, Fondrk MK, and Hartfelder K (2007) Variation in endocrine signaling underlies variation in social life history. *American Naturalist* 170: 37–46.

Arakane Y, Li B, Muthukrishnan S, Beeman RW, Kramer KJ, and Park Y (2008) Functional analysis of four neuropeptides, EH, ETH, CCAP, and bursicon, and their receptors in the adult ecdysis behavior of the red flour beetle, *Tribolium castaneum*. *Mechanisms of Development* 125: 984–995.

Evans JD and Wheeler DE (2001) Gene expression and the evolution of insect polyphenisms. *BioEssays* 23: 62–68.

Fahrbach SE and Robinson GE (1996) Juvenile hormone, behavioral maturation, and brain structure in the honey bee. *Developmental Neuroscience* 18: 102–114.

Klowden MJ (2007) *Physiological Systems in Insects*, 2nd edn. San Diego, CA: Elsevier.

Marco Antonio DS, Guidugli-Lazzarini KR, do Nascimento AM, Simões ZL, and Hartfelder K (2008) RNAi-mediated silencing of vitellogenin gene function turns honeybee (*Apis mellifera*) workers into extremely precocious foragers. *Naturwissenschaften* 95: 953–961.

O'Brien MA, Katahira EJ, Flanagan TR, Arnold LW, Haughton G, and Bollenbacher WE (1988) A monoclonal antibody to the insect prothoracicotropic hormone. *Journal of Neuroscience* 8: 3247–3257.

Scharrer B (1987) Insects as models in neuroendocrine research. *Annual Review of Entomology* 32: 1–16.

Simonet G, Poels J, Claeys I, et al. (2004) Neuroendocrinological and molecular aspects of insect reproduction. *Journal of Neuroendocrinology* 16: 649–659.

Truman JW (2005) Hormonal control of insect ecdysis: Endocrine cascades for coordinating behavior with physiology. *Vitamins and Hormones* 73: 1–30.

White KP, Rifkin SA, Hurban P, and Hogness DS (1999) Microarray analysis of *Drosophila* development during metamorphosis. *Science* 286: 2179–2184.

Whitfield CW, Ben-Shahar Y, Brillet C, et al. (2006) Genomic dissection of behavioral maturation in the honey bee. *Proceedings of the National Academy of Sciences of the United States of America* 103: 16068–16075.

Experimental Design: Basic Concepts

C. W. Kuhar, Cleveland Metroparks Zoo, Cleveland, OH, USA

Observational Research

Most research projects begin with reconnaissance observations. Researchers will watch animal behavior with the idea of forming or refining research questions, developing an ethogram, or defining specific behaviors. Ultimately, the focus of the research shifts from documenting occurrences to recording variables and making predictions about which variable or variables are predictive of certain behavior patterns.

In research paradigms, independent variables, or variables that are assumed to be predictive of behavior, are compared with dependent variables, which are the variables of interest, often rates or numbers of behaviors. In reconnaissance observations or observational research, independent variables are not manipulated but are instead observed or recorded (i.e., sex or age class of individuals in a wild group of chimpanzees). However, many observed independent variables are confounded, or interrelated, making it difficult to discern the true relationship between the independent variables and the dependent variable or the behavior of interest. For example, in male hippopotamus, dominance status, age, and weight may all be highly correlated. Picking just a single one of these variables may result in misleading conclusions unless the other variables are measured and controlled in an analysis.

A behavioral measure can be thought of as having two key components: a treatment component, which is the independent variable or variables impacting on the behavior, and an error component, which is composed of the uncontrolled or unmeasured variables or inherent variability that is also driving the behavioral measure. In observational research, it can be extremely challenging to determine whether it is the treatment component or the error component that is driving a behavioral measure because of lack of control or confounds. For this reason, many researchers turn to experimental methods.

Experimental Research

The goal of experimental research is to utilize experimental designs to tease out the effects of variables on behavior by controlling as many variables as possible. Experimental designs are protocols intended to manipulate and control independent variables so a small number of variables can be tested to determine what, if any, the effects on the behavior in question are.

Replications

Because the research questions involved in animal behavior research often apply to large populations of individuals, a single measurement from a single individual is rarely sufficient to make intelligent statements about the relationships between variables. As a result, multiple measurements, or replications, are collected and inferential statistics are used to test hypotheses about the relationship between the variables.

Replications can be generated in many different ways. The simplest and most common method is to take a sample of multiple individuals and randomly assign each individual to one of two experimental groups. This is called a between-group experimental design. In between-group designs, the groups may be naturally occurring, such as comparisons of males and females, or they may be manipulated as part of a controlled experiment, such as the administration of a drug to one group and not the other.

There are many types of experimental designs based on this model of assigning subjects to treatment groups. The most basic design includes a control group, which is not manipulated in any way, and a second group, called the treatment group, which is changed or manipulated in some manner. More complicated designs may include multiple treatment groups, or subgroups nested within a treatment.

Randomly assigning a group of subjects to one of two conditions is one way of achieving replication, but it is not the only method. Other methods may provide a better measure of the impact of the independent variable. For example, with random assignment to conditions, a great deal of variability remains within the replications because each subject is different. By assigning subjects to groups randomly, it is possible that more aggressive or healthier subjects end up in a certain group, thereby confounding the results. An alternative strategy for controlling this variation is to use an experimental design called a matched-pair design.

Using a matched-pair design, subjects are matched for characteristics such that they are more like each other. In this way, individuals of similar age, sex, weight, rearing history, etc., can be paired, thereby making comparisons of the treatment more accurate. For example, in a laboratory study of working memory in food-caching birds of differing ages, it may be that older birds may have more experience with the task and are more proficient. If the subjects are randomly assigned, more older birds may end up in the same treatment group and confound the results. However, if the subjects are matched for age, paired-up, and one bird from each pair is randomly assigned to each treatment, then both treatment groups will be equally represented across the age groups and age will no longer be a factor in the experiment. This experimental design helps to control some of the error component in the measure and to reduce the results being confounded.

An extreme version of the matched-pair design is a within-subject or repeated-measure design. In within-subject designs each subject is evaluated in both the control and the treatment conditions. This design has the least variability because the same subjects serve as their own control, further reducing the number of potentially confounding variables. This method may be more time consuming because the two groups cannot be tested simultaneously. However, it is a preferred experimental design because fewer animals are required in experimental settings to achieve the same statistical power.

Inferential Statistics

Inferential statistics are often used to compare the differences between the treatment groups. Inferential statistics use measurements from the sample of subjects in the experiment to compare the treatment groups and make generalizations about the larger population of subjects.

There are many types of inferential statistics and each is appropriate for a specific research design and sample characteristics. Researchers should consult the numerous texts on experimental design and statistics to find the right statistical test for their experiment. However, most inferential statistics are based on the principle that a test-statistic value is calculated on the basis of a particular formula. That value along with the degrees of freedom, a measure related to the sample size, and the rejection criteria are used to determine whether differences exist between the treatment groups. The larger the sample size, the more likely a statistic is to indicate that differences exist between the treatment groups. Thus, the larger the sample of subjects, the more powerful the statistic is said to be.

Virtually all inferential statistics have an important underlying assumption. Each replication in a condition is assumed to be independent. That is each value in a condition is thought to be unrelated to any other value in the sample. This assumption of independence can create a number of challenges for animal behavior researchers.

Experimental Challenges

A number of challenges exist for employing experimental designs in animal behavior research. First, it may be impossible to select subjects for experimental treatments

and manipulations. This is the reality in most field research. In these cases, researchers often employ quasi-experimental designs. These designs have many of the same components as true experimental designs, including comparison of two levels of an independent variable, but they lack much of the control seen in true experimental designs. While quasi-experimental designs are not perfect, they often provide more information than simple observations. Thus, while the results must be interpreted with caution, the information can be extremely valuable.

Another challenge for animal behavior research is that experiments are often plagued by small sample sizes. For example, captive animal facilities such as zoos and aquariums often house only a small number of a given species. Small sample sizes can result in statistical tests with extremely low power. When sample sizes are extremely small, the resulting statistics may not indicate the need to reject the null hypothesis when they should. This is called a Type II error. Type II errors may be even more likely when small sample sizes indicate the need to use nonparametric statistics because of the violation of additional assumptions of parametric statistics. Nonparametric statistics are typically less powerful than their parametric counterparts and may result in an even higher incidence of making a Type II error.

In an effort to avoid a Type II error by increasing sample sizes, researchers may be guilty of errors of pseudoreplication or pooling. Pseudoreplication is the incorrect use of replications for inferential statistics. Pseudoreplicated studies result in measurements that are not statistically independent of one another and may increase the potential for having a Type I error, rejecting the null hypothesis when it should not be rejected.

In its most egregious form, pseudoreplication involves the inclusion of multiple measurements from a single individual within a condition. For example, when comparing the number of offspring in litters of lion cubs from two different habitat types, only a single litter from each lioness should be used. Including multiple litters from a single female would bias the sample, particularly if that female produced much more or much less offspring than others in her habitat type, and give the impression of group differences when the differences may be attributed only to a single female. In addition to biasing the mean value for that habitat type, using multiple measures from a single female artificially inflates the degrees of freedom and violates the assumption of independence, a component of virtually all inferential statistics.

Another less conspicuous form of pseudoreplication is known as pooling. The pooling of data occurs when data from individuals in different groups are compiled into the same treatment group for an inferential statistic. This is most easily illustrated in captive environments. For example, if rates of aggressive behavior in chichlids were compared using groups of 3–5 fish from multiple tanks, each group of fish would have more in common, that is, space, tank-mates, and water, than fish in other groups. This violates the assumption of independence and increases the potential for a Type I error.

Solutions

The use of experimental design in animal behavior research is very discipline-dependent. Researchers working with large populations of rats or other laboratory animals may have the sample sizes and the control to use traditional designs when conducting experiments. Alternatively, researchers in other settings, such as zoos and aquariums may have limited numbers of subjects and may struggle to obtain good control for experiments and battle small samples sizes and the statistical problems that go along with them. Researchers studying animal behavior in the field may also have to fight with small sample sizes. Additionally, they may not be able to conduct true experiments due to a lack of ability to control variables or assign animals to treatment conditions.

However, despite the experimental design limitations that animal behavior research presents, there are still a number of options. First, researchers should understand the principles of experimental design and the limitations their particular research discipline presents. In situations where control and manipulation are possible, researchers should utilize any of the large number of resources on research design to employ one of the traditional experimental designs, including the completely randomized design, blocked designs, and Latin square designs. While these designs require relatively large amounts of manipulation and control, they are well documented as to their effectiveness.

When control and manipulation of subjects and variables is possible but small sample sizes create statistical challenges, researchers should avoid pseudoreplicating or pooling their data. If researchers have multiple data points per subject, a mean or aggregate value can be used or a multilevel statistical model can be employed. Finally, measures of effective size can be paired with standard statistical tests to provide supplemental information and prevent Type II errors. When experimental manipulations are not possible, researchers should employ quasi-experimental designs to generate valuable information but the results should be interpreted cautiously.

Overall, experimental designs are intended to control variables and determine the relationship between independent variables and behaviors of interest. Many of these methods are time-tested and when properly employed can be very effective. However, it is important to remember that the practical application of many of these designs can be challenging and modified versions of these designs can produce valuable information but the results should always be interpreted cautiously.

See also: Endocrinology and Behavior: Methods; Ethograms, Activity Profiles and Energy Budgets; Experiment, Observation, and Modeling in the Lab and Field; Measurement Error and Reliability; Neuroethology: Methods.

Further Reading

Cohen J (1988) *Statistical Power Analysis for the Behavioral Sciences,* 2nd edn. Hillsdale, NJ: Lawrence Erlbaum Associates, Inc.

Heffner RA, Butler MJ, and Reilly CK (1996) Pseudoreplication revisited. *Ecology* 77: 2558–2562.

Hinkelmann K and Kempthorne O (2007) *Design and Analysis of Experiments, Introduction to Experimental Design,* vol. 1. New York, NY: John Wiley & Sons, Inc.

Hurlbert SH (1984) Pseudoreplication and the design of ecological field experiments. *Ecological Monographs* 54: 187–211.

Kirk RE (1968) *Experimental Design: Procedures for the Behavioral Sciences.* Belmont, CA: Brooks/Cole.

Kuhar CW (2006) In the deep end: Pooling data and other statistical challenges of zoo and aquarium research. *Zoo Biology* 25: 339–352.

Lehner PN (1996) *Handbook of Ethological Methods,* 2nd edn. Cambridge: Cambridge University Press.

Martin P and Bateson P (2007) *Measuring Behaviour,* 3rd edn. Cambridge: Cambridge University Press.

Shadish WR, Cook TD, and Campbell DT (2002) *Experimental and Quasi-Experimental Designs for Generalized Causal Inference.* Boston, MA: Houghton-Mifflin.

Sheskin DJ (2007) *Handbook of Parametric and Nonparametric Statistical Procedures,* 4th edn. Boca Raton, FL: CRC Press.

Female Sexual Behavior and Hormones in Non-Mammalian Vertebrates

D. L. Maney, Emory University, Atlanta, GA, USA

Introduction

The hormonal control of sexual behavior has been studied far less in females than in males. This disparity may stem from the perception that, for many species, female behaviors leading to the union of egg and sperm consist primarily of remaining stationary and acquiescing to the male's advances. In reality, however, female sexual behavior is far more complex, and in most species, far from simply passive. Sexual attraction, courtship, and engaging in copulatory behaviors must be mutual, and the control of sexual behavior in the female is at least as complex and interesting as in the male. In this article, we will see that female behavior is affected by many hormones, among which the best studied and most ubiquitous by far is estradiol, an estrogen that is secreted by the ovaries and can also be synthesized in the brain.

Estrogens such as estradiol are the most ancient steroid hormones, and their role in reproduction has remained relatively unchanged throughout vertebrate evolution. Estradiol associated with reproduction is secreted by developing ovarian follicles under the influence of gonadotropins from the anterior pituitary (**Figure 1**). As the follicles and the eggs inside them mature, estradiol levels increase in the plasma until peaking at or just before ovulation. Temporally, therefore, estradiol is in a good position to coordinate reproductive behavior with the fertilization of eggs – and indeed, in most vertebrates, the peak of sexual behavior does occur at precisely the same time as the peak in estradiol. Note, however, that mating is not always coincident with the highest levels of estradiol; in some species, mating occurs well outside that window. Even in those species, however, removal of the ovary usually abolishes sexual behavior and estradiol treatment restores it, which demonstrates that although the concentration of estradiol in blood does not need to be greatly elevated in order to support sexual behavior, it does need to reach some threshold level. Most research

indicates that estradiol, which is also involved in vitellogenesis and other physiological and morphological traits associated with reproduction, primes the brain so that sexual behavior is facilitated by other factors that would not be effective without the priming. The other factors include reproductive hormones such as gonadotropin-releasing hormone, the gonadotropins, progesterone, and prostaglandins, which also peak at various times during ovarian maturation, ovulation, and oviposition. Each of these hormones, including estradiol, is thus well suited to facilitate reproductive behaviors, such as territoriality, nest building, and of course sexual behavior, which we define here as social interactions that lead to the union of gametes.

A founding pioneer of behavioral endocrinology, Frank Beach, divided sexual behaviors into three categories which, although they were developed specifically for mammals, are useful when considering the behaviors of all vertebrates. The first, *attractivity*, is defined as the stimulus value of female in evoking sexual responses from the male. Attractivity is often related to an olfactory signal such as the sexual pheromone of female rough-skinned newts (*Tarica granulosa*) or a visual signal such as ornamental coloration in female budgerigars (*Melopsittacus undulates*) and *Crotaphytus* lizards. Such signals are often most attractive when plasma levels of ovarian hormones are high. Since the attractivity of the female is in the eye of the beholder, in this case the male, and because most examples of attractivity are nonbehavioral in that they involve signaling with chemicals or pigments, this review focuses primarily on the second and the third categories of female sexual behavior: *proceptivity* and *receptivity*. Attractivity, particularly as it relates to visual and olfactory signals, is covered elsewhere in this series.

Proceptive behaviors occur in response to stimuli from the male, and like the signals that fall into the category of attractivity, serve to lessen the distance between him and the female. Note that the presence of a male can affect

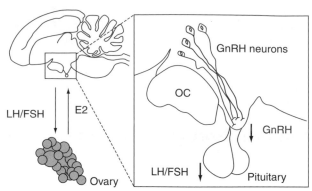

Figure 1 The hypothalamo–pituitary–gonadal (HPG) axis is nearly identical in birds and mammals. Gonadotropin-releasing hormone (GnRH) neurons project to the base of the brain where they release GnRH into the portal vasculature. From there, GnRH travels to the anterior pituitary to stimulate the release of luteinizing hormone (LH) and follicle-stimulating hormone (FSH) which act at the ovary to induce estradiol (E2) secretion and follicular development. E2 travels via the bloodstream back to the brain to stimulate reproductive behavior. OC, optic chiasm.

plasma estradiol concentrations, which in turn can affect the signals used in attraction; however, such responses are not normally considered proceptive. Rather, proceptive behaviors are those that are elicited on a more rapid time scale by signals from the male, and are normally directed to him specifically. They include behaviors as simple as approaching the male and as complex as multimodal courtship displays. *Receptive* behaviors are those that are both necessary and sufficient for the fertilization of eggs, and include the adoption of postures that facilitate copulation as well as the maintenance of contact with the male during copulation. With very few exceptions, both proceptive and receptive behaviors appear to depend universally on estradiol.

The Role of Ovarian Steroids in Proceptivity

Estradiol Is the Primary Ovarian Hormone Involved in Female Proceptive Behavior

During a proceptive response, the female reacts to stimuli from the male by attempting to bring him closer to her – either by attracting him or by approaching him herself. In the case of the latter, simply moving toward the male or spending time in his immediate vicinity is considered proceptive, and is in many cases under the control of ovarian steroids. In many species, particularly those that breed at night or in murky environments wherein the female may have difficulty locating a male, a male must advertise his position to a female by signaling to her. In midshipman fish (*Porichthys notatus*), for example, males

vibrate the muscles of their swim bladders to produce a distinctive hum. When a gravid female hears a male's hum, she orients toward it and approaches him – a behavior called 'phonotaxis.' Once she enters his nest, he stops humming and spawning begins. Only gravid females, in other words those ready to lay eggs, are attracted to the males' hums. Females that have already released their eggs do not respond to the hums, suggesting that the response may depend on reproductive condition and therefore have a hormonal basis.

The best-studied example of hormone-dependent phonotaxis occurs in female anurans. In many species, such as the Túngara frog (*Physalaemus pustulosus*) and the American toad (*Bufo americanus*), males emit loud vocalizations that are amplified by specialized structures called 'vocal sacs' and can be heard at a great distance. Because recordings of calls played from a speaker can induce female phonotaxis both in the field and in the lab, behavioral assays are a popular way to study it. Females do not perform this behavior unless they are gravid, so the early experiments had to be done using females collected as they began to mate with males. In an effort to circumvent this problem, it was discovered that injections of human chorionic gonadotropin induced phonotaxis in female African clawed frogs (*Xenopus laevis*), thereby showing some of the first evidence that this behavior has a hormonal basis. Human chorionic gonadotropin acts by mimicking the actions of gonadotropins on the ovary to induce the secretion of estradiol and progesterone. The induction of estradiol secretion may be the more important effect of human chorionic gonadotropin in the frog model; more recent research has demonstrated that experimental elevation of estradiol to breeding levels induces phonotaxis in ovariectomized females. Concurrent administration of progesterone does not increase phonotaxis any further, suggesting that estradiol alone is sufficient.

One of the best-studied proceptive behaviors in female nonmammalian vertebrates is the copulation solicitation display (CSD) of songbirds (**Figure 2**). Females can be seen and heard performing this display throughout the early part of the breeding season. During the display, the female raises her tail and head, quivers her wings dramatically, and gives a trill-like vocalization. CSD is usually performed in response to stimuli (such as song) from the male and functions as a signal to him that the female is ready to mate. Like phonotaxis in female anurans, CSD is strongly dependent on estradiol and is performed only by females with sufficiently elevated levels of this sex steroid. Also like phonotaxis in anurans, CSD can be elicited in laboratory-housed females in response to an audio recording of male courtship vocalizations. In wild-caught females of many species, endogenous plasma estradiol does not increase enough in captivity to support the behavior, and exogenous estradiol must be administered. In these species, the number of displays performed is

Figure 2 A female white-crowned sparrow performs a copulation solicitation display (CSD). This display signals to the male that she is in reproductive condition. Reprinted with permission from the *Journal of Neuroendocrinology* 10(8), cover material.

directly related, in a highly linear manner, to the dose of estradiol. In species that solicit and breed easily in captivity, such as canaries (*Serinus canaria*), the number of displays performed is not related to endogenous levels of plasma estradiol, but the latency to perform the display can be lengthened by blocking estradiol synthesis. No female of any species has been observed to perform CSD when not in breeding condition or otherwise treated with estradiol. It is also notable that progesterone has a clear inhibitory effect on CSD, which is perhaps not surprising given that in birds, progesterone levels are generally low until the first egg is laid and copulation ceases. Clearly, estradiol is the more important ovarian hormone governing proceptive responses to male vocalizations in the fish, frogs, and songbirds that have been studied.

Estradiol Acts Directly on Sensory Systems

By what neural mechanism does estradiol facilitate proceptive behavior? It is possible that estradiol acts directly on sensory structures to alter how the male's signal is processed or perceived. There is evidence, for example, that estradiol may facilitate phonotaxis by tuning peripheral auditory structures to the acoustic features of the signal. Electrophysiological recordings from the auditory nerve in midshipman fish have revealed that the female's peripheral auditory organ is better tuned to the male's hum during the breeding season than outside it. The inner ear of this species contains estrogen receptors, and the tuning shift toward the male hum can be induced outside the breeding season by treatment with estradiol. This shift in tuning parallels the female's behavioral response to the hum, which is strongest when her estradiol levels are

peaking. It will be interesting to determine whether the peripheral auditory organs in other species are also estradiol sensitive; this work is currently being carried out in frogs and songbirds.

Of course, in order to more completely understand the effects of estradiol on behavior, we must look inside the brain. Estradiol receptors are distributed widely in sensory areas as well as in areas more directly involved in social behavior. Sound-induced neural responses in these regions have been shown in some species to depend on reproductive state or season, implying hormonal regulation. For example, electrophysiological studies have shown that auditory response properties in the inferior colliculus, a major auditory integration center in the midbrain, appear to depend on reproductive state in several species of anuran. Hormonal modulation of neural responses can also be studied using the expression of immediate early genes, which are associated with new protein synthesis and thus allow the detection and labeling of neurons responding to stimuli. This technique allows quantification of sound-induced activity throughout the brain. In my laboratory, for example, we played male song or a less relevant sound, synthetic tones, to female white-throated sparrows (*Zonotrichia albicollis*) and compared the resulting expression of the immediate early gene Egr-1. We quantified Egr-1 responses not only in auditory areas, but also in an extensive network of brain regions involved more generally in social behavior. In females with normal breeding levels of estradiol, the Egr-1 response to song was higher than the response to tones in nearly every region we looked at (**Figure 3**). In birds with low estradiol, as they would have during the nonbreeding season, the responses to song and tones were indistinguishable in most regions. In other words, the Egr-1 responses were selective for male song only when estradiol was high. This result, together with the work done in fish and anurans, suggests that estradiol facilitates proceptive behavior by promoting the recognition of and attention to the signals that trigger it.

The Role of Ovarian Steroids in Receptivity

Estradiol

Feminine receptive behavior, whether it is the neck-bending posture of an anole, the crouch of a quail, the S-shaped posture of a killifish, or the knee flexion of a frog, serves one purpose – to maintain contact with the male and facilitate the union of eggs and sperm. In some species, mating involves the coordinated release of both types of gametes, and fertilization occurs externally. In species with internal fertilization, the sperm are usually transferred directly from the male to the female via an intromittent organ or cloacal contact. In some species,

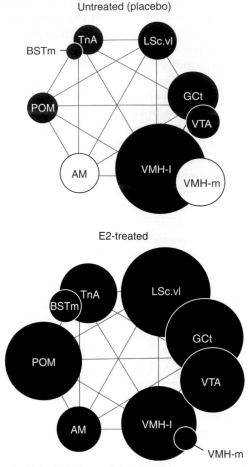

Figure 3 Estradiol (E2) modulates selective responses in interconnected brain regions important for social behavior. In this representation of the social behavior network, the brain regions or 'nodes' are shown as circles at each corner of a hexagon. The area of each circle corresponds to selective activation in that node by hearing song (the amount of Egr-1 induced by song, minus the amount induced by control sounds, or tones). Filled circles represent positive values and white denotes negative, that is, a larger response to tones than to song. In female white-throated sparrows, both the level and the selectivity of the response increase after treatment with E2. For abbreviations, see Reprinted from Maney DL, Goode CT, Lange HS, Sanford SE, and Solomon BL (2008) Estradiol modulates neural responses to song in a seasonal songbird. *Journal of Comparative Neurology* 511: 173–186, with permission from John Wiley & Sons, Inc.

such as urodele amphibians, the male deposits onto the substrate a spermatophore which the female picks up. Either way, in the majority of species eggs are ready and waiting to be fertilized either immediately or within a few days after mating. For this reason, it is highly efficient and effective for sexual behavior and ovulation to be under the same hormonal control. Female receptivity closely mirrors plasma estradiol in internally fertilizing fish as well as all anurans, most newts and lizards, many turtles, and all birds; in the species for which such

manipulations have been done, ovariectomy invariably inhibits receptivity and exogenous estradiol almost always restores it.

In a few species, mating appears to have become temporally dissociated from ovulation and peaking levels of plasma estradiol. For example, in sting rays (*Dasyatis sabina*), red-sided garter snakes (*Thamnophis sirtalis*), and green sea turtles (*Chelonia mydas*), mating occurs before ovarian development is well underway and the sperm can apparently be stored in the female's reproductive tract for months before ovulation. It is important to realize in these cases that although mating and the estradiol peak do not occur concurrently, estradiol does not have to be high to be involved in receptivity. The behavior can often be supported by relatively low levels of estradiol. Female skinks (*Niveoscincus ocellatus*) mate in the fall, store sperm all winter, and then ovulate in spring, yet in even in the fall vitellogenesis has clearly begun and plasma estradiol is elevated above basal levels. Even in red-sided garter snakes, which are sexually receptive at a time when their plasma estradiol levels are at an annual low, ovariectomy abolishes receptivity and estradiol treatment can restore it within minutes. Thus, although not much estradiol is needed, it is necessary to support receptivity. Because the frequency of sexual behavior is clearly not always correlated directly with plasma estradiol levels, it is thought that estradiol plays a priming, or permissive role, and that other hormones or neurotransmitters are also important.

Progesterone

In mammals, particularly rodents, estradiol-dependent feminine sexual behavior is often greatly facilitated by another important ovarian steroid, progesterone. In most vertebrate females, progesterone peaks either along with or somewhat after the peak in plasma estradiol. The role of progesterone has been investigated in nonmammalian vertebrate taxa, with mixed results. As noted earlier, progesterone does not seem important for proceptive behavior in that it does not enhance phonotaxis independent of estradiol, and it inhibits CSD. Some researchers, however, have found evidence that it may play a role in receptive behavior in some species. In ovariectomized African clawed frogs, for example, estradiol alone does not fully restore receptivity; treatment with progesterone is also required. In ovariectomized green anoles (*Anolis carolinensis*), estradiol treatment can fully restore receptivity but only at high doses. When progesterone is administered as well, much lower doses of estradiol are required, suggesting that estradiol primes the brain for the actions of progesterone. In many other species, however, such as some skinks and quail, progesterone does not restore receptivity any further after estradiol treatment.

Testosterone

Because plasma testosterone peaks in many female vertebrates around the same time as estradiol and progesterone, some researchers have hypothesized that it is involved in female sexual behavior. Those studying mammals, particularly primates, have often argued that testosterone is even more important than estradiol. This claim has been controversial, however, and there is little evidence supporting it from nonmammalian vertebrates. It is important to remember that particularly in the brain, testosterone can be converted into estradiol, so experiments manipulating testosterone may affect estradiol synthesis. When ovariectomized green anoles are treated with testosterone, feminine receptivity is restored; when they are treated with a drug that blocks the conversion of testosterone to estradiol, however, testosterone does not have the same effect. Similarly, in both green anoles and leopard geckos (*Eublepharis macularius*), receptivity is not restored by a form of testosterone that cannot be converted to estradiol. Thus, even if testosterone from the ovary does facilitate sexual behavior in female anoles, it most likely does so via conversion to estradiol in the brain.

Sexual Behavior and Nonsteroid Hormones

Prostaglandins

In many species, both proceptive and receptive behaviors appear to be stimulated by the presence of eggs in the reproductive tract. For example, in both midshipman fish and in anurans, only gravid females perform phonotaxis. After a female has released her eggs, she is much less likely to approach a speaker playing male courtship sounds. Do the eggs themselves act as a signal to facilitate sexual behavior? A large body of research, primarily on goldfish (*Cassius auratus*), indicates that receptivity is triggered by distention of the reproductive tract by ovulated eggs. Removing the ovulated eggs from the ovarian lumen will abolish spawning behavior, and the insertion of foreign objects will stimulate it. When it is distended, the goldfish ovary secretes nonsteroid fatty acid hormones called prostaglandins. Levels of the prostaglandin PGF2a increase many fold around the time of ovulation and decrease sharply once the eggs are released. Injection of PGF2a stimulates spawning within minutes in at least a dozen species of fish, and spawning can be blocked by treating with drugs that block prostaglandin synthesis. Although prostaglandins are thought to be rapidly metabolized, they can travel via the bloodstream from the reproductive tract to the brain and bind to receptors there. Injecting PGF2a directly into the brain of a female goldfish is many times more effective at stimulating receptivity than is systemic injection, suggesting that the brain is a primary target during prostaglandin-induced spawning. In another fish, the paradise fish (*Macropodus opercuhris*), PGF2a stimulates the female to approach a male, lead him to the nest, and display to him, demonstrating that prostaglandins can stimulate proceptive as well as receptive behaviors. Prostaglandins appear to have a similar function in anurans, stimulating both phonotaxis and receptive behavior.

Both estradiol and prostaglandins appear to act on the brain to facilitate sexual behavior, in fish and anurans, but whether both are required, or how they interact, is not well understood. Blocking prostaglandin synthesis can reduce expression of sexual behavior in estradiol-primed females, showing that prostaglandin action is necessary even when estradiol is high. Some of the work in goldfish has shown that prostaglandins cannot stimulate sexual behavior unless the ovary is present and active, which suggests that the dependence of one hormone on the other may be mutual. Other work, however, has revealed evidence that PGF2a can stimulate spawning and phonotaxis even in ovariectomized or otherwise nonovulatory females. These results are surprising given the nearly universal dependence of female sexual behavior on estradiol. The key to this mystery may lie in the fact that prostaglandin synthesis is related to estradiol and vice versa; each stimulates the synthesis of the other. Concentrations of the two are positively correlated in the brain, and applying prostaglandins to brain cells in culture activates enzymes in the estradiol synthetic pathway and increases estradiol production. Thus, treatment of females with prostaglandins may in fact increase estradiol within the brain even if the ovary has been removed. Because the two types of hormones are so closely linked, it is difficult to separate their actions and functions experimentally. They are likely to synergize with each other in the control of both proceptive and receptive behavior.

In snakes and lizards, prostaglandins inhibit receptivity – which at first seems curiously at odds with the strong stimulatory effect in fish and anurans. The mechanism underlying prostaglandin's effects, however, is remarkably similar to other taxa: stimulation of the reproductive tract, in this case not by eggs but by the male, causes prostaglandin secretion. In reptiles such as red-sided garter snakes and green anoles, females become unreceptive almost immediately after mating. The male's intromittent organ apparently stimulates the female's reproductive tract to produce prostaglandins (there may also be some prostaglandins in the male's seminal fluid), which travel to the brain where they rapidly and profoundly inhibit further mating. Treatment with exogenous PG2a, whether injected systemically or directly into the brain, facilitates the switch from receptive to unreceptive. So, although the secretion of prostaglandins from the stimulated reproductive tract appears to have been conserved across taxa, the behavioral result of prostaglandin action in the brain appears to be more plastic.

Gonadotropin-Releasing Hormone

At the top of the hypothalamic–pituitary–gonadal axis is a hormone known as gonadotropin-releasing hormone. Most vertebrates have at least two forms of it. One, abbreviated GnRH1, is synthesized primarily in the preoptic area of the hypothalamus, released into the bloodstream at the base of the brain, and controls ovarian estradiol production by stimulating gonadotropin release at the pituitary. In that sense, it is one of the most important hormones promoting ovarian development and oogenesis. Not all of the GnRH1 axons, however, terminate at the base of the brain to secrete their product into the bloodstream. In many amphibians, reptiles, and birds, large numbers of GnRH1 fibers project to brain regions associated with sexual behavior. GnRH1 is therefore in a position to affect reproductive physiology and behavior simultaneously. In female rough-skinned newts, for example, GnRH1 levels rise in the brain during courtship, and exogenous administration facilitates receptivity.

The second form of gonadotropin-releasing hormone, abbreviated GnRH2, is not thought to play a role in ovarian development but rather coordinates reproductive behavior and energy availability. As is the case for GnRH1, neuronal fibers carry GnRH2 to brain regions involved in sexual behavior. GnRH2 increases in the brain during spawning in goldfish and may synergize with prostaglandins. PGF2a increases GnRH2 levels in the brain, and exogenous GnRH2 enhances PGF2a-induced spawning behavior. Both forms of GnRH, when administered centrally in estradiol-treated female white-crowned sparrows (*Zonotrichia leucophrys*), enhance the CSD response to song playbacks. Although not a lot of research has been conducted on how these neuropeptides affect sexual behavior in nonmammalian vertebrates in general, the existing research confirms what is known for mammals – that the peptides play a facilitatory role.

Vasotocin

The neuropeptide vasotocin is involved in many types of social behaviors across vertebrate taxa. Most of what is known about its involvement in reproductive behavior has come from studying males, but some effects of vasotocin are clear in females. In killifish (*Fundulus heteroclitus*), for example, treatment with vasotocin induces spawning behavior in both sexes. In anurans, vasotocin increases receptive behavior and also facilitates phonotaxis, making females more likely to respond to male calls. In white-crowned sparrows, intracranial administration of vasotocin induces spontaneous CSD behavior. The auditory midbrain in both frogs and birds is innervated by neuronal fibers containing vasotocin and also has vasotocin receptors, suggesting that the neuropeptide may, like estradiol, affect sensory processing of courtship signals.

In addition to stimulating CSD, intracranial vasotocin in songbirds also stimulates singing, calling, and preening, which suggests a rather nonspecific effect on behavior. Vasotocin stimulates CSD, however, only in estradiol-treated females; in nonreproductive females not treated with estradiol, vasotocin induces all the other behaviors but not CSD. Other effects of vasotocin on proceptivity and receptivity also seem to depend on ovarian steroids; for example, phonotaxis in frogs is enhanced by vasotocin only in the spring when estradiol levels are high. The neural actions of vasotocin, like many other factors, may therefore depend on estrogen priming.

Conclusion

Estradiol is one of the most evolutionarily ancient hormones and has served to coordinate reproductive physiology with sexual behavior for close to 500 My. In females of all vertebrate taxa, normal sexual behavior cannot proceed without it. There are, of course, a small number of possible exceptions. In guppies (*Poecilia reticulata*), for example, even though receptivity in adult sexually experienced females depends on ovarian activity, young virgin females are always receptive – before the ovary is fully developed, and even if it has been removed. It is important to remember, however, that the dependence of behavior on steroid hormones is often difficult to study. First, the brain is itself a source of hormones. Removal of the ovaries does not necessarily remove all the estradiol. Second, because the actions of estradiol on behavior are permissive, low levels are often sufficient to prime the brain to respond to other factors such as prostaglandins, gonadotropin-releasing hormone, and vasotocin. Thus, sexual behavior is not always directly correlated with plasma estradiol levels, and may peak well outside the period wherein plasma estradiol concentrations are at their highest.

Although the interactions of steroid hormones with neuromodulators and neurotransmitters in the nonmammalian vertebrate brain are beginning to be understood for males, the same interactions are not well studied in females. Because females are more sensitive to environmental cues and because their reproduction is under tighter control than is male reproduction, studying them will provide valuable insight into phenomena such as steroid-dependent sensory processing, the neural and hormonal bases of adaptive mate choice, and the timing of reproduction. In addition, because the hormonal control of their sexual behavior is so well conserved, cross-species investigations can help us understand the evolution of hormone–behavior relationships.

See also: Forced or Aggressively Coerced Copulation; Male Sexual Behavior and Hormones in Non-Mammalian Vertebrates; Mammalian Female Sexual Behavior and

Hormones; Mate Choice in Males and Females; Mating Signals; Neural Control of Sexual Behavior; Seasonality: Hormones and Behavior; Sex Changing Organisms and Reproductive Behavior; Túngara Frog: A Model for Sexual Selection and Communication.

Further Reading

Adkins-Regan E (2005) *Hormones and Animal Social Behavior*. Princeton, NJ: Princeton University Press.

Beach F (1976) Sexual attractivity, proceptivity, and receptivity in female mammals. *Hormones & Behavior* 7: 105–138.

Goodson JL and Bass AH (2001) Social behavior functions and related anatomical characteristics of vasotocin/vasopressin systems in vertebrates. *Brain Research Reviews* 35: 246–265.

Guillette LJ, Dubois DH, and Cree A (1991) Prostaglandins, oviductal function, and parturient behavior in nonmammalian vertebrates. *American Journal of Physiology; Regulatory, Integrative and Comparative Physiology* 260: 854–861.

Liley NR and Stacey NE (1983) Hormones, pheromones, and reproductive behavior in fish. In: Hoar WS and Randall DJ (eds.) *Fish Physiology,* vol. IXB, pp. 1–63. London: Academic Press.

Maney DL, Goode CT, Lange HS, Sanford SE, and Solomon BL (2008) Estradiol modulates neural responses to song in a seasonal songbird. *Journal of Comparative Neurology* 511: 173–186.

Whittier JM and Tokarz RR (1992) Physiological regulation of sexual behavior in female reptiles. In: Gans C and Crews D (eds.) *Biology of the Reptilia,* vol. 18, pp. 24–69. Chicago: University of Chicago Press.

Field Techniques in Hormones and Behavior

L. Fusani, University of Ferrara, Ferrara, Italy

Introduction

About 100 hormones have been described to date, and many of them influence or are influenced by behavior. Well-controlled laboratory experiments have been providing basic knowledge about hormonal control of behavior and hormonal responses to behavioral interactions. However, laboratory settings and the use of domestic species set implicit limits on what can be studied. These limits do not concern only the reduced, altered behavioral repertoire of captive animals. In fact, captivity can have substantial effects on the endocrine systems. Thus, studying behavioral endocrinology in the field is a necessary complement to laboratory studies for understanding the complexity and variety of hormones and behavior relationships in animals. This is particularly true for complex behavioral interactions among individuals, which cannot be simulated in the laboratory, even in large enclosures. Studies in field behavioral endocrinology have focused mainly on hormonal systems that are capable of modulating behaviors associated with life history stages like reproduction, territoriality, and seasonal cycles. As a consequence, there is a large number of field studies of behavioral traits that are controlled by the hypothalamus–pituitary–gonadal axis and the hypothalamus–pituitary–adrenal axis. These two axes influence many key behavioral contexts including courtship, male and female sexual behavior, mating systems, territorial aggression, social rank, parental care, maternal effects, migration, parent–offspring conflicts, and pair bonding.

History

Traditionally the first study of behavioral endocrinology is reported to be that done by Arnold Adolph Berthold in 1846. It was well known that castration of young male cockerels prevents the development of male sexual traits including crowing and sexual behavior – an effect called caponization. Berthold showed that male typical behavior could be restored by reimplanting the testes in the same birds or in other castrated cockerels. Because the grafted testes did not build any tissue connection with the body of the host, Berthold concluded that some substances secreted by the testes – 'androgens' – were responsible for the activation of male sexual behavior. Soon afterwards, naturalists started to investigate the relationships between endocrine glands and behavior in free-living animals.

In fact, already in 1802 George Montagu had noted that songbirds sing more at the times of the year when their testes are larger. By the end of the 1940s, the most important androgen and estrogen hormones, testosterone and estradiol, had been isolated and methods for their synthesis had been developed. These discoveries opened new perspectives in endocrinology as they allowed studying how specific hormones influence behavior. The founders of behavioral endocrinology, Frank A. Beach, Daniel Lehrman, and William C. Young, worked on a number of domestic or laboratory animals, building the conceptual bases for later extending behavioral endocrinology to the field. Initially, field endocrinology involved shooting the animals to measure the size of their endocrine glands. In 1960, Rosalyn Yalow and Solomon Aaron Berson published a new method called Radioimmunoassy, which allowed the measurement of hormones in relatively small blood samples. The birth of 'modern' field behavioral endocrinology can be traced to 1975, when John C. Wingfield and Donald S. Farner published a method for measuring five different steroid hormones in small blood samples taken from free-living songbirds which could be released immediately after sampling. This allowed studying behavioral interactions between conspecifics and correlate behavioral differences with hormonal ones. More importantly, because animals did not have to be killed to measure hormonal parameters, it was possible to study time-dependent changes in hormones in individual animals that could be sampled repeatedly.

Major Issues in Field Behavioral Endocrinology

The benefits and limits of field compared to laboratory studies are a common denominator of all behavioral studies and are not discussed in detail here. Instead, this article focuses on conceptual issues specific to field behavioral endocrinology. Hormones influence the likelihood of the occurrence of behavior, and in turn behavioral interactions affect hormone concentrations. Hormone action depends on a number of regulatory factors such as hormone carrier molecules, hormone metabolism, hormone receptors, and hormone coactivators expression. Only few of these factors, however, such as blood concentration of hormone-carrier proteins, can be studied in free-living animals with minimally invasive methods. Thus, the large majority of field studies in behavioral endocrinology belong to two categories. The first category consists of correlation studies, in which the main question is whether variation in a given behavior or behavioral pattern is paralleled by variation in the circulating concentration of one or more hormones. This is one of the main methods used to ascertain if there exists a relationship between a hormone and a behavior. The second category includes experimental manipulations of the hormone and/or its action by means of treating the animals with the hormone or with drugs that interfere, for example, with its metabolism or its binding to the receptor. Changes in behavior following the hormonal manipulation are then recorded.

Measuring Hormones in Free-Living Animals

In principle, the measurement of hormones in free-living animals involves the same methods used in laboratory studies, among which the most important ones are various kinds of immunoassays (radio-, enzyme-, and fluorescence-immunoassay) and more recently high-performance liquid chromatography (HPLC), and mass-spectrometry combined with gas chromatography (MS–GC). Analytical protocols based on these assay methods were originally developed for measurements of hormone concentrations in bodily fluids such as plasma, serum, or urine. The peculiarities of field research, where the capture or handling of wild animals is not always possible or not desirable, has lead to the establishment of new protocols for measuring hormone concentrations in other tissues such as feces, hair, and skin. In most cases, hormones – or hormone metabolites – contained in these tissues need to be extracted and separated by the other components before they can be measured. Thus, new developments in hormone assays for field endocrinology concern mainly the methods used to purify the sample before the actual assay. However, in the case of excreta such as feces and urine, only the metabolites of the target hormone can be measured. In such cases, the establishment of new methods also involves the assessment of the relationships between the amount of hormone metabolites detected in the excreta and the actual circulating concentrations of the same hormone.

Capture and Handling

Tissue sampling for hormone measurement is often done when animals are caught for marking and/or measurement of morphological and physiological variables. This approach guarantees unambiguous individual identification and collection of determined amounts of samples. Because blood is to date the choice tissue for hormone measurement – hormones are by definition circulating factors – blood sampling of immobilized animals has been the most commonly used approach. There are, however, disadvantages in collecting samples for hormone measurement from caught animals. Depending on the species, traps, nets, or narcotic darts are used to temporarily immobilize the animals. All these methods induce a stress response that might affect the release of target hormones. For example, corticosteroids are well known to be affected by capture and handling within a short period of time, and when studying these hormones, it is imperative that the samples are taken within a few minutes of capture. Because corticosteroids are important modulators of a number of other hormonal systems such as the hypothalamus–pituitary–gonadal axes, rapid collection is always crucial with blood.

Sample Collection

Blood

Blood remains the choice tissue for hormone measurement for a series of reasons. Endocrinological research until the 1990s relied almost exclusively on blood concentrations of hormones. Thus, there is a large amount of information relating blood (or plasma/serum) hormone concentrations with behavioral, physiological, and morphological traits. This is not just the consequence of the lack of methods for measuring hormone concentrations in other tissues. Hormones are by definition those factors that are actively released into the circulation by the endocrine glands, and their presence in the feces or in the urine is the result of metabolism and excretion. Thus, blood concentrations represent the actual message – and the hormonal response – whereas fecal concentrations of hormone metabolites depend on a series of factors including food intake, metabolic rate, time required for the feces to be formed, etc. Hormone concentrations are rarely measured directly in the blood. Typically, red blood cells are separated by centrifugation to give plasma or by coagulation to give serum. Most assays are designed to measure plasma or serum concentrations. The main disadvantage of using blood to determine hormones is that the animals have to be caught and immobilized to collect the sample. Plasma samples are typically stored frozen until analysis, but recently other methods of conservation have been tested that do not depend on electricity or gas supplies.

Feces/urine

The development of methods to measure hormones in feces and urine has given a new pulse to field endocrinology in the last 20 years. These methods were developed during 1950s–1960s to evaluate hormone metabolism and for pharmaceutical studies, including the detection of anabolizing steroids illegally used by athletes. With the increased interest in conservation biology and the more stringent ethical rules for animal experimentation, however, the availability of noninvasive methods for collecting samples to use for hormone analysis has become of great help. Urine and feces in mammals or mixed fecal and urine droppings in birds and reptiles can be collected without disturbing the animals and as often as required. There are, however, several issues that need to be taken into account when working with this type of samples. Excreta typically contain metabolic products of the original hormones, not the hormone itself. Back calculating the original blood concentration of hormone from knowing the concentration of its metabolites in the excreta is very difficult if not impossible for it would require knowledge of several factors, such as number of metabolic pathways, their relative importance, that is, the time required to form and accumulate in the excreta, and the concentration of the excreta themselves. This means that although fecal or urine concentration can be compared between animals of the same species, comparison between species is problematic. The proper interpretation of fecal and urine hormone metabolites thus requires a rigorous validation of the method for each study species. In all cases, the validation procedure should make sure that the substance that is measured from urine, feces, or droppings is a true metabolite of the hormone in question. Without such a validation, any substance that cross-reacts with the antibody in the respective immunoassay may render the measurement meaningless. In addition, the concentration of the metabolite should vary in response to a specific stimulation and parallel changes in blood hormone concentrations. For example, the injection of gonadotropin-releasing hormone (GnRH) induces an increase in the blood concentration of testosterone, and a proper validation should demonstrate that the amount of testosterone metabolites in the excreta shows a similar increase – with a predictable delay.

Skin and skin derivatives

The measurement of hormones or hormone metabolites in skin derivates such as hair and feathers is a promising alternative to blood or excreta. Similar to the case of blood, collection of these samples typically requires the capture and immobilization of the subjects, although skin derivatives such as feathers are renewed regularly and can be collected in nests or tree holes. Skin derivatives accumulate hormones during their growth and thus provide a hormonal history of the animals during their growth. A limitation to the use of these types of tissues is that the concentrations of hormones tend to be relatively low and thus their determination requires either large amount of tissue or very sensitive methods. However, the potential applications of these techniques are many and they will probably receive much attention in the future.

Eggs

Hormones accumulate in eggs during their formation and their concentration will influence the development of the embryo and possibly the behavior of the adult. This is a particularly important phenomenon in species with large eggs (reptiles and birds) in which the entire development depends on reserves stored in the egg. The concentration of hormones in the egg has become the focus of a large number of studies on maternal effects.

Sample Storage and Transport

A major challenge in field studies in remote regions is the preparation and conservation of biological samples. Blood derivatives such as plasma or serum should be kept refrigerated or better frozen until assayed, which is not an easy task where electric power is not available. Recent tests with conservative agents such as ethanol have shown that at least steroid hormones can be stored at ambient temperature for relatively long periods of time. However, it is highly recommended that such methods are validated before they are used for other hormone categories.

Experimental Hormone Manipulation

The major challenges in field endocrinology are common to other research areas that deal with free-living animals: uncontrolled and often unpredictable environmental conditions, interference of conspecific or heterospecific individuals, and difficult planning of treatment and measurements. However, there is an additional problem with field endocrinology. Most of the available techniques and experimental protocols derive from laboratory studies and were often designed to address specific questions in specific contexts. For example, a large proportion of experiments in endocrinology were (are) based on the removal–replacement protocol: to investigate the function of a hormone, the gland producing the hormone is removed. In a subset of individuals, the hormone is then replaced by administering exogenous hormone. If the effects of the gland abduction are counteracted by the hormone replacement, the traits which were reduced by the removal and restored by the replacement are called hormone-dependent. This protocol has been of crucial importance in identifying the roles of most hormones and their implication for behavior. However, many experimental protocols that were derived from this basic one are not supported by a rationale in different conditions. For example, gonadectomy is rarely practiced in field studies for a series of reasons including complications of surgery and increased predation risk for experimental subjects. Thus, the role of testicular testosterone for controlling reproductive behaviors (such as courtship) is often tested by administering testosterone to intact animals. Such a treatment can elevate testosterone levels for a short time, but these effects are likely to be cancelled soon by feedback mechanisms that will reduce the production of endogenous testosterone. Thus, unless the treatment brings testosterone levels above the physiological range, which is not desirable, it is likely that it will loose its efficacy within a few days from the onset of the treatment. Therefore, any behavioral effect of the treatment should be registered within a few days to relate it to increased testosterone concentrations.

Treatment Methods

In the field, cases in which repeated administration of a hormone or a drug is possible are rare. Even if this was practically feasible, the repeated capture of the animal would have substantial consequences on its physiology – stress – and behavior. Therefore, most field studies that involve hormonal manipulation require a more or less permanent implant which guarantees a long-term release of the active principle. Among the most common methods for hormonal treatment are silicon tubing, osmotic pumps, and time-release pellets (**Figure 1**). All these methods have advantages and disadvantages which should be carefully evaluated.

Silicon tubing

Silicon tubing has been used extensively in endocrinological research and continues to be one of the favorite choices among researchers for its versatility of use. Tubing can be cut at any desired length, filled with lipophilic substances such as steroid hormones, and closed at both ends with silicon glue. The substance contained in

the tubing will slowly diffuse through the walls when submerged in a fluid, such as subcutaneous or internal fluids. The principle of use for lipophilic substances such as steroid hormones is that they pass through the wall of the tubing at a rate that depends on the substance and the thickness of the wall. Obviously, longer tubes of larger diameters will release the hormone at higher rates. The major advantage of using silicon tubing is that most lipophilic hormones and drugs can be delivered for long periods of time. The availability of tubing of different diameters with a length defined by the researcher allows great flexibility. The major disadvantage of the silastic tubing comes from the preparation procedure. Typically, the hormone is pushed into the tubing and tightly packed. However, the packing can differ considerably between operators and labs, which affects the replicability of the treatment.

Time-release pellets

The rationale of these devices is that of packing the drug/hormone to be released into an organic matrix that dissolves slowly when in contact with the bodily fluids. The main advantage of these pellets is that any type of substance in principle can be packed into an appropriate matrix, and because the pellets are machine-packed, the replicability is high and the variability low. On the other side, the pellets cannot be prepared by the researcher and are relatively expensive. Moreover, they should be properly tested when used for a new hormone/drug or taxa, to validate release rate and functionality.

Osmotic minipumps

These devices have become increasingly popular in laboratory studies and have been used in several field studies as well. The pump is filled with a solution of the drug to be delivered and implanted subcutaneously or intraperitoneally. The walls of the pump slowly absorb water from the bodily fluids and physically push out the solution contained in the pump at a given rate. The main advantage of this type of device is the reliable release rate and the replicability of the treatment between labs and operators. However, the pumps can be filled only with aqueous solutions and are relatively large and expensive. Moreover, because of their size and their composition, the osmotic pumps need to be removed at the end of the experiment, thus requiring capture, immobilization, and sedation of the animal at least twice.

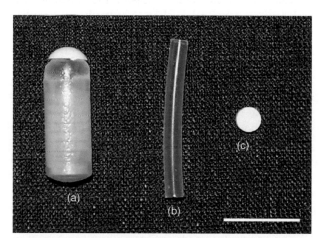

Figure 1 Some of the most common drug-releasing devices used for experimental studies on hormones and behavior. (a) Osmotic mini-pumps (ALZET, Durect Corporation, Cupertino, CA, USA). The internal reservoir is filled with the solution to be delivered. The external chamber has a high salt concentration and pulls water through the outer, semipermeable surface, pushing the drug solution out through the exit port. (b) Silastic Tubing (Dow Corning Corporation). The tubing can be cut at any desired length, filled with a lipophilic substance, and then closed at both ends with silicon glue. The substance diffuses slowly through the walls of the tubing. (c) Time-release pellets (Innovative Research of America, Sarasota, FL, USA). The hormone or drug is packed into an organic matrix which dissolves slowly when in contact with the bodily fluids. Scale = 10 mm.

Other treatment methods

Rapid technological advances in drug delivery and manipulation of gene expression offer a range of new techniques to study how alteration of hormonal action affects behavior. These include use of antisense olgonucleotides to interfere with gene translation, and viral vectors to induce gene expression. The application of these techniques in the field is a stimulating challenge for future studies.

Treatment Duration

Most methods for hormonal treatment in field studies rely on devices that release the hormonally active substance over a prolonged period of time, like those described in the previous section. There are two major aspects that need to be considered when performing long-term hormonal treatment. First, most methods can provide a continuous but seldom a constant release. Typically, the amount of released substance will be higher in the first days of treatment and decrease subsequently (**Figure 2**).

Thus, it is important to know the extent of this decrease if one desires to relate the frequency of a given behavior to a certain hormone concentration. Second, any hormonal manipulation will elicit a response by the endocrine system with consequent changes in the production of endogenous hormones. Thus, effects observed days or weeks after the beginning of the treatment could in fact reflect this endocrine response rather than the original treatment. This is an inescapable problem of any endocrinological manipulation, but is more severe in free-living animals which are typically intact, that is their endocrine glands are not surgically removed prior to the hormonal treatment. Therefore, it is important to conduct a thorough validation of the treatment (see section 'Treatment Validation') and to reduce the time lags between hormonal manipulation and the behavioral measure. For example, if the aim of the study is to test whether an increase in testosterone results

in an increase in aggression and the study is conducted by treating intact males with physiological doses of the hormone (which is highly recommended), it is important to record aggression within the first few days of the treatment. In the long term, the testosterone implant can lead to a regression of the testes and abolishment of the production of endogenous testosterone, which eventually might result in a decrease of testosterone levels below the initial, pretreatment values. Thus, aggression recorded weeks after the beginning of the treatment might be associated with a *decrease* in testosterone levels, even if the treatment is still in course.

Treatment Validation

A common problem encountered in designing field studies which involve hormonal manipulation is establishing the dose of the treatment. Sometimes an acute change in circulating hormone concentrations is desired, although the majority of studies in field behavioral endocrinology aim to modify for a relatively long time (hours, days, weeks) the hormonal profile of the experimental subjects. In both cases, a proper validation of the treatment is required especially when it is applied for the first time to the study species. The two main aspects to take into consideration are the resulting concentrations of target hormones and their variability over time.

Conclusions that can be drawn from the study will depend strongly on these two aspects. For example, several hormones are released in a pulsatile or circadian fashion. Examples are the gonadotrophin luteinizing hormone (LH) and the pineal hormone melatonin. To date, there are no devices readily available for long-term treatment which can reproduce time-dependent release. Thus, behavioral effects of the treatment will have to take into account that the circadian or pulsatile variation in circulating hormone concentration has been damped or even abolished by the treatment, which may lead to continuously high concentrations. The absolute concentration of the hormone is also very important. Endocrinologists often refer to the physiological range of concentrations, which is usually calculated as two standard deviations below and above the mean. However, such a definition can be misleading when dealing with species that show large seasonal or cyclical variations in hormone concentrations. For example, in birds androgen levels can be very low or undetectable during the molt and rise 100-fold at the beginning of the breeding season. Thus, a 'physiological' dose given to a molting individual can be very different from the concentrations that the animal experiences at another time of the year and during a particular life cycle stage.

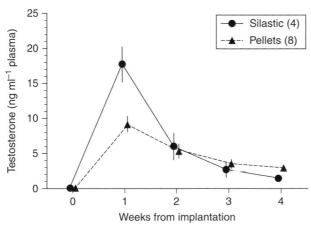

Figure 2 Concentration of testosterone in the plasma of female canaries following the subcutaneous implantation of time-release pellets or silicon tubing containing testosterone. Testosterone concentration decreased less sharply in pellet-implanted females compared to silicon-implanted ones. Data shown are the means ± the standard error of the mean; the numbers in the legend report the sample size. Redrawn from Fusani L (2008) Endocrinology in field studies: Problems and solutions for the experimental design. *General and Comparative Endocrinology* 157: 249–253.

See also: Experimental Approaches to Hormones and Behavior: Invertebrates; Hormones and Behavior: Basic Concepts.

Further Reading

Bradshaw D (2007) Environmental endocrinology. *General and Comparative Endocrinology* 152: 125–141.

Canoine V, Fusani L, Schlinger BA, and Hau M (2007) Low sex steroids, high steroid receptors: Increasing the sensitivity of the nonreproductive brain. *Developmental Neurobiology* 67: 57–67.

Costa DP and Sinervo B (2004) Field physiology: Physiological insights from animals in nature. *Annual Review of Physiology* 66: 209–238.

Fusani L (2008) Endocrinology in field studies: Problems and solutions for the experimental design. *General and Comparative Endocrinology* 157: 249–253.

Fusani L, Canoine V, Goymann W, Wikelski M, and Hau M (2005) Difficulties and special issues associated with field research in behavioral neuroendocrinology. *Hormones and Behavior* 48: 484–491.

Goldstein DL and Pinshow B (2006) Taking physiology to the field: using physiological approaches to answer questions about animals in their environments. *Physiological and Biochemical Zoology* 79: 237–241.

Goymann W (2005) Noninvasive monitoring of hormones in bird droppings – Physiological validation, sampling, extraction, sex differences, and the influence of diet on hormone metabolite levels. *Annals of the New York Academy of Sciences* 1046: 35–53.

Gwinner E, Roedl T, and Schwabl H (1994) Pair territoriality of wintering stonechats: Behaviour, function and hormones. *Behavioral Ecology and Sociobiology* 34: 321–327.

Ketterson ED and Nolan VJ (1999) Adaptation, exaptation, and constraint: A hormonal perspective. *American Naturalist* 154: 4–25.

Romero LM and Reed JM (2005) Collecting baseline corticosterone samples in the field: Is under 3 min good enough? *Comparative Biochemistry and Physiology. Part A: Molecular & Integrative Physiology* 140: 73–79.

Wikelski M and Cooke SJ (2006) Conservation physiology. *Trends in Ecology & Evolution* 21: 38–46.

Wingfield JC and Farner DS (1976) Avian endocrinology – Field investigations and methods. *Condor* 78: 570–573.

Wingfield JC and Farner DS (1993) Endocrinology of reproduction in wild species. In: Farner DS, King JR, and Parkes KC (eds.) *Avian Biology*, pp. 163–327. San Diego, CA: Academic Press.

Zera AJ, Zhao Z, and Kaliseck K (2007) Hormones in the field: Evolutionary endocrinology of juvenile hormone and ecdysteroids in field populations of the wing-dimorphic cricket *Gryllus firmus*. *Physiological and Biochemical Zoology* 80: 592–606.

Fight or Flight Responses

L. M. Romero, Tufts University, Medford, MA, USA

Introduction

When animals are faced with perceived or anticipated dangers from the environment, they initiate a stress response – a generic term for a bewildering suite of physiological and behavioral responses that are designed to help an animal survive these dangers (called stressors because they elicit a stress response). The two best-studied physiological responses are the release of glucocorticoid steroid hormones and the activation of the sympathetic nervous system. Discussion of the glucocorticoid response is presented in other articles and sympathetic activation, usually referred to as the fight-or-flight response, is the subject of this article.

The term fight-or-flight dates back to the early twentieth century and is an excellent brief description of the role of sympathetic nervous system activation. Fight-or-flight evokes immediacy. Life is hanging in the balance and there is no time for reproduction, mate selection, foraging, digestion, etc. At this precise instant, only survival is important. The fight-or-flight response is the first-line physiological mechanism for giving an animal its best chance for survival. If an animal mounts a fight-or-flight response, it suggests that the animal is reacting quickly, strongly, and immediately in order to survive. The sympathetic nervous system mediates these reactions.

Although many stressors can elicit a fight-or-flight response, the stressor most commonly associated with a fight-or-flight response is a predator attack. Predators can have numerous behavioral effects on their prey, including increasing vigilance, decreasing foraging times, changing how animals interact with other individuals (such as flocking or forming herds), and even altering where individual animals choose to live (i.e., changing prey distribution patterns). Predator presence, predator calls, and predator odors can all evoke a fight-or-flight response and even the threat of predation is sufficiently powerful that animals often react as if there were predators present even when there are none. These predator-induced changes in behavior have been studied for decades, but predator pressure can also change hormonal and physiological systems. Predation risk alters the hormonal and physiological regulation of reproduction and appears to have led to the evolution, in some bird species, of the restriction of sleep to one hemisphere of the brain at a time, an adaptation thought to allow the bird to maintain vigilance for predators at all times. The physiological changes associated with predation risk are so powerful that chronic exposure to predator cues is used as a laboratory model for studying human anxiety.

Clearly, the fight-or-flight response occupies a critical position in coping with a short-term stressor such as a predator attack. This article presents an overview of how the sympathetic nervous system regulates a fight-or-flight response and how that response might help an animal to survive.

Sympathetic Nervous System

The details of sympathetic activation are highly conserved among vertebrates. Although there are some species differences, the broad outlines are present in every species examined, from fish to mammals. This should highlight how central the fight-or-flight response is for survival.

Catecholamines

The activity of the sympathetic nervous system is primarily regulated by a class of hormones called catecholamines. Epinephrine and norepinephrine are the two primary catecholamines involved in the fight-or-flight response. They are equivalent to adrenaline and noradrenaline, epinephrine/norepinephrine being the names used in the United States and adrenaline/noradrenaline used in Europe. Epinephrine and norepinephrine are produced primarily in neurons (or modified neurons in the case of adrenal tissue). Synthesis begins with the amino acid tyrosine and ends with the rate-limiting conversion to norepinephrine by tyrosine hydroxylase. Epinephrine can then be converted from norepinephrine, using the enzyme phenylethanolamine N-methyltransferase (PNMT). PNMT is then the rate-limiting enzyme for epinephrine production. Both tyrosine hydroxylase and PNMT are often the targets for assays and in situ localizations to determine where, and potentially how much, epinephrine and norepinephrine are being produced.

Anatomy

The location of epinephrine release depends in part on the species. In mammals, epinephrine is primarily produced in the adrenal medulla – the center portion of the adrenal gland. The adrenal medulla is essentially a modified sympathetic ganglion where each secretory cell is a neuron without an axon. Epinephrine and norepinephrine are released from two different cell populations in the adrenal medulla. Another major source of norepinephrine is nerve terminals of the sympathetic nervous system. Most nonmammalian species, however, lack a well-defined adrenal medulla. In these species, cells that release epinephrine and norepinephrine are embedded in the wall of the kidneys. These cells are called chromaffin and are homologous to cells in the adrenal medulla of mammals. The physiological responses in mammals and nonmammals, however, appear to be essentially identical. When a stressor begins, epinephrine and norepinephrine are released from the adrenal medulla and norepinephrine is released from the sympathetic nerve terminals. Because the secretory cells are neurons, catecholamine release is very quick and effects can be seen in less than a second.

Catecholamines orchestrate the entire fight-or-flight response. The amount released, however, is very important. Sympathetic activation is not an all-or-nothing response and the strength of the response can be modulated to the needs of the moment. If too little response is released, the impact on target tissues will be insufficient; too much release, however, is often fatal. Consequently, the amount of epinephrine and norepinephrine released is usually carefully titrated to correspond to the severity of the stressor.

Physiological Effects

Once released, the catecholamines exert a number of effects throughout the body. Catecholamine-induced changes in the cardiovascular system have been known for almost a century. Catecholamines alter the delivery of nutrients, especially glucose and oxygen, to the brain, heart, lungs, and skeletal muscles, at the cost of the peripheral tissues. They do this by increasing cardiac output, increasing blood pressure, vasodilating arteries in skeletal muscle, vasoconstricting arteries in the kidney, gut, and skin, vasoconstricting veins in general, stimulating the lungs to dilate air passages, and initiating hyperventilation.

Although it might seem counterintuitive that catecholamines would exert such opposite effects as both vasodilation and vasoconstriction, the explanation resides in a difference in receptors. Catecholamine receptors come in both α- and β-varieties, both of which have two subforms (α_1 and α_2, β_1 and β_2). The α-receptors have higher affinity for norepinephrine, whereas the β-receptors have a higher affinity for epinephrine. Furthermore, each receptor type mediates different functions. For example, in arteries α_1-receptors mediate vasoconstriction, whereas β_2-receptors mediate vasodilation. However, the binding dynamics of α- and β-receptors described earlier apply to mammals. Birds have different

binding dynamics (both epinephrine and norepinephrine bind preferentially to β-receptors) and other taxa may show slight differences as well. Consequently, the physiology of catecholamine function can be richly varied in different species.

A second major effect of catecholamines is to increase the energy available to the muscles and brain, especially when glucose is being rapidly consumed during a fight or when fleeing. Catecholamines accomplish this by stimulating the liver to increase the production of glucose, specifically via glycogen breakdown, which is then released for delivery to peripheral tissues. The end result is a quick burst of glucose that can be used by muscles, the brain, and other essential tissues. Catecholamines also stimulate white adipose tissue to release free fatty acids that the liver can then use to produce more glucose. Once extra glucose becomes available in the blood stream, the final step in this process is to get the glucose into the cells that need it. Catecholamines also stimulate increased glucose uptake in these cells.

A third major effect of catecholamines is to regulate a number of effects in the skin. Catecholamines vasoconstrict blood vessels in the skin in order to shunt blood preferentially to internal organs, stimulate sweat production, and induce piloerection, the standing up of hairs in their follicles. Piloerection may serve two purposes: to enhance heat retention and to make the animal appear larger and fiercer to rivals and predators. Catecholamines may also regulate facultative changes in skin color in some species, especially to hide from predators.

Finally, the nervous system is also a major target for catecholamines. Their major effects in the brain are to increase attention and alertness. This leads to increased performance on cognitive tasks as well as a decrease in muscular and psychological fatigue. Catecholamines also cause the pupils to dilate, which aids in distance vision.

Increases in Heart Rate

Measuring the strength of the fight-or-flight response or even determining whether a fight-or-flight response is initiated, is often difficult. One problem is how fast the sympathetic nervous system is activated. Catecholamines are released into the blood so quickly that it is virtually instantaneous. Currently, one of the few techniques available is to measure changes in heart rate as an index of catecholamine release. Basic regulation of cardiac function is common across the vertebrates with epinephrine as the primary mediator of heart rate during exposure to a stressor. Epinephrine is released either directly from nerve terminals or indirectly from the adrenal and binds to β-receptors on the heart. Epinephrine results in an increase in heart rate (tachycardia) after most stressors. Furthermore, the degree of tachycardia depends upon the strength of the stressor – stronger stressors evoke higher increases in heart rate.

The diversity of stressors that can elicit tachycardia is impressive. Much of the work has been done under laboratory conditions, where stressors such as sounds, lighting conditions, novel odors, confinement, and abnormal social groups can stimulate increases in heart rate. In addition, a small but growing number of studies indicate that wild free-living animals increase heart rate in response to stressors such as human disturbance, social interactions, and capture and handling.

Social interactions are potent stressors that can result in robust tachycardia. However, animals can modulate these responses. For example, animals can habituate to intermittent social stressors, resulting over time in lower responses to equivalent social situations. Only the animal that wins the social encounter habituates, however. The loser retains, and even augments, its original tachycardia during subsequent encounters. In addition, the degree of tachycardia in the loser can depend upon the individual coping style of the animal. Animals have been shown to have either reactive or proactive coping styles when faced with novel situations. When faced with many, but not all, stressors, animals with reactive coping styles show a stronger activation of the sympathetic nervous system and thus a stronger tachycardia.

Although stressors induce tachycardia, the relationship can be reversed experimentally. The degree of tachycardia can be used to infer the strength of a stressor. If a stimulus evokes tachycardia, it is assumed to be a stressor, and if stressor evokes greater tachycardia than another stressor, then the stressor associated with the greater tachycardia is assumed to be the stronger stressor. There are several examples of this type of work. First, increased crowding evokes greater tachycardia, with the subsequent inference that crowding is stressful to these species. Second, increases in heart rate have also been used to show that many species can distinguish between familiar and unfamiliar conspecific calls, suggesting that vocalizations from neighbors are less stressful than vocalizations from strangers. Finally, animals can have robust increases in heart rate by simply watching agonistic interactions by other animals, even though they are not directly involved. This bystander effect suggests that social interactions, even those in which the individual is not directly involved, can elicit far more robust responses than other potentially dangerous stimuli.

In fact, tachycardia can be a more sensitive index of a fight-or-flight response than behavior. Animals often use behavior in order to avoid a costly physiological response – in other words, to avoid a fight-or-flight response. The result is that behavioral and physiological responses are often uncoupled. For example, an animal moving away from a disturbance will not necessarily be in the midst of a fight-or-flight response. In fact, it may be moving away specifically to avoid a potential stressor. Conversely, a number of studies have indicated that tachycardia can show a strong increase without any overt behavioral changes. Birds sitting in a nest, for example, may be acutely aware of a nearby predator, and show marked tachycardia, even though there is no outward change in behavior. Consequently, increases in heart rate are often better indicators of an underlying physiological fight-or-flight response than overt changes in behavior.

Heart rate can also be used to determine whether free-living wild animals are affected by putative anthropogenic stressors. Many things that humans do, such as building roads, ecotourism, wilderness sports, etc., are presumed to serve as potent stressors to wildlife. However, this assumption is rarely tested. Even if these activities change a species' settlement patterns or reproductive success, it is not necessarily true that the activities will induce a physiological fight-or-flight response. Implanted or attached heart rate transmitters, devices that collect heart rate data from freely behaving animals and transmit those data to a remote detection device, can be used to determine whether anthropogenic activities are, in fact, stressors. The evidence to date suggests that the impact of anthropogenic disturbances is more complex than originally thought. Many anthropogenic disturbances elicit robust fight-or-flight responses and thus can clearly be described as stressors, but other disturbances do not. The presence or absence of a fight-or-flight response may be an excellent diagnostic tool to determine what is, or is not, a stressor.

Finally, it should be remembered that an increase in heart rate, driven by the sympathetic nervous system, can extract a heavy price. For many years we have known that humans can go into sudden cardiac arrest and die because of severe emotional trauma. Although it is unknown whether this occurs in wild animals, it could be the mechanism underlying reports of trap death where wild animals spontaneously die for no apparent reason when captured. The fight-or-flight response is clearly necessary to escape from predators, but the increase in heart rate can create its own problems.

Decreases in Heart Rate

The uncoupling of tachycardia and behavior points out a weakness of the term fight-or-flight. Not all immediate emergency behaviors can be easily categorized as a fight response or a flight response. Flight generally evokes images of animals running/flying away from a predator. However, moving toward a predator may be a better strategy. Several studies indicate that moving toward an attacking predator, thereby reducing the predator's maneuvering time, can actually decrease predator success. In fact, often the most effective tactic is to freeze and not move at all: a tactic taken to its extreme in those species that feign death. Furthermore, the type of tactic employed often varies, both between individuals and within the same individual over time. Sometimes the individual flees and sometimes it freezes. These behaviors, freezing or moving towards a predator, are not generally considered to be part of the fight-or-flight response, yet are likely regulated by the same mechanisms.

Whether an animal flees or freezes when faced with a predator is not always predictable, but the choices are mutually exclusive with respective pluses and minuses. Fleeing immediately is the better choice when there is sufficient distance and speed to outrun the predator. The downside is that the animal also draws the predator's attention immediately and almost guarantees a chase. Freezing, on the other hand, may allow the animal to elude detection. This response is especially useful if the animal is not yet detected or has an asset, such as nearby young, that needs to stay hidden. The downside of this choice is that it can allow a predator to get lethally close. The decision whether to freeze or flee is complex, partially dependent upon the individual animal's predilection, its distance to a refuge, and the potential benefit of confusing a predator by being unpredictable.

When an animal chooses to freeze, however, there is a very different change in the sympathetic nervous system. In general, the response is bradycardia, not tachycardia. A classic example is the feigned death of species like the opossum. When faced with a predator, the animal will become limp and nonresponsive to poking and prodding. This behavioral response is accompanied by a marked bradycardia. Heart rate plummets regardless of the behavior of the predator. Bradycardia makes sense in this context – if the goal is to appear dead, then decreasing heart rate helps to damp any behavioral and/or physiological responses. Once the danger has passed, however, the classic sympathetic response resumes and a strong tachycardia ensues. Interestingly, freezing behavior is rarely, if ever, seen in captive animals. Sympathetic activation, with its associated tachycardia, appears to be the default response. Freezing, with its attendant bradycardia, appears to require a specific context that is absent in caged animals.

Seasonal Differences in the Fight-or-Flight Response

All animals seasonally adjust behavioral and physiological responses. Cardiovascular function and the underlying fight-or-flight response is no exception. For example, resting heart rate is lower during the winter than during the summer for many species. In general, seasonal changes are linked to a lower energetic demand resulting from a lower winter metabolism. However, there are also seasonal differences in sympathetic activation in response to a

stressor. The fight-or-flight response can be modulated depending upon the life-history stage. For example, animals can show a stronger fight-or-flight response to conspecific crowding when they are defending territories in the spring than when they are gregarious in the winter. The fight- or flight response can also be modulated depending upon the physiological state of the animal. When an animal is in a particularly energy-intensive period, such as molt or pregnancy, the fight- or flight response can be dramatically suppressed. The lack of response highlights that the magnitude, and perhaps even the presence, of a fight-or-flight response may depend upon the season and/or physiological state of the animal.

Seasonal and life-history-stage differences in the fight-or-flight response are not well studied. However, understanding how and when sympathetic responses are modulated should provide important insights into the survival benefits of the generalized fight-or-flight response. The modulation of sympathetic activation may be related to seasonal changes in the prevalence and severity of stressors. The end result would be an animal fine-tuning its fight-or-flight response in order to maximize effectiveness.

Sympathetic Responses During Chronic Stress

The fight-or flight response seems well-suited to help an animal cope with short-term emergency situations. If a stressor continues for a long time, however, or if a series of short-term stressors continues in rapid succession, many of the short-term emergency responses can themselves become damaging. When this occurs, it is called chronic stress. The constant and/or repeated initiation of the fight-or flight response can lead to profound disruption of the sympathetic nervous system. For example, chronic stress can lead to coronary heart disease in both humans and animals.

Many studies indicate that, over time, chronic stress leads to a lowering of the magnitude of heart rate elevations in response to a variety of stressors. In other words, chronic stress leads to a damping of the fight-or-flight response. The chronically stressed animal can no longer mount a robust sympathetic response to a novel stressor. This decrease is often interpreted as habituation to the stressor, but other data suggest that this might be too simple an explanation. Stores of both epinephrine and norepinephrine are depleted during chronic stress, which results in diminished release of both hormones. This appears to be the underlying mechanism that results in the attenuated fight-or flight response.

If the fight-or-flight response is indeed down-regulated during chronic stress, it would have tremendous fitness implications. An appropriate fight-or-flight response, necessary to survive stressors in the wild, would be compromised. This suggests that chronic stress greatly impacts on an animal's potential survival, especially in terms of evading predators.

In contrast, other studies show long-term increases in heart rate during chronic stress. The increase is in both basal heart rates, indicating chronic sympathetic overstimulation, and in the heart rate response to a stressor, that is the fight-or-flight response. The underlying mechanism appears to be increased synthesis and storage of catecholamines, which may allow the animal to respond stronger to a novel stressor.

Note that the two sets of studies come to completely opposite results. One set shows a decrease in catecholamines, leading to decreases in the fight-or-flight response, and the other set shows an increase in catecholamines, leading to increases in the fight-or-flight response. The reasons for this disparity are currently unknown, but a difference in individual coping styles is one candidate.

The impact of heart rate changes during chronic stress could also be ameliorated over the course of the day. In some models of chronic stress, the heart rates recover quickly once the chronic stress ends. This suggests that the chronic stress-induced changes are not long-lasting. Furthermore, most chronic stress models apply stressors only during a portion of the 24 h cycle (e.g., during the active period). The heart rate can often recover and even overcompensate during nonstress periods (e.g., during the sleep period). Because mounting a fight-or-flight response is costly energetically, the heart rate changes at night might be an attempt to balance the daily energy budget and compensate for the energy lost when responding to the chronic stress.

See also: Conservation and Anti-Predator Behavior; Ecology of Fear; Hormones and Breeding Strategies, Sex Reversal, Brood Parasites, Parthenogenesis; Stress, Health and Social Behavior; Trade-Offs in Anti-Predator Behavior; Vigilance and Models of Behavior.

Further Reading

Bohus B and Koolhaas JM (1993) Stress and the cardiovascular system: Central and peripheral physiological mechanisms. In: Stanford SC, Salmon P, and Gray JA (eds.) *Stress: From Synapse to Syndrome*, pp. 75–117. Boston, MA: Academic Press.

Cannon WB (1932) *The Wisdom of the Body*. New York, NY: W.W. Norton.

Goldstein DS (1987) Stress-induced activation of the sympathetic nervous-system. *Baillieres Clinical Endocrinology and Metabolism* 1: 253–278.

Reid SG, Bernier NJ, and Perry SF (1998) The adrenergic stress response in fish: Control of catecholamine storage and release. *Comparative Biochemistry and Physiology C: Toxicology & Pharmacology* 120: 1–27.

Stanford SC (1993) Monoamines in response and adaptation to stress. In: Stanford SC, Salmon P, and Gray JA (eds.) *Stress: From Synapse to Syndrome*, pp. 281–331. Boston, MA: Academic Press.

Steen JB, Gabrielsen GW, and Kanwisher JW (1988) Physiological aspects of freezing behavior in willow ptarmigan hens. *Acta Physiologica Scandinavica* 134: 299–304.

Young JB and Landsberg L (2001) Synthesis, storage, and secretion of adrenal medullary hormones: Physiology and pathophysiology. In: McEwen BS and Goodman HM (eds.) *Handbook of Physiology; Section 7: The Endocrine System; Volume IV: Coping with the Environment: Neural and Endocrine Mechanisms*, pp. 3–19. New York, NY: Oxford University Press.

Fish Migration

R. D. Grubbs and R. T. Kraus, Florida State University Coastal and Marine Laboratory, St. Teresa, FL, USA; George Mason University, Fairfax, VA, USA

Introduction

Application of the term, migration, to fishes generally adheres to the definition originally proposed by ecologist Walter Heape and refers to predictable movements between areas or habitats, related to resource availability, in which the migrants are compelled to return to their place of origin. This concept guided early classifications of fishes based on their migratory habits, but the prevalence of migration among the fishes remains difficult to quantify for three primary reasons. First, a wide range of definitions for migration are commonly used (both explicitly and implicitly) in current literature (see below for the concept adopted here). Second, the temporal and spatial characteristics of movements in fishes represent continuums and are therefore, not always amenable to classification. Alexander Meek recognized this fact nearly a century ago and asserted that the majority of fish species are migratory to some degree. Third, large gaps still exist in our knowledge of movement and life history patterns in many fish taxa. Due to these difficulties, attention to migration has focused on the 1% of fishes (~300 species) that make 'extensive' migrations. The distinction of 'extensive' migrations is arbitrary, and often biased toward species that are of economic importance and make long-distance migrations with seasonal periodicity. Migratory species (primarily, anchovies; shads and herrings; hakes and cods; and mackerels and tunas) account for most of the catch among fisheries worldwide and, consequently, also account for the greatest body of published research on fishes. Nevertheless, cyclical, to-and-fro movements exist in fishes over many spatial and temporal scales, and sometimes, across generations. Thus, it is important to recognize that small spatial and short temporal scale movements of one species may have the same adaptive significance (i.e., increased fitness) as the extensive migrations seen in other species. For example, many live-bearing poecilid fishes (e.g., guppies) have life spans less than 1 year and make migrations of several hundred meters that are extensive in the context of their life histories.

Though many definitions of migration exist, perhaps the most inclusive one defines migration as an *adaptive response to spatial changes in the availability of resources and/or mortality risk.* This definition applies to disparate taxa and yet recognizes migration as a singular phenomenon different from other types of movement. At the level of the individual, a hallmark of adaptive migration is uninterrupted movement in which reactions to stimuli, such as suitable habitats or forage, are temporarily suppressed. At the level of the population, migration is a hypothesis to explain seasonal or other cyclical changes in the distribution of individuals. Thus, migration is central along the continuum of movements represented by foraging and commuting (see section Vertical Migrations for an example) at one end and ranging or natal dispersal (i.e., metapopulation dynamics) at the other. Here, our intention is to introduce the major patterns and scales of migration in fishes, along with examples, some of which are not commonly cited. We first provide a short background outlining the historical context for considering the adaptive function of migration in fishes. We also emphasize insights that have been gained from the study of intrapopulation variability in migration. Finally, we examine the great diversity of migratory patterns found in fishes, applying traditional classifications and terminology. The key references and vocabulary in this summary should provide the reader with a synoptic understanding and productive directions for further study of migration in this incredibly diverse group of vertebrates.

Adaptive Function of Migrations

The evolution of migratory behavior requires interaction between multiple genetic determinants and environmental factors. The resulting suite of physiological and behavioral character states associated with migration thus has a complex explanation representing what some have termed a migratory syndrome. One prominent example is the preemptive changes in morphology, coloration, osmoregulation, growth rate, and rheotaxis that accompany smoltification (metamorphosis from a stream-dwelling to a marine form) and prepare juvenile salmon to leave natal streams and survive at sea. To be adaptive, such changes must provide distinct advantages in terms of reproductive opportunities, energetics, and/or survival of the individual that outweigh the costs of movement and risks of starvation, predation, and reproductive failure as well as the costs and risks of not migrating. Further, the adaptive advantage of migration may be condition-dependent such that an individual may have higher fitness by becoming a nonmigratory resident (see section Partial Migration).

At the level of the population, ontogenetic niche shift (ONS) theory provides great insights to the adaptive function of migration. ONS theory predicts the body size at which a change in niche (e.g., habitat, diet, and behavior) would provide an adaptive advantage by minimizing the ratio of mortality to growth. Interestingly, movement to a niche with higher mortality may be favored if there are sufficiently high opportunities for resource acquisition. More importantly, the ONS concept can be extended to reproductive stages as much as individual growth and cohort biomass are related to fecundity and fertility. For the vast majority of egg-laying fishes, fecundity increases approximately as a cubic function of length; therefore, at least in females, strong selection for minimizing the mortality-to-growth ratio should be expected. In practice, understanding how this dynamic operates can be far more difficult because the migration activity may impose significant energetic costs with attendant consequences for growth (or achievable size) in the alternative habitat and/or energy allocated to reproduction.

Recognizing that a migration involves numerous character states and evolves in response to a suite of factors, it is clear that the ultimate cause for migration can vary substantially across species. In the earliest works on animal movements, migrations were classified as alimentary, gametic, or climatic based upon endogenous considerations of the primary motivation for movement. With the application of the definition of migration we have adopted, this scheme provides an effective context for developing and testing hypotheses about migration, but we also note that refuge migrations are an additional category to include.

Gametic Migrations

Gametic migrations are movements that increase reproductive success of individuals by promoting gonad development, increasing sexual encounter rates, or increasing the survival of offspring. Gametic migrations are complex and highly evolved and the consequence of not migrating is often reproductive failure. In fishes, gametic migrations are often tied to geographical locations (e.g., larval retention areas) or a specific type of habitat. The most impressive and well-known gametic migrations in fishes are seen in those that migrate between freshwater and saltwater during specific life stages to reproduce (see section Diadromy).

Alimentary Migrations

Alimentary migrations are those that increase trophic success by allowing access to new or more abundant prey resources or by decreasing competition for available prey. A majority of the migratory patterns exhibited by fishes are alimentary in function and may occur over widely varying scales dependent upon life stage. Trophic movements, referred to as migrations, often occur on tidal, diel, lunar, and seasonal temporal scales as well as a wide range of horizontal and vertical spatial scales. Thus, the correlated changes in forage resources may be represented by short-term accessibility to intertidal habitats exploited by many drum species (family Sciaenidae) and also broad-scale synchrony between the movement of some planktivorous elasmobranchs (e.g., whale sharks and mantas) and the spawning episodes of corals and reef fishes.

Climatic Migrations

Climatic migrations are driven by physiological tolerances of individuals to environmental factors such as temperature or salinity. Climatic forcings are often the proximate causes for migrations that ultimately serve to increase foraging or reproductive success. Nonetheless, consideration of climatic migrations is critical because energetic costs and mortality risks are often greatly increased for individuals that fail to migrate. An important distinction for climatic migrations is that movement occurs despite the immediate presence of sufficient foraging opportunities in the initial habitat. For example, in many temperate estuaries, boreal species may temporarily fill the niches of temperate species that have migrated due to seasonal declines in water temperature.

Refuge Migrations

Refuge migration functions directly to decrease the risk of mortality from predation. Often, refuge habitats are inaccessible to predators due to physical (e.g., water depth) or physiological (e.g., salinity) tolerances. However, complex habitats may also serve as refuges even in the presence of predators by allowing concealment. In addition, juveniles of many species migrate to distinct nursery habitats that provide refuge but are also highly productive forage areas; therefore, migration has both refuge and alimentary purposes. Compared to the other categories, the refuge function of migration has probably received the least attention, though experiments to test proximate mechanisms of the refuge hypothesis might easily be constructed through manipulations of food supply.

It is important to emphasize that fish migrations often entail a combination of the above motivations, and when adaptive advantages are gained in more than one of these dimensions, we frequently observe a highly stereotyped life history pattern. This point is best illustrated by the large body of research on salmon species in which the downstream removal migration of juveniles provides a different adaptive advantage than the return migration of adults. Smoltification and downstream migration in salmon during a nonreproductive life stage are likely driven by higher productivity in coastal habitats. The advantages of shifting to a marine environment include increased forage and reduced intraspecific competition compared to stream habitats, fitting the paradigm of an alimentary migration. The return migration to spawn involves cessation of feeding and loss of energy reserves from upstream movement that are typically between 50% and 70%. Because the return of adults entails natal homing mechanisms, reproduction, and death (in some species), this migration is clearly gametic. As expected, we also find a similar degree of highly stereotyped life history complexity in other diadromous species (see section Diadromy).

Intrapopulation Variability

Most empirical insights about the interplay between the genetic architecture and phenotypic plasticity of migration have been gained through investigations of individual variability in the occurrence or magnitude of migration. This variability generally falls into two categories: *partial migration*, where only a fraction of individuals migrate, and *differential migration*, where the pattern of migration varies among population segments (e.g., between sexes or life stages). Examining this variability provides opportunities to learn how obligately migratory populations evolve or how some species can establish migratory populations from nonmigratory ones and vice versa. More importantly, when this intrapopulation variability persists, we are faced with a more complex challenge for conservation or resource management that demands an understanding of its adaptive significance.

Partial Migration

A growing body of research indicates that, for a given genetic architecture, migratory versus nonmigratory life history alternatives represent tactics dependent upon the condition of the individual. Condition-dependent migration presumes that the individual can somehow evaluate its own situation (resource acquisition rate, mortality risk, and/or competition) when 'deciding' (unconsciously) whether to migrate. Support for condition-dependent migration comes from information on other animals, and in fishes, manipulative experiments have demonstrated phenotypic plasticity where transplanted nonmigratory fish became migratory and where manipulated nutritional states determined migratory tendencies. The interesting question is why partial migration is a persistent phenomenon in many migratory fish populations.

Evolutionary biologist, Mart Gross, produced a classic example that illustrated how condition-dependent phenotypic plasticity could be involved in maintaining partial migration for *Oncorhynchus kisutch*, coho salmon. In Gross's study, most male coho salmon either matured precociously and remained resident in freshwater (jack) or became migratory and matured later at a larger size (hooknose). The conditions that lead to either life history alternative

appeared to be linked to parr body size, which is influenced by multiple factors (e.g., maternal effects, emergence time, growth-rate, intraspecific interactions, and hydrological conditions) such that any individual male presumably could adopt either tactic. Hooknose males typically exhibited a fighting behavior to gain access to females for spawning, whereas jacks typically exhibited a sneak-spawning tactic. More importantly, the breeding success of the jacks was dependent upon the frequency of hooknoses in the population and vice versa. Overall reproductive success was reduced when there were too many hooknose males (interference between males defending territory) or too many sneaker males, which require hooknose males in order to spawn successfully. This negative frequency-dependent selection leads to an evolutionarily stable strategy with both male life histories. Similarly, density-dependent (as opposed to frequency-dependent) effects on reproductive success or survival in a particular habitat could promote partially migratory populations. In addition, when there are multiple optimal habitats depending upon the particular condition and mortality risk, partial migration may include a gradient of tactics.

Differential Migration

Migration patterns that vary as a function of sex or age can be observed in both immature and reproductive life stages. In juvenile fishes, differential migrations are often a balance between obtaining a trophic advantage (alimentary) and seeking physical protection (refuge). Larger or older conspecifics may become migratory due to trophic requirements and size-based reductions in mortality risk. Alternatively, density-dependent competition in a habitat with low mortality risk may result in differential migration of inferior competitors. Considering the entire life cycle, a reduced tendency to migrate can provide a reproductive advantage by allowing some individuals to remain in close proximity to the spawning area and reduce energetic costs associated with migration or gain priority in the selection of reproductive habitats. When differential migration occurs at a mature life stage, requirements of parental care may have a dramatic influence on observed patterns. The most striking examples of sexual differences in migration occur in species that employ internal fertilization. Both sexes necessarily exhibit a gametic migration to a common geographical location for mating, but females (or in rare cases, brooding males) show an additional migratory phase to specific regions or habitats for parturition. As the gametic contribution of the males ends at mating, they do not participate in this migration phase. For example, sandbar sharks *Carcharhinus plumbeus*, in the western North Atlantic exhibit ontogenetically and sexually differential migrations that serve a variety of adaptive functions (**Figure 1**).

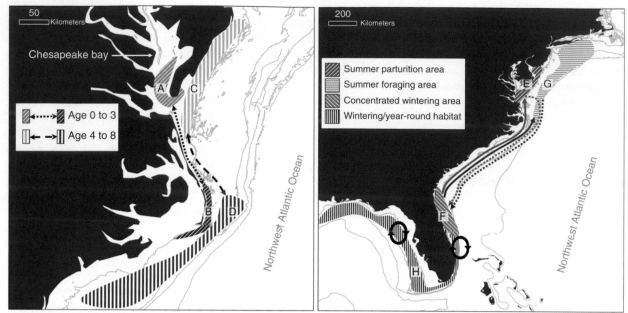

Figure 1   Major patterns of migration in the sandbar shark, *Carcharhinus plumbeus*, in the Northwest Atlantic Ocean. Sandbar sharks are born in warm-temperate estuaries (e.g., Chesapeake Bay), which serve as primary nurseries, providing forage and refuge from predation during the first summer. The juvenile sharks undergo a climatic migration south to wintering areas in the fall and a return alimentary/refuge migration to the natal estuary the following summer. This migration pattern is repeated annually until they are at least 7 years of age but is ontogenetically differential. Very young sharks (<3 years) use the estuary during summer (a) and overwinter in relatively shallow coastal waters (b). Older sharks (4–8 years) are larger and more vagile, and therefore are less susceptible to predation and can exploit additional habitats. The summer habitats used by these older juveniles are in nearshore coastal waters (c) adjacent to and north of the natal estuary and the wintering areas are progressively expanded southward and to deeper waters (d). Juvenile female sandbar sharks maintain this migratory pattern until about 12 years of age while juvenile males older than age 7 rarely return to the natal region, a sexual difference that is likely linked to sexually differential migration of adults. Adult females (>15 years) are philopatric, making gametic migrations to their natal estuaries for parturition on a 2-year cycle (e). This adult migration is not exhibited by male sandbar sharks. Many subadult and adult sandbar sharks of both sexes overwinter nearshore off the east coast of Florida (f), though some move into the Gulf of Mexico, concentrating along the edge of the continental shelf (h). Some sandbar sharks, especially nonpregnant (resting) females, remain in southern waters year-round while others, especially adult males, migrate to foraging areas along the edge of the continental shelf off the northeast coast of the United States during summer (g). This diagram is based primarily on the work of R. D. Grubbs and J. A. Musick (Virginia Institute of Marine Science).

Migratory Patterns

The Greek term *dromos* translates to 'running' or 'race' and is the root for describing the major migratory patterns in fishes. The terms *anadromy* (up-running) to describe migration from the sea into freshwater and *catadromy* (down-running) to describe migration from freshwater to the sea dates back nearly a century. These terms originally described migrations toward spawning areas; however, their use has broadened their use to refer to any inshore or upstream migration as anadromy and any offshore or downstream migration as catadromy. The wider use of the terms complicated their meanings, but ichthyologist George Myers brought some consistency to the nomenclature in the later 1940s by introducing the term *diadromy* (through-running) to include all true migrations between marine and freshwater habitats while preserving the original definitions of anadromy and catadromy. Myers also introduced the term *amphidromy* to describe diadromous fishes that migrate between freshwater and

seawater for purposes other than spawning. In addition, the terms *oceanodromy* and *potamodromy* (Greek *potamos* = river) were introduced to describe those migrations that take place wholly in seawater or freshwater, respectively.

Diadromy

Although diadromy is exhibited by only ~1% of fish species, ~27% of recent scientific literature on fish migration addresses this narrow topic. In part, the bias is due to a small number of diadromous species that support highly productive and valuable fisheries such as Pacific salmon or anguillid eels. Another related bias is that much more is known about the habits of diadromous species during the freshwater phase of life due to the relative difficulties of studying fish behavior in the open ocean. Sea-going habits of Pacific salmon are poorly understood and the specific marine spawning locations of anguillid eels are still unknown. An obvious predisposition for the evolution of diadromy is the ability to switch reversibly between hypo- and hyper-osmoregulation. In addition, since many diadromous species are semelparous (reproducing once before death; e.g., anadromous salmon species), we find elaborate sensory capabilities (rather than social learning mechanisms) adapted for natal homing, including olfactory imprinting and mechanisms for geomagnetic and celestial navigation.

Anadromy and Catadromy

The adaptive significance of migration likely has similar origins in anadromy and catadromy, and is linked to productivity differences between adjacent marine and freshwater ecosystems. As a percentage of diadromous species, anadromy predominates in the northern hemisphere where ocean habitats are more productive than adjacent freshwater habitats. Likewise, in the southern hemisphere where ocean and freshwater primary production rates are more similar, catadromy is more frequent. The advantages of a more productive environment are apparent in the faster growth rates of migratory versus nonmigratory individuals in many salmonid populations. Faster growth and larger size leads to earlier maturation and greater potential for gamete production. As catadromous and anadromous species are sympatric in some freshwater systems (e.g., American shad and American eel populations), an additional consideration for the ocean productivity hypothesis is ancestral legacies. Salmon populations introduced into southern hemisphere rivers retain anadromous behavior, and the ancestral origin of catadromous eels in the northern hemisphere is likely the tropical regions of the southern hemisphere.

Atlantic or Pacific salmon are the archetypal examples of anadromy, but temperate sea basses (Moronidae), alosine shad, and sturgeons are also well-known examples from the northern hemisphere. Within these, there is a wide range of migration distances. Some shad and salmon species migrate over 1000 km inland with an accompanying elevation change of 1 km. By comparison, anadromous migrations of some temperate sea basses do not extend beyond the influence of tides. Classic examples of catadromy are American and European eels, which have broad freshwater distributions as juveniles occupying nearly every catchment in subtropical, temperate, and boreal climates of the northern Atlantic. Both eel species undergo a striking metamorphosis (transforming from juvenile yellow eels to nonfeeding mature silver eels) and migrate to an unknown location in the Sargasso Sea where they spawn and die.

There is growing knowledge of the importance of partial migration in anadromy and catadromy which may manifest as multiple distinct and persistent migratory modes, referred to as *contingents* in the fisheries science literature. Temperate sea basses, such as white perch (*Morone americana*), typify this wide variation in partial

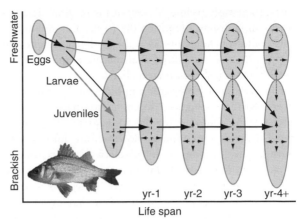

Figure 2 Life cycle diversity and migratory pathways of an anadromous temperate sea bass (*Morone americana*). White perch exhibit partial migration characterized by early life differences in habitat use that perpetuate into adulthood. Migratory adults make gametic migrations (dotted arrows) to common, spatially limited freshwater environments where nonassortive mating with nonmigratory individuals occurs. Seasonal refuge and foraging migrations are also observed as upstream–downstream (vertical dashed arrows) and littoral movements (horizontal dashed arrows). Eggs and larvae develop in tidal freshwater, and upon metamorphosis, juveniles may take up residence in the natal freshwater habitats or disperse to brackish habitats. This pattern is related to growth performance early in life with correlation between faster larval growth and nonmigratory behavior. The disparity in larval growth is reversed later in life when migratory adults exhibit faster growth rates and attain larger size. In addition, infrequent strong year-classes (gray arrows) generate a higher proportion of migratory individuals; therefore, the biomass productivity is primarily determined by the migratory fraction of the population. The relative reproductive contribution of nonmigratory individuals becomes increasingly important when conditions for early life survival are poor and ensures some reproductive success during such episodes. Some nonmigratory adults may become migratory later in life, and this may be due to higher than average growth performance of these individuals. This diagram is based primarily upon the works of R. T. Kraus and D. H. Secor (University of Maryland).

migration (see **Figure 2**). The diversity of individual migratory behaviors may also help to resolve the paradox of catadromy in anguillid eels in the northern hemisphere. There is evidence that eel catadromy may be facultative, and a few researchers have posited that freshwater eels make an insignificant reproductive contribution to the population. Instead, most population productivity appears to occur in more productive, coastal marine habitats, which supports the ocean productivity hypothesis of diadromy.

Amphidromy

The ocean productivity hypothesis does not fully explain the adaptive function of amphidromy. Amphidromy has been a source of much debate and confusion, and most recent discussions have restricted its definition to freshwater amphidromy only. While Myers, in his original description of amphidromy, identified the freshwater amphidromous gobies of the genus *Sicydium* as the representative example, he also recognized that marine amphidromy is probably more common than realized.

The pattern of freshwater amphidromy is well documented. Spawning, egg development, and hatching occur in freshwater, followed by the immediate transport of larvae through entrainment downstream into marine habitats. The juveniles spend a relatively short time in seawater and then actively migrate back into the streams or rivers where most somatic growth, as well as maturation and spawning, occur. Most fishes that exhibit this pattern are riverine species associated with oceanic islands. The streams associated with many oceanic islands are small high-velocity torrents that flow through complex and steep terrain. The upstream migrations of tiny juveniles of these species can be incredibly impressive. The endemic Hawaiian goby *Lentipes concolor* may climb vertical waterfalls and damp rock walls more than 100 m in height to reach the natal habitat. Fish migration expert, Robert McDowall, hypothesized freshwater amphidromy to be an island adaptation related to the colonization and recolonization of obligatory, spatially limited freshwater habitats after calamities such as volcanic eruption. Considering this, freshwater amphidromy lies at the ranging end of the migration spectrum, and is thus most important as a metapopulation phenomenon, an idea supported by recent analyses indicating genetic homogeneity among conspecific populations.

Marine amphidromy may be more common than previously realized because most amphidromous species appear to exhibit this behavior facultatively. Many marine coastally spawning species make significant migrations into freshwater habitats during early life and return to the sea as juveniles, supporting predictions that marine amphidromy may be more widespread than freshwater amphidromy. The distances and speed traveled often exceed the swimming capabilities of larvae, thus selective tidal stream transport (see section Vertical Migrations) is a key migration mechanism during early life. Transport of larvae into freshwater estuarine habitats may result in increased growth rates and reduced mortality due to increased feeding opportunities and turbidity, thus reducing recruitment variability. Subsequently, these species return to coastal marine habitats where reproduction and, for some, the majority of somatic growth occurs. These species are often estuarine-dependent and rarely move into nontidal freshwater habitats. North American examples include Atlantic menhaden, Atlantic needlefish, temperate drum species, and some flatfishes. As freshwater habitat use is facultative in these species, they are not typically classified as diadromous. Future research on fish migration should include reevaluating the condition of diadromy in these and other coastal and estuarine species.

Oceanodromy

Migrations occurring entirely at sea are not easily characterized because many forms operate on varied spatial and temporal scales, and some are not as regimented as diadromous migrations. One critical difference between diadromy and oceanodromy involves the role of rheotaxis and currents. Freshwater movements in diadromous migrations involve adaptations to deal with primarily unidimensional currents that vary only on temporal intensity. In contrast, current regimes in oceanic environments occur in three spatial dimensions simultaneously and are typically weaker and more variable than those in freshwater. Further, gyres and counter currents are common, and movement perpendicular to currents can promote physical entrainment and retention within a specific geographical area. These movements cost far less in terms of lifetime energy budgets than the upstream movements of diadromous migrations. It is not surprising, therefore, that the migratory patterns of oceandromous fishes initially appear less complex and regimented, but a closer examination reveals a complexity, which indicates that these patterns are no less significant to the evolutionary success of marine fishes.

Open ocean

Many oceanic migration patterns are linked to highly persistent oceanographic currents. Large pelagic fishes, such as tunas, are of high economic value and have garnered much attention from researchers. The migratory patterns of some species are highly regimented and typically include three fundamental component migrations to spawning areas, feeding areas, and wintering areas, a pattern described by fish ecologist, Roy Harden Jones, as a migration triangle. The migration of adults to spawning grounds is adaptive by increasing the encounter rates of reproductive individuals, and spawning grounds are located where pelagic eggs and larvae will be retained in areas that promote development, feeding, and growth. Spring and summer migrations of juveniles and adults to separate feeding grounds, often in productive waters that are physiologically intolerable in winter, serve an energetic function maximizing growth, maturation rate, and gonadal development. Juveniles typically alternate seasonally between feeding areas and wintering areas, while adults rotate between feeding areas, wintering areas, and spawning areas. This pattern of oceanic migration is much more prevalent for pelagic fishes that have feeding areas in temperate latitudes but spawn in subtropical latitudes (e.g., **Figure 3**, bluefin tuna – *Thunnus thynnus*). Many tropical pelagic fishes (e.g., yellowfin tuna, *Thunnus albacares*) have poorly defined gametic migrations because spawning may occur over broad areas or throughout the year whenever water temperature and nutritional state promote gonad development. Migrations in these species often facilitate

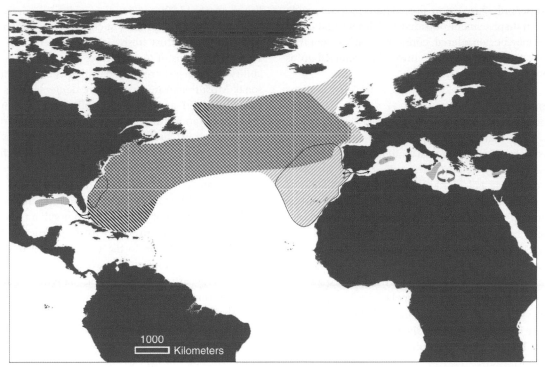

Figure 3 Ocean basin scale migration patterns of Atlantic bluefin tuna. Owing to the high metabolic rate and endothermy in this species, seasonal foraging areas (hashed areas) and individual foraging movements may span the entire Atlantic Ocean and a wide range of temperatures. Concordant with the concept of alimentary migrations, individual movements within these vast areas correlate strongly with temporal changes in food supply and are to some degree anticipatory because of predictable seasonal occurrences of these fishes throughout their range. Despite extensive mixing during seasonal feeding migrations, gametic migrations (black arrows) to principal spawning locations (green areas) isolate western Atlantic (red) from eastern Atlantic populations (blue). Juveniles in the Mediterranean Sea may remain there for multiple years until they migrate into the Atlantic. Partial migration may be occurring in populations of the Mediterranean Sea as size (and presumably growth) may determine the tendency to migrate. By comparison, there are no resident bluefin tuna in the Gulf of Mexico. Adults leave immediately after spawning, and juveniles migrate to the Atlantic during the first 3–6 months of life. Overwintering areas (black outlines) appear to be concentrated within limited subregions of the seasonal foraging range, but unlike the gametic migrations mixing between western and eastern populations can occur. The boundaries of these migration areas vary on decadal and interannual scales. The migration of Atlantic bluefin tuna has been revealed primarily through electronic tagging methods and the synthesis of information on the species' biology and population structure. Many researchers deserve credit here, and a good starting place for further reading is the work of J. R. Rooker (Texas A&M University).

arrival at discreet locations coinciding with concentrated forage that is temporally predictable (e.g., mesopelagic shrimp spawning on seamounts) or they follow major oceanographic features to remain in relatively productive areas (e.g., convergence zones) throughout the year.

Coastal ocean/estuaries

Coastal and estuarine fishes exhibit a wide array of migratory behaviors, including migratory circuits between wintering areas, feeding areas (and refuges), and spawning areas similar to the patterns described for the pelagic fishes. The sea-snail (*Liparis liparis*) a confusingly named marine fish, lives 1 year but has a highly regimented migration pattern. Spawning takes place in coastal marine waters, but juveniles migrate to inshore estuaries for trophic benefits and refuge. The following fall, sea-snails leave the estuaries on a gametic return migration to coastal

spawning areas with the specific habitats (demersal hydroids) necessary for egg deposition.

Reef fishes often have very precise diel and seasonal alimentary migrations, and regimented gametic migrations. Adults of some coral reef taxa migrate seasonally to discrete areas that aggregate spawners (e.g., groupers and snappers), while others move to specialized habitats to deposit demersal eggs (e.g., damselfishes). The larvae of many reef fishes settle in seagrass meadows and mangroves that serve as nursery habitats. With growth, the juveniles egress to intermediate patch reefs before ultimately migrating to adult habitats. Many reef fishes exhibit precise diel (crepuscular) movements between resting and foraging areas. Some tropical haemulids (grunts), for example, aggregate at specific sites on reefs or in mangrove habitats during the day but disperse in dendritic patterns at sunset into seagrass habitats to forage. At sunrise, they follow the same course, returning to the highly conserved aggregation sites. These movements do not result in displacement outside the fish's normal home range and may be considered 'commutes' rather than true migrations. However, they are invariably referred to as migrations in the literature.

Deep sea

The study of migration in deep sea fishes is in its infancy, but it is likely that many deep sea fishes exhibit highly developed gametic and alimentary migrations. While most abiotic factors (e.g., temperature, salinity, and dissolved oxygen) vary little below 1000 m, primary production of phytoplankton and the resulting vertical fluxes of carbon vary on multiple temporal scales, especially lunar and seasonal. Some bathyal fishes (species occurring 200–4000 m deep) such as orange roughy (*Hoplostethus atlanticus*) range widely through most of the year but make seasonal gametic migrations hundreds of kilometers to discrete spawning sites. Similarly, many large mesopelagic fishes that live most of the year on the lower continental slope migrate to bathymetric features such as the shelf/slope break to spawn (e.g., the gempylid, Thyrsites atun).

Littoral Migrations

Littoral migrations specifically refer to those occurring in the littoral (i.e., intertidal) zone; however, the term is often applied to any migrations occurring between inshore and offshore or between shallow and deep waters. By this expanded use, littoral migrations may occur in freshwater as well as marine habitats and with tidal to seasonal periodicities. Many fishes that inhabit rocky and muddy intertidal shores (e.g., gobies, blennies, and sculpins) migrate with the ebb and flood of tides, exploiting resources that are inaccessible to subtidal competitors while avoiding most marine predators. Thereby, these migrations serve alimentary and refuge functions. Abiotic factors (temperature, salinity, pH, and dissolved oxygen) vary dramatically in littoral habitats, and tidal or seasonal movements (climatic migrations) are often necessary for fishes to remain within physiological tolerances. Littoral migrations serving gametic functions are also common. Many atheriniform fishes (silversides and grunions) migrate inshore to spawn with fortnightly frequency. Spring high tides provide the only access to appropriate habitats for embryonic development. Silversides need access to rooted vegetation in intertidal areas for the attachment of filamentous eggs, while grunions lay eggs in the moist, warm sand above the high-tide line where abrasion is minimized.

Potamodromy

Less than 0.01% of all water on Earth is contained in freshwater lakes, rivers, and streams, yet ~40% of all fishes, nearly 12 000 species, live exclusively in freshwater. The migratory patterns of such a diverse fauna are not well described. Most freshwater ecosystems are strongly seasonal in terms of temperature, pH, and flow (or availability of water); therefore, spawning is also strongly seasonal. Most freshwater fishes have demersal eggs that are typically larger and higher in yolk content than marine fishes, adaptations that prevent loss of offspring to downstream transport and high predation in a spatially limited habitat. Spawning requires migrating to areas that meet specific habitat requirements (e.g., depth, substrate, temperature, and current speed) for the survival and development of eggs. Most known potamodromous migrations are freshwater analogs of anadromy and involve the upstream migration of adults to spawn (e.g., paddlefish, *Polyodon spathula*), an adaptation to the unidirectional flow of most freshwater environments. However, lateral migrations (littoral analogs) are also common, as many freshwater fishes (e.g., sunfishes, family Centrarchidae) migrate seasonally into shallow waters to tend nests for spawning. Flood plains that are inundated seasonally also provide critical spawning habitats for riverine species, for example, alligator gar (*Atractosteus spatula*) and Nile perch (*Lates niloticus*), providing areas of high productivity and forage for postlarvae and/or juveniles while affording protection from aquatic predators and damaging currents.

Vertical Migrations

Vertical migrations have received considerable attention by biological oceanographers but little consideration in discussions of fish migration. This is largely because the most dramatic vertical migrations correspond to tidal and diel cycles, and thus, may be considered commutes rather than true migrations. Considering that vertical movements are frequently called 'migrations' and often include changes in physical environments (temperature, depth, light, and pressure) equivalent to extreme horizontal migrations (e.g., between tropic and polar latitudes), they represent an important category of fish migration. Vertical migrations are commonly alimentary, exhibited with diel periodicity in freshwater (African cichlids), marine (cod, haddock, and herring), and diadromous (clupeids and juvenile salmonids) species. Fishes of mesopelagic boundary communities (MBC) in the open ocean are perhaps the best known. These communities, which include many taxa, spend daylight hours 500–1000 m deep, but migrate en masse to 0–200 m depths at sunset where they remain until sunrise when they return to deeper habitats. These migrations may be pelagic (in the water column) or demersal (following the bottom slope). The upward migration allows access to plankton concentrated in the photic zone and thermocline. Mesopelagic fishes that vertically migrate have higher metabolic rates, necessitating higher daily rations, than nonmigratory taxa. The return

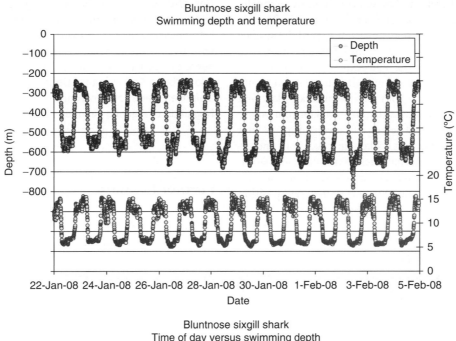

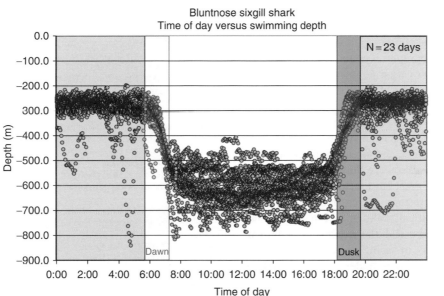

Figure 4 Diel vertical migration of an adult bluntnose sixgill shark (*Hexanchus griseus*) near Hawaii, as revealed by an electronic tag. The top chart shows the regimented vertical patterns in the swimming depth and temperature of the shark over a 14-day period. The bottom graph shows the vertical position of the shark pooled over a 23-day period relative to dawn and dusk. The sixgill shark spent daytime hours between 500 and 800 m deep, but followed the insular slope to depths of 200–300 m at sunset. Before sunrise, the shark returned to the deeper daytime depths. This diel migration is coincident with the movements of the mesopelagic boundary community (MBC). The sharks occupy a trophic position two or three levels higher than most of the MBC taxa. Direct predators of the MBC often have similar patterns of vertical movements. It is hypothesized that predators of MBC taxa are prey for the sixgill sharks. This migration thereby serves and trophic function. These data are from ongoing research by R. D. Grubbs to elucidate the life history and ecology of this and other deep sea elasmobranchs.

migration (deeper at sunrise) is linked to decreased predation rates and slower metabolism in darker, colder waters. Many mesopelagic fishes possess bioluminescent photophores which emit light of similar wavelength to downwelling light, thereby obliterating silhouettes. This mechanism of counter-illumination is a direct adaptation for vertical migration. Often, multiple trophic levels of

pelagic and demersal fishes undergo diel vertical migrations coincident or opposite those of the MBC (**Figure 4**).

Selective tidal stream transport (STST) is a specialized and highly evolved form of vertical migration exhibited by larvae and postsettlement juveniles of many coastal and estuarine species. These small fishes seek refuge and forage inshore or upstream but are unable to swim against tidal currents. Therefore, they selectively move vertically to harness tidal energy. The fish are demersal during ebb, seeking to maintain position, but swim toward the surface during flood where currents transport them upstream. The results are dramatic, and explain how relatively young (~1 month) larvae and juveniles in the drum family traverse brackish waters with net downstream flow and arrive in freshwater habitats that are 300 km from where they were spawned. The exact mechanism is not completely understood, but it likely involves both endogenous rhythms and external cues. A similar mechanism exists in some open ocean taxa that selectively use surface currents and deeper countercurrents to maintain position over productive habitats such as seamounts.

Concluding Remarks

The great diversity of fishes and the parallel diversity of migratory patterns present a fascinating and massive challenge for the conservation of biodiversity and the sustainable management of human food resources of great significance. Traditional conservation and resource management draw fixed geographical boundaries and develop benchmarks based upon closed populations, whereas the complex life history patterns of fishes demand a spatially explicit and sometimes adaptive approach, especially when the ecological boundaries are variable or unclear. In some cases, critical habitats may be occupied only briefly and at varying times, and for a majority of species, the major aspects of migration are simply unknown. Fisheries science has made great advances (albeit with only a few species) to deal with these issues, but they are often short-circuited by political realities. Given the driving forces of climate change and the expansion of human population, our success in facing these challenges will depend largely on our ability to understand how complex spatial life histories can adapt and evolve.

See also: Behavioral Endocrinology of Migration; Conservation and Animal Behavior; Evolution: Fundamentals; Fish Social Learning; Habitat Imprinting; Habitat Selection; Life Histories and Predation Risk; Migratory Connectivity; Optimal Foraging Theory: Introduction; Risk Allocation in Anti-Predator Behavior; Risk-Taking in Self-Defense; Trade-Offs in Anti-Predator Behavior; Vertical Migration of Aquatic Animals.

Further Reading

Dingle H and Drake VA (2007) What is migration? *Bioscience* 57: 113–121.

Gross MR (1991) Salmon breeding-behavior and life-history evolution in changing environments. *Ecology* 72: 1180–1186.

Harden-Jones FR (1968) *Fish Migrations*. London: Edward Arnold.

Heape W (1931) *Emigration, Migration and Nomadism*. Cambridge: W. Heffer.

McDowall RM (1988) *Diadromy in Fishes: Migrations Between Freshwater and Marine Environments*. Portland: Timber Press.

McDowall RM (2007) On amphidromy, a distinct form of diadromy in aquatic organisms. *Fish and Fisheries* 8: 1–13.

Meek A (1916) *The Migrations of Fish*. London: Edward Arnold.

Myers G (1949) Usage of anadromous, catadromous and allied terms for migratory fishes. *Copeia* 1949: 89–97.

Pearre S, Jr (2003) Eat and run? The hungersatiation hypothesis in vertical migration: History, evidence and consequences. *Biological Review* 78: 1–79.

Quinn TJ (2005) *The Behavior and Ecology of Pacific Salmon and Trout*. Seattle, WA: University of Washington Press.

Rooker JR, Alvarado-Bremer JR, Block BA, et al. (2007) Life history and stock structure of Atlantic bluefin tuna (*Thunnus thynnus*). *Reviews in Fisheries Science* 15: 265–310.

Secor DH (1999) Specifying divergent migrations in the concept of stock: The contingent hypothesis. *Fisheries Research* 43: 13–34.

Sinclair M (1988) *Marine Populations: An Essay on Population Regulation and Speciation*. Seattle, WA: University of Washington Press.

Tesch F-W (2003) *The Eel*. Oxford: Blackwell Science.

Werner EE and Gilliam JF (1984) The ontogenetic niche and species interactions in size-structured populations. *Annual Review of Ecology and Systematics* 15: 393–425.

Fish Social Learning

J.-G. J. Godin, Carleton University, Ottawa, ON, Canada

Introduction

To make appropriate decisions regarding where to live and reproduce, where to forage and which foods to eat, which potential predators to avoid and how to avoid them, and with whom to mate, animals need information about their alternatives. Individuals can use either personal (asocial) information that they acquire directly through experience with their environment or they can use social (public) information produced by other animals. Acquisition and use of public information can lead to social learning, defined as any process whereby the behavior of an individual (an observer) is altered as a result of the observer either observing the behavior of another individual (a demonstrator or model), or interacting with the demonstrator, or being exposed to its products.

In theory, whether individuals acquire and use asocial or social sources of information to make adaptive behavioral decisions depends on their relative availability, associated benefits and costs, and on whether individual and social learning conflict with one another.

Here, I provide an overview of social learning in fishes, some of the behavioral contexts and circumstances under which social learning occurs, and some of its evolutionary consequences. For more comprehensive recent reviews of social learning and related behavioral phenomena in fishes, the reader is referred to Brown and Laland (2003), Kendal et al. (2005), Brown et al. (2006).

Do Group Living and Social Networks Facilitate Social Learning?

Group living is ubiquitous among fishes. Social groups among fish are referred to as shoals or schools, nonrandom social associations of individuals commonly assorted by species, body length, sex, and (or) parasite load. Living in a social group can confer a number of benefits to individuals, including increased foraging efficiency and reduced risk of predation. However, life in social groups also has potential costs, such as increased competition for resources and risk of disease. Underlying the benefits associated with shoaling and schooling is the inadvertent social transmission and sharing of public information among individual members of a group about features of the external environment, including the movements of near neighbors.

Further, it has been shown recently in a couple of fish species (guppy, *Poecilia reticulata*; threespine stickleback, *Gasterosteus aculeatus*) that social networks exist within populations. Such networks may be formed by preferential social associations and repeated behavioral interactions between certain individuals in a population. Social networks can persist over extended periods. Both group living in general and social networks in particular are likely to facilitate social learning. However, neither is necessary for social learning to occur. Nonetheless, because fishes commonly live in social groups, possess the cognitive ability to recognize and remember the identity of other fish, and often preferentially associate with familiar individuals, we should expect social learning to be prevalent in fishes.

Migration and Habitat Choice

Most, if not all, fish species undergo a number of adaptive habitat changes during their lifetimes. Such habitat changes are generally achieved though directional movements (migrations) of individuals varying widely in distance traversed, orientation, and time scale. Such movements or migrations commonly occur in shoals or schools, which, as previously noted, provide opportunities for social learning to occur. Habitat choice, as an outcome of migratory movements, may be facilitated through social learning when individuals copy the observed prior habitat choice of others.

Using a generic agent-based model of grouping behavior, Couzin et al. recently showed that informed virtual individuals that had been programmed to prefer to move in a particular direction within mobile groups could influence the directional movements of naïve individuals within the group. As a result, and as is commonly observed in real fish schools in the wild, the entire group moved cohesively in the same direction. This rapid social transmission of information about preferred direction from informed individuals to neighboring naïve individuals within the group occurred without any overt signaling and without knowledge among group members about which individuals were informed.

Consistent with the results of Couzin et al.'s model is the observation from some laboratory studies that individual fish in aquaria can be trained to swim along specific routes to feeders or to specific locations at specific times of day to obtain food rewards, and that these informed individuals subsequently lead naïve individuals in their shoal to location where food is available. Initially-naïve observer fish learned a specific route to a foraging patch or to travel to a specific location to feed by observing the behavior of informed demonstrators. The information concerning movement was socially transmitted among shoal members, and the learned behavior was subsequently maintained in the absence of the trained demonstrator fish. If such conformity of movement behavior among shoal members is sufficiently strong and maintained for sufficient time through social learning, then as illustrated below, it may favor the evolution of local behavioral traditions.

The strongest evidence for a role of social learning in fish migration in nature comes from two field studies with coral-reef fishes. French grunts (*Haemulon flavolineatum*) in the Virgin Islands form daytime resting shoals at specific sites on coral reefs. At dusk and dawn, shoals migrate along regular routes from their resting sites to distant foraging sites. Juvenile grunts occasionally follow older and presumably more informed individuals into such foraging groups and, in doing so, can socially learn the specific migratory route of the group that they have joined. Helfman and colleagues transplanted juvenile grunts from other sites into resting shoals. They then allowed the transplants to follow their foster shoals along their specific migratory routes for 2 days. Next, they removed all the original members of the foster shoal from the reef. The transplanted juveniles continued daily use of the migratory routes of their foster shoal and returned to the reef resting site that the foster shoal had used. Juvenile grunts in a control treatment were transplanted to reef sites from which resting resident shoals had already been removed. These transplanted juveniles were thus not provided the opportunity to learn from a foster shoal. They did not adopt the latter's migratory routes or resting sites. Rather, they continued to use migratory routes appropriate to their original home resting site.

In a second field study, Warner found that blue head wrasse (*Thalassoma bifasciatum*) in Panama migrate to specific sites on reefs to spawn that remain constant over several generations. To determine whether social learning maintained the stable mating sites, Warner experimentally removed all wrasses from some reefs and replaced them with individuals collected at a different location. The transplanted wrasses rapidly established mating sites on their new home reefs different from those used by the original population and used these new sites consistently over several generations. Presumably, female wrasses learn the locations of mating sites from other, more experienced female conspecifics.

These two field studies provide compelling experimental evidence for a role of social learning in migration in general and specifically in the establishment and maintenance of local behavioral traditions or cultures in natural fish populations. However, whether social learning is implicated in the more spectacular long-distance, annual migrations of other fishes, such as salmon and eels, remains unknown and open for future investigation.

Antipredator Behavior

Fishes exhibit a wide range of behaviors in response to predators, including avoidance of dangerous habitats, immobility, hiding, fleeing, and shoaling. Selection for individuals to respond appropriately to predators is strong, because failure to do so is likely to result in death. Consequently, it has long been thought that fishes should inherently recognize and respond to their natural predators, and numerous studies on different fish species support this contention. However, an increasing body of research reveals that individual fish can assess the local risk of predation and adjust their antipredator behavior to the perceived level of threat. Evidence for a direct role of social learning in the development of such antipredator behavior in fishes is limited but accumulating.

Individual fish may learn about the presence or identity of a predator by observing the behavior of nearby conspecifics or heterospecifics without having directly experienced the predator themselves. For example, many fishes can individually learn to recognize and respond appropriately to a novel predator by associating visual or odor cues emitted by a predator with chemical alarm cues released from the skin of other fishes when they are frightened, injured, or captured by a predator. The antipredator response to the detection of alarm cues is referred to as a fright response. It has been shown experimentally that acquired fright responses of one or more individuals can rapidly be socially transmitted to nearby naïve fish, who in turn behave similarly without having either seen the predator or directly detected either its odor or alarm cues released by other fishes. Presumably, observer fish learn socially about the presence and identity of a predator and the level of risk it poses by associating the fright response of other fishes with the predator, and subsequently avoid any stimulus that elicits fright responses in others. Whether such socially-learned antipredator behavior represents an example of observational conditioning or some other underlying process remains uncertain.

There is little evidence that fishes learn socially about specific antipredator tactics (e.g., fleeing, hiding) by observing others. A notable exception is a recent study showing that initially naïve guppies learn an appropriate escape route in response to a simulated predation threat by observing and following demonstrator fish that have been trained previously to use a particular route. The observer guppies continued to use the learned escape route even after removal of their demonstrators and exhibited increased efficiency at escaping.

Foraging Behavior

Actively foraging animals are faced with the tasks of finding patchily-distributed food, deciding which patches to forage in, when to leave a food patch, and which prey to eat within a patch. Social learning can play a role both in finding food and in patch choice of fishes. However, surprisingly little information is available on whether social learning influences prey selection.

Early studies on fish shoaling behavior revealed that shoals of fish locate patchily-distributed food faster than do individuals. Once an individual in a group finds a food patch and begins foraging, other members of the group are quickly attracted to it by observing the foraging behavior of the finder. This phenomenon has been referred to as 'forage area copying,' which is probably a form of local enhancement. Moreover, there is evidence that individual foraging rates are enhanced when neighboring fish can see one another forage, suggesting social enhancement of foraging performance. Alternatively, enhanced foraging rates could result from per capita reduction in time spent in antipredator vigilance, and concomitant increase in time available for foraging, that commonly occur with increasing group size.

Theory suggests that use of public information in making foraging decisions should become more likely as either the costs associated with acquiring and using personal (asocial) information or the degree of uncertainty about food resources increases. Uncertainty may result from either lack of information, unreliable personal information or outdated personal information about food resources in the environment. Accumulating evidence, particularly from the research of Laland and his colleagues, reveals that fishes will preferentially use public information, and thus learn socially, under both of these circumstances (i.e., cost and uncertainty). For example, vulnerability to predation while foraging can increase the propensity to use social information. More generally, use of public information may depend on the costs associated with acquiring personal information. Threespine sticklebacks are better protected against predators than are ninespine sticklebacks (*Pungitius pungitius*), because threespines possess more extensive body armour. Accordingly, threespines are more willing than ninespines to swim in open water to sample for themselves the profitability of available food patches; ninespines are more likely to remain in the relative safety of vegetative cover and rather let other fish sample food patches and than use their public information about the relative profitability of the patches when choosing a patch. Using public information may be a less costly strategy than direct sampling for individuals that are relatively vulnerable to predation. Because personal information is generally more current and reliable than public information, all else being equal, individuals that are not particularly vulnerable to predators should prefer to use personal information to evaluate foraging patches.

Exploitation of public information about food resources is also expected, when foragers are uncertain about the relative quality of available food patches in the environment. In a recent study, van Bergen and colleagues experimentally manipulated both the reliability and recency of personal information regarding the profitability of two food patches that they presented to ninespine sticklebacks in an aquarium. As expected from theory, focal fish that had reliable, current personal information obtained through their own foraging ignored the observed foraging success of demonstrator fish at the feeders, but switched to using public information when their own personal information was unreliable or outdated. Individual fish do not indiscriminately copy the foraging behavior of others. Rather, they continually assess the relative costs and benefits of social and asocial learning and select the appropriate strategy.

Mating Behavior

Competition for mates, assortative (i.e., nonrandom) mating and mate preferences are widespread in fishes. Female mate choice is most common, but mutual mate choice and male choice also occur. Most evolutionary models of sexual selection assume that individual mating preferences are genetically based and inherited and that individuals choose mates independently of one another. However, considerable evidence from a wide range of taxa, including fishes, reveals both that the mate preferences of individuals can be flexible and that social experiences can influence mate-choice decisions.

Because mating is often a social phenomenon in vertebrates, public information associated with mating activities is readily available. Individuals can acquire social information from conspecifics about potential mates that can influence their subsequent choice of mates, as illustrated in the following section.

Mate-Choice Copying

Mate-choice copying is a form of nonindependent mate choice resulting from social learning, in which an individual gains information about potential mates by observing courtship and mating behaviors of nearby conspecifics. Mate choice copying is considered to have occurred if a focal individual's observation of a sexual interaction between a male and a female increases its likelihood of subsequently preferring the individual observed mating.

Questions as to (1) the advantages to an individual of copying the mate choice of another rather than assessing and choosing a mate based on personal information, (2) which individuals should be copied, and (3) when to copy and when to rely on personal experience have guided much recent research.

Potential benefits of mate-choice copying include a reduction in any costs associated with the search for and assessment of potential mates and an increase in the accuracy of mate assessment, particularly when the relative quality of potential mates is difficult to determine. Putative costs of mate-choice copying include: (1) acquisition of inaccurate or outdated information about potential mates from a demonstrator, (2) a risk of reduced fertility by mating with an individual who has recently mated with another, either through sperm depletion of males or sperm competition in females, and (3) increased risk of predation from spending time in the vicinity of a consorting pair whose behavior may attract the attention of predators. Reliance on social information about mates should theoretically be favored, when costs associated with independent mate choice are high and when discrimination between potential mates is uncertain or difficult.

A number of theoretical models have shown that, under certain conditions, a strategy of mate-choice copying can invade and be maintained in a population. Mate-choice copying should be favored and, therefore, most common in nonresource-based, polygynous/promiscuous mating systems, where some individuals have many mates and others few mates and the choosy sex (usually female) gains only gametes from the chosen sex (usually male). Evolutionary models have shown that mate-choice copying can have important implications for biological evolution. Copying the mate choice of others may increase variance in mating success in the population, increasing the probability that more matings are achieved by fewer individuals, and thus may influence the opportunity for sexual selection and the evolution of the traits preferred by the choosy sex. Further, depending on conditions, copying can either favor or constrain the spread of a novel trait in a population.

To date, mate-choice copying has been documented in at least eight species of fish, all of which exhibit a polygynous or promiscuous mating system. The most extensive evidence comes from research on the guppy and sailfin molly (*Poecilia latipinna*). The first strong experimental evidence for mate-choice copying came from Dugatkin's (1992) laboratory study of the guppy, a small poeciliid fish species native to Trinidad and adjacent islands. Using guppies descended from a natural population living in the Turure River in Trinidad, Dugatkin first showed that focal adult female guppies preferred to affiliate with males that they had previously observed courting a nearby female (to another male of similar body length and coloration seen alone). This initial result could have been explained by a number of behavioral mechanisms other than mate-choice copying. However, additional experiments systematically eliminated these alternative explanations for his initial findings.

Subsequent to Dugatkin's initial study with Turure River guppies, and using variations on his original experimental design, several studies have demonstrated mate-choice copying in guppies descended from two other natural populations in Trinidad, and some of the conditions under which it occurs have been elucidated. Researchers working with other fish species, particularly the sailfin molly, also have observed mate-choice copying. However, experiments using feral or guppies obtained from pet shops have found no evidence of mate-choice copying, and some species with resource-based mating systems, such as the threespine stickleback, may not copy the mate choices of others when given the opportunity to do so.

Despite the evidence for mate-choice copying in a number fish species, knowledge of both the fitness-related benefits and costs of copying to an individual and the prevalence of mate-choice copying in nature remains very limited. Indeed, there is no evidence in fishes unambiguously demonstrating that mate-choice copying increases the fitness of the copier. In theory, copying should benefit an individual if it increases the reliability of information about potential mates or an individual's ability to discriminate between them, thus reducing uncertainty about which potential mate to choose. Consistent with this proposition, female guppies are more likely to copy the mate choice of a nearby female when the difference in body coloration and body length of potential mates being assessed is small. When phenotypic differences between males are large, females do not copy. Rather, they choose males based on genetically-based preferences for more colorful and larger males. Further, the tendency to copy can be influenced by the amount or nature of the information gained from observing a sexual interaction between nearby males and females. Young female guppies are more likely to copy the mate choices of older model females, perhaps because the former are likely to be more experienced and, therefore, better able to assess male quality than younger females. In both guppies and sailfin mollies, focal females are more likely to copy the mate choices of conspecific demonstrator females, when they observe two rather than one demonstrator female interacting sexually with a particular male and when they observe sexual interactions between a demonstrator female and a male over long rather than short periods.

A further proposed benefit of copying the mate choice of others is a reduction in the costs associated with directly searching for and assessing potential mates. To date, only two studies, both with guppies, have investigated this putative benefit. Both failed to provide support for it. Neither experimentally varying the level of predation threat nor the hunger level of focal females increased reliance on public information. Clearly, more research is needed concerning this possible advantage of mate-choice copying.

Because alternative mechanisms can generate a mating pattern similar to that of copying, investigating the occurrence of mate-choice copying behavior in free-ranging fishes is difficult. Consequently, knowledge of mate-choice copying in fishes in the wild is currently limited to four experimental field studies, two on river-dwelling species (sailfin molly, guppy) and two on marine reef species (whitebelly damselfish, *Amblyglyphidodon leucogaster*, ocellated wrasse, *Symphodus ocellatus*). All four studies report results that are consistent with mate-choice copying. Additional field studies on other species are required to determine the prevalence of use of social information in making mate-choice decisions.

Social Eavesdropping on Male–Male Competitive Contests

Females not only copy the mate choices of nearby females, they can also gain at little cost information about the quality of potential mates by observing, or eavesdropping on, aggressive interactions between competing males and use this public information to make mate choices. Such social eavesdropping occurs when a focal individual, an eavesdropper, extracts social information from observing a signaling interaction between others in which the focal animal is taking no direct part. Because a male's ability to fight is a reliable indicator of his quality, eavesdropping females that are biased towards winners of aggressive encounters between males should increase the quality of the males with whom they mate. After observing two males interact aggressively in the laboratory, eavesdropping female Siamese fighting fish (*Betta splendens*) preferred to affiliate with the winner; focal females that were not allowed to eavesdrop on the aggressive interaction did not. A similar finding has been reported for male pipefish (*Syngnathus typhle*), a sex-role reversed species in which females are more aggressive than males and males are more parental than females. Male pipefish that observed display contests between females and then chose between contestants as potential mates preferred the more competitive females.

Prospects

Although individual learning in fishes is well established, experimental evidence for social learning in this taxon has accumulated only during the past two decades. Most work has been focused on the use of social information when foraging and mating, is based on just a few species, and is largely restricted to laboratory studies. Relatively little is known about the possible role of social learning in the development of antipredatory, migratory, habitat selection, communication, cooperative, and aggressive behaviors, or of the prevalence of social learning in natural populations.

Nonetheless, informed by theoretical models, there has been much progress in identifying conditions under which the use of social information and social learning is favored over the use of personal information in decision making in fishes. In brief, available evidence suggests that fishes prefer to use personally-acquired information, but will switch to acquiring and using social information when asocial learning is costly or when they are relatively uncertain about what to do. Because fish commonly live in shoals or schools and possess sensitive lateral line, visual and chemosensory systems, public information can be rapidly transmitted among members of a group, facilitating social learning.

Fishes constitute the most species-rich and ecologically-diverse group of vertebrates. Consequently, they are an ideal taxon in which to investigate the ecological conditions that favor the use of social learning. Such comparative studies should provide information bearing on the question of whether within-species facultative reliance on social information is an adaptive response to the demands of particular ecologies.

See also: Avian Social Learning; Culture; Imitation: Cognitive Implications; Mammalian Social Learning: Non-Primates; Social Learning: Theory.

Further Reading

Brown GE (2003) Learning about danger: Chemical alarm cues and local risk assessment in prey fishes. *Fish and Fisheries* 4: 227–234.

Brown C and Laland KN (2003) Social learning in fishes: A review. *Fish and Fisheries* 4: 280–288.

Brown C and Laland K (2006) Social learning in fishes. In: Brown C, Laland K, and Krause J (eds.) *Fish Cognition and Behavior*, pp. 186–202. Oxford: Blackwell.

Brown C, Laland K, and Krause J (eds.) (2006) *Fish Cognition and Behavior*. Oxford: Blackwell.

Couzin ID, James R, Mawdsley D, Croft DP, and Krause J (2006) Social organization and information transfer in schooling fishes. In: Brown C, Laland K, and Krause J (eds.) *Fish Cognition and Behavior*, pp. 166–185. Oxford: Blackwell.

Danchin E, Giraldeau L-A, Valone TJ, and Wagner RH (2004) Public information: From nosy neighbors to cultural evolution. *Science* 305: 487–491.

Dugatkin LA (1992) Sexual selection and imitation: Females copy the mate choice of others. *American Naturalist* 139: 1384–1389.

Godin J-GJ (ed.) (1997) *Behavioural Ecology of Teleost Fishes*. Oxford: Oxford University Press.

Kelley JL and Magurran AE (2003) Learned predator recognition and antipredator responses in fishes. *Fish and Fisheries* 4: 216–226.

Kendal RL, Coolen I, van Bergen Y, and Laland KN (2005) Tradeoffs in the adaptive use of social and asocial learning. *Advances in the Study of Behavior* 35: 333–380.

Krause J and Ruxton GD (2002) *Living in Groups*. Oxford: Oxford University Press.

Laland K (2008) Animal cultures. *Current Biology* 18: R366–R370.

Valone TJ (2007) From eavesdropping on performance to copying the behavior of others: A review of public information use. *Behavioral Ecology and Sociobiology* 62: 1–14.

Witte K (2006) Learning and mate choice. In: Brown C, Laland K, and Krause J (eds.) *Fish Cognition and Behavior*, pp. 70–95. Oxford: Blackwell Publishing Ltd.

Flexible Mate Choice

M. Ah-King, University of California, Los Angeles, CA, USA

Introduction

Traditionally, investigators and theorists have supposed that mate choice is directional and fixed within a species as well as static within individuals over time, so that the most extreme expression of a sexually selected trait should always be preferred. However, mate choice is dynamic and can select simultaneously for elaborate traits in the opposite sex, and also for behavioral displays and genital morphology. Most investigators of mate choice have worked on female choice, since females are regarded as the choosier sex. However, mate choice occurs in both sexes and this article includes examples of both. Males can benefit from choosing, either when mating is costly (sperm limitation, high predation risk, absence of competition, temporally high chances of fathering offspring) or when there is any variance in female reproductive output. In many species with indeterminate growth, males prefer larger females. There are also examples of species in which both females and males manifest mate choice simultaneously (mutual mate choice), as in *Drosophila pseudoobscura*, and broad-nosed pipefish (*Syngnathus typhle*). In this pipefish, females compete with each other for males that brood the embryos in foldings on their abdomen. Males are smaller than females and both sexes mate with multiple mates. Males prefer large, dominant, and ornamented females and females prefer large males with thick brood pouches. Mutual mate choice also occurs in, for example, the gregarious cockroach (*Blattella germanica*) and the three-spined stickleback (*Gasterosteus aculeatus*). Importantly, which sex is predominantly competitive does not determine whether that sex is discriminating when it comes to choosing mates, as competition and mate choice are not mutually exclusive processes. It is also possible, theoretically at least, that all individuals assess potential mates before acting as though they are choosy (rejecting some potential mates) or indiscriminate (accepting all potential mates as encountered) as posited by Gowaty and Hubbell (see section 'Theory').

Previously, there has been a strong tendency to assume that mate preferences are stable for a certain species and that an individual's preferences or assessments are consistent over time. Methods for predicting mate choice have often focused on the population level, for example, Potential Reproductive Rate Theory and Operational Sex Ratio Theory, and thus do not take individual variation into account. In such cases, investigators interpret individual variability in mate choice as statistical error. However, recent discoveries have revealed large variation in individual mate choice behavior in insects, birds, amphibians, and fish (**Table 1**). Mating preferences, thus, are flexible and change according to ecological conditions, in relation to social interactions and the state of the choosing individual, and experimental studies in flies, mice, and other species show that mate preferences correlate positively with offspring viability and the number of offspring surviving to reproductive age.

History

The first indications that mate choice might be flexible in response to environmental conditions came from studies of guppies. In 1970s, male guppies (*Poecilia reticulata*) from rivers with high predation pressure were found to have fewer and smaller color spots than those from low predation sites. Therefore, investigators suggested that male color patterns are the result of balancing selection pressures from female choice for brightly colored males and predation selecting for crypsis. This idea was later supported experimentally; female guppies change their mate choice behavior in response to perceived predation risk. These results are consistent also with other models of flexible mate choice behavior (see section 'Theory').

Theory

In a review in 1997, Jennions and Petrie pointed out the lack of studies of variation in female mate choice. They summarized theoretical and empirical studies of variation in mating preferences and put forward a theoretical framework for exploring variation in female mate choice (see **Figure 1**). Jennions and Petrie distinguished two aspects of mate choice, preference function, that is, the order in which potential mates are ranked, and choosiness, the willingness to invest time and energy in mate choice (assuming that mate assessment is costly in contrast to Gowaty and Hubbell's model, see below in this section). Animals may have innate preferences, but these preferences are not always realized in their choice of mate. Thus, variation between individuals in either preference function or choosiness results in flexible mate choice. Jennions and Petrie's review inspired a growing field of studies in variation of female mate choice. Lately, evidence of individually flexible mate choice is accumulating rapidly.

Changes in the environment favor flexibility in all kinds of behavior and should do so also in mate choice.

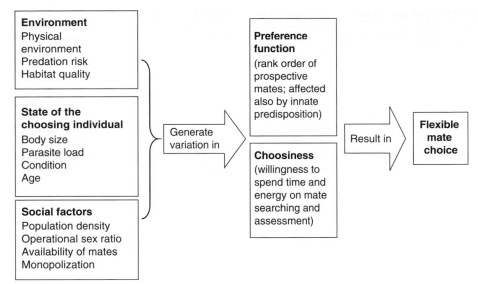

Figure 1 Factors that give rise to variation in mate choice according to the model by Jennions and Petrie (1997). Individuals can vary in mate choice because of differences in preference function (the order in which they rank potential mates) and choosiness (their willingness to invest in mate search) which result in flexible mate choice.

Indeed, models by Patricia Gowaty and Steve Hubbell show that fitness is enhanced when individuals are able to change from choosy to indiscriminate behavior dynamically as environmental and social conditions change. This model predicts that individuals, regardless of sex, should adjust mate choice dynamically and moment by moment to changing environmental and social conditions. Their model provides a justification for the idea that all individuals regardless of their sex assess likely fitness rewards from mating with alternative potential partners before expressing what we usually call choosy or indiscriminate behavior. In nature and in the laboratory, what we see are individuals that accept all or reject some. Gowaty and Hubbell's model says that even individuals that accept all base their decision on assessments of fitness rewards.

Mate choice decisions vary over different time scales, throughout the breeding season and over an individual's lifespan, thus, time constraints are important. As the need for breeding becomes more urgent, for example, when an individual comes closer to the end of the breeding season, the threshold for acceptable mates decreases to increase the number of possible mates. Gowaty and Hubbell's model is a theorem that predicts that individuals should accept or reject potential mates in response to variation in their survival probabilities, encounter rates, time before an individual can remate (latency), and distribution of fitness that would be conferred from mating with each potential opposite sex mate in the population (see **Figure** 2). An individual is expected to assess fitness differences of mating with potential mates and the time it has left to reproduce and respond adaptively to variable environmental cues. Thus, an individual should accept more potential mates (i.e., become less choosy) when it experiences

(1) lower survival probability (e.g., increased predation risk or enhanced parasite load), or (2) decreased encounter rates with potential mates (e.g., by reduced population density or increased competition), or (3) decreased latency (e.g., shortened reproductive rate), or (4) if the distribution of fitnesses conferred is more right-skewed (so that a larger proportion of potential mates in the population result in high fitness) (see **Figure** 2). The distribution of conferred fitnesses for all individuals in a population can be, for example, left-skewed (if many combinations result in low fitness) or right-skewed (if many combinations result in high fitness). This model assumes that individuals assess the fitness that would result from mating with potential mates. Furthermore, chance is important because, when potential mates or competitors face catastrophes, otherwise leave the population, or enter latency, encounter rates and survival probabilities are affected. Using this model, chance effects on fitness outcomes can also be discerned from other effects on variability in mate choice. Therefore, evolution of flexible mate choice is expected to occur through the evolution of sensitivity to environmental cues, the ability to assess potential mates, and the response so that mate choice can be adjusted adaptively.

Importantly, whether an individual gets to mate with its preferred mate or not affects offspring viability. When females and males get to mate with their preferred partners, their offspring have higher viability than when mated with potential mates they did not prefer. This viability enhancement has been shown in a number of species, in female grasshoppers (*Gryllus bimaculatus*) and both sexes in mice and fruit flies. These results suggest mate choice for adaptive gene combinations.

Table 1   Studies that have observed flexible mating behavior

Factor	Species, scientific name
Environmental factors	
Increased predation risk	*Females*
	Crickets, *Gryllus integer*
	Water striders, *Aquarius remigis*
	Sand gobies, *Pomatoschistus minutus*
	Guppies, *Poecilia reticulata*
	Tungara frogs, *Physalaemus pustulosus*
	Green swordtails, *Xiphophorus helleri*
	Males
	Broad-nosed pipefish, *Syngnathus typhle*
Habitat quality	*Females*
	Cockroach, *Nauphoeta cinerea*
	Males
	Cockroach, *Nauphoeta cinerea*
Demographic factors	
Increased density of opposite-sex conspecifics	*Females*
	Fruitfly, *Drosophila melanogaster*
	Guppies, *Poecilia reticulata*
	Butterflies, *Acraca encedon*
	Pill bugs, *Armadillidium vulgare*
	Males
	Pipefish, *Syngnathus typhle*
	Katydids
Changing OSR	*Females*
	Two-spotted gobies, *Gobiusculus flavescens*
	Katydids
	Males
	Two-spotted gobies, *G. flavescens*
	Broad-nosed pipefish, *Syngnathus typhle*
	Katydids
Increased guarding or territoriality	*Females*
	Mosquito fish, *Gambusia holbrooki*
	Eurasian dotterel, *Charadrius morinellus*
Increased territory homogeneity among males	*Females*
	Beaugregory damselfish, *Stegastes leucostictus*
State of the choosing individual	
Increased parasite load of the chooser	*Females*
	Upland bullies, *Gobiomorphus breviceps*
	Calopterygid damselfly, *Calopteryx haemorrhoidalis*
Increased age of chooser	*Females*
	House crickets, *Acheta domesticus*
	Tanzanian cockroaches, *Nauphoeta cinerea*
	Guppies, *Poecilia reticulata*

Continued

Table 1   Continued

Factor	Species, scientific name
Increased body condition of chooser	*Females*
	Wolf spider, *Schizocosa*
	Males
	Two-spotted goby, *Gobiusculus flavescens*
	Sail-fin molly, *Poecilia latipinna*
Relative attractiveness or availability of resources near the chooser	*Males*
	Beaugregory damselfish, *Stegastes leucostictus*
	Threespine sticklebacks, *Gasterosteus aculeatus*
	Common goby, *Pomatoschistus microps*
Experience	*Females*
	Field crickets, *Teleogryllus oceanicus*
	Bark beetles, *Ips pini*
	Males
	Drosophila paulistorum
	Red-sided garter snakes, *Thamnophis sirtalis parietalis*

This article demonstrates that mate choice is often flexible and adjusted to environmental conditions, that is, phenotypically plastic. Mate preferences, like other reproductive decisions, are dynamic and affected by both internal and external factors.

Environmental Factors

Environmental factors affect population density and habitat quality, which in turn affects possibilities for encounters with potential mates. Generally, it has been assumed that mating systems result from the spatial distribution of suitable breeding sites that determine males' opportunities to monopolize females. However, females often mate multiply and male distribution will also affect female sampling costs. Furthermore, physical properties of the environment such as light intensity will affect the efficacy of visual signaling, and water current affects the cost of mate sampling in fish. For example, under certain light conditions or in deeper waters, it is impossible to distinguish between certain colors. Therefore, attractiveness of different phenotypes may differ under different environmental conditions. Hence, variation in environmental conditions results in flexible mate choice.

Predation Risk

Under predation risk, individual survival probabilities decline and searching for mates becomes more costly;

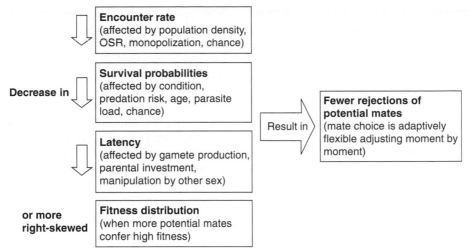

Figure 2 How environmental, social, and intrinsic factors and chance result in adaptively flexible mate choice behavior as modeled by Gowaty and Hubbell (2005, 2009). An individual, regardless of sex, will adjust its mate choice behavior according to experienced variation in encounter rate, survival probability, latency, and fitness distribution conferred from mating with each potential opposite sex mate in the population. A decrease in encounter rate, survival probability or latency or a more right-skewed fitness distribution (more potential mates in the population result in high fitness) is predicted to result in fewer rejections of potential mates. Likewise, an increase in either of these factors or a more left-skewed fitness distribution would result in more rejections. Environmental, social, and intrinsic factors as well as chance cause variation in encounter rate, survival probability, latency and fitness distribution, which result in adaptive flexible mate choice.

for either or both of these reasons, individuals should adjust their behavior. One way to reduce the increased cost under predation risk is to mate at random. By spending less time searching for mates and mating as potential mates are encountered, mate choice under predation risk often results in mating with lower quality mates or mates of more variable quality. For example, female sand gobies (*Pomatoschistus minutus*) prefer large and colorful males in the absence of predators, but become indiscriminate when presented with a visible predator. Likewise, broad-nosed pipefish males choose larger females in the absence of predators. With a predator present, males do not discriminate between large and small females.

Another way to reduce the risk of predation is to change preference to partners that are less conspicuous. This pattern has been shown in green swordtails (*Xiphophorus helleri*) and guppies.

Female green swordtails prefer males with long swords under predator-free conditions, but after having seen a video with a long-sworded male being eaten by a predator, females shifted their preference to males with shorter tails (**Figure 3**).

Responses to predation pressure are also dependent on the evolutionary history of a population. In experiments with guppies collected from two rivers in Trinidad differing in predation pressure, females showed preference for colorful males in the absence of predators. However, when exposed to predation risk, only the females from the high-predation site became indiscriminate with regard to male coloration. This experiment shows that female responses to individual males have evolved in response to predation

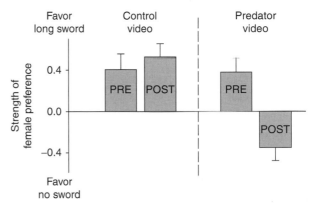

Figure 3 Individual green swordtails (*Xiphophorus helleri*) change their mating preferences from males with long swords to those with short swords after having seen a video with a long-sworded male being eaten by a predator. Reproduced from Johnson JB and Basolo AL (2003) Predator exposure alters female mate choice in the green swordtail. *Behavioral Ecology* 14: 614–625, with permission from Oxford University Press.

pressure. Animals change their mating behavior both in immediate response to predation risk and over evolutionary time as the sensitivity to predation risk differs between populations.

Habitat Quality

Habitat quality affects the distribution of individuals in space and influences the potential for resource acquisition. A poor habitat can also entail decreased survival. Interestingly, high-quality habitats sometimes render animals choosier and sometimes less choosy. A classical example of flexible mate choice due to changes in habitat quality occurs in katydid insects (undescribed species of Zaprochilinae). When pollen resources are scarce, females are dependent on nutrient-rich spermatophores for the production of eggs. Males produce spermatophores and under low resources take longer time to produce them, thus few males are available for mating. Thus, when pollen is scarce, females compete with each other for male partners. But during the season when pollen is abundant, males instead compete for females and females reject more males. Hence, in this case, abundant resources induce females to reject more potential mates and males to accept more potential mates.

Furthermore, mate choice is often based on both male phenotype and the quality of the territory. In heterogeneous habitats, female choice in pied flycatchers (*Ficedula hypoleuca*) was based on territory quality and unrelated to male phenotype. In contrast, in a homogenous environment, females preferred certain plumage patterns. Thus, in heterogeneous environments, large variation in territory quality overshadowed female preference for male plumage pattern.

Demographic Factors

Demographic factors, such as operational sex ratio (OSR), population density, and competition, influence the encounter rate and can result in flexible mate choice (see **Figure 2**).

Population Density and Operational Sex Ratio

When potential mates are abundant, encounter rates are usually high, and little time is lost during searching for a mate. When potential mates are few, or where there are few chances of finding a mate at all so that encounters are unlikely, an individual will gain more fitness by accepting most potential encountered mates. Changes in OSR are known to influence mate choice in a number of species. Male pipefish change their likelihood of accepting or rejecting potential mates in relation to OSR. When OSR is female-biased, males reject many smaller females as mates, but in male-biased OSR, males accept both small and large females as mates.

Two-spotted gobies (*Gobiusculus flavescens*) change the sex that predominantly compete and perform mate choice over the season. At the beginning of each summer, males compete among themselves for access to females, and the females often reject potential mates. At the end of the season, this pattern is reversed, with females competing for males and males often rejecting potential mates. This flexibility corresponds to a change in the adult sex ratio. At the beginning of the summer, there are plenty of males performing courtship to arriving females, but as the

season proceeds, breeding males become scarce, possibly dying. Later in the season, there are more females than males available for mating, and this overturned balance in available mates leads to female–female competition and male mate choice (**Figure 4**).

Mate preferences are adjusted to perceived fitness differences between potential mates, and therefore an individual's previous experience may change mating preferences (see **Figure 2**). Female bark beetles (*Ips pini*) are more prone to mate with intermediate-sized males when first presented with small rather than large males. Similarly, male red-sided garter snakes (*Thamnophis sirtalis parietalis*) adjust their mate choice criteria after exposure to large or small females.

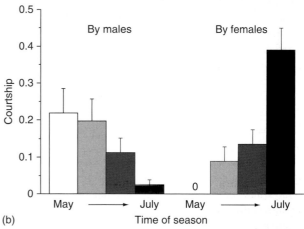

Figure 4 Change in the adult sex ratio can completely change which sex is choosy at the population level. In the two-spotted goby, females are choosy and males competitive in the beginning of the breeding season. Late in the season, males are scarce and choosy while females compete among each other and courtship males. (a) Two-spotted goby pair and (b) propensity to courtship by males and females is reversed over the season. Part (a) with permission from E. Forsgren. Part (b) reproduced from Forsgren E, Amundsen T, Borg AA, and Bjelvenmark J (2004) Unusually dynamic sex roles in a fish. *Nature* 429: 551–554, with permission from Nature Publishing Group.

Competition

Intrasexual competition and monopolization of mates may hinder females and males from exerting their mate preferences. It is well-known that male–male competition can reduce females' encounter rates and thus the fraction of potential mates they find acceptable. In the mosquito fish (*Gambusia holbrooki*), male guarding succeeds in restricting female encounters with other potential mates. Likewise, in the polyandric shorebird Eurasian dotterel (*Charadrius morinellus*), female competition restricts subdominant females to mate with dull males. However, female–female competition is important not only in species in which female competition is more common than male competition, but also in any situation where there is variation in male quality. Moreover, intrasexual competition may also make mate assessment easier.

State of the Choosing Individual

Mate choice is expected to be costly, but the costs have been difficult to measure directly. Instead, experimental manipulations and correlational studies show that individuals discriminate less when costs of mate choice increase. For example, swimming against a current, which is energetically costly, reduces preference for colorful male three-spined sticklebacks (*Gasterosteus aculeatus*).

If an individual is in bad condition, this can lower encounter rates (less energy to search for mates), lower survival probability, or prolong the latency to remating (see **Figure 2**). Individuals in good condition have more time available for mating and reproduction, and therefore their fitness may be favored by rejecting more potential mates. Variation in condition between individuals and within individuals over time will therefore lead to flexibility in mate choice.

Body Size

The prediction that higher quality individuals should reject more potential mates has received some support. In the sail-fin molly (*Poecilia latipinna*), large males are more energetic in courtship and reject more potential mates than small males. Similarly, large male two-spotted gobies accept more colorful females, while small males accept females without regard to female coloration.

Parasite Load

Parasites may decrease the condition and change the behavior of the infected host. Females reject heavily parasitized males, but what happens if the chooser is parasitized? When parasites decrease the body condition of an individual, the individual's survival probability declines, so that it has less time for mating and reproduction, and we would

expect it to accept more potential mates. This behavioral change has been shown in several species of fish.

One example is the upland bully (*Gobiomorphus breviceps*): parasitized females made fewer visits to potential mates before accepting a mate, and often ended up with a mate of smaller size. Likewise, in broad-nosed pipefish, male and female pipefish were infected experimentally with a parasite. Healthy males rejected infected females, whereas infected males accepted more potential mates.

One can interpret these results as being due to reduced choosiness of infected individuals, or due to a shift in infected individuals' thresholds for accepting potential mates because of reduced encounters or reduced survival.

Condition

Just as parasites may reduce an individual's physical condition and thus their survival probabilities, high-quality resources may increase their condition and thus affect the probability of accepting or rejecting potential mates. For example, female black field crickets (*Teleogryllus commodus*) fed on high-protein diet expressed stronger mate preference than those that were fed low-protein diet.

Age

As individuals grow older, their survival probability may decrease, and therefore, they should accept more potential mates as they age. Consistent with this prediction, older female house crickets (*Acheta domesticus*) accept more potential mates, as do Tanzanian cockroaches (*Nauphoeta cinerea*) and guppies (*Poecilia reticulata*).

Implications for Sexual Selection

As we have seen, variation in accepting and rejecting potential mates can result from a number of factors and mechanisms. By studying this variation, we may be able to understand better the variation in sexually selected traits, the maintenance of heritable variation in cues that induce acceptance or rejection of potential mates, and the evolutionary history of preferences and traits.

The effects of flexible acceptance and rejection of potential mates on sexual selection are not straightforward. Adaptive variation in factors affecting acceptance and rejection of potential mates might constrain the evolution of selected characters through a less intense selection pressure that leads to a lower degree of change in selected traits. If different phenotypes are successful breeders during different years, then preferences may change accordingly. One example is that traits correlated with acceptance of potential mates in lark buntings (*Calamosiza melanocorys*) shift between years. Male ornaments that females prefer change dramatically between years.

Additionally, the preferred traits correlate with high nesting success, suggesting that traits serving as fitness indicators switch between years. Adaptive variation, such as shifting between preferred traits, may reduce or even eliminate male trait evolution.

At the same time, condition-dependent mate choice is expected to result in high-quality individuals choosing highly ornamented mates, and the ornamented individuals gain even higher reproductive success by attracting not only more mates, but also mates in better condition. Condition-dependent mate preferences could therefore reinforce linkage disequilibrium between genes for ornament, preference, and condition, and could thereby result in stronger sexual selection.

One example that illustrates the multifaceted effects of flexible mate choice on sexual selection is how predation risk influences both signaling and the accept versus reject behavior of three-spined sticklebacks. Under predation risk, males develop less nuptial coloration, making it harder for females to discriminate among males on the basis of their colors. However, in this case, predation risk also reduces the number of males guarding territories so that only high-quality males are able to hold a territory. Even if females accept more potential mates under predation risk, they may still be able to mate with high-quality males since low-quality males are less prone to build nests.

There are a number of theoretical predictions of when flexible mate choice should be adaptive. Genetic complementarity, for example, choice of mates with complementary major histocompatibility complex (MHC) alleles, results in better resistance against parasites. Mate preferences based on genetic compatibility are often frequency dependent, changing over time, and therefore, they may maintain genetic polymorphism. Phenotypic compatibility might facilitate mating, for example, female red-groined toads (*Uperoleia laevigatga*) choose males that weigh 70% of their own weight. Larger males may be so heavy as to drown them, while smaller males may not provide enough sperm. Therefore the optimal mate choice is related to the female's own body size.

When there is a heritable component of variation in mate choice, it can lead to the evolution of mating preferences. Genetic models expect mating preferences and preferred traits to co-evolve. However, at times there might also be sexually antagonistic evolution, so that selection of a trait in one sex is coupled with deleterious effects of the other. Such sexually antagonistic pleiotropy will preserve variation in genetic quality.

Furthermore, mate choice may also differ between different social, ecological, or genetic contexts. Both males and females often mate multiply and the criteria for acceptance of a social partner and extra-pair partners may differ. For example, female pied flycatchers accept as primary mates males with good territories or an ornament signaling good parental abilities, but when females solicit extra

pair copulations, mate choice may be based on genetic qualities alone. Sometimes, it might even be advantageous to have a social partner of the same sex. In black swans, same-sexed male pairs have higher breeding success, are more aggressive, have larger territories, and share incubation time more evenly than opposite-sex pairs.

Multiple mating also makes cryptic female choice possible. Very little is known about the extent to which females determine the outcome of sperm competition, which also possibly varies with environmental, social, and intrinsic factors.

Quality assortative mating can occur when a low-quality individual expects a high-quality mate to be taken over by higher quality individuals. Assortative mating is thought to be a strong force in speciation and requires variation in acceptance and rejection of potential mates. Assortative mating has even been found to result in sympatric speciation in seahorses. Hence, examining processes leading to variation in mate preferences is important for understanding sexual selection and speciation processes that ultimately generate diversity in nature.

Some Current Questions

Flexible mate choice is a new field of study that has just started. At this moment, studies are accumulating on context- and condition-dependent mate choice. We still have much to learn about determinants of individuals' decisions to accept or reject potential mates, and why different individuals make different decisions. Temporal variation in mate preferences might also be rewarding to look for. Furthermore, mate choice and intrasexual competition interact. Mate choice for partners with good genes may result in offspring with increased fecundity and survival, but also for success in intersexual selection. Future studies will reveal more about these interactions.

We are beginning to understand mate choice as an integral part of life history. For example, female guppies either change preferences to duller males or become sexually unreceptive under predation pressure. Thus, animals can adjust their life histories in response to ecological factors to maximize lifetime reproductive success.

In the future, we will see further exploration of the genetic mechanisms of mate choice. Since the expression of sexually selected traits are context-dependent, benefits from choosing the most ornamented partner might differ between environments. Genes that are good in one environment might have a negative effect in another. One fruitful way to go would be to investigate the effects of genes and environment on mate preferences. Is there a reaction norm of mate choice?

It will be exciting to see further tests of Gowaty and Hubbell's predictions that individuals adjust their behavior in response to changing environmental, social, and intrinsic cues, on a moment-by-moment basis, resulting in adaptively flexible acceptance or rejection of particular potential mates and competitive behavior. Ultimately, we will be able to distinguish to what extent perceived typical differences between females and males are due to genetic sex-linked traits or an effect of ecological forces that individuals experience.

See also: Cryptic Female Choice; Mate Choice in Males and Females; Reproductive Behavior and Parasites: Invertebrates; Reproductive Behavior and Parasites: Vertebrates; Sexual Selection and Speciation; Social Selection, Sexual Selection, and Sexual Conflict.

Further Reading

Candolin U and Wong BBM (2008) Mate choice. In: Magnhagen C, Braithwaite V, Forsgren E, and Kapoor BG (eds.) *Fish Behaviour*, pp. 337–376. Enfield, NH: Science Publishers.

Cotton S, Small J, and Pomiankowski A (2006) Sexual selection and condition-dependent mate preferences. *Current Biology* 16: R755–R765.

Forsgren E, Amundsen T, Borg AA, and Bjelvenmark J (2004) Unusually dynamic sex roles in a fish. *Nature* 429: 551–554.

Godin JGJ and Briggs SE (1996) Female mate choice under predation risk in the guppy. *Animal Behaviour* 51: 117–130.

Gowaty PA and Hubbell SP (2005) Chance, time allocation, and the evolution of adaptively flexible sex role behavior. *Integrative Comparative Biology* 45: 931–944.

Gowaty PA and Hubbell SP (2009) Reproductive decisions under ecological constraints: It's about time. *Proceedings of the National Academy of Sciences* 106: 10017–10024.

Jennions MD and Petrie M (1997) Variation in mate choice and mating preferences: A review of causes and consequences. *Biological Reviews* 72: 283–327.

Johnson JB and Basolo AL (2003) Predator exposure alters female mate choice in the green swordtail. *Behavioral Ecology* 14: 614–625.

Qvarnström A (2001) Context-dependent genetic benefits from mate choice. *Trends in Ecology and Evolution* 16: 5–7.

Widemo F and Saether SA (1999) Beauty in the eye of the beholder: Causes and consequences of variation in mating preferences. *Trends in Ecology and Evolution* 14: 26–31.

Food Intake: Behavioral Endocrinology

T. Boswell, Newcastle University, Newcastle upon Tyne, UK

Introduction

Food intake is a complex behavior that serves several functions. Animals eat to acquire the energy, vitamins, minerals, and macronutrients necessary for survival. This is linked to the expression of food-seeking behaviors that involve learning and reward mechanisms to ensure the selection of foods that meet the body's needs while avoiding possible toxicity. On a daily basis, patterns of feeding are adjusted to an animal's way of life (e.g., whether nocturnal or diurnal), and eating is generally organized into bouts or meals. The availability of food is often unpredictable, both on a daily and a seasonal basis, and animals cope with this by storing excess energy in times of plenty that can be drawn upon when food is scarce. Vertebrates are able to match energy intake to expenditure very precisely over long periods in order to maintain what is called 'energy balance' or 'homeostasis.' An animal's life history may require food intake to be modulated on a seasonal basis, and adjustments must be made continuously to changes in the environment. Also, males and females are likely to have different feeding requirements, necessitating sex-specific controls over food intake. This study considers the neuroendocrine mechanisms that integrate external cues with internal physiological signals to control feeding behavior in vertebrates.

Neuroendocrine Control of Feeding Behavior by the Brain

Role of the Hypothalamus and Signaling Molecules

The importance of the hypothalamus in the regulation of food intake was established by investigations in the 1940s and 1950s showing stimulation or inhibition of feeding in rats following electrical stimulation or lesions in different hypothalamic regions. These findings were later extended to other vertebrates including birds and teleost fish. The initial interpretation of these findings was that two specific stimulatory and inhibitory feeding centers exist in the hypothalamus. This has been repeatedly challenged as it has become apparent that the regulation of feeding involves complex coordination and integration between neural networks throughout the brain. However, the experiments demonstrated the importance of the hypothalamus in vertebrates, and it has remained a central focus of research on the neuroendocrine regulation of

feeding, not least because it is an important site of production of many neural signaling molecules that exert potent effects on food intake.

A key development was the advances in protein chemistry in the 1980s that allowed small neural peptide signaling molecules, or neuropeptides, to be extracted from brain tissue and their amino acid sequence determined. This allowed them to be synthesized and be readily made available to researchers. The powerful influence on feeding of these molecules is exemplified by neuropeptide Y (NPY). The amount of food eaten after an NPY injection can be greater than the meal eaten after a rat has been deprived of food for 24 h, and even satiated animals will eat a normal-sized meal. Neuropeptides that influence feeding can be divided into those that exert stimulatory (orexigenic) effects when injected into the brain, and those that are inhibitory (anorexigenic). In addition to neuropeptides, many of which will be considered later, other signaling molecules that influence feeding throughout the brain include the classical neurotransmitters glutamate, catecholamines, serotonin, and GABA. Attention has also been focused recently on the endocannabinoid system. The appetite-stimulating properties of marijuana are well known, but the cannabinoid receptors in the brain that bind the psychoactive component of the drug were only identified in the 1990s. The natural ligands for these receptors – endocannabinoids – are derived from phospholipids. The endocannabinoid system acts to influence food intake both in the brain and in the body, and it interacts with many of the peptide-based signaling systems considered in the following lines.

Five key regions of the rat and mouse hypothalamus have been linked to the regulation of food intake. These are the arcuate, paraventricular (PVN), ventromedial (VMH), and dorsomedial (DMH) nuclei, together with the lateral hypothalamic area (LHA). Of these, the arcuate nucleus has been a particular focus of attention.

The Arcuate Nucleus: A Neural Network that Monitors and Responds to Energy Deficit

Free-living vertebrates will naturally experience periods of reduced food availability during which the energy needed for survival and reproduction can be drawn from body energy stores. Homeostatic mechanisms exist to monitor these stores and replace them when they are depleted. One of these mechanisms involves a network of neuropeptide-producing neurons within the arcuate nucleus of mammals. This is found in the basal hypothalamus, just above the pituitary gland. Its positioning is significant because this region has greater access to blood capillaries than most other parts of the brain, and is well placed for neurons to receive nutritional signals from blood-borne nutrients and metabolic hormones. The neuroendocrine signaling network within the arcuate nucleus is centered on the melanocortin system. This includes several melanocortin peptides encoded by a single pro-opiomelanocortin (POMC) precursor gene that exert their biological effects by interacting with melanocortin receptors. Melanocortin peptides are secreted by the pituitary gland in vertebrates. A well-known example is α-melanocyte-stimulating hormone (α-MSH) that stimulates pigment cells to modulate skin color in amphibians. By the 1980s, it became clear that melanocortin peptides were also synthesized within the brain. In addition to those encoded by the POMC gene, another important melanocortin system peptide is agouti-related peptide (AGRP), so called because the structurally similar agouti protein is responsible for producing characteristic yellow fur and obesity in the mutant line of *agouti* mice. Melanocortin receptors exist in several subtypes. The MC4R is of particular significance for regulation of food intake. Pharmacological studies indicate that it exerts an inhibitory influence. Two melanocortin system neuropeptides produced within the arcuate nucleus compete for access to the MC4R. AGRP acts as an antagonist, reversing the receptor's normal inhibitory influence and thereby stimulating feeding. In contrast, α-MSH is an agonist that promotes the MC4R's inhibitory effect.

The neurons synthesizing AGRP and α-MSH exist as two distinct cell groups within the arcuate nucleus. Individual AGRP neurons not only produce AGRP, but also NPY. Neurons transcribing the POMC gene produce α-MSH and, from another gene, also synthesize cocaine and amphetamine-regulated transcript (CART). CART, as its name suggests, was originally identified in rats administered cocaine or metamphetamine and has been linked to the anorexia induced by these drugs: its normal physiological role in the brain is to inhibit food intake. The appetite-stimulatory AGRP/NPY and inhibitory POMC/CART cell groups share reciprocal neural connections allowing them to influence each other's activity. A key feature of the neuronal network within the arcuate nucleus is the energy sensitivity of the neurons. They synthesize receptors for several metabolic hormones, as is discussed in the following section, and have access to blood-borne nutrients. This enables them to monitor the animal's energy status and make adjustments to maintain body energy balance. The network is activated following periods of energy deficiency that arise either because an animal is unable to feed, or because its energy expenditure exceeds its energy intake. Under such circumstances, AGRP/NPY neurons are activated (neural activity and

neuropeptide release and gene transcription are stimulated) and POMC/CART neurons inhibited.

In some working models of the hypothalamic control of feeding, the arcuate nucleus neurons responsible for receiving and integrating metabolic information from the blood are regarded as first-order neurons that send signals to second-order neurons that are responsible for modulating feeding behavior. Second-order neurons that are innervated by axons from AGRP/NPY and POMC/CART neurons in the rat brain include those in the PVN. This nucleus is believed to exert an inhibitory effect on feeding owing to the presence of neurons that synthesize the inhibitory MC4R as well as those producing feeding-inhibitory peptides such as corticotropin-releasing factor (CRF, also known as CRH) and thyrotropin-releasing hormone (TRH). In mammals and birds, α-MSH appears to induce its inhibitory effect on feeding by acting on CRF neurons. In contrast, second-order neurons in the LHA and perifornical area produce the feeding-stimulatory neuropeptides orexin A and B (also known as hypocretin 1 and 2), and melanin-concentrating hormone (MCH). Although the details of this signaling network were established in studies of laboratory rats, the first-order melanocortin system neurons showing sensitivity to energy deficit are present in other mammalian orders as well as in the neuroanatomical equivalents of the mammalian arcuate nucleus in several bird and teleost fish species. However, the organization of second-order neurons may differ among vertebrates because orexins and MCH appear ineffective in stimulating food intake in birds. The functional role of the arcuate nucleus network is to initiate and coordinate a behavioral and physiological response to restore lost energy stores. Many vertebrates will increase food intake when food becomes available following a period of fasting or food restriction, and this can be combined with a decreased body heat production and metabolic changes to promote the building of new body energy stores. The behavioral response to energy deficit of some species such as the Siberian hamster (*Phodopus sungorus*) is expressed primarily as an increase in food hoarding rather than increased feeding. Food hoards can be viewed as an external fat store. In this species, injections of AGRP and NPY into the brain preferentially stimulate hoarding behavior over increased food intake.

Control of Feeding and Digestion by the Brainstem

The brainstem is another important region for the control of food intake. Like the arcuate nucleus, this region is well placed to monitor blood-borne hormonal and metabolic signals owing to its proximity to the brain's ventricular system where the blood–brain barrier is incomplete. The nucleus of the solitary tract (NTS) is a particularly important center because it is connected both to the hypothalamus and also to sensory fibers of the vagus nerve that innervates the gut. It is therefore able to integrate feeding with digestive processes. Neurons in the NTS produce receptors for many of the neuropeptides and hormones involved in feeding regulation. A population of NTS cells also synthesizes the neuropeptide precursor molecule proglucagon that is processed to produce glucagon-like peptides 1 and 2 (GLP-1 and GLP-2), and oxyntomodulin, all of which decrease food intake when injected into the brain.

Regulation of Food Reward

It is well known that the presentation of palatable food, such as high-fat items, can cause food intake to be increased to a level greater than that required by an animal's immediate energy requirements. One important brain region for the regulation of reward is the midbrain dopamine system. In rats and mice, this consists of neurons containing dopamine within the ventral tegmental area (VTA) that connect with the nucleus accumbens. Selection of palatable food has been linked to signaling by opioid neuropeptides in the nucleus accumbens.

Hormones Influencing Food Intake

Adipose Tissue Hormones: Leptin

Given the importance of adipose tissue as a body energy store, it could be predicted that feedback systems exist to inform the brain of body fat content. This was encapsulated in the lipostat hypothesis proposed by Kennedy in the 1950s. However, the discovery that adipose tissue actively secretes signaling molecules known as 'adipokines' has been made only recently. The breakthrough in this area came in 1994 with the characterization of the gene that is mutated and causes extreme obesity in the *ob/ob* strain of mice studied since the 1950s. The protein product of the *ob* gene was given the name 'leptin' (derived from the Greek *leptos* for thin), and the *ob* gene is now known as 'the leptin gene.' Leptin is secreted by fat cells, circulates in the blood in proportion to body fat content, and is transported into the brain where it inhibits food intake and increases energy expenditure. Leptin received wide publicity at the time of its discovery as a potential treatment for clinical obesity in humans. However, it became clear that obese patients tend to be unresponsive to leptin treatment, a phenomenon known as 'leptin resistance.' While the clinical emphasis of early investigations focused on the effect of high circulating concentrations of leptin, it became clear that from an evolutionary perspective, decreasing levels of leptin are a more relevant signal to free-living animals that face fluctuations in food availability in their environment. Leptin concentrations in the blood fall during fasting or food restriction and signal the activation in the brain of

mechanisms to compensate for reduced energy intake, including increased foraging and feeding.

Much research has focused on the interaction between leptin and the arcuate nucleus cellular network described earlier. Both the AGRP/NPY and POMC/CART cell groups synthesize the leptin receptor. The hormone's inhibitory effect on food intake is signaled through activation of the POMC/CART neurons when the hormone is at high circulating concentrations. In fasting conditions, when leptin levels are low, the AGRP/NPY neurons are activated and feeding is stimulated. Leptin acts as a signal of body fat content (or 'adiposity') to coordinate an animal's feeding behavior and physiology in relation to its energetic state. As such, it impinges on facets of feeding behavior other than the feeding response to energy deficit. For example, leptin interacts with mechanisms governing the rewarding properties of food. In mice, the leptin receptor is produced in taste bud cells on the tongue, and leptin selectively inhibits responsiveness to sweet taste. Leptin appears to suppress the rewarding properties of food and, in rats, there is evidence that this is mediated by leptin signaling in the ventral tegmental area of the midbrain, richly supplied with dopamine neurons containing leptin receptors.

While rapid progress was made in uncovering leptin's role in regulating food intake and energy balance in mammals, it took 10 years for leptin genes to be identified in other vertebrate taxa. These have now been characterized in several species of teleost fish (which possess two leptin genes) and in amphibians. The effects of leptin on food intake in these groups appear to have been conserved during evolution, with leptin administration reducing food intake, and leptin gene transcription being sensitive to fasting. Whether leptin is present in birds is controversial. No unequivocal evidence exists for an avian leptin gene and it is absent from the sequenced chicken and zebra finch genomes. However, mammalian leptins inhibit food intake in birds, and a leptin receptor is present in the avian genome that shows functional signaling properties.

Peptide Signals from the Gut

Food intake in vertebrates tends to be episodic, with food being ingested in discrete meals owing to the development of satiation mechanisms that terminate a meal. The gut secretes a number of peptide satiety signals to influence meal size. The best known of these is cholecystokinin (CCK), which has been investigated since the 1970s. In rats, the octapeptide form of CCK, CCK-8, meets the criteria that have been established in the literature for a molecule to be considered a true physiological satiety signal. These are that it should reduce meal size when administered before a meal at doses in the physiological range, and that this should not occur as a result of illness; that it should be secreted as a result of food ingestion; and that meal size is increased when it is removed or its action antagonized. CCK is released from the small intestine in response to the presence of nutrients. Some of it enters the blood to influence digestive processes, but its inhibition of food intake is mediated by a local action on sensory fibers of the vagus nerve. The vagus nerve makes neural connections to the NTS in the brainstem. Leptin acting in the arcuate nucleus, and also possibly the brainstem, enhances the strength of the satiety signal mediated by CCK. Thus, the control of meal size is integrated in relation to the animal's overall energy status. The eight amino acids of CCK-8 are highly conserved between vertebrate taxa, and it inhibits food intake when injected into the body of teleost fish and also birds, where there is evidence that it signals via the vagus nerve as in mammals. Thus, the control of meal size by CCK appears to have been conserved during evolution.

Other gut peptides have also been implicated in the regulation of meal size in vertebrates. These include the amphibian peptide bombesin, structurally related to gastrin-releasing peptide (GRP) in other vertebrates, and several peptides derived from processing of the proglucagon precursor in the small intestine: GLP-1 and-2, and oxyntomodulin. An interesting evolutionary characteristic of bombesin/GRP, proglucagon-derived peptides, and CCK is that they are produced by neurons in the brain as well as in the gut in vertebrates. Their inhibitory effect on food intake by action in the gut is mirrored by a reduced feeding response after injection into the brain. Peptide YY (PYY) is another satiety factor produced by the small intestine. It is structurally related to NPY and, in common with other members of the pancreatic polypeptide family, interacts with NPY receptors. Its secretion increases following a meal and this is reflected in increased circulating PYY in the blood. The truncated form of the PYY peptide, PYY_{3-36}, inhibits food intake in rodents and humans, and this appears to be mediated by an action on the NPY Y2 receptor in arcuate nucleus neurons. The action of this signaling pathway on the regulation of food intake in other vertebrates has yet to be investigated.

In contrast to other gut peptides, ghrelin, discovered in rat stomach in 1999, stimulates food intake in mammals. Ghrelin was identified as the previously unknown natural ligand of a growth hormone secretagogue (GHS) receptor. Ghrelin is produced in the stomach and parts of the intestine, and is released into the blood during fasting. This pattern contrasts with decreased leptin levels during a fast. Ghrelin is believed to induce feeding in rodents by acting on neurons producing the GHS receptor in the arcuate nucleus and PVN, and also in the brainstem. In particular, ghrelin stimulates the AGRP/NPY neurons in the arcuate nucleus. Ghrelin also induces effects on feeding in the brain via a second population of ghrelin-producing cells located close to the arcuate nucleus. The hormone influences food intake in other vertebrates.

A stimulation of ghrelin production in response to fasting has been observed in birds, teleost fish, and an amphibian. In fish, ghrelin injections into the brain or body stimulate feeding as in mammals. However, in birds, a general inhibitory effect on food intake has been observed, the reasons for which are uncertain. The actions of ghrelin in promoting eating after fasting in mammals are complemented by other actions of the hormone in the brain. As for leptin, there is suggestive evidence that ghrelin acts on the midbrain dopamine system to influence the rewarding properties of food. There is also evidence for ghrelin acting in the hippocampus to modulate memory retention, and this has been linked to foraging behavior.

Pancreatic Hormones

Pancreatic hormones play a prominent role in the regulation of food intake. Prior to the discovery of leptin, insulin was the strongest candidate for a hormonal signal of body fat content. The function of insulin in regulating energy storage and feeding behavior is evolutionarily ancient, occurring in invertebrates. Although injections of insulin into the body in mammals provide a stimulus to eat, this occurs as a secondary effect of the hormone reducing concentrations of blood glucose. In its normal physiological role, insulin inhibits food intake in vertebrates. Insulin is secreted in proportion to body fat content and is transported into the brain where, in rodents, it interacts with insulin receptors produced in feeding-relating neurons, including AGRP/NPY neurons in the arcuate nucleus. The intracellular signaling pathways stimulated by insulin and leptin to regulate feeding overlap. As with leptin, POMC/CART neurons are stimulated by insulin, while AGRP/NPY neurons are inhibited by insulin at higher concentrations, and stimulated when insulin levels decrease during fasting. An inhibitory effect of insulin on feeding when injected into the brain has also been observed in birds. Like leptin and ghrelin, insulin regulates food reward in the midbrain dopamine system in rodents and also influences memory processes in the hippocampus. Insulin also acts like leptin in modulating the sensitivity of the feeding response to CCK.

Among the other pancreatic hormones, glucagon and amylin act as satiety signals to reduce meal size in mammals, teleost fish, and birds. Pancreatic polypeptide (PP) is structurally related to NPY and PYY. In mammals and in the chicken, PP is secreted after food ingestion and this is related to coordination of digestion. When injected into the body, PP reduces food intake in rodents and humans, but this has not been investigated in other vertebrates.

Adrenal Hormones: Glucocorticoids

Glucocorticoids play a fundamental role in the regulation of feeding, metabolism, and energy storage. The type of glucocorticoid secreted by the adrenal cortex in vertebrates varies. For example, the principal glucocorticoid in rats, mice, terrestrial amphibians, reptiles, and birds is corticosterone, whereas in primates and teleost fish, it is cortisol. In rats, removal of the adrenal gland (adrenalectomy) results in reduced food intake and can be reversed by administration of corticosterone. Glucocorticoids signal through two main types of glucocorticoid receptor – mineralocorticoid (MR), type I, and glucocorticoid (GR), type II receptors – and these are expressed in many brain areas, including the arcuate nucleus and the paraventricular nucleus. Administration of glucocorticoids stimulates food intake and foraging behavior in mammals, fish, and amphibians. In rats, corticosterone acts in the arcuate nucleus to stimulate synthesis of NPY in the AGRP/NPY neurons, a mechanism that comes into play in the adaptive response to fasting, when glucocorticoid concentrations are increased in birds and mammals. The effect is to promote foraging behavior and to increase food intake when food is found. A series of feedback loops between NPY, CRF, insulin, and glucocorticoids integrates food intake and energy storage or mobilization with the glucocorticoid response to stressors in the rat. The effects of glucocorticoids are closely linked to insulin secretion. Corticosterone stimulates insulin secretion that, in turn, reduces the corticosterone-mediated stimulation of food intake. The two hormones exert opposing effects in the body, with insulin being anabolic (promoting energy storage) and glucocorticoids catabolic (promoting mobilization of energy stores). Thus, fat will be stored when both glucocorticoids and insulin are high, whereas energy stores will be broken down when glucocorticoids are high and insulin low. The effect of glucocorticoids in promoting feeding is opposed by the inhibitory effects of CRF produced in the PVN. Situations of reduced food intake in response to stressors are linked to the action of CRF, because injections of CRF into the brain reduce feeding in all vertebrate classes.

Pituitary Hormones: Prolactin

Prolactin, secreted by the anterior pituitary gland, has been linked to the phase of increased food intake associated with maternal provisioning in vertebrates. In ring doves (*Streptopelia risoria*), male and female parent birds feed their young on crop milk, a situation resembling lactation in mammals. To support this, both sexes increase food intake during the posthatching period, when prolactin concentrations in the blood are highest. Prolactin stimulates food intake when injected into the brain of ring doves and has been linked to a stimulatory effect of the hormone on NPY neurons in the avian equivalent of the arcuate nucleus. A comparable situation applies during lactation in mammals when food intake is also increased. In lactating rats, the prolactin receptor is synthesized in NPY neurons in the DMH and NPY production is regulated in this nucleus by suckling and prolactin.

Sex Steroid Hormones

The effects of removal of the ovaries and testes on food intake are variable among vertebrates and even among mammals. However, evidence indicates that sex steroids interact with neuronal circuits controlling feeding to bring about sex differences in food intake. In rats and mice, food intake is reduced at the time when estrogen levels rise before ovulation. Removal of the ovaries (ovariectomy) results in an increase in food intake that is sustained until body mass stabilizes at a new, higher, level. Injections of estrogen into the brain or body of an ovariectomized rat decrease food intake. In contrast to the situation in females, testis removal in males decreases food intake and administration of testosterone reverses this. The respective increase and decrease in food intake after gonad removal are associated with increased meal size in females and decreased meal frequency in males. The effect on meal size in females has been linked to an interaction of estrogen with estrogen receptors in the brainstem to alter the sensitivity of the feeding response to CCK. Suggestive evidence also exists for regulatory effects of estrogen on the sensitivity of food intake to inhibition by leptin, which appear to differ between male and female rats.

Regulation of Seasonal Cycles of Food Intake and Fat Deposition

In many vertebrates, food intake is adjusted seasonally to enable animals to meet the demands of life history events such as reproduction, molt, migration, and hibernation. The best-studied model of seasonal food intake and fat deposition is the Siberian hamster. In this species, when maintained in the laboratory, exposing the animals to short days results in decreased food intake and loss of body mass. It might be predicted that these changes could be achieved by adjustments in the production of leptin and arcuate nucleus neuropeptides. However, blood levels of the feeding-inhibitory hormone leptin are paradoxically highest when food intake is highest under long days, while short day animals show greater leptin sensitivity. Thus, seasonally obese hamsters show the phenomenon of leptin resistance observed in obese humans. Similarly, synthesis of most of the hypothalamic neuropeptides is not adjusted between long days and short days in the direction expected to explain the seasonal difference in food intake. This can be explained by the operation of a 'sliding set point' for body mass. Thus, the function of leptin and the hypothalamic neuropeptide networks is to return energy stores to a homeostatic level in response to energy deficit. However, the homeostatic level at which food intake and body mass are maintained is adjusted on a seasonal basis in hamsters, a process known as involving a 'sliding set point' or 'rheostasis'. The use of DNA microarrays identified some of the genes that may regulate this process. These include genes involved in thyroid hormone metabolism, histamine and retinoic acid signaling, and VGF production. Implantation of tri-iodothyronine (T3) into the hypothalamus blocked short day weight loss.

See also: Behavioral Endocrinology of Migration; Caching; Digestion and Foraging; Hormones and Behavior: Basic Concepts; Hunger and Satiety; Parental Behavior and Hormones in Mammals; Parental Behavior and Hormones in Non-Mammalian Vertebrates; Taste: Vertebrates; Water and Salt Intake in Vertebrates: Endocrine and Behavioral Regulation.

Further Reading

Asarian L and Geary N (2006) Modulation of appetite by gonadal steroid hormones. *Philosophical Transactions of the Royal Society B* 361: 1251–1263.

Bellocchio L, Cervino C, Pasquali R, and Pagotto U (2008) The endocannabinoid system and energy metabolism. *Journal of Neuroendocrinology* 20: 850–857.

Boswell T.(in press) Molecular aspects of leptin in the chicken. In: Paolucci M.(ed.) *Leptin in Non-mammalian Vertebrates.* Kerala, India: Research Signpost, Transworld Research Network.

Dallman MF, La Fleur SE, Pecoraro NC, Gomez F, Houshyar H, and Akana SF (2004) Minireview: Glucocorticoids – food intake, abdominal obesity and wealthy nations in 2004. *Endocrinology* 145: 2633–2638.

Denver RJ (2009) Structure and functional evolution of vertebrate neuroendocrine stress systems. *Annals of the New York Academy of Sciences* 1163: 1–16.

Ebling FJP and Barrett P (2008) The regulation of seasonal changes in food intake and body weight. *Journal of Neuroendocrinology* 20: 827–833.

Figlewicz DP and Benoit SC (2009) Insulin, leptin and food reward: update 2008. *American Journal of Physiology* 296: R9–R19.

Gao Q and Horvath TL (2008) Neuronal control of energy homeostasis. *FEBS Letters* 582: 132–141.

Kaiya H, Miyazato M, Kangawa K, Peter RE, and Unniappan S (2008) Ghrelin: A multifunctional hormone in non-mammalian vertebrates. *Comparative Biochemistry and Physiology, Part A* 149: 109–128.

Morton GJ, Cummings DE, Baskin DG, Barsh GS, and Schwartz MW (2006) Central nervous system control of food intake and body weight. *Nature* 443: 289–295.

Richards MP and Proszkowiec-Weglarz M (2007) Mechanisms regulating feed intake, energy expenditure and body weight in poultry. *Poultry Science* 86: 1478–1490.

Sawchenko P (1998) Toward a new neurobiology of energy balance, appetite and obesity: The anatomists weigh in. *The Journal of Comparative Neurology* 402: 435–441.

Volkoff H, Canosa LF, Unniappan S, et al. (2005) Neuropeptides and the control of food intake in fish. *General and Comparative Endocrinology* 142: 3–19.

Woods SC (2004) Gastrointestinal satiety signals I. An overview of gastrointestinal signals that influence food intake. *American Journal of Physiology* 286: G7–G13.

Woodside B (2007) Prolactin and the hyperphagia of lactation. *Physiology & Behavior* 91: 375–382.

Food Signals

C. T. Snowdon, University of Wisconsin, Madison, WI, USA

Introduction

Many animals produce vocal signals in the context of discovering or ingesting food and such signals are of interest for a variety of reasons. First, do these calls actually refer to food suggesting that they may function as referential signals much as predator-specific calls do with each call type representing a specific predator type? Or do food-associated signals simply represent emotional expressions of elation with no true specificity to food? In the latter case, it may simply be that food is the most commonly observed context of elation, and signals may be mislabeled as food signals when in fact something different is communicated. Second, why should individuals discovering food call attention to others? The production of food-associated signals would seem to be a cost to the individual who locates the food if the individual subsequently must share food with others. Third, if food-associated signals are honest signals of food, what prevents individuals from cheating and failing to produce food-associated signals when out of view of others? Fourth, how does the production and usage of food-associated signals change with development? Are individuals able to produce signals appropriately at birth and use them in feeding contexts or are these signals learned? Finally, can food-associated signals be used by adults in teaching young how to forage or to learn which foods are appropriate to eat or to avoid? In one family of primates, the marmosets and tamarins (Callitrichids), food signals appear to play an important role in teaching. These species are cooperative breeders meaning that typically only one female in a group reproduces and all other group members assist with infant care, often to the detriment of their own reproduction.

Before addressing each of these questions, let us first look at some examples of food-associated signals. Chickens produce calls in response to food and they appear to call more frequently to highly preferred foods such as nuts than to less preferred foods such as peas. Other avian species with food-associated calls include sparrows, swallows, and ravens. Among nonprimate mammals, food-associated vocalizations are found in naked mole rats, dolphins, and greater spear-nosed bats. In non-human primates, calls associated with food have been observed in marmosets, tamarins, capuchin monkeys, spider monkeys, macaques, chimpanzees, and bonobos. Thus, calls given in conjunction with feeding are seen in a wide array of species.

Are Food-Associated Signals Functionally Referential?

So far I have been careful in saying 'food-associated signals' rather than 'food calls,' since it is important to know whether the calls are specific to foods or whether they communicate about something different that happens to be associated with feeding. Thus, the calls may serve primarily to promote social attraction, or may be used to attract mates or to signal status at feeding sites without directly communicating about food. It is equally important to demonstrate that the call structure differs from other calls in the repertoire if it is to have a unique function as a food call.

Signals are labeled as functionally referential when they are used predominantly in contexts where some external referent can be identified and when playback of calls in the absence of the hypothesized referent leads animals to respond as if the referent was present. Thus, in studies of chickens, food calls are given at higher rates to food items such as peanuts than to nonfood items such as peanut shells, but calls are still given to peanut shells. Toque macaques in Sri Lanka produce a certain type of call most often when an individual has found an abundant source of food such as a fig tree with ripe figs, but these monkeys also give the same calls on a sunny day at the end of the monsoon and to the first rain clouds at the end of the dry season, suggesting an elation component to the signal. Captive cotton-top tamarins give calls mainly to food, but as with toque macaques, 3% of the situations during which the calls were recorded did not involve food. However, when adult tamarins were offered small, manipulable objects the same size as food pieces, they did not vocalize suggesting that food rather than some other feature was most likely to elicit calls.

Tamarins also gave a greater number of vocalizations to highly preferred foods than to less preferred foods, supporting an emotional or motivational component to the calling. Similarly, chimpanzees give more calls to many items of fruit than to small or only single items of fruit and they gave more calls to a watermelon cut into 20 pieces than to a single intact watermelon suggesting that the ability to share food plays an important role in whether an animal calls or not. Initially, referential signals were thought to indicate something about the cognitive skills of animals, but signals may be primarily emotional from the perspective of the caller and still communicate about an external referent such as food.

Determining whether signals are truly related to food or to something else is critical in interpreting the results of studies of these calls. For example, an initial interpretation of the chimpanzee food calls was that chimpanzees suppressed calling to small numbers of items or nonsharable items, and that this was evidence of deception. Yet an alternative explanation might be that small amounts of food are not sufficient to motivate calling. One study on chickens found that males frequently 'food called' and that hens would approach the male, but nearly 50% of the time the male had no food and thus must be deceiving his mate. However, chickens and many other birds engage in tid-bitting behavior (food sharing) as part of courtship, and the calls that were labeled as 'food calls' may have served multiple functions, including courtship and affiliation. Before labeling a signal as a 'food signal,' it is necessary to evaluate alternative or additional functions so that results are not overinterpreted. This is best done experimentally by offering animals a variety of food items that vary in quality and amount as well as by testing with nonfood items of similar size and shape. Since one hypothesis of food calling may be simple elation, testing for calls in nonfeeding contexts where elation may be present is also important.

Relatively few playback studies have involved food-associated calls, but two independent field studies of two species of capuchin monkeys have found results suggestive of functional reference. Both the studies identified calls that were given mainly in the context of feeding and when these calls were played back, capuchin monkeys were more likely to orient to the speaker and to move in the direction of the speaker than to control sounds, suggesting that the monkeys reacted to the sounds as if food had been located.

Why Produce Food Signals?

It is puzzling that animals should vocalize when they discover food. If the food is also an animal, then vocalizations may allow the prey to escape. Indeed, even though animate prey are highly preferred foods for them, neither wild capuchin monkeys nor captive pygmy marmosets (**Figure 1**) vocalized to animate prey although they do vocalize to inanimate preferred foods.

Feeding competition is often cited as a major determinant of spacing patterns and mating systems in animals. If feeding competition is critical to individual survival and reproductive success, then it seems paradoxical that an individual should vocalize when it finds food. One conclusion would be that food calling should be more likely to occur when a food source is abundant beyond the ability of the discoverer to ingest all the food or if the spatial distribution of the food allows multiple individuals to eat

Figure 1 Pygmy marmoset in Amazon ingesting exudates. Courtesy of Pablo Yepez.

simultaneously. Indeed some studies find that food-related calling is more likely at large abundant patches than at small patches.

Another explanation is related to kin selection. If by calling when one locates food, one can lead kin (offspring or siblings) to a food resource, then even if there is some cost to the caller, its kin may benefit from the information about the location of food. Additionally, there may be social benefits beyond helping kin that make the production of food-associated signals adaptive. In some tamarin species, but not others, and in capuchin monkeys and chickens, the presence of conspecific group members leads to a greater probability of calling. Thus, if a rooster is tested alone with food, he is unlikely to vocalize, but if he is tested with a hen nearby, he will vocalize in the presence of food. The male's calling is not simply triggered by the presence of another animal alone, since if another rooster is present instead of a hen, the rooster exposed to the food will inhibit his calling.

In capuchin monkeys, both social status and the presence of an audience affected calling in the presence of food. In a study in captivity, high-ranking capuchin monkeys called less frequently over all, but called as often as lower ranking monkeys when they found food in the presence of all group members. Lower ranking monkeys called less often when alone than with one other partner or with the whole group present. In a field study of capuchin monkeys, food finders called less often in the season when food was scarce and less often to smaller amounts of food than to larger amounts, regardless of season. Females called with a significantly longer latency than males. There was a clear audience effect with animals of all social status calling with shorter latency when other group mates were close by. One conclusion from these results is that food calling is being used strategically by both chickens and capuchin monkeys with calling occurring only when a potential mate or other group

locate and process. In field studies of golden lion tamarins, juveniles were able to forage successfully on fruits and leaves but were much less successful with animate prey. Adults in the wild continued to give food transfer signals and offer prey to juveniles, but as juveniles became more skilled adults reduced signals and direct transfers. However, observers have noted adults giving food vocalizations that attract a juvenile, but with no obvious prey to transfer. When the juvenile approached the calling adult, it looked for prey in nearby areas and was often successful. This systematic use of food-related vocalizations with the reduction of adult calling and food transfers as infants and juveniles become more successful suggests that adults are sensitive to what the infants and juveniles know about food and how to obtain it. In the light of the earlier point that some monkeys do not give food-associated calls in the presence of animate prey, the use of food-associated vocalizations to juveniles only in the context of animate prey that juveniles find difficult to obtain is further support for the fact that this is some form of teaching.

Summary

There is evidence that at least some calls that are produced in feeding contexts are functionally referential, that is, they directly signal the presence of food. However, these calls also convey motivational information about food quality and quantity and may also be used in non-feeding contexts of excitement. Food-associated calls are often used strategically, given only in the presence of a potential mate or conspecific or to inanimate foods. There is some evidence that animals that fail to call when they find food are punished and that those that call honestly are likely to receive more food even if it must be shared with others. Finally, food-associated calls are used in more intense forms by cooperatively breeding primates in the context of food transfers to infants and the withdrawal of calling and food transfers by adults as infants acquire foraging skills is suggestive of teaching in these animals.

See also: Acoustic Signals; Communication and Hormones; Monkeys and Prosimians: Social Learning; Motivation and Signals; Parent–Offspring Signaling; Referential Signaling; Wintering Strategies: Moult and Behavior.

Further Reading

Caro TM and Hauser MD (1992) Is there teaching in nonhuman animals? *Quarterly Review of Biology* 67: 151–174.

Di Bitetti MS (2003) Food-associated calls of tufted capuchin monkeys (*Cebus apella nigritus*) are functionally referential signals. *Behaviour* 140: 565–592.

Di Bitetti MS (2005) Food-associated calls and audience effects in tufted capuchin monkeys, *Cebus apella nigritus*. *Animal Behaviour* 69: 911–919.

Elgar MA (1986) House sparrows establish foraging flocks by giving chirrup calls in the resources are divisible. *Animal Behavior* 34: 169–174.

Elowson AM, Tannenbaum PT, and Snowdon CT (1991) Food associated calls correlate with food preferences in cotton-top tamarins. *Animal Behaviour* 42: 931–937.

Evans CS (1997) Referential signals. *Perspectives in Ethology* 12: 99–143.

Gros-Louis J (2004) Responses of white-faced capuchins (*Cebus capucinus*) to naturalistic and experimentally presented food associated calls. *Journal of Comparative Psychology* 118: 396–402.

Hauser MD (1992) Costs of deception: Cheaters are punished in rhesus monkeys (*Macaca mulatta*). *Proceedings of the National Academy of Science USA* 89: 12137–12139.

Hauser MD, Teixidor P, Field L, and Flaherty R (1993) Food associated calls in chimpanzees, effects of food quality and divisibility. *Animal Behaviour* 45: 817–819.

Humle T and Snowdon CT (2008) Socially biased learning in the acquisition of a complex foraging task in captive cotton-top tamarins (*Saguinus oedipus*). *Animal Behaviour* 75: 267–277.

Joyce SM and Snowdon CT (2007) Developmental changes in food transfers in cotton-top tamarins (*Saguinus oedipus*). *American Journal of Primatology* 69: 1–11.

Marler P, Dufty A, and Pickert R (1986) Vocal communication in the domestic chicken: II. Is a caller sensitive to the presence and nature of a receiver? *Animal Behaviour* 34: 194–198.

Rapaport LG and Brown GR (2008) Social influences on foraging behavior in young nonhuman primates: Learning what, where, and how to eat. *Evolutionary Anthropology* 17: 189–201.

Roush RS and Snowdon CT (1999) The effects of social status on food-associated calling behaviour in captive cotton-top tamarins. *Animal Behaviour* 58: 1299–1305.

Foraging Modes

D. Raubenheimer, Massey University, Auckland, New Zealand

Introduction

It would be an exaggeration to say that in the evolution of foraging anything is possible, but not much of an exaggeration. Collectively, animals eat a huge diversity of different food types, from shoe polish (the fly *Megaselia scalaris*) to feces to the living tissues of other animals. Add to this the many ways that animals mix foods to compose their diets, and locate, capture, and process these foods, and the number of foraging strategies approaches the number of species. Furthermore, many species eat different diets at different stages in the life cycle, and in some, different individuals of the same age have different foraging adaptations and diets (called *resource polymorphisms*). The diversity of foraging strategies can therefore in theory exceed the number of species.

To help to understand this diversity, researchers have classified animals into 'foraging modes.' Although not perfect, these categories have been useful, because they help to impose some order on what would otherwise be an unstructured catalog of different cases. They also expose some important questions which help to build a more complete understanding of the diversity of foraging observed in nature. I return to these questions in the final section, after presenting an overview of the criteria that have been used for classifying animals according to their foraging mode.

Foraging Modes

Foraging involves an interrelated set of components, which can usefully be categorized into a series of roughly sequential steps. At the one extreme is the choice of the general category of foods that are eaten and the general habitat in which to forage. Within the chosen habitat, the animal needs to locate, capture, and ingest the food items, while avoiding hazards such as predators, physical defenses, and toxins. The swallowed foods are then further processed in the digestive tract to separate absorbable components from wastes. Many schemes exist for grouping animals into modes based on these aspects of foraging, with different schemes emphasizing different components (habitat selection, food choice, the pattern of foraging, food detection, food capture, and postingestive processing). These classifications are not, of course, mutually exclusive, because an animal can be classified separately according to different criteria or combinations of criteria

(**Figure 1**): it might, for example, be similar to some animals in the foods that it eats, but similar to others in the way that it processes them in the gut.

Habitat

A very broad categorization of foraging strategies concerns the habitat in which an animal forages. At the most general level, *terrestrial* foragers seek their foods on land, while *aquatic* consumers forage in water. If aquatic, foraging might take place in the sea (*marine*), lakes (*lacustrine*), or rivers and streams (*lotic*). Within these habitats, some animals feed on the bottom (*benthic*), while others feed in the open water (*pelagic* if marine, *limnetic* if in lakes). Terrestrial foragers are similarly classified according to a finer-scale distinction between types of habitats – for example, *arboreal* (in trees), *fossorial* (underground), *aerial* (in the air), desert, grasslands, etc. Within each of these habitats, foraging might take place during the day (*diurnal*), at night (*nocturnal*), at dawn (*matinal*), or dusk (*vespertine*). Many animals forage in more than one habitat type: *amphibious* foragers, such as crocodiles, forage on land and in water, while *crepuscular* animals are active both at dawn and dusk.

Some animals feed in one habitat and perform other activities (e.g., sleep, hide from predators, etc.) in another. For example, archer fish capture insects and other small invertebrates from outside their aquatic habitat by shooting them off overhanging vegetation using a jet of water forcefully expelled from a specialized mouth. In general, it is expected that cases where the two habitats are very different will be rare, because of the challenges involved in adapting to both. This is illustrated by the Australian spinifex hopping mouse (*Notomys alexis*), which burrows underground but uses *saltatorial* locomotion (hopping) to move through the arid open areas in which it forages. Research has shown that the adaptation of this species for saltatorial locomotion makes burrowing significantly more energetically expensive compared with specialized burrowers.

Foods

The most conspicuous, and in some respects the most important, aspect of foraging concerns the categories of foods that animals eat. For this reason, a good deal of attention has been paid to classifying animals into foraging modes according to their diet choice. These categories

Figure 1 Continued

Figure 1 Continued

Figure 1 A diversity of foraging modes. (a) New Zealand butterfish (*Odax pullus*) start life as carnivores, then become obligate algivorous herbivores – that is, they are life-history omnivores. In the herbivore stage, they use oral teeth to bite off pieces off macroalgae, which are then mechanically processed between the plates of a pharyngeal mill. Microbial populations in the hindgut ferment the food, but there is no specialized fermentation chamber. (b) Common dolphins (*Delphinus delphis*) are active marine predators, which feed both on fish (piscivorous) and squid (molluscivorous). They usually hunt in groups, using a range of cooperative hunting strategies. (c) New Zealand Weka (*Gallirallus australis*) are broad-scope terrestrial omnivores, eating a variety of plant- and animal-derived foods. They prefer forest and bush, but will forage across a wide range of habitats, including sandy beaches (as pictured). (d) The butterflyfish *Chaetodonton baronessa* is a specialist corallivore, which in some areas of its range feeds exclusively on a single species of coral (*Acropora hyacinthus*). (e) Blue mao mao (*Scorpis violacea*) are marine planktivores which often feed in large groups. (f) A parasitic *Cymothoa* isopod leaving the mouth of a recently dead butterfish (*O. pullus*). *C. exigua* are ectoparasites, which attach to the tongues of their hosts causing the organ to degenerate. They remain attached to the stub of the tongue, occupying the space formerly dedicated to the defunct organ. (g) Corals (here *Acropora* sp.) have nutritional symbioses with algal dinoflagellates. These unicellular algae provide the coral with photosynthesates, and in turn utilize waste nutrients excreted by the coral. (h) *Holothuria* sea cucumbers are detritivores, which pass large amounts of sediments through their guts extracting organic material and excreting the undigested sand (upper left). (i) A herbivorous black and white colobus monkey eating soil (geophagy). This common behavior is most likely a means to redress deficiencies in micronutrients, particularly sodium which is deficient in their folivorous diet. (j) *Acraea* butterflies in Uganda feeding on decaying feces of a mammalian predator. This behavior, known as 'puddling,' is a form of supplementary feeding believed to be targeted at obtaining micronutrients. (k) Scorpion fish (*Scorpaena* spp.) are cryptic ambush predators. (l) Australasian gannets (*Morus serrator*) are colonial-living central place foragers. They are visual-hunting, plunge pursuit predators, which hunt pelagic prey (fishes and squid). Here, a parent has just transferred a meal of fish to its unfledged chick. (m) New Zealand kakapo (*Strigops habroptila*) are solitary, flightless, nocturnal herbivores. They are central place foragers which eat a wide range of plant foods at night and nest during the day. Their eyes are not well developed, and it is likely that olfaction is an important sense used in foraging. (n) When eating fibrous leaves, kakapo (**Figure 1(m)**) selectively extract the juice and soft portions, discarding the less digestible, fibrous component as a 'chew.' Here, the chew is still attached to the plant. Kakapo are thus cytoplasm feeders. (o) Giant pandas eat large amounts of bamboo and selectively excrete the fibrous components. They are thus concentrate feeders. All photos by D. Raubenheimer.

are usually (but not always) assigned a name consisting of a description of the food with suffix '*vore*' (e.g., folivore = eater of leaves) or '*phage*' (e.g., phytophage = eater of plants). The '*vore*' versus '*phage*' distinction is of no particular significance: they both mean 'eater of,' but in Latin and Greek, respectively. In some cases, a feeding habit is described using Greek- and Latin-derived names interchangeably – for example, animals that eat plants are known both as 'herbivores' and 'phytophages.'

A common basis for denominating foraging modes is based (roughly) on the taxonomic group from which foods are drawn. Thus *cannibals* eat members of their own species, *bacterivores* eat bacteria, *fungivores* eat fungi, while *herbivores* eat plant-produced foods (**Figure 1(a)**), *carnivores* (sometimes called *faunivores*) (**Figure 1(b)**) eat animal-produced foods, and *omnivores* (**Figure 1(c)**) eat a combination of plant- and animal-derived foods. These categories are further split into subgroups: *graminivores* are herbivores that eat mainly grasses, while *piscivores* (**Figure 1(b)**) are carnivores that eat fish, *molluscivores* eat molluscs, *spongivores* eat sponges, *arthropodovores* eat arthropods, and *corallivores* (**Figure 1(d)**) eat corals. In some cases, there is further nesting within these groups: *entomophages* are arthropodovores that specifically eat insects, while *myrmecophages* and *lepidophages* are entomophages that eat ants and Lepidoptera, respectively. Other denominations are more specific about the parts of food organisms that are taken; for example, *folivores* eat leaves, *nectarivores* eat nectar, *xylophages* eat wood, *palinivores* eat pollen, *graminivores* eat seeds, *haematophages* eat blood, *mucophages* eat mucus, and *coprophages* eat feces.

A common distinction made for terrestrial herbivores is between *grazers* (which feed on grasses), *browsers* (which feed on dicotyledonous plants), and *mixed feeders* or *intermediate feeders* (which feed on both). Although grasses and dicotyledonous plants are taxonomically distinct, the grazer versus browser dichotomy for terrestrial mammalian herbivores is more about the properties of the foods than their taxonomic status. Thus, browsers are sometimes referred to as *concentrate feeders*, and grazers as *roughage selectors*, because grasses tend to contain lower concentrations of nutrients than do the dicotyledonous foods of browsers. However, since grasses are not always more fibrous than dicotyledonous leaves, a more refined, two-way, classification has been proposed, where animals are distinguished both according to the proportion of grass versus browse in the diet and the degree to which they feed selectivity within each category: *selective browsers* (i.e., concentrate selectors, e.g., pronghorn), *selective grazers* (e.g., bighorn sheep), *unselective browsers* (e.g., moose), and *unselective grazers* (e.g., elk). The *browser* versus *grazer* distinction is also used for marine animals, but the usage there has been more variable than for terrestrial herbivores. Some marine biologists consider grazers to be herbivores that feed by scraping or sucking and in so doing

ingest significant amounts of inorganic material, in contrast with browsers, which tear or bite pieces from more upright macroalgae and so rarely ingest inorganic material. Others use the term 'grazer' more generally to denote all marine herbivores whatever their feeding mode (including browsers, scrapers, particle feeders, etc.), and yet others do not restrict the use of these terms to herbivores. For example, the turtle-headed sea snake (*Emydocephalus annulatus*), which moves slowly through its environment feeding frequently on small, immobile, defenseless items (fish eggs), has been described as a browser, while some parrotfish are said to graze on the reproductive parts of corals. The criteria for distinguishing marine grazers and browsers are thus not consistently based on the properties of the foods, but sometimes on the mode of feeding.

Some groupings of foraging mode are based on more explicitly ecological criteria. For example, *planktivores* (e.g., **Figure 1(e)**) feed on plankton, a taxonomically mixed group (including plants, animals, bacteria, and archaea) which is defined by their ecological niche: small organisms that drift in the water column. An important ecologically inspired categorization of foraging modes emphasizes the trophic level from which foods are taken. In this classification, *herbivores* (e.g., **Figure 1(a)**) eat primary producers (plants), *carnivores* (**Figure 1(b)**) eat secondary or higher-order consumers (herbivores, omnivores, or other carnivores), and *omnivores* (**Figure 1(c)**) eat both plants and other animals. *Life-history omnivores* (**Figure 1(a)**) switch diets during development. Some ecologists distinguish between *strict predators* and *intraguild predators*, the former being predators that eat herbivores, and the latter predators that prey on other predators. In this approach, *trophic omnivores* are predators that feed both on herbivores and other predators, while *closed loop omnivores* eat both a resource and other consumers of that same resource.

A classification of foraging modes that is important both in the context of ecology and evolution is based on the nature of the biological interaction between the consumer and its foods. *Predators* kill their prey either in the act of or prior to eating them. *Parasites* (**Figure 1(f)**) take food from living hosts, which reduces the evolutionary fitness of the host but does not usually directly result in its death. Some parasites live inside the bodies of their hosts (*endoparasites*), while others live on their hosts (*ectoparasites*). *Parasitoids* spend a significant proportion of the life cycle living on or in the host, and ultimately kill it. *Commensals*, likewise, take food from other organisms, but at no net cost to those organisms. This is the case for some microbes living in the guts of animals, including humans, and also for some species of fish that swim beneath others eating their freshly excreted feces. *Mutualists* take food from another organism, but reciprocate by providing a nutritional or other benefit to the donor. Ants, for example, harvest honeydew from living aphids, and in return protect the aphids from predation; some invertebrates, including corals (**Figure 1(g)**), derive carbohydrates from symbiotic unicellular algae which, in turn, benefit from waste nutrients excreted by the corals. In contrast to these categories, *detritivores* (**Figure 1(h)**) feed on dead and decaying organic matter and are therefore of no direct functional relevance to their foods (although living microbes might contribute significantly to their nutritional gain). Some animals supplement their diet with nutrients from inorganic sources, including clay (a feeding habit known as *geophagy*) (**Figure 1(i)**), and many butterflies feed from nutrient-rich puddles or decaying organic matter (known as *puddling*) (**Figure 1(j)**). In this case, there is clearly no evolutionary significance for the food.

A widespread criterion for grouping animals into foraging modes is the breadth of foods eaten. Various terms have been used for this, the most common being *generalist* (wide range of foods) (**Figure 1(c)**) versus *specialist* (narrow range of foods) (**Figure 1(d)**). Other, more arcane terms include *monophage* (single food type), *oligophage* (a few food types), and *polyphage* (a wide range of foods). The term *stenophage* is sometimes used to refer to animals that have a narrow diet range (i.e., monophages and oligophages). It is important in characterizing diet breadth to distinguish between the range of foods eaten by individuals and the species as a whole. In some generalists, each individual mixes its diet from a wide range of foods (*individual generalists*), while in others the species as a whole has a diverse diet but each individual eats a subset of this diet (*population generalists*). One mechanism for this is *local specialization*, where individuals in different parts of the geographic range specialize on different foods. Another is *life-history omnivory* (as mentioned earlier in this section), in which animals switch diets at different stages in the life cycle (e.g., **Figure 1(a)**).

Foraging, Food Detection, and Food Capture

Among predators, the strategies of hunting and prey capture are widely used criteria for distinguishing foraging modes. At the most general level, *active predators* move in search of prey (**Figure 1(b)**), while *sit-and-wait predators* or *ambush predators* (**Figure 1(k)**) adopt an ambush position and wait for prey to come to them. Sit-and-wait predators can enhance the probability of encounter through the choice of ambush position, but many also actively lure prey. Examples of the latter include female angler fish, which have a modified dorsal spine that protrudes forward dangling a glowing lure above the mouth, and spiders that attract insects by mimicking the bright coloration of flowers or chemical sex attractants (pheromones).

Several variants of these strategies have been described. Active predators that spend a high percentage of their time moving are sometimes referred to as *widely foraging predators*, while those that have frequent short bursts of movement are called *saltatory* or *pause-travel*

foragers. Some biologists use the term *cruise predators* to describe animals that search actively but move slowly compared with other active predators. Other classifications distinguish between *pursuit predators*, which pursue mobile prey, and *close-quarter* predators which detect and capture prey over short distances. *Flush-pursuit* predators use conspicuous movements to startle their prey from hiding and then pursue them. Hunting strategies of snakes have been described as *browsing* (see previous section), *slow pursuit*, or *active pursuit* predation. Among the strategies of active pursuit by seabirds are *deep pursuit divers*, which use underwater flight (e.g., penguins), *foot-propelled divers*, which swim with their feet in pursuit of benthic or pelagic prey (e.g., cormorants), and *plunge pursuit predators* which capture prey using steep dives often from heights of up to 30 m (e.g., gannets; **Figure 1(l)**). Birds in the genus *Rhynchops* (skimmers) are described as *skim feeders*, for their behavior of flying low over water with the tip of the lower mandible submersed and seizing prey items on contact. Some species of whales (e.g., right and sei whales), which feed by swimming through patches of plankton with their mouths open and trapping the prey in a baleen filter, have also been described as skim feeders. Skim-feeding whales are contrasted with *gulp feeders*, which engulf large mouthfuls of water and use their tongue to force it through the baleen filter (e.g., humpback whales), and *bottom feeders* (e.g., the gray whale) which filter invertebrates from mouthfuls of sand and mud taken from the seabed. Four main modes of prey capture occur in fish: *biting*, where teeth are used to capture food (e.g., **Figure 1(a)**); *suction feeding*, where the food is sucked into the mouth; *ram feeding*, where the predator uses body propulsion to capture prey in their elongated jaws; and *filter feeding* (e.g., **Figure 1(e)**) where prey are filtered from the water using various mechanisms that are functionally equivalent to the baleen filter of the whales referred to above. Fish that feed in a manner equivalent to skim-feeding whales are sometimes referred to as *ram filterers*.

Animals that return to a central place (e.g., a nest, burrow, or sleeping site) between foraging trips are referred to as *central place foragers*. Where several such sites are used, this behavior is called *multiple central place foraging*. Many nesting birds are central place foragers when feeding young (**Figure 1(l)**). Some spider monkeys and Lapland longspur birds are examples of multiple central place foragers. Animals are referred to as *hoarders* if they collect a surplus of food when availability allows and store it for later consumption. The food might either be stored in one or a few large caches (*larder hoarders*, e.g., hamsters) or in multiple, dispersed sites (*scatter hoarders*, e.g., gray squirrels).

The principal sensory modality used to detect and distinguish prey is another criterion that has been used to categorize foraging modes. Thus, *visual* foragers (e.g., gannets, **Figure 1(l)**) principally use their eyes; *auditory* foragers (e.g., bat-eared foxes) use their ears, while *olfactory* foragers use their noses (e.g., rats) or tongues (e.g., some reptiles). Many predators combine the use of several sensory modalities in hunting. The American water shrew (*Sorex palustris*), for example, is a flush-pursuit predator that feeds on, among other foods, fish. Experiments have shown that on foraging dives it uses motion to detect moving prey, but also uses odor and tactile hairs (vibrissae) to inspect stationary objects. This is a highly effective combination for a flush-pursuit predator, because it exposes prey to detection both when sitting immobile in the substrate and when attempting to escape.

Finally, an important criterion for classification of foraging strategies concerns the social context in which animals forage. Some animals are *solitary foragers* (e.g., kakapo, **Figure 1(m)**), while others forage in small or large groups. Among the latter, groups may consist of individuals that cooperate in locating and subduing prey (*cooperative hunters* – **Figure 1(b)**), or merely comprise an assemblage of individually foraging animals (**Figure 1(e)**). Cooperative hunting enables predators to prey on animals that are larger, have greater endurance or are faster than would otherwise be possible, while group foraging has been demonstrated in several species to reduce the risk of the foragers falling prey to other animals.

Food Processing

The ability of animals to process foods after capture is an important consideration, because it influences all aspects of foraging from habitat choice to food selection. Foods are first harvested from the environment using various oral structures (teeth, mandibles, filters, sucking stylets, etc.), and exposed to digestive enzymes which help to extract nutrients and convert them to a form suitable for absorption. These enzymes are synthesized either by the animal (i.e., are *endogenous*) or by symbiotic microbes in the gut (*exogenous*). Nutrients are absorbed through the gut wall either passively by diffusion, actively by means of energy-driven nutrient transporters, or using a combination of passive and active mechanisms. In some cases, solid foods are swallowed whole (e.g., many predators), but many animals mechanically process foods to facilitate swallowing and increase the surface area exposed to digestive enzymes. This is commonly done using oral teeth or equivalent structures (e.g., arthropod mandibles), but some birds, reptiles, fish, and invertebrates use a muscular part of the gut called a *gastric mill* (also known as a gizzard) to grind food, sometimes with the aid of stones that are swallowed especially for this purpose. In some fish, a secondary set of jaws in the pharynx, called *pharyngeal jaws*, is used to crush and grind food (e.g., the New Zealand inanga, **Figure 1(j)**). There are many examples of how the strategies of mechanical and chemical breakdown of foods are adapted to the diet and lifestyle of animals, but in the context of foraging modes the most interesting aspect is the relationship between the rates at which nutrients are eaten and the efficiency with which they are extracted from the foods.

Overall nutrient gain is the product of the amounts of nutrients eaten and the efficiency with which they are extracted from the foods and retained by the animal. The rate of overall nutrient gain is often constrained by a trade-off between these parameters: an increase in either the feeding rate or the efficiency of digestion is offset by a decrease in the other. Both *extrinsic constraints* (ecological) and *intrinsic constraints* (i.e., properties of the animal) are responsible for this trade-off. An example of an extrinsic constraint is that high-quality foods that are easily digested, such as other animals, young leaves and seeds, tend to be less abundant and/or more difficult to find or capture than foods with a high proportion of poorly digestible material such as grasses, tree leaves, wood, and corals. An important intrinsic constraint is that gut size imposes a limit on the amount that can be eaten, and when filled to capacity the animal can increase its intake further only by passing the food through the gut more rapidly. In so doing, it will decrease the time available for digestion and absorption, and hence reduce the efficiency with which nutrients are extracted. Conversely, increased digestive efficiency requires prolonged exposure to digestive enzymes and this limits the rate of intake.

The trade-off between the quantity of a food type available and its digestibility has helped to explain many issues associated with the foraging strategies of animals. For example, herbivores tend to spend more time eating, retain foods for longer in the gut, and have larger guts than do carnivores, while the guts of omnivores tend to be intermediate. Large guts ease the limit on the rate of consumption by herbivores, and help to compensate for the low digestibility of plants by increasing the time that food is exposed to digestive enzymes and absorptive tissue of the gut. The smaller guts of carnivores, on the other hand, reduce the energetic costs of maintaining expensive gut tissue and possibly also improve agility and hence prey capture efficiency. Carnivores also tend to have larger brain size when scaled for body mass than do herbivores. Since both gut and brain tissues are energetically expensive to maintain, it has been suggested that the energetic savings due to the small guts of carnivores might help to fund their larger brains and the associated behavioral complexity.

The quantity–quality framework has also helped to make sense of differences in the distributions of feeding types among birds and mammals. Very few birds (3% of living species) eat leaves, probably because birds have a high metabolic rate, and hence higher energetic requirements than can readily be satisfied by a poor quality diet. Also, the larger guts needed for processing leaves are costly to maintain and transport, and even more costly to transport when filled with slowly digesting bulky foods. Not surprisingly, most folivorous birds are poor flyers and some, like the New Zealand kakapo (**Figure 1(m)**), have an unusually low metabolic rate. Kakapo also preprocess leaves to selectively ingest the soluble components and discard a fibrous 'chew' (**Figure 1(n)**). An interesting exception among birds is geese, which are both folivorous and exceptionally strong flyers. Studies have shown that these birds have biochemical adaptations that increase the rates at which enzymes digest and gut transporters absorb nutrients.

Diet quality varies not only between herbivores and carnivores, but also within these groups. For example, as discussed in the section 'Foods', some mammalian browsers tend to select high-quality plants and plant parts (i.e., they are 'concentrate selectors'), while others eat varying amounts of fibrous foods with lower concentration of readily extractable nutrients. A related classification is based on the ways that herbivores deal with the fibrous, structural components of plants. At the one extreme are *cytoplasm consumers*, which avoid dietary fiber through selective feeding (e.g., **Figure 1(n)**), or else eat large amounts of fibrous plants and selectively assimilate the cell contents while defecating the tough cell wall components more or less intact (e.g., the giant panda, **Figure 1(o)**). Other herbivores, called *cell wall consumers*, retain fibrous foods in the gut for long periods during which the cell walls are digested. Many herbivorous insects are cytoplasm consumers, whereas most mammalian herbivores are cell wall consumers. This distinction might have evolved because the large size and high metabolic rate of mammals makes it difficult to harvest cytoplasm at a rate that can satisfy their nutritional needs, whereas for smaller insects it is more challenging to retain enough food in the gut for the long periods needed to digest fiber. Exceptions are many wood-feeding insects, including termites, which are cell wall consumers (rely on cellulose digestion as a nutrient source), and among mammals giant pandas (**Figure 1(o)**), which are cytoplasm consumers (eat large amounts of fiber in their diet of bamboo, most of which is excreted in the feces).

Cell wall consumers tend to rely on specialized fermentative chambers in the gut which house fiber-digesting microbial symbionts. In some of these animals, known as *hindgut fermenters*, the chamber is located posterior to the small intestine, whereas *foregut fermenters* have a primary chamber located in the first part of the stomach (but almost always also have some fermentation in the hindgut). Both strategies have advantages and disadvantages. Having the fermentation chamber after the stomach and small intestine allows hindgut fermenters to absorb soluble nutrients before they are utilized or transformed by the microbes. On the other hand, amino acids that are released from microbes in posterior fermentation chambers do not make contact with the more anterior absorptive sites, and

so are lost in the feces. Animals with anterior fermentation chambers, by contrast, are able to absorb symbiont-derived amino acids as they are released from the chamber and move down the gut over the absorptive surfaces. Anterior chambers therefore help to increase nitrogen utilization efficiency, and they also enable microbes to detoxify some noxious plant compounds. The problem of losing microbial amino acids has been solved by some mammalian hindgut fermenters by eating their own feces (*coprophagy*), while some herbivorous birds have evolved the ability to absorb amino acids in the hindgut. Not all herbivores that ferment their foods do so in specialized chambers: the New Zealand butterfish (**Figure 1(a)**), for example, ferments algae in a relatively undifferentiated, tubular hindgut.

Finally, the strategies of digestion are sometimes classified using chemical reactor theory. Three types of chemical reactors have been recognized in animal digestive system. *Batch reactors* process ingested foods in large, discrete batches, which are poorly mixed. They may have only one opening, such that food is ingested and feces excreted through the same hole (e.g., cnidarians, such as sea anemones), but are also found in some animals that infrequently ingest large prey items and eject indigestible remains through the mouth (e.g., some predators, such as owls, regurgitate undigested hairs and bones). In *plug flow reactors*, the food passes along a tubular gut or gut section in the order that it was ingested, and as it does so nutrients get digested and absorbed. The small intestine of many animals most closely approximates this strategy. Plug flow reactors provide a high rate of digestion but are usually not particularly efficient, and so are best suited to animals with a readily digestible diet such as carnivores and nectar-feeding birds. This explains why the guts of such animals are dominated by the small intestine. In the third category of digestive strategy, the *continuous-flow stirred tank reactor*, food flows more or less continuously through a spherical reaction chamber which is kept well stirred. The fermentation chambers of foregut fermenters best approximate this strategy.

Priorities: The Dimensions of Foraging

So far, I have shown that many criteria have been used to classify animals into foraging modes – habitat, foods, sensory capabilities, hunting strategies, food processing, etc. The priority for the future, I believe, is to move ahead in understanding the ecological and evolutionary factors that drive this diversity. For this, attention needs to be focused on the classifications themselves and how animals are distributed in relation to these classifications. As I see it, there are three central questions, which are characterized geometrically in **Figure 2**: the 'dimensions,' 'distribution,' and 'cohesion' questions.

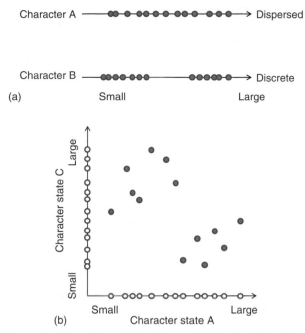

(a)

(b)

Figure 2 Dimensions, distributions, and cohesion in foraging modes. (a) Two dimensions are assessed (characters A and B), one of which shows a discrete distribution of cases and the other a dispersed distribution. (b) Bivariate distributions (filled circles) can be discrete, even when both the univariate component distributions (hollow circles) are continuous. The discrete bivariate distribution reveals cohesion between the two characters: animals with a high value for character A tend to have a low value for character B and vice versa.

The 'dimensions' question concerns the decision of which criteria to use in classifying the foraging mode of an animal (**Figure 2(a)**). For example, do we use the diet of the animal, or the mode of capture as well as the diet? If the diet, which aspect or aspects of the diet should we use: the size of the prey (microalgae vs. macroalgae), the food consistency (liquid vs. solid), or the trophic status of the prey (e.g., plant vs. animal)? Ultimately, this decision will depend on the specific interests of the researcher. It will, however, often be the case that carefully chosen combinations of variables will provide a better framework for addressing unresolved questions and revealing new ones than will any single dimension. We saw an example of this in the section 'Foods', in relation to the classification of mammalian herbivores into grazers versus browsers. In another example, the diets of insects have been usefully classified according to their consistency (liquid vs. solid) and their trophic status (plant vs. animal).

The 'distributions' question asks whether animals fall into discrete groups along an axis or whether they are dispersed on a continuum (**Figure 2(a)**). The digestive strategies of mammalian and avian cell wall consumers (see section 'Food Processing') provides an example of a discrete strategy, since these animals tend to either have an enlarged fermentative chamber in the foregut or in the hindgut, but not both. Some researchers believe that the searching behavior of lizards represents an example of a continuous strategy in which sit-and-wait and active predators occupy the two extremes. Sometimes, however, it can be misleading to categorize animals based on the distributions of a single dimension, because combining dimensions may give a very different picture (**Figure 2(b)**). This is illustrated in a recent study, where two measures of movement were integrated to examine the foraging strategies of lizards: percent time moving (PTM) and number of movements per minute (MPM). Results showed that while PTM and MPM are individually continuously distributed, lizards tend to occupy discrete regions of the space that is defined by both variables.

Examination of the distributions of two or more dimensions exposes a third question, concerning the 'cohesion' of traits related to foraging modes. This asks: to what extent are animals that are similar on some axes also similar on other axes? One example comes from studies of amphipods, which suggest that mobile foragers (mobility axis) might benefit more from mixed diets (diet breadth axis) than to do sedentary foragers. A second example is the association of a high proportion of leaves in the diet with large gut size (see section 'Food Processing').

Finally, an understanding of foraging modes requires that the question of cohesion is extended beyond traits that are immediately related to foraging, including also other aspects of the biology, life history, and ecology of animals. An example of this is the observation that in endothermic vertebrate herbivores diet quality correlates negatively with body size. This correlation exists because both nutrient requirements and gut size decrease with body size, but the slope of decrease is steeper for gut size than for nutrient requirements. Consequently, smaller animals have higher nutrient requirements per unit of gut tissue than do larger animals, and to satisfy these requirements they need to target readily digestible foods; in contrast, the relatively larger guts of bigger animals enables them to process poorer-quality foods. This allometric model has provided a framework within which to understand various adaptations that have evolved for partially circumventing the constraint that body size imposes on diet selection and allowing some smaller animals to capitalize on relatively abundant low-quality foods. One mechanism, which exists both in small mammals and birds, is to selectively retain the readily digestible components of the food and more rapidly excrete the larger, poorly digestible particles. In this way, smaller animals are able to assemble a diet of relatively high digestibility by eating poor-quality *foods*.

Body size has, similarly, been an important factor in understanding the diet choice of mammalian predators. Analyses show that a discrete transition exists in the relationship between the body size of mammalian predators and the size of their prey, such that animals of 25 kg or lighter take prey that are less than 50% of their body weight, whereas only larger predators prey on animals with body mass that approaches their own. The likely reason for this is that small prey such as insects provides an abundant food source for smaller predators, but larger predators cannot acquire these at a high enough rate to satisfy their higher absolute energy requirements. This study provides a good example of how dimensions can be combined (prey size and body size) to reveal interesting distributions (in the relationship between predator and prey body size) which provide new insight into cohesion between traits associated with foraging.

See also: Digestion and Foraging; Empirical Studies of Predator and Prey Behavior; Habitat Selection; Klepto-parasitism and Cannibalism; Optimal Foraging and Plant–Pollinator Co-Evolution; Optimal Foraging Theory: Introduction.

Further Reading

Clements KD and Raubenheimer D (2006) Feeding and nutrition. In: Evans DH (ed.) *The Physiology of Fishes,* 3rd edn., pp. 47–82. Gainesville, FL: CRC Press.

Cooper WE (2005) The foraging mode controversy: Both continuous variation and clustering of foraging movements occur. *Journal of Zoology* 267: 179–190.

Horn MH (1992) Herbivorous fishes: Feeding and digestive mechanisms. In: John DM, Hawkins SJ, and Price JH (eds.) *Plant-Animal Interactions in the Marine Benthos*, pp. 339–362. Oxford: Clarendon Press.

Hume ID (2002) Digestive strategies of mammals. *Acta Zoologica Sinica* 48: 1–19.

Illius AW and Gordon IJ (1993) Diet selection in mammalian herbivores: Constraints and tactics. In: Hughes RN (ed.) *Diet Selection: An Interdisciplinary Approach to Foraging Behaviour*, pp. 157–181. Oxford: Blackwell Scientific Publishers.

Karasov W and del Rio CM (2007) *Physiological Ecology: How Animals Process Energy, Nutrients and Toxins*. Princeton, NJ: Princeton University Press.

McWilliams SR (1999) Digestive strategies of avian herbivores. In: Adams NJ and Slotow RH (eds.) *Proceedings of the 22 International Ornithological Congress Durban*, pp. 2198–2207. Johannesburg: BirdLife South Africa.

Starck JM and Wang T (eds.) (2005) *Physiological and Ecological Adaptations to Feeding in Vertebrates*. Enfield, NH: Science Publishers.

Stephens DW, Brown JS, and Ydenberg RC (eds.) (2006) *Foraging*. Chicago, IL: University of Chicago Press.

Forced or Aggressively Coerced Copulation

P. A. Gowaty, University of California, Los Angeles, CA, USA; Smithsonian Tropical Research Institute, USA

© 2010 Elsevier Ltd. All rights reserved.

Definitions: Phenomenological and Functional Considerations

Forced or coerced copulation is here defined as any copulation that is not freely sought or accepted by one of the individuals copulating. This definition is sex-neutral, indicating that in theory individuals of either sex may be vulnerable to forced copulation, depending on the behavior, physiology, and morphology of gamete sharing. Forced copulation does not depend on observation of either insemination or fertilization. The definition is operational and depends only on observable variation in copulatory behavior of individuals of both sexes in a given population or species.

It is not always easy to clearly determine if a copulation is coerced or forced, so that investigators of non-human animal behavior justify claims of forced copulation based on the existence of resistance and comparison of a copulation within the context of 'normal' copulatory behavior for a given population or species. For example, in obligatory traumatic insemination that occurs in bedbugs or octopus, every insemination may appear to be forced, when it remains likely that some females may freely seek or accept insemination; thus, obligatory traumatic insemination is not always resisted by females and thus may not be forced. In many other species, females signal their receptivity to copulate with a particular male. Receptivity signals are sometimes subtle as in *Drosophila* where female acceptance is often indicated by standing still. In contrast, rejection signals are often dramatic, including motor patterns that are ritualized and stereotyped, such as kicking, biting, emitting alarm calls, or females moving the position of their ovipositors, running or flying away or attempting to run or fly away. If rejection and resistance signals are observable, investigators are justified in naming a copulation as 'forced' when males ignore female rejection and resistance signals and attempt to copulate. In some cases such as when males copulate with teneral females as described for *Drosophilia* species by T. Markow in American Naturalist, copulation is called 'forced' because females cannot resist as their exoskeletons are not yet hardened. Furthermore, these immature females are sometimes injured when males copulate with them.

An advantage of the operational definition used here is that it allows investigators to examine alternative functional explanations of adaptive significance, if any, of forced copulation (the function is not implied by the definition of the motor pattern). Thus, this definition differs from Randy Thornhill's insistence in an early study on the adaptive significance of forced copulation that to demonstrate forced copulation one must observe not only that females resist the copulation (phenomenological criterion), but also that a male that force-copulates a female thereby enhances his own fitness (functional criterion). A disadvantage of the definition used here is that it is biased against observing forced inseminations that are cryptic and difficult to observe as they are in garter snakes (Shine et al., 2003). In such cases, the extraordinary knowledge of particular species allows natural scientists to demonstrate coercive copulation.

Investigators of non-human animals, who observe that individuals being forced to copulate are often physically hurt or even killed in the process, are inclined to seek ultimate or functional explanations for escalating female rejections. Why do females resist copulation if there is a risk of injury or death from resistance? Because the effects of aggressive copulation are sometimes so great, it is hard not to imagine that the stakes are very high for females who give in to copulations they resist. Yet, in most cases, investigators do not know what the fitness costs of accepting a resisted copulation may be. Unless it is extremely rare, forced copulation and/or male aggression against females is likely to have some function for males who force copulate. But again, in most cases, it is not yet clear what the function(s) of forced copulations are for males (in section below 'Fitness Dynamics of Sexual Conflict'). It is also likely that the adaptive significance of female resistance is different from the adaptive significance of force by males. Aggression associated with so-called forced copulation is exaggerated, one wonders if it is reasonable to categorize the behavior as copulation. Are apparent forced copulations better characterized as male aggression against females? Thus, understanding the adaptive significance of female resistance to forced copulation is a crucial goal for investigators of animal behavior.

Is It Forced Copulation or Male Aggression Against Females?

A further complication arises when male aggression against females looks like forced copulation, something that happens when more than one male 'gangs up' on female bank swallow or female mallard. Gangs force the female to the ground, so that males can stand on the female's back and pummel her. When this happens, it is difficult, perhaps impossible, for observers to see cloacal contact (bank

759

swallows) or intromission of the penis (in mallards). Observers assume such interactions are forced copulation because males are in the copulatory position on the backs of females. When what looks like forced copulation occurs in the winter when fertilization is impossible as it often does in mallards, it is more likely that it is male aggression against females rather than a reproductive act. With or without intromission and insemination, males can hurt females when males are on females' backs, and indeed these interactions sometimes kill females. This moves the question away from why do males force-copulate to why are males are aggressive to females. As Barb Smuts and R. Smuts showed years ago, male aggression against females is common.

Rape Includes more than Forced Copulation

In contrast to forced copulation, by social convention (laws) in many democracies of the world, rape is a type of sexual assault that is violence against women, and whether it affects the biological fitness of the women who are raped (except when women are severely injured or killed) and the men who rape them is extraordinarily difficult to evaluate. In fact, decrements in the victim's Darwinian fitness or enhancements to the rapist's Darwinian fitness are not included in the criminalization of rape. Therefore, many in the social sciences and humanities, in law, and among the public-at-large argue that the adaptive significance of rape is beside the point. What is agreed upon in the social sciences is that rape is violence against women, and rape is widely considered a mechanism for the control of women's sexuality and reproductive behavior. Rape in humans is sometimes not physically forced; an example of rape that is not physically coerced is statutory rape that occurs when an adult has apparent consensual sex with a minor. Men most often are statutory rapists; nevertheless, women too have recently been convicted as statutory rapists, because they had apparent consensual sex with teenage boys. Because of the important differences between rape and forced copulations, those of us who study animal behavior agreed years ago to refer to 'forced copulation' in non-human animals, and to reserve the term 'rape' for humans.

Taxonomic Distribution of Forced Copulation

Forced copulation occurs in a wide variety of species including the great apes – chimpanzees, but not bonobos, more frequently in orangutans – as well as several species of old and new world monkeys including spider monkeys *Ateles belzebuth chamek*, quanacos *Lama quanicoe*, lizards including iguanas, *Iguana iguana*, marine iguanas, garter snakes, fish, crustacia including some spiders, and insects species including dragonflies such as *Calopteryx haemorrhoidalis*.

Forced copulations are common in duck species, and forced copulation is particularly well studied in mallards. Despite the widespread taxonomic distribution of forced copulation, it is far from ubiquitous, being absent from most species studied to date. For example, in some species such as *Anolis carolinensis*, in which investigators have specifically looked for evidence of coerced matings (those that are forced, associated with harassment or intimidation), forced copulations do not occur. In addition, under some ecological conditions, females are more likely to be force copulated than others; so that interpopulation variation in forced copulation exists. Thus, two questions arise, (1) why are some species more likely to exhibit forced copulations than other species? And, (2) why within populations and species are some females less vulnerable to forced copulation and other forms of male aggression than other females?

Fitness Dynamics of Forced Copulation

Two premises suggest that female resistance is a defensive response to some loss of fitness that would be associated with the copulation. The first is that selection should favor females able to control their own reproductive decisions, and the second is that female resistance to male control should evolve whenever male control of females' reproduction is costly to females as discussed in a review by P. Gowaty in 1997. Forced copulation may affect females' survival probabilities as well as reproductive success. Experiments on flies *Drosophila pseudoobscura*, cockroaches *Naupheta cinerea*, mice *Mus musculus*, fish, and mallards *Anas platyrynchos*, which tested the fitness effects for experimental females that were constrained to reproduce with males they did not prefer and females that were allowed to mate with a male they did prefer, showed that offspring viability and mother's productivity were significantly lower when females lacked control of the decision about with whom to mate. Despite some clarity on the benefits of resistance for females, the fitness benefits for males who force copulate are not yet clear. As Gowaty and Buschhaus hypothesized in 1998, males may increase their mating success, their immediate fertilization success, or they may gain kin-related benefits.

Fitness Dynamics Suggest that Sexual Conflict Results in Sexual Dialectics

When males seek and females reject copulations, it is hard not to think that the interests of the male and female are in conflict over the copulation. When the fitness interests of the sexes conflict, animal behaviorists expect dynamic interactions between the subjects. When females reject male solicitations for copulation, some males persist, so that, when they can, females will run or fly away. When females cannot escape by leaving, whether the female is force-copulated or not, will depend on her ability to retaliate by kicking, hitting, or biting the forcing male, as happens among dwarf chameleons, *Bradyodion pumilum*. If female resistance behavior fails, once inseminated, females may often expel the inseminate immediately. If forced insemination is successful, evolutionary biologists expect that female resistance physiology will evolve, so that many investigators now assume that once a force-copulated female is inseminated that females most often control what happens after that, something that Bill Eberhard described in his 1996 book called *Female Control*. Female reproductive physiology and morphology suggest very long histories of female control of inseminates, in that females can kill sperm, encapsulate them, or store sperm for later use. And, in response, mechanisms for the physiological manipulation of females have evolved among males including hormone mimics that act as physiological chastity belts, which may manipulate females into using the sperm of the inseminating male rather than another male's sperm. No one has yet asked whether females have the ability to inhibit the effects of manipulative male seminal peptides. However, as long as the fitness interests of the sexes conflict, evolutionary theory says that the dynamics of male attempts to control and female resistance will continue whenever any variation exists upon which selection can act.

Variation Between Females in Their Vulnerability to Forced Copulation

Females in different species experience different degrees of threat of sexual coercion and forced copulation, and in most species between-female variation in their vulnerability to control by others including forced-copulation is predictable. Variation in females' vulnerabilities to control of reproduction by others depends upon both intrinsic differences between females and ecological differences. For example, in *Iguana iguana*, a male force copulated a female who lacked a front leg and was therefore crippled, possibly unable to escape the forcing male by running away or fighting. Despite the fact that between-individual, within-population variation exists in females' abilities to avoid forced copulation, and despite the fact that variation among females is a key predictor of the outcome of forced copulation attempts as P. A. Gowaty has insisted, there are few studies in non-human animals that yet test the importance of between-female variation in vulnerabilities to forced copulation or in female resistance behavior during attempted forced copulation.

Females Resist Forced Copulation in Many Ways

Females at risk of forced copulation may solicit protective services from other males or from relatives to reduce their risk of forced copulation. During the season when sub-adult males are most likely to harass female Sumatran oranutans *Pongo pygmaeus abilii*, females solicit protective services of adult males from forced copulation by subadult males. Thus, females may trade protection from males by allowing protective males sexual access. This is not an uncommon pattern in species with larger males than females and male aggression against females. The potential trade of sexual access for protection was one of the key ideas in CODE hypothesis of Gowaty and Buschhaus.

Females may resist copulation attempts or other types of male harassment by hiding, wedging themselves between rocks as do marine iquanas, *Amblyrhynchus cristatus*, or flying away as adult *Drosophila* females do.

If males are able to force copulate females, females may eject the inseminate as do mosquitoes *Aedes aegypti*, some species of ducks and geese, razorbills *Alca torda*, northern gannets, penguins, humans and presumably the other great apes, and zebras. Or, they can denature sperm, sequester it, or store it for later use, possibilities that are common in insects in which females often have more than one type of organ for killing, sequestering, or storing sperm for later use.

Questions that have yet to be addressed include how variable females are within populations in their morphological, physiological, or behavioral resistance mechanisms. Given that it is this variation on which rests the likelihood of successful forced copulation, these are important questions that deserve much more research attention.

The Evolutionary Power of Female Variation

The selective force of female resistance may have favored the evolutionary loss of the intromittant organ in passerines. In species in which males have intromittant organs, females are more vulnerable to forced copulation than in species without. Consider birds. Ancestral birds had intromittant organs, but 97% of extant bird species lack intromittant organs. Copulation occurs when both the female and the male evert the second compartments of their cloacae through their vents into the air and are then touched together. During these usually very brief cloacal kisses lasting from 1 s to about 2 min, sperm are transferred from the male's cloacal surface to the female's. Because females are as active as males in these copulations, sperm transfer cannot take place without females' active cooperation. Therefore, many ornithologists and animal behaviorists consider forced copulation impossible in bird species that lack male intromittant organs. The fact that forced copulation is common in bird species with intromittant organs and much rarer, perhaps absent, in species without intromittant organs is consistent with the hypothesis that male intromittant organs are a means of male control of females' reproduction. The selection pressure favoring loss of intromittant organs in birds is much debated, but it is readily explained by a female control hypothesis. Jim Briskie's and Bob Montgomerie's argument is related to avian life histories, notably that females lay only a single egg at 1- or 2-day intervals, so females may readily abandon eggs fertilized by males they do not prefer without risking the loss of her entire clutch. Flexible females able to facultatively destroy an egg produced the selective pressure that favored males who could not force-copulate. A similar argument is based on the conclusions of experiments described earlier that showed that females constrained to mate with males they did not prefer had lower viability offspring then females not so constrained. If this is usually the case, the selection pressure favoring the loss of male intromittant organs would have been variation in the viability of offspring of females who were forced copulated and females who freely chose with whom to copulate. Thus variation in offspring viability would favor not just female resistance, but the loss of intromittant organs in males. Both the female life history constraint hypothesis and the offspring viability hypothesis are reasonable explanations for the loss of the male intromittant organ in most passerines.

Forced Copulation Favored Induced Ovulation in Canids

Induced ovulation may function as a guard against forced copulation and subsequent forced fertilization by males that females do not prefer. An observation inconsistent with induced ovulation as a resistance mechanism would be no differences in the likelihood that ovulation is induced by copulation with a male the female prefers compared to a male the female does not prefer.

Hypotheses Explaining Forced Copulation

All evolutionary hypotheses attempting to explain forced copulation assume that the fitness interests of the sexes conflict. Some hypotheses explaining forced copulation deal exclusively with why males attempt forced copulation. There have as yet been no strong inferential tests in any species designed to eliminate alternative predictions of these hypotheses.

The inferior male hypothesis of Thornhill explains forced copulation only in species in which females depend on males to provide resources necessary for reproduction such as food or nesting sites. It assumes that females are unable to provide themselves with resources necessary to reproduce. The hypothesis implicitly assumes that females cannot sequester or denature sperm once it is inseminated. It also assumes that the chief basis for female preferences for males is male resource accruing ability. This hypothesis predicts that males who are unable to provide females with resources for reproduction resort to forced copulation and

thereby gain some immediate enhancement to their reproductive success, such as immediate fertilization success. This hypothesis therefore cannot explain forced copulation in the many species in which males do not provide females with resources for reproduction; nor does it apply to species in which males that do supply resources to females also force copulate females.

The by-product hypothesis of Palmer says that forced copulation is not in itself fitness enhancing for males but is a by-product of other traits that do enhance male fitness. This hypothesis assumes near universal sex differences in behaviors associated with reproductive decisions: it assumes females are evolved to be choosy in contrast to males being evolved to be ardent, indiscriminate, and profligate about with whom they mate. This hypothesis says that male ardor occasionally slips into intimidation, threat, and forced copulation of females who reject them, so that this hypothesis predicts that all males may force copulate. It is notable in not predicting fitness variation among males that force copulate, but does predict that females will suffer reduced fitness relative to females who are not force-copulated.

The CODE hypothesis of Gowaty and Buschaus assumes that whenever females are not in control of their reproductive decisions, that female fitness is lower than when they are in control of their reproductive decisions. The CODE hypothesis assumes females can eject, denature, or sequester sperm; thus, it assumes that forced copulation is unlikely to result in immediate fertilization success for forcing males. It says that forced copulation is a type of extreme male aggression against females that conditions the future behavior of females – those who are actually force-copulated and those who only witness the force – for the benefit of the male and his male kin. This hypothesis says that forced copulation creates a dangerous environment for all females so that females are willing to trade future copulations for male protective services. The CODE hypothesis predicts that any male may force copulate. It predicts that male benefits of forced copulation accrue to all males in the population enhancing the probability that all males have a social mate who copulates exclusively with them; thus, the CODE hypothesis can account for the evolution of social monogamy in species with no male parental care. It says that because male aggression constrains female reproductive decisions, females have lower fitness when male aggression against females is higher than in contexts where male aggression against females is reduced or absent.

The killing time hypothesis of Gowaty and Hubbell does not assume that females have evolved to be necessarily choosy and males necessarily indiscriminate. It does assume that females can kill, sequester, or eject sperm of males they do not prefer. This hypothesis says that male aggression against females decreases female's survival probabilities. The killing time hypothesis is derived from the switchpoint theorem of Gowaty and Hubbell that demonstrated that individuals who trade off likely fitness gains and losses (from mating with this or that particular mate) against the time they have available for mating and reproduction will have higher fitness than individuals who do not trade off time and fitness. That it is says that individuals who have fixed reproductive decisions, such as only accept these, always reject others, will have lower fitness than individuals who flexibly adjust their decisions. Thus, the switchpoint theorem says that individuals who have more time available for mating and reproduction will reject more potential mates than individuals who have less time. The switchpoint theorem showed that whenever time available for mating is reduced, say from a very long search time for potential mates or from reduced individual survival probabilities, individuals will reduce their threshold for acceptance of potential mates. That is, when individuals have less time they will more likely accept a potential mate who will confer lower fitness. Thus, the killing time hypothesis assumes that aggressive or forced copulation reduces female survival probabilities and perhaps also their encounters with potential mates. In turn, females with reduced survival probabilities or reduced encounter probabilities will be manipulated into accept males as mates who confer on them lower fitness. Thus, aggressive or forced copulation may manipulate a female's reproductive decision to use a particular male's sperm. The killing time hypothesis says that male aggression against females is a manipulative mechanism that males may use to exploit females' preexisting biases so that such males persuade females to use rather than to kill his sperm. This hypothesis is the only quantitative hypothesis predicting the effects of aggressive or forced copulation.

Some Remaining Questions about the Natural History of Forced Copulation

There are many additional questions one could ask about forced copulation and aggression against females. Future studies should, as many of the investigators cited here did, thoroughly describe behavior of females and males when they think they are observing forced copulation. If observations are insufficient for definitive observation of intromission or insemination, investigators should consider the related question of why males are aggressive to females. What are the effects on female survival probabilities of forced copulation or male aggression against females? Are there effects also on females' subsequent encounter probabilities with potential mates? As in some of the cited studies here, investigators should carefully document the timing of forced copulations relative to the likelihood of fertilization. The identity of forcing males is not always easy to determine, but that should be a high priority in future studies of forced copulation. Are forcing males also the social partners of the females they aggress or force-copulate? Perhaps most important, because it is least known, investigators should document variation among females in their vulnerabilities to forced copulation. Of course, questions about how forced copulation affects fitness are extremely important for informing our understanding of the fitness effects of forced copulation and should be a component of most studies of forced copulation. There are a few studies of hormonal correlates in males of forced copulation, but so far, few or no studies of the proximate effects on females.

See also: Cryptic Female Choice; Infanticide; Invertebrates: The Inside Story of Post-Insemination, Pre-Fertilization Reproductive Interactions; Social Selection, Sexual Selection, and Sexual Conflict.

Further Reading

Brownmiller S (1975) *Against Our Will: Men, Women and Rape.* New York: Simon and Shuster.

Cluttonbrock TH and Parker GA (1995) Sexual coercion in animal societies. *Animal Behaviour* 49: 1345–1365.

Davis ES (2002) Male reproductive tactics in the mallard, *Anas platyrhynchos*: Social and hormonal mechanisms. *Behavioral Ecology and Sociobiology* 52: 224–231.

Dunn PO, Afton AD, Gloutney ML, and Alisauskas RT (1999) Forced copulation results in few extrapair fertilizations in Ross's and lesser snow geese. *Animal Behaviour* 57: 1071–1081.

Gowaty PA (1996) Battles of the sexes and origins of monogamy. In: Black JL (ed.) *Partnerships in Birds. Oxford Series in Ecology and Evolution,* pp. 21–52. Oxford: Oxford University Press.

Gowaty PA (1999) Extra-pair paternity and paternal care: Differential fitness among males via male exploitation of variation among females. In: Adams N and Slotow R (eds.) *Proceedings 22nd International Ornithological Congress*, Durban, University of Natal, pp. 2639–2656. Johannesburg: BirdLife South Africa.

Gowaty PA (2003) Power asymmetries between the sexes, mate preferences, and components of Fitness. In: Travis Cheryl (ed.) *Women, Evolution, and Power,* pp. 61–86. Boston, MA: MIT.

Gowaty PA and Buschhaus N (1998) Ultimate causation of aggressive and forced copulation in birds: Female resistance, the CODE hypothesis, and social monogamy. *American Zoologist* 38: 207–225.

Gowaty PA and Hubbell SP (2009) Reproductive decisions under ecological constraints: It's about time. *Proceedings of the National Academy of Sciences* 106(1): 10017–10024.

Low M, Castro I, and Berggren A (2005) Cloacal erection promotes vent apposition during forced copulation in the New Zealand stitchbird (hihi): Implications for copulation efficiency in other species. *Behavioral Ecology and Sociobiology* 58: 247–255.

Palmer CT (1991) Human rape: Adaptation or by-product? *Journal of Sex Research* 28: 365–386.

Persaud KN and Galef BG (2005) Female Japanese quail (*Coturnix japonica*) mated with males that harassed them are unlikely to lay fertilized eggs. *Journal of Comparative Psychology* 119: 440–446.

Shine R, Langkilde T, and Mason RT (2003) Cryptic forcible insemination: Male snakes exploit female physiology, anatomy, and behavior to obtain coercive matings. *American Naturalist* 162: 653–667.

Thornhill R (1980) Rape in Panorpa scorpionflies and a general rape hypothesis. *Animal Behaviour* 28: 52–59.

Future of Animal Behavior: Predicting Trends

L. C. Drickamer, Northern Arizona University, Flagstaff, AZ, USA
P. A. Gowaty, University of California, Los Angeles, CA, USA; Smithsonian Tropical Research Institute, USA

Introduction

Animal Behavior Is Biology and Biology Is Animal Behavior

Typical investigators of animal behavior are complete biologists, and sometimes physicists, chemists, or conceptual and mathematical theorists. Often investigators work simultaneously on wild subjects in field settings and on captive animals in laboratories. Most animal behaviorists are experimentalists; some rely entirely on observations of animals in the wild. Animal behaviorists increasingly rely on mathematical modeling to explore the limits of their hypotheses. It is no exaggeration to call animal behaviorists among the most broadly knowledgeable, highly disciplined of biological scientists: We have to be, because our questions are complex.

Pathways to the Future

As in any attempt to predict the future, the past and the present are starting points for our comments and conclusions. Themes – or our main predictions about the future – that wind through our review include:

1. Animal behavior will be thoroughly integrated across traditional lines of proximate and ultimate causation.
2. Some animal behaviorists will continue to have major influence not just as animal behaviorists but as evolutionary theorists.
3. More robust theory, new methods, better-trained investigators, and a broader, more inclusive understanding of hereditary mechanisms will influence the questions we ask and how we investigate them.
4. Social factors, as always, will continue to affect the conduct of research. How we deal with the vagaries of research funding, reviews (of papers, of grant proposals, or colleagues) as individuals and collaboratively will determine the vitality of the discipline of animal behavior.

In the following sections, we examine the classic tensions between ultimate and proximate causation of behavior to introduce some speculations about further integration in the future. We then explore the ways in which new methods will impact what we study and the importance of molecular studies of development, evolution, and gene expression to the fully integrated future we imagine. We conclude with a section on the emerging synthetic approach to behavior.

Tinbergen's Four Questions and the Future

Investigators studying animal behavior classically pursue what is known as 'Tinbergen's four questions': (1) What, if any, adaptive function does behavior accomplish? (2) What are the evolutionary patterns of behavior? How does behavior map to phylogenies? (3) What are the hormonal and neural mechanisms mediating behavior, and (4) How does development or ontogeny affect behavior? The first two questions fall into the category of ultimate questions and the last two questions fall into the category of proximate questions. These questions – deeply embedded in organismal, population, and evolutionary biology – stress the comprehensive and integrative nature of animal behavior done well.

We imagine a future for animal behavior that is fully integrated, rather than dissected into parts and neatly stacked in proximate and ultimate bins. Thus, it should be no surprise to today's animal behaviorists that the new evolutionary theorists are often animal behaviorists. The works of animal behaviorists like Mary Jane West-Eberhard and Marion Lamb and her colleagues are harbingers of an integrated theoretical future for all of biology in which the concept of the determinism of genes will finally give way to a fuller understanding of what influences phenotypes. Their work elaborates and explains the remarkable discoveries of the scientists studying evolutionary development who have relatively recently discovered that organisms as diverse as flies and mammals share most of their genes, and that so-called house-keeping genes are conserved from worms to great apes. These discoveries have propelled longstanding debates about the determination of phenotypes onto center stage, where evolutionary theorists like animal behaviorist West-Eberhard are leaders.

The remarkable discoveries of Frans de Waal and others like him interested in the evolution of morality, the significance of learning and culture to the biology of organisms and to the demography of their populations, suggests that animal behaviorists will continue to lead the full integration of biology, linking processes of individual and group selection for further understanding of the ultimate origins of developmental plasticity. No one is surprised that it is animal behaviorists who are elucidating the influence of learning and culture on individual and group phenotypes. But, historians of science will likely mark as significant the fact that the theorists of developmental plasticity, flexible phenotypes and epigenetics are animal behaviorists, just as the prominent evolutionary theorists of the last 40 years were. (In this context, we think of animal behaviorists/evolutionary theorists William D. Hamilton, Robert L. Trivers, John Maynard Smith, and David Sloan Wilson).

Dominance of Ultimate Questions

Shortly after the birth of sociobiology in the mid-1970s, behavioral ecology was born; several journals specifically devoted to behavioral ecology originated, and it seemed that this subdiscipline was ascending. During the last 35+ years, behavioral ecology has organized most of its questions around the adaptive significance of behavior. Behavioral ecologists ask questions about fitness variation in the present. Typically, they ask two kinds of questions: What is the adaptive significance, if any, of this or that trait? And, what traits evolve in response to this or that selection pressure?

Successes and Unresolved Challenges

Empirical studies testing the predictions of optimal foraging theory were spectacularly successful, as were empirical studies about the evolution of so-called helpers-at-the-nest. Other notable achievements include a vastly enhanced understanding of the evolution of sex allocation. We believe that the success of these three subdisciplines in behavioral ecology is associated with the existence of efficient theory (e.g., optimal foraging theory, kin selection theory, and Fisher's sex allocation theory) and the investigators' use of strong inference in observational and experimental protocols.

Yet, in other areas of behavioral ecology, what we learned in the 1980s is remarkably similar to today's continuing discoveries. For example, advances in our understanding of mating system evolution seem to be limited to small incremental achievements, rather than to definitive resolution of some outstanding questions. Consider what animal behaviorists have learned about social monogamy and extra-pair paternity. In the 1980s and today, we are able to conclude that in most socially monogamous species, some females, at least, and sometimes all females are genetically polyandrous. Nevertheless, we still seem unwilling to conclude what is consistent with most data: that the vast majority of males in these species are genetically monogamous as well as socially monogamous. Strong evidence in support of the hypothesis that some males in these species are genetically polygynous has seldom been observed or reported. Some reasons could be that (1) our tools may not be adequate to answer the question at hand about male fitness variation; or (2) we may need other hypotheses to explain mating systems.

Or, consider the state of sex role science, one of the most contentious areas in animal behavior. Most behavioral ecological theory in sex role science begins with assumptions about 'genes for coyness in females' and 'genes for profligate behavior in males.' There is no doubt that the intuitive theories from the early 1970s from Bob Trivers and Geoff Parker, which were affected by Angus Bateman's classic data on *Drosophila*, changed not only the questions we asked, but over the long run, led to an incredibly large number of new observations. Some of these observations and resulting predictions are consistent with their predictions, and importantly, some that are not supportive of those predictions. What is hard to explain is the failure of students of sexual selection to incorporate the insight of Stephen P. Hubbell and Leslie K. Johnson that demographic stochasticity may explain variances in fitness usually attributed to sexual selection. It is now clear that demographic stochasticity should be a component of efficient theory (see section 'Dialectical Cycles of Efficient Theory, Observation, and Revision') in sexual selection.

We believe that current state of the behavioral ecology of mating systems and sex roles is attributable to (1) paradigmatic dominance, (2) biases against alternative ideas, (3) bandwagon effects on what questions get asked and who gets to publish, and (4) the resultant narrowing of attention to a few questions. It might also be because most theory in mating systems and sex roles has been conceptual, graphical, and intuitive rather than efficient theory from first principles (see later).

Does this somewhat pessimistic view of some areas of behavioral ecology mean that behavioral ecology is dead? We don't think so. Rather, we believe that what we need is new strong, efficient theory from first principles (see later) to guide new investigations. Despite the fact that there is some new efficient theory in animal behavior, it remains for the future to know where additional efficient theory will lead us.

Mechanisms: Proximate Causation

Phenotypic Plasticity

Proximate causes are the mechanisms underlying behavior. Mechanisms tell us how behavior occurs. In terms of mechanisms, it is useful to think of an isosceles triangle (**Figure 1**) with the points labeled as nervous system, endocrine system, and immune system; all of the physiological processes that are the basis for behavior. Inside the triangle, we envision the genetics of the organism. A balloon surrounding the triangle represents all of the organism's behavior. Outside the circle are myriad ecological and life history traits of the animal, bridging to functional and evolutionary aspects of behavior.

There is a longstanding discussion concerning whether behavior is determined by genes or environmental factors.

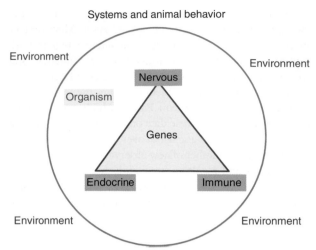

Figure 1 This depiction of the various elements that comprise the scope of work in animal behavior, build outward from the genes and physiological processes to the behavior of organisms in populations and the surrounding ecological conditions.

Today, we accept that all behaviors have a genetic basis, and that all behaviors are influenced by environmental factors. The work of Mary Jane West-Eberhard focused attention on developmental plasticity not just in behavior and physiology, but in morphology as well. This provides a much broader empirical and conceptual platform for a fuller integration of how other mechanisms of heredity along with genes influence phenotypes. In her big book published in 2003, West-Eberhard engages her readers in a long argument that enjoins us to the view that genes and environments are inextricably intertwined in the development of phenotypes. She says her book is about 'the dual nature of the phenotype – the undeniable fact that the phenotype is a product of both genotype and environment, and the equally undeniable fact that phenotypes evolve.' She goes on to say 'there is no escape from the conclusion that evolution can occur without genetic change' (West-Eberhard, 2003, p. 17). Mary Jane West-Eberhard's book is game changing, and it arrived just when gene-expression studies were beginning to be commonplace. The study of the proximate and ultimate causation of behavior will be profoundly influenced by West-Eberhard's work, and we refer readers to her germinal work.

Thus, **Figure 1** can be expanded to include epigenetic mechanisms of behavior as well as environmental variation that induces flexibility of individuals when they are confronted with different environmental or social circumstances.

Mechanistic Effects

What will research on internal mechanisms and developmental aspects of the study of behavior look like in the future? There are three major themes that we think will

emerge for the study of proximate mechanisms of behavior. First, little has been done with respect to interactions between immunology and behavior, and (**Figure 1**) the interactions of the immune, nervous, and endocrine systems affecting behavior. The likelihood of strong connections between immunology and behavior is obvious, as many studies in this collection demonstrate; this field is off and running and also wonderfully open avenue for future research.

Second, techniques involving genetic profiling will enable scientists to relate genes to the interactive effects of environmental factors on behavior. This is going to be true for both the initial development of individuals and, most notably, for changes in its phenotype during its lifetime. These studies may well be the center of animal behavior research in the next decade or two, linking ecology and internal mechanisms.

Third, we will, with modern field techniques, combined with laboratory analyses, gather new information on aspects of the physiology and related behavior of free-living organisms. Wingfield uses the integration of techniques to study annual cycles in birds with an emphasis on endocrine systems. This includes reproduction, migration, and overwintering and accompanying stress levels. His work also integrates theoretical and empirical techniques. Andy Bass has used singing fish to examine ways that phenotypic variation in vertebrate brain organization leads to adaptive behavioral phenotypes, with an emphasis on neural systems.

New Methods

Field and laboratory technologies and protocols from the last several decades, coupled with recent and future advances in the apparatus and analysis of data from these systems, will enhance the study of behavior. Simple radio tracking of specific animals has evolved to the use of satellites that can not only continuously record the animal's position, but can also provide information on various physiological parameters. Sophisticated remote camera systems, augmenting the earlier photos triggered by animals, can give us more than just a picture; these images can be used for gathering data on behavioral sequences and even interactions.

Another important advance was development of collection of samples for analysis of hormone levels and other chemical constituents in urine and feces. Animals need not be disrupted or handled, to obtain samples, reducing the confounding effects of capture and restraint. These are not immediately new techniques, but combined with other, newer technologies, they produce much more accurate and highly significant information.

A series of new laboratory technologies promises fresh insights into behavior and the underlying genetics and physiology. The first of these, with a longer than 30-year-old history, involves bringing samples to the laboratory from field settings and then translating the information back to the natural setting. Several of these techniques involve the use of DNA and other genetic material for assignment of parentage, determination of mating partners, and using gut contents to determine specific prey items consumed by predators. New uses for genetic information arise on a regular basis. The future holds real-time, mobile, genetic parentage devices that will result in immediate analysis in the field that will allow further observations and experiments on wild-living animals whose histories we can readily infer. Our work will go faster in the future.

The use of stable isotope technology enables scholars to examine, in detail, the specific locations where animals obtained their foods. Elaboration of detailed food chains and food webs is possible using such technology. The measurement of hormone levels from urine and feces through laboratory-based assays leads back to the interpretation of behaviors in relation to reproductive success, stressful environments, or social stress.

Several procedures with laboratory animals provide insights regarding neural aspects of behavior and genetics and behavior. Some of these techniques may, eventually, be applicable to animals in free-living situations. Techniques for brain imaging and measuring activity in specific brain areas when the animal is engaged in particular activities, which we do regularly with human subjects, and some non-human primates, are applicable to other animals as well. Molecular biology-based systems for genetic profiling are a good bet in terms of working out relationships between genetics, development, and phenotypic traits. As we gather information with genetic profiling, a technique, which shows us which genes 'light up' via tags we have applied, provides basic information on which genes are turning on and off, in what sequences, and on what sorts of time scales. While the uses for this technology are mainly in the basic medical sciences at present, there will be increased applications in animal behavior.

Other Developments: Applied Animal Behavior

There are some important problems involving animal behavior that are both new and expanding. Behavioral toxicology is most often associated with humans and involves a variety of pollutants and toxic substances that negatively impact behavior. Lead paint and brain development in children is an example. Animal models and direct effects of toxic substances on non-human animals is also a problem. The entire study of endocrine disrupters on physiology and behavior, which involves invertebrates such as snails, and many insect species, and vertebrates such as fish, birds, mammals, and humans, is based on human-made substances that now pollute our environment. Chemicals

that disrupt endocrine systems influence diverse behaviors including reproduction, parental care, activity level, intraspecific partner preferences, and others. Current and future studies will require additional attention to the behavioral consequences of the chemicals that we directly or indirectly place in our environment.

The mix of behavior, ecology, and disease, as exemplified by aspects of the some of the newly emerging viruses like hanta virus, is quite important, both for human health and for wildlife biology and conservation. Understanding the behavior of organisms like deer mice that carry the hanta virus will enable better understanding of how the disease is transmitted and perhaps some suggestions on how to control the spread of the virus. Another example involves plague, which is found in a number of western US rodent species, most notably in prairie dogs where it is fatal. Plague can infect humans and several other mammals; the Black Death in Europe in the fourteenth century was due to the same organism. Knowing more about prairie dog behavior, the nature of the rodent reservoirs for this disease, and possible vectors will be necessary to protect humans and various wildlife species.

We already have considerable progress to show for work on applied animal behavior as it concerns both domestic species and animals that are of interest in terms of conservation. Understanding the behavior of our pets provides a better experience for both the pet and its owner. Work on behavior of domestic livestock helps in terms of animal welfare concerns and with productivity. Conservation of a number of endangered and threatened species benefits from our knowledge of behavior. This is most evident when one examines the social and mating systems of particular species. Giving animals in captive breeding programs more choice in terms of mates leads to better success with production of progeny. Understanding some aspects of the behavior of animals like oryx, from both wild and captive studies, enables reintroductions of these animals in areas that they formerly occupied. We expect that new avenues of work, including a fuller understanding of topics like phenotypic flexibility mentioned earlier, will be important to these investigations of applied aspects of behavior. So too will new technologies and some of the protocols for collecting, for example, endocrine data, on free-living animals in seminatural settings.

Synthesis: The Wave of the Future
Dialectical Cycles of Efficient Theory, Observation, and Revision

The great flowering of animal behavior since the late 1960s came about, we think, because animal behaviorists formulated hypotheses and models that captured relevant details but provided an abstracted description of behavior that yielded novel predictions of yet unobserved phenomena. These efficient theories, in turn, motivated novel studies that fueled the observational and the experimental successes of the last 40 years. Think of the simplicity and elegance of Hamilton's kinselection theory, for example, an idea that has reached all areas of biology, explained the adaptive significance of social insects and helping in birds, and predicted the then unobserved phenomenon of kin recognition. We mention the spectacular advances that accrued from kin selection theory because we think the efficiencies of theories like Hamilton's represent the wave of the future in animal behavior. Hamilton's theory came from the iterative processes of induction and deduction; his focus on first principles yielded novel predictions and inspired new tests. As more animal behaviorists and their students recognize the value of the dialectic tension between developing efficient theories from first principles, assumption and prediction testing, and theory revision, we predict even greater progress in the science of animal behavior.

One of the great proponents of this dialectic between theory, prediction, and testing was John Platt, who in 1964 invited all scientists to pursue what he named strong inference. Strong inference is a simple but rigorous generalized protocol useful in experimental or observational science including animal behavior that allows for efficient testing and falsification of alternative hypotheses. One of the key insights associated with strong inference is the idea of a crucial prediction, one that if tested well in a carefully designed experiment or observational protocol, would simultaneously reject one hypothesis and support another. Platt and other proponents of strong inference argue that using strong inference reduces the likelihood of unintentional bias in favor of one hypothesis or another because no matter what the results, the scientific endeavor will move forward and investigators will have something to say. But, the strongest of Platt's insights was that using strong inference is efficient. Scientific progress is faster when investigators use strong inference.

By analogy to strong inference in hypothesis testing, we argue that animal behaviorists who embrace efficient theory will enhance the progress of our science. (We use theory here in the vernacular to mean a hypothesis or a model, i.e., we are not using it as philosophers do to indicate an accepted body of already-tested ideas.) The definition that we use says that efficient theories make fewer, simpler, and more fundamental assumptions, and simultaneously generate a greater number of testable predictions per free parameter, than do less efficient theories. Some efficient theories will begin as approximate theories that when confronted with data require revision; it is the process of confrontation between prediction and data that results in successive theoretical refinement and progress in science. As animal behaviorists become more conversant with efficiency in theory, as well as strong inference experiments, animal behavior will integrate ever faster.

Reaching New Horizons

The achievement of new levels of understanding concerning animal behavior will grow, in part from our own interactions as scientists. These collaborations will continue to expand beyond the current involvement of ecologists, psychologists, anthropologists, and zoologists. We already see molecular biologists, neurobiologists, engineers, computer scientists, sociologists, and others joining teams that work on problems in animal behavior. As we train students with crossdisciplinary emphases, we will see other disciplines involved with problems in animal behavior. With the depth and breadth of our explorations of behavior, the need arises for specialized expertise in order to work out both evolutionary mechanisms and the underlying genetic, developmental, and physiological mechanisms for behavior. Indeed, scholars trained to look at all aspects of phenotypic flexibility are needed in great numbers. The silos or narrow subdisciplinary bins that we have inhabited for the first three generations of animal behavior will disappear, to be replaced by new types of scholars with a different set of skills.

Another form of collaboration will involve a greater mixing of theoretical considerations with empirical studies. This has occurred in many areas of behavioral ecology in the past two decades. We look for that pattern of theoretical–empirical combinations to expand to other areas of behavior, particularly pieces of the puzzle that involve genetics, development, and physiological mechanisms underlying phenomena like phenotypic plasticity.

How will we get from our current state of affairs to these exciting new horizons? The process is transpiring even now – science always has forward momentum. First, we need to carefully consider the training that we provide for our students. Here again, the concept of subdisciplinary silos must be discarded. Older scholars in the field need to grasp the need for a different sort of training for their students. Early in the graduate experience, we need to incorporate exposure to a broad range of frameworks, both theoretical and empirical, and to a wide range of techniques now available for exploring animal behavior. No individual will take away all of the in-depth skills needed for studying behavior, but by exposing students to the range of models, testing paradigms, and techniques, they can be conversant with collaborators who do have a complete grasp of particular modes of thinking or use of certain techniques. This means that training students will involve much more than working in a single laboratory, honing skills specific to the project(s) supported by a major professor. We must rethink the training process and engage a wide range of faculty and current students to arrive at what likely will be several different models for solving the problem of providing proper training. Also, we must consider and make available the possibilities for current scholars of animal behavior (in its broadest sense) to become exposed to the new ways of thinking and new technologies. This can and should be a shared endeavor.

While the major focus will be on procedures for training students and current scholars, there are several other aspects of the future of animal behavior. First, journals should consider their mission, with particular attention to highlighting studies that contain new, more integrative approaches. These may involve mixes of subdisciplines like endocrinology and ecology, or development and ecology, or they could be reviews making strong connections between theoretical and empirical work. In a similar manner, we as practicing animal behaviorists need to assist, in a positive manner, the various granting agencies with a push to secure both additional funding and funding that is specifically directed at the sorts of studies and training we have suggested here.

Like most shifts in subfields of science, this will be an incremental process; nothing will transform overnight, or even in a few years. However, this is an exciting new pathway and we expect that as it gains momentum, it will flourish and give a new emphasis to the importance of the exploration and explanation of animal behavior. We suggest that targeted symposia, workshops, and plenary talks at national and international meetings can be both a training venue and a stimulus to excite interest in these new approaches.

Funding Animal Behavior Research: The Changing Sociology of Science

What animal behaviorists study makes those who work with Disney, the Discovery Channel, or Animal Planet wealthy. It is no mystery as to why students recruit to animal behavior in significant numbers. The lives of non-human animals, their behavior in the wild, and what they teach us about evolution are fascinating. Undergraduates turn on to biology in their first exposures to animal behavior. Enthusiastic graduate students compete fiercely for the few graduate fellowships typically available for the study of animal behavior.

Yet, increasingly there is less funding per capita and fewer rather than more jobs for animal behaviorists. We find this distressing given that we think animal behavior is the linchpin holding together the integration of biology. The behavior of animals is at the center of a continuum that begins with cells and biochemistry at one end, passes through tissues, organs, and physiology to the organism, and moves to the other end of the spectrum through populations, communities, and ecosystems. In effect, behavior is the set of tools with which an animal plays the evolutionary game. Thus, we contend that animal behavior should receive considerably more grant support. Instead, animal behaviorists are among the most poorly funded in biology at our universities.

Sources of funding for animal behavior research, such the National Science Foundation and National Institutes of Health in the United States, must realize and act on the importance of understanding the integrative importance of behavior. We are also hopeful that nongovernmental organizations who use the work generated by animal behaviorists can establish foundations to supply funds for work in this area of investigation. To increase the likelihood of enhanced funding from all sources, we need more spokespersons able to discuss the history and current state of animal behavior studies.

See also: Avoidance of Parasites; Bateman's Principles: Original Experiment and Modern Data For and Against; Beyond Fever: Comparative Perspectives on Sickness Behavior; Caching; Compensation in Reproduction; Conservation, Behavior, Parasites and Invasive Species; Differential Allocation; Digestion and Foraging; Ectoparasite Behavior; Evolution of Parasite-Induced Behavioral Alterations; Flexible Mate Choice; Foraging Modes; Habitat Selection; Helpers and Reproductive Behavior in Birds and Mammals; Hormones and Breeding Strategies, Sex Reversal, Brood Parasites, Parthenogenesis; Hunger and Satiety; Intermediate Host Behavior; Internal Energy Storage; Kleptoparasitism and Cannibalism; Mate Choice in Males and Females; Monogamy and Extra-Pair Parentage; Optimal Foraging and Plant–Pollinator Co-Evolution; Optimal Foraging Theory: Introduction; Parasite-Induced Behavioral Change: Mechanisms; Parasite-Modified Vector Behavior; Parasites and Sexual Selection; Patch Exploitation; Propagule Behavior and Parasite Transmission; Reproductive Behavior and Parasites: Invertebrates; Reproductive Behavior and Parasites: Vertebrates; Self-Medication: Passive Prevention and Active Treatment; Sex Allocation, Sex Ratios and Reproduction; Sexual Selection and Speciation; Social Behavior and Parasites; Social Selection, Sexual Selection, and Sexual Conflict; Wintering Strategies: Moult and Behavior.

Further Reading

Jablonka E and Lamb MJ (2006) *Evolution in Four Dimensions: Genetic, Epigenetic, Behavioral, and Symbolic Variation in the History of Life*. Cambridge: MIT Press.
Platt JR (1964) Science, strong inference. *Science* 146: 347–353.
Popper K (1959) *The Logic of Scientific Discovery*. London: Routledge.
West-Eberhard MJ (2003) *Developmental Plasticity and Evolution*. Oxford: Oxford University Press.

Relevant Websites

http://ehponline.org.members/1995/Suppl-6/needleman-full.html – Behavioral toxicology.
http://www.nidirh.org/medlineplus/GEP.html – Genetic profiling.
http://en.wikipedia.org/wiki/Neuroimaging – Neuroimaging.
http://www.ehponline.org/members/2006/114-3/focus.html – Epigenetics.

Glossary

11-ketotestosterone (11KT)   Potent androgenic steroid hormone in teleost fishes that is analogous to dihydrotestosterone in tetrapod vertebrates in terms of inducing the development of secondary sexual characters often associated with territoriality and courtship in large males.

Abiotic   Nonliving.

Absconding   In honeybees, an absconding colony leaves its nest and searches for a new nest site. Absconding in response to low food availability, parasites, or predation is more common in 'African' strains of *Apis mellifera* than in 'European' strains.

Absolute sensitivity   The lowest amount of light that can be perceived by an animal.

Acanthocephalan   A phylum of parasitic worms known as 'acanthocephalans,' 'thorny-headed worms,' or 'spiny-headed worms,' characterized by the presence of an evertable proboscis, armed with spines, which it uses to pierce and hold the gut wall of its host. Acanthocephalans typically have complex life cycles, involving a number of hosts, including invertebrates, fishes, amphibians, birds, and mammals. About 1150 species have been described.

Accessory gland   A gland associated with reproductive organs of either males or females and producing substances accompanying the sperms or eggs.

Accommodation   An optical adjustment made by the eye to focus an object at a given distance.

Acoustic startle response   The behavioral and/or physiological response of an individual to an unexpected acoustic stimulus such as the sound of a nearby predator.

Action component   Any behavior elicited on the part of an evaluator by the act of recognition.

Action-oriented representations   The idea that internal representations should describe the external world by depicting it in terms of the possible actions an animal can take.

Activational effects   A change in behavior and/or physiology that occurs in response to a hormonal signal and that disappears once the influence of the hormonal signal ends.

Active electrolocation   The ability of weakly electric fish to detect objects and orient in their environment based on their electric sense. For active electrolocation, fish generate a carrier signal (EOD), which is modulated in amplitude and phase by the environment, resulting in the projection of a modulated signal onto their electrosensory skin surface. By sampling, the thus projected electric image fish can gain information about the properties and the location of nearby objects.

Active space   The area in which a signal (or cue) can be detected from the source.

Actual conflict   Observed conflict over reproduction in a social group; actual conflict can be much lower than potential conflict.

Acute phase response   A rapid, systemic, innate immune response that includes heterothermy (fever or hypothermia), production of proinflammatory cytokines, synthesis of defensive and other immune regulatory proteins, and sickness behaviors.

Ad libitum sampling   Noting whenever something of interest occurs.

Adaptation   (1). At the level of evolution, a process, driven by natural selection, whereby species or populations become better suited to the environment. It occurs over generations and results in an increase in those genes that allow individuals in a population to better survive and reproduce in an environment. (2). At the individual level, the use of regulatory systems, with their behavioral and physiological components, in order to allow an individual to cope with its environmental conditions.

Adaptive demography   The composition of eusocial insect workers within a colony so that different sizes and/or ages enhance the efficiency of colony operations and fitness.

Adaptive flexibility   The ability of individuals to adjust behavior or physiology as ecological or social conditions change in ways that enhance their fitness.

Adaptive radiation   Evolutionary diversification of a lineage into multiple species or differentiated populations (radiation), in which natural selection in novel environments has played a prominent role (adaptive).

Adaptive response   Refers to flexible behavior that an individual uses to adjust to another type of behavior or situation. Adaptively flexible behavior allows an individual to enhance its reproductive success or survival.

Adaptive suicide   Individual mortality that enhances inclusive fitness by benefiting relatives.

Additive character optimization   A type of character coding that applies differential costs for transformations across character-states arranged in leaner order. For example, if character-states {0,1,2} are observed, and 1 is assessed to be of intermediate similarity, additive coding can be employed to apply this conclusion. Thus, transformations from 0 to 1 and from 1 to 2 would cost the same, but the cost of transforming 0 directly in 2 would be equal to the cost of transforming from 0 to 1 plus the cost of transforming from 1 to 2 (hence, additive).

Additive genetic variance   The genetic variance of a quantitative character associated with the average effect of substituting one allele for another. Additive genetic effects are the only strictly heritable genetic effects.

Adipose tissue   Tissues that serve as the principal storage sites for body fat.

Adrenal glands   Endocrine glands located on the kidneys, which play a role in water and electrolyte balance.

Adrenocorticotropic hormone (ACTH)   Small polypeptide hormone derived from a larger precursor (proopiomelanocortin) produced by the anterior pituitary gland that stimulates the adrenal cortex (inter-renal glands in nonmammalian species) to produce corticosteroids (primarily glucocorticoids).

Aestivation   Spending the summer in a dormant stage. It occurs in crustaceans, snails, amphibians, reptiles, and lungfishes.

Affect   Subjective feelings.

Affective states   Emotional state, that is, feelings.

Affiliative social relationship   Strong association between individuals, usually manifested by high rates of physical proximity to one another and nonaggressive social interactions.

Affordance learning   A form of observational learning in which the crucial information an observer acquires is about properties of objects manipulated by the model and the opportunities they 'afford,' the observer then exploits this information rather than imitating the model's actions.

'African' honeybee   Bees derived from *A. mellifera* ecotypes that evolved in Africa were introduced into Brazil in the 1950s. Fiercely defensive of their nest, these bees have caused public health problems, due to the dangers of massive stinging, and agricultural management problems as their range has increased to cover much of South America, all of Central America, Mexico, California, Arizona, Texas, and parts of the southern United States.

Age polyethism   A mechanism for division of labor in which individuals within a social group specialize in different tasks at different developmental stages or different ages.

Aggression   Overt, complex, social behavior with the intention of inflicting damage or status change upon another individual.

Aggression against females   A category of male aggression. Some aggression against females may be mistaken for forced copulation.

Agonism (adj., agonistic)   Aggressive behavior including responses to aggression such as flight and submission. Conflict resolution through a series of aggressive or submissive signals.

Alarm call   A chemical, auditory, or visual signal emitted in the presence of a predator that may serve one or more functions, including advertisement of perception to the predator, advertisement of signaler quality, and warning conspecifics.

Alarm pheromone   Pheromones released in response to threats. In honeybees, the alarm pheromones are associated with the sting.

Alarm reaction   A behavior induced by chemical stimuli that tend to bring the animal in a position where it is less exposed to predation.

Alarm substance   Substance(s) in the skin of fishes that induce alarm reactions.

Allele   One of several alternative forms (nucleotide sequences) of a gene.

Allelochemical   A chemical signal produced by an organism that influences the behavior or physiology of an organism of a different species.

Alliance   A close social bond between two or more adult individuals. Alliances often support each other during conflict and are more likely to share resources with each other than with other animals.

Allochthonous   Originating elsewhere; not native to a place.

Allometric   Describing the relationship between the size of an organism and the proportional size of its parts.

Allomone   A chemical produced by individuals of a species used in communication with other species; typically used in defense against predators, etc.

Alloparental care (Alloparenting, alloparents)   Care for infants and juveniles that mimics and substitutes for the parental behavior of a parent. Typically, the caregivers are kin and the social group is cohesive and related.

Allopatric (n. Allopatry)   Geographically separated; for example, populations on different islands with little or no movement between islands. Allopatric speciation is the development of isolating mechanisms while incipient species are separated by a geographic barrier.

Allostatic load   The cumulative wear and tear and energetic demand of daily and annual routines. Allostatic load can also include increased demands of poor habitat, injury and infection, human disturbance and life history stages, such as breeding, migration, etc.

Allostasis   An elaboration on the concept of homeostasis, where there is an emphasis on the fact that (1) what counts as an ideal physiological measure can change over time, and (2) numerous physiological systems may become activated in the body's attempt to solve a challenge to its equilibrium.

Allotype   The allorecognition phenotype of an individual. An allotype is the composite of an individual's allorecognition genes, or the expressed gene products of allorecognition loci that confer cue specificity.

Alternative reproductive tactics (ARTs)   Discontinuous variation in mating behavior within one sex of a species, often associated with morphological variation.

Alternative splicing   Different exons of an RNA transcript from a single gene are spliced together to produce different mRNA transcripts and thus different proteins.

Altricial   Relatively immobile young (usually birds or mammals) depend on parents for food and warmth.

Altruism   A behavior that is costly to the performer's fitness, but beneficial to others (evolutionary biology); helping behavior resulting from selfless concern for the positive well-being of other individuals (social science).

Altruistic punishment   The costly infliction of harm on another individual or group that produces net benefits for all the individuals in the group (social science).

Alzheimer's disease   Neural disease accompanied by cognitive dementia and occurrence of plaques and tangles in the brain.

Ammocoete   Premetamorphic larva of a lamprey.

Amoebic dysentery (or amoebiasis)   An infection of the intestine caused by *Entamoeba histolytica*, a unicellular protozoan parasite, which causes severe diarrhea. Infection occurs by consuming food or water contaminated with amoeba cysts.

Amplified fragment length polymorphism (AFLP)   Genetic variation found by cutting DNA strands with restriction enzymes and amplifying the resulting segments by using PCR (polymerase chain reaction).

Amplitude   Sound intensity, as determined by the magnitude of vibration by a sound-producing object. This physical property of sound is the primary determinant of our psychological experience of the loudness of sound.

Amplitude modulation   Changes in the amplitude of a sound over time. The process of modulation produces extra frequencies in the sound, called sidebands.

Amygdala   A brain region that (among other functions) plays a critical role in fear, anxiety, and aggression.

Anadromy   Migratory pattern of fish that hatch and develop in fresh water and then migrate to saltwater for adult development to return to fresh water and breed.

Analogy (Analogical reasoning)   (In the field of *logic*) A form of reasoning in which one thing can be inferred (see *Inference*) as similar to another thing in certain respects, on the basis of the known similarity between the two things in other respects.

Anautogeny   The adult female ectoparasite requires that protein be ingested, often in the form of a blood meal, in order to mature her eggs.

Androgen (pl. androgens)   A steroid hormone with 19 carbon atoms, so named because of their *andros* (male)-generating effects. Examples include testosterone, androstenedione, 5-α dihydrotestosterone, and 11-ketotestosterone. Although the testes are an abundant source of androgens, they can also be synthesized in other glands, including the adrenal gland and ovaries. Some androgens such as testosterone can be converted by the enzyme aromatase into estrogens. Responsible for the development and maintenance of male-typical characteristics, including development of secondary sex characteristics and behaviors.

Androgenic gland   A gland near the distal portion of the sperm duct in crustaceans that secretes androgenic gland hormone.

Angiotensin II   An octapeptide that plays a prominent role in the regulation of cardiovascular and body fluid homeostasis.

Angular acceptance function   Photoreceptors have a limited 'field of view.' The angular extent of visual space over which a receptor receives light is described by its angular acceptance angle or function.

Anhedonia   The inability to feel pleasure; a defining symptom of clinical depression.

Animal communication   A behavior in which an animal produces a signal, which conveys information and influences the behavior or physiology of another animal.

Anisogamy   Refers to the differences in size of the gametes of the two sexes: sperm are generally small and eggs are generally large.

Anorexia   A change in eating behavior characterized by markedly reduced appetite or a total aversion to food. Anorexia is a component of sickness behavior but may refer to a behavioral change apart from febrile illnesses, such as with food allergies or psychological stress.

Anosmic animals   Animals without the sense of smell.

Antagonistic pleiotropy   A single gene controls for more than one trait with at least one trait being beneficial to the organism's fitness and at least one being detrimental to the organism's fitness. In analogy, a certain maternal hormone may have beneficial or detrimental effects on different offspring traits.

Anthropocentrism   Regarding humans as the central element of the universe; interpreting reality exclusively in terms of human values and experience.

Anthropogenic   Related to or caused by human activities (e.g., human-induced).

Anthropomorphism   Attribution of human motivation, characteristics, or behavior to animals.

Antiaphrodisiacs   Compounds transferred by the male to the female during mating that are either synthesized by the male or sequestered from the environment that render the female unattractive to rival males.

Apical gland   A gland associated with the reproductive tract of the sea hare that produces a hormone that influences egg-laying behavior.

Aposematic signal   Traits of the prey that predators can detect prior to attack and that inform the predator that the prey is defended or otherwise unattractive to attack.

Aposematism   The correlation between conspicuous signals, particularly warning coloration, and the presence of defenses in prey.

Apparent competition   When the fitness of one species is indirectly lowered by the presence of another species because of a shared parasite or predator.

Appeasement   Post-conflict interaction directed from a bystander to the aggressor to reduce the risk of being attacked.

Appetitive behavior   Behaviors that increase the probability that a particular need is satisfied. In the case of food deprivation, appetitive behavior would increase the organism's chance of locating food.

Appetitive cue   A stimulus associated with a resource (such as a food item, a host plant, or a prospective mate) that an individual would normally respond to, but which is ignored when the individual is actively migrating.

Appetitive movement   Movements that an individual makes while searching for, or in response to, appetitive cues (also called 'Trivial Movements').

Appetitive sexual behavior   A phase of reproductive behavior during which male searches, orients toward, and courts a female in preparation for copulation.

Apyrase   A calcium-dependent enzyme, found in mosquito saliva, that catalyzes the hydrolysis of ATP to ADP and inorganic phosphate.

Arginine-vasopressin   Peptide hormone involved in osmoregulation in both sexes, aggression and affiliative behavior of males.

Arginine vasotocin   Nine amino acid neuropeptide that is the homolog of arginine vasopressin found in mammals. These hormones are released at the posterior pituitary gland, but also widely in the brain where they act as neuromodulators.

Armpit effect   A system of kin recognition in which individuals learn their own phenotypic cues and use them as a template for determining the kinship status of other individuals.

Arms race   A metaphor for predator–prey coevolution wherein adaptation proceeds in an escalation/ counterescalation dynamic that leads to ever exaggerated traits on both sides of the interaction.

Aromatase   The enzyme that catalyzes the transformation of androgens such as testosterone into estrogens such as estradiol.

Arthropod   Animals with an exoskeleton, a segmented body, and jointed appendages.

Artificial fruit   A device modeled on the natural problems animals deal with in opening difficult-to-process natural foods, such as fruits that needed cracking, peeling, and other forms of manipulation; typically designed to afford two or more successful opening techniques so that the fidelity of social learning about such alternatives can be objectively measured and compared.

Artificial neural network modeling   The mathematical modeling of biological nervous systems in order to simulate animal behavior and its evolution.

Asset protection   Organisms act to protect their expected future reproductive success (the asset here) from loss due to predation. Because predation would eliminate future reproduction, individuals that can expect great future success are predicted to take greater actions to protect it.

Association   (In psychology) The process of forming mental connections or bonds between sensations, ideas, or memories; two stimuli or events are associated when the experience of one leads to the effects of another, because of repeated pairing.

Associative class   A collection of objects or events signaling the same consequence or follow-up event; the members are grouped on the basis of a common association.

Assortative mating   A system in which individuals choose mates nonrandomly on the basis of a particular characteristic, selecting either mates more dissimilar to themselves than expected under random mate choice (negative assortative mating) or mates more similar to themselves than expected under random mate choice (positive assortative mating).

Asymmetric game   A subset of games in game theory, in which the differences between the contestants may affect their choice of strategies.

Attentional states   Perceptual states of the eyes that are either directed toward a viewer or directed away from them.

Attractivity   A female's ability to elicit sexual responses from a male.

Attractor   Mathematically, a set of values that a dynamical system maintains after a sufficient time, with 'sufficient' depending upon the system. An important property of an attractor is that the system returns to this set of values even after it has been slightly disturbed, that is, when it is moved a small distance away from the set of values.

Auditory template model   The model of vocal development which proposes that an animal is constrained only to copy the sounds that it hears which match a template with which it hatches or is born.

Autochthonous   Native to a place; indigenous.

Autogeny   A female ectoparasite that is able to mature a batch of eggs without an external protein meal.

Autonoesis   A special form of consciousness that allows us to be aware of being the author of the episodic memory and the episodic future imagined event.

Autotomous sting   When the sting easily tears from the body of the worker insect, sting autotomy is found in all honeybee species and some wasps.

Autotomy   The loss of a body part (generally a limb or tail) by an animal, generally as a means of escape when held by that body part.

Avoidance   The use of a habitat that has few associated natural enemies (i.e., enemy-free space).

***Avpr1a* gene**   A gene coding the arginine-vasopressin receptor V1aR.

Awareness   Refers to the ability to perceive or feel something; awareness can refer to a wide range of sensitivity and experience, from dim perceptions to detailed conscious experience.

Bacillus   A rod-shaped bacteria cell.

Bag-cells   Neurosecretory cells of the abdominal ganglion of the sea hare that secrete egg-laying hormone.

Balanced polymorphism   The condition of having two or more alleles maintained in a population as a result of opposing evolutionary forces.

Banding (ringing)   Placing an inscribed metal or colored plastic band (ring) on the 'leg' of a bird so that the movement of the bird can be determined when recovered.

Basolateral membrane   Basal and lateral surfaces of *enterocyte* epithelial cells of intestinal *mucosa*.

Bateman gradients   Are regression lines that show the relationship between the number of mates and reproductive success for each sex in a group or population. Sometimes, they are also referred to as 'sexual selection gradients.' Generally, the steeper the slope of the regression line, the more intense will be sexual selection on that respective sex.

Batesian mimicry   Mimicry of body coloration and patterning of a toxic species by another coexisting toxic species. Sharing of the same aposematic signal between a defended species (called 'the model') and an undefended species (called 'the mimic'), such that individuals of the mimic species gain an antipredatory advantage from the shared signal.

Bathyal   Associated with benthic habitats of the continental slope between 200 and 4000 m deep.

Bayesian   Relative to animal behavior, the assumption that animals continually use new information to change their expectations of the environment (and therefore change their behavioral decisions).

Bayesian information criterion   A measure of the fit of a model to the data combining the log-likelihood of the model with a penalty term, taking into account the complexity of the model. This measure can be used to select a model among several alternatives.

Beacon   A unique marker for a location, analogous to a sign post.

Beeswax   A complex mix of hydrocarbons produced from wax glands on the abdomen of honeybees and worked into the comb structure to form the bees' nest.

Behavior sampling   Observing a group of animals and recording each occurrence of a particular behavior along with the individuals who perform it.

Behavioral deficit   A change in a behavior as a result of a contaminant or other treatment, usually having a negative effect on the animal.

Behavioral ecology   The study of the evolution and adaptive significance of behavior in the framework of recent views on the levels of action of natural selection and the importance of kin selection. There is an underlying assumption that behavior is selected for individuals to maximize the representation of their genes in the gene pools of future generations.

Behavioral fever   An increase in temperature as a response to parasites or disease. In endotherms, fevers are physiological, but in ectotherms, they may be caused by basking or the production of heat through muscular contractions.

Behavioral hierarchy   A description of interactions among behaviors, taken two at a time, that indicates which of each pair of behaviors overrides the other. Such maps are generally unidirectional (behavior A overrides B, which overrides C, which overrides D), but can have branches ($A \geq B \geq C$ *or* D), or feedback ($A = B \geq C \geq A$). Typically, these maps can be modified by such factors as an animal's state (e.g., hungry, sleepy, reproductive) and age.

Behavioral strategy   In game theory, a player's complete plan of action in a game, taking all other players possible actions into account.

Behavioral tradition   Nongenetic, heritable differences in behavior among groups or populations with overlapping generations, which are socially transmitted within and between generations.

Behavioristic psychology   A branch of psychology. The goal is to use the animal in an effort to understand a process of interest. Such processes include the mechanisms of learning, the prediction and control of learned behavior, and motivational processes. Some behaviorists focus on the prediction and control of behavior.

Betweenness   Centrality based on the number of shortest paths between every pair of other group members on which the focal individual lies.

Bidirectional control procedure   Manipulation of the direction of movement of an object (e.g., screen or rod) by a demonstrator to determine if an observer will manipulate the object in the same direction.

Bidirectional sex change   An individual is capable of changing sex in both directions.

Binary   Having two states; communication codes having two alternative signals.

Binding globulin   A protein molecule that binds steroids in the bloodstream and prevents both hepatic metabolism and the hormone from binding to its receptors, thus keeping the hormone in circulation. In some cases, hormones bound to binding globulins are capable of binding to receptors specific to binding-globulin-bound hormones.

Binocular stereopsis   Animals with widely separated eyes can judge the relative distance of objects because objects at different distances are imaged on slightly different parts of the retina in the two eyes. This difference is called 'retinal disparity.'

Bioassay   An appraisal of the biological activity of a substance, performed by testing its effect on an organism and comparing the result with some agreed standard.

Biodiversity   The variety of life forms at all levels of a biological system, but most often referring to the number of species.

Biogenic amine   A neurotransmitter, such as serotonin or dopamine, that can regulate behavior.

Bioindicator   A species or attribute (morphology, behavior, reproductive success) of a species or population that can be used to assess the health and well-being of an animal or plant species, a population or an ecological community.

Biological fitness   An individual's ability to survive and reproduce.

Biological model   A conceptual or mathematical description of a biological phenomenon, which generally aims to facilitate comprehension and/or to make predictions.

Biologically active (or bioactive)   Describes a substance, usually a chemical, that acts upon or influences the bodily functions of an organism.

Biologically inspired robots   Robots that are inspired by principles and mechanisms of biological systems. Bioinspired robots often share certain detailed morphological features with their biological analogs, but this is not a requirement.

Bioluminescence   Light produced by living organisms.

Biomagnification   The ability of chemicals to increase in concentration with each step in the food chain. That is, when a large fish eats a smaller fish with a given level of a contaminant, it accumulates a higher level of that contaminant in its own tissues.

Biomimesis   Mimic or imitate biological systems by artificial means (adj: biomimetic).

Biotic   Living.

Biotype   An ill-defined term generally applied to a herbivore exhibiting a specific host plant association that is noteworthy for some reason.

Bit   Contraction of *binary* digit, the un*it* of information in the mathematical theory of communication.

Bitter pith chewing   A form of self-medication practised by chimpanzees in which an ill animal removes the outer bark and leaves of a plant, *Vernonia amygdalina*, to chew on the exposed, bitter pith. The pith has medicinal properties.

Blind to treatment   Refers to the investigator being unaware and unable to identify which animal has been treated (e.g., with a chemica) and which is a control (and has not).

Blood–brain barrier   The limited diffusion of substances from the bloodstream into the brain and cerebrospinal fluid. Small molecules and molecules that have active transport mechanisms can cross the blood brain barrier, whereas larger molecules without active transport mechanisms are prevented from crossing. The barrier makes the brain less susceptible to blood–borne substances.

Bonanza food source   A very (quantity-) rich food source that lasts for a long time.

Bouton   The enlarged terminus of a nerve cell that forms a connection, or synapse, with another nerve cell.

Bradycardia   A decrease in heart rate.

Brain size   The absolute mass or volume of the brain, based on measures of fresh tissues, brain images, or corrected endocranial volumes. The term 'relative brain size' usually refers to allometrically-corrected measures, most often residuals of log-transformed brain mass regressed against log-transformed body mass.

Branchiostegal membrane   The membrane deep to the gill operculum, connected to the small support bones of the gills (the branchiostegal bones).

Breeding dispersal   The distance between the breeding site of an individual in one year and the breeding site of the same individual in another year.

Breeding range   In migratory birds, the area in which populations reproduce.

Bridge whistle   A whistle used by animal trainers to immediately indicate to the animal that it has performed a behavior correctly. The use of the bridge whistle facilitates training if rewards cannot be given immediately after a task was performed, for example when the animal is far away from the trainer.

Bridging stimulus   A conditioned stimulus that signals the imminent delivery of reinforcement.

Broadband   A vocalization with a broad energy spectrum that often lacks sharp harmonic peaks and has an irregular pulse repetition rate.

Broodiness   Behavior of female poultry as they sit on and incubate a clutch of eggs.

Brood parasitism   Leaving eggs or young to be raised by a nonparent, usually heterospecific, host.

Brood reduction   Occurs when the number of chicks in a brood falls due to the death of one or more of them. Brood reduction is considered an adaptive process if it enhances the viability or survival of the remaining chicks or parental prospects for future reproduction.

Brush-border   Microvilli-covered apical surface of *enterocyte* epithelial cells of intestinal *mucosa*.

Budding   Mode of colony multiplication in which new colonies are founded by the departure of a relatively small force of workers accompanied by one or more queens.

Bumblefoot   Bacterial infection and inflammatory reaction of the foot.

By-product mutualism   Where two or more individuals benefit each other by investing in a cooperative behavior.

Calf   A young animal dependent on its mother.

Cameleon   A genetically engineered protein that consists of two fluorescent proteins on the N and C terminus with calmodulin and M13 domains in between. It is used to detect calcium ion concentration, using fluorescence resonance energy transfer (FRET). Cameleon remains in a linear form when no calcium ions are present. When calcium ions are present, calmodulin binds to calcium, which allows M13 to bind to calmodulin/calcium ion complex and leads to a change in confirmation (shape). The confirmation change induced by M13 and calmdulin/calcium binding pulls the two fluorescent proteins into close proximity that allows FRET to occur.

Camouflage   Concealment strategies that have evolved to reduce the chances of being detected or recognized by predators.

Candidate gene   A gene whose function suggests it may be involved in specifying variation in a quantitative trait.

Canid   Species that are dog-like, classified within the family Canidae in the order Carnivora.

Cannibalism   Feeding on conspecifics.

Capturing   A training technique that involves reinforcing a behavior that is offered spontaneously and in its final target form.

Carapace   The part of a crab's exoskeleton that covers the cephalothorax (the fused head and thorax segments).

Cardiovascular disease   Disease of the heart and/or blood vessels.

Carrier-mediated transport   Passage of glucose, amino acids, and other polar molecules through a cell membrane by 'carrier' or 'transporter' proteins in the cell membrane.

Carry-over effect   Nonfatal condition that transfers between periods of the annual cycle and influences an individual's performance through effects on the condition or timing.

Caste   Distinct social roles within a colony. Caste typically refers to reproductive caste (queen or worker), but may also refer to specialized groups within workers. These are persistent specializations in function or task.

Caste totipotency   The capacity of a social insect larva to develop into any caste within the colony.

Catadromy   The migratory pattern of fish that hatch and develop in saltwater, then migrate to fresh water for adult development, and then return to the sea and breed.

Catecholamines   The class of hormones that includes epinephrine and norepinephrine.

Categorical perception   Occurs when the continuous, variable, and confusable stimulation that reaches the sense organs is sorted by the mind into discrete, distinct categories the members of which somehow come to resemble one another more than they resemble members of other categories.

Category   (The representation of) A specifically defined, general or comprehensive division in a system of classification; often used synonymously with 'class' (see 'class').

Cation   An ion with more protons than electrons giving it a net positive charge.

Caudodorsal cells   Neurosecretory cells in a particular part of the pond snail brain that secrete a hormone that mediates egg-laying behavior.

Causal knowledge   Knowledge of causal structures or properties.

Causal properties   The properties of objects that dictate the possible ways in which they can interact with one another (e.g., solid objects cannot pass through one another).

Causal structure   The directionality of physical events (i.e., cause and effect).

cDNA library   A collection of cDNA molecules that have been inserted into host cells, typically bacteria or viruses, so that the individual cDNAs can be replicated in high numbers.

Ceilometers   A brilliant shaft of light projected on the base of the cloud layer for cloud height measurement. On misty nights, with low ceiling, birds were attracted to the light beams and collided with other birds and the ground. These devices are no longer used by the weather service.

Central pattern generator (CPG)   A neuron or neuronal circuit that produces an activity pattern that varies in time and space to produce a behavior without any need for sensory feedback. For instance, the CPG for walking in a mouse coordinates the four legs (variation in space) to produce a series of steps (variation in time), and the basic motor output pattern can be elicited in an isolated spinal cord from which all connections to sensory and motor structures have been eliminated. Although the best-studied CPGs are those that produce rhythmically patterned behaviors (e.g., walking, swimming, breathing), there are also CPGs for nonrhythmic behaviors (e.g., withdrawal, shortening, vomiting).

Centrality   A measure of an individual's structural importance in a group on the basis of its network position.

Cephalofoil   Flattened and lateral extensions of the head, typically used to describe the head of the hammerhead shark.

Cercaria (plural: cercariae)   A small free-living larval stage of the Trematoda which swims using a tail and does not feed, relying on stored glycogen for energy to find and infect the subsequent host, often a mammal.

Cerebral ganglia   Ganglia located in the head of arthropods. In insects, these are the supra and subesophageal ganglia ('brain' and SEG, respectively).

Cervical connectives   Large nerve bundles passing through an insect's neck; comparable to the spinal cord.

Cestodes   Class of parasitic flatworms, commonly called 'tapeworms,' that live in the digestive tract of vertebrates as adults and often in the bodies of various animals as juveniles.

CF–FM bats   Bats that emit a long constant frequency component terminated by a brief frequency modulated sweep for echolocation. CF–FM bats compensate for Doppler shifts in the echoes they receive.

c-fos   An immediate early gene that is expressed in cells relatively rapidly (e.g., within 30 minutes) in response to the experience of environmental stimuli or after engaging in a particular behavior. The expression of the mRNA for the c-fos gene or the protein product of this gene has been widely employed by behavioral neuroscientists to localize brain areas implicated in the expression of a given behavior.

Chagas disease   A tropical disease also known as American trypanomiasis. It is caused by the flagellate protozoan *Trypanosoea cruzi* and is transmitted by the assassin bug.

Chain migration   Where northern wintering populations breed in the northern portions of the breeding range and the southern wintering population breed in the southern parts of the range.

Channel   A communication system; the collection of alternative signals composing a communication system; the physical system conveying signals.

Channel capacity   The maximum amount of information a communication system is theoretically capable of transmitting.

Chappius effect   A phenomenon in which wavelength-specific absorption by ozone affects the spectral composition of atmospheric light; there is a relative reduction in the spectral region of 540–625 nm (yellow) and increases near 500 nm (blue-green) and 680 nm (red).

Character   A biologically transmitted attribute of a species; in behavioral phylogenetics, such attributes might include learned behaviors not encoded in the genome.

Character reconstruction   An illustration of the simplest (and putatively most likely) pattern of evolutionary changes of a trait as depicted on a phylogenetic tree.

Chase-away coevolution   A form of arms race dynamics in which reciprocal selection pushes one species to stay ahead (in terms of a trait value) of the other.

Cheating, cheater, cheat   A party in a social interaction that does not contribute its fair share. In the case of *Dictyostelium*, it would be a clone that contributed proportionally more to fertile spore cells than to sterile stalk cells during the social stage.

Chelae   Front legs of crustaceans that have been modified into claws.

Chemoreceptor   Sensillum that houses either olfactory or gustatory neurons.

Chemosensor   A sensory receptor that detects specific chemical stimuli in the environment.

Chimera   An organism that is made up of two genetically distinct lineages.

Choosy   Rejecting a particular encountered potential mate.

Chromatic aberration   Optical imperfection caused by light of different wavelengths being focused in different planes by a refractive element such as a lens.

Chromophore   Part or moiety of a molecule responsible for its color. In vertebrate visual pigments, the chromophore is either retinal or 3,4 dehydroretinal, aldehydes of vitamin A.

Chronesthesia   The subjective awareness of the passage of time, an ability that allows us to address our own personally experienced past.

Chronic stress   Either long-term exposure to a stressor or repeated exposure to an acute stressor.

Chronobiology   Chronobiology, which comes from 'chrono,' meaning time, and biology, is the field of science that deals with cyclic activities in organisms and their relations to time. It is the formal study of biological rhythms.

Circadian rhythm   The term 'circadian' comes from the words, 'circa,' which means about, and 'diem,' meaning day. Circadian rhythms are endogenously organized oscillations in biological processes that occur roughly with a period of about 24h and are sustained in constant conditions.

Circannual rhythm   Circannual rhythms are endogenously organized oscillations in biological processes that occur each year, such as the migration patterns of some birds.

Circumboreal   Occurring around the globe in the boreal, or northern regions.

Circumventricular organ   A brain structure lacking a blood–brain barrier.

***cis*-Regulatory regions**   DNA regions outside of the protein coding region of a gene involved in regulating transcription.

Class   A collection of things sharing a common attribute, characteristic, quality, or trait.

Classic foraging theory   A body of economic models concerned with prey choice and patch residence time characterized by the use of simple optimization that applies to cases without frequency-dependent payoffs and hence mostly nonsocial situations.

Classical (aka Pavlovian or Respondent) conditioning   A stimulus (the unconditioned stimulus) that normally elicits a response (e.g., altered respiration) is repeatedly paired with a stimulus that does not normally elicit the response (the conditioned stimulus, e.g., light). The CS and US eventually become associated, and the organism begins to produce the behavioral response to the CS alone.

Claustral founding   Colony foundation by a non-foraging queen or queens, in which energy to rear the first generation of workers comes entirely from queens stored body reserves.

Clever Hans effect   The artifact that occurs when animals, including humans, may be sensitive to cues from the experimenter or the environment of which the experimenter is unaware. Double blind designs are often used to minimize Clever Hans effects.

Cloaca   A single posterior opening of the gut to which the bladder and reproductive organs also join.

Cloacal protuberances   Seasonally variable, occurring during the breeding season, in male birds. The protuberances are from engorgement by sperm of the storage tubules around the cloaca.

Clone   A genetically identical population of cells.

Clustering coefficient (C)   The density of the subnetwork of a focal individual's neighbors; the number of edges between neighbors is divided by the maximal possible number of edges between them.

Cnidocytes   Stinging cells found in cnidarians (e.g., jellyfish) that contain cnidocysts that are fired out into potential predators, injecting venom.

Coalition formation   Agonistic acts that involve at least two aggressors simultaneously joining forces to direct aggression toward the same target; such acts of coalitionary support indicate short-term cooperation between coalition partners, whereas the relationship between two individuals that repeatedly join forces over long time periods is considered to be an alliance.

Cochlear nucleus   The first auditory nucleus in the brain that receives the projections from the auditory nerve. The projections from neurons in cochlear nucleus are then sent into the medulla as a series of parallel pathways that form the ascending auditory system.

Code word   In certain types of codes, an ordered collection of signals making up the smallest decodable unit.

Code   The way in which signals stand for their referents.

Coevolution   Evolution of organisms of two or more species in which each adapts to changes in the other.

Cofoundress   A female that founds a new colony in association with other females.

Cognition   Psychological mechanisms that process perceptual information to enable behavioral decisions to be made, for example, learning, memory, generalization, and categorization.

Cognitive control   Process in which one cognitive mechanism exerts inhibitory, excitatory, or supervisory influence on another cognitive process; executive control, executive function.

Cognitive imitation   Adopting a decision rule after observing another use of that rule.

Cognitive psychology   The study of the mind's function, including perception, attention, memory, imagery, and decision-making.

Collective behavior   A phrase to describe how interactions between individuals produce group-level patterns of behavior.

Collective decision-making   The selection of one from two or more options by a group of individuals in which all members contribute to the choice, rather than following the decision of a leader.

Collective detection   Transfer of information within a group from animals that detect predation threats to others that have not detected the threats directly. Collective detection assumes that once an individual in the group has detected a threat, conspicuous signals of detection, such as alarm calls or flushing, will rapidly alert all other group members.

Collective intelligence   A group of agents that together act as a single cognitive unit to solve problems, make decisions, and carry out other complex tasks. Natural examples include social insect colonies, fish schools, and bacterial aggregations. Artificial examples include robot collectives and decentralized computer algorithms. Also known as 'swarm intelligence.'

Collective robotics   The design of groups of autonomous artificial agents that cooperate to carry out tasks. This field is strongly inspired by examples of collective behavior in animal groups.

Colonial spider   Spiders living in individual webs or nests that are interconnected by silk threads.

Colony budding   Colony founding by a group of workers and one or more queens.

Colony collapse disorder   A syndrome of unknown origin and cause afflicting beekeepers with high rates of colony mortality.

Combinatorial neurons   Neurons that respond best to signals that have two frequencies, that are harmonically or nearly harmonically related.

Command neurons   Neurons that, when stimulated individually or in small groups, can elicit a complex behavior. By strict definition, to be called a command neuron, that neuron must be active whenever the behavior occurs (correlation), stimulation of the neuron must elicit the behavior (sufficiency), and elimination of the neuron must make it impossible to trigger the behavior by its normal sensory input (necessity). Because necessity is often difficult to test, neurons that show just correlation plus sufficiency are often called command neurons.

Commensalism   An interaction between species in which at least one species is not affected, although others benefit.

Common orientation   The phenomenon whereby multiple individuals flying at high altitudes and not in visual contact of each other all take up similar flight headings, which are usually closely aligned with either the downwind direction or a seasonally preferred compass direction.

Communication   The transfer of information from a sender to a receiver by means of signals.

Communication system   An evolved network involving signal givers that produce information containing messages intended for a particular set of signal receivers; both signalers and receivers experience a net gain in reproductive success from their actions and responses.

Communicative culture   A group-specific system of signals, responses to those signals, and preferences for the class of individuals toward which signals are directed, that is socially learned and transmitted across generations.

Comparative psychology   Defined in many different ways. One approach would be the study of a great variety of behavior in a variety of species with a goal of understanding the evolutionary history, adaptive significance, development, and immediate control of behavior.

Competition   In strict biological terms, competition occurs when a necessary resource is in short supply and the use of the resource by one party denies access to that resource by another. Note that the two parties do not even have to be aware of one another, as in scramble competition, where the first party uses the resource before the second arrives and with neither necessarily aware of the other. In more general discussions of social behavior, competition may be described as a striving to outperform another where both parties are aware of the other.

Complete dimorphism   A size distribution of workers composed only of large and small individuals, with no intermediates.

Complete migration   All populations leave the breeding range of the species and move in some cases considerable distances to occupy a nonbreeding range of the species.

Components of fitness   Measures of individual fitness. Reproductive success components include the number of mates, the number of eggs laid or offspring born, the number of offspring that survive to reproductive age, the number of offspring that produce grand-offspring, and the number of grand-offspring. Survival components include age at death or lifespan.

Compound eyes   Crabs, like insects, have eyes that are composed of many repeated units called *ommatidia*, each with a *facet lens* and a transparent *crystalline cone*, which together focus light onto a narrow, elongated light-sensitive structure called the *rhabdom*. The rhabdom consists of densely packed microvilli, protruding from eight *retinula cells*, the photoreceptors. The membranes of these microvilli contain visual pigment molecules, the transmembrane protein part of which is called *opsin*.

Screening pigments in special pigment cells and in retinula cells prevent stray light from reaching the rhabdom from any other direction except through the fact lens belonging to the same ommatidium (apposition compound eyes).

Comprehension learning   Where an animal comes to extract a novel meaning from a signal as a result of experience of the usage of signals by other individuals.

Computational neuroethology   The modeling of the neural basis of animal behavior, with an emphasis on the interaction of the simulated animal with its simulated environment.

Concentrated animal-feeding operations (CAFOs)   Agricultural facilities that house a number of large animals; these operations may release waste into the environment (see http://cfpub.epa.gov/npdes/home.cfm?program_id=7 for additional information).

Concept   An abstract or general idea inferred or derived from specific instances; a mental construct or representation or idea of something formed by (mentally) combining all its characteristics or particulars (synonymously used with a general notion, a scheme or a plan).

Concerted evolution   The changes in one brain region caused as a result of changes in other associated brain regions, often thought to be due to underlying developmental mechanisms.

Condition dependence   Expression of a trait or behavior depends on the state of the organism. Many possible state variables are possible including size, age, energetic reserves, immune function, nest quality, and presence of a mate.

Conduction velocity   The speed with which action potentials travel along an axon; increases with increasing axon diameter; in vertebrates, also increased by myelin around the axon; for unmyelinated axons, typically below 20 m s^{-1}.

Conflict outcome   The result of an actual conflict in terms of the amount of conflict in the colony and the winning party, if any. For example, in honeybees, the outcome of the conflict over caste fate is that selfish individuals lose because they lack means to successful selfish behavior.

Conflict resolution   Exchange of threat and submissive signals between individuals over ownership of resources.

Conformity   A term used to define a family of biases towards high levels of fidelity in social learning, most commonly the case of copying whichever of various options is being shown by the majority of other group members.

Confound   To mingle so that the causes cannot be distinguished or separated.

Confusion effect   A reduction in capture rate by predators attacking a group resulting from their inability to single out one prey from the group.

Consciousness   In a strict medical sense, it is the state of being awake. Often refers to the state of being aware of oneself or environment.

Consensus decision   When the members of a group choose between two or more mutually exclusive actions and reach a consensus, that is, they all 'agree' on the same action.

Conservation reintroduction/benign reintroduction   An attempt to establish a species for the purpose of conservation outside its recorded distribution but within an appropriate habitat and ecogeographical area.

Consistency index   The number of character states specified in the character matrix, divided by the number of character-state transformations appearing on a phylogeny in question. The index is widely used to measure how closely the data fit a given tree.

Consolation   Postconflict affiliation from a bystander to the recipient of aggression with a stress-reducing function for the recipient of aggression. Reassuring body contact provided by a bystander to a distressed party.

Conspecific sperm precedence   Disproportional fertilization of a female by conspecific over heterospecific sperm following mating with both a con- and a heterospecific.

Conspecific   Used as either an adjective or noun to refer to another member of the same species, as contrasted with *heterospecific*, referring to a member of another species.

Constrained parents   Individuals mated to partners they do not individually prefer.

Constraints on mate preferences   Social or ecological factors that reduce the likelihood or opportunity for individuals to mate with partners they individually prefer under the assumption that mate preferences predict offspring viability.

Consummatory sexual behavior   The terminal phase of a sexual behavior sequence during which male gametes are emitted so that they can fertilize oocytes produced by the female the male is mating with.

Contagion   The unconditioned release of a predisposed behavior in one animal by the performance of the same behavior in another animal.

Contaminant   A chemical that has the potential to cause adverse effects in plants or animals.

Context   The circumstances under which a decision is made, which could be including but is not limited to, number of alternative options, nature of alternative options, temporal and spatial information.

Context-specific behavior   A behavior that occurs in one situation, but not others; the context can be defined by a social setting like aggression or mating, an environmental cue like darkness, internal condition like reproductive state or hunger, ongoing activity like flight or walking, or other factors.

Contextual fear conditioning   Pavlovian conditioning can be used to study contextual learning in which the composite properties of an experimental apparatus (e.g., its configuration, odor, illumination) frequently accompanied by an acoustical cue, act as conditioned stimuli predicting a previously experienced foot shock. Rats receiving foot shocks will typically display conditioned freezing when placed in the apparatus the following day.

Continuous (all-occurrences) sampling or recording   Observational method in which an observer records all behavioral onsets, transitions, and interactions of a single, focal animal during an observation period.

Conventional signal   A signal whose meaning could, at least theoretically, be exchanged for another within the same repertoire.

Convergent evolution   The development of similar anatomical, physiological, behavioral, or cognitive traits that may have a similar function, in two or more distantly related species; for example insects, birds, and bats all have evolved wings enabling them to fly. The traits may evolve through similar selection pressures, such as finding and processing food. Convergent evolution (analogy) is different from evolution via shared ancestry (homology), in which traits evolve because they are present in closely related species with a shared ancestor.

Cooperation   A behavior which provides a benefit to another individual (recipient), and which is selected for because of its beneficial effect on the recipient (cf. altruism which is a special case of cooperation).

Cooperative breeding   A social system in which individuals help care for young that are not their own. The parental care givers may be other reproducing adults or reproductively mature but nonreproducing adults.

Co-option or exaptation   Co-option or exaptation is the use of an ancestral adaptation (gene-trait relationship) for a new function for which the adaptation did not originally evolve. Evolutionary co-option occurs when natural selection causes traits, including behavioral traits, to assume new functions, often in new contexts. Motor patterns that are elicited in new contexts ('co-opted') can subsequently become ritualized.

Copulation   Mating; the act of inserting the male reproductive organ into the female.

Copulation solicitation display   An estrogen-dependent courtship display performed by female songbirds.

Corm   Bulblike underground part of a plant stem.

Cornicles   A pair of small upright tubes found on the hind dorsal side of aphids that are used to excrete droplets of defensive compounds.

Corpora allata   Glands near the insect brain that secrete juvenile hormone.

Corpora cardiaca   Neurohemal organs near the insect brain that store and release prothoracicotropic hormone and other neuropeptides.

Corticoids   A class of C_{21} steroid hormones secreted primarily from the adrenal cortices. There are two main types of corticoids: glucocorticoids (e.g., cortisol and corticosterone) and mineralocorticoids (e.g., aldosterone).

Corticosterone   Glucocorticoid hormones found in birds, reptiles, and mammals.

Corticotropin-releasing factor (CRF)   Forty-one amino acid polypeptides produced in the hypothalamus and extrahypothalamic sites that stimulate the release of ACTH (all vertebrates studied) and TSH (nonmammalian vertebrates) by the anterior pituitary gland. CRF-like peptides play central roles in developmental, behavioral, and physiological responses to stressors.

Cortisol   Glucocorticoid hormone most commonly found in mammals.

Corvids   Members of the crow family, which includes the rooks, ravens, magpies, jackdaws, jays, and choughs as well as crows.

Cost–benefit analysis   Cost–benefit analysis as applied to animal behavior predicts that if a behavior is adaptive, the benefits of a behavior must exceed the costs of that behavior. These costs are typically measured in terms of energy, time, and survival or reproduction.

Counterconditioning   A respondent learning technique designed to replace an undesirable response with a more desirable one. Often used to reverse fear conditioning.

Courtship   A suite of behaviors by members of one sex to attract members of the other sex for the purposes of mating.

Crepuscular   Active during periods of twilight, that is dawn and dusk.

Criterion   A rule or test on which to base a decision.

Critical flicker fusion frequency   Frequency of a flickering light at which it is perceived as steady.

Crop   A pouch-like enlargement of a bird's gullet.

Cryophilic   Having an affinity for low temperature. In behavior, cryophilic refers to animals having a tendency to move toward lower temperature.

Crypsis   Defense strategies that have specifically evolved to reduce the probability of detection.

Cryptic female choice   A type of sexual selection that can occur if a female's morphological, behavioral, or physiological traits (for instance, triggering of oviposition, ovulation, sperm transport or storage, resistance to further mating, inhibition of sperm dumping soon after copulation, etc.)

consistently biases the chances that a particular subset of conspecific mates have of siring offspring, when she copulates with more than one male. This is the postcopulatory equivalent of Darwinian female choice.

Cryptochrome Flavoprotein ultraviolet-A receptor involved in circadian rhythm entrainment in plants, insects, and mammals.

Cue A change in the environment made by one animal that allows another animal to acquire information, but does not benefit the animal that produced it. A source of information that can be used during orientation (e.g., a landmark).

Cue bearer Any organism or object that carries a recognizable set of identity cues.

Cue calibration The process of comparing compass information (i.e., directional references) derived from multiple sensory cues, such as magnetic and celestial cues, and calibrating one compass with respect to another. This can lead to a hierarchy of sensory cues, in which one particular sensory cue is being used to calibrate all the others.

Cue readers Unintended receivers of signals (predators or parasites) using a signal to detect the location of a potential prey/host, but for which the information content of the signal is unimportant.

Culture (a) [as commonly used by biologists]: between-group variation in behavior that owes its existence at least in part to social learning processes; (b) [as commonly used by anthropologists]: 'the complex whole which includes knowledge, belief, art, law, morals, custom, and any other capabilities and habits acquired by man as a member of society' (Tylor, 1924, p. 1).

Cupula Gelatinous covering of the hair cells in a lateral line neuromast. The cupula forms the mechanical coupling between water movement and the displacement of the hair cell cilia.

Currency Any quantity that can be used to evaluate the costs and benefits of different behavioral acts.

Cutaneous receptors A cutaneous receptor is a type of sensory receptor found in the dermis or epidermis. They are a part of the somatosensory system. Cutaneous receptors include cutaneous mechanoreceptors, nociceptors (pain), and thermoreceptors (temperature).

Cysticercoids The larval stage of many tapeworms.

Cytokine The name literally refers to a 'moving cell,' but in this case, cytokines refer to protein and peptide molecules that act as a cell signals. Cytokines, which are secreted by immune cells that have encountered a pathogen, encompass a large and diverse family of protein and polypeptide regulators that are critical to the development and functioning of both innate and adaptive immune responses. Endogenous pyrogens, which evoke the fever reaction and sickness behavior, are a type of cytokine.

Cytoplasmic incompatibility Differences carried within the cytoplasm of an egg or sperm prevent the formation or lead to the degradation of the zygote due to an interaction with the cytoplasm and the nuclear genetic material. The cytoplasmic effect may be due to gene products existing in the cytoplasm or cytoplasm-associated endosymbiotic organisms.

Dance language A series of movements displayed by honeybees to recruit their nestmates to food or nest sites.

Darwinian fitness or fitness The capability of an individual of certain genotype to reproduce, which is usually equal to the proportion of the individual's genes in all the genes of the next generation.

De novo synthesis Produced by the organism; self-made.

Death feigning The assumption of a false catatonic state after being captured by a predator in which the animal appears rigid and lifeless; may function to convince the predator that no further attack is necessary, allowing the prey to escape (also called: letisimulation, thanatosis, death shamming, akinesis, hypnosis, and tonic immobility).

Deception The production of a signal that induces a receiver to behave in ways that reduce its reproductive success.

Decibel A measurement of sound amplitude. A decibel is the ratio of two pressures on a logarithmic scale: $dB = 20 \log (p_1/p_2)$, where p_1 is the sound being measured and p_2 is a reference pressure referred to the threshold of human hearing.

Decision algorithm A set of behavioral steps that ends with selection of one option from a choice set. The steps govern how an individual reacts to the options themselves, other aspects of the environment, and its own state. In a collective decision, they also govern interactions among group members.

Decision-making An outcome of cognitive processes, leading to the selection of one particular course of action (or option) among several alternatives.

Declarative In memory research, declarative memories are contrasted with nondeclarative (or implicit) memories; originally, declarative memories were those that could be explicitly talked about although today other properties may be used to characterize declarative memory; declarative memories are widely thought to depend on the temporal lobes of the brain. Nondeclarative memories control behavior without the awareness of the existence of stored information, for example, one can ride a bicycle without being able to state, in detail, how it is accomplished.

Decoding The process of extracting information from signals.

Defeminization   A component of the sexual differentiation process during which the capacity to display female-typical behaviors is lost or reduced.

Defense call   Auditory call given by an animal standing its ground in the face of an approaching predator that may mimic the call of a species that is threatening to the predator and function to deter its further attack.

Deflective markings   Patterning on the body of a prey type that produces a fitness advantage to the bearer by manipulating the point of predatory attack on the prey's body such that successful prey capture is less likely.

Degree (k)   The number of edges a focal animal has; in an unweighted network, this is the number of other animals with which the focal individual interacts; in a weighted network, this will reflect the strength or frequency of interactions; also called *connectivity*.

Degrees of freedom   In statistical analyses it is the number of independent pieces of information upon with a statistical value is based. This along with a statistical value and the rejection criteria determine the statistical significance of a test.

Deimatic signal   A sudden change in the appearance of prey that can cause a predator to delay (or even abandon) an attack.

Delayed gratification task   Experimental situation in which rewards accumulate over time, and decision makers can choose when to stop the accumulation.

Delay-tuned neurons   Neurons in the auditory system that respond most vigorously to two brief signals that have a particular temporal separation that mimic an emitted pulse and echo that returns from a particular distance.

Demersal   Living or occurring in habitats near the bottom or seafloor.

Demographic stochasticity   The fact that some individuals fail by chance to encounter potential mates or by chance die, processes that cause fluctuations in demographic parameters.

Demography   The size and age structure of a colony.

Dendrite   Peripheral extension of a sensory neuron on which the receptor proteins are located.

Dendritic   Fingerlike, branching as a tree from a single root.

Dense cored vesicles   Small, intracellular, membrane-enclosed sacs found in neuronal terminals. Also called 'granular vesicles.'

Denticles (placoid scales)   Small outgrowths, similar in structure to teeth, which cover the skin of many cartilaginous fish including sharks. Denticles of sharks are formed of dentine with dermal papillae located in the core. The shape of a denticle varies from species to species and can be used in identification.

Dependent founding   Initiation of a new colony that requires the aid of workers. It involves colony budding or fission.

Dependent variable   A variable that is presumed to be affected or controlled by one or many independent variables.

Depth perception   The ability of animals to see the world in three dimensions.

DES   Diethylstilbestrol is a strong estrogen that was used as a preventive treatment against miscarriage.

Desensitization   The mitigation of a response to a distressing stimulus by gradual and repeated exposure to that stimulus.

Desquamation   Physical loss of skin, scales, etc.

Developmental plasticity   Environmental variation induces variation in phenotypes among individuals within populations and sometimes within individuals.

Developmental psychology   Focused on the changes in behavior as the animal matures and the interplay between genes, environment, and the organism during ontogeny.

Dewlap   A fleshy and sometimes colorful patch of skin on the throat area of some lizards. Many species have muscles that allow the dewlap to be extended as part of displays.

Dialect   The situation where acoustic communication signals form a mosaic pattern of geographic variation, with individuals within a local population producing very similar signals that are separated by relatively sharp borders from those of the neighboring groups.

Diameter (d)   The largest distance between any two vertices in the network.

Diapause   A state of arrested behavior, growth, and development that occurs at one stage in the life cycle. Quiescence accompanied by decreased metabolic rate and other physiological processes.

Diel vertical migration (DVM)   Vertical movements at sunrise and sunset, commonly used by aquatic organisms to balance feeding and predator avoidance. DVM usually involves an ascent to shallow water at sunset and descent to deeper water at sunrise, often linked to temperature and light.

DIF   Differentiation inducing factor is a chlorinated alkyl phenone produced by strong cells that induces weaker cells to become stalk, not spore.

Differential allocation hypothesis   A hypothesis about selection on parents to allocate their parental resources differently to offspring depending on the relative attractiveness of mothers versus fathers.

Differential migration   When the timing or distance of migration is different for males and females, or for young and adults, or both sex and age differences.

Diffusion chain   An experimental design for studying the serial transmission of information from model to novice, typically used to assess fidelity, corruption, and other changes, along a chain of individuals.

Dilution effect   A decrease in predation risk due to the presence of alternative targets in a group when a predator cannot capture all group members during an attack.

Dimorphism   Having two different patterns, usually referring to physical features. Males and females differ in their color patterns or sizes.

Dipsogenic   Thirst provoking.

Direct benefits   Material benefits of mate choice that accrue directly to the choosing individual as a result of the choice, such as nutrients, territory quality, or parental care provided by the mate.

Direct fitness   Fitness achieved through direct reproduction of one's own offspring. Direct reproduction is one component of inclusive fitness.

Directionality   The ability to locate the source of a stimulus in space.

Directional selection   A form of selection in which more extreme phenotypes are favored over existing phenotypes, such as larger more colorful ornaments, resulting in progressive elaboration of the phenotype over evolutionary time.

Dispersal   Movement of individuals away from an existing population or away from the parent organism.

Displacement activities   Behaviors performed in an abnormal context and in response to a seemingly unrelated motivation.

Dissociated pattern of reproduction   An annual reproductive cycle in which expression of copulatory behaviors and fertilization are not synchronized with the period of maximal activity of the gonads.

Distal   Farther from a body midline – used to describe order of segments in an appendage (e.g., a hand is distal to a shoulder).

Distractor option   A member of a choice set that is unlikely to be chosen but which may influence preferences for other options. Distractor effects exemplify the irrational decision-making often seen in humans and other animals.

Distributed cognition   Distributed cognition is an interdisciplinary branch of cognitive science that holds that cognitive processes are not confined to the brains of animals, but extend across individuals and out into the environment. An animal's 'cognitive system' consists not of its brain alone, but of its brain, body, and environment (including other animals) acting in concert. It is closely connected to the concept of embodied cognition.

Disturbance and disturbance stimulus   Disturbance is a deviation in an animal's behavior from patterns occurring without human influences. A disturbance stimulus is a human-related presence or object (e.g., birdwatcher, motorized vehicle) or sound (e.g., seismic blast) that creates a disturbance.

Diurnal   Active primarily during the daytime.

Diurnal rhythm   A biological rhythm that is synchronized to the 24 h light-dark cycle.

Diversionary display   A display performed by a parent at the approach of a predator that poses a risk to vulnerable young. If successful, the display attracts the attention of the predator causing it to move toward the parent and away from the young. These displays, most commonly described in ground-nesting birds, but also found in stickleback, incorporate elements that seem to have been co-opted and ritualized.

Division of labor   A property of a social group in which different individuals specialize in different tasks.

DNA methylation   Chemical modification of individual cytosine nucleotides in DNA that alters gene transcription.

DNQX (6,7-Dinitroquinoxaline-2,3-dione)   An AMPA and kainate antagonist. It is used in neurobiology as a tool to block AMPA and kainate type ionotropic glutamate receptors.

Domain of danger   The space closer to a focal individual than to any other group members.

Dominance   The state of having high social status in a group, often won through aggressive encounters or threats of aggressive encounters with conspecifics. Dominance is often linked to increased acquisition of resources, including food, territories, and mates.

Dominance hierarchy   A dominance hierarchy describes predictable interactions among individuals, with one giving way to another in competition for resources. A linear dominance hierarchy is transitive.

Dominance–subordinance relations   In groups of animals some individuals dominate (i.e., are higher in the 'pecking order') others that become subordinate (lower in the 'pecking order'). These relationships may be stable over many days, weeks, months, or even years, whereas in other cases they can be changing constantly (e.g., as in large groups).

Dominant frequency   The highest amplitude frequency component in a harmonic sound.

Dominant individual   High-ranking individual within a social group. This individual often has primary access to the best resources, such as food and mating partners. Dominance is often (but not always) correlated with large size and fighting ability, but also the ability to form coalitions (friendships) with other individuals.

Dopamine   A neurotransmitter occurring in both vertebrates and invertebrates. Massive loss of DA neurons in the substantia nigra in humans results in Parkinson's disease whose main characteristic is the paucity of voluntary movements, or hypokinesia. It is also associated with the pleasure system of the mammalian brain.

Doppler shifts   The increase in the frequency of a returning echo due to the difference in velocity between a bat and its target.

Dorsal root ganglion   A nodule near the spinal cord that contains cell bodies of sensory neurons in spinal nerves.

Drone   A male honeybee.

dsRNA   Double-stranded RNA.

Duration eggs   Encapsulated eggs, which can dry out or freeze and hatch once conditions are favorable again. Often found in zooplankton, especially in temporary ponds.

Dynamical system   A mathematical description of how a system behaves as a function of time. The description consists of an equation or set of equations that define the system's current state, as well as its past and predicted trajectory. The goal in considering nervous systems as dynamical systems is to characterize their oscillatory or quasi-oscillatory behavior.

Eavesdropping   The use of a signal by an animal that is not the intended receiver of the signaler.

Ecdysis   The shedding of the old, overlying exoskeleton of an arthropod; a process necessary for growth.

Ecdysteroid   A general term for a family of steroid hormones known in insects and other invertebrates. In insects, it is known as a molting hormone during larval stages, but has many other nondevelopmental effects. In most insects, the primary ecdysteroid is 20-hydroxyecdysone.

Echo ranging   The measurement of distance between a bat and its target. The acoustic cue for ranging is the time interval between the emitted pulse and the returning echo.

Echolocation   The ability to use sound waves reflected from a surface to detect objects at a distance.

Ecological determinism   Similarities among closely related species that reflect the ecological selection pressures acting on species, rather than the phylogenetic relationships between species.

Ecological time scale   A time scale of the same order of magnitude as the life span of the organisms investigated. Measured in days, months, or years, as opposed to evolutionary time scale, which is measured in thousands or millions of years.

Ecotoxicology   The study of the effect of chemicals (toxicology) on the ecology of animals or plants.

Ectoparasitoid   A parasitoid with a life-history strategy where the larva develops outside the host body by attaching or embedding in the host's tissues.

Edge   A relationship between two components of a network where the two related components are vertices in the graph model representing the network; in a social network, these can be any sort of social relationship, such as social interactions or information transfer; also called a *tie* or *link*.

Education by master apprenticeship   This is a phrase coined to describe how chimpanzees acquire new behaviors through observational learning. It is characterized by the following four aspects: (1) a long-term affectionate bond between mother and infant, (2) the mother takes on the role of the 'model' who demonstrates specific behaviors in the correct context, (3), the infant has a strong motivation to copy the model's behavior, and (4) the mother is highly tolerant toward the infant.

Effective population size   The number of breeding individuals within an idealized population, mating at random, that would have the same amount of inbreeding or of random gene frequency drift as the population under consideration.

Efficient theory   These are theories built from first principles. They are often, but not always, expressed mathematically; have few assumptions and free parameters (those that cannot be derived from a model or hypothesis); describe nature in an approximate way and is used iteratively to approach an ever-better understanding of nature. Characteristically, efficient theory has considerably fewer input variables than output variables.

Egg dumping   Occurs when a female bird lays her egg or eggs in the nest of another female and leaves that other female to care for them.

Egg pod   A capsule which encloses the egg mass of grasshoppers and which is formed through the cementing of soil particles together by secretions of the ovipositing female.

Egress   To come out or exit.

Elasmobranch   The cartilaginous fishes of the subclass Elasmobranchii including the sharks, skates, rays, and their extinct relatives.

Electric organ discharge (EOD)   The electrical signal produced by the electric organs of electric fishes. Electric organ discharges create an electrical field around the fish that can be detected by electroreceptor organs in the skin. Electric organ discharges have three functions. Extremely strong discharges (hundreds of volts) of strongly electric fish such as electric eels (*Electrophorus electricus*), electric rays (*Torpedo* spp.), and strongly electric catfish (*Malapterurus electricus*) can stun prey or potential predators. Weak electric organ discharges (typically less than a volt) of South

American knifefishes (Gymnotiformes) or African Mormyriformes are used to detect objects and prey or to communicate with conspecifics in dark, murky waters at night.

Electrocommunication   The ability of weakly electric fish to emit and receive electrical signals for the purpose of communication. Electrocommunication is limited to aquatic environments where the electrical conductivity of the medium is sufficient to transmit electric signals.

Electromyographic activity   Product of the electrical activity of muscle, which normally generates an electric current only when contracting or when its nerve is stimulated. Electrical impulses are often recorded as an electromyogram (EMG).

Electro-olfactogram   An electrical recording of the voltage across the olfactory epithelium. This type of recording allows experimenters to detect the electrical responses of olfactory sensory cells to odors.

Electroreceptor organ   Lateral-line-derived epidermal sense organs consisting of electroreceptor cells and associative structures located on the head and the trunk of certain fishes. They direct the flow of electrical current through low-resistive canals or through loosely layered patches of epithelial cells to specialized receptor cells containing membrane-bound voltage-gated ion channels, which convert outside electrical signals into sizable membrane potentials and subsequent transmitter release.

Electroretinogram (ERG)   The massed electrical response of the retina recorded by extracellular electrodes on the retinal, or more usually, corneal surface.

Embodied cognition   Embodied cognition is an interdisciplinary branch of cognitive science that argues that cognitive processes emerge from the unique manner in which an animal's morphological structure and sensorimotor capacities allow it to successfully engage with its environment. It aims to capture the way in which an animal's brain, body, and world act in concert to produce adaptive behavior, and, as such, is closely allied to the concept of distributed cognition.

Emergence   When a behavioral response to a multimodal signal is entirely different from responses elicited by any single component.

Emergency life history stage   A syndrome of physiological and behavioral traits triggered by perturbations of the environment that are designed to allow the individual to cope with the perturbation in the best condition possible until it passes.

Emergent phenomenon   Complex biological event that itself is not the target of natural or sexual selection, but which arises as the collective result of many simpler events that are under direct selection pressure.

Emergent relations   Relations between classes or class members that arise through a process of association, generalization, or inference.

Emigration   Dispersal or migration of organisms away from an area.

Emotion   A physiological and psychological state that functions to increase the survival of the organism. Basic emotions include anger, disgust, fear, happiness, sadness, and surprise.

Emotional contagion   Automatic state matching as a result of perceived emotions in others.

Empathy   The ability to recognize or understand another's state of mind or feelings (i.e., emotions).

Empirical   Evidence that can be observed.

Emulation   Recreation of the results of the efforts of another animal.

Encapsulation   A physiological immune response in host insects where a parasitoid egg, or other foreign body, is coated or engulfed by specialized cells called plasmatocytes resulting in the death of the parasitoid egg.

Encoding   The process of endowing signals with information.

Endemic   Native or restricted to a certain area.

Endocrine disruptor   A compound that interferes with the endocrine system, typically by binding to a receptor and either stimulating the effects of the receptor's hormone or blocking those effects, rendering the receptor inert. A compound produced for use as insecticides, herbicides, fungicides, or industrial applications as well as plant-produced chemicals with biological activity in living systems due to similarities in the structural and functional characteristics of native hormones, resulting in interference of endocrine systems.

Endocrine gland   A ductless gland from which hormones are released into the blood system in response to specific physiological signals. These signals can result from internal or external stimuli.

Endogenous   Phenomena arising within an organism, such as a biological rhythm.

Endogenous metabolic marker   An endogenous indicator of changes in metabolic activity within a cell.

Endogenous oscillator   An oscillator that is reset by internal stimuli. An endogenous oscillator is self-sustaining (i.e., periodic output continues after the termination of periodic input).

Endogenous pyrogens   Endogenous refers to inside the body and pyrogen refers to the generation of heat, in this case the increase in body temperature associated with a fever. Endogenous pyrogens, now commonly referred to as cytokines, evoke sickness behavior along with fever. Endogenous pyrogens are released in the body upon exposure to bacteria, bacterial cell-wall lipopolysaccharides, and viruses.

Endogenous rhythm   Internally generated, and not dependent on (but may be modified by) an external stimulus. Usually applied to seasonal processes, such as gonad growth, and migration, or diurnal processes, such as sleep.

Endoparasitoids   A parasitoid with a life-history strategy where the larva develops within the host body.

Enemy-free space   A habitat (e.g., host plant) where the herbivore is exposed to reduced rates of predation and parasitism.

Enemy release hypothesis   Hypothesizes that nonnative species become invasive because they are free from the predation and parasitism pressures of their native region.

Energy balance   Physiological adjustment of energy intake and expenditure resulting in precise maintenance of body mass; also known as 'energy homeostasis.'

Enrichment   Any aspect of enclosure design or husbandry practice that increases behavioral opportunities and promotes physical and psychological well-being in captive animals.

Enterocytes   Epithelial cells comprising the innermost layer of the gut.

Entrain (entrainment)   To adjust a rhythm so that it synchronizes with an external cycle, for example, the entraining of the internal rhythm of an organism to a light/dark cycle.

Entropy   The average amount of information encoded by signals of a system, less than or (rarely) equal to the channel capacity.

Environmental signaling   The signaling effects of environmental chemicals that directly or indirectly lead to changes in physiological functions or behaviors through interference with endocrine or exocrine mechanisms.

Environmental task specialization   Task threshold is primarily determined by the environment. Workers vary in their behavior on the basis of the environment they have experienced particularly during larval feeding.

Eph–ephrin receptors   Eph and ephrin receptors are components of cell signaling pathways involved in animal development and axon guidance. Eph receptors are classified as receptor tyrosine kinases (RTKs) and form the largest subfamily of RTKs.

Epidermis   The outermost layer of cells acting as the organism's major barrier against the environment.

Epigenetic   Originally, the term 'epigenetic' was used in a broad sense to refer to the processes of development as an interaction of genes and their products to produce the phenotype. This original definition did not imply heritability. Its definition became narrower with epigenetic being viewed as any aspect other than DNA sequence that influences the development of an organism. Modern usage of the term in molecular biology refers to the heritability of a trait over cell generations in an individual or across generations of individuals without changes in underlying DNA sequence. Epigenetic changes are preserved during the cell cycle and remain stable over the course of an individual's lifetime. For example, methylation of DNA at a Cytosine followed by a Guanine (CphosphateG or CpG site) can epigenetically switch off the adjacent gene, which may then stay 'off' in ensuing generations.

Episodic memory   The ability to remember and reexperience specific personal happenings from the past.

Epistasis   Where multiple genes interact to influence a trait.

Epistemic acts   Acts which serve to change the cognitive demands of a task so as to make it easier to solve, but which do not move an animal closer to task completion.

Epistemic engineering   The manner in which animals change their environments in order to alter the nature of the informational environments, as a means of either reducing its own cognitive load or increasing that of its enemies and rivals.

Eradication   An attempt to completely remove exotic fauna or flora from an area.

ERαKO   Knockout mice lacking a functional estrogen receptor α.

ERβKO   Knockout mice lacking a functional estrogen receptor β.

Eruption   In migration studies, a massive emigration from a particular region.

Estivation (Aestivation)   A period of dormancy over the summer that allows animals to survive an extended period of high temperatures or drought.

Estradiol   An estrogen (hormone) secreted by the ovary; it binds to estrogen receptors in many tissues including the brain.

Estradiol 17β   The most important circulating estrogen in both teleost fishes and mammals, produced in the ovaries but also other tissues including brain through the action of the enzyme aromatase.

Estrogen   A steroid hormone with 18 carbons and an aromatic ring, so named because of their estrus-generating properties in female mammals. Examples include estradiol, estriol, and estrone. Estrogens are synthesized from androgens with the help of the enzyme aromatase.

Estrus (estrous)   The period during which a female is sexually attractive, proceptive, and receptive to males and is capable of conceiving.

Ethnic marker   A seemingly arbitrary cultural element that signals membership of a particular ethnic group.

Ethnopharmacology   The study of the pharmacologically active compounds in plants used by traditional societies pertaining to the health care of humans and their animals.

Ethogram   Inventory of behaviors of a species, with definitions.

Ethology   Approach to the study of behavior developed by European zoologists that emphasizes, but is not limited to, the study of the naturally occurring behavioral patterns of free-ranging animals with particular emphasis on evolution and adaptive significance but not to the exclusion of development and immediate causation.

Ethopharmacology   The study of the effects of drugs on the neurochemical mechanisms of behavior. According to some authors, ethopharmacology should include the biological variability and adaptive significance of behavior, thus explicitly relying on an evolutionary approach. Ethopharmacological studies of host–parasite interactions attempt to unravel the neuromodulatory mechanisms that underlie the behavioral alterations of a host induced by a manipulative parasite.

Euphotic zone   Upper water layer of a lake or ocean to which 1% sunlight penetrates.

Euryhaline   The ability to tolerate various salt concentrations, that is, describes water organisms that tolerate a wide range of salinity.

Eusocial   A classification of social organization with (1) reproductive suppression, (2) overlapping generations, and (3) cooperative care of young (e.g., naked mole rat).

Eusocial (eusociality)   Colonies of animals structured around in which the generations overlap and there is a division of reproductive labor with members of the older generation producing most or all of the offspring of the colony. In primitively eusocial species, the differentiation between the parental generation (queens) and their daughter workers is weak and the daughters may have the potential to reproduce. In highly eusocial species, the queen and workers are highly differentiated and workers typically lack the physical and physiological attributes required to mate and reproduce.

Eutherian mammals   Eutheria are a group of mammals consisting of placental mammals plus all extinct mammals that are more closely related to living placentals (such as humans) than to living marsupials (such as kangaroos). They are distinguished from noneutherians by various features of the feet, ankles, jaws, and teeth.

Evaluator   Any organism that evaluates a cue bearer and makes a behavioral decision regarding the cue bearer's identity.

Evo-devo   Evolutionary developmental biology, a field of biology that integrates studies of genetics, development, and evolution in order to understand the evolution of morphology and developmental processes.

Evolutionarily stable strategy (ESS)   A strategy that, if adopted by a population of players, cannot be invaded by any alternative strategy that is initially rare.

Evolutionary algorithms   Several computational techniques that use iterative progress to solve problems. Inspired by evolutionary processes, such as reproduction, mutation, recombination, and selection, the techniques are based on a population that evolves in a guided random search until the individuals who use the best solution or strategy take over.

Evolutionary game theory   Evolutionary game theory is an application of the mathematical theory of games to evolutionary biology contexts, arising from the realization that frequency-dependent fitness introduces a strategic aspect. A game defines fitness of players, which reflects not only strategy of the protagonist player but also strategy of other ones. Evolutionary game theory analyzes transition of strategists' frequency in the population according to the expected fitness of each strategist, which reflects the current relative frequencies of the strategists and the game rules.

Evolutionary psychology   The application of evolutionary principles to human behavior in which behavior is regarded as the product of mechanisms that evolved early in human history, possibly in the Pleistocene epoch, and may not be adaptive in the present environment. Thus, behavior need not be adaptive in the present environment. Often behavior is viewed as the product of relatively specialized modules in the brain.

Exogenous   Phenomena arising outside of an organism, such as the light–dark cycle.

Exogenous metabolic marker   An exogenous substance that, when introduced to an animal, can indicate changes in metabolic activity within a cell.

Exotherm   An animal that depends on external sources of heat to maintain its body temperature in a viable range, as contrasted with *endotherms*, which have physiological mechanisms to generate heat and reduce heat stress.

Exotic species   A species that was accidentally or deliberately transported to an area far from its native distribution range.

Expected group size   The group size that is predicted on the basis of a given hypothesis for the advantage to being in groups.

Explicit   In memory research, equivalent in meaning to declarative; contrasts with implicit and nondeclarative.

Exposure   Process or situation in which a substance in the environment, such as a chemical, gains entrance to an organism (through ingestion, inhalation, dermal, or injection).

Expression component   The production or acquisition of identity cues by a cue bearer.

External validity   How well results of a study can be generalized to other situations or conditions.

Extinction   Withholding or preventing reinforcement of a previously reinforced behavior with the goal of reducing the frequency of the behavior to baseline or eliminating it altogether.

Extracellular fluid   One of the major fluid compartments of the body comprising all fluid residing outside cells.

Extracellular recording   Monitoring the electrical activity of neurons with an electrode outside the cells; normally records the activity of many neurons simultaneously.

Extractive-foraging   Behavior aimed at accessing food embedded in a protective matrix (such as shells or spines), or that is otherwise inaccessible (such as termites in nests or insect larvae in tree holes).

Extra-pair copulations (EPCs)   Copulations with individual(s) other than a mate or social partner.

Extra-pair fertilizations (EPFs)   Fertilizations that occur when females copulate with males other than their social mate.

Extra-pair offspring (EPO)   Offspring obtained by extra-pair copulations.

Extra-pair paternity (EPP)   Occurs when a socially paired female reproduces with a male, who is not the social mate.

Extrinsic isolation   Low fitness of hybrids because of hybrid phenotypes not being adapted to the resources of either parental population.

Extrinsic marker   Tag or band affixed to an animal at the time of capture that yields data only when an individual is re-sighted or recaptured later on.

Extrinsic mortality   Mortality caused by extrinsic agents such as predators, diseases, and accidents independently of any risks taken for reproduction.

Exudate   An escape of fluid as a consequence of increased vascular permeability and inflammation.

Exuviae   The remains of a molted arthropod exoskeleton.

Facial nerve   The seventh (VII) of twelve paired cranial nerves. It emerges from the brainstem between the pons and the medulla, and controls the muscles of facial expression, and taste to the anterior regions of the tongue.

Facultative   Applies to organisms that can adopt alternative ways of living. More specifically, individual facultative migrants have the choice of whether to migrate or not.

False belief   A belief is a mental state representing knowledge about the state of the world, for example that food is hidden in a particular container. A false belief is a mental state that is contrary to reality, for example the food may have been moved without an individual witnessing the change, and therefore it will have a false belief about the location of the food. Understanding that others can have false beliefs has been suggested as the key test for theory of mind in children.

False workers   In termites, the majority of the individuals within a colony of wood-dwelling termites. They differ from the (true) workers of foraging termites as they are totipotent larvae that lack morphological differentiations. Correspondingly, they are less involved in truly altruistic working tasks, such as foraging, brood care, or building behaviors. Therefore, they may rather be regarded as large immatures that delay reproductive maturity ('hopeful reproductives').

Family group   A group of individuals that repeatedly interact, composed of one or both parents and their direct offspring; may or may not include other relatives as in 'extended family group.'

Fast mapping   A type of inference by exclusion used by psycholinguists to denote the ability of children to form quick and rough hypotheses about the meaning of a new word after only a single exposure.

Fear effects   Another term for nonconsumptive effects. This term should be avoided except in cases where it has been established that antipredator responses are driven by fear.

Fear scream   Loud, harsh auditory call emitted after being captured by a predator that may serve one or more functions, including mobbing, startling the predator, warning kin of danger, calling for help from conspecifics, and attracting other nearby predators to distract the captor (also called: distress call).

Feature learning   (In the process of categorization) The use or abstraction of common features; in contrast to feature analysis, this process is characterized by a continuous adaptation of the feature set or the feature weights in order to cope with the actual categorization task.

Fecundity   The reproductive capacity of an organism; the quantity of eggs, sperm, or offspring produced by an individual.

Fecundity selection   Selection generated by variation in the number of offspring produced among individuals of a population.

Feeling   A brain construct, involving at least perceptual awareness, associated with a life-regulating system, which is recognizable by the individual when it recurs and may change behavior or act as a reinforcer when learning.

Felid   Species that are docat-like, classified within the family Felidae in the order Carnivora.

Female control   Refers to the idea that in species with internal insemination and fertilization that females are likely to control the fate of sperm and the likelihood of fertilization by particular sperm.

Female resistance   Describes behavior, physiology, and morphology of females that decreases the likelihood that males will attempt to force them to copulate.

Fertility   The number of reproductive bouts for an individual female over a season or a lifetime.

Fertilization   Occurs when sperm enters an egg.

Fidelity   Faithfulness, usually applied to a locality or mate.

Finder's advantage   In the context of group feeding, it is the part of a clump of food that a finder gets to eat before the arrival of any other individuals at the patch.

Finder's share   The fraction of the total food patch that makes up the finder's advantage.

Fisher's sex-ratio theory   Sex-ratio argument predicted for diploid species that sex ratios should stabilize at 1:1 (female:male) because each offspring derives from the pairing of a female and a male, and each sex, thus, produces overall the same total number of offspring; any deviations from an even sex ratio are unstable because negative frequency-dependent selection gives the rarer sex a reproductive advantage over the more common sex, ultimately leading to equal sex ratios at the population level.

Fission   Mode of colony multiplication in which new colonies are founded by one colony dividing into two relatively equal halves.

Fission–fusion society   A society in which members belong to a single, permanent social group, but in which all group members are rarely observed together concurrently. Instead, individuals form temporary subgroups that change frequently in their size and composition, often in response to ecological variation.

Fitness   The relative capacity of an organism to survive and transmit its genotype to reproductive offspring.

Fixed threshold model   A model of task allocation in insect colonies that holds that individual workers vary in the level of stimulus required to undertake a particular task. Workers with a low threshold are likely to engage in the task. High-threshold workers will not.

Flank marking   A behavior in which an animal rubs its flanks on objects to deposit contact pheromones from scent glands located on or near the flanks.

Flexible individual phenotypes   Phenotypes that are induced by environmental variation; these often appear to enhance the instantaneous fitness of the individual.

Flight boundary layer   The narrow layer of the atmosphere closest to the surface within which the self-powered flight speed of an individual exceeds the mean wind speed; thus within this layer, the individual can control its direction and make headway against the wind.

Flight initiation distance   Distance separating a prey and an approaching predator when the prey begins to flee; synonyms: approach distance, flight distance, flush distance.

Flight zone   It is the animal's personal space. The size of the flight zone is determined by how wild or tame the animal is. Animals that are trained to lead have no flight zone.

Fluctuating asymmetry   Difference between the values of bilateral symmetric traits of the same individual which can be of either sign with respect to the body axis and is assumedly the product of problems during development.

Fluffing   The act of shaking and loosening the feathers.

Fluorescence resonance energy transfer (FRET) also known as Forster resonance energy transfer
A phenomenon in which nonradioactive transfer of energy occurs between donor and acceptor molecules when the two are in close proximity. The most common donor and acceptor pair used in molecular biology are CFP and YFP, respectively. When FRET occurs, CFP transfers its excited energy to YFP. As a result, YFP is observed instead of CFP fluorescence emission. An example of FRET application in neurobiology is using Cameleon, a genetically engineered protein, to detect temporal calcium activity inside a living cell.

Flyway   A flyway is the entire range of a migratory bird species (or groups of related species or distinct populations of a single species) through which it moves on an annual basis from the breeding grounds to nonbreeding areas, including intermediate resting and feeding places as well as the area within which the birds migrate.

FM bats   Bats that emit a brief pulse for echolocation where the frequencies of the emitted call sweep from high to low throughout the duration of the pulse.

Focal sampling   Observational method in which an observer focuses on a single individual during a sampling period.

Follicle-stimulating hormone (FSH)   Gonadotropin that supports spermatogenesis and oocyte development in the gonads; also responsible for production of the hormone, inhibin, by the gonad.

Food aversion learning   A form of associative learning in which an animal associates sensory cues from a food with some deleterious consequence of eating that food and subsequently avoids the food.

Forced copulation   Contrasts with copulation that individuals seek or freely accept. Most investigators infer that copulation is forced when it is preceded by aggression or the threat of aggression, including 'violent restraint.'

Forward genetics   A phenotype-driven mutant screen.

Forward masking   Reduction of perceptual sensitivity over a given time interval following the perception of a specific stimulus.

Foundress/cofoundress   Foundresses are females that are initiating a nest, or living, on a newly established nest before the emergence of the first offspring. If more than one foundress is present in a nest, they are called cofoundresses.

Fourier analysis   A type of time series analysis that involves fitting a series of sine waves to data. The analysis identifies the amount of strength or power associated with a set of periods.

Fovea   Specifically, a depression in the center of the retina of many vertebrates, providing high-resolution vision. More generally, areas of high visual acuity in vertebrate retinas are called 'area centralis' or 'visual streak.'

Framework   A simplified conceptual structure used to solve complex problems.

Frass   The waste product from an animal's digestive tract expelled during defecation (also known as fecal material, or feces).

Free choice profiling   An experimental methodology in which observers have complete freedom to choose their own descriptive terms and apply them to the observed behavior of animal subjects.

Free-running rhythm   Free-running rhythm refers to fluctuations in physiological or behavioral responses, with a period of about 24h, that recur in the absence of environmental cues.

Freeze tolerance   The ability of an animal to survive freezing of tissues.

Freezing   Remaining motionless upon detection of a predator in hopes of avoiding detection by the predator either through cryptic morphology or habitat cover.

Frequency-dependent selection   Selection that varies depending on trait frequency in the population.

Frequency of sound   The number of cycles of vibration per second of a sound-producing object, expressed in Hz (Hertz, or cycles per second). A good set of human ears can detect frequencies of 20 Hz–20 kHz (a kHz is a kilohertz, or 1000 cycles per second). This physical property of sound is the primary determinant of our psychological experience of sound pitch.

Frequency modulation   Cyclic changes in the frequency composition of a sound over time. The process of modulation produces extra frequencies in the sound, called sidebands.

Frontal cortex   A brain region that (among other functions) plays a key role in long-term planning, executive decision-making, and impulse control.

Functional activity mapping   An analysis of the patterns of neural activity, or its correlates, during the performance of a behavior or in response to a stimulus.

Functional class   A class defined by a common (inherent) function of its members.

Fundamental frequency (f_0)   The lowest frequency component in a harmonic sound.

Future planning   The ability to imagine and preexperience specific personal scenarios that might occur in the future.

GABA–γ   Aminobutyric acid is the chief inhibitory neurotransmitter in the mammalian CNS. The binding of GABA to its receptors causes the opening of ion channels to allow the flow of either negatively charged chloride ions into the cell, or positively charged potassium ions out of the cell, to produce an inhibition of the cell. Receptors to GABA are found in both the central and peripheral nervous systems of several invertebrate phyla. Insect GABA receptors show some similarities with vertebrate GABA receptors.

Gametes   A cell that fuses with another gamete during fertilization.

Game theoretic models   These are mathematical calculations of an individual's success (fitness) in making choices when their choice depends on the choices of others.

Game theory   A mathematical technique for choosing the best strategy given the likely choice of others.

Ganglion   The CNS of insects and other invertebrates comprises a ganglion – a processing center ('brain') – for each body segment connected to the ganglia of adjacent segments by bundles of axons called 'connectives.'

Gap junctions   Specialized intercellular complexes that directly connect the cytoplasm of two cells. Gap junctions allow various molecules and ions to pass freely between cells. Between two neurons, gap junctions form electrical synapses.

Gasterosteidae   Latin name for the family of stickleback fish.

Gating neurons   A type of *command neuron* that must be active during the whole time while a behavior takes place. This term was coined in the study of leech swimming activation to distinguish these neurons from *trigger neurons*, a class of command neurons that is active only for a short time when a behavior begins.

Gene chip   A commercial microarray.

Gene flow The transfer of alleles of genes from one population to another.

Gene regulation Relating to the activation (expression) of genes, including both transcription and translation.

Genetically effective population size The number of reproducing individuals in a randomly mating population; actual population size is usually larger than its genetically effective size owing to the presence of sexually immature or nonbreeding individuals.

Genetic complementarity The potential for traits on both sides of an ecological interaction to respond evolutionarily to reciprocal selection.

Genetic diversity The level of biodiversity within a species, in reference to its total existing number of genetic characteristics, which, importantly, provides the raw material for evolution and is critical for long-term sustainability of a population.

Genetic drift Chance variations in gene frequencies that result from random sampling error.

Genetic monogamy An exclusive mating relationship between a male and a female resulting in all offspring being genetically directly related to both partners.

Genetic polymorphism A portion of the genome that is represented by numerous distinct versions in the population. The more polymorphic a given locus is, the greater the number of distinct versions that will exist in the population. Genetic polymorphisms are based on sequence variation at specific loci.

Genetic relatedness The fraction of genes identical by descent between two individuals. Only the fraction of genes shared above background count. See piece on relatedness.

Genetic structure The array of alleles and genotype combinations in a population.

Genetic subdivision Reduced gene flow between populations allows them to differ in the presence and/or frequency of alleles as a result of random genetic drift or natural selection.

Genetic task specialization Task threshold is genetically influenced. Workers of particular parentage are more likely to engage in particular tasks.

Genic selection Selection within individual bodies between alleles at a locus.

Genomic imprinting Form of inheritance in which the expression of a gene depends upon the parent from which the gene is inherited. Because imprinting allows genes to be silenced when inherited from one sex and not the other, it provides a potential mechanism for achieving sex-specific expression. The imprint alters the chemical structure and expression (suppression) of the gene but not its nucleotide sequence. Thus, the imprint can be erased and an active gene can be passed down in the next generation.

Genomic library A collection of fragments of genomic DNA that have been inserted into host cells, typically bacteria or viruses, so that the individual fragments can be replicated in high numbers.

Genotype The genetic constitution of an organism or one of the loci within that organism.

Geocentric cue A cue based on information external to the organism.

Geographic mosaic Ecological interactions vary across space because of the specifics of biotic and abiotic local environments, leading to a spatial mosaic of coevolutionary intensity. Hotspots, where reciprocal selection is strong, and coldspots, where reciprocal selection is weak or absent, characterize the geographic mosaic.

Geolocator A daylight-level recorder affixed to an animal at capture that can be recovered at recapture up to one year later to estimate the latitude and longitude for each day the device was attached.

Geomagnetic field Magnetic field associated with the Earth. It is essentially dipolar (it has two poles), the northern and southern magnetic poles on the Earth's surface. Away from the surface, the field becomes distorted.

Geophagy The ingestion of soil particles which can reduce the potency of ingested toxins.

Geotaxis Directed movement with respect to Earth's gravitational field. Movement away from Earth is 'negative,' movement toward Earth is 'positive.'

Germinal vesicle breakdown Dissolution of the nuclear membrane that signals continuation of meiosis.

Ghost experiment An experiment in which the model who would normally produce some effect in the world is absent, the effect being produced instead by surreptitious ('ghostly') means, such as pulling fine fishing line, allowing a test of how much an observer will learn from this component of the display alone.

Gill operculum The hard flaps covering the gills of a fish.

Gilliam's rule The prediction that animals favor using patches that minimize the ratio of predation risk to either expected growth or foraging rates.

Giving-up density and time (GUD and GUT) Giving-up density is the amount of food or prey items still remaining in the patch, when a forager leaves it. Giving-up time is the length of time a forager will go without encountering a food item before it leaves a patch. Both are important metrics for testing predictions of the marginal value theorem.

Glossopharyngeal nerve The ninth (IX) of twelve pairs of cranial nerves. It exits the brainstem from the medulla, just rostral (closer to the nose) to the vagus nerve. The glossopharyngeal nerve is mostly sensory and is involved in tasting, swallowing, and salivary secretions.

GLU   Glutamic acid (glutamate) is the most common excitatory neurotransmitter in the mammalian brain. Receptors to GLU are found in both the central and peripheral nervous systems of several invertebrate phyla. Insect GLU receptors show some similarities with vertebrate GLU receptors.

Glucocorticoids (Glucocorticosteroids)   (1) A class of steroid hormones released from the adrenal gland, particularly in response to stress; these include cortisol and corticosterone; (2) a class of synthetic steroid hormones; these include prednisone, dexamethasone and triamcinolone.

G-matrix   A square and symmetrical matrix in which the main diagonal consists of the additive genetic variance for a series of traits, and the other elements are additive genetic covariances between pairs of traits. Additive genetic variances have values between 0 and $+1$, whereas additive genetic covariances can range between -1 and $+1$.

Gonadotropin-releasing hormone (GnRH)   One of several neuropeptides synthesized in the brain.

Gonadotropins   Peptide hormones released from the pituitary in response to gonadotropin-releasing hormone from the brain; they stimulate growth of the gonads and synthesis of gonadal steroids.

Gonochorism   A sexual pattern in which individuals mature as one sex and remain that sex.

Good genes hypotheses   Refer, collectively, to explanations of mate preferences based on information or cues about the genes in potential mates. Good genes hypotheses can refer to complementarity (dissimilarity), relative individual heterozygosity, or to traits that indicate the possession of particular genes.

Granivorous   A diet of mostly seeds.

Gravid   Ready to lay eggs, for example carrying ovulated eggs in the ovarian lumen or oviduct.

Green-beard gene   A gene that affects copies of itself via three effects: production of trait, recognition of the trait in others, and differential treatment based on that trait. Sometimes not considered as part of kin selection because benefits go not to relatives but to actual bearers of the gene.

Green leaf volatiles   A suite of chemicals released from many plants upon mechanical damage.

Gregarious   Tending to aggregate actively into groups or clusters.

Gregarization   Density-dependent behavioral phase change in locusts from mutual repulsion to attraction and aggregation.

Ground-reaction forces   The forces that are developed as an animal or robot walks by pushing against a substrate (positive) or absorb momentum (negative, braking forces).

Group foraging   The searching, handling, and consumption of food by animals in close spatial proximity, whether or not there are social interactions between them.

Group memory   Information that is stored in the properties of an entire group, rather than encoded in the nervous system of an individual animal. The distribution of honeybee waggle dancers across food sources, for example, encodes the colony's ranking of the value of these sources.

Group selection   Selection between assemblages of individuals.

Group size effect   The phenomenon that individual vigilance declines as group size increases. This is most often explained by individual adjustments to a reduced perceived predation risk.

Gustation   Sense of taste.

Gustatory receptor protein (GR)   7-transmembrane protein located on the dendrite membrane of a gustatory neuron; detects and binds specific chemicals such as sugars or minerals.

Gustatory receptor   Sensillum that houses gustatory neurons; usually a tip pore *sensillum trichodeum*.

Gymnotiform   Electric knifefish of the New World order Gymnotiformes comprising five families. All gymnotiforms are electrogenic. 'Gymnotid' refers to members of the family 'Gymnotidae' including the weakly electric genus *Gymnotus* and the strongly electric *Electrophorus* (electric eel).

Gyne   Gynes are young females who have the potential to become egg-laying foundresses.

Habituation   Often considered the most basic form of learning that is defined as a response decrease in the presence of repeated stimulation.

Hamilton's rule   Named after W.D. (Bill) Hamilton, it is an inequality ($rb-c > 0$) that predicts when a trait is favored by kin selection, where c is the fitness cost to the actor of performing the behavior, b is the benefit to the individual to which the behavior is directed, and r is a measure of the genetic relatedness between those individuals. An altruistic act by definition has positive c and positive b and so is more likely to be favored by natural selection when r is high, and requires r to be positive. A selfish act, such as cannibalizing a member of the same species, has negative c and negative b and so is more likely to be favored by natural selection when r is low, and especially when r is zero.

Handicap   A trait whose expression incurs a cost, such that the degree of trait expression reflects the quality or condition of the bearer, in the sense than only an individual of high quality or condition can afford the cost of expressing the trait. Handicaps are one type of indicator mechanism and comprise a subset of the various indirect benefit hypotheses for the evolution of sexual dimorphisms via mate choice.

Handicap principle A hypothesis to explain honest signaling that proposes that reliable signals must be costly to the signaler in a manner that an individual with less of that trait could not afford.

Haplodiploidy A genetic system in which females come from fertilized eggs and are diploid, while males come from unfertilized eggs and are haploid.

Haplometrosis The founding of a eusocial insect colony by a single queen.

Haplotype A set of alleles of closely linked loci that are usually inherited together.

Harassment Occurs when males attempt repeatedly to copulate and in so doing impose costs on females that supposedly induce females to submit to copulation attempts.

Harderian gland A gland found within the eye's orbit, which occurs in vertebrates that possess a nictitating membrane. In some animals, it secretes fluid that lubricates movement of the nictitating membrane.

Hardy–Weinberg law The foundation of population genetics; the law shows that in the absence of evolutionary forces genotype and allele frequencies are stable and related to each other algebraically.

Harmonic An integer multiple of the fundamental frequency of a sound (e.g., $2f$, $3f$, $4f$).

Harmonic sound A complex sound consisting of multiple frequencies (sine waves), all in integer relation with each other.

Hawk–dove game A game theory analysis of alternate strategies hawk (attack immediately) and dove (display and retreat if attacked).

Helpers/helpers-at-the-nest Individuals, especially birds that provide care for conspecific young that are not their own offspring.

Hematophagy The habit of feeding on blood.

Hemimetabolous Having no pupal stage in the transition from larva to adult.

Hemoglobin Oxygen-carrying component of red blood cells.

Hemolymph The circulatory fluid of insects and other invertebrates, comparable to vertebrate blood.

Herbicides Chemicals produced to kill plants/weeds; generally used in no-till agricultural operations where the previous planting and weeds are not removed prior to seeding the new crop.

Heritability A measure of the proportion of phenotypic variation that is due to genetic variation in a population.

Hermaphroditism A condition in which individuals have gonads of both sexes (testes and ovaries) either simultaneously or sequentially.

Heterochrony hypothesis Proposes that an early step in the evolution of eusociality is based on simple evolutionary modification of the timing of expression of maternal care behaviors, from postreproductively towards offspring, to prereproductively, towards sibs (see reproductive groundplan hypothesis).

Heterospecific An individual of a different species.

Heterozygosity The proportion of genetic loci in an organism that have different alleles.

Heuristic Is a 'rule of thumb,' educated guess or a general way to solve a problem. Often used to describe a method that rapidly leads to a solution that is good in most situations. In phylogenetics, heuristic procedures are common because exact solutions are either mathematically impossible or nearly so.

Hibernation Dormancy during the winter. The seasonal occurrence of profound physiological changes that include strongly reduced basal rates of metabolism, heartbeat, and respiration.

Hidden Markov model Extension of the Markov chain concept to the modeling of nonhomogeneous data. The model combines a hidden variable driven by a Markov chain and an observed variable. A different distribution of the visible variable is associated with each possible value of the hidden variable.

Hiding time Latency between entering and emerging from refuge; synonym: emergence time.

Higher-order conditioning This Pavlovian learning process has two phases. First-order conditioning results in the conditioned stimulus predicting the occurrence of the provocative unconditioned stimulus. Second-order conditioning involves exposing the animal to the first conditioned stimulus, which has now acquired provocative properties, in temporal association with a second, emotionally neutral conditioned stimulus. The second conditioned stimulus then becomes a predictor of both the first conditioned stimulus and the unconditioned stimulus (not employed in the second-order association) and acquires its emotionally provocative properties at even a lower level of intensity.

Highly eusocial Eusocial society in which there are developmentally distinct specializations where some individuals are specialized for reproduction, and others have developmental differences that preclude mating and make them totally/effectively sterile under normal circumstances.

High-speed video Allows very high time resolution for analyzing fast behaviors by using high frame rates (commonly 500–2000 frames per second); frame rate for normal video is 30 frames per second.

Hippocampus   A brain region that (among other functions), plays a critical role in learning and memory, especially spatial learning.

Historical contingency   Evolutionary changes in a characteristic are dependent on what is inherited from evolutionary ancestors and the extent to which a characteristic diverges from that historic phenotype in response to selection.

Holarctic   The northern continents of the world.

Holometabolous   Insects that undergo complete metamorphosis involving four life stages: egg, larva, pupa, and adult.

Homeostasis   The ability of or tendency for an organism or a cell to maintain ideal internal equilibrium by adjusting its physiological processes.

Homeostatic sleep regulation   A sleep regulatory mechanism that aims to keep sleep amounts unchanged over a certain period of time; for example, after an overnight sleep loss, the activation of homeostatic sleep-promoting mechanisms induces sleepiness and compensatory increases in sleep next day.

Home range   The geographic space that an individual or group utilizes over the course of a year or longer.

Homing   The ability of an animal to return to its specific territory, or home range.

Hominization   The process of human evolution. Humans (*Homo sapiens*) are a member of Hominoids, that is a group of primates, which include humans, chimpanzees, gorillas, orangutans, and gibbons.

Homolog   A gene that shares ancestry, and hence DNA sequence composition with a gene from another species.

Homology   Biological similarity due to ancestry. For example, bat wings and mammalian forelegs are homologous.

Homoplasy   Biological similarity not due to ancestry, such as convergence or parallelism, for example bat wings and insect wings.

Honest signal   A structure or behavior that conveys reliable information to a receiver.

Horizontal social influence   Social influence on behavior that occurs within a generational cohort; for example, among juveniles.

Hormone   A chemical signal produced by one gland or tissue in the body that influences the physiology of a remote tissue.

Host   An organism harboring another parasitic organism that provides nourishment and shelter for the developing parasite.

Host plant   A plant species naturally used by a herbivore for its life activities.

Host range   The suite of host plant species used by a herbivore.

Host record   Documentation from field observation that a particular herbivore naturally uses a particular plant as a host.

Host shift   An evolutionary change by a herbivore lineage from using one host plant to using another; implies the abandonment of the ancestral host.

HPG axis   Hypothalamo–pituitary–gonadal axis.

Hybridization   Nucleic acid hybridization, the annealing, or binding, of two complementary, single-stranded, nucleic acid molecules.

Hybrid vigor   The tendency of a crossbred individual to show qualities superior to those of both parents.

Hydrozoan   A class of cnidarians that includes colonial polyps such as *Hydractinia*, individual polyps such as *Hydra*, and a diverse array of jellyfish with complex life cycles that include an attached polypoid and swimming medusoid phase.

Hyperosmolarity   An abnormally high osmolarity. The osmotic concentration of a solution, normally expressed as osmoles of solute per liter of solution.

Hyperparasitoids   A type of parasitoid that uses other parasitoids as host insects (also known as secondary parasitoids).

Hyperphagia   Seasonal occurrence of excess eating to build up fat reserves.

Hyperpolarization   A change in a nerve cell's membrane potential that makes it more negative.

Hypertrophy   Growth and enlargement of tissues and organs without cell division.

Hypokinesia   Abnormally slow or diminished movement of an animal.

Hypophysectomy   Removal of the pituitary gland.

Hypothalamus   A small region in the forebrain, containing various substructures (nuclei, including the arcuate nucleus) that collectively play a role in hunger, satiety, thirst, temperature regulation, hormone release, autonomic control, and circadian rhythms.

Hypothetico-deductive method   Hypothesis testing in which a scientific hypothesis could be falsified by a test of a prediction of that hypothesis.

Hypoxia   The presence of a low oxygen environment.

Hysteresis   The dependence of a physical system's performance on its history, apparent in some emergent collective properties of animal groups. For example, the ability of a group of ants to form a pheromone recruitment trail may depend on whether it reached its current size by growth from a smaller size or reduction from a larger one.

Hysteria   Uncontrollable and potentially violent episodes of extreme nervousness.

Ideal despotic distribution   Expected spatial distribution of organisms that have perfect information on the relative quality of all available habitats and current residents of habitats can exclude others from entering.

Ideal free distribution (IFD)   Expected spatial distribution of organisms that have perfect information on the relative quality of all available habitats and can move freely among these habitats.

Idiobiont   A parasitoid life-history strategy where host development is arrested upon parasitism. Idiobiont parasitoids are typically ectoparasitoids that attack host eggs or pupae.

Imitation   The reproduction of the form of a behavior produced by another animal.

Immediate early genes   The first genes transcribed in a cell during a response to a stimulus; their protein products regulate the transcription of other genes.

Immigration   The arrival of new individuals from elsewhere.

Immunocompetence   The ability of the body to produce a normal immune response (i.e., antibody production and/or cell-mediated immunity) following exposure to an antigen, which might be an actual virus itself or an immunization shot. Immunocompetence is the opposite of immunodeficiency or immuno-incompetent or immuno-compromised.

Imposex   A form of sexual abnormality in gastropods where male sex organs such as the penis and vas deferens develop in ('imposed upon') a genetic female as a result of exposure to organotin.

Impulsivity   A preference for the less delayed outcome.

In situ hybridization   A process in which labeled DNA or RNA probes are used to localize specific DNA or RNA sequences in sections of tissue.

In vitro   Literally, 'in glass,' meaning a reaction, process, or experiment in a metaphorical test tube rather than in a living organism. As opposed to in vivo: Literally, 'in life,' meaning a reaction, process, or experiment in a living organism.

Inadvertent social information   Information generated as a by-product of the behavior of other individuals.

Inbreeding   Mating among close relatives.

Inclusive fitness   Calculated from an individual's own reproductive success plus his/her effects on the

reproductive success of his/her relatives, each one weighted by the appropriate coefficient of relatedness.

Inclusive fitness theory   A synonym of kin selection theory emphasizing inclusive fitness.

Independence from irrelevant alternatives   Principle of rational choice behavior. It describes the expectation that preference between a pair of options should be independent of the presence of inferior alternatives.

Independent founding   Initiation of a new colony by reproductives without the help of workers.

Independent variable   A variable that is presumed to affect or control the value of a dependent variable.

Indeterminate growth   Growth that is not terminated in contrast to determinate growth that stops once a genetically predetermined structure has completely formed.

Index   A signal whose reliability is maintained due to some physical constraint on their performance.

Indicator models   A subset of indirect benefit hypotheses proposing that extravagant traits evolve via mate choice because their expression indicates the quality or condition of the bearer, which is assumed to be heritable. A handicap is an example of an indicator mechanism.

Indifference point   A set of options between which agents are indifferent; that is, in preference tasks, they choose the options equally.

Indirect benefits   Genetic benefits of mate choice that accrue indirectly to the choosing individual in the form of improved genetic quality of its offspring.

Indirect environmental maternal effect   Indirect environmental effects occur when the mother's environment influences her own and in turn her offsprings' phenotype. With regard to hormone-mediated maternal effects, differences in the environment the mothers live in result in differences in hormonal signaling to the offspring.

Indirect fitness   Indirect fitness is one component of inclusive fitness. The effects of an individual on the fitness of other individuals weighted by their genetic relatedness.

Indirect genetic maternal effect   Indirect genetic maternal effects are influences on offspring phenotype due to differences in the genetic background of mothers. With regard to hormone-mediated maternal effects, genetic differences between mothers would result in, for example, the expression of certain genes that regulate hormone secretion.

Indirect reciprocity   An observer C witnesses an altruistic act by A toward B, and as a result, cooperates with A in the future.

Individual comparison   Direct comparison of two or more options by a single animal, allowing it to determine which option is best. Individual comparison is not necessary for a collective decision, which can emerge from interactions among individuals who have each assessed only some of the available options.

Individual- or agent-based models   Computer simulations which can be used to describe and predict the global (group or population) consequences of the local interactions of individuals.

Individual recognition   The ability to learn the phenotypes of other individuals in a population and to use that information to shape individual-specific behavioral responses during interactions.

Induced ovulation   Occurs when ovulation is tied directly to copulation or some other stimulus associated with copulation. It may have evolved as a guard against forced or coerced copulation.

Inducible defenses   Defenses that occur only when predators are present.

Induction of preference   When past experience with a plant increases the degree of preference for that plant relative to others.

Inequity aversion   An aversion to unequal distributions of resources.

Infanticide   Killing a young, relatively defenseless, member of the same species.

Infectious coryza   Acute or subacute bacterial respiratory infection in chicken, pheasant, and guinea fowl caused by *Avibacterium paragallinarum.*

Inference   (In the field of *logic*) The act of passing from one proposition, statement, or judgment considered as true to another the truth of which is believed to follow from that of the former.

Inference by exclusion   Choice of an undefined stimulus (i.e., a stimulus that does not already have a learned association with a category) over a defined one (i.e., a stimulus that is already associated) by excluding (logically rejecting) the latter, which leads to the emergence of an untrained association between the undefined stimulus and the category.

Inferential reasoning   The ability to associate a visible and an imagined event.

Inferior colliculus   The midbrain auditory nucleus where the projections from most lower centers converge and are integrated. The inferior colliculus is the nexus of the auditory system.

Infinitesimal model   A genetic model in which it is assumed that traits are determined by a large (infinite) number of loci, each with a very small (infinitesimal) effect.

Inflorescence   A group or cluster of flowers arranged on a stem.

Information   Data that, when acquired, reduces an animal's uncertainty about environmental or social conditions. A quantity in the mathematical theory of communication expressed in bits.

Information sharing   A foraging system in which all group members are instantly informed of each other's food discoveries as they search for their own food.

Information transferred   The average variety conveyed by a communicative act, less than or (commonly) equal to the entropy.

Initial phase   The first sexual phenotype seen in many protogynous species, often characterized by relatively drab colors and relatively low displays of aggression and courtship behavior.

Inka cell   Endocrine cells near the insect spiracles that secrete pre-ecdysis-triggering hormone and ecdysis-triggering hormone.

Innate behavior   A behavior that is not learnt, but inherited.

Innovation (sensu process)   A process that introduces novel behavioral variants into a population's repertoire and results in new or modified learned behavior. The introduction of a novel behavior by social learning is not considered innovation.

Innovation (sensu product)   A new or modified learned behavior not previously found in the population.

Insectivorous   A diet of mainly insects.

Insemination   Occurs when males ejaculate inside the copulatory organ of a female.

Insight   The view that problem solving occurs by sudden recognition of a solution, or 'ah-ha' experience, rather than by trial-and-error learning. It is characterized by a sudden shift in behavior with a smooth and error-free transformation, a shift before the reward is obtained, long-term retention, transfer to other, similar problems, and to appear based on a perceptual restructuring of the problem.

Instantaneous (point) sampling   Observational method in which an observer records behavior of an individual at preset intervals.

Instar   The growth stage between two successive molts.

Insulin resistance   A state in which fat cells and muscle become insensitive to insulin's signal to take up glucose from the circulation, thereby producing high blood glucose levels (hyperglycemia). This is often seen in obesity and can be a precursor to diabetes.

Integument   All components of the outer layer of an organism – includes skin, hair, feathers, scales, nails, horns, wattles, warts, etc.

Interaural time difference When sound comes from one side of the body, it reaches one ear before the other. This creates an interaural time difference (ITD) which is used to localize sound in the horizontal plane. When the ITD is zero, the source appears at the midpoint between the ears. When ITD is varied, the source shifts toward the ear at which the signal arrives earlier. ITDs depend upon head size and in some cases on an interaural canal. In general, animals with large heads have larger time differences available to them.

Interference A reversible decline in fitness with increasing competitor density.

Interleukin-1 This is one of the earliest described endogenous pyrogens or cytokines. IL-1 is also known as lymphocyte activating factor and mononuclear cell factor. IL-1 is actually composed of two distinct proteins, IL-1α and IL-1β.

Intermediate host A host which is used by a parasite during its life cycle, in which it may multiply asexually but not sexually.

Internal validity Suitability of the study design to answer the question. The extent to which an effect seen in a study can be attributed to a specific cause.

Interneuron A neuron which connects neurons to other neurons in neural circuitries and whose cell body lies in the CNS.

Interobserver reliability The extent to which two or more observers consistently score behavior in the same way.

Interommatidial angle The angle between the viewing directions of two neighboring ommatidia in compound eyes.

Intersex An individual carrying the sexual characteristics of both sexes.

Intersexual selection Selection arising from variance in mating success due to interactions between males and females, such as female preference for males with a particular trait or resource.

Interspecific competition Competition between individuals of two different species.

Intertemporal choice A choice between outcomes that yield benefits at different points in time.

Intimidation A type of male aggressive response to females' refusals to mate. It may increase the likelihood that a female will mate with a male in the future.

Intracellular fluid One of the major fluid compartments of the body comprising all fluid within cells.

Intracerebroventricular administration Injection of a substance into one of the cerebral ventricles. Drugs and hormones injected this way come into relatively direct access to the brain tissue.

Intraguild predation An interaction in which predator and prey compete for basal resources (e.g., top predators eating mesopredators as well as smaller prey eaten by mesopredators).

Intralocus sexual conflict A form of genomic conflict that occurs when males and females differ in their fitness optima for a shared trait that is coded by the same locus or set of loci. Intralocus sexual conflict arises from intrasexual genetic correlations that constrain sex-specific expression of the shared trait and it is resolved by the evolution of sex-linked inheritance or sex-limited gene expression and the subsequent evolution of sexual dimorphism.

Intraobserver reliability The extent to which an observer consistently scores behavior in the same way at successful time intervals.

Intrasexual selection An evolutionary process that favors traits which improve an individual's competitive ability against members of the same sex for access to mates. Selection arising from variance in mating success due to competitive interactions within one sex, such as male–male combat or territory defense for access to females.

Intrinsic isolation Low fitness of hybrids because of genetic incompatibilities.

Intrinsic markers Genetic material, stable isotopes, or other markers that are carried within the animal itself and require only a single capture to yield data.

Intromittant organs Male copulatory organs, which deposit sperm and other seminal fluids into the female reproductive tracts. In mammals, a very few birds (only 3% of species), lizards, and snakes, males have an intromittant organ called 'a penis.' In insects, a male's intromittant organ is called 'an eadeagus.'

Introspection Self-observation based on private mental processes; often thought to be limited to consideration of one's own conscious thoughts, feelings, and perceptions.

Invariant feature A feature (quantity or property or function) that remains unchanged under a transformation.

Invasion In migration studies, the same as irruption. More generally, the colonization of an area by a species formerly absent there.

Invasive species A nonnative species that spreads rapidly once established, with the potential to cause economic or environmental harm.

Inverse square law A mathematical formula describing the attenuation of sound as it propagates through an ideal environment. By the inverse square law, sound amplitude decreases by 6 dB per doubling of distance.

Ionospheric circulation Large-scale convection in the inner magnetosphere and the conjugated ionosphere.

Irruption   In migration studies, a massive immigration to a particular region. More generally, a form of migration in which the proportions of individuals that participate, and the distances they travel, vary greatly from year to year.

Isodar   The set of points on a plot of density of one species in different habitats at which the fitness payoffs for choosing between habitats are equal.

Isogamy   Refers to eggs and sperm that are similar, or approximately more similar, in size than is usually the case.

Isoleg   The set of points on a plot of densities of different species at which the fitness payoffs for using both (or multiple) habitats and using only one habitat are equal.

Isolume   A level of constant light intensity in the water column that is commonly represented as a line of points on a plot.

Iterated game   Contestants play a game such as the prisoner's dilemma many times, thus allowing a strategy to be contingent on past moves.

Iteroparity   The repeated or iterated cycles of reproduction, production of young, throughout the life cycle of an organism before it succumbs.

Jack   In salmon, a male that matures precociously and typically does not spend any time at sea; jacks are typically much smaller.

Juvenile hormone   A sesquiterpenoid insect hormone known to regulate many functions across insect taxa, including larval development, reproduction, and behavior.

Kairomone   Chemical signal molecule that is produced by one species and perceived by another species, resulting in altered physiology or behavior in the species perceiving the cue that benefits that species.

Kappa coefficient   An index of concordance that measures agreement between two observers in behavioral observation, taking into account the probability of agreement by chance alone.

Kendall's coefficient of concordance   A nonparametric method for measuring agreement among more than two observers in behavioral observation.

Kinematics   The characterization of a behavior in terms of the movements of the body. Most commonly, such studies involve a frame-by-frame analysis of films or videotapes of the behavior. Kinematic studies are often carried out to determine which muscles produce the movements underlying a behavior, so they are often accompanied by recording the tension or electrical activity generated by active muscles.

Kinesis   Behavior in which the organism does not move in a particular direction with reference to a stimulus but instead simply moves at an increasing or decreasing rate, or rate or turning, until it ends up farther from or closer to the object. (Contrast with taxis.)

Kinocilium   A special structure on the apex of hair cells located in the sensory epithelium of various vertebrate sensory receptors including electroreceptors.

Kin recognition   The ability to discriminate kin from nonkin, or the ability to make discriminations among kin based on degree of relatedness.

Kin selection   The process of selection as it acts through effects on relatives. Sometimes viewed as co-extensive with inclusive fitness, but sometimes viewed as excluding green-beard effects. See Hamilton's Rule.

Kleptoparasitic spiders   Spiders that live in webs of other species and steal prey from the host.

Koinobiont   A parasitoid life-history strategy where hosts continue to grow and develop after parasitism.

Labellum   Bottom part of the proboscis in flies, equipped with fine grooves to assist ingestion of liquid food.

Lag-sequential analysis   Method used for the identification of the most likely sequences of successive events appearing in a time series.

Lairage   European term for the stockyards that hold animals at a slaughter plant.

Larviposition   The act of depositing living larvae instead of eggs.

Larynx   A musculoskeletal structure that functions as a vocal organ among amphibians, reptiles, and mammals.

Laser ablation   In biology, a process of killing cells by irradiating them with a laser beam.

Latency   The amount of time until a behavior occurs. The delay between the onset of the stimulus and the beginning of the response (neural or behavioral).

Leaf swallowing   The slow and deliberate swallowing, one at a time without chewing, of whole leaves that are folded between tongue and palate, and pass through the gastrointestinal tract visibly unchanged. The behavior is known to occur in apes, some monkey species, other mammals and some birds.

Leapfrog migration   Where northern wintering populations breed in the southern portions of the breeding range and southern wintering populations breed in the northern parts of the range.

Leghorn   Breed of egg-type chickens that produce white-shelled eggs; named after the city of Leghorn, Italy, where they are considered to have originated; leghorns have provided the genetic foundation of most modern egg-type chicken strains.

Leishmaniasis   Caused by protozoan parasites in the genus *Leishmania* that are transmitted by sandflies. They can affect the skin, mucus membranes, or internal organs.

Lek Is an aggregation or cluster of male territories into arenas used for attracting, courting, and mating with females. Males that form leks provide only sperm and no other resource to the females. No lasting bonds are formed and males do not engage in any parental care.

Lek paradox The persistence of strong directional selection for exaggerated sexual ornaments or display despite the apparent lack of benefit for such choice, particularly in lek mating systems.

Lek polygyny A mating system in which individual males mate with multiple females during a breeding season and in which males aggregate at small, nonresource-containing display sites to attract females.

Leptin A type I cytokine secreted by fat cells that regulates food intake. Leptin acts on the brain to signal when the body has sufficient energy stores, thus inhibiting appetite (i.e., it is an 'adipostat'). However, leptin and its receptor are widely expressed, suggesting that leptin is much more than an 'adipostat,' and likely plays diverse roles in animal development.

Levels of organization A complex behavioral system can be broken into a hierarchy of components or networks based on their physical size and functional complexity. Causal influences operate in both a top-down and a bottom-up fashion with one-way causation characterizing the simplest interactions and two-way causation operating across multiple levels. The lowest level of organization for predator recognition is sensory input from the environment, followed by the processing of predator features in a down-stream hierarchical integration of predator features, yielding higher-order predator recognition and mediation of antipredator behavior.

Lexical syntax Structured rules for ordering semantically meaningful sound units such that their ordering carries additional meaning beyond that reflected in the units alone.

Life cycle of chemicals The passage of a compound through the environment beginning with the source of production and release; consideration of the physical/chemical properties and the migration of the chemical in various media including soil, water, and air including the production of metabolites and their activity in living systems.

Life cycle of organisms Consideration of all stages in the life of an individual with ontogeny, maturation, adult, and aging including reproductive strategy and lifespan as well as unique species characteristics.

Life-for-life relatedness Relatedness including a concept of relative sex-specific reproductive value. Life-for-life relatedness = Regression relatedness × (sex specific reproductive value of the recipient/sex-specific reproductive value of the actor).

Life history Characteristics of the growth and development of an organism, such as its length and timing of gestation, maturation (beginning, period, ending), reproductive period, and lifespan.

Life history stage (LHS) A syndrome of morphological, physiological, and behavioral traits associated with a specific process (e.g., reproduction, nonreproduction).

Life-history traits Features of the life cycle, with particular reference to survival and reproduction (e.g., age at first reproduction, fecundity, etc.)

Lignified When something has been made hard like wood as a result of the internal deposition of *lignin*, a substance related to cellulose that provides rigidity to plant cell walls.

Linear timing The hypothesis that psychological estimates of time are linearly related to physical time.

Linkage A phenomenon whereby two genes are spatially located close to each other on a chromosome so that crossing over rarely occurs between them during meiosis. Thus, two variants in the corresponding genes are said to be in 'linkage disequilibrium' when they tend to be coinherited. If one variant is in a gene that encodes a phenotype, the linked variant acts as a marker. This is the basis for linkage studies.

Lipophilic Having an affinity for, tending to combine with, or capable of, dissolving in lipids (fats).

Lipopolysaccharides (LPS) Large molecules consisting of a lipid and a polysaccharide that are found in the outer membrane of some bacteria. The molecules, referred to as endotoxins, cause the release of endogenous pyrogens that evoke a fever, resulting in sickness behavior in the animals exposed to them. An integral component of Gram-negative bacterial cell walls that induces an acute phase response in most vertebrates.

Local enhancement Attention drawn to the location where another animal is performing a response.

Local mate competition Theory that competition for mates is stronger between related males than between related females, reducing the relative value of males; thus, sex-ratio interests of queens and workers become more closely aligned and female-biased sex ratios are considered optimal for both parties.

Local resource enhancement hypothesis The idea that related females cooperate synergistically to enhance their joint reproduction, increasing the relative value of females; thus, sex ratios should be female-biased.

Locomotor system The way an animal moves from one location to another. In primates, locomotor systems include brachiation, vertical clinging and leaping, quadrupedality, knuckle-walking and bipedality.

Logistic A logistic function or logistic curve is the most common sigmoid curve. The initial stage is approximately exponential, then, as saturation begins, the rate of increase slows and approaches an asymptote.

Log-linear model   Model for the analysis of multiway contingency tables. The principle is to first consider all possible associations between a finite set of categorical variables, and then to remove nonsignificant associations.

Longitudinal   Correlational research study that involves repeated observations of the same items over long periods of time – often many decades.

Lophotrochozoa   A major subdivision of protostome animals that includes molluscs, annelids, bryozoans, brachiopods, and other less conspicuous animal phyla. The group is named for the presence fan-like feeding structures called 'lophophores' (in the bryozoans, phoronids, and brachiopods) and trochophore larval stages found in many of the group's members. The Lophotrochozoa can be contrasted with the other group of protostomes called the 'Ecdysozoa,' which includes arthropods, nematodes, and other animal phyla.

Lordosis   A female receptive behavior exhibited by many rodents and birds, highlighted by an immobile posture with arched back and raised rump and head.

Lumen   The space within the intestinal tube.

Luminance   An indicator of brightness.

Luteinizing hormone (LH)   Gonadotropin that stimulates gonadal production of steroid hormones and supports gamete production.

Lymphatic filariasis   A tropical parasitic disease caused by thread-like filarial nematode worms that live in the lymphatic system and cause lymphedema. The worms are transmitted by mosquitoes.

Lymphocyte   This type of white blood cell makes up 25–30% of white blood cells. Lymphocytes are concentrated in central lymphoid organs and tissues, such as the spleen, tonsils, and lymph nodes. Lymphocytes determine the specificity of the immune response to infectious microorganisms. The two broad categories of lymphocytes are the large granular lymphocytes and the small lymphocytes. Large, granular lymphocytes are known as the natural killer cells and the small lymphocytes are the T cells and B cells.

Macrocyst   The sexual, diploid stage of the *Dictyostelium* life cycle.

Macroevolution   Evolutionary change that is observed as differences between species, genera, or higher taxa.

Macronutrient   Those nutrients that are needed by the body in large amounts and potentially can be used as a source of energy (proteins, carbohydrates, and fats).

Macroparasite   A parasite that does not multiply inside its definitive host.

Macrophages   Literally meaning 'big eaters,' in actuality these are white blood cells dwelling within tissues that phagocytose or engulf cellular debris and bacteria.

Macrophytes   Aquatic vegetation with roots.

Magnetic compass   A compass that provides a direction bearing, or reference, based on the polarity or inclination of the Earth's magnetic field.

Magnetic inclination angle   The angle at which field lines of Earth's magnetic field intersect the surface of the Earth.

Magnetic intensity   The strength of a magnetic field.

Magnetic map   A map based on geographic variation in the Earth's magnetic field, which could be used to determine geographic position.

Magnetite (Fe_3O_4)   One of several types of biogenically produced iron oxides. Lustrous black, magnetic mineral, Fe_3O_4. It occurs in crystals of the cubic system. A cubic mineral and member of the *spinel* structure type.

Magnetoreception   The sensory detection and use of magnetic fields, particularly the Earth's magnetic field.

Magnetoreceptor   A sensory neuron that transduces magnetic stimuli into a neural (i.e., bioelectric) signal.

Major histocompatibility complex (MHC)   A genetic region (containing > 150 genes in humans) that plays an important role in autoimmunity and immune diversity in jawed vertebrates. MHC genes products mediate self/nonself recognition in vertebrate immune systems and are involved in tissue compatibility (histocompatibility).

Male harassment of females   A type of coercion that may not be immediately associated with copulation attempts.

Mandibular gland   A salivary gland on either side of the mouth, inside the lower jaw, that discharges saliva into the oral cavity.

Mantle   Soft extensions of the body wall that in many mollusks secrete a shell. It also forms a cavity that shelters the gills.

Marginal costs   The change in costs with a change in behavior. (In economics, marginal is synonymous with the derivative from calculus.)

Marginal value theorem (MVT)   A model within optimal foraging theory that predicts whether an animal should continue to exploit a given patch based on its current (marginal) value relative to the expected gain from moving to another patch.

Marker   A trait that signals a particular genotype.

Markov chain   Stochastic process in which the value taken by a random variable X at time t is explained by the values observed for the same variable at time $t-1$ (first-order model) and possibly at times $t-2$, $t-3$, ... (high-order model). Transition probabilities between different values are summarized as a transition matrix.

Mark-recapture method   A method commonly used to estimate population sizes which relies on recording individually distinctive traits or making individuals, and later using these traits or marks to recognize them in future encounters. In a closed population, the proportion of animals resighted in relation to newly encountered animals allows a calculation of population size. A set of methods for estimating one or more of abundance, survival, and recruitment by recording repeated sightings or captures of animals, some of which are identifiables from marks previously placed on them. Increasingly, natural marks are used, identified from photographs or DNA fingerprinting.

Masculinization   A component of the sexual differentiation process during which the capacity to display male-typical behaviors is acquired or enhanced.

Mate   A social associate and need not refer to an individual with which one copulates.

Mate assessment   Results from the process of evaluating potential mates; mate assessment determines an individual's preference function.

Mate choice   The decision made by an individual in selecting a partner for reproduction.

Mate-choice copying   A form of nonindependent mate choice whereby an individual chooses the same mate that it previously observed being chosen by another individual.

Maternal effects   Nongenetic influences of the mother's phenotype (including behavior) on an individual's phenotype, especially those with evolutionary consequences.

Maternal inheritance   Maternal inheritance describes the maternal inheritance of DNA and is distinct from maternal effect.

Maternal rank inheritance   The process by which juveniles (e.g., cercopithecine primates, spotted hyenas) attain positions in the dominance hierarchy adjacent to those of their mothers.

Mating system   The demographic pattern of breeding individuals within a group.

Matrigene   In diploids, the allele inherited from the mother.

Matriline   Individuals of two or more generations that are descended from the same female.

Matrotrophic   A form of gestation in which the developing offspring take nutrients directly from the mother's blood through specialized embryonic structures throughout the gestation period.

Maxillary palps   Sensory structures on the outer surface of the maxillae used for detecting food.

Mechanisms of heredity   Ways in which information physical or otherwise — are transferred between generations. Such mechanisms include genes, culture, learning, developmental systems, and epigenetics.

Mechanisms of sexual selection   Include behavioral and physiological interactions between individuals, whether male–male, female–female, or male–female that result in within-sex variance in some component of fitness.

Medulla (medulla oblongata)   The lower half of the brainstem. It contains the cardiac, respiratory, and vasomotor centers and deals with autonomic functions, such as breathing, heart rate, and blood pressure.

Melanin   A pigment that underlies the rusty coloration of the ventral feathers of the barn swallow and many other birds. Melanins are produced endogenously rather than acquired through diet like carotenoid pigments that add red and orange colors to the feathers of many other birds like house finches.

Melanophores   A pigment cell that contains melanin.

Melatonin   Melatonin is a hormone secreted by the pineal gland. Plasma levels of melatonin are low in the day and high in the night. It enables an organism to detect changes in the seasons because its expression mimics changes in day length.

Memory   The retention of information from prior experience.

Memory monitoring   The process of tracking or evaluating the contents of one's own memory.

Mental representation   The process of internalizing a referent (external stimulus) into specific mental content. The term can also be used to refer to the content itself.

Mental state   An unobservable, internal or cognitive representation of 'things' in the world (e.g., the perception of objects), the actions or plans required to interact with those objects (e.g., intentions and desires), and information about those objects' current structure, location, properties, etc. (e.g., knowledge). In theory of mind research, David Premack suggested that there are three important categories of mental states: Perceptual, Motivational, and Informational.

Mental time travel   The ability to travel backwards and forwards in the mind's eye in order to reminisce about the past and imagine future scenarios.

Mentalistic psychology   An approach in which the scientist attempts to understand the mental life of the animal. One posits experiences in the animal mind that are similar, at least in some respects, to those of human experience.

Mesoconsumer   Intermediate consumers (i.e., herbivores and mesopredators).

Mesocosm experiments   Experiments that achieve highly controlled manipulations by working at small spatial scales (e.g. experimental ecosystems in mesocosms). Because smaller and invertebrate consumers often comprise the highest trophic level). They often test general principles that potentially apply to large spatial scales where experimental tests are logistically more difficult (e.g., large vertebrates in vast landscapes).

Mesopelagic   Associated with the midwater oceanic zone between 200 and 1000 m depth, a zone characterized by dim light and a steep persistent thermocline.

Mesopredator   A carnivore occupying a mid-trophic level and at risk of predation from carnivores at higher trophic levels.

Message   A decodable collection of signals transmitted as a unit; also the meaning of a communication.

Metacognition   Thinking about thinking; the ability to reflect on or think about one's own thoughts, feelings, and knowledge.

Metamemory   Knowledge of the contents and function of one's own memory; memory monitoring.

Metamorphic climax   The final and most rapid phase of morphological change when thyroid activity is at its peak.

Metamorphosis   The change in form that occurs during the postembryonic lives of insects as they transition from early feeding stages to the adult reproductive stage.

Metapopulation   A group of semi-isolated populations that are linked through exchange of individuals such that the dynamics of each subpopulation are asynchronous. A series of populations connected by dispersal; the dynamics of metapopulations involve extinction and recolonization events.

Methylation   The addition of a methyl group to a molecule; in DNA methylation, methyl groups are attached to cytosine residues and can lead to changes in gene expression including gene silencing.

Microarray   A series of microscopic spots of DNA that are attached to a solid surface; the microarray is hybridized with cDNA or RNA in order to measure differences in gene expression between two samples.

Microevolution   Evolutionary change that takes place within a population. The direct or indirect genetic response to selection or drift.

Microparasite   A parasite that reproduces inside its host.

Microsatellites   Neutral segments of DNA consisting of repeating base pairs that show a high degree of intra- and inter-specific polymorphism.

Microspectrophotometry   Measurement of the spectral composition of light that is reflected or transmitted by materials, at a microscopic scale.

Microvilli   Microscopic cellular membranous protrusions that increase the surface area of epithelial cells and are involved in a wide variety of functions, including absorption and secretion.

Migration   A seasonal, usually two-way, movement from one habitat to another to avoid unfavorable climatic conditions and/or to seek more favorable energetic conditions.

Migration syndrome   The suite of coadapted morphological, physiological, and life-history traits that enable migration and that is underlain by a genetic complex that controls the development and expression of these traits.

Migratory connectivity   Geographic linking of populations between different periods of the annual cycle, including breeding, migration, and wintering.

Miraoidium (plural: miracidia)   A small free-living larval stage of the Trematoda which swims using cilia and does not feed, relying on glycogen stores to enable it to find and infect the subsequent parasite host, often a mollusc.

Mirror neurons   Brain cells that react similarly during one's own motor actions as those observed in others.

Mitochondrial DNA   An abundant single-stranded circular DNA molecule occurring in mitochondria and containing a few genes; the control region where DNA replication begins has especially high mutation rates and is valuable for population genetics studies.

Müllerian mimicry   Mimicry of body coloration, body patterning, and/or behavior of a toxic prey species by a nontoxic, coexisting species.

Müllerian ring   A group of species that are Müllerian mimics and have converged on the same aposematic signal.

Mobbing   A coordinated effort by a group (three or more) of prey in response to a predatory attack in which the prey approach, observe, harass, attack, and sometimes injure or kill the predator before it is able to attack.

Modal action pattern   An innate, relatively invariant series of behaviors, common to all members of a species, that are dependent on an external signal (sign stimulus) to trigger the sequence. Originally termed 'fixed action pattern,' George Barlow argued that because the motor pattern is not performed identically each time it is elicited 'modal action pattern' would be a more appropriate term for a recognizable motor pattern elicited by a sign stimulus.

Modules   Phenotypic units, often occurring in a repeating series that develop more or less independently of each other, which come together to form a larger whole.

Molt   The process of shedding the outer covering of the body.

Molt cycles   Replacement of skin, hair, feathers, scales, etc. usually is cyclic and occurs during restricted periods called 'molts.' There may be one to several molt cycles each year depending on the species.

Monoamines Important neural signaling molecules characterized by having an amino group connected to an aromatic ring. Important examples include serotonin, dopamine, and norepinephrine.

Monocularly With one eye only.

Monocytes This type of white blood cell changes into a macrophage. While they make up only 3–8% of all white blood cells, monocytes have two important functions related to the immune system; one is to replenish resident tissue macrophages that get used up engulfing bacteria and cell debris and the other is to move quickly to new sites of infection where they then differentiate into a new population of macrophages.

Monodomy An ant colony that occupies a single nest.

Monogamy Mating system in which males and females mate with a single partner during a particular breeding season. This typically reduces variance in mating success in both sexes, thereby limiting the opportunity for sexual selection.

Monogynous (monogyne, monogyny) Colonies having one queen.

Monomorphism Individuals of a prey species which are invariant in a specific trait, for example, color pattern.

Monophagous versus oligophagous versus polyphagous Whether a herbivore uses one versus several versus many plant taxa as hosts.

Moon watching A technique for studying nocturnal migration by observing through a 20–30× telescope birds as they pass before the disc of the moon.

Morgan's Canon States that we should not attribute behavior to higher cognitive abilities if it can be explained in terms of simpler processes.

Mormyrid African, weakly electric fishes of the family Mormyridae.

Morph A discontinuous class of morphological variation.

Morphological caste A mechanism for division of labor in which individuals vary in physical attributes, particularly size, with corresponding differences in the tasks they perform.

Morphological computation The idea that the physical body of an animal, interacting with its environment, can function in a manner that removes the need for direct neural control in the production of adaptive behavior.

Morphology The form, structure, and configuration of an organism. This includes aspects of outward appearance such as coloration as well as the form and structure of internal parts such as bones and organs.

Mosaic evolution The ability of selective pressures to produce independent changes in brain regions.

Mosquito control Many states have programs to control mosquitos, either with chemical spray or by altering habitat, such as cutting ditches in salt marshes to drain them so there is no habitat for the mosquitos to breed.

Motion parallax Motion parallax is a monocular depth cue that results from motion of the object or observer. Closer objects move farther across the visual field than distant ones.

Motoneuron (or motor neuron) A neuron located in the CNS that project its axon outside the CNS to innervated and control muscles.

Motor imitation Performing an action after seeing another perform that action.

Mucopolysaccharide Class of polysaccharide molecules, also known as 'glycosaminoglycans,' composed of amino sugars chemically linked into repeating units that give a linear unbranched polymeric compound.

Mucosa The innermost layer of the gastrointestinal tract (gut) that surrounds the lumen, or space within the intestinal tube. This layer of epithelial cells, known as *enterocytes*, comes in direct contact with food, and is the primary site of nutrient absorption.

Multifunctional neurons Neurons, particularly interneurons, that are active in – and presumably contribute to – several different behaviors.

Multiharmonic A vocalization with a nearly constant pulse repetition rate or fundamental frequency, and several prominent harmonics.

Multilevel selection theory Also known as levels of selection theory. Describes how variation in fitness can be partitioned into selection at multiple levels (e.g., between and within groups) to provide insight into how selection affects phenotypic evolution. For example, for social evolution, the balance of selection between and within social groups explains the evolution of sociality (note this can equivalently be described in terms of inclusive fitness/ kin selection theory).

Multimale groups More or less permanent social groups containing multiple, reproductively active adults of each sex.

Multimodal signal Signals produced in multiple sensory modes or channels at the same time.

Multiple messages When the individual components of multimodal signals each convey distinct information.

Multivalued Having more than two values; communication codes having three or more alternative signals.

Multivariate More than one-variable quantity; communication codes having two or more signals making up a decodable unit.

Mutant screen Organisms are exposed to a mutagenic substance and the offspring of the mutagenized organisms are then screened for mutant phenotypes.

Mutation   Any change in DNA sequence, typically caused by errors during DNA replication.

Mutual benefit/mutualism   A behavior performed by the actor that contributes to the lifetime fitness benefits of both the actor and recipient (evolutionary biology); mutualism refers to interspecies cooperation (evolutionary biology); a behavior that produces immediate benefits for both actor and recipient (social science).

Mutual gaze   Eye-to-eye contact is an important characteristic of early mother–infant relationships. Mothers look into the eyes of their infants, while the infants look back into their mothers.' This is called 'mutual gaze.' It is a truly unique feature shared by humans and chimpanzees.

Mutualism   Intra- or interspecific social interactions in which both parties benefit.

Mycophagy   Feeding on fungi.

Myelination   The development of a myelin sheath around sensory or motor neurones. Myelination improves the conduction speed of nerve impulses, enabling fast reactions and skilled movements to occur.

Narrow-sense sexual selection   Variance in fitness due entirely to variation in number of mates. It is often associated with exaggerated traits in males.

Nash equilibrium   A combination of strategies for the players of a game in which each player's strategy is the best response (i.e., one that maximizes expected payoffs) to the other players' strategies. Equilibrium point in a game at which no player can improve its payoff by changing its tactic unilaterally.

Natal   Related to ones birthplace.

Natal dispersal   The movement of an individual from birthplace to breeding place.

Natal homing   Tendency for an animal to return to reproduce in the same geographic area where it began life.

Natriorexigenic   That which provokes salt intake.

Natural selection   Nonrandom differential preservation of traits across generations, leading to changes in the distribution of traits in a population over time.

Nature–nurture controversy   Controversy over the relative importance of genetic factors (nature) and the environment (nurture) in the development of behavior. This is now regarded as supplanted by an epigenetic approach to development.

Necessity and sufficiency   Criteria required to prove causation; for instance, if a behavior disappears when a specific neuron is killed, that neuron is necessary for the behavior; if stimulation of only that single neuron elicits the behavior, it is sufficient; necessity and sufficiency can occur separately.

Necrophagy (adj. necrophagous)   Eating dead and/or decaying insects.

Necrophoresis   Movement toward dead organisms.

Nectar corridor   A series of populations of flowering plants that permit nectar-feeding bats to migrate from one area to another.

Nectarivore   An animal that eats nectar produced by flowering plants.

Negative frequency-dependent selection   A type of selection that favors rare polymorphisms in the population. Under this type of selection, the fitness of a given locus is inversely proportional to its prevalence in the population.

Negative punishment   The removal of a desirable outcome, or the opportunity for reinforcement, coincident with a behavior such that the future probability of that behavior is decreased.

Negative reinforcement   Increasing the future probability of a behavior by the removal of, or a decrease in the intensity of, an aversive stimulus.

Neighborhoods   All the groups of individuals that live within one fragment of habitat, more likely to interact with each other than with individuals from other areas; technically called a deme of a population.

Nematocytes   The stinging cells of cnidarians. These cells contain organelles called nematocysts, among the most complex intracellular structures known in animals. Nematocysts serve a variety of functions including feeding, defense, and locomotion. Nematocytes are found only in the phylum Cnidaria, although a few other noncnidarian groups possess superficially similar cells.

Nematodes (or roundworms)   Phylum of worms with an unsegmented body. Abundant in marine and freshwater habitats, in soil, and as parasites of plants and animals. Nematode species are very difficult to distinguish; over 80 000 have been described, of which over 15 000 are parasitic.

Nematomorpha   Commonly known as 'Horsehair worms' or 'Gordian worms,' parasitic animals that are morphologically and ecologically similar to nematode worms, hence the name. They range in size from 1 cm to 1 meter long, and 1–3 mm in diameter. The adult worms are free living, but the larvae are parasitic on beetles, cockroaches, Orthoptera, and crustaceans. About 326 species are known and a conservative estimate suggests that there may be about 2000 species worldwide.

Neonatal smiling   Human newborns are known to smile spontaneously with their eyes closed, a behavior known as 'neonatal smiling.'

Neophilia   A form of nonassociative learning in which novel things become more acceptable.

Neophobia Fear of novelty. A form of nonassociative learning in which novel things become less acceptable.

Neotenic reproductives In termites, wingless reproductives that develop within the natal colony *via* a single molt from any instar after the third larval instar. At this neotenic molt, their gonads grow and they develop some imaginal characters while maintaining an otherwise larval appearance; some characters, like wing pads, may regress. Neotenic reproductives are characterized by the absence of wings and usually by the lack of compound eyes. The cuticle is less sclerotized than in primary reproductives. They are subdivided into: (i) *replacement reproductives* if they develop after the death of the same-sex reproductive of a colony or (ii) *supplementary reproductives* if they develop in addition to other same-sex reproductive(s) already present within a colony.

Neoteny Persistence of juvenile characteristics into adulthood.

Neotropics An ecozone that includes Central and South America, the Mexican lowlands, the Caribbean islands, and southern Florida.

Nepotism The preferential treatment of relatives.

Nervous system maps A physical organization of neurons that corresponds to locations in the external world; analogous to a road map that depicts the locations of the real roads; can be sensory as in the mapping of touch sensation onto a body representation in the primate cortex or can be motor.

Nest defense Behavior by a parent that reduces the probability that a potential predator will hurt the parent's offspring; the parent may incur some cost of defense, including increased probability of injury or death.

Neural tracer Any substance that, when injected into brain tissue, is taken up by one part of a neuron and is transported to another part and can be used, therefore, to determine connections among brain regions.

Neuroendocrine General interactions between the nervous and endocrine systems; specific production of endocrine signaling molecules by neurons.

Neurohemal organ The enlarged endings of neurosecretory neurons that serve as a distinct storage and release site.

Neurohormone A hormone that is released into the blood from a neuron rather than from endocrine tissue.

Neuromast Functional unit of the lateral line, consisting of a cluster of hair cells with surrounding support cells, and an overlying gelatinous mass called 'a cupula.'

Neuromodulator A chemical messenger, typically a peptide, which is released from presynaptic terminals and acts on the postsynaptic membrane to modulate the responsiveness of the postsynaptic cells to the effects of the neurotransmitter.

Neuro-muscular junction Synapse between the motor neuron terminals and the muscle. In vertebrates, the signal passes through the neuromuscular junction via the neurotransmitter acetylcholine. In invertebrates, the transmitter is Glutamate.

Neuropeptide Peptides found in neural tissue acting as chemical signals to communicate information (such as endorphins, or some hormones like oxytocin and vasopressin).

Neurosteroids Some regions of the brain, especially those involved in territorial aggression, express all the enzymes needed to synthesize sex steroids such as testosterone and estradiol-17beta de novo from cholesterol. These neurosteroids are thought to act locally on neurons associated with aggressive behavior.

Neurotoxin A toxin that acts specifically on neurons usually but not exclusively by interacting with membrane proteins such as ion channels.

Neurotransmitter Chemicals (monoamines, ions, gases, hormones) that relay and modulate signals between a presynaptic neuron and a postsynaptic cell.

New York epigeneticists A group of animal psychologists that developed around T. C. Schneirla and was located primarily at the American Museum of Natural History and the Institute of Animal Behavior. They generally favored nurture over nature and a 'levels' view of evolution according to which only very limited generalizations can be made across well-defined taxonomic levels.

Niche conservatism Closely related species tending to occupy similar environments.

Niche displacement The removal of a species from its ecological and functional space in the environment. It is most often caused by a natural catastrophe, or by interspecific interactions like predation, competition, or mating interference.

Niche (ecological niche) The features of the environment that characterize an organism's position in the ecosystem, such as diet, preferred habitat, location within the habitat, and activity pattern. The ecological role of a species in an ecosystem encompassing abiotic, biotic, and geographical dimensions.

Nocturnal Active at night.

Nomenclature As subdiscipline of taxonomy, it is the naming of taxa, including species and higher level groups. In phylogenetic systematics, nomenclature must be tied to phylogeny. Formal rules governing the naming of animals are codified by the International Code of Zoological Nomenclature (ICZN).

Nonconsumptive effects The effect of a predator's presence on the survival and reproduction of prey, not due to direct killing. In essence, nonconsumptive effects are the costs of antipredator behavior.

Nonelemental learning　Associative forms of learning in which individual events are ambiguous and only logical combinations of them can be used to solve a discrimination problem.

Nongenomic effects of steroids　Effects of steroids on behavior or physiological responses that are not mediated by their binding to their well-characterized cognate intracellular receptors that normally results in a change in gene transcription. These effects are rather thought to come about via an interaction of the steroid with the cell membrane including the binding to membrane receptors of various sorts. These nongenomic effects are observed with much shorter latencies than the traditional genomic effects and by definition do not involve the induction of their biological effects via changes in gene transcription but via changes in protein state and second messenger systems.

Nonlinear timing　The hypothesis that psychological estimates of time are nonlinearly related to physical time.

Nonrapid-eye-movement sleep　One of the two basic forms of sleep in mammals and birds; in adult humans, it constitutes about 75% of total sleep time. It is characterized by high-amplitude, low-frequency brain waves, suppressed muscle tone, and decreased metabolic rate.

Nonredundant signals　When component modes in a multimodal signal contain distinctly different kinds of information, indicated by different responses of receivers to each mode.

Norm enforcement　The infliction of harm (including gossip, shunning, and ostracism as well as physical harm) on another individual for violations of social rules and conventions (social science).

Novelty response　Sudden acceleration of the rate of EOD emitted by a pulse-type electric fish caused by the sudden appearance of a novel sensory stimulus of any modality.

Noxious　Harmful or poisonous.

Nuclear species　A species that plays an important role in the formation and maintenance of a mixed-species group, usually leading the group.

Numerical ratio effect　When comparing a set of numerical values, one's ability to discriminate the values is based on both their magnitude and the difference between them (also known as Weber's law).

Nymph　Generally, a nymph is the juvenile stage of any hemimetabolous insect; it appears similar to an adult except it is smaller and lacks wing structures and developed reproductive organs. In termites, it refers to the preadult instars that perform nonreproductive tasks in the nest. However, since they are juveniles, they can later develop into either reproductive members of the colony or into sterile adult workers. Sometimes, this term is used interchangeably with larvae; however, there is some contention to this dichotomy.

Object　Something perceptible by one or more of the senses, especially by vision or touch; also a focus of attention, feeling, thought, or action.

Object movement reenactment　Reproducing the movement of an object manipulated by another animal.

Obligate　Applies to organisms that have to behave in a particular way to survive and whose behavior is innate. More specifically, individual obligate migrants migrate every year, and do not have the option of migrating or not, their behavior being genetically fixed.

Observational conditioning　Facilitation of the acquisition of a response due to the association between an object and secondary reinforcement (the observation of the other animal making contact with the object and obtaining a reinforcer).

Occasion setting　A learning situation in which a stimulus, the occasion setter, sets the occasion for when or where a predictive relationship applies. Contextual learning is closely related to occasion setting.

Occipital nerves　One or more nerves that originate in the brain and exit the posterior end of the skull through a foramen to innervate muscles that develop from occipital somites; considered a homolog of the hypoglossal nerve of tetrapods.

Occipital somites　Embryonic segments of mesoderm in all developing vertebrates that give rise to several skeletal muscles in the head including vocal/sonic muscles associated with the larynx, syrinx, and swimbladder.

Octavolateralis system　The group of sensory systems related to the eighth, and lateral line cranial nerves. It includes the sense organs of the inner ear, the lateral line, and the electrosense.

Octopamine　A neurotransmitter found in the CNS and elsewhere in all major classes of invertebrates. Its vertebrate equivalent is considered to be noradrenaline. It has been suggested that it plays a crucial role in the flight or flight reaction in insects. In particular, OA has been suspected of having a general effect on insect arousal.

Odiferous　Producing a pungent smell, often unpleasant.

Odometry　The measurement of distance traveled.

Odorant receptor　A protein molecule situated on the membrane of the sensory neuron that recognize a particular odorant (or a class of similar odorants).

Offspring viability　A measure of the relative health of offspring and/or their survival probability.

Offspring viability selection　Occurs when variation in the number of offspring surviving to reproductive age (productivity) differs between constrained and unconstrained parents.

Oil droplets Lipid globules located in the inner segment of the cone photoreceptor of many birds and reptiles that filter light at different wavelengths and decrease the overlap in sensitivity between cones.

Olfaction Sense of smell.

Omnivorous A diversified diet of plant and animal materials.

One-way migration Movement of an organism from a location where it develops to where it breeds without returning to the natal habitat before succumbing.

One–zero sampling A time sample that produces a proportion of periods in which the behavior occurred.

Ontogeny The development of an organism from fertilization through maturity and adulthood, also used to refer to the development of a particular trait over the same time.

Oocyst A zygote stage in the sporozoan life cycle that sporulates to form sporozoites.

Oogenesis Production of eggs.

Oogenesis-flight syndrome A kind of migration syndrome described by C. G. Johnson, and found in many insects, in which migratory activity is limited to the brief period of sexual immaturity of the adult stage that immediately follows metamorphosis to the adult form.

Ootheca An egg case; in cockroaches, a double row of eggs enclosed by a protective outer shell.

Opaque imitation A form of imitation in which the observer cannot see its own reproduction of the behavior that is observed (e.g., imitating a demonstrator who places his hand on his head).

Open diffusion An experimental design for studying the social diffusion of information, in which one or more individuals proficient in a novel action pattern is introduced into, or reunited with, a group of individuals and the potential spread of the action tracked.

Operant conditioning Associative forms of learning in which an individual learns the consequences of its own behavior. It is a form of conditioning in which the desired behavior or increasingly closer approximations to it are followed by a rewarding or reinforcing stimulus.

Operational sex ratio (OSR) The sex ratio among individuals ready to mate (i.e., being in operation).

Opportunistic breeder An organism that can breed at any time of year, as long as specific environmental conditions exist (can thus also be a continuous breeder under correct circumstances).

Opportunistic foragers Animals that feed on whatever is available, and can make use of new and novel food sources.

Opportunity cost The cost of choosing one option and foregoing the opportunity associated with another option.

Opportunity for selection The upper bound on the rate of evolutionary change in the mean of all phenotypes in a population, which is equal to the variance in relative fitness among members of the population divided by the squared average in fitness of those individuals.

Opsin The membrane-bound G-protein-coupled receptor protein found in photoreceptors in the retina, which when combined to the chromophore, forms a visual pigment.

Optic lobe The portion of the insect brain that processes visual input.

Optic tectum A portion of the vertebrate midbrain, which processes sensory information from the eyes. In mammals, the optic tectum is called 'the superior colliculus.'

Optimal foraging theory A body of theory that predicts behavior relative to maximizing or minimizing one or a set of goals.

Optimal group size A group size for which the net benefits of group members are at a maximum.

Optimality The cost–benefit approach has been extended to model when the benefit-to-cost ratio is maximized so that an individual should maximize the benefit of the behavior while simultaneously minimizing any costs associated with the behavior.

Optimal outbreeding Mating with animals that share, due to identity by descent, favorable gene combinations, while avoiding matings with first or second degree relatives (parents, sibs, offspring) that might expose deleterious lethal genetic combinations.

Optimization and trade-offs *Optimization* is a mathematical concept in which a function is either minimized or maximized given a restricted set of alternative inputs into the function. In behavioral ecology, it applies to predicting or interpreting behavioral decisions that maximize net fitness (e.g., lifetime reproductive success) in the face of conflicting demands, such as avoiding predation, which reduces feeding rates, and foraging, which increases exposure and risk of death by predation. *Trade-offs* are the outcomes of these decisions, such as greater safety at the cost of poorer energy stores or better energy stores at the cost of higher predation risk.

Optomotor response Innate behavior used to stabilize a moving image through movements of the eyes, head, or body.

Organizational effects Permanent changes in morphology, physiology, and/or neural circuitry dependent on hormone exposure during development.

Oropharynx Region including the oral cavity and pharynx.

Ortholog A similar gene in different species, thought to be derived from a common ancestor.

Oscillator An oscillator is a process that repeats periodically.

Osmoregulation   The homeostatic control (see homeostasis) of osmotic potential or water potential, resulting in the maintenance of a constant volume of body fluids.

Otolith   Also known as 'ear stones,' these calcium carbonate structures are attached to the sensory epithelium of subdivisions of the vertebrate inner ear that are known as the lagena, saccule, and utricle. Each subdivision may serve either a vestibular (balance) and/or an auditory (hearing) function.

Oviposition   Egg-laying.

Ovipositor   The valved egg-laying apparatus of a female insect.

Ovipositor valve   The blade-like paired structures comprising the ovipositor shaft.

Oxytocin   A peptide produced almost exclusively within the hypothalamus that is released from the posterior pituitary and from neural projections to numerous intra- and extrahypothalamic brain sites. I Involved in milk-let-down, mother–offspring, and pair-bond formation in females, and contraction of nonstriated muscles for example during parturition.

Paedomorphosis   Reproductive maturity is attained while in a larval or branchiate form.

Pain   An aversive sensation and a feeling associated with actual or potential tissue damage.

Pair bond   The temporary or permanent association formed between a female and male, potentially leading to breeding.

Palps   Lateral mouthparts of invertebrates.

Panmictic (panmixia)   When mating between individuals in a population occurs randomly.

Pan-pipes   A device designed to present a naturalistic challenge to a tool-using animal such as the chimpanzee (*Pan*). A blockage in the upper of two pipes traps a food item. In social learning experiments, the blockage is released by using the tool in either of two quite different ways, the spread of which through social learning can thus later be objectively recorded (see Whiten et al., 2005).

Paracellular solvent drag   Movement of small molecules from interior of intestine (*lumen*) to circulatory fluids by passing between *enterocyte* epithelial cells of small intestine.

Paracrine agent   A chemical messenger that is released into the extracellular fluid and diffuses to and acts on adjacent target cells without entering the systemic circulation.

Paradigm   A combination of methods used to investigate problems, or an overall model of scientific conclusions regarding a given subject (e.g., how a contaminant affects the behavior of an animal, including humans).

Paralog   A gene that duplicated from an ancestral gene.

Parasite   Something that lives in, with, or on another organism (the host) and obtains benefits from that organism. A parasite is detrimental to the host in varying degrees.

Parasite manipulation   The ability, shared by several parasite groups, to modify their hosts' behavior to their own advantage, generally through increased probability of transmission in parasitic cycles.

Parasite propagules   Life-cycle stages which enable dispersion, transmission between hosts and from which new organisms can develop.

Parasitic wasps   A number of families of wasps which lay their eggs inside or outside of the larvae, pupae, or adult-host arthropods. The eggs hatch and the wasp's larvae feed inside the host eventually killing it. The wasp's larvae then pupate inside the host and emerge as adult wasps.

Parasitoid   An organism that spends a significant portion of its life history attached to or within a single host organism that it ultimately kills.

Parasocial route   Social grouping that originates in aggregations of individuals, usually around a rich resource.

Parathyroid glands   Small endocrine glands in the neck which are involved in calcium homeostasis.

Parentage analysis   Are studies or experiments that allow the investigator to determine the parents of any given individual offspring. In modern times, this is done using molecular techniques involving DNA analyses using mostly microsatellites.

Parental distraction display   Any behavior by a parent that reduces the probability that a predator will harm the parent's offspring by means of drawing the predator's attention away from the offspring; may take the form of feigning injury, tail-flagging, explosive flight, or erratic or conspicuous running.

Parental investment (PI)   Any investment by a parent that increases offspring fitness, at the cost of investing in other offspring.

Parental manipulation   Proposes that offspring helping behavior, a fundamental characteristic of the evolution of eusociality, arises as a result of parents influencing offspring development and condition.

Parent–offspring conflict   The disparity in selective pressures arising because optima in parental investment differ between parents and offspring.

Parr   A juvenile salmon during the initial freshwater phase of life.

Parsimony   The fundamental scientific principle that assumptions (especially process assumptions) need not be inflated beyond what is necessary to explain the phenomenon. In phylogenetics, the optimal tree is one that summarizes the putative homologies in such a way that as many as possible are retained. That is, homology is maximized, and as a result, the minimum number of evolutionary changes necessary is preferred.

Parthenogenesis Development from an unfertilized egg.

Partial migration A situation in which some birds from a given breeding area migrate away for the nonbreeding season, while others remain in the breeding area year-round.

Passerine A bird belonging to the order Passeriformes, also referred to as 'perching birds.' Songbirds also belong to this group.

Patch A relatively homogeneous area that differs in some way from its surroundings.

Path integration Estimation of the current position relative to a starting location by integrating distances traveled and changes in direction throughout the journey.

Pathogen Any disease-causing agent, especially a microorganism.

Patience A preference for the more delayed outcome.

Patrigene In diploids, the allele inherited from the father.

Pavlovian conditioning This term is used interchangeably with the term 'classical conditioning.' It is a method of learning in which animals have inescapable exposure (one or more times) to an emotionally neutral stimulus, the conditioned stimulus, in temporal association with an innately provocative stimulus, unconditioned stimulus. Because of this association, the conditioned stimulus becomes a predictor of the occurrence of the unconditioned stimulus, and it typically acquires emotionally provocative properties similar to the unconditioned stimulus but at lower intensity.

Payoff matrix A mathematical description of the fitness benefits to one behavioral strategy when it plays other strategies.

PCR Polymerase chain reaction, a chemical reaction that utilizes a polymerase enzyme to replicate a target DNA sequence using primers that bind to the target DNA.

Pearson coefficient A measure of correlation between ordinal or ratio data; can be used as a measure of observer reliability.

Pecking The act of striking with the beak.

Pectoral girdle That part of the skeleton that connects the fins or limbs to the axial skeleton (homologous to the shoulder region in mammals).

Pelage Soft covering of a mammal such as hair, fur, or wool.

Penetrance A genetic term referring to the extent to which the effect of a gene is expressed.

Peptide hormones Small proteins, typically around 100 amino acids or shorter that are released from one tissue and have their action in another. These hormones differ slightly from species to species as the result of evolutionary changes in the DNA sequence, posttranslational processing, etc.

Perception Physical sensation interpreted in the light of experience; as a fundamental means of allowing an organism to process changes in its external environment it depends on, but is not equal, to *sensation* – the detection of a stimulus and the recognition that an event has occurred; it can be viewed as the process whereby sensory stimuli are translated into organized experience. In the human cognitive sciences, perception is the process of attaining *awareness* or understanding of sensory information.

Perception-action mechanism (PAM) Perception of another's state or situation activates neural representations of similar states or situations that the self has experienced.

Perception component The recognition and processing of cues and cue bearers by an evaluator.

Perceptual class A collection of items sharing perceptual properties, that is, arrays of features or elements defined in their own absolute values; thus class membership is solely based on similarity.

Period The amount of time taken to complete one cycle of a sinusoid is its period. Period is the reciprocal of frequency. Low-frequency sounds have long periods and high-frequency sounds have short periods. In chronobiology, period is the time it takes for a full oscillation or rhythm to occur.

Periodogram analysis A type of time series analysis that involves combining average response rate functions assuming different underlying periodic trends. The analysis identifies the underlying periods that minimize errors of prediction.

Perspective taking Being in a position to form a 'mental picture' of what another can see even when you cannot see it directly yourself.

Pesticides A range of chemicals produced to kill insects; many chemical forms exist some of which are endocrine active.

Pet A domestic or tamed animal that is individually identified, kept by a person or persons as a companion, and cared for with affection.

Phagocytosis The cellular consumption or elimination of foreign tissues, cells, or particles.

Phagomimicry Release of a chemical that induces feeding behavior toward the chemical (a false food stimulant) and not the animal that releases it.

Phagostimulant Anything that triggers feeding behavior.

Pharmaceuticals   Chemicals produced for the treatment of biomedical conditions.

Pharmacology   The science of the properties of drugs and their affects on the body.

Pharynx   The part of the neck and throat situated immediately posterior to (behind) the mouth and nasal cavity.

Phase   Temporal relationship between two rhythmic processes having the same frequency.

Phase angle   Measurement of phase, expressed as the time delay between two rhythmic processes, divided by the length of the common period and multiplied by 360°.

Phase (in locusts-solitarious, gregarious, or transiens)   A combination of traits defining morphological, physiological, and behavioral state of a locust.

Phase locking   The auditory system uses phase-locked spikes to encode the timing or phase of the auditory signal. Phase-locked neurons fire spikes at, or near, particular phase angles of sinusoidal waveform. Physiological experiments measure this spike phase with respect to the stimulus period. Spike phase is plotted in a period histogram and is used to calculate the statistic vector strength (r). Each spike defines a vector of unit length with a measured phase angle. The vectors characterizing the spikes are plotted on a unit circle and the mean vector calculated. The length of the mean vector provides a measure of the degree of synchronization.

Phase shift   A phase shift is a change in the timing, or phase, of an oscillation or rhythm in response to an external cue. A widely used manipulation in the study of biological rhythms. The event that is thought to reset timing (e.g., light-dark cycle in the case of circadian rhythms) is advanced or delayed. Gradual adjustment in response to a phase shift is a characteristic feature of an endogenous oscillator.

Phenology   The repetitive sequence of events of the life cycle of plants and animals that are affected by environmental conditions.

Phenomenology   One's subjective experience or the experience from the first-person point of view.

Phenotype   Any characteristic of an organism that is the result of that individual's genotype and the interaction of the genotype with the environment during development.

Phenotype matching   The ability to learn phenotypes of group members, such as littermates, and to extend that knowledge of phenotype to make discriminations among previously unmet animals.

Phenotypic flexibility   See phenotypic plasticity.

Phenotypic interface of coevolution   The traits that mediate ecological interactions between coevolving species, such as chemical defenses of prey and resistance to those compounds by predators.

Phenotypic plasticity   The capacity of an individual organism to produce different phenotypes (morphology, physiology, behavior, etc.) in response to different environmental inputs.

Pheromone   A chemical messenger produced by an organism that influences the behavior or physiology of another organism of the same species.

Phi coefficient   A measure of correlation between nominal data; can be used as a measure of observer reliability.

Philopatric reproduction   Breeding at the natal nest.

Philopatry   The tendency of an individual to remain or return to its birthplace.

Phonological syntax   Structured rules for constructing sequences of otherwise meaningless sound units.

Phonotaxis   Locomotion towards or away from a sound source.

Photic zone   The portion of the upper water column with sufficient light for photosynthesis to occur, typically reaches between 50 and 200 m depth in oceanic waters.

Photomechanic infrared receptor   A receptor rapidly dissipating infrared energy into a micromechanical event (i.e., a brief increase in internal pressure in the core of the receptor) which is measured by a mechanoreceptor.

Photoperiod   Length of day.

Photoperiodism   Changes in reproductive physiology and behavior in response to changing day length.

Photorefractoriness   A complete shutdown of the hypothalamic-pituitary-gonad axis that terminates the breeding phase.

Photorefractoriness in birds   Physiological state in which photoperiodic birds terminate reproduction during long day lengths. Signals the end of the breeding season.

Photorefractoriness in mammals   Physiological state in which photoperiodic mammals reactivate the HPG axis after prolonged exposure to short days. Spontaneous gonadal recrudescence occurs and the short days no longer inhibit reproduction.

Phototaxis   From photos (light) and taxis (movement). It refers to movement toward or away from a light source (positive or negative phototaxis, respectively).

Phylogenetic   Relating to, or based on, evolutionary history.

Phylogenetic signal   The extent to which similarities among closely related species, such as the form of communication they use, is dependent on the phylogenetic relationships between those species.

Phylogenetic systematics (also, phylogenetics, cladistics)   The particular method of systematics proposed by Willi Hennig. Phylogenetic systematics relies on two fundamental precepts: (1) only whole character-state transformations, in the form of synapomorphies, count as evidence of relationship; and (2) taxonomic names must be applied only to natural evolutionary groups (i.e., nomenclature is united with phylogeny).

Phylogeny   A genealogy of species that reflects their evolutionary relationships.

Physiological psychology   The study of mechanisms internal to the animal that affect and are affected by behavior. Included are studies of the nervous system, endocrine function, and other internal processes.

Phytohormones   Hormone-life chemicals produced by plants that have structural characteristics that allow them to interact with steroid hormone receptors in vertebrate physiological systems; for example soy phytoestrogens.

Phytophagous   Plant eating.

Pied Piper effect   The idea (now largely discounted) that in the northern hemisphere, northward movements of insects in the spring are facilitated by favorable winds, but that the progeny of these immigrants are then trapped at high latitudes as winter approaches, leading to mass fatality.

Piloerection   This term derives from 'pilo,' meaning hair, and refers to the erection of the hair of the skin. Piloerection starts when a stimulus such as cold or a frightening stimulus causes an involuntary contraction of the small muscles that attach to the base of the hairs deep in the hair follicles. Contraction of these muscles elevates the hair follicles above the rest of the skin so the hairs seem to stand on end.

Planktivorous   Feeding primarily on organisms that drift or possess insufficient motor capabilities to overcome currents (plankton).

Plant secondary metabolite   See secondary plant compound.

Playback studies   'Playbacks' can be broadly defined as the use of broadcast signals in any sensory modality with an accompanying bioassay to address questions concerning communication and animal behavior.

Pleiotropic   (see pleiotropy).

Pleiotropy   When a gene affects multiple traits.

Pleometrosis   The founding of a eusocial insect colony by several queens.

Poikilothermic   Having a body temperature that varies with the temperature of its surroundings.

Point of balance   It is a point at the animal's shoulder that handlers can use to control animal movement. When a person stands behind the point of balance, the animal

moves forward. When a person stands in front of the point of balance, the animal backs up.

Policing   Repression of selfish or competitive behavior (evolutionary biology); in social insects, inhibition of worker reproduction by aggression or destruction of eggs (evolutionary biology); impartial intervention in conflicts (social science); enforcement of societal norms and laws (social science).

Polyandry   One female has a breeding relationship with two or more males. In eusocial insects, a queen that has mated many times.

Polydomy   Of colonies having more than one nest each.

Polyembryonic   A form of reproductive in which one sexually produced embryo splits into many genetically identical offspring.

Polygamous   Having more than one partner or spouse.

Polygenic   Where several genes interact to influence a phenotype; each gene may have a varying degree of influence upon the phenotype.

Polygyny (polygynous)   Mating system in which some or all males in a population mate with more than one female per breeding season. This typically increases variance in male mating success, thereby generating sexual selection on males. In eusocial insects, a colony that has two or more queens.

Polymorphism   The existence of multiple forms within a population or species. The term can refer to morphology or alleles or physiology or behavior or any other kind of trait. In eusocial insects, size or shape variation in the worker caste.

Polymorphous class   A class in which no single feature is necessary or sufficient to determine class membership, but several features contribute to this to some degree.

Polyphagy (adj. polyphagous)   Eating many kinds of food, for example, many plant species from a range of families.

Polyphenism   Within a population, different phenotypes that arise from environmental rather than genetic causes.

Polyspermy   When more than one sperm enters the egg during fertilization.

Ponerine ants   The Ponerinae is a subfamily of ants (Hymenoptera, Formicidae). Some species within the Ponerinae are queenless, having lost the queen caste. A colony is headed by one or more mated workers that fulfill the queen's role and are sometimes known as gamergates.

Population density   The number of individuals within a specified unit of space.

Population dynamics   Study of short- and long-term changes in the size and age composition of populations, and the factors influencing those changes.

Porphyropsin   All visual pigments whose chromophore is 3,4 dehydroretinal.

Positional cloning   A technique used in molecular cloning that utilizes a set of unique genomic elements called 'genetic markers' that flank the gene of interest.

Positive punishment   The application of a stimulus immediately after a behavior which results in a reduction in the future probability of that behavior.

Positive reinforcement   The application of a stimulus coincident with or immediately after a behavior which results in an increase in the future probability of that behavior.

Postconflict bystander affiliation   Postconflict affiliative interaction between a conflict opponent and a bystander uninvolved in the conflict.

Postconflict quadratic affiliation   Postconflict affiliation between two bystanders.

Postcopulatory sexual competition   The term generally used to refer (somewhat imprecisely) to all events following the initiation of genital coupling.

Postmating-prezygotic isolation   Barriers between species or populations that result from mechanisms that prevent zygote formation after mating.

Postzygotic compensatory mechanisms   Flexible responses of constrained individuals of either sex that increase the likelihood that already produced zygotes will survive to reproductive age.

Postzygotic isolation   Barriers between species or populations that result from low fitness of hybrids.

Potential conflict   Differences in the reproductive optima of individuals or groups in a colony. For example, there is potential conflict over male production in a colony of eusocial Hymenoptera as each individual is more related to its own sons (0.5) than to the sons of the mother queen (0.25) or sister workers (full nephews 0.375, half nephews 0.125).

Potential reproductive rate (PRR)   Offspring production per unit time when unconstrained by mate availability.

Praying mantis   A predatory insect with prominent eyes and an elongated body; in the order Dictyoptera along with the cockroaches and termites.

Precedence effect   Psychophysical phenomenon in which two or more stimuli separated by a brief time interval are perceived as a single stimulus originating from the source of the first one.

Precocial   Mobile young (usually birds or mammals) that are dependent on parents for food and warmth.

Predation risk theory   The framework used to predict or interpret antipredator behavior, risk effects, and the behavioral component of trophic cascades. Fundamental to it is the assumption that prey maximize fitness (e.g., lifetime reproductive success) by making behavioral decisions that optimize trade-offs between predator avoidance and resource acquisition.

Predation sequence   The sequence of events that is necessary for a predator to kill one or more prey individuals, including search, encounter, hunting, and killing. Each of these four stages can include distinct substages.

Predator inspection   Alone or in groups, an approach toward a predator to observe and gain information about it that may function to deter attack by advertising that the predator has been detected; the behavior may also advertise ability to incur risk and escape.

Prediction   Statement of results of studies that could be performed.

Preening   The act of cleaning and trimming the feathers with the beak.

Preference function   The order in which an individual ranks potential mates.

Preference hierarchies   A ranking indicating the relative degree to which each of a set of alternative plants are preferred by a herbivore.

Premating isolation   Barriers between species or populations that result from mechanisms that prevent mating.

Premetamorphosis   Stage of amphibian larval development when the animal grows but little or no morphological change occurs; plasma thyroid hormone concentrations are low.

Preoptic area   A region of the brain just rostral to the optic chiasma where steroid action plays a key role in the activation of male sexual behavior in many vertebrate species.

Prepubescent   Prior to puberty.

Prezygotic compensatory mechanisms   Flexible or facultative responses of constrained individuals of either sex that increase the likelihood that their offspring survive to reproductive age.

Price equation   A mathematical statement of evolutionary change that partitions selection into a between- and within-group component.

Primary polygyny   Polygyny that arises through pleometrosis.

Primary predator–prey behaviors   Behaviors concerned with predators encountering prey, or prey avoiding predators, before any attack occurs.

Primary reproductive   The winged, founding members of a termite colony. A winged reproductive male and female found a new colony as the primary reproductives.

Primitively eusocial Eusocial society in which all individuals are capable of mating and reproducing, though behaviorally specialized reproductives occur.

Prisoner's dilemma In its simplest form, it is a two-player game in which players decide whether to cooperate (C) or defect (D). The relative sizes of the payoffs define the game, in that mutual cooperation pays more than mutual defection, but defecting while your partner cooperates provides the highest payoff, and cooperating while your partner defects provides the lowest payoff. The game captures both the temptation to defect and the low payoff for being a 'sucker' (cf. the 'tragedy of the commons,' which arises in a multiplayer version of this game).

PRKO Knockout mice lacking a functional progesterone receptor.

Probability matching In the study of foraging behavior, this refers to an animal's tendency to match its proportion of visits to a feeding site with the proportion of times that site produced food.

Probing motor acts (PMA) Characteristic behaviors composed of a series of swimming movements in close proximity to an object under investigation.

Problem-solving The use of novel means to reach a goal when direct means are unavailable.

Proboscis Central trunk-like mouthpart of insects that feed on liquid food.

Proceptivity Feminine behaviors that are evoked by stimuli from the male and which serve to reduce the distance between the female and the male. The extent to which a female initiates mating (i.e., a female's willingness and motivation to mate).

Producer and scrounger Behavioral alternatives for group foragers when a resource, for example, food, is found by one individual, the producer, and then exploited by one or more animals in the group, the scroungers. The term also describes a game theory model that applies to the two alternatives.

Production learning Where a signal is modified in form as a result of experience of the usage of signals by other individuals.

Productivity The number of offspring that survive to reproductive age.

Progesterone Steroid hormone produced mostly in gonads and the brain.

Progressive molt A molt characterizing the gradual development from egg via several instars into an adult. Associated with progressive molts is an increase in body size and morphological development. This is the default developmental program in all hemimetabolous and holometabolous insects.

Progressive provisioning Type of larval provisioning in which the larvae are fed throughout their development. In contrast to mass provisioning, in which all of the food necessary for larval development is amassed before laying an egg.

Prohormone Precursor to the active form of a hormone.

Prometamorphosis Stage of amphibian larval development when metamorphosis begins. Hindlimb growth and development is evident externally. The thyroid gland becomes active and secretes thyroid hormone in response to increasing plasma concentrations of pituitary thyrotropin (TSH).

Propagules Any structure that can give rise to a new individual. This could include sexually or asexually produced zygotes, embryos, larvae, seeds, or fragments or buds.

Propolis Plant resins collected by honeybees and used for sealing gaps and cracks in their nest.

Proprioception The ability to sense the position and location and orientation and movement of the body and its parts.

Prosocial behavior Tendency to help others even if this provides no immediate reward to the self.

Prosociality The tendency to help another in a situation where there are no personal gains, and little or no personal cost.

Prostaglandins Fatty acid hormones, such as PGF2a, that are secreted by the reproductive tract and ovary.

Protandry A sexual pattern in which individuals mature as males and can then later change functional sex to become female.

Proteome The set of proteins expressed by the entire genome of an organism under given environmental conditions at a given time. Proteomics is the large-scale study of the structure and function of this entire set of proteins, generally in a particular cell, tissue-type, or organ (such as the brain).

Prothoracic gland The molting gland of the insect that secretes ecdysone, the precursor of the active form of the molting hormone, 20-hydroxyecdysone.

Protogyny A sexual pattern in which individuals can mature as females and then later change functional sex to become secondary males. In *monandric* ('one male') protogyny, all secondary males first pass through a female stage. In *diandric* ('two males') protogyny, individuals can mature as either males or females and both can change from the initial phase (IP) to become the larger and typically colorful and aggressive terminal phase (TP) males.

Prototype The 'best' or most typical example of a category that corresponds to the average, or central tendency, of all of the exemplars that have been experienced; it serves as the basis or standard for other members of the same category.

Protozoan   Unicellular microorganisms among eukaryotes. Comprises flagellates, ciliates, sporozoans, amoebas, foraminifers.

Proximal   Closer to a body midline (opposite of distal).

Proximate causation   Explanations of an animal's behavior based on internal and external mediators of behavior including genetic underpinnings, epigenetic forces, maternal effects on physiology, morphology, and development. Questions about proximate causes are sometimes said to be about how animal behavior is expressed or about mechanisms of animal behavior.

Proximate factors   External stimuli (such as specific daylengths) which are used as cues by an animal to trigger preparation for breeding, migration, molt, or other events, or as time keepers to set their endogenous time programs at appropriate times of the year.

Pseudergate   In termites, an alternative technical term that can be found which distinguishes workers with a flexible development and options for direct reproduction from workers with restricted developmental trajectories. Pseudergates are the 'workers' of many lower termites (including wood-dwelling and foraging species) that have broad developmental options, generally including progressive, stationary, and regressive molts. Current use of this term often lacks the precision of its original definition for individuals that develop regressively from nymphal instars to 'worker' instars without wing buds.

Pseudopregnant   Reproductive condition in which a female shows external indicators of pregnancy but is not actually pregnant.

Pseudoreciprocity   The act of increasing another individual's fitness to acquire or enhance the by-product benefits obtained from that individual.

Pseudoreplication   A statistical error in which interrelated observations or measures are treated as though they are statistically independent.

Psychoneuroimmunology   A relatively new field in medicine that explores the ability of the nervous system and psychological states to influence immune defenses, and the ability of the immune system to influence the brain and behavior.

Pterygoid teeth   Small teeth on the roof of the mouth.

Ptilochronology   The study of growth bands in feathers that indicate condition or problems during feather molt in birds.

PTT   A platform transmitter terminal (PTT) sends an ultrahigh frequency (401.650 MHz) signal to satellites.

PTTs are attached to animals in order to track their movements.

Public good   A resource that is costly to produce and provides a benefit to all the individuals in the local group. Public goods systems are often open to exploitation by cheats who benefit, but do not pay the cost.

Public information   Cues produced by animals that can potentially be used by observer animals in making behavioral decisions.

Pulse repetition rate   The rate at which individual sound pulses are produced within a single call.

Punishment   A costly behavior that is negatively reciprocal (decreases harmful behavior in the recipient) (evolutionary biology); any stimulus that reduces the frequency of a behavior (social science); behavior correction and the enforcement of social norms, typically by impartial parties; see also Third-party punishment, Policing (social science).

Pupa   A life stage in some insects that undergo complete metamorphosis that results in the transition between the larval and adult stage.

Purging selection   Mechanisms eliminating deleterious genes from the population.

Pyrophilous insects   Species strongly attracted to burning or newly burned areas, and species that have their main occurrence in burned forests 0–3 years after the fire.

Quality of life   Well-being; a multidimensional, experiential continuum that comprises an array of affective states, broadly classifiably as relating to the states of comfort–discomfort and pleasure; often equated to welfare and well-being.

Quantitative trait   A continuous trait such as body mass that is influenced by many genes and the environment.

Quantitative trait locus (QTL)   A region of DNA that is associated with a particular quantitative trait, containing a gene or genes that influence that trait. Quantitative traits typically have continuous distributions rather than discrete states, and are influenced by several or many loci, each with relatively small or large effects on the expression of the trait.

Quasi-experimental design   An experimental design where a treatment variable may be manipulated but subjects within groups are not equated or randomly assigned.

Quasiparisitism   Occurs when the female that dumps eggs in another female's nest is the resident male's extra-pair partner and her dumping is assisted by that male.

Queen   Reproductive female in a eusocial insect society. She is developmentally and/or behaviorally disposed towards performing all reproductive function for a colony.

Questing   The behavior of ticks, involving an ascent on vegetation that allows for a maximum exposure of sensory receptors on the forelegs to stimuli from approaching hosts.

Quorum decision   A minimum number of individuals required to perform a specific behavior (such as choosing a direction of travel) that results in all of the other members of a group adopting this behavior.

Quorum sensing   A rule under which a social group member's execution of a particular act or behavioral transition is conditioned on the presence of a threshold number of fellow group members.

Radiotracking   The location and tracking of a radiomarked individual from a signal emitted frequently by the radio.

Rape   A legal term and includes other forms of sexual assault as well as forced copulation, including statutory rape, which may appear to be consensual copulation but with a minor; in this case women, not just men, can be rapists.

Rapid-eye-movement sleep   The other basic sleep form in mammals and birds. It is often called 'paradoxical sleep' because the brain activity resembles that of the awake brain. It is characterized by the complete inhibition of muscle tone and suppressed autonomic regulation of most homeostatic functions such as thermoregulation and blood pressure.

Rate of return   The ratio of the amount of food obtained to the time it took to procure the food.

Rationality   A set of consistency principles that decision-makers are expected to follow if they are attempting to maximize some currency such as utility or fitness. Fitness maximization by natural selection is expected to yield rationality, but many instances of irrational choice are known in humans and other animals. Property of individual choice is used both to describe the process of making a choice and to describe the behavioral outcome of choice.

Rayleigh scatter   Light scatter by particles smaller than the wavelength of light.

Reaction norm   A reaction norm describes the production of a range of phenotypes by a single genotype in response to a range of an environmental parameter. Different genotypes may produce different response trajectories in response to a gradient of an environmental parameter. Reaction norms resemble dose-response curves in physiology, for example the effects of a gradient in hormone concentrations. Dose-response relationships are not necessarily monotonic but can include thresholds or show maximal (minimal) effects at low and high doses or medium doses.

Reasoning   A form of logic-based thinking; the cognitive process of looking for reasons for beliefs, conclusions, actions, or feelings.

Receiver psychology   Sensory capabilities of the signal receiver that affect the detectability, discriminability, and/or memorability of signals, and play a role in the evolution of signal design.

Receptivity   Sexual behaviors that are necessary and sufficient for mating.

Reciprocal altruism   Where individual A pays a personal cost to help individual B with the expectation that B will return the favor.

Reciprocal selection   Positive feedback between selection by ecological enemies. Natural selection by predators on prey generates the evolution of increased defense, which in turn causes stronger selection by prey on predators to evolve greater exploitative abilities.

Reciprocity   Delayed exchange of benefits between parties.

Recognition signals   Signals that evolved to make a signaler distinctive.

Recombination   In evolutionary algorithms, a process of crossover that combines elements of existing solutions in order to create at the next generation a new solution, with some of the features of each 'parent solution.' It is analogous to biological crossover.

Reconciliation   Postconflict affiliative reunion between former opponents that restores their social relationship disturbed by the conflict.

Recruitment   Entry of progeny into a population as reproductive adults.

Red queen   Based on the quote from Lewis Carroll's Red Queen, 'It takes all the running you can do, to keep in the same place,' this metaphor describes a coevolutionary dynamic where frequencies of traits or genotypes of ecological enemies cycle through time so that as one type becomes common, it is disfavored and a rare type can spread through the population.

Redirected aggression   Postconflict aggressive interaction directed from the original recipient of aggression to a bystander uninvolved in the conflict.

Redirected behavior   The direction of some behavior, such as an act of aggression, away from the primary target and toward another, inappropriate target.

Redundancy reduction   The reduction in the overlap of information encoded by neurons in the nervous system.

Referent   The on model on which a signal is based.

Reflectance   The ratio of reflected to incident light on a given area (e.g., colored patch in the plumage).

Refraction   Change in direction of light caused by alteration of its velocity on obliquely entering a medium of different refractive index.

Refractive index A measure of the speed of light in a medium.

Refractive state The resting refractive state of an animal determines the point at which it is focused without having to expend any accommodative effort.

Regressive molt A molt that is characterized by a decrease in body size and/or regression of morphological development, generally a reduction of wing bud size in nymphal instars. This type of development is unique to termites.

Regularity A specific version of independence from irrelevant alternatives. It describes the expectation that the absolute preference for an option should never be increased by the addition of inferior options to the choice set.

Regurgitant A substance produced in the gut of an insect that is excreted from the mouth as a defensive secretion.

Reinforcement The evolution of premating isolation after secondary contact as a result of selection against hybrids or hybridization.

Reinforcement/supplementation Addition of individuals to an existing population of conspecifics.

Reintroduction An attempt to establish a species in an area which was once part of its historical range, but from which it has been extirpated or become extinct.

Relatedness asymmetries A group of individuals are more closely related with a certain group of individuals than others within a colony.

Relatedness, r Genetic similarity between individuals, in comparison with randomly chosen individuals in the population, that have a mean relatedness of zero by definition.

Relational class A class defined by relations between or among its members and going beyond any perceptual similarities or functional interconnections.

Relative risk An individual's risk of predation given the abundance of its type.

Relaxed selection This occurs when the sources of natural selection engendering physical or behavioral traits that promote fitness diminish markedly or are no longer present in the environment. In the case of predators, prey species might be separated from their former predators by their isolation on islands. In another context, climate change tolerated by prey might diminish contact with their predators that are intolerant to climate change and eventually disappear.

Reliability The percentage of signals of a particular type X that are accurately associated with a stimulus (X′).

REMI Restriction enzyme mediated integration (REMI) is an ingenious method of introducing single gene knockouts in a genome in a way that allows one to identify the actual gene that is knocked out. Used in *Dictyostelium*.

Repeatability Consistency between different measurements separated in time of a trait of a certain individual, used in population genetics as the upper limit of heritability.

Repertoire expansion A pattern of temporal polyethism in which workers increase the types of tasks they perform as they age.

Replication Using more than one observation per observational unit or subject per experimental treatment group.

Reproductive age The age at which an individual becomes receptive to mating the first time.

Reproductive character displacement The process of phenotypic evolution in a population caused by cross-species mating and which results in enhanced prezygotic reproductive isolation between sympatric species. Referred to as 'reinforcement' if postzygotic isolation is incomplete.

Reproductive compensation Refers to any flexible response of constrained individuals that increases the likelihood that their offspring will survive to reproductive age.

Reproductive division of labor Differentiation of individuals within a eusocial colony into those capable of reproducing, and functionally or physically sterile workers.

Reproductive effort The proportion of available time, nutrient or energy resources that an adult invests in current reproduction, usually detracting from those available for other functions.

Reproductive groundplan hypothesis (Originally described as ovarian groundplan hypothesis) Proposes that the evolution of eusociality is based on simple evolutionary modification of conserved reproductive and corresponding behavioral cycles so that during the course of social evolution, reproductive and nonreproductive behavioral and physiological components can be separated and used to build reproductive (queen) and nonreproductive (worker) phenotypes.

Reproductive isolation Reduced genetic exchange between populations via reduced interbreeding and lower fitness of hybrid offspring; speciation has occurred when reproductive isolation between populations is complete.

Reproductive skew Asymmetry in the distribution of direct reproduction among individuals within a social group.

Reproductive strategy An organism's relative investment, behaviorally and physiologically, in offspring, including reproduction and parental care.

Reproductive success (RS) Refers to the number of offspring an individual produces which survive and go on to reproduce in the next generation. Although 'life-time reproductive success' is the most accurate measure, logistically it is not always possible to obtain this measure.

Consequently, RS may be measured as number of eggs produced, number of young produced, number of young that fledge from the nest (e.g., birds) or survive to weaning (e.g., mammals), or number of young that survive to reproductive age.

Reproductive suppression A mature individual does not reproduce because of physiological mechanisms that inhibit production of gametes as a direct result of communication with conspecifics.

Reproductive value The expected reproduction of an individual from its current age onward, given that it has survived to that age. It changes with age, increasing at first and declining until death.

Residual reproductive value The number of offspring an individual is expected to produce during its remaining lifespan.

Resource competition A particular form of competition in which members of the same or different species compete for the same resource in an ecosystem (e.g., food, space).

Resource constraint hypothesis (Trivers–Willard effect) Colonies should invest more in the cheaper sex (i.e., males, which are generally smaller than females in Hymenoptera) when resources are limited.

Resource holding potential The relative fighting ability of a contestant.

Response blocking Also called *flooding* – The process of exposing a subject to constant, high levels of a distressing stimulus, while preventing escape from the situation, in an attempt to reduce or extinguish the distress produced by the stimulus.

Retinal disparity Difference between the images projected on the two retinas when looking at an object that serves as a binocular cue for the perception of depth.

Retinoscopy A technique used to obtain an objective measurement of the refractive state of the eye, in which a moving light is shone into an animal's eyes and the relative motion of the reflection is observed.

Reverse genetics A molecule-driven approach to understanding a phenotype.

Rheotaxis Orientation or response to current flow; moving upstream is positive and downstream is negative rheotaxis.

Rhinophores Tentacles in some gastropod mollusks that carry the olfactory organ.

Rhodopsin All visual pigments whose chromophore is retinal, but commonly (although erroneously) used to refer only to rod visual pigments.

$R_{male-male}$ Androgen responsiveness (i.e., the change in testosterone from baseline to maximal aggressive interactions) between territorial males.

R_{season} Seasonal androgen response, reflecting the increase from breeding baseline testosterone concentrations to maximum concentrations during specific parts of the breeding life-cycle stage, that is, during the phase of territory establishment or mate guarding.

Riparian Interface between terrestrial and aquatic ecosystem. When intact, riparian ecosystems limit soil runoff and are characterized by high biodiversity and thus are an important buffer zone.

Risk effects Nonconsumptive effects of predators on prey, namely the lost foraging opportunities and lower levels of growth and reproduction experienced by prey investing in antipredator behavior (also known as nonlethal effects). This term avoids the complication that prey that are not directly killed by a predator may in fact be consumed.

Risk history The frequency, intensity, and duration of predation risk events experienced by prey in the past.

Risk threshold The level of risk that must be exceeded for the prey to start reducing its antipredator behavior under the risk allocation hypothesis.

Ritualization Communicative behaviors used in social interactions that evolved from other behaviors with different functions. For example, when attacked an ancestor of the wolf might have flattened the ears, crouched, and tucked the tail to avoid injury; over time these behaviors evolved to communicate submission. Evolutionary modification of a motor pattern used in communication that is thought to improve signal function, often through increased stereotypy and exaggeration.

RNA interference (RNAi) A technique of molecular biology in which expression of a particular gene is silenced by introducing double-stranded RNA into a eukaryotic organism. RNA interference can provide conclusive proof that a particular gene influences behavior.

Roosting The act of perching to rest or sleep.

Round-trip migration A subcategory of migration, with seasonal to-and-fro movements between regular breeding and wintering sites, typical of many birds but rare in insects.

RT-PCR Reverse transcription PCR, PCR that is performed on DNA that was synthesized from RNA by a reverse transcriptase enzyme.

Rule learning The ability to infer rule information from a number of different examples connected by a logical operation 'if → then.'

Rules of thumb Simple measures that animals can use to approximate solutions to optimal foraging problems. An example would be using the number of prey items encountered to leave patches as predicted by the marginal value theorem.

Runaway selection A theoretical model for the evolution of extravagant traits based on female preference.

The model proposes that female preference for a male trait results in a genetic correlation between preference and trait, such that the trait evolves beyond the level favored by natural selection in a 'runaway' process fuelled by female preference. Also called the Fisher process in reference to Sir Ronald Fisher, who developed the theory.

Saccule   An otolithic subdivision of the inner ear in all vertebrates that has an auditory (hearing) function among many fishes.

Saprophagy   Feeding on dead materials.

Satellite transmitters   These tracking devices are larger than radio transmitters and emit signals that are detected by geosynchronous satellites; these devices carry substantial batteries or are solar powered and continue to transmit for relatively long periods of time (i.e., a year or more); they enable tracking to occur over substantial geographic distances.

Satiation   The feeling of fullness at the end of a meal.

Satiety   The persisting sensation of repletion that results from eating.

Scalar timing   The dominant theory of timing which assumes that the coefficient of variability (i.e., the standard deviation of time estimates divided by the mean of time estimates) is constant across a broad range of temporal estimates (i.e., a specific proposal of the linear timing hypothesis).

Scale-free power-law   A degree distribution described by $p(k) \approx k^{-\gamma}$; demonstrated by a straight line on a log–log plot.

Scan sampling   A type of instantaneous sampling in which a group of individuals is scanned at specified intervals and the behavior of each individual at that instant is recorded.

Scanning   Often synonymous to vigilance.

Scatter hoarding   Hoarding of individual food items in many different locations.

Schistosomiasis (or bilharzias)   A disease caused by a blood fluke of the genus *Schistosoma*, a type of flatworm parasite. The intermediate host is a snail, in which cercariae (larvae) develop and migrate out into water; the cercariae penetrate the skin of hosts which make contact with the water. Symptoms depend on species causing infection, but can include rash, fever, aching, cough, diarrhea, and liver and spleen enlargement.

Schnauzenorgan response   A twitching movement of the elongated chin (Schnauzenorgan) of *Gnathonemus petersii, an electric fish,* evoked by the sudden emergence of a novel object near the animal's head, which is detected through the active or passive electric sense.

Schreckstoffe   Chemical alarm signals released by aquatic injured conspecifics, which is used to warn animals about an imminent danger.

Sclerotized   The hardening of tissue.

Scolopidium   A multicellular sensory structure of arthropods used to detect stretch, vibration, or sound.

Scout   A member of a social group, such as an ant or bee colony, that searches for food sources, nest sites, or other targets of interest. It may exploit its discoveries by itself or recruit other group members to help.

Scramble competition   Organisms use up a common limiting resource but otherwise do not contest or harm each other.

Scrounging   A behavioral strategy that consists of exploiting a resource uncovered by some other individual's efforts.

Seasonal breeder   An organism that breeds only in specific seasons (i.e., not continuously).

Seasonal interaction   When events in one period of the annual cycle, such as timing or condition, of an animal to influence events in subsequent periods.

Seasonality   Changes in hormonal or behavioral status in response to change in seasons.

Secondary defenses   Traits of the prey that influence the action of the predator, subsequent to prey detection, in ways that benefit the prey. Compare with primary defenses that act prior to the predator detecting the prey.

Secondary plant compound   Molecules produced by plants, the presence of which is often characteristic of particular plant taxa and which appear not to be directly involved in primary metabolism.

Secondary polygyny   Polygyny that arises from monogyny, generally through queen adoption.

Secondary predator–prey behaviors   Behaviors concerned with predators capturing prey, or prey escaping from predators, during an attack.

Secondary reproductive   These are produced by many termite species; they are sexually capable individuals who do not have wings, and are capable of superceding sick, injured, or absent parental primary reproductives.

Secondary sexual character   A trait that differs between the sexes and is neither required for reproduction nor related to sex differences in ecology. Most such traits do not develop fully until sexual maturity, are expressed more strongly in males than in females, and are useless or costly for survival. Traits that do not differ between the sexes but share the other two qualities may also be referred to as secondary sexual characters (e.g., ornate plumage in sexually monomorphic birds).

Segregation distortion   Within-individual selection for one or another allele of a diploid body.

Selective attention   The cognitive processes of (selectively) concentrating on one aspect of the environment while ignoring others; consciously or unconsciously, the perceiving organism is focused on particular areas of the environment. This is determined by past experience and the skill being performed.

Selective differential   The difference in fitness between two or more subsets of a population subjected to different selective pressures with resulting differences in fitness.

Selective sweep   Recent and strong positive natural selection on a particular gene which leads to reduced variation in DNA sequence among individuals in a population.

Selective tidal stream transport (STST)   Vertical movements of aquatic organisms relative to tides; provides a mechanism for zooplankton and small nekton to move horizontally within and between estuaries and coastal regions.

Self-awareness (self-recognition)   Increased self-other distinction, oftentimes indicated by self-recognition in a mirror. Sensitivity to one's own thoughts and feelings; sometimes used to indicate the knowledge that one exists independent of other entities.

Self-control task   Experimental situation in which decision-makers must choose between smaller–sooner and larger–later options.

Selfish-herd effect   Bunching by foragers to decrease their relative domain of danger when facing predation threats.

Self-medicate   The use by animals of secondary plant compounds or other nonnutritional substances in preventing or treating diseases.

Self-organization   The idea that the development of complex structures and behaviors in a system can emerge from events taking place primarily within and through the system itself.

Self-propelled particle (SPP) models   Models of collective motion in which each group member is treated as a particle that responds to other group members within interaction zones. An individual moves toward or away from other individuals, or aligns itself with them, depending on which zone they occupy.

Semantic memory   The ability to acquire general factual knowledge about the world.

Semelparous (semelparity)   Reproducing once during a lifetime.

Semiclaustral founding   Colony founding procedure in which a founding queen or queens forage outside the brood cell to secure sufficient energy to rear the first generation of workers.

Semi-intact preparation   A piece of an animal, along with its nervous system, that produces a behavior or a component of a behavior. Such preparations are normally used primarily to allow access to the nervous system, but can also be used to eliminate sensory input or confounding inputs from other parts of the nervous system.

Semisociality   Social groups of same-generation adults and their offspring characterized by cooperative brood care (i.e., alloparental care occurs), and a reproductive division of labor, such that some individuals mainly reproduce while others mainly perform other tasks such as foraging and brood-care.

Senescence   The combination of biological processes of deterioration of organismic function in a living organism approaching an advanced age.

Sensillum   Hair-like structure that houses sensory neurons.

Sensitive phase   A stage of life during which the ability to learn is enhanced. Occurs most commonly early in life.

Sensory drive   The hypothesis that sensory systems and sensory conditions in the environment 'drive' evolution in particular directions.

Sensory environment   Multiple types of information – signals and cues from other animals and the physical environment – that may be perceived by an animal on the basis of its unique sensory capabilities (i.e., 'umwelt').

Sensory mode   The physical characteristics of signal production, on the basis of animal sense organs by which it is perceived (e.g., sound, patterns of light and color, vibration, etc.).

Sensory traps   In attempts to induce certain responses in other individuals, the use of stimuli whose effectiveness in inducing these responses evolved in a different context. In a sexual context, the male can produce a stimulus that elicits a particular female response; this female response exists because previous natural selection in another context favored such a response to the same (or a similar) stimulus.

Sentience   A general term for the ability to feel or perceive subjectively.

Sentinel   An individual in a group that remains vigilant and stands guard while other group members forage or carry out other activities (also called: sentry or guard).

Sentinel cells   A newly discovered cell that sweeps through a *Dictyostlium* slug mopping up toxins and bacteria, acting as a kidney, a liver, and an innate immune system.

Sequence divergence   Changes in the sequence of DNA bases in different populations or different species. Comparisons of the degree of sequence divergence are used to estimate how long ago the populations or species began to evolve independently.

Sequestering   Accumulation of a chemical in the integument or inner organs of an organism from an outside source (e.g., diet).

Serotonergic basal cells   Round cells at the base of the taste bud, which are immunoreactive to serotonin.

Serotonergic medications   Psychotropic medications that effectively increase the availability of the neurotransmitter serotonin in the brain.

Serotonin (5-HT)   A monoamine neurotransmitter that is derived from tryptophan. It is synthesized in the gut, pineal, and CNS. In the brain, 5-HT influences learning and memory as well as appetite, sleep, and muscle contraction.

Sex allocation   Sometimes used to refer to the process by, or the time at, which a parent bestows gender on offspring (see sex determination and sex allocation sequence, respectively), but more generally used to refer to how resources are apportioned to each gender (also referred to as *investment ratio*). Sex allocation can be thought of as an evolutionarily derived reproductive strategy of the parents and the sex ratio as one of its manifestations.

Sex allocation sequence   The order in which offspring of different gender are produced by a parent. Nonrandom sequences can, but do not always, imply parental control and can influence sex ratio variance.

Sex determination   The genetic basis of an individual's gender. There is an astonishing diversity of sex determination mechanisms among animals, often exerting a profound influence on reproductive behavior.

Sex-limited polymorphism   Occurrence of several discrete forms or morphs within one sex, but not the other sex.

Sex ratio   The proportion of individuals that are male, that is, males/(males + females). Sex ratios are sometimes given as the proportion females (this is not incorrect; there is no strict convention) and sometimes reported as the ratio of males to females, that is, males/females (termed sex ratio *sensu stricto*): this is not a recommended measure as it is not readily amenable to statistical analysis. The sampling unit may be indicated, for example, *population sex ratio*, *clutch sex ratio*, *parental sex ratio* (the sex ratio of offspring produced by a given parent or pair of parents). The developmental stage of offspring may also be indicated: *primary sex ratio* (the sex ratio at offspring production; this may be used to indicate the sex ratios at fertilization or at egg laying), *secondary sex ratio* (the sex ratio at some defined later stage of offspring development, for example, emergence or mating (adulthood)). Developmental mortality can mean that primary and secondary sex ratios are not equivalent.

Sex ratio variance   A measure of the diversity of sexual composition in groups of offspring (e.g., clutches, litters, etc.). Heterogametic sex determination (e.g., the XY system in mammals, the WZ system in birds) leads to the null expectation that distributions of group sex ratios conform to binomial variance. Deviations from the binomial expectation can, but do not necessarily, imply sex ratio control. Under haplodiploid sex determination, there is no particular null expectation of variance, but subbinomial variances have been observed in many haplodiploid species.

Sex-ratio conflict   Conflict between queens and workers over the investment into male versus female reproductives produced by the colony.

Sex role reversal   Occurs when males provide the majority of parental care, resulting in sexual selection on females, who can increase their reproductive success by obtaining additional mates.

Sex-role reversed species   Are those in which females compete for males and males choose among females. Typically, males take care of the young.

Sexual behavior   Behavioral interactions that facilitate the union of eggs and sperm.

Sexual coercion   Occurs when one sex, usually males, use force or the threat of force – forced copulation, harassment, intimidation, restriction of the movement of the other – to increase the probability that mating will occur.

Sexual conflict   Occurs whenever the fitness interests of individuals of different sexes conflict.

Sexual dialectics hypothesis   The idea that whenever the behavior and physiology of one sex decreases the fitness of the other, flexible individuals adaptively modify their behavior or physiology to resist the deleterious effects of interaction(s) with the other sex. Because control and resistance interactions are likely to be dynamic, changing during the lifetime of an individual, the sexual dialectics hypothesis predicts that individuals flexibly adjust resistance behavior in contemporary time.

Sexual dichromatism   A subset of sexual dimorphisms in which males and females of a species differ systematically in coloration or color pattern.

Sexual differentiation   In ontogeny, the anatomical and behavioral differentiation of males and females.

Sexual dimorphism   Refers to differences in morphology, behavior or physiology between males and females. Generally, more intense sexual selection results in greater sexual dimorphism.

Sexually antagonistic selection   A type of selection that is characterized by dynamic interactions – actions and reactions – between individuals of different sexes that can lead to a coevolutionary arms race.

Sexual reproduction   Reproduction involving gamete formation by meiosis and gamete fusion to form new individuals.

Sexual selection   Selection for traits that make individuals of one sex better able to compete for individuals of the opposite sex. As a consequence, some individuals have a mating advantage over other individuals of their own sex, such that there is nonrandom differential reproductive success among these individuals.

Sexual signals   Advertise the signaler's genetic or phenotypic quality in order to attract mates and deter rivals. Examples include conspicuous traits, such as bright colors and elaborate songs. Signals can be visual, acoustic, olfactory, tactile, or electric.

Sexual size dimorphism (SSD)   A subset of sexual dimorphisms in which males and females of a species differ systematically in body size.

Shaping   The procedure of reinforcing successive approximations of a desired behavior.

Short day breeder   An organism that enters full reproductive capability during short days of winter.

Sibling species   Anatomically similar species that are nonetheless reproductively isolated; in herbivorous insects, such species often use different host plants.

Sickness responses   The suite of adaptive behavioral and febrile reactions among vertebrate animals associated with the acute phase immune response that includes fever, iron withholding, reduced motivated behaviors such as food and water intake, and lack of sexual, parental, or other social interactions. These responses are critical to survival.

Sign   A signal; also anything that gives evidence or trace of something else; also a physical object, usually fixed in space, that is a signal when encountered by a receiver.

Sign stimulus   An external stimulus that elicits a specific motor pattern (modal or fixed action pattern).

Signal   A character or behavior that has evolved so as to provide information to other organisms.

Signal detection theory   A general model of the discrimination of signals from background noise that can be applied to data from psychophysical studies with animals and to situations where an animal must make a discrimination under conditions of uncertainty.

Signal dominance   When a multimodal signal generates a response in only one of its component modes in relation to other modes.

Signal enhancement   When receiver responses to redundant multimodal signals are increased in their intensity compared to unimodal signals.

Signal equivalence   When receiver responses to redundant multimodal signals are the same or equal to unimodal signals in their intensity (equivalence).

Signal independence   When the response to a multimodal signal includes the (different) responses to each of its unimodal components.

Signaling mode   The physical characteristics of a signal that enables it to be received by a specific type of sensory neuron in a receiver. Signaling modes include chemical, electric, sound, light, and vibration.

Signal parasite   An individual that exploits an existing communication system in a way that benefits itself at the expense of a signal giver or a signal receiver.

Signal redundancy   When individual components of a multimodal signal presented separately elicit the same response from a receiver and likely contain the same or similar kinds of information about the sender.

Significance level/criterion   In statistical analyses it is a criterion of probability below which a statistical test value is said to indicate a significant difference between populations.

Silkie   Asiatic breed of chickens characterized by fur-like plumage and dark blue flesh.

Simultaneous hermaphroditism   A sexual pattern characterized by individuals possessing both mature ovarian and spermatogenic tissue within the same functional gonad.

Single nucleotide polymorphism (SNP)   Variation in a DNA sequence that occurs when a single nucleotide – A, T, C, or G – varies between individuals of the same species.

Sinus gland   A neurohemal organ associated with the crustacean X-organ.

Siphon   Cylinder created by curling the edges of the mantle in some mollusks. It can be used to forcibly discharge the contents of the mantle cavity.

Sister groups   A pair of evolutionary lineages that share their most recent common ancestor and thus are necessarily equal in age.

Site fidelity (see philopatry).

Size constancy   The ability to determine the true size of objects despite viewing them at different distances when their images subtend various angles on the retina.

Skylight polarization   Due to scattering by particles in the earth's atmosphere, sunlight becomes polarized, with the light wave's electric field oscillating in one direction. The degree of polarization is maximal at 90°, relative to the direction of incident light.

Sloughing behavior   Specific behavior associated with sloughing off skin and associated structures such as hair, feathers, and scales. Often, this involves rhythmic movements to lift off old skin layers (e.g., in snakes), or movements allowing abrasion of skin with substrate (many birds and mammals) to break up and shed skin and its components.

Smoltification   The transformation or metamorphosis of anadromous salmonids from the parr to smolt stages, including changes in morphology, endocrinology, and behavior in preparation for saltwater entry. Some of these include increased plasma, thyroid hormone, and cortisol levels, as well as the deposition of guanine in the skin, giving the fish a silvery appearance. This impedes water loss, and with the increase of Na^+-ATPase pumps in the gills and gut, the osmoregulatory function improves as the fish enters the hyperosmotic conditions for seawater. Behaviorally, smolts leave the natal streams and migrate to open water.

Sneak spawning   Male reproductive behavior where an individual will attempt to fertilize eggs that are released during the courtship and spawning episode of another male–female pair; the individual is usually unable to defend a territory and court a female independently, and spends most of the time hiding to avoid agonistic encounters with territory-holding males.

Social cognition   Knowledge about group mates and social interactions.

Social (cooperative) spider   Spiders that share a nest, feed together, and have cooperative breeding.

Social cues   Products of the behavior of others that convey inadvertent social information.

Social dominance   The state of having high social status relative to other individuals, which react submissively during dyadic agonistic encounters. Dominant individuals have priority of access to resources over subordinate individuals.

Social eavesdropping   The extraction of social information by an individual (the eavesdropper) from a signaling interaction between other individuals (usually conspecifics) in which the eavesdropper takes no direct part.

Social facilitation or social enhancement   The effect of the mere presence of another animal on the production of a target response. The increase or initiation of a behavior already in one's behavioral repertoire when in the presence of others engaged in the same behavior.

Social foraging theory   A body of game theoretic models designed to analyze foraging decisions made under conditions of frequency-dependent payoffs.

Social hymenoptera   Meaning the eusocial Hymenoptera. Eusociality has evolved approximately nine times in the Hymenoptera, once in ants, three times in wasps, and approximately five times in bees.

Social influence   The effect of another animal on the production of a target response (e.g., contagion or social facilitation) that does not involve the acquisition of information about the to-be acquired response (e.g., imitation).

Social information   Information obtained by an individual from other animals in its social group.

Social insects (see eusocial)   Insects that live in groups in which some group members rear offspring that are not their own.

Social intelligence   An influential theory developed by Alison Jolly, Nick Humphrey, and others to explain the superior intelligence of primates, including humans. The theory is based on the idea that living in a complex social world requires cognitive abilities related to learning from others, forming social relationships in order to gain dominance, and deceiving others to gain resources normally unavailable to them. This extreme form has been named Machiavellian Intelligence.

Social interaction   A dynamic, changing sequence of social actions between individuals that modify their actions and reactions according to those of their interaction partner(s).

Sociality   Associations and interactions of individuals within a social group.

Social learning   Any process whereby the behavior of an individual is altered as a result of it either observing the behavior of another individual, interacting with it, or being exposed to its products.

Social learning strategy   An evolved psychological rule specifying under what circumstances an individual learns from others and/or from whom it learns.

Socially mediated learning   Learning that is influenced by presence and activity of conspecifics, also referred to as socially biased learning; the process by which social context contributes to learning.

Social mimicry   Imitation between species that associate among each other.

Social monogamy   A type of mating system in which one male and one female form a bonded pair for the purposes of reproduction. Typically, the pair will stay together and raise young together. However, both the male and/or the female may engage in copulations with other individuals from outside the bonded pair (EPCs).

Social network   Pattern of social connectedness, either through behavioral interactions or spatial proximity, between individuals in a population.

Social norm   A pattern of behavior that is accepted as being the normal way of behaving for a particular group of people, and to which all the group members are expected to conform.

Social organization   The size, demographic composition, and spatiotemporal coordination of individuals within a group.

Social (other-regarding) preferences   Behavior motivated out of concern for the effects it has on other individuals over and above material self-interest; these can be positive or negative (social science).

Social parasitism   The coexistence in the same colony of two species of social insects, one of which parasitizes the other.

Social selection   A type of natural selection characterized by nonrandom, differential reproductive success of individuals bearing some trait relevant to social interactions (either competitive or cooperative) for access to resources such as food, territories, allies, and mates.

Social structure   The pattern of relationships among individuals within groups, groups within demes (subpopulations), and demes within a population of a given species.

Social transmission   Transfer of information among individuals in a group or population, both within and between generations, through social learning or teaching.

Social transport   A form of recruitment used by certain ant species, in which one ant carries another to a destination, typically in a stereotyped posture. This is most commonly seen when colonies emigrate from one nest site to another.

Sociobiology   An extension of Darwinian theory and the evolutionary synthesis that developed during the 1960s and 1970s. The core principles were that natural selection works at the level of the individual or gene, not the population or species (still contested) and that the representation of one's genes in future generations could be achieved by facilitating the reproductive success of close relatives.

Sociomatrix   For a group with n members, an $n \times n$ matrix with each group member along the vertical and horizontal axes and each entry in the grid as the weight of the social relationship, if any, between the two intersecting individuals.

Soldier   Similar to workers, these are nonreproductive members of a colony (sometimes known, especially in ants, as 'major workers'). Unlike normal workers, however, they are generally larger, with specialized head structures, and primarily perform nest-defense tasks.

Solitarious   Living singly or in pairs; the term refers to behavioral, morphological, and physiological traits (especially for the solitarious phase of locusts).

Somatic   Pertaining to the body.

Somatic fusion   The process by which the nonreproductive tissues of two individuals join to form a single individual with a shared body (soma). In many taxa, this can occur either between clones or closely related individuals. Fusion between allogeneic (nonclonemates) organisms produces a genetically chimeric individual.

Somatic recombination   The process by which regions of the genome are physically edited in the nucleus, giving rise to novel genetic elements. Accounts of this process are rare and are known from a few systems where genetic diversity is of primary importance.

Somatic rejection   The process by which two individuals reject each other, often involving the formation of a physical barrier between them and the preservation of genetic individuality.

Somatosensory system   A diverse sensory system comprising the receptors and processing centers to produce the sensory modalities such as touch, temperature, proprioception (body position), and nociception (pain). The sensory receptors cover the skin and epithelia, skeletal muscles, bones and joints, internal organs, and the cardiovascular system.

Somatotropic axis   A group of hierarchically regulated hypothalamic, pituitary, and peripheral tissue hormones which are involved in the regulation of somatic growth.

Song   Loud, often complex sound usually produced by males of a species in defense of a breeding territory and/or to attract females.

Song control nuclei   Interconnected regions of the brain in songbirds that regulate the production and learning of song.

Sore footed   A type of lameness which is caused by pain in the animal's hoof.

Sparse coding   The representation of information in the nervous system by the activation of a relatively small set of neurons.

Spatial contrast sensitivity function   Plot of the contrast required to detect gratings of different spatial frequencies.

Spatial resolution (acuity)   The ability of an animal to perceive spatial detail.

Spawning   Oviposition, or the deposition of eggs, in water.

Spearman coefficient   A measure of correlation between ranked data; can be used as a measure of observer reliability.

Spectral sensitivity   The differential sensitivity of photoreceptors to different wavelengths of light.

Spectrogram   A display of the frequency components of a sound over time.

Speed/accuracy tradeoff   A fundamental decision-making constraint that captures the cost in time that must be paid to improve the accuracy with which the best available option can be chosen.

Sperm allocation   Refers to situations in which males that are running low on sperm will vary the amount of sperm in an ejaculate, so as to provide more sperm for some females and less for others. Generally, it is assumed that males will provide more sperm for females that are of higher quality or status.

Spermatheca   A small sac associated with the median oviduct of the female, in which sperm are stored following copulation.

Spermatophore   A sac produced by accessory glands of male insects and directly or indirectly transferred to the female, containing sperm and often proteinaceous material.

Sperm capacitation   Changes the spermatozoa undergo to become ready to interact with the ovum and hence able to fertilize.

Sperm competition   A type of sexual selection that can occur if a male or his seminal products directly reduce the changes that the sperm of other males which have mated with the same female have of fathering her offspring. This is the postcopulatory equivalent of male–male battles.

Sperm depletion   Refers to the fact that males may be limited in the number of sperm that they can produce per unit time and eventually they may run out of sperm. In such cases, males need a period of time to rebuild their sperm supplies.

Sperm precedence   An individual male's share of paternity when females mate with multiple partners.

Spherical aberration   Optical imperfection caused by light striking a refractive surface at different points being focused in different planes.

Spite   A behavior that reduces the lifetime fitness of the recipient while also reducing the fitness of the actor (evolutionary biology); harming behavior resulting from a desire for the suffering or misfortunes of another individual (social science).

Split sex ratios   Population-wide bimodal sex-ratio distributions with co-occurring colonies that specialize in the production of either male or female reproductives.

Sporozoites   A stage in the life cycle of apicomplexan protists that is produced by sporulation and invades host cells.

Stabilizing selection   A form of selection in which deviations from a main phenotype, such as changes to a conspecific call type, are selected against maintaining the same phenotype over evolutionary time. Contrast with directional selection.

Stable group size   A group size at which no individual can gain by unilaterally leaving or joining the group.

Stable isotopes   Nonradioactive forms of an element having an extra neutron; stable isotopes of carbon, nitrogen, and hydrogen, among others, are very useful for ecological and behavioral studies.

Stable supine posture   The ability of infants to lie on their backs on the ground or another surface is uniquely human. Chimpanzee and other non-human primate infants are unstable when they are laid on their backs – they move their limbs in an attempt to grasp and cling to something. From a developmental perspective stable supine posture enabled humans to become by far the most versatile and proficient tool users in nature. The stable supine posture provides the basis of tool use, face-to-face communication, and vocal exchange.

Stage 4 sleep   The deepest stage of NREMS. It is characteristic of the first half of the night in humans; our ability to enter this stage diminishes with aging.

Startle signal   See deimatic signal.

Starvation–predation risk trade-off   Animals must balance the time or effort they spend feeding to prevent themselves from starving, with the time or effort they spend looking out for predators to prevent themselves from being eaten. Any animal that spends all its time looking out for, or avoiding predators will starve to death. Any animal that spends all its time feeding may not ever starve, but is more likely to be caught by a predator.

State-dependent model   Models that use the techniques of stochastic dynamic optimization to predict animal behavior. Often used to model tradeoffs that animals face when having to decide between competing factors such as getting food and avoiding predators.

Stationary molt   An intermittent molt that is associated with a lack of increase in body size and morphological development. This type of development occurs in several insect species and is frequently associated with periods of food shortage, when a larva or nymph is not capable of passing a critical mass threshold in an instar. In some termites, it might also be linked to the wear of mandibles.

Statocyst   Inertial balance organ of aquatic invertebrates, consisting of a heavy mineral body (statolith) resting on a bed of mechanoreceptors that register the displacement of this body whenever its orientation relative to the direction of gravity changes.

Stereocilia   Nonmotile tufts of secretory microvilli on the free surface of cells. Thought to be a variant of microvilli and characterized by their length (distinguishing them from microvilli) and their lack of motility (distinguishing them from cilia).

Stereotypic behavior   Behavior that is repetitive, relatively invariant, and has no obvious goal or function.

Steroid hormone   A class of molecules that include the sex hormones and stress hormones from the adrenal cortex that share a common biosynthetic pathway.

Steroid receptors   Steroid hormones act largely through intracellular steroid receptor proteins that bind hormone and then function as 'ligand activated transcription factors' to regulate gene expression in target cells. Teleosts have multiple forms of both the estrogen and the androgen receptors.

Stimulant   A substance that quickens and enlivens the physiological and metabolic activity of the body.

Stimulus   (In physiology) Something that can elicit or evoke a physiological response in a (sensory) cell, a (sense) organ, or an organism; it can be internal or external; (in psychology) something that has an impact or an effect on an organism so that its behavior is modified in a detectable way.

Stimulus enhancement   The facilitation of an observer's response (e.g., through approach and manipulation) resulting from the pairing of an object with reinforcement.

Stimulus generalization and discrimination   When prey respond to the olfactory, auditory, or visual cues of a species which are similar to those of another species, prey are said to generalize their species recognition to these cues. If prey fail to respond or respond weakly to these cues because they are dissimilar, they are said to discriminate these cues from those of another species. Therefore, stimulus generalization and discrimination are reciprocal effects, with higher stimulus generalization indicating lower stimulus discrimination.

Stochastic dynamic optimization   A mathematical technique that predicts optimal behavior by having computers examine every possible set of behaviors. This produces a numerical rather than an analytical solution as found by the marginal value theorem or Gilliam's rule.

Stop-over habitats   Habitats along the migration routes of animals that allow them to feed and replenish fat stores before moving on.

Stotting   Vertical jumping in ungulates during flight away from a predator in which all four legs leave the ground at the same time, the legs being held straight while the animal is in the air; similar behaviors include pronking, spronking, bounding, and leaping; may function to deter further attack by a predator or distract the predator's attention away from vulnerable offspring.

Strategic design   Aspects of signals relating to its function, for example, brightness of plumage conveying male quality.

Strategy   A set of behavioral decisions that are highly heritable, associated with a particular genotype within the gene pool of a species.

Stratified squamous epithelium   An epithelium characterized by multiple layers of flat, scale-like cells called 'squamous cells.'

Stress   A descriptive label with varying meanings for the biological processes involved when an animal perceives a threat that challenges internal homeostasis (both motivational and physiological 'set points') and the behavioral and physiological adjustments that the organism undergoes to avoid or adapt to the stressor and return to homeostasis. An environmental effect on an individual that overtaxes its control systems and results in adverse consequences and eventually in reduced fitness.

Stress response   The physiological and behavioral responses to a sudden emergency situation.

Stressor   A challenge (whether physical or psychological) to homeostatic balance (see 'homeostasis').

Stress-response   The array of neural and endocrine adaptations that occur in the body in response to a stressful challenge.

Stretch activation   In some muscles, physical elongation by mechanical means can lead directly to contraction of the muscle, counteracting the induced stretch.

Striated muscle   Also known as 'skeletal muscles,' these muscles have alternating bands of overlap and nonoverlap between thick (myosin) and thin (actin) filaments, giving them a striated appearance.

Stridulation   The rubbing of skeletal elements against one another that is a common form of sound production in fishes and many insects.

Strong inference   A method in the cognitive structure and logic of scientific discovery in which investigators attempt to identify and test simultaneously alternative hypothetical-deductive hypotheses with crucial predictions. Crucial predictions are predictions about a phenomenon that are in opposite directions. If tested well with a crucial experiment, two hypotheses can be tested simultaneously and one hypothesis supported and another rejected.

Strong reciprocity   A propensity to reward others for cooperative, norm-abiding, behaviors coupled with a propensity to punish others for norm violations.

Stunning   A method that renders animals insensible to pain before slaughter.

Sublethal effects   Effects that are negative, but do not immediately kill the organisms, such as decreased ability to stand, walk, eat, or avoid predators.

Submission   Behavior that indicates a low probability of initiating aggressive behavior. A submissive individual, however, may respond to injurious aggression with aggression. Submissive individuals often terminate interactions by physical distancing.

Subordinance   The state of having low social status in a group, often because the individual was defeated in an aggressive encounter.

Subordinate   A low-ranking individual within the group that does not usually get access to resources. Subordinate individuals tend to be smaller and weaker and do not form close social networks.

Subsociality   Family groups consisting of parents and immature offspring, and are characterized by brood defense or brood provisioning by parents.

Subsocial route   Social grouping that originates from an extended family and restricted, or no dispersal of, young.

Suprachiasmatic nucleus (SCN)   The suprachiasmatic nucleus (SCN) of the hypothalamus is a bilateral structure that sits at the base of the mammalian brain. It serves as a central pacemaker and coordinates the daily rhythms with the environment. It also synchronizes endogenous circadian rhythms. The SCN can be divided into two areas, a ventral area containing cells that receive direct input from the retina, and a dorsal area which contains highly rhythmic cells that serve in output processes. Input from the ventral SCN synchronizes the rhythmic cells of the dorsal SCN.

Surprisal   In the mathematical theory of communication, the entropy or information associated with a particular signal.

Survivorship cost   Reduction in fitness in the form of decreased probability of survival.

Swimbladder   An anatomical structure comprising connective tissue, filled with a mixture of oxygen, carbon dioxide, and nitrogen (hence, also known as a 'gas bladder') that has multiple functions among fishes including control of buoyancy, sound production, and sound reception.

Symbol   Something – such as an object, picture, written word, a sound, or particular mark – that represents (or stands for) something else through association, resemblance, or convention, especially a material object used to represent something invisible.

Sympathetic nervous system   The branch of the peripheral nervous system, regulated by epinephrine and norepinephrine release, that orchestrates the immediate responses to a stressor.

Sympatric   Geographically overlapping; for example, populations on the same island with no barrier to movement between them.

Sympatric speciation   The development of isolating mechanisms while incipient species are within the same geographic area, specifically when individuals from each population are within cruising range of one another.

Synanthropic (synanthropy)   Describes a population of wild animals that lives near or within human settlements or anthropogenic habitats; usually implies some degree of dependence on humans or exploitation of human-derived resources.

Synapomorphy   A shared, derived character; the only valid character type for revealing phylogenetic affinities.

Synapse   The gap or junction between nerve cells.

Synaptic pruning   The reduction in the number and connectivity of synapses that may accompany development.

Synaptic transmission   The transmission of an electrical signal from one cell to another which occurs at the point of connection between these two nerve cells called synapse.

Syrinx   Musculoskeletal structure that functions as a vocal organ found among birds.

Systematics   The general field of researching inferring, and proposing the evolutionary relationships of organisms. One of the oldest fields of biology, its relevance today is stronger than ever, including multiple sources of data and methodological techniques.

Tachycardia   An increase in heart rate.

Tactical design   Aspects of signals relating to its effectiveness in transmission, for example, male songs with higher amplitude signals in certain frequencies get more female attention.

Tactics   A set of behavioral decisions for which the phenotype develops as a result of any combination of learned mechanisms (genetic heritability is unspecified).

Tail streamer   The elongated outer tail feathers (rectrices) of the swallow tail, giving the tail its forked appearance.

Tandem run   A form of recruitment used by certain ant species, in which one ant leads a single follower to a destination. The pair remain in contact by the exchange of pheromone signals from the leader and tactile signals from the follower.

Tangled bank theory   The idea that the world, and the challenges that it poses to organisms, is variable and complex. In such a world, the production of genetically variable offspring increases the chances of at least some of them being able to survive and reproduce.

Tapetum   Reflective layer in either the retinal pigment epithelium or choroid that reflects light not absorbed by the photoreceptors back through the retina, thus improving sensitivity in animals in low light levels.

Task   A behavior or set of behaviors that contribute to the work necessary for the function of a social group.

Task specialization   When an individual within a social group preferentially performs one task over other tasks being performed by that group.

Task threshold   The level of stimulus required to make a worker engage in a task.

Tastant   Chemical molecule that induces the sensation of taste, such as sugars or salts.

Tautologous   A circular logical argument in which the conclusion is included in the propositions.

Taxis   The movement of an organism in a particular direction with reference to a stimulus. A taxis usually involves the employment of one sense and a movement directly toward or away from the stimulus, or else the maintenance of a constant angle to it. (Contrast with kinesis.) See phototaxis and geotaxis as examples.

Taxonomy   The scientific discipline concerned with studies of taxa, including the subdisciplines of systematics and nomenclature.

Teaching Behavior modified by an experienced individual in the presence of a naïve individual, such that the naïve individual learns the behavior more quickly than it would otherwise and at some cost to the teacher.

Tegmen (pl. tegmina) A leathery, hardened forewing (usually of Orthopteroids).

Teleost Fish infraclass Teleostei within the ray-finned, bony fishes, excluding gars and bowfins. One of three infraclasses of ray-finned fishes (Actinopterygii) that includes most common fish.

Template The neurological or physical model against which cue bearers are compared and evaluated.

Temporal caste discretization A form of age-related division of labor in which workers form distinct age groups that have roles composed of sets of nonoverlapping tasks.

Temporal contrast sensitivity function Plot of the contrast required for detection of a light flickering at different frequencies.

Temporal discounting A decrease in the subjective value of a delayed benefit.

Temporal information Of, relating to, or involving an awareness of time.

Temporal information processing The sequence of computational steps that are hypothesized to occur while processing events that unfold in time.

Temporal polyethism A pattern of division of labor in eusocial insect colonies in which task performance is associated with worker age.

Temporal representation The internal format of stored information about events that unfold in time.

Temporal structure Describes the amplitude and frequency modulations of an acoustic waveform over time.

Tergal glands Glands on the dorsal surface of the abdomen; usually referring to those on males that entice females into position for copulatory engagement.

Terminal investment strategy Is a term sometimes used to refer to species in which the male usually, or always, is killed and cannibalized by the female during, or immediately after, copulation. The term implies that the male may be investing in its future offspring by providing food and nutrients (its own body) to the female.

Termites, higher Comprises only the termite species of the family Termitidae. They have bacterial gut symbionts only.

Termites, lower All termites with the exception of the Termitidae. Lower termites harbor bacteria and flagellates in their guts.

Territory Any defended space; can be for breeding, foraging, caring for young, or a combination.

Test of congruence A central component of phylogenetic systematics; it is the result of simultaneous analysis of characters. Given sufficient evidence, true synapomorphies will tend to reinforce one another guiding tree inference, and characters that do not in fact reveal phylogeny will be revealed as such. The test of congruence is therefore the primary tool in testing homology and identifying homoplasy, and it flows logically from the recognition that there is but one optimal phylogeny for a group of taxa.

Testosterone (T) Important androgenic steroid hormone in all classes of vertebrates; critically, this steroid often functions as a biosynthetic intermediate in estradiol or 11-ketotestosterone production.

Tethered flight A laboratory technique in which an insect is suspended by a wire or stick attached to its dorsal surface; with a wind blowing on the head, removal of foot (tarsal) contact triggers sustained flight.

Thanatosis An antipredator behavior in which the organism feigns death.

Thelytokous automixis A kind of parthenogenesis in which two gametes produced by meiosis fuse to produce a diploid female.

Thelytoky A form of parthenogenetic reproduction in which only female offspring are produced.

Theory of mind The ability to attribute mental (cognitive) states to others.

Thermocline A zone of rapidly changing temperature.

Thermolability See poikilothermic.

Thermoneutral ambient temperature An ambient temperature where the activities of heat-producing and the heat loss mechanisms are at a minimum level; the animal needs the least thermoregulatory effort to maintain its normal body temperature.

Thermoregulation The ability of an organism to keep its body temperature within certain boundaries, even when the temperature surrounding is very different. The regulation of body temperature.

Thiamine A water-soluble vitamin of the B complex (vitamin B_1), whose phosphate derivatives are involved in many cellular processes.

Third-party punishment Imposition of sanctions by an impartial observer on an individual for actions directed toward a third party; see also Policing punishment (social science).

Third-party relationships Relationships or interactions among conspecific group members in which the observer itself is not directly involved.

Threshold The lowest stimulus strength that reliably elicits a response; a low threshold means high sensitivity; the exact criterion for threshold differs among studies.

Thyroid hormones   Iodinated tyrosine residues produced in the thyroid gland. The gland mostly secretes thyroxine (T4) that is then converted in the blood or in target organs by deiodinases to tri-iodothyronine (T3). T3 is regarded as being the biologically active form. Biological effects of thyroid hormones include regulation of metabolism (temperature regulation), development, and behavioral effects.

Thyrotropin (TSH)   Glycoprotein hormone comprising two subunits produced by the anterior pituitary gland that stimulates the production of thyroid hormone by the thyroid gland.

Thyrotropin-releasing hormone (TRH)   Tripeptide produced in the hypothalamus and extrahypothalamic sites that stimulates the release of TSH by the anterior pituitary gland.

Time perception   The experience of time.

Time sampling   Behavior is sampled periodically at a specified sample point at the end of a specific sample interval.

Time series analysis   Events that unfold in time may be characterized by the periodic trends that make up the temporal structure of the events.

Timing   The general ability to keep track of time.

Tonotopy   The orderly mapping of frequency along the cochlea. The orderly arrangement of frequency is then preserved in each of the successively higher nuclei of the auditory system up to and including the auditory cortex.

Tool-use   Directing an unattached object towards one's self or another object (animate or inanimate) in order to achieve a goal.

Tool-user   A species that regularly uses tools in its natural environment.

Totipotent   In eusocial insects, having the ability to express either the reproductive queen or the nonreproductive worker phenotype.

Toxic   Substances that are poisonous and injurious or lethal to predators that attempt to consume it.

Toxicological effects   Direct effects of chemicals that interfere with physiological processes resulting in the deterioration of function and may ultimately cause organ and system failure.

Trade-off   The cost–benefit approach has been extended to model when this benefit-to-cost ratio is optimal, and states that an individual should maximize the benefit of the behavior while simultaneously minimizing any costs associated with the behavior. In other words, the benefit of any particular behavior should be considered with the costs associated with the behavior.

Trade-off theory   A theory to explain the emergence of symbolic representation in humans. At a certain point in human evolution, brain capacity reached a limit and in order to accumulate new functions, old functions needed to be lost. Consequently, humans may have lost much of their ability for olfactory processing and developed instead highly sensitive visual, auditory, and crossmodal functions. A similar scenario may be applied to the trade-off between memory and symbol use, where human memory capacity may have been sacrificed in exchange for enhanced symbolic capabilities.

Tradition   An enduring behavior pattern shared among members of a group that depends to a measurable degree on social contributions to learning.

Tragedy of the commons   A situation in which individuals would do better if they all cooperate, compared to them all defecting, but in which cooperation is unstable because each individual gains by selfishly pursuing their own short-term interests (cf. Prisoner's dilemma).

Trained losing and winning   The learning processes whereby an animal either acquires a stronger tendency to submit or yield to other individuals after losing previous agonistic encounters, or acquires a stronger tendency to attack or dominate other individuals after winning previous encounters. Modification of the tendency is in relation to other individuals generally, not limited to opponents involved in previous encounters.

Transcellular diffusion   Substances travel through the cell, passing through both the *apical membrane* and the *basolateral membrane*.

Transcription factor   A gene that directly affects the expression of another gene or genes.

Transcriptional   Relating to transcription, the process by which DNA is converted into messenger RNA.

Transcriptome   The set of all messenger RNA (mRNA) molecules produced in one cell or a population of cells, or in a given organism, under particular environmental conditions at a given time. Transcriptomics is the large-scale study of gene expression level (mRNAs) in a given cell population (such as brain cells), often using high-throughput techniques based on DNA microarray technology.

Transduction mechanism   In a sensory neuron the odorant receptor-ligand complex induces a series of cellular reactions that ultimately release action potentials in the axons.

Transiens   Transitional locust phase, from the solitarious to gregarious or vice versa.

Transition matrix   Squared matrix in which each row is a probability distribution. This is the fundamental element of a Markov chain.

Transitive inference   A form of reasoning in which given prior information a subject deduces a logical conclusion.

Specifically, the ordinal relation between two elements in a series must be inferred from information that establishes the relations of those two elements to a third.

Transitivity   A fundamental principle of rational choice behavior that applies specifically to binary choices. Preferences are transitive between the three options A, B, and C if A is preferred to B, B is preferred to C, and A is preferred to C.

Translational   Relating to translation, a process by which mRNA is converted into protein.

Translocation   Technique used in wildlife conservation, wherein wild individuals are captured from one location and transported and introduced to another part of their range, often with the purpose of re-establishing a local population which has become extirpated.

Transmission distance   Refers to the change of sound intensity with increasing distance relative to a reference point.

Transposable element   A mobile piece of DNA that can insert itself into the genome.

Trematodes   Groups of parasitic worms, commonly referred to as 'flukes.' Almost all trematodes infect mollusks as the first host in the life cycle, and most have a complex life cycle involving other hosts. Most trematodes are monoecious and alternately reproduce sexually and asexually. The two main exceptions to this are the Aspidogastrea, which have no asexual reproduction, and the schistosomes, which are dioecious. The Trematoda are estimated to include 18 000–24 000 species.

Triadic mother–infant–object relationships   It is also called 'social referencing.' Human infants often manipulate objects within a social context. Suppose that a human infant encounters a new toy. She may look up at the mother *before* touching it. The mother may nod or smile, and only then will the infant actually start manipulating the object. While playing with the toy, the infant may often show it to the mother while smiling. The mother may smile back at her child and give social praise.

Trigeminal nerve   The fifth cranial nerve in vertebrates, which is known to be both sensory and motor in function. The ophthalmic branch of the trigeminal has been shown to be sensitive to magnetic fields.

Trigger neurons   A class of command neurons whose short-lasting activation (e.g., less than a second) produces a long-lasting behavioral response (e.g., for tens of seconds). This term was coined in the study of leech swimming activation to distinguish these neurons from *gating neurons*, a class of command neurons that must be active during the whole time while a behavior takes place.

Tri-trophic level interactions   Interactions that take place between organisms at three different levels within a food chain, for example, a plant, herbivore, and a carnivore.

Trivial movement   See Appetitive movement.

Trophic cascades   The indirect effects of top predators on the population processes of plants and animal species at lower trophic levels, as mediated by the density and foraging behavior of intermediate consumers.

Trophic level   An organism's feeding position in a food web, with primary producers occupying the lowest level, herbivores the second, and carnivores occupying higher trophic levels.

Trophollaxis   Mouth-to-mouth transfer of food or other substances.

Tropic hormone   A hormone that modulates the secretion of another hormone.

True workers   In termites, workers in colonies of foraging termites. They can be considered altruistic individuals as they perform most tasks within a colony (e.g., foraging, brood care, and building behavior) except for reproduction and specialized defense. Although they sometimes, especially in lower termites, still have some reproductive options (for instance as neotenic reproductives), their morphological differentiations (especially their sclerotization) largely restrict their developmental capability. In functional terms, these true workers, often just called workers, are equivalent to the workers of the social Hymenoptera, even though the latter are imagoes, whereas the true workers here are preimaginal stages.

Trypanosomiasis   The name given to several diseases of vertebrates, including man, that are endemic in parts of Africa and the American continents. They are caused by protozoan parasites of the genus *Trypanosoma*.

Tuber   Enlarged area of a root (e.g., a sweet potato).

Two-action design   An experimental design used in social learning studies, in which each of two different actions on the same object is modeled in either of two different experimental conditions, permitting measurement of the extent to which observers match their later behavior to the alternative they witnessed.

Two-action procedure   The demonstration of a response in two distinctly different ways that results in the same effect on the environment (e.g., stepping on vs. pecking at a treadle).

Tympanum   Eardrum; a thin membrane that vibrates in response to sound.

Type I error   A statistical error in which the null hypothesis is rejected when it is, in fact, true.

Type II error   A statistical error in which the null hypothesis is not rejected when it is, in fact, not true.

Ultimate causation   Evolutionary explanations of animal behavior. Questions about ultimate causes of behavior are about why a behavior is expressed. Ultimate causes explain the adaptive significance of behavior.

Ultrasonic vocalization   A vocalization consisting only of frequencies higher than 20 kHz, that is, higher than the range of frequencies audible to human ears. Many species hear very high frequencies, well above the frequency range of human hearing.

Ultrasound   Sounds with frequencies above the limit of human hearing; normally considered to be 20 kHz and higher.

Unconstrained parents   Individuals mated to partners they do individually prefer.

Undertaking behavior   A behavioral routine found in social insects that involves collecting and removing the corpses of colony-mates from the nest.

Units   In extracellular multichannel recordings, investigators use mathematical techniques to separate differently sized and shaped action potentials from each other, calling each one a 'unit.' It is thought that these represent recordings from individual neurons. Because of the properties of extracellular recording, however, one cannot be absolutely certain that these are unique neurons. Hence, people who work in this area often use the less specific term 'unit.'

Univoltine   Having but a single generation a year.

Unpalatable   Unable to be eaten due to an unpleasant/noxious taste or toxicity.

Usage learning   Where an animal comes to use an existing signal in a new context as a result of experience of the usage of signals by other individuals.

Usurpation   Take over or adoption of nest, brood, and/or workers produced by other queens.

Vacuum activities   Behaviors, such as fly snapping, performed out of context, without an obvious stimulus.

Vagotomy   The transsection of the vagus nerve.

Vagus nerve   The tenth of the 12 pairs of cranial nerves, which originates in the brain stem and sends nerve fibers to the head, neck and viscera, including the lungs, heart, liver, and gastrointestinal tract. Most of the nerve fibers in the vagus nerve are sensory and the remainder are part of the parasympathetic nervous system. It contributes to the innervation of the viscera and conveys sensory information about the state of the body's organs to the central nervous system. The vagus is also called the *pneumogastric* nerve since it innervates both the lungs and the stomach.

Value   One of the alternative states of a variable; in communication, one of the alternative signals of a code.

Value of information   The fitness of an animal with access to information, contrasted to the fitness of an animal without access to the information.

Variable reinforcement schedule   In operant conditioning, the reinforcement of a desired behavior is given at random intervals.

Variance   A measure of statistical dispersion obtained by averaging the squared distance of its possible values from the expected value (mean). Whereas the mean is a way to describe the location of a distribution, the variance is a way to capture its scale or degree of being spread out.

Variance in fitness   A measure of deviation from mean fitness.

Variance in number of mates   Refers to a measure of the variation (deviation around the mean) in the number of mates obtained by different individuals of the same sex within a population. For example, in any given population, variance in number of mates is low when all individuals of one sex are able to obtain more or less the same number of mates. Variance in number of mates is high when some individuals mate with many members of the opposite sex, while others mate with very few or none.

Variance in reproductive success   Refers to a measure of the variation (deviation around the mean) in number of young produced by different individuals of the same sex within a population. For example, in any given population, variance in RS is low when all individuals of one sex produce more or less the same number of young. Variance in RS is high when some individuals produce most of the young, while others produce few or none.

Varroa mite   The mite species *Varroa destructor*, originally a pest of *Apis cerana* and now found on *A. mellifera*. A serious pest of honeybees and the cause of substantial colony mortality in *A. mellifera*.

Vasopressin   A peptide produced predominantly by magnocellular cells within hypothalamus, but also by centrally projecting neurons within the hypothalamus and amygdala.

Vasotocin   Peptide hormone secreted by the posterior pituitary; also released in the brain where it affects many social behaviors.

Veliger   One of the larval stages of some mollusks, including gastropods.

Venomous   Substances that are toxic and injure or kill animals, in most cases injected by biting or stinging.

Vent   External opening of the cloaca.

Ventricle   A cavity within the brain that is filled with cerebrospinal fluid. The cerebroventricular system comprises four ventricles: two lateral ventricles, the third ventricle, and the fourth ventricle. Cerebrospinal fluid flows from the lateral ventricles, to the third ventricle, then to the fourth ventricle before leaving the brain and entering the central canal of the spinal cord or into the subarachnoid space.

Vergence eye movements   Eye movements where the angle between the eyes changes.

Vertex   A component of a network with known relationships to others in the graph model representing the network; in a social network, this can be an individual animal or group; also called a *node* or *point*.

Vertical social influence   Influence by an individual on another from a different generation, such as a mother's influence on her offspring.

Vesicle   Knob-like structure on the terminal region of a nerve cell that stores and releases neurotransmitters. Also called synaptic vesicle.

Viability   Capacity for survival, more specifically used to mean a capacity for living, developing, or germinating under favorable conditions.

Viability selection   Selection generated by variation in survival among individuals of a population.

Vibrissae   Specialized hairs usually used for tactile sensation (singular: vibrissa).

Vicarious (or social) sampling   Gathering of information about the environment by observation of the behavior or products of the behavior of others.

Vigilance   Visual or auditory monitoring of the surroundings aimed at detecting threats related to predation. Vigilance can also be aimed at rivals or mates within the group.

Viral vector   A virus that is engineered to transport a specific DNA sequence into infected cells.

Viscera   The organs in the cavities of the body.

Visual acuity   Spatial visual resolution, the minimum angular separation between two objects that are perceived as different within the visual field.

Visual fields   Volume of space around an animal from which visual information can be obtained.

Vitellogenesis   Yolk deposition into the oocyte (egg).

Vitellogenin   Egg yolk precursor protein involved in regulation of behavioral maturation in social insects.

Viviparous   An animal giving birth to live young which have developed inside the body of the parent.

Vocal mimicry   Imitation by one species of sounds produced by another.

Vocal muscle   A vertebrate striated muscle used in sound production; also known as 'sonic muscle.'

Vocal production learning   Signals are modified in form as a result of experience with those of other individuals, leading to signals that are either similar or dissimilar to the model.

Vomeronasal organ (Jacobson's organ)   An accessory olfactory (odor-detecting) organ that is located in the roof of the mouth or nasal septum. The vomeronasal organ is particularly important for processing odors related to social signals.

Waders   Used in Europe and refers to shorebirds but used in North America with reference to herons and egrets.

Waiting game   A game in which both predator and prey need to decide for how long to wait when the prey entered a refuge that restricts its ability to collect information about the continued presence of the predator.

Wave refraction zone   The shallow area of ocean adjacent to a coastline where waves approaching the shore at an angle are redirected by interactions with the sea floor so that they approach directly toward shore.

Wavelength   The spatial distance between two consecutive cycles of a sine wave. Numerically, wavelength is the velocity of sound divided by its frequency. In a given medium, low-frequency sounds have long wavelengths and high-frequency sounds have short wavelengths.

Weakly electric fish   Electric fish with electric organs that produce very weak electric organ discharges that function in electrolocation and communication, but that are too weak to function in stunning predators or prey.

Weaning   The transition of young mammals from nursing to independent feeding, especially the parent's role in facilitating that transition.

Weber's law   A psychological law stating that one's ability to discriminate two quantities or intensities depends on the ratio between them.

Welfare   The health, happiness, and prosperity of an individual in its state as regards its attempts to cope with its environment; equated with 'well-being,' generally measured on a scale from very good to very poor.

Welfare illustrator grid   The assessment and two-dimensional illustration of welfare, designed to account for a temporal component and the cause of the animal's suffering.

'When' strategy   A social learning strategy specifying the circumstances under which individuals copy others.

'Who' strategy   A social learning strategy specifying from whom individuals learn.

Wild-type   The phenotypic composition of an organism as it occurs in nature.

Wing aspect ratio   The ratio of wing length to wing width; high aspect ratio wings permit fast, agile flight; low aspect ratio wings permit slow, maneuverable flight.

Wing polymorphism   Having more than one wing form within a population, for example, long-winged (migratory) and short-winged (nonmigratory) individuals may be found within the same population of many species of planthoppers (known as 'wing dimorphism').

Wintering area   In migratory birds, the area where populations spend the nonbreeding season, usually at lower latitudes.

Wintering dispersal   The distance between the wintering site of an individual in one year and its wintering site in another year.

Winter territory   A home range that an individual occupies and defends its boundaries against others (usually conspecifics but sometimes other species as well). This territory/home range may be held exclusively by the individual or as a pair or as a small group.

Wiring costs   The energetic costs associated with total length neural wiring (axons and dendrites).

Wisdom of crowds   The principle that the collective performance of a group of decision-makers can exceed that of a randomly chosen individual acting alone.

Within-pair offspring (WPC)   Offspring sired by the social father.

Within-sex variance in reproductive success   An operational definition of sexual selection.

Worker   Individual in a eusocial society that primarily performs all nonreproductive tasks in a colony. In primitively eusocial groups, this individual may be physically capable of reproduction; however, in highly eusocial groups, it is effectively sterile.

Xenoestrogens   Chemicals that are produced for agricultural, private, or industrial use that have estrogenic activity in living organisms.

X-organ   A group of neurosecretory neurons in the crustacean eyestalk that synthesize several peptide hormones.

Y-organ   The molting gland of crustaceans that usually secretes ecdysone, the precursor of the active form of the molting hormone, 20-hydroxyecdysone.

Zeitgeber   German word for 'time-giver'; an exogenous cue that entrains an endogenous biological rhythm.

Zoological psychology   A part of animal psychology that lies at the boundary between psychology and zoology. The approach is animal-centered in that the focus is primarily on studying the life of the animal rather than on asking arbitrary questions in a so-called animal model. The emphasis is often upon the natural behavioral repertoire of the animal rather than training the animal to engage in some arbitrary task.

Zoopharmacognosy   The study of how animals use medicinal substances. Interchangeably used by some with the term animal self-medication.

Zooplankton   Small pelagic organisms in aquatic ecosystems that form central part of the food web. They typically eat algae (phytoplankton) and are consumed by small (planktivorous) fish.

Zugunruhe   Migratory restlessness (hopping or hovering) in caged migratory birds often oriented with respect to seasonal directions of migration (e.g., northward in spring and southward in fall).

Zygote   A newly fertilized egg.